Common Formulas

Distance

$d = rt$

$d =$ distance traveled
$t =$ time
$r =$ rate

Temperature

$F = \dfrac{9}{5}C + 32$

$F =$ degrees Fahrenheit
$C =$ degrees Celsius

Simple Interest

$I = Prt$

$I =$ interest
$P =$ principal
$r =$ annual interest rate
$t =$ time in years

Compound Interest

$A = P\left(1 + \dfrac{r}{n}\right)^{nt}$

$A =$ balance
$P =$ principal
$r =$ annual interest rate
$n =$ compoundings per year
$t =$ time in years

Coordinate Plane: Midpoint Formula

Midpoint of line segment
joining (x_1, y_1) and (x_2, y_2)
$\left(\dfrac{x_1 + x_2}{2}, \dfrac{y_1 + y_2}{2}\right)$

Coordinate Plane: Distance Formula

$d =$ distance between
points (x_1, y_1) and
(x_2, y_2)
$d = \sqrt{(x_2 - x_1)^2 + (y_2 - y_1)^2}$

Quadratic Formula

Solutions of $ax^2 + bx + c = 0$

$x = \dfrac{-b \pm \sqrt{b^2 - 4ac}}{2a}$

Rules of Exponents

$a^0 = 1$

$a^m \cdot a^n = a^{m+n}$

$(ab)^m = a^m \cdot b^m$

$(a^m)^n = a^{mn}$

$\dfrac{a^m}{a^n} = a^{m-n}, \quad a \neq 0$

$\left(\dfrac{a}{b}\right)^m = \dfrac{a^m}{b^m}, \quad b \neq 0$

$a^{-n} = \dfrac{1}{a^n}, \quad a \neq 0$

$\left(\dfrac{a}{b}\right)^{-n} = \dfrac{b^n}{a^n}, \quad a \neq 0, \, b \neq 0$

Basic Rules of Algebra

Commutative Property of Addition

$a + b = b + a$

Commutative Property of Multiplication

$ab = ba$

Associative Property of Addition

$(a + b) + c = a + (b + c)$

Associative Property of Multiplication

$(ab)c = a(bc)$

Left Distributive Property

$a(b + c) = ab + ac$

Right Distributive Property

$(a + b)c = ac + bc$

Additive Identity Property

$a + 0 = a$

Multiplicative Identity Property

$a \cdot 1 = 1 \cdot a = a$

Additive Inverse Property

$a + (-a) = 0$

Multiplicative Inverse Property

$a \cdot \dfrac{1}{a} = 1, \quad a \neq 0$

Properties of Equality

Addition Property of Equality

If $a = b$, then $a + c = b + c$.

Multiplication Property of Equality

If $a = b$, then $ac = bc$.

Cancellation Property of Addition

If $a + c = b + c$, then $a = b$.

Cancellation Property of Multiplication

If $ac = bc$, and $c \neq 0$, then $a = b$.

Zero Factor Property

If $ab = 0$, then $a = 0$ or $b = 0$.

Intermediate Algebra

Third Edition

Ron Larson
The Pennsylvania State University
The Behrend College

Robert P. Hostetler
The Pennsylvania State University
The Behrend College

With the assistance of
David E. Heyd
The Pennsylvania State University
The Behrend College

Houghton Mifflin Company
Boston   New York

Sponsoring Editor: Jack Shira
Managing Editor: Cathy Cantin
Senior Associate Editor: Maureen Ross
Associate Editor: Laura Wheel
Assistant Editor: Carolyn Johnson
Supervising Editor: Karen Carter
Project Editor: Patty Bergin
Editorial Assistant: Christine E. Lee
Art Supervisor: Gary Crespo
Marketing Manager: Ros Kane
Senior Manufacturing Coordinator: Sally Culler
Composition and Art: Meridian Creative Group

We have included examples and exercises that use real-life data as well as technology output from a variety of software. This would not have been possible without the help of many people and organizations. Our wholehearted thanks go to all for their time and effort.

Trademark acknowledgment: TI is a registered trademark of Texas Instruments, Inc.

Printed in the U.S.A.

Library of Congress Catalog Card Number: 99-71986

ISBN: 0-395-97660-X

23456789–DOW–04 03 02 01 00

Contents

A Word from the Authors

Welcome to *Intermediate Algebra*, Third Edition. In this revision, we have continued to focus on developing students' proficiency and conceptual understanding of algebra. We hope you enjoy the Third Edition.

In response to intermediate algebra instructors, we have revised and reorganized the coverage of topics for the Third Edition. To improve the flow of the material, Chapter 2 "Graphs and Functions" now includes Section 2.4 "Equations of Lines" (formerly Section 7.1). Chapter 4 "Rational Expressions, Equations, and Functions" now includes Section 4.1 "Integer Exponents and Scientific Notation" (formerly Section 5.1). Chapter 7 has been renamed "Linear Models and Graphs of Nonlinear Models." "Variation" has been moved forward from Section 7.5 to Section 7.1, and "Graphs of Rational Functions" has been moved from Section 4.5 to Section 7.5.

In order to address the diverse needs and abilities of students, we offer a straightforward approach to the presentation of difficult concepts. In the Third Edition, the emphasis is on helping students learn a variety of techniques—symbolic, numeric, and visual—for solving problems. We are committed to providing students with a successful and meaningful course of study.

Our approach begins with Motivating the Chapter, a new feature that introduces each chapter. These multipart problems are designed to show students the relevance of algebra to the world around them. Each Motivating the Chapter feature is a real-life application that requires students to apply the concepts of the chapter in order to solve each part of the problem. Problem-solving and critical thinking skills are emphasized here and throughout the text in applications that appear in the examples and exercise sets.

To improve the usefulness of the text as a study tool, we added Objectives, which highlight the main concepts that students will learn throughout the section. Each objective is restated in the margin at the point where the concept is introduced, to help keep students focused as they read the section. The Chapter Summary was revised for the Third Edition to make it a more comprehensive and effective study tool. It now highlights the important mathematical vocabulary (Key Terms) and primary concepts (Key Concepts) of the chapter. For easy reference, the Key Terms are correlated to the chapter by page number and the Key Concepts by section number.

As students proceed through each chapter they have many opportunities to assess their understanding. They can check their progress after each section with the exercise sets (which are correlated to examples in the section), midway through the chapter with the Mid-Chapter Quiz, and at the end of the chapter with the Review Exercises (which are correlated to the sections) and the Chapter Test. The exercises and test items were carefully chosen and graded in difficulty to allow students to gain confidence as they progress. In addition, students can assess their understanding of previously learned concepts through the Integrated Review exercises that precede the section exercise sets and the Cumulative Tests that follow Chapters 3, 6, and 9.

In the Third Edition, we combined the Technology and Discovery features of the Second Edition. Technology Tips provide point-of-use instructions for using a graphing utility. Technology Discovery features encourage students to explore mathematical concepts with graphing utilities and scientific calculators. Both are highlighted and can easily be omitted without loss of continuity in coverage of material.

To show students the practical uses of algebra, we highlight the connections between the mathematical concepts and the real world in the multitude of applications found throughout the text. We believe that students can overcome their difficulties in mathematics if they are encouraged and supported throughout the learning process. Too often, students become frustrated and lose interest in the material when they cannot follow the text. With this in mind, every effort has been made to write a readable text that can be understood by every student. We hope that your students find our approach engaging and effective.

Ron Larson

Robert P. Hostetler

Features

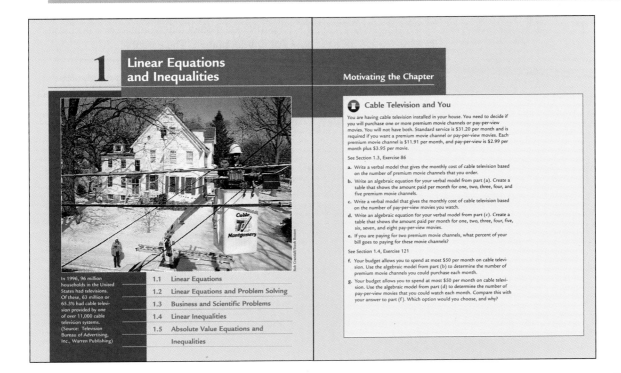

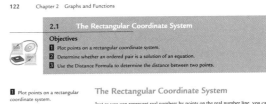

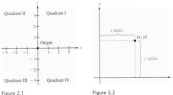

Chapter Opener   *New*

Every chapter opens with *Motivating the Chapter*. Each of these multipart problems incorporates the concepts presented in the chapter in the context of a single real-world application. *Motivating the Chapter* problems are correlated to sections and exercises and can be assigned as students work through the chapter or can be assigned as individual or group projects. The icon 🟠 identifies an exercise that relates back to *Motivating the Chapter*.

Section Opener   *New*

Every section begins with a list of learning objectives. Each objective is restated in the margin at the point where it is covered.

Historical Note

Historical notes featuring mathematicians or mathematical artifacts are included throughout the text.

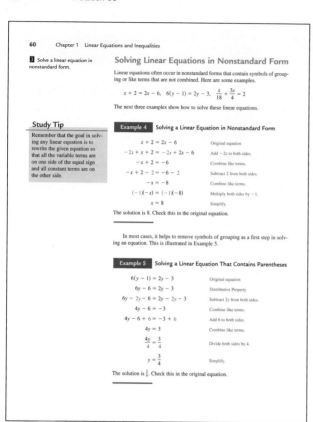

3 Solve a linear equation in nonstandard form.

Solving Linear Equations in Nonstandard Form

Linear equations often occur in nonstandard forms that contain symbols of grouping or like terms that are not combined. Here are some examples.

$$x + 2 = 2x - 6, \quad 6(y - 1) = 2y - 3, \quad \frac{x}{18} + \frac{3x}{4} = 2$$

The next three examples show how to solve these linear equations.

Study Tip

Remember that the goal in solving any linear equation is to rewrite the given equation so that all the variable terms are on one side of the equal sign and all constant terms are on the other side.

Example 4   Solving a Linear Equation in Nonstandard Form

$x + 2 = 2x - 6$	Original equation
$-2x + x + 2 = -2x + 2x - 6$	Add $-2x$ to both sides.
$-x + 2 = -6$	Combine like terms.
$-x + 2 - 2 = -6 - 2$	Subtract 2 from both sides.
$-x = -8$	Combine like terms.
$(-1)(-x) = (-1)(-8)$	Multiply both sides by -1.
$x = 8$	Simplify.

The solution is 8. Check this in the original equation.

In most cases, it helps to remove symbols of grouping as a first step in solving an equation. This is illustrated in Example 5.

Example 5   Solving a Linear Equation That Contains Parentheses

$6(y - 1) = 2y - 3$	Original equation
$6y - 6 = 2y - 3$	Distributive Property
$6y - 2y - 6 = 2y - 2y - 3$	Subtract 2y from both sides.
$4y - 6 = -3$	Combine like terms.
$4y - 6 + 6 = -3 + 6$	Add 6 to both sides.
$4y = 3$	Combine like terms.
$\dfrac{4y}{4} = \dfrac{3}{4}$	Divide both sides by 4.
$y = \dfrac{3}{4}$	Simplify.

The solution is $\frac{3}{4}$. Check this in the original equation.

Examples

Each example was carefully chosen to illustrate a particular mathematical concept or problem-solving technique. The examples cover a wide variety of problems and are titled for easy reference. Many examples include detailed, step-by-step solutions with side comments, which explain the key steps of the solution process.

Applications

A wide variety of real-life applications are integrated throughout the text in examples and exercises. These applications demonstrate the relevance of algebra in the real world. Many of the applications use current, real data. The icon indicates an example that involves a real-life application.

As a consumer today, you are presented almost daily with vast amounts of data given in various forms. Data are given in *numerical* form using lists and tables and in *graphical* form using scatter plots, lines, circle graphs, and bar graphs. Graphical forms are more visual and make wide use of Descartes's rectangular coordinate system to show the relationship between two variables. Today, Descartes's ideas are commonly used in virtually every scientific and business-related field.

Example 3   Representing Data Graphically

The population (in millions) of California from 1982 through 1997 is listed in the table. Plot these points on a rectangular coordinate system.   (Source: U.S. Bureau of the Census)

Year	1982	1983	1984	1985	1986	1987	1988	1989
Population	24.8	25.4	25.8	26.4	27.1	27.8	28.5	29.2

Year	1990	1991	1992	1993	1994	1995	1996	1997
Population	29.8	30.4	30.9	31.2	31.4	31.6	31.9	32.3

Solution

Begin by choosing which variable will be plotted on the horizontal axis and which will be plotted on the vertical axis. For these data, it seems natural to plot the years on the horizontal axis (which means that the population must be plotted on the vertical axis). Next, use the data in the table to form ordered pairs. For instance, the first three ordered pairs are (1982, 24.8), (1983, 25.4), and (1984, 25.8). All 16 points are shown in Figure 2.5. Note that the break in the x-axis indicates that the numbers between 0 and 1982 have been omitted. The break in the y-axis indicates that the numbers between 0 and 24 have been omitted.

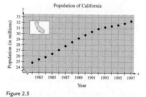

Figure 2.5

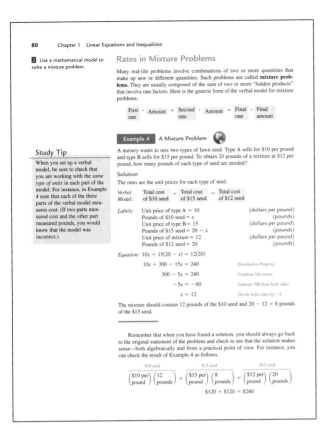

Problem Solving

This text provides many opportunities for students to sharpen their problem-solving skills. In both the examples and the exercises, students are asked to apply verbal, numerical, analytical, and graphical approaches to problem solving. In the spirit of the AMATYC and NCTM standards, students are taught a five-step strategy for solving applied problems, which begins with constructing a verbal model and ends with checking the answer.

Geometry

Coverage and integration of geometry in examples and exercises have been enhanced throughout the Third Edition.

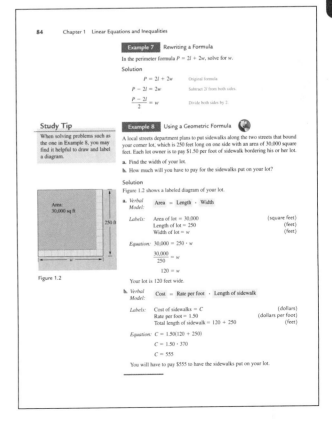

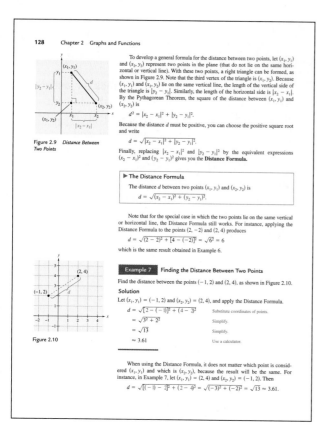

Definitions and Rules

All important definitions, rules, formulas, properties, and summaries of solution methods are highlighted for emphasis. Each of these features is also titled for easy reference.

Graphics

Visualization is a critical problem-solving skill. To encourage the development of this skill, students are shown how to use graphs to reinforce algebraic and numeric solutions and to interpret data. The numerous figures in examples and exercises throughout the text were computer generated for accuracy.

Technology Tips

Point-of-use instructions for using graphing utilities appear in the margins. They provide convenient reference for students using graphing technology. In addition, they encourage the use of graphing technology as a tool for visualization of mathematical concepts, for verification of other solution methods, and for facilitation of computations. The *Technology Tips* can easily be omitted without loss of continuity in coverage.

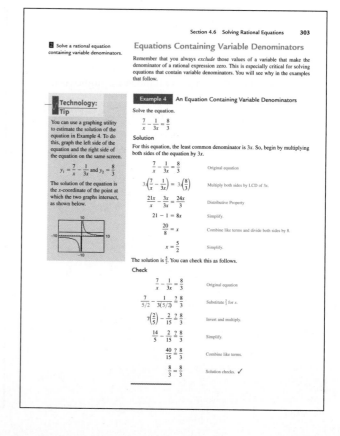

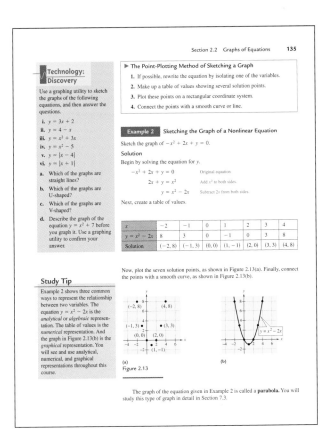

Technology Discovery

Utilizing the power of technology (scientific calculators and graphing utilities), *Technology Discovery* invites students to engage in active exploration of mathematical concepts and discovery of mathematical relationships. These activities encourage students to use their critical thinking skills and help them develop an intuitive understanding of theoretical concepts. *Technology Discovery* features can easily be omitted without loss of continuity of coverage.

Study Tips

Study Tips offer students specific point-of-use suggestions for studying algebra, as well as pointing out common errors and discussing alternative solution methods. They appear in the margins.

Discussing the Concept

Each section concludes with a *Discussing the Concept* feature. Designed as a section wrap-up activity to give students an opportunity to think, talk, and write about mathematics, each of these activities encourages students to synthesize the mathematical concepts presented in the section. *Discussing the Concept* can be assigned as an independent or collaborative activity or can be used as a basis for a class discussion.

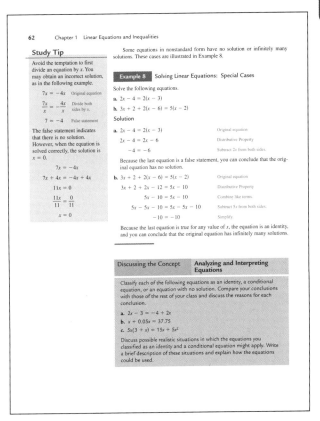

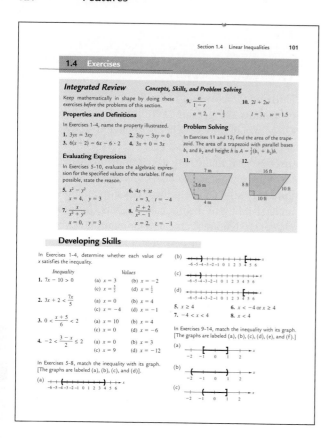

Integrated Review

Each exercise set (except in Chapter P) is preceded by *Integrated Review* exercises. These exercises are designed to help students keep up with concepts and skills learned in previous chapters. Answers to all *Integrated Review* problems are given in the back of the book.

Exercises

The exercise sets have been reorganized in the Third Edition. Each exercise set is grouped into three categories: *Developing Skills, Solving Problems*, and *Explaining Concepts*. The exercise sets offer a diverse variety of computational, conceptual, and applied problems to accommodate many teaching and learning styles. Designed to build competence, skill, and understanding, each exercise set is graded in difficulty to allow students to gain confidence as they progress. Detailed solutions to all odd-numbered exercises are given in the *Student Solutions Guide*, and answers to all odd-numbered exercises are given in the back of the book.

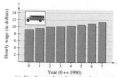

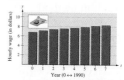

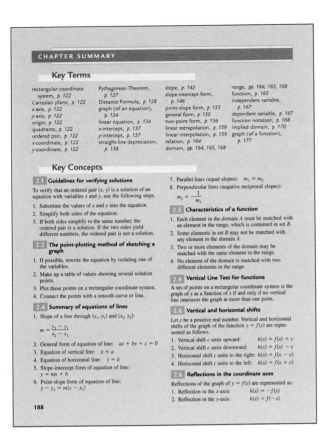

Chapter Summary

The *Chapter Summary* has been completely revised in the Third Edition. Designed to be an effective study tool for students preparing for exams, it highlights the *Key Terms* (referenced by page) and the *Key Concepts* (referenced by section) presented in the chapter.

Review Exercises

The *Review Exercises* at the end of each chapter have been reorganized in the Third Edition. They are grouped into two categories: *Reviewing Skills* and *Solving Problems*. Exercises under *Reviewing Skills* are correlated to sections in the chapter. The *Review Exercises* offer students additional practice in preparation for exams. Answers to all odd-numbered exercises are given in the back of the book.

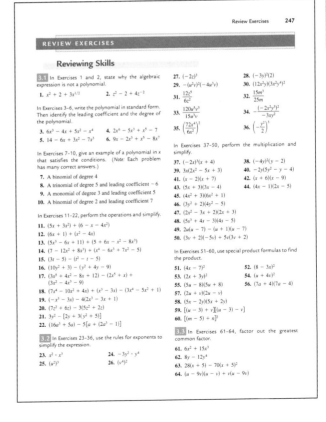

Mid-Chapter Quiz

Take this quiz as you would take a quiz in class. After you are done, check your work against the answers given in the back of the book.

1. Determine the degree and leading coefficient of the polynomial
$$3 - 2x + 4x^3 - 2x^4.$$

2. Explain why $2x - 3x^{1/2} + 5$ is not a polynomial.

In Exercises 3–18, perform the indicated operations and simplify.

3. Add $2t^3 + 3t^2 - 2$ to $t^3 + 9$.
4. $(3 - 7y) + (7y^2 + 2y - 3)$
5. $(7x^3 - 3x^2 + 1) - (x^2 - 2x^3)$
6. $(5 - u) - 2[3 - (u^2 + 1)]$
7. $(-5n^2)(-2n^3)$
8. $(-2x^3)^3(x^4)$
9. $\dfrac{6x^7}{(-2x^3)^3}$
10. $\left(\dfrac{4y^2}{5x}\right)^2$
11. $7y(4 - 3y)$
12. $(x - 7)(x + 3)$
13. $(4x - y)(6x - 5y)$
14. $2z(z + 5) - 7(z + 5)$
15. $(6r + 5)(6r - 5)$
16. $(2x - 3)^2$
17. $(x + 1)(x^2 - x + 1)$
18. $(x^2 - 3x + 2)(x^2 + 5x - 10)$

In Exercises 19–22, factor the expression completely.

19. $28a^2 - 21a$
20. $25 - 4x^2$
21. $z^3 + 3z^2 - 9z - 27$
22. $4y^3 - 32x^3$

23. Find all possible products of the form $(5x + m)(2x + n)$ such that $mn = 10$.
24. Find the area of the shaded portion of the figure.

25. An object is thrown downward from the top of a 100-foot building with an initial velocity of -5 feet per second. Use the position function $h(t) = -16t^2 - 5t + 100$ to find the height of the object when $t = 1$ and $t = 2$.

26. A manufacturer can produce and sell x T-shirts per week. The total cost (in dollars) for producing the T-shirts is given by $C = 5x + 2000$ and the total revenue is given by $R = 19x$. Find the profit obtained by selling 1000 T-shirts per week.

Chapter Test

Take this test as you would take a test in class. After you are done, check your work against the answers given in the back of the book.

1. Determine the quadrant in which the point (x, y) lies if $x > 0$ and $y < 0$.
2. Plot the points $(0, 5)$ and $(3, 1)$. Then find the distance between them.
3. Find the x- and y-intercepts of the graph of the equation $y = -3(x + 1)$.
4. Sketch the graph of the equation $y = |x - 2|$.
5. Find the slope (if possible) of the line passing through each pair of points.
   (a) $(-4, 7), (2, 3)$ (b) $(3, -2), (3, 6)$
6. Sketch the graph of the line passing through the point $(0, -6)$ with slope $m = \frac{4}{3}$.
7. Plot the x- and y-intercepts of the graph of $2x + 5y - 10 = 0$. Use the results to sketch the graph.
8. Write the equation $5x + 3y - 9 = 0$ in slope-intercept form. Find the slope of the line that is perpendicular to this line.
9. Find an equation of the line through the points $(25, -15)$ and $(75, 10)$.
10. Find an equation of the line with slope -2 that passes through the point $(2, -4)$.
11. Find an equation of the vertical line through the point $(-2, 4)$.
12. The graph of $y^2(4 - x) = x^3$ is shown at the left. Does the graph represent y as a function of x? Explain your reasoning.
13. Determine whether the relation represents a function. Explain.
    (a) $\{(2, 4), (-6, 3), (3, 3), (1, -2)\}$ (b) $\{(0, 0), (1, 5), (-2, 1), (0, -4)\}$
14. Evaluate $g(x) = x/(x - 3)$ for the indicated values.
    (a) $g(2)$ (b) $g(\frac{7}{2})$ (c) $g(x + 2)$
15. Find the domain of each function.
    (a) $h(t) = \sqrt{9 - t}$ (b) $f(x) = \dfrac{x + 1}{x - 4}$
16. Sketch the graph of the function $g(x) = \sqrt{2 - x}$.
17. Describe the transformation of the graph of $f(x) = x^2$ that would produce the graph of $g(x) = -(x - 2)^2 + 1$.
18. After 4 years, the value of a \$26,000 car will have depreciated to \$10,000. Write the value V of the car as a linear function of t, the number of years since the car was purchased. When will the car be worth \$16,000? Explain your reasoning.
19. Use the graph of $f(x) = |x|$ to write an equation for each graph.
    (a)   (b)   (c)

Figure for 12

Cumulative Test: Chapters P–3

Take this test as you would take a test in class. After you are done, check your work against the answers given in the back of the book.

1. Place the correct symbol ($<$, $>$, or $=$) between the two numbers.
   (a) -2 ___ 5 (b) $\frac{1}{3}$ ___ $\frac{1}{2}$ (c) $|2.3|$ ___ $-|-4.5|$
2. Write an algebraic expression for the statement, "The number n is tripled and the product is decreased by 8."

In Exercises 3–5, perform the operations and simplify.

3. (a) $t(3t - 1) - 2t(t + 4)$ (b) $3x(x^2 - 2) - x(x^2 + 5)$
4. (a) $(2a^2b)^3(-ab^2)^2$ (b) $\left(\dfrac{2x^4y^2}{4x^3y}\right)^2$
5. (a) $(2x + 1)(x - 5)$ (b) $[2 + (x - y)]^2$

In Exercises 6–8, solve the equations or inequalities.

6. (a) $12 - 5(3 - x) = x + 3$ (b) $1 - \dfrac{x + 2}{4} = \dfrac{7}{8}$
7. (a) $|3x - 5| = 7$ (b) $2t^2 - 5t - 3 = 0$
8. (a) $3(1 - x) > 6$ (b) $-12 \le 4x - 6 < 10$

9. Your annual automobile insurance premium is \$1225. Because of a driving violation, your premium is increased 15%. What is your new premium?
10. The triangles at the left are similar. Solve for x by using the fact that corresponding sides of similar triangles are proportional.
11. Solve $|x - 2| \ge 3$ and sketch its solution.
12. The revenue from selling x units of a product is $R = 12.90x$. The cost of producing x units is $C = 8.50x + 450$. To obtain a profit, the revenue must be greater than the cost. For what values of x will this product produce a profit? Explain your reasoning.
13. Determine whether the equation $x - y^3 = 0$ represents y as a function of x.
14. Find the domain of the function $f(x) = \sqrt{x - 2}$.
15. Given $f(x) = x^2 - 3x$, find (a) $f(4)$ and (b) $f(c + 3)$.
16. Find the slope of the line passing through $(-4, 0)$ and $(4, 6)$. Then find the distance between the points.
17. Determine the equation of a line through the point $(-2, 1)$ (a) parallel to $2x - y = 1$ and (b) perpendicular to $3x + 2y = 5$.

In Exercises 18 and 19, factor the polynomials.

18. (a) $3x^2 - 8x - 35$ (b) $9x^2 - 144$
19. (a) $y^3 - 3y^2 - 9y + 27$ (b) $8t^3 - 40t^2 + 50t$

In Exercises 20 and 21, graph the equation.

20. $4x + 3y - 12 = 0$ 21. $y = 1 - (x - 2)^2$

Figure for 10

Mid-Chapter Quiz

Each chapter contains a *Mid-Chapter Quiz*. This feature allows students to perform a self-assessment midway through the chapter. Answers to all *Mid-Chapter Quiz* exercises are given in the back of the book.

Chapter Test

Each chapter ends with a *Chapter Test*. This feature allows students to perform a self-assessment at the end of the chapter. Answers to all *Chapter Test* exercises are given in the back of the book.

Cumulative Test

The *Cumulative Tests* that follow Chapters 3, 6, and 9 provide a comprehensive self-assessment tool that helps students check their mastery of previously covered material. Answers to all *Cumulative Test* exercises are given in the back of the book.

Supplements

Intermediate Algebra, Third Edition, by Larson and Hostetler is accompanied by a comprehensive supplements package, which includes resources for both students and instructors. All items are keyed to the text.

Printed Resources

For the Student

Study and Solutions Guide by Gerry C. Fitch, Louisiana State University
(0-395-97662-6)
• Detailed, step-by-step solutions to all Integrated Review exercises and to all odd-numbered exercises in the section exercise sets and in the review exercises
• Detailed, step-by-step solutions to all Mid-Chapter Quiz, Chapter Test, and Cumulative Test questions

Graphing Calculator Keystroke Guide by Benjamin N. Levy and Laurel Technical Services
(0-395-87777-6)
• Keystroke instructions for the following graphing calculators: (Texas Instruments) *TI-80, TI-81, TI-82, TI-83, TI-85,* and *TI-92*; (Casio) *fx-7700GE, fx-9700GE,* and *CFX-9800G*; (Hewlett Packard) *HP-38G*; and (Sharp) *EL-9200/9300*
• Examples with step-by-step solutions
• Extensive graphics screen output
• Technology tips

For the Instructor

Instructor's Annotated Edition
(0-395-97663-4)
• Includes entire student edition
• Instructor's answer section, which includes answers to all even-numbered exercises, Technology Discovery boxes, Technology Tip boxes, and Discussing the Concept activities
• Annotations at point of use that offer strategies and suggestions for teaching the course and point out common student errors

Test Item File and Instructor's Resource Guide by Ann R. Kraus, The Pennsylvania State University, The Behrend College
(0-395-97661-8)
• Printed test bank with approximately 3300 test items, coded by level of difficulty
• Technology-required test items, coded for easy reference
• Chapter test forms with answer key
• Two final exams
• Transparency masters

- Notes to the instructor, which include information on standardized tests such as the Texas Academic Skills Program (TASP), the Florida College Level Academic Skills Test (CLAST), and the California State University Entry Level Mathematics (ELM) Exam. A list of skills covered by the test and the corresponding sections in the text where the topics are covered are also provided.
- Alternative assessment strategies

Media Resources

For Students and Instructors

Website (*www.hmco.com*)
Contains, but is not limited to, the following student and instructor resources:
- Study guide (for students), which includes section summaries, additional examples with solutions, and starter exercises with answers
- Chapter projects and additional real-life applications
- Geometry review
- ACE Algebra Tutor
- Graphing calculator programs
- Math Matters and Career Interviews

HM³ Tutor
(Instructor's version Windows: 0-618-04208-3)
This networkable, interactive tutorial software offers the following features:
- Algorithmically generated practice and quiz problems
- A variety of multiple-choice and free-response questions, varying in degree of difficulty
- Animated examples and interactivity within lessons
- Hints and full solutions available for every problem
- Integrated classroom management system (for instructors), which includes a syllabus builder and the capability to track and report student performance
- Non-networkable student version (Windows: 0-395-97656-1)

For the Student

Videotape Series by Dana Mosely
(0-395-97670-7)
- Comprehensive section-by-section coverage
- Detailed explanations of important concepts
- Numerous examples and applications, often illustrated by means of computer-generated animations
- Discussion of study skills

For the Instructor

Computerized Test Bank
(Windows: 0-395-97665-0; Macintosh: 0-395-97666-9)
- Test-generating software for IBM and Macintosh computers
- Approximately 3300 test items
- Also available as a printed test bank

Acknowledgments

We would like to thank the many people who have helped us prepare the Third Edition of this text. Their encouragement, criticisms, and suggestions have been invaluable to us.

Third Edition Reviewers

Beverly Broomwell, Suffolk County Community College; Connie L. Buller, Metropolitan Community College; David L. Byrd, Enterprise State Junior College; E. Judith Cantey, Jefferson State Community College; Kelly E. Champagne, Nicholls State University; Sally Copeland, Johnson County Community College; Fletcher Gross, University of Utah; William Hoard, Front Range Community College; Judith Kasabian, El Camino College; Harvey W. Lambert, University of Nevada–Reno; Carol McVey, Florence–Darlington Technical College; William Naegele, South Suburban College; Judith Pranger, Binghamton University; Scott Reed, College of Lake County; Eveline Robbins, Gainesville College; Joel Siegel, Sierra College; Allan Struck, California State University–Chico; Katherine R. Struve, Columbus State Community College; Marilyn Treder, Rochester Community and Technical College; Bettie Truitt, Black Hawk College; Christine Walker, Utah Valley State College; Maureen Watson, Nicholls State University; Matrid Whiddon, Edison Community College; Betsey S. Whitman, Framingham State College; George J. Witt, Glendale Community College.

We would also like to thank the staff of Larson Texts, Inc., and the staff of Meridian Creative Group, who assisted in proofreading the manuscript, preparing and proofreading the art package, and checking and typesetting the supplements.

On a personal level, we are grateful to our wives, Deanna Gilbert Larson and Eloise Hostetler, for their love, patience, and support. Also, a special thanks goes to R. Scott O'Neil.

If you have suggestions for improving this text, please feel free to write to us. Over the past two decades we have received many useful comments from both instructors and students, and we value these comments very much.

Ron Larson
Robert P. Hostetler

ACKNOWLEDGMENTS

How to Study Algebra

Your success in algebra depends on your active participation both in class and outside of class. Because the material you learn each day builds on the material you learned previously, it is important that you keep up with the course work every day and develop a clear plan of study. To help you learn how to study algebra, we have prepared a set of guidelines that highlight key study strategies.

Preparing for Class

The syllabus your instructor provides is an invaluable resource that outlines the major topics to be covered in the course. Use it to help you prepare. As a general rule, you should set aside two to four hours of study time for each hour spent in class. Being prepared is the first step toward success in algebra. Before class,

- ❑ Review your notes from the previous class.
- ❑ Read the portion of the text that will be covered in class.
- ❑ Use the objectives list at the beginning of each section to keep you focused on the main ideas presented in the section.
- ❑ Pay special attention to the definitions, rules, and concepts highlighted in boxes. Also, be sure you understand the meanings of mathematical symbols and of terms written in boldface type. Keep a vocabulary journal for easy reference.
- ❑ Read through the solved examples. Use the side comments given in the solution steps to help you follow the solution process. Also, read the *Study Tips* given in the margins.
- ❑ Make notes of anything you do not understand as you read through the text. If you still do not understand after your instructor covers the topic in question, ask questions before your instructor moves on to a new topic.
- ❑ If you are using technology in this course, read the *Technology Tips* and try the *Technology Discovery* exercises.

Keeping Up

Another important step toward success in algebra involves your ability to keep up with the work. It is very easy to fall behind, especially if you miss a class. To keep up with the course work, be sure to

- ❑ Attend every class. Bring your text, a notebook, and a pen or pencil. If you miss a class, get the notes from a classmate as soon as possible and review them carefully.
- ❑ Take notes in class. After class, read through your notes and add explanations so that your notes make sense to *you*.
- ❑ Reread the portion of the text that was covered in class. This time, work each example *before* reading through the solution.

- ❏ Do your homework as soon as possible, while concepts are still fresh in your mind.
- ❏ Use your notes from class, the text discussion, the examples, and the *Study Tips* as you do your homework. Many exercises are keyed to specific examples in the text for easy reference.

Getting Extra Help

It can be very frustrating when you do not understand concepts and are unable to complete homework assignments. However, there are many resources available to help you with your study of algebra.

- ❏ Your instructor may have office hours. If you are feeling overwhelmed and need help, make an appointment to discuss your difficulties with your instructor.
- ❏ Find a study partner or a study group. Sometimes it helps to work through problems with another person.
- ❏ Arrange to get regular assistance from a tutor. Many colleges have math resource centers available on campus.
- ❏ Consult one of the many ancillaries available with this text: the *Student Solutions Guide*, tutorial software, videotapes, and additional study resources available at our website at *www.hmco.com*.

Preparing for an Exam

The last step toward success in algebra lies in how you prepare for and complete exams. If you have followed the suggestions given above, then you are almost ready for exams. Do not assume that you can cram for the exam the night before: this seldom works. As a final preparation for the exam,

- ❏ Read the *Chapter Summary*, which is keyed to each section, and review all pertinent concepts and terms.
- ❏ Work through the *Review Exercises* if you need extra practice on material from a particular section.
- ❏ Take the *Mid-Chapter Quiz* and the *Chapter Test* as if you were in class. You should set aside at least one hour per test. Check your answers against the answers given in the back of the book.
- ❏ Review your notes and the portion of the text that will be covered on the exam.
- ❏ Avoid studying up until the last minute. This will only make you anxious.
- ❏ Once the exam begins, read through the directions and the entire exam before beginning. Work the problems that you know how to solve first, to avoid spending too much of the allotted time on any one problem. Time management is extremely important when taking an exam.
- ❏ If you finish early, use the remaining time to go over your work.
- ❏ When you get an exam back, review it carefully and go over your errors. Rework the problems you answered incorrectly. Discovering the mistakes you made will help you improve your test-taking ability.

P Prerequisites: Fundamentals of Algebra

Riley & Riley Photography/Tony Stone Images

The Universal Product Code first appeared on supermarket products in 1973. Since that time other types of bar codes have been used for identification purposes in factories, hospitals, libraries, and warehouses.

The Universal Product Code

Packaged products in the United States have a Universal Product Code (UPC) or bar code, as shown at the right. When the UPC is scanned and verified at a checkout counter, the manufacturer and the product are identified. Then a database in the store provides the retail price.

To verify a UPC, the first 11 digits are evaluated according to the algorithm given below. The result should equal the 12th digit, called the check digit.

First digit   Check digit

Algorithm

1. Add the numbers in the odd-numbered positions together. Multiply by 3.
2. Add the numbers in the even-numbered positions together.
3. Add the results of Steps 1 and 2.
4. Subtract the result of Step 3 from the next-highest multiple of 10.

Using the algorithm on the UPC at the right produces the following.

1. $(0 + 5 + 1 + 0 + 6 + 8) \times 3 = 60$
2. $2 + 2 + 5 + 4 + 5 = 18$
3. $60 + 18 = 78$
4. The next-highest multiple of 10 is 80. So, $80 - 78 = 2$, is the check digit.

Here are some types of questions you will be able to answer as you study this chapter. You will be asked to answer parts (a)–(c) in Section P.2, Exercise 135.

a. One student decides to combine Steps 1, 2, and 3 with the following results.

$(0 + 5 + 1 + 0 + 6 + 8) \times 3 + (2 + 2 + 5 + 4 + 5)$
$= 20 \times 3 + 18 = 20 \times 21 = 420$

The next-highest multiple of 10 is 430, so $430 - 420 = 10$.

What did the student do wrong?

b. Does a UPC of 0 76737 20012 9 check? Explain.

c. Does a UPC of 0 41800 48700 3 check? Explain.

You will be asked to answer parts (d) and (e) in Section P.4, Exercise 111.

d. The UPC below is missing the 10th-place digit. Write an algebraic expression for this UPC that represents the result of Step 3 of the algorithm.

0 43819 236*a*7 4

e. Evaluate the expression in part (e) and do Step 4 of the algorithm for the digits 0 through 9. What is the value of the missing digit? Could there be more than one value for this digit? Explain.

P.1   The Real Number System

Objectives

1 Review the set of real numbers and the subsets of real numbers.

2 Use the real number line to order real numbers.

3 Find the distance between two real numbers using the number line.

4 Determine the absolute value of a real number.

1 Review the set of real numbers and the subsets of real numbers.

Study Tip

In this text, whenever a mathematical term is introduced, the word will appear in boldface type. Be sure you understand the meaning of each new word—it is important that each word become part of your mathematical vocabulary. It may be helpful to keep a vocabulary journal.

Sets and Real Numbers

This chapter reviews the basic definitions, operations, and rules that form the fundamental concepts of algebra. The chapter begins with real numbers and their representation on the real number line. Sections P.2 and P.3 review operations and properties of real numbers, and Sections P.4 and P.5 review algebraic expressions.

The formal term that is used in mathematics to talk about a collection of objects is the word **set.** For instance, the set

$$\{1, 2, 3\}$$ A set with three members

contains the three numbers 1, 2, and 3. Note that the members of the set are enclosed in braces { }. Parentheses ( ) and brackets [ ] are used to represent other concepts.

The set of numbers that is used in arithmetic is called the set of **real numbers.** The term *real* distinguishes real numbers from *imaginary* or *complex* numbers—a type of number that you will study later in this text.

If all members of a set A are also members of a set B, then A is a **subset** of B. One of the most commonly used subsets of real numbers is the set of **natural numbers** or **positive integers.**

$$\{1, 2, 3, 4, \ldots\}$$ The set of positive integers

Note that the three dots indicate that the pattern continues. For instance, the set also contains the numbers 5, 6, 7, and so on.

Positive integers can be used to describe many quantities that you encounter in everyday life—for instance, you might be taking four classes this term, or you might be paying 240 dollars a month for rent. But even in everyday life, positive integers cannot describe some concepts accurately. For instance, you could have a zero balance in your checking account. To describe such a quantity you need to expand the set of positive integers to include **zero,** forming the set of **whole numbers.** To describe a quantity such as $-10°$ (10 degrees below zero), you need to expand the set of whole numbers to include **negative integers.** This expanded set is called the set of **integers.**

$$\underbrace{\{\ldots, -3, -2, -1, \overbrace{0, 1, 2, 3, \ldots}^{\text{Whole numbers}}\}}$$ The set of integers

Negative integers   Positive integers

The set of integers is also a *subset* of the set of real numbers.

Technology: Tip

You can use a calculator to round decimals. For instance, to round 0.2846 to three decimal places on a scientific calculator, enter

FIX 3 .2846 =

or, on a graphing utility, enter

round (.2846, 3) ENTER .

Consult the user's manual for your graphing utility for specific keystrokes or instructions. Without using a calculator, round 0.38174 to four decimal places. Verify your answer with a calculator.

Even with the set of integers, there are still many quantities in everyday life that you cannot describe accurately. The costs of many items are not in whole dollar amounts, but in parts of dollars, such as $1.19 or $39.98. You might work $8\frac{1}{2}$ hours, or you might miss the first *half* of a movie. To describe such quantities, you can expand the set of integers to include **fractions.** The expanded set is called the set of **rational numbers.** Formally, a real number is **rational** if it can be written as the ratio p/q of two integers, where $q \neq 0$ (the symbol $\neq$ means *does not equal*). For instance,

$$2 = \frac{2}{1}, \quad \frac{1}{3} = 0.333 \ldots, \quad \frac{1}{8} = 0.125, \quad \text{and} \quad \frac{125}{111} = 1.126126 \ldots$$

are rational numbers. The decimal representation of a rational number is either *terminating* or *repeating*. For instance, the decimal representation of $\frac{1}{4} = 0.25$ is terminating, and the decimal representation of

$$\frac{4}{11} = 0.363636 \ldots = 0.\overline{36}$$

is repeating. (The line over "36" indicates which digits repeat.) A real number that cannot be written as a ratio of two integers is **irrational.** For instance, the numbers

$$\sqrt{2} = 1.4142135\ldots \quad \text{and} \quad \pi = 3.1415926\ldots$$

are irrational.

The decimal representation of an irrational number neither terminates nor repeats. When you perform calculations using decimal representations of nonterminating decimals, you usually use a decimal approximation that has been **rounded** to a certain number of decimal places. The rounding rule used in this text is to round up if the succeeding digit is 5 or more and round down if the succeeding digit is 4 or less. For example, to one decimal place, 7.35 would *round up* to 7.4. Similarly, to two decimal places, 2.364 would *round down* to 2.36. Rounded to four decimal places, the decimal approximations of the rational number $\frac{2}{3}$ and the irrational number π are

$$\frac{2}{3} \approx 0.6667 \quad \text{and} \quad \pi \approx 3.1416.$$

The symbol $\approx$ means **is approximately equal to.** Figure P.1 shows several commonly used subsets of real numbers and their relationships to each other.

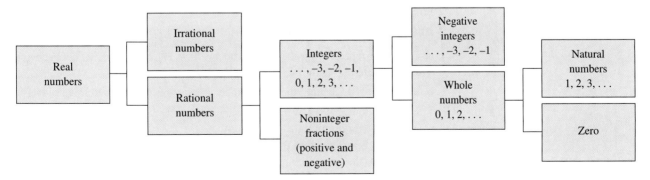

Figure P.1   *Subsets of the Set of Real Numbers*

2 Use the real number line to order real numbers.

The Real Number Line

The picture that represents the real numbers is called the **real number line.** It consists of a horizontal line with a point (the **origin**) labeled as 0. Numbers to the left of zero are **negative** and numbers to the right of 0 are **positive,** as shown in Figure P.2.

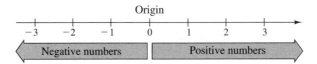

Figure P.2   *The Real Number Line*

The real number zero is neither positive nor negative. So, to describe a real number that might be positive *or* zero, you can use the term **nonnegative real number.**

Each point on the real number line corresponds to exactly one real number, and each real number corresponds to exactly one point on the real number line, as shown in Figure P.3. When you draw the point (on the real number line) that corresponds to a real number, you are **plotting** the real number.

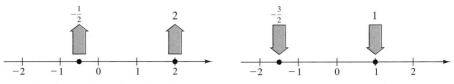

Each point on the real number line corresponds to a real number.

Each real number corresponds to a point on the real number line.

Figure P.3

Example 1   Plotting Points on the Real Number Line

Plot the points that represent the real numbers.

a. $-\dfrac{5}{3}$ **b.** 2.3   **c.** $\dfrac{9}{4}$ **d.** -0.3

Solution

All four points are shown in Figure P.4.

a. The point representing the real number $-\frac{5}{3} = -1.666 \ldots$ lies between -2 and -1, but closer to -2, on the real number line.

b. The point representing the real number 2.3 lies between 2 and 3, but closer to 2, on the real number line.

c. The point representing the real number $\frac{9}{4} = 2.25$ lies between 2 and 3, but closer to 2, on the real number line. Note that the point representing $\frac{9}{4}$ lies slightly to the left of the point representing 2.3.

d. The point representing the real number -0.3 lies between -1 and 0, but closer to 0, on the real number line.

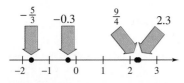

Figure P.4

The real number line provides a way of comparing any two real numbers. For instance, if you choose any two (different) numbers on the real number line, one of the numbers must be to the left of the other. You can describe this by saying that the number to the left is **less than** the number to the right, or that the number to the right is **greater than** the number to the left, as shown in Figure P.5.

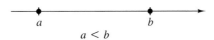

$$a < b$$

Figure P.5 *a is to the left of b.*

> ▶ Order on the Real Number Line
>
> If the real number a lies to the left of the real number b on the real number line, then a is **less than** b, which is written as
>
> $$a < b.$$
>
> This relationship can also be described by saying that b is **greater than** a and writing $b > a$. The symbol $a \leq b$ means that a is **less than or equal to** b, and the symbol $b \geq a$ means that b is **greater than or equal to** a. The symbols $<$, $>$, $\leq$, and $\geq$ are called **inequality symbols.**

When asked to **order** two numbers, you are simply being asked to say which of the two numbers is greater.

(a)

(b)

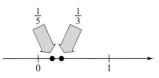

(c)

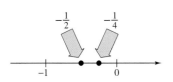

(d)
Figure P.6

Example 2   Ordering Real Numbers

Place the correct inequality symbol ($<$ or $>$) between the two numbers.

a. -4 ___ 0 **b.** -3 ___ -5 **c.** $\dfrac{1}{5}$ ___ $\dfrac{1}{3}$ **d.** $-\dfrac{1}{4}$ ___ $-\dfrac{1}{2}$

Solution

a. Because -4 lies to the left of 0 on the real number line, as shown in Figure P.6(a), you can say that -4 is *less than* 0, and write $-4 < 0$.

b. Because -3 lies to the right of -5 on the real number line, as shown in Figure P.6(b), you can say that -3 is *greater than* -5, and write $-3 > -5$.

c. Because $\frac{1}{5}$ lies to the left of $\frac{1}{3}$ on the real number line, as shown in Figure P.6(c), you can say that $\frac{1}{5}$ is *less than* $\frac{1}{3}$, and write $\frac{1}{5} < \frac{1}{3}$.

d. Because $-\frac{1}{4}$ lies to the right of $-\frac{1}{2}$ on the real number line, as shown in Figure P.6(d), you can say that $-\frac{1}{4}$ is *greater than* $-\frac{1}{2}$, and write $-\frac{1}{4} > -\frac{1}{2}$.

One effective way to order two fractions such as $5/12$ and $9/23$ is to compare their decimal equivalents. Because $\frac{5}{12} = 0.41\overline{6}$ and $\frac{9}{23} \approx 0.391$, you can write

$$\frac{5}{12} > \frac{9}{23}.$$

3 Find the distance between two real numbers using the number line.

Distance on the Real Line

Once you know how to represent real numbers as points on the real number line, it is natural to talk about the **distance between two real numbers.** Specifically, if a and b are two real numbers such that $a \le b$, then the distance between a and b is defined to be $b - a$.

> ▶ **Distance Between Two Real Numbers**
>
> If a and b are two real numbers such that $a \le b$, then the **distance between a and b** is given by
>
> (Distance between a and b) $= b - a$.

Note from this definition that if $a = b$, the distance between a and b is zero. If $a \ne b$, the distance between a and b is positive.

Example 3   Finding the Distance Between Two Real Numbers

Find the distance between each pair of real numbers.

a. -2 and 3   **b.** 0 and 4   **c.** -4 and 0   **d.** 1 and $-\dfrac{1}{2}$

Study Tip

Recall that when you subtract a negative number, as in Example 3(a), you add the opposite of the second number to the first. Because the opposite of -2 is 2, you add 2 to 3.

Solution

All four distances are shown in Figure P.7.

a. Because $-2 \le 3$, the distance between -2 and 3 is

$$3 - (-2) = 3 + 2 = 5. \qquad \text{Distance between } -2 \text{ and } 3$$

b. Because $0 \le 4$, the distance between 0 and 4 is

$$4 - 0 = 4. \qquad \text{Distance between } 0 \text{ and } 4$$

c. Because $-4 \le 0$, the distance between -4 and 0 is

$$0 - (-4) = 0 + 4 = 4. \qquad \text{Distance between } -4 \text{ and } 0$$

d. Because $-\frac{1}{2} \le 1$, the distance between 1 and $-\frac{1}{2}$ is

$$1 - \left(-\frac{1}{2}\right) = 1 + \frac{1}{2} = 1\frac{1}{2}. \qquad \text{Distance between } 1 \text{ and } -\tfrac{1}{2}$$

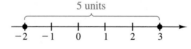

(a)

(b)

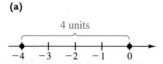

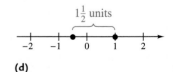

(c)

(d)

Figure P.7

4 Determine the absolute value of a real number.

Absolute Value

The distance between a real number a and 0 (the origin) is called the **absolute value** of a. Absolute value is denoted by double vertical bars, $|\quad|$. For example,

$$|5| = \text{“distance between 5 and 0”} = 5$$

and

$$|-8| = \text{“distance between } -8 \text{ and 0”} = 8.$$

Study Tip

Be sure you see from this definition that the absolute value of a real number is never negative. For instance, if $a = -3$, then $|-3| = -(-3) = 3$. Moreover, the only real number whose absolute value is zero is 0. That is, $|0| = 0$.

▶ Absolute Value of a Real Number

The **absolute value** of a real number a is the distance between a and 0 on the real number line.

1. If $a > 0$, then $|a| = a - 0 = a$.

2. If $a = 0$, then $|a| = 0 - 0 = 0$.

3. If $a < 0$, then $|a| = 0 - a = -a$.

Two real numbers are called **opposites** of each other if they lie the same distance from, but on opposite sides of, 0 on the real number line. For instance, -2 is the opposite of 2. (See Figure P.8.) Because *opposite* numbers lie the same distance from 0 on the real number line, they have the same absolute value. So, $|5| = 5$ and $|-5| = 5$. The use of a negative sign to denote the *opposite* of a number gives meaning to expressions such as $-(-3)$ and $-|-3|$, as follows.

$$-(-3) = (\text{opposite of } -3) = 3$$

$$-|-3| = \left(\text{opposite of } |-3|\right) = -3$$

Opposite numbers are also referred to as **additive inverses** because their sum is zero. For instance, $3 + (-3) = 0$, or, in general, $b + (-b) = 0$.

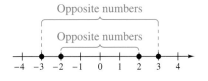

Opposite numbers

Opposite numbers

-4 -3 -2 -1 0 1 2 3 4

Figure P.8

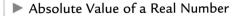

Example 4   Finding Absolute Values

Evaluate each expression.

a. $|-10|$ **b.** $\left|\dfrac{3}{4}\right|$ **c.** $|-3.2|$ **d.** $-|-6|$

Solution

a. $|-10| = 10$ The absolute value of -10 is 10.

b. $\left|\dfrac{3}{4}\right| = \dfrac{3}{4}$ The absolute value of $\frac{3}{4}$ is $\frac{3}{4}$.

c. $|-3.2| = 3.2$ The absolute value of -3.2 is 3.2.

d. $-|-6| = -(6) = -6$ The opposite of $|-6|$ is -6.

Note that part (d) does not contradict the fact that the absolute value of a number cannot be negative. The expression $-|-6|$ calls for the *opposite* of an absolute value and so it must be negative.

For any two real numbers a and b, exactly one of the following orders must be true.

$$a < b, \quad a = b, \quad \text{or} \quad a > b.$$

This property of real numbers is called the **Law of Trichotomy.** In words, this property tells you that if a and b are any two real numbers, then a is less than b, a is equal to b, or a is greater than b.

Example 5   Comparing Real Numbers

Place the correct symbol ($<$, $>$, or $=$) between each pair of real numbers.

a. $|-2|$ ___ 1   **b.** -4 ___ $-|-4|$

c. $|12|$ ___ $|-15|$ **d.** $|-3|$ ___ 5

e. 2 ___ $-|-2|$ **f.** $-|-3|$ ___ -3

Solution

a. $|-2| > 1$, because $|-2| = 2$ and 2 is greater than 1.
b. $-4 = -|-4|$, because $-|-4| = -4$ and -4 is equal to -4.
c. $|12| < |-15|$, because $|12| = 12$ and $|-15| = 15$, and 12 is less than 15.
d. $|-3| < 5$, because $|-3| = 3$ and 3 is less than 5.
e. $2 > -|-2|$, because $-|-2| = -2$ and 2 is greater than -2.
f. $-|-3| = -3$, because $-|-3| = -3$ and -3 is equal to -3.

When the distance between the two real numbers a and b was defined to be $b - a$, we specified that a was less than or equal to b. Using absolute value, you can generalize this definition. That is, if a and b are *any* two real numbers, then the distance between a and b is given by

$$(\text{Distance between } a \text{ and } b) = |a - b|.$$

For instance, the distance between -2 and 1 is given by

$$|-2 - 1| = |-3| = 3. \qquad \text{Distance between } -2 \text{ and } 1$$

Discussing the Concept   Comparing Real Numbers

For each of the following lists of real numbers, arrange the entries in increasing order. Compare your lists with those of others in your class and resolve any differences.

a. $3, -\sqrt{7}, \sqrt{2}, \pi, -3, 1, \frac{22}{7}, -\frac{15}{7}$

b. $\left|-\frac{3}{5}\right|, -\left|\frac{3}{5}\right|, \left|-\frac{5}{3}\right|, -\left|\frac{5}{3}\right|, \left|-\frac{3}{4}\right|, -\left|\frac{3}{4}\right|, \left|-\frac{4}{5}\right|, -\left|\frac{4}{5}\right|$

Developing Skills

In Exercises 1–4, which of the real numbers in the set are (a) natural numbers, (b) integers, (c) rational numbers, and (d) irrational numbers?

1. $\left\{-10, -\sqrt{5}, -\frac{2}{3}, -\frac{1}{4}, 0, \frac{5}{8}, 1, \sqrt{3}, 4, 2\pi, 6\right\}$

2. $\left\{-\frac{7}{2}, -\sqrt{6}, -\frac{\pi}{2}, -\frac{3}{8}, 0, \sqrt{15}, \frac{10}{3}, 8, 245\right\}$

3. $\left\{-3.5, -\sqrt{4}, -\frac{1}{2}, -0.\overline{3}, 0, 3, \sqrt{5}, 3\pi, 25.2\right\}$

4. $\left\{-\sqrt{25}, -\sqrt{6}, -0.\overline{1}, -\frac{5}{3}, 0, 0.85, 3, 110\right\}$

In Exercises 5–8, list all members of the set.

5. The integers between -5.8 and 3.2

6. The even integers between -2.1 and 10.5

7. The odd integers between 0 and 3π

8. All prime numbers between 0 and 25

In Exercises 9 and 10, plot the points that represent the real numbers on the real number line. See Example 1.

9. (a) 3 (b) $\frac{5}{2}$ (c) $-\frac{7}{2}$ (d) -5.2

10. (a) 8 (b) $\frac{4}{3}$ (c) -6.75 (d) $-\frac{9}{2}$

In Exercises 11–14, approximate the numbers and order them.

11.

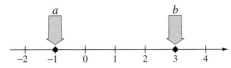

12.

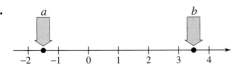

13.

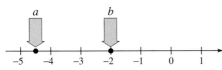

14.

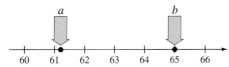

In Exercises 15–30, place the correct inequality symbol ($<$ or $>$) between the two numbers. See Example 2.

15. $2 \quad\quad 5$

16. $8 \quad\quad 3$

17. $10 \quad\quad 4$

18. $3.5 \quad\quad 8.5$

19. $-7 \quad\quad -2$

20. $-2 \quad\quad -5$

21. $-5 \quad\quad -2$

22. $-8 \quad\quad 3$

23. $\frac{1}{3} \quad\quad \frac{1}{4}$

24. $\frac{4}{5} \quad\quad 1$

25. $-\frac{5}{8} \quad\quad \frac{1}{2}$

26. $-\frac{3}{2} \quad\quad -\frac{5}{2}$

27. $-\frac{2}{3} \quad\quad -\frac{10}{3}$

28. $-\frac{5}{3} \quad\quad -\frac{3}{2}$

29. $2.75 \quad\quad \pi$

30. $-\pi \quad\quad -3.1$

In Exercises 31–42, find the distance between each pair of real numbers. See Example 3.

31. 4 and 10

32. 75 and 20

33. -12 and 7

34. -54 and 32

35. 18 and -32

36. 14 and -6

37. -8 and 0

38. 0 and 125

39. 0 and 35

40. -35 and 0

41. -6 and -9

42. -12 and -7

In Exercises 43–56, evaluate each expression. See Example 4.

43. $|10|$

44. $|62|$

45. $|-225|$

46. $|-14|$

47. $-|-85|$

48. $-|-36.5|$

49. $-|16|$

50. $-|-25|$

51. $-\left|-\frac{3}{4}\right|$

52. $-\left|\frac{3}{8}\right|$

53. $-|3.5|$

54. $|-1.4|$

55. $|-\pi|$

56. $-|\pi|$

In Exercises 57–64, place the correct symbol ($<$, $>$, or $=$) between the two real numbers. See Example 5.

57. $|-6| \quad\quad |2|$

58. $|-2| \quad\quad |2|$

59. $|47| \quad\quad |-27|$

60. $|150| \quad\quad |-310|$

61. $-|-16.8| \quad\quad -|16.8|$

62. $|12.5| \quad\quad -|-25|$

63. $\left|-\frac{3}{4}\right| \quad\quad -\left|\frac{4}{5}\right|$

64. $-\left|-\frac{7}{3}\right| \quad\quad -\left|\frac{1}{3}\right|$

In Exercises 65–74, find the opposite and the absolute value of the given number.

65. 34
66. 225
67. -160
68. -52
69. $-\frac{3}{11}$
70. $\frac{7}{32}$
71. $\frac{5}{4}$
72. $\frac{4}{3}$
73. 4.7
74. -0.4

In Exercises 75–86, plot the number and its opposite on the real number line. Determine the distance of each from 0.

75. 7
76. -3
77. -5
78. 6
79. $-\frac{3}{5}$
80. $\frac{7}{4}$
81. $\frac{5}{3}$
82. $-\frac{3}{4}$
83. -4.25
84. 3.5
85. 0.7
86. -1.2

In Exercises 87–94, write the statement using inequality notation.

87. x is negative.
88. y is more than 25.
89. x is nonnegative.
90. u is at least 16.
91. z is greater than 2 and no more than 10.
92. The tire pressure p is at least 30 pounds per square inch and no more than 35 pounds per square inch.
93. The price p is less than \$225.
94. The Dow Jones Average A will exceed 10,000.

Explaining Concepts

In Exercises 95–100, decide whether the statement is true or false. If the statement is false, give an example of a real number that makes the statement false.

95. Every integer is a rational number.
96. Every real number is either rational or irrational.
97. Every rational number is an integer.
98. Because $|a|$ is a nonnegative real number for any real number a, there exists no real number a such that $|a| = -a$.
99. If x and y are real numbers, $|x + y| = |x| + |y|$.
100. $\frac{1}{6} = 0.17$

101. Describe the difference between the set of natural numbers and the set of integers.
102. Describe the difference between the rational numbers 0.15 and $0.\overline{15}$.
103. Is there a difference between saying that a real number is positive and saying that a real number is nonnegative? Explain your answer.
104. Which real number lies farther from -4: -8 or 6? Explain your answer.
105. If you are given two real numbers a and b, how can you tell which is greater?

P.2 Operations with Real Numbers

Objectives

1 Review addition, subtraction, multiplication, and division of real numbers.

2 Write repeated multiplication in exponential form and evaluate an exponential expression.

3 Use order of operations to evaluate an expression.

4 Evaluate an expression using a calculator and order of operations.

1 Review addition, subtraction, multiplication, and division of real numbers.

Operations with Real Numbers

This section reviews the four basic operations of arithmetic: addition, subtraction, multiplication, and division.

▶ **Addition of Two Real Numbers**

1. To add two real numbers with *like signs*, add their absolute values and attach the common sign to the result.

2. To add two real numbers with *unlike signs*, subtract the smaller absolute value from the greater absolute value and attach the sign of the number with the greater absolute value.

The result of adding two real numbers is the **sum** of the two numbers, and the two real numbers are the **terms** of the sum.

Example 1   Adding Integers and Decimals

a. $-84 + 14 = -(84 - 14)$ Choose negative sign.

$ = -70$ Subtract absolute values.

b. $-138 + (-62) = -(138 + 62)$ Use common sign.

$ = -200$ Add absolute values.

c. $3.2 + (-0.4) = +(3.2 - 0.4)$ Choose positive sign.

$ = 2.8$ Subtract absolute values.

▶ **Subtraction of Two Real Numbers**

To **subtract** the real number b from the real number a, add the opposite of b to a. That is,

$$a - b = a + (-b).$$

The result of this subtraction is the **difference** of a and b.

Example 2 Subtracting Integers and Decimals

a. $-25 - (-27) = -25 + 27$ Add the opposite of -27.

$= +(27 - 25)$ Choose positive sign.

$= 2$ Subtract absolute values.

b. $-13.8 - 7.02 = -13.8 + (-7.02) = -(13.8 + 7.02) = -20.82$

Study Tip

Here is an alternative method for adding and subtracting fractions with unlike denominators.

$$\frac{a}{c} + \frac{b}{d} = \frac{ad + bc}{cd}$$

$$\frac{a}{c} - \frac{b}{d} = \frac{ad - bc}{cd}$$

For example,

$$\frac{1}{6} + \frac{3}{8} = \frac{1(8)}{6(8)} + \frac{3(6)}{8(6)}$$

$$= \frac{8}{48} + \frac{18}{48}$$

$$= \frac{8 + 18}{48}$$

$$= \frac{26}{48}$$

$$= \frac{13}{24}.$$

Note that an additional step is needed to simplify the fraction after the numerators have been added.

▶ **Addition and Subtraction of Fractions**

1. *Like Denominators:* The sum and difference of two fractions with like denominators are as follows.

$$\frac{a}{c} + \frac{b}{c} = \frac{a + b}{c} \qquad \frac{a}{c} - \frac{b}{c} = \frac{a - b}{c}$$

2. *Unlike Denominators:* To add or subtract two fractions with unlike denominators, first rewrite the fractions so that they have the same denominator and apply the first rule.

To find the **least common denominator (LCD)** for two or more fractions, find the **least common multiple (LCM)** of their denominators. For instance, the LCM of 6 and 8 is 24. To see this, consider all multiples of 6 (6, 12, 18, 24, 30, 36, 42, 48, . . .) and all multiples of 8 (8, 16, 24, 32, 40, 48, . . .). The numbers 24 and 48 are common multiples, and the number 24 is the smallest of the common multiples. To add $\frac{1}{6}$ and $\frac{3}{8}$, proceed as follows.

$$\frac{1}{6} + \frac{3}{8} = \frac{1(4)}{6(4)} + \frac{3(3)}{8(3)} = \frac{4}{24} + \frac{9}{24} = \frac{4 + 9}{24} = \frac{13}{24}$$

Example 3 Adding and Subtracting Fractions

a. $\dfrac{5}{17} + \dfrac{9}{17} = \dfrac{5 + 9}{17}$ Add numerators.

$= \dfrac{14}{17}$ Simplify.

b. $\dfrac{3}{8} - \dfrac{5}{12} = \dfrac{3(3)}{8(3)} - \dfrac{5(2)}{12(2)}$ Least common denominator is 24.

$= \dfrac{9}{24} - \dfrac{10}{24}$ Simplify.

$= \dfrac{9 - 10}{24}$ Subtract numerators.

$= -\dfrac{1}{24}$ Simplify.

Study Tip

A quick way to convert the mixed number $1\frac{4}{5}$ into the fraction $\frac{9}{5}$ is to multiply the whole number by the denominator of the fraction and add the result to the numerator, as follows.

$$1\frac{4}{5} = \frac{1(5) + 4}{5} = \frac{9}{5}$$

Example 4   Adding Mixed Numbers

$$1\frac{4}{5} + \frac{11}{7} = \frac{9}{5} + \frac{11}{7}$$ Write $1\frac{4}{5}$ as $\frac{9}{5}$.

$$= \frac{9(7)}{5(7)} + \frac{11(5)}{7(5)}$$ Least common denominator is 35.

$$= \frac{63}{35} + \frac{55}{35}$$ Simplify.

$$= \frac{63 + 55}{35}$$ Add numerators.

$$= \frac{118}{35}$$ Simplify.

Multiplication of two real numbers can be described as repeated addition. For instance, 7×3 can be described as $3 + 3 + 3 + 3 + 3 + 3 + 3$. Multiplication is denoted in a variety of ways. For instance,

$$7 \times 3, \quad 7 \cdot 3, \quad 7(3), \quad \text{and} \quad (7)(3)$$

all denote the product "7 times 3."

▶ **Multiplication of Two Real Numbers**

The result of multiplying two real numbers is their **product,** and each of the two numbers is a **factor** of the product.

1. To multiply two real numbers with *like signs*, find the product of their absolute values. The product is *positive*.

2. To multiply two real numbers with *unlike signs*, find the product of their absolute values, and attach a minus sign. The product is *negative*.

3. The product of zero and any other real number is zero.

Study Tip

To find the product of more than two numbers, first find the product of their absolute values. If there is an *even* number of negative factors, as in Example 5(c), the product is positive. If there is an *odd* number of negative factors, the product is negative.

Example 5   Multiplying Integers

Unlike signs

a. $-6 \cdot 9 = -54$ The product is negative.

Like signs

b. $(-5)(-7) = 35$ The product is positive.

Like signs

c. $5(-3)(-4)(7) = 420$ The product is positive.

Like signs

Study Tip

When operating with fractions, you should check to see whether your answers can be simplified by canceling (or dividing out) factors that are common to the numerator and denominator. For instance, the fraction $\frac{4}{6}$ can be written in simplified form as

$$\frac{4}{6} = \frac{\overset{1}{\cancel{2}} \cdot 2}{\underset{1}{\cancel{2}} \cdot 3} = \frac{2}{3}.$$

Note that canceling a common factor is the division of a number by itself, and that what remains is a factor of 1.

▶ **Multiplication of Two Fractions**

The product of the two fractions a/c and b/d is given by

$$\frac{a}{c} \cdot \frac{b}{d} = \frac{ab}{cd}.$$

Example 6   Multiplying Fractions

$$\left(-\frac{3}{8}\right)\left(\frac{11}{6}\right) = -\frac{3(11)}{8(6)}$$ Multiply numerators and denominators.

$$= -\frac{3(11)}{8(2)(3)}$$ Factor and cancel common factor.

$$= -\frac{11}{16}$$ Simplify.

Study Tip

Division by 0 is not defined because 0 has no reciprocal. If 0 had a reciprocal value b, then you would obtain the *false* result

$$\frac{1}{0} = b$$ The reciprocal of zero is b.

$$1 = b \cdot 0$$ Multiply both sides by 0.

$$1 = 0.$$ False result, $1 \neq 0$

▶ **Division of Two Real Numbers**

To divide the real number a by the nonzero real number b, multiply a by the **reciprocal** of b. That is,

$$a \div b = a \cdot \frac{1}{b}.$$

The result of dividing two real numbers is the **quotient** of the numbers. The number a is the **dividend** and the number b is the **divisor**. Using the symbols a/b or $\frac{a}{b}$, a is the **numerator** and b is the **denominator.**

Example 7   Division of Real Numbers

a. $-30 \div 5 = -30 \cdot \frac{1}{5}$ Invert divisor and multiply.

$$= -\frac{30}{5}$$ Multiply.

$$= -\frac{6 \cdot 5}{5}$$ Factor and cancel common factor.

$$= -6$$ Simplify.

b. $\frac{5}{16} \div 2\frac{3}{4} = \frac{5}{16} \div \frac{11}{4} = \frac{5}{16} \cdot \frac{4}{11} = \frac{5(4)}{16(11)} = \frac{5}{44}$

Positive Integer Exponents

Just as multiplication by a positive integer can be described as repeated addition, *repeated multiplication* can be written in what is called **exponential form.** Here is an example.

Repeated Multiplication   *Exponential Form*

$$\underbrace{7 \cdot 7 \cdot 7 \cdot 7}_{\text{4 factors of 7}} = 7^4$$

$$\underbrace{\left(-\frac{3}{4}\right)\left(-\frac{3}{4}\right)\left(-\frac{3}{4}\right)}_{\text{3 factors of } -\frac{3}{4}} = \left(-\frac{3}{4}\right)^3$$

▶ **Exponential Notation**

Let n be a positive integer and let a be a real number. Then the product of n factors of a is given by

$$a^n = \underbrace{a \cdot a \cdot a \cdots a}_{n \text{ factors}}.$$

In the exponential form a^n, a is the **base** and n is the **exponent.** Writing the exponential form a^n is called **"raising a to the nth power."**

When a number is raised to the *first* power, you do not need to write the exponent 1. For instance, write 5 rather than 5^1. Raising a number to the *second* power is called **squaring** the number. Raising a number to the *third* power is called **cubing** the number.

Example 8   Evaluating Exponential Expressions

a. $(-3)^4 = (-3)(-3)(-3)(-3) = 81$ Negative sign is part of the base.

b. $-3^4 = -(3)(3)(3)(3) = -81$ Negative sign is not part of the base.

c. $\left(\frac{2}{5}\right)^3 = \left(\frac{2}{5}\right)\left(\frac{2}{5}\right)\left(\frac{2}{5}\right) = \frac{8}{125}$

d. $(-2)^5 = (-2)(-2)(-2)(-2)(-2) = -32$

e. $(-2)^6 = (-2)(-2)(-2)(-2)(-2)(-2) = 64$

In parts (a) and (b) of Example 8, be sure you see the distinction between the expressions $(-3)^4$ and -3^4. Here is a similar example.

$$(-5)^2 = (-5)(-5) = 25 \qquad \text{Negative sign is part of the base.}$$

$$-5^2 = -(5)(5) = -25 \qquad \text{Negative sign is not part of the base.}$$

3 Use order of operations to evaluate an expression.

Order of Operations

One of your goals in studying this book is to learn to communicate about algebra by reading and writing information about numbers. One way to help avoid confusion when communicating algebraic ideas is to establish an **order of operations.** This is done by giving priorities to different operations. First priority is given to exponents, second priority is given to multiplication and division, and third priority is given to addition and subtraction. To distinguish between operations with the same priority, use the *Left-to-Right Rule*.

Study Tip

The order of operations for multiplication applies when multiplication is written with the symbols $\times$ or $\cdot$. When multiplication is implied by parentheses, it has a higher priority than the Left-to-Right Rule. For instance,

$$8 \div 4(2) = 8 \div 8 = 1$$

but

$$8 \div 4 \cdot 2 = 2 \cdot 2 = 4.$$

▶ **Order of Operations**

To evaluate an expression involving more than one operation, use the following order.

1. First do operations that occur within symbols of grouping.

2. Then evaluate powers.

3. Then do multiplications and divisions from left to right.

4. Finally, do additions and subtractions from left to right.

Example 9 Order of Operations Without Symbols of Grouping

a. $20 - 2 \cdot 3^2 = 20 - 2 \cdot 9 = 20 - 18 = 2$

b. $5 - 6 - 2 = (5 - 6) - 2 = -1 - 2 = -3$

c. $8 \div 2 \cdot 2 = (8 \div 2) \cdot 2 = 4 \cdot 2 = 8$

When you want to change the established order of operations, you must use parentheses or other symbols of grouping. Part (d) in the next example shows that a fraction bar acts like a symbol of grouping.

Example 10 Order of Operations with Symbols of Grouping

a. $7 - 3(4 - 2) = 7 - 3(2) = 7 - 6 = 1$

b. $4 - 3(2)^3 = 4 - 3(8) = 4 - 24 = -20$

c. $1 - [4 - (5 - 3)] = 1 - (4 - 2) = 1 - 2 = -1$

d. $\dfrac{2 \cdot 5^2 - 10}{3^2 - 4} = (2 \cdot 5^2 - 10) \div (3^2 - 4)$ Rewrite using parentheses.

$= (50 - 10) \div (9 - 4)$ Evaluate powers and multiply within symbols of grouping.

$= 40 \div 5$ Subtract within symbols of grouping.

$= 8$ Divide.

4 Evaluate an expression using a calculator and order of operations.

Calculators and Order of Operations

When using your own calculator, be sure that you are familiar with the use of each of the keys. For each of the calculator examples in the text, we will give two possible keystroke sequences: one for a standard *scientific* calculator, and one for a *graphing* calculator.

Study Tip

Be sure you see the difference between the change sign key $+/-$ and the subtraction key $-$ on a scientific calculator. Also notice the difference between the minus key $(-)$ and the subtraction key $-$ on a graphing calculator.

| Example 11 | Evaluating Expressions on a Calculator |

a. To evaluate the expression $7 - (5 \cdot 3)$, use the following keystrokes.

Keystrokes	Display	
7 $-$ $($ 5 $\times$ 3 $)$ $=$	-8	Scientific
7 $-$ $($ 5 $\times$ 3 $)$ ENTER	-8	Graphing

b. To evaluate the expression $(-3)^2 + 4$, use the following keystrokes.

Keystrokes	Display	
3 $+/-$ x^2 $+$ 4 $=$	13	Scientific
$($ $(-)$ 3 $)$ x^2 $+$ 4 ENTER	13	Graphing

c. To evaluate the expression $5/(4 + 3 \cdot 2)$ use the following keystrokes.

Keystrokes	Display	
5 $\div$ $($ 4 $+$ 3 $\times$ 2 $)$ $=$	0.5	Scientific
5 $\div$ $($ 4 $+$ 3 $\times$ 2 $)$ ENTER	.5	Graphing

Some calculators follow the established order of operations and some do not. To see whether yours does, try entering $30 - 5 \times 4$. If your calculator follows the established order of operations, it will display 10. If it does not, it will display 100.

The Granger Collection

Abacus

Next to fingers, the *abacus* is the oldest calculating device known. It can be used to add, subtract, multiply, divide, and calculate square and cube roots. Different forms of the abacus were used by Egyptians, Greeks, Romans, Hindus, and Chinese.

| Discussing the Concept | Evaluating Expressions |

Decide which of the following expressions are equal to 27 when you follow the standard order of operations.

a. $40 - 10 + 3$ **b.** $5^2 + \frac{1}{2} \cdot 4$

c. $8 \cdot 3 + 30 \div 2$ **d.** $75 \div 2 + 1 + 2$

e. $9 \cdot 4 - 18 \div 2$ **f.** $7 \cdot 4 - 4 - 5$

For the expressions that are not equal to 27, see if you can discover a way to insert symbols of grouping (parentheses, brackets, and absolute value symbols) that make the expression equal to 27. Discuss the value of symbols of grouping in mathematical communication.

P.2   Exercises

Developing Skills

In Exercises 1–30, evaluate the expression. See Examples 1–4.

1. $13 + 32$

2. $16 + 84$

3. $-13 + 32$

4. $-16 + 84$

5. $13 + (-32)$

6. $16 + (-84)$

7. $-7 - 15$

8. $-22 - 6$

9. $-13 + (-8)$

10. $-5 + (-52)$

11. $4 - 16 + (-8)$

12. $-15 + (-6) + 32$

13. $5.8 - 6.2 + 1.1$

14. $46.08 - 35.1 - 16.25$

15. $\frac{3}{8} + \frac{7}{8}$

16. $\frac{5}{6} + \frac{7}{6}$

17. $\frac{3}{4} - \frac{1}{4}$

18. $\frac{5}{9} - \frac{1}{9}$

19. $\frac{3}{5} + \left(-\frac{1}{2}\right)$

20. $\frac{5}{6} - \frac{3}{4}$

21. $\frac{5}{8} + \frac{1}{4} - \frac{5}{6}$

22. $\frac{3}{11} + \frac{-5}{2}$

23. $5\frac{3}{4} + 7\frac{3}{8}$

24. $8\frac{1}{2} - 4\frac{2}{3}$

25. $85 - |-25|$

26. $-36 + |-8|$

27. $-(-11.325) + |34.625|$

28. $|-16.25| - 54.78$

29. $-|-15.667| - 12.333$

30. $-\left|-15\frac{2}{3}\right| - 12\frac{1}{3}$

In Exercises 31–34, write the expression as a repeated addition.

31. $4(5)$

32. $5(2)$

33. $3(-4)$

34. $6(-2)$

In Exercises 35–38, write the expression as a multiplication problem.

35. $\frac{1}{4} + \frac{1}{4} + \frac{1}{4} + \frac{1}{4} + \frac{1}{4} + \frac{1}{4}$

36. $9 + 9 + 9 + 9$

37. $(-15) + (-15) + (-15) + (-15)$

38. $\left(-\frac{5}{22}\right) + \left(-\frac{5}{22}\right) + \left(-\frac{5}{22}\right)$

In Exercises 39–52, evaluate the product. See Examples 5 and 6.

39. $5(-6)$

40. $-7(3)$

41. $(-8)(-6)$

42. $(-4)(-7)$

43. $-6(12)$

44. $7(10)$

45. $\left(-\frac{5}{8}\right)\left(-\frac{4}{5}\right)$

46. $\left(\frac{10}{13}\right)\left(-\frac{3}{5}\right)$

47. $-\frac{3}{2}\left(\frac{8}{5}\right)$

48. $\left(-\frac{4}{7}\right)\left(-\frac{4}{5}\right)$

49. $\frac{1}{2}\left(\frac{1}{6}\right)$

50. $\frac{1}{3}\left(\frac{2}{3}\right)$

51. $-\frac{9}{8}\left(\frac{16}{27}\right)\left(\frac{1}{2}\right)$

52. $\frac{2}{3}\left(-\frac{18}{5}\right)\left(-\frac{5}{6}\right)$

In Exercises 53–66, evaluate the expression. See Example 7.

53. $\frac{-18}{-3}$

54. $-\frac{30}{-15}$

55. $-48 \div 16$

56. $-27 \div (-9)$

57. $63 \div (-7)$

58. $-72 \div 12$

59. $-\frac{4}{5} \div \frac{8}{25}$

60. $\frac{8}{15} \div \frac{32}{5}$

61. $\left(-\frac{1}{3}\right) \div \left(-\frac{5}{6}\right)$

62. $-\frac{11}{12} \div \frac{5}{24}$

63. $5\frac{3}{4} \div 2\frac{1}{8}$

64. $-3\frac{5}{6} \div -2\frac{2}{3}$

65. $4\frac{1}{8} \div 3\frac{3}{2}$

66. $26\frac{2}{3} \div 10\frac{5}{6}$

In Exercises 67–72, write the expression as a repeated multiplication problem.

67. 4^3

68. $(-6)^5$

69. $\left(-\frac{3}{4}\right)^4$

70. $\left(\frac{2}{3}\right)^3$

71. $(-0.8)^6$

72. $(0.67)^4$

In Exercises 73–78, write the expression using exponential notation.

73. $(-7) \cdot (-7) \cdot (-7)$

74. $(-4)(-4)(-4)(-4)(-4)(-4)$

75. $(-5)(-5)(-5)(-5)$

76. $\left(\frac{5}{8}\right) \cdot \left(\frac{5}{8}\right) \cdot \left(\frac{5}{8}\right) \cdot \left(\frac{5}{8}\right)$

77. $-(7 \cdot 7 \cdot 7)$

78. $-(5 \cdot 5 \cdot 5 \cdot 5 \cdot 5 \cdot 5)$

In Exercises 79–92, evaluate the expression. See Example 8.

79. $(-2)^4$

80. $(-3)^2$

81. $(-2)^3$

82. $(-3)^3$

83. -4^3

84. -3^4

85. $\left(\frac{4}{5}\right)^3$

86. $-\left(\frac{2}{3}\right)^4$

87. $-\left(-\frac{1}{2}\right)^5$ **88.** $\left(\frac{3}{4}\right)^3$

89. $(0.3)^3$ **90.** $(0.2)^4$

91. $5(-0.4)^3$ **92.** $-3(0.8)^2$

In Exercises 93–114, evaluate the expression. See Examples 9 and 10.

93. $16 - 6 - 10$ **94.** $18 - 12 + 4$

95. $24 - 5 \cdot 2^2$ **96.** $18 + 3^2 - 12$

97. $28 \div 4 + 3 \cdot 5$ **98.** $6 \cdot 7 - 6^2 \div 4$

99. $14 - 2(8 - 4)$ **100.** $21 - 5(7 - 5)$

101. $45 + 3(16 \div 4)$ **102.** $72 - 8(6^2 \div 9)$

103. $2 + [8 - (14 \div 2)]$

104. $18 - [4 + (17 - 12)]$

105. $5^2 - 2[9 - (18 - 8)]$ **106.** $8 \cdot 3^2 - 4(12 + 3)$

107. $5^3 + |-14 + 4|$ **108.** $|(-2)^5| - (25 + 7)$

109. $\dfrac{8 + 7}{12 - 15}$ **110.** $\dfrac{9 + 6(2)}{3 + 4}$

111. $\dfrac{4^2 - 5}{11} - 7$ **112.** $\dfrac{5^3 - 50}{-15} + 27$

113. $\dfrac{6 \cdot 2^2 - 12}{3^2 + 3}$ **114.** $\dfrac{7^2 - 2(11)}{5^2 + 8(-2)}$

In Exercises 115–120, evaluate the expression using a calculator. Round the result to two decimal places. See Example 11.

115. $5.6[13 - 2.5(-6.3)]$ **116.** $35(1032 - 4650)$

117. $5^6 - 3(400)$ **118.** $300(1.09)^{10}$

119. $\dfrac{500}{(1.055)^{20}}$ **120.** $5(100 - 3.6^4) \div 4.1$

Solving Problems

Circle Graphs In Exercises 121 and 122, find the unknown fractional part of the circle graph.

121.
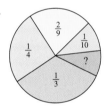

122.
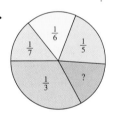

123. *Balance in an Account* During one month, you made the following transactions in your checking account.

Initial Balance:	$2618.68
Deposit:	$1236.45
Withdrawal:	$25.62
Withdrawal:	$455.00
Withdrawal:	$125.00
Withdrawal:	$715.95

Find the balance at the end of the month. (Disregard any interest that may have been earned.)

124. *Company Profit* The midyear financial statement of a company showed a profit of $1,415,322.62. At the close of the year, the financial statement showed a profit for the year of $916,489.26. Find the profit (or loss) of the company for the second 6 months of the year.

125. *Organizing Data* On Monday you purchased $500 worth of stock. The value of the stock during the remainder of the week is shown in the bar graph.

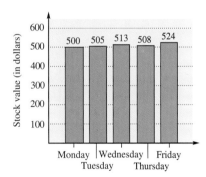

(a) Use the graph to complete the table.

Day	Daily gain or loss
Tuesday	
Wednesday	
Thursday	
Friday	

(b) Find the sum of the daily gains and losses. Interpret the result in the context of the problem. How could you determine this sum from the graph?

126. *Organizing Data* The annual profits for a company (in millions of dollars) for the years 1994 to 1999 are shown in the bar graph. Use the graph to create a table that shows the yearly gains or losses.

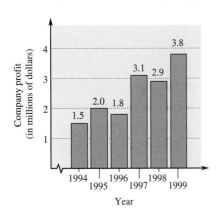

127. *Savings Plan*

(a) If you save $50 per month for 18 years, how much money will you set aside during the 18 years?

(b) If the money in part (a) is deposited in a savings account earning 9% interest compounded monthly, the total amount in the fund after 18 years will be

$$50\left[\left(1 + \frac{0.09}{12}\right)^{216} - 1\right]\left(1 + \frac{12}{0.09}\right).$$

Use a calculator to determine this amount.

(c) How much of the amount in part (b) is earnings from interest?

128. *Savings Plan*

(a) If you save $75 per month for 30 years, how much money will you set aside during the 30 years?

(b) If the money in part (a) is deposited in a savings account earning 8% interest compounded monthly, the total amount in the fund after 30 years will be

$$75\left[\left(1 + \frac{0.08}{12}\right)^{360} - 1\right]\left(1 + \frac{12}{0.08}\right).$$

Use a calculator to determine this amount.

(c) How much of the amount in part (b) is earnings from interest?

Area In Exercises 129–132, find the area of the figure. (The area of a rectangle is $A = lw$, and the area of a triangle is $A = \frac{1}{2}bh$.)

129.

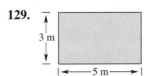

130.

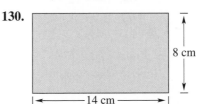

131.

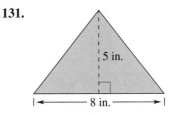

132.

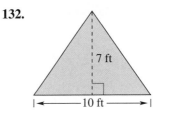

Volume In Exercises 133 and 134, use the following information. A bale of hay is a rectangular solid having the dimensions shown in the figure and weighing approximately 50 pounds. (The volume of a rectangular solid is $V = lwh$.)

133. Find the volume of a bale of hay in cubic feet if 1728 cubic inches equals 1 cubic foot.

134. Approximate the number of bales in a ton of hay. Then approximate the volume of a stack of baled hay that weighs 12 tons.

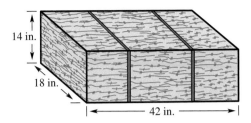

14 in.

18 in.

42 in.

Figure for 133 and 134

Explaining Concepts

135. Answer parts (a)–(c) of Motivating the Chapter on page 1.

True or False?   In Exercises 136–140, determine whether the statement is true or false. Explain your reasoning.

136. The reciprocal of every nonzero integer is an integer.

137. The reciprocal of every nonzero rational number is a rational number.

138. If a negative real number is raised to the 12th power, the result will be positive.

139. If a negative real number is raised to the 11th power, the result will be positive.

140. $a \div b = b \div a$

141. Can the sum of two real numbers be less than either number? If so, give an example.

142. Explain how to subtract one real number from another.

143. In your own words, describe the rules for determining the sign of the product or the quotient of two real numbers.

144. If $a > 0$, state the values of n such that
$$(-a)^n = -a^n.$$

145. In your own words, describe the established order of operations. Without these priorities, explain why the expression $6 - 5 - 2$ would be ambiguous.

Error Analysis   In Exercises 146–149, explain the error.

146. $\dfrac{2}{3} + \dfrac{3}{2} = \dfrac{2 + 3}{3 + 2} = 1$

147. $\dfrac{5 + 12}{5} = \dfrac{5 + 12}{5} = 12$

148. $\dfrac{28}{83} = \dfrac{28}{83} = \dfrac{2}{3}$

149. $3 \cdot 4^2 = 12^2$

Objectives

1 Identify and use the properties of real numbers.

2 Develop additional properties of real numbers.

1 Identify and use the properties of real numbers.

Basic Properties of Real Numbers

The following list summarizes the basic properties of addition and multiplication. Although the examples involve real numbers, you will learn later that these properties can also be applied to algebraic expressions.

Study Tip

The operations of subtraction and division are not listed at the right because they do not have many of the properties of real numbers. For instance, subtraction and division are not commutative or associative. To see this, consider the following.

$$4 - 3 \neq 3 - 4$$

$$15 \div 5 \neq 5 \div 15$$

$$8 - (6 - 2) \neq (8 - 6) - 2$$

$$20 \div (4 \div 2) \neq (20 \div 4) \div 2$$

▶ **Properties of Real Numbers**

Let a, b, and c represent real numbers, variables, or algebraic expressions.

Property	Example
Commutative Property of Addition: $a + b = b + a$	$3 + 5 = 5 + 3$
Commutative Property of Multiplication: $ab = ba$	$2 \cdot 7 = 7 \cdot 2$
Associative Property of Addition: $(a + b) + c = a + (b + c)$	$(4 + 2) + 3 = 4 + (2 + 3)$
Associative Property of Multiplication: $(ab)c = a(bc)$	$(2 \cdot 5) \cdot 7 = 2 \cdot (5 \cdot 7)$
Distributive Property: $a(b + c) = ab + ac$	$4(7 + 3) = 4 \cdot 7 + 4 \cdot 3$
$(a + b)c = ac + bc$	$(2 + 5)3 = 2 \cdot 3 + 5 \cdot 3$
$a(b - c) = ab - ac$	$6(5 - 3) = 6 \cdot 5 - 6 \cdot 3$
$(a - b)c = ac - bc$	$(7 - 2)4 = 7 \cdot 4 - 2 \cdot 4$
Additive Identity Property: $a + 0 = a$	$4 + 0 = 4$
Multiplicative Identity Property: $a \cdot 1 = 1 \cdot a = a$	$-5 \cdot 1 = 1 \cdot (-5) = -5$
Additive Inverse Property: $a + (-a) = 0$	$3 + (-3) = 0$
Multiplicative Inverse Property: $a \cdot \dfrac{1}{a} = 1, \quad a \neq 0$	$5 \cdot \dfrac{1}{5} = 1$

Example 1 Identifying Properties of Real Numbers

Name the property of real numbers that justifies the statement.

a. $4(a + 3) = 4 \cdot a + 4 \cdot 3$ **b.** $6 \cdot \dfrac{1}{6} = 1$

c. $-3 + (2 + b) = (-3 + 2) + b$

d. $(b + 8) + 0 = b + 8$

Solution

a. This is an example of the Distributive Property.

b. This is an example of the Multiplicative Inverse Property.

c. This is an example of the Associative Property of Addition.

d. This is an example of the Additive Identity Property, where $(b + 8)$ is an algebraic expression.

The properties of real numbers make up the third component of what is called a **mathematical system.** These three components are a *set of numbers* (Section P.1), *operations* with the set of numbers (Section P.2), and *properties* of the operations with the numbers (Section P.3).

Set of Numbers ⟹ Operations with the Numbers ⟹ Properties of the Operations

Be sure you see that the properties of real numbers can be applied to variables and algebraic expressions as well as to real numbers.

Example 2 Using the Properties of Real Numbers

Complete each statement using the specified property of real numbers.

a. Multiplicative Identity Property: $(4a)1 =$

b. Associative Property of Addition: $(b + 8) + 3 =$

c. Additive Inverse Property: $0 = 5c +$

d. Distributive Property: $4 \cdot b + 4 \cdot 5 =$

Solution

a. By the Multiplicative Identity Property, $(4a)1 = 4a$.

b. By the Associative Property of Addition, $(b + 8) + 3 = b + (8 + 3)$.

c. By the Additive Inverse Property, $0 = 5c + (-5c)$.

d. By the Distributive Property, $4 \cdot b + 4 \cdot 5 = 4(b + 5)$.

To help you understand each property of real numbers, try stating the properties in your own words. For instance, the Associative Property of Addition can be stated as follows: *When three real numbers are added, it makes no difference which two are added first.*

2 Develop additional properties of real numbers.

Additional Properties of Real Numbers

Once you have determined the basic properties (or *axioms*) of a mathematical system, you can go on to develop other properties. These additional properties are **theorems,** and the formal arguments that justify the theorems are **proofs.**

▶ **Additional Properties of Real Numbers**

Let a, b, and c be real numbers, variables, or algebraic expressions.

Properties of Equality

Addition Property of Equality: If $a = b$, then $a + c = b + c$.

Multiplication Property of Equality: If $a = b$, then $ac = bc$.

Cancellation Property of Addition: If $a + c = b + c$, then $a = b$.

Cancellation Property of Multiplication: If $ac = bc$ and $c \neq 0$, then $a = b$.

Properties of Zero

Multiplication Property of Zero: $0 \cdot a = 0$

Division Property of Zero: $\dfrac{0}{a} = 0, \quad a \neq 0$

Division by Zero Is Undefined: $\dfrac{a}{0}$ is undefined.

Properties of Negation

Multiplication by -1: $(-1)(a) = -a,$
                                          $(-1)(-a) = a$

Placement of Negative Signs: $(-a)(b) = -(ab) = (a)(-b)$

Product of Two Opposites: $(-a)(-b) = ab$

Study Tip

When you are using the properties of real numbers in actual applications, the process is usually less formal than it would appear from the list of properties on this page. For instance, the steps shown at the right are less formal than those shown in Examples 5 and 6 on page 26. The importance of the properties is that they can be used to justify the steps of a solution. They do not always need to be listed for *every* step of the solution.

In Section 1.1, you will see that the properties of equality are useful for solving equations, as shown below. Note that the Addition and Multiplication Properties of Equality can be used to subtract or divide both sides of an equation by the same nonzero quantity.

$5x + 4 = -2x + 18$	Original equation
$5x + 4 - 4 = -2x + 18 - 4$	Subtract 4 from both sides.
$5x = -2x + 14$	Simplify.
$5x + 2x = -2x + 2x + 14$	Add $2x$ to both sides.
$7x = 14$	Simplify.
$\dfrac{7x}{7} = \dfrac{14}{7}$	Divide both sides by 7.
$x = 2$	Simplify.

Each of the additional properties in the list on page 24 can be proved using the basic properties of real numbers. Examples 3 and 4 illustrate such proofs.

Example 3   Proof of the Cancellation Property of Addition

Prove that if $a + c = b + c$, then $a = b$.

Solution

Notice how each step is justified from the previous step by means of a property of real numbers.

$$a + c = b + c \qquad\qquad \text{Original equation}$$

$$(a + c) + (-c) = (b + c) + (-c) \qquad\qquad \text{Addition Property of Equality}$$

$$a + [c + (-c)] = b + [c + (-c)] \qquad\qquad \text{Associative Property of Addition}$$

$$a + 0 = b + 0 \qquad\qquad \text{Additive Inverse Property}$$

$$a = b \qquad\qquad \text{Additive Identity Property}$$

Example 4   Proof of a Property of Negation

Prove that $(-1)a = -a$.

Solution

At first glance, it is a little difficult to see what you are being asked to prove. However, a good way to start is to consider carefully the definitions of the three numbers in the equation.

$$a = \text{given real number}$$

$$-1 = \text{the additive inverse of } 1$$

$$-a = \text{the additive inverse of } a$$

Now, by showing that $(-1)a$ has the same properties as the additive inverse of a, you will be showing that $(-1)a$ must be the additive inverse of a.

$$(-1)a + a = (-1)a + (1)(a) \qquad\qquad \text{Multiplicative Identity Property}$$

$$= (-1 + 1)a \qquad\qquad \text{Distributive Property}$$

$$= (0)a \qquad\qquad \text{Additive Inverse Property}$$

$$= 0 \qquad\qquad \text{Multiplication Property of Zero}$$

Because $(-1)a + a = 0$, you can use the fact that $-a + a = 0$ to conclude that $(-1)a + a = -a + a$. From this, you can complete the proof as follows.

$$(-1)a + a = -a + a \qquad\qquad \text{Shown in first part of proof}$$

$$(-1)a = -a \qquad\qquad \text{Cancellation Property of Addition}$$

The list of additional properties of real numbers on page 24 forms a very important part of algebra. Knowing the names of the properties is useful, but knowing how to use each property is extremely important. The next two examples show how several of the properties can be used to solve equations. (You will study these techniques in detail in Section 1.1.)

Example 5   Applying the Properties of Real Numbers

In the solution of the equation $b + 2 = 6$, identify the property of real numbers that justifies each step.

Solution

$b + 2 = 6$	Original equation
$(b + 2) + (-2) = 6 + (-2)$	Addition Property of Equality
$b + [2 + (-2)] = 6 - 2$	Associative Property of Addition
$b + 0 = 4$	Additive Inverse Property
$b = 4$	Additive Identity Property

Example 6   Applying the Properties of Real Numbers

In the solution of the equation $3x = 15$, identify the property of real numbers that justifies each step.

Solution

$3x = 15$	Original equation
$\left(\dfrac{1}{3}\right)3x = \left(\dfrac{1}{3}\right)15$	Multiplication Property of Equality
$\left(\dfrac{1}{3} \cdot 3\right)x = 5$	Associative Property of Multiplication
$(1)(x) = 5$	Multiplicative Inverse Property
$x = 5$	Multiplicative Identity Property

Discussing the Concept   Error Analysis

One of your classmates argues that the following statement is true because of the Associative Property of Addition.

$$(7 - 3) + 4 = 7 - (3 + 4)$$

Discuss why this statement is incorrect. Can you explain how to alter the statement so that the Associative Property of Addition can be correctly applied?

P.3 Exercises

Developing Skills

In Exercises 1–28, name the property of real numbers that justifies the statement. See Example 1.

1. $3 + (-5) = -5 + 3$

2. $-5(7) = 7(-5)$

3. $25 - 25 = 0$

4. $5 + 0 = 5$

5. $6(-10) = -10(6)$

6. $2(6 \cdot 3) = (2 \cdot 6)3$

7. $7 \cdot 1 = 7$

8. $4 \cdot \frac{1}{4} = 1$

9. $25 + 35 = 35 + 25$

10. $(-4 \cdot 10) \cdot 8 = -4(10 \cdot 8)$

11. $3 + (12 - 9) = (3 + 12) - 9$

12. $(16 + 8) - 5 = 16 + (8 - 5)$

13. $(8 - 5)(10) = 8 \cdot 10 - 5 \cdot 10$

14. $7(9 + 15) = 7 \cdot 9 + 7 \cdot 15$

15. $(10 + 8) + 3 = 10 + (8 + 3)$

16. $(5 + 10)(8) = 8(5 + 10)$

17. $5(2a) = (5 \cdot 2)a$

18. $10(2x) = (10 \cdot 2)x$

19. $1 \cdot (5t) = 5t$

20. $10x \cdot \dfrac{1}{10x} = 1$

21. $3x + 0 = 3x$

22. $2x - 2x = 0$

23. $4 + (3 - x) = (4 + 3) - x$

24. $3(2 + x) = 3 \cdot 2 + 3x$

25. $3(6 + b) = 3 \cdot 6 + 3 \cdot b$

26. $(x + 1) - (x + 1) = 0$

27. $6(x + 3) = 6 \cdot x + 6 \cdot 3$

28. $(6 + x) - m = 6 + (x - m)$

In Exercises 29–38, use the property of real numbers to fill in the missing part of the statement. See Example 2.

29. Associative Property of Multiplication:
$3(6y) = $

30. Commutative Property of Addition:
$10 + (-6) = $

31. Commutative Property of Multiplication:
$15(-3) = $

32. Associative Property of Addition:
$6 + (5 + y) = $

33. Distributive Property:
$5(6 + z) = $

34. Distributive Property:
$(8 - y)(4) = $

35. Commutative Property of Addition:
$25 + (-x) = $

36. Additive Inverse Property:
$13x + (-13x) = $

37. Multiplicative Identity Property:
$(x + 8) \cdot 1 = $

38. Additive Identity Property:
$(8x) + 0 = $

In Exercises 39–46, give (a) the additive inverse and (b) the multiplicative inverse of the quantity.

39. 10

40. 18

41. -16

42. -52

43. $6z, \quad z \neq 0$

44. $2y, \quad y \neq 0$

45. $x + 1, \quad x \neq -1$

46. $y - 4, \quad y \neq 4$

In Exercises 47–54, rewrite the expression using the Associative Property of Addition or the Associative Property of Multiplication.

47. $(x + 5) - 3$

48. $(z + 6) + 10$

49. $32 + (4 + y)$

50. $15 + (3 - x)$

51. $3(4 \cdot 5)$

52. $(10 \cdot 8) \cdot 5$

53. $6(2y)$

54. $8(3x)$

In Exercises 55–62, rewrite the expression using the Distributive Property.

55. $20(2 + 5)$

56. $-3(4 - 8)$

57. $5(3x - 4)$

58. $6(2x + 5)$

59. $(x + 6)(-2)$

60. $(z - 10)(12)$

61. $-6(2y - 5)$

62. $-4(10 - b)$

In Exercises 63–66, the right side of the equation is *not* equal to the left side. Change the right side so that it *does* equal the left side.

63. $3(x + 5) \neq 3x + 5$

64. $4(x + 2) \neq 4x + 2$

65. $-2(x + 8) \neq -2x + 16$

66. $-9(x + 4) \neq -9x + 36$

In Exercises 67 and 68, use the basic properties of real numbers to prove the statement. See Examples 3 and 4.

67. If $ac = bc$ and $c \neq 0$, then $a = b$.

68. $(-1)(-a) = a$.

In Exercises 69–72, identify the property of real numbers that justifies each step. See Examples 5 and 6.

69.
$$x + 5 = 3 \qquad \text{Original equation}$$
$$(x + 5) + (-5) = 3 + (-5)$$
$$x + [5 + (-5)] = -2$$
$$x + 0 = -2$$
$$x = -2$$

70.
$$x - 8 = 20 \qquad \text{Original equation}$$
$$(x - 8) + 8 = 20 + 8$$
$$x + (-8 + 8) = 28$$
$$x + 0 = 28$$
$$x = 28$$

71.
$$2x - 5 = 6 \qquad \text{Original equation}$$
$$(2x - 5) + 5 = 6 + 5$$
$$2x + (-5 + 5) = 11$$
$$2x + 0 = 11$$
$$2x = 11$$
$$\tfrac{1}{2}(2x) = \tfrac{1}{2}(11)$$
$$\left(\tfrac{1}{2} \cdot 2\right)x = \tfrac{11}{2}$$
$$1 \cdot x = \tfrac{11}{2}$$
$$x = \tfrac{11}{2}$$

72.
$$3x + 4 = 10 \qquad \text{Original equation}$$
$$(3x + 4) + (-4) = 10 + (-4)$$
$$3x + [4 + (-4)] = 6$$
$$3x + 0 = 6$$
$$3x = 6$$
$$\tfrac{1}{3}(3x) = \tfrac{1}{3}(6)$$
$$\left(\tfrac{1}{3} \cdot 3\right)x = 2$$
$$1 \cdot x = 2$$
$$x = 2$$

Mental Math   In Exercises 73–78, use the Distributive Property to perform the required arithmetic mentally. For example, suppose you work in an industry where the wage is $14 per hour with "time and a half" for overtime. So, your hourly wage for overtime is

$$14(1.5) = 14\left(1 + \frac{1}{2}\right) = 14 + 7 = \$21.$$

73. $16(1.75) = 16\left(2 - \tfrac{1}{4}\right)$

74. $15\left(1\tfrac{2}{3}\right) = 15\left(2 - \tfrac{1}{3}\right)$

75. $7(62) = 7(60 + 2)$

76. $5(51) = 5(50 + 1)$

77. $9(6.98) = 9(7 - 0.02)$

78. $12(19.95) = 12(20 - 0.05)$

Solving Problems

79. *Geometry*   The figure shows two adjoining rectangles. Demonstrate the Distributive Property by filling in the blanks to express the total area of the two rectangles in two ways.

$$\left(\boxed{} + \boxed{}\right) = \boxed{} + \boxed{}$$

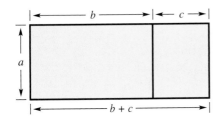

80. *Geometry*   The figure shows two adjoining rectangles. Demonstrate the "subtraction version" of the Distributive Property by filling in the blanks to express the area of the left rectangle in two ways.

$$\boxed{}\left(\boxed{} - \boxed{}\right) = \boxed{} - \boxed{}$$

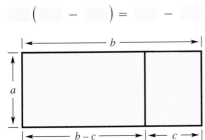

Dividends   In Exercises 81–84, the dividend paid per share of common stock by the Gillette Company for the years 1990 through 1997 is approximated by the model

Dividend per share $= 0.08t + 0.21$.

In this model, the dividend per share is measured in dollars and t represents the year, with $t = 0$ corresponding to 1990 (see figure).   (Source: The Gillette Company 1997 Annual Report)

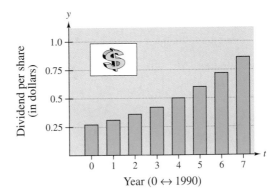

81. Use the graph to approximate the dividend paid in 1995.

82. Use the model to approximate the annual increase in the dividend paid per share.

83. Use the model to forecast the dividend per share in 2000.

84. In 1996, the actual dividend paid per share of common stock was $0.72. Compare this with the approximation given by the model.

Explaining Concepts

85. In your own words, give a verbal description of the Commutative Property of Addition.

86. What is the additive inverse of a real number? Give an example of the Additive Inverse Property.

87. What is the multiplicative inverse of a real number? Give an example of the Multiplicative Inverse Property.

88. Does every real number have a multiplicative inverse? Explain.

89. State the Multiplicative Property of Zero.

90. Explain how the Addition Property of Equality can be used to allow you to subtract the same number from both sides of an equation.

91. Suppose you define a new mathematical operation using the symbol $\odot$. This operation is defined as $a \odot b = 2 \cdot a + b$. Give examples to show that neither the Commutative Property nor the Associative Property is true for this operation.

Mid-Chapter Quiz

Take this quiz as you would take a quiz in class. After you are done, check your work against the answers given in the back of the book.

In Exercises 1 and 2, show each real number as a point on the real number line and place the correct inequality symbol ($<$ or $>$) between the real numbers.

1. -4.5 ___ -6 **2.** $\frac{3}{4}$ ___ $\frac{3}{2}$

In Exercises 3 and 4, evaluate the expression.

3. $|-3.2|$ **4.** $-|5.75|$

In Exercises 5 and 6, find the distance between the two real numbers.

5. -15 and 7 **6.** -10.5 and -6.75

In Exercises 7–16, evaluate the expression. Write fractions in simplified form.

7. $32 + (-18)$ **8.** $-10 - 12$

9. $\frac{3}{4} + \frac{7}{4}$ **10.** $\frac{2}{3} - \frac{1}{6}$

11. $(-12)(-4)$ **12.** $\left(-\frac{4}{5}\right)\left(\frac{15}{32}\right)$

13. $\frac{7}{12} \div \frac{5}{6}$ **14.** $\left(-\frac{3}{2}\right)^3$

15. $3 - 2^2 + 25 \div 5$ **16.** $\dfrac{18 - 2(3 + 4)}{6^2 - (12 \cdot 2 + 10)}$

In Exercises 17 and 18, name the property of real numbers that justifies the given statement.

17. (a) $8(u - 5) = 8 \cdot u - 8 \cdot 5$ (b) $10x - 10x = 0$

18. (a) $(7 + y) - z = 7 + (y - z)$ (b) $2x \cdot 1 = 2x$

19. At the beginning of a certain month, the initial balance in your checking account is $1522.76. During the month, you make the following transactions.

Withdrawal	$328.37
Withdrawal	$65.99
Withdrawal	$50.00
Deposit	$413.88

Find the balance at the end of the month. (Disregard any interest that may have been earned.)

20. If you deposit $30 in a retirement account twice each month, how much will you deposit in the account in 5 years?

21. Determine the unknown fractional part of the circle graph at the left. Explain how you were able to make this determination.

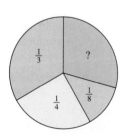

Figure for 21

P.4   Algebraic Expressions

Objectives

1 Identify the terms of an algebraic expression.

2 Use the rules of exponents to simplify an algebraic expression.

3 Simplify algebraic expressions by combining like terms and removing symbols of grouping.

4 Evaluate an algebraic expression by substituting a value for the variable.

1 Identify the terms of an algebraic expression.

Algebraic Expressions

One of the basic characteristics of algebra is the use of letters (or combinations of letters) to represent numbers. The letters used to represent the numbers are called **variables,** and combinations of letters and numbers are called **algebraic expressions.** Here are some examples.

$$3x, \quad x + 2, \quad \frac{x}{x^2 + 1}, \quad 2x - 3y, \quad 2x^3 - y^2$$

▶ **Algebraic Expression**

A collection of letters (called **variables**) and real numbers (called **constants**) combined using the operations of addition, subtraction, multiplication, and division is called an **algebraic expression.**

The **terms** of an algebraic expression are those parts that are separated by *addition.* For example, the algebraic expression $x^2 - 3x + 6$ has three terms: x^2, $-3x$, and 6. Note that $-3x$ is a term, rather than $3x$, because

$$x^2 - 3x + 6 = x^2 + (-3x) + 6. \qquad \text{Terms are separated by addition.}$$

The terms x^2 and $-3x$ are the **variable terms** of the expression, and 6 is the **constant term.** The numerical factor of a variable term is called the **coefficient.** For instance, the coefficient of the variable term $-3x$ is -3, and the coefficient of the variable term x^2 is 1.

Example 1   Identifying the Terms of an Algebraic Expression

Study Tip

Remember that subtraction can be written as addition. For instance, in Example 1(a) the expression $5x - \frac{1}{3}$ can be written as $5x + \left(-\frac{1}{3}\right)$ and so the terms of the expression are $5x$ and $-\frac{1}{3}$.

Algebraic Expression	*Terms*
a. $5x - \dfrac{1}{3}$	$5x, \quad -\dfrac{1}{3}$
b. $4y + 6x - 9$	$4y, \quad 6x, \quad -9$
c. $-x^3 + 4x^2 - 5x + 1$	$-x^3, \quad 4x^2, \quad -5x, \quad 1$
d. $\dfrac{1}{x} + 5x^4$	$\dfrac{1}{x}, \quad 5x^4$

2 Use the rules of exponents to simplify an algebraic expression.

Rules of Exponents

When multiplying two exponential expressions that have the *same* base, you add exponents. To see why this is true, consider the product $a^3 \cdot a^2$. Because the first expression represents $a \cdot a \cdot a$ and the second represents $a \cdot a$, it follows that the product of the two expressions is represented by $a \cdot a \cdot a \cdot a \cdot a$. That is,

$$a^3 \cdot a^2 = \underbrace{(a \cdot a \cdot a)}_{\substack{\text{Three} \\ \text{factors}}} \cdot \underbrace{(a \cdot a)}_{\substack{\text{Two} \\ \text{factors}}} = \underbrace{(a \cdot a \cdot a \cdot a \cdot a)}_{\substack{\text{Five} \\ \text{factors}}} = a^{3+2} = a^5$$

This and one other rule of exponents are summarized below.

> ### ▶ Rules of Exponents
>
> Let m and n be positive integers, and let a and b be real numbers, variables, or algebraic expressions.
>
> **1.** To multiply two exponential expressions that have the same base, add the exponents.
>
> $$a^m \cdot a^n = a^{m+n}$$
>
> **2.** To raise a product to a power, raise each factor to the power and multiply the results.
>
> $$(ab)^m = a^m b^m$$

The rules of exponents extend to three or more factors. For example,

$$a^m \cdot a^n \cdot a^k = a^{m+n+k}$$

and

$$(abc)^m = a^m b^m c^m.$$

The next two examples illustrate the use of these rules.

Study Tip

In the expression $x + 5$, the coefficient of x is understood to be 1. Similarly, the power (or exponent) of x is also understood to be 1. So,

$$x^4 \cdot x \cdot x^2 = x^{4+1+2} = x^7.$$

Note such occurrences in Example 2.

Example 2   Applying the Rules of Exponents

a. $-3^2 \cdot 3^4 \cdot 3 = -(3^2 \cdot 3^4 \cdot 3^1) = -3^{2+4+1} = -3^7$

b. $b^3 b^2 b = b^{3+2+1} = b^6$

c. $4^2 x^2 \cdot x = (4^2)(x^2 \cdot x) = 16x^3$

d. $(-3x^2y)(5xy)(2y^2) = (-3)(5)(2)(x^2 \cdot x)(y \cdot y \cdot y^2) = -30x^3 y^4$

Example 3   Applying the Rules of Exponents

a. $(3x)^3 = 3^3 \cdot x^3 = 27x^3$

b. $(-x^2)(x^2) = (-1)(x^2 \cdot x^2) = (-1)x^{2+2} = -x^4$

c. $(-x)^2(x^2) = x^2 \cdot x^2 = x^{2+2} = x^4$

d. $3x^2 \cdot (-5x)^3 = 3 \cdot x^2 \cdot (-5)^3 \cdot x^3 = 3(-125)x^{2+3} = -375x^5$

3 Simplify algebraic expressions by combining like terms and removing symbols of grouping.

Simplifying Algebraic Expressions

In an algebraic expression, two terms are said to be **like terms** if they are both constant terms or if they have the same variable factor. For example, the terms $4x$ and $-2x$ are like terms because they have the same variable factor x. Similarly, $2x^2y$, $-x^2y$, and $\frac{1}{2}(x^2y)$ are like terms because they have the same variable factor x^2y. Note that $4x^2y$ and $-x^2y^2$ are not like terms because their variable factors x^2y and x^2y^2 are different.

One of the most common uses of the basic rules of algebra is to rewrite an algebraic expression in a simpler form. One way to **simplify** an algebraic expression is to combine like terms.

Example 4   Combining Like Terms

Simplify each expression by combining like terms.

a. $2x + 3x - 4$ **b.** $-3 + 5 + 2y - 7y$ **c.** $5x + 3y - 4x$

Solution

a. $2x + 3x - 4 = (2 + 3)x - 4$ Distributive Property

$= 5x - 4$ Simplify.

b. $-3 + 5 + 2y - 7y = (-3 + 5) + (2 - 7)y$ Distributive Property

$= 2 - 5y$ Simplify.

c. $5x + 3y - 4x = 3y + 5x - 4x$ Commutative Property

$= 3y + (5x - 4x)$ Associative Property

$= 3y + (5 - 4)x$ Distributive Property

$= 3y + x$ Simplify.

As you gain experience with the rules of algebra, you may want to combine some of the steps in your work. For instance, you might feel comfortable listing only the following steps to solve Example 4(c).

$$5x + 3y - 4x = 3y + (5x - 4x) = 3y + x$$

Example 5   Combining Like Terms

a. $7x + 7y - 4x - y = (7x - 4x) + (7y - y)$ Group like terms.

$= 3x + 6y$ Combine like terms.

b. $2x^2 + 3x - 5x^2 - x = (2x^2 - 5x^2) + (3x - x)$ Group like terms.

$= -3x^2 + 2x$ Combine like terms.

c. $3xy^2 - 4x^2y^2 + 2xy^2 + (xy)^2$

$= (3xy^2 + 2xy^2) + (-4x^2y^2 + x^2y^2)$ Group like terms.

$= 5xy^2 - 3x^2y^2$ Combine like terms.

Another way to simplify an algebraic expression is to remove symbols of grouping. When removing symbols of grouping, remove the innermost symbols first and combine like terms. Then repeat the process for any remaining symbols of grouping.

A set of parentheses preceded by a *minus* sign can be removed by changing the sign of each term inside the parentheses. For instance,

$$3x - (2x - 7) = 3x - 2x + 7.$$

This is equivalent to using the Distributive Property with a multiplier of -1. That is,

$$3x - (2x - 7) = 3x + (-1)(2x - 7) = 3x - 2x + 7.$$

A set of parentheses preceded by a *plus* sign can be removed without changing the sign of each term inside the parentheses. For instance,

$$3x + (2x - 7) = 3x + 2x - 7.$$

Example 6   Removing Symbols of Grouping

Simplify each expression.

a. $3(x - 5) - (2x - 7)$ **b.** $-4(x^2 + 4) + x^2(x + 4)$

Solution

a. $3(x - 5) - (2x - 7) = 3x - 15 - 2x + 7$ Distributive Property

$$= (3x - 2x) + (-15 + 7)$$ Group like terms.

$$= x - 8$$ Combine like terms.

b. $-4(x^2 + 4) + x^2(x + 4) = -4x^2 - 16 + x^3 + 4x^2$ Distributive Property

$$= x^3 + (4x^2 - 4x^2) - 16$$ Group like terms.

$$= x^3 + 0 - 16$$ Combine like terms.

$$= x^3 - 16$$ Simplify.

Example 7   Removing Symbols of Grouping

a. $5x - 2x[3 + 2(x - 7)] = 5x - 2x(3 + 2x - 14)$ Distributive Property

$$= 5x - 2x(2x - 11)$$ Combine like terms.

$$= 5x - 4x^2 + 22x$$ Distributive Property

$$= -4x^2 + 27x$$ Combine like terms.

b. $-3x(5x^4) + (2x)^5 = -15x^5 + (2^5)(x^5)$ Properties of exponents

$$= -15x^5 + 32x^5$$ Simplify.

$$= 17x^5$$ Combine like terms.

4 Evaluate an algebraic expression by substituting a value for the variable.

Evaluating Algebraic Expressions

To **evaluate** an algebraic expression, substitute numerical values for each of the variables in the expression. Here are some examples.

Expression	Value of Variable	Substitute	Value of Expression
$3x + 2$	$x = 2$	$3(2) + 2$	$6 + 2 = 8$
$4x^2 + 2x - 1$	$x = -1$	$4(-1)^2 + 2(-1) - 1$	$4 - 2 - 1 = 1$
$2x(x + 4)$	$x = -2$	$2(-2)(-2 + 4)$	$2(-2)(2) = -8$

Note that you must substitute the value for *each* occurrence of the variable.

Example 8   Evaluating Algebraic Expressions

Evaluate each algebraic expression when $x = -2$ and $y = 5$.

a. $2y - 3x$ **b.** $5 + x^2$ **c.** $5 - x^2$

Solution

a. When $x = -2$ and $y = 5$, the expression $2y - 3x$ has a value of

$$2(5) - 3(-2) = 10 + 6 = 16.$$

b. When $x = -2$, the expression $5 + x^2$ has a value of

$$5 + (-2)^2 = 5 + 4 = 9.$$

c. When $x = -2$, the expression $5 - x^2$ has a value of

$$5 - (-2)^2 = 5 - 4 = 1.$$

Example 9   Evaluating Algebraic Expressions

Evaluate each algebraic expression when $x = 2$ and $y = -1$.

a. $y - x$ **b.** $|y - x|$ **c.** $|y^2 - x^2|$

Solution

a. When $x = 2$ and $y = -1$, the expression $y - x$ has a value of

$$(-1) - (2) = -1 - 2 = -3.$$

b. When $x = 2$ and $y = -1$, the expression $|y - x|$ has a value of

$$|(-1) - (2)| = |-3| = 3.$$

c. When $x = 2$ and $y = -1$, the expression $|y^2 - x^2|$ has a value of

$$|(-1)^2 - (2)^2| = |1 - 4| = |-3| = 3.$$

Craig Aurness/Corbis

In 1997, the mining industry in the
United States employed about
592,000 people.

Example 10   Using a Mathematical Model

From 1986 to 1997, the average hourly wage for miners in the United States can be modeled by

$$\text{Hourly wage} = 0.0006t^2 + 0.3t + 12.3, \quad 0 \le t \le 11$$

where $t = 0$ represents 1986. Create a table that shows the average hourly wages for these years.   (Source: U.S. Bureau of Labor Statistics)

Solution

To create a table of values that shows the average hourly wages for the years 1986 to 1997, evaluate the expression

$$0.0006t^2 + 0.3t + 12.3$$

for each integer value of t from $t = 0$ to $t = 11$.

Year	1986	1987	1988	1989	1990	1991
t	0	1	2	3	4	5
Hourly wage	$12.30	$12.60	$12.90	$13.21	$13.51	$13.82

Year	1992	1993	1994	1995	1996	1997
t	6	7	8	9	10	11
Hourly wage	$14.12	$14.43	$14.74	$15.05	$15.36	$15.67

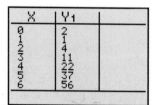

Discussing the Concept   Using a Graphing Utility

Most graphing utilities can evaluate an algebraic expression for several values of x and display the results in a table. For instance, to evaluate $2x^2 - 3x + 2$ when x is 0, 1, 2, 3, 4, 5, and 6, you can use the following steps.

1. Enter the expression into the graphing utility.
2. Set the minimum value of the table to 0.
3. Set the table step or table increment to 1.
4. Display the table.

The results are shown at the left. Consult the user's guide for your graphing utility for specific instructions. Then complete the table below for the expression $-4x^2 + 5x - 8$ when x is 0, 1, 2, 3, 4, 5, and 6.

x	0	1	2	3	4	5	6
$-4x^2 + 5x - 8$							

P.4   Exercises

Developing Skills

In Exercises 1-8, identify the terms of the algebraic expression. See Example 1.

1. $10x + 5$

2. $-16t^2 + 48$

3. $-3y^2 + 2y - 8$

4. $25z^3 - 4.8z^2$

5. $4x^2 - 3y^2 - 5x + 2y$

6. $14u^2 + 25uv - 3v^2$

7. $x^2 - 2.5x - \dfrac{1}{x}$

8. $\dfrac{3}{t^2} - \dfrac{4}{t} + 6$

In Exercises 9-12, determine the coefficient of the term.

9. $5y^3$

10. $4x^6$

11. $-\frac{3}{4}t^2$

12. $-8.4x$

In Exercises 13-16, identify the property of algebra that is illustrated by the equation.

13. $4 - 3x = -3x + 4$

14. $(10 + x) - y = 10 + (x - y)$

15. $-5(2x) = (-5 \cdot 2)x$

16. $(x - 2)(3) = 3(x - 2)$

In Exercises 17-20, use the given property to rewrite the expression.

17. Distributive Property

$5(x + 6) =$

18. Commutative Property of Multiplication

$5(x + 6) =$

19. Distributive Property

$6x + 6 =$

20. Commutative Property of Addition

$6x + 6 =$

In Exercises 21-24, write the expression as repeated multiplication.

21. $x^3 \cdot x^4$

22. $-2x^4$

23. $z^2 \cdot z^5$

24. $(-2x)^3$

In Exercises 25-28, write in exponential notation.

25. $(-5x)(-5x)(-5x)(-5x)$

26. $(-9t)(-9t)(-9t)(-9t)(-9t)(-9t)$

27. $(x \cdot x \cdot x)(y \cdot y \cdot y)$

28. $(y \cdot y \cdot y)(y \cdot y \cdot y \cdot y)$

In Exercises 29-54, simplify the expression. See Examples 2 and 3.

29. $-2^3 \cdot 2^4$

30. $-4^2 \cdot 4^5$

31. $x^5 \cdot x^7 \cdot x$

32. $u^3 \cdot u^5 \cdot u$

33. $3^3y^4 \cdot y^2$

34. $6^2x^3 \cdot x^5$

35. $(-4x)^2$

36. $(-4x)^3$

37. $-4(2x)^2$

38. $-2(-4x)^3$

39. $(-5z^2)^3$

40. $(-5z^3)^2$

41. $(2xy)(3x^2y^3)$

42. $(-5a^2b^3)(2ab^4)$

43. $(5y^2)(-y^4)(2y^3)$

44. $(3y)(2y^2)$

45. $-5z^4(-5z)^4$

46. $(-6n)(-3n^2)^2$

47. $(-2a^2)^3(-8a)$

48. $(-2a)^2(-2a)^2$

49. $(3uv)^2(-6u^3v)$

50. $(10x^2y)^3(2x^4y)$

51. $(x^n)^4$

52. $(a^3)^k$

53. $x^{n+1} \cdot x^3$

54. $y^{m-2} \cdot y^2$

In Exercises 55-66, simplify the expression by combining like terms. See Examples 4 and 5.

55. $3x + 4x$

56. $-2x^2 + 4x^2$

57. $9y - 5y + 4y$

58. $8y + 7y - y$

59. $3x - 2y + 5x + 20y$

60. $-2a + 4b - 7a - b$

61. $7x^2 - 2x - x^2$

62. $9y + y^2 - 6y$

63. $-3z^4 + 6z - z + 8 + z^4$

64. $-5y^3 + 3y - 6y^2 + 8y^3 + y - 4$

65. $2uv + 5u^2v^2 - uv - (uv)^2$

66. $7x^2y + 8x^2y^2 + 2xy^2 - (xy)^2$

In Exercises 67-70, use the Distributive Property to rewrite the expression.

67. $4(2x^2 + x - 3)$

68. $8(z^3 - 4z^2 + 2)$

69. $-3(6y^2 - y - 2)$

70. $-5(-x^2 + 2y + 1)$

In Exercises 71–86, simplify the expression. See Examples 6 and 7.

71. $10(x - 3) + 2x - 5$ **72.** $3(x + 1) + x - 6$

73. $-3(3y - 1) + 2(y - 5)$

74. $5(a + 6) - 4(2a - 1)$

75. $-3(y^2 - 2) + y^2(y + 3)$

76. $x(x^2 - 5) - 4(4 - x)$

77. $y^2(y + 1) + y(y^2 + 1)$ **78.** $x(x^2 + 3) - 3(x + 4)$

79. $3[2x - 4(x - 8)]$ **80.** $4[5 - 3(x^2 + 10)]$

81. $8x + 3x[10 - 4(3 - x)]$

82. $5y - y[9 + 6(y - 2)]$

83. $2[3(b - 5) - (b^2 + b + 3)]$

84. $5[3(z + 2) - (z^2 + z - 2)]$

85. $2x(5x^2) - 4x^3(x + 15)$

86. $5y^3(-3y)^2 - 4(x^2)^2(x - 1)$

In Exercises 87–100, evaluate the expression for the specified values of the variable(s). If not possible, state the reason. See Examples 8 and 9.

Expression	Values
87. $5 - 3x$	(a) $x = \frac{2}{3}$ (b) $x = 5$
88. $\frac{3}{2}x - 2$	(a) $x = 6$ (b) $x = -3$
89. $10 - 4x^2$	(a) $x = -1$ (b) $x = \frac{1}{2}$
90. $2x^2 + 5x - 3$	(a) $x = 2$ (b) $x = -3$

Expression	Values
91. $\dfrac{x}{x^2 + 1}$	(a) $x = 0$ (b) $x = 3$
92. $5 - \dfrac{3}{x}$	(a) $x = 0$ (b) $x = -6$
93. $3x + 2y$	(a) $x = 1,\ \ y = 5$
	(b) $x = -6,\ \ y = -9$
94. $6x - 5y$	(a) $x = -2,\ \ y = -3$
	(b) $x = 1,\ \ y = 1$
95. $x^2 - xy + y^2$	(a) $x = 2,\ \ y = -1$
	(b) $x = -3,\ \ y = -2$
96. $\dfrac{x}{x - y}$	(a) $x = 0,\ \ y = 10$
	(b) $x = 4,\ \ y = 4$
97. $|y - x|$	(a) $x = 2,\ \ y = 5$
	(b) $x = -2,\ \ y = -2$
98. $|x^2 - y|$	(a) $x = 0,\ \ y = -2$
	(b) $x = 3,\ \ y = 15$
99. Distance traveled: rt	(a) $r = 40,\ \ t = 5\frac{1}{4}$
	(b) $r = 35,\ \ t = 4$
100. Simple interest: Prt	(a) $P = \$5000,$
	$r = 0.085,\ \ t = 10$
	(b) $P = \$750,$
	$r = 0.07,\ \ t = 3$

Solving Problems

Geometry   In Exercises 101 and 102, find an expression for the area of the figure. Then evaluate the expression for the given value of the variable.

101. $b = 15$ **102.** $h = 12$

    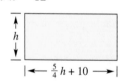

Geometry   In Exercises 103 and 104, find an expression for the area of the region. Then simplify the expression.

103.   **104.**

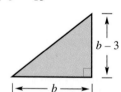   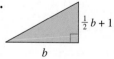

Using a Model   In Exercises 105 and 106, use the model, which approximates the annual sales (in millions of dollars) of exercise equipment in the United States from 1990 to 1996 (see figure).

Sales $= 193.89t + 1830.89, \quad 0 \le t \le 6$

In this formula, $t = 0$ represents 1990.   (Source: National Sporting Goods Association)

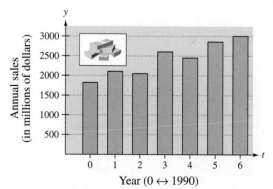

105. Graphically approximate the sales of exercise equipment in 1995. Then use the model to confirm your estimate algebraically.

106. Use the model and a calculator to complete the table showing the sales from 1990 to 1996.

Year	1990	1991	1992	1993
Forecast				

Year	1994	1995	1996
Forecast			

Using a Model   In Exercises 107 and 108, use the model, which approximates the median sale price (in thousands of dollars) for homes in the Midwest from 1990 to 1997 (see figure).

Sale price $= 5.9t + 106.0$, $0 \le t \le 7$

In this formula, $t = 0$ represents 1990.   (Source: U.S. Bureau of the Census, U.S. Department of Housing and Urban Development)

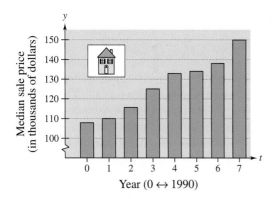

Median sale price (in thousands of dollars)

Year $(0 \leftrightarrow 1990)$

107. Graphically approximate the median sale price for homes in the Midwest in 1995. Then use the model to confirm your estimate algebraically.

108. Use the model to complete the table showing the median sale price for homes in the Midwest from 1990 to 1997.

Year	1990	1991	1992	1993
Price				

Year	1994	1995	1996	1997
Price				

109. *Geometry*   The roof shown in the figure is made up of two trapezoids and two triangles. Find the total area of the roof.

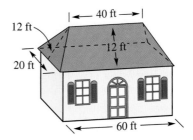

110. *Exploration*

(a) A convex polygon with n sides has

$$\frac{n(n-3)}{2}, \quad n \ge 4$$

diagonals. Verify the formula for a square, a pentagon, and a hexagon.

(b) Explain why the formula in part (a) will always yield a natural number

Explaining the Concepts

111. Answer parts (d) and (e) of Motivating the Chapter on page 1.

112. Explain the difference between terms and factors in an algebraic expression.

113. In your own words, explain how to combine like terms. Give an example.

114. State the procedure for removing nested symbols of grouping. Nested symbols of groupings are sym-

bols of groupings within symbols of groupings such as $5[1 + 4(3x + 2)]$.

115. Explain how the Distributive Property can be used to simplify the expression $5x + 3x$.

116. Explain how to rewrite $[x - (3 \cdot 4)] \div 5$ without using symbols of grouping.

117. Is it possible to evaluate $(x + 2)/(y - 3)$ when $x = 5$ and $y = 3$? Explain.

P.5   Constructing Algebraic Expressions

Objectives

1 Translate a verbal phrase into an algebraic expression.

2 Construct an algebraic expression with hidden products.

1 Translate a verbal phrase into an algebraic expression.

Translating Phrases

In this section, you will study ways to *construct* algebraic expressions. When you translate a verbal sentence or phrase into an algebraic expression, it helps to watch for key words and phrases that indicate the four different operations of arithmetic.

▶ **Translating Key Words and Phrases**

Key Words and Phrases	Verbal Description	Algebraic Expression
Addition: Sum, plus, greater than, increased by, more than, exceeds, total of	The sum of 5 and x Seven more than y	$5 + x$ $y + 7$
Subtraction: Difference, minus, less than, decreased by, subtracted from, reduced by, the remainder	b is subtracted from 4. Three less than z	$4 - b$ $z - 3$
Multiplication: Product, multiplied by, twice, times, percent of	Two times x	$2x$
Division: Quotient, divided by, ratio, per	The ratio of x and 8	$\dfrac{x}{8}$

Example 1   Translating Verbal Phrases

a. *Verbal Description:* Seven more than three times x
   Algebraic Expression: $3x + 7$

b. *Verbal Description:* Four less than the product of 6 and n
   Algebraic Expression: $6n - 4$

c. *Verbal Description:* The quotient of x and 3 decreased by 6
   Algebraic Expression: $\dfrac{x}{3} - 6$

Example 2   Translating Verbal Phrases

a. *Verbal Description:* Eight added to the product of 2 and *n*

  Algebraic Expression: $2n + 8$

b. *Verbal Description:* Four times the sum of *y* and 9

  Algebraic Expression: $4(y + 9)$

c. *Verbal Description:* The difference of *a* and 7, divided by 9

  Algebraic Expression: $\dfrac{a - 7}{9}$

  In Examples 1 and 2, the verbal description specified the name of the variable. In most real-life situations, however, the variables are not specified and it is your task to assign variables to the *appropriate* quantities.

Example 3   Translating Verbal Phrases

a. *Verbal Description:* The sum of 7 and a number

  Label: The number $= x$

  Algebraic Expression: $7 + x$

b. *Verbal Description:* Four decreased by the product of 2 and a number

  Label: The number $= x$

  Algebraic Expression: $4 - 2x$

  A good way to learn algebra is to do it forward and backward. For instance, the next example translates algebraic expressions into verbal form. Keep in mind that other key words could be used to describe the operations in each expression.

Example 4   Translating Expressions into Verbal Phrases

Without using a variable, write a verbal description for each expression.

a. $5x - 10$ **b.** $\dfrac{3 + x}{4}$

c. $2(3x + 4)$ **d.** $\dfrac{4}{x - 2}$

Solution

a. 10 less than the product of 5 and a number

b. The sum of 3 and a number, divided by 4

c. Twice the sum of 3 times a number and 4

d. Four divided by a number reduced by 2.

---

2 Construct an algebraic expression with hidden products.

Constructing Mathematical Models

Translating a verbal phrase into a mathematical model is critical in problem solving. The next four examples will demonstrate three steps for creating a mathematical model.

1. Construct a verbal model that represents the problem situation.
2. Assign labels to all quantities in the verbal model.
3. Construct a mathematical model (algebraic expression).

When verbal phrases are translated into algebraic expressions, products are often overlooked. Watch for hidden products in the next two examples.

Study Tip

Most real-life problems do not contain verbal expressions that clearly identify the arithmetic operations involved in the problem. You need to rely on past experience and the physical nature of the problem in order to identify the operations hidden in the problem statement.

Example 5   Constructing a Mathematical Model

A cash register contains x quarters. Write an algebraic expression for this amount of money in dollars.

Solution

Verbal Model: $\boxed{\text{Value of coin}} \cdot \boxed{\text{Number of coins}}$

Labels: Value of coin $= 0.25$ (dollars per quarter)
Number of coins $= x$ (quarters)

Expression: $0.25x$

Example 6   Constructing a Mathematical Model

A cash register contains n nickels and d dimes. Write an algebraic expression for this amount of money in cents.

Solution

Verbal Model: $\boxed{\text{Value of nickel}} \cdot \boxed{\text{Number of nickels}} + \boxed{\text{Value of dime}} \cdot \boxed{\text{Number of dimes}}$

Labels: Value of nickel $= 5$ (cents per nickel)
Number of nickels $= n$ (nickels)
Value of dime $= 10$ (cents per dime)
Number of dimes $= d$ (dimes)

Expression: $5n + 10d$

In Example 6, the final expression $5n + 10d$ is measured in cents. This makes "sense" in the following way.

$$\frac{5 \text{ cents}}{\text{nickel}} \cdot n \text{ nickels} + \frac{10 \text{ cents}}{\text{dime}} \cdot d \text{ dimes}$$

Note that the nickels and dimes "cancel," leaving cents as the unit of measure for each term. This technique is called *unit analysis,* and it can be very helpful in determining the final unit of measure.

Example 7   Constructing a Mathematical Model

Write an algebraic expression showing how far a person can ride a bicycle in t hours if the person travels at a constant rate of 12 miles per hour.

Solution

For this problem, use the formula (distance) = (rate)(time).

Verbal Model:	Rate · Time	
Labels:	Rate = 12	(miles per hour)
	Time = t	(hours)

Expression: $12t$

Using unit analysis, you can see that the expression in Example 7 has *miles* as its unit of measure.

$$12 \frac{\text{miles}}{\text{hour}} \cdot t \text{ hours}$$

When translating verbal phrases involving percents, be sure you write the percent *in decimal form*. For instance, 25% should be written as 0.25. Remember that when you find a percent of a number, you multiply. For instance, 25% of 78 is given by

$$0.25(78) = 19.5. \qquad \text{25\% of 78}$$

Example 8   Constructing a Mathematical Model

A person adds k liters of fluid containing 55% antifreeze to a car radiator. Write an algebraic expression that indicates how much antifreeze was added.

Solution

Verbal Model:	Percent antifreeze · Number of liters	
Labels:	Percent of antifreeze = 0.55	(in decimal form)
	Number of liters = k	(liters)

Expression: $0.55k$

Note that the algebraic expression uses the decimal form of 55%. That is, you compute with 0.55 rather than 55%.

Hidden operations are often involved when labels are assigned to *two* unknown quantities. For example, suppose two numbers add up to 18 and one of the numbers is assigned the variable x. What expression can you use to represent the second number? Let's try a specific case first, then apply it to a general case.

Specific case: If the first number is 7, the second number is $18 - 7 = 11$.

General case: If the first number is x, the second number is $18 - x$.

The strategy of using a *specific* case to help determine the general case is often helpful in applications. Observe the use of this strategy in the next example.

Example 9   Using Specific Cases to Model General Cases

a. A person's weekly salary is d dollars. Write an expression for the person's annual salary.

b. A person's annual salary is y dollars. Write an expression for the person's monthly salary.

Solution

a. *Specific Case:*   If the weekly salary is \$300, the annual salary is 52(300) dollars.

   General Case: If the weekly salary is d dollars, the annual salary is $52 \cdot d$ or $52d$ dollars.

b. *Specific Case:*   If the annual salary is \$24,000, the monthly salary is $24,000 \div 12$ dollars.

   General Case: If the annual salary is y dollars, the monthly salary is $y \div 12$ or $y/12$ dollars.

Study Tip

You can check that your algebraic expressions are correct for even, odd, or consecutive integers by substituting an integer for n. For instance, let $n = 5$ and you can see that $2n = 2(5) = 10$ is an even integer, $2n - 1 = 2(5) - 1 = 9$ is an odd integer, and $2n + 1 = 2(5) + 1 = 11$ is an odd integer.

In mathematics it is useful to know how to represent certain types of integers algebraically. For instance, consider the set $\{2, 4, 6, 8, \ldots\}$ of *even* integers. Because every even integer has 2 as a factor,

$$2 = 2 \cdot 1, \quad 4 = 2 \cdot 2, \quad 6 = 2 \cdot 3, \quad 8 = 2 \cdot 4, \ldots$$

it follows that any integer n multiplied by 2 is sure to be the *even* number $2n$. Moreover, if $2n$ is even, then $2n - 1$ and $2n + 1$ are sure to be *odd* integers.

Two integers are called **consecutive integers** if they differ by 1. For any integer n, its next two larger consecutive integers are $n + 1$ and $(n + 1) + 1$ or $n + 2$. So, you can denote three consecutive integers by n, $n + 1$, and $n + 2$. These results are summarized below.

▶ **Labels for Integers**

Let n represent an integer. Then even integers, odd integers, and consecutive integers can be represented as follows.

1. $2n$ denotes an *even* integer for $n = 1, 2, 3, \ldots$.

2. $2n - 1$ and $2n + 1$ denote *odd* integers for $n = 1, 2, 3, \ldots$.

3. $\{n, n + 1, n + 2, \ldots\}$ denotes a set of *consecutive* integers.

Example 10 Constructing a Mathematical Model

Write an expression for the following phrase.

The sum of two consecutive integers, the first of which is n

Solution

The first integer is n. The next consecutive integer is $n + 1$. So the sum of two consecutive integers is

$$n + (n + 1) = 2n + 1.$$

Sometimes an expression may be written directly from a diagram using a common geometric formula, as shown in the next example.

Example 11 A Geometric Application

For Figure P.9, write expressions for the following. Simplify your results.

a. The perimeter of the figure   **b.** The area of the figure

Solution

a. The perimeter P is

$$P = 3x + 3 + 2x + x + 2 + (x + 3)$$
$$= 7x + 8. \qquad \text{Combine like terms.}$$

b. Using the dotted line, there are two rectangles with areas A_1 and A_2, so the area of the figure is $A_1 + A_2$.

$$A_1 + A_2 = (2 \cdot x) + (3 \cdot 3x) \qquad \text{Substitute values for } A_1 \text{ and } A_2.$$
$$= 2x + 9x \qquad \text{Multiply.}$$
$$= 11x \qquad \text{Combine like terms.}$$

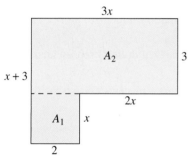

Figure P.9

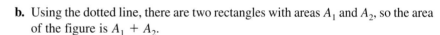

Discussing the Concept   **Avoiding Ambiguity**

When you write a verbal model or construct an algebraic expression, watch out for statements that may be interpreted in more than one way. For instance, the statement

"The sum of 4 and x divided by 5"

is ambiguous because it could mean

$$\frac{4 + x}{5} \quad \text{or} \quad 4 + \frac{x}{5}.$$

Write up at least three other statements like this that can be interpreted in two different ways. Try them out with other people in your class to see whether they are truly ambiguous.

P.5   Exercises

Developing Skills

In Exercises 1–24, translate the verbal phrase into an algebraic expression. See Examples 1–3.

1. The sum of 8 and a number n

2. Five more than a number n

3. The sum of 12 and twice a number n

4. The total of 25 and three times a number n

5. Six less than a number n

6. Fifteen decreased by three times a number n

7. Four times a number n minus 3

8. The product of a number y and 10 is decreased by 35.

9. One-third of a number n

10. Seven-fifths of a number n

11. The quotient of a number x and 6

12. The ratio of y to 3

13. Eight times the ratio of N and 5

14. Twenty times the ratio of x and 9

15. The number c is quadrupled and the product is increased by 10.

16. The number u is tripled and the product is increased by 250.

17. Thirty percent of the list price L

18. Forty percent of the cost C

19. The sum of a number and 5, divided by 10

20. The sum of 3 and four times a number x, divided by 8

21. The absolute value of the difference between a number and 5

22. The absolute value of the quotient of a number and 4

23. The product of 3 and the square of a number is decreased by 4.

24. The sum of 10 and one-fourth the square of a number

In Exercises 25–40, write a verbal description of the algebraic expression without using the variable. See Example 4.

25. $t - 2$

26. $x - 5$

27. $y + 50$

28. $2y + 3$

29. $3x + 2$

30. $4x - 5$

31. $\dfrac{z}{2}$

32. $\dfrac{y}{8}$

33. $\frac{4}{5}x$

34. $\frac{2}{3}t$

35. $8(x - 5)$

36. $-3(x + 2)$

37. $\dfrac{x + 10}{3}$

38. $\dfrac{x - 2}{3}$

39. $x(x + 7)$

40. $x^2 + 2$

In Exercises 41–64, write an algebraic expression that represents the specified quantity in the verbal statement, and simplify if possible. See Examples 5–10.

41. The amount of money (in dollars) represented by n quarters

42. The amount of money (in dollars) represented by x nickels

43. The amount of money (in dollars) represented by m dimes

44. The amount of money (in cents) represented by x nickels and y quarters

45. The amount of money (in cents) represented by m nickels and n dimes

46. The amount of money (in cents) represented by m dimes and n quarters

47. The distance traveled in t hours at an average speed of 55 miles per hour

48. The distance traveled in 5 hours at an average speed of r miles per hour

49. The time to travel 100 miles at an average speed of r miles per hour

50. The average rate of speed when traveling 360 miles in t hours

51. The amount of antifreeze in a cooling system containing y gallons of coolant that is 45% antifreeze

52. The amount of water in q quarts of a food product that is 65% water

53. The amount of wage tax due for a taxable income of I dollars that is taxed at the rate of 1.25%

54. The amount of sales tax on a purchase valued at L dollars if the tax rate is 6%

55. The sale price of a coat that has a list price of L dollars if it is a "20% off" sale

56. The total cost for a family to stay one night at a campground if the charge is $18 for the parents plus $3 for each of the *n* children

57. The total hourly wage for an employee when the base pay is $8.25 per hour plus 60 cents for each of *q* units produced per hour

58. The total hourly wage for an employee when the base pay is $11.65 per hour plus 80 cents for each of *q* units produced per hour

59. The sum of a number *n* and three times the number

60. The sum of three consecutive integers, the first of which is *n*

61. The sum of two consecutive odd integers, the first of which is $2n + 1$

62. The sum of two consecutive even integers, the first of which is $2n$

63. The product of two consecutive even integers, divided by 4

64. The difference of two consecutive integers, divided by 2

Solving Problems

Geometry   In Exercises 65–68, write an expression for the area of the region.

65.

66.

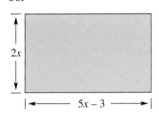

67.

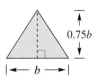

68.

In Exercises 69–72, write expressions for the perimeter and area of the region. Simplify the expressions. See Example 11.

69.

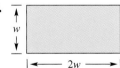

70.

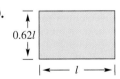

71.

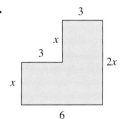

72.

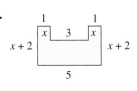

73. *Geometry*   Write an expression that represents the area of the top of the billiard table in the figure. What is the unit of measure for the area?

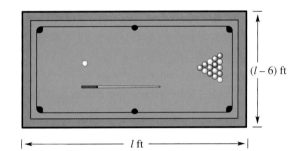

74. *Geometry*   Write an expression that represents the area of the advertising banner in the figure. What is the unit of measure for the area?

75. *Finding a Pattern*   Complete the table. The third row contains the differences between consecutive entries of the second row. Describe the pattern of the third row.

n	0	1	2	3	4	5
$5n - 3$						
Differences						

76. *Finding a Pattern*   Complete the table. The third row contains the differences between consecutive entries of the second row. Describe the pattern of the third row.

n	0	1	2	3	4	5
$3n + 1$						
Differences						

77. *Finding a Pattern*   Using the results of Exercises 75 and 76, guess the third-row difference that would result in a similar table if the algebraic expression were $an + b$.

78. *Think About It*   Find a and b such that the expression $an + b$ would yield the following table.

n	0	1	2	3	4	5
$an + b$	3	7	11	15	19	23

Explaining Concepts

79. The phrase *reduced by* implies what operation?

80. The word *ratio* indicates what operation?

81. Which are equivalent to $4x$?

    (a) x multiplied by 4

    (b) x increased by 4

    (c) the product of x and 4

    (d) the ratio of 4 and x

82. When each phrase is translated into an algebraic expression, is order important? Explain.

    (a) y is multiplied by 5

    (b) 5 is decreased by y

    (c) y divided by 5

    (d) the sum of 5 and y

83. When translating a statement into an algebraic expression, explain why it may be helpful to use a specific case before writing the expression.

84. If n is an integer, how are the integers $2n - 1$ and $2n + 1$ related? Explain.

Key Terms

set, *p.2*
real numbers, *p.2*
subset, *p.2*
natural numbers, *p.2*
integers, *p.2*
whole numbers, *p.2*
fractions, *p.3*
rational numbers, *p.3*
irrational numbers, *p.3*
real number line, *p.4*
origin, *p.4*

nonnegative real number, *p.4*
inequality symbols, *p.5*
absolute value, *p.7*
opposites, *p.7*
additive inverses, *p.7*
sum, *p.11*
difference, *p.11*
least common denominator, *p.12*
product, *p.13*

factor, *p.13*
reciprocal, *p.14*
quotient, *p.14*
dividend, *p.14*
divisor, *p.14*
numerator, *p.14*
denominator, *p.14*
exponential form, *p.15*
base, *p.15*
exponential notation, *p.15*

variables, *p.31*
algebraic expressions, *p.31*
variable terms, *p.31*
constant term, *p.31*
coefficient, *p.31*
like terms, *p.33*
simplify, *p.33*
consecutive integers, *p.44*

Key Concepts

P.1 Order on the real number line

If the real number a lies to the left of the real number b on the real number line, then a is less than b, which is written as $a < b$.

P.1 Distance between two real numbers

If a and b are two real numbers such that $a \leq b$, then the distance between a and b is given by

(Distance between a and b) $= b - a$.

P.2 Addition of two real numbers

1. To add two real numbers with like signs, add their absolute values and attach the common sign to the result.

2. To add two real numbers with unlike signs, subtract the smaller absolute value from the greater absolute value and attach the sign of the number with the greater absolute value.

P.2 Subtraction of two real numbers

To subtract the real number b from the real number a, add the opposite of b to a. That is,

$$a - b = a + (-b).$$

P.2 Addition and subtraction of fractions

1. Like Denominators: The sum and difference of two fractions with like denominators are as follows.

$$\frac{a}{c} + \frac{b}{c} = \frac{a + b}{c} \qquad \frac{a}{c} - \frac{b}{c} = \frac{a - b}{c}$$

2. Unlike Denominators: To add or subtract two fractions with unlike denominators, first rewrite the fractions so that they have the same denominator and then apply the first rule.

P.2 Multiplication of two real numbers

1. To multiply two real numbers with like signs, find the product of their absolute values. The product is positive.
2. To multiply two real numbers with unlike signs, find the product of their absolute values, and attach a minus sign. The product is negative.
3. The product of zero and any other number is zero.

P.2 Multiplication of two fractions

The product of the two fractions a/c and b/d is given by

$$\frac{a}{c} \cdot \frac{b}{d} = \frac{ab}{cd}.$$

P.2 Division of two real numbers

To divide the real number a by the nonzero real number b, multiply a by the reciprocal of b. That is,

$$a \div b = a \cdot \frac{1}{b}.$$

P.2 Order of operations

To evaluate an expression involving more than one operation, use the following order.

1. Do operations that occur within symbols of grouping.
2. Evaluate powers.
3. Do multiplications and divisions from left to right.
4. Do additions and subtractions from left to right.

49

Key Concepts   *(continued)*

P.3 Properties of real numbers

Let a, b, and c represent real numbers, variables, or algebraic expressions.

Commutative Property of Addition:

$a + b = b + a$

Commutative Property of Multiplication:

$ab = ba$

Associative Property of Addition:

$(a + b) + c = a + (b + c)$

Associative Property of Multiplication:

$(ab)c = a(bc)$

Distributive Property:

$a(b + c) = ab + ac \qquad a(b - c) = ab - ac$

$(a + b)c = ac + bc \qquad (a - b)c = ac - bc$

Additive Identity Property:

$a + 0 = a$

Multiplicative Identity Property:

$a \cdot 1 = 1 \cdot a = a$

Additive Inverse Property:

$a + (-a) = 0$

Multiplicative Inverse Property:

$a \cdot \dfrac{1}{a} = 1, \quad a \neq 0$

Addition Property of Equality:

If $a = b$, then $a + c = b + c$.

Multiplication Property of Equality:

If $a = b$, then $ac = bc$.

Cancellation Property of Addition:

If $a + c = b + c$, then $a = b$.

Cancellation Property of Multiplication:

If $ac = bc$ and $c \neq 0$, then $a = b$. $\quad 0 \cdot a = 0$

Multiplication Property of Zero:

Division Property of Zero: $\dfrac{0}{a} = 0, \quad a \neq 0$

Division by Zero Is Undefined: $\dfrac{a}{0}$ is undefined.

Multiplication by -1:

$(-1)(a) = -a, \quad (-1)(-a) = a$

Placement of Negative Signs:

$(-a)(b) = -(ab) = (a)(-b)$

Product of Two Opposites:

$(-a)(-b) = ab$

P.4 Rules of exponents

Let m and n be positive integers, and let a and b be real numbers, variables, or algebraic expressions.

1. To multiply two exponential expressions that have the same base, add the exponents.

   $a^m \cdot a^n = a^{m+n}$

2. To raise a product to a power, raise each factor to the power and multiply the results.

   $(ab)^m = a^m b^m$

P.5 Translating key words and phrases

Addition: sum, plus, greater than, increased by, more than, exceeds, total of

Subtraction: difference, minus, less than, decreased by, subtracted from, reduced by, the remainder

Multiplication: product, multiplied by, twice, times, percent of

Division: quotient, divided by, ratio, per

P.5 Labels for integers

Let n represent an integer. Then even integers, odd integers, and consecutive integers can be represented as follows.

1. $2n$ denotes an even integer for $n = 1, 2, 3, \ldots$.

2. $2n - 1$ and $2n + 1$ denote odd integers for $n = 1, 2, 3, \ldots$.

3. $\{n, n + 1, n + 2, \ldots\}$ denotes a set of consecutive integers.

REVIEW EXERCISES

Reviewing Skills

P.1 In Exercises 1–4, plot the real numbers on a real number line and place the correct inequality symbol (< or >) between them.

1. $-5 \quad\quad 3$

2. $-2 \quad\quad -8$

3. $-\frac{8}{5} \quad\quad -\frac{2}{5}$

4. $8.4 \quad\quad -3.2$

In Exercises 5–8, find the distance between each pair of real numbers.

5. 9 and -2

6. -7 and 4

7. -13.5 and -6.2

8. -8.4 and -0.3

In Exercises 9–12, determine the absolute value of the expression.

9. $|-5|$

10. $|6|$

11. $-|-7.2|$

12. $|-3.6|$

P.2 In Exercises 13–34, evaluate the expression. If it is not possible, state the reason. Write all fractions in simplified form.

13. $15 + (-4)$

14. $-12 + 3$

15. $340 - 115 + 5$

16. $-154 + 86 - 240$

17. $-63.5 + 21.7$

18. $14.35 - 10.3$

19. $\frac{4}{21} + \frac{7}{21}$

20. $\frac{21}{16} - \frac{13}{16}$

21. $-\frac{5}{6} + 1$

22. $\frac{21}{32} + \frac{11}{24}$

23. $8\frac{3}{4} - 6\frac{5}{8}$

24. $-2\frac{9}{10} + 5\frac{3}{20}$

25. $-7 \cdot 4$

26. $(-8)(-3)$

27. $120(-5)(7)$

28. $(-16)(-15)(-4)$

29. $\frac{3}{8} \cdot \left(-\frac{2}{15}\right)$

30. $\frac{5}{21} \cdot \frac{21}{5}$

31. $\frac{-56}{-4}$

32. $\frac{85}{0}$

33. $-\frac{7}{15} \div \left(-\frac{7}{30}\right)$

34. $-\frac{2}{3} \div \frac{4}{15}$

In Exercises 35–40, evaluate the exponential expression.

35. $(-6)^3$

36. $-(-3)^4$

37. -4^2

38. 2^5

39. $-\left(-\frac{1}{2}\right)^3$

40. $\left(-\frac{1}{3}\right)^3$

In Exercises 41–44, use the order of operations to evaluate the expression.

41. $120 - (5^2 \cdot 4)$

42. $45 - 45 \div 3^2$

43. $8 + 3[6^2 - 2(7 - 4)]$

44. $2^4 - [10 + 6(1 - 3)^2]$

P.3 In Exercises 45–50, name the property of real numbers that justifies the statement.

45. $13 - 13 = 0$

46. $7\left(\frac{1}{7}\right) = 1$

47. $7(9 + 3) = 7 \cdot 9 + 7 \cdot 3$

48. $15(4) = 4(15)$

49. $5 + (4 - y) = (5 + 4) - y$

50. $6(4z) = (6 \cdot 4)z$

In Exercises 51–54, identify the property of real numbers that is illustrated by the equation.

51. $(u - v)(2) = 2(u - v)$

52. $(x + y) + 0 = x + y$

53. $8(x - y) = 8 \cdot x - 8 \cdot y$

54. $x(yz) = (xy)z$

In Exercises 55–58, expand the expression by using the Distributive Property.

55. $-(-u + 3v)$

56. $-5(2x - 4y)$

57. $-y(3y - 10)$

58. $x(3x + 4y)$

P.4 In Exercises 59–64, use the rules of exponents to simplify the algebraic expression.

59. $x^2 \cdot x^3 \cdot x$

60. $6x^2 \cdot x^5$

61. $(xy)(-3x^2y^3)$

62. $(12x^2y)(3x^2y^4)$

63. $(5ab)(25a^3)$

64. $3uv(-2uv^2)^2$

In Exercises 65–68, simplify the algebraic expression by combining like terms.

65. $7x - 2x$

66. $25y + 32y$

67. $3u - 2v + 7v - 3u$

68. $7r - 4 - 9 + 3r$

In Exercises 69–74, simplify the algebraic expression using the Distributive Property.

69. $5(x - 4) + 10$

70. $15 - 7(z + 2)$

71. $3x - (y - 2x)$

72. $30x - (10x + 80)$

73. $3[b + 5(b - a)]$

74. $-2t[8 - (6 - t)] + 5t$

In Exercises 75 and 76, evaluate the algebraic expression for the specified values of the variable(s). If not possible, state the reason.

Expression	Values
75. $x^2 - 2x - 3$	(a) $x = 3$ (b) $x = 0$
76. $\dfrac{x}{y + 2}$	(a) $x = 0, \quad y = 3$
	(b) $x = 5, \quad y = -2$

P.5 In Exercises 77–80, translate the verbal phrase into an algebraic expression. (Let n represent the arbitrary real number.)

77. Two hundred decreased by three times a number

78. One hundred increased by the product of 15 and a number

79. The sum of the square of a number and 49

80. The absolute value of the sum of a number and 10, divided by 2

In Exercises 81–84, write a verbal description of the algebraic expression without using the variable.

81. $2y + 7$

82. $5u - 3$

83. $\dfrac{x - 5}{4}$

84. $-3(a - 10)$

In Exercises 85–88, construct an algebraic expression that represents the quantity given by the verbal statement.

85. The amount of income tax on a taxable income of I dollars when the tax rate is 18%

86. The distance traveled when you travel 8 hours at the average speed of r miles per hour

87. The area of a rectangle whose length is l inches and whose width is 5 units less than the length

88. The sum of three consecutive odd integers, the first of which is $2n + 1$

Solving Problems

Graphical Interpretation   In Exercises 89 and 90, use the figure, which shows the expenditures (in billions of dollars) for advertising for various media in 1996. (Source: McCann-Erickson, Inc.)

89. Determine the combined expenditures for advertising for the five media.

90. What is the difference between expenditures for television and for radio?

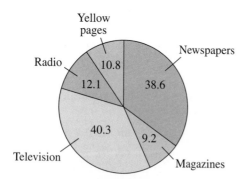

Numerical Interpretation   In Exercises 91–94, use the following information. The numbers (in millions) of revenue passengers enplaned in 1996 in the top six airports in the United States are Atlanta/Hartsfield (30.8), Chicago/O'Hare (30.5), Dallas/Ft. Worth (26.6), Denver International (15.2), Los Angeles International (22.7), and San Francisco International (15.2). (Source: U.S. Bureau of Transportation Statistics)

91. Find the volume difference between the airports with the greatest and smallest passenger volumes.

92. Determine the total number of passengers using the six airports.

93. Rank the airports from greatest to smallest passenger volume. Sketch a bar graph of the data.

94. There were approximately 381,000 aircraft departures from O'Hare in 1996. Approximate the average number of passengers per plane.

95. *Total Charge*   You purchase a product and make a down payment of $239 plus nine monthly payments of $45 each. What is the total amount you paid for the product?

96. *Total Charge*   You purchase a product and make a down payment of $387 plus 12 monthly payments of $68 each. What is the total amount you paid for the product?

Chapter Test

Take this test as you would take a test in class. After you are done, check your work against the answers given in the back of the book.

1. Place the correct symbol ($<$ or $>$) between the numbers.

  (a) $-\frac{5}{2}$ $-|-3|$ (b) $-\frac{2}{3}$ $-\frac{3}{2}$

2. Find the distance between -6.2 and 5.7.

In Exercises 3–10, evaluate the expression.

3. $-14 + 9 - 15$ **4.** $\frac{2}{3} + \left(-\frac{7}{6}\right)$

5. $-2(225 - 150)$ **6.** $(-3)(4)(-5)$

7. $\left(-\frac{7}{16}\right)\left(-\frac{8}{21}\right)$ **8.** $\frac{5}{18} \div \frac{15}{8}$

9. $\left(-\frac{3}{5}\right)^3$ **10.** $\dfrac{4^2 - 6}{5} + 13$

11. Name the property of real numbers demonstrated by the equation.

  (a) $(-3 \cdot 5) \cdot 6 = -3(5 \cdot 6)$ (b) $3y \cdot \dfrac{1}{3y} = 1$

12. Rewrite the expression $5(2x - 3)$ using the Distributive Property.

In Exercises 13–16, simplify the expression.

13. $(3x^2y)(-xy)^2$

14. $3x^2 - 2x - 5x^2 + 7x - 1$

15. $a(5a - 4) - 2(2a^2 - 2a)$

16. $4t - [3t - (10t + 7)]$

17. Explain the meaning of "evaluating an expression." Evaluate the expression $4 - (x + 1)^2$ for the given value of x.

  (a) $x = -1$ (b) $x = 3$

18. Translate the following statement into an algebraic expression.

  "The product of a number n and 5 is decreased by 8."

19. Write algebraic expressions for the perimeter and area of the rectangle shown at the left. Then simplify the expressions.

20. Write an algebraic expression for the sum of two consecutive even integers, the first of which is $2n$.

21. It is necessary to cut a 144-foot rope into nine pieces of equal length. What is the length of each piece?

22. A *cord* of wood is a pile 4 feet high, 4 feet wide, and 8 feet long. The volume of a rectangular solid is its length times its width times its height. Find the number of cubic feet in 5 cords of wood.

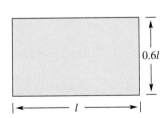

$0.6l$

l

Figure for 19

1 Linear Equations and Inequalities

Rob Crandall/Stock Boston

In 1996, 96 million households in the United States had televisions. Of these, 63 million or 65.3% had cable television provided by one of over 11,000 cable television systems. (Source: Television Bureau of Advertising, Inc., Warren Publishing)

Cable Television and You

You are having cable television installed in your house. You need to decide if you will purchase one or more premium movie channels or pay-per-view movies. You will not have both. Standard service is $31.20 per month and is required if you want a premium movie channel or pay-per-view movies. Each premium movie channel is $11.91 per month, and pay-per-view is $2.99 per month plus $3.95 per movie.

See Section 1.3, Exercise 86

a. Write a verbal model that gives the monthly cost of cable television based on the number of premium movie channels that you order.

b. Write an algebraic equation for your verbal model from part (a). Create a table that shows the amount paid per month for one, two, three, four, and five premium movie channels.

c. Write a verbal model that gives the monthly cost of cable television based on the number of pay-per-view movies you watch.

d. Write an algebraic equation for your verbal model from part (c). Create a table that shows the amount paid per month for one, two, three, four, five, six, seven, and eight pay-per-view movies.

e. If you are paying for two premium movie channels, what percent of your bill goes to paying for these movie channels?

See Section 1.4, Exercise 121

f. Your budget allows you to spend at most $50 per month on cable television. Use the algebraic model from part (b) to determine the number of premium movie channels you could purchase each month.

g. Your budget allows you to spend at most $50 per month on cable television. Use the algebraic model from part (d) to determine the number of pay-per-view movies that you could watch each month. Compare this with your answer to part (f). Which option would you choose, and why?

1.1   Linear Equations

Objectives

1 Check the solution of an equation.

2 Solve a linear equation in standard form.

3 Solve a linear equation in nonstandard form.

1 Check the solution of an equation.

Introduction

An **equation** is a statement that equates two mathematical expressions. Some examples are

$$x = 4, \quad 4x + 3 = 15, \quad 2x - 8 = 2(x - 4), \quad \text{and} \quad x^2 - 16 = 0.$$

Solving an equation involving a variable means finding all values of the variable for which the equation is true. Such values are **solutions** and are said to **satisfy** the equation. For instance, 3 is a solution of $4x + 3 = 15$ because $4(3) + 3 = 15$ is a true statement.

The **solution set** of an equation is the set of all solutions of the equation. Sometimes, an equation will have the set of all real numbers as its solution set. Such an equation is an **identity.** For instance, the equation

$$2x - 8 = 2(x - 4) \qquad \text{Identity}$$

is an identity because the equation is true for all real values of x. Try values such as 0, 1, -2, and 5 in this equation to see that each one is a solution.

An equation whose solution set is not the entire set of real numbers is called a **conditional equation.** For instance, the equation

$$x^2 - 16 = 0 \qquad \text{Conditional equation}$$

is a conditional equation because it has only two solutions, 4 and -4. Example 1 shows how to **check** whether a given value is a solution.

Study Tip

When checking a solution, we suggest you write a question mark over the equal sign to indicate that you are uncertain whether the "equation" is true for a given value of the variable.

Example 1   Checking a Solution of an Equation

Decide whether -3 is a solution of $-3x - 5 = 4x + 16$.

Solution

$$
\begin{aligned}
-3x - 5 &= 4x + 16 && \text{Original equation} \\
-3(-3) - 5 &\overset{?}{=} 4(-3) + 16 && \text{Substitute } -3 \text{ for } x. \\
9 - 5 &\overset{?}{=} -12 + 16 && \text{Simplify.} \\
4 &= 4 && \text{Solution checks. } \checkmark
\end{aligned}
$$

Because both sides turn out to be the same number, you can conclude that -3 is a solution of the original equation. Try checking to see whether -2 is a solution.

It is helpful to think of an equation as having two sides that are "in balance." Consequently, when you try to solve an equation, you must be careful to maintain that balance by performing the same operation(s) on both sides.

Two equations that have the same set of solutions are **equivalent equations.** For instance, the equations $x = 3$ and $x - 3 = 0$ are equivalent equations because both have only one solution—the number 3. When any one of the four techniques in the following list is applied to an equation, the resulting equation is equivalent to the original equation.

Kurt Gödel

(1906–1978)

In the late 1800s, a movement was begun to identify a complete set of axioms for each branch of mathematics from which all other propositions could be deduced. In 1931, Gödel, a faculty member at the University of Vienna, showed that this goal was unattainable. He proved that a complete set of axioms could never be identified for a branch of mathematics such that all of its propositions could be proven or disproven on the basis of those axioms. Although this closed one avenue of research, Gödel also pointed out new directions for the future.

▶ **Forming Equivalent Equations: Properties of Equality**

A given equation can be transformed into an *equivalent equation* using one or more of the following procedures.

	Original Equation	*Equivalent Equation*
1. *Simplify Each Side:* Remove symbols of grouping, combine like terms, or simplify fractions on one or both sides of the equation.	$4x - x = 8$	$3x = 8$
2. *Apply the Addition Property of Equality:* Add (or subtract) the same quantity to (from) *both* sides of the equation.	$x - 3 = 5$	$x = 8$
3. *Apply the Multiplication Property of Equality:* Multiply (or divide) *both* sides of the equation by the same nonzero quantity.	$3x = 12$	$x = 4$
4. *Interchange Sides:* Interchange the two sides of the equation.	$7 = x$	$x = 7$

When solving an equation, you can use any of the four techniques for forming equivalent equations to eliminate terms or factors in the equation. For example, to solve the equation

$$x + 4 = 2,$$

you need to remove the term $+4$ from the left side. This is accomplished by subtracting 4 from both sides.

$x + 4 = 2$	Original equation
$x + 4 - 4 = 2 - 4$	Subtract 4 from both sides.
$x + 0 = -2$	Combine like terms.
$x = -2$	Simplify.

Although this solution involved subtracting 4 from both sides, you could just as easily have added -4 to both sides. Both techniques are legitimate—which one you decide to use is a matter of personal preference.

2 Solve a linear equation in standard form.

Solving Linear Equations in Standard Form

The most common type of equation in one variable is a **linear equation.**

> ▶ **Definition of Linear Equation**
>
> A **linear equation** in one variable x is an equation that can be written in the standard form
>
> $$ax + b = c \qquad \text{Standard form}$$
>
> where a, b, and c are real numbers with $a \neq 0$.

A linear equation in one variable is also called a **first-degree equation** because its variable has an implied exponent of 1. Some examples of linear equations in the standard form $ax + b = c$ are $3x + 2 = 0$ and $5x - 4 = -6$.

Remember that to *solve* an equation in x means that you are to find the values of x that satisfy the equation. For a linear equation in the standard form

$$ax + b = c$$

the goal is to **isolate** x by rewriting the standard equation in the form

$$x = \text{(a number)}.$$

Beginning with the original equation, you write a sequence of equivalent equations, each having the same solution as the original equation.

Example 2   Solving a Linear Equation in Standard Form

Solve the equation $4x - 2 = 10$.

Solution

$4x - 2 = 10$	Original equation
$4x - 2 + 2 = 10 + 2$	Add 2 to both sides.
$4x = 12$	Combine like terms.
$\dfrac{4x}{4} = \dfrac{12}{4}$	Divide both sides by 4.
$x = 3$	Simplify.

It appears that the solution is 3. You can check this as follows.

Check

$4x - 2 = 10$	Original equation
$4(3) - 2 \stackrel{?}{=} 10$	Substitute 3 for x.
$12 - 2 \stackrel{?}{=} 10$	Simplify.
$10 = 10$	Solution checks. ✓

Study Tip

Be sure you see that solving an equation such as the one in Example 2 has two basic steps. The first step is to *find* the solution(s). The second step is to *check* that each solution you find actually satisfies the original equation. You can improve your accuracy in algebra by developing the habit of checking each solution.

You know that 3 is a solution of the equation in Example 2, but at this point you might be asking, "How can I be sure that the equation does not have other solutions?" The answer is that a linear equation in one variable always has *exactly one* solution. You can show this with the following steps.

$$ax + b = c \qquad \text{Original equation, with } a \neq 0$$

$$ax + b - b = c - b \qquad \text{Subtract } b \text{ from both sides.}$$

$$ax = c - b \qquad \text{Combine like terms.}$$

$$\frac{ax}{a} = \frac{c - b}{a} \qquad \text{Divide both sides by } a.$$

$$x = \frac{c - b}{a} \qquad \text{Simplify.}$$

It is clear that the last equation has only one solution, $x = (c - b)/a$. Because the last equation is equivalent to the given equation, you can conclude that every linear equation written in standard form in one variable has exactly one solution.

Example 3   Solving a Linear Equation in Standard Form

Solve the equation $2x - 3 = -5$.

Solution

$$2x - 3 = -5 \qquad \text{Original equation}$$

$$2x - 3 + 3 = -5 + 3 \qquad \text{Add 3 to both sides.}$$

$$2x = -2 \qquad \text{Combine like terms.}$$

$$\frac{2x}{2} = \frac{-2}{2} \qquad \text{Divide both sides by 2.}$$

$$x = -1 \qquad \text{Simplify.}$$

The solution is -1. Check this in the original equation, as follows.

Check

$$2x - 3 = -5 \qquad \text{Original equation}$$

$$2(-1) - 3 \stackrel{?}{=} -5 \qquad \text{Substitute } -1 \text{ for } x.$$

$$-2 - 3 \stackrel{?}{=} -5 \qquad \text{Simplify.}$$

$$-5 = -5 \qquad \text{Solution checks. } \checkmark$$

As you gain experience in solving linear equations, you will probably be able to perform some of the solution steps in your head. For instance, in Example 3, you might write only the following steps.

$$2x - 3 = -5 \qquad \text{Original equation}$$

$$2x = -2 \qquad \text{Add 3 to both sides.}$$

$$x = -1 \qquad \text{Divide both sides by 2.}$$

3 Solve a linear equation in nonstandard form.

Solving Linear Equations in Nonstandard Form

Linear equations often occur in nonstandard forms that contain symbols of grouping or like terms that are not combined. Here are some examples.

$$x + 2 = 2x - 6, \quad 6(y - 1) = 2y - 3, \quad \frac{x}{18} + \frac{3x}{4} = 2$$

The next three examples show how to solve these linear equations.

Study Tip

Remember that the goal in solving any linear equation is to rewrite the given equation so that all the variable terms are on one side of the equal sign and all constant terms are on the other side.

Example 4   Solving a Linear Equation in Nonstandard Form

$x + 2 = 2x - 6$	Original equation
$-2x + x + 2 = -2x + 2x - 6$	Add $-2x$ to both sides.
$-x + 2 = -6$	Combine like terms.
$-x + 2 - 2 = -6 - 2$	Subtract 2 from both sides.
$-x = -8$	Combine like terms.
$(-1)(-x) = (-1)(-8)$	Multiply both sides by -1.
$x = 8$	Simplify.

The solution is 8. Check this in the original equation.

In most cases, it helps to remove symbols of grouping as a first step in solving an equation. This is illustrated in Example 5.

Example 5   Solving a Linear Equation That Contains Parentheses

$6(y - 1) = 2y - 3$	Original equation
$6y - 6 = 2y - 3$	Distributive Property
$6y - 2y - 6 = 2y - 2y - 3$	Subtract $2y$ from both sides.
$4y - 6 = -3$	Combine like terms.
$4y - 6 + 6 = -3 + 6$	Add 6 to both sides.
$4y = 3$	Combine like terms.
$\dfrac{4y}{4} = \dfrac{3}{4}$	Divide both sides by 4.
$y = \dfrac{3}{4}$	Simplify.

The solution is $\frac{3}{4}$. Check this in the original equation.

If a linear equation contains fractions, we suggest that you first *clear the equation of fractions* by multiplying both sides of the equation by the least common denominator (LCD) of the fractions.

Example 6   Solving a Linear Equation That Contains Fractions

Solve the equation $\dfrac{x}{18} + \dfrac{3x}{4} = 2$.

Solution

$$\frac{x}{18} + \frac{3x}{4} = 2 \qquad \text{Original equation}$$

$$36\left(\frac{x}{18} + \frac{3x}{4}\right) = 36(2) \qquad \text{Multiply both sides by LCD of 36.}$$

$$36 \cdot \frac{x}{18} + 36 \cdot \frac{3x}{4} = 36(2) \qquad \text{Distributive Property}$$

$$2x + 27x = 72 \qquad \text{Simplify.}$$

$$29x = 72 \qquad \text{Combine like terms.}$$

$$\frac{29x}{29} = \frac{72}{29} \qquad \text{Divide both sides by 29.}$$

$$x = \frac{72}{29} \qquad \text{Simplify.}$$

The solution is $\frac{72}{29}$. Check this in the original equation.

The next example shows how to solve a linear equation involving decimals. The procedure is basically the same, but the arithmetic can be messier.

Example 7   Solving a Linear Equation Involving Decimals

Solve the equation $0.12x + 0.09(5000 - x) = 513$.

Solution

$$0.12x + 0.09(5000 - x) = 513 \qquad \text{Original equation}$$

$$0.12x + 450 - 0.09x = 513 \qquad \text{Distributive Property}$$

$$0.03x + 450 = 513 \qquad \text{Combine like terms.}$$

$$0.03x + 450 - 450 = 513 - 450 \qquad \text{Subtract 450 from both sides.}$$

$$0.03x = 63 \qquad \text{Combine like terms.}$$

$$\frac{0.03x}{0.03} = \frac{63}{0.03} \qquad \text{Divide both sides by 0.03.}$$

$$x = 2100 \qquad \text{Simplify.}$$

The solution is 2100. Check this in the original equation.

Study Tip

A different approach to Example 7 would be to begin by multiplying both sides of the equation by 100. This would clear the equation of decimals to produce

$$12x + 9(5000 - x) = 51,300.$$

Try solving this equation to see that you obtain the same solution.

Study Tip

Avoid the temptation to first divide an equation by x. You may obtain an incorrect solution, as in the following example.

$$7x = -4x \quad \text{Original equation}$$

$$\frac{7x}{x} = -\frac{4x}{x} \quad \text{Divide both sides by } x.$$

$$7 = -4 \quad \text{False statement}$$

The false statement indicates that there is no solution. However, when the equation is solved correctly, the solution is $x = 0$.

$$7x = -4x$$

$$7x + 4x = -4x + 4x$$

$$11x = 0$$

$$\frac{11x}{11} = \frac{0}{11}$$

$$x = 0$$

Some equations in nonstandard form have no solution or infinitely many solutions. These cases are illustrated in Example 8.

Example 8   Solving Linear Equations: Special Cases

Solve the following equations.

a. $2x - 4 = 2(x - 3)$

b. $3x + 2 + 2(x - 6) = 5(x - 2)$

Solution

a.
$2x - 4 = 2(x - 3)$	Original equation
$2x - 4 = 2x - 6$	Distributive Property
$-4 = -6$	Subtract $2x$ from both sides.

Because the last equation is a false statement, you can conclude that the original equation has no solution.

b.
$3x + 2 + 2(x - 6) = 5(x - 2)$	Original equation
$3x + 2 + 2x - 12 = 5x - 10$	Distributive Property
$5x - 10 = 5x - 10$	Combine like terms.
$5x - 5x - 10 = 5x - 5x - 10$	Subtract $5x$ from both sides.
$-10 = -10$	Simplify.

Because the last equation is true for any value of x, the equation is an identity, and you can conclude that the original equation has infinitely many solutions.

Discussing the Concept	Analyzing and Interpreting Equations

Classify each of the following equations as an identity, a conditional equation, or an equation with no solution. Compare your conclusions with those of the rest of your class and discuss the reasons for each conclusion.

a. $2x - 3 = -4 + 2x$

b. $x + 0.05x = 37.75$

c. $5x(3 + x) = 15x + 5x^2$

Discuss possible realistic situations in which the equations you classified as an identity and a conditional equation might apply. Write a brief description of these situations and explain how the equations could be used.

1.1   Exercises

Integrated Review   *Concepts, Skills, and Problem Solving*

Keep mathematically in shape by doing these exercises *before* the problems of this section.

Properties and Definitions

In Exercises 1–4, identify the property illustrated by the equation.

1. $5 + x = x + 5$ **2.** $10 \cdot \frac{1}{10} = 1$

3. $6(x - 2) = 6x - 6 \cdot 2$

4. $3 + (4 + x) = (3 + 4) + x$

Simplifying Expressions

In Exercises 5–10, evaluate the expression.

5. $4 - |-3|$ **6.** $-10 - (4 - 18)$

7. $\dfrac{3 - (5 - 20)}{4}$ **8.** $\dfrac{|3 - 18|}{3}$

9. $6\left(\frac{2}{15}\right)$ **10.** $\frac{7}{12} \div \frac{5}{16}$

Problem Solving

11. You plan to save $75 per month for 20 years. How much money will you set aside during the 20 years?

12. It is necessary to cut a 135-foot rope into 15 pieces of equal length. What is the length of each piece?

Developing Skills

In Exercises 1–8, decide whether each value of the variable is a solution of the equation. See Example 1.

	Equation	*Values*
1.	$3x - 7 = 2$	(a) $x = 0$
		(b) $x = 3$
2.	$5x + 9 = 4$	(a) $x = -1$
		(b) $x = 2$
3.	$x + 8 = 3x$	(a) $x = 4$
		(b) $x = -4$
4.	$10x - 3 = 7x$	(a) $x = 0$
		(b) $x = -1$
5.	$3x + 3 = 2(x - 4)$	(a) $x = -11$
		(b) $x = 5$
6.	$7x - 1 = 5(x + 5)$	(a) $x = 2$
		(b) $x = 13$
7.	$\frac{1}{4}x = 3$	(a) $x = -4$
		(b) $x = 12$
8.	$3(y + 2) = y - 5$	(a) $y = -\frac{3}{2}$
		(b) $y = -5.5$

In Exercises 9–12, identify the equation as a conditional equation, an identity, or an equation with no solution.

9. $3(x - 1) = 3x$

10. $2x + 8 = 6x$

11. $5(x + 3) = 2x + 3(x + 5)$

12. $\frac{2}{3}x + 4 = \frac{1}{3}x + 12$

In Exercises 13–16, determine whether the equation is linear. If not, state why.

13. $3x + 4 = 10$ **14.** $x^2 + 3 = 8$

15. $\dfrac{4}{x} - 3 = 5$ **16.** $3(x - 2) = 4x$

In Exercises 17–20, justify each step of the solution.

17. $3x + 15 = 0$ Original equation

$3x + 15 - 15 = 0 - 15$

$3x = -15$

$\dfrac{3x}{3} = \dfrac{-15}{3}$

$x = -5$

18. $\quad 7x - 21 = 0$ $\qquad$ Original equation

$$7x - 21 + 21 = 0 + 21$$

$$7x = 21$$

$$\frac{7x}{7} = \frac{21}{7}$$

$$x = 3$$

19. $\quad -2x + 5 = 12$ $\qquad$ Original equation

$$-2x + 5 - 5 = 12 - 5$$

$$-2x = 7$$

$$\frac{-2x}{-2} = \frac{7}{-2}$$

$$x = -\frac{7}{2}$$

20. $\quad 25 - 3x = 10$ $\qquad$ Original equation

$$25 - 3x + 3x = 10 + 3x$$

$$25 = 10 + 3x$$

$$25 - 10 = 10 + 3x - 10$$

$$15 = 3x$$

$$\frac{15}{3} = \frac{3x}{3}$$

$$5 = x$$

In Exercises 21–80, solve the equation and check the result. (If it has no solution, state the reason.) See Examples 2–8.

21. $x - 3 = 0$

22. $x + 8 = 0$

23. $3x = 12$

24. $-14x = 28$

25. $-6y = 4.2$

26. $0.5t = 7$

27. $6x + 4 = 0$

28. $8z - 10 = 0$

29. $-2u + 5 = 7$

30. $3 - 2y = 5$

31. $4x - 7 = -11$

32. $5y + 9 = -6$

33. $23x - 4 = 42$

34. $15x - 18 = 27$

35. $3t + 8 = -2$

36. $10 - 6x = -5$

37. $8 - 5t = 20 + t$

38. $3y + 14 = y + 20$

39. $4x - 5 = 2x - 1$

40. $8 - 7y = 5y - 4$

41. $7 - 8x = 13x$

42. $2s - 16 = 34s$

43. $4y - 3 = 4y$

44. $24 - 2x = x$

45. $-8t = -16t$

46. $4x = -12x$

47. $-9y - 4 = -9y$

48. $6a + 2 = 6a$

49. $8(x - 8) = 24$

50. $6(x + 2) = 30$

51. $-4(t + 2) = 0$

52. $8(z - 8) = 0$

53. $3(x - 4) = 7x + 6$

54. $-2(t + 3) = 9 - 5t$

55. $8x - 3(x - 2) = 12$

56. $12 = 6(y + 1) - 8y$

57. $5 - (2y - 4) = 15$

58. $26 - (3x - 10) = 6$

59. $12(x + 3) = 7(x + 3)$

60. $-5(x - 10) = 6(x - 10)$

61. $2(x + 7) - 9 = 5(x - 4)$

62. $4(2 - x) = -2(x + 7) - 2x$

63. $\dfrac{u}{5} = 10$

64. $-\dfrac{z}{2} = 7$

65. $t - \frac{2}{5} = \frac{3}{2}$

66. $z + \frac{1}{15} = -\frac{3}{10}$

67. $\dfrac{t}{5} - \dfrac{t}{2} = 1$

68. $\dfrac{t}{6} + \dfrac{t}{8} = 1$

69. $\dfrac{8x}{5} - \dfrac{x}{4} = -3$

70. $\dfrac{11x}{6} + \dfrac{1}{3} = 2x$

71. $\frac{1}{3}x + 1 = \frac{1}{12}x - 4$

72. $\frac{1}{9}x + \frac{1}{3} = \frac{11}{18}$

73. $\dfrac{25 - 4u}{3} = \dfrac{5u + 12}{4} + 6$

74. $\dfrac{8 - 3x}{4} - 4 = \dfrac{x}{6}$

75. $0.3x + 1.5 = 8.4$

76. $16.3 - 0.2x = 7.1$

77. $1.2(x - 3) = 10.8$

78. $6.5(1 - 2x) = 13$

79. $\frac{2}{3}(2x - 4) = \frac{1}{2}(x + 3) - 4$

80. $\frac{3}{4}(6 - x) = \frac{1}{3}(4x + 5) + 2$

Solving Problems

81. *Number Problem*   The sum of two consecutive integers is 251. Find the integers.

82. *Number Problem*   The sum of two consecutive integers is 137. Find the integers.

83. *Number Problem*   The sum of two consecutive even integers is 166. Find the integers.

84. *Number Problem*   The sum of two consecutive even integers is 626. Find the integers.

85. *Car Repair*   The bill for the repair of your car was $210. The cost for parts was $162. The cost for labor was $32 per hour. How many hours did the repair work take?

86. *Appliance Repair*   The bill for the repair of your refrigerator was $172. The cost for parts was $74. The cost for the service call and the first half hour of service was $50. The additional cost for labor was $16 per half hour. How many hours did the repair work take?

87. *Maximum Height of a Fountain*   Consider the fountain shown in the figure. The initial velocity of the stream of the water is 48 feet per second. The velocity v of the water at any time t (in seconds) is given by $v = 48 - 32t$. Find the time for a drop of water to travel from the base to the maximum height of the fountain. (*Hint:* The maximum height is reached when $v = 0$.)

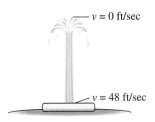

88. *Maximum Height of an Object*   The velocity v of an object projected vertically upward with an initial velocity of 64 feet per second is given by $v = 64 - 32t$, where t is time in seconds. When does the object reach its maximum height?

89. *Work-Rate Problem*   Two people can complete a task in t hours, where t must satisfy the equation

$$\frac{t}{10} + \frac{t}{15} = 1.$$

Find the required time t.

90. *Work-Rate Problem*   Two people can complete a task in t hours, where t must satisfy the equation

$$\frac{t}{12} + \frac{t}{20} = 1.$$

Find the required time t.

91. *Investigation*   The length of a rectangle is t times its width. So, the perimeter P is given by $P = 2w + 2(tw)$, where w is the width of the rectangle. The perimeter of the rectangle is 1200 meters.

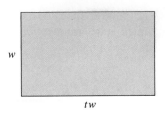

w

t w

Figure for 91

(a) Complete the table of widths, lengths, and areas of the rectangle for the specified values of t.

t	1	1.5	2	3	4	5
Width						
Length						
Area						

(b) Use the table to write a short paragraph describing the relationship among the width, length, and area of a rectangle that has a *fixed* perimeter.

92. *Geometry*   The length of a rectangle is 10 meters greater than its width. If the perimeter is 64 meters, find the dimensions of the rectangle.

93. *Using a Model*   The average annual expenditures per student for primary and secondary public schools in the United States from 1990 to 1997 can be approximated by the model

$$y = 207t + 4962, \quad 0 \le t \le 7$$

where y represents expenditures in dollars and t represents the year, with $t = 0$ corresponding to 1990 (see figure). According to this model, during which year did the expenditures first reach $5500? Explain how to answer the question graphically, numerically, and algebraically. (Source: National Education Association)

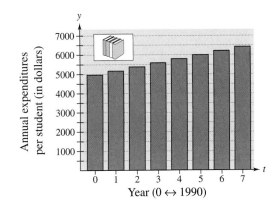

94. *Using a Model*   The number of cable television subscribers in the United States from 1990 to 1997 can be approximated by the model

$$y = 49.3 + 1.93t, \quad 0 \le t \le 7$$

where y represents the number of subscribers (in millions) and t represents the year, with $t = 0$ corresponding to 1990 (see figure). According to this model, during which year were there 57 million subscribers? Explain how to answer the question graphically, numerically, and algebraically.   (Source: Television & Cable Factbook)

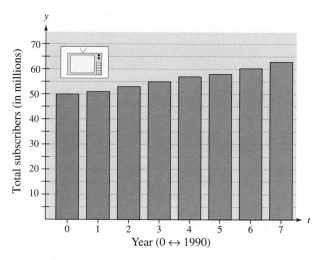

Figure for 94

Explaining Concepts

95. Explain the difference between a conditional equation and an identity.

96. Explain how you can decide whether a real number is a solution of an equation.

97. Explain the difference between evaluating an expression and solving an equation.

98. Give the standard form of a linear equation. Why is a linear equation sometimes called a first-degree equation?

99. What is meant by equivalent equations? Give an example of two equivalent equations.

100. In your own words, describe the steps used to transform an equation into an equivalent equation.

101. *True or False?*   Multiplying both sides of an equation by 0 yields an equivalent equation.

102. *True or False?*   Subtracting 0 from both sides of an equation yields an equivalent equation.

1.2   Linear Equations and Problem Solving

Objectives

1 Use mathematical modeling to write an algebraic equation representing a real-life situation.

2 Solve a percent problem using a percent equation.

3 Use a ratio to compare unit prices for particular products.

4 Solve a proportion.

1 Use mathematical modeling to write an algebraic equation representing a real-life situation.

Mathematical Modeling

In this section you will see how algebra can be used to solve problems that occur in real-life situations. This process is called **mathematical modeling,** and its basic steps are shown below.

Verbal Description ⟹ Verbal Model ⟹ Assign Labels ⟹ Algebraic Equation

Example 1   Mathematical Modeling

Write an algebraic equation that represents the following problem. Then solve the equation and answer the question.

You have accepted a job at an annual salary of $27,630. This salary includes a year-end bonus of $750. If you are paid twice a month, what will your gross pay be for each paycheck?

Solution

Because there are 12 months in a year and you will be paid twice a month, it follows that you will receive 24 paychecks during the year. Construct an algebraic equation for this problem, as follows. Begin with a verbal model, then assign labels, and finally form an algebraic equation.

Verbal Model: $\boxed{\text{Income for year}} = 24 \times \boxed{\text{Amount of each paycheck}} + \boxed{\text{Bonus}}$

Labels: Income for year = 27,630   (dollars)
Amount of each paycheck = x (dollars)
Bonus = 750   (dollars)

Equation: $27,630 = 24x + 750$

$27,630 - 750 = 24x + 750 - 750$ Subtract 750 from both sides.

$26,880 = 24x$ Combine like terms.

$\dfrac{26,880}{24} = \dfrac{24x}{24}$ Divide both sides by 24.

$1120 = x$ Simplify.

Each paycheck will be $1120. Check this in the original statement of the problem.

Study Tip

You could solve the problem in Example 1 *without* algebra by simply subtracting the bonus of $750 from the annual salary of $27,630 and dividing the result by 24 pay periods. The reason for listing this example is to allow you to practice writing algebraic versions of the problem-solving skills *you already possess.* Your goals in this section are to practice formulating problems by logical reasoning *and* to use this reasoning to write algebraic versions of the problems. Later, you will encounter more complicated problems in which algebra is a necessary part of the solution.

2 Solve a percent problem using a percent equation.

Percent Problems

Rates that describe increases, decreases, and discounts are often given as percents. **Percent** means *per hundred*, so 40% means 40 per hundred or, equivalently, $\frac{40}{100}$. The word "per" occurs in many other rates, such as price per ounce, miles per gallon, revolutions per minute, and cost per share. In applications involving percents, you need to convert the percent number to decimal or fraction form before performing any arithmetic operations. Some examples are listed below.

Percent	10%	$12\frac{1}{2}$%	20%	25%	$33\frac{1}{3}$%	50%	$66\frac{2}{3}$%	75%
Decimal	0.1	0.125	0.2	0.25	0.33...	0.5	0.66...	0.75
Fraction	$\frac{1}{10}$	$\frac{1}{8}$	$\frac{1}{5}$	$\frac{1}{4}$	$\frac{1}{3}$	$\frac{1}{2}$	$\frac{2}{3}$	$\frac{3}{4}$

> ∿ **Technology: Tip**
>
> Some scientific calculators and graphing utilities can convert percents to decimal and fraction forms. Consult the user's guide for your calculator or graphing utility for the proper keystrokes. Use your calculator or graphing utility to convert the following percents to decimal and fraction forms.
>
> **a.** 35%   **b.** 0.1%
>
> **c.** 8%   **d.** 96%
>
> **e.** 150%   **f.** 200%

The primary use of percents is to compare two numbers. For example, you can compare 3 and 6 by saying that 3 is 50% of 6. In this statement, 6 is the **base number,** and 3 is the number being compared to the base number. The following model, which is called the **percent equation,** is helpful.

Verbal Model: Compared number = Percent (decimal form) · Base number

Labels: Compared number = a
       Percent = p (decimal form)
       Base number = b

Equation: $a = p \cdot b$ Percent equation

Remember to convert p to a decimal value before multiplying by b.

Example 2   Solving a Percent Problem

The number 15.6 is 26% of what number?

Solution

Verbal Model: Compared number = Percent (decimal form) · Base number

Labels: Compared number = 15.6
       Percent = 0.26   (decimal form)
       Base number = b

Equation: $15.6 = 0.26b$

       $\dfrac{15.6}{0.26} = b$ Divide both sides by 0.26.

       $60 = b$ Simplify.

Check that 15.6 is 26% of 60 by multiplying 60 by 0.26 to get 15.6.

Example 3   Solving a Percent Problem

The number 28 is what percent of 80?

Solution

Verbal Model: $\dfrac{\text{Compared}}{\text{number}} = \dfrac{\text{Percent}}{\text{(decimal form)}} \cdot \dfrac{\text{Base}}{\text{number}}$

Labels: Compared number $= 28$
Percent $= p$ (decimal form)
Base number $= 80$

Equation: $28 = p(80)$

$$\frac{28}{80} = p$$

$$0.35 = p$$

So, 28 is 35% of 80. Check this solution by multiplying 80 by 0.35 to see that you obtain 28.

In most real-life applications, the base number b and the number a are much more disguised than in Examples 2 and 3. It sometimes helps to think of a as the "new" amount and b as the "original" amount.

Example 4   A Percent Application

A real estate agency receives a commission of $8092.50 for the sale of a $124,500 house. What percent commission is this?

Solution

A commission is a percent of the sale price paid to the agency for their services. To determine the percent commission, start with a verbal model.

Verbal Model: $\text{Commission} = \dfrac{\text{Percent}}{\text{(decimal form)}} \cdot \dfrac{\text{Sale}}{\text{price}}$

Labels: Commission $= 8092.50$ (dollars)
Percent $= p$ (decimal form)
Sale price $= 124,500$ (dollars)

Equation: $8092.50 = p(124,500)$

$$\frac{8092.50}{124,500} = p$$

$$0.065 = p$$

The real estate agency receives a commission of 6.5%. Check this solution by multiplying 124,500 by 0.065 to see that you obtain 8092.50.

3 Use a ratio to compare unit prices for particular products.

Ratios and Unit Prices

You know that a percent compares a number with 100. A **ratio** is a more generic rate form that compares one number with another. Specifically, if a and b have the same units of measure, then a/b is called the ratio of a to b. Note the *order* implied by a ratio. The ratio of a to b means a/b, whereas the ratio of b to a means b/a.

| Example 5 | Using a Ratio |

Find the ratio of 4 feet to 8 inches.

Solution

Because the units of feet and inches are not the same, you must first convert 4 feet into its equivalent in inches or convert 8 inches into its equivalent in feet. You can convert 4 feet to 48 inches (by multiplying by 12) to obtain

$$\frac{4 \text{ feet}}{8 \text{ inches}} = \frac{48 \text{ inches}}{8 \text{ inches}} = \frac{48}{8} = \frac{6}{1}.$$

Or, you can convert 8 inches to $\frac{8}{12}$ feet (by dividing by 12) to obtain

$$\frac{4 \text{ feet}}{8 \text{ inches}} = \frac{4 \text{ feet}}{\frac{8}{12} \text{ feet}} = 4 \div \frac{8}{12} = 4 \cdot \frac{12}{8} = \frac{6}{1}.$$

The **unit price** of an item is the quotient of the total price divided by the total units. That is, unit price = total price/total units. To state unit prices, use the word "per." For instance, the unit price for a particular brand of coffee might be 4.79 dollars *per* pound.

| Example 6 | Comparing Unit Prices |

Which is the better buy, a 12-ounce box of breakfast cereal for $2.79 or a 16-ounce box of the same cereal for $3.59?

Solution

The unit price for the smaller box is

$$\text{Unit price} = \frac{\text{total price}}{\text{total units}} = \frac{\$2.79}{12 \text{ ounces}} \approx \$0.2325 \text{ per ounce.}$$

The unit price for the larger box is

$$\text{Unit price} = \frac{\text{total price}}{\text{total units}} = \frac{\$3.59}{16 \text{ ounces}} \approx \$0.2244 \text{ per ounce.}$$

The larger box has a slightly smaller unit price, and so it is the better buy.

4 Solve a proportion.

Solving Proportions

A **proportion** is a statement that equates two ratios. For example, if the ratio of a to b is the same as the ratio of c to d, you can write the proportion as $a/b = c/d$. In typical problems, you know three of the values and need to find the fourth.

Example 7   Solving a Proportion in Geometry

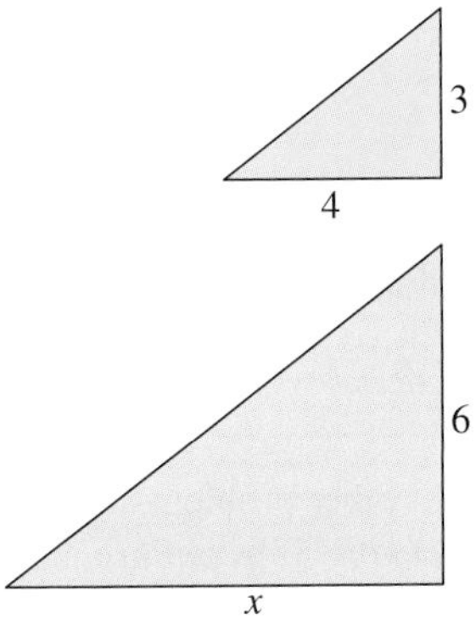

The triangles shown in Figure 1.1 are similar triangles. They have the same shape, but are different in size. Because they are similar triangles, their corresponding sides are proportional. Use this fact to find the length of the unknown side x of the bottom triangle.

Solution

$$\frac{4}{3} = \frac{x}{6}$$ Set up proportion.

$$6 \cdot \frac{4}{3} = 6 \cdot \frac{x}{6}$$ Multiply both sides by 6.

$$8 = x$$ Simplify.

So, $x = 8$. Check this in the original statement of the problem.

Figure 1.1

Example 8   An Application of Proportion

You are driving from Arizona to New York, a trip of 2750 miles. You begin the trip with a full tank of gas and after traveling 424 miles, you refill the tank for $22.00. Approximately how much should you expect to spend on gasoline for the entire trip?

Solution

Verbal Model: $$\frac{\text{Cost of gas for trip}}{\text{Cost of gas for tank}} = \frac{\text{Miles for trip}}{\text{Miles for tank}}$$

Labels:

Cost of gas for entire trip = x	(dollars)
Cost of gas for tank = 22	(dollars)
Miles for entire trip = 2750	(miles)
Miles for tank = 424	(miles)

Proportion: $$\frac{x}{22} = \frac{2750}{424}$$

$$x = 22 \cdot \left(\frac{2750}{424} \right)$$

$$x \approx 142.69$$

You should expect to spend approximately $142.69 for gasoline on the trip. Check this in the original statement of the problem.

The following list summarizes the strategy for modeling and solving a real-life problem.

▶ **Strategy for Solving Word Problems**

1. Ask yourself what you need to know to solve the problem. Then *write a verbal model* that will give you what you need to know.

2. *Assign labels* to each part of the verbal model—numbers to the known quantities and letters (or expressions) to the variable quantities.

3. Use the labels to *write an algebraic model* based on the verbal model.

4. *Solve* the resulting algebraic equation.

5. *Answer* the original question and *check* that your answer satisfies the original problem as stated.

In previous mathematics courses, you studied several other problem-solving strategies, such as *drawing a diagram*, *making a table*, *looking for a pattern*, and *solving a simpler problem*. Each of these strategies can also help you to solve problems in algebra.

Discussing the Concept	Checking the Sensibility of an Answer

When you solve a problem related to a real-life situation, you should always ask yourself whether your answer makes sense. Discuss why the following answers are suspicious and decide what a more reasonable answer might be.

a. A problem asks you to find the life expectancy of an American male born in 1970, and you use a formula to obtain a preliminary answer of 124 years.

b. A problem asks you to find the Celsius temperature equivalent to 80° Fahrenheit, and you obtain a preliminary answer of −24°C.

c. A problem asks you to find the diameter of a round household electrical extension cord, and you obtain a preliminary answer of 5.0 inches.

d. A problem asks you to find the floor area of a gymnasium, and you obtain a preliminary answer of 300 square feet.

1.2   Exercises

Integrated Review   *Concepts, Skills, and Problem Solving*

Keep mathematically in shape by doing these exercises *before* the problems of this section.

Properties and Definitions

1. In your own words, define an algebraic expression.

2. State the definition of the *terms* of an algebraic expression.

3. Complete the property of exponents:

   $a^m \cdot a^n =$ _____ .

4. Complete the property of exponents:

   $(ab)^m =$ _____ .

Algebraic Operations

In Exercises 5–10, perform the indicated operations.

5. $-360 + 120$

6. $5(57 - 33)$

7. $-\frac{4}{15} \cdot \frac{15}{16}$

8. $\frac{3}{8} \div \frac{5}{16}$

9. $(12 - 15)^3$

10. $\left(\frac{5}{8}\right)^2$

Problem Solving

In Exercises 11 and 12, find an expression for the perimeter of the figure. Simplify the expression.

11.

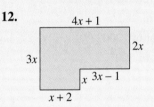

12.

Developing Skills

Mathematical Modeling   In Exercises 1–4, construct a verbal model and write an algebraic equation that represents the problem. Solve the equation. See Example 1.

1. Find a number such that the sum of the number and 30 is 82.

2. Find a number such that the difference of the number and 18 is 27.

3. You have accepted a job offer at an annual salary of $30,500. This salary includes a year-end bonus of $2300. If you are paid every 2 weeks, what will your gross pay be for each paycheck?

4. You have a job on an assembly line for which you are paid $10 per hour plus $0.75 per unit assembled. Find the number of units produced in an 8-hour day if your earnings for the day are $146.

In Exercises 5–10, complete the table showing the equivalent forms of various percents.

Percent	Parts out of 100	Decimal	Fraction
5. 30%			
6. 75%			
7.		0.075	
8.			$\frac{2}{3}$
9.			$\frac{1}{8}$
10.	100		

In Exercises 11–26, solve using a percent equation. See Examples 2 and 3.

11. What is 35% of 250?

12. What is 68% of 800?

13. What is 8.5% of 816?

14. What is $33\frac{1}{3}\%$ of 816?

15. What is 0.4% of 150,000?

16. What is 300% of 16?

17. 84 is 24% of what number?

18. 416 is 65% of what number?

19. 42 is $10\frac{1}{2}\%$ of what number?

20. 168 is 350% of what number?

21. 96 is 0.8% of what number?

22. 496 is what percent of 800?

23. 1650 is what percent of 5000?

24. 2.4 is what percent of 480?

25. 2100 is what percent of 1200?

26. 900 is what percent of 500?

In Exercises 27–34, write the verbal expression as a ratio. Use the same units in both the numerator and denominator, and simplify. See Example 5.

27. 120 meters to 180 meters

28. 12 ounces to 20 ounces

29. 36 inches to 48 inches

30. 125 centimeters to 2 meters

31. 40 milliliters to 1 liter

32. 1 pint to 1 gallon

33. 5 pounds to 24 ounces

34. 45 minutes to 2 hours

In Exercises 35–44, solve the proportion. See Example 7.

35. $\dfrac{x}{6} = \dfrac{2}{3}$

36. $\dfrac{y}{36} = \dfrac{6}{7}$

37. $\dfrac{t}{4} = \dfrac{3}{2}$

38. $\dfrac{5}{16} = \dfrac{x}{4}$

39. $\dfrac{5}{4} = \dfrac{t}{6}$

40. $\dfrac{7}{8} = \dfrac{x}{2}$

41. $\dfrac{y}{6} = \dfrac{y-2}{4}$

42. $\dfrac{a}{5} = \dfrac{a+4}{8}$

43. $\dfrac{y+1}{10} = \dfrac{y-1}{6}$

44. $\dfrac{z-3}{3} = \dfrac{z+8}{12}$

Solving Problems

45. *College Enrollment* Thirty-eight percent of the students enrolled at a college are freshmen. Determine the number of freshmen if the enrollment of the college is 3000.

46. *Pension Fund* Your employer withholds $6\frac{1}{2}\%$ of your gross income for your retirement. Determine the amount withheld each month if your gross monthly income is $3800.

47. *Passing Grade* There are 40 students in your class. On one test, 95% of the students received passing grades. How many students failed the test?

48. *Number of Eligible Voters* The news media reported that 7387 votes were cast in the last election and that this represented 63% of the eligible voters in the district. Assuming that this is true, how many eligible voters are in the district?

49. *Company Layoff* Because of slumping sales, a small company laid off 25 of its 160 employees. What percent of the work force was laid off?

50. *Monthly Rent* If you spend $748 of your monthly income of $3400 for rent, what percent of your monthly income is your monthly rent payment?

51. *Tip Rate* A customer left $10 for a meal that cost $8.45. Determine the tip rate.

52. *Tip Rate* A customer left $40 for a meal that cost $34.73. Determine the tip rate.

53. *Real Estate Commission* A real estate agency receives a commission of $12,250 for the sale of a $175,000 house. What percent commission is this?

54. *Real Estate Commission* A real estate agency receives a commission of $20,400 for the sale of a $240,000 house. What percent commission is this?

55. *Quality Control* A quality control engineer reported that 1.5% of a sample of parts were defective. The engineer found three defective parts. How large was the sample?

56. *Price Inflation*   A new van costs $29,750, which is approximately 115% of what a comparable van cost 3 years ago. What did it cost 3 years ago?

57. *Floor Space*   You are planning to build a tool shed, but you are undecided about the size. The two sizes you are considering are 12 feet by 15 feet and 16 feet by 20 feet. The floor space of the larger is what percent of the floor space of the smaller? The floor space of the smaller is what percent of the floor space of the larger?

58. *Geometry*   The floor of a room that measures 10 feet by 12 feet is partially covered by a circular rug with a radius of 4 feet (see figure). What percent of the floor is covered by the rug? (*Hint:* The area of a circle is $A = \pi r^2$.)

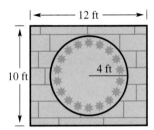

59. *Reading a Circle Graph*   The populations of six counties are shown in the circle graph. What percent of the total population is each county's population?

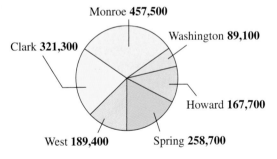

60. *Energy Use*   The circle graph shows the sources of the approximately 90.6 quadrillion British thermal units (Btu) of energy consumed in the United States in 1995. How many quadrillion Btu were obtained from coal?   (Source: Energy Information Administration)

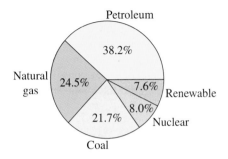

Figure for 60

Graphical Estimation   In Exercises 61–64, use the bar graph to answer the questions. The graph shows the per capita food consumption of selected meats for 1980, 1990, and 1995.   (Source: U.S. Department of Agriculture)

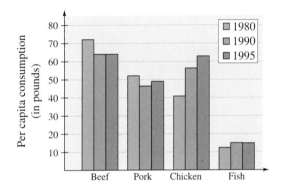

61. Approximate the decrease in the per capita consumption of beef from 1980 to 1995. Use this estimate to approximate the percent decrease.

62. Approximate the increase in the per capita consumption of chicken from 1980 to 1995. Use this estimate to approximate the percent increase.

63. Approximate the total number of pounds of pork consumed in 1990 if the population of the United States was approximately 250 million.

64. Of the four categories of meats shown in the table, what percent of the meat diet was met by fish in 1995?

65. *State Income Tax*   You have $12.50 of state tax withheld from your paycheck per week when your gross pay is $625. Find the ratio of tax to gross pay.

66. *Price-Earnings Ratio*   The ratio of the price of a stock to its earnings is called the **price-earnings ratio.** Find the price-earnings ratio of a stock that sells for $56.25 per share and earns $6.25 per share.

67. *Compression Ratio* The **compression ratio** of a cylinder is the ratio of its expanded volume to its compressed volume (see figure). The expanded volume of one cylinder of a small diesel engine is 425 cubic centimeters, and its compressed volume is 20 cubic centimeters. Find the compression ratio of this engine.

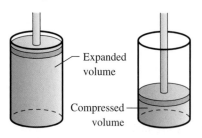

68. *Gear Ratio* The **gear ratio** of two gears is the number of teeth in one gear to the number of teeth in the other gear. If two gears in a gearbox have 60 teeth and 40 teeth (see figure), find the gear ratio.

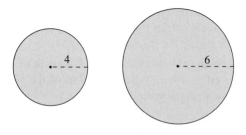

69. *Geometry* Find the ratio of the areas of the two circles in the figure. (*Hint:* The area of a circle is $A = \pi r^2$.)

70. *Geometry* Find the ratio of the areas of the two triangles in the figure.

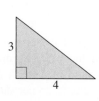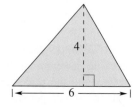

Unit Prices In Exercises 71–74, find the unit price (in dollars per ounce) of the product.

71. A 20-ounce can of pineapple for 95¢

72. A 64-ounce bottle of juice for $1.29

73. A 1-pound, 4-ounce loaf of bread for $1.69

74. An 18-ounce box of cereal for $3.49

Comparison Shopping In Exercises 75–78, use unit prices to determine the better buy. See Example 6.

75. (a) A $14\frac{1}{2}$-ounce bag of chips for $2.32

  (b) A $5\frac{1}{2}$-ounce bag of chips for $0.99

76. (a) A $10\frac{1}{2}$-ounce package of cookies for $1.79

  (b) A 16-ounce package of cookies for $2.39

77. (a) A 4-ounce tube of toothpaste for $1.69

  (b) A 6-ounce tube of toothpaste for $2.39

78. (a) A 2-pound package of hamburger for $3.49

  (b) A 3-pound package of hamburger for $5.29

Geometry In Exercises 79–82, solve for the length x of the side of the triangle by using the fact that corresponding sides of similar triangles are proportional. See Example 7.

79.

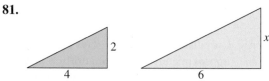

80.

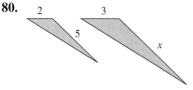

81.

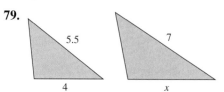

82.

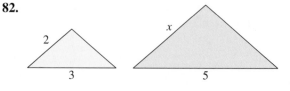

83. *Geometry*   A man who is 6 feet tall walks directly toward the tip of the shadow of a tree. When the man is 75 feet from the tree, he starts forming his own shadow beyond the shadow of the tree (see figure). Find the height of the tree if the length of the shadow of the tree beyond this point is 11 feet.

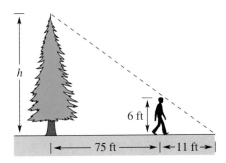

84. *Geometry*   Find the length of the shadow of a man who is 6 feet tall and is standing 15 feet from a streetlight that is 20 feet high (see figure).

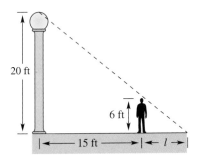

85. *Fuel Usage*   A tractor uses 5 gallons of diesel fuel to plow for 105 minutes. Assuming conditions remain the same, determine the number of gallons of fuel used in 6 hours.

86. *Spring Length*   A force of 32 pounds stretches a spring 6 inches. Determine the number of pounds of force required to stretch it 9 inches.

87. *Property Tax*   The tax on a property with an assessed value of $110,000 is $1650. Find the tax on a property with an assessed value of $160,000.

88. *Recipe Proportions*   Three cups of flour are required to make one batch of cookies. How many cups are required to make $3\frac{1}{2}$ batches?

89. *Quality Control*   A quality control engineer finds one defective unit in a sample of 75. At this rate, what is the expected number of defective units in a shipment of 200,000?

90. *Public Opinion Poll*   In a public opinion poll, 870 people from a sample of 1500 indicate they will vote for a certain candidate. Assuming this poll to be a correct indicator of the electorate, how many votes can the candidate expect to receive from 80,000 votes cast?

Explaining Concepts

91. Explain the meaning of the word *percent*.

92. Explain how to change percents to decimals and decimals to percents. Give examples.

93. Is it true that $\frac{1}{2}\% = 50\%$? Explain.

94. Define the term *ratio*. Give an example of a ratio.

95. You are told that the ratio of the number of boys to the number of girls in a class is 2 to 1. Is this sufficient information to determine the number of students in the class? Explain your reasoning.

96. During a year of financial difficulties, your company reduces your salary by 7%. What percent increase in this reduced salary is required to raise your salary to the amount it was prior to the reduction? Why isn't the percent increase the same as the percent of the reduction?

97. In your own words, describe the meaning of *mathematical modeling*. Give an example.

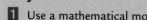

1.3   Business and Scientific Problems

Objectives

1 Use a mathematical model to solve business-related problems.

2 Use a mathematical model to solve a mixture problem.

3 Use a mathematical model to solve classic rate problems.

4 Use a formula to solve an application problem.

---

1 Use a mathematical model to solve business-related problems.

Rates in Business Problems

Many business problems can be represented by mathematical models involving the sum of a fixed term and a variable term. The variable term is often a *hidden product* in which one of the factors is a percent or some other type of rate. Watch for these occurrences in the discussions and examples that follow.

The **markup** on a consumer item is the difference between the **cost** a retailer pays for an item and the **price** at which the retailer sells the item. A verbal model for this relationship is as follows.

$$\boxed{\text{Selling price}} \;=\; \boxed{\text{Cost}} \;+\; \boxed{\text{Markup}} \qquad \text{Markup is a hidden product.}$$

The markup is the hidden product of the **markup rate** and the cost.

$$\boxed{\text{Markup}} \;=\; \boxed{\text{Markup rate}} \;\cdot\; \boxed{\text{Cost}}$$

Example 1   Finding the Markup Rate

A clothing store sells a pair of jeans for $42. If the cost of the jeans is $16.80, what is the markup rate?

Solution

Verbal Model: $\boxed{\text{Selling price}} = \boxed{\text{Cost}} + \boxed{\text{Markup}}$

Labels:
Selling price = 42   (dollars)
Cost = 16.80   (dollars)
Markup rate = p (percent in decimal form)
Markup = $p(16.80)$ (dollars)

Equation:

$$42 = 16.8 + p(16.8)$$

$$42 - 16.8 = p(16.8) \qquad \text{Subtract 16.8 from both sides.}$$

$$25.2 = p(16.8) \qquad \text{Combine like terms.}$$

$$\frac{25.2}{16.8} = p \qquad \text{Divide both sides by 16.8.}$$

$$1.5 = p \qquad \text{Simplify.}$$

Because $p = 1.5$, it follows that the markup rate is 150%. Check this in the original statement of the problem.

In 1874, Levi Strauss designed the first pair of blue jeans. Today, billions of pairs of jeans are sold each year throughout the world.

Study Tip

Although markup and discount are similar, it is important to remember that markup is based on cost and discount is based on list price.

The model for a **discount** is similar to that for a markup.

| Selling price | = | List price | − | Discount | Discount is a hidden product.

The discount is the hidden product of the **discount rate** and the list price.

| **Example 2** | Finding the Discount and the Discount Rate |

A compact disc player is marked down from its list price of $820 to a sale price of $574. What is the discount rate?

Solution

Verbal Model: | Discount | = | Discount rate | · | List price |

Labels: Discount = 820 − 574 = 246   (dollars)
       List price = 820   (dollars)
       Discount rate = p (percent in decimal form)

Equation: $246 = p(820)$

       $\dfrac{246}{820} = p$

       $0.30 = p$

The discount rate is 30%. Check this in the original statement of the problem.

In Example 3, the price of labor is a hidden product.

| **Example 3** | Finding the Hours of Labor |

An auto repair bill of $338 lists $170 for parts and the rest for labor. If it took 6 hours to repair the auto, what is the hourly rate for labor?

Solution

Verbal Model: | Total bill | = | Price of parts | + | Price of labor |

Labels: Total bill = 338   (dollars)
       Price of parts = 170   (dollars)
       Hours of labor = 6   (hours)
       Hourly rate for labor = x (dollars per hour)
       Price of labor = $6x$ (dollars)

Equation: $338 = 170 + 6x$

       $168 = 6x$ Subtract 170 from both sides.

       $\dfrac{168}{6} = x$ Divide both sides by 6.

       $28 = x$ Simplify.

The hourly rate for labor is $28 per hour. Check this in the original problem.

---

2 Use a mathematical model to solve a mixture problem.

Rates in Mixture Problems

Many real-life problems involve combinations of two or more quantities that make up new or different quantities. Such problems are called **mixture problems.** They are usually composed of the sum of two or more "hidden products" that involve rate factors. Here is the generic form of the verbal model for mixture problems.

$$\boxed{\text{First rate}} \cdot \boxed{\text{Amount}} + \boxed{\text{Second rate}} \cdot \boxed{\text{Amount}} = \boxed{\text{Final rate}} \cdot \boxed{\text{Final amount}}$$

| Example 4 | A Mixture Problem | |

A nursery wants to mix two types of lawn seed. Type A sells for \$10 per pound and type B sells for \$15 per pound. To obtain 20 pounds of a mixture at \$12 per pound, how many pounds of each type of seed are needed?

Study Tip

When you set up a verbal model, be sure to check that you are working with the *same type of units* in each part of the model. For instance, in Example 4 note that each of the three parts of the verbal model measures cost. (If two parts measured cost and the other part measured pounds, you would know that the model was incorrect.)

Solution

The rates are the unit prices for each type of seed.

Verbal Model: $\boxed{\text{Total cost of \$10 seed}} + \boxed{\text{Total cost of \$15 seed}} = \boxed{\text{Total cost of \$12 seed}}$

Labels:
Unit price of type A = 10   (dollars per pound)
Pounds of \$10 seed = x (pounds)
Unit price of type B = 15   (dollars per pound)
Pounds of \$15 seed = $20 - x$ (pounds)
Unit price of mixture = 12   (dollars per pound)
Pounds of \$12 seed = 20   (pounds)

Equation: $10x + 15(20 - x) = 12(20)$

$10x + 300 - 15x = 240$ Distributive Property

$300 - 5x = 240$ Combine like terms.

$-5x = -60$ Subtract 300 from both sides.

$x = 12$ Divide both sides by -5.

The mixture should contain 12 pounds of the \$10 seed and $20 - 12 = 8$ pounds of the \$15 seed.

Remember that when you have found a solution, you should always go back to the original statement of the problem and check to see that the solution makes sense—both algebraically and from a practical point of view. For instance, you can check the result of Example 4 as follows.

$$\overbrace{\left(\frac{\$10 \text{ per}}{\text{pound}}\right)\left(\frac{12}{\text{pounds}}\right)}^{\$10 \text{ seed}} + \overbrace{\left(\frac{\$15 \text{ per}}{\text{pound}}\right)\left(\frac{8}{\text{pounds}}\right)}^{\$15 \text{ seed}} = \overbrace{\left(\frac{\$12 \text{ per}}{\text{pound}}\right)\left(\frac{20}{\text{pounds}}\right)}^{\$12 \text{ seed}}$$

$$\$120 + \$120 = \$240$$

3 Use a mathematical model to solve classic rate problems.

Classic Rate Problems

Time-dependent problems such as distance traveled at a given speed and work done at a specified rate are classic types of **rate problems.** The distance-rate-time problem fits the verbal model

$$\boxed{\text{Distance}} = \boxed{\text{Rate}} \cdot \boxed{\text{Time}}.$$

For instance, if you travel at a constant (or average) rate of 55 miles per hour for 45 minutes, the total distance you travel is given by

$$\left(55\,\frac{\text{miles}}{\text{hour}}\right)\!\left(\frac{45}{60}\,\text{hour}\right) = 41.25\text{ miles.}$$

As with all problems involving applications, be sure to check that the units in the verbal model make sense. For instance, in this problem the rate is given in *miles per hour*. Therefore, in order for the solution to be given in *miles*, you must convert the time (from minutes) to *hours*. In the model, you can think of the two "hours" as canceling, as follows.

$$\left(55\,\frac{\text{miles}}{\text{hour}}\right)\!\left(\frac{45}{60}\,\text{hour}\right) = 41.25\text{ miles}$$

Study Tip

To convert minutes to hours, use the fact that there are 60 minutes in each hour. So, 45 minutes is equivalent to $\frac{45}{60}$ of 1 hour. In general, x minutes is $x/60$ of 1 hour.

Example 5   Distance-Rate-Time Problem

If you ride your bike at an average rate of 18 kilometers per hour, how long will it take you to ride 30 kilometers?

Solution

Verbal Model: $\boxed{\text{Distance}} = \boxed{\text{Rate}} \cdot \boxed{\text{Time}}$

Labels: Distance = 30   (kilometers)
            Rate = 18   (kilometers per hour)
            Time = t (hours)

Equation: $30 = 18t$

$$\frac{30}{18} = t$$

$$\frac{5}{3} = t$$

It will take you $1\frac{2}{3}$ hours (or 1 hour and 40 minutes). You can check this in the original statement of the problem, as follows.

Check

$$\left(18\,\frac{\text{kilometers}}{\text{hour}}\right)\!\left(\frac{5}{3}\,\text{hours}\right) = 30\text{ kilometers}$$

In work-rate problems, the **rate of work** is the *reciprocal* of the time needed to do the entire job. For instance, if it takes 5 hours to complete a job, then the per hour work rate is $\frac{1}{5}$. In general,

$$\text{Per hour work rate} = \frac{1}{\text{Total hours to complete a job}}.$$

The next example involves two rates of work and so fits the model for solving *mixture* problems.

Example 6 Work-Rate Problem

Consider two machines in a paper manufacturing plant. Machine 1 can produce 2000 pounds of paper in 4 hours. Machine 2 is newer and can produce 2000 pounds of paper in $2\frac{1}{2}$ hours. How long will it take the two machines working together to produce 2000 pounds of paper?

Solution

Verbal Model:

$$\boxed{\text{Work done}} = \boxed{\text{Portion done by machine 1}} + \boxed{\text{Portion done by machine 2}}$$

Labels:

Work done by both machines = 1   (job)

Time for each machine = t (hours)

Per hour work rate for machine 1 = $\frac{1}{4}$ (job per hour)

Per hour work rate for machine 2 = $\frac{2}{5}$ (job per hour)

Equation:

$$1 = \left(\frac{1}{4}\right)(t) + \left(\frac{2}{5}\right)(t) \qquad \text{Rate · time + rate · time}$$

$$1 = \left(\frac{1}{4} + \frac{2}{5}\right)(t) \qquad \text{Distributive Property}$$

$$1 = \left(\frac{13}{20}\right)(t) \qquad \text{Simplify.}$$

$$\frac{1}{13/20} = t \qquad \text{Divide both sides by } \frac{13}{20}.$$

$$\frac{20}{13} = t \qquad \text{Simplify.}$$

It would take $\frac{20}{13}$ hours (or about 1.5 hours) for both machines to complete the job. Check this solution in the original statement of the problem.

Note in Example 6 that the "2000 pounds" of paper was unnecessary information. We simply represented the 2000 pounds as one job. This type of unnecessary information in an applied problem is sometimes called a *red herring*.

[4] Use a formula to solve an application problem.

Formulas

Many common types of geometric, scientific, and investment problems use ready-made equations called **formulas.** Knowing formulas such as those in the following list will help you translate and solve a wide variety of real-life problems involving perimeter, area, volume, temperature, interest, and distance.

▶ **Common Formulas for Area, Perimeter, and Volume**

Square	*Rectangle*	*Circle*	*Triangle*
$A = s^2$	$A = lw$	$A = \pi r^2$	$A = \frac{1}{2} bh$
$P = 4s$	$P = 2l + 2w$	$C = 2\pi r$	$P = a + b + c$

	Rectangular Solid	*Circular Cylinder*	*Sphere*
Cube			
$V = s^3$	$V = lwh$	$V = \pi r^2 h$	$V = \frac{4}{3} \pi r^3$

▶ **Miscellaneous Common Formulas**

Temperature: F = degrees Fahrenheit, C = degrees Celsius

$$F = \frac{9}{5} C + 32$$

Simple Interest: I = interest, P = principal, r = interest rate, t = time

$$I = Prt$$

Distance: d = distance traveled, r = rate, t = time

$$d = rt$$

Technology: Tip

You can use a graphing utility to solve simple interest problems by using the program found at our website *www.hmco.com.* Use the program and the Guess, Check, and Revise method to find P when I = \$3330, r = 6%, and t = 3 years.

When working with applied problems, you often need to rewrite one of the common formulas, as shown in the next example.

Example 7   Rewriting a Formula

In the perimeter formula $P = 2l + 2w$, solve for w.

Solution

$$P = 2l + 2w \qquad \text{Original formula}$$

$$P - 2l = 2w \qquad \text{Subtract } 2l \text{ from both sides.}$$

$$\frac{P - 2l}{2} = w \qquad \text{Divide both sides by 2.}$$

Study Tip

When solving problems such as the one in Example 8, you may find it helpful to draw and label a diagram.

Example 8   Using a Geometric Formula

A local streets department plans to put sidewalks along the two streets that bound your corner lot, which is 250 feet long on one side with an area of 30,000 square feet. Each lot owner is to pay $1.50 per foot of sidewalk bordering his or her lot.

a. Find the width of your lot.

b. How much will you have to pay for the sidewalks put on your lot?

Solution

Figure 1.2 shows a labeled diagram of your lot.

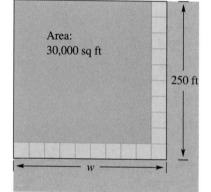

Area: 30,000 sq ft

250 ft

w

Figure 1.2

a. *Verbal Model:*   Area = Length · Width

Labels: Area of lot = 30,000   (square feet)
Length of lot = 250   (feet)
Width of lot = w (feet)

Equation: $30,000 = 250 \cdot w$

$$\frac{30,000}{250} = w$$

$$120 = w$$

Your lot is 120 feet wide.

b. *Verbal Model:*   Cost = Rate per foot · Length of sidewalk

Labels: Cost of sidewalks = C (dollars)
Rate per foot = 1.50   (dollars per foot)
Total length of sidewalk = 120 + 250   (feet)

Equation: $C = 1.50(120 + 250)$

$$C = 1.50 \cdot 370$$

$$C = 555$$

You will have to pay $555 to have the sidewalks put on your lot.

Example 9   Simple Interest

A deposit of $8000 earned $300 in interest in 6 months.

a. What was the annual interest rate for this account?

b. At this rate, how long would it take to earn $800 in total interest?

Solution

a. *Verbal*
Model: Interest $=$ Principal $\cdot$ Rate $\cdot$ Time

  Labels: Interest $= 300$ (dollars)
          Principal $= 8000$ (dollars)
          Time $= \frac{1}{2}$ (year)
          Annual interest rate $= r$ (percent in decimal form)

  Equation: $300 = 8000(r)\left(\dfrac{1}{2}\right)$

          $\dfrac{300}{4000} = r$

          $0.075 = r$

  The annual interest rate is $r = 0.075$ (or 7.5%).

b. Using the same verbal model as in part (a) with t representing time, you obtain the following equation.

$$800 = 8000(0.075)(t)$$

$$\frac{800}{8000(0.075)} = t$$

$$\frac{4}{3} = t$$

So, it would take $\frac{4}{3}$ years, or $\frac{4}{3} \times 12 = 16$ months.

Discussing the Concept   **Translating a Formula**

Use the information provided in the following statement to write a mathematical formula for the 10-second pulse count.

"The target heart rate is the heartbeat rate a person should have during aerobic exercise to get the full benefit of the exercise for cardio-vascular conditioning . . . Using the American College of Sports Medicine Method to calculate one's target heart rate, an individual should subtract his or her age from 220, then multiply by the desired intensity level (as a percent—sedentary persons may want to use 60% and highly fit individuals may want to use 85 to 95%) of the workout. Then divide the answer by 6 for a 10-second pulse count. (The 10-second pulse count is useful for checking whether the target heart rate is being achieved during the workout. One can easily check one's pulse—at the wrist or side of the neck—counting the number of beats in 10 seconds.)"   (Source: Aerobic Fitness Association of America)

Calculate your own 10-second pulse count.

1.3   Exercises

Integrated Review   *Concepts, Skills, and Problem Solving*

Keep mathematically in shape by doing these exercises *before* the problems of this section.

Properties and Definitions

1. What is the sign of the sum $(-7) + (-3)$? State the rule used.

2. What is the sign of the sum $-7 + 3$? State the rule used.

3. What is the sign of the product $(-6)(-2)$? State the rule used.

4. What is the sign of the product $6(-2)$? State the rule used.

Solving Equations

In Exercises 5–10, solve the equation.

5. $2x - 5 = x + 9$ 6. $6x + 8 = 8 - 2x$

7. $2x + \dfrac{3}{2} = \dfrac{3}{2}$ 8. $-\dfrac{x}{10} = 1000$

9. $-0.35x = 70$ 10. $0.60x = 24$

Problem Solving

11. The length of a relay race is 2.5 miles. The last change of runners occurs at the 1.8-mile marker. How far does the last person run?

12. During the months of January, February, and March, a farmer bought $34\frac{1}{3}$ tons, $18\frac{1}{5}$ tons, and $25\frac{5}{6}$ tons of soybeans, respectively. Find the total amount of soybeans purchased during the first quarter of the year.

Developing Skills

In Exercises 1–8, find the missing quantities. (Assume the markup rate is a percent based on the cost.) See Example 1.

	Cost	Selling Price	Markup	Markup Rate
1.	$45.97	$64.33		
2.	$84.20	$113.67		
3.		$250.80	$98.80	
4.		$603.72	$184.47	
5.		$26,922.50	$4672.50	
6.		$16,440.50	$3890.50	
7.	$225.00			85.2%
8.	$732.00			$33\frac{1}{3}\%$

In Exercises 9–16, find the missing quantities. (Assume the discount rate is a percent based on the list price.) See Example 2.

	List Price	Sale Price	Discount	Discount Rate
9.	$49.95	$25.74		
10.	$119.00	$79.73		
11.	$300.00		$189.00	
12.	$345.00		$134.55	
13.	$95.00			65%
14.		$15.92		20%
15.		$893.10	$251.90	
16.		$257.32	$202.18	

Solving Problems

17. *Mathematical Modeling*   The selling price of a jacket in a department store is $85. The cost of the jacket to the store is $62.95. What is the markup?

18. *Mathematical Modeling*   A shoe store sells a pair of shoes for $35. The cost of the shoes to the store is $18.75. What is the markup?

19. *Mathematical Modeling*   A jewelry store sells a pair of earrings for $25. The cost of the earrings to the store is $15. What is the markup rate?

20. *Mathematical Modeling*   A department store sells a sweater for $60. The cost of the sweater to the store is $35. What is the markup rate?

21. *Mathematical Modeling*   A shoe store sells a pair of athletic shoes for $75. The shoes go on sale for $50. What is the discount?

22. *Mathematical Modeling*   A bakery sells a dozen rolls for $1.75. You can buy day-old rolls for $0.75. What is the discount?

23. *Mathematical Modeling*   An auto store sells a pair of car mats for $20. On sale, the car mats sell for $16. What is the discount rate?

24. *Mathematical Modeling*   A department store sells a beach towel for $14. On sale, the beach towel sells for $10. What is the discount rate?

25. *Long-Distance Rates*   The weekday rate for a telephone call is $0.75 for the first minute plus $0.55 for each additional minute. Determine the length of a call that cost $5.15. What would have been the cost of the call if it had been made during the weekend, when there is a 60% discount?

26. *Insurance Premiums*   The annual insurance premium for a policyholder is $862. Find the annual premium if the policyholder must pay a 20% surcharge because of an accident.

27. *Tire Cost*   An auto store gives the list price of a tire as $79.42. During a promotional sale, the store is selling four tires for the price of three. The store needs a markup on cost of 10% during the sale. What is the cost to the store of each tire?

28. *Price per Pound*   The produce manager of a supermarket pays $22.60 for a 100-pound box of bananas. From past experience, the manager estimates that 10% of the bananas will spoil before they are sold. At what price per pound should the bananas be sold to give the supermarket an average markup rate on cost of 30%?

29. *Amount Financed*   A customer bought a lawn tractor for $4450 plus 6% sales tax. Find the amount of the sales tax and the total bill. Find the amount financed if a down payment of $1000 was made.

30. *Weekly Pay*   The weekly salary of an employee is $375 plus a 6% commission on the employee's total sales. Find the weekly pay for a week in which the sales are $5500.

31. *Labor Charges*   An auto repair bill of $216.37 lists $136.37 for parts and the rest for labor. If the labor rate is $32 per hour, how many hours did it take to repair the auto?

32. *Labor Charges*   An appliance repair store charges $50 for the first $\frac{1}{2}$ hour of a service call. For each additional $\frac{1}{2}$ hour of labor, there is a charge of $18. Find the length of a service call for which the charge is $104.

33. *Mathematical Modeling*   The bill for the repair of an automobile is $380. Included in this bill is a charge of $275 for parts, and the remainder of the bill is for labor. If the charge for labor is $35 per hour, how many hours were spent in repairing the automobile?

34. *Mathematical Modeling*   The bill for the repair of an automobile is $648. Included in this bill is a charge of $315 for parts, and the remainder of the bill is for labor. If it took 9 hours to repair the automobile, what was the charge per hour for labor?

Mixture Problem   In Exercises 35–38, determine the number of units of solutions 1 and 2 needed to obtain the desired amount and concentration of the final solution.

Concentration of Solution 1	Concentration of Solution 2	Concentration of Final Solution	Amount of Final Solution
35. 20%	60%.	40%	100 gal
36. 50%	75%	60%	10 L
37. 15%	60%	45%	24 qt
38. 60%	80%	75%	55 gal

39. *Seed Mixture*   A nursery wants to mix two types of lawn seed. One type sells for $12 per pound and the other type sells for $20 per pound. To obtain 100 pounds of a mixture at $14 per pound, how many pounds of each type of seed are needed?

40. *Nut Mixture*   A grocer mixes two kinds of nuts costing $3.88 per pound and $4.88 per pound to make 100 pounds of a mixture costing $4.13 per pound. How many pounds of each kind of nut are put into the mixture?

41. *Ticket Sales*   Ticket sales for a play total $2200. There are three times as many adult tickets sold as children's tickets. The prices of the tickets for adults and children are $6 and $4, respectively. Find the number of children's tickets sold.

42. *Ticket Sales*   Ticket sales for a spaghetti dinner total $1350. There are four times as many adult tickets sold as children's tickets. The prices of the tickets for adults and children are $6 and $3, respectively. Find the number of children's tickets sold.

43. *Antifreeze Coolant*   The cooling system on a truck contains 5 gallons of coolant that is 40% antifreeze. How much must be withdrawn and replaced with 100% antifreeze to bring the coolant in the system to 50% antifreeze?

44. *Fuel Mixture*   You mix gasoline and oil to obtain $2\frac{1}{2}$ gallons of mixture for an engine. The mixture is 40 parts gasoline and 1 part two-cycle oil. How much gasoline must be added to bring the mixture to 50 parts gasoline and 1 part oil?

Distance   In Exercises 45–50, determine the unknown distance, rate, or time. See Example 5.

	Distance, d	Rate, r	Time, t
45.		650 mi/hr	$3\frac{1}{2}$ hr
46.		45 ft/sec	10 sec
47.	1000 km	110 km/hr	
48.	250 ft	32 ft/sec	
49.	1000 ft		$\frac{3}{2}$ sec
50.	385 mi		7 hr

51. *Time*   You ride your bike at an average of 12 miles per hour. How long will it take you to ride 30 miles?

52. *Time*   You ride your bike at an average of 8 miles per hour. How long will it take you to ride 12 miles?

53. *Distance*   Two planes leave an airport at approximately the same time and fly in opposite directions. How far apart are the planes after $1\frac{1}{3}$ hours if their speeds are 480 miles per hour and 600 miles per hour?

54. *Distance*   Two trucks leave a depot at approximately the same time and travel the same route. How far apart are the trucks after $4\frac{1}{2}$ hours if their average speeds are 52 miles per hour and 56 miles per hour?

55. *Time*   Determine the time for a space shuttle to travel a distance of 5000 miles in orbit when its speed is 17,000 miles per hour (see figure).

56. *Speed*   Determine the time for light to travel from the sun to the earth if the distance between the sun and the earth is 93,000,000 miles and the speed of light is 186,282.369 miles per second (see figure).

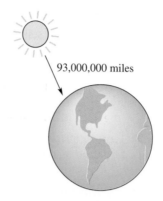

93,000,000 miles

57. *Time*   On the first part of a 317-mile trip, a sales representative averaged 58 miles per hour. The sales representative averaged only 52 miles per hour on the remainder of the trip because of an increased volume of traffic (see figure). Find the amount of driving time at each speed if the total time was 5 hours and 45 minutes.

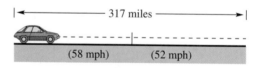

317 miles

(58 mph)   (52 mph)

58. *Time*   Two cars start at a given point and travel in the same direction at average speeds of 30 miles per hour and 45 miles per hour. How much time must elapse before the two cars are 5 miles apart?

59. *Work Rate*   Determine the work rate for each task.

(a) A printer can print 8 pages per minute.

(b) A machine shop can produce 30 units in 8 hours.

60. *Work-Rate Problem*   You can complete a typing project in 5 hours, and a friend estimates that it would take him 8 hours.

(a) What fractional part of the task can be accomplished by each person in 1 hour?

(b) If you both work on the project, in how many hours can it be completed?

61. *Work-Rate Problem* You can mow a lawn in 3 hours, and your friend can mow it in 4 hours.

 (a) What fractional part of the lawn can each of you mow in 1 hour?

 (b) How long will it take both of you to mow the lawn working together?

62. *Work-Rate Problem* It takes 30 minutes for a pump to empty a water tank. A larger pump can empty the tank in half the time. If both pumps were operating, how long would it take to empty the tank?

In Exercises 63–68, solve for the specified variable. See Example 7.

63. *Ohm's Law* Solve for R in

$$E = IR.$$

64. *Simple Interest* Solve for r in

$$A = P + Prt.$$

65. *Discount* Solve for L in

$$S = L - rL.$$

66. *Markup* Solve for C in

$$S = C + rC.$$

67. *Free-Falling Body* Solve for a in

$$h = 48t + \frac{1}{2}at^2.$$

68. *Area of a Trapezoid* Solve for b in

$$A = \frac{1}{2}(a + b)h.$$

Geometry In Exercises 69 and 70, use the closed rectangular box shown in the figure.

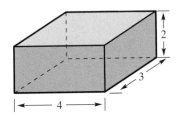

69. Find the volume of the box.

70. Find the surface area of the box. (*Hint:* The surface area is the combined area of the six surfaces.)

71. *Geometry* Find the volume of the circular cylinder shown in the figure.

$3\frac{1}{2}$ cm

12 cm

72. *Geometry* Find the volume of the sphere shown in the figure.

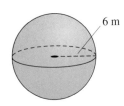

6 m

73. *Geometry* A rectangular picture frame has a perimeter of 3 feet. The width of the frame is 0.62 times its height. Find the height of the frame.

74. *Geometry* A rectangular stained-glass window has a perimeter of 18 feet. The height of the window is 1.25 times its width. Find the width of the window.

75. *Geometry* A "Slow Moving Vehicle" sign has the shape of an equilateral triangle. The sign has a perimeter of 129 centimeters. Find the length of each side. Include a labeled diagram with your model.

76. *Geometry* The length of a rectangle is three times its width. The perimeter of the rectangle is 64 inches. Find the dimensions of the rectangle.

77. *Simple Interest* Find the interest on a $5000 bond that pays an annual percentage rate of $9\frac{1}{2}\%$ for 6 years.

78. *Simple Interest* Find the annual interest rate on a certificate of deposit that accumulated $400 interest in 2 years on a principal of $2500.

79. *Simple Interest* Find the principal required to earn $500 in interest in 2 years if the annual interest rate is 7%.

80. *Simple Interest* You borrow $36,000 for 6 months. You agree to pay back the principal and the interest (finance charges) in one lump sum. What will be the amount of the payment if the interest rate is 13%?

81. *Simple Interest*   An inheritance of $40,000 is divided into two investments earning 8% and 10% simple interest. (The 10% investment has a greater risk.) What is the smallest amount that can be invested in the 10% fund if the total annual interest from both investments is at least $3500?

82. *Simple Interest*   An investment of $7000 is divided into two accounts earning 5% and 7% simple interest. (The 7% investment has a greater risk.) What is the smallest amount that can be invested in the 7% account if the total annual interest from both investments is at least $400?

83. *Average Wage*   The average hourly wage for bus drivers at public schools in the United States from 1990 through 1997 can be approximated by

$$y = 9.24 + 0.307t, \quad 0 \le t \le 7$$

where y represents the hourly wage (in dollars) and t represents the year, with $t = 0$ corresponding to 1990 (see figure).   (Source: Educational Research Service)

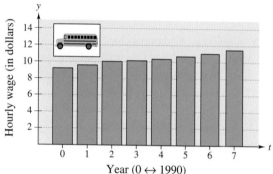

Year (0 ↔ 1990)

(a) Use the graph to determine the year when the average hourly wage was $10.15. Would the result be the same if you used the model? Explain.

(b) What was the average annual hourly raise for bus drivers during this 8-year period? Explain how you determined your answer.

84. *Average Wage*   The average hourly wage for cafeteria workers at public schools in the United States from 1990 through 1997 can be approximated by the linear model

$$y = 6.88 + 0.209t, \quad 0 \le t \le 7$$

where y represents the hourly wage (in dollars) and t represents the year, with $t = 0$ corresponding to 1990 (see figure).   (Source: Educational Research Service)

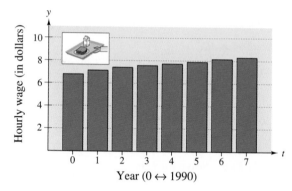

Year (0 ↔ 1990)

(a) Use the graph to determine the year when the average hourly wage was $7.72. Would the result be the same if you used the model? Explain.

(b) What was the average annual hourly raise for cafeteria workers during this 8-year period? Explain how you determined your answer.

85. *Comparing Wage Increases*   Use the information given in Exercises 83 and 84 to determine which of the two groups' average salaries was increasing at a greater annual rate during the 8-year period from 1990 to 1997.

Explaining Concepts

86. Answer parts (a)–(e) of Motivating the Chapter on page 55.

87. Explain the difference between markup rate and markup.

88. Explain how to find the sale price of an item when you are given the list price and the discount rate.

89. If it takes you t hours to complete a task, what portion of the task can you complete in 1 hour?

90. If the sides of a square are doubled, does the perimeter double? Explain.

91. If the sides of a square are doubled, does the area double? Explain.

92. If you forget the formula for the volume of a right circular cylinder, how can you derive it?

The symbol 🔻 indicates an exercise that relates to the Motivating the Chapter feature at the beginning of the chapter.

Mid-Chapter Quiz

Take this quiz as you would take a quiz in class. After you are done, check your work against the answers given in the back of the book.

In Exercises 1–8, solve the equation and check the result. (If it is not possible, state the reason.)

1. $4x + 3 = 11$

2. $-3(z - 2) = 0$

3. $2(y + 3) = 18 - 4y$

4. $5t + 7 = 7(t + 1) - 2t$

5. $\dfrac{1}{4}x + 6 = \dfrac{3}{2}x - 1$

6. $\dfrac{u}{4} + \dfrac{u}{3} = 1$

7. $\dfrac{4 - x}{5} + 5 = \dfrac{5}{2}$

8. $0.2x + 0.3 = 1.5$

In Exercises 9 and 10, solve the equation and round your answer to two decimal places. (A calculator may be helpful.)

9. $3x + \frac{11}{12} = \frac{5}{16}$

10. $0.42x + 6 = 5.25x - 0.80$

11. Explain how to write the decimal 0.45 as a fraction and as a percent.

12. 500 is 250% of what number?

13. Find the unit price (in dollars per ounce) of a 12-ounce box of cereal that sells for $2.35.

14. A quality control engineer for a manufacturer finds one defective unit in a sample of 300. At this rate, what is the expected number of defective units in a shipment of 600,000?

15. A store is offering a discount of 25% on a computer with a list price of $1750. A mail-order catalog has the same machine for $1250 plus $24.95 for shipping. Which is the better buy?

16. Last week you earned $616. Your regular hourly wage is $12.25 for the first 40 hours, and your overtime hourly wage is $18. How many hours of over-time did you work?

17. Fifty gallons of a 30% acid solution is obtained by combining solutions that are 25% acid and 50% acid. How much of each solution is required?

18. On the first part of a 300-mile trip, a sales representative averaged 62 miles per hour. The sales representative averaged 46 miles per hour on the remainder of the trip because of an increased volume of traffic. Find the amount of driving time at each speed if the total time was 6 hours.

19. You can paint a room in 6 hours, and your friend can paint it in 8 hours. How long will it take both of you to paint the room?

20. The accompanying figure shows three squares. The perimeters of squares I and II are 20 inches and 32 inches, respectively. Find the area of square III.

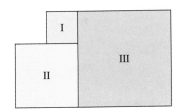

Figure for 20

1.4   Linear Inequalities

Objectives

1 Sketch the graph of an inequality.

2 Identify the properties of inequalities that can be used to create equivalent inequalities.

3 Solve a linear inequality.

4 Solve a compound inequality.

5 Solve an application problem involving inequalities.

---

1 Sketch the graph of an inequality.

Intervals on the Real Number Line

In this section you will study **algebraic inequalities,** which are inequalities that contain one or more variable terms. Some examples are

$$x \leq 4, \quad x \geq -3, \quad x + 2 < 7, \quad \text{and} \quad 4x - 6 < 3x + 8.$$

As with an equation, you **solve** an inequality in the variable x by finding all values of x for which the inequality is true. Such values are called **solutions** and are said to **satisfy** the inequality. The set of all solutions of an inequality is the **solution set** of the inequality. The **graph** of an inequality is obtained by plotting its solution set on the real number line. Often, these graphs are intervals—either bounded or unbounded.

▶ **Bounded Intervals on the Real Number Line**

Let a and b be real numbers such that $a < b$. The following intervals on the real number line are called **bounded intervals.** The numbers a and b are the **endpoints** of each interval. A bracket indicates that an endpoint is included in the interval, and a parenthesis indicates that the endpoint is excluded.

Notation	Interval Type	Inequality	Graph
$[a, b]$	Closed	$a \leq x \leq b$	
(a, b)	Open	$a < x < b$	
$[a, b)$		$a \leq x < b$	
$(a, b]$		$a < x \leq b$	

The **length** of the interval $[a, b]$ is the distance, $b - a$, between its endpoints. The lengths of $[a, b]$, (a, b), $[a, b)$, and $(a, b]$ are the same. The reason that these four types of intervals are called "bounded" is that each has a finite length. An interval that *does not* have a finite length is **unbounded** (or **infinite**).

John Wallis
(1616–1703)
Wallis, an English mathematician, introduced the symbol ∞ for infinity. He was an early historian of mathematics, and his *Treatise on Algebra* (1685) brought the work of Thomas Harriot (1560–1621) to mathematicians. Harriot's work included research on the theory of equations. Aside from his contributions to mathematics, Wallis was an expert in cryptography and devised a system for teaching deaf-mutes.

▶ **Unbounded Intervals on the Real Number Line**

Let a and b be real numbers. The following intervals on the real number line are called **unbounded intervals.**

Notation	Interval Type	Inequality	Graph
$[a, \infty)$		$a \leq x$	
(a, ∞)	Open	$a < x$	
$(-\infty, b]$		$x \leq b$	
$(-\infty, b)$	Open	$x < b$	
$(-\infty, \infty)$	Entire real line		

The symbols ∞ (**positive infinity**) and −∞ (**negative infinity**) do not represent real numbers. They are simply convenient symbols used to describe the unboundedness of an interval such as $(1, \infty)$.

Example 1   Graphs of Inequalities

Sketch the graph of each inequality.

a. $-3 < x \leq 1$ **b.** $0 < x < 2$ **c.** $-3 < x$ **d.** $x \leq 2$

Solution

a. The graph of $-3 < x \leq 1$ is a bounded interval.

b. The graph of $0 < x < 2$ is a bounded interval.

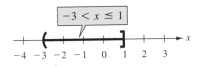

c. The graph of $-3 < x$ is an unbounded interval.

d. The graph of $x \leq 2$ is an unbounded interval.

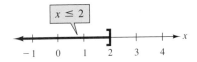

2 Identify the properties of inequalities that can be used to create equivalent inequalities.

Properties of Inequalities

Solving a linear inequality is much like solving a linear equation. To isolate the variable, you make use of **properties of inequalities.** These properties are similar to the properties of equality, but there are two important exceptions: *when both sides of an inequality are (1) multiplied by a negative number or (2) divided by a negative number, the direction of the inequality symbol must be reversed.* Here is an example.

$$-2 < 5 \qquad \text{Original inequality}$$

$$(-3)(-2) > (-3)(5) \qquad \text{Multiply both sides by } -3 \text{ and reverse the inequality.}$$

$$6 > -15 \qquad \text{Simplify.}$$

Two inequalities that have the same solution set are **equivalent inequalities.** The following list describes operations that can be used to create equivalent inequalities.

▶ **Properties of Inequalities**

1. *Addition and Subtraction Properties*

Adding the same quantity to, or subtracting the same quantity from, both sides of an inequality produces an equivalent inequality.

If $a < b$, then $a + c < b + c$.

If $a < b$, then $a - c < b - c$.

2. *Multiplication and Division Properties: Positive Quantities*

Multiplying or dividing both sides of an inequality by a *positive* quantity produces an equivalent inequality.

If $a < b$ and c is positive, then $ac < bc$.

If $a < b$ and c is positive, then $\dfrac{a}{c} < \dfrac{b}{c}$.

3. *Multiplication and Division Properties: Negative Quantities*

Multiplying or dividing both sides of an inequality by a *negative* quantity produces an equivalent inequality in which the inequality symbol is reversed.

If $a < b$ and c is negative, then $ac > bc$.   Reverse inequality

If $a < b$ and c is negative, then $\dfrac{a}{c} > \dfrac{b}{c}$.   Reverse inequality

4. *Transitive Property*

Consider three quantities for which the first quantity is less than the second, and the second is less than the third. It follows that the first quantity must be less than the third quantity.

If $a < b$ and $b < c$, then $a < c$.

These properties remain true if the symbols $<$ and $>$ are replaced by $\leq$ and $\geq$. Moreover, a, b, and c can represent real numbers, variables, or expressions. Note that you cannot multiply or divide both sides of an inequality by zero.

3 Solve a linear inequality.

Solving a Linear Inequality

An inequality in one variable is a **linear inequality** if it can be written in one of the following forms.

$$ax + b \leq 0, \quad ax + b < 0, \quad ax + b \geq 0, \quad ax + b > 0$$

The solution set of a linear inequality can be written in set notation. For the solution $x > 1$, the set notation is $\{x | x > 1\}$ and is read "the set of all x such that x is greater than 1."

As you study the following examples, pay special attention to the steps in which the inequality symbol is reversed. *Remember that when you multiply or divide an inequality by a negative number, you must reverse the inequality symbol.*

Study Tip

Checking the solution set of an inequality is not as simple as checking the solution set of an equation. (There are usually too many x-values to substitute back into the original inequality.) You can, however, get an indication of the validity of a solution set by substituting a few convenient values of x. For instance, in Example 2, try checking that $x = 0$ satisfies the original inequality, whereas $x = 4$ does not.

Example 2   Solving a Linear Inequality

$x + 6 < 9$	Original inequality
$x + 6 - 6 < 9 - 6$	Subtract 6 from both sides.
$x < 3$	Combine like terms.

The solution set consists of all real numbers that are less than 3. The solution set in interval notation is $(-\infty, 3)$ and in set notation is $\{x | x < 3\}$. The graph is shown in Figure 1.3.

Figure 1.3

Example 3   Solving a Linear Inequality

$8 - 3x \leq 20$	Original inequality
$8 - 8 - 3x \leq 20 - 8$	Subtract 8 from both sides.
$-3x \leq 12$	Combine like terms.
$\dfrac{-3x}{-3} \geq \dfrac{12}{-3}$	Divide both sides by -3 and reverse the inequality symbol.
$x \geq -4$	Simplify.

The solution set in interval notation is $[-4, \infty)$ and in set notation is $\{x | x \geq -4\}$. The graph is shown in Figure 1.4.

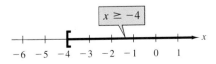

Figure 1.4

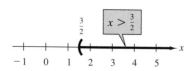

Figure 1.5

Study Tip

An inequality can be cleared of fractions in the same way an equation can be cleared of fractions—by multiplying both sides by the LCD. This is shown in Example 5.

Example 4   Solving a Linear Inequality

$9x - 4 > 5x + 2$	Original inequality
$9x - 4 + 4 > 5x + 2 + 4$	Add 4 to both sides.
$9x > 5x + 6$	Combine like terms.
$9x - 5x > 5x + 6 - 5x$	Subtract $5x$ from both sides.
$4x > 6$	Combine like terms.
$\dfrac{4x}{4} > \dfrac{6}{4}$	Divide both sides by 4.
$x > \dfrac{3}{2}$	Simplify.

The solution set consists of all real numbers that are greater than $\frac{3}{2}$. The solution set in interval notation is $\left(\frac{3}{2}, \infty\right)$ and in set notation is $\left\{x \mid x > \frac{3}{2}\right\}$. The graph is shown in Figure 1.5.

Example 5   Solving a Linear Inequality

$\dfrac{2x}{3} + 12 < \dfrac{x}{6} + 18$	Original inequality
$6 \cdot \left(\dfrac{2x}{3} + 12\right) < 6 \cdot \left(\dfrac{x}{6} + 18\right)$	Multiply both sides by LCD of 6.
$4x + 72 < x + 108$	Distributive Property
$4x - x < 108 - 72$	Subtract x and 72 from both sides.
$3x < 36$	Combine like terms.
$x < 12$	Divide both sides by 3.

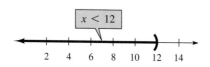

Figure 1.6

The solution set consists of all real numbers that are less than 12. The solution set in interval notation is $(-\infty, 12)$ and in set notation is $\{x \mid x < 12\}$. The graph is shown in Figure 1.6.

4 Solve a compound inequality.

Solving a Compound Inequality

Two inequalities joined by the word *and* or the word *or* constitute a **compound inequality.** When two inequalities are joined by the word *and*, the solution set consists of all real numbers that satisfy *both* inequalities. The solution set for the compound inequality $-4 \le 5x - 2$ *and* $5x - 2 < 7$ can be written more simply as the **double inequality**

$$-4 \le 5x - 2 < 7.$$

A compound inequality formed by the word *and* is called **conjunctive** and is the only kind that has the potential to form a double inequality. A compound inequality joined by the word *or* is called **disjunctive** and cannot be reformed into a double inequality.

Example 6   Solving a Double Inequality

Solve the double inequality $-7 \leq 5x - 2 < 8$.

Solution

$$-7 \leq 5x - 2 < 8 \qquad \text{Original inequality}$$

$$-7 + 2 \leq 5x - 2 + 2 < 8 + 2 \qquad \text{Add 2 to all three parts.}$$

$$-5 \leq 5x < 10 \qquad \text{Combine like terms.}$$

$$\frac{-5}{5} \leq \frac{5x}{5} < \frac{10}{5} \qquad \text{Divide each part by 5.}$$

$$-1 \leq x < 2 \qquad \text{Simplify.}$$

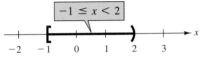

Figure 1.7

The solution set consists of all real numbers that are greater than or equal to -1 and less than 2. The interval notation for the solution set is $[-1, 2)$. The graph is shown in Figure 1.7.

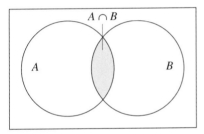

Intersection of two sets

Union of two sets
Figure 1.8

The double inequality in Example 6 could have been solved in two parts, as follows.

$$-7 \leq 5x - 2 \qquad \text{and} \qquad 5x - 2 < 8$$

$$-5 \leq 5x \qquad\qquad\qquad 5x < 10$$

$$-1 \leq x \qquad\qquad\qquad x < 2$$

The solution set consists of all real numbers that satisfy *both* inequalities. In other words, the solution set is the set of all values of x for which $-1 \leq x < 2$.

Compound inequalities can be written using *set notation*. In set notation, the word *and* is represented by the symbol ∩, which is read as **intersection.** The word *or* is represented by the symbol ∪, which is read as **union.** A graphical representation is shown in Figure 1.8. If A and B are sets, then x is in $A \cap B$ if it is in both A *and* B. Similarly, x is in $A \cup B$ if it is in A *or* B, or possibly both.

Example 7   Writing a Solution Set Using Union

A solution set is shown on the number line in Figure 1.9.

a. Write the solution set as a compound inequality.

b. Write the solution set using set notation and union.

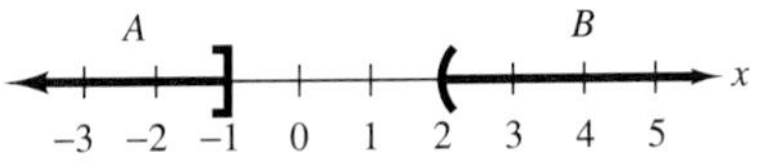

Solution

a. As a compound inequality, you can write the solution set as $x \leq -1$ *or* $x > 2$.

b. Using set notation, you can write the left interval as $A = \{x | x \leq -1\}$ and the right interval as $B = \{x | x > 2\}$. So, using the union symbol, the entire solution set can be written as $A \cup B$.

Figure 1.9

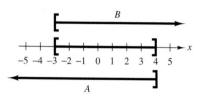

Figure 1.10

Example 8   Writing a Solution Set Using Intersection

Write the compound inequality $-3 \leq x \leq 4$ using set notation and intersection.

Solution

Consider the two sets $A = \{x | x \leq 4\}$ and $B = \{x | x \geq -3\}$. These two sets overlap, as shown on the number line in Figure 1.10. The compound inequality $-3 \leq x \leq 4$ consists of all numbers that are in $x \leq 4$ *and* $x \geq -3$, which means that it can be written as $A \cap B$.

Example 9   Solving a Conjunctive Inequality

Solve the compound inequality $-1 \leq 2x - 3$ and $2x - 3 < 5$.

Solution

Begin by writing the conjunctive inequality as a double inequality.

$$-1 \leq 2x - 3 < 5 \qquad \text{Double inequality}$$

$$-1 + 3 \leq 2x - 3 + 3 < 5 + 3 \qquad \text{Add 3 to all three parts.}$$

$$2 \leq 2x < 8 \qquad \text{Combine like terms.}$$

$$\frac{2}{2} \leq \frac{2x}{2} < \frac{8}{2} \qquad \text{Divide each part by 2.}$$

$$1 \leq x < 4 \qquad \text{Solution set}$$

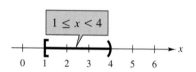

Figure 1.11

The solution set is $1 \leq x < 4$ or, in set notation, $\{x | 1 \leq x < 4\}$. The graph of the solution set is shown in Figure 1.11.

Example 10   Solving a Disjunctive Inequality

Solve the compound inequality

$$-3x + 6 \leq 2 \qquad \text{or} \qquad -3x + 6 \geq 7.$$

Solution

$-3x + 6 \leq 2$	or $\qquad -3x + 6 \geq 7$	Original inequality
$-3x + 6 - 6 \leq 2 - 6$	$-3x + 6 - 6 \geq 7 - 6$	Subtract 6 from all parts.
$-3x \leq -4$	$-3x \geq 1$	Combine like terms.
$\dfrac{-3x}{-3} \geq \dfrac{-4}{-3}$	$\dfrac{-3x}{-3} \leq \dfrac{1}{-3}$	Divide all parts by -3 and reverse both inequality symbols.
$x \geq \dfrac{4}{3}$	$x \leq -\dfrac{1}{3}$	Simplify.

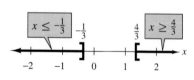

Figure 1.12

The solution set is $x \leq -\frac{1}{3}$ *or* $x \geq \frac{4}{3}$ or, in set notation, $\left\{x | x \leq -\frac{1}{3} \text{ or } x \geq \frac{4}{3}\right\}$. The graph of the solution set is shown in Figure 1.12.

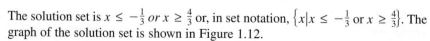

5 Solve an application problem involving inequalities.

Applications

Linear inequalities in real-life problems arise from statements that involve phrases such as "at least," "no more than," "minimum value," and so on. Study the meanings of the key phrases in the next example.

Example 11   Translating Verbal Statements

Verbal Statement	Inequality	
a. x is at most 3.	$x \leq 3$	"at most" means "less than or equal to."
b. x is no more than 3.	$x \leq 3$	
c. x is at least 3.	$x \geq 3$	"at least" means "greater than or equal to."
d. x is no less than 3.	$x \geq 3$	
e. x is more than 3.	$x > 3$	
f. x is less than 3.	$x < 3$	
g. x does not exceed 3.	$x \leq 3$	
h. x is at least 2, but less than 7.	$2 \leq x < 7$	
i. x is greater than 2, but no more than 7.	$2 < x \leq 7$	

To solve real-life problems involving inequalities, you can use the same "verbal-model approach" you use with equations.

Example 12   Finding the Maximum Width of a Package

An overnight delivery service will not accept any package whose combined length and girth (perimeter of a cross section) exceeds 132 inches. Suppose that you are sending a rectangular package that has square cross sections. If the length of the package is 68 inches, what is the maximum width of the sides of its square cross sections?

Solution

Begin by making a sketch. In Figure 1.13, notice that the length of the package is 68 inches, and each side is x inches wide because the package has a square cross section.

Verbal Model: Length + Girth ≤ 132 inches

Labels: Width of a side $= x$ (inches)
             Length $= 68$ (inches)
             Girth $= 4x$ (inches)

Inequality: $68 + 4x \leq 132$

$$4x \leq 64$$

$$x \leq 16$$

The width of each side of the package must be less than or equal to 16 inches.

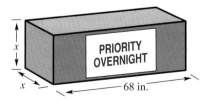

Figure 1.13

Example 13   Comparing Costs

A subcompact car can be rented from Company A for $240 per week with no extra charge for mileage. A similar car can be rented from Company B for $100 per week plus 25 cents for each mile driven. How many miles must you drive in a week so that the rental fee for Company A is less than that for Company B?

Solution

Verbal
Model: | Weekly cost for Company B | > | Weekly cost for Company A |

Labels: Number of miles driven in one week $= m$ (miles)
            Weekly cost for Company A $= 240$ (dollars)
            Weekly cost for Company B $= 100 + 0.25m$ (dollars)

Inequality: $100 + 0.25m > 240$

$$0.25m > 140$$

$$m > 560$$

The car from Company A is cheaper if you drive more than 560 miles in a week. A table helps confirm this conclusion.

Miles driven	520	530	540	550	560	570
Company A	$240.00	$240.00	$240.00	$240.00	$240.00	$240.00
Company B	$230.00	$232.50	$235.00	$237.50	$240.00	$242.50

Discussing the Concept   Problem Posing

Suppose you own a small business and must choose between two carriers for long-distance telephone service. Create realistic data for cost of the first minute of a call and cost per additional minute for each carrier, and decide what question(s) would be most helpful to ask when making such a choice. Solve the problem you created. Write a short memo to your company's business manager outlining the situation, explaining your mathematical solution, and summarizing your recommendations.

1.4   Exercises

Integrated Review   *Concepts, Skills, and Problem Solving*

Keep mathematically in shape by doing these exercises *before* the problems of this section.

Properties and Definitions

In Exercises 1–4, name the property illustrated.

1. $3yx = 3xy$

2. $3xy - 3xy = 0$

3. $6(x - 2) = 6x - 6 \cdot 2$

4. $3x + 0 = 3x$

Evaluating Expressions

In Exercises 5–10, evaluate the algebraic expression for the specified values of the variables. If not possible, state the reason.

5. $x^2 - y^2$

　　$x = 4, \quad y = 3$

6. $4s + st$

　　$s = 3, \quad t = -4$

7. $\dfrac{x}{x^2 + y^2}$

　　$x = 0, \quad y = 3$

8. $\dfrac{z^2 + 2}{x^2 - 1}$

　　$x = 2, \quad z = -1$

9. $\dfrac{a}{1 - r}$

　　$a = 2, \quad r = \frac{1}{2}$

10. $2l + 2w$

　　$l = 3, \quad w = 1.5$

Problem Solving

In Exercises 11 and 12, find the area of the trapezoid. The area of a trapezoid with parallel bases b_1 and b_2 and height h is $A = \frac{1}{2}(b_1 + b_2)h$.

11.

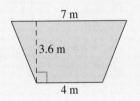

12.

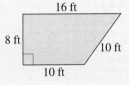

Developing Skills

In Exercises 1–4, determine whether each value of x satisfies the inequality.

Inequality		*Values*	
1. $7x - 10 > 0$	(a) $x = 3$	(b) $x = -2$	
	(c) $x = \frac{5}{2}$	(d) $x = \frac{1}{2}$	

2. $3x + 2 < \dfrac{7x}{5}$　　(a) $x = 0$　　(b) $x = 4$

　　　　　　　　　　(c) $x = -4$　　(d) $x = -1$

3. $0 < \dfrac{x + 5}{6} < 2$　　(a) $x = 10$　　(b) $x = 4$

　　　　　　　　　　(c) $x = 0$　　(d) $x = -6$

4. $-2 < \dfrac{3 - x}{2} \le 2$　　(a) $x = 0$　　(b) $x = 3$

　　　　　　　　　　(c) $x = 9$　　(d) $x = -12$

In Exercises 5–8, match the inequality with its graph. [The graphs are labeled (a), (b), (c), and (d)].

(a)

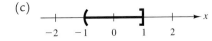

(b)

(c)

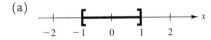

(d)

5. $x \ge 4$　　　　　　**6.** $x < -4$ or $x \ge 4$

7. $-4 < x < 4$　　　　**8.** $x < 4$

In Exercises 9–14, match the inequality with its graph. [The graphs are labeled (a), (b), (c), (d), (e), and (f).]

(a)

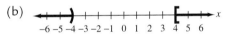

(b)

(c)

(d)

(e)

(f)

9. $-1 < x \le 2$

10. $-1 < x \le 1$

11. $-1 \le x \le 1$

12. $-1 < x < 2$

13. $-2 \le x < 1$

14. $-2 < x < 1$

In Exercises 15–28, sketch the graph of the inequality. See Example 1.

15. $x \le 2$

16. $x > -6$

17. $x > 3.5$

18. $x \le -2.5$

19. $-5 < x \le 3$

20. $-1 < x \le 5$

21. $4 > x \ge 1$

22. $9 \ge x \ge 3$

23. $\frac{3}{2} \ge x > 0$

24. $-\frac{15}{4} < x < -\frac{5}{2}$

25. $x < -5$ or $x \ge -1$

26. $x \le -4$ or $x > 0$

27. $x \le 3$ or $x > 7$

28. $x \le -1$ or $x \ge 1$

29. Write an inequality equivalent to $5 - \frac{1}{3}x > 8$ by multiplying both sides by -3.

30. Write an inequality equivalent to $5 - \frac{1}{3}x > 8$ by adding $\frac{1}{3}x$ to both sides.

In Exercises 31–78, solve the inequality and sketch the solution on the real number line. See Examples 2–6, 9, and 10.

31. $x - 4 \ge 0$

32. $x + 1 < 0$

33. $x + 7 \le 9$

34. $z - 4 > 0$

35. $2x < 8$

36. $3x \ge 12$

37. $-9x \ge 36$

38. $-6x \le 24$

39. $-\frac{3}{4}x < -6$

40. $-\frac{1}{5}x > -2$

41. $5 - x \le -2$

42. $1 - y \ge -5$

43. $2x - 5 > 9$

44. $3x + 4 \le 22$

45. $5 - 3x < 7$

46. $12 - 5x > 5$

47. $3x - 11 > -x + 7$

48. $21x - 11 \le 6x + 19$

49. $-3x + 7 < 8x - 13$

50. $6x - 1 > 3x - 11$

51. $\frac{x}{4} > 2 - \frac{x}{2}$

52. $\frac{x}{6} - 1 \le \frac{x}{4}$

53. $\frac{x - 4}{3} + 3 \le \frac{x}{8}$

54. $\frac{x + 3}{6} + \frac{x}{8} \ge 1$

55. $\frac{3x}{5} - 4 < \frac{2x}{3} - 3$

56. $\frac{4x}{7} + 1 > \frac{x}{2} + \frac{5}{7}$

57. $0 < 2x - 5 < 9$

58. $-6 \le 3x - 9 < 0$

59. $8 < 6 - 2x \le 12$

60. $-10 \le 4 - 7x < 10$

61. $-1 < -\frac{x}{6} < 1$

62. $-2 < -\frac{1}{2}s \le 0$

63. $-3 < \frac{2x - 3}{2} < 3$

64. $0 \le \frac{x - 5}{2} < 4$

65. $1 > \frac{x - 4}{-3} > -2$

66. $-\frac{2}{3} < \frac{x - 4}{-6} \le \frac{1}{3}$

67. $2x - 4 \le 4$ and $2x + 8 > 6$

68. $8 - 3x > 5$ and $x - 5 \ge -10$

69. $7 + 4x < -5 + x$ and $2x + 10 \le -2$

70. $9 - x \le 3 + 2x$ and $3x - 7 \le -22$

71. $6 - \frac{x}{2} > 1$ or $\frac{5}{4}x - 6 \ge 4$

72. $\frac{x}{3} - 2 \ge 1$ or $5 + \frac{3}{4}x \le -4$

73. $7x + 11 < 3 + 4x$ or $\frac{5}{2}x - 1 \ge 9 - \frac{3}{2}x$

74. $3x + 10 \le -x - 6$ or $\frac{x}{2} + 5 < \frac{5}{2}x - 4$

75. $-3(y + 10) \ge 4(y + 10)$

76. $2(4 - z) \ge 8(1 + z)$

77. $-4 \le 2 - 3(x + 2) < 11$

78. $16 < 4(y + 2) - 5(2 - y)$

In Exercises 79–84, write the solution set as a compound inequality. Then write the solution using set notation and union or intersection. See Example 7.

79.

80.

81.

82.

83.

84.

In Exercises 85–90, write the compound inequality using set notation and union or intersection. See Example 8.

85. $-7 \leq x < 0$

86. $2 < x < 8$

87. $x < -5$ or $x > 3$

88. $x \geq -1$ or $x < -6$

89. $-\frac{9}{2} < x \leq -\frac{3}{2}$

90. $x < 0$ or $x \geq \frac{2}{3}$

In Exercises 91–96, rewrite the statement using inequality notation. See Example 11.

91. x is nonnegative.

92. y is more than -2.

93. z is at least 2.

94. m is at least 4.

95. n is at least 10, but no more than 16.

96. x is at least 450, but no more than 500.

In Exercises 97–102, write a verbal description of the inequality.

97. $x \geq \frac{5}{2}$

98. $t < 4$

99. $3 \leq y < 5$

100. $-4 \leq t \leq 4$

101. $0 < z \leq \pi$

102. $-2 < x \leq 5$

Solving Problems

103. *Travel Budget*   A student group has $4500 budgeted for a field trip. The cost of transportation for the trip is $1900. To stay within the budget, all other costs C must be no more than what amount?

104. *Monthly Budget*   You have budgeted $1800 per month for your total expenses. The cost of rent per month is $600 and the cost of food is $350. To stay within your budget, all other costs C must be no more than what amount?

105. *Comparing Average Temperatures*   Miami's average temperature is greater than the average temperature in Washington, DC, and the average temperature in Washington, DC is greater than the average temperature in New York City. How does the average temperature in Miami compare with the average temperature in New York City?

106. *Comparing Elevations*   The elevation (above sea level) of San Francisco is less than the elevation of Dallas, and the elevation of Dallas is less than the elevation of Denver. How does the elevation of San Francisco compare with the elevation of Denver?

107. *Operating Costs*   A utility company has a fleet of vans. The annual operating cost per van is

$C = 0.35m + 2900,$

where m is the number of miles traveled by a van in a year. What is the maximum number of miles that will yield an annual operating cost that is less than $12,000?

108. *Operating Costs*   A fuel company has a fleet of trucks. The annual operating cost per truck is

$C = 0.58m + 7800,$

where m is the number of miles traveled by a truck in a year. What is the maximum number of miles that will yield an annual operating cost that is less than $25,000?

Profit   In Exercises 109 and 110, the revenue R for selling x units and the cost C of producing x units of an item are given. In order to obtain a profit, the revenue must be greater than the cost. For what values of x will this item produce a profit?

109. $R = 89.95x$
$C = 61x + 875$

110. $R = 105.45x$
$C = 78x + 25,850$

111. *Long-Distance Charges*   The cost of an international long-distance telephone call is $0.96 for the first minute and $0.75 for each additional minute. If the total cost of the call cannot exceed $5, find the interval of time that is available for the call.

112. *Long-Distance Charges*   The cost of an international long-distance telephone call is $1.45 for the first minute and $0.95 for each additional minute. If the total cost of the call cannot exceed $15, find the interval of time that is available for the call.

113. *Geometry*   The perimeter of the rectangle in the figure must be at least 36 centimeters and not more than 64 centimeters. Find the interval for x.

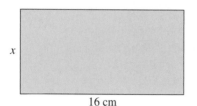

16 cm

114. *Geometry*   The perimeter of the rectangle in the figure must be at least 100 meters and not more than 120 meters. Find the interval for x.

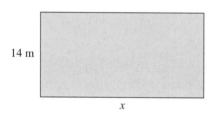

14 m

x

115 Four times a number n must be at least 12 and no more than 30. What interval contains this number?

116. Determine all real numbers n such that $\frac{1}{3}n$ must be more than 7.

117. *Hourly Wage*   Your company requires you to select one of two payment plans. One plan pays a straight $12.50 per hour. The second plan pays $8.00 per hour plus $0.75 per unit produced per hour. Write an inequality for the number of units that must be produced per hour so that the second option yields the greater hourly wage. Solve the inequality.

118. *Monthly Wage*   Your company requires you to select one of two payment plans. One plan pays a straight $3000 per month. The second plan pays $1000 per month plus a commission of 4% of your gross sales. Write an inequality for the gross sales per month for which the second option yields the greater monthly wage. Solve the inequality.

Air Pollutant Emissions   In Exercises 119 and 120, use the following equation, which models the air pollutant emissions of carbon monoxide in the continental United States from 1987 to 1995 (see figure).

$$y = 5.890 - 0.276t, \quad -3 \le t \le 5$$

In this model, y represents the amount of pollutant in parts per million and t represents the year, with $t = 0$ corresponding to 1990.   (Source: U.S. Environmental Protection Agency)

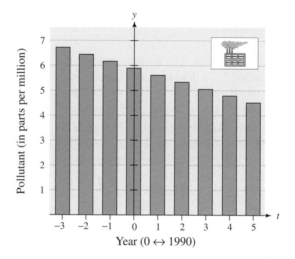

Pollutant (in parts per million)

Year ($0 \leftrightarrow 1990$)

119. During which years was the air pollutant emission of carbon monoxide greater than 6 parts per million?

120. During which years was the air pollutant emission of carbon monoxide less than 5 parts per million?

Explaining Concepts

121. Answer parts (f) and (g) of Motivating the Chapter on page 55.

122. Is adding -5 to both sides of an inequality the same as subtracting 5 from both sides? Explain.

123. Is dividing both sides of an inequality by 5 the same as multiplying both sides by $\frac{1}{5}$? Explain.

124. Give an example of a linear inequality that has an unbounded solution set.

125. Describe any differences between properties of equalities and properties of inequalities.

126. Give an example of "reversing an inequality symbol."

127. If $-5 \le t < 8$, then $-t$ must be in what interval?

128. If $-3 \le x \le 10$, then $-x$ must be in what interval?

1.5   Absolute Value Equations and Inequalities

Objectives

1 Solve an absolute value equation.

2 Solve an inequality involving absolute value.

1 Solve an absolute value equation.

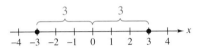

Figure 1.14

Solving Equations Involving Absolute Value

Consider the **absolute value equation**

$$|x| = 3.$$

The only solutions of this equation are -3 and 3, because these are the only two real numbers whose distance from zero is 3. (See Figure 1.14.) In other words, the absolute value equation $|x| = 3$ has exactly two solutions: $x = -3$ and $x = 3$.

> ▶ **Solving an Absolute Value Equation**
>
> Let x be a variable or a variable expression and let a be a real number such that $a \geq 0$. The solutions of the equation $|x| = a$ are given by $x = -a$ and $x = a$. That is,
>
> $$|x| = a \implies x = -a \quad \text{or} \quad x = a.$$

Example 1   Solving Absolute Value Equations

Solve each absolute value equation.

a. $|x| = 10$

b. $|x| = 0$

c. $|y| = -1$

Solution

a. This equation is equivalent to the two linear equations

$$x = -10 \quad \text{and} \quad x = 10. \qquad \text{Equivalent linear equations}$$

So, the absolute value equation has two solutions: -10 and 10.

b. This equation is equivalent to the two linear equations

$$x = 0 \quad \text{and} \quad x = 0. \qquad \text{Equivalent linear equations}$$

Because both equations are the same, you can conclude that the absolute value equation has only one solution: 0.

c. This absolute value equation has *no solution* because it is not possible for the absolute value of a real number to be negative.

Study Tip

The strategy for solving absolute value equations is to *rewrite* the equation in *equivalent forms* that can be solved by previously learned methods. This is a common strategy in mathematics. That is, when you encounter a new type of problem, you try to rewrite the problem so that it can be solved by techniques you already know.

Example 2   Solving Absolute Value Equations

Solve $|3x + 4| = 10$.

Solution

$$|3x + 4| = 10 \qquad \text{Original equation}$$

$$3x + 4 = -10 \quad \text{or} \quad 3x + 4 = 10 \qquad \text{Equivalent equations}$$

$$3x + 4 - 4 = -10 - 4 \qquad 3x + 4 - 4 = 10 - 4 \qquad \text{Subtract 4 from both sides.}$$

$$3x = -14 \qquad\qquad\qquad 3x = 6 \qquad \text{Combine like terms.}$$

$$x = -\frac{14}{3} \qquad\qquad\qquad x = 2 \qquad \text{Divide both sides by 3.}$$

Check

$$|3x + 4| = 10 \qquad\qquad |3x + 4| = 10$$

$$\left|3\left(-\tfrac{14}{3}\right) + 4\right| \overset{?}{=} 10 \qquad\qquad |3(2) + 4| \overset{?}{=} 10$$

$$|-14 + 4| \overset{?}{=} 10 \qquad\qquad |6 + 4| \overset{?}{=} 10$$

$$|-10| = 10 \checkmark \qquad\qquad |10| = 10 \checkmark$$

When solving absolute value equations, remember that it is possible that they have no solution. For instance, the equation $|3x + 4| = -10$ has no solution because the absolute value of a real number cannot be negative. Do not make the mistake of trying to solve such an equation by writing the "equivalent" linear equations as $3x + 4 = -10$ and $3x + 4 = 10$. These equations have solutions, but they are both extraneous.

The equation in the next example is not given in the **standard form**

$$|ax + b| = c, \quad c \geq 0.$$

Notice that the first step in solving such an equation is to write it in standard form.

Example 3   An Absolute Value Equation in Nonstandard Form

Solve $|2x - 1| + 3 = 8$.

Solution

$$|2x - 1| + 3 = 8 \qquad \text{Original equation}$$

$$|2x - 1| = 5 \qquad \text{Standard form}$$

$$2x - 1 = -5 \quad \text{or} \quad 2x - 1 = 5 \qquad \text{Equivalent equations}$$

$$2x = -4 \qquad\qquad 2x = 6 \qquad \text{Add 1 to both sides.}$$

$$x = -2 \qquad\qquad x = 3 \qquad \text{Divide both sides by 2.}$$

The solutions are -2 and 3. Check these in the original equation.

If two algebraic expressions are equal in absolute value, they must either be *equal* to each other or be the *opposites* of each other. So, you can solve equations of the form

$$|ax + b| = |cx + d|$$

by forming the two linear equations

Expressions equal

Expressions opposite

$$ax + b = cx + d \quad \text{and} \quad ax + b = -(cx + d).$$

Example 4   Solving an Equation Involving Two Absolute Values

Solve $|3x - 4| = |7x - 16|$.

Solution

$$|3x - 4| = |7x - 16| \qquad \text{Original equation}$$

$$3x - 4 = 7x - 16 \quad \text{or} \quad 3x - 4 = -(7x - 16) \qquad \text{Equivalent equations}$$

$$3x = 7x - 12 \qquad\qquad 3x - 4 = -7x + 16$$

$$-4x = -12 \qquad\qquad\quad 10x = 20$$

$$x = 3 \qquad\qquad\qquad\quad x = 2 \qquad \text{Solutions}$$

The solutions are 3 and 2. Check these in the original equation.

Study Tip

When solving equations of the form

$$|ax + b| = |cx + d|$$

it is possible that one of the resulting equations will not have a solution. Note this occurrence in Example 5.

Example 5   Solving an Equation Involving Two Absolute Values

Solve $|x + 5| = |x + 11|$.

Solution

By equating the expression $(x + 5)$ to the opposite of $(x + 11)$, you obtain

$$x + 5 = -(x + 11) \qquad \text{Equivalent equation}$$

$$x + 5 = -x - 11 \qquad \text{Distributive Property}$$

$$x = -x - 16 \qquad \text{Subtract 5 from both sides.}$$

$$2x = -16 \qquad \text{Add } x \text{ to both sides.}$$

$$x = -8. \qquad \text{Divide both sides by 2.}$$

However, by setting the two expressions equal to each other, you obtain

$$x + 5 = x + 11 \qquad \text{Equivalent equation}$$

$$x = x + 6 \qquad \text{Subtract 5 from both sides.}$$

$$0 = 6 \qquad \text{Subtract } x \text{ from both sides.}$$

which is a false statement. So, the original equation has only one solution: -8. Check this solution in the original equation.

2 Solve an inequality involving absolute value.

Solving Inequalities Involving Absolute Value

To see how to solve inequalities involving absolute value, consider the following comparisons.

$\lvert x \rvert = 2$	$\lvert x \rvert < 2$	$\lvert x \rvert > 2$
$x = -2$ and $x = 2$	$-2 < x < 2$	$x < -2$ or $x > 2$

These comparisons suggest the following rule for solving inequalities involving absolute value.

▶ **Solving an Absolute Value Inequality**

Let x be a variable or an algebraic expression and let a be a real number such that $a > 0$.

1. The solutions of $\lvert x \rvert < a$ are all values of x that lie between $-a$ and a. That is,

$$\lvert x \rvert < a \quad \text{if and only if} \quad -a < x < a.$$

2. The solutions of $\lvert x \rvert > a$ are all values of x that are *less than* $-a$ or *greater than* a. That is,

$$\lvert x \rvert > a \quad \text{if and only if} \quad x < -a \text{ or } x > a.$$

These rules are also valid if $<$ is replaced by $\leq$ and $>$ is replaced by $\geq$.

Example 6   Solving an Absolute Value Inequality

Solve $\lvert x - 5 \rvert < 2$.

Solution

$\lvert x - 5 \rvert < 2$	Original inequality
$-2 < x - 5 < 2$	Equivalent double inequality
$-2 + 5 < x - 5 + 5 < 2 + 5$	Add 5 to all three parts.
$3 < x < 7$	Combine like terms.

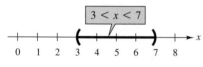

Figure 1.15

The solution set consists of all real numbers that are greater than 3 and less than 7. The interval notation for this solution set is $(3, 7)$. The graph of this solution set is shown in Figure 1.15.

To verify the solution of an absolute value inequality, you need to check values in the solution set and outside of the solution set. For instance, in Example 6 you can check that $x = 4$ is in the solution set and that $x = 2$ and $x = 8$ are not in the solution set.

Study Tip

In Example 6, note that an absolute value inequality of the form $|x| < a$ (or $|x| \le a$) can be solved with a double inequality. An inequality of the form $|x| > a$ (or $|x| \ge a$) cannot be solved with a double inequality. Instead, you must solve two separate inequalities, as demonstrated in Example 7.

Example 7 Solving an Absolute Value Inequality

Solve $|3x - 4| \ge 5$.

Solution

$|3x - 4| \ge 5$	Original inequality

$$3x - 4 \le -5 \quad \text{or} \quad 3x - 4 \ge 5 \qquad \text{Equivalent inequalities}$$

$$3x - 4 + 4 \le -5 + 4 \qquad 3x - 4 + 4 \ge 5 + 4 \qquad \text{Add 4 to all parts.}$$

$$3x \le -1 \qquad\qquad 3x \ge 9 \qquad\qquad \text{Combine like terms.}$$

$$\frac{3x}{3} \le \frac{-1}{3} \qquad\qquad \frac{3x}{3} \ge \frac{9}{3} \qquad\qquad \text{Divide both sides by 3.}$$

$$x \le -\frac{1}{3} \qquad\qquad x \ge 3 \qquad\qquad \text{Simplify.}$$

The solution set consists of all real numbers that are less than or equal to $-\frac{1}{3}$ *or* greater than or equal to 3. The interval notation for the solution set is $\left(-\infty, -\frac{1}{3}\right] \cup [3, \infty)$. The graph is shown in Figure 1.16.

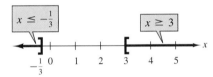

Figure 1.16

Technology: Tip

Most graphing utilities can sketch the graph of an absolute value inequality. The graph of the inequality

$$|x - 5| < 2$$

is shown below. Notice that the graph occurs as an interval *above* the x-axis. Consult the user's guide for your graphing utility for the steps to graph an inequality.

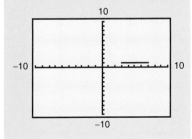

Example 8 Solving an Absolute Value Inequality

$$\left|2 - \frac{x}{3}\right| \le 0.01 \qquad\qquad \text{Original inequality}$$

$$-0.01 \le 2 - \frac{x}{3} \le 0.01 \qquad\qquad \text{Equivalent double inequality}$$

$$-2.01 \le -\frac{x}{3} \le -1.99 \qquad\qquad \text{Subtract 2 from all three parts.}$$

$$6.03 \ge x \ge 5.97 \qquad\qquad \text{Multiply all three parts by } -3 \text{ and reverse both inequality symbols.}$$

$$5.97 \le x \le 6.03 \qquad\qquad \text{Solution set in standard form}$$

The solution set consists of all real numbers that are greater than or equal to 5.97 *and* less than or equal to 6.03. The interval notation for the solution set is $[5.97, 6.03]$. The graph is shown in Figure 1.17.

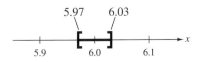

Figure 1.17

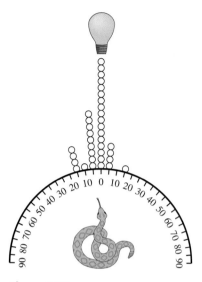

Figure 1.18

Example 9 Creating a Model

To test the accuracy of a rattlesnake's "pit-organ sensory system," a biologist blindfolded a rattlesnake and presented the snake with a warm "target." Of 36 strikes, the snake was on target 17 times. In fact, the snake was within 5 degrees of the target for 30 of the strikes. Let A represent the number of degrees by which the snake is off target. Then $A = 0$ represents a strike that is aimed directly at the target. Positive values of A represent strikes to the right of the target and negative values of A represent strikes to the left of the target. Use the diagram shown in Figure 1.18 to write an absolute value inequality that describes the interval in which the 36 strikes occurred.

Solution

From the diagram, you can see that the snake was never off by more than 15 degrees in either direction. As a compound inequality, this can be represented by

$$-15 \le A \le 15.$$

As an absolute value inequality, the interval in which the strikes occurred can be represented by

$$|A| \le 15.$$

Discussing the Concept   Error Analysis

Suppose you are teaching a class in algebra and one of your students hands in the following solution.

$$|3x - 4| \ge -5$$

$$3x - 4 \le -5 \quad \text{or} \quad 3x - 4 \ge 5$$

$$3x - 4 + 4 \le -5 + 4 \qquad 3x - 4 + 4 \ge 5 + 4$$

$$3x \le -1 \qquad\qquad 3x \ge 9$$

$$\frac{3x}{3} \le \frac{-1}{3} \qquad\qquad \frac{3x}{3} \ge \frac{9}{3}$$

$$x \le -\frac{1}{3} \qquad\qquad x \ge 3$$

What is wrong with this solution? What could you say to help your students avoid this type of error?

1.5   Exercises

Integrated Review   *Concepts, Skills, and Problem Solving*

Keep mathematically in shape by doing these exercises *before* the problems of this section.

Properties and Definitions

1. If n is an integer, how do the numbers $2n$ and $2n + 1$ differ? Explain.

2. Are $-2x^4$ and $(-2x)^4$ equal? Explain.

3. Explain how to write $\frac{35}{14}$ in simplified form.

4. Explain how to divide $\frac{4}{5}$ by $\frac{z}{3}$.

Order of Real Numbers

In Exercises 5–10, place the correct inequality symbol ($<$ or $>$) between the two real numbers.

5. $-3.2, 2$

6. $-3.2, -4.1$

7. $-\frac{3}{4}, -5$

8. $-\frac{1}{5}, -\frac{1}{3}$

9. $\pi, -3$

10. $6, \frac{13}{2}$

Problem Solving

In Exercises 11 and 12, determine whether there is more or less than a \$500 variance between the budgeted amount and the actual expense.

11. Wages
Budgeted: \$162,700
Actual: \$163,356

12. Taxes
Budgeted: \$42,640
Actual: \$42,335

Developing Skills

In Exercises 1–4, determine whether the value is a solution of the equation.

Equation	*Value*
1. $\|4x + 5\| = 10$	$x = -3$
2. $\|2x - 16\| = 10$	$x = 3$
3. $\|6 - 2w\| = 2$	$w = 4$
4. $\left\|\frac{1}{2}t + 4\right\| = 8$	$t = 6$

In Exercises 5–8, transform the absolute value equation into two linear equations.

5. $\|x - 10\| = 17$

6. $\|7 - 2t\| = 5$

7. $\|4x + 1\| = \frac{1}{2}$

8. $\|22k + 6\| = 9$

In Exercises 9–46, solve the equation. (Some of the equations have no solution.) See Examples 1–5.

9. $\|x\| = 4$

10. $\|x\| = 3$

11. $\|t\| = -45$

12. $\|s\| = 16$

13. $\|h\| = 0$

14. $\|x\| = -82$

15. $\|5x\| = 15$

16. $\left\|\frac{1}{3}x\right\| = 2$

17. $\|x - 16\| = 5$

18. $\|z - 100\| = 100$

19. $\|2s + 3\| = 25$

20. $\|7a + 6\| = 8$

21. $\|32 - 3y\| = 16$

22. $\|3 - 5x\| = 13$

23. $\|3x + 4\| = -16$

24. $\|20 - 5t\| = 50$

25. $\|4 - 3x\| = 0$

26. $\|3x - 2\| = -5$

27. $\left\|\frac{2}{3}x + 4\right\| = 9$

28. $\left\|3 - \frac{4}{5}x\right\| = 1$

29. $\|0.32x - 2\| = 4$

30. $\|2 - 1.054x\| = 2$

31. $\|5x - 3\| + 8 = 22$

32. $\|6x - 4\| - 7 = 3$

33. $\|3x + 9\| - 12 = -8$

34. $\|5 - 2x\| + 10 = 6$

35. $-2\|7 - 4x\| = -16$

36. $4\|5x + 1\| = 24$

37. $3\|2x - 5\| + 4 = 7$

38. $2\|4 - 3x\| - 6 = -2$

39. $\|x + 8\| = \|2x + 1\|$

40. $\|10 - 3x\| = \|x + 7\|$

41. $\|x + 2\| = \|3x - 1\|$

42. $\|x - 2\| = \|2x - 15\|$

43. $\|45 - 4x\| = \|32 - 3x\|$

44. $\|5x + 4\| = \|3x + 25\|$

45. $\|4x - 10\| = 2\|2x + 3\|$

46. $3\|2 - 3x\| = \|9x + 21\|$

Think About It In Exercises 47 and 48, write an absolute value equation that represents the verbal statement.

47. The distance between x and 5 is 3.

48. The distance between t and -2 is 6.

In Exercises 49–52, determine whether each x-value is a solution of the inequality.

Inequality	*Values*
49. $|x| < 3$	(a) $x = 2$ (b) $x = -4$
	(c) $x = 4$ (d) $x = -1$
50. $|x| \le 5$	(a) $x = -7$ (b) $x = -4$
	(c) $x = 4$ (d) $x = 9$
51. $|x - 7| \ge 3$	(a) $x = 9$ (b) $x = -4$
	(c) $x = 11$ (d) $x = 6$
52. $|x - 3| > 5$	(a) $x = 16$ (b) $x = 3$
	(c) $x = -2$ (d) $x = -3$

In Exercises 53–56, transform the absolute value inequality into a double inequality or two separate inequalities.

53. $|y + 5| < 3$
54. $|6x + 7| \le 5$
55. $|7 - 2h| \ge 9$
56. $|8 - x| > 25$

In Exercises 57–60, sketch a graph that shows the real numbers that satisfy the statement.

57. All real numbers greater than -2 *and* less than 5

58. All real numbers greater than or equal to 3 *and* less than 10

59. All real numbers less than or equal to 4 *or* greater than 7

60. All real numbers less than -6 *or* greater than or equal to 6

In Exercises 61–94, solve the inequality. See Examples 6–8.

61. $|y| < 4$
62. $|x| < 6$
63. $|x| \ge 6$
64. $|y| \ge 4$
65. $|2x| < 14$
66. $|4z| \le 9$
67. $\left|\dfrac{y}{3}\right| \le 3$
68. $\left|\dfrac{t}{2}\right| < 4$

69. $|y - 2| \le 4$
70. $|x - 3| \le 6$
71. $|x + 6| > 10$
72. $|x - 4| \ge 3$
73. $|2x - 1| \le 7$
74. $|3x + 4| < 2$
75. $|6t + 15| \ge 30$
76. $|3t + 1| > 5$
77. $|2 - 5x| > -8$
78. $|8 - 7x| < -6$
79. $|3x + 10| < -1$
80. $|4x - 5| > -3$
81. $\dfrac{|x + 2|}{10} \le 8$
82. $\dfrac{|s - 3|}{5} > 4$
83. $\dfrac{|y - 16|}{4} < 30$
84. $\dfrac{|a + 6|}{2} \ge 16$
85. $\left|\dfrac{z}{10} - 3\right| > 8$
86. $\left|\dfrac{x}{8} + 1\right| < 0$
87. $|0.2x - 3| < 4$
88. $|1.5t - 8| \le 16$
89. $\left|6 - \dfrac{3}{5}x\right| \le 0.4$
90. $\left|3 - \dfrac{x}{4}\right| > 0.15$
91. $-2|3x + 6| < 4$
92. $-4|2x - 7| > -12$
93. $\left|9 - \dfrac{x}{2}\right| - 7 \le 4$
94. $\left|8 - \dfrac{2}{3}x\right| + 6 \ge 10$

In Exercises 95–100, use a graphing utility to solve the inequality.

95. $|3x + 2| < 4$
96. $|2x - 1| \le 3$
97. $|2x + 3| > 9$
98. $|7r - 3| > 11$
99. $|x - 5| + 3 \le 5$
100. $|a + 1| - 4 < 0$

In Exercises 101–104, match the inequality with its graph. [The graphs are labeled (a), (b), (c), and (d).]

(a)

(b)

(c)

(d)

101. $|x - 4| \le 4$
102. $|x - 4| < 1$
103. $\frac{1}{2}|x - 4| > 4$
104. $|2(x - 4)| \ge 4$

The symbol ▦ indicates an exercise in which you are to use a graphing utility.

In Exercises 105–108, write an absolute value inequality that represents the interval.

105.

106.

107.

108.

In Exercises 109–112, write an absolute value inequality that represents the verbal statement.

109. The set of all real numbers x whose distance from 0 is less than 3.

110. The set of all real numbers x whose distance from 0 is more than 2.

111. The set of all real numbers x whose distance from 5 is more than 6.

112. The set of all real numbers x whose distance from 16 is less than 5.

Solving Problems

113. *Temperature* The operating temperature of an electronic device must satisfy the inequality

$$|t - 72| < 10$$

where t is given in degrees Fahrenheit. Sketch the graph of the solution of the inequality.

114. *Time Study* A time study was conducted to determine the length of time required to perform a particular task in a manufacturing process. The times required by approximately two thirds of the workers in the study satisfied the inequality

$$\left| \frac{t - 15.6}{1.9} \right| < 1$$

where t is time in minutes. Sketch the graph of the solution of the inequality.

115. *Accuracy of Measurements* In woodshop class, you must cut several pieces of wood to within $\frac{3}{16}$ inch of the teacher's specifications. Let $(s - x)$ represent the difference between the specification s and the measured length x of a cut piece.

(a) Write an absolute value inequality that describes the values of x that are within specifications.

(b) The length of one piece of wood is specified to be $s = 5\frac{1}{8}$ inches. Describe the acceptable lengths for this piece.

116. *Think About It* When you buy a 16-ounce bag of chips, you probably expect to get *precisely* 16 ounces. Suppose the actual weight w (in ounces) of a "16-ounce" bag of chips is given by $|w - 16| \leq \frac{1}{2}$. If you buy four 16-ounce bags, what is the greatest amount you can expect to get? What is the least? Explain.

Explaining Concepts

117. Give a graphical description of the absolute value of a real number.

118. Give an example of an absolute value equation that has only one solution.

119. In your own words, explain how to solve an absolute value equation. Illustrate your explanation with an example.

120. Describe the solution of the inequality $|x| > 3$.

121. The graph of the inequality $|x - 3| < 2$ can be described as *all real numbers that are within 2 units of 3*. Give a similar description of $|x - 4| < 1$.

122. The graph of the inequality $|y - 1| > 3$ can be described as *all real numbers that are more than 3 units from 1*. Give a similar description of $|y + 2| > 4$.

123. Complete $|x - 3| \leq \quad$ so that the solution is $0 \leq x \leq 6$.

124. Complete $|2x - 6| \leq \quad$ so that the solution is $0 \leq x \leq 6$.

Key Terms

equation, *p. 56*
solutions (equation), *p. 56*
solution set, *p. 56*
identity, *p. 56*
conditional equation, *p. 56*
equivalent equations, *p. 57*
linear equation, *p. 58*
mathematical modeling, *p. 67*
percent, *p. 68*

percent equation, *p. 68*
ratio, *p. 70*
unit price, *p. 70*
proportion, *p. 71*
markup, *p. 78*
markup rate, *p. 78*
discount, *p. 79*
discount rate, *p. 79*
rate of work, *p. 82*
algebraic inequalities, *p. 92*
solutions (inequality), *p. 92*

solution set of an inequality, *p. 92*
graph of an inequality, *p. 92*
bounded intervals, *p. 92*
endpoints, *p. 92*
length of [*a*, *b*], *p. 92*
unbounded (infinite) intervals, *p. 93*
positive infinity, *p. 93*
negative infinity, *p. 93*

equivalent inequalities, *p. 94*
linear inequality, *p. 95*
compound inequality, *p. 96*
intersection, *p. 97*
union, *p. 97*
absolute value equation, *p. 105*
standard form (absolute value equation), *p. 106*

Key Concepts

1.1 Forming equivalent equations: properties of equality

A given equation can be transformed into an equivalent equation using one or more of the following procedures.

1. Simplify each side.
2. Apply the Addition Property of Equality.
3. Apply the Multiplication Property of Equality.
4. Interchange sides.

1.2 Strategy for solving word problems

1. Ask yourself what you need to know to solve the problem. Then write a verbal model that will give you what you need to know.
2. Assign labels to each part of the verbal model—numbers to the known quantities and letters (or expressions) to the variable quantities.
3. Use the labels to write an algebraic model based on the verbal model.
4. Solve the resulting algebraic equation.
5. Answer the original question and check that your answer satisfies the original problem as stated.

1.3 Common formulas

See the summary boxes on page 83.

1.4 Properties of inequalities

1. Addition/subtraction properties

   If $a < b$, then $a + c < b + c$.
   If $a < b$, then $a - c < b - c$.

2. Multiplication/division properties: positive quantities

   If $a < b$ and c is positive, then $ac < bc$.

   If $a < b$ and c is positive, then $\dfrac{a}{c} < \dfrac{b}{c}$.

3. Multiplication/division properties: negative quantities

   If $a < b$ and c is negative, then $ac > bc$.

   If $a < b$ and c is negative, then $\dfrac{a}{c} > \dfrac{b}{c}$.

4. Transitive property

   If $a < b$ and $b < c$, then $a < c$.

1.5 Solving an absolute value equation

Let x be a variable or a variable expression and let a be a real number such that $a \geq 0$. The solutions of the equation $|x| = a$ are given by $x = -a$ and $x = a$. That is,

$$|x| = a \implies x = -a \quad \text{or} \quad x = a.$$

1.5 Solving an absolute value inequality

Let x be a variable or an algebraic expression and let a be a real number such that $a > 0$.

1. The solutions of $|x| < a$ are all values of x that lie between $-a$ and a. That is,

   $$|x| < a \quad \text{if and only if} \quad -a < x < a.$$

2. The solutions of $|x| > a$ are all values of x that are less than $-a$ or greater than a. That is,

   $$|x| > a \quad \text{if and only if} \quad x < -a \text{ or } x > a.$$

These rules are also valid if $<$ is replaced by $\leq$ and $>$ is replaced by $\geq$.

REVIEW EXERCISES

1.1 In Exercises 1–4, determine whether each value of the variable is a solution of the equation.

Equation	Values
1. $45 - 7x = 3$	(a) $x = 3$ (b) $x = 6$
2. $3(3 - x) = -x$	(a) $x = \frac{9}{2}$ (b) $x = -\frac{2}{3}$
3. $\frac{x}{7} + \frac{x}{5} = 1$	(a) $x = \frac{35}{12}$ (b) $x = -\frac{2}{35}$
4. $\frac{t + 2}{6} = \frac{7}{2}$	(a) $t = -12$ (b) $t = 19$

In Exercises 5–26, solve the linear equation and check the result. (Some of the equations have no solution.)

5. $x + 2 = 5$ **6.** $x - 7 = 3$

7. $-3x = 36$ **8.** $11x = 44$

9. $-\frac{1}{8}x = 3$ **10.** $\frac{1}{10}x = 5$

11. $5x + 4 = 19$ **12.** $3 - 2x = 9$

13. $17 - 7x = 3$ **14.** $3 + 6x = 51$

15. $7x - 5 = 3x + 11$ **16.** $9 - 2x = 4x - 7$

17. $3(2y - 1) = 9 + 3y$ **18.** $-2(x + 4) = 2x - 7$

19. $4y - 6(y - 5) = 2$ **20.** $7x + 2(7 - x) = 8$

21. $4(3x - 5) = 6(2x + 3)$

22. $8(x - 2) = 3(x - 2)$ **23.** $\frac{4}{5}x - \frac{1}{10} = \frac{3}{2}$

24. $\frac{1}{4}s + \frac{3}{8} = \frac{5}{2}$ **25.** $1.4t + 2.1 = 0.9t$

26. $2.5x - 6.2 = 3.7x - 5.8$

1.2 In Exercises 27 and 28, complete the table.

27.

Percent	Parts out of 100	Decimal	Fraction
87%			

28.

Percent	Parts out of 100	Decimal	Fraction
			$\frac{1}{6}$

In Exercises 29–34, solve the percent problem using the percent equation.

29. What is 130% of 50?

30. What is 0.4% of 7350?

31. 645 is $21\frac{1}{2}\%$ of what number?

32. 498 is 83% of what number?

33. 250 is what percent of 200?

34. 162.5 is what percent of 6500?

In Exercises 35–38, express as a ratio. (Use the same units for the numerator and denominator.)

35. 16 feet to 4 yards

36. 3 quarts to 5 pints

37. 45 seconds to 5 minutes

38. 3 meters to 150 centimeters

In Exercises 39–42, solve the proportion.

39. $\frac{7}{8} = \frac{y}{4}$ **40.** $\frac{x}{16} = \frac{5}{12}$

41. $\frac{b}{6} = \frac{5 + b}{15}$ **42.** $\frac{x + 1}{3} = \frac{x - 1}{2}$

1.3 In Exercises 43–46, find the missing quantities. (Assume the markup rate is a percent based on the cost.)

	Cost	Selling Price	Markup	Markup Rate
43.	$99.95	$149.93		
44.	$23.50	$31.33		
45.		$125.85	$44.13	
46.		$895.00	$223.75	

In Exercises 47–50, find the missing quantities. (Assume the discount rate is a percent based on the list price.)

	List Price	Sale Price	Discount	Discount Rate
47.	$71.95	$53.96		
48.	$559.95	$279.98		
49.	$1995.50		$598.65	
50.	$39.00		$15.60	

In Exercises 51–54, solve the equation for the specified variable.

51. Solve for x in $2x - 7y + 4 = 0$.

52. Solve for v in $\frac{2}{3}u - 4v = 2v + 3$.

53. Solve for h in $V = \pi r^2 h$.

54. Solve for h in $S = 2\pi r^2 + 2\pi rh$.

1.4 In Exercises 55–72, solve the linear inequality and sketch the solution on the real number line.

55. $x - 5 \leq -1$

56. $x + 8 > 5$

57. $-5x < 30$

58. $-11x \geq 44$

59. $5x + 3 > 18$

60. $3x - 11 \leq 7$

61. $8x + 1 \geq 10x - 11$

62. $12 - 3x < 4x - 2$

63. $\frac{1}{3} - \frac{1}{2}y < 12$

64. $\frac{x}{4} - 2 < \frac{3x}{8} + 5$

65. $-6 \leq 2x + 8 < 4$

66. $-13 \leq 3 - 4x < 13$

67. $5 > \frac{x + 1}{-3} > 0$

68. $12 \geq \frac{x - 3}{2} > 1$

69. $5x - 4 < 6$ and $3x + 1 > -8$

70. $6 - 2x \leq 1$ or $10 - 4x > -6$

71. $-4(3 - 2x) \leq 3(2x - 6)$

72. $3(2 - y) \geq 2(1 + y)$

In Exercises 73–76, write an inequality for the given statement.

73. z is no more than 10.

74. x is nonnegative.

75. y is at least 7 but less than 14.

76. The volume V is less than 27 cubic feet.

1.5 In Exercises 77–84, solve the absolute value equation.

77. $|x| = 6$

78. $|x| = -4$

79. $|4 - 3x| = 8$

80. $|2x + 3| = 7$

81. $|5x + 4| - 10 = -6$

82. $|x - 2| - 2 = 4$

83. $|3x - 4| = |x + 2|$

84. $|5x + 6| = |2x - 1|$

In Exercises 85–92, solve the absolute value inequality.

85. $|x - 4| > 3$

86. $|t + 3| > 2$

87. $|3x| > 9$

88. $\left|\frac{t}{3}\right| < 1$

89. $|2x - 7| < 15$

90. $|5x - 1| < 9$

91. $|b + 2| - 6 > 1$

92. $|2y - 1| + 4 < -1$

In Exercises 93 and 94, use a graphing utility to solve the inequality.

93. $|2x - 5| \geq 1$

94. $|5(1 - x)| \leq 25$

In Exercises 95 and 96, write an absolute value inequality that represents the interval.

95.

96.

Solving Problems

97. *Number Problem* Find two consecutive positive integers whose sum is 147.

98. *Number Problem* Find two consecutive even integers whose sum is 74.

99. *Real Estate Commission* A real estate agency receives a commission of $9000 for the sale of a $150,000 house. What percent commission is this?

100. *Quality Control* A quality control engineer reported that 1.6% of a sample of parts were defective. The engineer found six defective parts. How large was the sample?

101. *Comparison Shopping* A mail-order catalog has a plant stand with a list price of $24.95 plus $6.95 for shipping and handling. A local department store has the same plant stand for $35.95. The department store has a special 30% off sale. Which is the better buy?

102. *Pension Fund* Your employer withholds $216 of your gross income each month for your retirement. Determine the percent of your total monthly gross income of $3200 that is withheld for retirement.

103. *Sales Tax* The state sales tax on an item you purchase is $7\frac{1}{4}\%$. If the item costs $34, how much sales tax will you pay?

104. *Property Tax* The tax on a property with an assessed value of $105,000 is $1680. Find the tax on a property with an assessed value of $125,000.

105. *Recipe Enlargement* One and one-half cups of milk are required to make one batch of pudding. How much milk is required to make $2\frac{1}{2}$ batches?

106. *Map Scale* One-third inch represents 50 miles on a map. Approximate the distance between two cities that are $3\frac{1}{4}$ inches apart on the map.

107. *Fuel Mixture* The gasoline-to-oil ratio for a lawn mower engine is 50 to 1. Determine the amount of gasoline required if $\frac{1}{2}$ pint of oil is used in producing the mixture.

Similar Triangles In Exercises 108 and 109, solve for the length x by using the fact that corresponding sides of similar triangles are proportional.

108.

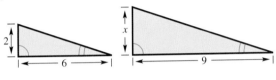

109.

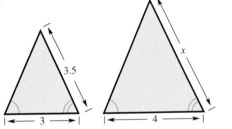

110. *Geometry* You want to measure the height of a flagpole. To do this, you measure the flagpole's shadow and find that it is 30 feet long. You also measure the height of a 5-foot lamp post and find its shadow to be 3 feet long (see figure). How tall is the flagpole?

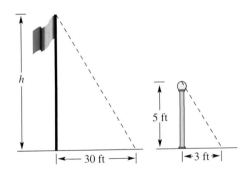

111. *Geometry* You want to measure the height of a silo. To do this, you measure the silo's shadow and find that it is 20 feet long. You are 6 feet tall and your shadow is $1\frac{1}{2}$ feet long. How tall is the silo?

112. *Revenue Increase* The revenues for a corporation, in millions of dollars, in the years 1999 and 2000 were $4521.4 and $4679.0, respectively. Determine the percent increase in revenue from 1999 to 2000.

113. *Price Increase* The manufacturer's suggested retail price for a certain truck model is $25,750. Estimate the price of a comparably equipped truck for the next model year if it is projected that truck prices will increase by $5\frac{1}{2}\%$.

114. *Retail Price* A camera that costs a retailer $259.95 is marked up by 35%. Find the price to the consumer.

115. *Markup Rate* A calculator selling for $175.00 costs the retailer $95.00. Find the markup rate.

116. *Sale Price* The list price of a coat is $259. Find the sale price of the coat if the price is reduced by 25%.

117. *Comparison Shopping* A mail-order catalog has an attaché case with a list price of $99.97 plus $4.50 for shipping and handling. A local department store has the same attaché case for $125.95. The department store has a special 20% off sale. Which is the better buy?

118. *Sales Goal* The weekly salary of an employee is $150 plus a 6% commission on total sales. The employee needs a minimum salary of $650 per week. How much must be sold to produce this salary?

119. *Mixture Problem* Determine the number of liters of a 30% saline solution and the number of liters of a 60% saline solution that are required to make 10 liters of a 50% saline solution.

120. *Mixture Problem* Determine the number of gallons of a 25% alcohol solution and the number of gallons of a 50% alcohol solution that are required to make 8 gallons of a 40% alcohol solution.

121. *Distance* Determine the distance an Air Force jet can travel in $2\frac{1}{3}$ hours if its average speed is 1200 miles per hour.

122. *Time* Determine the time for a bus to travel 330 miles if its average speed is 52 miles per hour.

123. *Speed* A truck driver traveled at an average speed of 48 miles per hour on a 100-mile trip to pick up a load of freight. On the return trip with the truck fully loaded, the average speed was 40 miles per hour. Find the average speed for the round trip.

124. *Speed*   For 2 hours of a 400-mile trip, your average speed is only 40 miles per hour. Determine the average speed that must be maintained for the remainder of the trip if you want the average speed for the entire trip to be 50 miles per hour.

125. *Work-Rate Problem*   Find the time for two people working together to complete a task if it takes them 4.5 hours and 6 hours working individually.

126. *Work-Rate Problem*   Find the time for two people working together to complete half a task if it takes them 8 hours and 10 hours to complete the entire task working individually.

127. *Simple Interest*   Find the total simple interest you will earn on a $1000 corporate bond that matures in 4 years and has an 8.5% interest rate.

128. *Simple Interest*   Find the annual simple interest rate on a certificate of deposit that pays $37.50 per year in interest on a principal of $500.

129. *Simple Interest*   Find the principal required to have an annual interest income of $20,000 if the annual simple interest rate on the principal is 9.5%.

130. *Simple Interest*   A corporation borrows 3.25 million dollars for 2 years to modernize one of its manufacturing facilities. If it pays an annual simple interest rate of 12%, what is the total principal and interest that must be repaid?

131. *Simple Interest*   An inheritance of $50,000 is divided between two investments earning 8.5% and 10% simple interest. (The 10% investment has a greater risk.) What is the smallest amount that can be invested in the 10% fund if the total annual interest from both investments must be at least $4700?

132. *Simple Interest*   You invest $1000 in a certificate of deposit that has an annual simple interest rate of 7%. After 6 months, the interest is computed and added to the principal. During the second 6 months, the interest is computed using the original investment plus the interest earned during the first 6 months. What is the total interest earned during the first year of the investment?

133. *Geometry*   The area of the rectangle in the figure is 48 square inches. Find x.

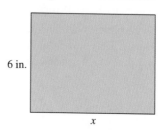

6 in.

x

134. *Geometry*   The perimeter of the rectangle in the figure is 110 feet. Find its dimensions.

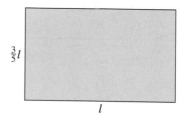

$\frac{3}{5}l$

l

135. *Geometry*   The perimeter of the rectangle in the figure must be at least 50 feet and not more than 100 feet. Find the interval for x.

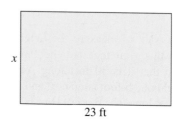

x

23 ft

136. *Long-Distance Charges*   The cost of an international long-distance telephone call is $0.99 for the first minute and $0.49 for each additional minute. If the total cost of the call cannot exceed $7.50, find the interval of time that is available for the call.

137. *Temperature*   The storage temperature of a computer must satisfy the inequality

$$|t - 78.3| < 38.3$$

where t is given in degrees Fahrenheit. Sketch the graph of the inequality.

138. *Temperature*   The operating temperature of a computer must satisfy the inequality

$$|t - 77| < 27$$

where t is given in degrees Fahrenheit. Sketch the graph of the inequality.

Chapter Test

Take this test as you would take a test in class. After you are done, check your work against the answers given in the back of the book.

In Exercises 1–4, solve the equation.

1. $6x - 5 = 19$

2. $5x - 6 = 7x - 12$

3. $15 - 7(1 - x) = 3(x + 8)$

4. $\dfrac{2x}{3} = \dfrac{x}{2} + 4$

5. What is 27% of 3200?

6. 1200 is what percent of 800?

7. A store is offering a 20% discount on all items in its inventory. Find the list price on a tractor that has a sale price of $6400.

8. Which of the packages at the left is a better buy? Explain your reasoning.

9. The tax on a property with an assessed value of $90,000 is $1200. Estimate the tax on a property with an assessed value of $110,000.

10. The bill (including parts and labor) for the repair of a home appliance was $165. The cost for parts was $85. How many hours were spent in repairing the appliance if the cost of labor was $16 per half hour?

11. Two solutions—10% concentration and 40% concentration—are mixed to make 100 liters of a 30% solution. Determine the numbers of liters of the 10% solution and the 40% solution that are required.

12. Two cars start at a given time and travel in the same direction at average speeds of 40 miles per hour and 55 miles per hour. How much time must elapse before the two cars are 10 miles apart?

13. Find the principal required to earn $300 in simple interest in 2 years if the annual interest rate is 7.5%.

14. Solve each absolute value equation.

　　a. $|2x + 6| = 16$　　**b.** $|3x - 5| = |6x - 1|$　　**c.** $|9 - 4x| - 10 = 1$

15. Solve each inequality and sketch its solution.

　　a. $3x + 12 \geq -6$　　**b.** $1 + 2x > 7 - x$

　　c. $0 \leq \dfrac{1 - x}{4} < 2$　　**d.** $-7 < 4(2 - 3x) \leq 20$

16. Solve each absolute value inequality.

　　a. $|x - 3| \leq 2$　　**b.** $|5x - 3| > 12$　　**c.** $\left|\dfrac{x}{4} + 2\right| < 0.2$

17. Rewrite the statement "t is at least 8" using inequality notation.

18. A utility company has a fleet of vans. The annual operating cost per van is

　　$C = 0.37m + 2700$

where m is the number of miles traveled by a van in a year. What is the maximum number of miles that will yield an annual operating cost that is less than or equal to $11,950?

$2.49   12 oz

$2.99   15 oz

Figure for 8

2 Graphs and Functions

David Ximeno Tejada/Tony Stone Images

In 1997, over 8.2 million new cars were sold. The average price of a domestic new car was $18,580 while the average price of an import new car was $29,296. (Source: U.S. Bureau of Economic Analysis)

Automobile Depreciation

You are a sales representative for a pharmaceutical company. You have just paid $15,900 for a new car that you will use for traveling between clients. After 3 years, the car is expected to have a value of $10,200. Because you will use the car for business purposes, its depreciation is deductible from your taxable income. Let x represent the time, with $x = 0$ corresponding to the time of purchase and $x = 3$ to the time 3 years later. Let y represent the value of the car. Using the straight-line method of depreciation, answer the following.

See Section 2.3, Exercise 85

a. Plot the value of the car when new and the value 3 years later on a rectangular coordinate system. Connect the points with a straight line.

b. Calculate the slope of the line.

c. In the context of this problem, what is the meaning of the slope?

See Section 2.4, Exercise 82

d. Make up a table of values showing the expected value of the car over the first 4 years ($x = 0, 1, 2, 3,$ and 4) of its use.

e. Write a linear equation that models this depreciation problem.

f. Does a fifth-year value ($x = 5$) of $7000 fit this model? Explain.

See Section 2.5, Exercise 91

g. Write the linear equation from part (e) as a function of x. What is the value of the car after 7 years?

h. Why might straight-line depreciation not be a fair model for automobile depreciation? Include a sketch of a graph that might better represent automobile depreciation.

i. From a practical perspective, describe the *domain* and the *range* of the linear function in part (e).

2.1   The Rectangular Coordinate System

Objectives

1. Plot points on a rectangular coordinate system.
2. Determine whether an ordered pair is a solution of an equation.
3. Use the Distance Formula to determine the distance between two points.

---

1 Plot points on a rectangular coordinate system.

Corbis-Bettmann

René Descartes

(1596–1650)

Descartes was a French mathematician, philosopher, and scientist. He is sometimes called the father of modern philosophy, and his phrase "I think, therefore I am" has been quoted often. In mathematics, Descartes is known as the father of analytic geometry. Prior to Descartes's time, geometry and algebra were separate mathematical studies. It was Descartes's introduction of the rectangular coordinate system that brought the two studies together.

The Rectangular Coordinate System

Just as you can represent real numbers by points on the real number line, you can represent ordered pairs of real numbers by points in a plane. This plane is called a **rectangular coordinate system** or the **Cartesian plane,** after the French mathematician René Descartes.

A rectangular coordinate system is formed by two real number lines intersecting at a right angle, as shown in Figure 2.1. The horizontal number line is usually called the **x-axis,** and the vertical number line is usually called the **y-axis.** (The plural of axis is *axes.*) The point of intersection of the two axes is called the **origin,** and the axes separate the plane into four regions called **quadrants.**

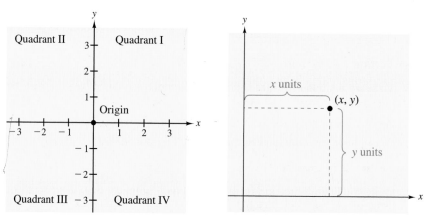

Figure 2.1   Figure 2.2

Each point in the plane corresponds to an **ordered pair** (x, y) of real numbers x and y, called the **coordinates** of the point. The first number (or **x-coordinate**) tells how far to the left or right the point is from the vertical axis, and the second number (or **y-coordinate**) tells how far up or down the point is from the horizontal axis, as shown in Figure 2.2.

A positive x-coordinate implies that the point lies to the *right* of the vertical axis; a negative x-coordinate implies that the point lies to the *left* of the vertical axis; and an x-coordinate of zero implies that the point lies *on* the vertical axis. Similar statements can be made about y-coordinates. A positive y-coordinate implies that the point lies *above* the horizontal axis; a negative y-coordinate implies that the point lies *below* the horizontal axis; and a y-coordinate of zero implies that the point lies *on* the horizontal axis.

Locating a given point in a plane is called **plotting** the point. Example 1 shows how this is done.

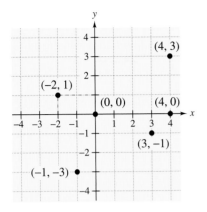

Figure 2.3

Example 1   Plotting Points on a Rectangular Coordinate System

Plot the points $(-2, 1)$, $(4, 0)$, $(3, -1)$, $(4, 3)$, $(0, 0)$, and $(-1, -3)$ on a rectangular coordinate system.

Solution

The point $(-2, 1)$ is 2 units to the *left* of the vertical axis and 1 unit *above* the horizontal axis.

$$\underset{\substack{\text{2 units to the left of}\\\text{the vertical axis}}}{} \quad \underset{\substack{\text{1 unit above the}\\\text{horizontal axis}}}{}$$

$$(-2, 1)$$

Similarly, the point $(4, 0)$ is 4 units to the *right* of the vertical axis and *on* the horizontal axis. (It is on the horizontal axis because its *y*-coordinate is 0.) The other four points can be plotted in a similar way, as shown in Figure 2.3.

In Example 1, you were given the coordinates of several points and asked to plot the points on a rectangular coordinate system. Example 2 looks at the reverse problem. That is, you are given points on a rectangular coordinate system and are asked to determine their coordinates.

Example 2   Finding Coordinates of Points

Determine the coordinates of each of the points shown in Figure 2.4.

Solution

Point *A* lies 2 units to the *right* of the vertical axis and 1 unit *below* the horizontal axis. So, point *A* must be given by the ordered pair $(2, -1)$. The coordinates of the other four points can be determined in a similar way. The results are summarized as follows.

Point	Position	Coordinates
A	2 units *right*, 1 unit *down*	$(2, -1)$
B	1 unit *left*, 5 units *up*	$(-1, 5)$
C	4 units *right*, 0 units *up* or *down*	$(4, 0)$
D	3 units *left*, 2 units *down*	$(-3, -2)$
E	2 units *right*, 4 units *up*	$(2, 4)$
F	1 unit *left*, 2 units *up*	$(-1, 2)$

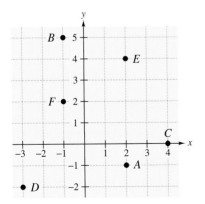

Figure 2.4

In Example 2, note that point $A(2, -1)$ and point $F(-1, 2)$ are different points. The order in which the numbers appear in an ordered pair is important.

As a consumer today, you are presented almost daily with vast amounts of data given in various forms. Data are given in *numerical* form using lists and tables and in *graphical* form using scatter plots, lines, circle graphs, and bar graphs. Graphical forms are more visual and make wide use of Descartes's rectangular coordinate system to show the relationship between two variables. Today, Descartes's ideas are commonly used in virtually every scientific and business-related field.

| Example 3 | Representing Data Graphically | |

The population (in millions) of California from 1982 through 1997 is listed in the table. Plot these points on a rectangular coordinate system. (Source: U.S. Bureau of the Census)

Year	1982	1983	1984	1985	1986	1987	1988	1989
Population	24.8	25.4	25.8	26.4	27.1	27.8	28.5	29.2

Year	1990	1991	1992	1993	1994	1995	1996	1997
Population	29.8	30.4	30.9	31.2	31.4	31.6	31.9	32.3

Solution

Begin by choosing which variable will be plotted on the horizontal axis and which will be plotted on the vertical axis. For these data, it seems natural to plot the years on the horizontal axis (which means that the population must be plotted on the vertical axis). Next, use the data in the table to form ordered pairs. For instance, the first three ordered pairs are (1982, 24.8), (1983, 25.4), and (1984, 25.8). All 16 points are shown in Figure 2.5. Note that the break in the *x*-axis indicates that the numbers between 0 and 1982 have been omitted. The break in the *y*-axis indicates that the numbers between 0 and 24 have been omitted.

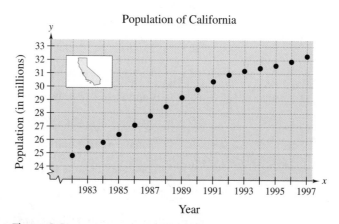

Figure 2.5

2 Determine whether an ordered pair is a solution of an equation.

Ordered Pairs as Solutions

In Example 3, the relationship between the year and the population was given by a **table of values.** In mathematics, the relationship between the variables x and y is often given by an equation. From the equation, you can construct your own table of values.

Example 4   Constructing a Table of Values

Construct a table of values for $y = 3x + 2$. Then plot the solution points on a rectangular coordinate system. Choose x-values of $-3, -2, -1, 0, 1, 2$, and 3.

Solution

For each x-value, you must calculate the corresponding y-value. For example, if you choose $x = 1$, then the y-value is

$$y = 3(1) + 2 = 5.$$

The ordered pair $(x, y) = (1, 5)$ is a **solution point** (or **solution**) of the equation.

Choose x	Calculate y from $y = 3x + 2$	Solution point
$x = -3$	$y = 3(-3) + 2 = -7$	$(-3, -7)$
$x = -2$	$y = 3(-2) + 2 = -4$	$(-2, -4)$
$x = -1$	$y = 3(-1) + 2 = -1$	$(-1, -1)$
$x = 0$	$y = 3(0) + 2 = 2$	$(0, 2)$
$x = 1$	$y = 3(1) + 2 = 5$	$(1, 5)$
$x = 2$	$y = 3(2) + 2 = 8$	$(2, 8)$
$x = 3$	$y = 3(3) + 2 = 11$	$(3, 11)$

Once you have constructed a table of values, you can get a visual idea of the relationship between the variables x and y by plotting the solution points on a rectangular coordinate system, as shown in Figure 2.6.

In many places throughout this course, you will see that approaching a problem in different ways can help you understand the problem better. In Example 4, for instance, solutions of an equation are arrived at in three ways.

▶ **Three Approaches to Problem Solving**

1. Algebraic Approach   Use algebra to find several solutions.

2. Numerical Approach   Construct a table that shows several solutions.

3. Graphical Approach   Draw a graph that shows several solutions.

Technology: Discovery

In the table of values in Example 4, successive x-values differ by 1. How do the successive y-values differ? Use the table feature of your graphing utility to create a table of values for the following equations. If successive x-values differ by 1, how do the successive y-values differ?

a. $y = x + 2$

b. $y = 2x + 2$

c. $y = 4x + 2$

d. $y = -x + 2$

Describe the pattern.

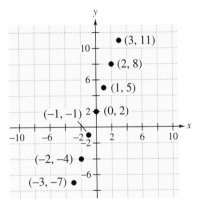

Figure 2.6

> ▶ **Guidelines for Verifying Solutions**
>
> To verify that an ordered pair (x, y) is a solution of an equation with variables x and y, use the following steps.
>
> **1.** Substitute the values of x and y into the equation.
>
> **2.** Simplify both sides of the equation.
>
> **3.** If both sides simplify to the same number, the ordered pair is a solution. If the two sides yield different numbers, the ordered pair is not a solution.

Example 5   Verifying Solutions of an Equation

Which of the ordered pairs are solutions of $x^2 - 2y = 6$?

a. $(2, 1)$ **b.** $(0, -3)$ **c.** $(-2, -5)$ **d.** $\left(1, -\frac{5}{2}\right)$

Solution

a. For the ordered pair $(2, 1)$, substitute $x = 2$ and $y = 1$ into the equation.

$$(2)^2 - 2(1) \overset{?}{=} 6 \qquad \text{Substitute 2 for } x \text{ and 1 for } y.$$

$$2 \neq 6 \qquad \text{Is not a solution} \; ✗$$

Because the substitution does not satisfy the given equation, you can conclude that the ordered pair $(2, 1)$ *is not* a solution of the given equation.

b. For the ordered pair $(0, -3)$, substitute $x = 0$ and $y = -3$ into the equation.

$$(0)^2 - 2(-3) \overset{?}{=} 6 \qquad \text{Substitute 0 for } x \text{ and } -3 \text{ for } y.$$

$$6 = 6 \qquad \text{Is a solution} \; ✓$$

Because the substitution satisfies the given equation, you can conclude that the ordered pair $(0, -3)$ *is* a solution of the given equation.

c. For the ordered pair $(-2, -5)$, substitute $x = -2$ and $y = -5$ into the equation.

$$(-2)^2 - 2(-5) \overset{?}{=} 6 \qquad \text{Substitute } -2 \text{ for } x \text{ and } -5 \text{ for } y.$$

$$14 \neq 6 \qquad \text{Is not a solution} \; ✗$$

Because the substitution does not satisfy the given equation, you can conclude that the ordered pair $(-2, -5)$ *is not* a solution of the given equation.

d. For the ordered pair $\left(1, -\frac{5}{2}\right)$, substitute $x = 1$ and $y = -\frac{5}{2}$ into the equation.

$$(1)^2 - 2\left(-\tfrac{5}{2}\right) \overset{?}{=} 6 \qquad \text{Substitute 1 for } x \text{ and } -\tfrac{5}{2} \text{ for } y.$$

$$6 = 6 \qquad \text{Is a solution} \; ✓$$

Because the substitution satisfies the given equation, you can conclude that the ordered pair $\left(1, -\frac{5}{2}\right)$ *is* a solution of the given equation.

3 Use the Distance Formula to determine the distance between two points.

The Distance Formula

You know from Section P.1 of the Prerequisites chapter that the distance d between two points a and b on the real number line is simply

$$d = |b - a|.$$

The same "absolute value rule" is used to find the distance between two points that lie on the same *vertical or horizontal line* in the coordinate plane, as shown in Example 6.

> **Example 6**   Finding Horizontal and Vertical Distances

a. Find the distance between the points $(2, -2)$ and $(2, 4)$.

b. Find the distance between the points $(-3, -2)$ and $(2, -2)$.

Solution

a. Because the x-coordinates are equal, you can visualize a vertical line through the points $(2, -2)$ and $(2, 4)$, as shown in Figure 2.7. The distance between these two points is the absolute value of the difference of their y-coordinates.

$$\text{Vertical distance} = |4 - (-2)| \qquad \text{Subtract } y\text{-coordinates.}$$

$$= 6 \qquad \text{Evaluate absolute value.}$$

b. Because the y-coordinates are equal, you can visualize a horizontal line through the points $(-3, -2)$ and $(2, -2)$, as shown in Figure 2.7. The distance between these two points is the absolute value of the difference of their x-coordinates.

$$\text{Horizontal distance} = |2 - (-3)| \qquad \text{Subtract } x\text{-coordinates.}$$

$$= 5 \qquad \text{Evaluate absolute value.}$$

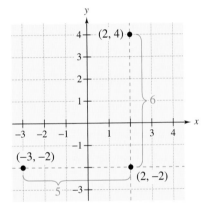

Figure 2.7

In Figure 2.7, note that the horizontal distance between the points $(-3, -2)$ and $(2, -2)$ is the absolute value of the difference of the x-coordinates, and the vertical distance between the points $(2, -2)$ and $(2, 4)$ is the absolute value of the difference of the y-coordinates.

The technique applied in Example 6 can be used to develop a general formula for finding the distance between two points in the plane. This general formula will work for any two points, even if they do not lie on the same vertical or horizontal line. To develop the formula, you use the **Pythagorean Theorem,** which states that for a right triangle, the hypotenuse c and sides a and b are related by the formula

$$a^2 + b^2 = c^2 \qquad \text{Pythagorean Theorem}$$

as shown in Figure 2.8. (The converse is also true. That is, if $a^2 + b^2 = c^2$, then the triangle is a right triangle.)

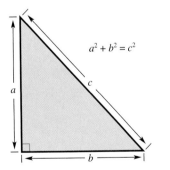

Figure 2.8   *Pythagorean Theorem*

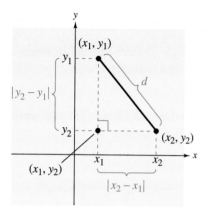

Figure 2.9 *Distance Between Two Points*

To develop a general formula for the distance between two points, let (x_1, y_1) and (x_2, y_2) represent two points in the plane (that do not lie on the same horizontal or vertical line). With these two points, a right triangle can be formed, as shown in Figure 2.9. Note that the third vertex of the triangle is (x_1, y_2). Because (x_1, y_1) and (x_1, y_2) lie on the same vertical line, the length of the vertical side of the triangle is $|y_2 - y_1|$. Similarly, the length of the horizontal side is $|x_2 - x_1|$. By the Pythagorean Theorem, the square of the distance between (x_1, y_1) and (x_2, y_2) is

$$d^2 = |x_2 - x_1|^2 + |y_2 - y_1|^2.$$

Because the distance d must be positive, you can choose the positive square root and write

$$d = \sqrt{|x_2 - x_1|^2 + |y_2 - y_1|^2}.$$

Finally, replacing $|x_2 - x_1|^2$ and $|y_2 - y_1|^2$ by the equivalent expressions $(x_2 - x_1)^2$ and $(y_2 - y_1)^2$ gives you the **Distance Formula.**

▶ **The Distance Formula**

The distance d between two points (x_1, y_1) and (x_2, y_2) is

$$d = \sqrt{(x_2 - x_1)^2 + (y_2 - y_1)^2}.$$

Note that for the special case in which the two points lie on the same vertical or horizontal line, the Distance Formula still works. For instance, applying the Distance Formula to the points $(2, -2)$ and $(2, 4)$ produces

$$d = \sqrt{(2 - 2)^2 + [4 - (-2)]^2} = \sqrt{6^2} = 6$$

which is the same result obtained in Example 6.

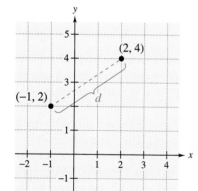

Figure 2.10

Example 7 Finding the Distance Between Two Points

Find the distance between the points $(-1, 2)$ and $(2, 4)$, as shown in Figure 2.10.

Solution

Let $(x_1, y_1) = (-1, 2)$ and $(x_2, y_2) = (2, 4)$, and apply the Distance Formula.

$$d = \sqrt{[2 - (-1)]^2 + (4 - 2)^2} \qquad \text{Substitute coordinates of points.}$$
$$= \sqrt{3^2 + 2^2} \qquad \text{Simplify.}$$
$$= \sqrt{13} \qquad \text{Simplify.}$$
$$\approx 3.61 \qquad \text{Use a calculator.}$$

When using the Distance Formula, it does not matter which point is considered (x_1, y_1) and which is (x_2, y_2), because the result will be the same. For instance, in Example 7, let $(x_1, y_1) = (2, 4)$ and $(x_2, y_2) = (-1, 2)$. Then

$$d = \sqrt{[(-1) - 2]^2 + (2 - 4)^2} = \sqrt{(-3)^2 + (-2)^2} = \sqrt{13} \approx 3.61.$$

The Distance Formula has many applications in mathematics. For instance, the next example shows how you can use the Distance Formula and the converse of the Pythagorean Theorem to verify that three points form the vertices of a right triangle.

Example 8	An Application of the Distance Formula

Show that the points $(1, 2)$, $(3, 1)$, and $(4, 3)$ are vertices of a right triangle.

Solution

The three points are plotted in Figure 2.11. Using the Distance Formula, you can find the lengths of the three sides of the triangle.

$$d_1 = \sqrt{(3 - 1)^2 + (1 - 2)^2} = \sqrt{4 + 1} = \sqrt{5}$$
$$d_2 = \sqrt{(4 - 3)^2 + (3 - 1)^2} = \sqrt{1 + 4} = \sqrt{5}$$
$$d_3 = \sqrt{(4 - 1)^2 + (3 - 2)^2} = \sqrt{9 + 1} = \sqrt{10}$$

Because

$$d_1{}^2 + d_2{}^2 = 5 + 5$$
$$= 10$$
$$= d_3{}^2$$

you can conclude from the converse of the Pythagorean Theorem that the triangle is a right triangle.

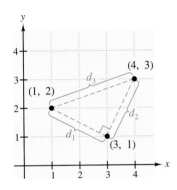

Figure 2.11

Discussing the Concept	Determining Collinearity

Three or more points are **collinear** if they all lie on the same line. Use the following steps to determine if the set of points

$$\{A(3, 1), B(5, 4), C(9, 10)\}$$

and the set of points

$$\{A(2, 2), B(4, 3), C(5, 4)\}$$

are collinear.

a. For each set of points, use the Distance Formula to find the distances from A to B, from B to C, and from A to C. What relationship exists among these distances for each set of points?

b. Plot each set of points on a rectangular coordinate system. Do all the points of either set appear to lie on the same line?

c. Compare your conclusions from part (a) with the conclusions you made from the graphs in part (b). Make a general statement about how to use the Distance Formula to determine collinearity.

2.1 Exercises

Integrated Review   Concepts, Skills, and Problem Solving

Keep mathematically in shape by doing these exercises *before* the problems of this section.

Properties and Definitions

1. Is $3x = 7$ a linear equation? Explain.
Is $x^2 + 3x = 2$ a linear equation? Explain.

2. Explain how to check if $x = 3$ is a solution of the equation $5x - 4 = 11$.

Simplifying Expressions

In Exercises 3–10, simplify the expression.

3. $6x(2x^2)$

4. $3t^2 \cdot t^4$

5. $-(-3x^2)^3(2x^4)$

6. $(4x^3y^2)(-2xy^3)$

7. $4 - 3(2x + 1)$

8. $5(x + 2) - 4(2x - 3)$

9. $24\left(\dfrac{y}{3} + \dfrac{y}{6}\right)$

10. $0.12x + 0.05(2000 - 2x)$

Problem Solving

11. You can mow a lawn in 4 hours, and your friend can mow it in 5 hours. What fractional part of the lawn can each of you mow in 1 hour? Working together, how long will it take to mow the lawn?

12. A truck driver traveled at an average speed of 50 miles per hour on a 200-mile trip. On the return trip with the truck fully loaded, the average speed was 42 miles per hour. Find the average speed for the round trip.

Developing Skills

In Exercises 1–8, plot the points on a rectangular coordinate system. See Example 1.

1. $(4, 3), (-5, 3), (3, -5)$

2. $(-2, 5), (-2, -5), (3, 5)$

3. $(-8, -2), (6, -2), (6, 5)$

4. $(0, 4), (0, 0), (3, 0)$

5. $\left(\frac{5}{2}, -2\right), \left(-2, \frac{1}{4}\right), \left(\frac{3}{2}, -\frac{7}{2}\right)$

6. $\left(-\frac{2}{3}, 3\right), \left(\frac{1}{4}, -\frac{5}{4}\right), \left(-5, -\frac{7}{4}\right)$

7. $\left(\frac{3}{2}, 1\right), (4, -3), \left(-\frac{4}{3}, \frac{7}{3}\right)$

8. $(-3, -5), \left(\frac{9}{4}, \frac{3}{4}\right), \left(\frac{5}{2}, -2\right)$

In Exercises 9–12, determine the coordinates of the points. See Example 2.

9.

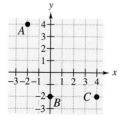

10.

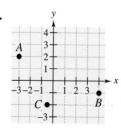

11.

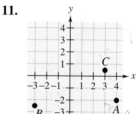

12.

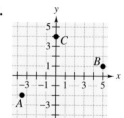

In Exercises 13–20, plot the points and connect them with line segments in such a way as to form the figure. (*Note:* A *rhombus* is a parallelogram whose sides are all of the same length.)

13. *Square:* $(0, 6), (3, 3), (0, 0), (-3, 3)$

14. *Rectangle:* $(7, 0), (9, 1), (4, 6), (6, 7)$

15. *Triangle:* $(-1, 2), (2, 0), (3, 5)$

16. *Triangle:* $(1, 4), (0, -1), (5, 9)$

17. *Parallelogram:* $(4, 0), (6, -2), (0, -4), (-2, -2)$

18. *Parallelogram:* $(-1, 1), (0, 4), (4, -2), (5, 1)$

19. *Rhombus:* $(0, 0), (3, 2), (2, 3), (5, 5)$

20. *Rhombus:* $(-3, -3), (-2, -1), (-1, -2), (0, 0)$

In Exercises 21–28, find the coordinates of the point.

21. The point is located 5 units to the left of the y-axis and 2 units above the x-axis.

22. The point is located 10 units to the right of the y-axis and 4 units below the x-axis.

23. The point is located 3 units to the right of the y-axis and 2 units below the x-axis.

24. The point is located 2 units to the right of the y-axis and 5 units above the x-axis.

25. The coordinates of the point are equal, and the point is located in the third quadrant 10 units to the left of the y-axis.

26. The coordinates of the point are equal in magnitude and opposite in sign, and the point is located 7 units to the right of the y-axis.

27. The point is on the positive x-axis 10 units from the origin.

28. The point is on the negative y-axis 5 units from the origin.

In Exercises 29–36, determine the quadrant of the point without plotting it.

29. $(-3, -5)$

30. $(4, -2)$

31. $\left(3, -\frac{5}{8}\right)$

32. $\left(-\frac{5}{11}, -\frac{3}{8}\right)$

33. $(200, 1365.6)$

34. $(-6.2, 8.05)$

35. $(x, y), \quad x > 0, y < 0$

36. $(x, y), \quad x > 0, y > 0$

In Exercises 37–42, determine the quadrants in which the point could be located. (x and y are real numbers.)

37. $(x, 4)$

38. $(-10, y)$

39. $(-3, y)$

40. $(x, 5)$

41. $(x, y), \quad xy > 0$

42. $(x, y), \quad xy < 0$

In Exercises 43–46, plot the points whose coordinates are given in the table. See Example 3.

43. *Exam Score*   The table gives the time x in hours invested in concentrated study for five different algebra exams and the resulting score y.

x	5	2	3	6.5	4
y	81	71	88	92	86

44. *Net Sales*   The net sales y (in billions of dollars) of Wal-Mart for the years 1991 through 1997 are given in the table. The time in years is given by x. (Source: Wal-Mart 1998 Annual Report)

x	1991	1992	1993	1994	1995	1996	1997
y	32.6	43.9	55.5	67.3	82.5	93.6	104.9

45. *Average Temperature*   The table gives the average temperature y (in degrees Fahrenheit) for Duluth, Minnesota for each month of the year, with $x = 1$ representing January.   (Source: NOAA)

x	1	2	3	4	5	6
y	6.3	12.0	22.9	38.3	50.3	59.4

x	7	8	9	10	11	12
y	65.3	63.2	54.0	44.2	28.2	13.8

46. *Fuel Efficiency*   The table gives the speed x of a car in miles per hour and the approximate fuel efficiency y in miles per gallon.

x	50	55	60	65	70
y	28	26.4	24.8	23.4	22

Shifting a Graph   In Exercises 47 and 48, the figure is shifted to a new location in the plane. Find the coordinates of the vertices of the figure in its new location.

47.

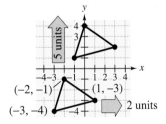

48.

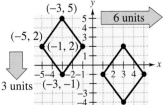

In Exercises 49–52, complete the table of values. Then plot the solution points on a rectangular coordinate system. See Example 4.

49.

x	-2	0	2	4	6
$y = 5x - 1$					

50.

x	-2	0	2	4	6
$y = \frac{3}{4}x + 2$					

51.

x	-4	$\frac{2}{5}$	4	8	12
$y = -\frac{5}{2}x + 4$					

52.

x	-6	-3	0	$\frac{3}{4}$	10
$y = \frac{4}{3}x - \frac{1}{3}$					

In Exercises 53 and 54, use the table feature of a graphing utility to complete the table of values.

53.

x	-2	0	2	4	6
$y = 4x^2 + x - 2$					

54.

x	-2	0	2	4	6
$y = |3x - 4| + 1$					

In Exercises 55–60, determine whether the ordered pairs are solutions of the equation. See Example 5.

55. $x^2 + 3y = -5$
  (a) $(3, -2)$
  (b) $(-2, -3)$
  (c) $(3, -5)$
  (d) $(4, -7)$

56. $y^2 - 4x = 8$
  (a) $(0, 6)$
  (b) $(-4, 2)$
  (c) $(-1, 3)$
  (d) $(7, 6)$

57. $4y - 2x + 1 = 0$
  (a) $(0, 0)$
  (b) $\left(\frac{1}{2}, 0\right)$
  (c) $\left(-3, -\frac{7}{4}\right)$
  (d) $\left(1, -\frac{3}{4}\right)$

58. $5x - 2y + 50 = 0$
  (a) $(-10, 0)$
  (b) $(-5, 5)$
  (c) $(0, 25)$
  (d) $(20, -2)$

59. $y = \frac{7}{8}x + 3$
  (a) $\left(\frac{8}{7}, 4\right)$
  (b) $(8, 10)$
  (c) $(0, 0)$
  (d) $(-16, 14)$

60. $y = \frac{5}{8}x - 2$
  (a) $(0, 0)$
  (b) $(8, 3)$
  (c) $(-16, -7)$
  (d) $\left(-\frac{8}{5}, 3\right)$

In Exercises 61–68, plot the points and find the distance between them. State whether the points lie on a horizontal or vertical line. See Example 6.

61. $(3, -2), (3, 5)$
62. $(-2, 8), (-2, 1)$
63. $(3, 2), (10, 2)$
64. $(-120, -2), (130, -2)$
65. $\left(-3, \frac{3}{2}\right), \left(-3, \frac{9}{4}\right)$
66. $\left(\frac{3}{4}, 1\right), \left(\frac{3}{4}, -10\right)$
67. $\left(-4, \frac{1}{3}\right), \left(\frac{5}{2}, \frac{1}{3}\right)$
68. $\left(\frac{1}{2}, \frac{7}{8}\right), \left(\frac{11}{2}, \frac{7}{8}\right)$

In Exercises 69–78, find the distance between the points. See Example 7.

69. $(3, 7), (4, 5)$
70. $(5, 2), (8, 3)$
71. $(1, 3), (5, 6)$
72. $(3, 10), (15, 5)$
73. $(-3, 0), (4, -3)$
74. $(0, -5), (2, -8)$
75. $(-2, -3), (4, 2)$
76. $(-5, 4), (10, -3)$
77. $(1, 3), (3, -2)$
78. $\left(\frac{1}{2}, 1\right), \left(\frac{3}{2}, 2\right)$

In Exercises 79–82, show that the points are vertices of a right triangle. See Example 8.

79.

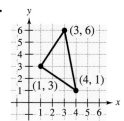

80.

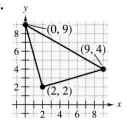

81.

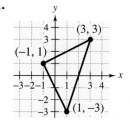

82.

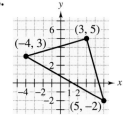

In Exercises 83–86, use the Distance Formula to determine whether the three points lie on the same line.

83. $(2, 3), (2, 6), (6, 3)$
84. $(2, 4), (-1, 6), (-3, 1)$
85. $(8, 3), (5, 2), (2, 1)$
86. $(2, 4), (1, 1), (0, -2)$

In Exercises 87–90, plot the points and the midpoint of the line segment joining the points. The coordinates of the midpoint of the line segment joining the points (x_1, y_1) and (x_2, y_2) are

$$\text{Midpoint} = \left(\frac{x_1 + x_2}{2}, \frac{y_1 + y_2}{2}\right).$$

87. $(-2, 0), (4, 8)$

88. $(-3, -2), (7, 2)$

89. $(1, 6), (6, 3)$

90. $(2, 7), (9, -1)$

Solving Problems

91. *Numerical Interpretation*   The cost C for producing x units is given by $C = 28x + 3000$. Use a table to help write a paragraph that describes the relationship between x and C.

92. *Numerical Interpretation*   When an employee produces x units per hour, the hourly wage y is given by $y = 0.75x + 8$. Use a table to help write a paragraph that describes the relationship between x and y.

93. *Housing Construction*   A house is 30 feet wide and the ridge of the roof is 7 feet above the tops of the walls (see figure). The rafters overhang the edges of the walls by 2 feet. How long are the rafters?

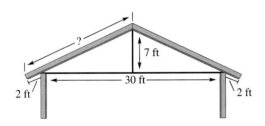

94. *Housing Construction*   Determine the length of the handrail over the stairs shown in the figure.

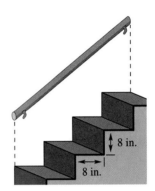

Geometry   In Exercises 95 and 96, find the perimeter of the triangle having the given vertices.

95. $(-2, 0), (0, 5), (1, 0)$

96. $(-5, -2), (-1, 4), (3, -1)$

Explaining Concepts

97. Discuss the significance of the word *ordered* when referring to an ordered pair (x, y).

98. When plotting the point (x, y), what does the x-coordinate measure? What does the y-coordinate measure?

99. What is the x-coordinate of any point on the y-axis? What is the y-coordinate of any point on the x-axis?

100. Explain why the ordered pair $(-3, 4)$ is not a solution point of the equation $y = 4x + 15$.

101. When plotting points on the rectangular coordinate system, is it true that the scales on the x- and y-axes must be the same? Explain.

102. State the Pythagorean Theorem and give examples of its use.

103. *Conjecture*   Plot the points $(2, 1), (-3, 5)$, and $(7, -3)$ on a rectangular coordinate system. Then change the sign of the x-coordinate of each point and plot the three new points on the same rectangular coordinate system. What conjecture can you make about the location of a point when the sign of the x-coordinate is changed?

104. *Conjecture*   Plot the points $(2, 1), (-3, 5)$, and $(7, -3)$ on a rectangular coordinate system. Then change the sign of the y-coordinate of each point and plot the three new points on the same rectangular coordinate system. What conjecture can you make about the location of a point when the sign of the y-coordinate is changed?

2.2   Graphs of Equations

Objectives

1 Sketch the graph of an equation using the point-plotting method.

2 Find and use x- and y-intercepts as aids to sketching a graph.

3 Use a pattern to write an equation for an application problem, and sketch its graph.

1 Sketch the graph of an equation using the point-plotting method.

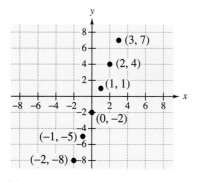

(a)

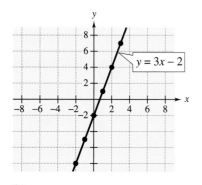

(b)

Figure 2.12

The Graph of an Equation

In Section 2.1, you saw that the solutions of an equation in x and y can be represented by points on a rectangular coordinate system. The set of *all* solutions of an equation is called its **graph.** In this section, you will study a basic technique for sketching the graph of an equation—the **point-plotting method.**

| Example 1 | Sketching the Graph of an Equation |

Sketch the graph of $3x - y = 2$.

Solution

To begin, solve the equation for y.

$$3x - y = 2 \qquad \text{Original equation}$$

$$-y = -3x + 2 \qquad \text{Subtract } 3x \text{ from both sides.}$$

$$y = 3x - 2 \qquad \text{Divide both sides by } -1.$$

Next, create a table of values. The choice of x-values to use in the table is somewhat arbitrary. However, the more x-values you choose, the easier it will be to recognize a pattern.

x	-2	-1	0	1	2	3
$y = 3x - 2$	-8	-5	-2	1	4	7
Solution	$(-2, -8)$	$(-1, -5)$	$(0, -2)$	$(1, 1)$	$(2, 4)$	$(3, 7)$

Now, plot the points, as shown in Figure 2.12(a). It appears that all six points lie on a line, so complete the sketch by drawing a line through the points, as shown in Figure 2.12(b).

The equation in Example 1 is an example of a **linear equation** in two variables—it is of first degree in both variables, and its graph is a straight line. By drawing a line through the plotted points, we are implying that every point on this line is a solution point of the given equation.

2 Find and use *x*- and *y*-intercepts as aids to sketching a graph.

Intercepts: An Aid to Sketching Graphs

Two types of solution points that are especially useful are those having zero as the *x*-coordinate and those having zero as the *y*-coordinate. Such points are called **intercepts** because they are the points at which the graph intersects, respectively, the *y*- and *x*-axes.

Study Tip

When creating a table of values for a graph, choose a span of *x*-values that lie at least 1 unit to the left and right of the intercepts of the graph. This helps to give a more complete view of the graph.

▶ **Definition of Intercepts**

The point $(a, 0)$ is called an **x-intercept** of the graph of an equation if it is a solution point of the equation. To find the *x*-intercepts, let $y = 0$ and solve the equation for *x*.

The point $(0, b)$ is called a **y-intercept** of the graph of an equation if it is a solution point of the equation. To find the *y*-intercepts, let $x = 0$ and solve the equation for *y*.

| Example 4 | Finding the Intercepts of a Graph |

Find the intercepts and sketch the graph of $y = 2x - 3$.

Solution

Find the *x*-intercept by letting $y = 0$ and solving for *x*.

$$y = 2x - 3 \qquad \text{Original equation}$$
$$0 = 2x - 3 \qquad \text{Substitute 0 for } y.$$
$$3 = 2x \qquad \text{Add 3 to both sides.}$$
$$\tfrac{3}{2} = x \qquad \text{Solve for } x.$$

Find the *y*-intercept by letting $x = 0$ and solving for *y*.

$$y = 2x - 3 \qquad \text{Original equation}$$
$$y = 2(0) - 3 \qquad \text{Substitute 0 for } x.$$
$$y = -3 \qquad \text{Solve for } y.$$

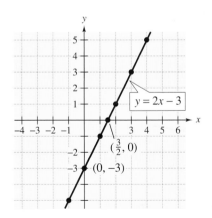

Figure 2.15

So, the graph has one *x*-intercept, which occurs at the point $\left(\tfrac{3}{2}, 0\right)$, and one *y*-intercept, which occurs at the point $(0, -3)$. To sketch the graph of the equation, create a table of values. (Include the intercepts in the table.) Finally, using the solution points given in the table, sketch the graph of the equation, as shown in Figure 2.15.

x	-1	0	1	$\frac{3}{2}$	2	3	4
$y = 2x - 3$	-5	-3	-1	0	1	3	5
Solution	$(-1, -5)$	$(0, -3)$	$(1, -1)$	$\left(\frac{3}{2}, 0\right)$	$(2, 1)$	$(3, 3)$	$(4, 5)$

3 Use a pattern to write an equation for an application problem, and sketch its graph.

Real-Life Application of Graphs

Newspapers and news magazines frequently use graphs to show real-life relationships between variables. Example 5 shows how such a graph can help you visualize the concept of **straight-line depreciation.**

| **Example 5** | Straight-Line Depreciation: Finding the Pattern | |

Your small business buys a new printing press for $65,000. For income tax purposes, you decide to depreciate the printing press over a 10-year period. At the end of the 10 years, the salvage value of the printing press is expected to be $5000. Find an equation that relates the depreciated value of the printing press to the number of years since it was purchased. Then sketch the graph of the equation.

Solution

The total depreciation over the 10-year period is $65,000 − 5000 = $60,000. Because the same amount is depreciated each year, it follows that the annual depreciation is $60,000/10 = $6000. So, after 1 year, the value of the printing press is

$$\text{Value after 1 year} = \$65,000 - (1)6000 = \$59,000.$$

By similar reasoning, you can see that the values after 2, 3, and 4 years are

$$\text{Value after 2 years} = \$65,000 - (2)6000 = \$53,000$$

$$\text{Value after 3 years} = \$65,000 - (3)6000 = \$47,000$$

$$\text{Value after 4 years} = \$65,000 - (4)6000 = \$41,000.$$

Let y represent the value of the printing press after t years and follow the pattern determined for the first 4 years to obtain

$$y = 65,000 - 6000t.$$

A sketch of the graph of this equation is shown in Figure 2.16.

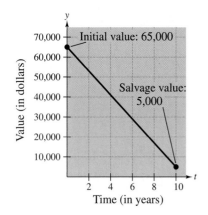

Figure 2.16   *Straight-Line Depreciation*

| Discussing the Concept | **Straight-Line Depreciation** |

In Example 5, suppose that you depreciated the printing press over 8 years instead of 10 years. Write an equation that represents the depreciated value of the printing press during the 8-year period. Then graph both depreciation models on the same rectangular coordinate system and compare the results. What are the advantages and disadvantages of each model?

2.2   Exercises

Integrated Review   *Concepts, Skills, and Problem Solving*

Keep mathematically in shape by doing these exercises *before* the problems of this section.

Properties and Definitions

1. If $t - 3 > 7$ and c is an algebraic expression, then what is the relationship between $t - 3 + c$ and $7 + c$?

2. If $t - 3 < 7$ and $c < 0$, then what is the relationship between $(t - 3)c$ and $7c$?

3. Complete the Multiplicative Inverse Property:
   $y(1/y) =$.

4. Name the property illustrated by $u + v = v + u$.

Simplifying Expressions

In Exercises 5–10, solve the inequality.

5. $2x + 3 \geq 5$

6. $5 - 3x > 14$

7. $-4 < 10x + 1 < 6$

8. $-2 \leq 1 - 2x \leq 2$

9. $-3 \leq -\dfrac{x}{2} \leq 3$

10. $-5 < x - 25 < 5$

Problem Solving

11. The price of a new van is approximately 112% of what it was 3 years ago. What was the approximate price 3 years ago if the current price is \$32,500?

12. Your employer withholds $3\frac{1}{2}\%$ of your gross income for medical insurance coverage. Determine the amount withheld each month if your gross monthly income is \$3100.

Developing Skills

In Exercises 1–6, match the equation with its graph. [The graphs are labeled (a), (b), (c), (d), (e), and (f).]

(a)
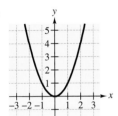

(b)
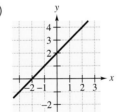

(c)
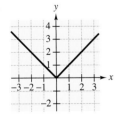

(d)
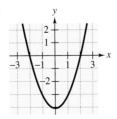

(e)
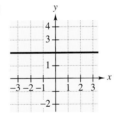

(f)
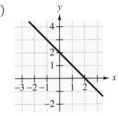

1. $y = 2$
2. $y = 2 + x$
3. $y = 2 - x$
4. $y = x^2$
5. $y = x^2 - 4$
6. $y = |x|$

In Exercises 7–30, sketch the graph of the equation. See Examples 1–3.

7. $y = 3x$

8. $y = -2x$

9. $y = 4 - x$

10. $y = -x + 2$

11. $2x - y = 3$

12. $3x - y = -2$

13. $3x + 2y = 2$

14. $2y - x = 4$

15. $y = -x^2$

16. $y = x^2$

17. $y = x^2 - 4$ **18.** $y = 1 - x^2$

19. $-x^2 - 3x + y = 0$ **20.** $-x^2 + x + y = 0$

21. $x^2 - 2x - y = 1$ **22.** $x^2 + 3x - y = 4$

23. $y = |x|$ **24.** $y = -|x|$

25. $y = |x| + 3$ **26.** $y = |x| - 1$

27. $y = |x + 3|$ **28.** $y = |x - 1|$

29. $y = -x^3$ **30.** $y = x^3$

In Exercises 31–44, find the x- and y-intercepts (if any) of the graph of the equation. See Example 4.

31. $y = 6x - 3$ **32.** $y = x + 2$

33. $y = \frac{3}{4}x + 15$ **34.** $y = 12 - \frac{2}{5}x$

35. $x + 2y = 10$ **36.** $3x - 2y + 12 = 0$

37. $4x - y + 3 = 0$ **38.** $2x + 3y - 8 = 0$

39. $y = |x| - 1$ **40.** $y = |x| + 4$

41. $y = |x + 2|$ **42.** $y = |x - 4|$

43. $y = |x - 1| - 3$ **44.** $y = |x + 3| - 1$

In Exercises 45–48, graphically estimate the x- and y-intercepts of the graph.

45. $2x + 3y = 6$ **46.** $3x - y + 9 = 0$

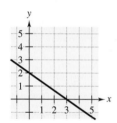   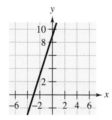

47. $y = x^2 + 3$ **48.** $x = 3$

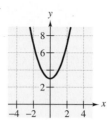   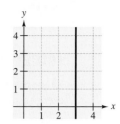

In Exercises 49–54, use a graphing utility to sketch a graph of the equation. Estimate the x- and y-intercepts (if any).

49. $y = 4x - 6$ **50.** $y = 3x + 12$

51. $y = (x - 1)(x - 6)$ **52.** $y = (x + 2)(x - 3)$

53. $y = |4x + 6| - 2$ **54.** $y = |2x - 4| + 1$

In Exercises 55–76, sketch the graph of the equation and show the coordinates of three solution points (including intercepts).

55. $y = 3 - x$ **56.** $y = x - 3$

57. $y = 2x - 3$ **58.** $y = -4x + 8$

59. $4x + y = 3$ **60.** $y - 2x = -4$

61. $2x - 3y = 6$ **62.** $3x + 4y = 12$

63. $x + 5y = 10$ **64.** $5x - y = 10$

65. $y = x^2 - 9$ **66.** $y = 9 - x^2$

67. $y = 1 - x^2$ **68.** $y = x^2 - 4$

69. $y = x(x - 2)$ **70.** $y = -x(x + 4)$

71. $y = |x| - 3$ **72.** $y = |x| + 2$

73. $y = |x + 2|$ **74.** $y = |x - 3|$

75. $y = -|x| + |x + 1|$ **76.** $y = |x| + |x - 2|$

Solving Problems

77. *Straight-Line Depreciation*   A manufacturing plant purchases a new molding machine for $225,000. The depreciated value y after t years is given by

$$y = 225{,}000 - 20{,}000t, \quad 0 \le t \le 8.$$

Sketch a graph of this model.

78. *Straight-Line Depreciation*   A manufacturing plant purchases a new computer system for $20,000. The depreciated value y after t years is given by

$$y = 20{,}000 - 3000t, \quad 0 \le t \le 6.$$

Sketch a graph of this model.

79. *Straight-Line Depreciation*   Your company purchases a new delivery van for $40,000. For tax purposes, the van will be depreciated over a 7-year period. At the end of 7 years, the value of the van is expected to be $5000. Find an equation that relates the depreciated value of the van to the number of years since it was purchased. Then sketch the graph of the equation.

80. *Straight-Line Depreciation*   Your company purchases a new limousine for $55,000. For tax purposes, the limousine will be depreciated over a 10-year period. At the end of 10 years, the value of the limousine is expected to be $10,000. Find an equation that relates the depreciated value of the limousine to the number of years since it was purchased. Then sketch the graph of the equation.

81. *Hooke's Law*   The force F (in pounds) to stretch a spring x inches from its natural length is given by

$$F = \frac{4}{3}x, \quad 0 \le x \le 12.$$

(a) Use the model to complete the table.

x	0	3	6	9	12
F					

(b) Sketch a graph of the model.

(c) Determine the required change in F if x is doubled. Explain your reasoning.

82. *Comparing Data with a Model*   The numbers of farms N (in thousands) in the United States for the years 1990 through 1996 are given in the table. (Source: U.S. Department of Agriculture)

Year	1990	1991	1992	1993	1994	1995	1996
N	2146	2117	2108	2083	2065	2072	2063

A model for these data is

$$N = -13.6t + 2134$$

where t is time in years, with $t = 0$ corresponding to 1990.

(a) Sketch a graph of the model and plot the data in the table on the same graph.

(b) How well does the model represent the data? Explain your reasoning.

(c) Use the model to predict the number of farms in the year 2000.

(d) Explain why this model may not be accurate in the future.

83. *Misleading Graphs*   Graphs can help you visualize relationships between two variables, but they can also be misused to imply results that are not correct. The two graphs below represent the *same* data points. Why do the graphs appear different? Identify ways in which this could be misleading.

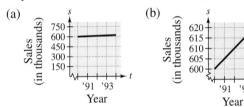

84. *Exploration*   Graph the equations $y = x^2 + 1$ and $y = -(x^2 + 1)$ on the same set of coordinate axes. Explain how the graph of an equation changes when the expression for y is multiplied by -1. Justify your answer by giving additional examples.

Explaining Concepts

85. Define the graph of an equation.

86. How many solution points make up the graph of $y = 2x - 1$? Explain.

87. Explain how to find the intercepts of a graph. Give examples.

88. An equation gives the relationship between profit y and time t. Profit has been decreasing at a slower rate than in the past. Is it possible to sketch the graph of such an equation? If so, sketch a representative graph.

89. The graph represents the distance d in miles that a person drives during a 10-minute trip from home to work.

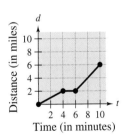

(a) How far is the person's home from the person's place of work? Explain.

(b) Describe the trip for time $4 < t < 6$. Explain.

(c) During what time interval was the person's speed greatest? Explain.

2.3 Slope and Graphs of Linear Equations

Objectives

1 Determine the slope of a line through two points.

2 Write a linear equation in slope-intercept form and use it to sketch the graph of the equation.

3 Use slope to determine whether two lines are parallel, perpendicular, or neither.

1 Determine the slope of a line through two points.

The Slope of a Line

The **slope** of a nonvertical line is the number of units the line rises or falls vertically for each unit of horizontal change from left to right. For example, the line in Figure 2.17 rises 2 units for each unit of horizontal change from left to right, and we say that this line has a slope of $m = 2$.

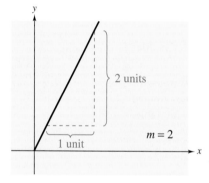

Figure 2.17   Figure 2.18

> ▶ **Definition of the Slope of a Line**
>
> The **slope** m of the nonvertical line passing through the points (x_1, y_1) and (x_2, y_2) is
>
> $$m = \frac{y_2 - y_1}{x_2 - x_1} = \frac{\text{change in } y}{\text{change in } x} = \frac{\text{rise}}{\text{run}}$$
>
> where $x_1 \neq x_2$ (see Figure 2.18).

When the formula for slope is used, the *order of subtraction* is important. Given two points on a line, you are free to label either of them (x_1, y_1) and the other (x_2, y_2). However, once this is done, you must form the numerator and denominator using the same order of subtraction.

$$m = \underbrace{\frac{y_2 - y_1}{x_2 - x_1}}_{\text{Correct}} \qquad m = \underbrace{\frac{y_1 - y_2}{x_1 - x_2}}_{\text{Correct}} \qquad \underbrace{m = \frac{y_2 - y_1}{x_1 - x_2}}_{\text{Incorrect}}$$

| Example 1 | Finding the Slope of a Line Through Two Points |

Find the slope of the line passing through each pair of points.

a. $(1, 2)$ and $(4, 5)$ **b.** $(1, 4)$ and $(3, 4)$

c. $(-1, 4)$ and $(2, 1)$ **d.** $(3, 1)$ and $(3, 3)$

Solution

a. Let $(x_1, y_1) = (1, 2)$ and $(x_2, y_2) = (4, 5)$.

$$m = \frac{y_2 - y_1}{x_2 - x_1} \qquad \begin{aligned} &\text{Difference in } y\text{-values} \\ &\text{Difference in } x\text{-values} \end{aligned}$$

$$= \frac{5 - 2}{4 - 1}$$

$$= 1$$

b. Let $(x_1, y_1) = (1, 4)$ and $(x_2, y_2) = (3, 4)$.

$$m = \frac{y_2 - y_1}{x_2 - x_1} \qquad \begin{aligned} &\text{Difference in } y\text{-values} \\ &\text{Difference in } x\text{-values} \end{aligned}$$

$$= \frac{4 - 4}{3 - 1}$$

$$= \frac{0}{2}$$

$$= 0$$

c. The slope of the line through $(-1, 4)$ and $(2, 1)$ is

$$m = \frac{1 - 4}{2 - (-1)} = \frac{-3}{3} = -1.$$

d. The slope of the line through $(3, 1)$ and $(3, 3)$ is

$$m = \frac{3 - 1}{3 - 3} = \frac{2}{0}.$$

Because division by zero is not defined, the slope of a vertical line is not defined.

The graphs of the four lines are shown in Figure 2.19.

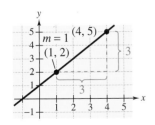

(a) Positive Slope

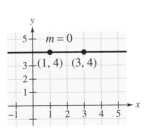

(b) Zero Slope

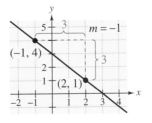

(c) Negative Slope

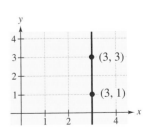

(d) Slope is undefined.

Figure 2.19

From the slopes of the lines shown in Figure 2.19, you can make the following generalizations about the slope of a line.

> ▶ **Slope of a Line**
>
> **1.** A line with positive slope ($m > 0$) *rises* from left to right.
>
> **2.** A line with negative slope ($m < 0$) *falls* from left to right.
>
> **3.** A line with zero slope ($m = 0$) is *horizontal.*
>
> **4.** A line with undefined slope is *vertical.*

Example 2   Using Slope to Describe Lines

Describe the line through each pair of points.

a. $(2, -1), (2, 3)$ **b.** $(-2, 4), (3, 1)$ **c.** $(1, 3), (4, 3)$ **d.** $(-1, 1), (2, 5)$

Solution

a. Because the slope is undefined, the line is vertical.

$$m = \frac{3 - (-1)}{2 - 2} = \frac{4}{0}$$ Undefined slope (See Figure 2.20(a).)

b. Because the slope is negative, the line falls from left to right.

$$m = \frac{1 - 4}{3 - (-2)} = -\frac{3}{5} < 0$$ Negative slope (See Figure 2.20(b).)

c. Because the slope is zero, the line is horizontal.

$$m = \frac{3 - 3}{4 - 1} = \frac{0}{3} = 0$$ Zero slope (See Figure 2.20(c).)

d. Because the slope is positive, the line rises from left to right.

$$m = \frac{5 - 1}{2 - (-1)} = \frac{4}{3} > 0$$ Positive slope (See Figure 2.20(d).)

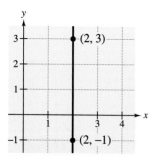

(a) Vertical line
     Undefined slope

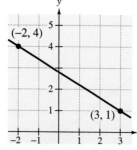

(b) Line falls
     Negative slope

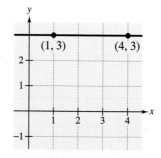

(c) Horizontal line
     Zero slope

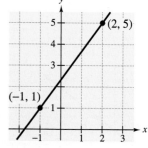

(d) Line rises
     Positive slope

Figure 2.20

Any two points on a nonvertical line can be used to calculate its slope. This is demonstrated in the next example.

Example 3   Finding the Slope of a Line

Sketch the graph of the line given by $2x + 3y = 6$. Then find the slope of the line. (Choose two different pairs of points on the line and show that the same slope is obtained from either pair.)

Solution

Begin by solving the equation for y.

$$2x + 3y = 6 \qquad \text{Original equation}$$

$$3y = -2x + 6 \qquad \text{Subtract } 2x \text{ from both sides.}$$

$$y = \frac{-2x + 6}{3} \qquad \text{Divide both sides by 3.}$$

$$y = -\frac{2}{3}x + 2 \qquad \text{Simplify.}$$

Then construct a table of values, as shown below.

x	-3	0	3	6
$y = -\frac{2}{3}x + 2$	4	2	0	-2
Solution point	$(-3, 4)$	$(0, 2)$	$(3, 0)$	$(6, -2)$

From the solution points shown in the table, sketch the graph of the line (see Figure 2.21). To calculate the slope of the line using two different sets of points, first use the points $(-3, 4)$ and $(0, 2)$, as shown in Figure 2.21(a), and obtain

$$m = \frac{2 - 4}{0 - (-3)} = -\frac{2}{3}.$$

Next, use the points $(3, 0)$ and $(6, -2)$, as shown in Figure 2.21(b), and obtain

$$m = \frac{-2 - 0}{6 - 3} = -\frac{2}{3}.$$

Try some other pairs of points on the line to see that you obtain a slope of $m = -\frac{2}{3}$ regardless of which two points you use.

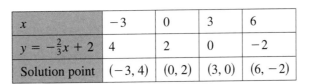

(a)

(b)

Figure 2.21

Technology: Tip   Setting the viewing window on a graphing utility affects the appearance of a line's slope. When you are using a graphing utility, you cannot judge whether a slope is steep or shallow *unless* you use a square setting. See Appendix A for more information on setting a viewing window.

2 Write a linear equation in slope-intercept form and use it to sketch the graph of the equation.

Slope as a Graphing Aid

You have seen that, before creating a table of values for an equation, you should first solve the equation for y. When you do this for a linear equation, you obtain some very useful information. Consider the results of Example 3.

$$2x + 3y = 6 \qquad \text{Original equation}$$

$$3y = -2x + 6 \qquad \text{Subtract } 2x \text{ from both sides.}$$

$$y = -\frac{2}{3}x + 2 \qquad \text{Divide both sides by 3.}$$

Observe that the coefficient of x is the slope of the graph of this equation (see Example 3). Moreover, the constant term, 2, gives the y-intercept of the graph.

$$y = -\frac{2}{3}x + \boxed{2}$$

Slope y-intercept $(0, 2)$

This form is called the **slope-intercept form** of the equation of the line.

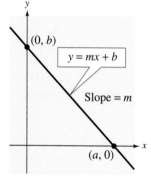

Figure 2.22

▶ **Slope-Intercept Form of the Equation of a Line**

The graph of the equation

$$y = mx + b$$

is a line whose slope is m and whose y-intercept is $(0, b)$. (See Figure 2.22.)

Example 4   Slope and y-Intercept of a Line

Find the slope and y-intercept of the graph of the equation

$$4x - 5y = 15.$$

Solution

Begin by writing the equation in slope-intercept form, as follows.

$$4x - 5y = 15 \qquad \text{Original equation}$$

$$-4x + 4x - 5y = -4x + 15 \qquad \text{Add } -4x \text{ to both sides.}$$

$$-5y = -4x + 15 \qquad \text{Combine like terms.}$$

$$y = \frac{-4x + 15}{-5} \qquad \text{Divide both sides by } -5.$$

$$y = \frac{4}{5}x - 3 \qquad \text{Slope-intercept form}$$

From the slope-intercept form, you can see that $m = \frac{4}{5}$ and $b = -3$. So, the slope of the graph of the equation is $\frac{4}{5}$ and the y-intercept is $(0, -3)$.

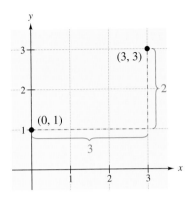

(a)

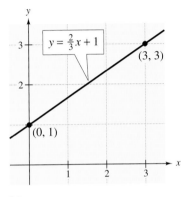

(b)

Figure 2.23

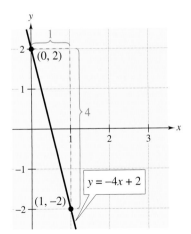

Figure 2.24

So far, you have been plotting several points to sketch the equation of a line. However, now that you can recognize equations of lines (linear equations), you don't have to plot as many points—two points are enough. (You might remember from geometry that *two points are all that are necessary to determine a line.*)

Example 5   Using the Slope and *y*-Intercept to Sketch a Line

Use the slope and *y*-intercept to sketch the graph of

$$y = \frac{2}{3}x + 1.$$

Solution

The equation is already in slope-intercept form.

$$y = mx + b$$

$$y = \frac{2}{3}x + 1 \qquad\qquad \text{Slope-intercept form}$$

So, the slope of the line is $m = \frac{2}{3}$ and the *y*-intercept is $(0, b) = (0, 1)$. Now you can sketch the graph of the line as follows. First, plot the *y*-intercept. Then, using a slope of $\frac{2}{3}$

$$m = \frac{2}{3} = \frac{\text{change in } y}{\text{change in } x},$$

locate a second point on the line by moving 3 units to the right and 2 units up (or 2 units up and 3 units to the right), as shown in Figure 2.23(a). Finally, obtain the graph by drawing a line through the two points, as shown in Figure 2.23(b).

Example 6   Using the Slope and *y*-Intercept to Sketch a Line

Use the slope and *y*-intercept to sketch the graph of

$$12x + 3y = 6.$$

Solution

Begin by writing the equation in slope-intercept form.

$12x + 3y = 6$	Original equation
$3y = -12x + 6$	Subtract $12x$ from both sides.
$y = \dfrac{-12x + 6}{3}$	Divide both sides by 3.
$y = -4x + 2$	Slope-intercept form

So, the slope of the line is $m = -4$ and the *y*-intercept is $(0, b) = (0, 2)$. Now you can sketch the graph of the line as follows. First, plot the *y*-intercept. Then, using a slope of -4, locate a second point on the line by moving 1 unit to the right and 4 units down (or 4 units down and 1 unit to the right). Finally, obtain the graph by drawing a line through the two points, as shown in Figure 2.24.

3 Use slope to determine whether two lines are parallel, perpendicular, or neither.

Parallel and Perpendicular Lines

You know from geometry that two lines in a plane are *parallel* if they do not intersect. What this means in terms of their slopes is suggested by Example 7.

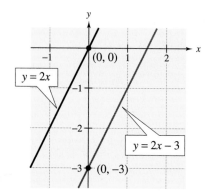

Figure 2.25

Example 7 Lines That Have the Same Slope

On the same set of axes, sketch the lines given by

$$y = 2x \text{ and } y = 2x - 3.$$

Solution

For the line given by $y = 2x$ the slope is $m = 2$ and the y-intercept is $(0, 0)$. For the line given by $y = 2x - 3$ the slope is also $m = 2$ and the y-intercept is $(0, -3)$. The graphs of these two lines are shown in Figure 2.25.

In Example 7, notice that the two lines have the same slope *and* appear to be parallel. The following rule states that this is always the case. That is, two (nonvertical) lines are parallel *if and only if* they have the same slope.

▶ **Parallel Lines**

Two distinct nonvertical lines are parallel if and only if they have the same slope.

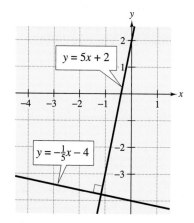

Figure 2.26

The phrase "if and only if" in this rule is used in mathematics as a way to write two statements in one. The first statement says that *if two distinct nonvertical lines have the same slope, they must be parallel.* The second statement says that *if two distinct nonvertical lines are parallel, they must have the same slope.*

Another rule from geometry is that two lines in a plane are *perpendicular* if they intersect at right angles. In terms of their slopes, this means that two nonvertical lines are perpendicular if their slopes are negative reciprocals of each other. For instance, the negative reciprocal of 5 is $-\frac{1}{5}$, so the lines

$$y = 5x + 2 \quad \text{and} \quad y = -\frac{1}{5}x - 4$$

are perpendicular to each other, as shown in Figure 2.26.

▶ **Perpendicular Lines**

Consider two nonvertical lines whose slopes are m_1 and m_2. The two lines are perpendicular if and only if their slopes are *negative reciprocals* of each other. That is,

$$m_1 = -\frac{1}{m_2}, \quad \text{or equivalently,} \quad m_1 \cdot m_2 = -1.$$

Example 8   Parallel or Perpendicular?

Are the following pairs of lines parallel, perpendicular, or neither?

a. $y = -2x + 4$, $y = \frac{1}{2}x + 1$

b. $y = \frac{1}{3}x + 2$, $y = \frac{1}{3}x - 3$

Solution

a. The first line has a slope of

$$m_1 = -2,$$

and the second line has a slope of

$$m_2 = \frac{1}{2}.$$

Because these slopes are negative reciprocals of each other, the two lines must be perpendicular, as shown in Figure 2.27.

b. Each of these two lines has a slope of $m = \frac{1}{3}$. So, the two lines must be parallel, as shown in Figure 2.28.

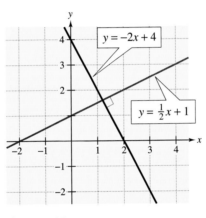

Figure 2.27

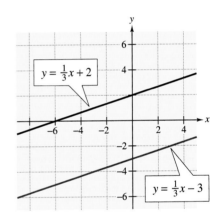

Figure 2.28

Discussing the Concept   **Interpreting Slope**

Write a function for the given verbal model. Identify the slope and then interpret the slope in the real-life setting.

1. $\dfrac{\text{Total pay}}{\text{per hour}} = \dfrac{\text{Piecework}}{\text{rate}} \cdot \dfrac{\text{Number}}{\text{of pieces}} + \dfrac{\text{Fixed}}{\text{hourly rate}}$

2. $\dfrac{\text{Total}}{\text{cost}} = \dfrac{\text{Tax}}{\text{rate}} \cdot \dfrac{\text{List}}{\text{price}} + \dfrac{\text{List}}{\text{price}}$

2.3   Exercises

Integrated Review   *Concepts, Skills, and Problem Solving*

Keep mathematically in shape by doing these exercises *before* the problems of this section.

Properties and Definitions

1. When two equations such as $2x - 3 = 5$ and $2x = 8$ have the same set of solutions, the equations are called _____.

2. Use the Addition Property of Equality to fill in the blank.

$$12x - 5 = 13$$
$$12x = 13 + \underline{}$$

Solving Equations

In Exercises 3–10, solve the equation.

3. $x + \dfrac{x}{2} = 4$

4. $\dfrac{1}{3}x + 1 = 10$

5. $-4(x - 5) = 0$

6. $\frac{3}{8}x + \frac{3}{4} = 2$

7. $8(x - 14) = 32$

8. $12(3 - x) = 5 - 7(2x + 1)$

9. $-(2x + 8) + \frac{1}{3}(6x + 5) = 0$

10. $(1 + r)500 = 550$

Problem Solving

11. The cost for an international telephone call is $1.10 for the first minute and $0.45 for each additional minute. If the total cost of the call cannot exceed $11, find the interval of time that is available for the call.

12. A fuel company has a fleet of trucks. The annual operating cost per truck is $C = 0.65m + 4500$, where m is the number of miles traveled by a truck in a year. What is the maximum number of miles that will yield an annual operating cost that is less than $20,000?

Developing Skills

In Exercises 1–6, estimate the slope of the line from its graph.

1.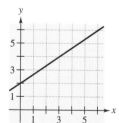

2.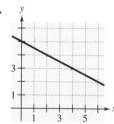

3.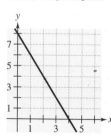

4.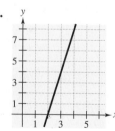

5.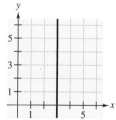

6.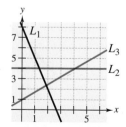

In Exercises 7 and 8, identify the line that has the specified slope m.

7. (a) $m = \frac{3}{4}$

   (b) $m = 0$

   (c) $m = -3$

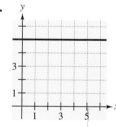

8. (a) $m = -\frac{5}{2}$

(b) m is undefined.

(c) $m = 2$

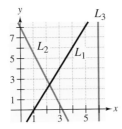

In Exercises 9–24, plot the points and, if possible, find the slope of the line passing through them. Then describe the line. See Examples 1 and 2.

9. $(0, 0), (7, 5)$

10. $(0, 0), (-3, -4)$

11. $(0, 0), (5, -4)$

12. $(0, 0), (-2, 1)$

13. $(-4, 3), (-2, 5)$

14. $(7, 1), (4, -5)$

15. $(-5, -3), (-5, 4)$

16. $(9, 2), (-9, 2)$

17. $(2, -5), (7, -5)$

18. $(-3, 4), (-3, 8)$

19. $\left(\frac{3}{4}, 2\right), \left(5, -\frac{5}{2}\right)$

20. $\left(\frac{1}{2}, -1\right), \left(3, \frac{2}{3}\right)$

21. $\left(\frac{3}{4}, \frac{1}{4}\right), \left(-\frac{3}{2}, \frac{1}{8}\right)$

22. $\left(-\frac{3}{2}, -\frac{1}{2}\right), \left(\frac{5}{8}, \frac{1}{2}\right)$

23. $(2.5, -2), (4.75, 5.25)$

24. $(0, 4.5), (3, 4.5)$

In Exercises 25–30, sketch the graph of the line. Then find the slope of the line. See Example 3.

25. $y = 2x - 1$

26. $y = 3x + 2$

27. $y = -\frac{1}{2}x + 4$

28. $y = \frac{3}{4}x - 5$

29. $4x + 5y = 10$

30. $3x - 2y = 8$

In Exercises 31 and 32, determine a value for x such that the line through the points has the given slope.

31. $(4, 5), (x, 7)$

$m = -\frac{2}{3}$

32. $(x, -2), (5, 0)$

$m = \frac{3}{4}$

In Exercises 33 and 34, determine a value for y such that the line through the points has the given slope.

33. $(-3, y), (9, 3)$

$m = \frac{3}{2}$

34. $(-3, 20), (2, y)$

$m = -6$

In Exercises 35–42, a point on a line and the slope of the line are given. Find two additional points on the line.

35. $(5, 2)$

$m = 0$

36. $(-4, 3)$

m is undefined.

37. $(3, -4)$

$m = 3$

38. $(-1, -5)$

$m = 2$

39. $(0, 3)$

$m = -1$

40. $(-2, 6)$

$m = -3$

41. $(-5, 0)$

$m = \frac{4}{3}$

42. $(-1, 1)$

$m = -\frac{3}{4}$

In Exercises 43–50, write the equation in slope-intercept form.

43. $6x - 3y = 9$

44. $2x + 4y = 16$

45. $4y - x = -4$

46. $3x - 2y = -10$

47. $2x + 5y - 3 = 0$

48. $8x - 6y + 1 = 0$

49. $y = \frac{1}{2}x + 2$

50. $y = -\frac{2}{3}x + 4$

In Exercises 51–56, find the slope and y-intercept of the graph of the equation. See Example 4.

51. $y = 3x - 2$

52. $y = 4 - 2x$

53. $3y - 2x = 3$

54. $4x + 8y = -1$

55. $5x + 3y - 2 = 0$

56. $6y - 5x + 18 = 0$

In Exercises 57–64, write the equation in slope-intercept form, and then use the slope and y-intercept to sketch the graph of the equation. See Examples 5 and 6.

57. $3x - y - 2 = 0$

58. $x - y - 5 = 0$

59. $x + y = 0$

60. $x - y = 0$

61. $3x + 2y - 2 = 0$

62. $x - 2y - 2 = 0$

63. $x - 4y + 2 = 0$

64. $8x + 6y - 3 = 0$

In Exercises 65–70, sketch the graph of a line through the point $(3, 2)$ having the given slope.

65. $m = 3$

66. $m = \frac{3}{2}$

67. $m = -\frac{1}{3}$

68. $m = 0$

69. m is undefined.

70. $m = -\frac{2}{3}$

In Exercises 71–74, plot the x- and y-intercepts and sketch the graph of the line.

71. $2x - y + 4 = 0$

72. $3x + 5y + 15 = 0$

73. $-5x + 2y - 20 = 0$

74. $3x - 5y - 15 = 0$

In Exercises 75–78, are the lines parallel, perpendicular, or neither? See Examples 7 and 8.

75. L_1: $y = \frac{1}{2}x - 2$

L_2: $y = \frac{1}{2}x + 3$

76. L_1: $y = 3x - 2$

L_2: $y = 3x + 1$

77. L_1: $y = \frac{3}{4}x - 3$

L_2: $y = -\frac{4}{3}x + 1$

78. L_1: $y = -\frac{2}{3}x - 5$

L_2: $y = \frac{3}{2}x + 1$

Solving Problems

79. *Road Grade*   When driving down a mountain road, you notice warning signs indicating a "12% down-grade." This means that the slope of the road is $-\frac{12}{100}$. Over a particular stretch of the road, your elevation drops 2000 feet (see figure). What is the horizontal change in your position?

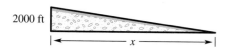

80. *Slope of a Ramp*   A loading dock ramp rises 4 feet above the ground. The ramp has a slope of $\frac{1}{10}$. What is the length of the ramp?

81. *Height of an Attic*   The slope, or pitch, of a roof (see figure) is such that it rises (or falls) 3 feet for every 4 feet of horizontal distance. Determine the maximum height of the attic of the house if the house is 30 feet wide.

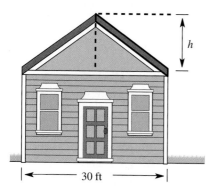

Figure for 81 and 82

82. *Height of an Attic*   The slope, or pitch, of a roof (see figure) is such that it rises (or falls) 4 feet for every 5 feet of horizontal distance. Determine the maximum height of the attic of the house if the house is 30 feet wide.

83. *Tuition Costs*   The average annual amount of tuition and fees y paid by an in-state student attending a 4-year college in the United States from 1990 to 1996 can be approximated by the model

$$y = 192.64t + 2015.79, \quad 0 \le t \le 6$$

where t represents the year, with $t = 0$ corresponding to 1990.   (Source: U.S. National Center for Education Statistics)

(a) Use the equation to complete the table.

t	0	1	2	3	4	5	6
y							

(b) Graph the equation on a rectangular coordinate system.

(c) On the average, how much did tuition and fees increase each year from 1990 to 1996? How does this relate to the slope of the graph?

(d) If this increase continued at the current rate, predict the amount of tuition and fees that would be paid in the year 2005.

84. *Simple Interest*   An inheritance of $8000 is invested in two different accounts. One account pays 6% simple interest and the other pays $7\frac{1}{2}\%$ simple interest.

(a) If x dollars is invested in the account paying 6%, how much is invested in the account paying $7\frac{1}{2}\%$?

(b) Use the result of part (a) to write the annual interest y in terms of x.

(c) Use a graphing utility to graph the equation in part (b).

(d) Explain why the slope of the line in part (b) is negative.

Explaining Concepts

85. Answer parts (a)–(c) of Motivating the Chapter on page 121.

86. Can any pair of points on a line be used to calculate the slope of the line? Explain.

87. In your own words, give interpretations of a negative slope, a zero slope, and a positive slope.

88. The slopes of two lines are -3 and $\frac{3}{2}$. Which is steeper? Explain.

89. In the form $y = mx + b$, what does m represent? What does b represent?

90. What is the relationship between the x-intercept of the graph of the line $y = mx + b$ and the solution to the equation $mx + b = 0$? Explain.

91. Is it possible for two lines with positive slopes to be perpendicular to each other? Explain.

Mid-Chapter Quiz

Take this quiz as you would take a quiz in class. After you are done, check your work against the answers given in the back of the book.

1. Determine the quadrants in which the point $(x, 4)$ must be located if x is a real number. Explain your reasoning.

2. Find the coordinates of the point that lies 10 units to the right of the y-axis and 3 units below the x-axis.

3. Determine whether the following ordered pairs are solution points of the equation $4x - 3y = 10$.

   (a) $(2, 1)$ (b) $(1, -2)$ (c) $(2.5, 0)$ (d) $\left(2, -\frac{2}{3}\right)$

In Exercises 4 and 5, plot the points on a rectangular coordinate system and find the distance between them.

4. $(-1, 5), (3, 2)$

5. $(-3, -2), (2, 10)$

6. Find the x- and y-intercepts of the graph of the equation $6x - 8y + 48 = 0$.

In Exercises 7–12, sketch the graph of the equation and show the coordinates of three solution points (including intercepts).

7. $y = 2x - 3$

8. $3x + y - 6 = 0$

9. $y = 6x - x^2$

10. $y = x^2 - 4$

11. $y = |x| + 1$

12. $y = |x - 2| - 3$

In Exercises 13–16, determine the slope of the line through the two points, if possible. Then describe the line.

13. $(5, 3), (5, -2)$

14. $(-3, 8), (7, 8)$

15. $(3, 0), (6, 5)$

16. $(-1, 4), (5, -6)$

In Exercises 17–19, write the linear equation in slope-intercept form. State the slope and y-intercept and use them to sketch the graph of the equation.

17. $3x + 6y = 6$ 18. $-2x + y = 8$ 19. $x - 2y = 4$

In Exercises 20–22, determine whether the lines are parallel, perpendicular, or neither.

20. $y = 3x + 2, y = -\frac{1}{3}x - 4$

21. $y = 2x - 3, y = -2x - 3$

22. $y = 4x + 3, y = \frac{1}{2}(8x + 5)$

23. Your company purchases a new printing press for $85,000. For tax purposes, the printing press will be depreciated over a 10-year period. At the end of 10 years, the salvage value of the printing press is expected to be $4000. Find an equation that relates the depreciated value of the printing press to the number of years since it was purchased. Then sketch the graph of the equation.

2.4   Equations of Lines

Objectives

1 Write an equation of a line using the point-slope form.

2 Write the equation of a horizontal, vertical, parallel, or perpendicular line.

3 Use a linear model to solve an application problem.

1 Write an equation of a line using the point-slope form.

The Point-Slope Equation of a Line

In Sections 2.1 through 2.3, you have been studying analytic (or coordinate) geometry. Analytic geometry uses a coordinate plane to give visual representations of algebraic concepts, such as equations or functions.

There are two basic types of problems in analytic geometry.

1. Given an equation, sketch its graph.

Algebra ⟹ Geometry

2. Given a graph, write its equation.

Geometry ⟹ Algebra

In Section 2.3, you worked primarily with the first type of problem. In this section, you will study the second type. Specifically, you will learn how to write the equation of a line when you are given its slope and a point on the line. Before we give a general formula for doing this, consider the following example.

Example 1   Writing an Equation of a Line

Write an equation of the line that has a slope of $\frac{4}{3}$ and passes through the point $(-2, 1)$.

Solution

Begin by sketching the line, as shown in Figure 2.29. You know that the slope of a line is the same through any two points on the line. So, to find an equation of the line, let (x, y) represent *any* point on the line. Using the representative point (x, y) and the point $(-2, 1)$, it follows that the slope of the line is

$$m = \frac{y - 1}{x - (-2)}.$$

⟸ Difference in y-values
⟸ Difference in x-values

Because the slope of the line is $m = \frac{4}{3}$, this equation can be rewritten as follows.

$$\frac{4}{3} = \frac{y - 1}{x + 2} \qquad \text{Slope formula}$$

$$4(x + 2) = 3(y - 1) \qquad \text{Cross-multiply.}$$

$$4x + 8 = 3y - 3 \qquad \text{Distributive Property}$$

$$4x - 3y = -11 \qquad \text{Subtract 8 and } 3y \text{ from both sides.}$$

An equation of the line is $4x - 3y = -11$.

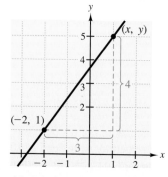

Figure 2.29

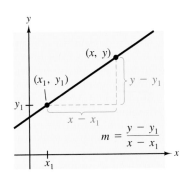

Figure 2.30

The procedure in Example 1 can be used to derive a *formula* for the equation of a line, given its slope and a point on the line. In Figure 2.30, let (x_1, y_1) be a given point on the line whose slope is m. If (x, y) is any *other* point on the line, it follows that

$$\frac{y - y_1}{x - x_1} = m.$$

This equation in variables x and y can be rewritten in the form

$$y - y_1 = m(x - x_1)$$

which is called the **point-slope form** of the equation of a line.

> ▶ **Point-Slope Form of the Equation of a Line**
>
> The **point-slope form** of the equation of the line that passes through the point (x_1, y_1) and has a slope of m is
>
> $$y - y_1 = m(x - x_1).$$

Example 2   The Point-Slope Form of the Equation of a Line

Write an equation of the line that passes through the point $(2, -3)$ and has a slope of -2.

Solution

Use the point-slope form with $(x_1, y_1) = (2, -3)$ and $m = -2$.

$$y - y_1 = m(x - x_1) \qquad \text{Point-slope form}$$
$$y - (-3) = -2(x - 2) \qquad \text{Substitute } y_1 = -3,\ x_1 = 2,\ \text{and}\ m = -2.$$
$$y + 3 = -2x + 4 \qquad \text{Simplify.}$$
$$y = -2x + 1 \qquad \text{Subtract 3 from both sides.}$$

So, an equation of the line is $y = -2x + 1$. Note that this is the slope-intercept form of the equation. The graph of the line is shown in Figure 2.31.

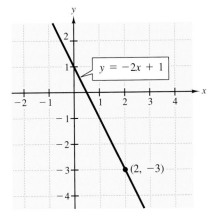

Figure 2.31

In Example 2, note that we conclude by saying that $y = -2x + 1$ is "an" equation of the line rather than saying it is "the" equation of the line. The reason for this is that every equation can be written in many equivalent forms. For instance,

$$y = -2x + 1, \quad 2x + y = 1, \quad \text{and} \quad 2x + y - 1 = 0$$

are all equations of the line in Example 2. The first of these equations $(y = -2x + 1)$ is the slope-intercept form

$$y = mx + b \qquad \text{Slope-intercept form}$$

and it provides the most information about the line. The last of these equations $(2x + y - 1 = 0)$ is the **general form** of the equation of a line.

$$ax + by + c = 0 \qquad \text{General form}$$

The point-slope form can be used to find the equation of a line passing through two points (x_1, y_1) and (x_2, y_2). First, use the formula for the slope of a line passing through two points.

$$m = \frac{y_2 - y_1}{x_2 - x_1}$$

Then, substitute this value for m into the point-slope form to obtain the equation

$$y - y_1 = \frac{y_2 - y_1}{x_2 - x_1}(x - x_1). \qquad \text{Two-point form}$$

This is sometimes called the **two-point form** of the equation of a line.

Example 3   An Equation of a Line Passing Through Two Points

Write the general form of the equation of the line that passes through the points $(4, 2)$ and $(-2, 3)$.

Solution

Let $(x_1, y_1) = (4, 2)$ and $(x_2, y_2) = (-2, 3)$. Then apply the formula for the slope of a line passing through two points, as follows.

$$m = \frac{y_2 - y_1}{x_2 - x_1}$$

$$= \frac{3 - 2}{-2 - 4}$$

$$= -\frac{1}{6}$$

Now, using the point-slope form, you can find the equation of the line.

$$y - y_1 = m(x - x_1) \qquad \text{Point-slope form}$$

$$y - 2 = -\frac{1}{6}(x - 4) \qquad \text{Substitute } y_1 = 2, \; x_1 = 4, \text{ and } m = -\frac{1}{6}.$$

$$6(y - 2) = -(x - 4) \qquad \text{Multiply both sides by 6.}$$

$$6y - 12 = -x + 4 \qquad \text{Distributive Property}$$

$$x + 6y - 12 = 4 \qquad \text{Add } x \text{ to both sides.}$$

$$x + 6y - 16 = 0 \qquad \text{Subtract 4 from both sides.}$$

The general form of the equation of the line is $x + 6y - 16 = 0$. The graph of this line is shown in Figure 2.32.

**Technology:
Tip**

A program that uses the two-point form to find the equation of a line is available at our website *www.hmco.com*. Programs for several models of calculators are available.

The program prompts for the coordinates of the two points and then outputs the slope and the y-intercept of the line that passes through the two points. Verify Example 3 using this program.

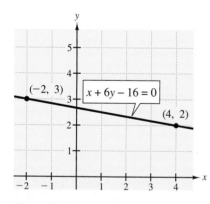

Figure 2.32

In Example 3, it does not matter which of the two points is labeled (x_1, y_1) and which is labeled (x_2, y_2). Try switching these labels to

$$(x_1, y_1) = (-2, 3) \quad \text{and} \quad (x_2, y_2) = (4, 2)$$

and reworking the problem to see that you obtain the same equation.

Other Equations of Lines

2 Write the equation of a horizontal, vertical, parallel, or perpendicular line.

From the slope-intercept form of the equation of a line, you can see that a horizontal line ($m = 0$) has an equation of the form

$$y = (0)x + b \quad \text{or} \quad y = b. \qquad \text{Horizontal line}$$

This is consistent with the fact that each point on a horizontal line through $(0, b)$ has a y-coordinate of b, as shown in Figure 2.33. Similarly, each point on a vertical line through $(a, 0)$ has an x-coordinate of a, as shown in Figure 2.34. So, a vertical line has an equation of the form

$$x = a. \qquad \text{Vertical line}$$

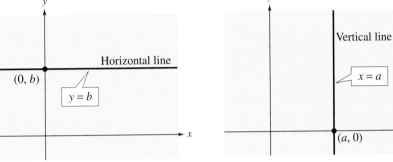

Figure 2.33

Figure 2.34

> **Example 4**   Writing Equations of Horizontal and Vertical Lines
>
> Write an equation for each line.
>
> **a.** Vertical line through $(-2, 4)$
>
> **b.** Horizontal line through $(0, -2)$
>
> **c.** Line passing through $(-2, 3)$ and $(3, 3)$
>
> **d.** Line passing through $(-1, 2)$ and $(-1, 3)$
>
> **Solution**
>
> **a.** Because the line is vertical and passes through the point $(-2, 4)$, you know that every point on the line has an x-coordinate of -2. So, the equation is
>
> $$x = -2.$$
>
> **b.** Because the line is horizontal and passes through the point $(0, -2)$, you know that every point on the line has a y-coordinate of -2. So, the equation of the line is
>
> $$y = -2.$$
>
> **c.** Because both points have the same y-coordinate, the line through $(-2, 3)$ and $(3, 3)$ is horizontal. So, its equation is $y = 3$.
>
> **d.** Because both points have the same x-coordinate, the line through $(-1, 2)$ and $(-1, 3)$ is vertical. So, its equation is $x = -1$.

In Section 2.3, you learned that parallel lines have the same slope and perpendicular lines have slopes that are negative reciprocals of each other. Use these facts to find an equation of a line parallel or perpendicular to a given line.

Example 5   Parallel and Perpendicular Lines

Write an equation of the line that passes through the point $(3, -2)$ and is (a) parallel and (b) perpendicular to the line $x - 4y = 6$, as shown in Figure 2.35.

Solution

By writing the given line in slope-intercept form, $y = \frac{1}{4}x - \frac{3}{2}$, you can see that it has a slope of $\frac{1}{4}$. So, a line parallel to it must also have a slope of $\frac{1}{4}$ and a line perpendicular to it must have a slope of -4.

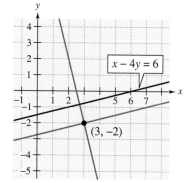

a.

$$y - y_1 = m(x - x_1)$$ Point-slope form

$$y - (-2) = \frac{1}{4}(x - 3)$$ Substitute $y_1 = -2$, $x_1 = 3$, and $m = \frac{1}{4}$.

$$y = \frac{1}{4}x - \frac{11}{4}$$ Equation of parallel line

b.

$$y - y_1 = m(x - x_1)$$ Point-slope form

$$y - (-2) = -4(x - 3)$$ Substitute $y_1 = -2$, $x_1 = 3$, and $m = -4$.

$$y = -4x + 10$$ Equation of perpendicular line

Figure 2.35

The equation of a vertical line cannot be written in slope-intercept form because the slope of a vertical line is undefined. However, *every* line has an equation that can be written in the general form $ax + by + c = 0$ where a and b are not *both* zero.

Study Tip

Keep in mind that the *general form* of an equation is commonly used to write linear equation problems or to express answers. The *slope-intercept form* gives the most information about the graph of a line. And the *point-slope form* is most often used to create an equation for a line.

▶ Summary of Equations of Lines

1. Slope of a line through (x_1, y_1) and (x_2, y_2): $m = \dfrac{y_2 - y_1}{x_2 - x_1}$

2. General form of equation of line: $ax + by + c = 0$

3. Equation of vertical line: $x = a$

4. Equation of horizontal line: $y = b$

5. Slope-intercept form of equation of line: $y = mx + b$

6. Point-slope form of equation of line: $y - y_1 = m(x - x_1)$

7. Parallel lines (equal slopes): $m_1 = m_2$

8. Perpendicular lines (negative reciprocal slopes): $m_2 = -\dfrac{1}{m_1}$

3 Use a linear model to solve an application problem.

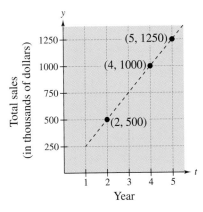

Figure 2.36

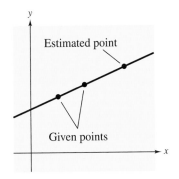

Linear Extrapolation

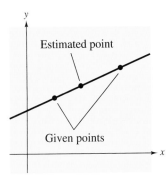

Linear Interpolation
Figure 2.37

Application

Example 6   Total Sales

The total sales of a new computer software company were $500,000 for the second year and $1,000,000 for the fourth year. Using only this information, what would you estimate the total sales to be during the fifth year?

Solution

To solve this problem, use a *linear model*, with y representing the total sales (in thousands of dollars) and t representing the year. That is, in Figure 2.36, let $(2, 500)$ and $(4, 1000)$ be two points on the line representing the total sales for the company. The slope of the line passing through these points is

$$m = \frac{1000 - 500}{4 - 2} = 250.$$

Now, using the point-slope form, the equation of the line is

$$y - y_1 = m(t - t_1) \qquad \text{Point-slope form}$$

$$y - 500 = 250(t - 2) \qquad \text{Substitute } y_1 = 500, \ t_1 = 2, \text{ and } m = 250.$$

$$y = 250t. \qquad \text{Linear model for sales}$$

Finally, estimate the total sales during the fifth year ($t = 5$) to be

$$y = 250(5) = \$1250 \text{ thousand} = \$1,250,000.$$

The estimation method illustrated in Example 6 is called **linear extrapolation.** Note in Figure 2.37 that for linear extrapolation, the estimated point lies to the right of the given points. When the estimated point lies *between* two given points, the procedure is called **linear interpolation.**

Discussing the Concept	Mathematical Modeling

You are asked to make sense of the following set of data, in which y represents the total expenditures (in billions of dollars) for public elementary and secondary schools in the United States and x represents the year from 1990 to 1997, with $x = 0$ corresponding to 1990. (Source: National Education Association)

x	0	1	2	3	4	5	6	7
y	210	227	237	249	262	277	292	308

Plot the ordered pairs on a rectangular coordinate system. This is known as a **scatter plot.** Draw a line through the points that seems to have the "best fit." Do you think a linear model would represent the data well? If so, find an equation for your best-fitting line. Interpret the meaning of the slope in the context of the data. Use the model to predict the public school expenditures in the year 2005 (assuming the trend continues).

2.4   Exercises

Integrated Review   Concepts, Skills, and Problem Solving

Keep mathematically in shape by doing these exercises *before* the problems of this section.

Properties and Definitions

1. State the definition of the ratio of the real number a to the real number b.

2. $\dfrac{4}{5} = \dfrac{12}{u}$ is a statement of equality of two ratios. What is this statement of equality called?

Solving Percent Problems

In Exercises 3–10, solve the percent problem.

3. What is $7\frac{1}{2}\%$ of 25?

4. What is 150% of 6000?

5. 225 is what percent of 150?

6. 93 is what percent of 600?

7. What percent of 240 is 160?

8. 12% of what number is 42?

9. 0.5% of what number is 400?

10. 48% of what number is 132?

Problem Solving

11. The ratio of cement to sand in a 90-pound bag of dry mix is 1 to 4. Find the number of pounds of sand in the bag.

12. The velocity v of an object projected vertically upward with an initial velocity of 96 feet per second is given by $v = 96 - 32t$, where t is time in seconds and air resistance is neglected. Find the time when the maximum height ($v = 0$) of the object is attained.

Developing Skills

In Exercises 1–4, match the equation with its graph. [The graphs are labeled (a), (b), (c), and (d).]

1. $y = \frac{2}{3}x + 2$

2. $y = \frac{2}{3}x - 2$

3. $y = -\frac{3}{2}x + 2$

4. $y = -3x + 2$

(a)

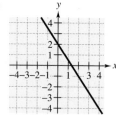

(b)

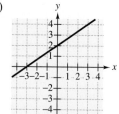

(c)

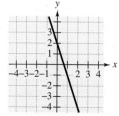

(d)

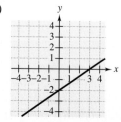

In Exercises 5–10, write an equation of the line with the specified slope that passes through the given point. See Example 1.

5. $(2, 5)$
   $m = 3$

6. $(1, -4)$
   $m = -2$

7. $(-3, 1)$
   $m = -\frac{1}{2}$

8. $(6, 9)$
   $m = \frac{2}{3}$

9. $\left(\frac{3}{4}, -1\right)$
   $m = \frac{4}{5}$

10. $\left(-2, \frac{3}{2}\right)$
    $m = -\frac{1}{6}$

In Exercises 11–26, use the point-slope form of the equation of a line to write an equation of the line with the specified slope passing through the given point. See Example 2.

11. $(0, 0)$
    $m = -\frac{1}{2}$

12. $(0, 0)$
    $m = \frac{2}{3}$

13. $(0, -4)$
    $m = 3$

14. $(0, 2)$
    $m = 3$

15. $(0, 6)$
    $m = -\frac{3}{4}$

16. $(0, -3)$
    $m = \frac{1}{3}$

17. $(-2, 8)$

$m = -2$

18. $(4, -1)$

$m = 3$

19. $(-4, -7)$

$m = \frac{5}{4}$

20. $(6, -8)$

$m = -\frac{2}{3}$

21. $\left(-2, \frac{7}{2}\right)$

$m = -4$

22. $\left(1, -\frac{3}{2}\right)$

$m = 1$

23. $\left(\frac{3}{4}, \frac{5}{2}\right)$

$m = \frac{4}{3}$

24. $\left(-\frac{5}{2}, \frac{1}{2}\right)$

$m = -\frac{2}{5}$

25. $(2, -1)$

$m = 0$

26. $(-8, 5)$

$m = 0$

In Exercises 27–42, write the general form of the equation of the line passing through the points. See Example 3.

27. $(0, 0), (2, 3)$

28. $(0, 0), (3, -5)$

29. $(0, 4), (4, 0)$

30. $(0, -2), (2, 0)$

31. $(-2, 3), (4, 0)$

32. $(1, -2), (2, 8)$

33. $(-5, 2), (5, -2)$

34. $(5, 4), (3, 5)$

35. $\left(\frac{3}{2}, 3\right), \left(\frac{9}{2}, 4\right)$

36. $\left(4, \frac{7}{3}\right), \left(-1, \frac{1}{3}\right)$

37. $\left(10, \frac{1}{2}\right), \left(\frac{3}{2}, \frac{7}{4}\right)$

38. $\left(-4, \frac{3}{5}\right), \left(\frac{3}{4}, -\frac{2}{5}\right)$

39. $(5, 9), (8, -1.4)$

40. $(2, -8), (6, 2.3)$

41. $(2, 0.6), (8, -4.2)$

42. $(-5, 0.6), (3, -3.4)$

In Exercises 43–46, write the equation of the line passing through the two points in slope-intercept form.

43. $(-2, 2), (4, 5)$

44. $(0, 10), (5, 0)$

45. $(-2, 3), (4, 3)$

46. $(-6, -3), (4, 3)$

In Exercises 47–52, write an equation for each line. See Example 4.

47. Vertical line through $(-1, 5)$

48. Vertical line through $(2, -3)$

49. Horizontal line through $(-4, 6)$

50. Horizontal line through $(0, -3)$

51. Line through $(-7, 2)$ and $(-7, -1)$

52. Line through $(6, 4)$ and $(-9, 4)$

In Exercises 53–62, write an equation of the line that passes through the point and is (a) parallel and (b) perpendicular to the given line. See Example 5.

53. $(2, 1)$

$6x - 2y = 3$

54. $(-3, 4)$

$x + 6y = 12$

55. $(-5, 4)$

$5x + 4y = 24$

56. $(6, -4)$

$3x + 10y = 24$

57. $(3, 7)$

$4x - y - 3 = 0$

58. $(-5, -10)$

$2x + 5y - 12 = 0$

59. $\left(\frac{2}{3}, \frac{4}{3}\right)$

$x - 5 = 0$

60. $\left(\frac{5}{8}, \frac{9}{4}\right)$

$-5x + 4y = 0$

61. $(-1, 2)$

$y + 5 = 0$

62. $(3, -4)$

$x - 10 = 0$

In Exercises 63–66, write an equation of the line with intercepts $(a, 0)$ and $(0, b)$ where the equation is given by

$$\frac{x}{a} + \frac{y}{b} = 1, \ a \neq 0, \ b \neq 0.$$

63. x-intercept: $(3, 0)$

y-intercept: $(0, 2)$

64. x-intercept: $(-6, 0)$

y-intercept: $(0, 2)$

65. x-intercept: $\left(-\frac{5}{6}, 0\right)$

y-intercept: $\left(0, -\frac{7}{3}\right)$

66. x-intercept: $\left(-\frac{8}{3}, 0\right)$

y-intercept: $(0, -4)$

Solving Problems

67. *Cost*   The cost C (in dollars) of producing x units of a certain product is given in the table. Find a linear model to represent the data. Estimate the cost of producing 400 units.

x	0	50	100	150	200
C	5000	6000	7000	8000	9000

68. *Temperature Conversion*   The relationship between the Fahrenheit F and Celsius C temperature scales is given in the table. Find a linear model to represent the data. Estimate the Celsius temperature when the Fahrenheit temperature is 72 degrees.

F	41	50	59	68	77
C	5	10	15	20	25

69. *Total Sales* The total sales for a new camera equipment store were $200,000 for the second year and $500,000 for the fifth year. Find a linear model to represent the data. Estimate the total sales for the sixth year.

70. *Total Sales* The total sales for a new sportswear store were $150,000 for the third year and $250,000 for the fifth year. Find a linear model to represent the data. Estimate the total sales for the sixth year.

71. *Sales Commission* The salary for a sales representative is $1500 per month plus a commission of total monthly sales. The table gives the relationship between the salary S and total monthly sales M. Write an equation of the line giving the salary S in terms of the monthly sales M. What is the commission rate?

M	0	1000	2000	3000	4000
S	1500	1530	1560	1590	1620

72. *Reimbursed Expenses* A sales representative is reimbursed $125 per day for lodging and meals plus an amount per mile driven. The table below gives the relationship between the daily cost C to the company and the number of miles driven x. Write an equation giving the daily cost C to the company in terms of x, the number of miles driven. How much is the sales representative reimbursed per mile?

x	50	100	150	200	250
C	141	157	173	189	205

73. *Discount Price* A store is offering a 30% discount on all items in its inventory.

(a) Write an equation of the line giving the sale price S for an item in terms of its list price L.

(b) Use the equation in part (a) to find the sale price of an item that has a list price of $135.

74. *Reimbursed Expenses* A sales representative is reimbursed $150 per day for lodging and meals plus $0.34 per mile driven.

(a) Write an equation of the line giving the daily cost C to the company in terms of x, the number of miles driven.

(b) Use a graphing utility to graph the line in part (a) and graphically estimate the daily cost to the company if the representative drives 230 miles. Confirm your estimate algebraically.

(c) Use the graph in part (b) to estimate graphically the number of miles driven if the daily cost to the company is $200. Confirm your estimate algebraically.

75. *Straight-Line Depreciation* A small business purchases a photocopier for $7400. After 4 years, its depreciated value will be $1500.

(a) Assuming straight-line depreciation, write an equation of the line giving the value V of the copier in terms of time t in years.

(b) Use the equation in part (a) to find the value of the copier after 2 years.

76. *Straight-Line Depreciation* A business purchases a van for $27,500. After 5 years, its depreciated value will be $12,000.

(a) Assuming straight-line depreciation, write an equation of the line giving the value V of the van in terms of time t in years.

(b) Use the equation in part (a) to find the value of the van after 2 years.

77. *College Enrollment* A small college had an enrollment of 1500 students in 1990. During the next 10 years, the enrollment increased by approximately 60 students per year.

(a) Write an equation of the line giving the enrollment N in terms of the year t. (Let $t = 0$ correspond to the year 1990.)

(b) (*Linear Extrapolation*) Use the equation in part (a) to predict the enrollment in the year 2005.

(c) (*Linear Interpolation*) Use the equation in part (a) to estimate the enrollment in 1995.

78. *Soft Drink Sales* When soft drinks sold for $0.80 per can at football games, approximately 6000 cans were sold. When the price was raised to $1.00 per can, the demand dropped to 4000. Assume the relationship between the price p and demand x is linear.

(a) Write an equation of the line giving the demand x in terms of the price p.

(b) (*Linear Extrapolation*) Use the equation in part (a) to predict the number of cans of soft drinks sold if the price is raised to $1.10.

(c) (*Linear Interpolation*) Use the equation in part (a) to predict the number of cans of soft drinks sold if the price is $0.90.

79. *Data Analysis*   The table gives the expected number of years of life E for a person of age A. (Source: U.S. National Center for Health Statistics)

A	Birth	10	20	40	60	80
E	75.8	66.6	56.9	38.3	21.1	8.3

(a) Plot the points.

(b) Use a ruler to sketch the "best-fitting" line through the points.

(c) Find an equation of the line sketched in part (b).

(d) Use the equation in part (c) to estimate the life expectancy of a person who is 30 years old.

80. *Data Analysis*   An instructor gives 20-point quizzes and 100-point exams in a mathematics course. The average quiz and test scores for six students are given as ordered pairs (x, y) where x is the average quiz score and y is the average test score. The ordered pairs are $(18, 87)$, $(10, 55)$, $(19, 96)$, $(16, 79)$, $(13, 76)$, and $(15, 82)$.

(a) Plot the points.

(b) Use a ruler to sketch the "best-fitting" line through the points.

(c) Find an equation of the line sketched in part (b).

(d) Use the equation in part (c) to estimate the average test score for a person with an average quiz score of 17.

81. *Depth Markers*   A swimming pool is 40 feet long, 20 feet wide, 4 feet deep at the shallow end, and 9 feet deep at the deep end. Position the side of the pool on a rectangular coordinate system as shown in the figure and find an equation of the line representing the edge of the inclined bottom of the pool. Use this equation to determine the distances from the deep end at which markers must be placed to indicate each 1-foot change in the depth of the pool.

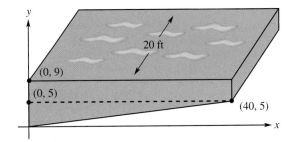

Explaining Concepts

82. Answer parts (d)–(f) of Motivating the Chapter on page 121.

83. Can any pair of points on a line be used to determine the equation of the line? Explain.

84. Write the point-slope form, the slope-intercept form, and the general form of an equation of a line.

85. In the equation $y = 3x + 5$, what does the 3 represent? What does the 5 represent?

86. In the equation of a vertical line, the variable y is missing. Explain why.

2.5   Relations and Functions

Objectives

1 Identify the domain and range of a relation.

2 Determine if a relation is a function by inspection.

3 Use function notation and evaluate a function.

4 Identify the domain and range of a function.

1 Identify the domain and range of a relation.

Relations

Many everyday occurrences involve two quantities that are paired or matched with each other by some rule of correspondence. The mathematical term for such a correspondence is a **relation.**

▶ Definition of a Relation

A **relation** is any set of ordered pairs. The set of first components in the ordered pairs is the **domain** of the relation and the set of second components is the **range** of the relation.

Example 1   Analyzing a Relation

Find the domain and range of the relation $\{(0, 1), (1, 3), (2, 5), (3, 5), (0, 3)\}$.

Solution

The domain is the set of all first components of the relation, and the range is the set of all second components.

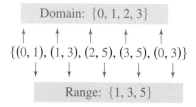

A graphical representation of this relation is given in Figure 2.38.

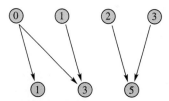

Figure 2.38

2 Determine if a relation is a function by inspection.

Functions

In modeling real-life situations, you will work with a special type of relation called a function. A **function** is a relation in which no two ordered pairs have the same first component and different second components. For instance, $(2, 3)$ and $(2, 4)$ could not be ordered pairs of a function.

▶ **Definition of a Function**

A **function** f from a set A to a set B is a rule of correspondence that assigns to each element x in the set A exactly one element y in the set B.

The set A is called the **domain** (or set of inputs) of the function f, and the set B contains the **range** (or set of outputs) of the function.

The rule of correspondence for a function establishes a set of "input-output" ordered pairs of the form (x, y), where x is an input and y is the corresponding output. In some cases, the rule may generate only a finite set of ordered pairs, whereas in other cases the rule may generate an infinite set of ordered pairs.

Example 2 Input-Output Ordered Pairs for Functions

Write a set of ordered pairs that represents the rule of correspondence.

a. Winner of the Super Bowl in 1996, 1997, 1998, and 1999

b. The squares of all positive integers less than 7

c. The squares of all real numbers

d. The equation $y = x - 2$

Solution

a. For the function that pairs the year from 1996 to 1999 with the winner of the Super Bowl, each ordered pair is of the form (year, winner).

{(1996, Cowboys), (1997, Packers), (1998, Broncos), (1999, Broncos)}

b. For the function that pairs the positive integers that are less than 7 with their squares, each ordered pair is of the form (n, n^2).

$\{(1, 1), (2, 4), (3, 9), (4, 16), (5, 25), (6, 36)\}$

c. For the function that pairs each real number with its square, each ordered pair is of the form (x, x^2).

{All points (x, x^2), where x is a real number}

d. For the function given by $y = x - 2$, each ordered pair is of the form $(x, x - 2)$.

{All points $(x, x - 2)$, where x is a real number}

In Example 2, the sets in parts (a) and (b) have only finite numbers of ordered pairs, whereas the sets in parts (c) and (d) have an infinite number of ordered pairs.

A function has certain characteristics that distinguish it from a relation. To determine whether a relation is a function, use the following list of characteristics of a function.

▶ **Characteristics of a Function**

1. Each element in the domain A must be matched with an element in the range, which is contained in the set B.

2. Some elements in set B may not be matched with any element in the domain A.

3. Two or more elements of the domain may be matched with the same element in the range.

4. No element of the domain is matched with two different elements in the range.

Example 3   Test for Functions Represented by Ordered Pairs

Let $A = \{a, b, c\}$ and let $B = \{1, 2, 3, 4, 5\}$. Which relations represent a function from A to B?

a. $\{(a, 2), (b, 3), (c, 4)\}$ **b.** $\{(a, 4), (b, 5)\}$ **c.** $\{(a, 5), (b, 5), (c, 2), (a, 1)\}$

d.
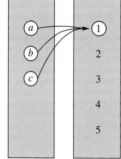

e.
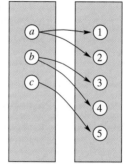

Solution

a. This set of ordered pairs *does* represent a function from A to B. Each element of A is matched with exactly one element of B.

b. This set of ordered pairs *does not* represent a function from A to B. Not every element of A is matched with an element of B.

c. This set of ordered pairs *does not* represent a function from A to B. The element a in A is matched with two elements, 1 and 5, in B.

d. This diagram *does* represent a function from A to B. It does not matter that each element of A is matched with the same element in B.

e. This diagram *does not* represent a function from A to B. The element a in A is matched with *two* elements, 1 and 2, in B. This is also true of b.

Representing functions by sets of ordered pairs is a common practice in the study of *discrete mathematics*, which deals mainly with finite sets of data or with finite subsets of the set of real numbers. In algebra, however, it is more common to represent functions by equations or formulas involving two variables. For instance, the equation

$$y = x^2 \qquad \text{Squaring function}$$

represents the variable y as a function of the variable x. The variable x is the **independent variable** and the variable y is the **dependent variable.** In this context, the domain of the function is the set of all *allowable* real values for the independent variable x, and the range of the function is the *resulting* set of all values taken on by the dependent variable y.

Example 4   Testing for Functions Represented by Equations

Which of the equations represent y as a function of x?

a. $y = x^2 + 1$ **b.** $x - y^2 = 2$ **c.** $-2x + 3y = 4$

Solution

a. For the equation

$$y = x^2 + 1$$

there corresponds just one value of y for each value of x. For instance, when $x = 1$, the value of y is

$$y = 1^2 + 1$$
$$= 2$$

So, y *is* a function of x.

b. By writing the equation $x - y^2 = 2$ in the form

$$y^2 = x - 2$$

you can see that there corresponds *two* values of y for some values of x. For instance, when $x = 3$,

$$y^2 = 3 - 2$$
$$y^2 = 1$$
$$y = 1 \quad \text{or} \quad y = -1$$

So, the solution points $(3, 1)$ and $(3, -1)$ show that y *is not* a function of x.

c. By writing the equation $-2x + 3y = 4$ in the form

$$y = \tfrac{2}{3}x + \tfrac{4}{3}$$

you can see that there corresponds just one value of y for each value of x. For instance, when $x = 2$, the value of y is $\tfrac{4}{3} + \tfrac{4}{3} = \tfrac{8}{3}$. So, y *is* a function of x.

An equation that defines y as a function of x may or may not also define x as a function of y. For instance, the equation in part (a) of Example 4 does not define x as a function of y, but the equation in part (c) does.

3 Use function notation and evaluate a function.

Function Notation

When an equation is used to represent a function, it is convenient to name the function so that it can be easily referenced. For example, the function $y = x^2 + 1$ in Example 4(a) can be given the name "f" and written in **function notation** as

$$f(x) = x^2 + 1.$$

> ▶ **Function Notation**
>
> In the notation $f(x)$:
>
> > f is the **name** of the function,
> >
> > x is the **domain** (or input) value, and
> >
> > $f(x)$ is a **range** (or output) value y for a given x.
>
> The symbol $f(x)$ is read as *the value of f at x* or simply *f of x.*

The process of finding the value of $f(x)$ for a given value of x is called **evaluating a function.** This is accomplished by substituting a given x-value (input) into the equation to obtain the value of $f(x)$ (output). Here is an example.

Function	*x-Value*	*Function Value*
$f(x) = 3 - 4x$	$x = -1$	$f(-1) = 3 - 4(-1) = 3 + 4 = 7$

Although f is often used as a convenient function name and x as the independent variable, you can use other letters. For instance, the equations

$$f(x) = 2x^2 + 5, \quad f(t) = 2t^2 + 5, \quad \text{and} \quad g(s) = 2s^2 + 5$$

all define the same function. In fact, the letters used are simply "placeholders," and this same function is well described by the form

$$f(\quad) = 2(\quad)^2 + 5$$

where the parentheses are used in place of a letter. To evaluate $f(-2)$, simply place -2 in each set of parentheses, as follows.

$$f(-2) = 2(-2)^2 + 5$$
$$= 2(4) + 5$$
$$= 8 + 5$$
$$= 13$$

When evaluating a function, you are not restricted to substituting only numerical values into the parentheses. For instance, the value of $f(3x)$ is

$$f(3x) = 2(3x)^2 + 5$$
$$= 2(9x^2) + 5$$
$$= 18x^2 + 5.$$

Example 5   Evaluating a Function

Let $g(x) = 3x - 4$ and find the following.

a. $g(1)$ **b.** $g(-2)$ **c.** $g(y)$ **d.** $g(x + 1)$ **e.** $g(x) + g(1)$

Solution

a. Replacing x by 1 produces $g(1) = 3(1) - 4 = 3 - 4 = -1$.

b. Replacing x by -2 produces $g(-2) = 3(-2) - 4 = -6 - 4 = -10$.

c. Replacing x by y produces $g(y) = 3(y) - 4 = 3y - 4$.

d. Replacing x by $(x + 1)$ produces $g(x + 1) = 3(x + 1) - 4 = 3x + 3 - 4 = 3x - 1$.

e. Using the result of part (a),

$$g(x) + g(1) = (3x - 4) + (-1) = 3x - 4 - 1 = 3x - 5.$$

Study Tip

Note that

$$g(x + 1) \neq g(x) + g(1).$$

In general, $g(a + b)$ is not equal to $g(a) + g(b)$.

Sometimes a function is defined by more than one equation, each of which is given a portion of the domain. Such a function is called a **piecewise-defined function.** To evaluate a piecewise-defined function f for a given value of x, first determine the portion of the domain in which the x-value lies and then use the corresponding equation to evaluate f. This is illustrated in Example 6.

Example 6   A Piecewise-Defined Function

Let $f(x) = \begin{cases} x^2 + 1, & \text{if } x < 0 \\ x - 2, & \text{if } x \geq 0 \end{cases}$. Find the following.

a. $f(-1)$ **b.** $f(0)$ **c.** $f(-2)$ **d.** $f(-3) + f(4)$

Solution

a. Because $x = -1 < 0$, use $f(x) = x^2 + 1$ to obtain

$$f(-1) = (-1)^2 + 1 = 1 + 1 = 2.$$

b. Because $x = 0 \geq 0$, use $f(x) = x - 2$ to obtain

$$f(0) = (0) - 2 = -2.$$

c. Because $x = -2 < 0$, use $f(x) = x^2 + 1$ to obtain

$$f(-2) = (-2)^2 + 1 = 4 + 1 = 5.$$

d. Because $x = -3 < 0$, use $f(x) = x^2 + 1$ to obtain

$$f(-3) = (-3)^2 + 1 = 9 + 1 = 10.$$

Because $x = 4 \geq 0$, use $f(x) = x - 2$ to obtain

$$f(4) = (4) - 2 = 2.$$

So, $f(-3) + f(4) = 10 + 2 = 12$.

4 Identify the domain and range of a function.

Finding the Domain and Range of a Function

The domain of a function may be explicitly described along with the function, or it may be *implied* by the expression used to define the function. The **implied domain** is the set of all real numbers (inputs) that yield real number values for the function. For instance, the function given by

$$f(x) = \frac{1}{x-3}$$ Domain: all $x \neq 3$

has an implied domain that consists of all real values of x other than $x = 3$. The value $x = 3$ is excluded from the domain because division by zero is undefined. Another common type of implied domain is that used to avoid even roots of negative numbers. For instance, the function given by

$$f(x) = \sqrt{x}$$ Domain: all $x \geq 0$

is defined only for $x \geq 0$. So, its implied domain is the interval $[0, \infty)$. More will be said about the domains of square root functions in Chapter 5.

Example 7   Finding the Domain of a Function

Find the domain of each function.

a. $f(x) = \sqrt{2x - 6}$ **b.** $g(x) = \dfrac{4x}{(x-1)(x+3)}$

Solution

a. The domain of f consists of all x such that $2x - 6 \geq 0$. Solving this inequality yields

$2x - 6 \geq 0$	Original equation
$2x \geq 6$	Add 6 to both sides.
$x \geq 3$	Divide both sides by 2.

So, the domain consists of all real numbers x such that $x \geq 3$.

b. The domain of g consists of all x such that the denominator is not equal to zero. The denominator will be equal to zero when either factor of the denominator is zero.

First Factor

$x - 1 = 0$	Set the first factor equal to zero.
$x = 1$	Add 1 to both sides.

Second Factor

$x + 3 = 0$	Set the second factor equal to zero.
$x = -3$	Subtract 3 from both sides.

So, the domain consists of all real numbers x such that $x \neq 1$ and $x \neq -3$.

| Example 8 | Finding the Domain and Range of a Function |

Find the domain and range of each function.

a. f: $\{(-3, 0), (-1, 2), (0, 4), (2, 4), (4, -1)\}$

b. Area of a circle: $A = \pi r^2$

Solution

a. The domain of f consists of all first coordinates in the set of ordered pairs. The range consists of all second coordinates in the set of ordered pairs. So, the domain is

$$\text{Domain} = \{-3, -1, 0, 2, 4\}$$

and the range is

$$\text{Range} = \{0, 2, 4, -1\}.$$

b. For the area of a circle, you must choose nonnegative values for the radius r. So, the domain is the set of all real numbers r such that $r \geq 0$. The range is therefore the set of all real numbers A such that $A \geq 0$.

Note in Example 8(b) that the domain of a function can be implied by a physical context. For instance, from the equation $A = \pi r^2$, we would have no strictly mathematical reason to restrict r to positive values. However, because we know that this function represents the area of a circle, we conclude that the radius must be positive.

| Discussing the Concept | Determining Relationships That Are Functions |

Compile a list of statements describing relationships in everyday life. For each statement, identify the dependent and independent variables and discuss whether the statement *is* a function or *is not* a function and why. Here are two examples.

a. In the statement, "The number of ceramic tiles required to floor a kitchen is a function of the floor's area," the dependent variable is the required number of ceramic tiles and the independent variable *is* the area of the floor. This statement *is* a mathematically correct use of the word "function" because for each possible floor area there corresponds exactly one number of tiles needed to do the job.

b. In the statement, "Interest rates are a function of economic conditions," the dependent variable is interest rates and the independent variable is economic conditions. This statement *is not* a mathematically correct use of the word "function" because "economic conditions" is ambiguous; it is difficult to tell if one set of economic conditions would always result in the same interest rates.

2.5 Exercises

Integrated Review   *Concepts, Skills, and Problem Solving*

Keep mathematically in shape by doing these exercises *before* the problems of this section.

Properties and Definitions

1. If $a < b$ and $b < c$, then what is the relationship between a and c? Name this property of inequalities.

2. Demonstrate the Multiplicative Property of Equality for the equation $9x = 36$.

3. Use inequality notation to write the statement, "y is no more than 45."

4. Use inequality notation to write the statement, "x is at least 15."

Simplifying Expressions

In Exercises 5–10, simplify the expression.

5. $6y - 3x + 3x - 10y$

6. $8(x - 2) - 3(x - 2)$

7. $\frac{2}{3}t - \frac{5}{8} + \frac{5}{6}t$

8. $\frac{3}{8}x - \frac{1}{12}x + 8$

9. $3x^2 - 5x + 3 + 28x - 33x^2$

10. $4x^3 - 3x^2y + 4xy^2 + 15x^2y + y^3$

Problem Solving

11. Two and one-half cups of flour are required to make one batch of cookies. How many cups are required to make $3\frac{1}{2}$ batches?

12. The gasoline-to-oil ratio of a two-cycle engine is 32 to 1. Determine the amount of gasoline required to produce a mixture that has $\frac{1}{2}$ pint of oil.

Developing Skills

In Exercises 1–4, give the domain and the range of the relation. Then draw a graphical representation of the relation. See Example 1.

1. $\{(-2, 0), (0, 1), (1, 4), (0, -1)\}$

2. $\{(3, 10), (4, 5), (6, -2), (8, 3)\}$

3. $\{(0, 0), (4, -3), (2, 8), (5, 5), (6, 5)\}$

4. $\{(-3, 6), (-3, 2), (-3, 5)\}$

In Exercises 5–10, write a set of ordered pairs that represents the rule of correspondence. See Example 2.

5. In a given week, a salesperson travels a distance d in t hours at an average speed of 50 miles per hour. The travel times for each day are 3 hours, 2 hours, 8 hours, 6 hours, and $\frac{1}{2}$ hour.

6. A court stenographer translates and types a court record of w words for t minutes at a rate of 60 words per minute. The amounts of time spent for each page of the court record are 8 minutes, 10 minutes, 7.5 minutes, and 4 minutes.

7. The cubes of all positive integers less than 8

8. The cubes of all integers greater than -2 and less than 5

9. The winners of the World Series from 1995 to 1998

10. The men inaugurated as president of the United States in 1977, 1981, 1985, 1989, 1993, and 1997

In Exercises 11–22, decide whether the relation is a function.

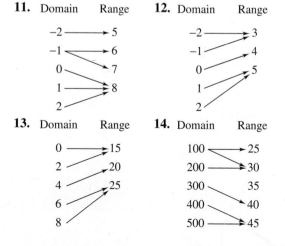

11. Domain   Range
    $-2 \longrightarrow 5$
    $-1 \longrightarrow 6$
    $0 \longrightarrow 7$
    $1 \longrightarrow 8$
    2

12. Domain   Range
    $-2 \longrightarrow 3$
    $-1 \longrightarrow 4$
    $0 \longrightarrow 5$
    1
    2

13. Domain   Range
    $0 \longrightarrow 15$
    $2 \longrightarrow 20$
    $4 \longrightarrow 25$
    6
    8

14. Domain   Range
    $100 \longrightarrow 25$
    $200 \longrightarrow 30$
    $300 \quad 35$
    $400 \longrightarrow 40$
    $500 \longrightarrow 45$

15. Domain   Range

16. Domain   Range

| | (Percent of single women |
| (Year) | in the labor force) |

1992 ──────────→ 66.2
1993 ──────────→ 66.7
1994 ──────────→ 66.8
1995 ──────────→ 67.1
1996 ─────

17. Domain   Range

60 Minutes
CBS ───────→ Chicago Hope
Dan Rather
20/20
ABC ───────→ The Practice
Peter Jennings

18.   Domain   Range

60 Minutes
Chicago Hope ───────→ CBS
Dan Rather
20/20
The Practice ───────→ ABC
Peter Jennings

19.

| Input value | 0 | 1 | 2 | 3 | 4 |
| Output value | 0 | 1 | 4 | 9 | 16 |

20.

| Input value | 0 | 1 | 2 | 1 | 0 |
| Output value | 1 | 8 | 12 | 15 | 20 |

21.

| Input value | 4 | 7 | 9 | 7 | 4 |
| Output value | 2 | 4 | 6 | 8 | 10 |

22.

| Input value | 0 | 2 | 4 | 6 | 8 |
| Output value | 5 | 5 | 5 | 5 | 5 |

In Exercises 23 and 24, determine which sets of ordered pairs represent a function from *A* to *B*. See Example 3.

23. $A = \{0, 1, 2, 3\}$ and $B = \{-2, -1, 0, 1, 2\}$
   (a) $\{(0, 1), (1, -2), (2, 0), (3, 2)\}$
   (b) $\{(0, -1), (2, 2), (1, -2), (3, 0), (1, 1)\}$
   (c) $\{(0, 0), (1, 0), (2, 0), (3, 0)\}$
   (d) $\{(0, 2), (3, 0), (1, 1)\}$

24. $A = \{1, 2, 3\}$ and $B = \{9, 10, 11, 12\}$
   (a) $\{(1, 10), (3, 11), (3, 12), (2, 12)\}$
   (b) $\{(1, 10), (2, 11), (3, 12)\}$
   (c) $\{(1, 10), (1, 9), (3, 11), (2, 12)\}$
   (d) $\{(3, 9), (2, 9), (1, 12)\}$

In Exercises 25–28, show that both ordered pairs are solutions of the equation and explain why this implies that *y* is not a function of *x*.

25. $x^2 + y^2 = 25, \ (0, 5), (0, -5)$
26. $x^2 + 4y^2 = 16, \ (0, 2), (0, -2)$
27. $|y| = x + 2, \ (1, 3), (1, -3)$
28. $|y - 2| = x, \ (2, 4), (2, 0)$

In Exercises 29–34, explain why the equation represents *y* as a function of *x*. See Example 4.

29. $y = 10x + 12$
30. $y = 3 - 8x$
31. $3x + 7y - 2 = 0$
32. $x - 9y + 3 = 0$
33. $y = x(x - 10)$
34. $y = (x + 2)^2 + 3$

In Exercises 35–40, fill in the blank and simplify.

35. $f(x) = 3x + 5$
   (a) $f(2) = 3(\quad) + 5$
   (b) $f(-2) = 3(\quad) + 5$
   (c) $f(k) = 3(\quad) + 5$
   (d) $f(k + 1) = 3(\quad) + 5$

36. $f(x) = 6 - 2x$
   (a) $f(3) = 6 - 2(\quad)$
   (b) $f(-4) = 6 - 2(\quad)$
   (c) $f(n) = 6 - 2(\quad)$
   (d) $f(n - 2) = 6 - 2(\quad)$

37. $f(x) = 3 - x^2$

    (a) $f(0) = 3 - (\quad)^2$

    (b) $f(-3) = 3 - (\quad)^2$

    (c) $f(m) = 3 - (\quad)^2$

    (d) $f(2t) = 3 - (\quad)^2$

38. $f(x) = \sqrt{x + 8}$

    (a) $f(1) = \sqrt{(\quad) + 8}$

    (b) $f(-4) = \sqrt{(\quad) + 8}$

    (c) $f(h) = \sqrt{(\quad) + 8}$

    (d) $f(h - 8) = \sqrt{(\quad) + 8}$

39. $f(x) = \dfrac{x}{x + 2}$

    (a) $f(3) = \dfrac{(\quad)}{(\quad) + 2}$

    (b) $f(-4) = \dfrac{(\quad)}{(\quad) + 2}$

    (c) $f(s) = \dfrac{(\quad)}{(\quad) + 2}$

    (d) $f(s - 2) = \dfrac{(\quad)}{(\quad) + 2}$

40. $f(x) = \dfrac{2x}{x - 7}$

    (a) $f(2) = \dfrac{2(\quad)}{(\quad) - 7}$

    (b) $f(-3) = \dfrac{2(\quad)}{(\quad) - 7}$

    (c) $f(t) = \dfrac{2(\quad)}{(\quad) - 7}$

    (d) $f(t + 5) = \dfrac{2(\quad)}{(\quad) - 7}$

In Exercises 41–56, evaluate the function as indicated, and simplify. See Examples 5 and 6.

41. $f(x) = 12x - 7$

    (a) $f(3)$ (b) $f\left(\frac{3}{2}\right)$

    (c) $f(a) + f(1)$ (d) $f(a + 1)$

42. $f(x) = 3 - 7x$

    (a) $f(-1)$ (b) $f\left(\frac{1}{2}\right)$

    (c) $f(t) + f(-2)$ (d) $f(2t - 3)$

43. $g(x) = 2 - 4x + x^2$

    (a) $g(4)$ (b) $g(0)$

    (c) $g(2y)$ (d) $g(4) + g(6)$

44. $h(x) = x^2 - 2x$

    (a) $h(2)$ (b) $h(0)$

    (c) $h(1) - h(-4)$ (d) $h(4t)$

45. $f(x) = \sqrt{x + 5}$

    (a) $f(-1)$ (b) $f(4)$

    (c) $f(z - 5)$ (d) $f(5z)$

46. $h(x) = \sqrt{2x - 3}$

    (a) $h(4)$ (b) $h(2)$

    (c) $h(4n)$ (d) $h(n + 2)$

47. $g(x) = 8 - |x - 4|$

    (a) $g(0)$ (b) $g(8)$

    (c) $g(16) - g(-1)$ (d) $g(x - 2)$

48. $g(x) = \dfrac{|x + 1|}{x + 1}$

    (a) $g(2)$ (b) $g\left(-\frac{1}{3}\right)$

    (c) $g(-4)$ (d) $g(3) + g(-5)$

49. $f(x) = \dfrac{3x}{x - 5}$

    (a) $f(0)$ (b) $f\left(\frac{5}{3}\right)$

    (c) $f(2) - f(-1)$ (d) $f(x + 4)$

50. $f(x) = \dfrac{x + 2}{x - 3}$

    (a) $f(-3)$ (b) $f\left(-\frac{3}{2}\right)$

    (c) $f(4) + f(8)$ (d) $f(x - 5)$

51. $f(x) = \begin{cases} x + 8, & \text{if } x < 0 \\ 10 - 2x, & \text{if } x \geq 0 \end{cases}$

    (a) $f(4)$ (b) $f(-10)$

    (c) $f(0)$ (d) $f(6) - f(-2)$

52. $f(x) = \begin{cases} -x, & \text{if } x \leq 0 \\ 6 - 3x, & \text{if } x > 0 \end{cases}$

    (a) $f(0)$ (b) $f\left(-\frac{3}{2}\right)$

    (c) $f(4)$ (d) $f(-2) + f(25)$

53. $h(x) = \begin{cases} 4 - x^2, & \text{if } x \leq 2 \\ x - 2, & \text{if } x > 2 \end{cases}$

    (a) $h(2)$ (b) $h\left(-\frac{3}{2}\right)$

    (c) $h(5)$ (d) $h(-3) + h(7)$

54. $f(x) = \begin{cases} x^2, & \text{if } x < 1 \\ x^2 - 3x + 2, & \text{if } x \geq 1 \end{cases}$

   (a) $f(1)$ (b) $f(-1)$

   (c) $f(2)$ (d) $f(-3) + f(3)$

55. $f(x) = 2x + 5$

   (a) $\dfrac{f(x+2) - f(2)}{x}$ (b) $\dfrac{f(x-3) - f(3)}{x}$

56. $f(x) = 3x + 4$

   (a) $\dfrac{f(x+1) - f(1)}{x}$ (b) $\dfrac{f(x-5) - f(5)}{x}$

In Exercises 57–68, find the domain of the function. See Example 7.

57. $f(x) = 5 - 2x$ **58.** $h(x) = 4x - 3$

59. $f(x) = \dfrac{2x}{x - 3}$ **60.** $g(x) = \dfrac{x + 5}{x + 4}$

61. $f(t) = \dfrac{t + 3}{t(t + 2)}$ **62.** $g(s) = \dfrac{s - 2}{(s - 6)(s - 10)}$

63. $g(x) = \sqrt{x + 4}$ **64.** $f(x) = \sqrt{2 - x}$

65. $f(x) = \sqrt{2x - 1}$ **66.** $G(x) = \sqrt{8 - 3x}$

67. $f(t) = |t - 4|$ **68.** $f(x) = |x + 3|$

In Exercises 69–76, find the domain and range of the function. See Example 8.

69. $f: \{(0, 0), (2, 1), (4, 8), (6, 27)\}$

70. $f: \{(-3, 4), (-1, 3), (2, 0), (5, -2)\}$

71. $f: \left\{\left(-3, -\frac{17}{2}\right), \left(-1, -\frac{5}{2}\right), (4, 2), (10, 11)\right\}$

72. $f: \left\{\left(\frac{1}{2}, 4\right), \left(\frac{3}{4}, 5\right), (1, 6), \left(\frac{5}{4}, 7\right)\right\}$

73. Circumference of a circle: $C = 2\pi r$

74. Area of a square of side s: $A = s^2$

75. Area of a circle with radius r: $A = \pi r^2$

76. Volume of a sphere with radius r: $V = \frac{4}{3}\pi r^3$

Solving Problems

77. *Geometry*   Express the perimeter P of a square as a function of the length x of one of its sides.

78. *Geometry*   Express the surface area S of a cube as a function of the length x of one of its edges.

79. *Geometry*   Express the volume V of a cube as a function of the length x of one of its edges.

80. *Geometry*   Express the length L of the diagonal of a square as a function of the length x of one of its sides.

81. *Distance*   A plane is flying at a speed of 230 miles per hour. Express the distance d traveled by the plane as a function of time t in hours.

82. *Cost*   The inventor of a new game believes that the variable cost for producing the game is $1.95 per unit and the fixed costs are $8000. Write the total cost C as a function of x, the number of games produced.

83. *Geometry*   An open box is to be made from a square piece of material 24 inches on a side by cutting equal squares from the corners and turning up the sides (see figure). Write the volume V of the box as a function of x.

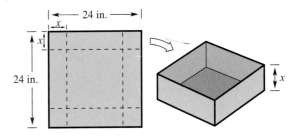

Figure for 83

84. *Geometry*   Strips of width x are cut from the four sides of a square that is 32 inches on a side (see figure). Write the area A of the remaining square as a function of x.

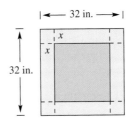

85. *Geometry* Strips of width x are cut from two adjacent sides of a square that is 32 inches on a side (see figure). Write the area A of the remaining square as a function of x.

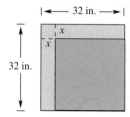

86. *Profit* The marketing department of a business has determined that the profit for selling x units of a product is approximated by the model

$$P(x) = 50\sqrt{x} - 0.5x - 500.$$

Find (a) $P(1600)$ and (b) $P(2500)$.

87. *Safe Load* A solid rectangular beam has a height of 6 inches and a width of 4 inches. The safe load S of the beam with the load at the center is a function of its length L and is approximated by the model

$$S(L) = \frac{128{,}160}{L},$$

where S is measured in pounds and L is measured in feet. Find (a) $S(12)$ and (b) $S(16)$.

88. *Wages* A wage earner is paid $12.00 per hour for regular time and time-and-a-half for overtime. The weekly wage function is

$$W(h) = \begin{cases} 12h, & 0 < h \le 40 \\ 18(h - 40) + 480, & h > 40, \end{cases}$$

where h represents the number of hours worked in a week.

(a) Evaluate $W(30)$, $W(40)$, $W(45)$, and $W(50)$.

(b) Could you use values of h for which $h < 0$ in this model? Why or why not?

Data Analysis In Exercises 89 and 90, use the graph, which shows the numbers of students (in millions) enrolled at all levels in public and private schools in the United States. (Source: U.S. National Center for Education Statistics)

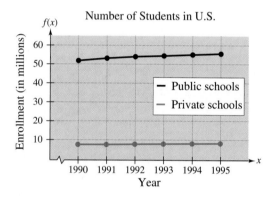

89. Is the public school enrollment a function of the year? Is the private school enrollment a function of the year? Explain.

90. Let $f(x)$ represent the number of public school students in year x. Estimate $f(1993)$.

Explaining Concepts

91. Answer parts (g)–(i) of Motivating the Chapter on page 121.

In Exercises 92 and 93, determine whether the statements use the word *function* in ways that are *mathematically* correct.

92. (a) The sales tax on a purchased item is a function of the selling price.

(b) Your score on the next algebra exam is a function of the number of hours you study the night before the exam.

93. (a) The amount in your savings account is a function of your salary.

(b) The speed at which a free-falling baseball strikes the ground is a function of the height from which it was dropped.

94. Explain the difference between a relation and a function.

95. Is every relation a function? Explain.

96. In your own words, explain the meanings of *domain* and *range*.

97. Describe an advantage of function notation.

2.6   Graphs of Functions

Objectives

1 Sketch the graph of a function on a rectangular coordinate system.

2 Identify the graphs of basic functions.

3 Use the Vertical Line Test to determine if a graph represents a function.

4 Identify transformations of the graph of a function and sketch their graphs.

---

1 Sketch the graph of a function on a rectangular coordinate system.

The Graph of a Function

Consider a function f whose domain and range are the set of real numbers. The **graph** of f is the set of ordered pairs $(x, f(x))$, where x is in the domain of f.

$x = x$-coordinate of the ordered pair

$f(x) = y$-coordinate of the ordered pair

Figure 2.39 shows a typical graph of such a function.

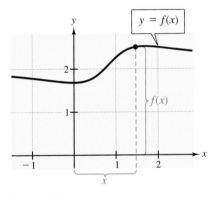

Figure 2.39

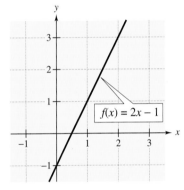

Figure 2.40

Example 1   Sketching the Graph of a Function

Sketch the graph of $f(x) = 2x - 1$.

Solution

Another way to write this function is $y = 2x - 1$. Using the methods you learned in Sections 2.2 and 2.3, you can sketch the graph of the function, as shown in Figure 2.40.

In Example 1, the (implied) domain of the function is the set of all real numbers. When writing the equation of a function, we sometimes choose to restrict its domain by writing a condition to the right of the equation. For instance, the domain of the function

$$f(x) = 4x + 5, \quad x \geq 0$$

is the set of all nonnegative real numbers (all $x \geq 0$).

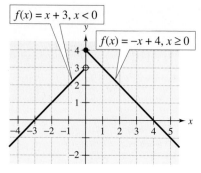

Figure 2.41

| Example 2 | Sketching the Graph of a Piecewise Function |

Sketch the graph of $f(x) = \begin{cases} x + 3, & x < 0 \\ -x + 4, & x \ge 0 \end{cases}$.

Solution

Begin by graphing $f(x) = x + 3$ for $x < 0$, as shown in Figure 2.41. You will recognize that this is the graph of the line $y = x + 3$ with the restriction that the x-values are negative. Because $x = 0$ is not in the domain, the right endpoint of the line is an open dot. Then graph $f(x) = -x + 4$ for $x \ge 0$ on the same set of coordinate axes, as shown in Figure 2.41. This is the graph of the line $y = -x + 4$ with the restriction that the x-values are nonnegative. Because $x = 0$ is in the domain, the left endpoint of the line is a solid dot.

2 Identify the graphs of basic functions.

Graphs of Basic Functions

To become good at sketching the graphs of functions, it helps to be familiar with the graphs of some basic functions. The functions shown in Figure 2.42, and variations of them, occur frequently in applications.

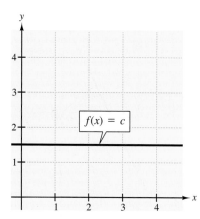

(a) Constant function

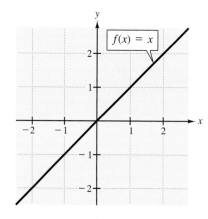

(b) Identity function

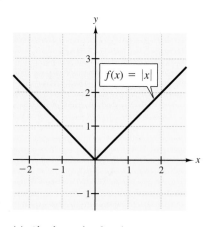

(c) Absolute value function

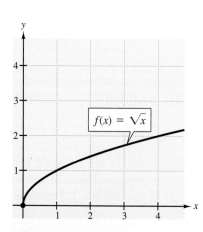

(d) Square root function

Figure 2.42

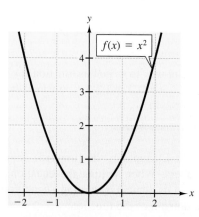

(e) Squaring function

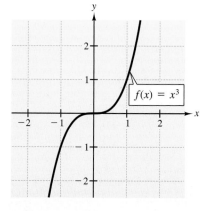

(f) Cubing function

3 Use the Vertical Line Test to determine if a graph represents a function.

The Vertical Line Test

By the definition of a function, at most one y-value corresponds to a given x-value. This implies that any vertical line can intersect the graph of a function at most once.

Study Tip

The Vertical Line Test provides you with an easy way to determine whether an equation represents y as a function of x. If the graph of an equation has the property that no vertical line intersects the graph at two (or more) points, then the equation represents y as a function of x. On the other hand, if you can find a vertical line that intersects the graph at two (or more) points, then the equation does not represent y as a function of x, because there are two (or more) values of y that correspond to certain values of x.

▶ **Vertical Line Test for Functions**

A set of points on a rectangular coordinate system is the graph of y as a function of x if and only if no vertical line intersects the graph at more than one point.

| Example 3 | Using the Vertical Line Test |

Decide whether each equation represents y as a function of x.

a. $y = x^2 - 3x + \frac{1}{4}$

b. $x = y^2 - 1$

c. $x = y^3$

Solution

a. From the graph of the equation in Figure 2.43(a), you can see that every vertical line intersects the graph at most once. So, by the Vertical Line Test, the equation does represent y as a function of x.

b. From the graph of the equation in Figure 2.43(b), you can see that a vertical line intersects the graph twice. So, by the Vertical Line Test, the equation does not represent y as a function of x.

c. From the graph of the equation in Figure 2.43(c), you can see that every vertical line intersects the graph at most once. So, by the Vertical Line Test, the equation does represent y as a function of x.

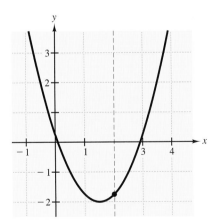

(a) Graph of a function of x.
Vertical line intersects once.

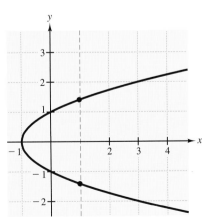

(b) Not a graph of a function of x.
Vertical line intersects twice.

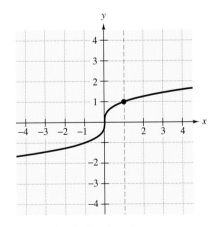

(c) Graph of a function of x.
Vertical line intersects once.

Figure 2.43

Transformations of Graphs of Functions

4 Identify transformations of the graph of a function and sketch their graphs.

Many functions have graphs that are simple transformations of the basic graphs shown in Figure 2.42. The following list summarizes the various types of **vertical** and **horizontal shifts** of the graphs of functions.

> ▶ **Vertical and Horizontal Shifts**
>
> Let c be a positive real number. **Vertical** and **horizontal shifts** of the graph of the function $y = f(x)$ are represented as follows.
>
> 1. Vertical shift c units **upward:** $\qquad\qquad h(x) = f(x) + c$
>
> 2. Vertical shift c units **downward:** $\qquad\quad h(x) = f(x) - c$
>
> 3. Horizontal shift c units to the **right:** $\qquad h(x) = f(x - c)$
>
> 4. Horizontal shift c units to the **left:** $\qquad\; h(x) = f(x + c)$

Note that for a vertical transformation the addition of a positive number c yields a shift upward (in the positive direction) and the subtraction of a positive number c yields a shift downward (in the negative direction). For a horizontal transformation the addition of a positive number c yields a shift to the left (in the negative direction) and the subtraction of a positive number c yields a shift to the right (in the positive direction).

Example 4 Shifts of the Graphs of Functions

Use the graph of $f(x) = x^2$ to sketch the graph of each function.

a. $g(x) = x^2 - 2$ **b.** $h(x) = (x + 3)^2$

Solution

a. Relative to the graph of $f(x) = x^2$, the graph of $g(x) = x^2 - 2$ represents a shift of 2 units *downward*, as shown in Figure 2.44.

b. Relative to the graph of $f(x) = x^2$, the graph of $h(x) = (x + 3)^2$ represents a shift of 3 units to the *left*, as shown in Figure 2.45.

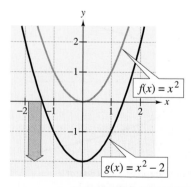

Vertical Shift: Two Units Downward
Figure 2.44

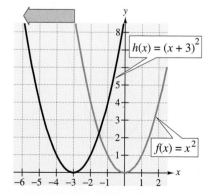

Horizontal Shift: Three Units Left
Figure 2.45

Some graphs can be obtained from *combinations* of vertical and horizontal shifts, as shown in part (b) of the next example.

Example 5   Shifts of the Graphs of Functions

Use the graph of $f(x) = x^3$ to sketch the graph of each function.

a. $g(x) = x^3 + 2$

b. $h(x) = (x - 1)^3 + 2$

Solution

a. Relative to the graph of $f(x) = x^3$, the graph of $g(x) = x^3 + 2$ represents a shift of 2 units *upward*, as shown in Figure 2.46.

b. Relative to the graph of $f(x) = x^3$, the graph of $h(x) = (x - 1)^3 + 2$ represents a shift of 1 unit to the *right*, followed by a shift of 2 units *upward*, as shown in Figure 2.47.

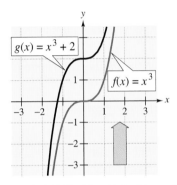

Vertical Shift: Two Units Upward

Figure 2.46

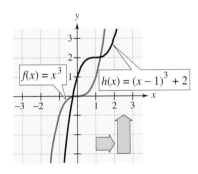

Horizontal Shift: One Unit Right
Vertical Shift: Two Units Upward

Figure 2.47

The second basic type of transformation is a **reflection.** For instance, if you imagine that the *x*-axis represents a mirror, then the graph of

$$h(x) = -x^2$$

is the mirror image (or reflection) of the graph of

$$f(x) = x^2,$$

as shown in Figure 2.48.

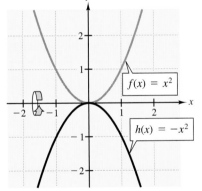

Figure 2.48   *Reflection*

▶ **Reflections in the Coordinate Axes**

Reflections of the graph of $y = f(x)$ are represented as follows.

1. Reflection in the *x*-axis: $h(x) = -f(x)$

2. Reflection in the *y*-axis: $h(x) = f(-x)$

Example 6   Reflections of the Graphs of Functions

Use the graph of $f(x) = \sqrt{x}$ to sketch the graph of each function.

a. $g(x) = -\sqrt{x}$

b. $h(x) = \sqrt{-x}$

Solution

a. Relative to the graph of $f(x) = \sqrt{x}$, the graph of $g(x) = -\sqrt{x} = -f(x)$ represents a *reflection in the x-axis*, as shown in Figure 2.49.

b. Relative to the graph of $f(x) = \sqrt{x}$, the graph of $h(x) = \sqrt{-x} = f(-x)$ represents a *reflection in the y-axis*, as shown in Figure 2.50.

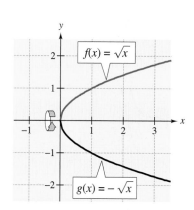

Reflection in x-Axis
Figure 2.49

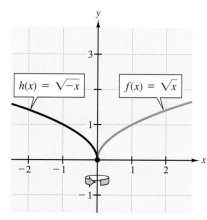

Reflection in y-Axis
Figure 2.50

Discussing the Concept   Constructing Transformations

Use a graphing utility to graph $f(x) = x^2 + 2$. Decide how to alter this function to produce each of the following transformation descriptions. Graph each transformation on the same screen with f; confirm that the transformation moved f as described.

a. The graph of f shifted to the left 3 units.

b. The graph of f shifted downward 5 units.

c. The graph of f shifted upward 1 unit.

d. The graph of f shifted to the right 2 units.

2.6   Exercises

Integrated Review   *Concepts, Skills, and Problem Solving*

Keep mathematically in shape by doing these exercises *before* the problems of this section.

Properties and Definitions

1. $8x \cdot \dfrac{1}{8x} = 1$

2. $3x + 0 = 3x$

3. $-4(x + 10) = -4 \cdot x + (-4)(10)$

4. $5 + (-3 + x) = (5 - 3) + x$

Simplifying Expressions

In Exercises 5–10, simplify the expression.

5. $5x^4(x^2)$

6. $3(x + 1)^2(x + 1)^3$

7. $(-4t)^3$

8. $-(-2x)^4$

9. $(u^2 v)^4$

10. $(3a^2 b)^2 (2b^3)$

Problem Solving

11. A department store is offering a discount of 20% on a sewing machine with a list price of \$239.95. A mail-order catalog has the same machine for \$188.95 plus \$4.32 for shipping. Which is the better bargain?

12. The annual automobile insurance premium for a policyholder is normally \$739. However, after having an accident, the policyholder was charged an additional 30%. What is the new annual premium?

Developing Skills

In Exercises 1–28, sketch the graph of the function. Then determine its domain and range. See Examples 1 and 2.

1. $f(x) = 2x - 7$

2. $f(x) = 3 - 2x$

3. $g(x) = \frac{1}{2}x^2$

4. $h(x) = \frac{1}{4}x^2 - 1$

5. $f(x) = -(x - 1)^2$

6. $g(x) = (x + 2)^2 + 3$

7. $h(x) = x^2 - 6x + 8$

8. $f(x) = -x^2 - 2x + 1$

9. $C(x) = \sqrt{x} - 1$

10. $Q(x) = 4 - \sqrt{x}$

11. $f(t) = \sqrt{t - 2}$

12. $h(x) = \sqrt{4 - x}$

13. $G(x) = 8$

14. $H(x) = -4$

15. $g(s) = s^3 + 1$

16. $f(x) = x^3 - 4$

17. $f(x) = |x + 3|$

18. $g(x) = |x - 1|$

19. $K(s) = |s - 4| + 1$

20. $Q(t) = 1 - |t + 1|$

21. $f(x) = 6 - 3x, \quad 0 \le x \le 2$

22. $f(x) = \frac{1}{3}x - 2, \quad 6 \le x \le 12$

23. $h(x) = x^3, \quad -2 \le x \le 2$

24. $h(x) = 6x - x^2, \quad 0 \le x \le 6$

25. $h(x) = \begin{cases} 2x + 3, & \text{if } x < 0 \\ 3 - x, & \text{if } x \ge 0 \end{cases}$

26. $f(x) = \begin{cases} x + 6, & \text{if } x < 0 \\ 6 - 2x, & \text{if } x \ge 0 \end{cases}$

27. $f(x) = \begin{cases} -x, & \text{if } x \le 0 \\ x^2 - 4x, & \text{if } x > 0 \end{cases}$

28. $h(x) = \begin{cases} 4 - x^2, & \text{if } x \le 2 \\ x - 2, & \text{if } x > 2 \end{cases}$

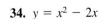 In Exercises 29–32, use a graphing utility to graph the function and find its domain and range.

29. $g(x) = 1 - x^2$

30. $f(x) = |x + 1|$

31. $f(x) = \sqrt{x - 2}$

32. $h(t) = \sqrt{4 - t^2}$

In Exercises 33–40, use the Vertical Line Test to determine whether y is a function of x. See Example 3.

33. $y = \frac{1}{3}x^3$

34. $y = x^2 - 2x$

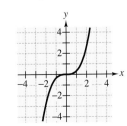

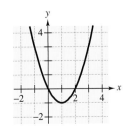

35. $y = (x + 2)^2$

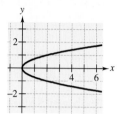

36. $x - 2y^2 = 0$

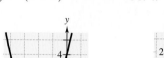

37. $x^2 + y^2 = 16$

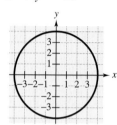

38. $y = x^3 - 3$

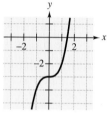

39. $y^2 = x$

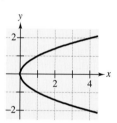

40. $|y| = x$

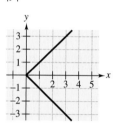

In Exercises 41–44, sketch a graph of the equation. Use the Vertical Line Test to determine whether y is a function of x.

41. $3x - 5y = 15$

42. $y = x^2 + 2$

43. $y^2 = x + 1$

44. $x = y^4$

In Exercises 45–48, match the function with its graph. [The graphs are labeled (a), (b), (c), and (d).]

(a)

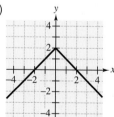

(b)

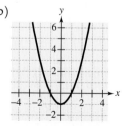

(c)

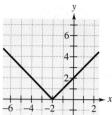

(d)

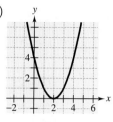

45. $f(x) = x^2 - 1$

46. $f(x) = (x - 2)^2$

47. $f(x) = 2 - |x|$

48. $f(x) = |x + 2|$

In Exercises 49 and 50, select the viewing window that shows the most complete graph of the function.

49. $f(x) = -(x^2 - 20x + 50)$

Xmin = 0	Xmin = 0	Xmin = 15
Xmax = 10	Xmax = 20	Xmax = 30
Xscl = 1	Xscl = 2	Xscl = 2
Ymin = 0	Ymin = -10	Ymin = -10
Ymax = 30	Ymax = 60	Ymax = 60
Yscl = 2	Yscl = 6	Yscl = 5

50. $f(x) = x^4 - 10x^3$

Xmin = -10	Xmin = -5	Xmin = -3
Xmax = 10	Xmax = 5	Xmax = 12
Xscl = 1	Xscl = 1	Xscl = 1
Ymin = -10	Ymin = -10	Ymin = -1200
Ymax = 10	Ymax = 20	Ymax = 400
Yscl = 1	Yscl = 2	Yscl = 100

In Exercises 51 and 52, identify the transformation of f and sketch a graph of the function h. See Examples 4–6.

51. $f(x) = x^2$

(a) $h(x) = x^2 + 2$

(b) $h(x) = x^2 - 4$

(c) $h(x) = (x + 2)^2$

(d) $h(x) = (x - 4)^2$

(e) $h(x) = -x^2$

(f) $h(x) = -x^2 + 4$

(g) $h(x) = (x - 3)^2 + 1$

(h) $h(x) = -(x + 2)^2 - 3$

52. $f(x) = x^3$

(a) $h(x) = x^3 + 3$

(b) $h(x) = x^3 - 5$

(c) $h(x) = (x - 3)^3$

(d) $h(x) = (x + 2)^3$

(e) $h(x) = (-x)^3$

(f) $h(x) = -x^3$

(g) $h(x) = 2 - (x - 1)^3$

(h) $h(x) = (x + 2)^3 - 3$

In Exercises 53–58, identify the transformation of the graph of $f(x) = |x|$ and use a graphing utility to graph h.

53. $h(x) = |x - 5|$

54. $h(x) = |x + 3|$

55. $h(x) = |x| - 5$

56. $h(x) = |-x|$

57. $h(x) = -|x|$

58. $h(x) = 5 - |x|$

In Exercises 59–66, use the graph of $f(x) = x^2$ to write a function that represents the graph.

59.

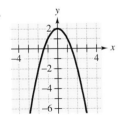

60.

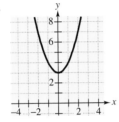

61.

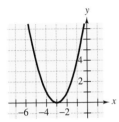

62.

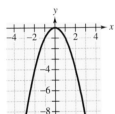

63.

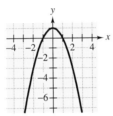

64.

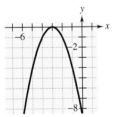

65.

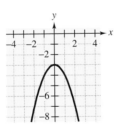

66.

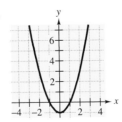

In Exercises 67–72, use the graph of $f(x) = \sqrt{x}$ to write a function that represents the graph.

67.

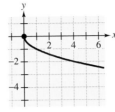

68.

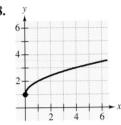

69.

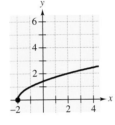

70.

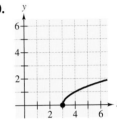

71.

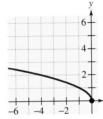

72.

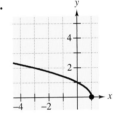

73. Use the graph of f to sketch the graphs.

   (a) $y = f(x) + 2$ (b) $y = -f(x)$

   (c) $y = f(x - 2)$ (d) $y = f(x + 2)$

   (e) $y = f(x) - 1$ (f) $y = f(-x)$

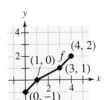

Figure for 73

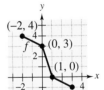

Figure for 74

74. Use the graph of f to sketch the graphs.

   (a) $y = f(x) - 1$ (b) $y = f(x + 1)$

   (c) $y = f(x - 1)$ (d) $y = -f(x - 2)$

   (e) $y = f(-x)$ (f) $y = f(x) + 2$

Solving Problems

75. *Automobile Engines* The percent x of antifreeze that prevents an automobile's engine coolant from freezing down to P degrees Fahrenheit is modeled by

$$P = 26.0 - 0.0242x^2, \quad 20 \le x \le 60.$$

(Source: Standard Handbook for Mechanical Engineers)

(a) Use a graphing utility to graph the model over the specified domain.

(b) Use the graph to approximate the value of x that yields protection against freezing at $-25°$ F.

76. *Profit* The profit P when x units of a product are sold is given by $P(x) = 0.47x - 100$ for x in the interval $0 \le x \le 1000$.

(a) Use a graphing utility to graph the profit function over the specified domain.

(b) Approximately how many units must be sold for the company to break even $(P = 0)$?

(c) Approximately how many units must be sold for the company to make a profit of $300?

77. *Geometry* The perimeter of a rectangle (see figure) is 200 meters.

(a) Show algebraically that the area of the rectangle is given by $A = x(100 - x)$, where x is its length.

(b) Use a graphing utility to graph the area function.

(c) Use the graph to determine the value of x that yields the largest value of A. Interpret the result.

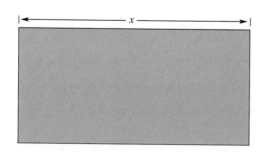

x

78. *Geometry* The length and width of a rectangular flower garden are 40 feet and 30 feet, respectively (see figure). A walkway of width x surrounds the garden.

(a) Write the outside perimeter y of the walkway as a function of x.

(b) Use a graphing utility to graph the function for the perimeter.

(c) Determine the slope of the graph in part (b). For each additional 1-foot increase in the width of the walkway, determine the increase in its outside perimeter.

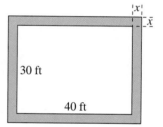

30 ft

40 ft

79. *Civilian Population of the United States* For 1950 through 1998, the civilian population P (in thousands) of the United States can be modeled by

$$P(t) = -5.46t^2 + 2665.56t + 153,363$$

where $t = 0$ represents 1950. (Source: U.S. Bureau of the Census)

(a) Use a graphing utility to graph the function over the appropriate domain.

(b) In the transformation of the population function

$$P_1(t) = -5.46(t + 20)^2 + 2665.56(t + 20) + 153,363,$$

$t = 0$ corresponds to what calendar year? Explain.

(c) Use a graphing utility to graph P_1 over the appropriate domain.

80. *Graphical Reasoning* An electronically controlled thermostat in a home is programmed to lower the temperature automatically during the night. The temperature T, in degrees Fahrenheit, is given in terms of t, the time on a 24-hour clock (see figure).

(a) Explain why T is a function of t.

(b) Find $T(4)$ and $T(15)$.

(c) Suppose the thermostat were reprogrammed to produce a temperature H where $H(t) = T(t - 1)$. Explain how this would change the temperature in the house.

(d) Suppose the thermostat were reprogrammed to produce a temperature H where $H(t) = T(t) - 1$. Explain how this would change the temperature in the house.

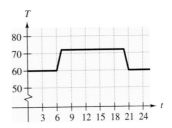

Figure for 80

Explaining Concepts

81. Explain the change in the range of the function $f(x) = 2x$ if the domain is changed from $[0, 2]$ to $[0, 4]$.

82. In your own words, explain how to use the Vertical Line Test.

83. Describe the four types of shifts of the graph of a function.

84. Describe the relationship between the graphs of $f(x)$ and $g(x) = -f(x)$.

85. Describe the relationship between the graphs of $f(x)$ and $g(x) = f(-x)$.

86. Describe the relationship between the graphs of $f(x)$ and $g(x) = f(x - 2)$.

Key Terms

rectangular coordinate system, *p. 122*
Cartesian plane, *p. 122*
x-axis, *p. 122*
y-axis, *p. 122*
origin, *p. 122*
quadrants, *p. 122*
ordered pair, *p. 122*
x-coordinate, *p. 122*
y-coordinate, *p. 122*

Pythagorean Theorem, *p. 127*
Distance Formula, *p. 128*
graph (of an equation), *p. 134*
linear equation, *p. 134*
x-intercept, *p. 137*
y-intercept, *p. 137*
straight-line depreciation, *p. 138*

slope, *p. 142*
slope-intercept form, *p. 146*
point-slope form, *p. 155*
general form, *p. 155*
two-point form, *p. 156*
linear extrapolation, *p. 159*
linear interpolation, *p. 159*
relation, *p. 164*
domain, *pp. 164, 165, 168*

range, *pp. 164, 165, 168*
function, *p. 165*
independent variable, *p. 167*
dependent variable, *p. 167*
function notation, *p. 168*
implied domain, *p. 170*
graph (of a function), *p. 177*

Key Concepts

2.1 Guidelines for verifying solutions

To verify that an ordered pair (x, y) is a solution of an equation with variables x and y, use the following steps.

1. Substitute the values of x and y into the equation.
2. Simplify both sides of the equation.
3. If both sides simplify to the same number, the ordered pair is a solution. If the two sides yield different numbers, the ordered pair is not a solution.

2.2 The point-plotting method of sketching a graph

1. If possible, rewrite the equation by isolating one of the variables.
2. Make up a table of values showing several solution points.
3. Plot these points on a rectangular coordinate system.
4. Connect the points with a smooth curve or line.

2.4 Summary of equations of lines

1. Slope of a line through (x_1, y_1) and (x_2, y_2):

$$m = \frac{y_2 - y_1}{x_2 - x_1}$$

2. General form of equation of line: $ax + by + c = 0$
3. Equation of vertical line: $x = a$
4. Equation of horizontal line: $y = b$
5. Slope-intercept form of equation of line:
$y = mx + b$
6. Point-slope form of equation of line:
$y - y_1 = m(x - x_1)$

7. Parallel lines (equal slopes): $m_1 = m_2$
8. Perpendicular lines (negative reciprocal slopes):

$$m_2 = -\frac{1}{m_1}$$

2.5 Characteristics of a function

1. Each element in the domain A must be matched with an element in the range, which is contained in set B.
2. Some elements in set B may not be matched with any element in the domain A.
3. Two or more elements of the domain may be matched with the same element in the range.
4. No element of the domain is matched with two different elements in the range.

2.6 Vertical Line Test for functions

A set of points on a rectangular coordinate system is the graph of y as a function of x if and only if no vertical line intersects the graph at more than one point.

2.6 Vertical and horizontal shifts

Let c be a positive real number. Vertical and horizontal shifts of the graph of the function $y = f(x)$ are represented as follows.

1. Vertical shift c units upward: $h(x) = f(x) + c$
2. Vertical shift c units downward: $h(x) = f(x) - c$
3. Horizontal shift c units to the right: $h(x) = f(x - c)$
4. Horizontal shift c units to the left: $h(x) = f(x + c)$

2.6 Reflections in the coordinate axes

Reflections of the graph of $y = f(x)$ are represented as:
1. Reflection in the x-axis: $h(x) = -f(x)$
2. Reflection in the y-axis: $h(x) = f(-x)$

REVIEW EXERCISES

Reviewing Skills

2.1 In Exercises 1 and 2, plot the points on a rectangular coordinate system.

1. $(0, -3), \left(\frac{5}{2}, 5\right), (-2, -4)$
2. $\left(1, -\frac{3}{2}\right), \left(-2, 2\frac{3}{4}\right), (5, 10)$

In Exercises 3 and 4, plot the points and connect them with line segments to form the indicated polygon.

3. *Right triangle:* $(1, 1), (12, 9), (4, 20)$
4. *Parallelogram:* $(0, 0), (7, 1), (8, 4), (1, 3)$

In Exercises 5–8, determine the quadrant(s) in which the point is or could be located.

5. $(2, -6)$ **6.** $(-4.8, -2)$
7. $(4, y)$ **8.** $(x, y), \quad xy > 0$

In Exercises 9 and 10, determine whether each ordered pair is a solution of the equation.

9. $y = 4 - \frac{1}{2}x$
  (a) $(4, 2)$ (b) $(-1, 5)$
  (c) $(-4, 0)$ (d) $(8, 0)$

10. $3x - 2y + 18 = 0$
  (a) $(3, 10)$ (b) $(0, 9)$
  (c) $(-4, 3)$ (d) $(-8, 0)$

In Exercises 11–14, use the Distance Formula to determine the distance between the points.

11. $(4, 3), (4, 8)$ **12.** $(2, -5), (6, -5)$
13. $(-5, -1), (1, 2)$ **14.** $(-2, 10), (3, -2)$

2.2 In Exercises 15–18, match the equation with its graph. [The graphs are labeled (a), (b), (c), and (d).]

(a)
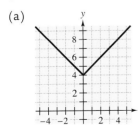

(b)
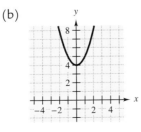

(c)
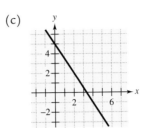

(d)
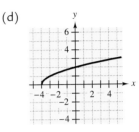

15. $y = 5 - \frac{3}{2}x$ **16.** $y = x^2 + 4$
17. $y = |x| + 4$ **18.** $y = \sqrt{x + 4}$

In Exercises 19–26, sketch the graph of the equation using the point-plotting method.

19. $y = 6 - \frac{1}{3}x$ **20.** $y = \frac{3}{4}x - 2$
21. $3y - 2x - 3 = 0$ **22.** $3x + 4y + 12 = 0$
23. $y = x^2 - 1$ **24.** $y = (x - 2)^2$
25. $y = |x| - 2$ **26.** $y = |x - 3|$

In Exercises 27–34, find the *x*- and *y*-intercepts of the graph.

27. $y = 4x - 6$ **28.** $y = 3x + 9$
29. $7x - 2y = -14$ **30.** $5x + 4y = 10$
31. $y = |x - 5|$ **32.** $y = |x| + 4$
33. $y = |2x + 1| - 5$ **34.** $y = |3 - 6x| - 15$

In Exercises 35–40, use a graphing utility to graph the equation. Approximate any intercepts.

35. $y = (x - 3)^2 - 3$ **36.** $y = \frac{1}{4}(x - 2)^3$
37. $y = -|x - 4| - 7$ **38.** $y = 3 - |x - 3|$
39. $y = \sqrt{3 - x}$ **40.** $y = x - 2\sqrt{x}$

2.3 In Exercises 41–46, determine the slope of the line through the points.

41. $(-1, 1), (6, 3)$ **42.** $(-2, 5), (3, -8)$
43. $(-1, 3), (4, 3)$ **44.** $(7, 2), (7, 8)$
45. $(0, 6), (8, 0)$ **46.** $(0, 0), \left(\frac{7}{2}, 6\right)$

In Exercises 47 and 48, find t such that the three points are collinear. (*Note:* Collinear means that the points lie on the same line.)

47. $(-3, -3), (0, t), (1, 3)$ **48.** $(2, 1), (1, t), (8, 3)$

In Exercises 49–54, a point on a line and the slope of the line are given. Find two additional points on the line.

49. $(2, -4)$
$m = -3$

50. $\left(-4, \frac{1}{2}\right)$
$m = 2$

51. $(3, 1)$
$m = \frac{5}{4}$

52. $\left(-3, -\frac{3}{2}\right)$
$m = -\frac{1}{3}$

53. $(3, 7)$
m is undefined.

54. $(7, -2)$
$m = 0$

In Exercises 55–58, write the equation of the line in slope-intercept form and sketch the line.

55. $5x - 2y - 4 = 0$ **56.** $x - 3y - 6 = 0$

57. $x + 2y - 2 = 0$ **58.** $y - 6 = 0$

In Exercises 59–64, are the lines parallel, perpendicular, or neither?

59. $L_1: y = \frac{3}{2}x + 1$
$L_2: y = \frac{2}{3}x - 1$

60. $L_1: y = 2x - 5$
$L_2: y = 2x + 3$

61. $L_1: y = \frac{3}{2}x - 2$
$L_2: y = -\frac{2}{3}x + 1$

62. $L_1: y = -0.3x - 2$
$L_2: y = 0.3x + 1$

63. $L_1: 2x - 3y - 5 = 0$
$L_2: x + 2y - 6 = 0$

64. $L_1: 4x + 3y - 6 = 0$
$L_2: 3x - 4y - 8 = 0$

2.4 In Exercises 65–72, write an equation of the line passing through the point with the specified slope.

65. $(1, -4)$
$m = 2$

66. $(-5, -5)$
$m = 3$

67. $(-1, 4)$
$m = -4$

68. $(5, -2)$
$m = -2$

69. $\left(\frac{5}{2}, 4\right)$
$m = -\frac{2}{3}$

70. $\left(-2, -\frac{4}{3}\right)$
$m = \frac{3}{2}$

71. $(-6, 5)$
$m = 0$

72. $(7, 8)$
m is undefined.

In Exercises 73–78, write an equation of the line passing through the two points using the point-slope form.

73. $(-6, 0), (0, -3)$ **74.** $(0, 10), (6, 10)$

75. $(-2, -3), (4, 6)$ **76.** $(-10, 2), (4, -7)$

77. $\left(\frac{4}{3}, \frac{1}{6}\right), \left(4, \frac{7}{6}\right)$ **78.** $\left(\frac{5}{2}, 0\right), \left(\frac{5}{2}, 5\right)$

In Exercises 79–82, write equations of the lines through the point that are (a) parallel and (b) perpendicular to the given line.

79. $\left(\frac{3}{5}, -\frac{4}{5}\right)$
$3x + y = 2$

80. $(-1, 5)$
$2x + 4y = 1$

81. $(12, 1)$
$5x = 3$

82. $\left(\frac{3}{8}, 3\right)$
$4x - 3y = 12$

2.5 In Exercises 83–86, determine whether the relation is a function.

83. Domain   Range

84. Domain   Range

85. Domain   Range

86. Domain   Range

In Exercises 87–94, evaluate the function for the specified values of the independent variable and simplify when possible.

87. $f(x) = 4 - \frac{5}{2}x$
(a) $f(-10)$ (b) $f\left(\frac{2}{5}\right)$
(c) $f(t) + f(-4)$ (d) $f(x + h)$

88. $h(x) = x(x - 8)$
(a) $h(8)$ (b) $h(10)$
(c) $h(-3) + h(4)$ (d) $h(4t)$

89. $f(t) = \sqrt{5 - t}$
(a) $f(-4)$ (b) $f(5)$
(c) $f(3)$ (d) $f(5z)$

90. $g(x) = \dfrac{|x + 4|}{4}$
(a) $g(0)$ (b) $g(-8)$
(c) $g(2) - g(-5)$ (d) $g(x - 2)$

91. $f(x) = \begin{cases} -3x, & \text{if } x \le 0 \\ 1 - x^2, & \text{if } x > 0 \end{cases}$

   (a) $f(2)$ (b) $f\left(-\frac{2}{3}\right)$

   (c) $f(1)$ (d) $f(4) - f(3)$

92. $h(x) = \begin{cases} x^3, & \text{if } x \le 1 \\ (x - 1)^2 + 1, & \text{if } x > 1 \end{cases}$

   (a) $h(2)$ (b) $h\left(-\frac{1}{2}\right)$

   (c) $h(0)$ (d) $h(4) - h(3)$

93. $f(x) = 3 - 2x$

   (a) $\dfrac{f(x + 2) - f(2)}{x}$ (b) $\dfrac{f(x - 3) - f(3)}{x}$

94. $f(x) = 7x + 10$

   (a) $\dfrac{f(x + 1) - f(1)}{x}$ (b) $\dfrac{f(x - 5) - f(5)}{x}$

In Exercises 95–98, identify the domain of the function.

95. $h(x) = 4x^2 - 7$ **96.** $g(s) = \dfrac{s + 1}{(s - 1)(s + 5)}$

97. $f(x) = \sqrt{5 - 2x}$ **98.** $f(x) = |x - 6| + 10$

2.6 In Exercises 99–108, sketch the graph of the function.

99. $y = 4 - (x - 3)^2$ **100.** $h(x) = 9 - (x - 2)^2$

101. $y = \sqrt{x} + 2$ **102.** $f(t) = \sqrt{t - 2}$

103. $y = 8 - |x|$ **104.** $f(x) = |x + 1| - 2$

105. $g(x) = 6 - 3x, \quad -2 \le x < 4$

106. $h(x) = x(4 - x), \quad 0 \le x \le 4$

107. $f(x) = \begin{cases} 2 - (x - 1)^2, & \text{if } x < 1 \\ 2 + (x - 1)^2, & \text{if } x \ge 1 \end{cases}$

108. $f(x) = \begin{cases} 2x, & \text{if } x \le 0 \\ x^2 + 1, & \text{if } x > 0 \end{cases}$

In Exercises 109–112, use the Vertical Line Test to determine if the graph represents y as a function of x.

109. $9y^2 = 4x^3$ **110.** $y = 4x^3 - x^4$

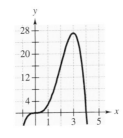

111. $y = x^2(x - 3)$ **112.** $x^3 + y^3 - 6xy = 0$

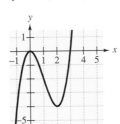
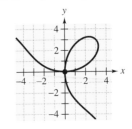

In Exercises 113–116, identify the transformation of the graph of $f(x) = \sqrt{x}$ and sketch the graph of h.

113. $h(x) = -\sqrt{x}$

114. $h(x) = \sqrt{x} + 3$

115. $h(x) = \sqrt{x - 1}$

116. $h(x) = 1 - \sqrt{x + 4}$

In Exercises 117–120, use the graph of $f(x) = x^2$ to write a function that represents the graph.

117.   **118.**

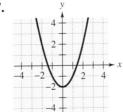
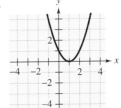

119.   **120.**

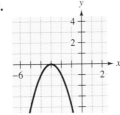
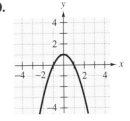

Solving Problems

121. *Slope of a Ramp*   A loading dock ramp rises 3 feet above the ground. The ramp has a slope of $\frac{1}{12}$. What is the length of the ramp?

122. *Road Grade*   When driving down a mountain road, you notice a warning sign indicating a "12% downgrade." This means that the slope of the road is $-\frac{12}{100}$. Over a particular stretch of road, your elevation drops 1500 feet. What is the horizontal change in your position?

123. *Straight-Line Depreciation*   You purchase new commercial washing machines for your laundromat for $20,000. For tax purposes, the washing machines will be depreciated over a 7-year period. At the end of 7 years, the value of the washing machines is expected to be $6000. Find an equation that relates the depreciated value of the washing machines to the number of years since they were purchased. Then sketch the graph of the equation.

124. *Straight-Line Depreciation*   You purchase new commercial clothes dryers for your laundromat for $8000. For tax purposes, the dryers will be depreciated over a 10-year period. At the end of 10 years, the value of the dryers is expected to be $3000. Find an equation that relates the depreciated value of the dryers to the number of years since they were purchased. Then sketch the graph of the equation.

Rocker Arm Construction   In Exercises 125 and 126, consider the rocker arm shown in the figure.

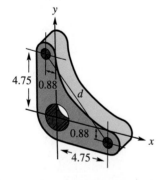

125. Find an equation of the line through the centers of the two small bolt holes in the rocker arm.

126. Find the distance between the centers of the two small bolt holes in the rocker arm.

127. *Wire Length*   A wire 150 inches long is to be cut into four pieces to form a rectangle whose shortest side has a length of x. Express the area A of the rectangle as a function of x. What is the domain of the function?

128. *Wire Length*   A wire 100 inches long is to be cut into four pieces to form a rectangle whose shortest side has a length of x. Express the area A of the rectangle as a function of x. What is the domain of the function?

129. *Velocity of a Ball*   The velocity of a ball thrown upward from ground level is given by $v = -32t + 80$, where t is time in seconds and v is velocity in feet per second.

  (a) Find the velocity when $t = 2$.

  (b) Find the time when the ball reaches its maximum height. (*Hint:* Find the time when $v = 0$.)

  (c) Find the velocity when $t = 3$.

Chapter Test

Take this test as you would take a test in class. After you are done, check your work against the answers given in the back of the book.

1. Determine the quadrant in which the point (x, y) lies if $x > 0$ and $y < 0$.

2. Plot the points $(0, 5)$ and $(3, 1)$. Then find the distance between them.

3. Find the x- and y-intercepts of the graph of the equation $y = -3(x + 1)$.

4. Sketch the graph of the equation $y = |x - 2|$.

5. Find the slope (if possible) of the line passing through each pair of points.
   (a) $(-4, 7), (2, 3)$ (b) $(3, -2), (3, 6)$

6. Sketch the graph of the line passing through the point $(0, -6)$ with slope $m = \frac{4}{3}$.

7. Plot the x- and y-intercepts of the graph of $2x + 5y - 10 = 0$. Use the results to sketch the graph.

8. Write the equation $5x + 3y - 9 = 0$ in slope-intercept form. Find the slope of the line that is perpendicular to this line.

9. Find an equation of the line through the points $(25, -15)$ and $(75, 10)$.

10. Find an equation of the line with slope -2 that passes through the point $(2, -4)$.

11. Find an equation of the vertical line through the point $(-2, 4)$.

12. The graph of $y^2(4 - x) = x^3$ is shown at the left. Does the graph represent y as a function of x? Explain your reasoning.

13. Determine whether the relation represents a function. Explain.
    (a) $\{(2, 4), (-6, 3), (3, 3), (1, -2)\}$ (b) $\{(0, 0), (1, 5), (-2, 1), (0, -4)\}$

14. Evaluate $g(x) = x/(x - 3)$ for the indicated values.
    (a) $g(2)$ (b) $g\left(\frac{7}{2}\right)$ (c) $g(x + 2)$

15. Find the domain of each function.
    (a) $h(t) = \sqrt{9 - t}$ (b) $f(x) = \dfrac{x + 1}{x - 4}$

16. Sketch the graph of the function $g(x) = \sqrt{2 - x}$.

17. Describe the transformation of the graph of $f(x) = x^2$ that would produce the graph of $g(x) = -(x - 2)^2 + 1$.

18. After 4 years, the value of a $26,000 car will have depreciated to $10,000. Write the value V of the car as a linear function of t, the number of years since the car was purchased. When will the car be worth $16,000? Explain your reasoning.

19. Use the graph of $f(x) = |x|$ to write an equation for each graph.
    (a)   (b)   (c)

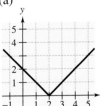
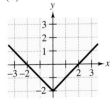
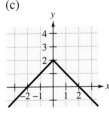

Figure for 12

3

Polynomials and Factoring

Steve Bly/Tony Stone Images

In 1997, over 1.6 million farms were operated by an individual or family, 169,000 farms were operated by a partnership and 84,000 farms were operated by a corporation. (Source: U.S. Dept. of Agriculture, National Agricultural Statistics Service)

A Storage Bin for Drying Grain

A rectangular grain bin has a width that is 5 feet greater than its height and a length that is 2 feet less than three times its height. A screen in the shape of a pyramid is centered at the base of the bin (see figure). Air is pumped into the bin through the screen pyramid in order to dry the grain. The screen pyramid has a height that is 3 feet less than the height of the bin, a width that is 6 feet less than the width of the bin, and a length that is twice the height of the pyramid.

See Section 3.2, Exercise 137

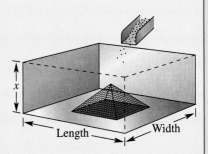

a. Write the dimensions, in feet, of the bin in terms of its height x. Write a polynomial function $V_B(x)$ that represents the volume of the rectangular grain bin before the screen pyramid is inserted.

b. Write the dimensions, in feet, of the screen pyramid in terms of x. Write a polynomial function $V_P(x)$ that represents the volume of the screen pyramid. [The formula for the volume of a pyramid is $V = \frac{1}{3}$ (area of base)(height).]

c. Write a polynomial function $V_S(x)$ that represents the volume of grain that can be stored in the bin when the screen pyramid is in place.

See Section 3.5, Exercise 99

d. If the opening for airflow (the base of the screen pyramid) must be 30 square feet, find the dimensions of the pyramid.

e. Under the conditions of part (d), calculate the maximum volume of grain that can be dried in this bin.

f. Under the conditions of part (d), what is the domain of the volume function $V_S(x)$ from part (c)?

3.1 Adding and Subtracting Polynomials

Objectives

1 Identify the leading coefficient and the degree of a polynomial.

2 Add and subtract polynomials using a vertical format and a horizontal format.

3 Use function notation to represent polynomials in application problems.

1 Identify the leading coefficient and the degree of a polynomial.

Basic Definitions

A **polynomial in x** is an algebraic expression whose terms are all of the form ax^k, where a is any real number and k is a nonnegative integer. The following *are not* polynomials for the reasons stated.

- The expression $2x^{-1} + 5$ is not a polynomial because the exponent in $2x^{-1}$ is not nonnegative.
- The expression $x^3 + 3x^{1/2}$ is not a polynomial because the exponent in $3x^{1/2}$ is not an integer.

▶ **Definition of a Polynomial in x**

Let $a_n, \ldots, a_2, a_1, a_0$ be real numbers and let n be a *nonnegative integer*. A **polynomial in x** is an expression of the form

$$a_n x^n + a_{n-1} x^{n-1} + \cdots + a_2 x^2 + a_1 x + a_0$$

where $a_n \neq 0$. The polynomial is of **degree n,** and the number a_n is the **leading coefficient.** The number a_0 is the **constant term.**

In the term ax^k, a is the **coefficient** and k is the **degree** of the term. Note that the degree of the term ax is 1, and the degree of the constant term is 0. Because a polynomial is an algebraic *sum*, the coefficients take on the signs between the terms. For instance,

$$x^3 - 4x^2 + 3 = (1)x^3 + (-4)x^2 + (0)x + 3$$

has coefficients 1, -4, 0, and 3. A polynomial that is written in order of descending powers of the variable is said to be in **standard form.** A polynomial with only one term is a **monomial.** Polynomials with two unlike terms are called **binomials,** and those with three unlike terms are called **trinomials.**

Example 1 Identifying Leading Coefficients and Degrees

	Polynomial	Standard Form	Degree	Leading Coefficient
a.	$5x^2 - 2x^7 + 4 - 2x$	$-2x^7 + 5x^2 - 2x + 4$	7	-2
b.	$16 - 8x^3$	$-8x^3 + 16$	3	-8
c.	10	10	0	10
d.	$5 + x^4 - 6x^3$	$x^4 - 6x^3 + 5$	4	1

2   Add and subtract polynomials using a vertical format and a horizontal format.

Adding and Subtracting Polynomials

To add two polynomials, combine like terms. This can be done in either a horizontal or a vertical format, as shown in Examples 2 and 3.

Technology: Tip

You can use a graphing utility to check the results of polynomial operations. For instance, in Example 2(a), you can graph the equations

$$y = (2x^3 + x^2 - 5) + (x^2 + x + 6)$$

and

$$y = 2x^3 + 2x^2 + x + 1$$

on the same screen. The fact that the graphs coincide, as shown below, confirms that the two polynomials are equivalent.

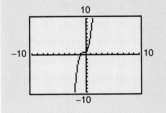

Example 2   Adding Polynomials Horizontally

a. $(2x^3 + x^2 - 5) + (x^2 + x + 6)$ Given polynomials

$\quad = (2x^3) + (x^2 + x^2) + (x) + (-5 + 6)$ Group like terms.

$\quad = 2x^3 + 2x^2 + x + 1$ Combine like terms.

b. $(3x^2 + 2x + 4) + (3x^2 - 6x + 3) + (-x^2 + 2x - 4)$

$\quad = (3x^2 + 3x^2 - x^2) + (2x - 6x + 2x) + (4 + 3 - 4)$

$\quad = 5x^2 - 2x + 3$

Example 3   Using a Vertical Format to Add Polynomials

Use a vertical format to find the sum.

$$(5x^3 + 2x^2 - x + 7) + (3x^2 - 4x + 7) + (-x^3 + 4x^2 - 8)$$

Solution

To use a vertical format, align the terms of the polynomials by their degrees.

$$
\begin{array}{r}
5x^3 + 2x^2 - \ x + 7 \\
3x^2 - 4x + 7 \\
-x^3 + 4x^2 \qquad - 8 \\
\hline
4x^3 + 9x^2 - 5x + 6
\end{array}
$$

To subtract one polynomial from another, *add the opposite*. You can do this by changing the sign of each term of the polynomial that is being subtracted and then adding the resulting like terms.

Example 4   Subtracting Polynomials Horizontally

$(3x^3 - 5x^2 + 3) - (x^3 + 2x^2 - x - 4)$ Given polynomials

$\quad = (3x^3 - 5x^2 + 3) + (-x^3 - 2x^2 + x + 4)$ Add the opposite.

$\quad = (3x^3 - x^3) + (-5x^2 - 2x^2) + (x) + (3 + 4)$ Group like terms.

$\quad = 2x^3 - 7x^2 + x + 7$ Combine like terms.

Be especially careful to use the correct signs when subtracting one polynomial from another. One of the most common mistakes in algebra is to forget to change signs correctly when subtracting one expression from another. Here is an example.

Wrong sign

$(x^2 - 2x + 3) - (x^2 + 2x - 2) \neq x^2 - 2x + 3 - x^2 + 2x - 2$ Common error

Wrong sign

The error illustrated above is forgetting to change two of the signs in the polynomial that is being subtracted. Remember to add the *opposite* of *every* term of the subtracted polynomial.

Example 5 Using a Vertical Format to Subtract Polynomials

Use a vertical format to find the difference.

$$(4x^4 - 2x^3 + 5x^2 - x + 8) - (3x^4 - 2x^3 + 3x - 4)$$

Solution

$$
\begin{array}{ll}
(4x^4 - 2x^3 + 5x^2 - x + 8) & 4x^4 - 2x^3 + 5x^2 - x + 8 \\
-(3x^4 - 2x^3 + 3x - 4) & -3x^4 + 2x^3 - 3x + 4 \\
\hline
& x^4 + 5x^2 - 4x + 12
\end{array}
$$

Example 6 Combining Polynomials

Use a horizontal format to perform the operations.

a. $(2x^2 - 7x + 2) - (4x^2 + 5x - 1) + (-x^2 + 4x + 4)$
b. $(-x^2 + 4x - 3) - [(4x^2 - 3x + 8) - (-x^2 + x + 7)]$

Solution

a. $(2x^2 - 7x + 2) - (4x^2 + 5x - 1) + (-x^2 + 4x + 4)$

$= 2x^2 - 7x + 2 - 4x^2 - 5x + 1 - x^2 + 4x + 4$

$= (2x^2 - 4x^2 - x^2) + (-7x - 5x + 4x) + (2 + 1 + 4)$

$= -3x^2 - 8x + 7$

b. $(-x^2 + 4x - 3) - [(4x^2 - 3x + 8) - (-x^2 + x + 7)]$

$= (-x^2 + 4x - 3) - (4x^2 - 3x + 8 + x^2 - x - 7)$

$= (-x^2 + 4x - 3) - [(4x^2 + x^2) + (-3x - x) + (8 - 7)]$

$= (-x^2 + 4x - 3) - (5x^2 - 4x + 1)$

$= -x^2 + 4x - 3 - 5x^2 + 4x - 1$

$= (-x^2 - 5x^2) + (4x + 4x) + (-3 - 1)$

$= -6x^2 + 8x - 4$

3 Use function notation to represent polynomials in application problems.

Galileo Galilei

(1564–1642)

Galileo believed that all objects, regardless of their weight (or mass), accelerate at the same rate. This law of falling bodies was not proven until the invention of the vacuum tube in the 1650s.

Applications

Function notation can be used to represent polynomials in a single variable. Such notation is useful for evaluating polynomials and is used in applications involving polynomials, as shown in Example 7.

There are many applications that involve polynomials. One commonly used second-degree polynomial is called a **position function.** This polynomial is a function of time and has the form

$$h(t) = -16t^2 + v_0 t + s_0 \qquad \text{Position function}$$

where the height h is measured in feet and the time t is measured in seconds.

This position function gives the height (above ground) of a free-falling object. The coefficient of t, v_0, is the **initial velocity** of the object, and the constant term s_0 is the **initial height** of the object. If the initial velocity is positive, the object was projected upward (at $t = 0$), and if the initial velocity is negative, the object was projected downward.

Example 7   Finding the Height of a Free-Falling Object

An object is thrown downward from the top of a 200-foot building. The initial velocity is -10 feet per second. Use the position function

$$h(t) = -16t^2 - 10t + 200$$

to find the height of the object when $t = 1$, $t = 2$, and $t = 3$. (See Figure 3.1.)

Solution

When $t = 1$, the height of the object is

$$h(1) = -16(1)^2 - 10(1) + 200 \qquad \text{Substitute 1 for } t.$$
$$= -16 - 10 + 200 \qquad \text{Simplify.}$$
$$= 174 \text{ feet.}$$

When $t = 2$, the height of the object is

$$h(2) = -16(2)^2 - 10(2) + 200 \qquad \text{Substitute 2 for } t.$$
$$= -64 - 20 + 200 \qquad \text{Simplify.}$$
$$= 116 \text{ feet.}$$

When $t = 3$, the height of the object is

$$h(3) = -16(3)^2 - 10(3) + 200 \qquad \text{Substitute 3 for } t.$$
$$= -144 - 30 + 200 \qquad \text{Simplify.}$$
$$= 26 \text{ feet.}$$

Use your calculator to determine the height of the object in Example 7 when $t = 3.2368$. What can you conclude?

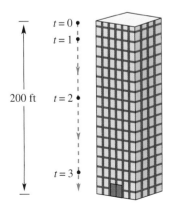

Figure 3.1

In 1995, there were about 16.5 million sole proprietorships (businesses owned by a single person or family) in the United States. Most of these were small businesses with 15 or fewer employees.

A second commonly used mathematical model that often involves polynomials is called a **profit equation.**

Example 8   Finding a Profit

A small manufacturing company can produce and sell x flashlights per week. The total cost (in dollars) for producing x flashlights is given by

$$C = 2x + 900,$$

and the total revenue from selling x flashlights is given by

$$R = 6x.$$

Find the profit obtained by selling 700 flashlights per week.

Solution

Verbal Model:
$$\boxed{\text{Profit}} \;=\; \boxed{\text{Revenue}} \;-\; \boxed{\text{Cost}}$$

Labels:
  Weekly profit $= P$ (dollars)
  Weekly revenue $R = 6x$ (dollars)
  Weekly cost $C = 2x + 900$ (dollars)
  Number of flashlights $x = 700$ (flashlights)

Equation:
$$P = R - C$$
$$P(x) = 6x - (2x + 900)$$
$$P(700) = 6(700) - [2(700) + 900]$$
$$= 4200 - (1400 + 900)$$
$$= \$1900$$

So, the total weekly profit from selling 700 flashlights is

$$P(700) = \$1900.$$

Discussing the Concept   Problem Posing

Create a problem involving cost and revenue equations and requiring a profit equation for the production and sale of a product of your choice. Be sure to describe your cost and revenue situations such that the profit is negative when no units are sold and positive when 200 units are sold. Solve your problem and interpret the solution. Exchange your problem for that of another student, and solve each other's problems.

3.1   Exercises

Integrated Review   *Concepts, Skills, and Problem Solving*

Keep mathematically in shape by doing these exercises *before* the problems of this section.

Properties and Definitions

1. The product of two real numbers is -96 and one of the factors is 12. What is the sign of the other factor?

2. Determine the sum of the digits of 576. Because this sum is divisible by 9, the number 576 is divisible by what number?

3. *True or False?* -6^2 is positive.

4. *True or False?* $(-6)^2$ is positive.

Solving Inequalities

In Exercises 5–10, solve the inequality.

5. $2x - 12 \geq 0$ **6.** $7 - 3x < 4 - x$

7. $-2 < 4 - 2x < 10$

8. $4 \leq x + 5 < 8$

9. $|x - 3| < 2$

10. $|x - 5| > 3$

Problem Solving

11. The tax on a property with an assessed value of \$145,000 is \$2400. Find the tax on a property with an assessed value of \$90,000.

12. A car uses 7 gallons of gasoline for a trip of 200 miles. How many gallons would be used on a trip of 325 miles? (Assume there is no change in the fuel efficiency.)

Developing Skills

In Exercises 1–12, write the polynomial in standard form, and find its degree and leading coefficient. See Example 1.

1. $10x - 4$ **2.** $5t + 3$

3. $3x^2 + 2 - x$ **4.** $3x^2 + 8 - 4x$

5. $5 - 3y^4 + y^5 - 2y^3$ **6.** $8z - 16z^2 + 6z^4$

7. $-3x^3 - 2x^2 - 3$ **8.** $6t + 4t^5 - t^2 + 3$

9. -4 **10.** 28

11. $v_0 t - 16t^2$ (v_0 is constant.)

12. $48 - \frac{1}{2}at^2$ (a is constant.)

In Exercises 13–18, determine whether the polynomial is a monomial, a binomial, or a trinomial.

13. $12 - 5y^2$ **14.** $-6y + 3 + y^3$

15. $x^3 + 2x^2 - 4$ **16.** t^3

17. 5 **18.** $25 - 2u^2$

In Exercises 19–22, give an example of a polynomial in x that satisfies the conditions. (*Note:* Each problem has many correct answers.)

19. A monomial of degree 3

20. A trinomial of degree 4 and leading coefficient -2

21. A binomial of degree 2 and leading coefficient 8

22. A monomial of degree 0

In Exercises 23–26, state why the expression is not a polynomial.

23. $y^{-3} - 2$ **24.** $x^3 - 4x^{1/3}$

25. $\dfrac{8}{x}$ **26.** $\dfrac{2}{x - 4}$

In Exercises 27–42, use a horizontal format to find the sum. See Example 2.

27. $5 + (2 + 3x)$ **28.** $(6 - 2x) + 4x$

29. $(2x^2 - 3) + (5x^2 + 6)$

30. $(3x + 1) + (6x - 1)$

31. $(5y + 6) + (4y^2 - 6y - 3)$

32. $(3x^3 - 2x + 8) + (3x - 5)$

33. $(2 - 8y) + (-2y^4 + 3y + 2)$

34. $(z^3 + 6z - 2) + (3z^2 - 6z)$

35. $(8 - t^4) + (5 + t^4)$

36. $(y^5 - 4y) + (3y - y^5) + (y^5 - 5)$

37. $(x^2 - 3x + 8) + (2x^2 - 4x) + 3x^2$

38. $(3a^2 + 5a) + (7 - a^2 - 5a) + (2a^2 + 8)$

39. $\left(\frac{2}{3}x^3 - 4x + 1\right) + \left(-\frac{3}{5} + 7x - \frac{1}{2}x^3\right)$

40. $\left(2 - \frac{1}{4}y^2 + y^4\right) + \left(\frac{1}{3}y^4 - \frac{3}{2}y^2 - 3\right)$

41. $(6.32t - 4.51t^2) + (7.2t^2 + 1.03t - 4.2)$

42. $(0.13x^4 - 2.25x - 1.63) +$
$\quad (5.3x^4 + 1.76x^2 + 1.29x)$

In Exercises 43–50, use a vertical format to find the sum. See Example 3.

43. $\quad 5x^2 - 3x + 4$
$\quad\quad \underline{-3x^2 \quad\quad\; - 4}$

44. $\quad 3x^4 - 2x^2 - 9$
$\quad\quad \underline{-5x^4 + \;\; x^2}$

45. $(4x^3 - 2x^2 + 8x) + (4x^2 + x - 6)$

46. $(4x^3 + 8x^2 - 5x + 3) + (x^3 - 3x^2 - 7)$

47. $(5p^2 - 4p + 2) + (-3p^2 + 2p - 7)$

48. $(16 - 32t) + (64 + 48t - 16t^2)$

49. $(2.5b - 3.6b^2) + (7.1 - 3.1b - 2.4b^2) + 6.6b^2$

50. $(1.7y^3 - 6.2y^2 + 5.9) + (2.2y + 6.7y^2 - 3.5y^3)$

In Exercises 51–62, use a horizontal format to find the difference. See Example 4.

51. $(4 - y^3) - (4 + y^3)$

52. $(5y^4 - 2) - (3y^4 + 2)$

53. $(3x^2 - 2x + 1) - (2x^2 + x - 1)$

54. $(5q^2 - 3q + 5) - (4q^2 - 3q - 10)$

55. $(6t^3 - 12) - (-t^3 + t - 2)$

56. $(-10s^2 - 5) - (2s^2 + 6s)$

57. $\left(\frac{1}{4}y^2 - 5y\right) - \left(12 + 4y - \frac{3}{2}y^2\right)$

58. $\left(12 - \frac{2}{3}x + \frac{1}{2}x^2\right) - \left(x^3 + 3x^2 - \frac{1}{6}x\right)$

59. $(10.4t^4 - 0.23t^5 + 1.3t^2) -$
$\quad (2.6 - 7.35t + 6.7t^2 - 9.6t^5)$

60. $(u^3 - 9.75u^2 + 0.12u - 3) -$
$\quad (0.7u^3 - 6.9u^2 - 4.83)$

61. Subtract $3x^3 - (x^2 + 5x)$ from $x^3 - 3x$.

62. Subtract $y^4 - (y^2 - 8y)$ from $y^2 + 3y^4$.

In Exercises 63–68, use a vertical format to find the difference. See Example 5.

63. $\quad x^2 - \;\, x + 3$
$\quad\quad \underline{- \quad\quad\; (x - 2)}$

64. $\quad 3t^4 - 5t^2$
$\quad\quad \underline{-(-t^4 + 2t^2 - 14)}$

65. $(25 - 15x - 2x^3) - (12 - 13x + 2x^3)$

66. $(4x^2 + 5x - 6) - (2x^2 - 4x + 5)$

67. $(6x^4 - 3x^7 + 4) - (8x^7 + 10x^5 - 2x^4 - 12)$

68. $(13x^3 - 9x^2 + 4x - 5) - (5x^3 + 7x + 3)$

In Exercises 69–84, perform the operations. See Example 6.

69. $-(2x^3 - 3) + (4x^3 - 2x)$

70. $(2x^2 + 1) - (x^2 - 2x + 1)$

71. $(4x^5 - 10x^3 + 6x) - (8x^5 - 3x^3 + 11) +$
$\quad (4x^5 + 5x^3 - x^2)$

72. $(15 - 2y + y^2) + (3y^2 - 6y + 1) -$
$\quad (4y^2 - 8y + 16)$

73. $(5y^2 - 2y) - [(y^2 + y) - (3y^2 - 6y + 2)]$

74. $(p^3 + 4) - [(p^2 + 4) + (3p - 9)]$

75. $(8x^3 - 4x^2 + 3x) -$
$\quad [(x^3 - 4x^2 + 5) + (x - 5)]$

76. $(5x^4 - 3x^2 + 9) -$
$\quad [(2x^4 + x^3 - 7x^2) - (x^2 + 6)]$

77. $3(4x^2 - 1) + (3x^3 - 7x^2 + 5)$

78. $(x^3 - 2x^2 - x) - 5(2x^3 + x^2 - 4x)$

79. $2(t^2 + 12) - 5(t^2 + 5) + 6(t^2 + 5)$

80. $-10(v + 2) + 8(v - 1) - 3(v - 9)$

81. $15v - 3(3v - v^2) + 9(8v + 3)$

82. $9(7x^2 - 3x + 3) - 4(15x + 2) - (3x^2 - 7x)$

83. $5s - [6s - (30s + 8)]$

84. $3x^2 - 2[3x + (9 - x^2)]$

Graphical Reasoning In Exercises 85 and 86, use a graphing utility to graph the expressions for y_1 and y_2. What conclusion can you make?

85. $y_1 = (x^3 - 3x^2 - 2) - (x^2 + 1)$
$\quad y_2 = x^3 - 4x^2 - 3$

86. $y_1 = \left(\frac{1}{2}x^3 + 2x\right) + (x^3 - x^2 - x + 1)$
$\quad y_2 = \frac{3}{2}x^3 - x^2 + x + 1$

In Exercises 87 and 88, $f(x) = 4x^3 - 3x^2 + 7$ and $g(x) = 9 - x - x^2 - 5x^3$. Find $h(x)$.

87. $h(x) = f(x) + g(x)$

88. $h(x) = f(x) - g(x)$

Solving Problems

Free-Falling Object In Exercises 89–92, find the height in feet of a free-falling object at the specified times using the position function. Then write a paragraph that describes the vertical path of the object.

89. $h(t) = -16t^2 + 64$

(a) $t = 0$ (b) $t = \frac{1}{2}$

(c) $t = 1$ (d) $t = 2$

90. $h(t) = -16t^2 + 256$

(a) $t = 0$ (b) $t = 1$

(c) $t = \frac{5}{2}$ (d) $t = 4$

91. $h(t) = -16t^2 + 80t + 50$

(a) $t = 0$ (b) $t = 2$

(c) $t = 4$ (d) $t = 5$

92. $h(t) = -16t^2 + 96t$

(a) $t = 0$ (b) $t = 2$

(c) $t = 3$ (d) $t = 6$

Free-Falling Object In Exercises 93–96, use the position function to determine whether the free-falling object was dropped, was thrown upward, or was thrown downward. Also determine the height in feet of the object at time $t = 0$.

93. $h(t) = -16t^2 + 100$ **94.** $h(t) = -16t^2 + 50t$

95. $h(t) = -16t^2 - 24t + 50$

96. $h(t) = -16t^2 + 32t + 300$

97. *Free-Falling Object* An object is thrown upward from the top of a 200-foot building (see figure). The initial velocity is 40 feet per second. Use the position function

$h(t) = -16t^2 + 40t + 200$

to find the height of the object when $t = 1$, $t = 2$, and $t = 3$.

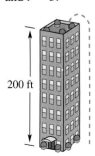

200 ft

Figure for 97

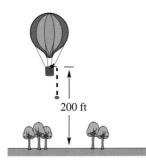

200 ft

Figure for 98

98. *Free-Falling Object* An object is dropped from a hot-air balloon that is 200 feet above the ground (see figure). Use the position function

$h(t) = -16t^2 + 200$

to find the height of the object when $t = 1$, $t = 2$, and $t = 3$.

99. *Profit* A manufacturer can produce and sell x radios per week. The total cost (in dollars) for producing the radios is given by

$C = 8x + 15{,}000$

and the total revenue is given by

$R = 14x.$

Find the profit obtained by selling 5000 radios per week.

100. *Profit* A manufacturer can produce and sell x golf clubs per week. The total cost (in dollars) for producing the golf clubs is given by

$C = 12x + 8000$

and the total revenue is given by

$R = 17x.$

Find the profit obtained by selling 10,000 golf clubs per week.

Geometry In Exercises 101 and 102, find the perimeter of the figure.

101.

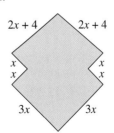

$2x + 4$ $2x + 4$

x x

x x

$3x$ $3x$

102.

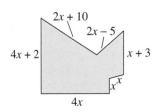

$2x + 10$

$2x - 5$

$4x + 2$ $x + 3$

x

x

$4x$

Geometry In Exercises 103 and 104, find a polynomial expression that represents the area of the entire region.

103. **104.**

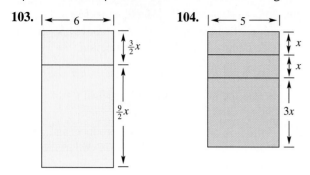

Geometry In Exercises 105 and 106, find the area of the shaded portion of the figure.

105. **106.**

107. *Consumption of Milk* The per capita consumptions (average consumptions per person) of all beverage milks (including whole milk) and whole milk (alone) in the United States from 1986 to 1995 can be approximated by the two polynomial models

$$y = 231.06 + 0.009t - 0.095t^2 \qquad \text{Beverage milks}$$

$$y = 171.17 - 11.415t + 0.325t^2. \qquad \text{Whole milk}$$

In these models, y represents the average consumption per person in gallons and t represents the year, with $t = 6$ corresponding to 1986. (Source: U.S. Department of Agriculture)

(a) Find a polynomial model that represents the per capita consumption of all beverage milks other than whole milk during the time period.

(b) During the given period, the per capita consumptions of beverage milks and whole milk were decreasing (see figure). Use a graphing utility to graph the model from part (a). Was the per capita consumption of beverage milks other than whole milk also decreasing over this period? Explain.

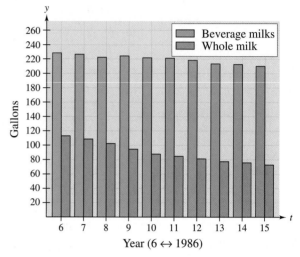

Figure for 107

108. *Stopping Distance* The total stopping distance of an automobile is the distance traveled during the driver's reaction time plus the distance traveled after the brakes are applied. In an experiment, these distances were measured (in feet) when the automobile was traveling at x miles per hour. The distance traveled during the reaction time was $R = 1.1x$, and the braking distance was $B = 0.14x^2 - 4.43x + 58.40$.

(a) Determine the polynomial that represents the total stopping distance T.

(b) Use a graphing utility to graph R, B, and T in the same viewing rectangle.

(c) Use the graph to estimate the total stopping distance when $x = 30$ and $x = 60$.

Explaining Concepts

109. Explain the difference between the degree of a term of a polynomial in x and the degree of a polynomial.

110. What algebraic operations separate terms of a polynomial? What operation separates factors of a term?

111. Give an example of combining like terms.

112. Can two third-degree polynomials be added to produce a second-degree polynomial? If so, give an example.

113. Is every trinomial a second-degree polynomial? Explain.

114. Describe the method for subtracting polynomials.

3.2   Multiplying Polynomials

Objectives

1 Use the rules for exponents to simplify an expression.

2 Use the FOIL Method and the Distributive Property to multiply polynomials.

3 Use special product formulas to multiply two binomials.

4 Use polynomials to find the area and volume of a box.

1 Use the rules for exponents to simplify an expression.

Rules for Exponents

You know from Chapter P that exponents are used to denote repeated multiplication. The next example uses repeated multiplication to illustrate the basic rules for operating with exponential expressions.

Example 1   Illustrating the Rules for Exponents

a. *Product Rule:*   To multiply exponential expressions that have the *same base*, add exponents.

$$x^2 \cdot x^3 = \underbrace{x \cdot x}_{2 \text{ factors}} \cdot \underbrace{x \cdot x \cdot x}_{3 \text{ factors}} = \underbrace{x \cdot x \cdot x \cdot x \cdot x}_{5 \text{ factors}} = x^{2+3} = x^5$$

b. *Quotient Rule:*   To divide exponential expressions that have the *same base*, subtract exponents.

$$\frac{x^4}{x^2} = \frac{\overbrace{x \cdot x \cdot x \cdot x}^{4 \text{ factors}}}{\underbrace{x \cdot x}_{2 \text{ factors}}} = x^{4-2} = x^2$$

c. *Power of a Product:*   To raise the product of two factors to the *same power*, raise each factor to the power and multiply.

$$(3x)^3 = \underbrace{3x \cdot 3x \cdot 3x}_{3 \text{ factors}} = \underbrace{3 \cdot 3 \cdot 3}_{3 \text{ factors}} \cdot \underbrace{x \cdot x \cdot x}_{3 \text{ factors}} = 3^3 \cdot x^3 = 27x^3$$

d. *Power of a Quotient:*   To raise the quotient of two expressions to a power, raise each expression to the power and divide.

$$\left(\frac{x}{3}\right)^3 = \frac{x}{3} \cdot \frac{x}{3} \cdot \frac{x}{3} = \frac{\overbrace{x \cdot x \cdot x}^{3 \text{ factors}}}{\underbrace{3 \cdot 3 \cdot 3}_{3 \text{ factors}}} = \frac{x^3}{3^3} = \frac{x^3}{27}$$

e. *Power of a Power:*   To raise an exponential expression to a power, multiply the powers.

$$(x^3)^2 = \underbrace{(x \cdot x \cdot x)}_{3 \text{ factors}} \cdot \underbrace{(x \cdot x \cdot x)}_{3 \text{ factors}} = \underbrace{(x \cdot x \cdot x \cdot x \cdot x \cdot x)}_{6 \text{ factors}} = x^{3 \cdot 2} = x^6$$

▶ **Summary of Rules for Exponents**

Let m and n be positive integers, and let a and b represent real numbers, variables, or algebraic expressions.

	Rule	*Example*
1.	Product Rule	
	$a^m \cdot a^n = a^{m+n}$	$x^3(x^2) = x^{3+2} = x^5$
2.	Quotient Rule	
	$\dfrac{a^m}{a^n} = a^{m-n}, \quad m > n, \quad a \neq 0$	$\dfrac{x^4}{x^2} = x^{4-2} = x^2$
3.	Power of a Product	
	$(ab)^m = a^m \cdot b^m$	$(2x)^3 = 2^3(x^3) = 8x^3$
4.	Power of a Quotient	
	$\left(\dfrac{a}{b}\right)^m = \dfrac{a^m}{b^m}, \quad b \neq 0$	$\left(\dfrac{x}{2}\right)^3 = \dfrac{x^3}{2^3} = \dfrac{x^3}{8}$
5.	Power of a Power	
	$(a^m)^n = a^{mn}$	$(x^2)^3 = x^{2 \cdot 3} = x^6$

The next example shows how the rules for exponents can be used to simplify expressions that involve products, powers, and quotients.

Study Tip

In some cases an expression is simplified using two or more rules in a single step. For instance, in Example 2(d)

$$\frac{(x^2)^3}{(2y)^3} = \frac{x^{2 \cdot 3}}{2^3 y^3}$$

the Power of a Power Rule is used in the numerator

$$(x^2)^3 = x^{2 \cdot 3}$$

and the Power of a Product Rule is used in the denominator

$$(2y)^3 = 2^3 y^3.$$

Example 2 **Applying Rules for Exponents**

Simplify each expression.

a. $(x^2 y^4)(3x)$ **b.** $(-2y^2)^3$ **c.** $\dfrac{14a^5b^3}{7a^2b^2}$ **d.** $\left(\dfrac{x^2}{2y}\right)^3$ **e.** $\dfrac{x^{n+2}y^{3n}}{x^2 y^n}$

Solution

a. $(x^2 y^4)(3x) = 3(x^2 \cdot x)(y^4)$

$\qquad\qquad\quad = 3(x^{2+1})(y^4)$

$\qquad\qquad\quad = 3x^3 y^4$

b. $(-2y^2)^3 = (-2)^3(y^2)^3$

$\qquad\qquad = -8y^{2 \cdot 3}$

$\qquad\qquad = -8y^6$

c. $\dfrac{14a^5b^3}{7a^2b^2} = \dfrac{14}{7}(a^{5-2})(b^{3-2}) = 2a^3b$

d. $\left(\dfrac{x^2}{2y}\right)^3 = \dfrac{(x^2)^3}{(2y)^3} = \dfrac{x^{2 \cdot 3}}{2^3 y^3} = \dfrac{x^6}{8y^3}$

e. $\dfrac{x^{n+2}y^{3n}}{x^2 y^n} = x^{(n+2)-2}y^{3n-n} = x^n y^{2n}$

2 Use the FOIL Method and the Distributive Property to multiply polynomials.

Multiplying Polynomials

The simplest type of polynomial multiplication involves a monomial multiplier. The product is obtained by direct application of the Distributive Property. For instance, to multiply the monomial $3x$ by the polynomial $(2x^2 - 5x + 3)$, multiply *each* term of the polynomial by $3x$.

$$(3x)(2x^2 - 5x + 3) = (3x)(2x^2) - (3x)(5x) + (3x)(3)$$

$$= 6x^3 - 15x^2 + 9x$$

Example 3   Finding Products with Monomial Multipliers

Multiply the polynomial by the monomial.

a. $(2x - 7)(-3x)$ **b.** $4x^2(3x - 2x^3 + 1)$ **c.** $(-x)(5x^2 - x)$

Solution

a. $(2x - 7)(-3x) = 2x(-3x) - 7(-3x)$ Distributive Property

$$= -6x^2 + 21x$$ Rules for exponents

b. $4x^2(3x - 2x^3 + 1)$

$$= 4x^2(3x) - 4x^2(2x^3) + 4x^2(1)$$ Distributive Property

$$= 12x^3 - 8x^5 + 4x^2$$ Rules for exponents

$$= -8x^5 + 12x^3 + 4x^2$$ Standard form

c. $(-x)(5x^2 - x) = (-x)(5x^2) - (-x)(x)$ Distributive Property

$$= -5x^3 + x^2$$ Rules for exponents

To multiply two *binomials*, you can use both (left and right) forms of the Distributive Property. For example, if you treat the binomial $(2x + 7)$ as a single quantity, you can multiply $(3x - 2)$ by $(2x + 7)$ as follows.

$$(3x - 2)(2x + 7) = 3x(2x + 7) - 2(2x + 7)$$

$$= (3x)(2x) + (3x)(7) - (2)(2x) - 2(7)$$

$$= 6x^2 + 21x - 4x - 14$$

| Product of First terms | Product of Outer terms | Product of Inner terms | Product of Last terms |

$$= 6x^2 + 17x - 14$$

With practice you should be able to multiply two binomials without writing out all of the steps shown above. In fact, the four products in the boxes above suggest that the product of two binomials can be written in just one step, which is called the **FOIL Method.** Note that the words *first, outer, inner,* and *last* refer to the positions of the terms in the original product, as shown at the left.

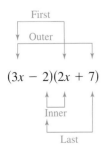

The FOIL Method

Example 4   Multiplying Binomials (FOIL Method)

Use the FOIL Method to multiply the binomials.

a. $(x - 3)(x + 3)$ **b.** $(3x + 4)(2x + 1)$

Solution

$$\text{a.} \quad (x - 3)(x + 3) = \overset{F}{\overbrace{x^2}} + \overset{O}{\overbrace{3x}} - \overset{I}{\overbrace{3x}} - \overset{L}{9}$$

$$= x^2 - 9 \qquad\qquad \text{Combine like terms.}$$

$$\text{b.} \quad (3x + 4)(2x + 1) = \overset{F}{\overbrace{6x^2}} + \overset{O}{\overbrace{3x}} + \overset{I}{\overbrace{8x}} + \overset{L}{4} = 6x^2 + 11x + 4$$

When multiplying polynomials that have three or more terms, use the same basic principle as for multiplying monomials and binomials. That is, *each term of one polynomial must be multiplied by each term of the other polynomial.* This can be done using either a horizontal or a vertical format.

Study Tip

When multiplying two polynomials, it is best to write each in standard form before using either the horizontal or the vertical format.

Example 5   Multiplying Polynomials (Horizontal Format)

Use a horizontal format to multiply the polynomials.

$$(4x^2 - 3x - 1)(2x - 5)$$

Solution

$$(4x^2 - 3x - 1)(2x - 5)$$

$$= (4x^2 - 3x - 1)(2x) - (4x^2 - 3x - 1)(5) \qquad \text{Distributive Property}$$

$$= 8x^3 - 6x^2 - 2x - (20x^2 - 15x - 5) \qquad \text{Distributive Property}$$

$$= 8x^3 - 6x^2 - 2x - 20x^2 + 15x + 5 \qquad \text{Subtract (change signs).}$$

$$= 8x^3 - 26x^2 + 13x + 5 \qquad \text{Combine like terms.}$$

Example 6   Multiplying Polynomials (Vertical Format)

Write the polynomials in standard form and use a vertical format to multiply.

$$(4x^2 + x - 2)(5 + 3x - x^2)$$

Solution

$$
\begin{array}{r}
4x^2 + \;\; x - \;\; 2 \qquad\qquad \text{Standard form}\\
\times \qquad\quad -x^2 + 3x + \;\; 5 \qquad\qquad \text{Standard form}\\
\hline
20x^2 + 5x - 10 \qquad \Longleftarrow\;\; 5(4x^2 + x - 2)\\
12x^3 + \;\; 3x^2 - 6x \qquad\qquad\quad \Longleftarrow\;\; 3x(4x^2 + x - 2)\\
-4x^4 - \;\;\; x^3 + 2x^2 \qquad\qquad\qquad\quad \Longleftarrow\;\; -x^2(4x^2 + x - 2)\\
\hline
-4x^4 + 11x^3 + 25x^2 - \;\; x - 10
\end{array}
$$

3 Use special product formulas to multiply two binomials.

Special Products

Some binomial products, such as that in Example 4(a), have special forms that occur frequently in algebra.

▶ **Special Products**

Let u and v be real numbers, variables, or algebraic expressions. Then the following formulas are true.

Special Product	*Example*
Sum and Difference of Two Terms	
$(u + v)(u - v) = u^2 - v^2$	$(3x - 4)(3x + 4) = 9x^2 - 16$
Square of a Binomial	
$(u + v)^2 = u^2 + 2uv + v^2$	$(2x + 5)^2 = 4x^2 + 2(2x)(5) + 25$
	$\qquad\qquad = 4x^2 + 20x + 25$
$(u - v)^2 = u^2 - 2uv + v^2$	$(x - 6)^2 = x^2 - 2(x)(6) + 36$
	$\qquad\qquad = x^2 - 12x + 36$

When squaring a binomial, note that the resulting middle term is always *twice* the product of the two terms.

Study Tip

A frequent error in calculating special products is to forget the middle term when squaring a binomial, $(x + y)^2$. Use $x = 3$ and $y = 2$ to verify the following statements.

1. $(x + y)^2 = x^2 + 2xy + y^2$

2. $(x + y)^2 \neq x^2 + y^2$

3. $(x - y)^2 = x^2 - 2xy + y^2$

4. $(x - y)^2 \neq x^2 - y^2$

Example 7   Product of the Sum and Difference of Two Terms

a. $(3x - 2)(3x + 2) = (3x)^2 - 4^2$ $(u + v)(u - v) = u^2 - v^2$

$\qquad\qquad\qquad\quad = 9x^2 - 16$ Rules for exponents

b. $(6 + 5x)(6 - 5x) = 6^2 - (5x)^2 = 36 - 25x^2$

Example 8   Squaring a Binomial

$(2x - 7)^2 = (2x)^2 - 2(2x)(7) + 7^2$ Square of a binomial

$\qquad\quad = 4x^2 - 28x + 49$ Rules for exponents

Example 9   Cubing a Binomial

$(x - 4)^3 = (x - 4)^2(x - 4)$ Rules for exponents

$\qquad\quad = (x^2 - 8x + 16)(x - 4)$ Square of a binomial

$\qquad\quad = x^2(x - 4) - 8x(x - 4) + 16(x - 4)$ Distributive Property

$\qquad\quad = x^3 - 4x^2 - 8x^2 + 32x + 16x - 64$ Distributive Property

$\qquad\quad = x^3 - 12x^2 + 48x - 64$ Combine like terms.

4 Use polynomials to find the area and volume of a box.

Application

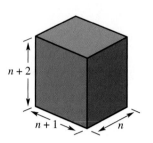

Figure 3.2

Example 10   Finding Area and Volume

The closed box shown in Figure 3.2 has sides whose lengths (in inches) are consecutive integers.

a. Find a polynomial function $V(n)$ that describes the volume of the box.

b. What is the volume if the length of the shortest side is 4 inches?

c. Write a polynomial function $A(n)$ for the area of the base of the box.

d. Write a polynomial function for the area of the base if its length and width increase by 3. That is, find $A(n + 3)$.

Solution

a. *Verbal Model:*

| Volume | = | Length | · | Width | · | Height |

Labels:
Length $= n$	(inches)
Width $= n + 1$	(inches)
Height $= n + 2$	(inches)
Volume $= V$	(cubic inches)

Equation:
$$V(n) = n(n + 1)(n + 2)$$
$$= n(n^2 + 3n + 2)$$
$$= n^3 + 3n^2 + 2n$$

b. If $n = 4$, the volume of the box is

$$V(4) = (4)^3 + 3(4)^2 + 2(4) = 64 + 48 + 8 = 120 \text{ cubic inches.}$$

c. Using the length and width from part (a), the area can be found as follows.

$$A(n) = (\text{length})(\text{width}) = n(n + 1) = n^2 + n$$

d. $A(n + 3) = (n + 3)^2 + (n + 3) = n^2 + 6n + 9 + n + 3 = n^2 + 7n + 12$

| **Discussing the Concept** | **Investigating a Demand Function** |

A company determines that the number of units of a product that it can sell depends on the price of the product. A model for this relationship is

$$p = 24 - 0.001x$$

where p is the price (in dollars) and x is the number of units.

a. Sketch the graph of this model and use the result to describe the relationship between the price and the number of units sold.

b. The revenue obtained from selling x units is given by $R = xp$. Sketch the graph of the revenue function and describe the relationship between the number of units sold and the revenue.

3.2   Exercises

Integrated Review   *Concepts, Skills, and Problem Solving*

Keep mathematically in shape by doing these exercises *before* the problems of this section.

Properties and Definitions

1. Relative to the x- and y-axes, explain the meaning of each coordinate of the point $(-2, 3)$.

2. A point lies 3 units from the x-axis and 4 units from the y-axis. Give the ordered pair for such a point in each quadrant.

Evaluating Expressions

In Exercises 3–6, find the missing coordinate of the solution point.

3. $y = \frac{3}{5}x + 4$

$(15, \quad)$

4. $y = 3 - \frac{5}{9}x$

$(12, \quad)$

5. $y = 5.5 - 0.95x$

$(\quad, -1)$

6. $y = 3 + 0.2x$

$(\quad, 4.4)$

In Exercises 7–10, evaluate the function.

7. $f(x) = \frac{1}{3}x^2$

(a) $f(6)$

(b) $f\left(\frac{3}{4}\right)$

8. $f(x) = 3 - 2x$

(a) $f(5)$

(b) $f(x + 3) - f(3)$

9. $g(x) = \dfrac{x}{x + 10}$

(a) $g(5)$

(b) $g(c - 6)$

10. $h(x) = \sqrt{x - 4}$

(a) $h(16)$

(b) $h(t + 3)$

Graphing

In Exercises 11 and 12, graph the function.

11. $g(x) = 7 - \frac{3}{2}x$

12. $h(x) = |3 - x|$

Developing Skills

In Exercises 1–10, use the expression to illustrate a rule for exponents. See Example 1.

1. $t^3 \cdot t^4$

2. $z^2 \cdot z^3$

3. $(-5x)^5$

4. $(2y)^3$

5. $(u^4)^2$

6. $(x^2)^3$

7. $\dfrac{x^6}{x^4}$

8. $\dfrac{y^5}{y^3}$

9. $\left(\dfrac{y}{5}\right)^4$

10. $\left(\dfrac{3}{t}\right)^5$

In Exercises 11–30, simplify (if possible) the expression. See Example 2.

11. (a) $-3x^3 \cdot x^5$ (b) $(-3x)^2 \cdot x^5$

12. (a) $5^2y^4 \cdot y^2$ (b) $(5y)^2 \cdot y^4$

13. (a) $(-5z^2)^3$ (b) $(-5z^4)^2$

14. (a) $(-5z^3)^2$ (b) $(-5z)^4$

15. (a) $(u^3v)(2v^2)$ (b) $(-4u^4)(u^5v)$

16. (a) $(6xy^7)(-x)$ (b) $(x^5y^3)(2y^3)$

17. (a) $5u^2 \cdot (-3u^6)$ (b) $(2u)^4(4u)$

18. (a) $(3y)^3(2y^2)$ (b) $3y^3 \cdot 2y^2$

19. (a) $-(m^5n)^3(-m^2n^2)^2$ (b) $(-m^5n)(m^2n^2)$

20. (a) $-(m^3n^2)(mn^3)$ (b) $-(m^3n^2)^2(-mn^3)$

21. (a) $\dfrac{27m^5n^6}{9mn^3}$ (b) $\dfrac{-18m^3n^6}{-6mn^3}$

22. (a) $\dfrac{28x^2y^3}{2xy^2}$ (b) $\dfrac{24xy^2}{8y}$

23. (a) $\left(\dfrac{3x}{4y}\right)^2$ (b) $\left(\dfrac{5u}{3v}\right)^3$

24. (a) $\left(\dfrac{2a}{3y}\right)^5$ (b) $-\left(\dfrac{2a}{3y}\right)^2$

25. (a) $-\dfrac{(-3x^2y)^3}{9x^2y^2}$ (b) $-\dfrac{(-2xy^3)^2}{6y^2}$

26. (a) $\dfrac{(-3xy)^3}{9xy^2}$ (b) $\dfrac{(-3xy)^4}{-3(xy)^2}$

27. (a) $\left[\dfrac{(-5u^3v)^2}{10u^2v}\right]^2$ (b) $\left[\dfrac{-5(u^3v)^2}{10u^2v}\right]^2$

28. (a) $\left[\dfrac{(3x^2)(2x)^2}{(-2x)(6x)}\right]^2$ (b) $\left[\dfrac{(3x^2)(2x)^4}{(-2x)^2(6x)}\right]^2$

29. (a) $\dfrac{x^{2n+4}\,y^{4n}}{x^5\,y^{2n+1}}$ (b) $\dfrac{x^{6n}\,y^{n-7}}{x^{4n+2}\,y^5}$

30. (a) $\dfrac{x^{3n}\,y^{2n-1}}{x^n\,y^{n+3}}$ (b) $\dfrac{x^{4n-6}\,y^{n+10}}{x^{2n-5}\,y^{n-2}}$

In Exercises 31–44, simplify the expression. See Example 3.

31. $(-2a^2)(-8a)$

32. $(-6n)(3n^2)$

33. $2y(5 - y)$

34. $5z(2z - 7)$

35. $4x(2x^2 - 3x + 5)$

36. $3y(-3y^2 + 7y - 3)$

37. $-2x^2(5 + 3x^2 - 7x^3)$

38. $-3a^2(8 - 2a - a^2)$

39. $-x^3(x^4 - 2x^3 + 5x - 6)$

40. $-y^4(7y^3 - 4y^2 + y - 4)$

41. $-3x(-5x)(5x + 2)$

42. $4t(-3t)(t^2 - 1)$

43. $u^2v(3u^4 - 5u^2v + 6uv^3)$

44. $ab^3(2a - 9a^2b + 3b)$

In Exercises 45–62, multiply using the FOIL Method. See Example 4.

45. $(x + 2)(x + 4)$

46. $(x - 5)(x - 3)$

47. $(x - 6)(x + 5)$

48. $(x + 7)(x - 1)$

49. $(x - 4)(x - 4)$

50. $(x - 6)(x + 6)$

51. $(2x - 3)(x + 5)$

52. $(3x + 1)(x - 4)$

53. $(5x - 2)(2x - 6)$

54. $(4x + 7)(3x + 7)$

55. $(8 - 3x^2)(4x + 1)$

56. $(6x^2 + 2)(9 - 2x)$

57. $\left(4y - \frac{1}{3}\right)(12y + 9)$

58. $\left(5t - \frac{3}{4}\right)(2t - 16)$

59. $(2x + y)(3x + 2y)$

60. $(2x - y)(3x - 2y)$

61. $(2t - 1)(t + 1) + (2t - 5)(t - 1)$

62. $(s - 3t)(s + t) - (s - 3t)(s - t)$

In Exercises 63–74, use a horizontal format to perform the multiplication. See Example 5.

63. $(x - 1)(x^2 - 4x + 6)$

64. $(z + 2)(z^2 - 4z + 4)$

65. $(3a + 2)(a^2 + 3a + 1)$

66. $(2t + 3)(t^2 - 5t + 1)$

67. $(2u^2 + 3u - 4)(4u + 5)$

68. $(2x^2 - 5x + 1)(3x - 4)$

69. $(x^3 - 3x + 2)(x - 2)$

70. $(x^2 + 4)(x^2 - 2x - 4)$

71. $(5x^2 + 2)(x^2 + 4x - 1)$

72. $(2x^2 - 3)(2x^2 - 2x + 3)$

73. $(t^2 + t - 2)(t^2 - t + 2)$

74. $(y^2 + 3y + 5)(2y^2 - 3y - 1)$

In Exercises 75–82, use a vertical format to perform the multiplication. See Example 6.

75. $\begin{array}{r} 7x^2 - 14x + 9 \\ \times \quad 4x^3 + 3 \\ \hline \end{array}$

76. $\begin{array}{r} 4x^4 - 6x^2 + 9 \\ \times \quad 2x^2 + 3 \\ \hline \end{array}$

77. $(u - 2)(2u^2 + 5u + 3)$

78. $(z - 2)(z^2 + z + 1)$

79. $(-x^2 + 2x - 1)(2x + 1)$

80. $(2s^2 - 5s + 6)(3s - 4)$

81. $(t^2 + t - 2)(t^2 - t + 2)$

82. $(y^2 + 3y + 5)(2y^2 - 3y - 1)$

In Exercises 83–112, use a special product formula to perform the multiplication. See Examples 7 and 8.

83. $(x + 2)(x - 2)$

84. $(x - 5)(x + 5)$

85. $(x - 7)(x + 7)$

86. $(x + 1)(x - 1)$

87. $(2 + 7y)(2 - 7y)$

88. $(4 + 3z)(4 - 3z)$

89. $(6 - 4x)(6 + 4x)$

90. $(8 - 3x)(8 + 3x)$

91. $(2a + 5b)(2a - 5b)$

92. $(5u + 12v)(5u - 12v)$

93. $(6x - 9y)(6x + 9y)$

94. $(8x - 5y)(8x + 5y)$

95. $\left(2x - \frac{1}{4}\right)\left(2x + \frac{1}{4}\right)$

96. $\left(\frac{2}{3}x + 7\right)\left(\frac{2}{3}x - 7\right)$

97. $(0.2t + 0.5)(0.2t - 0.5)$

98. $(4a - 0.1b)(4a + 0.1b)$

99. $(x + 5)^2$

100. $(x + 9)^2$

101. $(x - 10)^2$

102. $(u - 7)^2$

103. $(2x + 5)^2$

104. $(3x + 8)^2$

105. $(6x - 1)^2$

106. $(5 - 3z)^2$

107. $(2x - 7y)^2$

108. $(2x + 5y)^2$

109. $[(x + 2) + y]^2$

110. $[(x - 4) - y]^2$

111. $[u - (v - 3)][u + (v - 3)]$

112. $[z + (y + 1)][z - (y + 1)]$

In Exercises 113–116, simplify the expression. See Example 9.

113. $(x + 3)^3$

114. $(y - 2)^3$

115. $(u + v)^3$

116. $(u - v)^3$

In Exercises 117–120, use a graphing utility to graph the expressions for y_1 and y_2. What conclusion can you make? Verify the conclusion analytically.

117. $y_1 = (x + 1)(x^2 - x + 2)$

$y_2 = x^3 + x + 2$

118. $y_1 = (x - 3)^2$

$y_2 = x^2 - 6x + 9$

119. $y_1 = (2x - 3)(x + 2)$

$y_2 = 2x^2 + x - 6$

120. $y_1 = \left(x + \frac{1}{2}\right)\left(x - \frac{1}{2}\right)$

$y_2 = x^2 - \frac{1}{4}$

121. Given the function $f(x) = x^2 - 2x$, find and simplify each of the following.

(a) $f(t - 3)$

(b) $f(2 + h) - f(2)$

122. Given the function $f(x) = 2x^2 - 5x + 4$, find and simplify each of the following.

(a) $f(y + 2)$

(b) $f(1 + h) - f(1)$

Solving Problems

123. A closed box has sides of lengths n, $n + 2$, and $n + 4$ inches. (See figure.)

(a) Find a polynomial function $V(n)$ that describes the volume of the box.

(b) What is the volume if the length of the shortest side is 2 inches?

(c) Write a polynomial function $A(n)$ for the area of the base of the box.

(d) Write a polynomial function for the area of the base if the length and width increase by 4. Show that the polynomial function is $A(n + 4)$.

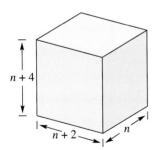

124. A closed box has sides of lengths $2n - 2$, $2n + 2$, and $2n$ inches. (See figure.)

(a) Find a polynomial function $V(n)$ that describes the volume of the box.

(b) What is the volume if the length of the shortest side is 6 inches?

(c) Write a polynomial function $A(n)$ for the area of the base of the box.

(d) Write a polynomial function for the area of the base if the length and width increase by 2. Show that the area of the base is not $A(n + 4)$.

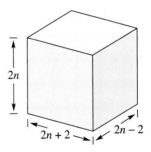

Figure for 124

Geometry   In Exercises 125–128, write an expression that represents the area of the shaded portion of the figure. Then simplify the expression.

125.

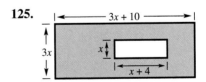

126.

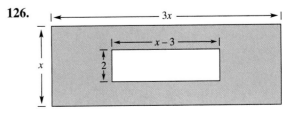

127.

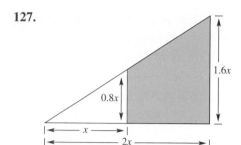

128.

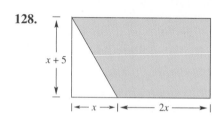

129. *Geometry*   The length of a rectangle is $1\frac{1}{2}$ times its width w. Find (a) the perimeter and (b) the area of the rectangle.

130. *Geometry*   The base of a triangle is $3x$ and its height is $x + 5$. Find the area A of the triangle.

131. *Compound Interest*   After 2 years, an investment of $1000 compounded annually at interest rate r will yield an amount $1000(1 + r)^2$. Find this product.

132. *Compound Interest*   After 2 years, an investment of $1000 compounded annually at an interest rate of 9.5% will yield an amount $1000(1 + 0.095)^2$. Find this product.

Geometric Modeling   In Exercises 133 and 134, use the area model to write two different expressions for the total area. Then equate the two expressions and name the algebraic property that is illustrated.

133.

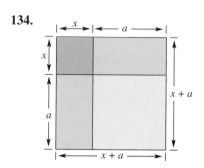

134.

135. *Finding a Pattern*   Perform the multiplications.

(a) $(x - 1)(x + 1)$

(b) $(x - 1)(x^2 + x + 1)$

(c) $(x - 1)(x^3 + x^2 + x + 1)$

From the pattern formed by these products, can you predict the result of $(x - 1)(x^4 + x^3 + x^2 + x + 1)$?

136. *Verification*   Use the FOIL Method to verify the following.

(a) $(x + y)^2 = x^2 + 2xy + y^2$

(b) $(x - y)^2 = x^2 - 2xy + y^2$

(c) $(x - y)(x + y) = x^2 - y^2$

Explaining Concepts

137. Answer parts (a)–(c) of Motivating the Chapter on page 195.

138. Write, from memory, the rules for exponents.

139. Discuss the difference between the expressions $(2x)^3$ and $2x^3$.

140. Give an example of how to use the Distributive Property to multiply two binomials.

141. Explain the meaning of each letter of FOIL as it relates to multiplying two binomials.

142. What is the degree of the product of two polynomials of degrees m and n?

143. *True or False?*   Decide whether the statement is true or false. If false, give an example.

(a) The product of two monomials is a monomial.

(b) The product of two binomials is a binomial.

3.3   Factoring Polynomials

Objectives

1 Factor the greatest common monomial factor from a polynomial.

2 Factor a polynomial by grouping terms.

3 Factor the difference of two squares and factor the sum and difference of two cubes.

4 Factor polynomials completely by repeated factoring.

---

1 Factor the greatest common monomial factor from a polynomial.

Common Monomial Factors

In Section 3.2, you studied ways of multiplying polynomials. In this section, you will study the reverse process—**factoring polynomials.** Here is an example.

Use Distributive Property to multiply.   *Use Distributive Property to factor.*

$$3x(4 - 5x) = 12x - 15x^2 \qquad 12x - 15x^2 = 3x(4 - 5x)$$

Notice that factoring changes a *sum of terms* into a *product of factors.*

To be efficient in factoring expressions, you need to understand the concept of the *greatest common factor* of two (or more) integers or terms. Recall from arithmetic that every integer can be factored into a product of prime numbers. The **greatest common factor** of two or more integers is the greatest integer that is a factor of each number.

Example 1   Finding the Greatest Common Factor

Find the greatest common factor of $6x^5$, $30x^4$, and $12x^3$.

Solution

From the factorizations

$$6x^5 = 2 \cdot 3 \cdot x \cdot x \cdot x \cdot x \cdot x = (6x^3)(x^2)$$

$$30x^4 = 2 \cdot 3 \cdot 5 \cdot x \cdot x \cdot x \cdot x = (6x^3)(5x)$$

$$12x^3 = 2 \cdot 2 \cdot 3 \cdot x \cdot x \cdot x = (6x^3)(2)$$

you can conclude that the greatest common factor is $6x^3$.

Consider the three terms given in Example 1 as terms of the polynomial

$$6x^5 + 30x^4 + 12x^3.$$

The common factor, $6x^3$, of these terms is the **greatest common monomial factor** of the polynomial. When you use the Distributive Property to remove this factor from each term of the polynomial, you are **factoring out** the greatest common monomial factor.

$$6x^5 + 30x^4 + 12x^3 = 6x^3(x^2) + 6x^3(5x) + 6x^3(2)$$ Factor each term.

$$= 6x^3(x^2 + 5x + 2)$$ Factor out common monomial factor.

> ► **Greatest Common Monomial Factor**
>
> If a polynomial in x with integer coefficients has a greatest common monomial factor of the form ax^n, the following statements must be true.
>
> 1. The coefficient a must be the greatest integer that *divides* each of the coefficients in the polynomial.
>
> 2. The variable factor x^n is the highest-powered variable factor that *is common to* all terms of the polynomial.

Example 2   The Greatest Common Monomial Factor

Factor the polynomial $24x^3 - 32x^2$.

Solution

For the terms $24x^3$ and $32x^2$, 8 is the greatest integer factor of 24 and 32 and x^2 is the highest-powered variable factor common to x^3 and x^2. So, the greatest common monomial factor of $24x^3$ and $32x^2$ is $8x^2$.

$$24x^3 - 32x^2 = (8x^2)(3x) - (8x^2)(4)$$
$$= 8x^2(3x - 4)$$

The greatest common monomial factor of a polynomial is usually considered to have a positive coefficient. However, sometimes it is convenient to factor a negative number out of a polynomial. You can see how this is done in the next example.

Study Tip

Whenever factoring a polynomial, remember that you can check your results by multiplying. That is, if you multiply the factors, you should obtain the original polynomial.

Example 3   A Negative Common Monomial Factor

Factor the polynomial $-3x^2 + 12x - 18$ in two ways.

a. Factor out 3.   **b.** Factor out -3.

Solution

a. To factor out the common monomial factor of 3, write the following.

$$-3x^2 + 12x - 18 = 3(-x^2) + 3(4x) + 3(-6)$$
$$= 3(-x^2 + 4x - 6)$$

Check this result by multiplying.

b. To factor out the common monomial factor of -3, write the following.

$$-3x^2 + 12x - 18 = -3(x^2) + (-3)(-4x) + (-3)(6)$$
$$= -3(x^2 - 4x + 6)$$

Check this result by multiplying.

2 Factor a polynomial by grouping terms.

Factoring by Grouping

Some polynomials have common factors that are not simple monomials. For instance, the polynomial

$$x^2(2x - 3) + 4(2x - 3)$$

has the common *binomial* factor $(2x - 3)$. Factoring out this common factor produces

$$x^2(2x - 3) + 4(2x - 3) = (2x - 3)(x^2 + 4).$$

This type of factoring is part of a procedure called **factoring by grouping.**

Example 4   Common Binomial Factors

Factor $5x^2(6x - 5) - 2(6x - 5)$.

Solution
Each of the terms of this polynomial has a binomial factor of $(6x - 5)$. Factoring this binomial out of each term produces the following.

$$5x^2(6x - 5) - 2(6x - 5) = (6x - 5)(5x^2 - 2)$$

In Example 4, the given polynomial was already grouped so that it was easy to determine the common binomial factor. In practice, you will have to do the grouping as well as the factoring.

Study Tip

You should put a polynomial in standard form before trying to factor by grouping. Then group and remove a common monomial factor from the first two terms and the last two terms. Finally, if possible, factor out the common binomial factor.

Example 5   Factoring By Grouping

Factor the polynomials by grouping.

a. $x^3 - 5x^2 + x - 5$ **b.** $4x^3 + 3x - 8x^2 - 6$

Solution

a. $x^3 - 5x^2 + x - 5 = (x^3 - 5x^2) + (x - 5)$ Group terms.

$$= x^2(x - 5) + 1(x - 5)$$ Factor grouped terms.

$$= (x - 5)(x^2 + 1)$$ Common binomial factor

b. $4x^3 + 3x - 8x^2 - 6 = 4x^3 - 8x^2 + 3x - 6$ Standard form

$$= (4x^3 - 8x^2) + (3x - 6)$$ Group terms.

$$= 4x^2(x - 2) + 3(x - 2)$$ Factor grouped terms.

$$= (x - 2)(4x^2 + 3)$$ Common binomial factor

You can *check* to see that you have factored the expression correctly by multiplying out the factors and comparing the result with the original expression.

3 Factor the difference of two squares and factor the sum and difference of two cubes.

Factoring Special Products

Some polynomials have special forms that you should learn to recognize so that you can factor them easily. One of the easiest special polynomial forms to recognize and to factor is the form $u^2 - v^2$, called a **difference of two squares.** This form arises from the special product $(u + v)(u - v)$ in Section 3.2.

> ▶ Difference of Two Squares
>
> Let u and v be real numbers, variables, or algebraic expressions. Then the expression $u^2 - v^2$ can be factored as follows.
>
> $$u^2 - v^2 = (u + v)(u - v)$$
>
> Difference   Opposite signs

To recognize perfect squares, look for coefficients that are squares of integers and for variables raised to *even* powers.

Example 6   Factoring the Difference of Two Squares

Factor the difference of two squares.

a. $x^2 - 64$ **b.** $49x^2 - 81$

Solution

a. $x^2 - 64 = x^2 - 8^2$ Write as difference of two squares.

$\qquad\qquad = (x + 8)(x - 8)$ Factored form

b. $49x^2 - 81 = (7x)^2 - 9^2$ Write as difference of two squares.

$\qquad\qquad = (7x + 9)(7x - 9)$ Factored form

Remember that the rule $u^2 - v^2 = (u + v)(u - v)$ applies to polynomials or expressions in which u and v are themselves expressions.

Example 7   Factoring the Difference of Two Squares

Factor $(x + 2)^2 - 9$.

Solution

$(x + 2)^2 - 9 = (x + 2)^2 - 3^2$ Write as difference of two squares.

$\qquad\qquad = [(x + 2) + 3][(x + 2) - 3]$ Factored form

$\qquad\qquad = (x + 5)(x - 1)$ Simplify.

To check this result, write the original polynomial in standard form. Then multiply the factored form to see that you obtain the same standard form.

▶ **Sum and Difference of Two Cubes**

Let u and v be real numbers, variables, or algebraic expressions. Then the expressions $u^3 + v^3$ and $u^3 - v^3$ can be factored as follows.

Like signs

1. $u^3 + v^3 = (u + v)(u^2 - uv + v^2)$

Unlike signs

Like signs

2. $u^3 - v^3 = (u - v)(u^2 + uv + v^2)$

Unlike signs

Example 8   Factoring Sums and Differences of Cubes

Factor each polynomial.

a. $x^3 - 125$ **b.** $8y^3 + 1$ **c.** $y^3 - 27x^3$

Solution

a. This polynomial is the difference of two cubes because x^3 is the cube of x and 125 is the cube of 5.

$$x^3 - 125 = x^3 - 5^3 \qquad \text{Difference of two cubes}$$
$$= (x - 5)(x^2 + 5x + 5^2) \qquad \text{Factored form}$$
$$= (x - 5)(x^2 + 5x + 25) \qquad \text{Simplify.}$$

b. This polynomial is the sum of two cubes because $8y^3$ is the cube of $2y$ and 1 is the cube of 1.

$$8y^3 + 1 = (2y)^3 + 1^3 \qquad \text{Sum of two cubes}$$
$$= (2y + 1)[(2y)^2 - (2y)(1) + 1^2] \qquad \text{Factored form}$$
$$= (2y + 1)(4y^2 - 2y + 1) \qquad \text{Simplify.}$$

c. $y^3 - 27x^3 = y^3 - (3x)^3 \qquad \text{Difference of two cubes}$

$$= (y - 3x)[y^2 + 3xy + (3x)^2] \qquad \text{Factored form}$$
$$= (y - 3x)(y^2 + 3xy + 9x^2) \qquad \text{Simplify.}$$

You can check the results of Example 8(a) by multiplying, as follows.

$$(x - 5)(x^2 + 5x + 25) = x(x^2 + 5x + 25) - 5(x^2 + 5x + 25)$$
$$= x(x^2) + x(5x) + x(25) - 5(x^2) - 5(5x) - 5(25)$$
$$= x^3 + 5x^2 + 25x - 5x^2 - 25x - 125$$
$$= x^3 - 125$$

4 Factor polynomials completely by repeated factoring.

Factoring Completely

Sometimes the difference of two squares can be hidden by the presence of a common monomial factor. Remember that with *all* factoring techniques, you should first remove any common monomial factors.

Example 9   Factoring Completely

Factor $125x^2 - 80$ completely.

Solution

Because both terms have a common factor of 5, begin by factoring 5 from the expression.

$$125x^2 - 80 = 5(25x^2 - 16) \qquad \text{Factor out common monomial factor.}$$
$$= 5[(5x)^2 - 4^2] \qquad \text{Write as difference of two squares.}$$
$$= 5(5x + 4)(5x - 4) \qquad \text{Factored form}$$

The polynomial in Example 9 is said to be **completely factored** because none of its factors can be further factored using integer coefficients.

Study Tip

The sum of two squares, such as $9m^2 + 1$ in Example 10(b), cannot be factored further using integer coefficients. Such polynomials are called **prime** with respect to the integers. Some other prime polynomials are $x^2 + 4$ and $4x^2 + 9$.

Example 10   Factoring Completely

Factor the polynomials completely.

a. $x^4 - y^4$ **b.** $81m^4 - 1$

Solution

a. $x^4 - y^4 = (x^2 + y^2)(x^2 - y^2)$ Factor as difference of two squares.

$\qquad\qquad = (x^2 + y^2)(x^2 - y^2)$ Find second difference of two squares.

$\qquad\qquad = (x^2 + y^2)(x + y)(x - y)$ Factored completely

b. $81m^4 - 1 = (9m^2 + 1)(9m^2 - 1)$ Factor as difference of two squares.

$\qquad\qquad = (9m^2 + 1)(9m^2 - 1)$ Find second difference of two squares.

$\qquad\qquad = (9m^2 + 1)(3m + 1)(3m - 1)$ Factored completely

Discussing the Concept   Position Functions

Write position functions for the height of an object thrown upward from ground level at initial velocities (in feet per second) of (a) 80, (b) 15, and (c) 44. Completely factor the polynomials and demonstrate their equivalence to the original polynomials.

3.3   Exercises

Integrated Review   Concepts, Skills, and Problem Solving

Keep mathematically in shape by doing these exercises *before* the problems of this section.

Properties and Definitions

1. In your own words, define a function of x.

2. State the definitions of the domain and range of a function of x.

3. Sketch a graph for which y is not a function of x.

4. Sketch a graph for which y is a function of x.

Graphing

In Exercises 5–10, graph the equation. Use the Vertical Line Test to determine whether y is a function of x.

5. $y = 6 - \frac{2}{3}x$ 6. $y = \frac{5}{2}x - 4$

7. $2y - 4x + 3 = 0$ 8. $3x + 2y + 12 = 0$
9. $|y| - x = 0$ 10. $|y| = 2 - x$

Graphs and Models

11. A manufacturer purchases a new computer system for \$175,000. The depreciated value y after t years is given by

$$y = 175{,}000 - 30{,}000t, \quad 0 \le t \le 5.$$

Sketch the graph of the function over its domain.

12. The length of a rectangle is x centimeters and its perimeter is 500 centimeters.

    (a) Express the area A of the rectangle as a function of x.

    (b) Graph the area function in part (a).

Developing Skills

In Exercises 1–10, find the greatest common factor of the expressions. See Example 1.

1. 48, 90, 96   2. 36, 150, 100

3. $3x^2, 12x$ 4. $27x^4, 18x^3$

5. $30z^2, -12z^3$ 6. $-45y, 150y^3$

7. $28b^2, 14b^3, 42b^5$

8. $16x^2y, 84xy^2, 36x^2y^2$

9. $42(x + 8)^2, 63(x + 8)^3$

10. $66(3 - y), 44(3 - y)^2$

In Exercises 11–30, factor out the greatest common monomial factor. (Some of the polynomials may have no common monomial factor other than 1.) See Example 2.

11. $8z - 8$ 12. $5x + 5$

13. $4u + 10$ 14. $-15t - 10$

15. $24x^2 - 18$ 16. $14z^3 + 21$

17. $2x^2 + x$ 18. $-a^3 - 4a$

19. $21u^2 - 14u$ 20. $36y^4 + 24y^2$

21. $11u^2 + 9$

22. $16 - 3y^3$

23. $28x^2 + 16x - 8$

24. $9 - 27y - 15y^2$

25. $3x^2y^2 - 15y$

26. $4uv + 6u^2v^2$

27. $15xy^2 - 3x^2y + 9xy$

28. $4x^2 - 2xy + 3y^2$

29. $14x^4y^3 + 21x^3y^2 + 9x^2$

30. $17x^5y^3 - xy^2 + 34y^2$

In Exercises 31–38, factor a negative real number out of the polynomial and then write the polynomial factor in standard form. See Example 3.

31. $10 - x$ 32. $32 - x^4$

33. $7 - 14x$ 34. $15 - 5x$

35. $16 + 4x - 6x^2$ 36. $12x - 6x^2 - 18$

37. $y - 3y^3 - 2y^2$ 38. $-2t^3 + 4t^2 + 7$

In Exercises 39–42, fill in the blank to complete the factored expression.

39. $2y - \frac{3}{5} = \frac{1}{5}(\quad)$ 40. $3z + \frac{3}{8} = \frac{1}{8}(\quad)$

41. $\frac{3}{2}x + \frac{5}{4} = \frac{1}{4}(\quad)$ 42. $\frac{1}{3}x - \frac{5}{6} = \frac{1}{6}(\quad)$

In Exercises 43–52, factor the polynomial by factoring out the greatest common binomial factor. See Example 4.

43. $2y(y - 3) + 5(y - 3)$

44. $7t(s + 9) - 6(s + 9)$

45. $5x(3x + 2) - 3(3x + 2)$

46. $6(4t - 3) - 5t(4t - 3)$

47. $2(7a + 6) - 3a^2(7a + 6)$

48. $4(5y - 12) + 3y^2(5y - 12)$

49. $8t^3(4t - 1)^2 + 3(4t - 1)^2$

50. $2y^2(y^2 + 6)^3 + 7(y^2 + 6)^3$

51. $(x - 5)(4x + 9) - (3x + 4)(4x + 9)$

52. $(3x + 7)(2x - 1) + (x - 6)(2x - 1)$

In Exercises 53–64, factor the expression by grouping. See Example 5.

53. $x^2 + 25x + x + 25$

54. $x^2 - 7x + x - 7$

55. $y^2 - 6y + 2y - 12$

56. $y^2 + 3y + 4y + 12$

57. $x^3 + 2x^2 + x + 2$

58. $t^3 - 11t^2 + t - 11$

59. $3a^3 - 12a^2 - 2a + 8$

60. $3s^3 + 6s^2 + 5s + 10$

61. $z^4 - 2z + 3z^3 - 6$

62. $4u^4 - 6u - 2u^3 + 3$

63. $5x^3 - 10x^2y + 7xy^2 - 14y^3$

64. $10u^4 - 8u^2v^3 - 12v^4 + 15u^2v$

In Exercises 65–86, factor the difference of two squares. See Examples 6 and 7.

65. $x^2 - 64$

66. $y^2 - 144$

67. $1 - a^2$

68. $16 - b^2$

69. $16y^2 - 9$

70. $9z^2 - 25$

71. $81 - 4x^2$

72. $49 - 64x^2$

73. $4z^2 - y^2$

74. $9u^2 - v^2$

75. $36x^2 - 25y^2$

76. $100a^2 - 49b^2$

77. $u^2 - \frac{1}{16}$

78. $v^2 - \frac{9}{25}$

79. $\frac{4}{9}x^2 - \frac{16}{25}y^2$

80. $\frac{1}{4}x^2 - \frac{36}{49}y^2$

81. $(x - 1)^2 - 16$

82. $(x - 3)^2 - 4$

83. $81 - (z + 5)^2$

84. $36 - (y - 6)^2$

85. $(2x + 5)^2 - (x - 4)^2$

86. $(3y - 1)^2 - (x + 6)^2$

In Exercises 87–98, factor the sum or difference of cubes. See Example 8.

87. $x^3 - 8$

88. $t^3 - 27$

89. $y^3 + 64$

90. $z^3 + 125$

91. $8t^3 - 27$

92. $27s^3 + 64$

93. $27u^3 + 1$

94. $64v^3 - 125$

95. $64a^3 + b^3$

96. $m^3 - 8n^3$

97. $x^3 + 27y^3$

98. $u^3 + 125v^3$

In Exercises 99–108, factor completely. See Examples 9 and 10.

99. $8 - 50x^2$

100. $8y^2 - 18$

101. $8x^3 + 64$

102. $a^3 - 16a$

103. $y^4 - 81$

104. $u^4 - 16$

105. $3x^4 - 300x^2$

106. $6x^5 + 30x^3$

107. $6x^6 - 48y^6$

108. $2u^6 + 54v^6$

In Exercises 109 and 110, factor the expression. (Assume $n > 0$.)

109. $4x^{2n} - 25$

110. $81 - 16y^{4n}$

Graphical Reasoning   In Exercises 111–114, use a graphing utility to graph y_1 and y_2 in the same viewing rectangle. What can you conclude by comparing the two graphs?

111. $y_1 = 3x - 6$
   $y_2 = 3(x - 2)$

112. $y_1 = x^3 - 2x^2$
   $y_2 = x^2(x - 2)$

113. $y_1 = x^2 - 4$
   $y_2 = (x + 2)(x - 2)$

114. $y_1 = x(x + 1) - 4(x + 1)$
   $y_2 = (x + 1)(x - 4)$

Think About It   In Exercises 115 and 116, show all the different groupings that can be used to factor completely the polynomial. Carry out the various factorizations to show that they yield the same result.

115. $3x^3 + 4x^2 - 3x - 4$

116. $6x^3 - 8x^2 + 9x - 12$

Solving Problems

Revenue   The revenue from selling x units of a product at a price of p dollars per unit is given by $R = xp$. In Exercises 117 and 118, factor the expression for revenue and determine an expression that gives the price in terms of x.

117. $R = 800x - 0.25x^2$ **118.** $R = 1000x - 0.4x^2$

119. *Simple Interest*   The total amount of money accrued from a principal of P invested at $r\%$ simple interest for t years is $P + Prt$. Factor this expression.

120. *Chemical Reaction*   The rate of change of a chemical reaction is given by $kQx - kx^2$, where Q is the amount of the original substance, x is the amount of substance formed, and k is a constant of proportionality. Factor this expression.

121. *Geometry*   The area of a rectangle of length l is given by $45l - l^2$. Factor this expression to determine the width of the rectangle.

122. *Geometry*   The area of a rectangle of width w is given by $32w - w^2$. Factor this expression to determine the length of the rectangle.

123. *Surface Area*   The surface area of a rectangular solid of height h and square base with edge of length x is $2x^2 + 4xh$. Factor this expression.

124. *Surface Area*   The surface area of a right circular cylinder is $S = \pi r^2 + 2\pi rh$ (see figure). Factor this expression.

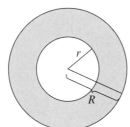

Figure for 124   Figure for 125

125. *Product Design*   A washer on the drive train of a car has an inside radius of r centimeters and an outside radius of R centimeters (see figure). Find the area of one of the flat surfaces of the washer and express the area in factored form.

126. *Geometry*   The cube shown in the figure is formed by solids I, II, III, and IV.

(a) Explain how you could determine the following expressions for volume.

	Volume
Entire cube	a^3
Solid I	$a^2(a - b)$
Solid II	$ab(a - b)$
Solid III	$b^2(a - b)$
Solid IV	b^3

(b) Add the volumes of solids I, II, and III. Factor the result to show that the total volume can be expressed as $(a - b)(a^2 + ab + b^2)$.

(c) Explain why the total volume of solids I, II, and III can also be expressed as $a^3 - b^3$. Then explain how the figure can be used as a geometric model for the *difference of two cubes* factoring pattern.

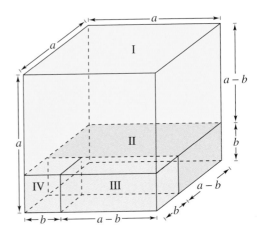

Explaining Concepts

127. Explain what is meant by saying that a polynomial is in factored form.

128. How do you check your result after factoring a polynomial?

129. Describe the method of finding the greatest common factor of two or more integers.

130. Explain how the word *factor* can be used as a noun or as a verb.

131. Give an example of using the Distributive Property to factor a polynomial.

132. Give an example of a polynomial that is prime with respect to the integers.

Mid-Chapter Quiz

Take this quiz as you would take a quiz in class. After you are done, check your work against the answers given in the back of the book.

1. Determine the degree and leading coefficient of the polynomial

$3 - 2x + 4x^3 - 2x^4$.

2. Explain why $2x - 3x^{1/2} + 5$ is not a polynomial.

In Exercises 3–18, perform the indicated operations and simplify.

3. Add $2t^3 + 3t^2 - 2$ to $t^3 + 9$.

4. $(3 - 7y) + (7y^2 + 2y - 3)$

5. $(7x^3 - 3x^2 + 1) - (x^2 - 2x^3)$

6. $(5 - u) - 2[3 - (u^2 + 1)]$

7. $(-5n^2)(-2n^3)$

8. $(-2x^2)^3(x^4)$

9. $\dfrac{6x^7}{(-2x^2)^3}$

10. $\left(\dfrac{4y^2}{5x}\right)^2$

11. $7y(4 - 3y)$

12. $(x - 7)(x + 3)$

13. $(4x - y)(6x - 5y)$

14. $2z(z + 5) - 7(z + 5)$

15. $(6r + 5)(6r - 5)$

16. $(2x - 3)^2$

17. $(x + 1)(x^2 - x + 1)$

18. $(x^2 - 3x + 2)(x^2 + 5x - 10)$

In Exercises 19–22, factor the expression completely.

19. $28a^2 - 21a$

20. $25 - 4x^2$

21. $z^3 + 3z^2 - 9z - 27$

22. $4y^3 - 32x^3$

23. Find all possible products of the form $(5x + m)(2x + n)$ such that $mn = 10$.

24. Find the area of the shaded portion of the figure.

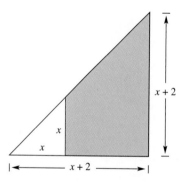

25. An object is thrown downward from the top of a 100-foot building with an initial velocity of -5 feet per second. Use the position function $h(t) = -16t^2 - 5t + 100$ to find the height of the object when $t = 1$ and $t = 2$.

26. A manufacturer can produce and sell x T-shirts per week. The total cost (in dollars) for producing the T-shirts is given by $C = 5x + 2000$ and the total revenue is given by $R = 19x$. Find the profit obtained by selling 1000 T-shirts per week.

3.4   Factoring Trinomials

Objectives

1 Recognize and factor perfect square trinomials.

2 Factor trinomials of the form $x^2 + bx + c$ and $ax^2 + bx + c$.

3 Factor trinomials of the form $ax^2 + bx + c$ by grouping.

4 Factor polynomials using the guidelines for factoring.

---

1 Recognize and factor perfect square trinomials.

Perfect Square Trinomials

A **perfect square trinomial** is the square of a binomial. For instance,

$$x^2 + 6x + 9 = (x + 3)^2$$

is the square of the binomial $(x + 3)$. Perfect square trinomials come in two forms, one in which the middle term is positive and the other in which it is negative.

> ▶ **Perfect Square Trinomials**
>
> Let u and v represent real numbers, variables, or algebraic expressions.
>
> **1.** $u^2 + 2uv + v^2 = (u + v)^2$ **2.** $u^2 - 2uv + v^2 = (u - v)^2$
>
>   Same sign   Same sign

To recognize a perfect square trinomial, remember that the first and last terms must be perfect squares and positive, and the middle term must be twice the product of u and v. (The middle term can be positive or negative.)

Example 1   Factoring Perfect Square Trinomials

a. $x^2 - 4x + 4 = x^2 - 2(2x) + 2^2 = (x - 2)^2$

b. $16y^2 + 24y + 9 = (4y)^2 + 2(4y)(3) + 3^2 = (4y + 3)^2$

c. $9x^2 - 30xy + 25y^2 = (3x)^2 - 2(3x)(5y) + (5y)^2 = (3x - 5y)^2$

Example 2   Removing a Common Monomial Factor First

a. $3x^2 - 30x + 75 = 3(x^2 - 10x + 25)$ Factor out common monomial factor.

$\qquad\qquad\qquad = 3(x - 5)^2$ Factor perfect square trinomial.

b. $16y^3 + 80y^2 + 100y = 4y(4y^2 + 20y + 25)$ Factor out common monomial factor.

$\qquad\qquad\qquad\qquad = 4y(2y + 5)^2$ Factor perfect square trinomial.

---

2 Factor trinomials of the form $x^2 + bx + c$ and $ax^2 + bx + c$.

Factoring Trinomials

To factor a trinomial of the form $x^2 + bx + c$, consider the following.

$$(x + m)(x + n) = x^2 + nx + mx + mn$$
$$= x^2 + \underbrace{(m + n)}x + \underbrace{mn}$$

Sum of   Product
terms   of terms

$$= x^2 + \boxed{b}\,x + \boxed{c}$$

From this, you can see that to factor a trinomial $x^2 + bx + c$ into a product of two binomials, you must find *factors of c whose sum is b*. There are many different techniques for factoring trinomials. The most common is to use "*Guess, Check, and Revise*" with mental math.

Example 3   Factoring a Trinomial

Factor $x^2 + 3x - 4$.

Solution

You need to find two numbers whose product is -4 and whose sum is 3. Using mental math, you can determine that the numbers are 4 and -1.

The product of 4 and -1 is -4.

$$x^2 + 3x - 4 = (x + 4)(x - 1)$$

The sum of 4 and -1 is 3.

Example 4   Factoring Trinomials

Factor each trinomial.

a. $x^2 - 2x - 8$

b. $x^2 - 5x + 6$

Solution

a. You need to find two numbers whose product is -8 and whose sum is -2.

The product of -4 and 2 is -8.

$$x^2 - 2x - 8 = (x - 4)(x + 2)$$

The sum of -4 and 2 is -2.

b. You need to find two numbers whose product is 6 and whose sum is -5.

The product of -3 and -2 is 6.

$$x^2 - 5x + 6 = (x - 3)(x - 2)$$

The sum of -3 and -2 is -5.

Study Tip

Use a list to help you find the two numbers with the required product and sum. For Example 4(a):

Factors of -8	Sum
$1, -8$	-7
$-1, 8$	7
$2, -4$	-2
$-2, 4$	2

Because -2 is the required sum, the correct factorization is

$$x^2 - 2x - 8 = (x - 4)(x + 2).$$

When factoring a trinomial of the form $x^2 + bx + c$, if you have trouble finding two factors of c whose sum is b, try making a list of all the distinct pairs of factors. For instance, consider the trinomial

$$x^2 - 2x - 24.$$

For this trinomial, $c = -24$ and $b = -2$. So, you need two factors of -24 whose sum is -2. Here is the complete list.

Factors of -24	*Sum of Factors*	
$(1)(-24)$	$1 - 24 = -23$	
$(-1)(24)$	$-1 + 24 = 23$	
$(2)(-12)$	$2 - 12 = -10$	
$(-2)(12)$	$-2 + 12 = 10$	
$(3)(-8)$	$3 - 8 = -5$	
$(-3)(8)$	$-3 + 8 = 5$	
$(4)(-6)$	$4 - 6 = -2$	⇐ Correct choice
$(-4)(6)$	$-4 + 6 = 2$	

With experience, you will be able *mentally* to narrow this list down to only two or three possibilities whose sums can then be tested to determine the correct factorization. Here are some suggestions for narrowing down the list.

▶ **Guidelines for Factoring** $x^2 + bx + c$

1. If c is *positive*, its factors have like signs that match the sign of b.

2. If c is *negative*, its factors have different signs.

3. If $|b|$ is small relative to $|c|$, first try those factors of c that are closest to each other in absolute value.

4. If $|b|$ is near $|c|$, first try those factors of c that are farthest from each other in absolute value.

Study Tip

With *any* factoring problem, remember that you can check your result by multiplying. For instance, in Example 5, you can check the result by multiplying $(x - 18)$ by $(x + 1)$ to see that you obtain $x^2 - 17x - 18$.

Remember that not all trinomials are factorable using integers. For instance, $x^2 - 2x - 4$ is not factorable using integers because there is no pair of factors of -4 whose sum is -2.

Example 5   Factoring a Trinomial

Factor $x^2 - 17x - 18$.

Solution

You need to find two numbers whose product is -18 and whose sum is -17. Because $|b| = |-17| = 17$ and $|c| = |-18| = 18$ are close in value, choose factors of -18 that are farthest from each other.

The product of -18 and 1 is -18.

$$x^2 - 17x - 18 = (x - 18)(x + 1)$$

The sum of -18 and 1 is -17.

To factor a trinomial whose leading coefficient is not 1, use the following pattern.

Factors of a

$$ax^2 + bx + c = (\quad x + \quad)(\quad x + \quad)$$

Factors of c

The goal is to find a combination of factors of a and c such that the outer and inner products add up to the middle term bx.

Example 6   Factoring a Trinomial of the Form $ax^2 + bx + c$

Factor $4x^2 + 5x - 6$.

Solution

First, observe that $4x^2 + 5x - 6$ has no common monomial factor. The leading coefficient 4 factors as $(1)(4)$ or as $(2)(2)$. The constant term -6 factors as $(-1)(6)$, $(1)(-6)$, $(-2)(3)$, or $(2)(-3)$. A test of the many possibilities is shown below.

Study Tip

If the original trinomial has no common monomial factor, its binomial factors cannot have common monomial factors. So, in Example 6, you do not have to test factors, such as $(4x - 6)$, that have a common factor of 2.

Factors	$O + I$	
$(x + 1)(4x - 6)$	$-6x + 4x = -2x$	$-2x$ does not equal $5x$.
$(x - 1)(4x + 6)$	$6x - 4x = 2x$	$2x$ does not equal $5x$.
$(x + 6)(4x - 1)$	$-x + 24x = 23x$	$23x$ does not equal $5x$.
$(x - 6)(4x + 1)$	$x - 24x = -23x$	$-23x$ does not equal $5x$.
$(x - 2)(4x + 3)$	$3x - 8x = -5x$	$-5x$ does not equal $5x$.
$(x + 2)(4x - 3)$	$-3x + 8x = 5x$	$5x$ equals $5x$. ✓
$(2x + 1)(2x - 6)$	$-12x + 2x = -10x$	$-10x$ does not equal $5x$.
$(2x - 1)(2x + 6)$	$12x - 2x = 10x$	$10x$ does not equal $5x$.
$(2x + 2)(2x - 3)$	$-6x + 4x = -2x$	$-2x$ does not equal $5x$.
$(2x - 2)(2x + 3)$	$6x - 4x = 2x$	$2x$ does not equal $5x$.
$(x + 3)(4x - 2)$	$-2x + 12x = 10x$	$10x$ does not equal $5x$.
$(x - 3)(4x + 2)$	$2x - 12x = -10x$	$-10x$ does not equal $5x$.

So, the correct factorization is

$$4x^2 + 5x - 6 = (x + 2)(4x - 3).$$

Check this result by multiplying $(x + 2)$ by $(4x - 3)$.

In Example 6, because the polynomial $4x^2 + 5x - 6$ has no common monomial factor, which of the factors listed did not need to be tested?

> ▶ **Guidelines for Factoring** $ax^2 + bx + c$
>
> **1.** First, factor out any common monomial factor.
>
> **2.** Because the resulting trinomial has no common monomial factor, don't use any binomial factors that have a common monomial factor.
>
> **3.** If the middle-term test $(O + I)$ yields the opposite of b, switch the signs of the factors of c.

Example 7   Factoring a Trinomial of the Form $ax^2 + bx + c$

Factor $2x^2 - x - 21$.

Solution

For this trinomial, $a = 2$, which factors as $(1)(2)$, and $c = -21$, which factors as $(1)(-21)$, $(-1)(21)$, $(3)(-7)$, or $(-3)(7)$. Because b is small, avoid the large factors of -21, and test the smaller ones.

Factors	$O + I$	
$(2x + 3)(x - 7)$	$-14x + 3x = -11x$	$-11x$ does not equal $-x$.
$(2x + 7)(x - 3)$	$-6x + 7x = x$	x does not equal $-x$.

Because $(2x + 7)(x - 3)$ results in a middle term that is the opposite of the correct term, you need only switch the signs of the factors of c to obtain the correct factorization.

$$2x^2 - x - 21 = (2x - 7)(x + 3). \qquad \text{Correct factorization}$$

Check this result by multiplying.

Example 8   Factoring a Trinomial of the Form $ax^2 + bx + c$

Factor $6x^2 + 19x + 10$.

Solution

For this trinomial, $a = 6$, which factors as $(1)(6)$ or $(2)(3)$, and $c = 10$, which factors as $(1)(10)$ or $(2)(5)$. Because the trinomial has no common monomial factor, you do not have to test binomial factors that have a common monomial factor.

Factors	$O + I$	
$(x + 10)(6x + 1)$	$x + 60x = 61x$	$61x$ does not equal $19x$.
$(x + 2)(6x + 5)$	$5x + 12x = 17x$	$17x$ does not equal $19x$.
$(2x + 1)(3x + 10)$	$20x + 3x = 23x$	$23x$ does not equal $19x$.
$(2x + 5)(3x + 2)$	$4x + 15x = 19x$	$19x$ equals $19x$. ✓

So, the correct factorization is

$$6x^2 + 19x + 10 = (2x + 5)(3x + 2).$$

Example 9	Factoring Completely

Factor $8x^3 - 60x^2 + 28x$.

Solution

Begin by factoring out the common monomial factor $4x$.

$$8x^3 - 60x^2 + 28x = 4x(2x^2 - 15x + 7)$$

Now, for the new trinomial $2x^2 - 15x + 7$, $a = 2$ and $c = 7$. The possible factorizations of this trinomial are as follows.

Factors	$O + I$	
$(2x - 7)(x - 1)$	$-2x - 7x = -9x$	$-9x$ does not equal $-15x$.
$(2x - 1)(x - 7)$	$-14x - x = -15x$	$-15x$ equals $-15x$. ✓

So, the complete factorization of the original trinomial is

$$8x^3 - 60x^2 + 28x = 4x(2x^2 - 15x + 7) = 4x(2x - 1)(x - 7).$$

Check this result by multiplying.

When factoring a trinomial with a negative leading coefficient, we suggest that you first factor -1 out of the trinomial.

Study Tip

If the leading coefficient is negative, you may find it helpful to factor out -1, as shown in Example 10.

Example 10	A Trinomial with a Negative Leading Coefficient

Factor $-3x^2 + 16x + 35$.

Solution

Begin by factoring out -1.

$$-3x^2 + 16x + 35 = (-1)(3x^2 - 16x - 35)$$

For the new trinomial $3x^2 - 16x - 35$, $a = 3$ and $c = -35$. Some possible factorizations of this trinomial are as follows.

Factors	$O + I$	
$(3x - 1)(x + 35)$	$105x - x = 104x$	$104x$ does not equal $-16x$.
$(3x - 35)(x + 1)$	$3x - 35x = -32x$	$-32x$ does not equal $-16x$.
$(3x - 7)(x + 5)$	$15x - 7x = 8x$	$8x$ does not equal $-16x$.
$(3x - 5)(x + 7)$	$21x - 5x = 16x$	$16x$ does not equal $-16x$.

Because $(3x - 5)(x + 7)$ resulted in a middle term that is the opposite of the correct term, you need only switch the signs of the factors of c to obtain the correct factorization.

$(3x + 5)(x - 7)$	$-21x + 5x = -16x$	$-16x$ equals $-16x$. ✓

So, the correct factorization is

$$-3x^2 + 16x + 35 = (-1)(3x + 5)(x - 7) = (3x + 5)(-x + 7).$$

3 Factor trinomials of the form $ax^2 + bx + c$ by grouping.

Factoring Trinomials by Grouping (Optional)

So far in this section, you have been using "Guess, Check, and Revise" to factor trinomials. An alternative technique is to use factoring by grouping to factor a trinomial. For instance, suppose you rewrote the trinomial $2x^2 + 7x - 15$ as

$$2x^2 + 7x - 15 = 2x^2 + 10x - 3x - 15.$$

Then, by grouping the first two terms and the third and fourth terms, you could factor the polynomial as follows.

$$
\begin{aligned}
2x^2 + 7x - 15 &= 2x^2 + (10x - 3x) - 15 && \text{Rewrite middle term.} \\
&= (2x^2 + 10x) - (3x + 15) && \text{Group terms.} \\
&= 2x(x + 5) - 3(x + 5) && \text{Factor groups.} \\
&= (2x - 3)(x + 5) && \text{Distributive Property}
\end{aligned}
$$

The key to this method of factoring is knowing how to rewrite the middle term. In general, *to factor a trinomial $ax^2 + bx + c$ by grouping, choose factors of the product ac that add up to b and use these factors to rewrite the middle term.*

Example 11   Factoring a Trinomial by Grouping

Use factoring by grouping to factor the trinomial.

$$3x^2 + 5x - 2$$

Solution

For the trinomial $3x^2 + 5x - 2$, $ac = 3(-2) = -6$, which has factors 6 and -1 that add up to 5. So, rewrite the middle term as $5x = 6x - x$. This produces the following.

$$
\begin{aligned}
3x^2 + 5x - 2 &= 3x^2 + (6x - x) - 2 && \text{Rewrite middle term.} \\
&= (3x^2 + 6x) - (x + 2) && \text{Group terms.} \\
&= 3x(x + 2) - (x + 2) && \text{Factor groups.} \\
&= (x + 2)(3x - 1) && \text{Distributive Property}
\end{aligned}
$$

So, the trinomial factors as

$$3x^2 + 5x - 2 = (x + 2)(3x - 1).$$

Check this result by multiplying, as follows.

Check

$$
\begin{aligned}
(x + 2)(3x - 1) &= x(3x - 1) + 2(3x - 1) && \text{Distributive Property} \\
&= 3x^2 - x + 6x - 2 && \text{Distributive Property} \\
&= 3x^2 + 5x - 2 && \text{Combine like terms. } \checkmark
\end{aligned}
$$

What do you think of this optional technique? Some people think that it is more efficient than the trial-and-error process, especially when the coefficients a and c have many factors.

4 Factor polynomials using the guidelines for factoring.

Summary of Factoring

▶ **Guidelines for Factoring Polynomials**

1. Factor out any common factors.

2. Factor according to one of the special polynomial forms: difference of squares, sum or difference of cubes, or perfect square trinomials.

3. Factor trinomials using the methods for $a = 1$ and $a \neq 1$.

4. Factor by grouping—for polynomials with four terms.

5. Check to see if the factors themselves can be factored further.

6. Check the results by multiplying the factors.

Example 12 Factoring Polynomials

Factor each polynomial completely.

a. $3x^2 - 108$ **b.** $4x^3 - 32x^2 + 64x$

c. $6x^3 + 27x^2 - 15x$ **d.** $x^3 - 3x^2 - 4x + 12$

Solution

a. $3x^2 - 108 = 3(x^2 - 36)$ Factor out common factor.

$= 3(x + 6)(x - 6)$ Difference of two squares

b. $4x^3 - 32x^2 + 64x = 4x(x^2 - 8x + 16)$ Factor out common factor.

$= 4x(x - 4)^2$ Perfect square trinomial

c. $6x^3 + 27x^2 - 15x = 3x(2x^2 + 9x - 5)$ Factor out common factor.

$= 3x(2x - 1)(x + 5)$ Factor.

d. $x^3 - 3x^2 - 4x + 12 = (x^3 - 3x^2) + (-4x + 12)$ Group terms.

$= x^2(x - 3) - 4(x - 3)$ Factor out common factors.

$= (x - 3)(x^2 - 4)$ Distributive Property

$= (x - 3)(x + 2)(x - 2)$ Difference of two squares

Discussing the Concept   **Creating a Test**

Create five factoring problems that you think represent a fair test of a person's factoring skills. Discuss how it is possible to *create* polynomials that are factorable. Exchange problems with another person in your class. Do each other's problems, then check each other's work.

3.4   Exercises

Integrated Review   Concepts, Skills, and Problem Solving

Keep mathematically in shape by doing these exercises *before* the problems of this section.

Properties and Definitions

1. Explain why a function of x cannot have two y-intercepts.

2. What is the leading coefficient of the polynomial $5t - 3t^2 + 6t^3 - 4$?

In Exercises 3 and 4, write an inequality involving absolute values to represent the verbal statement.

3. The set of all real numbers x whose distance from 0 is less than 5.

4. The set of all real numbers x whose distance from 6 is more than 3.

Slope

In Exercises 5–10, plot the points on a rectangular coordinate system and find the slope (if possible) of the line passing through the points.

5. $(-3, 2), (5, -4)$

6. $(2, 8), (7, -3)$

7. $\left(\frac{5}{2}, \frac{7}{2}\right), \left(\frac{7}{3}, -2\right)$

8. $\left(-\frac{9}{4}, -\frac{1}{4}\right), \left(-3, \frac{9}{2}\right)$

9. $(6, 4), (6, -3)$

10. $(-4, 5), (7, 5)$

Problem Solving

11. You borrow $12,000 for 6 months. You agree to pay back the principal and interest (finance charges) in one lump sum. What will be the amount of the payment if the interest rate is 12%?

12. A truck driver traveled at an average speed of 54 miles per hour on a 100-mile trip. On the return trip with the truck fully loaded, the average speed was 45 miles per hour. Find the average speed for the round trip.

Developing Skills

In Exercises 1–20, factor the trinomial. See Examples 1 and 2.

1. $x^2 + 4x + 4$

2. $z^2 + 6z + 9$

3. $a^2 - 12a + 36$

4. $y^2 - 14y + 49$

5. $25y^2 - 10y + 1$

6. $4z^2 + 28z + 49$

7. $9b^2 + 12b + 4$

8. $4x^2 - 4x + 1$

9. $u^2 + 8uv + 16v^2$

10. $x^2 - 14xy + 49y^2$

11. $36x^2 - 60xy + 25y^2$

12. $4y^2 + 20yz + 25z^2$

13. $5x^2 + 30x + 45$

14. $4x^2 - 32x + 64$

15. $2x^3 + 24x^2 + 72x$

16. $3u^3 - 48u^2 + 192u$

17. $20v^4 - 60v^3 + 45v^2$

18. $-18y^3 - 12y^2 - 2y$

19. $\frac{1}{4}x^2 - \frac{2}{3}x + \frac{4}{9}$

20. $\frac{1}{9}x^2 + \frac{8}{15}x + \frac{16}{25}$

In Exercises 21–24, find all values of b such that the expression is a perfect square trinomial.

21. $x^2 + bx + 81$

22. $x^2 + bx + \frac{9}{16}$

23. $4x^2 + bx + 9$

24. $16x^2 + bxy + 25y^2$

In Exercises 25–28, find a real number c such that the expression is a perfect square trinomial.

25. $x^2 + 8x + c$

26. $x^2 + 12x + c$

27. $y^2 - 6y + c$

28. $z^2 - 20z + c$

In Exercises 29–36, find the missing factor.

29. $x^2 + 5x + 4 = (x + 4)(\quad)$

30. $a^2 + 2a - 8 = (a + 4)(\quad)$

31. $y^2 - y - 20 = (y + 4)(\quad)$

32. $y^2 + 6y + 8 = (y + 4)(\quad)$

33. $x^2 - 2x - 24 = (x + 4)(\quad)$

34. $x^2 + 7x + 12 = (x + 4)(\quad)$

35. $z^2 - 6z + 8 = (z - 4)(\quad)$

36. $z^2 + 2z - 24 = (z - 4)(\quad)$

In Exercises 37–50, factor the trinomial. See Examples 3–5.

37. $x^2 + 4x + 3$ **38.** $x^2 + 7x + 10$

39. $x^2 - 5x + 6$ **40.** $x^2 - 10x + 24$

41. $y^2 + 7y - 30$ **42.** $m^2 - 3m - 10$

43. $t^2 - 4t - 21$ **44.** $x^2 + 4x - 12$

45. $x^2 - 20x + 96$ **46.** $y^2 - 35y + 300$

47. $x^2 - 2xy - 35y^2$ **48.** $u^2 + 5uv + 6v^2$

49. $x^2 + 30xy + 216y^2$ **50.** $a^2 - 21ab + 110b^2$

In Exercises 51–56, find all values of b for which the trinomial can be factored.

51. $x^2 + bx + 18$ **52.** $x^2 + bx + 14$

53. $x^2 + bx - 21$ **54.** $x^2 + bx - 7$

55. $x^2 + bx + 35$ **56.** $x^2 + bx - 38$

In Exercises 57–60, find two values of c for which the trinomial can be factored.

57. $x^2 + 6x + c$ **58.** $x^2 + 9x + c$

59. $x^2 - 3x + c$ **60.** $x^2 - 12x + c$

In Exercises 61–66, find the missing factor.

61. $5x^2 + 18x + 9 = (x + 3)(\quad)$

62. $5x^2 + 19x + 12 = (x + 3)(\quad)$

63. $5a^2 + 12a - 9 = (a + 3)(\quad)$

64. $5c^2 + 11c - 12 = (c + 3)(\quad)$

65. $2y^2 - 3y - 27 = (y + 3)(\quad)$

66. $3y^2 - y - 30 = (y + 3)(\quad)$

In Exercises 67–92, factor the trinomial, if possible. (Some of the trinomials may be prime.) See Examples 6–10.

67. $3x^2 + 4x + 1$ **68.** $5x^2 + 7x + 2$

69. $7x^2 + 15x + 2$ **70.** $3x^2 + 8x + 5$

71. $2x^2 - 9x + 9$ **72.** $2t^2 - 13t + 20$

73. $6x^2 - 11x + 3$ **74.** $4y^2 - 5y - 9$

75. $3t^2 - 4t - 10$ **76.** $2z^2 + 3z + 8$

77. $6b^2 + 19b - 7$ **78.** $10x^2 - 24x - 18$

79. $18y^2 + 35y + 12$ **80.** $20x^2 + x - 12$

81. $-2x^2 - x + 6$ **82.** $-6x^2 + 5x - 6$

83. $1 - 11x - 60x^2$ **84.** $2 + 5x - 12x^2$

85. $6x^2 - 3x - 84$ **86.** $12x^2 + 32x - 12$

87. $60y^3 + 35y^2 - 50y$ **88.** $12x^2 + 42x^3 - 54x^4$

89. $10a^2 + 23ab + 6b^2$ **90.** $6u^2 - 5uv - 4v^2$

91. $24x^2 - 14xy - 3y^2$ **92.** $10x^2 + 9xy - 9y^2$

In Exercises 93–98, factor the trinomial by grouping. See Example 11.

93. $3x^2 + 10x + 8$ **94.** $2x^2 + 9x + 9$

95. $6x^2 + x - 2$ **96.** $6x^2 - x - 15$

97. $15x^2 - 11x + 2$ **98.** $12x^2 - 28x + 15$

In Exercises 99–114, factor completely. See Example 12.

99. $3x^4 - 12x^3$

100. $20y^2 - 45$

101. $10t^3 + 2t^2 - 36t$

102. $16z^3 - 56z^2 + 49z$

103. $54x^3 - 2$

104. $3t^3 - 24$

105. $27a^3b^4 - 9a^2b^3 - 18ab^2$

106. $8m^3n + 20m^2n^2 - 48mn^3$

107. $x^3 + 2x^2 - 16x - 32$

108. $x^3 - 7x^2 - 4x + 28$

109. $36 - (z + 3)^2$

110. $(x + 7y)^2 - 4a^2$

111. $x^2 - 10x + 25 - y^2$

112. $a^2 - 2ab + b^2 - 16$

113. $x^8 - 1$

114. $x^4 - 16y^4$

Graphical Verification   In Exercises 115–118, use a graphing utility to graph the two equations on the same screen. What can you conclude?

115. $y_1 = x^2 + 6x + 9$

$\quad y_2 = (x + 3)^2$

116. $y_1 = 4x^2 - 4x + 1$

$\quad y_2 = (2x - 1)^2$

117. $y_1 = x^2 + 2x - 3$

$\quad y_2 = (x - 1)(x + 3)$

118. $y_1 = 3x^2 - 8x - 16$

$\quad y_2 = (3x + 4)(x - 4)$

Geometric Factoring Models   In Exercises 119–122, match the geometric factoring model with the correct factoring formula. [The models are labeled (a), (b), (c), and (d).]

(a)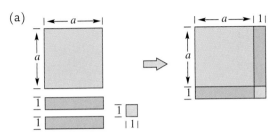

(b)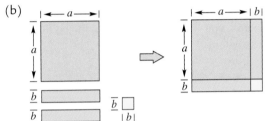

(c)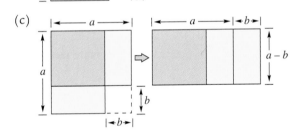

(d)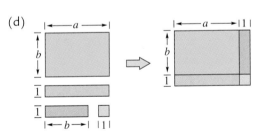

119. $a^2 - b^2 = (a + b)(a - b)$

120. $a^2 + 2a + 1 = (a + 1)^2$

121. $a^2 + 2ab + b^2 = (a + b)^2$

122. $ab + a + b + 1 = (a + 1)(b + 1)$

Geometry   In Exercises 123 and 124, write, in factored form, an expression for the shaded portion of the figure.

123.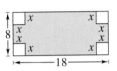

124.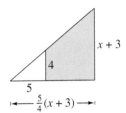

125. *Number Problem*   Let n be an integer.

(a) Factor $8n^3 - 8n$ so as to verify that it represents three consecutive even integers. (*Hint:* Show that each factor has a common factor of 2.)

(b) If $n = 10$, what are the three integers?

126. *Number Problem*   Let n be an integer.

(a) Factor $8n^3 + 12n^2 - 2n - 3$ so as to verify that it represents three consecutive odd integers.

(b) If $n = 15$, what are the three integers?

Explaining Concepts

127. In your own words, explain how you would factor $x^2 - 5x + 6$.

128. Give an example of a prime trinomial.

129. Explain how you can check the factors of a trinomial. Give an example.

130. *Error Analysis*   Identify the error.

$$9x^2 - 9x - 54 = (3x + 6)(3x - 9)$$
$$= 3(x + 2)(x - 3)$$

131. Is $x(x + 2) - 2(x + 2)$ in completely factored form? If not, show the complete factorization.

132. Is $(2x - 4)(x + 1)$ in completely factored form? If not, show the complete factorization.

3.5 Solving Polynomial Equations

Objectives

1 Use the Zero-Factor Property to solve an equation.

2 Use factoring to solve a quadratic equation.

3 Solve a polynomial equation by factoring.

4 Solve an application problem by factoring.

1 Use the Zero-Factor Property to solve an equation.

The Zero-Factor Property

You have spent the first part of this chapter developing skills for simplifying and factoring polynomials. In this section, you will use these skills with the following **Zero-Factor Property** to solve polynomial equations.

Study Tip

The Zero-Factor Property is just another way of saying that the only way the product of two or more factors can be zero is if one (or more) of the factors is zero.

▶ **Zero-Factor Property**

Let a and b be real numbers, variables, or algebraic expressions. If a and b are factors such that

$$ab = 0$$

then $a = 0$ or $b = 0$. This property also applies to three or more factors.

The Zero-Factor Property is the primary property for solving equations in algebra. For instance, to solve the equation

$$(x - 2)(x + 3) = 0 \qquad \text{Original equation}$$

you can use the Zero-Factor Property to conclude that either $(x - 2)$ or $(x + 3)$ equals zero. Setting the first factor equal to zero implies that $x = 2$ is a solution.

$$x - 2 = 0 \quad \Longrightarrow \quad x = 2 \qquad \text{First solution}$$

Similarly, setting the second factor equal to zero implies that $x = -3$ is a solution.

$$x + 3 = 0 \quad \Longrightarrow \quad x = -3 \qquad \text{Second solution}$$

So, the equation $(x - 2)(x + 3) = 0$ has exactly two solutions: 2 and -3. You can check these solutions by substituting them into the original equation.

Check

$$(x - 2)(x + 3) = 0 \qquad \text{Original equation}$$
$$(2 - 2)(2 + 3) \overset{?}{=} 0 \qquad \text{Substitute 2 for } x.$$
$$(0)(5) = 0 \qquad \text{First solution checks.} \checkmark$$
$$(-3 - 2)(-3 + 3) \overset{?}{=} 0 \qquad \text{Substitute } -3 \text{ for } x.$$
$$(-5)(0) = 0 \qquad \text{Second solution checks.} \checkmark$$

2 Use factoring to solve a quadratic equation.

Solving Quadratic Equations by Factoring

A **quadratic equation** in general form is an equation of the form

$$ax^2 + bx + c = 0.$$

Here are some examples.

$$x^2 - x - 6 = 0, \qquad 3x^2 + 2x - 1 = 0, \qquad \text{and} \qquad x^2 + 3x = 0$$

In the next four examples, note how you can combine your factoring skills with the Zero-Factor Property to solve quadratic equations.

Technology: Tip

You can use a graphing utility to find the solutions of an equation. For instance, to find the solutions of the equation in Example 1, graph the equation

$$y = x^2 - x - 12$$

as shown below. The two solutions correspond to the x-intercepts of the graph. Use the root or zero feature of the graphing utility to find the x-intercepts: $x = -3$ and $x = 4$. Consult the user's guide of your graphing utility for keystrokes.

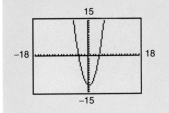

Example 1   Using Factoring to Solve a Quadratic Equation

Solve $x^2 - x - 12 = 0$.

Solution

First, check to see that the right side of the equation is zero. Next, factor the left side of the equation. Finally, apply the Zero-Factor Property to find the solutions.

$x^2 - x - 12 = 0$	Original equation
$(x + 3)(x - 4) = 0$	Factor left side of equation.
$x + 3 = 0$	Set 1st factor equal to 0.
$x = -3$	Subtract 3 from both sides.
$x - 4 = 0$	Set 2nd factor equal to 0.
$x = 4$	Add 4 to both sides.

The equation has two solutions: -3 and 4.

Check

$x^2 - x - 12 = 0$	Original equation
$(-3)^2 - (-3) - 12 \overset{?}{=} 0$	Substitute -3 for x.
$9 + 3 - 12 \overset{?}{=} 0$	Simplify.
$0 = 0$	Solution checks. ✓

Check

$x^2 - x - 12 = 0$	Original equation
$(4)^2 - (4) - 12 \overset{?}{=} 0$	Substitute 4 for x.
$16 - 4 - 12 \overset{?}{=} 0$	Simplify.
$0 = 0$	Solution checks. ✓

Factoring and the Zero-Factor Property allow you to solve a quadratic equation by converting it into two *linear* equations, which you already know how to solve. This is a common strategy in algebra—breaking down a given problem into simpler parts, each of which can be solved by previously learned methods.

If the Zero-Factor Property is to be used, a polynomial equation *must* be written in **general form.** That is, the polynomial must be on one side of the equation and zero must be the only term on the other side. For instance, to write $x^2 - 3x = 18$ in general form, subtract 18 from both sides of the equation.

$$x^2 - 3x = 18 \qquad\qquad \text{Original equation}$$

$$x^2 - 3x - 18 = 18 - 18 \qquad\qquad \text{Subtract 18 from both sides.}$$

$$x^2 - 3x - 18 = 0 \qquad\qquad \text{General form}$$

To solve this equation, factor the left side as $(x + 3)(x - 6)$ and then form the linear equations $x + 3 = 0$ and $x - 6 = 0$. The solutions of these two linear equations are -3 and 6, respectively. The general strategy for solving a quadratic equation by factoring is summarized in the following guidelines.

▶ **Guidelines for Solving Quadratic Equations**

1. Write the quadratic equation in general form.

2. Factor the left side of the equation.

3. Set each factor with a variable equal to zero.

4. Solve each linear equation.

5. Check each solution in the original equation.

Example 2   Solving a Quadratic Equation by Factoring

Solve $3x^2 + 5x = 12$.

Solution

$$3x^2 + 5x = 12 \qquad\qquad \text{Original equation}$$

$$3x^2 + 5x - 12 = 0 \qquad\qquad \text{Write in general form.}$$

$$(3x - 4)(x + 3) = 0 \qquad\qquad \text{Factor left side of equation.}$$

$$3x - 4 = 0 \quad\Longrightarrow\quad x = \frac{4}{3} \qquad\qquad \text{Set 1st factor equal to 0 and solve for } x.$$

$$x + 3 = 0 \quad\Longrightarrow\quad x = -3 \qquad\qquad \text{Set 2nd factor equal to 0 and solve for } x.$$

The solutions are $\frac{4}{3}$ and -3. Check these solutions in the original equation.

Be sure you see that the Zero-Factor Property can be applied only to a product that is equal to *zero*. For instance, you cannot conclude from the equation

$$x(x - 3) = 10$$

that $x = 10$ and $x - 3 = 10$ yield solutions. Instead, you must first write the equation in general form and then factor the left side, as follows.

$$x^2 - 3x - 10 = 0 \quad\Longrightarrow\quad (x - 5)(x + 2) = 0$$

Now, from the factored form you can see that the solutions are 5 and -2.

In Examples 1 and 2, the original equations each involved a second-degree (quadratic) polynomial and each had *two different* solutions. You will sometimes encounter second-degree polynomial equations that have only one (repeated) solution. This occurs when the left side of the equation is a perfect square trinomial, as shown in Example 3.

Example 3   A Quadratic Equation with a Repeated Solution

Solve $x^2 - 6x + 11 = 2$.

Solution

$x^2 - 6x + 11 = 2$	Original equation
$x^2 - 6x + 9 = 0$	Write in general form.
$(x - 3)^2 = 0$	Factor left side of equation.
$x - 3 = 0 \implies x = 3$	Set factor equal to 0 and solve for x.

Note that even though the left side of this equation has two factors, the factors are the same. So, the only solution of the equation is 3.

Check

$x^2 - 6x + 11 = 2$	Original equation
$(3)^2 - 6(3) + 11 \overset{?}{=} 2$	Substitute 3 for x.
$9 - 18 + 11 \overset{?}{=} 2$	Simplify.
$2 = 2$	Solution checks. ✓

Example 4   Solving a Polynomial Equation

Solve $(x + 3)(x + 6) = 4$.

Solution

Begin by multiplying the factors on the left side.

$(x + 3)(x + 6) = 4$	Original equation
$x^2 + 9x + 18 = 4$	Multiply factors.
$x^2 + 9x + 14 = 0$	Write in general form.
$(x + 2)(x + 7) = 0$	Factor left side of equation.
$x + 2 = 0 \implies x = -2$	Set 1st factor equal to 0 and solve for x.
$x + 7 = 0 \implies x = -7$	Set 2nd factor equal to 0 and solve for x.

The equation has two solutions: -2 and -7.

Check

$$(x + 3)(x + 6) = 4 \qquad\qquad (x + 3)(x + 6) = 4$$
$$(-2 + 3)(-2 + 6) \overset{?}{=} 4 \qquad (-7 + 3)(-7 + 6) \overset{?}{=} 4$$
$$(1)(4) = 4 \checkmark \qquad\qquad (-4)(-1) = 4 \checkmark$$

3 Solve a polynomial equation by factoring.

Solving Polynomial Equations by Factoring

Example 5 Solving a Polynomial Equation with Three Factors

Solve $3x^3 = 15x^2 + 18x$.

Solution

$3x^3 = 15x^2 + 18x$	Original equation
$3x^3 - 15x^2 - 18x = 0$	Write in general form.
$3x(x^2 - 5x - 6) = 0$	Factor out common factor.
$3x(x - 6)(x + 1) = 0$	Factor.
$3x = 0 \implies x = 0$	Set 1st factor equal to 0.
$x - 6 = 0 \implies x = 6$	Set 2nd factor equal to 0.
$x + 1 = 0 \implies x = -1$	Set 3rd factor equal to 0.

Check the three solutions 0, 6, and -1 in the original equation.

Notice that the equation in Example 5 is a third-degree equation and has three solutions. This is not a coincidence. In general, a polynomial equation can have *at most* as many solutions as its degree. For instance, a second-degree equation can have zero, one, or two solutions, and no more. Notice that the equation in Example 6 is a fourth-degree equation and has four solutions.

Example 6 Solving a Polynomial Equation with Four Factors

Solve $x^4 + x^3 - 4x^2 - 4x = 0$.

Solution

$x^4 + x^3 - 4x^2 - 4x = 0$	Original equation
$x(x^3 + x^2 - 4x - 4) = 0$	Factor out common factor.
$x[(x^3 + x^2) + (-4x - 4)] = 0$	Group terms.
$x[x^2(x + 1) - 4(x + 1)] = 0$	Factor grouped terms.
$x[(x + 1)(x^2 - 4)] = 0$	Distributive Property
$x(x + 1)(x + 2)(x - 2) = 0$	Difference of two squares
$x = 0 \implies x = 0$	
$x + 1 = 0 \implies x = -1$	
$x + 2 = 0 \implies x = -2$	
$x - 2 = 0 \implies x = 2$	

Check the four solutions 0, -1, -2, and 2 in the original equation.

---

4 Solve an application problem by factoring.

Applications

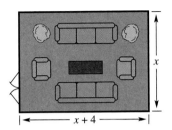

Figure 3.3

Example 7   Geometry

A rectangular room has an area of 192 square feet. The length of the room is 4 feet more than its width, as shown in Figure 3.3. Find the dimensions of the room.

Solution

Verbal Model:

$$\boxed{\text{Length}} \cdot \boxed{\text{Width}} = \boxed{\text{Area}}$$

Labels: Width $= x$ (feet)
            Length $= x + 4$ (feet)
            Area $= 192$ (square feet)

Equation: $(x + 4)x = 192$

$$x^2 + 4x - 192 = 0$$

$$(x + 16)(x - 12) = 0$$

$$x = -16 \quad \text{or} \quad x = 12$$

Because the negative solution does not make sense, choose the positive solution $x = 12$. When the width of the room is 12 feet, the length of the room is

Length $= x + 4 = 12 + 4 = 16$ feet.

So, the room is 12 feet by 16 feet. Check this solution in the original statement of the problem.

Example 8   A Falling-Body Problem

The height of a rock dropped into a well that is 64 feet deep above the water level is given by the position function $h(t) = -16t^2 + 64$, where the height is measured in feet and the time t is measured in seconds. (See Figure 3.4.) How long will it take the rock to hit the water at the bottom of the well?

Solution

In Figure 3.4, note that the water level of the well corresponds to a height of 0 feet. So, substitute a height of 0 for $h(t)$ in the equation and solve for t.

$$0 = -16t^2 + 64 \qquad \text{Substitute 0 for } h(t).$$

$$16t^2 - 64 = 0 \qquad \text{Write in general form.}$$

$$16(t^2 - 4) = 0 \qquad \text{Factor out common factor.}$$

$$16(t + 2)(t - 2) = 0 \qquad \text{Difference of two squares}$$

$$t = -2 \quad \text{or} \quad t = 2 \qquad \text{Solutions}$$

Because a time of -2 seconds does not make sense in this problem, choose the positive solution $t = 2$, and conclude that the rock hits the water 2 seconds after it is dropped. Check this solution in the original statement of the problem.

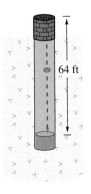

Figure 3.4

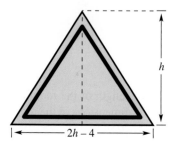

Figure 3.5

Example 9   Geometry

A triangular roadside sign has a base that is to be 4 feet less than twice its height, as shown in Figure 3.5. A local zoning ordinance restricts the area of signs to a maximum of 24 square feet. Find the base and height of the largest sign that meets the zoning ordinance.

Solution

Verbal Model: $\frac{1}{2} \cdot$ Base $\cdot$ Height $=$ Area

Labels: Height $= h$ (feet)
             Base $= 2h - 4$ (feet)
             Area $= 24$ (square feet)

Equation: $\frac{1}{2}(2h - 4)(h) = 24$

$$h^2 - 2h = 24$$

$$h^2 - 2h - 24 = 0$$

$$(h + 4)(h - 6) = 0$$

$$h = -4 \quad \text{or} \quad h = 6$$

Because the height must be positive, discard -4 as a solution and choose $h = 6$. So, the height of the sign is 6 feet, and the base is

Base $= 2h - 4$ Equation for base

$= 2(6) - 4$ Substitute 6 for h.

$= 8$ feet.   Simplify.

Check this solution in the original statement of the problem.

Discussing the Concept   Approximating a Solution

The polynomial equation

$$x^3 - x - 3 = 0$$

cannot be solved algebraically using any of the techniques described in this book. It does, however, have one solution that is a real number.

a. *Graphical Solution:* Use a graphing utility to estimate the solution.

b. *Numerical Solution:* Use a graphing utility to create a table such as the one below and estimate the solution.

x	1.0	1.1	1.2	1.3	1.4	1.5	1.6	1.7	1.8	1.9	2.0
$x^3 - x - 3$	-3										3

3.5   Exercises

Integrated Review   Concepts, Skills, and Problem Solving

Keep mathematically in shape by doing these exercises *before* the problems of this section.

Properties and Definitions

In Exercises 1–4, name the property illustrated.

1. $3uv - 3uv = 0$ **2.** $5z \cdot 1 = 5z$

3. $2s(1 - s) = 2s - 2s^2$ **4.** $(3x)y = 3(xy)$

Solving Equations

In Exercises 5–10, solve the equation.

5. $4 - \frac{1}{2}x = 6$ **6.** $500 - 0.75x = 235$

7. $4(x - 3) - (4x + 5) = 0$

8. $12(3 - x) = 5 - 7(2x + 1)$

9. $\dfrac{12 + x}{4} = 13$ **10.** $8(t - 24) = 0$

Problem Solving

11. The cost of producing x units is $C = 12 + 8x$. The revenue for selling x units is $R = 16x - \frac{1}{4}x^2$, where $0 \le x \le 20$. The profit is $P = R - C$.

   (a) Perform the subtraction required to find the polynomial representing profit.

   (b) Use a graphing utility to graph the polynomial representing profit.

   (c) Determine the profit when $x = 16$ units are produced and sold.

12. An object is dropped from a construction project 576 feet above the ground. Find the time t for the object to reach the ground by solving the equation $-16t^2 + 576 = 0$.

Developing Skills

In Exercises 1–12, use the Zero-Factory Property to solve the equation.

1. $2x(x - 8) = 0$ **2.** $z(z + 6) = 0$

3. $(y - 3)(y + 10) = 0$ **4.** $(s - 16)(s + 15) = 0$

5. $25(a + 4)(a - 2) = 0$ **6.** $17(t - 3)(t + 8) = 0$

7. $(2t + 5)(3t + 1) = 0$

8. $(5x - 3)(x - 8) = 0$

9. $4x(2x - 3)(2x + 25) = 0$

10. $\frac{1}{5}x(x - 2)(3x + 4) = 0$

11. $(x - 3)(2x + 1)(x + 4) = 0$

12. $(y - 39)(2y + 7)(y + 12) = 0$

In Exercises 13–66, solve the equation by factoring. See Examples 1–6.

13. $5y - y^2 = 0$ **14.** $3x^2 + 9x = 0$

15. $9x^2 + 15x = 0$ **16.** $4x^2 - 6x = 0$

17. $x(x + 2) - 10(x + 2) = 0$

18. $x(x - 15) + 3(x - 15) = 0$

19. $u(u - 3) + 3(u - 3) = 0$

20. $x(x + 10) - 2(x + 10) = 0$

21. $x^2 - 25 = 0$ **22.** $x^2 - 121 = 0$

23. $3y^2 - 48 = 0$ **24.** $25z^2 - 100 = 0$

25. $x^2 - 3x - 10 = 0$ **26.** $x^2 - x - 12 = 0$

27. $x^2 - 10x + 24 = 0$ **28.** $20 - 9x + x^2 = 0$

29. $4x^2 + 15x = 25$ **30.** $14x^2 + 9x = -1$

31. $7 + 13x - 2x^2 = 0$ **32.** $11 + 32y - 3y^2 = 0$

33. $m^2 - 8m + 18 = 2$ **34.** $a^2 + 4a + 10 = 6$

35. $x^2 + 16x + 57 = -7$

36. $x^2 - 12x + 21 = -15$

37. $4z^2 - 12z + 15 = 6$

38. $16t^2 + 48t + 40 = 4$

39. $x(x - 5) = 36$ **40.** $s(s + 4) = 96$

41. $y(y + 6) = 72$ **42.** $x(x - 4) = 12$

43. $t(2t - 3) = 35$ **44.** $3u(3u + 1) = 20$

45. $(a + 2)(a + 5) = 10$

46. $(x - 8)(x - 7) = 20$

47. $(x - 4)(x + 5) = 10$

48. $(u - 6)(u + 4) = -21$

49. $(t - 2)^2 = 16$ **50.** $(s + 4)^2 = 49$

51. $9 = (x + 2)^2$

52. $1 = (y + 3)^2$

53. $x^3 - 19x^2 + 84x = 0$

54. $x^3 + 18x^2 + 45x = 0$

55. $6t^3 = t^2 + t$

56. $3u^3 = 5u^2 + 2u$

57. $z^2(z + 2) - 4(z + 2) = 0$

58. $16(3 - u) - u^2(3 - u) = 0$

59. $a^3 + 2a^2 - 9a - 18 = 0$

60. $x^3 - 2x^2 - 4x + 8 = 0$

61. $c^3 - 3c^2 - 9c + 27 = 0$

62. $v^3 + 4v^2 - 4v - 16 = 0$

63. $x^4 - 3x^3 - x^2 + 3x = 0$

64. $x^4 + 2x^3 - 9x^2 - 18x = 0$

65. $8x^4 + 12x^3 - 32x^2 - 48x = 0$

66. $9x^4 - 15x^3 - 9x^2 + 15x = 0$

Graphical Reasoning In Exercises 67–70, determine the x-intercepts and explain how the x-intercepts of the graph correspond to the solutions of the polynomial equation when $y = 0$.

67. $y = x^2 - 9$

68. $y = x^2 - 4x + 4$

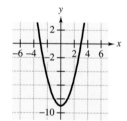
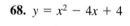

69. $y = x^2 - 2x - 3$

70. $y = x^3 - 3x^2 - x + 3$

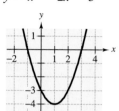
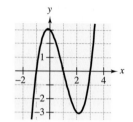

In Exercises 71–78, use a graphing utility to graph the equation and find any x-intercepts of the graph. Verify algebraically that any x-intercepts are solutions of the polynomial equation when $y = 0$.

71. $y = x^2 - 6x$

72. $y = x^2 - 11x + 28$

73. $y = x^2 - 8x + 12$

74. $y = (x - 2)^2 - 9$

75. $y = 2x^2 + 5x - 12$

76. $y = x^3 - 4x$

77. $y = 2x^3 - 5x^2 - 12x$

78. $y = 2 + x - 2x^2 - x^3$

79. Let a and b be real numbers such that $a \neq 0$. Find the solutions of $ax^2 + bx = 0$.

80. Let a be a nonzero real number. Find the solutions of $ax^2 - ax = 0$.

Think About It In Exercises 81 and 82, find a quadratic equation with the given solutions.

81. $x = -3, \quad x = 5$

82. $x = 1, \quad x = 6$

Solving Problems

83. *Number Problem* The sum of a positive number and its square is 240. Find the number.

84. *Number Problem* The sum of a positive number and its square is 72. Find the number.

85. *Number Problem* Find two consecutive positive integers whose product is 132.

86. *Geometry* The length of a rectangle is $2\frac{1}{4}$ times its width. Find the dimensions of the rectangle if its area is 900 square inches.

87. *Geometry* The rectangular floor of a storage shed has an area of 330 square feet. The length of the floor is 7 feet more than its width (see figure). Find the dimensions of the floor.

Figure for 87

88. *Geometry* The outside dimensions of a picture frame are 28 centimeters and 20 centimeters (see figure). The area of the exposed part of the picture is 468 square centimeters. Find the width w of the frame.

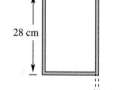

Figure for 88

89. *Geometry*   A triangle has an area of 48 square inches. Find the base and height of the triangle if the height is $1\frac{1}{2}$ times the base.

90. *Geometry*   The height of a triangle is 4 inches less than its base. Find the base and height of the triangle if its area is 70 square inches.

91. *Geometry*   An open box is to be made from a piece of material that is 5 meters long and 4 meters wide. The box is made by cutting squares of dimension x from the corners and turning up the sides, as shown in the figure. The volume V of a rectangular solid is the product of its length, width, and height.

(a) Show algebraically that the volume of the box is given by $V = (5 - 2x)(4 - 2x)x$.

(b) Determine the values of x for which $V = 0$. Determine an appropriate domain for the function V in the context of this problem.

(c) Complete the table.

x	0.25	0.50	0.75	1.00	1.25	1.50	1.75
V							

(d) Use the table to determine x if $V = 3$. Verify the result algebraically.

(e) Use a graphing utility to graph the volume function. Use the graph to approximate the value of x that yields the box of greatest volume.

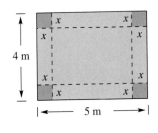
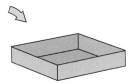

92. *Geometry*   An open box with a square base is to be constructed from 880 square inches of material. What should the dimensions of the base be if the height of the box is 6 inches?   (*Hint:* The surface area is given by $S = x^2 + 4xh$.)

93. *Free-Falling Object*   An object is dropped from a weather balloon 6400 feet above the ground. Find the time t for the object to reach the ground by solving the equation $-16t^2 + 6400 = 0$.

94. *Free-Falling Object*   An object is thrown upward from a height of 64 feet with an initial velocity of 48 feet per second. Find the time t for the object to reach the ground by solving the equation $-16t^2 + 48t + 64 = 0$.

95. *Break-Even Analysis*   The revenue R from the sale of x units of a product is given by $R = 90x - x^2$. The cost of producing x units of the product is given by $C = 200 + 60x$. How many units of the product must be produced and sold in order to break even?

96. *Break-Even Analysis*   The revenue R from the sale of x units of a product is given by $R = 60x - x^2$. The cost of producing x units of the product is given by $C = 75 + 40x$. How many units of the product must be produced and sold in order to break even?

97. *Investigation*   Solve the equation $2(x + 3)^2 + (x + 3) - 15 = 0$ in two ways.

(a) Let $u = x + 3$, and solve the resulting equation for u. Then find the values of x that satisfy the given equation.

(b) Expand and collect like terms in the given equation, and solve the resulting equation for x.

(c) Which method is easier? Explain.

98. *Investigation*   Solve each equation using both methods described in Exercise 97.

(a) $3(x + 6)^2 - 10(x + 6) - 8 = 0$

(b) $8(x + 2)^2 - 18(x + 2) + 9 = 0$

Explaining Concepts

99. Answer parts (d)–(f) of Motivating the Chapter on page 195.

100. Give an example of how the Zero-Factor Property can be used to solve a quadratic equation.

101. *True or False?*   If $(2x - 5)(x + 4) = 1$, then $2x - 5 = 1$ or $x + 4 = 1$. Explain.

102. Is it possible for a quadratic equation to have only one solution? Explain.

103. What is the maximum number of solutions of an nth-degree polynomial equation? Give an example of a third-degree equation that has only one real number solution.

Key Terms

polynomial in x, *p. 196*
degree n, *p. 196*
leading coefficient, *p. 196*
constant term, *p. 196*
standard form, *p. 196*
monomial, *p. 196*

binomial, *p. 196*
trinomial , *p. 196*
FOIL Method, *p. 207*
factoring polynomials,
    p. 215

greatest common
    monomial factor,
    p. 215
factoring by grouping,
    p. 217

completely factored,
    p. 220
perfect square trinomial,
    p. 225
quadratic equation, *p. 237*

Key Concepts

3.2 Summary of rules for exponents

1. Product Rule: $a^m \cdot a^n = a^{m+n}$

2. Quotient Rule: $\dfrac{a^m}{a^n} = a^{m-n}, \quad m > n, \quad a \neq 0$

3. Power of a Product: $(ab)^m = a^m \cdot b^m$

4. Power of a Quotient: $\left(\dfrac{a}{b}\right)^m = \dfrac{a^m}{b^m}, \quad b \neq 0$

5. Power of a Power: $(a^m)^n = a^{mn}$

3.2 Special products

Let u and v be real numbers, variables, or algebraic expressions. Then the following formulas are true.

1. Sum and Difference of Two Terms:

   $(u + v)(u - v) = u^2 - v^2$

2. Square of a Binomial: $(u \pm v)^2 = u^2 \pm 2uv + v^2$

3.3 Difference of two squares

Let u and v be real numbers, variables, or algebraic expressions. Then the expression $u^2 - v^2$ can be factored as follows: $u^2 - v^2 = (u + v)(u - v)$.

3.3 Sum and difference of two cubes

Let u and v be real numbers, variables, or algebraic expressions. Then the expressions $u^3 \pm v^3$ can be factored as follows: $u^3 \pm v^3 = (u \pm v)(u^2 \mp uv + v^2)$.

3.4 Perfect square trinomials

Let u and v be real numbers, variables, or algebraic expressions.

   $u^2 \pm 2uv + v^2 = (u \pm v)^2$

3.4 Guidelines for factoring $x^2 + bx + c$

See page 227 for factoring guidelines.

3.4 Guidelines for factoring $ax^2 + bx + c$

See page 229 for factoring guidelines.

3.4 Guidelines for factoring polynomials

1. Factor out any common factors.

2. Factor according to one of the special polynomial forms: difference of squares, sum or difference of cubes, or perfect square trinomials.

3. Factor trinomials using the methods for $a = 1$ and $a \neq 1$.

4. Factor by grouping—for polynomials with four terms.

5. Check to see if the factors themselves can be factored further.

6. Check the results by multiplying the factors.

3.5 Zero-Factor Property

Let a and b be real numbers, variables, or algebraic expressions. If a and b are factors such that $ab = 0$, then $a = b$ or $b = 0$. This property also applies to three or more factors.

3.5 Guidelines for solving quadratic equations

1. Write the quadratic equation in general form.

2. Factor the left side of the equation.

3. Set each factor with a variable equal to zero.

4. Solve each linear equation.

5. Check each solution in the original equation.

REVIEW EXERCISES

Reviewing Skills

3.1 In Exercises 1 and 2, state why the algebraic expression is not a polynomial.

1. $x^2 + 2 + 3x^{1/2}$ **2.** $z^2 - 2 + 4z^{-2}$

In Exercises 3–6, write the polynomial in standard form. Then identify the leading coefficient and the degree of the polynomial.

3. $6x^3 - 4x + 5x^2 - x^4$ **4.** $2x^6 - 5x^3 + x^5 - 7$

5. $14 - 6x + 3x^2 - 7x^3$ **6.** $9x - 2x^3 + x^5 - 8x^7$

In Exercises 7–10, give an example of a polynomial in x that satisfies the conditions. (*Note:* Each problem has many correct answers.)

7. A binomial of degree 4

8. A trinomial of degree 5 and leading coefficient -6

9. A monomial of degree 3 and leading coefficient 5

10. A binomial of degree 2 and leading coefficient 7

In Exercises 11–22, perform the operations and simplify.

11. $(5x + 3x^2) + (6 - x - 4x^2)$

12. $(6x + 1) + (x^2 - 4x)$

13. $(5x^3 - 6x + 11) + (5 + 6x - x^2 - 8x^3)$

14. $(7 - 12x^2 + 8x^3) + (x^4 - 6x^3 + 7x^2 - 5)$

15. $(3t - 5) - (t^2 - t - 5)$

16. $(10y^2 + 3) - (y^2 + 4y - 9)$

17. $(3x^5 + 4x^2 - 8x + 12) - (2x^5 + x) + (3x^2 - 4x^3 - 9)$

18. $(7x^4 - 10x^2 + 4x) + (x^3 - 3x) - (3x^4 - 5x^2 + 1)$

19. $(-x^3 - 3x) - 4(2x^3 - 3x + 1)$

20. $(7z^2 + 6z) - 3(5z^2 + 2z)$

21. $3y^2 - [2y + 3(y^2 + 5)]$

22. $(16a^3 + 5a) - 5[a + (2a^3 - 1)]$

3.2 In Exercises 23–36, use the rules for exponents to simplify the expression.

23. $x^2 \cdot x^3$ **24.** $-3y^2 \cdot y^4$

25. $(u^2)^3$ **26.** $(v^4)^2$

27. $(-2z)^3$ **28.** $(-3y)^2(2)$

29. $-(u^2v)^2(-4u^3v)$ **30.** $(12x^2y)(3x^2y^4)^2$

31. $\dfrac{12z^5}{6z^2}$ **32.** $\dfrac{15m^3}{25m}$

33. $\dfrac{120u^5v^3}{15u^3v}$ **34.** $-\dfrac{(-2x^2y^3)^2}{-3xy^2}$

35. $\left(\dfrac{72x^4}{6x^2}\right)^2$ **36.** $\left(-\dfrac{y^2}{2}\right)^3$

In Exercises 37–50, perform the multiplication and simplify.

37. $(-2x)^3(x + 4)$ **38.** $(-4y)^2(y - 2)$

39. $3x(2x^2 - 5x + 3)$ **40.** $-2y(5y^2 - y - 4)$

41. $(x - 2)(x + 7)$ **42.** $(x + 6)(x - 9)$

43. $(5x + 3)(3x - 4)$ **44.** $(4x - 1)(2x - 5)$

45. $(4x^2 + 3)(6x^2 + 1)$

46. $(3y^2 + 2)(4y^2 - 5)$

47. $(2x^2 - 3x + 2)(2x + 3)$

48. $(5s^3 + 4s - 3)(4s - 5)$

49. $2u(u - 7) - (u + 1)(u - 7)$

50. $(3v + 2)(-5v) + 5v(3v + 2)$

In Exercises 51–60, use special product formulas to find the product.

51. $(4x - 7)^2$ **52.** $(8 - 3x)^2$

53. $(2x + 3y)^2$ **54.** $(u + 4v)^2$

55. $(5u - 8)(5u + 8)$ **56.** $(7a + 4)(7a - 4)$

57. $(2u + v)(2u - v)$

58. $(5x - 2y)(5x + 2y)$

59. $[(u - 3) + v][(u - 3) - v]$

60. $[(m - 5) + n]^2$

3.3 In Exercises 61–64, factor out the greatest common factor.

61. $6x^2 + 15x^3$

62. $8y - 12y^4$

63. $28(x + 5) - 70(x + 5)^2$

64. $(u - 9v)(u - v) + v(u - 9v)$

In Exercises 65–68, factor the polynomial by grouping.

65. $v^3 - 2v^2 - v + 2$

66. $y^3 + 4y^2 - y - 4$

67. $t^3 + 3t^2 + 3t + 9$

68. $x^3 + 7x^2 + 3x + 21$

In Exercises 69–74, factor the difference of two squares.

69. $x^2 - 36$

70. $b^2 - 900$

71. $9a^2 - 100$

72. $16y^2 - 49$

73. $(u + 6)^2 - 81$

74. $(y - 3)^2 - 16$

In Exercises 75–78, factor the sum or difference of cubes.

75. $u^3 - 1$

76. $t^3 - 125$

77. $8x^3 + 27$

78. $27x^3 + 64$

3.4 In Exercises 79–82, factor the perfect square trinomial.

79. $x^2 - 18x + 81$

80. $y^2 + 16y + 64$

81. $4s^2 + 40st + 100t^2$

82. $u^2 - 10uv + 25v^2$

In Exercises 83–88, factor the trinomial.

83. $x^2 + 2x - 35$

84. $x^2 - 12x + 32$

85. $2x^2 - 7x + 6$

86. $5x^2 + 11x - 12$

87. $18x^2 + 27x + 10$

88. $12x^2 - 13x - 14$

In Exercises 89–98, factor the expression completely.

89. $4a - 64a^3$

90. $3b + 27b^3$

91. $8x(2x - 3) - 4(2x - 3)$

92. $x^3 + 3x^2 - 4x - 12$

93. $\frac{1}{4}x^2 + xy + y^2$

94. $x^2 - \frac{2}{3}x + \frac{1}{9}$

95. $4u^2 - 28u + 49$

96. $3x^3 + 23x^2 - 8x$

97. $x^2 - 10x + 25 - y^2$

98. $u^6 - 8v^6$

3.5 In Exercises 99–108, solve the polynomial equation.

99. $3s^2 - 2s - 8 = 0$

100. $4t^2 - 12t = -9$

101. $v^2 - 100 = 0$

102. $x^2 - 25x = -150$

103. $10x(x - 3) = 0$

104. $3x(4x + 7) = 0$

105. $z(5 - z) + 36 = 0$

106. $(x + 3)^2 - 25 = 0$

107. $2y^4 + 2y^3 - 24y^2 = 0$

108. $b^3 - 6b^2 - b + 6 = 0$

In Exercises 109 and 110, use a graphing utility to graph the equation and find any x-intercepts of the graph. Verify algebraically that any x-intercepts are solutions of the polynomial equation when $y = 0$.

109. $y = x^2 - 10x + 21$

110. $y = x^3 - 6x^2$

Think About It In Exercises 111–114, construct a polynomial equation with integer coefficients that has the specified solutions.

111. $5, -9$

112. $\frac{5}{2}, 3$

113. $0, -\frac{3}{2}, 2$

114. $-3, 3, 5$

Solving Problems

115. *Comparing Models* The table gives population projections (in millions) for the United States for selected years from 2000 to 2050. It gives three series of projections, which are the lowest P_L, middle P_M, and highest P_H. (Source: U.S. Bureau of the Census)

Year	2000	2010	2020	2030	2040	2050
P_L	271.2	281.5	288.8	291.1	287.7	282.5
P_M	274.6	297.7	322.7	347.0	370.0	393.9
P_H	278.1	314.6	357.7	405.1	458.4	518.9

In the following models for the data, $t = 0$ corresponds to 2000.

$$P_L = -0.022t^2 + 1.33t + 270.71$$

$$P_M = 2.386t + 274.857$$

$$P_H = 0.028t^2 + 3.40t + 278.18$$

(a) Use a graphing utility to plot the data and graph the models on the same screen.

(b) Find $(P_L + P_H)/2$. Use a graphing utility to graph this polynomial and state which graph from part (a) it is most like. Does this seem reasonable? Explain.

(c) Find $P_H - P_L$ and use a graphing utility to graph this model. Explain why it is increasing.

116. *Profit*   A manufacturer can produce and sell x book bags per week. The total cost (in dollars) for producing the book bags is given by

$C = 12x + 3000$

and the total revenue is given by

$R = 20x.$

Find the profit obtained by selling 1200 book bags per week.

117. *Profit*   A manufacturer can produce and sell x notepads per week. The total cost (in dollars) for producing the notepads is given by

$C = 0.5x + 1000$

and the total revenue is given by

$R = 1.1x.$

Find the profit obtained by selling 5000 notepads per week.

118. *Probability*   The probability of three successes in five trials of an experiment is given by $10p^3(1 - p)^2$. Find this product.

Geometry   In Exercises 119 and 120, write expressions for the perimeter and area of the region, and then simplify.

119.
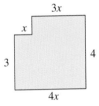

120.
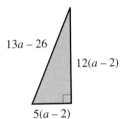

Geometry   In Exercises 121 and 122, find the area of the shaded portion of the figure.

121.
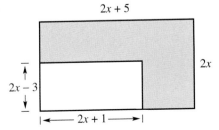

122.
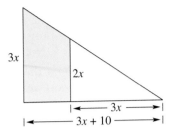

123. *Geometry*   A rectangle has a length of l units and a width of $l - 5$ units. Find (a) the perimeter and (b) the area of the rectangle.

124. *Geometry*   The width of a rectangle is $\frac{3}{4}$ of its length. Find the dimensions of the rectangle if its area is 432 square inches.

125. *Geometry*   The perimeter of a rectangular storage lot at a car dealership is 800 feet. The lot is surrounded by fencing that costs $15 per foot for the front side and $10 per foot for the remaining three sides. Find the dimensions of the storage lot if the total cost of the fencing was $9500.

126. *Free-Falling Object*   An object is thrown upward from a height of 48 feet with an initial velocity of 32 feet per second. Find the time t for the object to reach the ground by solving the following equation.

$-16t^2 + 32t + 48 = 0$

127. *Free-Falling Object*   An object is thrown upward from a height of 32 feet with an initial velocity of 16 feet per second. Find the time t for the object to reach the ground by solving the following equation.

$-16t^2 + 16t + 32 = 0$

128. *Number Problem*   Find two consecutive positive odd integers whose product is 195.

129. *Number Problem*   Find two consecutive positive even integers whose product is 224.

Chapter Test

Take this test as you would take a test in class. After you are done, check your work against the answers given in the back of the book.

1. Determine the degree and leading coefficient of $-5.2x^3 + 3x^2 - 8$.

2. Explain why the following expression is not a polynomial.

$$\frac{4}{x^2 + 2}$$

In Exercises 3–9, perform the operations and simplify.

3. (a) $(5a^2 - 3a + 4) + (a^2 - 4)$ (b) $(16 - y^2) - (16 + 2y + y^2)$

4. (a) $-2(2x^4 - 5) + 4x(x^3 + 2x - 1)$ (b) $4t - [3t - (10t + 7)]$

5. (a) $(-2u^2v)^3(3v^2)$ (b) $3(5x)(2xy)^2$

6. (a) $2y\left(\dfrac{y}{4}\right)^2$ (b) $\dfrac{(-3x^2y)^4}{6x^2}$

7. (a) $-3x(x - 4)$ (b) $(2x - 3y)(x + 5y)$

8. (a) $(x - 1)[2x + (x - 3)]$ (b) $(2s - 3)(3s^2 - 4s + 7)$

9. (a) $(4x - 3)^2$ (b) $[4 - (a + b)][4 + (a + b)]$

In Exercises 10–15, factor the expression completely.

10. $18y^2 - 12y$ **11.** $v^2 - \dfrac{16}{9}$

12. $x^3 - 3x^2 - 4x + 12$ **13.** $9u^2 - 6u + 1$

14. $6x^2 - 26x - 20$ **15.** $x^3 + 27$

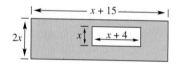

Figure for 18

In Exercises 16 and 17, solve the equation.

16. $(y + 2)^2 - 9 = 0$ **17.** $12 + 5y - 3y^2 = 0$

18. Find the area of the shaded region in the figure.

19. The length of a rectangle is $1\frac{1}{2}$ times its width. Find the dimensions of the rectangle if its area is 54 square centimeters.

20. The height $h(t)$ of a free-falling object is given by

$$h(t) = -16t^2 - 40t + 144$$

where h is measured in feet and t is measured in seconds. The object is projected downward when $t = 0$. How long does it take to hit the ground?

21. The height of a triangle is 4 feet more than twice its base. Find the base and height of the triangle if its area is 35 square feet.

Cumulative Test: Chapters P–3

Take this test as you would take a test in class. After you are done, check your work against the answers given in the back of the book.

1. Place the correct symbol ($<$, $>$, or $=$) between the two numbers.

   (a) -2 ____ 5 (b) $\frac{1}{3}$ ____ $\frac{1}{2}$ (c) $|2.3|$ ____ $-|-4.5|$

2. Write an algebraic expression for the statement, "The number n is tripled and the product is decreased by 8."

In Exercises 3–5, perform the operations and simplify.

3. (a) $t(3t - 1) - 2t(t + 4)$ (b) $3x(x^2 - 2) - x(x^2 + 5)$

4. (a) $(2a^2b)^3(-ab^2)^2$ (b) $\left(\dfrac{2x^4y^2}{4x^3y}\right)^2$

5. (a) $(2x + 1)(x - 5)$ (b) $[2 + (x - y)]^2$

In Exercises 6–8, solve the equations or inequalities.

6. (a) $12 - 5(3 - x) = x + 3$ (b) $1 - \dfrac{x + 2}{4} = \dfrac{7}{8}$

7. (a) $|3x - 5| = 7$ (b) $2t^2 - 5t - 3 = 0$

8. (a) $3(1 - x) > 6$ (b) $-12 \le 4x - 6 < 10$

9. Your annual automobile insurance premium is $1225. Because of a driving violation, your premium is increased 15%. What is your new premium?

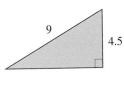

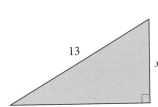

Figure for 10

10. The triangles at the left are similar. Solve for x by using the fact that corresponding sides of similar triangles are proportional.

11. Solve $|x - 2| \ge 3$ and sketch its solution.

12. The revenue from selling x units of a product is $R = 12.90x$. The cost of producing x units is $C = 8.50x + 450$. To obtain a profit, the revenue must be greater than the cost. For what values of x will this product produce a profit? Explain your reasoning.

13. Determine whether the equation $x - y^3 = 0$ represents y as a function of x.

14. Find the domain of the function $f(x) = \sqrt{x - 2}$.

15. Given $f(x) = x^2 - 3x$, find (a) $f(4)$ and (b) $f(c + 3)$.

16. Find the slope of the line passing through $(-4, 0)$ and $(4, 6)$. Then find the distance between the points.

17. Determine the equation of a line through the point $(-2, 1)$ (a) parallel to $2x - y = 1$ and (b) perpendicular to $3x + 2y = 5$.

In Exercises 18 and 19, factor the polynomials.

18. (a) $3x^2 - 8x - 35$ (b) $9x^2 - 144$

19. (a) $y^3 - 3y^2 - 9y + 27$ (b) $8t^3 - 40t^2 + 50t$

In Exercises 20 and 21, graph the equation.

20. $4x + 3y - 12 = 0$ 21. $y = 1 - (x - 2)^2$

4 Rational Expressions, Equations, and Functions

Daemmrich/The Image Works

In 1997, approximately 17% (103,600) of all boats sold in the United States were canoes. (Source: National Marine Manufacturers Association)

Motivating the Chapter

A Canoe Trip

You and a friend are planning a canoe trip on a river. You want to travel 10 miles upstream and 10 miles back downstream during daylight hours. You know that in still water you are able to paddle the canoe at an average speed of 5 miles per hour. While traveling upstream your average speed will be 5 miles per hour minus the speed of the current, and while traveling downstream your average speed will be 5 miles per hour plus the speed of the current.

See Section 4.4, Exercise 109

a. Write an expression that represents the time it will take to travel upstream in terms of the speed, x (in miles per hour), of the current. Write an expression that represents the time it will take to travel downstream in terms of the speed of the current.

b. Write a function f for the entire time (in hours) of the trip in terms of x.

c. Write the rational function f as a single fraction.

See Section 4.6, Exercise 95

d. What is the speed of the current if the time of the trip is $6\frac{1}{4}$ hours? Explain.

e. If the speed of the current is 4 miles per hour, can you and your friend make the trip during 12 hours of daylight? Explain.

4.1   Integer Exponents and Scientific Notation

Objectives

1 Rewrite an exponential expression involving negative exponents.

2 Write a very large or a very small number in scientific notation.

1 Rewrite an exponential expression involving negative exponents.

Integer Exponents

So far in the text, all exponents you have worked with have been positive integers. In this section, the definition of an exponent is extended to include zero and negative integers. If a is a real number such that $a \neq 0$, then a^0 is defined as 1. Moreover, if m is an integer, then a^{-m} is defined as the reciprocal of a^m.

> ▶ **Definitions of Zero Exponents and Negative Exponents**
>
> Let a and b be real numbers such that $a \neq 0$ and $b \neq 0$, and let m be an integer.
>
> **1.** $a^0 = 1$ **2.** $a^{-m} = \dfrac{1}{a^m}$ **3.** $\left(\dfrac{a}{b}\right)^{-m} = \left(\dfrac{b}{a}\right)^m$

These definitions are consistent with the properties of exponents given in Section 3.2. For instance, consider the following.

$$x^0 \cdot x^m = x^{0+m} = x^m = 1 \cdot x^m$$

$(x^0 \text{ is the same as } 1)$

Example 1   Zero Exponents and Negative Exponents

Evaluate each expression.

a. 3^0 **b.** 3^{-2} **c.** $\left(\frac{3}{4}\right)^{-1}$ **d.** $2^{-1} - 1$

Solution

a. $3^0 = 1$ Definition of zero exponent

b. $3^{-2} = \dfrac{1}{3^2} = \dfrac{1}{9}$ Definition of negative exponent

c. $\left(\dfrac{3}{4}\right)^{-1} = \left(\dfrac{4}{3}\right)^1 = \dfrac{4}{3}$ Definition of negative exponent

d. $2^{-1} - 1 = \dfrac{1}{2} - 1 = -\dfrac{1}{2}$ Definition of negative exponent

---

The following rules of exponents are valid for all integer exponents, including integer exponents that are zero or negative. (The first five properties were listed in Section 3.2.)

▶ **Summary of Rules of Exponents**

Let m and n be integers, and let a and b represent real numbers, variables, or algebraic expressions.

Product and Quotient Rules   *Example*

1. $a^m \cdot a^n = a^{m+n}$ $x^4(x^3) = x^{4+3} = x^7$

2. $\dfrac{a^m}{a^n} = a^{m-n}, \quad a \neq 0$ $\dfrac{x^3}{x} = x^{3-1} = x^2, \quad x \neq 0$

Power Rules

3. $(ab)^m = a^m \cdot b^m$ $(3x)^2 = 3^2(x^2) = 9x^2$

4. $\left(\dfrac{a}{b}\right)^m = \dfrac{a^m}{b^m}, \quad b \neq 0$ $\left(\dfrac{x}{3}\right)^2 = \dfrac{x^2}{3^2} = \dfrac{x^2}{9}$

5. $(a^m)^n = a^{mn}$ $(x^3)^3 = x^{3\cdot3} = x^9$

Zero and Negative Exponent Rules

6. $a^{-m} = \dfrac{1}{a^m}, \quad a \neq 0$ $x^{-2} = \dfrac{1}{x^2}, \quad x \neq 0$

7. $a^0 = 1, \quad a \neq 0$ $(x^2 + 1)^0 = 1$

8. $\left(\dfrac{a}{b}\right)^{-m} = \left(\dfrac{b}{a}\right)^m, \quad a \neq 0, \ b \neq 0$ $\left(\dfrac{x}{3}\right)^{-2} = \left(\dfrac{3}{x}\right)^2 = \dfrac{3^2}{x^2} = \dfrac{9}{x^2}$

Study Tip

As you become accustomed to working with negative exponents, you will probably not write as many steps as are shown in Example 2. For instance, to rewrite a fraction involving exponents, you might use the following simplified rule. *To move a factor from the numerator to the denominator or vice versa, change the sign of its exponent.* You can apply this rule to the expression in Example 2(c) by "moving" the factor x^{-2} to the numerator and changing the exponent to 2. That is,

$$\frac{3}{x^{-2}} = 3x^2.$$

Example 2   **Using Rules of Exponents**

a. $2x^{-1} = 2(x^{-1}) = 2\left(\dfrac{1}{x}\right) = \dfrac{2}{x}$ Use Negative Exponent Rule and simplify.

b. $(2x)^{-1} = \dfrac{1}{(2x)^1} = \dfrac{1}{2x}$ Use Negative Exponent Rule and simplify.

c. $\dfrac{3}{x^{-2}} = \dfrac{3}{\left(\dfrac{1}{x^2}\right)}$ Negative Exponent Rule

$\qquad = 3\left(\dfrac{x^2}{1}\right)$ Invert divisor and multiply.

$\qquad = 3x^2$ Simplify.

d. $\dfrac{1}{(3x)^{-2}} = \dfrac{1}{\left[\dfrac{1}{(3x)^2}\right]}$ Use Negative Exponent Rule.

$\qquad = \dfrac{1}{\left(\dfrac{1}{(9x^2)}\right)}$ Use Power Rule and simplify.

$\qquad = (1)\left(\dfrac{9x^2}{1}\right)$ Invert divisor and multiply.

$\qquad = 9x^2$ Simplify.

Example 3   Using Rules of Exponents

Rewrite each expression using only positive exponents. (For each expression, assume that $x \neq 0$ and $y \neq 0$.)

a. $(-5x^{-3})^2$ **b.** $-\left(\dfrac{7x}{y^2}\right)^{-2}$ **c.** $\dfrac{12x^2y^{-4}}{6x^{-1}y^2}$

Solution

a. $(-5x^{-3})^2 = (-5)^2(x^{-3})^2$ Power of a Product

$\qquad\qquad\quad\;\; = 25x^{-6}$ Power of a Power

$\qquad\qquad\quad\;\; = \dfrac{25}{x^6}$ Negative Exponent Rule

b. $-\left(\dfrac{7x}{y^2}\right)^{-2} = -\left(\dfrac{y^2}{7x}\right)^2$ Negative Exponent Rule

$\qquad\qquad\quad\;\; = -\dfrac{(y^2)^2}{(7x)^2}$ Power of a Quotient

$\qquad\qquad\quad\;\; = -\dfrac{y^4}{49x^2}$ Power of a Power

c. $\dfrac{12x^2y^{-4}}{6x^{-1}y^2} = 2(x^{2-(-1)})(y^{-4-2})$ Quotient Rule

$\qquad\qquad\quad\;\; = 2x^3y^{-6}$ Simplify.

$\qquad\qquad\quad\;\; = \dfrac{2x^3}{y^6}$ Negative Exponent Rule

Example 4   Using Rules of Exponents

Rewrite each expression using only positive exponents. (For each expression, assume that $x \neq 0$ and $y \neq 0$.)

a. $\left(\dfrac{8x^{-1}y^4}{4x^3y^2}\right)^{-3}$ **b.** $\dfrac{3xy^0}{x^2(5y)^0}$

Solution

a. $\left(\dfrac{8x^{-1}y^4}{4x^3y^2}\right)^{-3} = \left(\dfrac{2y^2}{x^4}\right)^{-3}$ Simplify.

$\qquad\qquad\quad\;\; = \left(\dfrac{x^4}{2y^2}\right)^3$ Negative Exponent Rule

$\qquad\qquad\quad\;\; = \dfrac{x^{12}}{2^3y^6}$ Power of a Quotient

$\qquad\qquad\quad\;\; = \dfrac{x^{12}}{8y^6}$ Simplify.

b. $\dfrac{3xy^0}{x^2(5y)^0} = \dfrac{3x(1)}{x^2(1)} = \dfrac{3}{x}$ Zero Exponent Rule

2 Write a very large or a very small number in scientific notation.

Scientific Notation

Exponents provide an efficient way of writing and computing with very large (or very small) numbers. For instance, a drop of water contains more than 33 billion billion molecules—that is, 33 followed by 18 zeros.

$$33,000,000,000,000,000,000$$

It is convenient to write such numbers in **scientific notation.** This notation has the form $c \times 10^n$, where $1 \leq c < 10$ and n is an integer. So, the number of molecules in a drop of water can be written in scientific notation as

$$3.3 \times 10,000,000,000,000,000,000 = 3.3 \times 10^{19}.$$

The *positive* exponent 19 indicates that the number being written in scientific notation is *large* (10 or more) and that the decimal point has been moved 19 places. A *negative* exponent in scientific notation indicates that the number is *small* (less than 1).

Example 5   Writing Scientific Notation

Write each real number in scientific notation.

a. 0.0000684   **b.** 937,200,000   **c.** 15,700

Solution

a. $0.0000684 = 6.84 \times 10^{-5}$ Small number ➡ negative exponent

    Five places

b. $937,200,000.0 = 9.372 \times 10^8$ Large number ➡ positive exponent

    Eight places

c. $15,700.0 = 1.57 \times 10^4$ Large number ➡ positive exponent

    Four places

Example 6   Writing Decimal Notation

Convert each number from scientific notation to decimal notation.

a. 2.486×10^2 **b.** 1.81×10^{-6} **c.** 6.28×10^5

Solution

a. $2.486 \times 10^2 = 248.6$ Positive exponent ➡ large number

    Two places

b. $1.81 \times 10^{-6} = 0.00000181$ Negative exponent ➡ small number

    Six places

c. $6.28 \times 10^5 = 628,000.0$ Positive exponent ➡ large number

    Five places

Example 7   Using Scientific Notation

Convert the factors to scientific notation and then evaluate

$$\frac{(2{,}400{,}000{,}000)(0.00000345)}{(0.00007)(3800)}.$$

Solution

$$\frac{(2{,}400{,}000{,}000)(0.00000345)}{(0.00007)(3800)} = \frac{(2.4 \times 10^9)(3.45 \times 10^{-6})}{(7.0 \times 10^{-5})(3.8 \times 10^3)}$$

$$= \frac{(2.4)(3.45)(10^3)}{(7)(3.8)(10^{-2})}$$

$$= \frac{(8.28)(10^5)}{26.6}$$

$$\approx 0.3112782(10^5)$$

$$= 31{,}127.82$$

Technology: Tip

Most scientific calculators and graphing utilities automatically switch to scientific notation when they are showing large (or small) numbers that exceed the display range.

To *enter* numbers in scientific notation, your calculator should have an exponential entry key labeled [EE] or [EXP]. Consult the user's guide for your graphing utility for instructions on keystrokes and how numbers in scientific notation are displayed.

Example 8   Using Scientific Notation with a Calculator

Use a scientific or graphing calculator to find the following.

a. $65{,}000 \times 3{,}400{,}000{,}000$ **b.** $0.000000348 \div 870$

Solution

a. 6.5 [EXP] 4 [×] 3.4 [EXP] 9 [=]   Scientific
    6.5 [EE] 4 [×] 3.4 [EE] 9 [ENTER]   Graphing

The calculator display should read [2.21E 14], which implies that

$$(6.5 \times 10^4)(3.4 \times 10^9) = 2.21 \times 10^{14} = 221{,}000{,}000{,}000{,}000.$$

b. 3.48 [EXP] 7 [+/−] [÷] 8.7 [EXP] 2 [=]   Scientific
    3.48 [EE] [(−)] 7 [÷] 8.7 [EE] 2 [ENTER]   Graphing

The calculator display should read [4.0E −10], which implies that

$$\frac{3.48 \times 10^{-7}}{8.7 \times 10^2} = 4.0 \times 10^{-10} = 0.0000000004.$$

Discussing the Concept	Developing a Mathematical Method

Develop an easy-to-use method for converting numbers to and from scientific notation. Demonstrate the method using one of the following.

a. 0.0000042   **b.** 293,600,000,000

c. 3.1×10^{-6} **d.** 5.12×10^{11}

4.1   Exercises

Integrated Review   *Concepts, Skills, and Problem Solving*

Keep mathematically in shape by doing these exercises *before* the problems of this section.

Properties and Definitions

1. In your own words, describe the graph of an equation.
2. Describe the point-plotting method for graphing an equation.
3. Find the coordinates of two points on the graph of $g(x) = \sqrt{x-2}$.
4. Describe the procedure for finding the x- and y-intercepts of the graph of an equation.

Simplifying Expressions

In Exercises 5–8, simplify the expression. (Assume the denominator is not zero.)

5. $(7x^2)(2x^3)$

6. $(y^2z^3)(z^2)^4$

7. $\dfrac{a^4b^2}{ab^2}$

8. $(x+2)^4 \div (x+2)^3$

Graphing Equations

In Exercises 9–12, use a graphing utility to graph the function. Identify any intercepts.

9. $f(x) = 5 - 2x$

10. $h(x) = \frac{1}{2}x + |x|$

11. $g(x) = x^2 - 4x$

12. $f(x) = 2\sqrt{x+1}$

Developing Skills

In Exercises 1–30, evaluate the expression. See Example 1.

1. 5^{-2}

2. 2^{-4}

3. -10^{-3}

4. -20^{-2}

5. $(-3)^0$

6. 25^0

7. $\dfrac{1}{4^{-3}}$

8. $\dfrac{1}{-8^{-2}}$

9. $\dfrac{1}{(-2)^{-5}}$

10. $-\dfrac{1}{6^{-2}}$

11. $\left(\frac{2}{3}\right)^{-1}$

12. $\left(\frac{4}{5}\right)^{-3}$

13. $\left(\frac{3}{16}\right)^{0}$

14. $\left(-\frac{5}{8}\right)^{-2}$

15. $27 \cdot 3^{-3}$

16. $4^2 \cdot 4^{-3}$

17. $\dfrac{3^4}{3^{-2}}$

18. $\dfrac{5^{-1}}{5^2}$

19. $\dfrac{10^3}{10^{-2}}$

20. $\dfrac{10^{-5}}{10^{-6}}$

21. $(4^2 \cdot 4^{-1})^{-2}$

22. $(5^3 \cdot 5^{-4})^{-3}$

23. $(2^{-3})^2$

24. $(-4^{-1})^{-2}$

25. $2^{-3} + 2^{-4}$

26. $4 - 3^{-2}$

27. $\left(\frac{3}{4} + \frac{5}{8}\right)^{-2}$

28. $\left(\frac{1}{2} - \frac{2}{3}\right)^{-1}$

29. $(5^0 - 4^{-2})^{-1}$

30. $(32 + 4^{-3})^0$

In Exercises 31–70, rewrite the expression using only positive exponents, and simplify. (Assume that any variables in the expression are nonzero.) See Examples 2–4.

31. $y^4 \cdot y^{-2}$

32. $x^{-2} \cdot x^{-5}$

33. $z^5 \cdot z^{-3}$

34. $t^{-1} \cdot t^{-6}$

35. $7x^{-4}$

36. $3y^{-3}$

37. $(4x)^{-3}$

38. $(5u)^{-2}$

39. $\dfrac{1}{x^{-6}}$

40. $\dfrac{4}{y^{-1}}$

41. $\dfrac{8a^{-6}}{6a^{-7}}$

42. $\dfrac{6u^{-2}}{15u^{-1}}$

43. $\dfrac{(4t)^0}{t^{-2}}$

44. $\dfrac{(5u)^{-4}}{(5u)^0}$

45. $(2x^2)^{-2}$

46. $(4a^{-2})^{-3}$

47. $(-3x^{-3}y^2)(4x^2y^{-5})$

48. $(5s^5t^{-5})(-6s^{-2}t^4)$

49. $(3x^2y^{-2})^{-2}$

50. $(-4y^{-3}z)^{-3}$

51. $\left(\dfrac{x}{10}\right)^{-1}$

52. $\left(\dfrac{4}{z}\right)^{-2}$

53. $\dfrac{6x^3y^{-3}}{12x^{-2}y}$

54. $\dfrac{2y^{-1}z^{-3}}{4yz^{-3}}$

55. $\left(\dfrac{3u^2v^{-1}}{3^3u^{-1}v^3}\right)^{-2}$

56. $\left(\dfrac{5^2x^3y^{-3}}{125xy}\right)^{-1}$

57. $\left(\dfrac{a^{-2}}{b^{-2}}\right)\left(\dfrac{b}{a}\right)^3$

58. $\left(\dfrac{a^{-3}}{b^{-3}}\right)\left(\dfrac{b}{a}\right)^3$

59. $(2x^3y^{-1})^{-3}(4xy^{-6})$

60. $(ab)^{-2}(a^2b^2)^{-1}$

61. $u^4(6u^{-3}v^0)(7v)^0$

62. $x^5(3x^0y^4)(7y)^0$

63. $[(x^{-4}y^{-6})^{-1}]^2$

64. $[(2x^{-3}y^{-2})^2]^{-2}$

65. $\dfrac{(2a^{-2}b^4)^3b}{(10a^3b)^2}$

66. $\dfrac{(5x^2y^{-5})^{-1}}{2x^{-5}y^4}$

67. $(u + v^{-2})^{-1}$

68. $x^{-2}(x^2 + y^2)$

69. $\dfrac{a + b}{ba^{-1} - ab^{-1}}$

70. $\dfrac{u^{-1} - v^{-1}}{u^{-1} + v^{-1}}$

In Exercises 71–84, write the number in scientific notation. See Example 5.

71. 3,600,000

72. 98,100,000

73. 47,620,000

74. 956,300,000

75. 0.00031

76. 0.00625

77. 0.0000000381

78. 0.0007384

79. *Land Area of Earth:* 57,500,000 square miles

80. *Ocean Area of Earth:* 139,400,000 square miles

81. *Light Year:* 9,461,000,000,000,000 kilometers

82. *Thickness of Soap Bubble:* 0.0000001 meter

83. *Relative Density of Hydrogen:* 0.0000899 grams per milliliter.

84. *One Micron (Millionth of Meter):* 0.00003937 inch

In Exercises 85–94, write the number in decimal notation. See Example 6.

85. 6×10^7

86. 5.05×10^{12}

87. 1.359×10^{-7}

88. 8.6×10^{-9}

89. *1997 Merrill Lynch Revenues:* $\$3.17 \times 10^{10}$

90. *Number of Air Sacs in Lungs:* 3.5×10^8

91. *Interior Temperature of Sun:* 1.3×10^7 degrees Celsius

92. *Width of Air Molecule:* 9.0×10^{-9} meter

93. *Charge of Electron:* 4.8×10^{-10} electrostatic unit

94. *Width of Human Hair:* 9.0×10^{-4} meter

In Exercises 95–104, evaluate without a calculator. See Example 7.

95. $(2 \times 10^9)(3.4 \times 10^{-4})$

96. $(6.5 \times 10^6)(2 \times 10^4)$

97. $(5 \times 10^4)^2$

98. $(4 \times 10^6)^3$

99. $\dfrac{3.6 \times 10^{12}}{6 \times 10^5}$

100. $\dfrac{2.5 \times 10^{-3}}{5 \times 10^2}$

101. $(4,500,000)(2,000,000,000)$

102. $(62,000,000)(0.0002)$

103. $\dfrac{64,000,000}{0.00004}$

104. $\dfrac{72,000,000,000}{0.00012}$

In Exercises 105–112, evaluate with a calculator. Write the answer in scientific notation, $c \times 10^n$, with c rounded to two decimal places. See Example 8.

105. $\dfrac{(0.0000565)(2,850,000,000,000)}{0.00465}$

106. $\dfrac{(3,450,000,000)(0.000125)}{(52,000,000)(0.000003)}$

107. $\dfrac{1.357 \times 10^{12}}{(4.2 \times 10^2)(6.87 \times 10^{-3})}$

108. $\dfrac{(3.82 \times 10^5)^2}{(8.5 \times 10^4)(5.2 \times 10^{-3})}$

109. $72,400 \times 2,300,000,000$

110. $(8.67 \times 10^4)^7$

111. $\dfrac{(5,000,000)^3(0.000037)^2}{(0.005)^4}$

112. $\dfrac{(6,200,000)(0.005)^3}{(0.00035)^5}$

Solving Problems

113. *Distance to the Sun* The distance from the earth to the sun is approximately 93 million miles. Write this distance in scientific notation.

114. *Copper Electrons* A cube of copper with an edge of 1 centimeter has approximately 8×10^{22} free electrons. Write this real number in decimal form.

115. *Light Year* One light year (the distance light can travel in one year) is approximately 9.45×10^{15} meters. Approximate the time for light to travel from the sun to the earth if that distance is approximately 1.49×10^{11} meters.

116. *Distance to a Star*   Determine the distance (in meters) to the star Alpha Andromeda if it is 90 light years from the earth (see figure). See Exercise 115 for the definition of a light year.

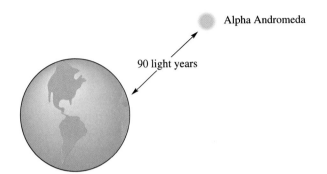

117. *Mass of Earth and Sun*   The masses of the earth and sun are approximately 5.975×10^{24} and 1.99×10^{30}, respectively. The mass of the sun is approximately how many times that of the earth?

118. *Metal Expansion*   When the temperature of an iron steam pipe 200 feet long is increased $75°$ C, the length of the pipe will increase by an amount

$$75(200)(10 \times 10^{-6}).$$

Find this amount of increase in length.

119. *Federal Debt*   In June 1998, the estimated population of the United States was 270 million people, and the estimated federal debt was 5506 billion dollars. Use these two numbers to determine the amount each person would have had to pay (per capita debt) to eliminate the debt. (Source: U.S. Bureau of the Census and U.S. Office of Management and Budget)

120. *Kepler's Third Law*   In 1619, Johannes Kepler, a German astronomer, discovered that the period T (in years) of each planet in our solar system is related to the planet's mean distance R (in astronomical units) from the sun by the equation

$$\frac{T^2}{R^3} = k.$$

Test Kepler's equation for the nine planets in our solar system, using the table. What value do you get for k for each planet? Are the values of k all approximately the same? (Astronomical units relate a planet's period and mean distance to the earth's period and mean distance.)

Planet	T	R
Mercury	0.241	0.387
Venus	0.615	0.723
Earth	1.000	1.000
Mars	1.881	1.523
Jupiter	11.861	5.203
Saturn	29.457	9.541
Uranus	84.008	19.190
Neptune	164.784	30.086
Pluto	248.350	39.508

Explaining Concepts

121. In $(3x)^4$, what is $3x$ called? What is 4 called?

122. Discuss any differences between $(-2x)^{-4}$ and $-2x^{-4}$.

123. In your own words, describe how you can "move" a factor from the numerator to the denominator or vice versa.

124. Is 32.5×10^5 in scientific notation? Explain.

125. When is scientific notation an efficient way of writing and computing real numbers?

4.2   Rational Expressions and Functions

Objectives

1. Determine the domain of a rational function.

2. Simplify a rational expression using the Cancellation Rule.

---

1 Determine the domain of a rational function.

The Domain of a Rational Function

Algebra may seem simpler when you realize that it consists primarily of a basic set of operations, including addition, subtraction, multiplication, division, factoring, and simplifying, that are applied to different types of algebraic expressions. For instance, in Chapter 3 you applied these operations to *polynomials*. In this chapter, you will apply these operations to *rational expressions*. Like polynomials, rational expressions can be used to describe functions. Such functions are called **rational functions.**

> ▶ **Definition of a Rational Function**
>
> Let $u(x)$ and $v(x)$ be polynomials. The function
>
> $$f(x) = \frac{u(x)}{v(x)}$$
>
> is a **rational function.** The **domain** of f is the set of all real numbers for which $v(x) \neq 0$.

Study Tip

Every polynomial is also a rational expression because you can consider the denominator to be 1. The domain of every polynomial is the set of all real numbers. That is,

Domain $= (-\infty, \infty)$.

Technology: Discovery

Use a graphing utility to graph the equation

$$y = \frac{4}{x - 2}.$$

Then use the trace or table feature of the utility to determine the behavior of the graph near $x = 2$. Graph the equations that correspond to parts (b) and (c) of Example 1. How does each of these graphs differ from the graph of $y = 4/(x - 2)$?

Example 1   Finding the Domain of a Rational Function

Find the domains of the following rational functions.

a. $f(x) = \dfrac{4}{x - 2}$ **b.** $g(x) = \dfrac{2x + 5}{8}$ **c.** $h(x) = \dfrac{3x - 1}{x^2 - 2x - 3}$

Solution

a. The denominator is zero when $x - 2 = 0$ or $x = 2$. So, the domain is all real values of x such that $x \neq 2$. In interval notation, you can write the domain as

Domain $= (-\infty, 2) \cup (2, \infty)$.

b. The denominator, 8, is never zero, so the domain is the set of *all* real numbers. In interval notation, you can write the domain as

Domain $= (-\infty, \infty)$.

c. The denominator is zero when $x^2 - 2x - 3 = 0$. Solving this equation by factoring, you find that the denominator is zero when $x = 3$ or when $x = -1$. So, the domain is all real values of x such that $x \neq 3$ and $x \neq -1$. In interval notation, you can write the domain as

Domain $= (-\infty, -1) \cup (-1, 3) \cup (3, \infty)$.

In applications involving rational functions, it is often necessary to place restrictions on the domain besides those that make the denominator zero. To indicate such a restriction, write the domain to the right of the fraction. For instance, the domain of the rational function

$$f(x) = \frac{x^2 + 20}{x + 4}, \qquad x > 0$$

is the set of positive real numbers, as indicated by the inequality $x > 0$. Note that the normal domain of this function would be all real values of x such that $x \neq -4$. However, because "$x > 0$" is listed to the right of the function, the domain is restricted by this inequality.

Example 2   An Application Involving a Restricted Domain

You have started a small manufacturing business. The initial investment for the business is \$120,000. The cost of each unit that you manufacture is \$15. So, your total cost of producing x units is

$$C = 15x + 120,000. \qquad \text{Cost function}$$

Your average cost per unit depends on the number of units produced. For instance, the average cost per unit $\overline{C}$ for producing 100 units is

$$\overline{C} = \frac{15(100) + 120,000}{100} \qquad \text{Substitute 100 for } x.$$

$$= \$1215. \qquad \text{Average cost per unit for 100 units}$$

The average cost per unit decreases as the number of units increases. For instance, the average cost per unit $\overline{C}$ for producing 1000 units is

$$\overline{C} = \frac{15(1000) + 120,000}{1000} \qquad \text{Substitute 1000 for } x.$$

$$= \$135. \qquad \text{Average cost per unit for 1000 units}$$

In general, the average cost of producing x units is

$$\overline{C} = \frac{15x + 120,000}{x}. \qquad \text{Average cost per unit for } x \text{ units}$$

What is the domain of this rational function?

Solution

If you were considering this function from only a mathematical point of view, you would say that the domain is all real values of x such that $x \neq 0$. However, because this function is a mathematical model representing a real-life situation, you must consider which values of x make sense in real life. For this model, the variable x represents the number of units that you produce. Assuming that you cannot produce a fractional number of units, you conclude that the domain is the set of positive integers. That is,

$$\text{Domain} = \{1, 2, 3, 4, \ldots\}.$$

2 Simplify a rational expression using the Cancellation Rule.

Simplifying Rational Expressions

As with numerical fractions, a rational expression is said to be **simplified** or **in reduced form** if its numerator and denominator have no factors in common (other than ±1). To simplify fractions, you can apply the following rule.

▶ **Cancellation Rule for Fractions**

Let u, v, and w represent numbers, variables, or algebraic expressions such that $v \neq 0$ and $w \neq 0$. Then the following Cancellation Rule is valid.

$$\frac{u\cancel{w}}{v\cancel{w}} = \frac{u}{v}$$

Be sure you see that this Cancellation Rule allows you to cancel only factors, not terms. For instance, consider the following.

$$\frac{2 \cdot 2}{2(x + 5)}$$ You can cancel common factor 2.

$$\frac{3 + x}{3 + 2x}$$ You cannot cancel common term 3.

Using the Cancellation Rule to simplify a rational expression requires two steps: (1) completely factor the numerator and denominator and (2) apply the Cancellation Rule to cancel any *factors* that are common to both the numerator and denominator. So, your success in simplifying rational expressions actually lies in your ability to *completely factor* the polynomials in both the numerator and denominator.

The Granger Collection

Karl Weierstrass

(1815–1897)

In the 19th century, mathematicians were expanding their knowledge of calculus and laying the foundation for complex numbers. At that time, the properties of real numbers had not been finalized. Teaching at the University of Berlin, Weierstrass recognized the need for a logical foundation for the real number system. His work contributed much to the formal real number system that forms the foundation of modern algebra.

Example 3   Simplifying a Rational Expression

Simplify $\dfrac{2x^3 - 6x}{6x^2}$.

Solution

First note that the domain of the rational expression is all real values of x such that $x \neq 0$. Then, completely factor both the numerator and denominator.

$$\frac{2x^3 - 6x}{6x^2} = \frac{2x(x^2 - 3)}{2x(3x)}$$ Factor numerator and denominator.

$$= \frac{2\cancel{x}(x^2 - 3)}{2\cancel{x}(3x)}$$ Cancel common factor $2x$.

$$= \frac{x^2 - 3}{3x}$$ Simplified form

In simplified form, the domain of the rational expression is the same as that of the original expression, all real values of x such that $x \neq 0$.

Example 4   Simplifying a Rational Expression

Simplify $\dfrac{x^2 + 2x - 15}{3x - 9}$.

Solution

The domain of the rational expression is all real values of x such that $x \neq 3$.

$$\frac{x^2 + 2x - 15}{3x - 9} = \frac{(x + 5)(x - 3)}{3(x - 3)} \qquad \text{Factor numerator and denominator.}$$

$$= \frac{(x + 5)(x - 3)}{3(x - 3)} \qquad \text{Cancel common factor } (x - 3).$$

$$= \frac{x + 5}{3}, \ x \neq 3 \qquad \text{Simplified form}$$

Technology: Tip

Use the table feature of your graphing utility to compare the two functions in Example 4.

$$y_1 = \frac{x^2 + 2x - 15}{3x - 9}$$

$$y_2 = \frac{x + 5}{3}$$

Set the increment value of the table to 1 and compare the values at $x = 0, 1, 2, 3, 4,$ and 5. Next set the increment value to 0.1 and compare the values at $x = 2.8, 2.9, 3.0, 3.1,$ and 3.2. From the table you can see that the functions differ only at $x = 3$. This shows why $x \neq 3$ must be written as part of the reduced form of the original expression.

Canceling common factors from the numerator and denominator of a rational expression can change its domain. For instance, in Example 4 the domain of the original expression is all real values of x such that $x \neq 3$. So, the original expression is equal to the simplified expression for all real numbers *except* 3.

Example 5   Simplifying a Rational Expression

Simplify $\dfrac{x^3 - 16x}{x^2 - 2x - 8}$.

Solution

The domain of the rational expression is all real values of x such that $x \neq -2$ and $x \neq 4$.

$$\frac{x^3 - 16x}{x^2 - 2x - 8} = \frac{x(x^2 - 16)}{(x + 2)(x - 4)} \qquad \text{Partially factor.}$$

$$= \frac{x(x + 4)(x - 4)}{(x + 2)(x - 4)} \qquad \text{Factor completely.}$$

$$= \frac{x(x + 4)(x - 4)}{(x + 2)(x - 4)} \qquad \text{Cancel common factor } (x - 4).$$

$$= \frac{x(x + 4)}{x + 2}, \ x \neq 4 \qquad \text{Simplified form}$$

In this text, when simplifying a rational expression, we follow the convention of listing *by the simplified expression* all values of x that must be specifically excluded from the domain in order to make the domains of the simplified and original expressions agree. For instance, in Example 5 the restriction $x \neq 4$ must be listed with the simplified expression in order to make the two domains agree. Note that the value of -2 is excluded from both domains, so it is not necessary to list this value.

Example 6 Simplification Involving a Change of Sign

Simplify $\dfrac{2x^2 - 9x + 4}{12 + x - x^2}$.

Solution

The domain of the rational expression is all real values of x such that $x \neq -3$ and $x \neq 4$.

$$\frac{2x^2 - 9x + 4}{12 + x - x^2} = \frac{(2x - 1)(x - 4)}{(4 - x)(3 + x)} \qquad \text{Factor numerator and denominator.}$$

$$= \frac{(2x - 1)(x - 4)}{-(x - 4)(3 + x)} \qquad (4 - x) = -(x - 4)$$

$$= \frac{(2x - 1)(x - 4)}{-(x - 4)(3 + x)} \qquad \text{Cancel common factor } (x - 4).$$

$$= -\frac{2x - 1}{3 + x}, \quad x \neq 4 \qquad \text{Simplified form}$$

The simplified form is equivalent to the original expression for all values of x except 4. Note that -3 is excluded from the domains of both the original and simplified expressions.

In Example 6, be sure you see that when dividing the numerator and denominator by the common factor of $(x - 4)$, you keep the negative sign. In the simplified form of the fraction, we usually like to move the negative sign out in front of the fraction. However, this is a personal preference. All of the following forms are legitimate.

$$-\frac{2x - 1}{3 + x} = \frac{-(2x - 1)}{3 + x} = \frac{2x - 1}{-3 - x} = \frac{2x - 1}{-(3 + x)}$$

In the next three examples, the Cancellation Rule is used to simplify rational expressions that involve more than one variable.

Example 7 A Rational Expression Involving Two Variables

Simplify $\dfrac{3xy + y^2}{2y}$.

Solution

The domain of the rational expression is all real values of y such that $y \neq 0$.

$$\frac{3xy + y^2}{2y} = \frac{y(3x + y)}{2y} \qquad \text{Factor numerator and denominator.}$$

$$= \frac{y(3x + y)}{2y} \qquad \text{Cancel common factor } y.$$

$$= \frac{3x + y}{2}, \quad y \neq 0 \qquad \text{Simplified form}$$

| Example 8 | A Rational Expression Involving Two Variables |

Simplify $\dfrac{2x^2 + 2xy - 4y^2}{5x^3 - 5xy^2}$.

Solution

The domain of the rational expression is all real numbers such that $x \neq 0$ and $x \neq \pm y$.

$$\frac{2x^2 + 2xy - 4y^2}{5x^3 - 5xy^2} = \frac{2(x - y)(x + 2y)}{5x(x - y)(x + y)}$$ Factor numerator and denominator.

$$= \frac{2(x - y)(x + 2y)}{5x(x - y)(x + y)}$$ Cancel common factor $(x - y)$.

$$= \frac{2(x + 2y)}{5x(x + y)}, \quad x \neq y$$ Simplified form

| Example 9 | A Rational Expression Involving Two Variables |

Simplify $\dfrac{4x^2y - y^3}{2x^2y - xy^2}$.

Solution

The domain of the rational expression is all real numbers such that $x \neq 0$, $y \neq 0$, and $y \neq 2x$.

$$\frac{4x^2y - y^3}{2x^2y - xy^2} = \frac{(2x - y)(2x + y)y}{(2x - y)xy}$$ Factor numerator and denominator.

$$= \frac{(2x - y)(2x + y)y}{(2x - y)xy}$$ Cancel common factors $(2x - y)$ and y.

$$= \frac{2x + y}{x}, \quad y \neq 0, \ y \neq 2x$$ Simplified form

As you study the examples and work the exercises in this chapter, keep in mind that you are *rewriting expressions in simpler forms.* You are not solving equations. Equal signs are used in the steps of the simplification process only to indicate that the new form of the expression is *equivalent* to the previous one.

Discussing the Concept   Error Analysis

Suppose you are the instructor of an algebra course. One of your students turns in the following incorrect solutions. Find the errors, discuss the student's misconceptions, and construct correct solutions.

a. $\dfrac{3x^2 + 5x - 4}{x} = 3x + 5 - 4 = 3x + 1$

b. $\dfrac{x^2 + 7x}{x + 7} = \dfrac{x^2}{x} + \dfrac{7x}{7} = x + x = 2x$

4.2 Exercises

Integrated Review   *Concepts, Skills, and Problem Solving*

Keep mathematically in shape by doing these exercises *before* the problems of this section.

Properties and Definitions

1. Define the slope of the line through the points (x_1, y_1) and (x_2, y_2).

2. Make a statement about the slope m of the line for each of the following.

    (a) The line rises from left to right.

    (b) The line falls from left to right.

    (c) The line is horizontal.

    (d) The line is vertical.

Simplifying Expressions

In Exercises 3–8, simplify the expression.

3. $2(x + 5) - 3 - (2x - 3)$

4. $3(y + 4) + 5 - (3y + 5)$

5. $4 - 2[3 + 4(x + 1)]$

6. $5x + x[3 - 2(x - 3)]$

7. $\left(\dfrac{5}{x^2}\right)^2$

8. $-\dfrac{(2u^2v)^2}{-3uv^2}$

Problem Solving

9. Determine the number of gallons of a 30% solution that must be mixed with a 60% solution to obtain 20 gallons of a 40% solution.

10. A suit sells for \$375 during a 25% storewide clearance sale. What was the original price of the suit?

Developing Skills

In Exercises 1–20, find the domain of the expression. See Example 1.

1. $\dfrac{5}{x - 8}$

2. $\dfrac{9}{x - 13}$

3. $\dfrac{7x}{x + 4}$

4. $\dfrac{2y}{6 - y}$

5. $\dfrac{x^2 + 9}{4}$

6. $\dfrac{y^2 - 3}{7}$

7. $x^4 - 2x^2 - 5$

8. $t^3 - 4t^2 + 1$

9. $\dfrac{x}{x^2 + 4}$

10. $\dfrac{4x}{x^2 + 16}$

11. $\dfrac{y - 4}{y(y + 3)}$

12. $\dfrac{z + 2}{z(z - 4)}$

13. $\dfrac{5t}{t^2 - 16}$

14. $\dfrac{x}{x^2 - 4}$

15. $\dfrac{y + 5}{y^2 - 3y}$

16. $\dfrac{t - 6}{t^2 + 5t}$

17. $\dfrac{8x}{x^2 - 5x + 6}$

18. $\dfrac{3t}{t^2 - 2t - 3}$

19. $\dfrac{u^2}{3u^2 - 2u - 5}$

20. $\dfrac{y + 5}{4y^2 - 5y - 6}$

In Exercises 21–26, evaluate the function as indicated. If not possible, state the reason.

21. $f(x) = \dfrac{4x}{x + 3}$

    (a) $f(1)$ (b) $f(-2)$
    (c) $f(-3)$ (d) $f(0)$

22. $f(x) = \dfrac{x - 10}{4x}$

    (a) $f(10)$ (b) $f(0)$
    (c) $f(-2)$ (d) $f(12)$

23. $g(x) = \dfrac{x^2 - 4x}{x^2 - 9}$

    (a) $g(0)$ (b) $g(4)$
    (c) $g(3)$ (d) $g(-3)$

24. $g(t) = \dfrac{t - 2}{2t - 5}$

    (a) $g(2)$ (b) $g\left(\frac{5}{2}\right)$
    (c) $g(-2)$ (d) $g(0)$

25. $h(s) = \dfrac{s^2}{s^2 - s - 2}$

    (a) $h(10)$ (b) $h(0)$
    (c) $h(-1)$ (d) $h(2)$

26. $f(x) = \dfrac{x^3 + 1}{x^2 - 6x + 9}$

(a) $f(-1)$ (b) $f(3)$

(c) $f(-2)$ (d) $f(2)$

In Exercises 27–32, describe the domain. See Example 2.

27. *Geometry*   A rectangle of length x inches has an area of 500 square inches. The perimeter P of the rectangle is given by

$$P = 2\left(x + \frac{500}{x}\right).$$

28. *Cost*   The cost C in millions of dollars for the government to seize $p\%$ of a certain illegal drug as it enters the country is given by

$$C = \frac{528p}{100 - p}.$$

29. *Inventory Cost*   The inventory cost I when x units of a product are ordered from a supplier is given by

$$I = \frac{0.25x + 2000}{x}.$$

30. *Average Cost*   The average cost $\overline{C}$ for a manufacturer to produce x units of a product is given by

$$\overline{C} = \frac{1.35x + 4570}{x}.$$

31. *Pollution Removal*   The cost C in dollars of removing $p\%$ of the air pollutants in the stack emission of a utility company is given by the rational function

$$C = \frac{80,000p}{100 - p}.$$

32. *Video Rental*   The average cost of a movie video rental $\overline{M}$ when you consider the cost of purchasing a video cassette recorder and renting x movie videos at \$2.49 per movie is

$$\overline{M} = \frac{150 + 2.49x}{x}.$$

In Exercises 33–40, complete the statement.

33. $\dfrac{5(\quad)}{6(x + 3)} = \dfrac{5}{6}, \quad x \neq -3$

34. $\dfrac{7(\quad)}{15(x - 10)} = \dfrac{7}{15}, \quad x \neq 10$

35. $\dfrac{3x(x + 16)^2}{2(\quad)} = \dfrac{x}{2}, \quad x \neq -16$

36. $\dfrac{25x^2(x - 10)}{12(\quad)} = \dfrac{5x}{12}, \quad x \neq 10, \quad x \neq 0$

37. $\dfrac{(x + 5)(\quad)}{3x^2(x - 2)} = \dfrac{x + 5}{3x}, \quad x \neq 2$

38. $\dfrac{(3y - 7)(\quad)}{y^2 - 4} = \dfrac{3y - 7}{y + 2}, \quad y \neq 2$

39. $\dfrac{8x(\quad)}{x^2 - 3x - 10} = \dfrac{8x}{x - 5}, \quad x \neq -2$

40. $\dfrac{(3 - z)(\quad)}{z^3 + 2z^2} = \dfrac{3 - z}{z^2}, \quad z \neq -2$

In Exercises 41–78, simplify the expression. See Examples 3–9.

41. $\dfrac{5x}{25}$

42. $\dfrac{32y}{24}$

43. $\dfrac{12y^2}{2y}$

44. $\dfrac{15z^3}{15z^3}$

45. $\dfrac{18x^2y}{15xy^4}$

46. $\dfrac{16y^2z^2}{60y^5z}$

47. $\dfrac{3x^2 - 9x}{12x^2}$

48. $\dfrac{8x^3 + 4x^2}{20x}$

49. $\dfrac{x^2(x - 8)}{x(x - 8)}$

50. $\dfrac{a^2b(b - 3)}{b^3(b - 3)^2}$

51. $\dfrac{2x - 3}{4x - 6}$

52. $\dfrac{y^2 - 81}{2y - 18}$

53. $\dfrac{5 - x}{3x - 15}$

54. $\dfrac{x^2 - 36}{6 - x}$

55. $\dfrac{a + 3}{a^2 + 6a + 9}$

56. $\dfrac{u^2 - 12u + 36}{u - 6}$

57. $\dfrac{x^2 - 7x}{x^2 - 14x + 49}$

58. $\dfrac{z^2 + 22z + 121}{3z + 33}$

59. $\dfrac{y^3 - 4y}{y^2 + 4y - 12}$

60. $\dfrac{x^2 - 7x}{x^2 - 4x - 21}$

61. $\dfrac{x^3 - 4x}{x^2 - 5x + 6}$

62. $\dfrac{x^4 - 25x^2}{x^2 + 2x - 15}$

63. $\dfrac{3x^2 - 7x - 20}{12 + x - x^2}$

64. $\dfrac{2x^2 + 3x - 5}{7 - 6x - x^2}$

65. $\dfrac{2x^2 + 19x + 24}{2x^2 - 3x - 9}$

66. $\dfrac{2y^2 + 13y + 20}{2y^2 + 17y + 30}$

67. $\dfrac{15x^2 + 7x - 4}{25x^2 - 16}$

68. $\dfrac{56z^2 - 3z - 20}{49z^2 - 16}$

69. $\dfrac{3xy^2}{xy^2 + x}$

70. $\dfrac{x + 3x^2y}{3xy + 1}$

71. $\dfrac{y^2 - 64x^2}{5(3y + 24x)}$

72. $\dfrac{x^2 - 25z^2}{x + 5z}$

73. $\dfrac{5xy + 3x^2y^2}{xy^3}$

74. $\dfrac{4u^2v - 12uv^2}{18uv}$

75. $\dfrac{u^2 - 4v^2}{u^2 + uv - 2v^2}$

76. $\dfrac{x^2 + 4xy}{x^2 - 16y^2}$

77. $\dfrac{3m^2 - 12n^2}{m^2 + 4mn + 4n^2}$

78. $\dfrac{x^2 + xy - 2y^2}{x^2 + 3xy + 2y^2}$

In Exercises 79–82, explain how you can show that the two expressions are not equivalent.

79. $\dfrac{x - 4}{4} \neq x - 1$

80. $\dfrac{x - 4}{x} \neq -4$

81. $\dfrac{3x + 2}{4x + 2} \neq \dfrac{3}{4}$

82. $\dfrac{1 - x}{2 - x} \neq \dfrac{1}{2}$

In Exercises 83 and 84, complete the table. What can you conclude?

83.

x	-2	-1	0	1	2	3	4
$\dfrac{x^2 - x - 2}{x - 2}$							
$x + 1$							

84.

x	-2	-1	0	1	2	3	4
$\dfrac{x^2 + 5x}{x}$							
$x + 5$							

Solving Problems

Geometry   In Exercises 85 and 86, find the ratio of the area of the shaded portion to the total area of the figure.

85.

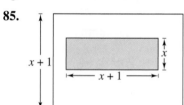

86.

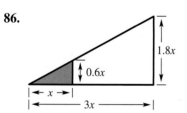

87. *Average Cost*   A machine shop has a setup cost of $2500 for the production of a new product. The cost for labor and material in producing each unit is $9.25.

(a) Write the total cost C as a function of x, the number of units produced.

(b) Write the average cost per unit $\overline{C} = C/x$ as a function of x, the number of units produced.

(c) Determine the domain of the function in part (b).

(d) Find the value of $\overline{C}(100)$.

88. *Average Cost*   A greeting card company has an initial investment of $60,000. The cost of producing one dozen cards is $6.50.

(a) Write the total cost C as a function of x, the number of cards in dozens produced.

(b) Write the average cost per dozen $\overline{C} = C/x$ as a function of x, the number of cards in dozens produced.

(c) Determine the domain of the function in part (b).

(d) Find the value of $\overline{C}(11,000)$.

89. *Distance Traveled*   A van starts on a trip and travels at an average speed of 45 miles per hour. Three hours later, a car starts on the same trip and travels at an average speed of 60 miles per hour.

(a) Find the distance each vehicle has traveled when the car has been on the road for t hours.

(b) Use the result of part (a) to write the distance between the van and the car as a function of t.

(c) Write the ratio of the distance the car has traveled to the distance the van has traveled as a function of t.

90. *Distance Traveled*   A car starts on a trip and travels at an average speed of 55 miles per hour. Two hours later, a second car starts on the same trip and travels at an average speed of 65 miles per hour.

(a) Find the distance each vehicle has traveled when the second car has been on the road for t hours.

(b) Use the result of part (a) to write the distance between the first car and the second car as a function of t.

(c) Write the ratio of the distance the second car has traveled to the distance the first car has traveled as a function of t.

91. *Geometry*   One swimming pool is circular and another is rectangular. The rectangular pool's width is three times its depth. Its length is 6 feet more than its width. The circular pool has a diameter that is twice the width of the rectangular pool, and it is 2 feet deeper. Find the ratio of the circular pool's volume to the rectangular pool's volume.

92. *Geometry*   A circular pool has a radius five times its depth. A rectangular pool has the same depth as the circular pool. Its width is 4 feet more than three times its depth and its length is 2 feet less than six times its depth. Find the ratio of the rectangular pool's volume to the circular pool's volume.

Cost of Medicare   In Exercises 93 and 94, use the following polynomial models, which give the total annual cost of Medicare C (in billions of dollars) and the U.S. population enrolled in Medicare P (in millions) from 1990 through 1996 (see figures).

$C = 107.30 + 15.09t$

$P = 34.26 + 0.65t$

In these models, t represents the year, with $t = 0$ corresponding to 1990. (Source: U.S. Health Care Financing Administration)

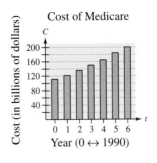

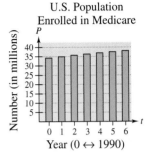

93. Find a rational model that represents the average cost of Medicare per person enrolled during the years 1990 to 1996.

94. Use the model found in Exercise 93 to complete the table showing the average cost of Medicare per person enrolled.

Year	1990	1991	1992	1993
Average cost				

Year	1994	1995	1996
Average cost			

Explaining Concepts

95. Define the term *rational expression*.

96. Give an example of a rational function whose domain is the set of all real numbers.

97. How do you determine whether a rational expression is in simplified form?

98. Can you cancel common terms from the numerator and denominator of a rational expression? Explain.

99. Explain the error in the following.

$$\frac{2x^2}{x^2+4} = \frac{2x^2}{x^2+4} = \frac{2}{1+4} = \frac{2}{5}$$

100. Is the following statement true? Explain.

$$\frac{6x-5}{5-6x} = -1$$

4.3   Multiplying and Dividing Rational Expressions

Objectives

1 Multiply rational expressions and simplify.

2 Divide rational expressions and simplify.

3 Simplify a complex fraction.

---

1 Multiply rational expressions and simplify.

Multiplying Rational Expressions

The rule for multiplying rational expressions is the same as the rule for multiplying numerical fractions.

$$\frac{3}{4} \cdot \frac{7}{6} = \frac{21}{24} = \frac{\cancel{3} \cdot 7}{\cancel{3} \cdot 8} = \frac{7}{8}$$

That is, you *multiply numerators, multiply denominators, and write the new fraction in simplified form.*

▶ **Multiplying Rational Expressions**

Let u, v, w, and z be real numbers, variables, or algebraic expressions such that $v \neq 0$ and $z \neq 0$. Then the product of u/v and w/z is given by

$$\frac{u}{v} \cdot \frac{w}{z} = \frac{uw}{vz}.$$

In order to recognize common factors, write the numerators and denominators in factored form, as demonstrated in Example 1.

Example 1   Multiplying Rational Expressions

Multiply the rational expressions.

$$\frac{4x^3y}{3xy^4} \cdot \frac{-6x^2y^2}{10x^4}$$

Solution

$$\frac{4x^3y}{3xy^4} \cdot \frac{-6x^2y^2}{10x^4} = \frac{(4x^3y) \cdot (-6x^2y^2)}{(3xy^4) \cdot (10x^4)} \qquad \text{Multiply numerators and denominators.}$$

$$= \frac{-24x^5y^3}{30x^5y^4} \qquad \text{Simplify.}$$

$$= \frac{-4(6)(x^5)(y^3)}{5(6)(x^5)(y^3)(y)} \qquad \text{Factor and cancel.}$$

$$= -\frac{4}{5y}, \quad x \neq 0 \qquad \text{Simplified form}$$

Example 2   Multiplying Rational Expressions

Multiply the rational expressions.

$$\frac{x}{5x^2 - 20x} \cdot \frac{x - 4}{2x^2 + x - 3}$$

Solution

$$\frac{x}{5x^2 - 20x} \cdot \frac{x - 4}{2x^2 + x - 3}$$

$$= \frac{x \cdot (x - 4)}{(5x^2 - 20x) \cdot (2x^2 + x - 3)}$$ Multiply numerators and denominators.

$$= \frac{x(x - 4)}{5x(x - 4)(x - 1)(2x + 3)}$$ Factor.

$$= \frac{x(x - 4)}{5x(x - 4)(x - 1)(2x + 3)}$$ Cancel common factors.

$$= \frac{1}{5(x - 1)(2x + 3)}, \quad x \neq 0, \ x \neq 4$$ Simplified form

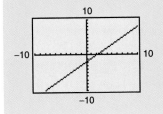

Technology: Tip

You can use a graphing utility to check your results when multiplying rational expressions. For instance, in Example 3, try graphing the equations

$$y_1 = \frac{4x^2 - 4x}{x^2 + 2x - 3} \cdot \frac{x^2 + x - 6}{4x}$$

and

$$y_2 = x - 2$$

on the same screen. If the two graphs coincide, as shown below, you can conclude that the two functions are equivalent.

Example 3   Multiplying Rational Expressions

Multiply the rational expressions.

$$\frac{4x^2 - 4x}{x^2 + 2x - 3} \cdot \frac{x^2 + x - 6}{4x}$$

Solution

$$\frac{4x^2 - 4x}{x^2 + 2x - 3} \cdot \frac{x^2 + x - 6}{4x}$$

$$= \frac{4x(x - 1)(x + 3)(x - 2)}{(x - 1)(x + 3)(4x)}$$ Factor and multiply.

$$= \frac{4x(x - 1)(x + 3)(x - 2)}{(x - 1)(x + 3)(4x)}$$ Cancel common factors.

$$= x - 2, \quad x \neq 0, \ x \neq 1, \ x \neq -3$$ Simplified form

The rule for multiplying fractions can be extended to cover products involving expressions that are not in fractional form. To do this, rewrite the nonfractional expression as a fraction whose denominator is 1. Here is a simple example.

$$\frac{x + 3}{x - 2} \cdot (5x) = \frac{x + 3}{x - 2} \cdot \frac{5x}{1}$$

$$= \frac{(x + 3)(5x)}{x - 2}$$

$$= \frac{5x(x + 3)}{x - 2}$$

In the next example, note how to divide out a factor that differs only in sign. The Distributive Property is used in the step in which $(y - x)$ is rewritten as $(-1)(x - y)$.

Example 4 Multiplying Rational Expressions

Multiply the rational expressions.

$$\frac{x - y}{y^2 - x^2} \cdot \frac{x^2 - xy - 2y^2}{3x - 6y}$$

Solution

$$\frac{x - y}{y^2 - x^2} \cdot \frac{x^2 - xy - 2y^2}{3x - 6y}$$

$$= \frac{x - y}{(y + x)(y - x)} \cdot \frac{(x - 2y)(x + y)}{3(x - 2y)} \qquad \text{Factor.}$$

$$= \frac{x - y}{(y + x)(-1)(x - y)} \cdot \frac{(x - 2y)(x + y)}{3(x - 2y)} \qquad \text{Factor: } (y - x) = -1(x - y).$$

$$= \frac{(x - y)(x - 2y)(x + y)}{(y + x)(-1)(x - y)(3)(x - 2y)} \qquad \text{Multiply.}$$

$$= \frac{(x - y)(x - 2y)(x + y)}{(x + y)(-1)(x - y)(3)(x - 2y)} \qquad \text{Cancel common factors.}$$

$$= -\frac{1}{3}, \ x \neq y, \ x \neq -y, \ x \neq 2y \qquad \text{Simplified form}$$

The rule for multiplying rational expressions can be extended to cover products of three or more fractions, as shown in Example 5.

Example 5 Multiplying Three Rational Expressions

Multiply the rational expressions.

$$\frac{x^2 - 3x + 2}{x + 2} \cdot \frac{3x}{x - 2} \cdot \frac{2x + 4}{x^2 - 5x}$$

Solution

$$\frac{x^2 - 3x + 2}{x + 2} \cdot \frac{3x}{x - 2} \cdot \frac{2x + 4}{x^2 - 5x}$$

$$= \frac{(x - 1)(x - 2)(3)(x)(2)(x + 2)}{(x + 2)(x - 2)(x)(x - 5)} \qquad \text{Factor and multiply.}$$

$$= \frac{(x - 1)(x - 2)(3)(x)(2)(x + 2)}{(x + 2)(x - 2)(x)(x - 5)} \qquad \text{Cancel common factors.}$$

$$= \frac{6(x - 1)}{x - 5}, \ x \neq 0, \ x \neq 2, \ x \neq -2 \qquad \text{Simplified form}$$

2 Divide rational expressions and simplify.

Dividing Rational Expressions

To divide two rational expressions, multiply the first fraction by the *reciprocal* of the second. That is, simply *invert the divisor and multiply.* For instance, to perform the following division

$$\frac{x}{x + 3} \div \frac{4}{x - 1}$$

invert the fraction $4/(x - 1)$ and multiply, as follows.

$$\frac{x}{x + 3} \div \frac{4}{x - 1} = \frac{x}{x + 3} \cdot \frac{x - 1}{4} \qquad \text{Invert divisor and multiply.}$$

$$= \frac{x(x - 1)}{(x + 3)(4)} \qquad \text{Multiply numerators and denominators.}$$

$$= \frac{x(x - 1)}{4(x + 3)} \qquad \text{Simplify.}$$

▶ **Dividing Rational Expressions**

Let u, v, w, and z be real numbers, variables, or algebraic expressions such that $v \neq 0$, $w \neq 0$, and $z \neq 0$. The quotient of u/v and w/z is

$$\frac{u}{v} \div \frac{w}{z} = \frac{u}{v} \cdot \frac{z}{w} = \frac{uz}{vw}.$$

Example 6   Dividing Rational Expressions

Perform the division.

$$\frac{2x}{3x - 12} \div \frac{x^2 - 2x}{x^2 - 6x + 8}$$

Solution

$$\frac{2x}{3x - 12} \div \frac{x^2 - 2x}{x^2 - 6x + 8}$$

$$= \frac{2x}{3x - 12} \cdot \frac{x^2 - 6x + 8}{x^2 - 2x} \qquad \text{Invert divisor and multiply.}$$

$$= \frac{(2)(x)(x - 2)(x - 4)}{(3)(x - 4)(x)(x - 2)} \qquad \text{Factor and multiply.}$$

$$= \frac{(2)(x)(x - 2)(x - 4)}{(3)(x - 4)(x)(x - 2)} \qquad \text{Cancel common factors.}$$

$$= \frac{2}{3}, \ x \neq 0, \ x \neq 2, \ x \neq 4 \qquad \text{Simplified form}$$

Remember that the original expression is equivalent to $2/3$ except for $x = 0$, $x = 2$, and $x = 4$.

3 Simplify a complex fraction.

Complex Fractions

Problems involving division of two rational expressions are sometimes written as **complex fractions.** A complex fraction is one that has a fraction in its numerator or denominator, or both. The rules for dividing fractions still apply in such cases.

Example 7 Simplifying a Complex Fraction

$$\frac{\left(\dfrac{x^2 + 2x - 3}{x - 3}\right)}{4x + 12} = \frac{\left(\dfrac{x^2 + 2x - 3}{x - 3}\right)}{\left(\dfrac{4x + 12}{1}\right)} \qquad \text{Rewrite denominator.}$$

$$= \frac{x^2 + 2x - 3}{x - 3} \cdot \frac{1}{4x + 12} \qquad \text{Invert divisor and multiply.}$$

$$= \frac{(x - 1)(x + 3)}{(x - 3)(4)(x + 3)} \qquad \text{Factor.}$$

$$= \frac{(x - 1)(x + 3)}{(x - 3)(4)(x + 3)} \qquad \text{Cancel common factor.}$$

$$= \frac{x - 1}{4(x - 3)}, \ x \neq -3 \qquad \text{Simplified form}$$

Note that in Example 7 the domain of the complex fraction is restricted by the two denominators in the expression, $x - 3$ and $4x + 12$. So, the domain of the original expression is all real values of x such that $x \neq 3$ and $x \neq -3$.

Discussing the Concept   Using a Table

Complete the following table for the given values of x.

x	60	100	1000	10,000	100,000	1,000,000
$\dfrac{x - 10}{x + 10}$						
$\dfrac{x + 50}{x - 50}$						
$\dfrac{x - 10}{x + 10} \cdot \dfrac{x + 50}{x - 50}$						

What kind of pattern do you see? Try to explain what is going on. Can you see why?

4.3   Exercises

Integrated Review   *Concepts, Skills, and Problem Solving*

Keep mathematically in shape by doing these exercises *before* the problems of this section.

Properties and Definitions

1. Explain how to factor the difference of two squares $9t^2 - 4$.

2. Explain how to factor the perfect square trinomial $4x^2 - 12x + 9$.

3. Explain how to factor the sum of two cubes $8x^3 + 64$.

4. Factor $3x^2 + 13x - 10$, and explain how you can prove that your answer is correct.

Algebraic Operations

In Exercises 5–10, factor the expression completely.

5. $5x - 20x^2$

6. $64 - (x - 6)^2$

7. $15x^2 - 16x - 15$

8. $16t^2 + 8t + 1$

9. $y^3 - 64$

10. $8x^3 + 1$

Graphs

In Exercises 11 and 12, sketch the graphs of the lines through the given point with the indicated slopes. Make the sketches on the same set of coordinate axes.

Point	Slopes	
11. $(2, -3)$	(a) 0	(b) undefined
	(c) 2	(d) $-\frac{1}{3}$
12. $(-1, 4)$	(a) 2	(b) -1
	(c) $\frac{1}{2}$	(d) undefined

Developing Skills

In Exercises 1 and 2, evaluate the function as indicated. If not possible, state the reason.

Expression

1. $f(x) = \dfrac{x - 10}{4x}$

Values
(a) $f(10)$ (b) $f(0)$
(c) $f(-2)$ (d) $f(12)$

2. $g(x) = \dfrac{x^2 - 4x}{x^2 - 9}$

(a) $g(0)$ (b) $g(4)$
(c) $g(3)$ (d) $g(-3)$

In Exercises 3–10, complete the statement.

3. $\dfrac{7x^2}{3y(\quad)} = \dfrac{7}{3y}, \quad x \neq 0$

4. $\dfrac{14x(x - 3)^2}{(x - 3)(\quad)} = \dfrac{2x}{x - 3}, \quad x \neq 3$

5. $\dfrac{3x(x + 2)^2}{(x - 4)(\quad)} = \dfrac{3x}{x - 4}, \quad x \neq -2$

6. $\dfrac{(x + 1)^3}{x(\quad)} = \dfrac{x + 1}{x}, \quad x \neq -1$

7. $\dfrac{3u(\quad)}{7v(u + 1)} = \dfrac{3u}{7v}, \quad u \neq -1$

8. $\dfrac{(3t + 5)(\quad)}{5t^2(3t - 5)} = \dfrac{3t + 5}{t}, \quad t \neq \dfrac{5}{3}$

9. $\dfrac{13x(\quad)}{4 - x^2} = \dfrac{13x}{x - 2}, \quad x \neq -2$

10. $\dfrac{x^2(\quad)}{x^2 - 10x} = \dfrac{x^2}{10 - x}, \quad x \neq 0$

In Exercises 11–40, multiply and simplify. See Examples 1–5.

11. $\dfrac{45}{28} \cdot \dfrac{77}{60}$

12. $24\left(-\dfrac{7}{18}\right)$

13. $7x \cdot \dfrac{9}{14x}$

14. $\dfrac{6}{5a} \cdot (25a)$

15. $\dfrac{8s^3}{9s} \cdot \dfrac{6s^2}{32s}$

16. $\dfrac{3x^4}{7x} \cdot \dfrac{8x^2}{9}$

17. $16u^4 \cdot \dfrac{12}{8u^2}$

18. $25x^3 \cdot \dfrac{8}{35x}$

19. $\dfrac{8}{3 + 4x} \cdot (9 + 12x)$

20. $(6 - 4x) \cdot \dfrac{10}{3 - 2x}$

21. $\dfrac{8u^2v}{3u + v} \cdot \dfrac{u + v}{12u}$

22. $\dfrac{1 - 3xy}{4x^2y} \cdot \dfrac{46x^4y^2}{15 - 45xy}$

23. $\dfrac{12 - r}{3} \cdot \dfrac{3}{r - 12}$

24. $\dfrac{8 - z}{8 + z} \cdot \dfrac{z + 8}{z - 8}$

25. $\dfrac{(2x - 3)(x + 8)}{x^3} \cdot \dfrac{x}{3 - 2x}$

26. $\dfrac{x + 14}{x^3(10 - x)} \cdot \dfrac{x(x - 10)}{5}$

27. $\dfrac{4r - 12}{r - 2} \cdot \dfrac{r^2 - 4}{r - 3}$

28. $\dfrac{5y - 20}{5y + 15} \cdot \dfrac{2y + 6}{y - 4}$

29. $\dfrac{2t^2 - t - 15}{t + 2} \cdot \dfrac{t^2 - t - 6}{t^2 - 6t + 9}$

30. $\dfrac{y^2 - 16}{y^2 + 8y + 16} \cdot \dfrac{3y^2 - 5y - 2}{y^2 - 6y + 8}$

31. $(x^2 - 4y^2) \cdot \dfrac{xy}{(x - 2y)^2}$

32. $(u - 2v)^2 \cdot \dfrac{u + 2v}{u - 2v}$

33. $\dfrac{x^2 + 2xy - 3y^2}{(x + y)^2} \cdot \dfrac{x^2 - y^2}{x + 3y}$

34. $\dfrac{(x - 2y)^2}{x + 2y} \cdot \dfrac{x^2 + 7xy + 10y^2}{x^2 - 4y^2}$

35. $\dfrac{x + 5}{x - 5} \cdot \dfrac{2x^2 - 9x - 5}{3x^2 + x - 2} \cdot \dfrac{x^2 - 1}{x^2 + 7x + 10}$

36. $\dfrac{t^2 + 4t + 3}{2t^2 - t - 10} \cdot \dfrac{t}{t^2 + 3t + 2} \cdot \dfrac{2t^2 + 4t^3}{t^2 + 3t}$

37. $\dfrac{9 - x^2}{2x + 3} \cdot \dfrac{4x^2 + 8x - 5}{4x^2 - 8x + 3} \cdot \dfrac{6x^4 - 2x^3}{8x^2 + 4x}$

38. $\dfrac{16x^2 - 1}{4x^2 + 9x + 5} \cdot \dfrac{5x^2 - 9x - 18}{x^2 - 12x + 36} \cdot \dfrac{12 + 4x - x^2}{4x^2 - 13x + 3}$

39. $\dfrac{x^3 + 3x^2 - 4x - 12}{x^3 - 3x^2 - 4x + 12} \cdot \dfrac{x^2 - 9}{x}$

40. $\dfrac{xu - yu + xv - yv}{xu + yu - xv - yv} \cdot \dfrac{xu + yu + xv + yv}{xu - yu - xv + yv}$

In Exercises 41–60, divide and simplify. See Examples 6 and 7.

41. $-\dfrac{5}{12} \div \dfrac{45}{32}$

42. $-\dfrac{7}{15} \div \left(-\dfrac{14}{25}\right)$

43. $x^2 \div \dfrac{3x}{4}$

44. $\dfrac{u}{10} \div u^2$

45. $\dfrac{7xy^2}{10u^2v} \div \dfrac{21x^3}{45uv}$

46. $\dfrac{25x^2y}{60x^3y^2} \div \dfrac{5x^4y^3}{16x^2y}$

47. $\dfrac{3(a + b)}{4} \div \dfrac{(a + b)^2}{2}$

48. $\dfrac{x^2 + 9}{5(x + 2)} \div \dfrac{x + 3}{5(x^2 - 4)}$

49. $\dfrac{(x^3y)^2}{(x + 2y)^2} \div \dfrac{x^2y}{(x + 2y)^3}$

50. $\dfrac{x^2 - y^2}{2x^2 - 8x} \div \dfrac{(x - y)^2}{2xy}$

51. $\dfrac{\left(\dfrac{x^2}{12}\right)}{\left(\dfrac{5x}{18}\right)}$

52. $\dfrac{\left(\dfrac{3u^2}{6v^3}\right)}{\left(\dfrac{u}{3v}\right)}$

53. $\dfrac{\left(\dfrac{25x^2}{x - 5}\right)}{\left(\dfrac{10x}{5 + 4x - x^2}\right)}$

54. $\dfrac{\left(\dfrac{5x}{x + 7}\right)}{\left(\dfrac{10}{x^2 + 8x + 7}\right)}$

55. $\dfrac{16x^2 + 8x + 1}{3x^2 + 8x - 3} \div \dfrac{4x^2 - 3x - 1}{x^2 + 6x + 9}$

56. $\dfrac{9x^2 - 24x + 16}{x^2 + 10x + 25} \div \dfrac{6x^2 - 5x - 4}{2x^2 + 3x - 35}$

57. $\dfrac{x^2 + 3x - 2x - 6}{x^2 - 4} \div \dfrac{x + 3}{x^2 + 4x + 4}$

58. $\dfrac{t^3 + t^2 - 9t - 9}{t^2 - 5t + 6} \div \dfrac{t^2 + 6t + 9}{t - 2}$

59. $\dfrac{\left(\dfrac{x^2 - 3x - 10}{x^2 - 4x + 4}\right)}{\left(\dfrac{21 + 4x - x^2}{x^2 - 5x - 14}\right)}$

60. $\dfrac{\left(\dfrac{x^2 + 5x + 6}{4x^2 - 20x + 25}\right)}{\left(\dfrac{x^2 - 5x - 24}{4x^2 - 25}\right)}$

In Exercises 61–68, perform the operations and simplify. (In Exercises 67 and 68, n is a positive integer.)

61. $\left[\dfrac{x^2}{9} \cdot \dfrac{3(x + 4)}{x^2 + 2x}\right] \div \dfrac{x}{x + 2}$

62. $\left(\dfrac{x^2 + 6x + 9}{x^2} \cdot \dfrac{2x + 1}{x^2 - 9}\right) \div \dfrac{4x^2 + 4x + 1}{x^2 - 3x}$

63. $\left[\dfrac{xy + y}{4x} \div (3x + 3)\right] \div \dfrac{y}{3x}$

64. $\dfrac{3u^2 - u - 4}{u^2} \div \dfrac{3u^2 + 12u + 4}{u^4 - 3u^3}$

65. $\dfrac{2x^2 + 5x - 25}{3x^2 + 5x + 2} \cdot \dfrac{3x^2 + 2x}{x + 5} \div \left(\dfrac{x}{x + 1}\right)^2$

66. $\dfrac{t^2 - 100}{4t^2} \cdot \dfrac{t^3 - 5t^2 - 50t}{t^4 + 10t^3} \div \dfrac{(t - 10)^2}{5t}$

67. $x^3 \cdot \dfrac{x^{2n} - 9}{x^{2n} + 4x^n + 3} \div \dfrac{x^{2n} - 2x^n - 3}{x}$

68. $\dfrac{x^{n+1} - 8x}{x^{2n} + 2x^n + 1} \cdot \dfrac{x^{2n} - 4x^n - 5}{x} \div x^n$

In Exercises 69–72, use a graphing utility to graph the two equations on the same screen. Use the graphs to verify that the expressions are equivalent. Verify the results algebraically.

69. $y_1 = \dfrac{3x + 2}{x} \cdot \dfrac{x^2}{9x^2 - 4}$

$y_2 = \dfrac{x}{3x - 2}, \quad x \neq 0, \quad x \neq -\dfrac{2}{3}$

70. $y_1 = \dfrac{x^2 - 10x + 25}{x^2 - 25} \cdot \dfrac{x + 5}{2}$

$y_2 = \dfrac{x - 5}{2}, \quad x \neq \pm 5$

71. $y_1 = \dfrac{3x + 15}{x^4} \div \dfrac{x + 5}{x^2}$

$y_2 = \dfrac{3}{x^2}, \quad x \neq -5$

72. $y_1 = (x^2 + 6x + 9) \cdot \dfrac{3}{2x(x + 3)}$

$y_2 = \dfrac{3(x + 3)}{2x}, \quad x \neq -3$

Solving Problems

Geometry   In Exercises 73 and 74, write an expression for the area of the shaded region. Then simplify.

73.

	$\dfrac{2w + 3}{3}$	$\dfrac{2w + 3}{3}$	$\dfrac{2w + 3}{3}$
$\dfrac{w}{2}$		▓	
$\dfrac{w}{2}$			

74.

	$\dfrac{2w - 1}{2}$	$\dfrac{2w - 1}{2}$
$\dfrac{w}{3}$		
$\dfrac{w}{3}$	░	
$\dfrac{w}{3}$		

Probability   In Exercises 75–78, consider an experiment in which a marble is tossed into a rectangular box with dimensions x centimeters by $2x + 1$ centimeters. The probability that the marble will come to rest in the unshaded portion of the box is equal to the ratio of the unshaded area to the total area of the figure. Find the probability in simplified form.

75.

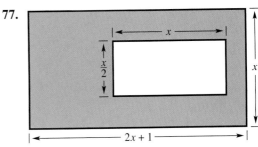

76.

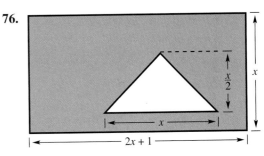

77.

78.

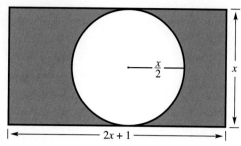

79. *Photocopy Rate*   A photocopier produces copies at a rate of 20 pages per minute.

(a) Determine the time required to copy 1 page.

(b) Determine the time required to copy x pages.

(c) Determine the time required to copy 35 pages.

80. *Pumping Rate*   The rate for a pump is 15 gallons per minute.

(a) Determine the time required to pump 1 gallon.

(b) Determine the time required to pump x gallons.

(c) Determine the time required to pump 130 gallons.

81. *Analyzing Data*   The number N (in thousands) of subscribers to a cellular telephone service and the annual revenue R (in millions of dollars) generated by subscribers in the United States for the period 1990 through 1996 can be modeled by

$$N = 6357 + 1070t^2 \quad \text{and} \quad R = 6115.2 + 590.7t^2$$

where t is time in years, with $t = 0$ representing 1990.   (Source: Cellular Telecommunications Industry Association)

(a) Use a graphing utility to graph the two models.

(b) Find a model for the average monthly bill per subscriber. *(Note:* Modify the revenue function from years to months.)

(c) Use the model in part (b) to complete the table.

Year, t	0	2	4	6
Monthly bill				

(d) The number of subscribers and the revenue were increasing over the last few years, and yet the average monthly bill was decreasing. Explain how this is possible.

Explaining Concepts

82. In your own words, explain how to divide rational expressions.

83. Explain how to divide a rational expression by a polynomial.

84. Define the term *complex fraction*. Give an example and show how to simplify the fraction.

85. *Error Analysis*   Describe the error.

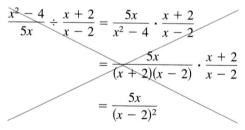

$$\frac{x^2 - 4}{5x} \div \frac{x + 2}{x - 2} = \frac{5x}{x^2 - 4} \cdot \frac{x + 2}{x - 2}$$

$$= \frac{5x}{(x + 2)(x - 2)} \cdot \frac{x + 2}{x - 2}$$

$$= \frac{5x}{(x - 2)^2}$$

Mid-Chapter Quiz

Take this quiz as you would take a quiz in class. After you are done, check your work against the answers given in the back of the book.

In Exercises 1–4, rewrite the expression using only positive exponents, and simplify. (Assume that any variables in the expression are nonzero.)

1. $(t^3)^{-4}(3t^3)$

2. $(3x^2y^{-1})(4x^{-2}y)^{-2}$

3. $\dfrac{10u^{-2}}{15u}$

4. $\dfrac{(10x)^0 x^{-2}}{(x^2)^{-1}}$

5. Write each number in scientific notation: (a) 13,400,000; (b) 0.00075.

6. Evaluate each expression without using a calculator.

(a) $(3 \times 10^3)^4$ (b) $\dfrac{3.2 \times 10^4}{16 \times 10^7}$

7. Determine the domain of $\dfrac{y + 2}{y(y - 4)}$.

8. Evaluate $h(x) = (x^2 - 9)/(x^2 - x - 2)$ for the indicated values of x. If it is not possible, state the reason.

(a) $h(-3)$ (b) $h(0)$ (c) $h(-1)$ (d) $h(5)$

In Exercises 9–14, write the expression in reduced form.

9. $\dfrac{9y^2}{6y}$

10. $\dfrac{8u^3v^2}{36uv^3}$

11. $\dfrac{4x^2 - 1}{x - 2x^2}$

12. $\dfrac{(z + 3)^2}{2z^2 + 5z - 3}$

13. $\dfrac{7ab + 3a^2b^2}{a^2b}$

14. $\dfrac{2mn^2 - n^3}{2m^2 + mn - n^2}$

In Exercises 15–20, perform the operations and simplify your answer.

15. $\dfrac{11t^2}{6} \cdot \dfrac{9}{33t}$

16. $(x^2 + 2x) \cdot \dfrac{5}{x^2 - 4}$

17. $\dfrac{4}{3(x - 1)} \cdot \dfrac{12x}{6(x^2 + 2x - 3)}$

18. $\dfrac{5u}{3(u + v)} \cdot \dfrac{2(u^2 - v^2)}{3v} \div \dfrac{25u^2}{18(u - v)}$

19. $\dfrac{\left(\dfrac{9t^2}{3 - t}\right)}{\left(\dfrac{6t}{t - 3}\right)}$

20. $\dfrac{\left(\dfrac{10}{x^2 + 2x}\right)}{\left(\dfrac{15}{x^2 + 3x + 2}\right)}$

21. You start a business with a setup cost of $6000. The cost of material for producing each unit of your product is $10.50.

(a) Write an algebraic function that gives the average cost per unit when x units are produced. Explain your reasoning.

(b) Find the average cost per unit when $x = 500$ units are produced.

4.4   Adding and Subtracting Rational Expressions

Objectives

1 Add or subtract rational expressions with like denominators and simplify.

2 Add or subtract rational expressions with unlike denominators and simplify.

3 Simplify a complex fraction.

1 Add or subtract rational expressions with like denominators and simplify.

Adding or Subtracting with Like Denominators

As with numerical fractions, the procedure used to add or subtract two rational expressions depends on whether the expressions have *like* or *unlike* denominators. To add or subtract two rational expressions with *like* denominators, simply combine their numerators and place the result over the common denominator.

▶ **Adding or Subtracting with Like Denominators**

If u, v, and w are real numbers, variables, or variable expressions, and $w \neq 0$, the following rules are valid.

1. $\dfrac{u}{w} + \dfrac{v}{w} = \dfrac{u+v}{w}$ Add fractions with like denominators.

2. $\dfrac{u}{w} - \dfrac{v}{w} = \dfrac{u-v}{w}$ Subtract fractions with like denominators.

Example 1   Adding and Subtracting with Like Denominators

a. $\dfrac{x}{4} + \dfrac{5-x}{4} = \dfrac{x+(5-x)}{4} = \dfrac{5}{4}$

b. $\dfrac{7}{2x-3} - \dfrac{3x}{2x-3} = \dfrac{7-3x}{2x-3}$

Example 2   Subtracting Rational Expressions and Simplifying

$$\frac{x}{x^2-2x-3} - \frac{3}{x^2-2x-3} = \frac{x-3}{x^2-2x-3} \qquad \text{Subtract.}$$

$$= \frac{x-3}{(x-3)(x+1)} \qquad \text{Factor.}$$

$$= \frac{(x-3)(1)}{(x-3)(x+1)} \qquad \text{Cancel common factor.}$$

$$= \frac{1}{x+1}, \quad x \neq 3 \qquad \text{Simplified form}$$

Study Tip

After adding or subtracting two (or more) rational expressions, check the resulting fraction to see if it can be simplified, as illustrated in Example 2.

The rules for adding and subtracting rational expressions with like denominators can be extended to cover sums and differences involving three or more rational expressions, as illustrated in Example 3.

Example 3　Combining Three Rational Expressions

$$\frac{x^2 - 26}{x - 5} - \frac{2x + 4}{x - 5} + \frac{10 + x}{x - 5}$$

$$= \frac{(x^2 - 26) - (2x + 4) + (10 + x)}{x - 5} \qquad \text{Write numerator over common denominator.}$$

$$= \frac{x^2 - 26 - 2x - 4 + 10 + x}{x - 5} \qquad \text{Distributive Property}$$

$$= \frac{x^2 - x - 20}{x - 5} \qquad \text{Simplify.}$$

$$= \frac{(x - 5)(x + 4)}{x - 5} \qquad \text{Factor and cancel common factor.}$$

$$= x + 4, \quad x \neq 5 \qquad \text{Simplified form}$$

---

2 Add or subtract rational expressions with unlike denominators and simplify.

Adding or Subtracting with Unlike Denominators

To add or subtract rational expressions with *unlike* denominators, you must first rewrite each expression using the **least common multiple (LCM)** of the denominators of the individual expressions. The least common multiple of two (or more) polynomials is the simplest polynomial that is a multiple of each of the original polynomials. This means that the LCM must contain all the *different* factors in the polynomials and each of these factors must be repeated the maximum number of times it occurs in any one of the polynomials.

Example 4　Finding Least Common Multiples

a. The least common multiple of

$$6x = 2 \cdot 3 \cdot x, \quad 2x^2 = 2 \cdot x \cdot x, \quad \text{and} \quad 9x^3 = 3 \cdot 3 \cdot x \cdot x \cdot x$$

is $2 \cdot 3 \cdot 3 \cdot x \cdot x \cdot x = 18x^3$.

b. The least common multiple of

$$x^2 - x = x(x - 1) \quad \text{and} \quad 2x - 2 = 2(x - 1)$$

is $2x(x - 1)$.

c. The least common multiple of

$$3x^2 + 6x = 3x(x + 2) \quad \text{and} \quad x^2 + 4x + 4 = (x + 2)^2$$

is $3x(x + 2)^2$.

To add or subtract rational expressions with *unlike* denominators, you must first rewrite the rational expressions so that they have *like* denominators. The like denominator that you use is the least common multiple of the original denominators and is called the **least common denominator (LCD)** of the original rational expressions. Once the rational expressions have been written with like denominators, you can simply add or subtract these rational expressions using the rules given at the beginning of this section.

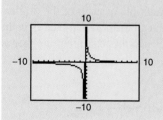

Example 5 Adding with Unlike Denominators

Add the rational expressions: $\dfrac{7}{6x} + \dfrac{5}{8x}$.

Solution

The least common denominator of $6x$ and $8x$ is $24x$, so the first step is to rewrite each fraction with this denominator.

$$\frac{7}{6x} + \frac{5}{8x} = \frac{7(4)}{6x(4)} + \frac{5(3)}{8x(3)} \qquad \text{Rewrite fractions using LCD of } 24x.$$

$$= \frac{28}{24x} + \frac{15}{24x} \qquad \text{Like denominators}$$

$$= \frac{28 + 15}{24x} \qquad \text{Add fractions.}$$

$$= \frac{43}{24x} \qquad \text{Simplified form}$$

Example 6 Subtracting with Unlike Denominators

Subtract the rational expressions: $\dfrac{3}{x - 3} - \dfrac{5}{x + 2}$.

Solution

The least common denominator is $(x - 3)(x + 2)$.

$$\frac{3}{x - 3} - \frac{5}{x + 2}$$

$$= \frac{3(x + 2)}{(x - 3)(x + 2)} - \frac{5(x - 3)}{(x - 3)(x + 2)} \qquad \text{Rewrite fractions using LCD of } (x - 3)(x + 2).$$

$$= \frac{3x + 6}{(x - 3)(x + 2)} - \frac{5x - 15}{(x - 3)(x + 2)} \qquad \text{Distributive Property}$$

$$= \frac{(3x + 6) - (5x - 15)}{(x - 3)(x + 2)} \qquad \text{Subtract fractions.}$$

$$= \frac{3x + 6 - 5x + 15}{(x - 3)(x + 2)} \qquad \text{Distributive Property}$$

$$= \frac{-2x + 21}{(x - 3)(x + 2)} \qquad \text{Simplified form}$$

Study Tip

In Example 7, the factors in the denominator are $x^2 - 4 = (x + 2)(x - 2)$ and $2 - x$. Because

$$(2 - x) = (-1)(x - 2),$$

the original addition problem can be written as a subtraction problem.

Example 7   Adding with Unlike Denominators

$$\frac{6x}{x^2 - 4} + \frac{3}{(2 - x)}$$

$$= \frac{6x}{(x + 2)(x - 2)} + \frac{3}{(-1)(x - 2)} \qquad \text{Factor.}$$

$$= \frac{6x}{(x + 2)(x - 2)} - \frac{3(x + 2)}{(x + 2)(x - 2)} \qquad \begin{array}{l}\text{Rewrite fractions}\\ \text{using LCD of}\\ (x + 2)(x - 2).\end{array}$$

$$= \frac{6x}{(x + 2)(x - 2)} - \frac{3x + 6}{(x + 2)(x - 2)} \qquad \text{Distributive Property}$$

$$= \frac{6x - (3x + 6)}{(x + 2)(x - 2)} \qquad \text{Subtract.}$$

$$= \frac{6x - 3x - 6}{(x + 2)(x - 2)} \qquad \text{Distributive Property}$$

$$= \frac{3x - 6}{(x + 2)(x - 2)} \qquad \text{Simplify.}$$

$$= \frac{3(x - 2)}{(x + 2)(x - 2)} \qquad \begin{array}{l}\text{Factor and cancel}\\ \text{common factor.}\end{array}$$

$$= \frac{3}{x + 2}, \quad x \ne 2 \qquad \text{Simplified form}$$

Example 8   Combining Three Rational Expressions

$$\frac{2x - 5}{6x + 9} - \frac{4}{2x^2 + 3x} + \frac{1}{x}$$

$$= \frac{(2x - 5)(x)}{3(2x + 3)(x)} - \frac{(4)(3)}{x(2x + 3)(3)} + \frac{3(2x + 3)}{(x)(3)(2x + 3)} \qquad \begin{array}{l}\text{Rewrite fractions}\\ \text{using LCD of}\\ 3x(2x + 3).\end{array}$$

$$= \frac{2x^2 - 5x}{3x(2x + 3)} - \frac{12}{3x(2x + 3)} + \frac{6x + 9}{3x(2x + 3)} \qquad \text{Distributive Property}$$

$$= \frac{2x^2 - 5x - 12 + 6x + 9}{3x(2x + 3)} \qquad \text{Combine numerators.}$$

$$= \frac{2x^2 + x - 3}{3x(2x + 3)} \qquad \text{Simplify.}$$

$$= \frac{(x - 1)(2x + 3)}{3x(2x + 3)} \qquad \text{Factor.}$$

$$= \frac{(x - 1)(2x + 3)}{3x(2x + 3)} \qquad \text{Cancel common factor.}$$

$$= \frac{x - 1}{3x}, \quad x \ne -\frac{3}{2} \qquad \text{Simplified form}$$

3 Simplify a complex fraction.

Complex Fractions

Complex fractions can have numerators or denominators that are the sums or differences of fractions. To simplify a complex fraction, first combine its numerator and its denominator into single fractions. Then divide by inverting the divisor and multiplying.

Example 9   Simplifying a Complex Fraction

Simplify $\dfrac{\left(\dfrac{x}{4} + \dfrac{3}{2}\right)}{\left(2 - \dfrac{3}{x}\right)}$.

Solution

$$\frac{\left(\dfrac{x}{4} + \dfrac{3}{2}\right)}{\left(2 - \dfrac{3}{x}\right)} = \frac{\left(\dfrac{x}{4} + \dfrac{6}{4}\right)}{\left(\dfrac{2x}{x} - \dfrac{3}{x}\right)} \qquad \text{Find least common denominators.}$$

$$= \frac{\left(\dfrac{x + 6}{4}\right)}{\left(\dfrac{2x - 3}{x}\right)} \qquad \begin{array}{l}\text{Add fractions in numerator}\\\text{and denominator.}\end{array}$$

$$= \frac{x + 6}{4} \cdot \frac{x}{2x - 3} \qquad \text{Invert divisor and multiply.}$$

$$= \frac{x(x + 6)}{4(2x - 3)}, \quad x \neq 0 \qquad \text{Simplified form}$$

Another way to simplify the complex fraction given in Example 9 is to multiply the numerator and denominator by the least common denominator of *every* fraction in the numerator and denominator. For this fraction, notice what happens when we multiply the numerator and denominator by $4x$.

$$\frac{\left(\dfrac{x}{4} + \dfrac{3}{2}\right)}{\left(2 - \dfrac{3}{x}\right)} = \frac{\left(\dfrac{x}{4} + \dfrac{3}{2}\right)}{\left(2 - \dfrac{3}{x}\right)} \cdot \frac{4x}{4x} \qquad \begin{array}{l}\text{Multiply numerator and denominator}\\\text{by LCD of } 4x.\end{array}$$

$$= \frac{\dfrac{x}{4}(4x) + \dfrac{3}{2}(4x)}{2(4x) - \dfrac{3}{x}(4x)} \qquad \text{Distributive Property}$$

$$= \frac{x^2 + 6x}{8x - 12} \qquad \text{Simplify.}$$

$$= \frac{x(x + 6)}{4(2x - 3)}, \quad x \neq 0 \qquad \text{Simplified form}$$

Example 10   Simplifying a Complex Fraction

Simplify $\dfrac{\left(\dfrac{2}{x+2}\right)}{\left(\dfrac{1}{x+2}+\dfrac{2}{x}\right)}$.

Solution

The least common denominator of the fractions is

$$x(x+2).$$

Multiplying each fraction by this LCD yields

$$\frac{\left(\dfrac{2}{x+2}\right)}{\left(\dfrac{1}{x+2}+\dfrac{2}{x}\right)}=\frac{\left(\dfrac{2}{x+2}\right)(x)(x+2)}{\dfrac{1}{x+2}(x)(x+2)+\dfrac{2}{x}(x)(x+2)}$$

Multiply numerator and denominator by LCD of $x(x+2)$.

$$=\frac{2x}{x+2(x+2)}$$

Simplify.

$$=\frac{2x}{3x+4},\quad x\neq-2, x\neq 0.$$

Simplified form

The factors of the LCD cannot be equal to zero in order for the expression to be defined. So, for the reduced form of the expression you must specify that $x\neq-2$ and $x\neq 0$.

Discussing the Concept   **Comparing Two Methods**

Evaluate each of the following expressions at the given value of the variable in two different ways: (1) combine and simplify the rational expressions first and then evaluate the simplified expression at the given variable value, and (2) substitute the given value of the variable first and then simplify the resulting expression. Do you get the same result with each method? Discuss which method you prefer and why. List any advantages and/or disadvantages of each method.

a. $\dfrac{1}{m-4}-\dfrac{1}{m+4}+\dfrac{3m}{m^2-16}$, $m=2$

b. $\dfrac{x-2}{x^2-9}+\dfrac{3x+2}{x^2-5x+6}$, $x=4$

c. $\dfrac{3y^2+16y-8}{y^2+2y-8}-\dfrac{y-1}{y-2}+\dfrac{y}{y+4}$, $y=3$

4.4   Exercises

Integrated Review   Concepts, Skills, and Problem Solving

Keep mathematically in shape by doing these exercises *before* the problems of this section.

Properties and Definitions

1. Write the equation $5y - 3x - 4 = 0$ in the following forms.

(a) Slope-intercept form

(b) Point-slope form (many correct answers)

2. Explain how you can visually determine the sign of the slope of a line by observing its graph.

Simplifying Expressions

In Exercises 3–10, perform the multiplication and simplify.

3. $-6x(10 - 7x)$

4. $(2 - y)(3 + 2y)$

5. $(11 - x)(11 + x)$

6. $(4 - 5z)(4 + 5z)$

7. $(x + 1)^2$

8. $t(t^2 + 1) - t(t^2 - 1)$

9. $(x - 2)(x^2 + 2x + 4)$

10. $t(t - 4)(2t + 3)$

Creating Expressions

In Exercises 11 and 12, find expressions for the perimeter and area of the region. Simplify the expressions.

11.   **12.**

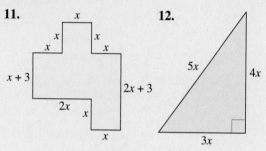

Developing Skills

In Exercises 1–18, combine and simplify. See Examples 1–3.

1. $\frac{5}{8} + \frac{7}{8}$

2. $\frac{7}{12} - \frac{5}{12}$

3. $\frac{5x}{8} - \frac{7x}{8}$

4. $\frac{7y}{12} + \frac{9y}{12}$

5. $\frac{2}{3a} - \frac{11}{3a}$

6. $\frac{6x}{13} - \frac{7x}{13}$

7. $\frac{x}{9} - \frac{x + 2}{9}$

8. $\frac{4 - y}{4} + \frac{3y}{4}$

9. $\frac{z^2}{3} + \frac{z^2 - 2}{3}$

10. $\frac{10x^2 + 1}{3} - \frac{10x^2}{3}$

11. $\frac{2x + 5}{3} + \frac{1 - x}{3}$

12. $\frac{16 + z}{5z} - \frac{11 - z}{5z}$

13. $\frac{3y}{3} - \frac{3y - 3}{3} - \frac{7}{3}$

14. $\frac{-16u}{9} - \frac{27 - 16u}{9} + \frac{2}{9}$

15. $\frac{3y - 22}{y - 6} - \frac{2y - 16}{y - 6}$

16. $\frac{5x - 1}{x + 4} + \frac{5 - 4x}{x + 4}$

17. $\frac{2x - 1}{x(x - 3)} + \frac{1 - x}{x(x - 3)}$

18. $\frac{7s - 5}{2s + 5} + \frac{3(s + 10)}{2s + 5}$

In Exercises 19–30, find the least common multiple of the expressions. See Example 4.

19. $5x^2, 20x^3$

20. $14t^2, 42t^5$

21. $9y^3, 12y$

22. $44m^2, 10m$

23. $15x^2, 3(x + 5)$

24. $6x^2, 15x(x - 1)$

25. $63z^2(z + 1), 14(z + 1)^4$

26. $18y^3, 27y(y - 3)^2$

27. $8t(t + 2), 14(t^2 - 4)$ **28.** $2y^2 + y - 1, 4y^2 - 2y$

29. $6(x^2 - 4), 2x(x + 2)$

30. $t^3 + 3t^2 + 9t, 2t^2(t^2 - 9)$

In Exercises 31–36, find the missing algebraic expression that makes the two fractions equivalent.

31. $\frac{7x^2}{4a(\quad)} = \frac{7}{4a}, \quad x \neq 0$

32. $\frac{3y(x - 3)^2}{(x - 3)(\quad)} = \frac{21y}{x - 3}$

33. $\frac{5r(\quad)}{3v(u + 1)} = \frac{5r}{3v}, \quad u \neq -1$

34. $\dfrac{(3t + 5)()}{10t^2(3t - 5)} = \dfrac{3t + 5}{2t}, \quad t \neq \dfrac{5}{3}$

35. $\dfrac{7y()}{4 - x^2} = \dfrac{7y}{x - 2}, \quad x \neq -2$

36. $\dfrac{4x^2()}{x^2 - 10x} = \dfrac{4x^2}{10 - x}, \quad x \neq 0$

In Exercises 37–44, find the least common denominator of the two fractions and rewrite each fraction using the least common denominator.

37. $\dfrac{n + 8}{3n - 12}, \dfrac{10}{6n^2}$

38. $\dfrac{8s}{(s + 2)^2}, \dfrac{3}{s^3 + s^2 - 2s}$

39. $\dfrac{2}{x^2(x - 3)}, \dfrac{5}{x(x + 3)}$

40. $\dfrac{5t}{2t(t - 3)^2}, \dfrac{4}{t(t - 3)}$

41. $\dfrac{v}{2v^2 + 2v}, \dfrac{4}{3v^2}$

42. $\dfrac{4x}{(x + 5)^2}, \dfrac{x - 2}{x^2 - 25}$

43. $\dfrac{x - 8}{x^2 - 25}, \dfrac{9x}{x^2 - 10x + 25}$

44. $\dfrac{3y}{y^2 - y - 12}, \dfrac{y - 4}{y^2 + 3y}$

In Exercises 45–78, perform the operation and simplify. See Examples 5–8.

45. $\dfrac{5}{4x} - \dfrac{3}{5}$

46. $\dfrac{10}{b} + \dfrac{1}{10b}$

47. $\dfrac{7}{a} + \dfrac{14}{a^2}$

48. $\dfrac{1}{6u^2} - \dfrac{2}{9u}$

49. $\dfrac{20}{x - 4} + \dfrac{20}{4 - x}$

50. $\dfrac{15}{2 - t} - \dfrac{7}{t - 2}$

51. $\dfrac{3x}{x - 8} - \dfrac{6}{8 - x}$

52. $\dfrac{1}{y - 6} + \dfrac{y}{6 - y}$

53. $25 + \dfrac{10}{x + 4}$

54. $\dfrac{100}{x - 10} - 8$

55. $\dfrac{3x}{3x - 2} + \dfrac{2}{2 - 3x}$

56. $\dfrac{y}{5y - 3} - \dfrac{3}{3 - 5y}$

57. $-\dfrac{1}{6x} + \dfrac{1}{6(x - 3)}$

58. $\dfrac{3}{t(t + 1)} + \dfrac{4}{t}$

59. $\dfrac{x}{x + 3} - \dfrac{5}{x - 2}$

60. $\dfrac{1}{x + 4} - \dfrac{1}{x + 2}$

61. $\dfrac{3}{x + 1} - \dfrac{2}{x}$

62. $\dfrac{5}{x - 4} - \dfrac{3}{x}$

63. $\dfrac{3}{x - 5} + \dfrac{2}{x + 5}$

64. $\dfrac{7}{2x - 3} + \dfrac{3}{2x + 3}$

65. $\dfrac{4}{x^2} - \dfrac{4}{x^2 + 1}$

66. $\dfrac{2}{y^2 + 2} + \dfrac{1}{2y^2}$

67. $\dfrac{x}{x^2 - 9} + \dfrac{3}{x^2 - 5x + 6}$

68. $\dfrac{x}{x^2 - x - 30} - \dfrac{1}{x + 5}$

69. $\dfrac{4}{x - 4} + \dfrac{16}{(x - 4)^2}$

70. $\dfrac{3}{x - 2} - \dfrac{1}{(x - 2)^2}$

71. $\dfrac{y}{x^2 + xy} - \dfrac{x}{xy + y^2}$

72. $\dfrac{5}{x + y} + \dfrac{5}{x^2 - y^2}$

73. $\dfrac{4}{x} - \dfrac{2}{x^2} + \dfrac{4}{x + 3}$

74. $\dfrac{5}{2} - \dfrac{1}{2x} - \dfrac{3}{x + 1}$

75. $\dfrac{3u}{u^2 - 2uv + v^2} + \dfrac{2}{u - v} - \dfrac{u}{u - v}$

76. $\dfrac{1}{x - y} - \dfrac{3}{x + y} + \dfrac{3x - y}{x^2 - y^2}$

77. $\dfrac{x + 2}{x - 1} - \dfrac{2}{x + 6} - \dfrac{14}{x^2 + 5x - 6}$

78. $\dfrac{x}{x^2 + 15x + 50} + \dfrac{7}{x + 10} - \dfrac{x - 1}{x + 5}$

In Exercises 79 and 80, use a graphing utility to graph the two equations on the same screen. Use the graphs to verify that the expressions are equivalent. Verify the results algebraically.

79. $y_1 = \dfrac{2}{x} + \dfrac{4}{x - 2}, y_2 = \dfrac{6x - 4}{x(x - 2)}$

80. $y_1 = 3 - \dfrac{1}{x - 1}, y_2 = \dfrac{3x - 4}{x - 1}$

In Exercises 81–96, simplify the complex fraction. See Examples 9 and 10.

81. $\dfrac{\dfrac{1}{2}}{\left(3 + \dfrac{1}{x}\right)}$

82. $\dfrac{\dfrac{2}{3}}{\left(4 - \dfrac{1}{x}\right)}$

83. $\dfrac{\left(\dfrac{4}{x} + 3\right)}{\left(\dfrac{4}{x} - 3\right)}$

84. $\dfrac{\left(\dfrac{1}{t} - 1\right)}{\left(\dfrac{1}{t} + 1\right)}$

85. $\dfrac{\left(16x - \dfrac{1}{x}\right)}{\left(\dfrac{1}{x} - 4\right)}$

86. $\dfrac{\left(\dfrac{36}{y} - y\right)}{6 + y}$

87. $\dfrac{\left(3 + \dfrac{9}{x - 3}\right)}{\left(4 + \dfrac{12}{x - 3}\right)}$

88. $\dfrac{\left(x + \dfrac{2}{x - 3}\right)}{\left(x + \dfrac{6}{x - 3}\right)}$

89. $\dfrac{\left(\dfrac{3}{x^2} + \dfrac{1}{x}\right)}{\left(2 - \dfrac{4}{5x}\right)}$

90. $\dfrac{\left(16 - \dfrac{1}{x^2}\right)}{\left(\dfrac{1}{4x^2} - 4\right)}$

91. $\dfrac{\left(\dfrac{y}{x} - \dfrac{x}{y}\right)}{\left(\dfrac{x + y}{xy}\right)}$

92. $\dfrac{\left(x - \dfrac{2y^2}{x - y}\right)}{x - 2y}$

93. $\dfrac{\left(1 - \dfrac{1}{y}\right)}{\left(\dfrac{1 - 4y}{y - 3}\right)}$

94. $\dfrac{\left(\dfrac{x + 1}{x + 2} - \dfrac{1}{x}\right)}{\left(\dfrac{2}{x + 2}\right)}$

95. $\dfrac{\left(\dfrac{x}{x - 3} - \dfrac{2}{3}\right)}{\left(\dfrac{10}{3x} + \dfrac{x^2}{x - 3}\right)}$

96. $\dfrac{\left(\dfrac{1}{2x} - \dfrac{6}{x + 5}\right)}{\left(\dfrac{x}{x - 5} + \dfrac{1}{x}\right)}$

In Exercises 97 and 98, use the function to find and simplify the expression for

$$\frac{f(2 + h) - f(2)}{h}.$$

97. $f(x) = \dfrac{1}{x}$

98. $f(x) = \dfrac{x}{x - 1}$

In Exercises 99 and 100, use a graphing utility to complete the table. Comment on the domains and equivalence of the expressions.

99.

x	$\dfrac{\left(1 - \dfrac{1}{x}\right)}{\left(1 - \dfrac{1}{x^2}\right)}$	$\dfrac{x}{x + 1}$
-3		
-2		
-1		
0		
1		
2		
3		

100.

x	$\dfrac{\left(1 + \dfrac{4}{x} + \dfrac{4}{x^2}\right)}{\left(1 - \dfrac{4}{x^2}\right)}$	$\dfrac{x + 2}{x - 2}$
-3		
-2		
-1		
0		
1		
2		
3		

Solving Problems

101. *Work Rate*   After working together for t hours on a common task, two workers have completed fractional parts of the job equal to $t/4$ and $t/6$. What fractional part of the task has been completed?

102. *Work Rate*   After working together for t hours on a common task, two workers have completed fractional parts of the job equal to $t/3$ and $t/5$. What fractional part of the task has been completed?

103. *Average of Two Numbers*   Determine the average of the two real numbers $x/4$ and $x/6$.

104. *Average of Three Numbers*   Determine the average of the three real numbers x, $x/2$, and $x/3$.

105. *Equal Parts*   Find two real numbers that divide the real number line between $x/5$ and $x/3$ into three equal parts (see figure).

Figure for 105

106. *Monthly Payment*   The approximate annual percentage rate r of a monthly installment loan is

$$r = \frac{\left[\dfrac{24(NM - P)}{N}\right]}{\left(P + \dfrac{MN}{12}\right)}$$

where N is the total number of payments, M is the monthly payment, and P is the amount financed.

(a) Approximate the annual percentage rate for a 4-year car loan of $10,000 that has monthly payments of $300.

(b) Simplify the expression for the annual percentage rate r, and then rework part (a).

107. *Parallel Resistance* When two resistors are connected in parallel (see figure), the total resistance is

$$\frac{1}{\left(\dfrac{1}{R_1} + \dfrac{1}{R_2}\right)}.$$

Simplify this complex fraction.

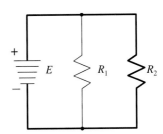

108. *Using Two Models* From 1990 through 1996, the circulations of morning and evening daily newspapers M and E (in millions) can be approximated by

$$M = 41.1 + 0.61t \quad \text{and} \quad E = \frac{20.675 - 1.675t}{1 - 0.023t}$$

where $t = 0$ represents 1990. (Source: Editor & Publisher Co.)

(a) Write an expression for the total daily circulation T. Simplify the result.

(b) Use a graphing utility to graph the functions M, E, and T on the same screen.

(c) Morning circulation is increasing over the time period, whereas evening circulation is decreasing. Use the graphs in part (b) to discuss the change in total circulation.

(d) Use the result of part (a) to approximate the total circulation in the United States in 1991.

Explaining Concepts

109. Answer parts (a)–(c) of Motivating the Chapter on page 253.

110. In your own words, describe how to add or subtract rational expressions with like denominators.

111. In your own words, describe how to add or subtract rational expressions with unlike denominators.

112. Is it possible for the least common denominator of two fractions to be the same as one of the fraction's denominators? If so, give an example.

113. *Error Analysis* Describe the error.

$$\frac{x - 1}{x + 4} - \frac{4x - 11}{x + 4} = \frac{x - 1 - 4x - 11}{x + 4}$$
$$= \frac{-3x - 12}{x + 4} = \frac{-3(x + 4)}{x + 4}$$
$$= -3$$

114. *Error Analysis* Describe the error.

$$\frac{2}{x} - \frac{3}{x + 1} + \frac{x + 1}{x^2}$$
$$= \frac{2x(x + 1) - 3x^2 + (x + 1)^2}{x^2(x + 1)}$$
$$= \frac{2x^2 + x - 3x^2 + x^2 + 1}{x^2(x + 1)}$$
$$= \frac{x + 1}{x^2(x + 1)} = \frac{1}{x^2}$$

4.5   Dividing Polynomials

Objectives

1 Divide a polynomial by a monomial and write in simplest form.

2 Use long division to divide a polynomial by a second polynomial.

3 Use synthetic division to divide a polynomial by a polynomial of the form $x - k$.

4 Use synthetic division to factor a polynomial.

1 Divide a polynomial by a monomial and write in simplest form.

Dividing a Polynomial by a Monomial

To divide a polynomial by a monomial, *reverse* the procedure used to add (or subtract) two rational expressions. Here is an example.

$$2 + \frac{1}{x} = \frac{2x}{x} + \frac{1}{x} = \frac{2x + 1}{x}$$ Add fractions.

$$\frac{2x + 1}{x} = \frac{2x}{x} + \frac{1}{x} = 2 + \frac{1}{x}$$ Divide by monomial.

> ### ▶ Dividing a Polynomial by a Monomial
>
> Let u, v, and w be real numbers, variables, or algebraic expressions such that $w \neq 0$.
>
> **1.** $\dfrac{u + v}{w} = \dfrac{u}{w} + \dfrac{v}{w}$ **2.** $\dfrac{u - v}{w} = \dfrac{u}{w} - \dfrac{v}{w}$

When dividing a polynomial by a monomial, remember to write the resulting expressions in simplest form, as illustrated in Example 1.

Example 1   Dividing a Polynomial by a Monomial

Perform the division and simplify.

$$\frac{12x^2 - 20x + 8}{4x}$$

Solution

$$\frac{12x^2 - 20x + 8}{4x} = \frac{12x^2}{4x} - \frac{20x}{4x} + \frac{8}{4x}$$ Divide each term in the numerator by $4x$.

$$= \frac{3(4x)(x)}{4x} - \frac{5(4x)}{4x} + \frac{2(4)}{4x}$$ Cancel common factors.

$$= 3x - 5 + \frac{2}{x}$$ Simplified form

2 Use long division to divide a polynomial by a second polynomial.

Long Division

In Section 4.3, you learned how to divide one polynomial by another by factoring and canceling common factors. For instance, you can divide $(x^2 - 2x - 3)$ by $(x - 3)$ as follows.

$$(x^2 - 2x - 3) \div (x - 3) = \frac{x^2 - 2x - 3}{x - 3} \qquad \text{Write as fraction.}$$

$$= \frac{(x + 1)(x - 3)}{x - 3} \qquad \text{Factor numerator.}$$

$$= \frac{(x + 1)(x - 3)}{x - 3} \qquad \text{Cancel common factor.}$$

$$= x + 1, \quad x \neq 3 \qquad \text{Simplified form}$$

This procedure works well for polynomials that factor easily. For those that do not, you can use a more general procedure that follows a "long division algorithm" similar to the algorithm used for dividing positive integers. We review that procedure in Example 2.

| **Example 2** | Long Division Algorithm for Positive Integers |

Use the long division algorithm to divide 6584 by 28.

Solution

$$
\begin{array}{r}
\text{Think } \frac{65}{28} \approx 2. \\
\text{Think } \frac{98}{28} \approx 3. \\
\text{Think } \frac{144}{28} \approx 5.
\end{array}
$$

$$
\begin{array}{r}
235 \\
28\overline{)6584} \\
\underline{56} \qquad \text{Multiply } 2 \cdot 28. \\
98 \qquad \text{Subtract and bring down 8.} \\
\underline{84} \qquad \text{Multiply } 3 \cdot 28. \\
144 \qquad \text{Subtract and bring down 4.} \\
\underline{140} \qquad \text{Multiply } 5 \cdot 28. \\
4 \qquad \text{Remainder}
\end{array}
$$

So, you have

$$6584 \div 28 = 235 + \frac{4}{28}$$

$$= 235 + \frac{1}{7}.$$

In Example 2, the numerator 6584 is the **dividend,** 28 is the **divisor,** 235 is the **quotient,** and 4 is the **remainder.**

To divide polynomials, begin by dividing the first term of the divisor into the first term of the dividend. Repeat this process for each subsequent division.

Example 3   Long Division Algorithm for Polynomials

Use the long division algorithm to divide $x^2 + 2x + 4$ by $x - 1$.

Solution

Think $x^2/x = x.$

Think $3x/x = 3.$

$$
\begin{array}{r}
x + 3 \\
x - 1 \overline{)\, x^2 + 2x + 4} \\
\underline{x^2 -\ x} \qquad\qquad \\
3x + 4 \\
\underline{3x - 3} \\
7
\end{array}
$$

Multiply $x(x - 1)$.

Subtract and bring down 4.

Multiply $3(x - 1)$.

Subtract.

Study Tip

When using the long division algorithm for polynomials, be sure that both the divisor and the dividend are written in standard form before beginning the division process.

The remainder is a fractional part of the divisor, so you can write

$$
\underbrace{\frac{x^2 + 2x + 4}{x - 1}}_{\text{Divisor}} = \overbrace{x + 3}^{\text{Quotient}} + \underbrace{\frac{7}{x - 1}}_{\text{Divisor}}.
$$

Dividend   Quotient   Remainder

Technology: Tip

You can check the result of a division problem *algebraically* by multiplying the result by the divisor. The product will be the dividend. You can check the result *graphically* with a graphing utility by comparing the graphs of the original quotient and the simplified form. The graphical check for Example 4 is shown below. Because the graphs coincide, it follows that the expressions are equivalent.

Example 4   Writing in Standard Form Before Dividing

Divide $-13x^3 + 10x^4 + 8x - 7x^2 + 4$ by $3 - 2x$.

Solution

First write the divisor and dividend in standard polynomial form.

$$
\begin{array}{r}
-\ 5x^3 -\ x^2 + 2x - 1 \\
-2x + 3 \overline{)\, 10x^4 - 13x^3 - 7x^2 + 8x + 4} \\
\underline{10x^4 - 15x^3} \qquad\qquad\qquad\qquad \\
2x^3 - 7x^2 \\
\underline{2x^3 - 3x^2} \\
-\ 4x^2 + 8x \\
\underline{-\ 4x^2 + 6x} \\
2x + 4 \\
\underline{2x - 3} \\
7
\end{array}
$$

Multiply $-5x^3(-2x + 3)$.

Subtract and bring down $-7x^2$.

Multiply $-x^2(-2x + 3)$.

Subtract and bring down $8x$.

Multiply $2x(-2x + 3)$.

Subtract and bring down 4.

Multiply $(-1)(-2x + 3)$.

This shows that

$$
\underbrace{\frac{10x^4 - 13x^3 - 7x^2 + 8x + 4}{-2x + 3}}_{\text{Divisor}} = \overbrace{-5x^3 - x^2 + 2x - 1}^{\text{Quotient}} + \underbrace{\frac{7}{-2x + 3}}_{\text{Divisor}}.
$$

Dividend   Quotient   Remainder

When the dividend is missing some powers of x, the long divison algorithm requires that you account for the missing powers by inserting zeros, as shown in Example 5.

Example 5   Accounting for Missing Powers of x

Divide $x^3 - 2$ by $x - 1$.

Solution

To account for the missing x^2- and x-terms, insert $0x^2$ and $0x$.

$$
\begin{array}{r}
x^2 + x + 1 \\
x - 1 \overline{)\, x^3 + 0x^2 + 0x - 2} \\
\underline{x^3 - x^2} \\
x^2 + 0x \\
\underline{x^2 - x} \\
x - 2 \\
\underline{x - 1} \\
- 1
\end{array}
$$

Insert $0x^2$ and $0x$.
Multiply $x^2(x - 1)$.
Subtract and bring down $0x$.
Multiply $x(x - 1)$.
Subtract and bring down -2.
Multiply $(1)(x - 1)$.
Subtract.

So, you have

$$\frac{x^3 - 2}{x - 1} = x^2 + x + 1 - \frac{1}{x - 1}.$$

In each of the long division examples so far, the divisor has been a first-degree polynomial. The long division algorithm works just as well with polynomial divisors of degree 2 or more, as shown in Example 6.

Example 6   A Second-Degree Divisor

Divide $x^4 + 6x^3 + 6x^2 - 10x - 3$ by $x^2 + 2x - 3$.

Solution

$$
\begin{array}{r}
x^2 + 4x + 1 \\
x^2 + 2x - 3 \overline{)\, x^4 + 6x^3 + 6x^2 - 10x - 3} \\
\underline{x^4 + 2x^3 - 3x^2} \\
4x^3 + 9x^2 - 10x \\
\underline{4x^3 + 8x^2 - 12x} \\
x^2 + 2x - 3 \\
\underline{x^2 + 2x - 3} \\
0
\end{array}
$$

Multiply $x^2(x^2 + 2x - 3)$.
Subtract and bring down $-10x$.
Multiply $4x(x^2 + 2x - 3)$.
Subtract and bring down -3.
Multiply $(1)(x^2 + 2x - 3)$.
Subtract.

Because the remainder is 0, $x^2 + 2x - 3$ **divides evenly** into $x^4 + 6x^3 + 6x^2 - 10x - 3$. That is,

$$\frac{x^4 + 6x^3 + 6x^2 - 10x - 3}{x^2 + 2x - 3} = x^2 + 4x + 1.$$

3 Use synthetic division to divide a polynomial by a polynomial of the form $x - k$.

Synthetic Division

There is a nice shortcut for division by polynomials of the form $x - k$. It is called **synthetic division** and is outlined for a third-degree polynomial as follows.

> ▶ **Synthetic Division of a Third-Degree Polynomial**
>
> Use synthetic division to divide $ax^3 + bx^2 + cx + d$ by $x - k$, as follows.
>
>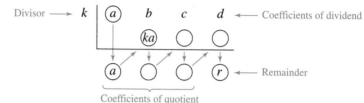
>
> *Vertical Pattern:* Add terms.
> *Diagonal Pattern:* Multiply by k.

Example 7 Using Synthetic Division

Use synthetic division to divide $x^3 + 3x^2 - 4x - 10$ by $x - 2$.

Solution

The coefficients of the dividend form the top row of the synthetic division tableau. Because you are dividing by $(x - 2)$, write 2 at the top left of the tableau. To begin the algorithm, bring down the first coefficient. Then multiply this coefficient by 2, write the result in the second row of the tableau, and add the two numbers in the second column. By continuing this pattern, you obtain the following tableau.

Divisor ⟶ 2 | 1   3   -4   -10 ⟵ Coefficients of dividend
            | 2   10   12
            ─────────────────────
              1   5   6   ② ⟵ Remainder
            Coefficients of quotient

The bottom row of the tableau shows the coefficients of the quotient. So, the quotient is

$$(1)x^2 + (5)x + (6)$$

and the remainder is 2. So, the result of the division problem is

$$\frac{x^3 + 3x^2 - 4x - 10}{x - 2} = x^2 + 5x + 6 + \frac{2}{x - 2}.$$

Notice that in synthetic division the quotient polynomial is always one degree *less* than the dividend polynomial.

4 Use synthetic division to factor a polynomial.

Factoring and Division

If the remainder in a synthetic division problem turns out to be zero, you can conclude that the divisor divides *evenly* into the dividend. When this happens, you know that the original polynomial can be factored as two polynomials of lesser degrees.

Example 8   Factoring a Polynomial

Completely factor $x^3 - 7x + 6$. Use the fact that $x - 1$ is one of the factors.

Solution

Because you are given one of the factors, divide this factor into the given polynomial using synthetic division.

$$
\begin{array}{r|rrrr}
1 & 1 & 0 & -7 & 6 \\
  &   & 1 & 1 & -6 \\
\hline
  & 1 & 1 & -6 & 0
\end{array}
$$

Because the remainder is zero, the divisor divides evenly into the dividend and you obtain

$$\frac{x^3 - 7x + 6}{x - 1} = x^2 + x - 6.$$

From this result, you can factor the original polynomial as follows.

$$
\begin{aligned}
x^3 - 7x + 6 &= (x - 1)(x^2 + x - 6) \\
&= (x - 1)(x + 3)(x - 2)
\end{aligned}
$$

Discussing the Concept   **Investigating Polynomials and Their Factors**

Use a graphing utility to graph the following polynomials in the same viewing rectangle using the standard setting. Use the zero or root feature to find the *x*-intercepts. What can you conclude about the polynomials? Verify your conclusion algebraically.

a. $y = (x - 4)(x - 2)(x + 1)$
b. $y = (x^2 - 6x + 8)(x + 1)$
c. $y = x^3 - 5x^2 + 2x + 8$

Now use your graphing utility to graph the function

$$f(x) = \frac{x^3 - 5x^2 + 2x + 8}{x - 2}.$$

Use the zero or root feature to find the *x*-intercepts. Why does this function have only two *x*-intercepts? To what other function does the graph of $f(x)$ appear to be equivalent? What is the difference between the two graphs?

4.5   Exercises

Integrated Review   *Concepts, Skills, and Problem Solving*

Keep mathematically in shape by doing these exercises *before* the problems of this section.

Properties and Definitions

1. Explain how to write the fraction $120y/90$ in simplified form.

2. Write an algebraic expression that represents the product of two consecutive odd integers, the first of which is $2n + 1$.

3. Write an algebraic expression that represents the sum of two consecutive odd integers, the first of which is $2n + 1$.

4. Write an algebraic expression that represents the product of two consecutive even integers, the first of which is $2n$.

Solving Equations

In Exercises 5–10, solve the equation.

5. $3(2 - x) = 5x$

6. $125 - 50x = 0$

7. $8y^2 - 50 = 0$

8. $t^2 - 8t = 0$

9. $x^2 + x - 42 = 0$

10. $x(10 - x) = 25$

Models and Graphs

11. You receive a monthly salary of $1500 plus a commission of 12% of sales. Find a model for the monthly wages y as a function of sales x. Graph the model.

12. In the year 2000, a college had an enrollment of 3500 students. If enrollment is projected to increase by 60 students per year, find a model for the enrollment N as a function of time t in years. (Let $t = 0$ represent the year 2000.) Graph the function for the years 2000 through 2010.

Developing Skills

In Exercises 1–14, perform the division. See Example 1.

1. $\dfrac{6z + 10}{2}$

2. $\dfrac{9x + 12}{3}$

3. $\dfrac{10z^2 + 4z - 12}{4}$

4. $\dfrac{4u^2 + 8u - 24}{16}$

5. $(7x^3 - 2x^2) \div x$

6. $(6a^2 + 7a) \div a$

7. $\dfrac{m^4 + 2m^2 - 7}{m}$

8. $\dfrac{l^2 - 8l + 4}{-l}$

9. $\dfrac{50z^3 + 30z}{-5z}$

10. $\dfrac{18c^4 - 24c^2}{-6c}$

11. $\dfrac{8z^3 + 3z^2 - 2z}{2z}$

12. $\dfrac{6x^4 + 8x^3 - 18x^2}{3x^2}$

13. $(5x^2y - 8xy + 7xy^2) \div 2xy$

14. $(-14s^4t^2 + 7s^2t^2 - 18t) \div 2s^2t$

In Exercises 15–52, perform the division. See Examples 2–6.

15. $\dfrac{x^2 - 8x + 15}{x - 3}$

16. $\dfrac{t^2 - 18t + 72}{t - 6}$

17. $(x^2 + 15x + 50) \div (x + 5)$

18. $(y^2 - 6y - 16) \div (y + 2)$

19. Divide $x^2 - 5x + 8$ by $x - 2$.

20. Divide $x^2 + 10x - 9$ by $x - 3$.

21. Divide $21 - 4x - x^2$ by $3 - x$.

22. Divide $5 + 4x - x^2$ by $1 + x$.

23. $\dfrac{5x^2 + 2x + 3}{x + 2}$

24. $\dfrac{2x^2 + 5x + 2}{x + 4}$

25. $\dfrac{12x^2 + 17x - 5}{3x + 2}$

26. $\dfrac{8x^2 + 2x + 3}{4x - 1}$

27. $(12t^2 - 40t + 25) \div (2t - 5)$

28. $(15 - 14u - 8u^2) \div (5 + 2u)$

29. Divide $2y^2 + 7y + 3$ by $2y + 1$.

30. Divide $10t^2 - 7t - 12$ by $2t - 3$.

31. $\dfrac{x^3 - 2x^2 + 4x - 8}{x - 2}$ **32.** $\dfrac{x^3 + 4x^2 + 7x + 28}{x + 4}$

33. $\dfrac{2x^3 - 5x^2 + x - 6}{x - 3}$ **34.** $\dfrac{5x^3 + 3x^2 + 12x + 20}{x + 1}$

35. $(2x + 9) \div (x + 2)$ **36.** $(12x - 5) \div (2x + 3)$

37. $\dfrac{x^2 + 16}{x + 4}$ **38.** $\dfrac{y^2 + 8}{y + 2}$

39. $\dfrac{6z^2 + 7z}{5z - 1}$ **40.** $\dfrac{8y^2 - 2y}{3y + 5}$

41. $\dfrac{16x^2 - 1}{4x + 1}$ **42.** $\dfrac{81y^2 - 25}{9y - 5}$

43. $\dfrac{x^3 + 125}{x + 5}$ **44.** $\dfrac{x^3 - 27}{x - 3}$

45. $(x^3 + 4x^2 + 7x + 6) \div (x^2 + 2x + 3)$

46. $(2x^3 + 2x^2 - 2x - 15) \div (2x^2 + 4x + 5)$

47. $(4x^4 - 3x^2 + x - 5) \div (x^2 - 3x + 2)$

48. $(8x^5 + 6x^4 - x^3 + 1) \div (2x^3 - x^2 - 3)$

49. Divide $x^6 - 1$ by $x - 1$.

50. Divide x^3 by $x - 1$.

51. $x^5 \div (x^2 + 1)$ **52.** $x^4 \div (x - 2)$

In Exercises 53–56, simplify the expression.

53. $\dfrac{4x^4}{x^3} - 2x$ **54.** $\dfrac{15x^3y}{10x^2} + \dfrac{3xy^2}{2y}$

55. $\dfrac{8u^2v}{2u} + \dfrac{3(uv)^2}{uv}$ **56.** $\dfrac{x^2 + 2x - 3}{x - 1} - (3x - 4)$

In Exercises 57–68, use synthetic division to perform the division. See Example 7.

57. $\dfrac{x^3 - 5x^2 + 3x - 4}{x - 2}$ **58.** $\dfrac{x^3 + 6x^2 + 8x - 2}{x + 3}$

59. $\dfrac{x^3 + 3x^2 - 1}{x + 4}$ **60.** $\dfrac{x^4 - 4x^2 + 6}{x - 4}$

61. $\dfrac{x^4 - 4x^3 + x + 10}{x - 2}$ **62.** $\dfrac{2x^5 - 3x^3 + x}{x - 3}$

63. $\dfrac{5x^3 - 6x^2 + 8}{x - 4}$ **64.** $\dfrac{5x^3 + 6x + 8}{x + 2}$

65. $\dfrac{10x^4 - 50x^3 - 800}{x - 6}$ **66.** $\dfrac{x^5 - 13x^4 - 120x + 80}{x + 3}$

67. $\dfrac{0.1x^2 + 0.8x + 1}{x - 0.2}$ **68.** $\dfrac{x^3 - 0.8x + 2.4}{x + 0.1}$

In Exercises 69–78, completely factor the polynomial given one of its factors. See Example 8.

	Polynomial	*Factor*
69.	$x^3 - 13x + 12$	$x - 3$
70.	$x^3 + x^2 - 32x - 60$	$x + 5$
71.	$6x^3 - 13x^2 + 9x - 2$	$x - 1$
72.	$9x^3 - 3x^2 - 56x - 48$	$x - 3$
73.	$9x^3 + 45x^2 - 4x - 20$	$x + 5$
74.	$4x^3 + 8x^2 - 25x - 50$	$x + 2$
75.	$x^4 + 7x^3 + 3x^2 - 63x - 108$	$x + 3$
76.	$x^4 - 6x^3 - 8x^2 + 96x - 128$	$x - 4$
77.	$15x^2 - 2x - 8$	$x - \frac{4}{5}$
78.	$18x^2 - 9x - 20$	$x + \frac{5}{6}$

In Exercises 79 and 80, find the constant c such that the denominator will divide evenly into the numerator.

79. $\dfrac{x^3 + 2x^2 - 4x + c}{x - 2}$ **80.** $\dfrac{x^4 - 3x^2 + c}{x + 6}$

In Exercises 81–84, use a graphing utility to graph the two equations on the same screen. Use the graphs to verify that the expressions are equivalent. Verify the results algebraically.

81. $y_1 = \dfrac{x + 4}{2x}$

$y_2 = \dfrac{1}{2} + \dfrac{2}{x}$

82. $y_1 = \dfrac{x^2 + 2}{x + 1}$

$y_2 = x - 1 + \dfrac{3}{x + 1}$

83. $y_1 = \dfrac{x^3 + 1}{x + 1}$

$y_2 = x^2 - x + 1, \quad x \neq -1$

84. $y_1 = \dfrac{x^3}{x^2 + 1}$

$y_2 = x - \dfrac{x}{x^2 + 1}$

Think About It   In Exercises 85 and 86, perform the division assuming that n is a positive integer.

85. $\dfrac{x^{3n} + 3x^{2n} + 6x^n + 8}{x^n + 2}$

86. $\dfrac{x^{3n} - x^{2n} + 5x^n - 5}{x^n - 1}$

Think About It In Exercises 87 and 88, the divisor, quotient, and remainder are given. Find the dividend.

	Divisor	Quotient	Remainder
87.	$x - 6$	$x^2 + x + 1$	-4
88.	$x + 3$	$x^3 + x^2 - 4$	8

Finding a Pattern In Exercises 89 and 90, complete the table for the given function. The first row is completed for Exercise 89. What conclusion can you draw as you compare the values of $f(k)$ with the remainders? (Use synthetic division to find the remainders.)

89. $f(x) = x^3 - x^2 - 2x$

90. $f(x) = 2x^3 - x^2 - 2x + 1$

k	$f(k)$	Divisor $(x - k)$	Remainder
-2	-8	$x + 2$	-8
-1			
0			
$\frac{1}{2}$			
1			
2			

Table for 89 and 90

Solving Problems

91. *Geometry* The area of a rectangle is $2x^3 + 3x^2 - 6x - 9$. Find its width if its length is $2x + 3$.

92. *Geometry* A rectangular house has a volume of

$$x^3 + 55x^2 + 650x + 2000$$

cubic feet (the space in the attic is not included). The height of the house is $x + 5$ feet (see figure). Find the number of square feet of floor space *on the first floor* of the house.

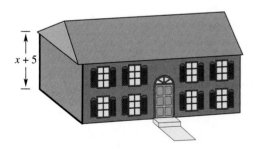

Geometry In Exercises 93 and 94, you are given the expression for the volume of the solid shown. Find the expression for the missing dimension.

93. $V = x^3 + 18x^2 + 80x + 96$

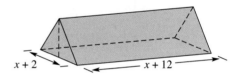

94. $V = h^4 + 3h^3 + 2h^2$

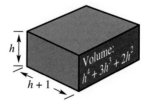

Explaining Concepts

95. *Error Analysis* Describe the error.

$$\frac{6x + 5y}{x} = \frac{6x + 5y}{x} = 6 + 5y$$

96. Create a polynomial division problem and identify the dividend, divisor, quotient, and remainder.

97. Explain what is meant for a divisor to *divide evenly* into a dividend.

98. Explain how you can check polynomial division.

99. *True or False?* If the divisor divides evenly into the dividend, the divisor and quotient are factors of the dividend. Explain.

100. For synthetic division, what form must the divisor have?

4.6 Solving Rational Equations

Objectives

1 Solve a rational equation containing constant denominators.

2 Solve a rational equation containing variable denominators.

3 Solve an application problem involving a rational equation.

---

1 Solve a rational equation containing constant denominators.

Equations Containing Constant Denominators

In Section 1.1, you studied a strategy for solving equations that contain fractions with *constant* denominators. We review that procedure here because it is the basis for solving more general equations involving fractions. Recall from Section 1.1 that you can "clear an equation of fractions" by multiplying both sides of the equation by the least common denominator (LCD) of the fractions in the equation. Note how this is done in the next three examples.

> **Example 1** An Equation Containing Constant Denominators

Solve $\dfrac{3}{5} = \dfrac{x}{2} + 1$.

Solution

The least common denominator of the two fractions is 10, so begin by multiplying both sides of the equation by 10.

$$\frac{3}{5} = \frac{x}{2} + 1 \qquad \text{Original equation}$$

$$10\left(\frac{3}{5}\right) = 10\left(\frac{x}{2} + 1\right) \qquad \text{Multiply both sides by LCD of 10.}$$

$$6 = 5x + 10 \qquad \text{Simplify.}$$

$$-4 = 5x \qquad \text{Subtract 10 from both sides.}$$

$$-\frac{4}{5} = x \qquad \text{Divide both sides by 5.}$$

The solution is $-\frac{4}{5}$. You can check this as follows.

Check

$$\frac{3}{5} \overset{?}{=} \frac{-4/5}{2} + 1 \qquad \text{Substitute } -\tfrac{4}{5} \text{ for } x \text{ in the original equation.}$$

$$\frac{3}{5} \overset{?}{=} -\frac{4}{5} \cdot \frac{1}{2} + 1 \qquad \text{Invert and multiply.}$$

$$\frac{3}{5} = -\frac{2}{5} + 1 \qquad \text{Solution checks.} \checkmark$$

Example 2 An Equation Containing Constant Denominators

Solve $\dfrac{x-3}{6} = 7 - \dfrac{x}{12}$.

Solution

$$\dfrac{x-3}{6} = 7 - \dfrac{x}{12}$$ Original equation

$$12\left(\dfrac{x-3}{6}\right) = 12\left(7 - \dfrac{x}{12}\right)$$ Multiply both sides by LCD of 12.

$$2x - 6 = 84 - x$$ Distribute and simplify.

$$3x - 6 = 84$$ Add x to both sides.

$$3x = 90$$ Add 6 to both sides.

$$x = 30$$ Divide both sides by 3.

Check

$$\dfrac{30-3}{6} \overset{?}{=} 7 - \dfrac{30}{12}$$ Substitute 30 for x in the original equation.

$$\dfrac{27}{6} = \dfrac{42}{6} - \dfrac{15}{6}$$ Solution checks. ✓

Example 3 An Equation Containing Constant Denominators

Solve $\dfrac{x+2}{6} - \dfrac{x-4}{8} = \dfrac{2}{3}$.

Solution

$$\dfrac{x+2}{6} - \dfrac{x-4}{8} = \dfrac{2}{3}$$ Original equation

$$24\left(\dfrac{x+2}{6} - \dfrac{x-4}{8}\right) = 24\left(\dfrac{2}{3}\right)$$ Multiply both sides by LCD of 24.

$$4(x+2) - 3(x-4) = 8(2)$$ Distribute and simplify.

$$4x + 8 - 3x + 12 = 16$$ Distributive Property

$$x + 20 = 16$$ Combine like terms.

$$x = -4$$ Subtract 20 from both sides.

Check

$$\dfrac{-4+2}{6} - \dfrac{-4-4}{8} \overset{?}{=} \dfrac{2}{3}$$ Substitute -4 for x in the original equation.

$$-\dfrac{1}{3} + 1 = \dfrac{2}{3}$$ Solution checks. ✓

2 Solve a rational equation containing variable denominators.

Equations Containing Variable Denominators

Remember that you always *exclude* those values of a variable that make the denominator of a rational expression zero. This is especially critical for solving equations that contain variable denominators. You will see why in the examples that follow.

Technology: Tip

You can use a graphing utility to estimate the solution of the equation in Example 4. To do this, graph the left side of the equation and the right side of the equation on the same screen.

$$y_1 = \frac{7}{x} - \frac{1}{3x} \text{ and } y_2 = \frac{8}{3}$$

The solution of the equation is the x-coordinate of the point at which the two graphs intersect, as shown below.

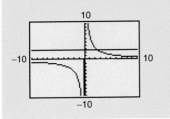

Example 4   An Equation Containing Variable Denominators

Solve the equation.

$$\frac{7}{x} - \frac{1}{3x} = \frac{8}{3}$$

Solution

For this equation, the least common denominator is $3x$. So, begin by multiplying both sides of the equation by $3x$.

$$\frac{7}{x} - \frac{1}{3x} = \frac{8}{3} \qquad \text{Original equation}$$

$$3x\left(\frac{7}{x} - \frac{1}{3x}\right) = 3x\left(\frac{8}{3}\right) \qquad \text{Multiply both sides by LCD of } 3x.$$

$$\frac{21x}{x} - \frac{3x}{3x} = \frac{24x}{3} \qquad \text{Distributive Property}$$

$$21 - 1 = 8x \qquad \text{Simplify.}$$

$$\frac{20}{8} = x \qquad \text{Combine like terms and divide both sides by 8.}$$

$$x = \frac{5}{2} \qquad \text{Simplify.}$$

The solution is $\frac{5}{2}$. You can check this as follows.

Check

$$\frac{7}{x} - \frac{1}{3x} = \frac{8}{3} \qquad \text{Original equation}$$

$$\frac{7}{5/2} - \frac{1}{3(5/2)} \overset{?}{=} \frac{8}{3} \qquad \text{Substitute } \tfrac{5}{2} \text{ for } x.$$

$$7\left(\frac{2}{5}\right) - \frac{2}{15} \overset{?}{=} \frac{8}{3} \qquad \text{Invert and multiply.}$$

$$\frac{14}{5} - \frac{2}{15} \overset{?}{=} \frac{8}{3} \qquad \text{Simplify.}$$

$$\frac{40}{15} \overset{?}{=} \frac{8}{3} \qquad \text{Combine like terms.}$$

$$\frac{8}{3} = \frac{8}{3} \qquad \text{Solution checks.} \checkmark$$

Throughout the text, we have emphasized the importance of checking solutions. Up to this point, the main reason for checking has been to make sure that you did not make errors in the solution process. In the next example you will see that there is another reason for checking solutions in the *original* equation. That is, even with no mistakes in the solution process, it can happen that a "trial solution" does not satisfy the original equation. This type of "solution" is called **extraneous.** An extraneous solution of an equation does not, by definition, satisfy the original equation, and therefore *must not* be listed as an actual solution.

Example 5   An Equation with No Solution

Solve the equation.

$$\frac{5x}{x-2} = 7 + \frac{10}{x-2}$$

Solution

The least common denominator for this equation is $x - 2$. So, begin by multiplying both sides of the equation by $x - 2$.

$$\frac{5x}{x-2} = 7 + \frac{10}{x-2}$$

   Original equation

$$(x-2)\left(\frac{5x}{x-2}\right) = (x-2)\left(7 + \frac{10}{x-2}\right)$$

   Multiply both sides by $x - 2$.

$$5x = 7(x-2) + 10, \quad x \neq 2$$

   Distribute and simplify.

$$5x = 7x - 14 + 10$$

   Distributive Property

$$5x = 7x - 4$$

   Combine like terms.

$$-2x = -4$$

   Subtract $7x$ from both sides.

$$x = 2$$

   Divide both sides by -2.

At this point, the solution appears to be 2. However, by performing the following check, you will see that this "trial solution" is extraneous.

Check

$$\frac{5x}{x-2} = 7 + \frac{10}{x-2}$$

   Original equation

$$\frac{5(2)}{2-2} \overset{?}{=} 7 + \frac{10}{2-2}$$

   Substitute 2 for x.

$$\frac{10}{0} \overset{?}{=} 7 + \frac{10}{0}$$

   Solution does not check. ✗

Because the check results in *division by zero*, 2 is extraneous. Therefore, the original equation has no solution.

Notice that 2 is excluded from the domains of the two fractions in the original equation in Example 5. You may find it helpful when solving these types of equations to list the domain restrictions before beginning the solution process.

| Example 6 | An Equation Containing Variable Denominators |

Solve $\dfrac{4}{x-2} + \dfrac{3x}{x+1} = 3$.

Solution

The domain is all real values of x such that $x \neq 2$ and $x \neq -1$. The least common denominator is $(x-2)(x+1)$.

$$\frac{4}{x-2} + \frac{3x}{x+1} = 3$$

$$(x-2)(x+1)\left(\frac{4}{x-2} + \frac{3x}{x+1}\right) = 3(x-2)(x+1)$$

$$4(x+1) + 3x(x-2) = 3(x^2 - x - 2), \quad x \neq 2, x \neq -1$$

$$4x + 4 + 3x^2 - 6x = 3x^2 - 3x - 6$$

$$3x^2 - 2x + 4 = 3x^2 - 3x - 6$$

$$x = -10$$

The solution is -10. Check this in the original equation.

_____

So far in this section, each of the equations has had one solution or no solution. The equation in the next example has two solutions.

| Example 7 | An Equation That Has Two Solutions |

Solve $\dfrac{3x}{x+1} = \dfrac{12}{x^2-1} + 2$.

Solution

The domain is all real values of x such that $x \neq 1$ and $x \neq -1$. The least common denominator is $(x+1)(x-1) = x^2 - 1$.

$$\frac{3x}{x+1} = \frac{12}{x^2-1} + 2 \qquad \text{Original equation}$$

$$(x^2-1)\left(\frac{3x}{x+1}\right) = (x^2-1)\left(\frac{12}{x^2-1} + 2\right) \qquad \begin{array}{l}\text{Multiply both sides by} \\ \text{LCD of } x^2 - 1.\end{array}$$

$$(x-1)(3x) = 12 + 2(x^2-1), \quad x \neq \pm 1 \qquad \text{Simplify.}$$

$$3x^2 - 3x = 12 + 2x^2 - 2 \qquad \text{Distributive Property}$$

$$x^2 - 3x - 10 = 0 \qquad \begin{array}{l}\text{Subtract } 2x^2 \text{ and } 10 \\ \text{from both sides.}\end{array}$$

$$(x+2)(x-5) = 0 \qquad \text{Factor.}$$

$$x + 2 = 0 \quad \Longrightarrow \quad x = -2 \qquad \text{Set } x + 2 \text{ equal to } 0.$$

$$x - 5 = 0 \quad \Longrightarrow \quad x = 5 \qquad \text{Set } x - 5 \text{ equal to } 0.$$

The solutions are -2 and 5. Check these in the original equation.

_____

Technology: Discovery

Use a graphing utility to graph the equation

$$y = \frac{3x}{x+1} - \frac{12}{x^2-1} - 2.$$

Then use the zoom and trace features of the utility to determine the x-intercepts. How do the x-intercepts compare with the solutions to Example 7? What can you conclude?

3 Solve an application problem involving a rational equation.

Applications

Example 8   Average Cost

A manufacturing plant can produce x units of a certain item for $26 per unit *plus* an initial investment of $80,000. How many units must be produced to have an average cost of $30 per unit?

Solution

Verbal Model:	Average cost per unit	=	Total cost	÷	Number of units

Labels: Number of units = x (units)
            Average cost per unit = 30   (dollars per unit)
            Total cost = $26x + 80{,}000$ (dollars)

Equation: $30 = \dfrac{26x + 80{,}000}{x}$

$$30x = 26x + 80{,}000, \quad x \neq 0$$

$$4x = 80{,}000$$

$$x = 20{,}000$$

The plant should produce 20,000 units.

Example 9   A Work-Rate Problem

With only the cold water valve open, it takes 8 minutes to fill the tub of a washer. With both the hot and cold water valves open, it takes only 5 minutes. How long will it take the tub to fill with only the hot water valve open?

Solution

Verbal Model:	Rate for cold water	+	Rate for hot water	=	Rate for warm water

Labels: Rate for cold water = $\dfrac{1}{8}$ (tub per minute)

            Rate for hot water = $\dfrac{1}{t}$ (tub per minute)

            Rate for warm water = $\dfrac{1}{5}$ (tub per minute)

Equation: $\dfrac{1}{8} + \dfrac{1}{t} = \dfrac{1}{5}$

$$5t + 40 = 8t$$

$$40 = 3t$$

$$\dfrac{40}{3} = t$$

So, it takes $13\frac{1}{3}$ minutes to fill the tub with hot water.

Example 10   Batting Average

In this year's playing season, a baseball player has been at bat 140 times and has hit the ball safely 35 times. So, the "batting average" for the player is $35/140 = .250$. How many consecutive times must the player hit safely to obtain a batting average of .300?

Solution

| *Verbal Model:* | $\dfrac{\text{Batting}}{\text{average}} = \dfrac{\text{Total}}{\text{hits}} \div \dfrac{\text{Total times}}{\text{at bat}}$ |

Labels: Current times at bat $= 140$
Current hits $= 35$
Additional consecutive hits $= x$

Equation:
$$.300 = \frac{x + 35}{x + 140}$$

$$.300(x + 140) = x + 35$$

$$.3x + 42 = x + 35$$

$$7 = 0.7x$$

$$10 = x$$

The player must hit safely the next 10 times at bat. After that, the batting average will be $45/150 = .300$.

Examples 8, 9, and 10 are types of application problems that you have seen earlier in the text. The difference now is that the variable appears in the denominator of a rational expression. When determining the domain of a real-life problem you must also consider the context of the problem. For instance, in Example 10 the additional times at bat could not be a negative number. The problem implies that the domain be all real numbers greater than or equal to 0.

Discussing the Concept   Interpreting Average Cost

You buy sand in bulk for a construction project. You find that the total cost of your order depends on the weight of the order. The total cost C in dollars is given by $C = 100 + 50x - 0.2x^2$, $1 \le x \le 50$, where x is the weight in thousands of pounds. Construct a rational function representing the average cost per thousand pounds. Because of cost constraints, you can proceed with the project only if the average cost of the sand is less than $50 per thousand pounds. What is the smallest order you can place and still proceed with the project?

4.6 Exercises

Integrated Review — *Concepts, Skills, and Problem Solving*

Keep mathematically in shape by doing these exercises *before* the problems of this section.

Properties and Definitions

In Exercises 1 and 2, determine the quadrants in which the point must be located.

1. $(-2, y)$, y is a real number.

2. $(x, 3)$, x is a real number.

3. Give the positions of points whose y-coordinates are 0.

4. Find the coordinates of the point 9 units to the right of the y-axis and 6 units below the x-axis.

Solving Inequalities

In Exercises 5–10, solve the inequality.

5. $7 - 3x > 4 - x$

6. $2(x + 6) - 20 < 2$

7. $|x - 3| < 2$

8. $|x - 5| > 3$

9. $\left|\frac{1}{4}x - 1\right| \geq 3$

10. $\left|2 - \frac{1}{3}x\right| \leq 10$

Problem Solving

11. A jogger leaves a given point on a fitness trail running at a rate of 6 miles per hour. Five minutes later a second jogger leaves from the same location running at 8 miles per hour. How long will it take the second runner to overtake the first, and how far will each have run at that point?

12. An inheritance of $24,000 is invested in two bonds that pay 7.5% and 9% simple interest. The annual interest is $1935. How much is invested in each bond?

Developing Skills

In Exercises 1–4, determine whether the values of x are solutions to the equation.

Equation	Values

1. $\frac{x}{3} - \frac{x}{5} = \frac{4}{3}$ (a) $x = 0$ (b) $x = -1$
                    (c) $x = \frac{1}{8}$ (d) $x = 10$

2. $x = 4 + \frac{21}{x}$ (a) $x = 0$ (b) $x = -3$
                    (c) $x = 7$ (d) $x = -1$

3. $\frac{x}{4} + \frac{3}{4x} = 1$ (a) $x = -1$ (b) $x = 1$
                    (c) $x = 3$ (d) $x = -3$

4. $5 - \frac{1}{x - 3} = 2$ (a) $x = \frac{10}{3}$ (b) $x = -\frac{1}{3}$
                    (c) $x = 0$ (d) $x = 1$

In Exercises 5–18, solve the equation. See Examples 1–3.

5. $\frac{x}{6} - 1 = \frac{2}{3}$

6. $\frac{y}{8} + 7 = -\frac{1}{2}$

7. $\frac{z + 2}{3} = 4 - \frac{z}{12}$

8. $\frac{x - 5}{5} + 3 = -\frac{x}{4}$

9. $\frac{2y - 9}{6} = 3y - \frac{3}{4}$

10. $\frac{4x - 2}{7} - \frac{5}{14} = 2x$

11. $\frac{4t}{3} = 15 - \frac{t}{6}$

12. $\frac{x}{3} + \frac{x}{6} = 10$

13. $\frac{5y - 1}{12} + \frac{y}{3} = -\frac{1}{4}$

14. $\frac{z - 4}{9} - \frac{3z + 1}{18} = \frac{3}{2}$

15. $\frac{h + 2}{5} - \frac{h - 1}{9} = \frac{2}{3}$

16. $\frac{u - 2}{6} + \frac{2u + 5}{15} = 3$

17. $\frac{x + 5}{4} - \frac{3x - 8}{3} = \frac{4 - x}{12}$

18. $\frac{2x - 7}{10} - \frac{3x + 1}{5} = \frac{6 - x}{5}$

In Exercises 19–58, solve the equation. (Check for extraneous solutions.) See Examples 4–7.

19. $\frac{9}{25 - y} = -\frac{1}{4}$

20. $\frac{2}{u + 4} = \frac{5}{8}$

21. $5 - \frac{12}{a} = \frac{5}{3}$

22. $\frac{6}{b} + 22 = 24$

23. $\dfrac{4}{x} - \dfrac{7}{5x} = -\dfrac{1}{2}$

24. $\dfrac{5}{3} = \dfrac{6}{7x} + \dfrac{2}{x}$

25. $\dfrac{12}{y+5} + \dfrac{1}{2} = 2$

26. $\dfrac{7}{8} - \dfrac{16}{t-2} = \dfrac{3}{4}$

27. $\dfrac{5}{x} = \dfrac{25}{3(x+2)}$

28. $\dfrac{10}{x+4} = \dfrac{15}{4(x+1)}$

29. $\dfrac{8}{3x+5} = \dfrac{1}{x+2}$

30. $\dfrac{500}{3x+5} = \dfrac{50}{x-3}$

31. $\dfrac{3}{x+2} - \dfrac{1}{x} = \dfrac{1}{5x}$

32. $\dfrac{12}{x+5} + \dfrac{5}{x} = \dfrac{20}{x}$

33. $\dfrac{1}{2} = \dfrac{18}{x^2}$

34. $\dfrac{1}{4} = \dfrac{16}{z^2}$

35. $\dfrac{32}{t} = 2t$

36. $\dfrac{20}{u} = \dfrac{u}{5}$

37. $x + 1 = \dfrac{72}{x}$

38. $\dfrac{48}{x} = x - 2$

39. $1 = \dfrac{16}{y} - \dfrac{39}{y^2}$

40. $x - \dfrac{24}{x} = 5$

41. $\dfrac{2x}{3x-10} - \dfrac{5}{x} = 0$

42. $\dfrac{x+42}{x} = x$

43. $\dfrac{2x}{5} = \dfrac{x^2-5x}{5x}$

44. $\dfrac{3x}{4} = \dfrac{x^2+3x}{8x}$

45. $\dfrac{2}{6q+5} - \dfrac{3}{4(6q+5)} = \dfrac{1}{28}$

46. $\dfrac{10}{x(x-2)} + \dfrac{4}{x} = \dfrac{5}{x-2}$

47. $\dfrac{4}{2x+3} + \dfrac{17}{5x-3} = 3$ **48.** $\dfrac{1}{x-1} + \dfrac{3}{x+1} = 2$

49. $\dfrac{2}{x-10} - \dfrac{3}{x-2} = \dfrac{6}{x^2-12x+20}$

50. $\dfrac{5}{x+2} + \dfrac{2}{x^2-6x-16} = \dfrac{-4}{x-8}$

51. $\dfrac{x+3}{x^2-9} + \dfrac{4}{3-x} - 2 = 0$

52. $1 - \dfrac{6}{4-x} = \dfrac{x+2}{x^2-16}$

53. $\dfrac{x}{x-2} + \dfrac{3x}{x-4} = \dfrac{-2(x-6)}{x^2-6x+8}$

54. $\dfrac{2(x+1)}{x^2-4x+3} + \dfrac{6x}{x-3} = \dfrac{3x}{x-1}$

55. $\dfrac{2(x+7)}{x+4} - 2 = \dfrac{2x+20}{2x+8}$

56. $\dfrac{2x^2-5}{x^2-4} + \dfrac{6}{x+2} = \dfrac{4x-7}{x-2}$

57. $\dfrac{x}{2} = \dfrac{2 - \dfrac{3}{x}}{1 - \dfrac{1}{x}}$ **58.** $\dfrac{2x}{3} = \dfrac{1 + \dfrac{2}{x}}{1 + \dfrac{1}{x}}$

In Exercises 59–62, (a) use the graph to determine any x-intercepts of the graph, and (b) set $y = 0$ and solve the resulting rational equation to confirm the result of part (a).

59. $y = \dfrac{x+2}{x-2}$ **60.** $y = \dfrac{2x}{x+4}$

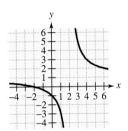 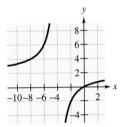

61. $y = x - \dfrac{1}{x}$ **62.** $y = x - \dfrac{2}{x} - 1$

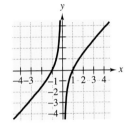 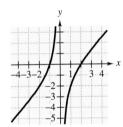

In Exercises 63–70, (a) use a graphing utility to graph the equation and determine any x-intercepts of the graph, and (b) set $y = 0$ and solve the resulting rational equation to confirm the result of part (a).

63. $y = \dfrac{x-4}{x+5}$ **64.** $y = \dfrac{1}{x} - \dfrac{3}{x+4}$

65. $y = \dfrac{1}{x} + \dfrac{4}{x-5}$ **66.** $y = 20\left(\dfrac{2}{x} - \dfrac{3}{x-1}\right)$

67. $y = (x+1) - \dfrac{6}{x}$ **68.** $y = \dfrac{x^2-4}{x}$

69. $y = (x-1) - \dfrac{12}{x}$ **70.** $y = \dfrac{x}{2} - \dfrac{4}{x} - 1$

Solving Problems

71. *Number Problem*   Find a number such that the sum of the number and its reciprocal is $\frac{65}{8}$.

72. *Number Problem*   Find a number such that the sum of two times the number and three times its reciprocal is $\frac{97}{4}$.

73. *Wind Speed*   A plane has a speed of 300 miles per hour in still air. Find the speed of the wind if the plane travels a distance of 680 miles with a tail wind in the same time it takes to travel 520 miles into a head wind.

74. *Average Speed*   During the first part of a 6-hour trip, you travel 240 miles at an average speed of r miles per hour. For the next 72 miles of the trip, you increase your speed by 10 miles per hour. What are your two average speeds?

75. *Speed*   One person runs 2 miles per hour faster than a second person. The first person runs 5 miles in the same time the second person runs 4 miles. Find the speed of each person.

76. *Speed*   The speed of a commuter plane is 150 miles per hour slower than that of a passenger jet. The commuter plane travels 450 miles in the same time the jet travels 1150 miles. Find the speed of each plane.

77. *Speed*   A boat travels at a speed of 20 miles per hour in still water. It travels 48 miles upstream and then returns to the starting point in a total of 5 hours. Find the speed of the current.

78. *Speed*   You traveled 72 miles in a certain time period. If you had traveled 6 miles per hour faster, the trip would have taken 10 minutes less time. What was your speed?

79. *Partnership Costs*   A group plans to start a new business that will require $240,000 for start-up capital. The individuals in the group will share the cost equally. If two additional people joined the group, the cost per person would decrease by $4000. How many people are presently in the group?

80. *Partnership Costs*   A group of people agree to share equally in the cost of a $150,000 endowment to a college. If they could find four more people to join the group, each person's share of the cost would decrease by $6250. How many people are presently in the group?

81. *Partnership Costs*   Some partners buy a piece of property for $78,000 by sharing the cost equally. To ease the financial burden, they look for three additional partners to reduce the cost per person by $1300. How many partners are presently in the group?

82. *Population Growth*   A biologist introduces 100 insects into a culture. The population P of the culture is approximated by the model

$$P = \frac{500(1 + 3t)}{5 + t}$$

where t is the time in hours. Find the time required for the population to increase to 1000 insects.

83. *Pollution Removal*   The cost C in dollars of removing $p\%$ of the air pollutants in the stack emission of a utility company is modeled by

$$C = \frac{120,000p}{100 - p}.$$

(a) Use a graphing utility to graph the model. Use the result to estimate graphically the percent of stack emission that can be removed for $680,000.

(b) Use the model to determine algebraically the percent of stack emission that can be removed for $680,000.

84. *Average Cost*   The average cost for producing x units of a product is given by

$$\text{Average cost} = 1.50 + \frac{4200}{x}.$$

Determine the number of units that must be produced to have an average cost of $2.90.

Work-Rate Problem   In Exercises 85 and 86, complete the table by finding the time required for two individuals to complete a task. The first two columns in the table give the times required for the two individuals to complete the task working alone. (Assume that when they work together their individual rates do not change.)

85.

Person #1	Person #2	Together
6 hours	6 hours	
3 minutes	5 minutes	
5 hours	$2\frac{1}{2}$ hours	

86.

Person #1	Person #2	Together
4 days	4 days	
$5\frac{1}{2}$ hours	3 hours	
a days	b days	

87. *Work-Rate Problem*   One landscaper works $1\frac{1}{2}$ times as fast as another landscaper. Find their individual times if it takes them 9 hours working together to complete a certain job.

88. *Work-Rate Problem*   Assume that the slower of the two landscapers in Exercise 87 is given another job after 4 hours. The faster of the two must work an additional 10 hours to complete the job. Find the individual times.

89. *Swimming Pool*   The flow rate for one pipe is $1\frac{1}{4}$ times that of another pipe. A swimming pool can be filled in 5 hours using both pipes. Find the time required to fill the pool using only the pipe with the slower flow rate.

90. *Swimming Pool*   Assume the pipe with the faster flow rate in Exercise 89 is shut off after 1 hour and it takes an additional 10 hours to fill the pool. Find the filling time for each pipe.

Computer and Data Processing Services   In Exercises 91 and 92, use the following model, which approximates the total revenue y (in billions of dollars) from computer and data services in the United States for the years 1990 through 1995.

$$y = \frac{87,709 - 1236t}{1000 - 93t}, \quad 0 \le t \le 5$$

In this model, $t = 0$ represents 1990.   (Source: Current Business Reports)

91. Evaluate the model for each year and compare the results with the actual data shown in the table.

Year	1990	1991	1992	1993	1994	1995
Revenue	88.3	94.4	104.7	116.8	133.1	152.2

92. (a) Use a graphing utility to graph the model. Use the graph to forecast when revenue will exceed $250 billion.

(b) Explain why the model fails after the year 2000.

93. *Construction*   The unit for determining the size of a nail is the *penny*. For example, 8d represents an 8-penny nail. The number N of finishing nails per pound can be modeled by

$$N = 43.4 + \frac{9353}{x^2}$$

where x is the size of the nail.   (Source: Standard Handbook for Mechanical Engineers)

(a) What is the domain of the function?

(b) Use a graphing utility to graph the function.

(c) Use the graph to determine the size of the finishing nail if there are 135 nails per pound.

94. *Think About It*   It is important to distinguish between equations and expressions. In parts (a) through (d), if the exercise is an equation, solve it; if it is an expression, simplify it.

(a) $\dfrac{16}{x^2 - 16} + \dfrac{x}{2x - 8} = \dfrac{1}{2}$

(b) $\dfrac{16}{x^2 - 16} + \dfrac{x}{2x - 8} + \dfrac{1}{2}$

(c) $\dfrac{5}{x + 3} + \dfrac{5}{3} + 3$

(d) $\dfrac{5}{x + 3} + \dfrac{5}{3} = 3$

Explaining Concepts

95. Answer parts (d) and (e) of Motivating the Chapter on page 253.

96. (a) Explain the difference between an equation and an expression.

(b) Compare the use of a common denominator in solving rational equations with its use in adding or subtracting rational expressions.

97. Describe how to solve a rational equation.

98. Define the term *extraneous solution*. How do you identify an extraneous solution?

99. Describe the steps that can be used to transform an equation into an equivalent equation.

100. Explain how you can use a graphing utility to estimate the solution of a rational equation.

101. When can you use cross-multiplication to solve a rational equation? Explain.

Key Terms

zero exponents, *p. 254*
negative exponents, *p. 254*
scientific notation, *p. 257*
rational function, *p. 262*
domain of a rational
   function, *p. 262*

simplified form, *p. 264*
complex fraction, *p. 276*
least common multiple,
   p. 283
least common
   denominator, *p. 284*

dividend, *p. 293*
divisor, *p. 293*
quotient, *p. 293*
remainder, *p. 293*
synthetic division, *p. 296*
extraneous solution, *p. 304*

Key Concepts

4.1 Summary of rules of exponents

Let m and n be integers, and let a and b represent real numbers, variables, or algebraic expressions.
Product and Quotient Rules

1. $a^m \cdot a^n = a^{m+n}$ 2. $\dfrac{a^m}{a^n} = a^{m-n}, \quad a \neq 0$

Power Rules

3. $(ab)^m = a^m \cdot b^m$ 4. $\left(\dfrac{a}{b}\right)^m = \dfrac{a^m}{b^m}, \quad b \neq 0$

5. $(a^m)^n = a^{mn}$

Zero and Negative Exponent Rules

6. $a^{-m} = \dfrac{1}{a^m}, \quad a \neq 0$

7. $a^0 = 1, \quad a \neq 0$

8. $\left(\dfrac{a}{b}\right)^{-m} = \left(\dfrac{b}{a}\right)^m, \quad a \neq 0, \ b \neq 0$

4.2 Cancellation Rule for fractions

Let u, v, and w represent numbers, variables, or algebraic expressions such that $v \neq 0$ and $w \neq 0$. Then the following Cancellation Rule is valid.

$$\frac{u\acute{w}}{v\acute{w}} = \frac{u}{v}$$

4.3 Multiplying rational expressions

Let u, v, w, and z be real numbers, variables, or algebraic expressions such that $v \neq 0$ and $z \neq 0$. Then the product of u/v and w/z is given by

$$\frac{u}{v} \cdot \frac{w}{z} = \frac{uw}{vz}.$$

4.3 Dividing rational expressions

Let u, v, w, and z be real numbers, variables, or algebraic expressions such that $v \neq 0$, $w \neq 0$, and $z \neq 0$. Then the quotient of u/v and w/z is given by

$$\frac{u}{v} \div \frac{w}{z} = \frac{u}{v} \cdot \frac{z}{w} = \frac{uz}{vw}.$$

4.4 Adding or subtracting with like denominators

If u, v, and w are real numbers, variables, or variable expressions, and $w \neq 0$, the following rules are valid.

1. $\dfrac{u}{w} + \dfrac{v}{w} = \dfrac{u+v}{w}$ 2. $\dfrac{u}{w} - \dfrac{v}{w} = \dfrac{u-v}{w}$

4.4 Adding or subtracting with unlike denominators

Rewrite the rational expressions with like denominators by finding the least common denominator. Then add or subtract as with like denominators.

4.5 Dividing a polynomial by a monomial

Let u, v, and w be real numbers, variables, or algebraic expressions such that $w \neq 0$.

1. $\dfrac{u+v}{w} = \dfrac{u}{w} + \dfrac{v}{w}$ 2. $\dfrac{u-v}{w} = \dfrac{u}{w} - \dfrac{v}{w}$

4.6 Solving rational equations

1. To remove a fraction in a rational equation, multiply both sides by the least common denominator of all fractions in the equation.

2. Exclude those values of a variable that make the denominator of a rational expression zero.

3. Check your solutions to determine if any are extraneous solutions.

REVIEW EXERCISES

Reviewing Skills

4.1 In Exercises 1–8, evaluate the expression.

1. $(2^3 \cdot 3^2)^{-1}$

2. $(2^{-2} \cdot 5^2)^{-2}$

3. $\left(\dfrac{2}{5}\right)^{-3}$

4. $\left(\dfrac{1}{3^{-2}}\right)^2$

5. $(6 \times 10^3)^2$

6. $(3 \times 10^{-3})(8 \times 10^7)$

7. $\dfrac{3.5 \times 10^7}{7 \times 10^4}$

8. $\dfrac{1}{(6 \times 10^{-3})^2}$

In Exercises 9 and 10, write the number in scientific notation.

9. 0.0000538

10. 30,296,000,000

In Exercises 11 and 12, write the number in decimal form.

11. 4.833×10^8

12. 2.74×10^{-4}

In Exercises 13–22, simplify the expression.

13. $(6y^4)(2y^{-3})$

14. $4(-3x)^{-3}$

15. $\dfrac{4x^{-2}}{2x}$

16. $\dfrac{15t^5}{24t^{-3}}$

17. $(x^3y^{-4})^2$

18. $5yx^0$

19. $\dfrac{t^{-5}}{t^{-2}}$

20. $\dfrac{a^5 \cdot a^{-3}}{a^{-2}}$

21. $\left(\dfrac{y}{3}\right)^{-3}$

22. $(2x^2y^4)^4(2x^2y^4)^{-4}$

4.2 In Exercises 23-26, determine the domain of the rational function.

23. $f(y) = \dfrac{3y}{y - 8}$

24. $g(t) = \dfrac{t + 4}{t + 12}$

25. $g(u) = \dfrac{u}{u^2 - 7u + 6}$

26. $f(x) = \dfrac{x - 12}{x(x^2 - 16)}$

In Exercises 27–34, simplify the rational expression using the Cancellation Rule.

27. $\dfrac{6x^4y^2}{15xy^2}$

28. $\dfrac{2(y^3z)^2}{28(yz^2)^2}$

29. $\dfrac{5b - 15}{30b - 120}$

30. $\dfrac{4a}{10a^2 + 26a}$

31. $\dfrac{9x - 9y}{y - x}$

32. $\dfrac{x + 3}{x^2 - x - 12}$

33. $\dfrac{x^2 - 5x}{2x^2 - 50}$

34. $\dfrac{x^2 + 3x + 9}{x^3 - 27}$

4.3 In Exercises 35–50, multiply or divide the rational expressions.

35. $3x(x^2y)^2$

36. $2b(-3b)^3$

37. $\dfrac{24x^4}{15x}$

38. $\dfrac{8u^2v}{6v}$

39. $\dfrac{7}{8} \cdot \dfrac{2x}{y} \cdot \dfrac{y^2}{14x^2}$

40. $\dfrac{15(x^2y)^3}{3y^3} \cdot \dfrac{12y}{x}$

41. $\dfrac{60z}{z + 6} \cdot \dfrac{z^2 - 36}{5}$

42. $\dfrac{1}{6}(x^2 - 16) \cdot \dfrac{3}{x^2 - 8x + 16}$

43. $\dfrac{u}{u - 3} \cdot \dfrac{3u - u^2}{4u^2}$

44. $x^2 \cdot \dfrac{x + 1}{x^2 - x} \cdot \dfrac{(5x - 5)^2}{x^2 + 6x + 5}$

45. $\dfrac{\left(\dfrac{6}{x}\right)}{\left(\dfrac{2}{x^3}\right)}$

46. $\dfrac{0}{\left(\dfrac{5x^2}{2y}\right)}$

47. $25y^2 \div \dfrac{xy}{5}$

48. $\dfrac{6}{z^2} \div 4z^2$

49. $\dfrac{x^2 - 7x}{x + 1} \div \dfrac{x^2 - 14x + 49}{x^2 - 1}$

50. $\left(\dfrac{6x}{y^2}\right)^2 \div \left(\dfrac{3x}{y}\right)^3$

In Exercises 51 and 52, simplify the complex fraction.

51. $\dfrac{\left(\dfrac{6x^2}{x^2+2x-35}\right)}{\left(\dfrac{x^3}{x^2-25}\right)}$

52. $\dfrac{\left[\dfrac{24-18x}{(2-x)^2}\right]}{\left(\dfrac{60-45x}{x^2-4x+4}\right)}$

4.4 In Exercises 53–64, add or subtract the rational expressions and simplify.

53. $\dfrac{4}{9}-\dfrac{11}{9}$

54. $-\dfrac{3}{8}+\dfrac{7}{6}-\dfrac{1}{12}$

55. $\dfrac{15}{16}-\dfrac{5}{24}-1$

56. $\dfrac{2(3y+4)}{2y+1}+\dfrac{3-y}{2y+1}$

57. $\dfrac{1}{x+5}+\dfrac{3}{x-12}$

58. $\dfrac{2}{x-10}+\dfrac{3}{4-x}$

59. $5x+\dfrac{2}{x-3}-\dfrac{3}{x+2}$

60. $4-\dfrac{4x}{x+6}+\dfrac{7}{x-5}$

61. $\dfrac{6}{x}-\dfrac{6x-1}{x^2+4}$

62. $\dfrac{5}{x+2}+\dfrac{25-x}{x^2-3x-10}$

63. $\dfrac{5}{x+3}-\dfrac{4x}{(x+3)^2}-\dfrac{1}{x-3}$

64. $\dfrac{8}{y}-\dfrac{3}{y+5}+\dfrac{4}{y-2}$

In Exercises 65–68, simplify the complex fraction.

65. $\dfrac{3t}{\left(5-\dfrac{2}{t}\right)}$

66. $\dfrac{\left(x-3+\dfrac{2}{x}\right)}{\left(1-\dfrac{2}{x}\right)}$

67. $\dfrac{\left(\dfrac{1}{a^2-16}-\dfrac{1}{a}\right)}{\left(\dfrac{1}{a^2+4a}+4\right)}$

68. $\dfrac{\left(\dfrac{1}{x^2}-\dfrac{1}{y^2}\right)}{\left(\dfrac{1}{x}+\dfrac{1}{y}\right)}$

In Exercises 69–72, use a graphing utility to graph the two functions on the same screen. Use the graphs to verify that the expressions are equivalent. Verify the results algebraically.

69. $y_1=\dfrac{x^2+6x+9}{x^2}\cdot\dfrac{x^2-3x}{x+3}$

$y_2=\dfrac{x^2-9}{x},\quad x\neq-3$

70. $y_1=\dfrac{1}{x}-\dfrac{3}{x+3}$

$y_2=\dfrac{3-2x}{x(x+3)}$

71. $y_1=\dfrac{\left(\dfrac{1}{x}-\dfrac{1}{2}\right)}{2x}$

$y_2=\dfrac{2-x}{4x^2}$

72. $y_1=\dfrac{x^3-2x^2-7}{x-2}$

$y_2=x^2-\dfrac{7}{x-2}$

4.5 In Exercises 73–80, perform the indicated division.

73. $(4x^3-x)\div 2x$

74. $(10x+15)\div(5x)$

75. $\dfrac{6x^3+2x^2-4x+2}{3x-1}$

76. $\dfrac{4x^4-x^3-7x^2+18x}{x-2}$

77. $\dfrac{x^4-3x^2+2}{x^2-1}$

78. $\dfrac{x^4-4x^3+3x}{x^2-1}$

79. $\dfrac{x^5-3x^4+x^2+6}{x^3-2x^2+x-1}$

80. $\dfrac{x^6+4x^5-3x^2+5x}{x^3+x^2-4x+3}$

In Exercises 81–84, use synthetic division to divide.

81. $\dfrac{x^3 + 7x^2 + 3x - 14}{x + 2}$

82. $\dfrac{x^4 - 2x^3 - 15x^2 - 2x + 10}{x - 5}$

83. $(x^4 - 3x^2 - 25) \div (x - 3)$

84. $(2x^3 + 5x - 2) \div \left(x + \frac{1}{2}\right)$

4.6 In Exercises 85–100, solve the equation.

85. $\dfrac{3x}{8} = -15 + \dfrac{x}{4}$

86. $\dfrac{t + 1}{6} = \dfrac{1}{2} - 2t$

87. $8 - \dfrac{12}{t} = \dfrac{1}{3}$

88. $5 + \dfrac{2}{x} = \dfrac{1}{4}$

89. $\dfrac{2}{y} - \dfrac{1}{3y} = \dfrac{1}{3}$

90. $\dfrac{7}{4x} - \dfrac{6}{8x} = 1$

91. $r = 2 + \dfrac{24}{r}$

92. $\dfrac{2}{x} - \dfrac{x}{6} = \dfrac{2}{3}$

93. $8\left(\dfrac{6}{x} - \dfrac{1}{x + 5}\right) = 15$

94. $\dfrac{3}{y + 1} - \dfrac{8}{y} = 1$

95. $\dfrac{4x}{x - 5} + \dfrac{2}{x} = -\dfrac{4}{x - 5}$

96. $\dfrac{2x}{x - 3} - \dfrac{3}{x} = 0$

97. $\dfrac{12}{x^2 + x - 12} - \dfrac{1}{x - 3} = -1$

98. $\dfrac{3}{x - 1} + \dfrac{6}{x^2 - 3x + 2} = 2$

99. $\dfrac{5}{x^2 - 4} - \dfrac{6}{x - 2} = -5$

100. $\dfrac{3}{x^2 - 9} + \dfrac{4}{x + 3} = 1$

In Exercises 101 and 102, (a) use a graphing utility to determine any x-intercepts of the graph of the equation, and (b) set $y = 0$ and solve the resulting equation to confirm the result algebraically.

101. $y = \dfrac{1}{x} - \dfrac{1}{2x + 3}$

102. $y = \dfrac{x}{4} - \dfrac{2}{x} - \dfrac{1}{2}$

Solving Problems

103. *Geometry* A rectangle with a width of w inches has an area of 36 square inches. The perimeter of the rectangle is given by

$$P = 2\left(w + \frac{36}{w}\right).$$

Describe the domain of the function.

104. *Average Cost* The average cost $\overline{C}$ for a manufacturer to produce x units of a product is given by

$$\overline{C} = \frac{15,000 + 0.75x}{x}.$$

Describe the domain of the function.

105. *Average Speed* You drive 56 miles on a service call for your company. On the return trip, which takes 10 minutes less than the original trip, your average speed is 8 miles per hour faster. What is your average speed on the return trip?

106. *Average Speed* You drive 220 miles to see a friend. On the return trip, which takes 20 minutes less than the original trip, your average speed is 5 miles per hour faster. What is your average speed on the return trip?

107. *Batting Average* In this year's playing season, a baseball player has been at bat 150 times and has hit the ball safely 45 times. So, the batting average for the player is

$$\frac{45}{150} = .300.$$

How many consecutive times must the player hit safely to obtain a batting average of .400?

108. *Batting Average* In this year's playing season, a softball player has been at bat 75 times and has hit the ball safely 23 times. So, the batting average for this player is

$$\frac{23}{75} \approx .307.$$

How many consecutive times must the player hit safely to obtain a batting average of .350?

109. *Forming a Partnership* A group of people agree to share equally in the cost of a $60,000 piece of machinery. If they could find two more people to join the group, each person's share of the cost would decrease by $5000. How many people are presently in the group?

110. *Forming a Partnership* An individual is planning to start a small business that will require $28,000 before any income can be generated. Because it is difficult to borrow for new ventures, the individual wants a group of friends to divide the cost equally for a future share of the profit. The person has found some investors, but three more are needed so that the price per person will be $1200 less. How many investors are currently in the group?

111. *Work-Rate Problem* Suppose that in 12 minutes your supervisor can complete a task that you require 15 minutes to complete. Determine the time required to complete the task if you work together.

112. *Work-Rate Problem* Suppose that in 21 minutes your supervisor can complete a task that you require 24 minutes to complete. Determine the time required to complete the task if you work together.

113. *Population of Fish* The Parks and Wildlife Commission introduces 80,000 fish into a large lake. The population (in thousands) is

$$N = \frac{20(4 + 3t)}{1 + 0.05t}, \quad t > 0$$

where t is time in years.

(a) Find the population when t is 5, 10, and 25.

(b) In how many years will the population reach 752,000?

Chapter Test

Take this test as you would take a test in class. After you are done, check your work against the answers given in the back of the book.

In Exercises 1–6, simplify the expressions.

1. $2^{-2} + 2^{-3}$

2. $\dfrac{6.3 \times 10^{-3}}{2.1 \times 10^2}$

3. $(5a^{-3})(2a^2)$

4. $\dfrac{r^2 s^{-3}}{r^5 s^2}$

5. $(x^2 y^{-3})^4$

6. $(3x^2 y^2)^3 (2x^{-2}y)^2$

7. Write 0.000032 in scientific notation.

8. Write 3.04×10^7 in decimal notation.

9. Find the domain of $\dfrac{3y}{y^2 - 25}$.

10. Find the least common denominator of $\dfrac{3}{x^2}$, $\dfrac{x}{x - 3}$, and $\dfrac{2x}{x^3(x + 3)}$.

11. Simplify each rational expression. (a) $\dfrac{2 - x}{3x - 6}$ (b) $\dfrac{2a^2 - 5a - 12}{5a - 20}$

In Exercises 12–23, perform the operation and simplify.

12. $\dfrac{4z^3}{5} \cdot \dfrac{25}{12z^2}$

13. $\dfrac{y^2 + 8y + 16}{2(y - 2)} \cdot \dfrac{8y - 16}{(y + 4)^3}$

14. $(4x^2 - 9) \cdot \dfrac{2x + 3}{2x^2 - x - 3}$

15. $\dfrac{(2xy^2)^3}{15} \div \dfrac{12x^3}{21}$

16. $\dfrac{\left(\dfrac{3x}{x + 2}\right)}{\left(\dfrac{12}{x^3 + 2x^2}\right)}$

17. $\dfrac{\left(9x - \dfrac{1}{x}\right)}{\left(\dfrac{1}{x} - 3\right)}$

18. $2x + \dfrac{1 - 4x^2}{x + 1}$

19. $\dfrac{5x}{x + 2} - \dfrac{2}{x^2 - x - 6}$

20. $\dfrac{3}{x} - \dfrac{5}{x^2} + \dfrac{2x}{x^2 + 2x + 1}$

21. $\dfrac{4}{x + 1} + \dfrac{4x}{x + 1}$

22. $\dfrac{t^4 + t^2 - 6t}{t^2 - 2}$

23. $\dfrac{2x^4 - 15x^2 - 7}{x - 3}$

In Exercises 24–26, solve the equation.

24. $\dfrac{3}{h + 2} = \dfrac{1}{8}$

25. $\dfrac{2}{x + 5} - \dfrac{3}{x + 3} = \dfrac{1}{x}$

26. $\dfrac{1}{x + 1} + \dfrac{1}{x - 1} = \dfrac{2}{x^2 - 1}$

27. One painter works $1\frac{1}{2}$ times as fast as another. Find their individual times for painting a room if it takes them 4 hours working together.

5

Radicals and Complex Numbers

William Taufic/The Stock Market

Greenhouses represented 57% (approximately 466 million square feet) of the total covered growing area used in floriculture crop production in 1997. (Source: U.S. Department of Agriculture)

Motivating the Chapter

Building a Greenhouse

You are building a greenhouse in the form of a half cylinder. The volume of the greenhouse is to be approximately 35,350 cubic feet.

See Section 5.1, Exercise 151

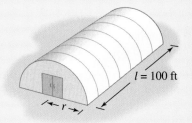

$l = 100$ ft

a. The formula for the radius (in feet) of a half cylinder is

$$r = \sqrt{\frac{2V}{\pi l}}$$

where V is the volume (in cubic feet) and l is the length (in feet). Find the radius of the greenhouse and round your result to the nearest whole number. Use this value of r in parts (b)–(d).

b. Beams to hold the sprinkler system are to be placed across the building. The formula for the height h at which the beams are to be placed is

$$h = \sqrt{r^2 - \left(\frac{a}{2}\right)^2}$$

where a is the length of the beam. Rewrite h as a function of a.

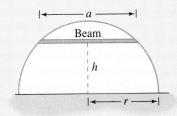

Cross Section of Greenhouse

c. Suppose the length of each beam is $a = 25$ feet. Find the height h at which the beams should be placed.

d. The equation from part (b) can be rewritten as

$$a = 2\sqrt{r^2 - h^2}.$$

Suppose that the height is $h = 8$ feet. What is the length a of each beam?

See Section 5.4, Exercise 103

e. The cost of building the greenhouse is estimated to be $25,000. The money to pay for the greenhouse was invested in an interest-bearing account 10 years ago at an annual percentage rate of 7%. The amount of money earned can be found using the formula

$$r = \left(\frac{A}{P}\right)^{1/n} - 1,$$

where r is the annual percentage rate (in decimal form), A is the amount in the account after 10 years, P is the initial deposit, and n is the number of years. What initial deposit P would have generated enough money to cover the building cost of $25,000?

5.1   Radicals and Rational Exponents

Objectives

1 Determine the *n*th root of a number and evaluate a radical expression.

2 Use the rules of exponents to evaluate an expression with a rational exponent.

3 Use a calculator to evaluate a radical expression.

4 Determine the domain of a radical function.

---

1 Determine the *n*th root of a number and evaluate a radical expression.

Roots and Radicals

The **square root** of a number is defined as one of its two equal factors. For example, 5 is a square root of 25 because 5 is one of the two equal factors of 25. In a similar way, a **cube root** of a number is one of its three equal factors.

Number	Equal Factors	Root	Type
$9 = 3^2$	$3 \cdot 3$	3	Square root
$25 = (-5)^2$	$(-5)(-5)$	-5	Square root
$-27 = (-3)^3$	$(-3)(-3)(-3)$	-3	Cube root
$64 = (4)^3$	$4 \cdot 4 \cdot 4$	4	Cube root
$16 = 2^4$	$2 \cdot 2 \cdot 2 \cdot 2$	2	Fourth root

> ▶ **Definition of *n*th Root of a Number**
>
> Let a and b be real numbers and let n be an integer such that $n \geq 2$. If
>
> $$a = b^n$$
>
> then b is an ***n*th root of *a*.** If $n = 2$, the root is a **square root,** and if $n = 3$, the root is a **cube root.**

Some numbers have more than one *n*th root. For example, both 5 and -5 are square roots of 25 because $25 = 5^2$ and $25 = (-5)^2$. To avoid ambiguity about which root of a number you are talking about, the **principal *n*th root** of a number is defined in terms of a radical symbol $\sqrt[n]{}$.

Study Tip

"Having the same sign as a" means that the principal *n*th root of a is positive if a is positive and negative if a is negative. For example, $\sqrt{4} = 2$ and $\sqrt[3]{-8} = -2$.

> ▶ **Principal *n*th Root of a Number**
>
> Let a be a real number that has at least one (real number) *n*th root. The **principal *n*th root of *a*** is the *n*th root that has the same sign as a, and it is denoted by the **radical symbol**
>
> $$\sqrt[n]{a}. \qquad \text{Principal } n\text{th root}$$
>
> The positive integer n is the **index** of the radical, and the number a is the **radicand.** If $n = 2$, omit the index and write $\sqrt{a}$ rather than $\sqrt[2]{a}$.

| Example 1 | Finding Roots of a Number |

Find the roots.

a. $\sqrt{36}$ **b.** $-\sqrt{36}$ **c.** $\sqrt{-4}$ **d.** $\sqrt[3]{8}$ **e.** $\sqrt[3]{-8}$

Solution

a. $\sqrt{36} = 6$ because $6 \cdot 6 = 6^2 = 36$.

b. $-\sqrt{36} = -6$ because $6 \cdot 6 = 6^2 = 36$.

c. $\sqrt{-4}$ is not real because there is no real number that when multiplied by itself yields -4.

d. $\sqrt[3]{8} = 2$ because $2 \cdot 2 \cdot 2 = 2^3 = 8$.

e. $\sqrt[3]{-8} = -2$ because $(-2)(-2)(-2) = (-2)^3 = -8$.

Study Tip

To remember the properties of *n*th roots in which *a* is negative, consider the following visual scheme.

$\sqrt[odd]{\text{negative number}}$ is negative.

$\sqrt[even]{\text{negative number}}$ is *not* a real number.

▶ **Properties of *n*th Roots**

Property	*Example*
1. If *a* is a positive real number and *n* is *even*, then *a* has exactly two (real) *n*th roots, which are denoted by $\sqrt[n]{a}$ and $-\sqrt[n]{a}$.	The two real square roots of 81 are $\sqrt{81} = 9$ and $-\sqrt{81} = -9$.
2. If *a* is any real number and *n* is *odd*, then *a* has only one (real) *n*th root, which is denoted by $\sqrt[n]{a}$.	$\sqrt[3]{27} = 3$ $\sqrt[3]{-64} = -4$
3. If *a* is a negative real number and *n* is *even*, then *a* has no (real) *n*th root.	$\sqrt{-64}$ is not a real number.

Integers such as 1, 4, 9, 16, 49, and 81 are called **perfect squares** because they have integer square roots. Similarly, integers such as 1, 8, 27, 64, and 125 are called **perfect cubes** because they have integer cube roots.

| Example 2 | Classifying Perfect Squares and Perfect Cubes |

State whether the number is a perfect square, a perfect cube, both, or neither.

a. 81   **b.** -125 **c.** 64   **d.** 32

Solution

a. 81 is a perfect square because $9^2 = 81$. It is not a perfect cube.

b. -125 is a perfect cube because $(-5)^3 = -125$. It is not a perfect square.

c. 64 is a perfect square because $8^2 = 64$, and it is also a perfect cube because $4^3 = 64$.

d. 32 is not a perfect square or a perfect cube. (It is a perfect 5th power because $2^5 = 32$.)

Raising a number to the nth power and taking the principal nth root of a number can be thought of as *inverse* operations. Here are some examples.

$$\left(\sqrt{4}\right)^2 = (2)^2 = 4 \quad \text{and} \quad \sqrt{2^2} = \sqrt{4} = 2$$

$$\left(\sqrt[3]{27}\right)^3 = (3)^3 = 27 \quad \text{and} \quad \sqrt[3]{3^3} = \sqrt[3]{27} = 3$$

$$\left(\sqrt[4]{16}\right)^4 = (2)^4 = 16 \quad \text{and} \quad \sqrt[4]{2^4} = \sqrt[4]{16} = 2$$

$$\left(\sqrt[5]{-243}\right)^5 = (-3)^5 = -243 \quad \text{and} \quad \sqrt[5]{(-3)^5} = \sqrt[5]{-243} = -3$$

▶ **Inverse Properties of nth Powers and nth Roots**

Let a be a real number, and let n be an integer such that $n \geq 2$.

Property	*Example*
1. If a has a principal nth root, then $\left(\sqrt[n]{a}\right)^n = a.$	$\left(\sqrt{5}\right)^2 = 5$
2. If n is *odd*, then $\sqrt[n]{a^n} = a.$	$\sqrt[3]{5^3} = 5$
If n is *even*, then $\sqrt[n]{a^n} = |a|.$	$\sqrt{(-5)^2} = |-5| = 5$

Example 3   Evaluating nth Roots of nth Powers

Evaluate each radical expression.

a. $\sqrt[3]{4^3}$ **b.** $\sqrt[3]{(-2)^3}$ **c.** $\left(\sqrt{7}\right)^2$
d. $\sqrt{(-3)^2}$ **e.** $\sqrt{-3^2}$

Solution

a. Because the index of the radical is odd, you can write

$$\sqrt[3]{4^3} = 4.$$

b. Because the index of the radical is odd, you can write

$$\sqrt[3]{(-2)^3} = -2.$$

c. Using the inverse property of powers and roots, you can write

$$\left(\sqrt{7}\right)^2 = 7.$$

d. Because the index of the radical is even, you must include absolute value signs, and write

$$\sqrt{(-3)^2} = |-3| = 3.$$

e. Because $\sqrt{-3^2} = \sqrt{-9}$ is an even root of a negative number, its value is not a real number.

2 Use the rules of exponents to evaluate an expression with a rational exponent.

Rational Exponents

> ▶ **Definition of Rational Exponents**
>
> Let a be a real number, and let n be an integer such that $n \geq 2$. If the principal nth root of a exists, we define $a^{1/n}$ to be
>
> $$a^{1/n} = \sqrt[n]{a}.$$
>
> If m is a positive integer that has no common factor with n, then
>
> $$a^{m/n} = (a^{1/n})^m = \left(\sqrt[n]{a}\right)^m \quad \text{and} \quad a^{m/n} = (a^m)^{1/n} = \sqrt[n]{a^m}.$$

Study Tip

The numerator of a rational exponent denotes the *power* to which the base is raised, and the denominator denotes the *root* to be taken.

$$\underbrace{\overbrace{a^{m}}^{\text{Power}}{}^{/n}}_{} = \left(\sqrt[n]{a}\right)^m$$

It does not matter in which order the two operations are performed, provided the nth root exists. Here is an example.

$8^{2/3} = \left(\sqrt[3]{8}\right)^2 = 2^2 = 4$ Cube root, then second power

$8^{2/3} = \sqrt[3]{8^2} = \sqrt[3]{64} = 4$ Second power, then cube root

The rules of exponents that we listed in Section 4.1 also apply to rational exponents (provided the roots indicated by the denominators exist). We relist those rules here, with different examples.

∿ Technology: Discovery

Use a calculator to evaluate the expressions below.

$$\frac{3.4^{4.6}}{3.4^{3.1}} \quad \text{and} \quad 3.4^{1.5}$$

How are these two expressions related? Use your calculator to verify some of the other rules of exponents.

> ▶ **Summary of Rules of Exponents**
>
> Let r and s be rational numbers, and let a and b be real numbers, variables, or algebraic expressions.
>
> | *Product and Quotient Rules* | *Example* |
> |---|---|
> | **1.** $a^r \cdot a^s = a^{r+s}$ | $4^{1/2}(4^{1/3}) = 4^{5/6}$ |
> | **2.** $\dfrac{a^r}{a^s} = a^{r-s}, \quad a \neq 0$ | $\dfrac{x^2}{x^{1/2}} = x^{2-(1/2)} = x^{3/2}$ |
>
> | *Power Rules* | |
> |---|---|
> | **3.** $(ab)^r = a^r \cdot b^r$ | $(2x)^{1/2} = 2^{1/2}(x^{1/2})$ |
> | **4.** $\left(\dfrac{a}{b}\right)^r = \dfrac{a^r}{b^r}, \quad b \neq 0$ | $\left(\dfrac{x}{3}\right)^{2/3} = \dfrac{x^{2/3}}{3^{2/3}}$ |
> | **5.** $(a^r)^s = a^{rs}$ | $(x^3)^{1/2} = x^{3/2}$ |
>
> | *Zero and Negative Exponent Rules* | |
> |---|---|
> | **6.** $a^{-r} = \dfrac{1}{a^r}, \quad a \neq 0$ | $4^{-3/2} = \dfrac{1}{4^{3/2}} = \dfrac{1}{(2)^3} = \dfrac{1}{8}$ |
> | **7.** $a^0 = 1$ | $(3x)^0 = 1$ |
> | **8.** $\left(\dfrac{a}{b}\right)^{-r} = \left(\dfrac{b}{a}\right)^r, \quad a \neq 0, \ b \neq 0$ | $\left(\dfrac{x}{4}\right)^{-1/2} = \left(\dfrac{4}{x}\right)^{1/2} = \dfrac{2}{x^{1/2}}$ |

Example 4   Evaluating Expressions with Rational Exponents

Use rules of exponents to rewrite each expression in simpler form.

a. $8^{4/3}$ **b.** $(4^2)^{3/2}$ **c.** $25^{-3/2}$

d. $\left(\frac{64}{125}\right)^{2/3}$ **e.** $-9^{1/2}$ **f.** $(-9)^{1/2}$

Solution

a. $8^{4/3} = (8^{1/3})^4 = (\sqrt[3]{8})^4 = 2^4 = 16$ Root is 3. Power is 4.

b. $(4^2)^{3/2} = 4^{2\cdot(3/2)} = 4^{6/2} = 4^3 = 64$ Root is 2. Power is 3.

c. $25^{-3/2} = \frac{1}{25^{3/2}} = \frac{1}{(\sqrt{25})^3} = \frac{1}{5^3} = \frac{1}{125}$ Root is 2. Power is 3.

d. $\left(\frac{64}{125}\right)^{2/3} = \frac{64^{2/3}}{125^{2/3}} = \frac{(\sqrt[3]{64})^2}{(\sqrt[3]{125})^2} = \frac{4^2}{5^2} = \frac{16}{25}$ Root is 3. Power is 2.

e. $-9^{1/2} = -\sqrt{9} = -(3) = -3$ Root is 2. Power is 1.

f. $(-9)^{1/2} = \sqrt{-9}$ is not a real number.   Root is 2. Power is 1.

In parts (e) and (f) of Example 4, be sure that you see the distinction between the expressions $-9^{1/2}$ and $(-9)^{1/2}$.

Example 5   Using Rules of Exponents

Rewrite each expression using rational exponents.

a. $x\sqrt[4]{x^3}$ **b.** $\frac{\sqrt[3]{x^2}}{\sqrt{x^3}}$ **c.** $\sqrt[3]{x^2 y}$

Solution

a. $x\sqrt[4]{x^3} = x(x^{3/4}) = x^{1+(3/4)} = x^{7/4}$

b. $\frac{\sqrt[3]{x^2}}{\sqrt{x^3}} = \frac{x^{2/3}}{x^{3/2}} = x^{(2/3)-(3/2)} = x^{-5/6} = \frac{1}{x^{5/6}}$

c. $\sqrt[3]{x^2 y} = (x^2 y)^{1/3} = (x^2)^{1/3} y^{1/3} = x^{2/3} y^{1/3}$

Example 6   Using Rules of Exponents

Use rules of exponents to rewrite each expression in simpler form.

a. $\sqrt{\sqrt[3]{x}}$ **b.** $\frac{(2x-1)^{4/3}}{\sqrt[3]{2x-1}}$

Solution

a. $\sqrt{\sqrt[3]{x}} = \sqrt{x^{1/3}} = (x^{1/3})^{1/2} = x^{(1/3)(1/2)} = x^{1/6}$

b. $\frac{(2x-1)^{4/3}}{\sqrt[3]{2x-1}} = \frac{(2x-1)^{4/3}}{(2x-1)^{1/3}} = (2x-1)^{(4/3)-(1/3)} = (2x-1)^{3/3} = 2x-1$

3 Use a calculator to evaluate a radical expression.

Radicals and Calculators

There are two methods of evaluating radicals on most calculators. For square roots, you can use the *square root key* $\boxed{\sqrt{}}$. For other roots, you should first convert the radical to exponential form and then use the *exponential key* $\boxed{y^x}$ or $\boxed{\wedge}$.

Example 7   Evaluating Roots with a Calculator

Evaluate the following. Round the result to three decimal places.

a. $\sqrt{5}$ **b.** $\sqrt[5]{25}$ **c.** $\sqrt[3]{-4}$ **d.** $(8)^{1/2}$ **e.** $(1.4)^{-2/5}$

Solution

a. 5 $\boxed{\sqrt{}}$ Scientific

$\boxed{\sqrt{}}$ 5 $\boxed{\text{ENTER}}$ Graphing

The display is 2.236068. Rounded to three decimal places, $\sqrt{5} \approx 2.236$.

b. First rewrite the expression as $\sqrt[5]{25} = 25^{1/5}$. Then use one of the following keystroke sequences.

25 $\boxed{y^x}$ $\boxed{(}$ 1 $\boxed{\div}$ 5 $\boxed{)}$ $\boxed{=}$ Scientific

25 $\boxed{\wedge}$ $\boxed{(}$ 1 $\boxed{\div}$ 5 $\boxed{)}$ $\boxed{\text{ENTER}}$ Graphing

The display is 1.9036539. Rounded to three decimal places, $\sqrt[5]{25} \approx 1.904$.

c. If your calculator does not have a cube root key, use the fact that

$$\sqrt[3]{-4} = \sqrt[3]{(-1)(4)} = \sqrt[3]{-1}\sqrt[3]{4} = -\sqrt[3]{4}$$

and attach the negative sign of the radicand as the last keystroke.

4 $\boxed{y^x}$ $\boxed{(}$ 1 $\boxed{\div}$ 3 $\boxed{)}$ $\boxed{=}$ $\boxed{+/-}$ Scientific

$\boxed{\sqrt[3]{}}$ $\boxed{(-)}$ 4 $\boxed{\text{ENTER}}$ Graphing

The display is −1.5874011. Rounded to three decimal places, $\sqrt[3]{-4} \approx -1.587$.

d. 8 $\boxed{y^x}$ $\boxed{(}$ 1 $\boxed{\div}$ 2 $\boxed{)}$ $\boxed{=}$ Scientific

8 $\boxed{\wedge}$ $\boxed{(}$ 1 $\boxed{\div}$ 2 $\boxed{)}$ $\boxed{\text{ENTER}}$ Graphing

The display is 2.8284271. Rounded to three decimal places, $(8)^{1/2} \approx 2.828$.

e. 1.4 $\boxed{y^x}$ $\boxed{(}$ 2 $\boxed{\div}$ 5 $\boxed{+/-}$ $\boxed{)}$ $\boxed{=}$ Scientific

1.4 $\boxed{\wedge}$ $\boxed{(}$ $\boxed{(-)}$ 2 $\boxed{\div}$ 5 $\boxed{)}$ $\boxed{\text{ENTER}}$ Graphing

The display is 0.8740752. Rounded to three decimal places, $(1.4)^{-2/5} \approx 0.874$.

Some calculators have a cube root key or submenu command. If your calculator does, try using it to evaluate the expression in Example 7(c).

4 Determine the domain of a radical function.

Radical Functions

The **domain** of the radical function $f(x) = \sqrt[n]{x}$ is the set of all real numbers such that x has a principal nth root.

Technology: Discovery

Consider the function $f(x) = x^{2/3}$.

a. What is the domain of the function?

b. Use your graphing utility to graph the following, in order.

$y_1 = x^{(2 \div 3)}$

$y_2 = (x^2)^{1/3}$ Power, then root

$y_3 = (x^{1/3})^2$ Root, then power

c. Are the graphs all the same? Are their domains all the same?

d. On your graphing utility, which of the forms properly represent the function $f(x) = x^{m/n}$?

$y_1 = x^{(m \div n)}$

$y_2 = (x^m)^{1/n}$

$y_3 = (x^{1/n})^m$

e. Explain how the domains of $f(x) = x^{2/3}$ and $g(x) = x^{-2/3}$ differ.

▶ **Domain of a Radical Function**

Let n be an integer that is greater than or equal to 2.

1. If n is odd, the domain of $f(x) = \sqrt[n]{x}$ is the set of all real numbers.

2. If n is even, the domain of $f(x) = \sqrt[n]{x}$ is the set of all nonnegative real numbers.

Example 8 Finding the Domain of a Radical Function

Describe the domain of each function.

a. $f(x) = \sqrt{x}$

b. $f(x) = \sqrt[3]{x}$

c. $f(x) = \sqrt{x^2}$

d. $f(x) = \sqrt{x^3}$

Solution

a. The domain of $f(x) = \sqrt{x}$ is the set of all nonnegative real numbers. For instance, 2 is in the domain, but -2 is not because $\sqrt{-2}$ is not a real number.

b. The domain of $f(x) = \sqrt[3]{x}$ is the set of all real numbers because for any real number x, the expression $\sqrt[3]{x}$ is a real number.

c. The domain of $f(x) = \sqrt{x^2}$ is the set of all real numbers because for any real number x, the expression x^2 is a nonnegative real number.

d. The domain of $f(x) = \sqrt{x^3}$ is the set of all nonnegative real numbers. For instance, 1 is in the domain, but -1 is not because $\sqrt{(-1)^3} = \sqrt{-1}$ is not a real number.

Discussing the Concept **Describing Domains and Ranges**

Discuss the domain and range of each of the following functions. Use a graphing utility to verify your conclusions.

a. $y = x^{3/2}$ **b.** $y = x^2$

c. $y = x^{1/3}$ **d.** $y = \left(\sqrt{x}\right)^2$

e. $y = x^{-4/5}$ **f.** $y = \sqrt[3]{x^2}$

5.1 Exercises

Integrated Review Concepts, Skills, and Problem Solving

Keep mathematically in shape by doing these exercises *before* the problems of this section.

Properties and Definitions

In Exercises 1–4, complete the property of exponents.

1. $a^m \cdot a^n =$ **2.** $(ab)^m =$

3. $(a^m)^n =$ **4.** $\dfrac{a^m}{a^n} =$, if $m > n$

Solving Equations

In Exercises 5–10, solve for y.

5. $3x + y = 4$

6. $2x + 3y = 2$

Developing Skills

In Exercises 1–8, find the root if it exists. See Example 1.

1. $\sqrt{64}$

2. $-\sqrt{100}$

3. $-\sqrt{49}$

4. $\sqrt{-25}$

5. $\sqrt[3]{-8}$

6. $\sqrt[3]{-64}$

7. $\sqrt{-1}$

8. $-\sqrt[3]{1}$

In Exercises 9–14, complete the statement. See Example 2.

9. Because $7^2 = 49$,   is a square root of 49.

10. Because $24.5^2 = 600.25$,   is a square root of 600.25.

11. Because $4.2^3 = 74.088$,   is a cube root of 74.088.

12. Because $6^4 = 1296$,   is a fourth root of 1296.

13. Because $45^2 = 2025$, 45 is called the   of 2025.

14. Because $12^3 = 1728$, 12 is called the   of 1728.

In Exercises 15–44, evaluate each radical expression without using a calculator. If not possible, state the reason. See Example 3.

15. $\sqrt{8^2}$

16. $-\sqrt{10^2}$

17. $\sqrt{(-10)^2}$

18. $\sqrt{(-12)^2}$

19. $\sqrt{-9^2}$

20. $\sqrt{-12^2}$

21. $-\sqrt{\left(\frac{2}{3}\right)^2}$

22. $\sqrt{\left(\frac{3}{4}\right)^2}$

23. $\sqrt{-\left(\frac{3}{10}\right)^2}$

24. $\sqrt{\left(-\frac{3}{5}\right)^2}$

25. $\left(\sqrt{5}\right)^2$

26. $-\left(\sqrt{10}\right)^2$

27. $-\left(\sqrt{23}\right)^2$

28. $\left(-\sqrt{18}\right)^2$

29. $\sqrt[3]{(5)^3}$

30. $\sqrt[3]{(-2)^3}$

31. $\sqrt[3]{10^3}$

32. $\sqrt[3]{4^3}$

33. $-\sqrt[3]{(-6)^3}$

34. $-\sqrt[3]{9^3}$

35. $\sqrt[3]{\left(-\frac{1}{4}\right)^3}$

36. $-\sqrt[3]{\left(\frac{1}{5}\right)^3}$

37. $\left(\sqrt[3]{11}\right)^3$

38. $\left(\sqrt[3]{-6}\right)^3$

39. $\left(-\sqrt[3]{24}\right)^3$

40. $\left(\sqrt[3]{21}\right)^3$

41. $\sqrt[4]{3^4}$

42. $\sqrt[5]{(-2)^5}$

43. $-\sqrt[4]{-5^4}$

44. $-\sqrt[4]{2^4}$

In Exercises 45–48, determine whether the square root is a rational or irrational number.

45. $\sqrt{6}$

46. $\sqrt{\frac{9}{16}}$

47. $\sqrt{900}$

48. $\sqrt{72}$

In Exercises 49–54, fill in the missing description.

Radical Form	Rational Exponent Form
49. $\sqrt{16} = 4$	
50. $\sqrt[4]{81} = 3$	
51. $\sqrt[3]{27^2} = 9$	
52.	$125^{1/3} = 5$
53.	$256^{3/4} = 64$
54.	$27^{2/3} = 9$

In Exercises 55–74, evaluate without using a calculator. See Example 4.

55. $25^{1/2}$

56. $49^{1/2}$

57. $-36^{1/2}$

58. $-121^{1/2}$

59. $-(16)^{3/4}$

60. $-(125)^{2/3}$

61. $32^{-2/5}$

62. $81^{-3/4}$

63. $(-27)^{-2/3}$

64. $(-243)^{-3/5}$

65. $\left(\frac{8}{27}\right)^{2/3}$

66. $\left(\frac{256}{625}\right)^{1/4}$

67. $\left(\frac{121}{9}\right)^{-1/2}$

68. $\left(\frac{27}{1000}\right)^{-4/3}$

69. $(3^3)^{2/3}$

70. $(8^2)^{3/2}$

71. $-(4^4)^{3/4}$

72. $(-2^3)^{5/3}$

73. $\left(\frac{1}{5^3}\right)^{-2/3}$

74. $\left(\frac{4}{6^2}\right)^{-3/2}$

In Exercises 75–94, rewrite the expression using rational exponents. See Example 5.

75. $\sqrt{t}$

76. $\sqrt[3]{x}$

77. $x\sqrt[4]{x^3}$

78. $t\sqrt[5]{t^2}$

79. $u^2\sqrt[3]{u}$

80. $y\sqrt[4]{y^2}$

81. $s^4\sqrt{s^5}$

82. $n^3\sqrt[4]{n^6}$

83. $\dfrac{\sqrt{x}}{\sqrt{x^3}}$

84. $\dfrac{\sqrt[3]{x^2}}{\sqrt[3]{x^4}}$

85. $\dfrac{\sqrt[4]{t}}{\sqrt{t^5}}$

86. $\dfrac{\sqrt[3]{x^4}}{\sqrt{x^3}}$

87. $\sqrt[3]{x^2} \cdot \sqrt[3]{x^7}$

88. $\sqrt[5]{z^3} \cdot \sqrt[5]{z^2}$

89. $\sqrt[4]{y^3} \cdot \sqrt[3]{y}$

90. $\sqrt[6]{x^5} \cdot \sqrt[3]{x^4}$

91. $\sqrt[4]{x^3 y}$

92. $\sqrt[3]{u^4 v^2}$

93. $z^2\sqrt{y^5 z^4}$

94. $x^2\sqrt[3]{xy^4}$

In Exercises 95–116, simplify the expression. See Example 6.

95. $3^{1/4} \cdot 3^{3/4}$

96. $2^{2/5} \cdot 2^{3/5}$

97. $(2^{1/2})^{2/3}$

98. $(4^{1/3})^{9/4}$

99. $\dfrac{2^{1/5}}{2^{6/5}}$

100. $\dfrac{5^{-3/4}}{5}$

101. $(c^{3/2})^{1/3}$

102. $(k^{-1/3})^{3/2}$

103. $\dfrac{18y^{4/3}z^{-1/3}}{24y^{-2/3}z}$

104. $\dfrac{a^{3/4} \cdot a^{1/2}}{a^{5/2}}$

105. $(3x^{-1/3}y^{3/4})^2$

106. $(-2u^{3/5}v^{-1/5})^3$

107. $\left(\dfrac{x^{1/4}}{x^{1/6}}\right)^3$

108. $\left(\dfrac{3m^{1/6}n^{1/3}}{4n^{-2/3}}\right)^2$

109. $\sqrt{\sqrt[4]{y}}$

110. $\sqrt[3]{\sqrt{2x}}$

111. $\sqrt[4]{\sqrt{x^3}}$

112. $\sqrt[5]{\sqrt[3]{y^4}}$

113. $\dfrac{(x+y)^{3/4}}{\sqrt[4]{x+y}}$

114. $\dfrac{(a-b)^{1/3}}{\sqrt[3]{a-b}}$

115. $\dfrac{(3u-2v)^{2/3}}{\sqrt{(3u-2v)^3}}$

116. $\dfrac{\sqrt[4]{2x+y}}{(2x+y)^{3/2}}$

In Exercises 117–130, use a calculator to approximate the quantity accurate to four decimal places. If not possible, state the reason. See Example 7.

117. $\sqrt{73}$

118. $\sqrt{-532}$

119. $315^{2/5}$

120. $962^{2/3}$

121. $1698^{-3/4}$

122. $382.5^{-3/2}$

123. $\sqrt[4]{342}$

124. $\sqrt[3]{159}$

125. $\sqrt[3]{545^2}$

126. $\sqrt[5]{-35^3}$

127. $\dfrac{8 - \sqrt{35}}{2}$

128. $\dfrac{-5 + \sqrt{3215}}{10}$

129. $\dfrac{3 + \sqrt{17}}{9}$

130. $\dfrac{7 - \sqrt{241}}{12}$

In Exercises 131–136, describe the domain of the function. See Example 8.

131. $f(x) = 3\sqrt{x}$

132. $h(x) = \sqrt[4]{x}$

133. $g(x) = \dfrac{2}{\sqrt[4]{x}}$

134. $g(x) = \dfrac{10}{\sqrt[3]{x}}$

135. $f(x) = \sqrt{-x}$

136. $f(x) = \sqrt[3]{x^4}$

In Exercises 137–140, describe the domain of the function algebraically. Use a graphing utility to graph the function. Did the graphing utility omit part of the domain? If so, complete the graph by hand.

137. $y = \dfrac{5}{\sqrt[4]{x^3}}$

138. $y = 4\sqrt[3]{x}$

139. $g(x) = 2x^{3/5}$

140. $h(x) = 5x^{2/3}$

In Exercises 141–144, perform the multiplication. Use a graphing utility to confirm your result.

141. $x^{1/2}(2x - 3)$

142. $x^{4/3}(3x^2 - 4x + 5)$

143. $y^{-1/3}(y^{1/3} + 5y^{4/3})$

144. $(x^{1/2} - 3)(x^{1/2} + 3)$

Solving Problems

Mathematical Modeling   In Exercises 145 and 146, use the formula for the *declining balances method*

$$r = 1 - \left(\frac{S}{C}\right)^{1/n}$$

to find the depreciation rate r. In the formula, n is the useful life of the item (in years), S is the salvage value (in dollars), and C is the original cost (in dollars).

145. A $75,000 truck depreciates over an 8-year period, as shown in the graph. Find r.

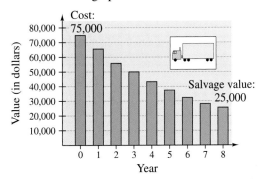

146. A $125,000 printing press depreciates over a 10-year period, as shown in the graph. Find r.

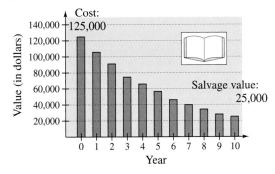

147. *Geometry*   Find the dimensions of a piece of carpet for a classroom with 529 square feet of floor space, assuming the floor is square.

148. *Geometry*   Find the dimensions of a square mirror with an area of 1024 square inches.

149. *Geometry*   The length of a diagonal of a rectangular solid of length l, width w, and height h is

$$\sqrt{l^2 + w^2 + h^2}.$$

Approximate to two decimal places the length of the diagonal of the solid shown in the figure.

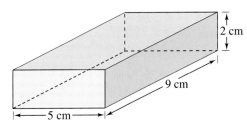

150. *Velocity of a Stream*   A stream of water moving at a rate of v feet per second can carry particles of size $0.03\sqrt{v}$ inches.

(a) Find the particle size that can be carried by a stream flowing at the rate of $\frac{3}{4}$ foot per second. Approximate to three decimal places.

(b) Find the particle size that can be carried by a stream flowing at the rate of $\frac{3}{16}$ foot per second. Approximate to three decimal places.

Explaining Concepts

151. Answer parts (a)–(d) of Motivating the Chapter on page 319.

152. In your own words, define the nth root of a number.

153. Define the *radicand* and the *index* of a radical.

154. If n is even, what must be true about the radicand for the nth root to be a real number? Explain.

155. Is it true that $\sqrt{2} = 1.414$? Explain.

156. Given a real number x, state the conditions on n for each of the following.

(a) $\sqrt[n]{x^n} = x$ (b) $\sqrt[n]{x^n} = |x|$

157. *Investigation*   Find all possible "last digits" of perfect squares. (For instance, the last digit of 81 is 1 and the last digit of 64 is 4.) Is it possible that 4,322,788,986 is a perfect square?

5.2 Simplifying Radical Expressions

Objectives

1 Use the Multiplication and Division Properties of Radicals to simplify a radical expression.

2 Use rationalization techniques to simplify a radical expression.

3 Use the Distributive Property to add and subtract like radicals.

4 Use the Pythagorean Theorem in an application problem.

---

1 Use the Multiplication and Division Properties of Radicals to simplify a radical expression.

Simplifying Radicals

In this section, you will study ways to simplify and combine radicals. For instance, the expression $\sqrt{12}$ can be simplified as

$$\sqrt{12} = \sqrt{4 \cdot 3} = \sqrt{4}\sqrt{3} = 2\sqrt{3}.$$

This rewritten form is based on the following rules for multiplying and dividing radicals.

Study Tip

The Multiplication and Division Properties of Radicals can be shown to be true by converting the radicals to exponential form and using the rules of exponents on page 323.

Using Property 3

$$\sqrt[n]{uv} = (uv)^{1/n}$$

$$= u^{1/n}v^{1/n}$$

$$= \sqrt[n]{u}\sqrt[n]{v}$$

Using Property 4

$$\sqrt[n]{\frac{u}{v}} = \left(\frac{u}{v}\right)^{1/n}$$

$$= \frac{u^{1/n}}{v^{1/n}} = \frac{\sqrt[n]{u}}{\sqrt[n]{v}}$$

▶ **Multiplying and Dividing Radicals**

Let u and v be real numbers, variables, or algebraic expressions. If the nth roots of u and v are real, the following properties are true.

1. $\sqrt[n]{uv} = \sqrt[n]{u}\sqrt[n]{v}$ Multiplication Property of Radicals

2. $\sqrt[n]{\dfrac{u}{v}} = \dfrac{\sqrt[n]{u}}{\sqrt[n]{v}}, \quad v \neq 0$ Division Property of Radicals

You can use the Multiplication Property of Radicals to *simplify* square root expressions by finding the largest perfect square factor and removing it from the radical, as follows.

$$\sqrt{48} = \sqrt{16 \cdot 3} = \sqrt{16}\sqrt{3} = 4\sqrt{3}$$

This simplification process is called **removing perfect square factors from the radical.**

Example 1   Removing Constant Factors from Radicals

Simplify each radical by removing as many factors as possible.

a. $\sqrt{75}$ **b.** $\sqrt{72}$ **c.** $\sqrt{162}$

Solution

a. $\sqrt{75} = \sqrt{25 \cdot 3} = \sqrt{25}\sqrt{3} = 5\sqrt{3}$ 25 is a perfect square factor of 75.

b. $\sqrt{72} = \sqrt{36 \cdot 2} = \sqrt{36}\sqrt{2} = 6\sqrt{2}$ 36 is a perfect square factor of 72.

c. $\sqrt{162} = \sqrt{81 \cdot 2} = \sqrt{81}\sqrt{2} = 9\sqrt{2}$ 81 is a perfect square factor of 162.

When removing *variable* factors from a square root radical, remember that it is not valid to write $\sqrt{x^2} = x$ *unless* you happen to know that x is nonnegative. Without knowing anything about x, the only way you can simplify $\sqrt{x^2}$ is to include absolute value signs when you remove x from the radical.

$$\sqrt{x^2} = |x| \qquad\qquad \text{Restricted by absolute value signs}$$

When simplifying the expression $\sqrt{x^3}$, it is not necessary to include absolute value signs because the domain of this expression does not include negative numbers.

$$\sqrt{x^3} = \sqrt{x^2(x)} = x\sqrt{x} \qquad\qquad \text{Restricted by domain of radical}$$

Example 2	Removing Variable Factors from Radicals

Simplify the radical expression.

a. $\sqrt{25x^2}$ **b.** $\sqrt{12x^3}, \quad x \ge 0$ **c.** $\sqrt{144x^4}$ **d.** $\sqrt{72x^3y^2}$

Solution

a. $\sqrt{25x^2} = \sqrt{5^2x^2} = \sqrt{5^2}\sqrt{x^2} = 5|x|$ $\qquad \sqrt{x^2} = |x|$

b. $\sqrt{12x^3} = \sqrt{2^2x^2(3x)} = 2x\sqrt{3x}$ $\qquad \sqrt{2^2}\sqrt{x^2} = 2x, \quad x \ge 0$

c. $\sqrt{144x^4} = \sqrt{12^2(x^2)^2} = 12x^2$ $\qquad \sqrt{12^2}\sqrt{(x^2)^2} = 12|x^2| = 12x^2$

d. $\sqrt{72x^3y^2} = \sqrt{6^2x^2y^2(2x)}$ $\qquad \sqrt{6^2}\sqrt{x^2}\sqrt{y^2} = 6|x||y|$

$\qquad\qquad\quad = 6|x||y|\sqrt{2x}$

In the same way that perfect squares can be removed from square root radicals, perfect nth powers can be removed from nth root radicals.

Study Tip

To find the perfect nth root factor of 486 in Example 3(c), you can write the prime factorization of the number.

$$486 = 2 \cdot 3 \cdot 3 \cdot 3 \cdot 3 \cdot 3$$
$$= 2 \cdot 3^5$$

From its prime factorization you can see that 3^5 is a fifth root factor of 486.

$$\sqrt[5]{486} = \sqrt[5]{2 \cdot 3^5}$$
$$= \sqrt[5]{3^5}\sqrt[5]{2}$$
$$= 3\sqrt[5]{2}$$

Example 3	Removing Factors from Radicals

Simplify the radical expressions.

a. $\sqrt[3]{40}$ **b.** $\sqrt[4]{x^5}, \quad x \ge 0$ **c.** $\sqrt[5]{486x^7}$ **d.** $\sqrt[3]{128x^3y^5}$

Solution

a. $\sqrt[3]{40} = \sqrt[3]{8(5)}$ $\qquad\qquad \sqrt[3]{2^3} = 2$

$\qquad\quad = \sqrt[3]{2^3(5)}$

$\qquad\quad = 2\sqrt[3]{5}$

b. $\sqrt[4]{x^5} = \sqrt[4]{x^4(x)}$ $\qquad\qquad \sqrt[4]{x^4} = x, \quad x \ge 0$

$\qquad\quad = x\sqrt[4]{x}$

c. $\sqrt[5]{486x^7} = \sqrt[5]{243x^5(2x^2)}$

$\qquad\qquad = \sqrt[5]{3^5x^5(2x^2)}$

$\qquad\qquad = 3x\sqrt[5]{2x^2}$ $\qquad\qquad \sqrt[5]{3^5}\sqrt[5]{x^5} = 3x$

d. $\sqrt[3]{128x^3y^5} = \sqrt[3]{64x^3y^3(2y^2)}$

$\qquad\qquad = \sqrt[3]{4^3x^3y^3(2y^2)}$

$\qquad\qquad = 4xy\sqrt[3]{2y^2}$ $\qquad\qquad \sqrt[3]{4^3}\sqrt[3]{x^3}\sqrt[3]{y^3} = 4xy$

2 Use rationalization techniques to simplify a radical expression.

Rationalization Techniques

Removing factors from radicals is only one of three techniques used to simplify radicals. The three techniques are summarized as follows.

> ▶ **Simplifying Radical Expressions**
>
> A radical expression is said to be in simplest form if all three of the following are true.
>
> **1.** All possible nth powered factors have been removed from each radical.
>
> **2.** No radical contains a fraction.
>
> **3.** No denominator of a fraction contains a radical.

To meet the last two conditions, you can use a technique called **rationalizing the denominator.** This involves multiplying both the numerator and denominator by a *rationalizing factor* that creates a perfect nth power in the denominator.

Example 4 Rationalizing the Denominator

Simplify the expression.

a. $\sqrt{\dfrac{3}{5}}$ **b.** $\dfrac{4}{\sqrt[3]{9}}$ **c.** $\dfrac{8}{3\sqrt{18}}$

Solution

a. $\sqrt{\dfrac{3}{5}} = \dfrac{\sqrt{3}}{\sqrt{5}} = \dfrac{\sqrt{3}}{\sqrt{5}} \cdot \dfrac{\sqrt{5}}{\sqrt{5}} = \dfrac{\sqrt{15}}{\sqrt{25}} = \dfrac{\sqrt{15}}{5}$ Multiply by $\sqrt{5}/\sqrt{5}$ to create a perfect square in the denominator.

b. $\dfrac{4}{\sqrt[3]{9}} = \dfrac{4}{\sqrt[3]{9}} \cdot \dfrac{\sqrt[3]{3}}{\sqrt[3]{3}} = \dfrac{4\sqrt[3]{3}}{\sqrt[3]{27}} = \dfrac{4\sqrt[3]{3}}{3}$ Multiply by $\sqrt[3]{3}/\sqrt[3]{3}$ to create a perfect cube in the denominator.

c. $\dfrac{8}{3\sqrt{18}} = \dfrac{8}{3\sqrt{18}} \cdot \dfrac{\sqrt{2}}{\sqrt{2}} = \dfrac{8\sqrt{2}}{3\sqrt{36}} = \dfrac{8\sqrt{2}}{3(6)} = \dfrac{4\sqrt{2}}{9}$

Example 5 Rationalizing the Denominator

Simplify the expression.

a. $\sqrt{\dfrac{8x}{12y^5}}$ **b.** $\sqrt[3]{\dfrac{54x^6y^3}{5z^2}}$

Solution

a. $\sqrt{\dfrac{8x}{12y^5}} = \sqrt{\dfrac{(4)(2)x}{(4)(3)y^5}} = \sqrt{\dfrac{2x}{3y^5}} = \dfrac{\sqrt{2x}}{\sqrt{3y^5}} \cdot \dfrac{\sqrt{3y}}{\sqrt{3y}} = \dfrac{\sqrt{6xy}}{\sqrt{9y^6}} = \dfrac{\sqrt{6xy}}{3y^3}$

b. $\sqrt[3]{\dfrac{54x^6y^3}{5z^2}} = \dfrac{\sqrt[3]{(3^3)(2)(x^6)(y^3)}}{\sqrt[3]{5z^2}} \cdot \dfrac{\sqrt[3]{25z}}{\sqrt[3]{25z}} = \dfrac{3x^2y\sqrt[3]{50z}}{\sqrt[3]{5^3z^3}} = \dfrac{3x^2y\sqrt[3]{50z}}{5z}$

Study Tip

When rationalizing a denominator, remember that for square roots you want a perfect square in the denominator, for cube roots you want a perfect cube, and so on.

3 Use the Distributive Property to add and subtract like radicals.

Adding and Subtracting Radicals

Two or more radical expressions are *alike* if they have the same radicand and the same index. For instance, $\sqrt{2}$ and $3\sqrt{2}$ are alike, but $\sqrt{3}$ and $\sqrt[3]{3}$ are not alike. Two radical expressions that are alike can be added or subtracted by adding or subtracting their coefficients. *Before* concluding that two radicals cannot be combined, you should first rewrite them in simplest form. This is illustrated in Example 6(c).

Example 6   Combining Radicals

a. $\sqrt{7} + 5\sqrt{7} - 2\sqrt{7} = (1 + 5 - 2)\sqrt{7}$ Distributive Property

$\phantom{\sqrt{7} + 5\sqrt{7} - 2\sqrt{7}} = 4\sqrt{7}$ Simplify.

b. $6\sqrt{x} - \sqrt[3]{4} - 5\sqrt{x} + 2\sqrt[3]{4} = 6\sqrt{x} - 5\sqrt{x} - \sqrt[3]{4} + 2\sqrt[3]{4}$ Group like terms.

$\phantom{6\sqrt{x} - \sqrt[3]{4} - 5\sqrt{x} + 2\sqrt[3]{4}} = (6 - 5)\sqrt{x} + (-1 + 2)\sqrt[3]{4}$ Distributive Property

$\phantom{6\sqrt{x} - \sqrt[3]{4} - 5\sqrt{x} + 2\sqrt[3]{4}} = \sqrt{x} + \sqrt[3]{4}$ Simplify.

c. $3\sqrt[3]{x} + \sqrt[3]{8x} = 3\sqrt[3]{x} + 2\sqrt[3]{x}$ $\sqrt[3]{8} = \sqrt[3]{2^3} = 2$

$\phantom{3\sqrt[3]{x} + \sqrt[3]{8x}} = (3 + 2)\sqrt[3]{x}$ Distributive Property

$\phantom{3\sqrt[3]{x} + \sqrt[3]{8x}} = 5\sqrt[3]{x}$ Simplify.

Example 7   Simplifying Before Combining Radicals

a. $\sqrt{45x} + 3\sqrt{20x} = 3\sqrt{5x} + 6\sqrt{5x} = 9\sqrt{5x}$

b. $5\sqrt{x^3} - x\sqrt{4x} = 5x\sqrt{x} - 2x\sqrt{x} = 3x\sqrt{x}$

c. $\sqrt[3]{54y^5} + 4\sqrt[3]{2y^2} = 3y\sqrt[3]{2y^2} + 4\sqrt[3]{2y^2} = (3y + 4)\sqrt[3]{2y^2}$

In some instances, it may be necessary to rationalize denominators before combining radicals.

Example 8   Rationalizing Denominators Before Simplifying

$$\sqrt{7} - \frac{5}{\sqrt{7}} = \sqrt{7} - \left(\frac{5}{\sqrt{7}} \cdot \frac{\sqrt{7}}{\sqrt{7}}\right)$$ Multiply by $\sqrt{7}/\sqrt{7}$ to create a perfect square in the denominator.

$$= \sqrt{7} - \frac{5\sqrt{7}}{7}$$ Simplify.

$$= \left(1 - \frac{5}{7}\right)\sqrt{7}$$ Distributive Property

$$= \frac{2}{7}\sqrt{7}$$ Simplify.

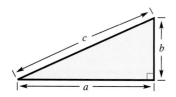

4 Use the Pythagorean Theorem in an application problem.

Application of Radicals

A common use of radicals occurs in applications involving right triangles. Recall that a right triangle is one that contains a right (or 90°) angle, as shown in Figure 5.1. The relationship among the three sides of a right triangle is described by the **Pythagorean Theorem,** which says that if a and b are the lengths of the legs and c is the length of the hypotenuse, then

$$c = \sqrt{a^2 + b^2}. \qquad \text{Pythagorean Theorem}$$

For instance, if $a = 6$ and $b = 9$, then

$$c = \sqrt{6^2 + 9^2} = \sqrt{117} = \sqrt{9}\sqrt{13} = 3\sqrt{13}.$$

Figure 5.1

Example 9   An Application of the Pythagorean Theorem

A softball diamond has the shape of a square with 60-foot sides (see Figure 5.2). The catcher is 5 feet behind home plate. How far does the catcher have to throw to reach second base?

Solution

In Figure 5.2, let x be the hypotenuse of a right triangle with 60-foot legs. So, by the Pythagorean Theorem, you have the following.

$$x = \sqrt{60^2 + 60^2} \qquad \text{Pythagorean Theorem}$$

$$x = \sqrt{7200} \qquad \text{Simplify.}$$

$$x = \sqrt{3600 \cdot 2} \qquad \text{3600 is a perfect square factor of 7200.}$$

$$x = 60\sqrt{2} \qquad \text{Simplify.}$$

$$x \approx 84.9 \text{ feet}$$

So, the distance from home plate to second base is approximately 84.9 feet. Because the catcher is 5 feet behind home plate, the catcher must make a throw of

$$x + 5 \approx 84.9 + 5 = 89.9 \text{ feet.}$$

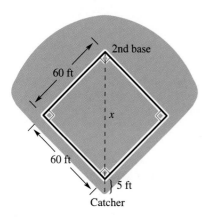

Figure 5.2

Discussing the Concept   Error Analysis

Suppose you are an algebra instructor and one of your students hands in the following work. Find and correct the errors, and discuss how you can help your student avoid such errors in the future.

a. $7\sqrt{3} + 4\sqrt{2} = 11\sqrt{5}$

b. $3\sqrt[3]{k} - 6\sqrt{k} = -3\sqrt{k}$

5.2   Exercises

Integrated Review   *Concepts, Skills, and Problem Solving*

Keep mathematically in shape by doing these exercises *before* the problems of this section.

Properties and Definitions

1. Explain how to determine the half-plane satisfying $x - y > -3$.

2. Describe the difference between the graphs of $3x + 4y \leq 4$ and $3x + 4y < 4$.

In Exercises 3–8, factor the expression completely.

3. $-x^3 + 3x^2 - x + 3$

4. $4t^2 - 169$

5. $x^2 - 3x + 2$

6. $2x^2 + 5x - 7$

7. $11x^2 + 6x - 5$

8. $4x^2 - 28x + 49$

Problem Solving

9. Twelve hundred tickets were sold for a theater production and the receipts for the performance were \$21,120. The tickets for adults and students sold for \$20 and \$12.50, respectively. How many of each kind of ticket were sold?

10. A quality control engineer for a certain buyer found two defective units in a sample of 75. At that rate, what is the expected number of defective units in a shipment of 10,000 units?

Developing Skills

In Exercises 1–20, simplify the radical. See Example 1.

1. $\sqrt{20}$

2. $\sqrt{27}$

3. $\sqrt{50}$

4. $\sqrt{125}$

5. $\sqrt{96}$

6. $\sqrt{84}$

7. $\sqrt{216}$

8. $\sqrt{147}$

9. $\sqrt{1183}$

10. $\sqrt{1176}$

11. $\sqrt{0.04}$

12. $\sqrt{0.25}$

13. $\sqrt{0.0072}$

14. $\sqrt{0.0027}$

15. $\sqrt{2.42}$

16. $\sqrt{9.8}$

17. $\sqrt{\frac{15}{4}}$

18. $\sqrt{\frac{5}{36}}$

19. $\sqrt{\frac{13}{25}}$

20. $\sqrt{\frac{15}{36}}$

In Exercises 21–56, simplify the expression. See Examples 2 and 3.

21. $\sqrt{9x^5}$

22. $\sqrt{64x^3}$

23. $\sqrt{48y^4}$

24. $\sqrt{32x}$

25. $\sqrt{117y^6}$

26. $\sqrt{160x^8}$

27. $\sqrt{120x^2y^3}$

28. $\sqrt{125u^4v^6}$

29. $\sqrt{192a^5b^7}$

30. $\sqrt{363x^{10}y^9}$

31. $\sqrt[3]{48}$

32. $\sqrt[3]{81}$

33. $\sqrt[3]{112}$

34. $\sqrt[4]{112}$

35. $\sqrt[3]{40x^5}$

36. $\sqrt[3]{54z^7}$

37. $\sqrt[4]{324y^6}$

38. $\sqrt[5]{160x^8}$

39. $\sqrt[3]{x^4y^3}$

40. $\sqrt[3]{a^5b^6}$

41. $\sqrt[4]{3x^4y^2}$

42. $\sqrt[4]{128u^4v^7}$

43. $\sqrt[5]{32x^5y^6}$

44. $\sqrt[3]{16x^4y^5}$

45. $\sqrt[3]{\frac{35}{64}}$

46. $\sqrt[4]{\frac{5}{16}}$

47. $\sqrt[5]{\frac{15}{243}}$

48. $\sqrt[3]{\frac{1}{1000}}$

49. $\sqrt[5]{\frac{32x^2}{y^5}}$

50. $\sqrt[3]{\frac{16z^3}{y^6}}$

51. $\sqrt[3]{\frac{54a^4}{b^9}}$

52. $\sqrt[4]{\frac{3u^2}{16v^8}}$

53. $\sqrt{\frac{32a^4}{b^2}}$

54. $\sqrt{\frac{18x^2}{z^6}}$

55. $\sqrt[4]{(3x^2)^4}$

56. $\sqrt[5]{96x^5}$

In Exercises 57–78, rationalize the denominator and simplify further, if possible. See Examples 4 and 5.

57. $\sqrt{\frac{1}{3}}$

58. $\sqrt{\frac{1}{5}}$

59. $\frac{1}{\sqrt{7}}$

60. $\frac{1}{\sqrt{15}}$

61. $\frac{12}{\sqrt{3}}$

62. $\frac{5}{\sqrt{10}}$

63. $\sqrt[4]{\dfrac{5}{4}}$

64. $\sqrt[3]{\dfrac{9}{25}}$

65. $\dfrac{6}{\sqrt[3]{32}}$

66. $\dfrac{10}{\sqrt[5]{16}}$

67. $\dfrac{1}{\sqrt{y}}$

68. $\sqrt{\dfrac{5}{c}}$

69. $\sqrt{\dfrac{4}{x}}$

70. $\sqrt{\dfrac{4}{x^3}}$

71. $\dfrac{1}{\sqrt{2x}}$

72. $\dfrac{5}{\sqrt{8x^5}}$

73. $\dfrac{6}{\sqrt{3b^3}}$

74. $\dfrac{1}{\sqrt{xy}}$

75. $\sqrt[3]{\dfrac{2x}{3y}}$

76. $\sqrt[3]{\dfrac{20x^2}{9y^2}}$

77. $\dfrac{a^3}{\sqrt[3]{ab^2}}$

78. $\dfrac{3u^2}{\sqrt[4]{8u^3}}$

In Exercises 79–92, combine the radical expressions, if possible. See Examples 6 and 7.

79. $3\sqrt{2} - \sqrt{2}$

80. $6\sqrt{5} - 2\sqrt{5}$

81. $12\sqrt{8} - 3\sqrt[3]{8}$

82. $4\sqrt{32} + 7\sqrt{32}$

83. $\sqrt[4]{3} - 5\sqrt[4]{7} - 12\sqrt[4]{3}$

84. $9\sqrt[3]{17} + 7\sqrt[3]{2} - 4\sqrt[3]{17} + \sqrt[3]{2}$

85. $2\sqrt[3]{54} + 12\sqrt[3]{16}$

86. $4\sqrt[4]{48} - \sqrt[4]{243}$

87. $5\sqrt{9x} - 3\sqrt{x}$

88. $3\sqrt{x+1} + 10\sqrt{x+1}$

89. $\sqrt{25y} + \sqrt{64y}$

90. $\sqrt[3]{16t^4} - \sqrt[3]{54t^4}$

91. $10\sqrt[3]{z} - \sqrt[3]{z^4}$

92. $5\sqrt[3]{24u^2} + 2\sqrt[3]{81u^5}$

In Exercises 93–98, perform the addition or subtraction and simplify your answer. See Example 8.

93. $\sqrt{5} - \dfrac{3}{\sqrt{5}}$

94. $\sqrt{10} + \dfrac{5}{\sqrt{10}}$

95. $\sqrt{20} - \sqrt{\dfrac{1}{5}}$

96. $\dfrac{x}{\sqrt{3x}} + \sqrt{27x}$

97. $\sqrt{2x} - \dfrac{3}{\sqrt{2x}}$

98. $\sqrt{\dfrac{4}{3x^3}} + \sqrt{3x^3}$

In Exercises 99–102, place the correct inequality symbol (<, >, or =) between the numbers.

99. $\sqrt{7} + \sqrt{18}$ $\sqrt{7+18}$

100. $\sqrt{10} - \sqrt{6}$ $\sqrt{10-6}$

101. 5 $\sqrt{3^2 + 2^2}$

102. 5 $\sqrt{3^2 + 4^2}$

In Exercises 103–106, find the length of the hypotenuse of the right triangle. See Example 9.

103.

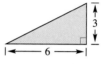

104.

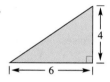

105.

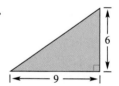

106.

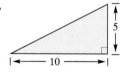

Solving Problems

107. *Geometry*   The foundation of a house is 40 feet long and 30 feet wide. The height of the attic is 5 feet (see figure).

(a) Use the Pythagorean Theorem to find the length of the hypotenuse of the right triangle formed by the roof line.

(b) Use the result of part (a) to determine the total area of the roof.

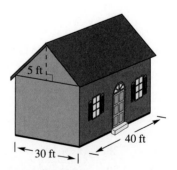

Figure for 107

108. *Geometry*   The four corners are cut from a 4-foot-by-8-foot sheet of plywood, as shown in the figure. Find the perimeter of the remaining piece of plywood.

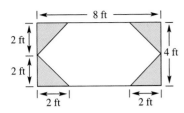

109. *Vibrating String*   The frequency f in cycles per second of a vibrating string is given by

$$f = \frac{1}{100} \sqrt{\frac{400 \times 10^6}{5}}.$$

Use a calculator to approximate this number. (Round the result to two decimal places.)

110. *Period of a Pendulum*   The period T in seconds of a pendulum (see figure) is given by

$$T = 2\pi \sqrt{\frac{L}{32}}$$

where L is the length of the pendulum in feet. Find the period of a pendulum whose length is 4 feet. (Round the result to two decimal places.)

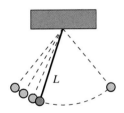

111. *Mathematical Modeling*   The average salary S (in thousands of dollars) of public school teachers for the years 1992 through 1997 in the United States is modeled by

$$S = 34.7 + 1.6t - 2.4\sqrt{t}, \quad 2 \le t \le 7$$

where t is years, with $t = 0$ corresponding to 1990. (Source: Educational Research Service)

(a) Use a graphing utility to graph the model over the specified domain.

(b) Use a graphing utility to predict the year when the average salary will reach \$48,000.

112. *Calculator Experiment*   Enter any positive real number into your calculator and find its square root. Then repeatedly take the square root of the result.

$$\sqrt{x}, \quad \sqrt{\sqrt{x}}, \quad \sqrt{\sqrt{\sqrt{x}}}, \dots$$

What real number does the display appear to be approaching?

113. Square the real number $5/\sqrt{3}$ and note that the radical is eliminated from the denominator. Is this equivalent to rationalizing the denominator? Why or why not?

114. *The Square Root Spiral*   The square root spiral (see figure) is formed by a sequence of right triangles, each with a side whose length is 1. Let r_n be the length of the hypotenuse of the nth triangle.

(a) Each leg of the first triangle has a length of 1 unit. Use the Pythagorean Theorem to show that $r_1 = \sqrt{2}$.

(b) Find r_2, r_3, r_4, r_5, and r_6.

(c) What can you conclude about r_n?

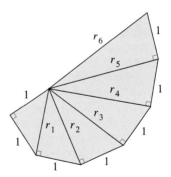

Explaining Concepts

115. Give an example of multiplying two radicals.

116. Describe the three conditions that characterize a simplified radical expression.

117. Is $\sqrt{2} + \sqrt{18}$ in simplest form? Explain.

118. Describe the steps you would use to simplify $1/\sqrt{3}$.

119. For what values of x is $\sqrt{x^2} \ne x$? Explain.

120. Explain what it means for two radical expressions to be alike.

5.3 Multiplying and Dividing Radical Expressions

Objectives

1 Use the Distributive Property or the FOIL Method to multiply radical expressions.

2 Determine the product of conjugates.

3 Simplify a quotient involving radicals by rationalizing the denominator.

1 Use the Distributive Property or the FOIL Method to multiply radical expressions.

Multiplying Radical Expressions

You can multiply radical expressions by using the Distributive Property or the FOIL Method. In both procedures, you also make use of the Multiplication Property of Radicals from Section 5.2,

$$\sqrt[n]{a}\sqrt[n]{b} = \sqrt[n]{ab},$$

where a and b are real numbers whose nth roots are also real numbers.

Example 1   Multiplying Radical Expressions

Find the products and simplify.

a. $\sqrt{6} \cdot \sqrt{3}$ **b.** $\sqrt[3]{5} \cdot \sqrt[3]{16}$

Solution

a. $\sqrt{6} \cdot \sqrt{3} = \sqrt{6 \cdot 3} = \sqrt{18} = \sqrt{9 \cdot 2} = 3\sqrt{2}$

b. $\sqrt[3]{5} \cdot \sqrt[3]{16} = \sqrt[3]{5 \cdot 16} = \sqrt[3]{80} = \sqrt[3]{8 \cdot 10} = 2\sqrt[3]{10}$

Example 2   Multiplying Radical Expressions

Find the products and simplify.

a. $\sqrt{3}\left(2 + \sqrt{5}\right)$ **b.** $\sqrt{2}\left(4 - \sqrt{8}\right)$ **c.** $\sqrt{6}\left(\sqrt{12} - \sqrt{3}\right)$

Solution

a. $\sqrt{3}\left(2 + \sqrt{5}\right) = 2\sqrt{3} + \sqrt{3}\sqrt{5}$ Distributive Property

$\qquad\qquad\qquad = 2\sqrt{3} + \sqrt{15}$ Multiplication Property of Radicals

b. $\sqrt{2}\left(4 - \sqrt{8}\right) = 4\sqrt{2} - \sqrt{2}\sqrt{8}$ Distributive Property

$\qquad\qquad\qquad = 4\sqrt{2} - \sqrt{16}$ Multiplication Property of Radicals

$\qquad\qquad\qquad = 4\sqrt{2} - 4$ Simplify.

c. $\sqrt{6}\left(\sqrt{12} - \sqrt{3}\right) = \sqrt{6}\sqrt{12} - \sqrt{6}\sqrt{3}$ Distributive Property

$\qquad\qquad\qquad = \sqrt{72} - \sqrt{18}$ Multiplication Property of Radicals

$\qquad\qquad\qquad = 6\sqrt{2} - 3\sqrt{2}$ Find perfect square factors.

$\qquad\qquad\qquad = 3\sqrt{2}$ Simplify.

In Example 2, the Distributive Property was used to multiply radical expressions. In Example 3, note how the FOIL Method can be used to multiply binomial radical expressions.

Example 3   Using the FOIL Method

$$\overbrace{}^{F} \quad \overbrace{}^{0} \quad \overbrace{}^{I} \quad \overset{L}{\vert}$$

a. $\left(2\sqrt{7} - 4\right)\left(\sqrt{7} + 1\right) = 2\left(\sqrt{7}\right)^2 + 2\sqrt{7} - 4\sqrt{7} - 4$ FOIL Method

$$= 2(7) + (2 - 4)\sqrt{7} - 4$$ Combine like radicals.

$$= 10 - 2\sqrt{7}$$ Combine like terms.

b. $\left(3 - \sqrt{x}\right)\left(1 + \sqrt{x}\right) = 3 + 3\sqrt{x} - \sqrt{x} - \left(\sqrt{x}\right)^2$

$$= 3 + 2\sqrt{x} - x, \quad x \geq 0$$ Combine like radicals.

Conjugates

2 Determine the product of conjugates.

The expressions $3 + \sqrt{6}$ and $3 - \sqrt{6}$ are called **conjugates** of each other. Notice that they differ only by the sign between the terms. The product of two conjugates is the difference of two squares, which is given by the special product formula $(a + b)(a - b) = a^2 - b^2$. Now, in addition to the FOIL Method, you can also use special product formulas to multiply certain binomial radical expressions.

Example 4   Using a Special Product Formula

Find the conjugate of the expression and multiply the number by its conjugate.

a. $\left(2 - \sqrt{5}\right)$ **b.** $\left(\sqrt{x} - 2\right)$ **c.** $\left(\sqrt{3} + \sqrt{x}\right)$

Solution

a. The conjugate of $\left(2 - \sqrt{5}\right)$ is $\left(2 + \sqrt{5}\right)$.

$$\left(2 - \sqrt{5}\right)\left(2 + \sqrt{5}\right) = 2^2 - \left(\sqrt{5}\right)^2$$ Special product formula

$$= 4 - 5$$

$$= -1$$

b. The conjugate of $\left(\sqrt{x} - 2\right)$ is $\left(\sqrt{x} + 2\right)$.

$$\left(\sqrt{x} - 2\right)\left(\sqrt{x} + 2\right) = \left(\sqrt{x}\right)^2 - (2)^2$$ Special product formula

$$= x - 4, \quad x \geq 0$$

c. The conjugate of $\left(\sqrt{3} + \sqrt{x}\right)$ is $\left(\sqrt{3} - \sqrt{x}\right)$.

$$\left(\sqrt{3} + \sqrt{x}\right)\left(\sqrt{3} - \sqrt{x}\right) = \left(\sqrt{3}\right)^2 - \left(\sqrt{x}\right)^2$$ Special product formula

$$= 3 - x, \quad x \geq 0$$

3 Simplify a quotient involving radicals by rationalizing the denominator.

Dividing Radical Expressions

To simplify a *quotient* involving radicals, we rationalize the denominator. For single-term denominators, you can use the rationalizing process described in Section 5.2. To rationalize a denominator involving two terms, multiply both the numerator and denominator by the *conjugate of the denominator.*

Example 5   Simplifying Quotients Involving Radicals

a. $\dfrac{\sqrt{3}}{1 - \sqrt{5}} = \dfrac{\sqrt{3}}{1 - \sqrt{5}} \cdot \dfrac{1 + \sqrt{5}}{1 + \sqrt{5}}$ Multiply numerator and denominator by conjugate of denominator.

$= \dfrac{\sqrt{3}\left(1 + \sqrt{5}\right)}{1^2 - \left(\sqrt{5}\right)^2}$ Special product formula

$= \dfrac{\sqrt{3} + \sqrt{15}}{1 - 5}$ Simplify.

$= \dfrac{\sqrt{3} + \sqrt{15}}{-4}$ Simplify.

b. $\dfrac{4}{2 - \sqrt{3}} = \dfrac{4}{2 - \sqrt{3}} \cdot \dfrac{2 + \sqrt{3}}{2 + \sqrt{3}}$ Multiply numerator and denominator by conjugate of denominator.

$= \dfrac{4\left(2 + \sqrt{3}\right)}{2^2 - \left(\sqrt{3}\right)^2}$ Special product formula

$= \dfrac{8 + 4\sqrt{3}}{4 - 3}$ Simplify.

$= 8 + 4\sqrt{3}$ Simplify.

Example 6   Simplifying Quotients Involving Radicals

$\dfrac{5\sqrt{2}}{\sqrt{7} + \sqrt{2}} = \dfrac{5\sqrt{2}}{\sqrt{7} + \sqrt{2}} \cdot \dfrac{\sqrt{7} - \sqrt{2}}{\sqrt{7} - \sqrt{2}}$ Multiply numerator and denominator by conjugate of denominator.

$= \dfrac{5\left(\sqrt{14} - \sqrt{4}\right)}{\left(\sqrt{7}\right)^2 - \left(\sqrt{2}\right)^2}$ Special product formula

$= \dfrac{5\left(\sqrt{14} - 2\right)}{7 - 2}$ Simplify.

$= \dfrac{5\left(\sqrt{14} - 2\right)}{5}$ Cancel common factor.

$= \sqrt{14} - 2$ Simplest form

Example 7   Dividing Radical Expressions

Perform the division and simplify.

a. $6 \div \left(\sqrt{x} - 2 \right)$ **b.** $\left(2 - \sqrt{3} \right) \div \left(\sqrt{6} + \sqrt{2} \right)$

Solution

a. $\dfrac{6}{\sqrt{x} - 2} = \dfrac{6}{\sqrt{x} - 2} \cdot \dfrac{\sqrt{x} + 2}{\sqrt{x} + 2}$ Multiply numerator and denominator by conjugate of denominator.

$= \dfrac{6\left(\sqrt{x} + 2 \right)}{\left(\sqrt{x} \right)^2 - 2^2}$ Special product formula

$= \dfrac{6\sqrt{x} + 12}{x - 4}, \quad x \geq 0$ Simplify.

b. $\dfrac{2 - \sqrt{3}}{\sqrt{6} + \sqrt{2}} = \dfrac{2 - \sqrt{3}}{\sqrt{6} + \sqrt{2}} \cdot \dfrac{\sqrt{6} - \sqrt{2}}{\sqrt{6} - \sqrt{2}}$ Multiply numerator and denominator by conjugate of denominator.

$= \dfrac{2\sqrt{6} - 2\sqrt{2} - \sqrt{18} + \sqrt{6}}{\left(\sqrt{6} \right)^2 - \left(\sqrt{2} \right)^2}$ Special product formula

$= \dfrac{3\sqrt{6} - 2\sqrt{2} - 3\sqrt{2}}{6 - 2}$ Simplify.

$= \dfrac{3\sqrt{6} - 5\sqrt{2}}{4}$ Simplify.

Discussing the Concept   The Golden Section

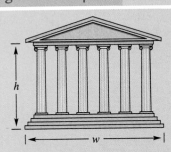

The ratio of the width of the Temple of Hephaestus to its height is approximately

$$\frac{w}{h} \approx \frac{2}{\sqrt{5} - 1}.$$

This number is called the **golden section.** Early Greeks believed that the most aesthetically pleasing rectangles were those whose sides had this ratio.

a. Rationalize the denominator for this number. Approximate your answer, rounded to two decimal places.

b. Use the Pythagorean Theorem, a straightedge, and a compass to construct a rectangle whose sides have the golden section as their ratio.

5.3   Exercises

Integrated Review   Concepts, Skills, and Problem Solving

Keep mathematically in shape by doing these exercises *before* the problems of this section.

Properties and Definitions

In Exercises 1–4, use $x^2 + bx + c = (x + m)(x + n)$.

1. $mn =$

2. If $c > 0$, then what must be true about the signs of m and n?

3. If $c < 0$, then what must be true about the signs of m and n?

4. If m and n have like signs, then $m + n =$ _____ .

Equations of Lines

In Exercises 5–10, find an equation of the line through the two points.

5. $(-1, -2), (3, 6)$ **6.** $(1, 5), (6, 0)$

7. $(6, 3), (10, 3)$

8. $(4, -2), (4, 5)$

9. $(\frac{4}{3}, 8), (5, 6)$

10. $(7, 4), (10, 1)$

Models

In Exercises 11 and 12, translate the phrase into an algebraic expression.

11. The time to travel 360 miles if the average speed is r miles per hour

12. The perimeter of a rectangle of length L and width $L/3$

Developing Skills

In Exercises 1–42, multiply and simplify. See Examples 1–3.

1. $\sqrt{2} \cdot \sqrt{8}$

2. $\sqrt{6} \cdot \sqrt{18}$

3. $\sqrt{3} \cdot \sqrt{6}$

4. $\sqrt{5} \cdot \sqrt{10}$

5. $\sqrt[3]{12} \cdot \sqrt[3]{6}$

6. $\sqrt[3]{9} \cdot \sqrt[3]{9}$

7. $\sqrt[4]{8} \cdot \sqrt[4]{6}$

8. $\sqrt[4]{54} \cdot \sqrt[4]{3}$

9. $\sqrt{5}(2 - \sqrt{3})$

10. $\sqrt{11}(\sqrt{5} - 3)$

11. $\sqrt{2}(\sqrt{20} + 8)$

12. $\sqrt{7}(\sqrt{14} + 3)$

13. $\sqrt{6}(\sqrt{12} - \sqrt{3})$

14. $\sqrt{10}(\sqrt{5} + \sqrt{6})$

15. $\sqrt{2}(\sqrt{18} - \sqrt{10})$

16. $\sqrt{5}(\sqrt{15} + \sqrt{5})$

17. $\sqrt{y}(\sqrt{y} + 4)$

18. $\sqrt{x}(5 - \sqrt{x})$

19. $\sqrt{a}(4 - \sqrt{a})$

20. $\sqrt{z}(\sqrt{z} + 5)$

21. $\sqrt[3]{4}(\sqrt[3]{2} - 7)$

22. $\sqrt[3]{9}(\sqrt[3]{3} + 2)$

23. $(\sqrt{3} + 2)(\sqrt{3} - 2)$

24. $(3 - \sqrt{5})(3 + \sqrt{5})$

25. $(\sqrt{5} + 3)(\sqrt{3} - 5)$

26. $(\sqrt{7} + 6)(\sqrt{2} + 6)$

27. $(\sqrt{20} + 2)^2$

28. $(4 - \sqrt{20})^2$

29. $(\sqrt[3]{6} - 3)(\sqrt[3]{4} + 3)$

30. $(\sqrt[3]{9} + 5)(\sqrt[3]{5} - 5)$

31. $(10 + \sqrt{2x})^2$

32. $(5 - \sqrt{3v})^2$

33. $(9\sqrt{x} + 2)(5\sqrt{x} - 3)$

34. $(16\sqrt{u} - 3)(\sqrt{u} - 1)$

35. $(3\sqrt{x} - 5)(3\sqrt{x} + 5)$

36. $(7 - 3\sqrt{3t})(7 + 3\sqrt{3t})$

37. $(\sqrt[3]{2x} + 5)^2$

38. $(\sqrt[3]{3x} - 4)^2$

39. $(\sqrt[3]{y} + 2)(\sqrt[3]{y^2} - 5)$

40. $(\sqrt[3]{2y} + 10)(\sqrt[3]{4y^2} - 10)$

41. $(\sqrt[3]{t} + 1)(\sqrt[3]{t^2} + 4\sqrt[3]{t} - 3)$

42. $(\sqrt{x} - 2)(\sqrt{x^3} - 2\sqrt{x^2} + 1)$

In Exercises 43–48, complete the statement.

43. $5x\sqrt{3} + 15\sqrt{3} = 5\sqrt{3}()$

44. $x\sqrt{7} - x^2\sqrt{7} = x\sqrt{7}()$

45. $4\sqrt{12} - 2x\sqrt{27} = 2\sqrt{3}()$

46. $5\sqrt{50} + 10y\sqrt{8} = 5\sqrt{2}()$

47. $6u^2 + \sqrt{18u^3} = 3u()$

48. $12s^3 - \sqrt{32s^4} = 4s^2()$

In Exercises 49–62, find the conjugate of the expression. Then multiply the expression by its conjugate. See Example 4.

49. $2 + \sqrt{5}$

50. $\sqrt{2} - 9$

51. $\sqrt{11} - \sqrt{3}$

52. $\sqrt{10} + \sqrt{7}$

53. $\sqrt{15} + 3$

54. $\sqrt{11} + 3$

55. $\sqrt{x} - 3$

56. $\sqrt{t} + 7$

57. $\sqrt{2u} - \sqrt{3}$

58. $\sqrt{5a} + \sqrt{2}$

59. $2\sqrt{2} + \sqrt{4}$

60. $4\sqrt{3} + \sqrt{2}$

61. $\sqrt{x} + \sqrt{y}$

62. $3\sqrt{u} + \sqrt{3v}$

In Exercises 63–66, simplify the expression.

63. $\dfrac{4 - 8\sqrt{x}}{12}$

64. $\dfrac{-3 + 27\sqrt{2y}}{18}$

65. $\dfrac{-2y + \sqrt{12y^3}}{8y}$

66. $\dfrac{-t^2 - \sqrt{2t^3}}{3t}$

In Exercises 67–70, evaluate the function.

67. $f(x) = x^2 - 6x + 1$
   (a) $f(2 - \sqrt{3})$
   (b) $f(3 - 2\sqrt{2})$

68. $g(x) = x^2 + 8x + 11$
   (a) $g(-4 + \sqrt{5})$
   (b) $g(-4\sqrt{2})$

69. $f(x) = x^2 - 2x - 1$
   (a) $f(1 + \sqrt{2})$
   (b) $f(\sqrt{4})$

70. $g(x) = x^2 - 4x + 1$
   (a) $g(1 + \sqrt{5})$
   (b) $g(2 - \sqrt{3})$

In Exercises 71–94, rationalize the denominator of the expression and simplify. See Examples 5–7.

71. $\dfrac{6}{\sqrt{2} - 2}$

72. $\dfrac{8}{\sqrt{7} + 3}$

73. $\dfrac{7}{\sqrt{3} + 5}$

74. $\dfrac{5}{9 - \sqrt{6}}$

75. $\dfrac{3}{2\sqrt{10} - 5}$

76. $\dfrac{4}{3\sqrt{5} - 1}$

77. $\dfrac{2}{\sqrt{6} + \sqrt{2}}$

78. $\dfrac{10}{\sqrt{9} + \sqrt{5}}$

79. $\dfrac{9}{\sqrt{3} - \sqrt{7}}$

80. $\dfrac{12}{\sqrt{5} + \sqrt{8}}$

81. $(\sqrt{7} + 2) \div (\sqrt{7} - 2)$

82. $(5 - \sqrt{3}) \div (3 + \sqrt{3})$

83. $(\sqrt{x} - 5) \div (2\sqrt{x} - 1)$

84. $(2\sqrt{t} + 1) \div (2\sqrt{t} - 1)$

85. $\dfrac{3x}{\sqrt{15} - \sqrt{3}}$

86. $\dfrac{5y}{\sqrt{12} + \sqrt{10}}$

87. $\dfrac{2t^2}{\sqrt{5} - \sqrt{t}}$

88. $\dfrac{5x}{\sqrt{x} - \sqrt{2}}$

89. $\dfrac{8a}{\sqrt{3a} + \sqrt{a}}$

90. $\dfrac{7z}{\sqrt{5z} - \sqrt{z}}$

91. $\dfrac{3(x - 4)}{x^2 - \sqrt{x}}$

92. $\dfrac{6(y + 1)}{y^2 + \sqrt{y}}$

93. $\dfrac{\sqrt{u} + v}{\sqrt{u - v} - \sqrt{u}}$

94. $\dfrac{z}{\sqrt{u + z} - \sqrt{u}}$

In Exercises 95–98, use a graphing utility to graph the functions on the same screen. Use the graphs to verify that the functions are equivalent. Verify your results algebraically.

95. $y_1 = \dfrac{10}{\sqrt{x} + 1}$

$y_2 = \dfrac{10(\sqrt{x} - 1)}{x - 1}$

96. $y_1 = \dfrac{4x}{\sqrt{x} + 4}$

$y_2 = \dfrac{4x(\sqrt{x} - 4)}{x - 16}$

97. $y_1 = \dfrac{2\sqrt{x}}{2 - \sqrt{x}}$

$y_2 = \dfrac{2(2\sqrt{x} + x)}{4 - x}$

98. $y_1 = \dfrac{\sqrt{2x} + 6}{\sqrt{2x} - 2}$

$y_2 = \dfrac{x + 6 + 4\sqrt{2x}}{x - 2}$

Rationalizing Numerators   In the study of calculus, students sometimes rewrite an expression by rationalizing the numerator. (*Note:* The results will not be in simplest radical form.) In Exercises 99–102, rationalize the numerator.

99. $\dfrac{\sqrt{2}}{7}$

100. $\dfrac{\sqrt{10}}{5}$

101. $\dfrac{\sqrt{7} + \sqrt{3}}{5}$

102. $\dfrac{\sqrt{x} + 6}{\sqrt{2}}$

Solving Problems

103. *Strength of a Wooden Beam* The rectangular cross section of a wooden beam cut from a log of diameter 24 inches (see figure) will have maximum strength if its width w and height h are given by

$$w = 8\sqrt{3} \quad \text{and} \quad h = \sqrt{24^2 - \left(8\sqrt{3}\right)^2}.$$

Find the area of the rectangular cross section and express the area in simplest form.

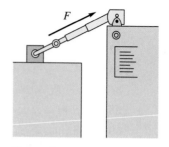

Figure for 105

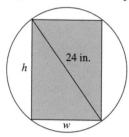

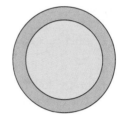

Figure for 103   Figure for 104

104. *Geometry* The areas of the circles in the figure are 15 square centimeters and 20 square centimeters. Find the ratio of the radius of the small circle to the radius of the large circle.

105. *Force to Move a Block* The force required to slide a steel block weighing 500 pounds across a milling machine is

$$\frac{500k}{\dfrac{1}{\sqrt{k^2+1}} + \dfrac{k^2}{\sqrt{k^2+1}}}$$

where k is the friction constant (see figure). Simplify this expression.

106. *Wind Chill* The term "wind chill" goes back to the Antarctic explorer Paul A. Siple, who coined it in a 1939 dissertation, *Adaptation of the Explorer to the Climate of Antarctica.* During the 1940s, Siple and Charles F. Passel conducted experiments on the time needed to freeze water in a plastic cylinder that was exposed to the elements. They found that the time depends on the initial temperature of the water, the outside temperature, and the wind speed. The National Weather Service uses the following formula for determining "wind chill."

$$T_{wc} = 0.0817\left(3.71\sqrt{v} + 5.81 - 0.25v\right)(T - 91.4) + 91.4$$

where T is the air temperature in degrees Fahrenheit, T_{wc} is the wind chill, and v is the wind speed in miles per hour. Use the formula to determine the wind chill for each combination of air temperature and wind speed given in the table.

v \ T	0°	5°	10°	15°	20°	25°
10 mi/hr						
20 mi/hr						
30 mi/hr						
40 mi/hr						

Explaining Concepts

107. Multiply $\sqrt{3}\left(1 - \sqrt{6}\right)$. State an algebraic property to justify each step.

108. Describe the differences and similarities of using the FOIL Method with polynomial expressions and with radical expressions.

109. Multiply $3 - \sqrt{2}$ by its conjugate. Explain why the result has no radicals.

110. Is the number $3/\left(1 + \sqrt{5}\right)$ in simplest form? If not, explain the steps for writing it in simplest form.

Mid-Chapter Quiz

Take this quiz as you would take a quiz in class. After you are done, check your work against the answers given in the back of the book.

In Exercises 1–4, evaluate the expression.

1. $\sqrt{225}$ **2.** $\sqrt[4]{\frac{81}{16}}$ **3.** $64^{1/2}$ **4.** $(-27)^{2/3}$

In Exercises 5–8, simplify the expression.

5. $\sqrt{27x^2}$ **6.** $\sqrt[4]{81x^6}$ **7.** $\sqrt{\frac{4u^3}{9}}$ **8.** $\sqrt[3]{\frac{16}{u^6}}$

In Exercises 9 and 10, combine the radical expressions, if possible.

9. $\sqrt{200y} - 3\sqrt{8y}$ **10.** $6x\sqrt[3]{5x^2} + 2\sqrt[3]{40x^4}$

In Exercises 11–18, simplify the radical expression.

11. $\sqrt{8}\left(3 + \sqrt{32}\right)$

12. $\left(\sqrt{50} - 4\right)\sqrt{2}$

13. $\left(\sqrt{6} + 3\right)\left(4\sqrt{6} - 7\right)$

14. $\left(9 + 2\sqrt{3}\right)\left(2 + 7\sqrt{3}\right)$

15. $\dfrac{\sqrt{7}}{1 + \sqrt{3}}$

16. $\dfrac{6\sqrt{2}}{2\sqrt{2} - 4}$

17. $4 \div \left(\sqrt{6} + 3\right)$

18. $\left(4\sqrt{2} - 2\sqrt{3}\right) \div \left(\sqrt{2} + \sqrt{6}\right)$

In Exercises 19 and 20, write the conjugate of the number. Find the product of the number and its conjugate.

19. $1 + \sqrt{4}$ **20.** $\sqrt{10} - 5$

21. Explain why $\sqrt{5^2 + 12^2} \neq 17$. Determine the correct value of the radical.

22. The four corners are cut from an $8\frac{1}{2}$-inch-by-11-inch sheet of paper, as shown in the figure at the left. Find the perimeter of the remaining piece of paper.

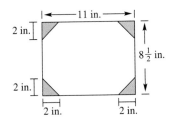

Figure for 22

(figure labels: 11 in., 2 in., $8\frac{1}{2}$ in., 2 in., 2 in., 2 in.)

5.4   Solving Radical Equations

Objectives

1 Solve a radical equation by raising both sides to the *n*th power.

2 Solve an application problem involving a radical equation.

1 Solve a radical equation by raising both sides to the *n*th power.

Solving Radical Equations

Solving equations involving radicals is somewhat like solving equations that contain fractions—you try to get rid of the radicals and obtain a polynomial equation. Then you solve the polynomial equation using the standard procedures. The following property plays a key role.

> ▶ **Raising Both Sides of an Equation to the *n*th Power**
>
> Let *u* and *v* be real numbers, variables, or algebraic expressions, and let *n* be a positive integer. If $u = v$, then it follows that
>
> $$u^n = v^n.$$
>
> This is called **raising both sides of an equation to the *n*th power.**

To use this property to solve an equation, first try to isolate one of the radicals on one side of the equation.

Technology: Tip

To use a graphing utility to check the solution in Example 1, sketch the graph of

$$y = \sqrt{x} - 8$$

as shown below. Notice that the graph crosses the *x*-axis when $x = 64$, which confirms the solution that was obtained algebraically.

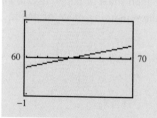

Example 1   Solving an Equation Having One Radical

Solve $\sqrt{x} - 8 = 0$.

Solution

$\sqrt{x} - 8 = 0$	Original equation
$\sqrt{x} = 8$	Isolate radical.
$(\sqrt{x})^2 = 8^2$	Square both sides.
$x = 64$	Simplify.

Check

$\sqrt{x} - 8 = 0$	Original equation
$\sqrt{64} - 8 \stackrel{?}{=} 0$	Substitute 64 for *x*.
$8 - 8 = 0$	Solution checks. ✓

So, the equation has one solution: $x = 64$.

Example 2   Solving an Equation Having One Radical

Solve $\sqrt{3x} + 6 = 0$.

Solution

$\sqrt{3x} + 6 = 0$	Original equation
$\sqrt{3x} = -6$	Isolate radical.
$\left(\sqrt{3x}\right)^2 = (-6)^2$	Square both sides.
$3x = 36$	Simplify.
$x = 12$	Divide both sides by 3.

Check

$\sqrt{3x} + 6 = 0$	Original equation
$\sqrt{3(12)} + 6 \overset{?}{=} 0$	Substitute 12 for x.
$6 + 6 \neq 0$	Solution does not check. ✗

The solution $x = 12$ is extraneous. So, the equation has no solution. You can also check this graphically, as shown in Figure 5.3.

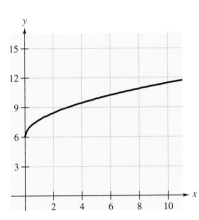

Figure 5.3

As you can see from Example 2, checking solutions of a radical equation is especially important because raising both sides of an equation to the *n*th power to remove the radical often introduces *extraneous* solutions.

Example 3   Solving an Equation Having One Radical

Solve $\sqrt[3]{2x + 1} - 2 = 3$.

Solution

$\sqrt[3]{2x + 1} - 2 = 3$	Original equation
$\sqrt[3]{2x + 1} = 5$	Isolate radical.
$\left(\sqrt[3]{2x + 1}\right)^3 = 5^3$	Cube both sides.
$2x + 1 = 125$	Simplify.
$2x = 124$	Subtract 1 from both sides.
$x = 62$	Divide both sides by 2.

Check

$\sqrt[3]{2x + 1} - 2 = 3$	Original equation
$\sqrt[3]{2(62) + 1} - 2 \overset{?}{=} 3$	Substitute 62 for x.
$\sqrt[3]{125} - 2 \overset{?}{=} 3$	Simplify.
$5 - 2 = 3$	Solution checks. ✓

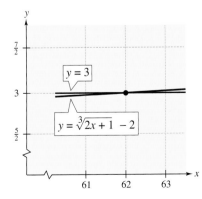

Figure 5.4

So, the equation has one solution: $x = 62$. You can also check the solution graphically, as shown in Figure 5.4.

Technology: Tip

In Example 4, you can graphically check the solution of the equation by "graphing the left side and right side on the same screen." That is, by graphing the equations

$$y = \sqrt{5x + 3}$$

and

$$y = \sqrt{x + 11}$$

on the same screen, as shown below, you can see that the two graphs intersect when $x = 2$.

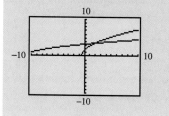

Example 4 Solving an Equation Having Two Radicals

Solve $\sqrt{5x + 3} = \sqrt{x + 11}$.

Solution

$$\sqrt{5x + 3} = \sqrt{x + 11} \qquad \text{Original equation}$$

$$\left(\sqrt{5x + 3}\right)^2 = \left(\sqrt{x + 11}\right)^2 \qquad \text{Square both sides.}$$

$$5x + 3 = x + 11 \qquad \text{Simplify.}$$

$$5x = x + 8 \qquad \text{Subtract 3 from both sides.}$$

$$4x = 8 \qquad \text{Subtract } x \text{ from both sides.}$$

$$x = 2 \qquad \text{Divide both sides by 4.}$$

Check

$$\sqrt{5x + 3} = \sqrt{x + 11} \qquad \text{Original equation}$$

$$\sqrt{5(2) + 3} \overset{?}{=} \sqrt{2 + 11} \qquad \text{Substitute 2 for } x.$$

$$\sqrt{13} = \sqrt{13} \qquad \text{Solution checks. } \checkmark$$

So, the equation has one solution: $x = 2$.

Example 5 Solving an Equation Having Two Radicals

Solve $\sqrt[4]{3x} + \sqrt[4]{2x - 5} = 0$.

Solution

$$\sqrt[4]{3x} + \sqrt[4]{2x - 5} = 0 \qquad \text{Original equation}$$

$$\sqrt[4]{3x} = -\sqrt[4]{2x - 5} \qquad \text{Isolate radicals.}$$

$$\left(\sqrt[4]{3x}\right)^4 = \left(-\sqrt[4]{2x - 5}\right)^4 \qquad \text{Raise both sides to 4th power.}$$

$$3x = 2x - 5 \qquad \text{Simplify.}$$

$$x = -5 \qquad \text{Subtract } 2x \text{ from both sides.}$$

Check

$$\sqrt[4]{3x} + \sqrt[4]{2x - 5} = 0 \qquad \text{Original equation}$$

$$\sqrt[4]{3(-5)} + \sqrt[4]{2(-5) - 5} \overset{?}{=} 0 \qquad \text{Substitute } -5 \text{ for } x.$$

$$\sqrt[4]{-15} + \sqrt[4]{-15} \neq 0 \qquad \text{Solution does not check. } \times$$

The solution does not check because it yields fourth roots of negative radicands. So, this equation has no solution. Try checking this graphically. If you graph both sides of the equation, you will discover that they do not intersect.

In the next example you will see that squaring both sides results in a quadratic equation. Remember that you must check the solutions in the *original* radical equation.

Example 6   An Equation That Converts to a Quadratic Equation

Solve $\sqrt{x} + 2 = x$.

Solution

$\sqrt{x} + 2 = x$	Original equation
$\sqrt{x} = x - 2$	Isolate radical.
$\left(\sqrt{x}\right)^2 = (x - 2)^2$	Square both sides.
$x = x^2 - 4x + 4$	Simplify.
$-x^2 + 5x - 4 = 0$	General form
$(-1)(x - 4)(x - 1) = 0$	Factor.
$x - 4 = 0 \implies x = 4$	Set 1st factor equal to 0.
$x - 1 = 0 \implies x = 1$	Set 2nd factor equal to 0.

Check

$\sqrt{x} + 2 = x$	Original equation	$\sqrt{x} + 2 = x$	Original equation
$\sqrt{4} + 2 \overset{?}{=} 4$	Substitute 4 for x.	$\sqrt{1} + 2 \overset{?}{=} 1$	Substitute 1 for x.
$2 + 2 \overset{?}{=} 4$	Simplify.	$1 + 2 \overset{?}{=} 1$	Simplify.
$4 = 4$	Solution checks. ✓	$3 = 1$	Solution does not check. ✗

From the check you can see that $x = 1$ is an extraneous solution. So, the only solution to the equation is $x = 4$.

When an equation contains two radicals, it may not be possible to isolate both. In such cases, you may have to raise both sides of the equation to a power at *two* different stages in the solution.

Example 7   Repeatedly Squaring Both Sides of an Equation

Solve $\sqrt{3t + 1} = 2 - \sqrt{3t}$.

Solution

$\sqrt{3t + 1} = 2 - \sqrt{3t}$	Original equation
$\left(\sqrt{3t + 1}\right)^2 = \left(2 - \sqrt{3t}\right)^2$	Square both sides (1st time).
$3t + 1 = 4 - 4\sqrt{3t} + 3t$	Simplify.
$-3 = -4\sqrt{3t}$	Isolate radical.
$(-3)^2 = \left(-4\sqrt{3t}\right)^2$	Square both sides (2nd time).
$9 = 16(3t)$	Simplify.
$\dfrac{3}{16} = t$	Divide both sides by 48.

The solution is $t = \frac{3}{16}$. Check this in the original equation.

2 Solve an application problem
involving a radical equation.

Applications

Example 8   An Application Involving Electricity

The amount of power consumed by an electrical appliance is given by

$$I = \sqrt{\frac{P}{R}}$$

where I is the current measured in amps, R is the resistance measured in ohms, and P is the power measured in watts. Find the power used by an electric heater for which $I = 10$ amps and $R = 16$ ohms.

Solution

$$I = \sqrt{\frac{P}{R}}$$ Original equation

$$10 = \sqrt{\frac{P}{16}}$$ Substitute for I and R.

$$10^2 = \left(\sqrt{\frac{P}{16}}\right)^2$$ Square both sides.

$$100 = \frac{P}{16}$$ Simplify.

$$1600 = P$$ Multiply both sides by 16.

Check

$$10 \overset{?}{=} \sqrt{\frac{1600}{16}}$$ Substitute 10 for I, 16 for R, and 1600 for P in the original equation.

$$10 \overset{?}{=} \sqrt{100}$$ Simplify.

$$10 = 10$$ Solution checks. ✓

So, the solution is $P = 1600$ watts.

An alternative way to solve the problem in Example 8 would be first to solve the equation for P.

$$I = \sqrt{\frac{P}{R}}$$ Original equation

$$I^2 = \left(\sqrt{\frac{P}{R}}\right)^2$$ Square both sides.

$$I^2 = \frac{P}{R}$$ Simplify.

$$I^2R = P$$ Multiply both sides by R.

At this stage, you can substitute the known values of I and R to obtain

$$P = (10)^2 16 = 1600.$$

| Example 9 | The Velocity of a Falling Object | |

The velocity of a free-falling object can be determined from the equation

$$v = \sqrt{2gh}$$

where v is the velocity measured in feet per second, $g = 32$ feet per second per second, and h is the distance (in feet) the object has fallen. Find the height from which a rock has been dropped if it strikes the ground with a velocity of 50 feet per second.

Solution

$v = \sqrt{2gh}$	Original equation
$50 = \sqrt{2(32)h}$	Substitute for v and g.
$50^2 = \left(\sqrt{64h}\right)^2$	Square both sides.
$2500 = 64h$	Simplify.
$39 \approx h$	Divide both sides by 64.

Check

Because the value of h was rounded in the solution, the check will not result in an equality. The expressions on each side of the equal sign will be approximately equal to each other.

$v = \sqrt{2gh}$	Original equation
$50 \stackrel{?}{\approx} \sqrt{2(32)(39)}$	Substitute 50 for v, 32 for g, and 39 for h.
$50 \stackrel{?}{\approx} \sqrt{2496}$	Simplify.
$50 \approx 49.96$	Solution checks. ✓

So, the height from which the rock has been dropped is approximately 39 feet.

Discussing the Concept   An Experiment

Without using a stopwatch, you can find the length of time an object has been falling by using the following equation from physics

$$t = \sqrt{\frac{h}{384}}$$

where t is the length of time in seconds and h is the distance in inches the object has fallen. How far does an object fall in 0.25 second? in 0.10 second?

Use this equation to test how long it takes members of your group to catch a falling ruler. Hold the ruler vertically while another student holds his or her hands near the lower end of the ruler ready to catch it. Before releasing the ruler, record the mark on the ruler closest to the top of the catcher's hands. Release the ruler. After it has been caught, again note the mark closest to the top of the catcher's hands. (The difference between these two measurements is h.) Which member of your class reacts most quickly?

5.4   Exercises

Integrated Review   *Concepts, Skills, and Problem Solving*

Keep mathematically in shape by doing these exercises *before* the problems of this section.

Properties and Definitions

1. Explain how you determine the domain of the function

$$f(x) = \frac{4}{(x + 2)(x - 3)}.$$

2. Explain the excluded value $(x \neq -3)$ in the following equation.

$$\frac{2x^2 + 5x - 3}{x^2 - 9} = \frac{2x - 1}{x - 3}, \quad x \neq 3, \quad x \neq -3$$

Simplifying Expressions

In Exercises 3–6, simplify the expression. (Assume that no variable is equal to 0.)

3. $(-3x^2y^3)^2 \cdot (4xy^2)$

4. $(x^2 - 3xy)^0$

5. $\dfrac{64r^2s^4}{16rs^2}$

6. $\left(\dfrac{3x}{4y^3}\right)^2$

In Exercises 7–10, perform the operation and simplify.

7. $\dfrac{x + 13}{x^3(3 - x)} \cdot \dfrac{x(x - 3)}{5}$

8. $\dfrac{x + 2}{5x + 15} \cdot \dfrac{x - 2}{5(x - 3)}$

9. $\dfrac{2x}{x - 5} - \dfrac{5}{5 - x}$

10. $\dfrac{3}{x - 1} - 5$

Graphs

In Exercises 11 and 12, graph the function and identify any intercepts.

11. $y = 2x - 3$

12. $y = -\frac{3}{4}x + 2$

Developing Skills

In Exercises 1–4, determine whether the values of x are solutions of the radical equation.

Equation	*Values of x*

1. $\sqrt{x} - 10 = 0$ (a) $x = -4$ (b) $x = -100$
  (c) $x = \sqrt{10}$ (d) $x = 100$

2. $\sqrt{3x} - 6 = 0$ (a) $x = \frac{2}{3}$ (b) $x = 2$
  (c) $x = 12$ (d) $x = -\frac{1}{3}\sqrt{6}$

3. $\sqrt[3]{x - 4} = 4$ (a) $x = -60$ (b) $x = 68$
  (c) $x = 20$ (d) $x = 0$

4. $\sqrt[4]{2x} + 2 = 6$ (a) $x = 128$ (b) $x = 2$
  (c) $x = -2$ (d) $x = 0$

In Exercises 5–52, solve the equation and check your solution(s) in the original equation. (Some of the equations have no solution.) See Examples 1–7.

5. $\sqrt{x} = 20$

6. $\sqrt{x} = 5$

7. $\sqrt{x} = 3$

8. $\sqrt{t} = 4$

9. $\sqrt[3]{z} = 3$

10. $\sqrt[4]{x} = 2$

11. $\sqrt{y} - 7 = 0$

12. $\sqrt{t} - 13 = 0$

13. $\sqrt{u} + 13 = 0$

14. $\sqrt{y} + 15 = 0$

15. $\sqrt{x} - 8 = 0$

16. $\sqrt{x} - 10 = 0$

17. $\sqrt{10x} = 30$

18. $\sqrt{8x} = 6$

19. $\sqrt{-3x} = 9$

20. $\sqrt{-4y} = 4$

21. $\sqrt{5t} - 2 = 0$

22. $6 - \sqrt{8x} = 0$

23. $\sqrt{3y + 1} = 4$

24. $\sqrt{3 - 2x} = 2$

25. $\sqrt{4 - 5x} = -3$

26. $\sqrt{2t - 7} = -5$

27. $\sqrt{3y + 5} - 3 = 4$

28. $\sqrt{5z - 2} + 7 = 10$

29. $5\sqrt{x + 2} = 8$

30. $2\sqrt{x + 4} = 7$

31. $\sqrt{3x + 2} + 5 = 0$

32. $\sqrt{1 - x} + 10 = 4$

33. $\sqrt{x + 3} = \sqrt{2x - 1}$

34. $\sqrt{3t + 1} = \sqrt{t + 15}$

35. $\sqrt{3y - 5} - 3\sqrt{y} = 0$

36. $\sqrt{2u + 10} - 2\sqrt{u} = 0$

37. $\sqrt[3]{3x - 4} = \sqrt[3]{x + 10}$

38. $2\sqrt[3]{10 - 3x} = \sqrt[3]{2 - x}$

39. $\sqrt[3]{2x + 15} - \sqrt[3]{x} = 0$

40. $\sqrt[4]{2x} + \sqrt[4]{x + 3} = 0$

41. $\sqrt{x^2 + 5} = x + 3$ **42.** $\sqrt{x^2 - 4} = x - 2$

43. $\sqrt{2x} = x - 4$ **44.** $\sqrt{x} = x - 6$

45. $\sqrt{8x + 1} = x + 2$ **46.** $\sqrt{3x + 7} = x + 3$

47. $\sqrt{z + 2} = 1 + \sqrt{z}$

48. $\sqrt{2x + 5} = 7 - \sqrt{2x}$

49. $\sqrt{2t + 3} = 3 - \sqrt{2t}$

50. $\sqrt{x} + \sqrt{x + 2} = 2$

51. $\sqrt{x + 5} - \sqrt{x} = 1$

52. $\sqrt{x + 3} - \sqrt{x - 1} = 1$

In Exercises 53–60, solve the equation and check your solution(s) in the original equation.

53. $t^{3/2} = 8$ **54.** $v^{2/3} = 25$

55. $3y^{1/3} = 18$ **56.** $2x^{3/4} = 54$

57. $(x + 4)^{2/3} = 4$ **58.** $(u - 2)^{4/3} = 81$

59. $(2x + 5)^{1/3} + 3 = 0$ **60.** $(x - 6)^{3/2} - 27 = 0$

In Exercises 61–70, use a graphing utility to graph each side of the equation on the same screen. Use the graphs to approximate the solution(s).

61. $\sqrt{x} = 2(2 - x)$ **62.** $\sqrt{2x + 3} = 4x - 3$

63. $\sqrt{x^2 + 1} = 5 - 2x$ **64.** $\sqrt{8 - 3x} = x$

65. $\sqrt{x + 3} = 5 - \sqrt{x}$ **66.** $\sqrt[3]{5x - 8} = 4 - \sqrt[3]{x}$

67. $4\sqrt[3]{x} = 7 - x$ **68.** $\sqrt[3]{x + 4} = \sqrt{6 - x}$

69. $\sqrt{15 - 4x} = 2x$ **70.** $\dfrac{4}{\sqrt{x}} = 3\sqrt{x} - 4$

In Exercises 71–76, match the function with its graph. [The graphs are labeled (a), (b), (c), (d), (e), and (f).]

(a)
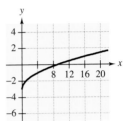

(b)
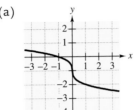

(c)
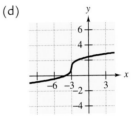

(d)
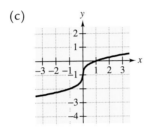

(e)
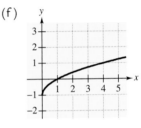

(f)
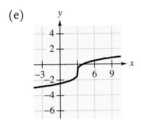

71. $f(x) = \sqrt[3]{x} - 1$ **72.** $f(x) = \sqrt[3]{x - 3} - 1$

73. $f(x) = \sqrt[3]{x + 3} + 1$ **74.** $f(x) = -\sqrt[3]{x} - 1$

75. $f(x) = \sqrt{x} - 1$ **76.** $f(x) = \sqrt{x} - 3$

Solving Problems

Geometry   In Exercises 77–80, find the length x of the unknown side of the right triangle. (Round your answer to two decimal places.)

77.
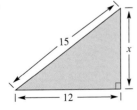

78.
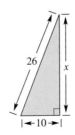

79.
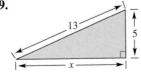

80.
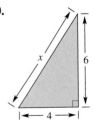

81. *Drawing a Diagram*   The screen of a computer monitor has a diagonal of 13.75 inches and a width of 8.25 inches. Draw a diagram of the computer monitor and find the length of the screen.

82. *Drawing a Diagram*   A basketball court is 50 feet wide and 94 feet long. Draw a diagram and find the length of the diagonal of the court.

83. *Geometry*   A house has a basement floor with dimensions 26 feet by 32 feet. The gas hot water heater and furnace are diagonally across the basement from where the natural gas line enters the house. Find the length of the gas line across the basement.

84. *Geometry*   A guy wire on a 100-foot-tall radio tower is attached to the top of the tower and to an anchor 50 feet from the base of the tower. Determine the length of the guy wire.

85. *Geometry*   A ladder is 17 feet long and the bottom of the ladder is 8 feet from the wall of a house. Determine the height at which the top of the ladder rests against the wall.

86. *Geometry*   A 10-foot plank is used to brace a basement wall during construction of a home. The plank is nailed to the wall 6 feet above the floor. Find the slope of the plank.

87. *Geometry*   Determine the length and width of a rectangle with a perimeter of 92 inches and a diagonal of 34 inches.

88. *Geometry*   Determine the length and width of a rectangle with a perimeter of 68 inches and a diagonal of 26 inches.

89. *Geometry*   The surface area of a cone is given by

$$S = \pi r \sqrt{r^2 + h^2}$$

as shown in the figure. Solve the equation for h.

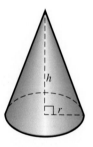

90. *Geometry*   Write a function that gives the radius of a circle in terms of the circle's area A. Use a graphing utility to graph this function.

Height of an Object   In Exercises 91 and 92, use the formula $t = \sqrt{d/16}$, which gives the time t in seconds for a free-falling object to fall d feet.

91. A construction worker drops a nail and observes it strike a water puddle after approximately 2 seconds. Estimate the height from which the nail was dropped.

92. A construction worker drops a nail and observes it strike a water puddle after approximately 3 seconds. Estimate the height from which the nail was dropped.

Free-Falling Object   In Exercises 93–96, use the equation for the velocity of a free-falling object ($v = \sqrt{2gh}$), as described in Example 9.

93. An object is dropped from a height of 50 feet. Find the velocity of the object when it strikes the ground.

94. An object is dropped from a height of 200 feet. Find the velocity of the object when it strikes the ground.

95. An object that was dropped strikes the ground with a velocity of 60 feet per second. Find the height from which the object was dropped.

96. An object that was dropped strikes the ground with a velocity of 120 feet per second. Find the height from which the object was dropped.

Length of a Pendulum In Exercises 97 and 98, use the equation of the time t in seconds for a pendulum of length L feet to go through one complete cycle (its period). The equation is $t = 2\pi\sqrt{L/32}$.

97. How long is the pendulum of a grandfather clock with a period of 1.5 seconds (see figure)?

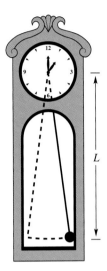

98. How long is the pendulum of a mantle clock with a period of 0.75 second?

99. *Demand for a Product* The demand equation for a certain product is

$$p = 50 - \sqrt{0.8(x - 1)}$$

where x is the number of units demanded per day and p is the price per unit. Find the demand if the price is $30.02.

100. *Airline Passengers* An airline offers daily flights between Chicago and Denver. The total monthly cost of the flights is

$$C = \sqrt{0.2x + 1}, \quad 0 \le x$$

where C is measured in millions of dollars and x is measured in thousands of passengers (see figure). The total cost of the flights for a certain month is 2.5 million dollars. Approximately how many passengers flew that month?

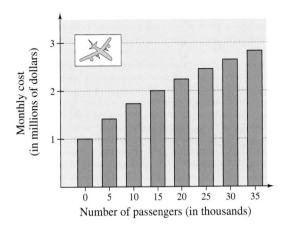

Number of passengers (in thousands)

101. *Federal Grants-in-Aid* The total amount spent on federal grants-in-aid F (in billions of dollars) in the United States for the years 1990 through 1997 is modeled by

$$F = 133.5 + 9.3t + 18.0\sqrt{t}, \quad 0 \le t \le 7$$

where t is time in years, with $t = 0$ corresponding to 1990. (Source: U.S. Office of Management and Budget)

(a) Use a graphing utility to graph the function.

(b) In what year did the grants total approximately 225 billion dollars?

102. *Exploration* The solution of the equation $x + \sqrt{x - a} = b$ is $x = 20$. Find a and b. (There are many correct answers.)

Explaining Concepts

103. Answer part (e) of Motivating the Chapter on page 319.

104. In your own words, describe the steps that can be used to solve a radical equation.

105. Does raising both sides of an equation to the nth power always yield an equivalent equation? Explain.

106. One reason for checking a solution in the original equation is to discover errors that were made when solving the equation. Describe another reason.

107. *Error Analysis* Describe the error.

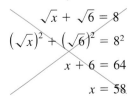

$$\sqrt{x} + \sqrt{6} = 8$$
$$\left(\sqrt{x}\right)^2 + \left(\sqrt{6}\right)^2 = 8^2$$
$$x + 6 = 64$$
$$x = 58$$

5.5   Complex Numbers

Objectives

1 Write the square root of a negative number in *i*-form and perform operations on numbers in *i*-form.

2 Determine the equality of two complex numbers.

3 Add, subtract, and multiply with complex numbers.

4 Use complex conjugates to divide complex numbers.

1 Write the square root of a negative number in *i*-form and perform operations on numbers in *i*-form.

The Imaginary Unit *i*

In Section 5.1, you learned that a negative number has no *real* square root. For instance, $\sqrt{-1}$ is not real because there is no real number x such that $x^2 = -1$. Thus, as long as you are dealing only with real numbers, the equation

$$x^2 = -1$$

has no solution. To overcome this deficiency, mathematicians have expanded the set of numbers, using the **imaginary unit *i*,** defined as

$$i = \sqrt{-1}. \qquad \text{Imaginary unit}$$

This number has the property that $i^2 = -1$. So, the imaginary unit *i* is a solution of the equation $x^2 = -1$.

▶ **The Square Root of a Negative Number**

Let c be a positive real number. Then the square root of $-c$ is given by

$$\sqrt{-c} = \sqrt{c(-1)} = \sqrt{c}\sqrt{-1} = \sqrt{c}\,i.$$

When writing $\sqrt{-c}$ in the **i-form,** $\sqrt{c}\,i$, note that i is outside the radical.

Example 1   Writing Numbers in *i*-Form

a. $\sqrt{-36} = \sqrt{36(-1)} = \sqrt{36}\sqrt{-1} = 6i$

b. $\sqrt{-\dfrac{16}{25}} = \sqrt{\dfrac{16}{25}(-1)} = \sqrt{\dfrac{16}{25}}\sqrt{-1} = \dfrac{4}{5}i$

c. $\sqrt{-5} = \sqrt{5(-1)} = \sqrt{5}\sqrt{-1} = \sqrt{5}\,i$

d. $\sqrt{-54} = \sqrt{54(-1)} = \sqrt{54}\sqrt{-1} = 3\sqrt{6}\,i$

e. $\dfrac{\sqrt{-48}}{\sqrt{-3}} = \dfrac{\sqrt{48}\sqrt{-1}}{\sqrt{3}\sqrt{-1}} = \dfrac{\sqrt{48}\,i}{\sqrt{3}\,i} = \sqrt{\dfrac{48}{3}} = \sqrt{16} = 4$

f. $\dfrac{\sqrt{-18}}{\sqrt{2}} = \dfrac{\sqrt{18}\sqrt{-1}}{\sqrt{2}} = \dfrac{\sqrt{18}\,i}{\sqrt{2}} = \sqrt{\dfrac{18}{2}}\,i = \sqrt{9}\,i = 3i$

Technology: Discovery

Use a calculator to evaluate the following radicals. Does one result in an error message? Explain why.

a. $\sqrt{121}$

b. $\sqrt{-121}$

c. $-\sqrt{121}$

To perform operations with square roots of negative numbers, you must *first* write the numbers in *i*-form. Once the numbers are written in *i*-form, you add, subtract, and multiply as follows.

$$ai + bi = (a + b)i \qquad\qquad \text{Addition}$$

$$ai - bi = (a - b)i \qquad\qquad \text{Subtraction}$$

$$(ai)(bi) = ab(i^2) = ab(-1) = -ab \qquad\qquad \text{Multiplication}$$

Example 2   Operations with Square Roots of Negative Numbers

Perform the operation.

a. $\sqrt{-9} + \sqrt{-49}$ **b.** $\sqrt{-32} - 2\sqrt{-2}$

Solution

a. $\sqrt{-9} + \sqrt{-49} = \sqrt{9}\sqrt{-1} + \sqrt{49}\sqrt{-1}$ Property of radicals

$$= 3i + 7i \qquad\qquad \text{Write in } i\text{-form.}$$

$$= 10i \qquad\qquad \text{Simplify.}$$

b. $\sqrt{-32} - 2\sqrt{-2} = \sqrt{32}\sqrt{-1} - 2\sqrt{2}\sqrt{-1}$ Property of radicals

$$= 4\sqrt{2}\,i - 2\sqrt{2}\,i \qquad\qquad \text{Write in } i\text{-form.}$$

$$= 2\sqrt{2}\,i \qquad\qquad \text{Simplify.}$$

Example 3   Multiplying Square Roots of Negative Numbers

Find each product.

a. $\sqrt{-15}\sqrt{-15}$ **b.** $\sqrt{-5}\left(\sqrt{-45} - \sqrt{-4}\right)$

Solution

a. $\sqrt{-15}\sqrt{-15} = \left(\sqrt{15}\,i\right)\left(\sqrt{15}\,i\right)$ Write in *i*-form.

$$= \left(\sqrt{15}\right)^2 i^2 \qquad\qquad \text{Multiply.}$$

$$= 15(-1) \qquad\qquad \text{Definition of } i$$

$$= -15 \qquad\qquad \text{Simplify.}$$

b. $\sqrt{-5}\left(\sqrt{-45} - \sqrt{-4}\right) = \sqrt{5}\,i\left(3\sqrt{5}\,i - 2i\right)$ Write in *i*-form.

$$= \left(\sqrt{5}\,i\right)\left(3\sqrt{5}\,i\right) - \left(\sqrt{5}\,i\right)(2i) \qquad\qquad \text{Distributive Property}$$

$$= 3(5)(-1) - 2\sqrt{5}(-1) \qquad\qquad \text{Multiply.}$$

$$= -15 + 2\sqrt{5} \qquad\qquad \text{Simplify.}$$

When multiplying square roots of negative numbers, be sure to write them in *i*-form *before multiplying.* If you do not, you can obtain incorrect answers. For instance, in Example 3(a) be sure you see that

$$\sqrt{-15}\sqrt{-15} \neq \sqrt{(-15)(-15)} = \sqrt{225} = 15.$$

Study Tip

When performing operations with numbers in *i*-form, you sometimes need to be able to evaluate powers of the imaginary unit *i*. The first several powers of *i* are as follows.

$$i^1 = i$$

$$i^2 = -1$$

$$i^3 = i(i^2) = i(-1) = -i$$

$$i^4 = (i^2)(i^2) = (-1)(-1) = 1$$

$$i^5 = i(i^4) = i(1) = i$$

$$i^6 = (i^2)(i^4) = (-1)(1) = -1$$

$$i^7 = (i^3)(i^4) = (-i)(1) = -i$$

$$i^8 = (i^4)(i^4) = (1)(1) = 1$$

Note how the pattern of values $i, -1, -i$, and 1 repeats itself for powers greater than 4.

2 Determine the equality of two complex numbers.

Complex Numbers

A number of the form $a + bi$, where a and b are real numbers, is called a **complex number.**

> ▶ **Definition of a Complex Number**
>
> If a and b are real numbers, the number $a + bi$ is a **complex number,** and it is said to be written in **standard form.** If $b = 0$, the number $a + bi = a$ is a real number. If $b \neq 0$, the number $a + bi$ is called an **imaginary number.** A number of the form bi, where $b \neq 0$, is called a **pure imaginary number.**

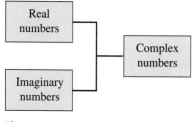

Figure 5.5

A number cannot be both real and imaginary. For instance, the numbers -2, 0, 1, $\frac{1}{2}$, and $\sqrt{2}$ are real numbers (but they are *not* imaginary numbers), and the numbers $-3i$, $2 + 4i$, and $-1 + i$ are imaginary numbers (but they are *not* real numbers). The diagram shown in Figure 5.5 further illustrates the relationship among real, complex, and imaginary numbers.

Two complex numbers $a + bi$ and $c + di$, in standard form, are equal if and only if $a = c$ and $b = d$.

Example 4   Equality of Two Complex Numbers

a. Are the complex numbers $\sqrt{9} + \sqrt{-48}$ and $3 - 4\sqrt{3}i$ equal?

b. Find values of x and y such that the equation is valid.

$$3x - \sqrt{-25} = -6 + 3yi$$

Solution

a. Begin by writing the first number in standard form.

$$\sqrt{9} + \sqrt{-48} = \sqrt{3^2} + \sqrt{4^2(3)(-1)} \qquad \text{Factor.}$$

$$= 3 + 4\sqrt{3}i \qquad \text{Write in } i\text{-form.}$$

From this form, you can see that the two numbers are not equal because they have imaginary parts that differ in sign.

b. Begin by writing the left side of the equation in standard form.

$$3x - \sqrt{-25} = -6 + 3yi \qquad \text{Original equation}$$

$$3x - 5i = -6 + 3yi \qquad \text{Both sides in standard form}$$

For these two numbers to be equal, their real parts must be equal to each other and their imaginary parts must be equal to each other.

Real Parts	*Imaginary Parts*
$3x = -6$	$3y = -5$
$x = -2$	$y = -\frac{5}{3}$

So, $x = -2$ and $y = -\frac{5}{3}$.

3 Add, subtract, and multiply with complex numbers.

Courtesy of Dr. Benoit Mandelbrot

Benoit Mandelbrot

(1924–)

Until very recently it was thought that shapes in nature, such as clouds, coastlines, and mountain ranges, could not be described in mathematical terms. In the 1970s, Mandelbrot discovered that many of these shapes do have patterns in their irregularity—they are made up of smaller parts that are scaled-down versions of the shapes themselves. Computers using mathematical terms with complex numbers are able to generate the larger images. Mandelbrot coined the term *fractals* for these shapes and for the geometry used to describe them.

Operations with Complex Numbers

The real number a is called the **real part** of the complex number $a + bi$, and the number bi is called the **imaginary part** of the complex number. To add or subtract two complex numbers, we add (or subtract) the real and imaginary parts separately. This is similar to combining like terms of a polynomial.

$$(a + bi) + (c + di) = (a + c) + (b + d)i \qquad \text{Addition of complex numbers}$$

$$(a + bi) - (c + di) = (a - c) + (b - d)i \qquad \text{Subtraction of complex numbers}$$

Example 5   Adding and Subtracting Complex Numbers

a. $(3 - i) + (-2 + 4i) = (3 - 2) + (-1 + 4)i = 1 + 3i$

b. $3i + (5 - 3i) = 5 + (3 - 3)i = 5$

c. $4 - (-1 + 5i) + (7 + 2i) = [4 - (-1) + 7] + (-5 + 2)i = 12 - 3i$

d. $(6 + 3i) + \left(2 - \sqrt{-8}\right) - \sqrt{-4} = (6 + 3i) + \left(2 - 2\sqrt{2}i\right) - 2i$

$$= (6 + 2) + \left(3 - 2\sqrt{2} - 2\right)i$$

$$= 8 + \left(1 - 2\sqrt{2}\right)i$$

Note in part (b) that the sum of two complex numbers can be a real number.

The Commutative, Associative, and Distributive Properties of real numbers are also valid for complex numbers.

Example 6   Multiplying Complex Numbers

a. $(7i)(-3i) = -21i^2$ \hfill Multiply.

$$= -21(-1) = 21 \qquad i^2 = -1$$

b. $(1 - i)\left(\sqrt{-9}\right) = (1 - i)(3i)$ \hfill Write in i-form.

$$= (1)(3i) - (i)(3i) \qquad \text{Distributive Property}$$

$$= 3i - 3(i^2) \qquad \text{Simplify.}$$

$$= 3i - 3(-1) = 3 + 3i \qquad i^2 = -1$$

c. $(2 - i)(4 + 3i) = 8 + 6i - 4i - 3i^2$ \hfill FOIL Method

$$= 8 + 6i - 4i - 3(-1) \qquad i^2 = -1$$

$$= 11 + 2i \qquad \text{Combine like terms.}$$

d. $(3 + 2i)(3 - 2i) = 9 - 6i + 6i - 4i^2$ \hfill FOIL Method

$$= 9 - 6i + 6i - 4(-1) \qquad i^2 = -1$$

$$= 9 + 4 = 13 \qquad \text{Combine like terms.}$$

4 Use complex conjugates to divide complex numbers.

Complex Conjugates

In Example 6(d), note that the product of two complex numbers can be a real number. This occurs with pairs of complex numbers of the form $a + bi$ and $a - bi$, called **complex conjugates.** In general, the product of complex conjugates has the following form.

$$(a + bi)(a - bi) = a^2 - (bi)^2$$

$$= a^2 - b^2 i^2$$

$$= a^2 - b^2(-1)$$

$$= a^2 + b^2$$

Here are some examples.

Complex Number	Complex Conjugate	Product
$4 - 5i$	$4 + 5i$	$4^2 + 5^2 = 41$
$3 + 2i$	$3 - 2i$	$3^2 + 2^2 = 13$
$-2 = -2 + 0i$	$-2 = -2 - 0i$	$(-2)^2 + 0^2 = 4$
$i = 0 + i$	$-i = 0 - i$	$0^2 + 1^2 = 1$

Complex conjugates are used to divide one complex number by another. To do this, multiply the numerator and denominator by the *complex conjugate of the denominator,* as shown in Example 7.

Example 7 Division of Complex Numbers

a. $\dfrac{2 - i}{4i} = \dfrac{2 - i}{4i} \cdot \dfrac{(-i)}{(-i)}$ Multiply numerator and denominator by complex conjugate of denominator.

$$= \dfrac{-2i + i^2}{-4i^2}$$ Multiply fractions.

$$= \dfrac{-2i + (-1)}{-4(-1)}$$ $i^2 = -1$

$$= -\dfrac{2i + 1}{4}$$ Simplify.

b. $\dfrac{5}{3 - 2i} = \dfrac{5}{3 - 2i} \cdot \dfrac{3 + 2i}{3 + 2i}$ Multiply numerator and denominator by complex conjugate of denominator.

$$= \dfrac{5(3 + 2i)}{(3 - 2i)(3 + 2i)}$$ Multiply fractions.

$$= \dfrac{5(3 + 2i)}{3^2 + 2^2}$$ Product of complex conjugates

$$= \dfrac{15 + 10i}{13}$$ Simplify.

$$= \dfrac{15}{13} + \dfrac{10}{13}i$$ Standard form

Example 8   Division of Complex Numbers

Divide $2 + 3i$ by $4 - 2i$.

Solution

$$\frac{2 + 3i}{4 - 2i} = \frac{2 + 3i}{4 - 2i} \cdot \frac{4 + 2i}{4 + 2i}$$ Multiply numerator and denominator by complex conjugate of denominator.

$$= \frac{8 + 16i + 6i^2}{4^2 + 2^2}$$ Multiply fractions.

$$= \frac{8 + 16i + 6(-1)}{20}$$ $i^2 = -1$

$$= \frac{2 + 16i}{20}$$ Combine like terms.

$$= \frac{2}{20} + \frac{16i}{20}$$ Standard form

$$= \frac{1}{10} + \frac{4}{5}i$$ Simplify.

Example 9   Verifying a Complex Solution of an Equation

Show that $x = 2 + i$ is a solution of the equation $x^2 - 4x + 5 = 0$.

Solution

$$x^2 - 4x + 5 = 0$$ Original equation

$$(2 + i)^2 - 4(2 + i) + 5 \overset{?}{=} 0$$ Substitute $2 + i$ for x.

$$4 + 4i + i^2 - 8 - 4i + 5 \overset{?}{=} 0$$ Expand.

$$i^2 + 1 \overset{?}{=} 0$$ Combine like terms.

$$(-1) + 1 \overset{?}{=} 0$$ $i^2 = -1$

$$0 = 0$$ Solution checks. ✓

So, $2 + i$ is a solution of the original equation.

Discussing the Concept   Prime Polynomials

The polynomial $x^2 + 1$ is prime *with respect to the integers*. It is not, however, prime *with respect to the complex numbers*. Show how $x^2 + 1$ can be factored using complex numbers.

5.5   Exercises

Integrated Review   *Concepts, Skills, and Problem Solving*

Keep mathematically in shape by doing these exercises *before* the problems of this section.

Properties and Definitions

1. In your own words, describe how to multiply

$$\frac{3t}{5} \cdot \frac{8t^2}{15}.$$

2. In your own words, describe how to divide

$$\frac{3t}{5} \div \frac{8t^2}{15}.$$

3. In your own words, describe how to add

$$\frac{3t}{5} + \frac{8t^2}{15}.$$

4. What is the value of $\dfrac{t-5}{5-t}$? Explain.

Simplifying Expressions

In Exercises 5–10, simplify the expression.

5. $\dfrac{x^2}{2x+3} \div \dfrac{5x}{2x+3}$

6. $\dfrac{x-y}{5x} \div \dfrac{x^2-y^2}{x^2}$

7. $\dfrac{\dfrac{9}{x}}{\left(\dfrac{6}{x}+2\right)}$

8. $\dfrac{\left(1+\dfrac{2}{x}\right)}{\left(x-\dfrac{4}{x}\right)}$

9. $\dfrac{\left(\dfrac{4}{x^2-9}+\dfrac{2}{x-2}\right)}{\left(\dfrac{1}{x+3}+\dfrac{1}{x-3}\right)}$

10. $\dfrac{\left(\dfrac{1}{x+1}+\dfrac{1}{2}\right)}{\left(\dfrac{3}{2x^2+4x+2}\right)}$

Problem Solving

11. Find two real numbers that divide the real number line between $x/2$ and $4x/3$ into three equal parts.

12. When two capacitors with capacitances C_1 and C_2, respectively, are connected in series, the equivalent capacitance is given by

$$\frac{1}{\left(\dfrac{1}{C_1}+\dfrac{1}{C_2}\right)}.$$

Simplify this complex fraction.

Developing Skills

In Exercises 1–20, write the number in *i*-form. See Example 1.

1. $\sqrt{-4}$

2. $\sqrt{-9}$

3. $-\sqrt{-144}$

4. $\sqrt{-49}$

5. $\sqrt{-\frac{4}{25}}$

6. $-\sqrt{-\frac{36}{121}}$

7. $\sqrt{-0.09}$

8. $\sqrt{-0.0004}$

9. $\sqrt{-8}$

10. $\sqrt{-75}$

11. $\sqrt{-27}$

12. $\sqrt{-80}$

13. $\sqrt{-7}$

14. $\sqrt{-15}$

15. $\dfrac{\sqrt{-12}}{\sqrt{-3}}$

16. $\dfrac{\sqrt{-45}}{\sqrt{-5}}$

17. $\dfrac{\sqrt{-20}}{\sqrt{4}}$

18. $\dfrac{\sqrt{72}}{\sqrt{-2}}$

19. $\sqrt{-\frac{18}{64}}$

20. $\sqrt{-\frac{8}{25}}$

In Exercises 21–42, perform the operation(s) and write the result in standard form. See Examples 2 and 3.

21. $\sqrt{-16} + \sqrt{-36}$

22. $\sqrt{-25} - \sqrt{-9}$

23. $\sqrt{-50} - \sqrt{-8}$

24. $\sqrt{-500} + \sqrt{-45}$

25. $\sqrt{-48} + \sqrt{-12} - \sqrt{-27}$

26. $\sqrt{-32} - \sqrt{-18} + \sqrt{-50}$

27. $\sqrt{-8}\sqrt{-2}$

28. $\sqrt{-25}\sqrt{-6}$

29. $\sqrt{-18}\sqrt{-3}$

30. $\sqrt{-7}\sqrt{-7}$

31. $\sqrt{-0.16}\sqrt{-1.21}$

32. $\sqrt{-0.49}\sqrt{-1.44}$

33. $\sqrt{-3}\left(\sqrt{-3}+\sqrt{-4}\right)$

34. $\sqrt{-12}\left(\sqrt{-3}-\sqrt{-12}\right)$

35. $\sqrt{-5}\left(\sqrt{-16} - \sqrt{-10}\right)$

36. $\sqrt{-24}\left(\sqrt{-9} + \sqrt{-4}\right)$

37. $\sqrt{-2}\left(3 - \sqrt{-8}\right)$ **38.** $\sqrt{-9}\left(1 + \sqrt{-16}\right)$

39. $\left(\sqrt{-16}\right)^2$ **40.** $\left(\sqrt{-2}\right)^2$

41. $\left(\sqrt{-4}\right)^3$ **42.** $\left(\sqrt{-5}\right)^3$

In Exercises 43-50, determine a and b. See Example 4.

43. $3 - 4i = a + bi$

44. $-8 + 6i = a + bi$

45. $5 - 4i = (a + 3) + (b - 1)i$

46. $-10 + 12i = 2a + (5b - 3)i$

47. $-4 - \sqrt{-8} = a + bi$

48. $\sqrt{-36} - 3 = a + bi$

49. $(a + 5) + (b - 1)i = 7 - 3i$

50. $(2a + 1) + (2b + 3)i = 5 + 12i$

In Exercises 51-66, perform the operation(s) and write the result in standard form. See Example 5.

51. $(4 - 3i) + (6 + 7i)$

52. $(-10 + 2i) + (4 - 7i)$

53. $(-4 - 7i) + (-10 - 33i)$

54. $(15 + 10i) - (2 + 10i)$

55. $13i - (14 - 7i)$

56. $(-21 - 50i) + (21 - 20i)$

57. $(30 - i) - (18 + 6i) + 3i^2$

58. $(4 + 6i) + (15 + 24i) - (1 - i)$

59. $6 - (3 - 4i) + 2i$

60. $22 + (-5 + 8i) + 10i$

61. $\left(\frac{4}{3} + \frac{1}{3}i\right) + \left(\frac{5}{6} + \frac{7}{6}i\right)$

62. $(0.05 + 2.50i) - (6.2 + 11.8i)$

63. $15i - (3 - 25i) + \sqrt{-81}$

64. $(-1 + i) - \sqrt{2} - \sqrt{-2}$

65. $8 - \left(5 - \sqrt{-63}\right) + (4 - 5i)$

66. $\left(7 - \sqrt{-96}\right) - (-8 + 10i) - 3i$

In Exercises 67-96, perform the operation and write the result in standard form. See Example 6.

67. $(3i)(12i)$ **68.** $(-5i)(4i)$

69. $(3i)(-8i)$ **70.** $(-2i)(-10i)$

71. $(-6i)(-i)(6i)$ **72.** $(10i)(12i)(-3i)$

73. $(-3i)^3$ **74.** $(8i)^2$

75. $(-3i)^2$ **76.** $(2i)^4$

77. $-5(13 + 2i)$ **78.** $10(8 - 6i)$

79. $4i(-3 - 5i)$ **80.** $-3i(10 - 15i)$

81. $(9 - 2i)\left(\sqrt{-4}\right)$ **82.** $(11 + 3i)\left(\sqrt{-25}\right)$

83. $\sqrt{-20}\left(6 + 2\sqrt{5}i\right)$

84. $\sqrt{-24}\left(-3\sqrt{6} - 4i\right)$

85. $(4 + 3i)(-7 + 4i)$ **86.** $(3 + 5i)(2 + 15i)$

87. $(-7 + 7i)(4 - 2i)$ **88.** $(3 + 5i)(2 - 15i)$

89. $\left(-2 + \sqrt{-5}\right)\left(-2 - \sqrt{-5}\right)$

90. $\left(-3 - \sqrt{-12}\right)\left(4 - \sqrt{-12}\right)$

91. $(3 - 4i)^2$ **92.** $(7 + i)^2$

93. $(2 + 5i)^2$ **94.** $(8 - 3i)^2$

95. $(2 + i)^3$ **96.** $(3 - 2i)^3$

In Exercises 97-108, multiply the number by its conjugate.

97. $2 + i$ **98.** $3 + 2i$

99. $-2 - 8i$ **100.** $10 - 3i$

101. $5 - \sqrt{6}i$ **102.** $-4 + \sqrt{2}i$

103. $10i$ **104.** 20

105. $1 + \sqrt{-3}$ **106.** $-3 - \sqrt{-5}$

107. $1.5 + \sqrt{-0.25}$ **108.** $3.2 - \sqrt{-0.04}$

In Exercises 109-118, perform the operation and write the result in standard form. See Examples 7 and 8.

109. $\dfrac{20}{2i}$ **110.** $\dfrac{1 + i}{3i}$

111. $\dfrac{4}{1 - i}$ **112.** $\dfrac{20}{3 + i}$

113. $\dfrac{-12}{2 + 7i}$ **114.** $\dfrac{15}{2(1 - i)}$

115. $\dfrac{4i}{1 - 3i}$ **116.** $\dfrac{17i}{5 + 3i}$

117. $\dfrac{2 + 3i}{1 + 2i}$ **118.** $\dfrac{4 - 5i}{4 + 5i}$

In Exercises 119-122, perform the operation and write the result in standard form.

119. $\dfrac{1}{1 - 2i} + \dfrac{4}{1 + 2i}$ **120.** $\dfrac{3i}{1 + i} + \dfrac{2}{2 + 3i}$

121. $\dfrac{i}{4 - 3i} - \dfrac{5}{2 + i}$ **122.** $\dfrac{1 + i}{i} - \dfrac{3}{5 - 2i}$

In Exercises 123–126, determine whether each number is a solution of the equation. See Example 9.

123. $x^2 + 2x + 5 = 0$

    (a) $x = -1 + 2i$ (b) $x = -1 - 2i$

124. $x^2 - 4x + 13 = 0$

    (a) $x = 2 - 3i$ (b) $x = 2 + 3i$

125. $x^3 + 4x^2 + 9x + 36 = 0$

    (a) $x = -4$ (b) $x = -3i$

126. $x^3 - 8x^2 + 25x - 26 = 0$

    (a) $x = 2$ (b) $x = 3 - 2i$

127. *Cube Roots* The principal cube root of 125, $\sqrt[3]{125}$, is 5. Evaluate the expression x^3 for each of the following values of x.

    (a) $\dfrac{-5 + 5\sqrt{3}\,i}{2}$ (b) $\dfrac{-5 - 5\sqrt{3}\,i}{2}$

128. *Cube Roots* The principal cube root of 27, $\sqrt[3]{27}$, is 3. Evaluate the expression x^3 for each of the following values of x.

    (a) $\dfrac{-3 + 3\sqrt{3}\,i}{2}$ (b) $\dfrac{-3 - 3\sqrt{3}\,i}{2}$

129. *Pattern Recognition* Use the results of Exercises 127 and 128 to list possible cube roots of (a) 1, (b) 8, and (c) 64. Verify your results algebraically.

130. *Algebraic Properties* Consider the complex number $1 + 5i$.

    (a) Find the additive inverse of the number.

    (b) Find the multiplicative inverse of the number.

In Exercises 131–134, perform the operations.

131. $(a + bi) + (a - bi)$ **132.** $(a + bi)(a - bi)$

133. $(a + bi) - (a - bi)$

134. $(a + bi)^2 + (a - bi)^2$

Explaining Concepts

135. Define the imaginary unit i.

136. Explain why the equation $x^2 = -1$ does not have real number solutions.

137. *Error Analysis* Describe the error.

$$\sqrt{-3}\sqrt{-3} = \sqrt{(-3)(-3)} = \sqrt{9} = 3$$

138. *True or False?* Some numbers are both real and imaginary. Explain.

139. Find the product of $3 - 2i$ and its complex conjugate.

140. Describe the methods for adding, subtracting, multiplying, and dividing complex numbers.

Key Terms

square root, *p. 320*
cube root, *p. 320*
*n*th root of *a*, *p. 320*
principal *n*th root of *a*,
   p. 320
radical symbol, *p. 320*

index, *p. 320*
radicand, *p. 320*
rationalizing the
   denominator, *p. 332*
Pythagorean Theorem,
   p. 334

conjugates, *p. 339*
imaginary unit *i*, *p. 356*
i-form, *p. 356*
complex number, *p. 358*
imaginary number, *p. 358*

real part, *p. 359*
imaginary part, *p. 359*
complex conjugates,
   p. 360

Key Concepts

5.1 Properties of *n*th roots

1. If a is a positive real number and n is even, then a has exactly two (real) nth roots, which are denoted by $\sqrt[n]{a}$ and $-\sqrt[n]{a}$.

2. If a is any real number and n is odd, then a has only one (real) nth root, which is denoted by $\sqrt[n]{a}$.

3. If a is a negative real number and n is even, then a has no (real) nth root.

5.1 Inverse properties of *n*th powers and *n*th roots

Let a be a real number, and let n be an integer such that $n \geq 2$.

1. If a has a principal nth root, then $\left(\sqrt[n]{a}\right)^n = a$.

2. If n is odd, then $\sqrt[n]{a^n} = a$.
   If n is even, then $\sqrt[n]{a^n} = |a|$.

5.1 Domain of a radical function

Let n be an integer that is greater than or equal to 2.

1. If n is odd, the domain of $f(x) = \sqrt[n]{x}$ is the set of all real numbers.

2. If n is even, the domain of $f(x) = \sqrt[n]{x}$ is the set of all nonnegative real numbers.

5.1 Rules of exponents

Let r and s be rational numbers, and let a and b be real numbers, variables, or algebraic expressions.

1. $a^r \cdot a^s = a^{r+s}$

2. $\dfrac{a^r}{a^s} = a^{r-s}, \quad a \neq 0$

3. $(ab)^r = a^r \cdot b^r$

4. $\left(\dfrac{a}{b}\right)^r = \dfrac{a^r}{b^r}, \quad b \neq 0$

5. $(a^r)^s = a^{rs}$

6. $a^{-r} = \dfrac{1}{a^r}, \quad a \neq 0$

7. $a^0 = 1$

8. $\left(\dfrac{a}{b}\right)^{-r} = \left(\dfrac{b}{a}\right)^r, \quad a \neq 0, b \neq 0$

5.2 Multiplying and dividing radicals

Let u and v be real numbers, variables, or algebraic expressions. If the nth roots of u and v are real, the following properties are true.

1. $\sqrt[n]{uv} = \sqrt[n]{u}\,\sqrt[n]{v}$

2. $\sqrt[n]{\dfrac{u}{v}} = \dfrac{\sqrt[n]{u}}{\sqrt[n]{v}}, \quad v \neq 0$

5.2 Simplifying radical expressions

A radical expression is said to be in simplest form if all three of the following are true.

1. All possible nth powered factors have been removed from each radical.

2. No radical contains a fraction.

3. No denominator of a fraction contains a radical.

5.4 Raising both sides of an equation to the *n*th power

Let u and v be real numbers, variables, or algebraic expressions, and let n be a positive integer. If $u = v$, then it follows that $u^n = v^n$. This is called raising both sides of an equation to the nth power.

5.5 The square root of a negative number

Let c be a positive real number. Then the square root of $-c$ is given by $\sqrt{-c} = \sqrt{c(-1)} = \sqrt{c}\sqrt{-1} = \sqrt{c}\,i$. When writing $\sqrt{-c}$ in the i-form, $\sqrt{c}\,i$, note that the i is outside the radical.

REVIEW EXERCISES

Reviewing Skills

5.1 In Exercises 1–16, evaluate the radical expression.

1. $\sqrt{49}$ **2.** $\sqrt{64}$

3. $-\sqrt{81}$ **4.** $\sqrt{-16}$

5. $\sqrt[3]{-8}$ **6.** $\sqrt[3]{-27}$

7. $-\sqrt[3]{64}$ **8.** $-\sqrt[3]{125}$

9. $\sqrt{(1.2)^2}$ **10.** $\sqrt{(0.4)^2}$

11. $\sqrt{\left(\frac{5}{6}\right)^2}$ **12.** $\sqrt{\left(\frac{8}{15}\right)^2}$

13. $\sqrt[3]{-\left(\frac{1}{5}\right)^3}$ **14.** $-\sqrt[3]{\left(-\frac{27}{64}\right)^3}$

15. $\sqrt{-2^2}$ **16.** $\sqrt{-4^2}$

In Exercises 17–20, fill in the missing description.

Radical Form	Rational Exponent Form
17. $\sqrt{49}=7$	
18. $\sqrt[3]{0.125}=0.5$	
19.	$216^{1/3}=6$
20.	$16^{1/4}=2$

In Exercises 21–28, use the rules of exponents to evaluate the expression.

21. $27^{4/3}$ **22.** $16^{3/4}$

23. $-(5^2)^{3/2}$ **24.** $(-9)^{5/2}$

25. $8^{-4/3}$ **26.** $243^{-2/5}$

27. $-\left(\frac{27}{64}\right)^{2/3}$ **28.** $\left(-\frac{8}{125}\right)^{1/3}$

In Exercises 29–40, use the rules of exponents to simplify the expression.

29. $x^{3/4}\cdot x^{-1/6}$ **30.** $a^{2/3}\cdot a^{3/5}$

31. $z\sqrt[3]{z^2}$ **32.** $x^2\sqrt[4]{x^3}$

33. $\dfrac{\sqrt[4]{x^3}}{\sqrt{x^4}}$ **34.** $\dfrac{\sqrt{x^3}}{\sqrt[3]{x^2}}$

35. $\sqrt[3]{a^3b^2}$ **36.** $\sqrt[5]{x^6y^2}$

37. $\sqrt[4]{\sqrt{x}}$ **38.** $\sqrt{\sqrt[3]{x^4}}$

39. $\dfrac{(3x+2)^{2/3}}{\sqrt[3]{3x+2}}$ **40.** $\dfrac{\sqrt[5]{3x+6}}{(3x+6)^{4/5}}$

In Exercises 41–44, evaluate the expression. Round the result to two decimal places.

41. $75^{-3/4}$ **42.** $510^{5/3}$

43. $\sqrt{13^2-4(2)(7)}$ **44.** $\dfrac{-3.7+\sqrt{15.8}}{2(2.3)}$

In Exercises 45–48, use a graphing utility to graph the function. Give the domain of the function.

45. $y=3\sqrt[3]{2x}$ **46.** $y=\dfrac{10}{\sqrt[4]{x^2+1}}$

47. $g(x)=4x^{3/4}$ **48.** $h(x)=\frac{1}{2}x^{4/3}$

5.2 In Exercises 49–58, simplify the radical expression.

49. $\sqrt{360}$ **50.** $\sqrt{\frac{50}{9}}$

51. $\sqrt{75u^5v^4}$ **52.** $\sqrt{24x^3y^4}$

53. $\sqrt{0.25x^4y}$ **54.** $\sqrt{0.16s^6t^3}$

55. $\sqrt[4]{64a^2b^5}$ **56.** $\sqrt{36x^3y^2}$

57. $\sqrt[3]{48a^3b^4}$ **58.** $\sqrt[4]{32u^4v^5}$

In Exercises 59–64, rationalize the denominator and simplify further, if possible.

59. $\sqrt{\frac{5}{6}}$ **60.** $\sqrt{\frac{3}{20}}$

61. $\dfrac{3}{\sqrt{12x}}$ **62.** $\dfrac{4y}{\sqrt{10z}}$

63. $\dfrac{2}{\sqrt[3]{2x}}$ **64.** $\sqrt[3]{\dfrac{16t}{s^2}}$

In Exercises 65–76, perform the operations and simplify.

65. $2\sqrt{7}-5\sqrt{7}+4\sqrt{7}$

66. $3\sqrt{5}-7\sqrt{5}+2\sqrt{5}$

67. $3\sqrt{40}-10\sqrt{90}$

68. $9\sqrt{50}-5\sqrt{8}+\sqrt{48}$

69. $5\sqrt{x}-\sqrt[3]{x}+9\sqrt{x}-8\sqrt[3]{x}$

70. $\sqrt{3x}-\sqrt[4]{6x^2}+2\sqrt[4]{6x^2}-4\sqrt{3x}$

71. $10\sqrt[4]{y+3}-3\sqrt[4]{y+3}$

72. $5\sqrt[3]{x-3}+4\sqrt[3]{x-3}$

73. $\sqrt{25x} + \sqrt{49x} - \sqrt[3]{8x}$

74. $\sqrt[3]{81x^4} + \sqrt[3]{24x^4} - \sqrt{3x}$

75. $\sqrt{5} - \dfrac{3}{\sqrt{5}}$ **76.** $\dfrac{4}{\sqrt{2}} + 3\sqrt{2}$

5.3 In Exercises 77–88, multiply the radical expressions and simplify.

77. $\sqrt{15} \cdot \sqrt{20}$ **78.** $\sqrt{42} \cdot \sqrt{21}$

79. $\sqrt{5}(\sqrt{10} + 3)$ **80.** $\sqrt{6}(\sqrt{24} - 8)$

81. $\sqrt{10}(\sqrt{2} + \sqrt{5})$ **82.** $\sqrt{12}(\sqrt{6} - \sqrt{8})$

83. $(2\sqrt{3} + 7)(\sqrt{6} - 2)$

84. $(2 - 4\sqrt{3})(7 + \sqrt{3})$

85. $(\sqrt{5} + 6)^2$ **86.** $(4 - 3\sqrt{2})^2$

87. $(\sqrt{3} - \sqrt{x})(\sqrt{3} + \sqrt{x})$

88. $(2 + 3\sqrt{5})(2 - 3\sqrt{5})$

In Exercises 89–96, simplify the quotient.

89. $\dfrac{3}{1 - \sqrt{2}}$ **90.** $\dfrac{\sqrt{5}}{\sqrt{10} + 3}$

91. $\dfrac{3\sqrt{8}}{2\sqrt{2} + \sqrt{3}}$ **92.** $\dfrac{7\sqrt{6}}{\sqrt{3} - 4\sqrt{2}}$

93. $\dfrac{\sqrt{2} - 1}{\sqrt{3} - 4}$ **94.** $\dfrac{3 + \sqrt{3}}{5 - \sqrt{3}}$

95. $(\sqrt{x} + 10) \div (\sqrt{x} - 10)$

96. $(3\sqrt{s} + 4) \div (\sqrt{s} + 2)$

In Exercises 97–100, use a graphing utility to graph the functions on the same screen. Use the graphs to verify that the expressions are equivalent. Verify the results algebraically.

97. $y_1 = \sqrt{\dfrac{5}{2x}}$ **98.** $y_1 = \dfrac{x}{1 + \sqrt{x}}$

$y_2 = \dfrac{\sqrt{10x}}{2x}$ $y_2 = \dfrac{x(1 - \sqrt{x})}{1 - x}$

99. $y_1 = 5\sqrt{x} - 2\sqrt{x}$ **100.** $y_1 = -2\sqrt{9x} + 10\sqrt{x}$

$y_2 = 3\sqrt{x}$ $y_2 = 4\sqrt{x}$

5.4 In Exercises 101–116, solve the given equation.

101. $\sqrt{y} = 15$ **102.** $\sqrt{x} - 3 = 0$

103. $\sqrt{3x} + 9 = 0$ **104.** $\sqrt{4x} + 6 = 9$

105. $\sqrt{2(a - 7)} = 14$ **106.** $\sqrt{5(4 - 3x)} = 10$

107. $\sqrt[3]{5x - 7} - 3 = -1$

108. $\sqrt[4]{2x + 3} + 4 = 5$

109. $\sqrt[3]{5x + 2} - \sqrt[3]{7x - 8} = 0$

110. $\sqrt[4]{9x - 2} - \sqrt[4]{8x} = 0$

111. $\sqrt{2(x + 5)} = x + 5$

112. $y - 2 = \sqrt{y + 4}$

113. $\sqrt{v - 6} = 6 - v$

114. $\sqrt{5t} = 1 + \sqrt{5(t - 1)}$

115. $\sqrt{1 + 6x} = 2 - \sqrt{6x}$

116. $\sqrt{2 + 9b} + 1 = 3\sqrt{b}$

5.5 In Exercises 117–122, write the complex number in *i*-form.

117. $\sqrt{-48}$

118. $\sqrt{-0.16}$

119. $10 - 3\sqrt{-27}$

120. $3 + 2\sqrt{-500}$

121. $\frac{3}{4} - 5\sqrt{-\frac{3}{25}}$

122. $-0.5 + 3\sqrt{-1.21}$

In Exercises 123–130, perform the operation and write the result in standard form.

123. $\sqrt{-81} + \sqrt{-36}$

124. $\sqrt{-49} + \sqrt{-1}$

125. $\sqrt{-121} - \sqrt{-84}$

126. $\sqrt{-169} - \sqrt{-4}$

127. $\sqrt{-5}\sqrt{-5}$

128. $\sqrt{-24}\sqrt{-6}$

129. $\sqrt{-10}(\sqrt{-4} - \sqrt{-7})$

130. $\sqrt{-5}(\sqrt{-10} + \sqrt{-15})$

In Exercises 131–134, find x and y such that the two complex numbers are equal.

131. $4x - \sqrt{-36} = 8 - 2yi$

132. $5x + \sqrt{-81} = 25 + 3yi$

133. $24 + \sqrt{-5y} = 6x + 25i$

134. $10 - \sqrt{-4y} = 2x - 16i$

In Exercises 135–142, add, subtract, or multiply the complex numbers.

135. $(-4 + 5i) - (-12 + 8i)$

136. $(-8 + 3i) - (6 + 7i)$

137. $(3 - 8i) + (5 + 12i)$

138. $(-6 + 3i) + (-1 + i)$

139. $(4 - 3i)(4 + 3i)$ **140.** $(12 - 5i)(2 + 7i)$

141. $(6 - 5i)^2$ **142.** $(2 - 9i)^2$

In Exercises 143–148, use complex conjugates to perform the division.

143. $\dfrac{7}{3i}$ **144.** $\dfrac{4}{5i}$

145. $\dfrac{4i}{2 - 8i}$ **146.** $\dfrac{5i}{2 + 9i}$

147. $\dfrac{3 - 5i}{6 + i}$ **148.** $\dfrac{2 + i}{1 - 9i}$

Solving Problems

149. *Perimeter*   The four corners are cut from an $8\frac{1}{2}$-inch-by-14-inch sheet of paper (see figure). Find the perimeter of the remaining piece of paper.

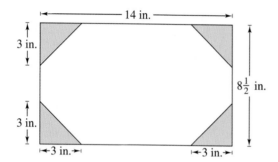

150. *Geometry*   Determine the length and width of a rectangle with a perimeter of 84 inches and a diagonal of 30 inches.

151. *Length of a Pendulum*   The time t in seconds for a pendulum of length L in feet to go through one complete cycle (its period) is

$$t = 2\pi\sqrt{\frac{L}{32}}.$$

How long is the pendulum of a grandfather clock with a period of 1.3 seconds?

152. *Height of a Bridge*   The time t in seconds for a free-falling object to fall d feet is given by

$$t = \sqrt{\frac{d}{16}}.$$

A child drops a pebble from a bridge and observes it strike the water after approximately 4 seconds. Estimate the height of the bridge.

Power   In Exercises 153–156, use the equation

$$I = \sqrt{\frac{P}{R}}$$

to find the amount of power P (in watts) consumed by an electrical device that operates with a current I (in amps) and a resistance R (in ohms).

153. $I = 5$ amps, $R = 20$ ohms

154. $I = 10$ amps, $R = 20$ ohms

155. $I = 15$ amps, $R = 40$ ohms

156. $I = 15$ amps, $R = 20$ ohms

157. *Velocity of a Falling Object*   The velocity of a free-falling object can be determined from the equation

$$v = \sqrt{2gh}$$

where v is the velocity (in feet per second), $g = 32$ feet per second per second, and h is the distance (in feet) the object has fallen. Find the height from which a rock has been dropped if it strikes the ground with a velocity of 25 feet per second.

Chapter Test

Take this test as you would take a test in class. After you are done, check your work against the answers given in the back of the book.

In Exercises 1 and 2, evaluate the expressions without using a calculator.

1. (a) $16^{3/2}$

(b) $\sqrt{5}\sqrt{20}$

2. (a) $27^{-2/3}$

(b) $\sqrt{2}\sqrt{18}$

In Exercises 3–5, simplify the expressions.

3. (a) $\left(\dfrac{x^{1/2}}{x^{1/3}}\right)^2$

(b) $5^{1/4} \cdot 5^{7/4}$

4. (a) $\sqrt{\dfrac{32}{9}}$

(b) $\sqrt[3]{24}$

5. (a) $\sqrt{24x^3}$

(b) $\sqrt[4]{16x^5y^8}$

6. In your own words, explain the meaning of "rationalize" and demonstrate by rationalizing the denominator in the expression $\dfrac{3}{\sqrt{6}}$.

7. Combine: $5\sqrt{3x} - 3\sqrt{75x}$

8. Multiply and simplify: $\sqrt{5}\left(\sqrt{15x} + 3\right)$

9. Expand: $\left(4 - \sqrt{2x}\right)^2$

10. Factor: $7\sqrt{27} + 14y\sqrt{12} = 7\sqrt{3}\left(\quad\right)$

In Exercises 11–13, solve the equation.

11. $\sqrt{3y} - 6 = 3$

12. $\sqrt{x^2 - 1} = x - 2$

13. $\sqrt{x} - x + 6 = 0$

In Exercises 14 and 15, find x and y such that the two complex numbers are equal.

14. $3x + \sqrt{-4y} = 12 + 40i$

15. $27 - \sqrt{-16y} = 9x - 4i$

In Exercises 16–19, perform the operation and simplify.

16. $(2 + 3i) - \sqrt{-25}$

17. $(2 - 3i)^2$

18. $\sqrt{-16}\left(1 + \sqrt{-4}\right)$

19. $(3 - 2i)(1 + 5i)$

20. Divide $5 - 2i$ by i. Write the result in standard form.

21. The velocity v (in feet per second) of an object is given by $v = \sqrt{2gh}$, where $g = 32$ feet per second per second and h is the distance (in feet) the object has fallen. Find the height from which a rock has been dropped if it strikes the ground with a velocity of 80 feet per second.

6 Quadratic Equations and Inequalities

Tom Bean/The Stock Market

The footbridge over the dells of the Eau Claire River is part of the Ice Age Trail, a 1000-mile state scenic trail in Marathon County, Wisconsin.

Height of a Falling Object

You drop a rock from a bridge 100 feet above a river. The height h (in feet) of the rock at any time t (in seconds) is

$$h = -16t^2 + v_0 t + h_0$$

where v_0 is the initial velocity (in feet per second) of the rock and h_0 is the initial height.

See Section 6.1, Exercise 137

a. Suppose you drop the rock ($v_0 = 0$ ft/sec). How long will it take to hit the water? What method did you use to solve the quadratic equation? Explain why you used that method.

b. Suppose you throw the rock straight up with an initial velocity of 32 feet per second. Find the time(s) when h is 100 feet. What method did you use to solve this quadratic equation? Explain why you used this method.

See Section 6.3, Exercise 103

c. Suppose you throw the rock straight up with an initial velocity of 32 feet per second. Find the time when h is 50 feet. What method did you use to solve this quadratic equation? Explain why you used this method.

d. You move to a lookout point that is 84 feet above the river. If you throw the rock straight up at the same rate as when you were 100 feet above the river, would you expect it to reach the water in less time? Verify your conclusion algebraically.

See Section 6.5, Exercise 116

e. You throw a rock straight up with an initial velocity of 32 feet per second from a height of 100 feet. During what interval of time is the height greater than 52 feet?

6.1 Factoring and Extracting Square Roots

Objectives

1. Solve a quadratic equation by factoring.
2. Solve a quadratic equation by extracting square roots.
3. Solve a quadratic equation with complex solutions by extracting complex square roots.
4. Use substitution to solve an equation of quadratic form.

1 Solve a quadratic equation by factoring.

Corbis-Bettmann

Pierre de Fermat

(1601–1665)

Fermat's Last Theorem states that the equation $x^n + y^n = z^n$ has no solution when $x, y,$ and z are nonzero integers and $n > 2$. In 1637, Fermat wrote in the margin of a book that he had discovered a proof of this theorem; however, his proof has never been found. On June 23, 1993, 356 years later, a 200-page proof was presented at a gathering of mathematicians at Cambridge University in England by an American mathematician, Andrew Wiles.

Solving Quadratic Equations by Factoring

In this chapter, you will study methods for solving quadratic equations and equations of quadratic form. To begin, let's review the method of factoring that you studied in Section 3.5.

Remember that the first step in solving a quadratic equation by factoring is to write the equation in general form. Next, factor the left side. Finally, set each factor equal to zero and solve for x. It is important to check each solution in the original equation.

Example 1 Solving Quadratic Equations by Factoring

a.

$$x^2 + 5x = 24 \qquad \text{Original equation}$$
$$x^2 + 5x - 24 = 0 \qquad \text{General form}$$
$$(x + 8)(x - 3) = 0 \qquad \text{Factor.}$$
$$x + 8 = 0 \implies x = -8 \qquad \text{Set 1st factor equal to 0.}$$
$$x - 3 = 0 \implies x = 3 \qquad \text{Set 2nd factor equal to 0.}$$

b.

$$3x^2 = 4 - 11x \qquad \text{Original equation}$$
$$3x^2 + 11x - 4 = 0 \qquad \text{General form}$$
$$(3x - 1)(x + 4) = 0 \qquad \text{Factor.}$$
$$3x - 1 = 0 \implies x = \tfrac{1}{3} \qquad \text{Set 1st factor equal to 0.}$$
$$x + 4 = 0 \implies x = -4 \qquad \text{Set 2nd factor equal to 0.}$$

c.

$$9x^2 + 12 = 3 + 12x + 5x^2 \qquad \text{Original equation}$$
$$4x^2 - 12x + 9 = 0 \qquad \text{General form}$$
$$(2x - 3)(2x - 3) = 0 \qquad \text{Repeated factor}$$
$$2x - 3 = 0 \implies x = \frac{3}{2} \qquad \text{Set factor equal to 0.}$$

When the two solutions of a quadratic equation are identical, they are called a **double** or **repeated solution.** This occurred in Example 1(c).

2 Solve a quadratic equation by extracting square roots.

Extracting Square Roots

Consider the following equation, where $d > 0$ and u is an algebraic expression.

$$u^2 = d \qquad \text{Original equation}$$

$$u^2 - d = 0 \qquad \text{General form}$$

$$(u + \sqrt{d})(u - \sqrt{d}) = 0 \qquad \text{Factor.}$$

$$u + \sqrt{d} = 0 \implies u = -\sqrt{d} \qquad \text{Set 1st factor equal to 0.}$$

$$u - \sqrt{d} = 0 \implies u = \sqrt{d} \qquad \text{Set 2nd factor equal to 0.}$$

Because the solutions differ only in sign, they can be written together using a "plus or minus sign"

$$u = \pm \sqrt{d}.$$

This form of the solution is read as "u is equal to plus or minus the square root of d." Solving an equation of the form $u^2 = d$ *without* going through the steps of factoring is called **extracting square roots.**

Technology: Tip

To graphically check solutions of an equation written in general form, graph the left side of the equation and locate its x-intercepts. For instance, in Example 2(b), write the equation as

$$(x - 2)^2 - 10 = 0$$

and then sketch the graph of

$$y = (x - 2)^2 - 10$$

as shown below. From the graph, you can determine that the x-intercepts are approximately 5.16 and -1.16.

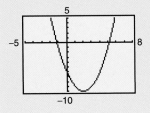

▶ **Extracting Square Roots**

The equation $u^2 = d$, where $d > 0$, has exactly two solutions:

$$u = \sqrt{d} \qquad \text{and} \qquad u = -\sqrt{d}.$$

These solutions can also be written as $u = \pm \sqrt{d}$.

Example 2   Extracting Square Roots

a. $3x^2 = 15 \qquad \text{Original equation}$

$x^2 = 5 \qquad \text{Divide both sides by 3.}$

$x = \pm \sqrt{5} \qquad \text{Extract square roots.}$

The solutions are $\sqrt{5}$ and $-\sqrt{5}$. Check these in the original equation.

b. $(x - 2)^2 = 10 \qquad \text{Original equation}$

$x - 2 = \pm \sqrt{10} \qquad \text{Extract square roots.}$

$x = 2 \pm \sqrt{10} \qquad \text{Add 2 to both sides.}$

The solutions are $2 + \sqrt{10} \approx 5.16$ and $2 - \sqrt{10} \approx -1.16$.

c. $(3x - 6)^2 - 8 = 0 \qquad \text{Original equation}$

$(3x - 6)^2 = 8 \qquad \text{Add 8 to both sides.}$

$3x - 6 = \pm 2\sqrt{2} \qquad \text{Extract square roots and rewrite } \sqrt{8} \text{ as } 2\sqrt{2}.$

$3x = 6 \pm 2\sqrt{2} \qquad \text{Add 6 to both sides.}$

$x = \dfrac{6 \pm 2\sqrt{2}}{3} \qquad \text{Divide both sides by 3.}$

The solutions are $\left(6 + 2\sqrt{2}\right)/3 \approx 2.94$ and $\left(6 - 2\sqrt{2}\right)/3 \approx 1.06$.

3 Solve a quadratic equation with complex solutions by extracting complex square roots.

Quadratic Equations with Complex Solutions

Prior to Section 5.5, the only solutions to find were real numbers. But now that you have studied complex numbers, it makes sense to look for other types of solutions. For instance, although the quadratic equation $x^2 + 1 = 0$ has no solutions that are real numbers, it does have two solutions that are complex numbers: i and $-i$. To check this, substitute i and $-i$ for x.

$$(i)^2 + 1 = -1 + 1 = 0 \qquad \text{Solution checks. ✓}$$

$$(-i)^2 + 1 = -1 + 1 = 0 \qquad \text{Solution checks. ✓}$$

One way to find complex solutions of a quadratic equation is to extend the *extracting square roots* technique to cover the case where d is a negative number.

> ▶ **Extracting Complex Square Roots**
>
> The equation $u^2 = d$, where $d < 0$, has exactly two solutions:
>
> $$u = \sqrt{|d|}\,i \qquad \text{and} \qquad u = -\sqrt{|d|}\,i.$$
>
> These solutions can also be written as $u = \pm\sqrt{|d|}\,i$.

Example 3	Extracting Complex Square Roots

a. $x^2 + 8 = 0$ Original equation

$$x^2 = -8 \qquad \text{Subtract 8 from both sides.}$$

$$x = \pm\sqrt{8}\,i \qquad \text{Extract complex square roots.}$$

$$x = \pm 2\sqrt{2}\,i \qquad \text{Simplify.}$$

The solutions are $2\sqrt{2}\,i$ and $-2\sqrt{2}\,i$. Check these in the original equation.

b. $(x - 4)^2 = -3$ Original equation

$$x - 4 = \pm\sqrt{-3} \qquad \text{Extract complex square roots.}$$

$$x - 4 = \pm\sqrt{3}\,i \qquad \text{Write in } i\text{-form.}$$

$$x = 4 \pm \sqrt{3}\,i \qquad \text{Add 4 to both sides.}$$

The solutions are $4 + \sqrt{3}\,i$ and $4 - \sqrt{3}\,i$. Check these in the original equation.

c. $2(3x - 5)^2 + 32 = 0$ Original equation

$$2(3x - 5)^2 = -32 \qquad \text{Subtract 32 from both sides.}$$

$$(3x - 5)^2 = -16 \qquad \text{Divide both sides by 2.}$$

$$3x - 5 = \pm 4i \qquad \text{Extract complex square roots.}$$

$$3x = 5 \pm 4i \qquad \text{Add 5 to both sides.}$$

$$x = \frac{5 \pm 4i}{3} \qquad \text{Divide both sides by 3.}$$

The solutions are $(5 + 4i)/3$ and $(5 - 4i)/3$. Check these in the original equation.

Technology: Tip

When graphically checking solutions, it is important to realize that only the real solutions appear as x-intercepts—the complex solutions cannot be estimated from the graph. For instance, in Example 3(a), the graph of

$$y = x^2 + 8$$

has no x-intercepts. This agrees with the fact that both of its solutions are complex numbers.

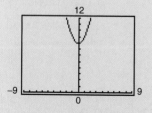

4 Use substitution to solve an equation of quadratic form.

Equations of Quadratic Form

Both the factoring and extraction of square roots methods can be applied to nonquadratic equations that are of **quadratic form.** An equation is said to be of quadratic form if it has the form

$$au^2 + bu + c = 0$$

where u is an algebraic expression. Here are some examples.

Equation	Written in Quadratic Form
$x^4 + 5x^2 + 4 = 0$	$(x^2)^2 + 5(x^2) + 4 = 0$
$x - 5\sqrt{x} + 6 = 0$	$(\sqrt{x})^2 - 5(\sqrt{x}) + 6 = 0$
$2x^{2/3} + 5x^{1/3} - 3 = 0$	$2(\sqrt[3]{x})^2 + 5(\sqrt[3]{x}) - 3 = 0$
$18 + 2x^2 + (x^2 + 9)^2 = 8$	$(x^2 + 9)^2 + 2(x^2 + 9) - 8 = 0$

To solve an equation of quadratic form, it helps to make a substitution and rewrite the equation in terms of u, as demonstrated in Examples 4 and 5.

Example 4   Solving an Equation of Quadratic Form

Solve $x^4 - 13x^2 + 36 = 0$.

Solution

Begin by writing the original equation in quadratic form, as follows.

$x^4 - 13x^2 + 36 = 0$	Original equation
$(x^2)^2 - 13(x^2) + 36 = 0$	Write in quadratic form.

Next, let $u = x^2$ and substitute u into the equation written in quadratic form. Then, factor and solve the equation.

$u^2 - 13u + 36 = 0$	Substitute u for x^2.
$(u - 4)(u - 9) = 0$	Factor.
$u - 4 = 0 \quad\Longrightarrow\quad u = 4$	Set 1st factor equal to 0.
$u - 9 = 0 \quad\Longrightarrow\quad u = 9$	Set 2nd factor equal to 0.

At this point you have found the "u-solutions." To find the "x-solutions," replace u by x^2 and solve for x.

$$u = 4 \quad\Longrightarrow\quad x^2 = 4 \quad\Longrightarrow\quad x = \pm 2$$
$$u = 9 \quad\Longrightarrow\quad x^2 = 9 \quad\Longrightarrow\quad x = \pm 3$$

The solutions are 2, -2, 3, and -3. Check these in the original equation.

Be sure you see in Example 4 that the u-solutions of 4 and 9 represent only a temporary step. They are not solutions of the original equation and cannot be substituted into the original equation.

Technology: Tip

You may find it helpful to graph the equation with a graphing utility before you begin. The graph will indicate the number of real solutions an equation has. For instance, the graph shown below is from the equation in Example 4. You can see from the graph that there are four real solutions.

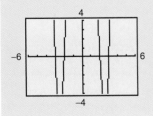

Example 5	Solving an Equation of Quadratic Form

Solve $x - 5\sqrt{x} + 6 = 0$.

Solution

This equation is of quadratic form with $u = \sqrt{x}$.

$$x - 5\sqrt{x} + 6 = 0 \qquad \text{Original equation}$$

$$\left(\sqrt{x}\right)^2 - 5\left(\sqrt{x}\right) + 6 = 0 \qquad \text{Write in quadratic form.}$$

$$u^2 - 5u + 6 = 0 \qquad \text{Substitute } u \text{ for } \sqrt{x}.$$

$$(u - 2)(u - 3) = 0 \qquad \text{Factor.}$$

$$u - 2 = 0 \quad\Longrightarrow\quad u = 2 \qquad \text{Set 1st factor equal to 0.}$$

$$u - 3 = 0 \quad\Longrightarrow\quad u = 3 \qquad \text{Set 2nd factor equal to 0.}$$

Now, using the u-solutions of 2 and 3, you obtain the following x-solutions.

$$u = 2 \quad\Longrightarrow\quad \sqrt{x} = 2 \quad\Longrightarrow\quad x = 4$$

$$u = 3 \quad\Longrightarrow\quad \sqrt{x} = 3 \quad\Longrightarrow\quad x = 9$$

Example 6	Surface Area of a Sphere

The surface area of a sphere of radius r is given by $S = 4\pi r^2$. If the surface area of a softball is $144/\pi$ square inches, find the diameter d of the softball.

Solution

$$\frac{144}{\pi} = 4\pi r^2 \qquad \text{Substitute } 144/\pi \text{ for } S.$$

$$\frac{36}{\pi^2} = r^2 \quad\Longrightarrow\quad \pm\sqrt{\frac{36}{\pi^2}} = r \qquad \begin{array}{l}\text{Divide both sides by } 4\pi \\ \text{and extract square roots.}\end{array}$$

Choosing the positive root, you get $r = 6/\pi$, and so the diameter of the softball is

$$d = 2r = 2\left(\frac{6}{\pi}\right) = \frac{12}{\pi} \approx 3.82 \text{ inches.}$$

Discussing the Concept	Analyzing Solutions of Quadratic Equations

Use a graphing utility to graph each of the following equations. How many times does the graph of each equation cross the x-axis?

a. $y = 2x^2 + x - 15$ **b.** $y = (3x - 1)^2 + 3$

Now set each equation equal to zero and solve the resulting equations. How many of each type of solution (real or complex) does each equation have? Summarize the relationship between the number of x-intercepts in the graph of a quadratic equation and the number and type of roots found algebraically.

6.1   Exercises

Integrated Review   Concepts, Skills, and Problem Solving

Keep mathematically in shape by doing these exercises *before* the problems of this section.

Properties and Definitions

1. Identify the leading coefficient in $5t - 3t^3 + 7t^2$. Explain.

2. State the degree of the product $(y^2 - 2)(y^3 + 7)$. Explain.

3. Sketch a graph for which y is not a function of x. Explain why it is not a function.

4. Sketch a graph for which y is a function of x. Explain why it is a function.

Simplifying Expressions

In Exercises 5–10, simplify the expression.

5. $(x^3 \cdot x^{-2})^{-3}$

6. $(5x^{-4}y^5)(-3x^2y^{-1})$

7. $\left(\dfrac{2x}{3y}\right)^{-2}$

8. $\left(\dfrac{7u^{-4}}{3v^{-2}}\right)\left(\dfrac{14u}{6v^2}\right)^{-1}$

9. $\dfrac{6u^2v^{-3}}{27uv^3}$

10. $\dfrac{-14r^4s^2}{-98rs^2}$

Problem Solving

11. The number N of prey t months after a natural predator is introduced into the test area is inversely proportional to the square root of $t + 1$. If $N = 300$ when $t = 0$, find N when $t = 8$.

12. The travel time between two cities is inversely proportional to the average speed. If a train travels between two cities in 2 hours at an average speed of 58 miles per hour, how long would it take at an average speed of 72 miles per hour? What does the constant of proportionality measure in this problem?

Developing Skills

In Exercises 1–20, solve the equation by factoring. See Example 1.

1. $x^2 - 12x + 35 = 0$

2. $x^2 + 15x + 44 = 0$

3. $x^2 + x - 72 = 0$

4. $x^2 - 2x - 48 = 0$

5. $x^2 + 4x = 45$

6. $x^2 - 7x = 18$

7. $x^2 - 12x + 36 = 0$

8. $x^2 + 60x + 900 = 0$

9. $9x^2 + 24x + 16 = 0$

10. $8x^2 - 10x + 3 = 0$

11. $4x^2 - 12x = 0$

12. $25y^2 - 75y = 0$

13. $u(u - 9) - 12(u - 9) = 0$

14. $16x(x - 8) - 12(x - 8) = 0$

15. $3x(x - 6) - 5(x - 6) = 0$

16. $3(4 - x) - 2x(4 - x) = 0$

17. $(y - 4)(y - 3) = 6$

18. $(6 + u)(1 - u) = 10$

19. $2x(3x + 2) = 5 - 6x^2$

20. $(2z + 1)(2z - 1) = -4z^2 - 5z + 2$

In Exercises 21–42, solve the quadratic equation by extracting square roots. See Example 2.

21. $x^2 = 64$

22. $z^2 = 169$

23. $6x^2 = 54$

24. $5t^2 = 125$

25. $25x^2 = 16$

26. $9z^2 = 121$

27. $\dfrac{y^2}{2} = 32$

28. $\dfrac{x^2}{6} = 24$

29. $4x^2 - 25 = 0$

30. $16y^2 - 121 = 0$

31. $4u^2 - 225 = 0$

32. $16x^2 - 1 = 0$

33. $(x + 4)^2 = 169$

34. $(y - 20)^2 = 625$

35. $(x - 3)^2 = 0.25$

36. $(x + 2)^2 = 0.81$

37. $(x - 2)^2 = 7$

38. $(x + 8)^2 = 28$

39. $(2x + 1)^2 = 50$

40. $(3x - 5)^2 = 48$

41. $(4x - 3)^2 - 98 = 0$

42. $(5x + 11)^2 - 300 = 0$

In Exercises 43-64, solve the equation by extracting complex square roots. See Example 3.

43. $z^2 = -36$

44. $x^2 = -9$

45. $x^2 + 4 = 0$

46. $y^2 + 16 = 0$

47. $9u^2 + 17 = 0$

48. $4v^2 + 9 = 0$

49. $(t - 3)^2 = -25$

50. $(x + 5)^2 = -81$

51. $(3z + 4)^2 + 144 = 0$

52. $(2y - 3)^2 + 25 = 0$

53. $(2x + 3)^2 = -54$

54. $(6y - 5)^2 = -8$

55. $9(x + 6)^2 = -121$

56. $4(x - 4)^2 = -169$

57. $(x - 1)^2 = -27$

58. $(2x + 3)^2 = -54$

59. $(x + 1)^2 + 0.04 = 0$

60. $(x - 3)^2 + 2.25 = 0$

61. $\left(c - \frac{2}{3}\right)^2 + \frac{1}{9} = 0$

62. $\left(u + \frac{5}{8}\right)^2 + \frac{49}{16} = 0$

63. $\left(x + \frac{7}{3}\right)^2 = -\frac{38}{9}$

64. $\left(y - \frac{5}{6}\right)^2 = -\frac{4}{5}$

In Exercises 65–80, find all real and complex solutions of the equation.

65. $2x^2 - 5x = 0$

66. $3t^2 + 6t = 0$

67. $2x^2 + 5x - 12 = 0$

68. $3x^2 + 8x - 16 = 0$

69. $x^2 - 900 = 0$

70. $y^2 - 225 = 0$

71. $x^2 + 900 = 0$

72. $y^2 + 225 = 0$

73. $\frac{2}{3}x^2 = 6$

74. $\frac{1}{3}x^2 = 4$

75. $(x - 5)^2 - 100 = 0$

76. $(y + 12)^2 - 400 = 0$

77. $(x - 5)^2 + 100 = 0$

78. $(y + 12)^2 + 400 = 0$

79. $(x + 2)^2 + 18 = 0$

80. $(x + 2)^2 - 18 = 0$

In Exercises 81–90, use a graphing utility to graph the function. Use the graph to approximate any x-intercepts. Set $y = 0$ and solve the resulting equation. Compare the result with the x-intercepts of the graph.

81. $y = x^2 - 9$

82. $y = 5x - x^2$

83. $y = x^2 - 2x - 15$

84. $y = 9 - 4(x - 3)^2$

85. $y = 4 - (x - 3)^2$

86. $y = 4(x + 1)^2 - 9$

87. $y = 2x^2 - x - 6$

88. $y = 4x^2 - x - 14$

89. $y = 3x^2 - 8x - 16$

90. $y = 5x^2 + 9x - 18$

In Exercises 91–96, use a graphing utility to graph the function and observe that the graph has no x-intercepts. Set $y = 0$ and solve the resulting equation. Identify the type of roots of the equation.

91. $y = x^2 + 7$

92. $y = x^2 + 5$

93. $y = (x - 1)^2 + 1$

94. $y = (x + 2)^2 + 3$

95. $y = (x + 3)^2 + 5$

96. $y = (x - 2)^2 + 3$

In Exercises 97–100, solve for y in terms of x. Let f and g be those functions where f represents the positive square root and g the negative square root. Use a graphing utility to sketch the graphs of f and g in the same viewing rectangle.

97. $x^2 + y^2 = 4$

98. $x^2 - y^2 = 4$

99. $x^2 + 4y^2 = 4$

100. $x - y^2 = 0$

In Exercises 101–120, solve the equation of quadratic form. (Find all real *and* complex solutions.) See Examples 4 and 5.

101. $x^4 - 5x^2 + 4 = 0$

102. $x^4 - 10x^2 + 25 = 0$

103. $x^4 - 5x^2 + 6 = 0$

104. $x^4 - 11x^2 + 30 = 0$

105. $x^4 - 3x^2 - 4 = 0$

106. $x^4 - x^2 - 6 = 0$

107. $(x^2 - 4)^2 + 2(x^2 - 4) - 3 = 0$

108. $(x^2 - 1)^2 + (x^2 - 1) - 6 = 0$

109. $x - 7\sqrt{x} + 10 = 0$

110. $x - 11\sqrt{x} + 24 = 0$

111. $x^{2/3} - x^{1/3} - 6 = 0$

112. $x^{2/3} + 3x^{1/3} - 10 = 0$

113. $2x^{2/3} - 7x^{1/3} + 5 = 0$

114. $3x^{2/3} + 8x^{1/3} + 5 = 0$

115. $x^{2/5} - 3x^{1/5} + 2 = 0$

116. $x^{2/5} + 5x^{1/5} + 6 = 0$

117. $2x^{2/5} - 7x^{1/5} + 3 = 0$

118. $2x^{2/5} + 3x^{1/5} + 1 = 0$

119. $\frac{1}{x^2} - \frac{3}{x} + 2 = 0$

120. $3\left(\frac{x}{x + 1}\right)^2 + 7\left(\frac{x}{x + 1}\right) - 6 = 0$

Think About It In Exercises 121–126, find a quadratic equation having the given solutions.

121. $5, -2$

122. $-2, 3$

123. $1 + \sqrt{2}, 1 - \sqrt{2}$

124. $-3 + \sqrt{5}, -3 - \sqrt{5}$

125. $5i, -5i$

126. $2i, -2i$

Solving Problems

Free-Falling Object   In Exercises 127–130, find the time required for an object to reach the ground when it is dropped from a height of s_0 feet. The height h (in feet) is given by

$$h = -16t^2 + s_0$$

where t measures time in seconds from the time when the object is released.

127. $s_0 = 256$
128. $s_0 = 48$
129. $s_0 = 128$
130. $s_0 = 500$

131. *Free-Falling Object*   The height h (in feet) of an object thrown vertically upward from a tower 144 feet tall is given by

$$h = 144 + 128t - 16t^2$$

where t measures the time in seconds from the time when the object is released. How long does it take for the object to reach the ground?

132. *Revenue*   The revenue R (in dollars) when x units of a product are sold is given by

$$R = x\left(120 - \frac{1}{2}x\right).$$

Determine the number of units that must be sold to produce a revenue of $7000.

Compound Interest   The amount A after 2 years when a principal of P dollars is invested at percentage rate r compounded annually is given by

$$A = P(1 + r)^2.$$

In Exercises 133 and 134, find r.

133. $P = \$1500$, $A = \$1685.40$

134. $P = \$5000$, $A = \$5724.50$

National Health Expenditures   In Exercises 135 and 136, use the following model, which gives the national expenditures for health care in the United States from 1990 through 1996.

$$y = (26.6 + t)^2, \quad 0 \le t \le 6$$

In this model, y represents the expenditures (in billions of dollars) and t represents the year, with $t = 0$ corresponding to 1990 (see figure).   (Source: U.S. Health Care Financing Administration)

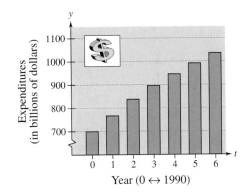

Year ($0 \leftrightarrow 1990$)

135. Analytically determine the year when expenditures were approximately $892 billion. Graphically confirm the result.

136. Analytically determine the year when expenditures were approximately $1000 billion. Graphically confirm the result.

Explaining Concepts

137. Answer parts (a) and (b) of Motivating the Chapter on page 371.

138. For a quadratic equation $ax^2 + bx + c = 0$, where a, b, and c are real numbers with $a \neq 0$, explain why b and c can equal 0, but a cannot.

139. Explain the Zero-Factor Property and how it can be used to solve a quadratic equation.

140. Is it possible for a quadratic equation to have only one solution? If so, give an example.

141. *True or False?*   The only solution of the equation $x^2 = 25$ is $x = 5$. Explain.

142. Describe the steps in solving a quadratic equation by extracting square roots.

143. Describe the procedure for solving an equation in quadratic form. Give an example.

6.2   Completing the Square

Objectives

1 Rewrite a quadratic equation in completed square form.

2 Solve a quadratic equation by completing the square.

---

1 Rewrite a quadratic equation in completed square form.

Constructing Perfect Square Trinomials

Consider the quadratic equation

$$(x - 2)^2 = 10.$$ Completed square form

You know from Example 2(b) in the preceding section that this equation has two solutions: $2 + \sqrt{10}$ and $2 - \sqrt{10}$. Suppose you had been given the equation in its general form

$$x^2 - 4x - 6 = 0.$$ General form

How would you solve this equation if you were given only the general form? You could try factoring, but after attempting to do so you would find that the left side of the equation is not factorable (using integer coefficients).

In this section, you will study a technique for rewriting an equation in a completed square form. This technique is called **completing the square.** Note that prior to completing the square, the coefficient of the second-degree term must be 1.

> ▶ Completing the Square
>
> To **complete the square** for the expression $x^2 + bx$, add $(b/2)^2$, which is the square of half the coefficient of x. Consequently,
>
> $$x^2 + bx + \left(\frac{b}{2}\right)^2 = \left(x + \frac{b}{2}\right)^2.$$

Example 1   Creating a Perfect Square Trinomial

What term should be added to $x^2 - 8x$ so that it becomes a perfect square trinomial?

Solution

For this expression, the coefficient of the x-term is -8. Divide this term by 2, and square the result to obtain $(-4)^2 = 16$. This is the term that should be added to the expression to make it a perfect square trinomial.

$$x^2 - 8x + 16 = x^2 - 8x + (-4)^2$$ Add 16 to the expression.

$$= (x - 4)^2$$ Completed square form

2 Solve a quadratic equation by completing the square.

Solving Equations by Completing the Square

When completing the square to solve an equation, remember that it is essential to *preserve the equality*. Thus, when you add a constant term to one side of the equation, you must be sure to add the same constant to the other side of the equation.

Study Tip

Completing the square can be used to solve *any* quadratic equation. However, sometimes it is easier to factor an equation than to complete the square. For instance, the equation in Example 2 could easily be factored as $x(x + 12) = 0$. But remember, not all equations are factorable. Don't spend a lot of time trying to factor when you know that completing the square will work.

| **Example 2** | Completing the Square: Leading Coefficient Is 1 |

Solve $x^2 + 12x = 0$.

Solution

$$x^2 + 12x = 0 \qquad \text{Original equation}$$

$$x^2 + 12x + (6)^2 = 36 \qquad \text{Add } \left(\tfrac{12}{2}\right)^2 = 36 \text{ to both sides.}$$

$$\underset{\text{(half)}^2}{\underline{\qquad}}$$

$$(x + 6)^2 = 36 \qquad \text{Completed square form}$$

$$x + 6 = \pm\sqrt{36} \qquad \text{Extract square roots.}$$

$$x = -6 \pm 6 \qquad \text{Subtract 6 from both sides.}$$

$$x = -6 + 6 \quad \text{or} \quad x = -6 - 6 \qquad \text{Separate solutions.}$$

$$x = 0 \qquad\qquad x = -12 \qquad \text{Solutions}$$

The solutions are 0 and -12. Check these in the original equation.

| **Example 3** | Completing the Square: Leading Coefficient Is 1 |

Solve $x^2 - 6x + 7 = 0$.

Solution

$$x^2 - 6x + 7 = 0 \qquad \text{Original equation}$$

$$x^2 - 6x = -7 \qquad \text{Subtract 7 from both sides.}$$

$$x^2 - 6x + (-3)^2 = -7 + 9 \qquad \text{Add } \left(-\tfrac{6}{2}\right)^2 = 9 \text{ to both sides.}$$

$$\underset{\text{(half)}^2}{\underline{\qquad}}$$

$$(x - 3)^2 = 2 \qquad \text{Completed square form}$$

$$x - 3 = \pm\sqrt{2} \qquad \text{Extract square roots.}$$

$$x = 3 \pm \sqrt{2} \qquad \text{Add 3 to both sides.}$$

$$x = 3 + \sqrt{2} \quad \text{or} \quad x = 3 - \sqrt{2} \qquad \text{Separate solutions.}$$

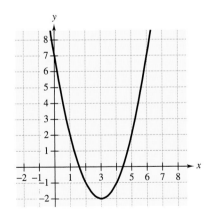

Figure 6.1

The solutions are $3 + \sqrt{2}$ and $3 - \sqrt{2}$. Check these in the original equation. Also try checking the solutions graphically, as shown in Figure 6.1.

If the leading coefficient of a quadratic expression is not 1, you must divide both sides of the equation by this coefficient *before* completing the square. This process is demonstrated in Example 4.

Example 4 A Leading Coefficient That Is Not 1

Solve $2x^2 - x - 2 = 0$.

Solution

$$2x^2 - x - 2 = 0 \qquad \text{Original equation}$$

$$2x^2 - x = 2 \qquad \text{Add 2 to both sides.}$$

$$x^2 - \frac{1}{2}x = 1 \qquad \text{Divide both sides by 2.}$$

$$x^2 - \frac{1}{2}x + \left(-\frac{1}{4}\right)^2 = 1 + \frac{1}{16} \qquad \text{Add } \left(-\frac{1}{4}\right)^2 = \frac{1}{16} \text{ to both sides.}$$

$$\left(x - \frac{1}{4}\right)^2 = \frac{17}{16} \qquad \text{Completed square form}$$

$$x - \frac{1}{4} = \pm\frac{\sqrt{17}}{4} \qquad \text{Extract square roots.}$$

$$x = \frac{1}{4} \pm \frac{\sqrt{17}}{4} \qquad \text{Add } \frac{1}{4} \text{ to both sides.}$$

The solutions are $\frac{1}{4}\left(1 + \sqrt{17}\right)$ and $\frac{1}{4}\left(1 - \sqrt{17}\right)$.

Example 5 A Quadratic Equation with Complex Solutions

Solve $x^2 - 4x + 8 = 0$.

Solution

$$x^2 - 4x + 8 = 0 \qquad \text{Original equation}$$

$$x^2 - 4x = -8 \qquad \text{Subtract 8 from both sides.}$$

$$x^2 - 4x + (-2)^2 = -8 + 4 \qquad \text{Add } (-2)^2 = 4 \text{ to both sides.}$$

$$(x - 2)^2 = -4 \qquad \text{Completed square form}$$

$$x - 2 = \pm 2i \qquad \text{Extract complex square roots.}$$

$$x = 2 \pm 2i \qquad \text{Add 2 to both sides.}$$

The solutions are $2 + 2i$ and $2 - 2i$. The first of these is checked below.

Check

$$x^2 - 4x + 8 = 0 \qquad \text{Original equation}$$

$$(2 + 2i)^2 - 4(2 + 2i) + 8 \overset{?}{=} 0 \qquad \text{Substitute } 2 + 2i \text{ for } x.$$

$$4 + 8i - 4 - 8 - 8i + 8 \overset{?}{=} 0 \qquad \text{Simplify.}$$

$$0 = 0 \qquad \text{Solution checks. } \checkmark$$

Figure 6.2

| Example 6 | Dimensions of a Cereal Box |

A cereal box has a volume of 441 cubic inches. Its height is 12 inches and its base has the dimensions x by $x + 7$. (See Figure 6.2.) Find the dimensions of the base in inches.

Solution

$$lwh = V$$ Formula for volume of a rectangular box

$$(x + 7)(x)(12) = 441$$ Substitute 441 for V, $x + 7$ for length, x for width, and 12 for height.

$$12x^2 + 84x = 441$$ Multiply factors.

$$x^2 + 7x = \frac{441}{12}$$ Divide both sides by 12.

$$x^2 + 7x + \left(\frac{7}{2}\right)^2 = \frac{147}{4} + \frac{49}{4}$$ Add $\left(\frac{7}{2}\right)^2 = \frac{49}{4}$ to both sides.

$$\left(x + \frac{7}{2}\right)^2 = \frac{196}{4}$$ Completed square form

$$x + \frac{7}{2} = \pm\sqrt{49}$$ Extract square roots.

$$x = -\frac{7}{2} \pm 7$$ Subtract $\frac{7}{2}$ from both sides.

Choosing the positive root, you get

$$x = -\frac{7}{2} + 7 = \frac{7}{2} = 3.5 \text{ inches}$$ Width of base

and

$$x + 7 = \frac{7}{2} + 7 = \frac{21}{2} = 10.5 \text{ inches.}$$ Length of base

---

| Discussing the Concept | Error Analysis |

Suppose you teach an algebra class and one of your students hands in the following solution. Find and correct the error(s). Discuss how to explain the error(s) to your student.

1. Solve $x^2 + 6x - 13 = 0$ by completing the square.

$$x^2 + 6x = 13$$

$$x^2 + 6x + \left(\frac{6}{2}\right)^2 = 13$$

$$(x + 3)^2 = 13$$

$$x + 3 = \pm\sqrt{13}$$

$$x = -3 \pm \sqrt{13}$$

6.2 Exercises

Integrated Review *Concepts, Skills, and Problem Solving*

Keep mathematically in shape by doing these exercises *before* the problems of this section.

Properties and Definitions

In Exercises 1–4, complete the property of exponents and/or simplify.

1. $(ab)^4 =$
2. $(a^r)^s =$

3. $\left(\dfrac{a}{b}\right)^{-r} =$, $a \neq 0, b \neq 0$

4. $a^{-r} =$, $a \neq 0$

Solving Equations

In Exercises 5–8, solve the equation.

5. $\dfrac{4}{x} - \dfrac{2}{3} = 0$

6. $2x - 3[1 + (4 - x)] = 0$

7. $3x^2 - 13x - 10 = 0$

8. $x(x - 3) = 40$

Graphing

In Exercises 9–12, graph the function.

9. $g(x) = \frac{2}{3}x - 5$ **10.** $h(x) = 5 - \sqrt{x}$

11. $f(x) = \dfrac{4}{x + 2}$ **12.** $f(x) = 2x + |x - 1|$

Developing Skills

In Exercises 1–16, add a term to the expression so that it becomes a perfect square trinomial. See Example 1.

1. $x^2 + 8x +$
2. $x^2 + 12x +$

3. $y^2 - 20y +$
4. $y^2 - 2y +$

5. $x^2 - 16x +$
6. $x^2 + 18x +$

7. $t^2 + 5t +$
8. $u^2 + 7u +$

9. $x^2 - 9x +$
10. $y^2 - 11y +$

11. $a^2 - \frac{1}{3}a +$
12. $y^2 + \frac{4}{3}y +$

13. $y^2 - \frac{3}{5}y +$
14. $x^2 - \frac{6}{5}x +$

15. $r^2 - 0.4r +$
16. $s^2 + 4.6s +$

In Exercises 17–34, solve the quadratic equation (a) by completing the square and (b) by factoring. See Examples 2–4.

17. $x^2 - 20x = 0$ **18.** $x^2 + 32x = 0$

19. $x^2 + 6x = 0$ **20.** $t^2 - 10t = 0$

21. $y^2 - 5y = 0$ **22.** $t^2 - 9t = 0$

23. $t^2 - 8t + 7 = 0$ **24.** $y^2 - 8y + 12 = 0$

25. $x^2 + 2x - 24 = 0$ **26.** $x^2 + 12x + 27 = 0$

27. $x^2 + 7x + 12 = 0$ **28.** $z^2 + 3z - 10 = 0$

29. $x^2 - 3x - 18 = 0$ **30.** $t^2 - 5t - 36 = 0$

31. $2x^2 - 14x + 12 = 0$ **32.** $3x^2 - 3x - 6 = 0$

33. $4x^2 + 4x - 15 = 0$ **34.** $3x^2 - 13x + 12 = 0$

In Exercises 35–72, solve the quadratic equation by completing the square. Give the solutions in exact form and in decimal form rounded to two decimal places. (The solutions may be complex numbers.) See Examples 2–5.

35. $x^2 - 4x - 3 = 0$ **36.** $x^2 - 6x + 7 = 0$

37. $x^2 + 4x - 3 = 0$ **38.** $x^2 + 6x + 7 = 0$

39. $u^2 - 4u + 1 = 0$ **40.** $a^2 - 10a - 15 = 0$

41. $x^2 + 2x + 3 = 0$ **42.** $x^2 - 6x + 12 = 0$

43. $x^2 - 10x - 2 = 0$ **44.** $x^2 + 8x - 4 = 0$

45. $y^2 + 20y + 10 = 0$ **46.** $y^2 + 6y - 24 = 0$

47. $t^2 + 5t + 3 = 0$ **48.** $u^2 - 9u - 1 = 0$

49. $v^2 + 3v - 2 = 0$ **50.** $z^2 - 7z + 9 = 0$

51. $-x^2 + x - 1 = 0$ **52.** $1 - x - x^2 = 0$

53. $x^2 - 7x + 12 = 0$ **54.** $y^2 + 5y + 9 = 0$

55. $x^2 - \frac{2}{3}x - 3 = 0$ **56.** $x^2 + \frac{4}{5}x - 1 = 0$

57. $v^2 + \frac{3}{4}v - 2 = 0$ **58.** $u^2 - \frac{2}{3}u + 5 = 0$

59. $2x^2 + 8x + 3 = 0$ **60.** $3x^2 - 24x - 5 = 0$

61. $3x^2 + 9x + 5 = 0$ **62.** $5x^2 - 15x + 7 = 0$

63. $4y^2 + 4y - 9 = 0$ **64.** $4z^2 - 3z + 2 = 0$

65. $5x^2 - 3x + 10 = 0$ **66.** $7x^2 + 4x + 3 = 0$

67. $x(x - 7) = 2$ **68.** $2x\left(x + \dfrac{4}{3}\right) = 5$

69. $0.5t^2 + t + 2 = 0$ **70.** $0.1x^2 + 0.5x = -0.2$

71. $0.1x^2 + 0.2x + 0.5 = 0$

72. $0.02x^2 + 0.10x - 0.05 = 0$

In Exercises 73–78, find the real solutions.

73. $\dfrac{x}{2} - \dfrac{1}{x} = 1$ **74.** $\dfrac{x}{2} + \dfrac{5}{x} = 4$

75. $\dfrac{x^2}{4} = \dfrac{x + 1}{2}$ **76.** $\dfrac{x^2 + 2}{24} = \dfrac{x - 1}{3}$

77. $\sqrt{2x + 1} = x - 3$ **78.** $\sqrt{3x - 2} = x - 2$

In Exercises 79–86, use a graphing utility to graph the function. Use the graph to approximate any x-intercepts of the graph. Set $y = 0$ and solve the resulting equation. Compare the result with the x-intercepts of the graph.

79. $y = x^2 + 4x - 1$ **80.** $y = x^2 + 6x - 4$

81. $y = x^2 - 2x - 5$ **82.** $y = 2x^2 - 6x - 5$

83. $y = \tfrac{1}{3}x^2 + 2x - 6$ **84.** $y = \tfrac{1}{2}x^2 - 3x + 1$

85. $y = -x^2 - x + 3$ **86.** $y = \sqrt{x} - x + 2$

Solving Problems

87. *Geometric Modeling*

(a) Find the area of the two adjoining rectangles and large square in the figure.

(b) Find the area of the small square region in the lower right-hand corner of the figure and add it to the area found in part (a).

(c) Find the dimensions and the area of the entire figure after adjoining the small square in the lower right-hand corner of the figure. Note that you have shown geometrically the technique of completing the square.

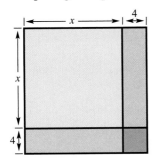

88. *Completing the Square*   Repeat Exercise 87 for the model shown below.

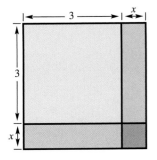

89. *Geometry*   Find the dimensions of the triangle in the figure if its area is 12 square centimeters.

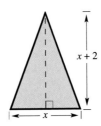

90. *Geometry*   The area of the rectangle in the figure is 160 square feet. Find the rectangle's dimensions.

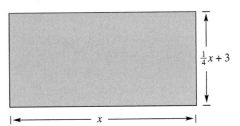

91. *Fencing In a Corral*   You have 200 meters of fencing to enclose two adjacent rectangular corrals (see figure). The total area of the enclosed region is 1400 square meters. What are the dimensions of each corral? (The corrals are the same size.)

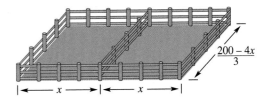

92. *Geometry* An open box with a rectangular base of x inches by $x + 4$ inches has a height of 6 inches (see figure). Find the dimensions of the box if its volume is 840 cubic inches.

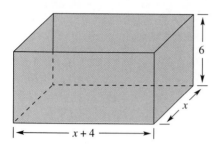

93. *Cutting Across the Lawn* On the sidewalk, the distance from the dormitory to the cafeteria is 400 meters (see figure). By cutting across the lawn, the walking distance is shortened to 300 meters. How long is each part of the L-shaped sidewalk?

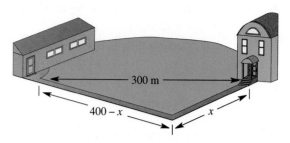

Figure for 93

94. *Revenue* The revenue R from selling x units of a certain product is

$$R = x\left(50 - \frac{1}{2}x\right).$$

Find the number of units that must be sold to produce a revenue of $1218.

95. *Revenue* The revenue R from selling x units of a certain product is

$$R = x\left(100 - \frac{1}{10}x\right).$$

Find the number of units that must be sold to produce a revenue of $12,000.

Explaining Concepts

96. What is a perfect square trinomial?

97. What term must be added to $x^2 + 5x$ to complete the square? Explain how you found the term.

98. Explain the use of extracting square roots when solving a quadratic equation by the method of completing the square.

99. Is it possible for a quadratic equation to have no real number solution? If so, give an example.

100. When using the method of completing the square to solve a quadratic equation, what is the first step if the leading coefficient is not 1? Is the resulting equation equivalent to the given equation? Explain.

101. *True or False?* If you solve a quadratic equation by completing the square and obtain solutions that are rational numbers, then you could have solved the equation by factoring. Explain.

102. Consider the following quadratic equation.

$$(x - 1)^2 = d$$

(a) What value(s) of d will produce a quadratic equation that has exactly one (repeated) solution?

(b) Describe the value(s) of d that will produce two different solutions, both of which are rational numbers.

(c) Describe the value(s) of d that will produce two different solutions, both of which are irrational numbers.

(d) Describe the value(s) of d that will produce two different solutions, both of which are complex numbers.

6.3   The Quadratic Formula

Objectives

1 Derive the Quadratic Formula by completing the square for a general quadratic equation.

2 Use the Quadratic Formula to solve a quadratic equation.

3 Determine the type of solution to a quadratic equation using the discriminant.

1 Derive the Quadratic Formula by completing the square for a general quadratic equation.

The Quadratic Formula

A fourth technique for solving a quadratic equation involves the **Quadratic Formula.** This formula is derived by completing the square for a general quadratic equation.

$$ax^2 + bx + c = 0 \qquad \text{General form, } a \neq 0$$

$$ax^2 + bx = -c \qquad \text{Subtract } c \text{ from both sides.}$$

$$x^2 + \frac{b}{a}x = -\frac{c}{a} \qquad \text{Divide both sides by } a.$$

$$x^2 + \frac{b}{a}x + \left(\frac{b}{2a}\right)^2 = -\frac{c}{a} + \left(\frac{b}{2a}\right)^2 \qquad \text{Add } \left(\frac{b}{2a}\right)^2 \text{ to both sides.}$$

$$\left(x + \frac{b}{2a}\right)^2 = \frac{b^2 - 4ac}{4a^2} \qquad \text{Simplify.}$$

$$x + \frac{b}{2a} = \pm\sqrt{\frac{b^2 - 4ac}{4a^2}} \qquad \text{Extract square roots.}$$

$$x = -\frac{b}{2a} \pm \frac{\sqrt{b^2 - 4ac}}{2|a|} \qquad \text{Subtract } \frac{b}{2a} \text{ from both sides.}$$

$$x = \frac{-b \pm \sqrt{b^2 - 4ac}}{2a} \qquad \text{Simplify.}$$

Study Tip

The Quadratic Formula is one of the most important formulas in algebra, and you should memorize it. We have found that it helps to try to memorize a verbal statement of the rule. For instance, you might try to remember the following verbal statement of the Quadratic Formula: "Minus b, plus or minus the square root of b squared minus $4ac$, all divided by $2a$."

▶ **The Quadratic Formula**

The solutions of $ax^2 + bx + c = 0$, $a \neq 0$, are given by the **Quadratic Formula**

$$x = \frac{-b \pm \sqrt{b^2 - 4ac}}{2a}.$$

The expression inside the radical, $b^2 - 4ac$, is called the **discriminant.**

1. If $b^2 - 4ac > 0$, the equation has two real solutions.

2. If $b^2 - 4ac = 0$, the equation has one (repeated) real solution.

3. If $b^2 - 4ac < 0$, the equation has no real solutions.

2 Use the Quadratic Formula to solve a quadratic equation.

Solving Equations by the Quadratic Formula

When using the Quadratic Formula, remember that *before* the formula can be applied, you must first write the quadratic equation in general form in order to determine the values of a, b, and c.

> **Example 1** The Quadratic Formula: Two Distinct Solutions

$$x^2 + 6x = 16 \qquad \text{Original equation}$$

$$x^2 + 6x - 16 = 0 \qquad \text{Write in general form.}$$

$$x = \frac{-b \pm \sqrt{b^2 - 4ac}}{2a} \qquad \text{Quadratic Formula}$$

$$x = \frac{-6 \pm \sqrt{6^2 - 4(1)(-16)}}{2(1)} \qquad \text{Substitute: } a = 1, b = 6, c = -16.$$

$$x = \frac{-6 \pm \sqrt{100}}{2} \qquad \text{Simplify.}$$

$$x = \frac{-6 \pm 10}{2} \qquad \text{Simplify.}$$

$$x = 2 \quad \text{or} \quad x = -8 \qquad \text{Solutions}$$

The solutions are 2 and -8. Check these in the original equation.

Study Tip

In Example 1, the solutions are rational numbers, which means that the equation could have been solved by factoring. Try solving the equation by factoring.

> **Example 2** The Quadratic Formula: Two Distinct Solutions

$$-x^2 - 4x + 8 = 0 \qquad \text{Leading coefficient is negative.}$$

$$x^2 + 4x - 8 = 0 \qquad \text{Multiply both sides by } -1.$$

$$x = \frac{-b \pm \sqrt{b^2 - 4ac}}{2a} \qquad \text{Quadratic Formula}$$

$$x = \frac{-4 \pm \sqrt{4^2 - 4(1)(-8)}}{2(1)} \qquad \text{Substitute: } a = 1, b = 4, c = -8.$$

$$x = \frac{-4 \pm \sqrt{48}}{2} \qquad \text{Simplify.}$$

$$x = \frac{-4 \pm 4\sqrt{3}}{2} \qquad \text{Simplify.}$$

$$x = \frac{2(-2 \pm 2\sqrt{3})}{2} \qquad \text{Factor numerator.}$$

$$x = \frac{2(-2 \pm 2\sqrt{3})}{2} \qquad \text{Cancel common factor.}$$

$$x = -2 \pm 2\sqrt{3} \qquad \text{Solutions}$$

The solutions are $-2 + 2\sqrt{3}$ and $-2 - 2\sqrt{3}$. Check these in the original equation.

Study Tip

If the leading coefficient of a quadratic equation is negative, we suggest that you begin by multiplying both sides of the equation by -1, as shown in Example 2. This will produce a positive leading coefficient, which is less cumbersome to work with.

Study Tip

Example 3 could have been solved as follows, without dividing both sides by 2 in the first step.

$$x = \frac{-(-24) \pm \sqrt{(-24)^2 - 4(18)(8)}}{2(18)}$$

$$x = \frac{24 \pm \sqrt{576 - 576}}{36}$$

$$x = \frac{24 \pm 0}{36}$$

$$x = \frac{2}{3}$$

While the result is the same, dividing both sides by 2 simplifies the equation before the Quadratic Formula is applied and so allows you to work with smaller numbers.

Example 3   The Quadratic Formula: One Repeated Solution

$$18x^2 - 24x + 8 = 0$$ Original equation

$$9x^2 - 12x + 4 = 0$$ Divide both sides by 2.

$$x = \frac{-b \pm \sqrt{b^2 - 4ac}}{2a}$$ Quadratic Formula

$$x = \frac{-(-12) \pm \sqrt{(-12)^2 - 4(9)(4)}}{2(9)}$$ Substitute 9 for a, -12 for b, and 4 for c.

$$x = \frac{12 \pm \sqrt{144 - 144}}{18}$$ Simplify.

$$x = \frac{12 \pm \sqrt{0}}{18}$$ Simplify.

$$x = \frac{2}{3}$$ Solution

The only solution is $\frac{2}{3}$. Check this in the original equation.

Note in the next example how the Quadratic Formula can be used to solve a quadratic equation that has complex solutions.

Example 4   The Quadratic Formula: Complex Solutions

$$2x^2 - 4x + 5 = 0$$ Original equation

$$x = \frac{-b \pm \sqrt{b^2 - 4ac}}{2a}$$ Quadratic Formula

$$x = \frac{-(-4) \pm \sqrt{(-4)^2 - 4(2)(5)}}{2(2)}$$ Substitute 2 for a, -4 for b, and 5 for c.

$$x = \frac{4 \pm \sqrt{-24}}{4}$$ Simplify.

$$x = \frac{4 \pm 2\sqrt{6}i}{4}$$ Write in i-form.

$$x = \frac{2(2 \pm \sqrt{6}i)}{2 \cdot 2}$$ Factor numerator and denominator.

$$x = \frac{2(2 \pm \sqrt{6}i)}{2 \cdot 2}$$ Cancel common factor.

$$x = \frac{2 \pm \sqrt{6}i}{2}$$ Solutions

The solutions are $\frac{1}{2}(2 + \sqrt{6}i)$ and $\frac{1}{2}(2 - \sqrt{6}i)$. Check these in the original equation.

3 Determine the type of solution to a quadratic equation using the discriminant.

The Discriminant

The radicand in the Quadratic Formula, $b^2 - 4ac$, is called the discriminant because it allows you to "discriminate" among different types of solutions.

Study Tip

From Examples 1–4, you can see that equations with rational or repeated solutions could have been solved by factoring. A quick calculation of the discriminant will help you decide which solution method to use to solve a quadratic equation.

1. Use factoring if

$$b^2 - 4ac \text{ is } \begin{cases} \text{zero} \\ \text{or a} \\ \text{perfect square} \end{cases}.$$

2. Use completing the square or the Quadratic Formula if

$$b^2 - 4ac \text{ is } \begin{cases} \text{negative} \\ \text{or not} \\ \text{a perfect square} \end{cases}.$$

▶ **Using the Discriminant**

Let a, b, and c be rational numbers such that $a \neq 0$. The discriminant of the quadratic equation $ax^2 + bx + c = 0$ is given by $b^2 - 4ac$, and can be used to classify the solutions of the equation as follows.

Discriminant: $b^2 - 4ac$	Solution Types
1. Perfect square	Two distinct rational solutions (Example 1)
2. Positive nonperfect square	Two distinct irrational solutions (Example 2)
3. Zero	One repeated rational solution (Example 3)
4. Negative number	Two distinct imaginary solutions (Example 4)

Example 5 Using the Discriminant

Determine the type of solution for each quadratic equation.

a. $x^2 - x + 2 = 0$ b. $2x^2 - 3x - 2 = 0$
c. $x^2 - 2x + 1 = 0$ d. $x^2 - 2x - 1 = 9$

Solution

Equation	Discriminant	Solution Types
a. $x^2 - x + 2 = 0$	$b^2 - 4ac = (-1)^2 - 4(1)(2)$ $= 1 - 8$ $= -7$	Two distinct imaginary solutions
b. $2x^2 - 3x - 2 = 0$	$b^2 - 4ac = (-3)^2 - 4(2)(-2)$ $= 9 + 16$ $= 25$	Two distinct rational solutions
c. $x^2 - 2x + 1 = 0$	$b^2 - 4ac = (-2)^2 - 4(1)(1)$ $= 4 - 4$ $= 0$	One repeated rational solution
d. $x^2 - 2x - 1 = 9$	$b^2 - 4ac = (-2)^2 - 4(1)(-10)$ $= 4 + 40$ $= 44$	Two distinct irrational solutions

Technology: Discovery

Use a graphing utility to graph the equations below.

a. $y = x^2 - x + 2$

b. $y = 2x^2 - 3x - 2$

c. $y = x^2 - 2x + 1$

d. $y = x^2 - 2x - 10$

Describe the solution type of each equation and check your results with those shown in Example 5. Why do you think the discriminant is used to determine solution types?

▶ Summary of Methods for Solving Quadratic Equations

Method	*Example*
1. Factoring	$3x^2 + x = 0$
	$x(3x + 1) = 0 \implies x = 0 \quad \text{and} \quad 3x + 1 = 0$
2. Extracting square roots	$(x + 2)^2 = 9$
	$x + 2 = \pm 3 \implies x = -2 + 3 = 1 \quad \text{and} \quad x = -2 - 3 = -5$
3. Completing the square	$x^2 + 6x = 3$
	$x^2 + 6x + \left(\frac{1}{2} \cdot 6\right)^2 = 3 + \left(\frac{1}{2} \cdot 6\right)^2$
	$(x + 3)^2 = 12 \implies x = -3 \pm \sqrt{12}$
4. Using the Quadratic Formula	$3x^2 - 2x + 2 = 0 \implies x = \dfrac{-(-2) \pm \sqrt{(-2)^2 - 4(3)(2)}}{2(3)} = \dfrac{1 \pm \sqrt{5}i}{3}$

Technology:
Tip

A graphing utility program that uses the Quadratic Formula to solve quadratic equations can be found at our website, *www.hmco.com*. Programs are available for several current models of graphing utilities.

Example 6   Using a Calculator with the Quadratic Formula

Solve $1.2x^2 - 17.8x + 8.05 = 0$.

Solution

Using the Quadratic Formula, you can write

$$x = \frac{-(-17.8) \pm \sqrt{(-17.8)^2 - 4(1.2)(8.05)}}{2(1.2)}.$$

To evaluate these solutions, begin by calculating the square root.

$$17.8 \boxed{+/-} \boxed{x^2} \boxed{-} 4 \boxed{\times} 1.2 \boxed{\times} 8.05 \boxed{=} \boxed{\sqrt{}} \qquad \text{Scientific}$$

$$\boxed{\sqrt{}} \boxed{(} \boxed{(} \boxed{(-)} 17.8 \boxed{)} \boxed{x^2} \boxed{-} 4 \boxed{\times} 1.2 \qquad \text{Graphing}$$

$$\boxed{\times} 8.05 \boxed{)} \boxed{\text{ENTER}}$$

The display for either of these keystroke sequences should be 16.67932852. Storing this result and using the recall key, we find the following two solutions.

$$x \approx \frac{17.8 + 16.67932852}{2.4} \approx 14.366 \qquad \text{Add stored value.}$$

$$x \approx \frac{17.8 - 16.67932852}{2.4} \approx 0.467 \qquad \text{Subtract stored value.}$$

Discussing the Concept   **Problem Posing**

Suppose you are writing a quiz that covers quadratic equations. Write four quadratic equations, including one with solutions $x = \frac{5}{3}$ and $x = -2$ and one with solutions $x = 4 \pm \sqrt{3}$, and instruct students to use any of the four solution methods: factoring, extracting square roots, completing the square, and using the Quadratic Formula. Trade quizzes with a class member and check each other's work.

6.3   Exercises

Integrated Review   *Concepts, Skills, and Problem Solving*

Keep mathematically in shape by doing these exercises *before* the problems of this section.

Properties and Definitions

In Exercises 1 and 2, rewrite the expression using the specified property, where a and b are nonnegative real numbers.

1. Multiplication Property: $\sqrt{ab} = \underline{\hspace{1cm}}$

2. Division Property: $\sqrt{\dfrac{a}{b}} = \underline{\hspace{1cm}}$

3. Is $\sqrt{72}$ in simplest form? Explain.

4. Is $10/\sqrt{5}$ in simplest form? Explain.

Simplifying Expressions

In Exercises 5–10, perform the operation and simplify the expression.

5. $\sqrt{128} + 3\sqrt{50}$ **6.** $3\sqrt{5}\sqrt{500}$

7. $\left(3 + \sqrt{2}\right)\left(3 - \sqrt{2}\right)$ **8.** $\left(3 + \sqrt{2}\right)^2$

9. $\dfrac{8}{\sqrt{10}}$ **10.** $\dfrac{5}{\sqrt{12} - 2}$

Problem Solving

11. Determine the length and width of a rectangle with a perimeter of 50 inches and a diagonal of $5\sqrt{13}$ inches.

12. The demand equation for a certain product is given by

$$p = 75 - \sqrt{1.2(x - 10)}$$

where x is the number of units demanded per day and p is the price per unit. Find the demand if the price is \$59.90.

Developing Skills

In Exercises 1–4, write in general form.

1. $2x^2 = 7 - 2x$ **2.** $7x^2 + 15x = 5$

3. $x(10 - x) = 5$ **4.** $x(3x + 8) = 15$

In Exercises 5–16, solve the quadratic equation (a) by using the Quadratic Formula and (b) by factoring. See Examples 1–4.

5. $x^2 - 11x + 28 = 0$ **6.** $x^2 - 12x + 27 = 0$

7. $x^2 + 6x + 8 = 0$ **8.** $x^2 + 9x + 14 = 0$

9. $4x^2 + 4x + 1 = 0$ **10.** $9x^2 + 12x + 4 = 0$

11. $4x^2 + 12x + 9 = 0$ **12.** $9x^2 - 30x + 25 = 0$

13. $6x^2 - x - 2 = 0$ **14.** $10x^2 - 11x + 3 = 0$

15. $x^2 - 5x - 300 = 0$ **16.** $x^2 + 20x - 300 = 0$

In Exercises 17–46, solve the quadratic equation by using the Quadratic Formula. (Find all real *and* complex solutions.) See Examples 1–4.

17. $x^2 - 2x - 4 = 0$ **18.** $x^2 - 2x - 6 = 0$

19. $t^2 + 4t + 1 = 0$ **20.** $y^2 + 6y + 4 = 0$

21. $x^2 + 6x - 3 = 0$ **22.** $x^2 + 8x - 4 = 0$

23. $x^2 - 10x + 23 = 0$ **24.** $u^2 - 12u + 29 = 0$

25. $2x^2 + 3x + 3 = 0$ **26.** $2x^2 - x + 1 = 0$

27. $3v^2 - 2v - 1 = 0$ **28.** $4x^2 + 6x + 1 = 0$

29. $2x^2 + 4x - 3 = 0$ **30.** $2x^2 + 3x + 3 = 0$

31. $9z^2 + 6z - 4 = 0$ **32.** $8y^2 - 8y - 1 = 0$

33. $-4x^2 - 6x + 3 = 0$

34. $-5x^2 - 15x + 10 = 0$

35. $8x^2 - 6x + 2 = 0$ **36.** $6x^2 + 3x - 9 = 0$

37. $-4x^2 + 10x + 12 = 0$

38. $-15x^2 - 10x + 25 = 0$

39. $9x^2 = 1 + 9x$ **40.** $7x^2 = 3 - 5x$

41. $3x - 2x^2 = 4 - 5x^2$

42. $x - x^2 = 1 - 6x^2$ **43.** $x^2 - 0.4x - 0.16 = 0$

44. $x^2 + 0.6x - 0.41 = 0$

45. $2.5x^2 + x - 0.9 = 0$

46. $0.09x^2 - 0.12x - 0.26 = 0$

In Exercises 47–56, use the discriminant to determine the type of solutions of the quadratic equation. See Example 5.

47. $x^2 + x + 1 = 0$

48. $x^2 + x - 1 = 0$

49. $2x^2 - 5x - 4 = 0$

50. $10x^2 + 5x + 1 = 0$

51. $5x^2 + 7x + 3 = 0$

52. $3x^2 - 2x - 5 = 0$

53. $4x^2 - 12x + 9 = 0$

54. $2x^2 + 10x + 6 = 0$

55. $3x^2 - x + 2 = 0$

56. $9x^2 - 24x + 16 = 0$

In Exercises 57–74, solve the quadratic equation by the most convenient method. (Find all real *and* complex solutions.)

57. $z^2 - 169 = 0$

58. $t^2 = 144$

59. $5y^2 + 15y = 0$

60. $7u^2 + 49u = 0$

61. $25(x - 3)^2 - 36 = 0$

62. $9(x + 4)^2 + 16 = 0$

63. $2y(y - 18) + 3(y - 18) = 0$

64. $4y(y + 7) - 5(y + 7) = 0$

65. $x^2 + 8x + 25 = 0$

66. $x^2 - 3x - 4 = 0$

67. $x^2 - 24x + 128 = 0$

68. $y^2 + 21y + 108 = 0$

69. $3x^2 - 13x + 169 = 0$

70. $2x^2 - 15x + 225 = 0$

71. $18x^2 + 15x - 50 = 0$

72. $14x^2 + 11x - 40 = 0$

73. $1.2x^2 - 0.8x - 5.5 = 0$

74. $2x^2 + 8x + 4.5 = 0$

In Exercises 75–82, use a graphing utility to graph the function. Use the graph to approximate any x-intercepts of the graph. Set $y = 0$ and solve the resulting equation. Compare the result with the x-intercepts of the graph.

75. $y = 3x^2 - 6x + 1$

76. $y = x^2 + x + 1$

77. $y = -(4x^2 - 20x + 25)$

78. $y = x^2 - 4x + 3$

79. $y = 5x^2 - 18x + 6$

80. $y = 15x^2 + 3x - 105$

81. $y = -0.04x^2 + 4x - 0.8$

82. $y = 3.7x^2 - 10.2x + 3.2$

In Exercises 83–86, use a graphing utility to determine the number of real solutions of the quadratic equation. Verify your answer using the discriminant.

83. $2x^2 - 5x + 5 = 0$

84. $2x^2 - x - 1 = 0$

85. $\frac{1}{5}x^2 + \frac{6}{5}x - 8 = 0$

86. $\frac{1}{3}x^2 - 5x + 25 = 0$

In Exercises 87–90, solve the equation.

87. $\dfrac{2x^2}{5} - \dfrac{x}{2} = 1$

88. $\dfrac{x^2 - 9x}{6} = \dfrac{x - 1}{2}$

89. $\sqrt{x + 3} = x - 1$

90. $\sqrt{2x - 3} = x - 2$

Think About It   In Exercises 91–94, describe the values of c such that the equation has (a) two real number solutions, (b) one real number solution, and (c) two complex number solutions.

91. $x^2 - 6x + c = 0$

92. $x^2 - 12x + c = 0$

93. $x^2 + 8x + c = 0$

94. $x^2 + 2x + c = 0$

Solving Problems

95. *Geometry*   A rectangle has a width of x inches, a length of $x + 6.3$ inches, and an area of 58.14 square inches. Find its dimensions.

96. *Geometry*   A rectangle has a length of $x + 1.5$ inches, a width of x inches, and an area of 18.36 square inches. Find its dimensions.

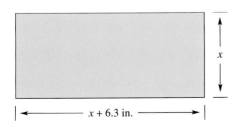

x

$x + 6.3$ in.

97. *Free-Falling Object*   A ball is thrown vertically upward at a velocity of 40 feet per second from a bridge that is 50 feet above the level of the water (see figure). The height h (in feet) of the ball at time t (in seconds) after it is thrown is

$$h = -16t^2 + 40t + 50.$$

(a) Find the time when the ball is again 50 feet above the water.

(b) Find the time when the ball strikes the water.

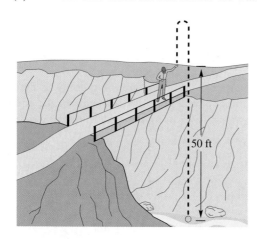

50 ft

98. *Free-Falling Object*   A ball is thrown vertically upward at a velocity of 20 feet per second from a bridge that is 40 feet above the level of the water. The height h (in feet) of the ball at time t (in seconds) after it is thrown is

$$h = -16t^2 + 20t + 40.$$

(a) Find the time when the ball is again 40 feet above the water.

(b) Find the time when the ball strikes the water.

99. *Aerospace Employment*   The following model approximates the number of people employed in the aerospace industry in the United States from 1990 through 1996.

$$y = 831.3 - 85.71t + 3.452t^2, \quad 0 \le t \le 6$$

In this model, y represents the number employed in the aerospace industry (in thousands) and t represents the year, with $t = 0$ corresponding to 1990. (Source: U.S. Department of Commerce)

(a) Use a graphing utility to graph the model.

(b) Use the graph in part (a) to find the year in which there were approximately 750,000 employed in the aerospace industry in the United States. Verify your answer algebraically.

(c) Use the model to estimate the number employed in the aerospace industry in 1997.

100. *Cellular Phone Subscribers*   The numbers of cellular phone subscribers (in millions) in the United States for the years 1989 through 1996 can be modeled by

$$s = 0.84t^2 + 1.51t + 4.70, \quad -1 \le t \le 6$$

where $t = 0$ corresponds to 1990. (Source: Cellular Telecommunications Industry Association)

(a) Use a graphing utility to graph the model.

(b) Use the model to determine the year in which the cellular phone companies had 10 million subscribers. Verify your answer algebraically.

101. *Exploration*   Determine the solutions x_1 and x_2 of each quadratic equation. Use the values of x_1 and x_2 to fill in the boxes.

Equation	x_1, x_2	$x_1 + x_2$	$x_1 x_2$
(a) $x^2 - x - 6 = 0$			
(b) $2x^2 + 5x - 3 = 0$			
(c) $4x^2 - 9 = 0$			
(d) $x^2 - 10x + 34 = 0$			

102. *Think About It*   Consider a general quadratic equation $ax^2 + bx + c = 0$ whose solutions are x_1 and x_2. Use the results of Exercise 101 to determine a relationship among the coefficients a, b, and c, and the sum $(x_1 + x_2)$ and product $(x_1 x_2)$ of the solutions.

Explaining Concepts

103. Answer parts (c) and (d) of Motivating the Chapter on page 371.

104. State the Quadratic Formula *in words*.

105. What is the discriminant of $ax^2 + bx + c = 0$? How is the discriminant related to the number and type of solutions of the equation?

106. Explain how completing the square can be used to develop the Quadratic Formula.

107. Summarize the four methods for solving a quadratic equation.

Mid-Chapter Quiz

Take this quiz as you would take a quiz in class. After you are done, check your work against the answers given in the back of the book.

In Exercises 1–8, solve the quadratic equation by the specified method.

1. Factoring:

$2x^2 - 72 = 0$

2. Factoring:

$2x^2 + 3x - 20 = 0$

3. Extracting square roots:

$t^2 = 12$

4. Extracting square roots:

$(u - 3)^2 - 16 = 0$

5. Completing the square:

$s^2 + 10s + 1 = 0$

6. Completing the square:

$2y^2 + 6y - 5 = 0$

7. Quadratic Formula:

$x^2 + 4x - 6 = 0$

8. Quadratic Formula:

$6v^2 - 3v - 4 = 0$

In Exercises 9–16, solve the equation by the most convenient method. (Find all the real *and* complex solutions.)

9. $x^2 + 5x + 7 = 0$

10. $36 - (t - 4)^2 = 0$

11. $x(x - 10) + 3(x - 10) = 0$

12. $x(x - 3) = 10$

13. $4b^2 - 12b + 9 = 0$

14. $3m^2 + 10m + 5 = 0$

15. $x - 2\sqrt{x} - 24 = 0$

16. $x^4 + 7x^2 + 12 = 0$

In Exercises 17 and 18, use a graphing utility to graph the function. Use the graph to approximate any x-intercepts of the graph. Set $y = 0$ and solve the resulting equation. Write a paragraph comparing the results of your algebraic and graphical solutions.

17. $y = \frac{1}{2}x^2 - 3x - 1$

18. $y = x^2 + 0.45x - 4$

19. The revenue R from selling x units of a certain product is given by

$R = x(20 - 0.2x).$

Find the number of units that must be sold to produce a revenue of $500.

20. The perimeter of a rectangle with sides x and $100 - x$ is 200 meters. Its area A is given by $A = x(100 - x)$. Determine the dimensions of the rectangle if its area is 2275 square meters.

6.4   Applications of Quadratic Equations

Objectives

1 Use a quadratic equation to solve an application problem.

1 Use a quadratic equation to solve an application problem.

Applications of Quadratic Equations

Example 1   An Investment Problem

A car dealer bought a fleet of cars from a car rental agency for a total of $120,000. By the time the dealer had sold all but four of the cars, at an average profit of $2500 each, the original investment of $120,000 had been regained. How many cars did the dealer sell, and what was the average price per car?

Solution

Although this problem is stated in terms of average price and average profit per car, we can use a model that assumes that each car sold for the same price.

Verbal Model: $\boxed{\text{Selling price per car}} = \boxed{\text{Cost per car}} + \boxed{\text{Profit per car}}$

Labels:
Number of cars sold $= x$ (cars)
Number of cars bought $= x + 4$ (cars)
Selling price per car $= \dfrac{120{,}000}{x}$ (dollars per car)
Cost per car $= \dfrac{120{,}000}{x + 4}$ (dollars per car)
Profit per car $= 2500$ (dollars per car)

Equation:

$$\frac{120{,}000}{x} = \frac{120{,}000}{x + 4} + 2500$$

$$120{,}000(x + 4) = 120{,}000x + 2500x(x + 4), \quad x \neq 0, \, x \neq -4$$

$$120{,}000x + 480{,}000 = 120{,}000x + 2500x^2 + 10{,}000x$$

$$0 = 2500x^2 + 10{,}000x - 480{,}000$$

$$0 = x^2 + 4x - 192$$

$$0 = (x - 12)(x + 16)$$

$$x - 12 = 0 \implies x = 12$$

$$x + 16 = 0 \implies x = -16$$

Choosing the positive value, it follows that the dealer sold 12 cars at an average price of $\frac{1}{12}(120{,}000) = \$10{,}000$ per car. Check this result in the original statement of the problem.

| Example 2 | Geometry | |

A picture is 6 inches taller than it is wide and has an area of 216 square inches. What are the dimensions of the picture?

Solution

Begin by drawing a diagram, as shown in Figure 6.3.

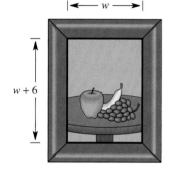

Figure 6.3

Verbal Model: Area of picture $=$ Width $\cdot$ Height

Labels: Picture width $= w$ (inches)
            Picture height $= w + 6$ (inches)
            Area $= 216$ (square inches)

Equation: $216 = w(w + 6)$

$$0 = w^2 + 6w - 216$$

$$0 = (w + 18)(w - 12)$$

$$w + 18 = 0 \implies w = -18$$

$$w - 12 = 0 \implies w = 12$$

Of the two possible solutions, choose the positive value of w and conclude that the picture is $w = 12$ inches wide and $w + 6$ or 18 inches tall. Check these dimensions in the original statement of the problem.

| Example 3 | An Interest Problem | |

The formula

$$A = P(1 + r)^2$$

represents the amount of money A in an account in which P dollars is deposited for 2 years at an annual interest rate of r (in decimal form). Find the interest rate if a deposit of $6000 increases to $6933.75 over a 2-year period.

Solution

$$A = P(1 + r)^2 \qquad \text{Given formula}$$

$$6933.75 = 6000(1 + r)^2 \qquad \text{Substitute for } A \text{ and } P.$$

$$1.155625 = (1 + r)^2 \qquad \text{Divide both sides by 6000.}$$

$$\pm 1.075 = 1 + r \qquad \text{Extract square roots.}$$

$$0.075 = r \qquad \text{Choose positive solution.}$$

The annual interest rate is $r = 0.075 = 7.5\%$.

Check

$$A = P(1 + r)^2 \qquad \text{Given formula}$$

$$6933.75 \overset{?}{=} 6000(1 + 0.075)^2 \qquad \text{Substitute 6933.75 for } A, \text{ 6000 for } P, \text{ and 0.075 for } r.$$

$$6933.75 \overset{?}{=} 6000(1.155625) \qquad \text{Simplify.}$$

$$6933.75 = 6933.75 \qquad \text{Solution checks.} \checkmark$$

Example 4 Reduced Rates

A ski club chartered a bus for a ski trip at a cost of $720. In an attempt to lower the bus fare per skier, the club invited nonmembers to go along. When four nonmembers joined the trip, the fare per skier decreased by $6.00. How many club members are going on the trip?

Solution

Verbal Model: $\boxed{\text{Cost per skier}} \cdot \boxed{\text{Number of skiers}} = \boxed{\$720}$

Labels:
Number of ski club members $= x$ (people)
Number of skiers $= x + 4$ (people)
Original cost per skier $= \dfrac{720}{x}$ (dollars)
New cost per skier $= \dfrac{720}{x} - 6.00$ (dollars)

Equation:

$$\left(\frac{720}{x} - 6.00\right)(x + 4) = 720$$

$$\left(\frac{720 - 6x}{x}\right)(x + 4) = 720$$

$$(720 - 6x)(x + 4) = 720x, \quad x \neq 0$$

$$720x - 6x^2 - 24x + 2880 = 720x$$

$$-6x^2 - 24x + 2880 = 0$$

$$x^2 + 4x - 480 = 0$$

$$(x + 24)(x - 20) = 0$$

$$x + 24 = 0 \quad \Longrightarrow \quad x = -24$$

$$x - 20 = 0 \quad \Longrightarrow \quad x = 20$$

Choosing the positive value of x implies that there are 20 ski club members. Check this solution in the original equation, as follows.

Check

Number of Skiers	Cost per Skier
20	$\dfrac{720}{20} = \$36.00$
24	$\dfrac{720}{24} = \$30.00$

From these two calculations, you can see that the difference in cost per skier is $36.00 - \$30.00 = \$6.00$.

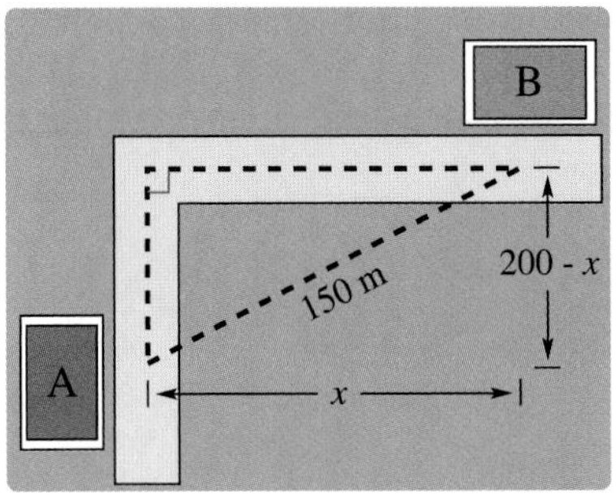

Example 5   An Application Involving the Pythagorean Theorem

An L-shaped sidewalk from building A to building B on a college campus is 200 meters long, as shown in Figure 6.4. By cutting diagonally across the grass, students shorten the walking distance to 150 meters. What are the lengths of the two legs of the sidewalk?

Figure 6.4

Solution

Verbal Model: $a^2 + b^2 = c^2$ Pythagorean Theorem

Labels: Length of one leg $= x$ (meters)
Length of other leg $= 200 - x$ (meters)
Length of diagonal $= 150$ (meters)

Equation: $x^2 + (200 - x)^2 = (150)^2$

$$2x^2 - 400x + 40{,}000 = 22{,}500$$

$$2x^2 - 400x + 17{,}500 = 0$$

$$x^2 - 200x + 8750 = 0$$

By the Quadratic Formula, you can find the solutions as follows.

$$x = \frac{200 \pm \sqrt{(-200)^2 - 4(1)(8750)}}{2(1)}$$

$$= \frac{200 \pm \sqrt{5000}}{2}$$

$$= \frac{200 \pm 50\sqrt{2}}{2}$$

$$= \frac{2(100 \pm 25\sqrt{2})}{2}$$

$$= 100 \pm 25\sqrt{2}$$

Both solutions are positive, and it does not matter which one you choose. If you let

$$x = 100 + 25\sqrt{2} \approx 135.4 \text{ meters},$$

the length of the other leg is

$$200 - x \approx 200 - 135.4$$

$$\approx 64.6 \text{ meters}.$$

In Example 5, notice that you obtain the same dimensions if you choose the other value of x. That is, if the length of one leg is

$$x = 100 - 25\sqrt{2} \approx 64.6 \text{ meters},$$

the length of the other leg is

$$200 - x \approx 200 - 64.6$$

$$\approx 135.4 \text{ meters}.$$

Example 6   Work-Rate Problem

An office contains two copy machines. Machine B is known to take 12 minutes longer than Machine A to copy the company's monthly report. Using both machines together, it takes 8 minutes to reproduce the report. How long would it take each machine alone to reproduce the report?

Solution

Verbal Model:

Work done by machine A	+	Work done by machine B	=	1 complete job

Rate for A	·	Time for both	+	Rate for B	·	Time for both	=	1

Labels:

Time for machine A $= t$ (minutes)

Rate for machine A $= \dfrac{1}{t}$ (job per minute)

Time for machine B $= t + 12$ (minutes)

Rate for machine B $= \dfrac{1}{t + 12}$ (job per minute)

Time for both machines $= 8$ (minutes)

Rate for both machines $= \frac{1}{8}$ (job per minute)

Equation:

$$\frac{1}{t}(8) + \frac{1}{t + 12}(8) = 1$$

$$8\left(\frac{1}{t} + \frac{1}{t + 12}\right) = 1$$

$$8\left[\frac{t + 12 + t}{t(t + 12)}\right] = 1$$

$$8t(t + 12)\left[\frac{2t + 12}{t(t + 12)}\right] = t(t + 12)$$

$$8(2t + 12) = t^2 + 12t$$

$$16t + 96 = t^2 + 12t$$

$$0 = t^2 - 4t - 96$$

$$0 = (t - 12)(t + 8)$$

$$t - 12 = 0 \quad \Longrightarrow \quad t = 12$$

$$t + 8 = 0 \quad \Longrightarrow \quad t = -8$$

Choose the positive value for t and find that

Time for machine A $= t = 12$ minutes

Time for machine B $= t + 12 = 24$ minutes.

Check these solutions in the original equation.

Example 7   The Height of a Model Rocket

A model rocket is projected straight upward from ground level according to the height equation $h = -16t^2 + 192t$, $t \geq 0$, where h is the height in feet and t is the time in seconds. (a) After how many seconds will the height be 432 feet? (b) When will the rocket hit the ground?

Solution

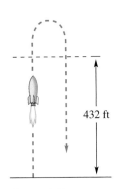

432 ft

Figure 6.5

a.

$h = -16t^2 + 192t$	Original equation
$432 = -16t^2 + 192t$	Substitute 432 for h.
$16t^2 - 192t + 432 = 0$	Standard form
$t^2 - 12t + 27 = 0$	Divide both sides by 16.
$(t - 3)(t - 9) = 0$	Factor.
$t - 3 = 0 \implies t = 3$	Set 1st factor equal to 0.
$t - 9 = 0 \implies t = 9$	Set 2nd factor equal to 0.

The rocket attains a height of 432 feet at two different times—once (going up) after 3 seconds, and again (coming down) after 9 seconds. (See Figure 6.5.)

b. To find the time it takes for the rocket to hit the ground, let the height be 0.

$$0 = -16t^2 + 192t$$

$$0 = t^2 - 12t$$

$$0 = t(t - 12)$$

$$t = 0 \quad \text{or} \quad t = 12$$

The rocket will hit the ground after 12 seconds. (Note that the time of $t = 0$ seconds corresponds to the time of lift-off.)

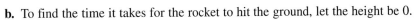

Discussing the Concept   Analyzing Quadratic Functions

Use a graphing utility to graph

$$y_1 = 3x^2 + 2x - 1$$

and

$$y_2 = -x^2 + 5x + 4.$$

For each function, use the zoom and trace features to find either the maximum or minimum function value. Discuss other methods that you could use to find these values.

6.4   Exercises

Properties and Definitions

1. Define the slope of the line through the points (x_1, y_1) and (x_2, y_2).

2. Give the following forms of an equation of a line.

(a) Slope-intercept form

(b) Point-slope form

(c) General form

(d) Horizontal line

Equations of Lines

In Exercises 3–10, find the general form of the equation of the line through the two points.

3. $(0, 0), (4, -2)$ **4.** $(0, 0), (100, 75)$

5. $(-1, -2), (3, 6)$ **6.** $(1, 5), (6, 0)$

7. $\left(\frac{3}{2}, 8\right), \left(\frac{11}{2}, \frac{5}{2}\right)$ **8.** $(0, 2), (7.3, 15.4)$

9. $(0, 8), (5, 8)$ **10.** $(-3, 2), (-3, 5)$

Problem Solving

11. A group of people agree to share equally in the cost of a $250,000 endowment to a college. If they could find two more people to join the group, each person's share of the cost would decrease by $6250. How many people are presently in the group?

12. A boat travels at a speed of 18 miles per hour in still water. It travels 35 miles upstream and then returns to the starting point in a total of 4 hours. Find the speed of the current.

Solving Problems

1. *Selling Price*   A store owner bought a case of eggs for $21.60. By the time all but 6 dozen of the eggs had been sold at a profit of $0.30 per dozen, the original investment of $21.60 had been regained. How many dozen eggs did the owner sell, and what was the selling price per dozen?

2. *Selling Price*   A manager of a computer store bought several computers of the same model for $27,000. When all but three of the computers had been sold at a profit of $750 per computer, the original investment of $27,000 had been regained. How many computers were sold, and what was the selling price of each?

3. *Selling Price*   A storeowner bought a case of video games for $480. By the time he had sold all but eight of them at a profit of $10 each, the original investment of $480 had been regained. How many video games were sold, and what was the selling price of each game?

4. *Selling Price*   A math club bought a case of sweatshirts for $850 to sell as a fundraiser. By the time all but 16 sweatshirts had been sold at a profit of $8 per sweatshirt, the original investment of $850 had been regained. How many sweatshirts were sold, and what was the selling price of each sweatshirt?

Dimensions of a Rectangle   In Exercises 5–14, complete the table of widths, lengths, perimeters, and areas of rectangles.

	Width	Length	Perimeter	Area
5.	$0.75l$	l	42 in.	
6.	w	$1.5w$	40 m	
7.	w	$2.5w$		250 ft^2
8.	w	$1.5w$		216 cm^2
9.	$\frac{1}{3}l$	l		192 in.^{2}
10.	$\frac{3}{4}l$	l		2700 in.^{2}
11.	w	$w + 3$	54 km	
12.	$l - 6$	l	108 ft	
13.	$l - 20$	l		12,000 m^2
14.	w	$w + 5$		500 ft^2

15. *Geometry*   A picture frame is 4 inches taller than it is wide and has an area of 192 square inches. What are the dimensions of the picture frame?

16. *Geometry*   The top of a coffee table is 3 feet longer than it is wide and has an area of 10 square feet. What are the dimensions of the top of the coffee table?

17. *Geometry*   The height of a triangle is 8 inches less than its base. The area of the triangle is 192 square inches. Find the dimensions of the triangle.

18. *Geometry*   The height of a triangle is 25 inches greater than its base. The area of the triangle is 625 square inches. Find the dimensions of the triangle.

19. *Lumber Storage Area*   A retail lumberyard plans to store lumber in a rectangular region adjoining the sales office (see figure). The region will be fenced on three sides and the fourth side will be bounded by the wall of the office building. Find the dimensions of the region if 350 feet of fencing is available and the area of the region is 12,500 square feet.

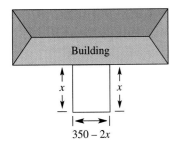

20. *Fencing the Yard*   You have 100 feet of fencing. Do you have enough to enclose a rectangular region whose area is 630 square feet? Is there enough to enclose a circular area of 630 square feet? Explain.

21. *Fencing the Yard*   A family built a fence around three sides of their property (see figure). In total, they used 550 feet of fencing. By their calculations, the lot is 1 acre (43,560 square feet). Is this correct? Explain your answer.

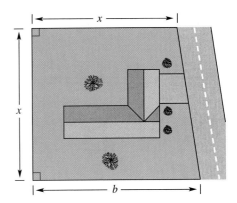

Figure for 21

22. *Geometry*   Your home is on a square lot. To add more space to your yard, you purchase an additional 20 feet along the side of the property (see figure). The area of the lot is now 25,500 square feet. What are the dimensions of the new lot?

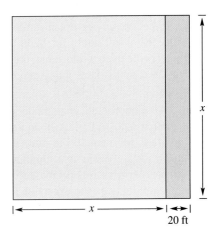

23. *Open Conduit*   An open-topped rectangular conduit for carrying water in a manufacturing process is made by folding up the edges of a sheet of aluminum 48 inches wide (see figure). A cross section of the conduit must have an area of 288 square inches. Find the width and height of the conduit.

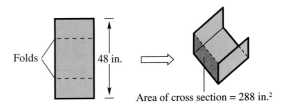

Compound Interest In Exercises 24–29, find the interest rate r. Use the formula $A = P(1 + r)^2$, where A is the amount after 2 years in an account earning r percent (in decimal form) compounded annually, and P is the original investment. See Example 3.

24. $P = \$10,000$
$A = \$11,990.25$

25. $P = \$3000$
$A = \$3499.20$

26. $P = \$500$
$A = \$572.45$

27. $P = \$250$
$A = \$280.90$

28. $P = \$6500$
$A = \$7370.46$

29. $P = \$8000$
$A = \$8420.20$

30. *Reduced Ticket Price* A service organization paid $210 for a block of tickets to a ball game. The block contained three more tickets than the organization needed for its members. By inviting three more people to attend (and share in the cost), the organization lowered the price per ticket by $3.50. How many people are going to the game?

31. *Reduced Ticket Price* A service organization buys a block of tickets for a ball game for $240. After eight more people decide to go to the game, the price per ticket is decreased by $1. How many people are going to the game?

32. *Reduced Fare* A science club charters a bus to attend a science fair at a cost of $480. In an attempt to lower the bus fare per person, the club invites nonmembers to go along. When two nonmembers join the trip, the fare per person is decreased by $1. How many people are going on the excursion?

33. *Venture Capital* Eighty thousand dollars is needed to begin a small business. The cost will be divided equally among the investors. Some have made a commitment to invest. If three more investors are found, the amount required from each would decrease by $6000. How many have made a commitment to invest in the business?

34. *Dimensions of a Rectangle* The perimeter of a rectangle is 102 inches and the length of the diagonal is 39 inches. Find the dimensions of the rectangle.

35. *Delivery Route* You are asked to deliver pizza to offices B and C in your city (see figure), and you are required to keep a log of all the mileages between stops. You forget to look at the odometer at stop B, but after getting to stop C you record the total distance traveled from the pizza shop as 18 miles. The return distance from C to A is 16 miles. If the route

approximates a right triangle, estimate the distance from A to B.

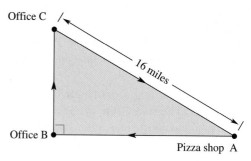

Figure for 35

36. *Shortcut* An L-shaped sidewalk from building A to building B on a high school campus is 100 yards long, as shown in the figure. By cutting diagonally across the grass, students shorten the walking distance to 80 yards. What are the lengths of the two legs of the sidewalk?

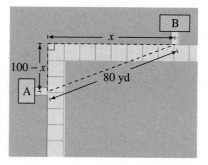

37. *Dimensions of a Rectangle* An adjustable rectangular form has minimum dimensions of 3 meters by 4 meters. The length and width can be expanded by equal amounts x (see figure).

(a) Write the length d of the diagonal as a function of x. Use a graphing utility to graph the function. Use the graph to approximate the value of x when $d = 10$ meters.

(b) Find x algebraically when $d = 10$.

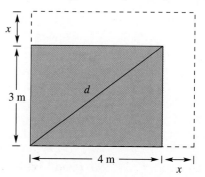

38. *Solving Graphically and Numerically*   A meteorologist is positioned 100 feet from the point where a weather balloon is launched (see figure). The instrument package lifted vertically by the balloon transmits data to the meteorologist.

(a) Write the distance d between the balloon and the meteorologist as a function of the height h of the balloon.

(b) Use a graphing utility to graph the function in part (a). Use the graph to approximate the value of h when $d = 200$ feet.

(c) Complete the following table.

h	0	100	200	300
d				

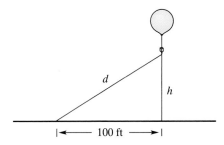

39. *Work-Rate Problem*   Working together, two people can complete a task in 5 hours. Working alone, how long would it take each to do the task if one person took 2 hours longer than the other?

40. *Work-Rate Problem*   An office contains two printers. Machine B is known to take 3 minutes longer than Machine A to produce the company's monthly financial report. Using both machines together, it takes 6 minutes to produce the report. How long would it take each machine to produce the report?

41. *Work-Rate Problem*   A builder works with two plumbing companies. Company A is known to take 3 days longer than Company B to do the plumbing in a particular style of house. Using both companies, it takes 4 days. How long would it take to do the plumbing using each company individually?

42. *Work-Rate Problem*   Working together, two people can complete a task in 6 hours. Working alone, one person takes 2 hours longer than the other. How long would it take each to do the task alone?

Free-Falling Object   In Exercises 43–46, find the time necessary for an object to fall to ground level from an initial height of h_0 feet if its height h at any time t (in seconds) is given by

$$h = h_0 - 16t^2.$$

43. $h_0 = 144$ **44.** $h_0 = 625$

45. $h_0 = 1454$ (height of the Sears Tower)

46. $h_0 = 984$ (height of the Eiffel Tower)

47. *Height of a Baseball*   The height h in feet of a baseball hit 3 feet above the ground is given by

$$h = 3 + 75t - 16t^2$$

where t is time in seconds. Find the time when the ball hits the ground in the outfield.

48. *Hitting Baseballs*   You are hitting baseballs. When tossing the ball into the air, your hand is 5 feet above the ground (see figure). You hit the ball when it falls back to a height of 4 feet. If you toss the ball with an initial velocity of 25 feet per second, the height h of the ball t seconds after leaving your hand is given by

$$h = 5 + 25t - 16t^2.$$

How much time will pass before you hit the ball?

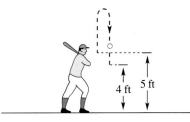

49. *Height of a Model Rocket*   A model rocket is projected straight upward from ground level according to the height equation $h = -16t^2 + 160t$, where h is the height of the rocket in feet and t is the time in seconds.

(a) After how many seconds will the height be 336 feet?

(b) When will the rocket hit the ground?

50. *Height of a Tennis Ball*   A tennis ball is tossed vertically upward from a height of 5 feet according to the height equation $h = -16t^2 + 21t + 5$, where h is the height of the tennis ball in feet and t is the time in seconds.

(a) After how many seconds will the height be 11 feet?

(b) When will the tennis ball hit the ground?

Number Problems   In Exercises 51–56, find two positive integers that satisfy the given requirement.

51. The product of two consecutive integers is 240.

52. The product of two consecutive integers is 1122.

53. The product of two consecutive even integers is 224.

54. The product of two consecutive even integers is 528.

55. The product of two consecutive odd integers is 483.

56. The product of two consecutive odd integers is 255.

57. *Air Speed*   An airline runs a commuter flight between two cities that are 720 miles apart. If the average speed of the planes could be increased by 40 miles per hour, the travel time would be decreased by 12 minutes. What air speed is required to obtain this decrease in travel time?

58. *Average Speed*   A truck traveled the first 100 miles of a trip at one speed and the last 135 miles at an average speed of 5 miles per hour less. If the entire trip took 5 hours, what was the average speed for the first part of the trip?

59. *Speed*   A small business uses a minivan to make deliveries. The cost per hour for fuel for the van is $C = v^2/600$, where v is the speed in miles per hour. The driver is paid $5 per hour. Find the speed if the cost for wages and fuel for a 110-mile trip is $20.39.

60. *Distance*   Find any points on the line $y = 14$ that are 13 units from the point $(1, 2)$.

61. *Geometry*   The area of an ellipse is given by $A = \pi ab$ (see figure). For a certain ellipse, it is required that $a + b = 20$.

(a) Show that $A = \pi a(20 - a)$.

(b) Complete the table.

a	4	7	10	13	16
A					

(c) Find two values of a such that $A = 300$.

(d) Use a graphing utility to graph the area function.

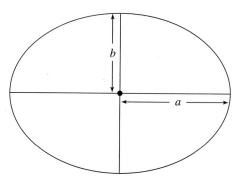

Figure for 61

62. *Data Analysis*   For the years 1990 through 1996, the sales s (in millions of dollars) of snowmobiles in the United States can be approximated by the model

$$s = 12.88t^2 + 43.86t + 300.83, \ 0 \le t \le 6$$

where t is time in years, with $t = 0$ corresponding to 1990. (Source: National Sporting Goods Association)

(a) Use a graphing utility to graph the model over the specified domain.

(b) In which year were sales approximately $400 million?

Explaining Concepts

63. In your own words, describe guidelines for solving word problems.

64. Describe the strategies that can be used to solve a quadratic equation.

65. *Unit Analysis*   Describe the units of the product.

$$\frac{9 \text{ dollars}}{\text{hour}} \cdot (20 \text{ hours})$$

66. *Unit Analysis*   Describe the units of the product.

$$\frac{20 \text{ feet}}{\text{minute}} \cdot \frac{1 \text{ minute}}{60 \text{ seconds}} \cdot (45 \text{ seconds})$$

67. Give an example of a quadratic equation that has only one repeated solution.

68. Give an example of a quadratic equation that has two imaginary solutions.

6.5   Quadratic and Rational Inequalities

Objectives

1. Determine test intervals for a polynomial inequality.
2. Use test intervals to solve a quadratic inequality.
3. Use test intervals to solve a rational inequality.
4. Use an inequality to solve an application problem.

1 Determine test intervals for a polynomial inequality.

Finding Test Intervals

When working with polynomial inequalities, it is important to realize that the value of a polynomial can change sign only at its **zeros**. That is, a polynomial can change signs only at the x-values that make the value of the polynomial zero. For instance, the first-degree polynomial $x + 2$ has a zero at -2, and it changes sign at that zero. You can picture this result on the real number line, as shown in Figure 6.6.

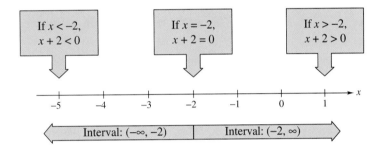

Figure 6.6

Note in Figure 6.6 that the zero of the polynomial partitions the real number line into two **test intervals**. The value of the polynomial is negative for every x-value in the first test interval $(-\infty, -2)$, and it is positive for every x-value in the second test interval $(-2, \infty)$. You can use the same basic approach to determine the test intervals for any polynomial.

> ▶ Finding Test Intervals for a Polynomial
>
> 1. Find all real zeros of the polynomial, and arrange the zeros in increasing order. The zeros of a polynomial are called its **critical numbers**.
>
> 2. Use the critical numbers of the polynomial to determine its test intervals.
>
> 3. Choose a representative x-value in each test interval and evaluate the polynomial at that value. If the value of the polynomial is negative, the polynomial will have negative values for *every* x-value in the interval. If the value of the polynomial is positive, the polynomial will have positive values for *every* x-value in the interval.

2 Use test intervals to solve a quadratic inequality.

Quadratic Inequalities

The concepts of critical numbers and test intervals can be used to solve nonlinear inequalities, as demonstrated in Examples 1, 2, and 4.

Example 1   Solving a Quadratic Inequality

Solve the inequality $x^2 - 5x < 0$.

Solution

First find the *critical numbers* for $x^2 - 5x < 0$ by finding the solutions of the equation $x^2 - 5x = 0$.

$$x^2 - 5x = 0 \qquad \text{Corresponding equation}$$

$$x(x - 5) = 0 \qquad \text{Factor.}$$

$$x = 0, x = 5 \qquad \text{Critical numbers}$$

This implies that the test intervals are

$$(-\infty, 0), \quad (0, 5), \quad \text{and} \quad (5, \infty).$$

To test an interval, choose a convenient number in the interval and compute the sign of $x^2 - 5x$.

Interval	x-Value	Polynomial Value	Conclusion
$(-\infty, 0)$	$x = -1$	$(-1)^2 - 5(-1) = 6$	Positive
$(0, 5)$	$x = 1$	$(1)^2 - 5(1) = -4$	Negative
$(5, \infty)$	$x = 6$	$(6)^2 - 5(6) = 6$	Positive

From this you can conclude that the value of the polynomial is positive for all x-values in $(-\infty, 0)$ and $(5, \infty)$, and negative for all x-values in $(0, 5)$. This implies that the solution of the inequality $x^2 - 5x < 0$ is the interval $(0, 5)$, as shown in Figure 6.7.

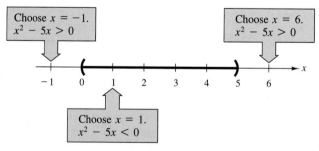

Figure 6.7

In Example 1, note that you would have used the same basic procedure if the inequality symbol had been $\leq$, $>$, or $\geq$. For instance, from Figure 6.7, you can see that the solution set of the inequality $x^2 - 5x \geq 0$ consists of the union of the half-open intervals $(-\infty, 0]$ and $[5, \infty)$, which is written as $(-\infty, 0] \cup [5, \infty)$.

Technology: Tip

Most graphing utilities can sketch the graph of the solution set of a quadratic inequality. Consult the user's guide for your graphing utility. Notice that the solution set for the quadratic inequality

$$x^2 - 5x < 0$$

shown below appears as a horizontal line above the x-axis.

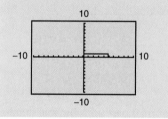

Just as in solving quadratic *equations*, the first step in solving a quadratic *inequality* is to write the inequality in **general form,** with the polynomial on the left and zero on the right. Factorization of the polynomial then shows the critical numbers, as demonstrated in Example 2.

| Example 2 | Solving a Quadratic Inequality |

Solve the inequality $2x^2 + 5x \ge 12$.

Solution

Begin by writing the inequality in general form.

$$2x^2 + 5x \ge 12 \qquad \text{Original inequality}$$

$$2x^2 + 5x - 12 \ge 0 \qquad \text{Write in general form.}$$

Next, find the critical numbers for $2x^2 + 5x - 12 \ge 0$ by finding the solutions to the equation $2x^2 + 5x - 12 = 0$.

$$2x^2 + 5x - 12 = 0 \qquad \text{Original equation}$$

$$(x + 4)(2x - 3) = 0 \qquad \text{Factor.}$$

$$x = -4, \, x = \frac{3}{2} \qquad \text{Critical numbers}$$

This implies that the test intervals are $(-\infty, -4)$, $\left(-4, \frac{3}{2}\right)$, and $\left(\frac{3}{2}, \infty\right)$. To test an interval, choose a convenient number in the interval and compute the sign of $2x^2 + 5x - 12$.

Interval	x-Value	Polynomial Value	Conclusion
$(-\infty, -4)$	$x = -5$	$2(-5)^2 + 5(-5) - 12 = 13$	Positive
$\left(-4, \frac{3}{2}\right)$	$x = 0$	$2(0)^2 + 5(0) - 12 = -12$	Negative
$\left(\frac{3}{2}, \infty\right)$	$x = 2$	$2(2)^2 + 5(2) - 12 = 6$	Positive

From this you can conclude that the value of the polynomial is positive for all x-values in $(-\infty, -4)$ and $\left(\frac{3}{2}, \infty\right)$, and negative for all x-values in $\left(-4, \frac{3}{2}\right)$. This implies that the solution set of the inequality $2x^2 + 5x - 12 \ge 0$ is $\left(-\infty, -4\right] \cup \left[\frac{3}{2}, \infty\right)$, as shown in Figure 6.8.

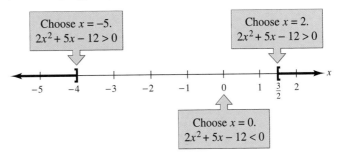

Choose $x = -5$.
$2x^2 + 5x - 12 > 0$

Choose $x = 2$.
$2x^2 + 5x - 12 > 0$

Choose $x = 0$.
$2x^2 + 5x - 12 < 0$

Figure 6.8

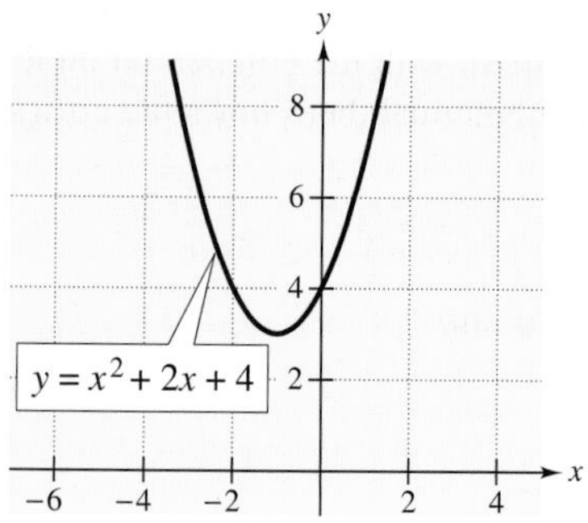

(a)

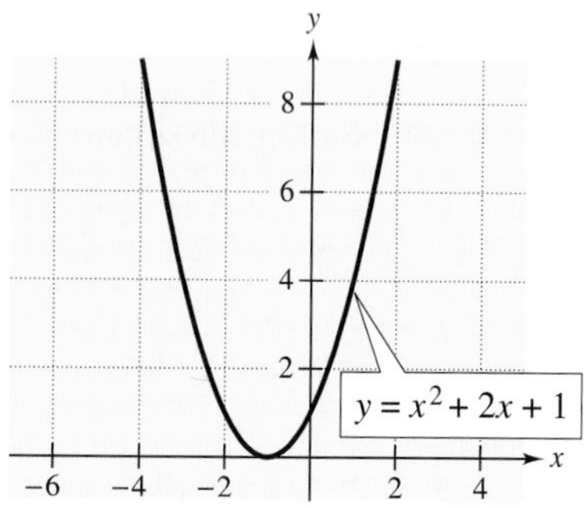

(b)

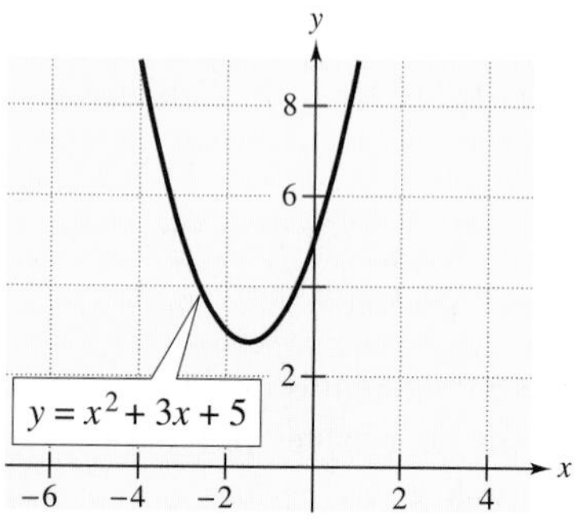

(c)

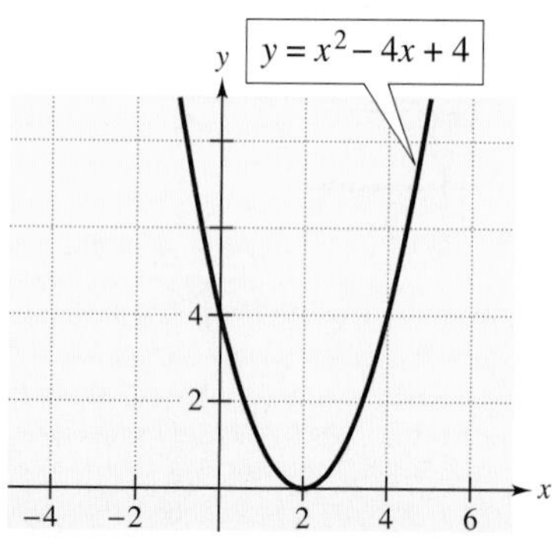

(d)

Figure 6.9

The solutions of the quadratic inequalities in Examples 1 and 2 consist, respectively, of a single interval and the union of two intervals. When solving the exercises for this section, you should be on the watch for some unusual solution sets, as illustrated in Example 3.

Example 3   Unusual Solution Sets

Solve each inequality to verify that the given solution set is correct.

a. The solution set of the quadratic inequality

$$x^2 + 2x + 4 > 0$$

consists of the entire set of real numbers, $(-\infty, \infty)$. This is true because the value of the quadratic $x^2 + 2x + 4$ is positive for every real value of x, as shown in Figure 6.9(a).

b. The solution set of the quadratic inequality

$$x^2 + 2x + 1 \leq 0$$

consists of the single number $\{-1\}$. This is true because $x^2 + 2x + 1 = (x + 1)^2$ has just one critical number, $x = -1$, and it is the only value that satisfies the inequality. [See Figure 6.9(b).]

c. The solution set of the quadratic inequality

$$x^2 + 3x + 5 < 0$$

is empty. This is true because the value of the quadratic $x^2 + 3x + 5$ is not less than zero for any value of x, as shown in Figure 6.9(c).

d. The solution set of the quadratic inequality

$$x^2 - 4x + 4 > 0$$

consists of all real numbers *except* the number 2. In interval notation, this solution set can be written as $(-\infty, 2) \cup (2, \infty)$. [See Figure 6.9(d).]

Remember that checking the solution set of an inequality is not as straight-forward as checking the solutions of an equation, because inequalities tend to have infinitely many solutions. Even so, we suggest that you check several x-values in your solution set to confirm that they satisfy the inequality. Also try checking x-values that are not in the solution set to verify that they do not satisfy the inequality.

For instance, the solution of $x^2 - 5x < 0$ is $(0, 5)$. Try checking some numbers in this interval to verify that they satisfy the inequality. Then check some numbers outside the interval to verify that they do not satisfy the inequality.

3 Use test intervals to solve a rational inequality.

Rational Inequalities

The concepts of critical numbers and test intervals can be extended to inequalities involving rational expressions. To do this, use the fact that the value of a rational expression can change sign only at its *zeros* (the x-values for which its numerator is zero) and its *undefined values* (the x-values for which its denominator is zero). These two types of numbers make up the **critical numbers** of a rational inequality. For instance, the critical numbers of the inequality

$$\frac{x - 2}{(x - 1)(x + 3)} < 0$$

are $x = 2$ (the numerator is zero), and $x = 1$ and $x = -3$ (the denominator is zero). From these three critical numbers you can see that the inequality has *four* test intervals.

$$(-\infty, -3), \quad (-3, 1), \quad (1, 2), \quad \text{and} \quad (2, \infty)$$

Study Tip

When solving a rational inequality, you should begin by writing the inequality in general form, with the rational expression (as a single fraction) on the left and zero on the right. For instance, the first step in solving

$$\frac{2x}{x + 3} < 4$$

is to write it as

$$\frac{2x}{x + 3} < 4$$

$$\frac{2x}{x + 3} - 4 < 0$$

$$\frac{2x - 4(x + 3)}{x + 3} < 0$$

$$\frac{-2x - 12}{x + 3} < 0.$$

Try solving this inequality. You should find that the solution set is $(-\infty, -6) \cup (-3, \infty)$.

Example 4   Solving a Rational Inequality

Solve the inequality $\dfrac{x}{x - 2} > 0$.

Solution

The numerator is zero when $x = 0$ and the denominator is zero when $x = 2$. So, the two critical numbers are 0 and 2, which implies that the test intervals are

$$(-\infty, 0), \quad (0, 2), \quad \text{and} \quad (2, \infty).$$

To test an interval, choose a convenient number and compute the sign of $x/(x - 2)$.

Interval	x-Value	Rational Expression Value	Conclusion
$(-\infty, 0)$	$x = -1$	$(-1)/(-1 - 2) = \frac{1}{3}$	Positive
$(0, 2)$	$x = 1$	$(1)/(1 - 2) = -1$	Negative
$(2, \infty)$	$x = 3$	$(3)/(3 - 2) = 3$	Positive

From this you can conclude that the value of the rational expression is positive for all x-values in $(-\infty, 0)$ and $(2, \infty)$, and negative for all x-values in $(0, 2)$. This implies that the solution set of the inequality $x/(x - 2) > 0$ is $(-\infty, 0) \cup (2, \infty)$, as shown in Figure 6.10.

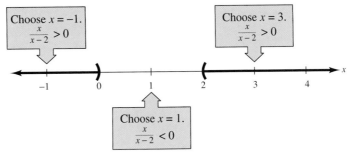

Figure 6.10

4 Use an inequality to solve an application problem.

Application

| Example 5 | The Height of a Projectile |

A projectile is fired straight up from ground level with an initial velocity of 256 feet per second, as shown in Figure 6.11, so that its height at any time t is given by

$$h = -16t^2 + 256t$$

where the height h is measured in feet and the time t is measured in seconds. During what interval of time will the height of the projectile exceed 960 feet?

Solution

To solve this problem, find the values of t for which h is greater than 960.

$$-16t^2 + 256t > 960 \qquad \text{Original inequality}$$

$$-16t^2 + 256t - 960 > 0 \qquad \text{Subtract 960 from both sides.}$$

$$t^2 - 16t + 60 < 0 \qquad \begin{array}{l}\text{Divide both sides by } -16 \text{ and}\\ \text{reverse the inequality.}\end{array}$$

$$(t - 6)(t - 10) < 0 \qquad \text{Factor.}$$

So, the critical numbers are $t = 6$ and $t = 10$. A test of the intervals $(-\infty, 6)$, $(6, 10)$, and $(10, \infty)$ shows that the solution interval is $(6, 10)$. So, the height of the object will exceed 960 feet for values of t that are greater than 6 seconds and less than 10 seconds.

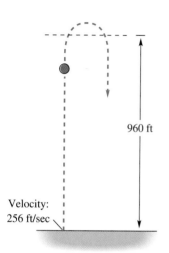

960 ft

Velocity:
256 ft/sec

Figure 6.11

Discussing the Concept   Graphing an Inequality

You can use a graph on a rectangular coordinate system as an alternative method for solving an inequality. For instance, to solve the inequality in Example 1

$$x^2 - 5x < 0$$

you can sketch the graph of $y = x^2 - 5x$. Using a graphing utility, you can obtain the graph shown below. From the graph, you can see that the only part of the curve that lies below the x-axis is the portion for which $0 < x < 5$. So, the solution of $x^2 - 5x < 0$ is $0 < x < 5$. Try using this graphing approach to solve the following inequalities.

a. $x^2 + 4x > 0$ **b.** $x^2 - 16 \le 0$

c. $x^2 - 3x - 18 \le 0$ **d.** $3x^2 + 5x + 4 \ge 0$

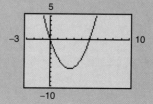

6.5   Exercises

Integrated Review   Concepts, Skills, and Problem Solving

Keep mathematically in shape by doing these exercises *before* the problems of this section.

Properties and Definitions

1. Is 36.82×10^8 written in scientific notation? Explain.

2. The numbers $n_1 \times 10^2$ and $n_2 \times 10^4$ are written in scientific notation. The product of these two numbers must lie in what interval? Explain.

Simplifying Expressions

In Exercises 3–8, factor the expression.

3. $6u^2v - 192v^2$ **4.** $5x^{2/3} - 10x^{1/3}$

5. $x(x - 10) - 4(x - 10)$

6. $x^3 + 3x^2 - 4x - 12$

7. $16x^2 - 121$ **8.** $4x^3 - 12x^2 + 16x$

Mathematical Modeling

In Exercises 9–12, find a mathematical model for the area of the figure.

9.   **10.**

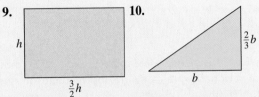

11.   **12.**

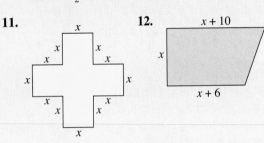

Developing Skills

In Exercises 1–10, find the critical numbers.

1. $x(2x - 5)$ **2.** $5x(x - 3)$

3. $4x^2 - 81$ **4.** $9y^2 - 16$

5. $x(x + 3) - 5(x + 3)$ **6.** $y(y - 4) - 3(y - 4)$

7. $x^2 - 4x + 3$ **8.** $3x^2 - 2x - 8$

9. $4x^2 - 20x + 25$ **10.** $4x^2 - 4x - 3$

In Exercises 11–20, determine the intervals for which the polynomial is entirely negative and entirely positive.

11. $x - 4$ **12.** $3 - x$ **13.** $3 - \frac{1}{2}x$ **14.** $\frac{2}{3}x - 8$

15. $2x(x - 4)$ **16.** $7x(3 - x)$

17. $4 - x^2$ **18.** $x^2 - 9$

19. $x^2 - 4x - 5$ **20.** $2x^2 - 4x - 3$

In Exercises 21–60, solve the inequality and sketch the graph of the solution on the real number line. (Some of the inequalities have no solution.) See Examples 1–3.

21. $2x + 6 \geq 0$ **22.** $5x - 20 < 0$

23. $-\frac{3}{4}x + 6 < 0$ **24.** $3x - 2 \geq 0$

25. $3x(x - 2) < 0$ **26.** $2x(x - 6) > 0$

27. $3x(2 - x) \geq 0$ **28.** $2x(6 - x) > 0$

29. $x^2 > 4$ **30.** $z^2 \leq 9$

31. $x^2 + 3x \leq 10$ **32.** $t^2 - 4t > 12$

33. $u^2 + 2u - 2 > 1$ **34.** $t^2 - 15t + 50 < 0$

35. $x^2 + 4x + 5 < 0$ **36.** $x^2 + 6x + 10 > 0$

37. $x^2 + 2x + 1 \geq 0$ **38.** $y^2 - 5y + 6 > 0$

39. $x^2 - 4x + 2 > 0$ **40.** $-x^2 + 8x - 11 \leq 0$

41. $x^2 - 6x + 9 \geq 0$ **42.** $x^2 + 8x + 16 < 0$

43. $u^2 - 10u + 25 < 0$ **44.** $y^2 + 16y + 64 \leq 0$

45. $3x^2 + 2x - 8 \leq 0$ **46.** $2t^2 - 3t - 20 \geq 0$

47. $-6u^2 + 19u - 10 > 0$ **48.** $4x^2 - 4x - 63 < 0$

49. $-2u^2 + 7u + 4 < 0$ **50.** $-3x^2 - 4x + 4 \leq 0$

51. $4x^2 + 28x + 49 \leq 0$ **52.** $9x^2 - 24x + 16 \geq 0$

53. $(x - 5)^2 < 0$ **54.** $(y + 3)^2 \geq 0$

55. $6 - (x - 5)^2 < 0$ **56.** $(y + 3)^2 - 6 \geq 0$

57. $16 \leq (u + 5)^2$ **58.** $25 \geq (x - 3)^2$

59. $x(x - 2)(x + 2) > 0$ **60.** $x^2(x - 2) \leq 0$

In Exercises 61–68, use a graphing utility to graph the function and solve the inequality.

61. $y = x^2 - 6x, \quad y < 0$

62. $y = 2x^2 + 5x, \quad y > 0$

63. $y = 0.5x^2 + 1.25x - 3, \quad y > 0$

64. $y = \frac{1}{3}x^2 - 3x, \quad y < 0$

65. $y = x^2 + 4x + 4, \quad y \geq 9$

66. $y = x^2 - 6x + 9, \quad y < 16$

67. $y = 9 - 0.2(x - 2)^2, \quad y < 4$

68. $y = 8x - x^2, \quad y > 12$

In Exercises 69–72, determine the critical numbers of the rational expression and locate them on the real number line.

69. $\dfrac{5}{x - 3}$

70. $\dfrac{-6}{x + 2}$

71. $\dfrac{2x}{x + 5}$

72. $\dfrac{x - 2}{x - 10}$

In Exercises 73–94, solve the rational inequality. As part of your solution, include a graph that shows the test intervals. See Example 4.

73. $\dfrac{5}{x - 3} > 0$

74. $\dfrac{3}{4 - x} > 0$

75. $\dfrac{-5}{x - 3} > 0$

76. $\dfrac{-3}{4 - x} > 0$

77. $\dfrac{x}{x - 3} < 0$

78. $\dfrac{x}{2 - x} < 0$

79. $\dfrac{x + 3}{x - 4} \leq 0$

80. $\dfrac{z - 1}{z + 3} < 0$

81. $\dfrac{y - 4}{y + 6} < 0$

82. $\dfrac{u + 3}{u + 7} \leq 0$

83. $\dfrac{y - 3}{2y - 11} \geq 0$

84. $\dfrac{x + 5}{3x + 2} \geq 0$

85. $\dfrac{x + 2}{4x + 6} \leq 0$

86. $\dfrac{u - 6}{3u - 5} \leq 0$

87. $\dfrac{3(u - 3)}{u + 1} < 0$

88. $\dfrac{2(4 - t)}{4 + t} > 0$

89. $\dfrac{6}{x - 4} > 2$

90. $\dfrac{1}{x + 2} > -3$

91. $\dfrac{4x}{x + 2} < -1$

92. $\dfrac{6x}{x - 4} < 5$

93. $\dfrac{x - 1}{x - 3} \leq 2$

94. $\dfrac{x + 4}{x - 5} \geq 10$

In Exercises 95–102, use a graphing utility to solve the rational inequality.

95. $\dfrac{1}{x} - x > 0$

96. $\dfrac{1}{x} - 4 < 0$

97. $\dfrac{x + 6}{x + 1} - 2 < 0$

98. $\dfrac{x + 12}{x + 2} - 3 \geq 0$

99. $\dfrac{6x - 3}{x + 5} < 2$

100. $\dfrac{3x - 4}{x - 4} < -5$

101. $x + \dfrac{1}{x} > 3$

102. $4 - \dfrac{1}{x^2} > 1$

Graphical Analysis In Exercises 103–106, use a graphing utility to graph the function. Use the graph to approximate the values of x that satisfy the specified inequalities.

Equation	Inequalities	
103. $y = \dfrac{3x}{x - 2}$	(a) $y \leq 0$	(b) $y \geq 6$
104. $y = \dfrac{2(x - 2)}{x + 1}$	(a) $y \leq 0$	(b) $y \geq 8$
105. $y = \dfrac{2x^2}{x^2 + 4}$	(a) $y \geq 1$	(b) $y \leq 2$
106. $y = \dfrac{5x}{x^2 + 4}$	(a) $y \geq 1$	(b) $y \geq 0$

Solving Problems

107. *Height of a Projectile* A projectile is fired vertically upward from ground level with an initial velocity of 128 feet per second, so that its height at any time t is given by $h = -16t^2 + 128t$ where the height h is measured in feet and the time t is measured in seconds. During what interval of time will the height of the projectile exceed 240 feet?

108. *Height of a Projectile* A projectile is fired vertically upward from ground level with an initial velocity of 88 feet per second, so that its height at any time t is given by $h = -16t^2 + 88t$ where the height h is measured in feet and the time t is measured in seconds. During what interval of time will the height of the projectile exceed 50 feet?

109. *Annual Interest Rate*   You are investing $1000 in a certificate of deposit for 2 years and you want the interest for that time period to exceed $150. The interest is compounded annually. What interest rate should you have? [*Hint:* Solve the inequality $1000(1 + r)^2 > 1150$.]

110. *Annual Interest Rate*   You are investing $500 in a certificate of deposit for 2 years and you want the interest for that time to exceed $50. The interest is compounded annually. What interest rate should you have? [*Hint:* Solve the inequality $500(1 + r)^2 > 550$.]

111. *Company Profits*   The revenue and cost equations for a product are given by

$$R = x(50 - 0.0002x)$$
$$C = 12x + 150{,}000$$

where R and C are measured in dollars and x represents the number of units sold (see figure). How many units must be sold to obtain a profit of at least $1,650,000?

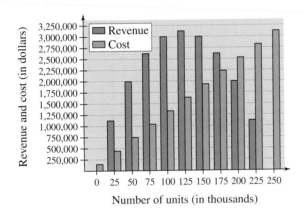

112. *Geometry*   A rectangular playing field with a perimeter of 100 meters is to have an area of at least 500 square meters. Within what bounds must the length of the field lie?

113. *Geometry*   You have 64 feet of fencing to enclose a rectangular region. Determine the interval for the length such that the area will exceed 240 square feet.

114. *Average Cost*   The cost of producing x units of a product is $C = 3000 + 0.75x$, $x > 0$.

(a) Write the average cost $\overline{C} = C/x$ as a function of x.

(b) Use a graphing utility to graph the average cost function in part (a). Determine the horizontal asymptote of the graph.

(c) How many units must be produced if the average cost per unit is to be less than $2?

115. *Data Analysis*   The temperature T (in degrees Fahrenheit) of a metal in a laboratory experiment was recorded every 2 minutes for a period of 16 minutes. The table gives the experimental data, where t is the time in minutes.

t	0	2	4	6	8
T	250	290	338	410	498

t	10	12	14	16
T	560	530	370	160

A model for these data is

$$T = \frac{244.20 - 13.23t}{1 - 0.13t + 0.005t^2}.$$

(a) Use a graphing utility to plot the data and graph the model.

(b) Use the graph to approximate the times when the temperature was at least $400°$F.

Explaining Concepts

116. Answer part (e) of Motivating the Chapter on page 371.

117. Explain the change in an inequality when both sides are multiplied by a negative real number.

118. Give a verbal description of the intervals $(-\infty, 5] \cup (10, \infty)$.

119. Define the term *critical number* and explain its use in solving quadratic inequalities.

120. In your own words, describe the procedure for solving quadratic inequalities.

121. Give an example of a quadratic inequality that has no real solution.

122. Explain the distinction between the critical numbers of a quadratic inequality and those of a rational inequality.

Key Terms

double or repeated
  solution, *p. 372*
quadratic form, *p. 375*

discriminant, *p. 387*
test intervals, *p. 407*

critical numbers, *p. 407*

Key Concepts

6.1 Extracting square roots

The equation $u^2 = d$, where $d > 0$, has exactly two solutions: $u = \sqrt{d}$ and $u = -\sqrt{d}$.

6.1 Extracting complex square roots

The equation $u^2 = d$, where $d < 0$, has exactly two solutions: $u = \sqrt{|d|}\,i$ and $u = -\sqrt{|d|}\,i$.

6.2 Completing the square

To complete the square for the expression $x^2 + bx$, add $(b/2)^2$, which is the square of half the coefficient of x. Consequently, $x^2 + bx + (b/2)^2 = (x + b/2)^2$.

6.3 The Quadratic Formula

The solutions of $ax^2 + bx + c = 0$, $a \neq 0$, are given by the Quadratic Formula $x = \left(-b \pm \sqrt{b^2 - 4ac}\right)/(2a)$. The expression inside the radical, $b^2 - 4ac$, is called the discriminant.

1. If $b^2 - 4ac > 0$, the equation has two real solutions.
2. If $b^2 - 4ac = 0$, the equation has one (repeated) real solution.
3. If $b^2 - 4ac < 0$, the equation has no real solutions.

6.3 Using the discriminant

The discriminant of the quadratic equation

$$ax^2 + bx + c = 0, \quad a \neq 0$$

can be used to classify the solutions of the equation as follows.

Discriminant	Solution Types
1. Perfect square	Two distinct rational solutions
2. Positive nonperfect square	Two distinct irrational solutions
3. Zero	One repeated rational solution
4. Negative number	Two distinct imaginary solutions

6.3 Summary of methods for solving quadratic equations

1. Factoring
2. Extracting square roots
3. Completing the square
4. Using the Quadratic Formula

6.5 Finding test intervals for inequalities

1. For a polynomial function, find all the real zeros. For a rational function, find all the real zeros and those x-values for which the function is undefined.

2. Arrange the numbers found in step 1 in increasing order. These numbers are called critical numbers.

3. Use the critical numbers to determine the test intervals.

4. Choose a representative x-value in each test interval and evaluate the function at that value. If the value of the function is negative, the function will have negative values for every x-value in the interval. If the value of the function is positive, the function will have positive values for every x-value in the interval.

REVIEW EXERCISES

Reviewing Skills

6.1 In Exercises 1–10, solve the quadratic equation by factoring.

1. $x^2 + 12x = 0$

2. $u^2 - 18u = 0$

3. $4y^2 - 1 = 0$

4. $2z^2 - 72 = 0$

5. $4y^2 + 20y + 25 = 0$

6. $x^2 + \frac{8}{3}x + \frac{16}{9} = 0$

7. $2x^2 - 2x - 180 = 0$

8. $15x^2 - 30x - 45 = 0$

9. $6x^2 - 12x = 4x^2 - 3x + 18$

10. $10x - 8 = 3x^2 - 9x + 12$

In Exercises 11–22, solve the quadratic equation by extracting square roots. Find all real and complex solutions.

11. $4x^2 = 10,000$

12. $2x^2 = 98$

13. $y^2 - 12 = 0$

14. $y^2 - 8 = 0$

15. $(x - 16)^2 = 400$

16. $(x + 3)^2 = 900$

17. $z^2 = -121$

18. $u^2 = -36$

19. $y^2 + 50 = 0$

20. $x^2 + 48 = 0$

21. $(y + 4)^2 + 18 = 0$

22. $(x - 2)^2 + 24 = 0$

In Exercises 23–30, solve the equation of quadratic form.

23. $x^4 - 4x^2 - 5 = 0$

24. $x^4 - 10x^2 + 9 = 0$

25. $x - 4\sqrt{x} + 3 = 0$

26. $x + 2\sqrt{x} - 3 = 0$

27. $(x^2 - 2x)^2 - 4(x^2 - 2x) - 5 = 0$

28. $\left(\sqrt{x} - 2\right)^2 + 2\left(\sqrt{x} - 2\right) - 3 = 0$

29. $x^{2/3} + 3x^{1/3} - 28 = 0$

30. $x^{2/5} + 4x^{1/5} + 3 = 0$

6.2 In Exercises 31–38, solve the equation by completing the square. (Find all real and complex solutions.)

31. $x^2 - 6x - 3 = 0$

32. $x^2 + 12x + 6 = 0$

33. $x^2 - 3x + 3 = 0$

34. $u^2 - 5u + 6 = 0$

35. $y^2 - \frac{2}{3}y + 2 = 0$

36. $t^2 + \frac{1}{2}t - 1 = 0$

37. $2y^2 + 10y + 3 = 0$

38. $3x^2 - 2x + 2 = 0$

6.3 In Exercises 39–46, use the Quadratic Formula to solve the equation. Find all real and complex solutions.

39. $y^2 + y - 30 = 0$

40. $x^2 - x - 72 = 0$

41. $2y^2 + y - 21 = 0$

42. $2x^2 - 3x - 20 = 0$

43. $5x^2 - 16x + 2 = 0$

44. $3x^2 + 12x + 4 = 0$

45. $0.3t^2 - 2t + 5 = 0$

46. $-u^2 + 2.5u + 3 = 0$

In Exercises 47–54, determine the type of solution of the quadratic equation using the discriminant.

47. $x^2 + 4x + 4 = 0$

48. $y^2 - 26y + 169 = 0$

49. $s^2 - s - 20 = 0$

50. $r^2 - 5r - 45 = 0$

51. $3t^2 + 17t + 10 = 0$

52. $7x^2 + 3x - 18 = 0$

53. $v^2 - 6v + 21 = 0$

54. $9y^2 + 1 = 0$

6.5 In Exercises 55–64, solve the inequality and graph its solution on the real number line.

55. $5x(7 - x) > 0$

56. $-2x(x - 10) \le 0$

57. $16 - (x - 2)^2 \le 0$

58. $(x - 5)^2 - 36 > 0$

59. $2x^2 + 3x - 20 < 0$

60. $3x^2 - 2x - 8 > 0$

61. $\dfrac{x + 3}{2x - 7} \ge 0$

62. $\dfrac{2x - 9}{x - 1} \le 0$

63. $\dfrac{2x - 2}{x + 6} + 2 < 0$

64. $\dfrac{3x + 1}{x - 2} > 4$

Solving Problems

65. *Selling Price* A car dealer bought a fleet of cars from a car rental agency for a total of $80,000. By the time the dealer had sold all but four of the cars, at an average profit of $1000 each, the original investment of $80,000 had been regained. How many cars did the dealer sell, and what was the average price per car?

66. *Selling Price* A manager of a computer store bought several computers of the same model for $27,000. When all but five of the computers had been sold at a profit of $900 per computer, the original investment of $27,000 had been regained. How many computers were sold, and what was the selling price of each computer?

67. *Geometry* The length of a rectangle is 12 inches greater than its width. The area of the rectangle is 108 square inches. Find the dimensions of the rectangle.

68. *Geometry* Find the dimensions of a triangle if its height is 4 centimeters less than its base and its area is 240 square centimeters.

69. *Compound Interest* You want to invest $20,000 for 2 years at an annual interest rate of r (in decimal form). Interest on the account is compounded annually. Find the interest rate if a deposit of $20,000 increases to $21,424.50 over a 2-year period.

70. *Compound Interest* You want to invest $35,000 for 2 years at an annual interest rate of r (in decimal form). Interest on the account is compounded annually. Find the interest rate if a deposit of $35,000 increases to $38,955.88 over a 2-year period.

71. *Reduced Fare* A college wind ensemble charters a bus at a cost of $360 to attend a concert. In an attempt to lower the bus fare per person, the ensemble invites nonmembers to go along. When eight nonmembers join the trip, the fare is decreased by $1.50 per person. How many people are going on the excursion?

72. *Think About It* When six nonmembers go along on the excursion described in Exercise 71, the fare is decreased by $16. Describe how it is possible to have fewer nonmembers and a greater decrease in the fare.

73. *Decreased Price* A Little League baseball team obtains a block of tickets for a ball game for $96. After three more people decide to go to the game, the price per ticket is decreased by $1.60. How many people are going to the game?

74. *Shortcut* A corner lot has an L-shaped sidewalk along its sides. The total length of the sidewalk is 51 feet. By cutting diagonally across the lot, the walking distance is shortened to 39 feet. What are the lengths of the two legs of the sidewalk?

75. *Shortcut* Two buildings are connected by an L-shaped protected walkway. The distance between buildings via the walkway is 140 feet. By cutting diagonally across the grass, the walking distance is shortened to 100 feet. What are the lengths of the two legs of the walkway?

76. *Work-Rate Problem* Working together, two people can complete a task in 6 hours. Working alone, how long would it take each to do the task if one person takes 3 hours longer than the other?

77. *Work-Rate Problem* Working together, two people can complete a task in 10 hours. Working alone, how long would it take each to do the task if one person takes 2 hours longer than the other?

78. *Vertical Motion* The height h in feet of an object above the ground is

$$h = 200 - 16t^2, \quad t \geq 0$$

where t is time in seconds.

(a) After how many seconds will the height be 164 feet?

(b) Find the time when the object strikes the ground.

79. *Vertical Motion* The height h in feet of an object above the ground is

$$h = -16t^2 + 64t + 192, \quad t > 0$$

where t is time in seconds.

(a) After how many seconds will the height be 256 feet?

(b) Find the time when the object strikes the ground.

80. *Average Cost*   The cost of producing x units of a product is

$$C = 100,000 + 0.9x,$$

and so the average cost per unit is

$$\overline{C} = C/x.$$

Find the number of units that must be produced if $\overline{C} < 2$.

81. *Average Cost*   The cost of producing x units of a product is

$$C = 50,000 + 1.2x,$$

and so the average cost per unit is

$$\overline{C} = C/x.$$

Find the number of units that must be produced if $\overline{C} < 5$.

82. *Annual Interest Rate*   You are investing $3000 in a certificate of deposit for 2 years and you want the interest for that period of time to exceed $370. The interest is compounded annually. What interest rate should you have? [*Hint:* Solve the inequality $3000(1 + r)^2 > 3370$.]

83. *Height of a Projectile*   A projectile is fired straight up from ground level with an initial velocity of 312 feet per second. Its height at any time t is given by

$$h = -16t^2 + 312t,$$

where the height h is measured in feet and the time t is measured in seconds. During what interval of time will the height of the projectile exceed 1200 feet?

Chapter Test

Take this test as you would take a test in class. After you are done, check your work against the answers given in the back of the book.

In Exercises 1–6, solve the equation by the specified method.

1. Factoring:

$x(x + 5) - 10(x + 5) = 0$

2. Factoring:

$8x^2 - 21x - 9 = 0$

3. Extracting square roots:

$(x - 2)^2 = 0.09$

4. Extracting square roots:

$(x + 3)^2 + 81 = 0$

5. Completing the square:

$2x^2 - 6x + 3 = 0$

6. Quadratic Formula:

$2y(y - 2) = 7$

In Exercises 7 and 8, solve the equation of quadratic form.

7. $x - 5\sqrt{x} + 4 = 0$

8. $x^4 + 6x^2 - 16 = 0$

9. Find the discriminant and explain how it can be used to determine the type of solutions of the quadratic equation $5x^2 - 12x + 10 = 0$.

10. Find a quadratic equation having the solutions -4 and 5.

In Exercises 11–14, solve the inequality and sketch its solution.

11. $16 \leq (x - 2)^2$

12. $2x(x - 3) < 0$

13. $\dfrac{3u + 2}{u - 3} \leq 2$

14. $\dfrac{3}{x - 2} > 4$

15. The width of a rectangle is 8 feet less than its length. The area of the rectangle is 240 square feet. Find the dimensions of the rectangle.

16. An English club chartered a bus for a trip to a Shakespearean festival. The cost of the bus was $1250. To lower the per person cost of the bus, non-members were invited. When 10 nonmembers joined the trip, the fare per person decreased by $6.25. How many club members were going on the trip?

17. An object is dropped from a height of 75 feet. Its height h (in feet) at any time t is given by $h = -16t^2 + 75$, where the time t is measured in seconds. Find the time required for the object to fall to a height of 35 feet.

18. The revenue R for a chartered bus trip is given by $R = -\frac{1}{20}(n^2 - 240n)$, where n is the number of passengers and $80 \leq n \leq 160$. How many passengers will produce a maximum revenue? Explain your reasoning.

19. A projectile is fired straight up from ground level with an initial velocity of 288 feet per second. Its height at any time t is given by $h = -16t^2 + 288t$, where the height h is measured in feet and the time t is measured in seconds. During what time interval will the height of the projectile exceed 1040 feet?

Cumulative Test: Chapters 4–6

Take this test as you would take a test in class. After you are done, check your work against the answers given in the back of the book.

1. Simplify $\left(\dfrac{2x^{-4}y^3}{3x^5y^{-3}z^0}\right)^{-2}$.

2. Evaluate $(4 \times 10^3)^2$ without the aid of a calculator.

3. Divide $4x^4 - 6x^3 + x - 4$ by $2x - 1$.

In Exercises 4–13, perform the operations and/or simplify.

4. $\dfrac{x^2 + 8x + 16}{18x^2} \cdot \dfrac{2x^4 + 4x^3}{x^2 - 16}$

5. $\dfrac{2}{x} - \dfrac{x}{x^3 + 3x^2} + \dfrac{1}{x + 3}$

6. $\dfrac{\left(\dfrac{x}{y} - \dfrac{y}{x}\right)}{\left(\dfrac{x - y}{xy}\right)}$

7. $\sqrt{-2}\left(\sqrt{-8} + 3\right)$

8. $(3 - 4i)^2$

9. $\left(\dfrac{t^{1/2}}{t^{1/4}}\right)^2$

10. $10\sqrt{20x} + 3\sqrt{125x}$

11. $\left(\sqrt{2x} - 3\right)^2$

12. $\dfrac{6}{\sqrt{10} - 2}$

13. $\dfrac{1 - 2i}{4 + i}$

In Exercises 14–19, solve the equation.

14. $\dfrac{1}{x} + \dfrac{4}{10 - x} = 1$

15. $\dfrac{x - 3}{x} + 1 = \dfrac{x - 4}{x - 6}$

16. $\sqrt{x} - x + 12 = 0$

17. $\sqrt{5 - x} + 10 = 11$

18. $(x - 5)^2 + 50 = 0$

19. $3x^2 + 6x + 2 = 0$

20. The volume V of a right circular cylinder is $V = \pi r^2 h$. The two cylinders in the figure have equal volumes. Write r_2 as a function of r_1.

21. The four corners are cut from a 12-inch-by-12-inch piece of glass, as shown in the figure. Find the perimeter of the remaining piece of glass.

22. Use a graphing utility to graph the equation $y = x^2 - 6x - 8$. Use the graph to approximate any x-intercepts of the graph. Set $y = 0$ and solve the resulting equation. Compare the results with the x-intercepts of the graph.

23. Find a quadratic equation having the solutions -2 and 6.

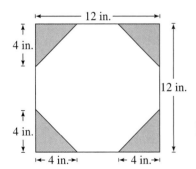

Figure for 20

Figure for 21

7 Linear Models and Graphs of Nonlinear Models

Mark Richard/PhotoEdit

In 1996, the computer and office equipment industry employed 259,000 people. Of these, 96,000 were production workers. (Source: Electronic Industries Association)

Motivating the Chapter

 ## Factory Shipments of Computers

The values (in billions of dollars) of factory shipments of computers and peripheral equipment for the years 1991 through 1996 are listed in the table. (Source: Electronic Industries Association)

Year	1991	1992	1993	1994	1995	1996
Value	50.1	51.9	54.8	59.3	73.6	78.7

A quadratic model for the value of the shipments is

$$y = 1.11x^2 - 1.69x + 50.5, \qquad 1 \leq x \leq 6$$

where y is the value of the shipments in billions of dollars and x represents the year, with $x = 1$ corresponding to 1991.

A rational model for the value of the shipments is

$$y = \frac{48.4 - 4.79x}{1 - 0.13x}, \qquad 1 \leq x \leq 6$$

where y is the value of the shipments in billions of dollars and x represents the year, with $x = 1$ corresponding to 1991.

See Section 7.3, Exercise 106

a. Does the graph of the quadratic model open upward or downward? Determine the vertex and the axis of the graph. Is the vertex a minimum value or a maximum value?

b. Use a graphing utility to draw a scatter plot of the data and the quadratic model in the same viewing window. Does the model appear to be accurate? Explain.

See Section 7.5, Exercise 81

c. Determine the domain of the graph of the rational model. What is the x-intercept of the graph? What are the horizontal and vertical asymptotes of the graph?

d. Use a graphing utility to draw a scatter plot of the data and the rational model in the same viewing window. Does the model appear to be accurate? Explain.

e. The domain for both models is $1 \leq x \leq 6$. Do you think the models are accurate for the years before 1991 and after 1996? If you were to estimate the value of the shipments of computers and peripheral equipment in 1998, which model would you use? Do you think this is a good estimate? Explain.

Objectives

1. Solve an application problem involving direct variation.

2. Solve an application problem involving inverse variation.

3. Solve an application problem involving joint variation.

1 Solve an application problem involving direct variation.

Direct Variation

In the mathematical model for **direct variation,** y is a *linear* function of x. Specifically, $y = kx$.

Study Tip

To use this mathematical model in applications involving direct variation, you are usually given specific values of x and y, which then enable you to find the value of the constant k.

▶ Direct Variation

The following statements are equivalent. The number k is the **constant of proportionality.**

1. y **varies directly** as x.

2. y is **directly proportional** to x.

3. $y = kx$ for some constant k.

Example 1 Direct Variation

Assume that the total revenue R (in dollars) obtained from selling x units of a product is directly proportional to the number of units sold. When 10,000 units are sold, the total revenue is $142,500.

a. Find a model that relates the total revenue R to the number of units sold x.

b. Find the total revenue obtained from selling 12,000 units.

Solution

a. Because the total revenue is directly proportional to the number of units sold, the linear model is $R = kx$. To find the value of the constant k, substitute 142,500 for R and 10,000 for x

$$142{,}500 = k(10{,}000) \qquad \text{Substitute for } R \text{ and } x.$$

which implies that $k = 142{,}500/10{,}000 = 14.25$. So, the equation relating the total revenue to the total number of units sold is

$$R = 14.25x. \qquad \text{Direct variation model}$$

The graph of this equation is shown in Figure 7.1.

b. When $x = 12{,}000$, the total revenue is

$$R = 14.25(12{,}000) = \$171{,}000.$$

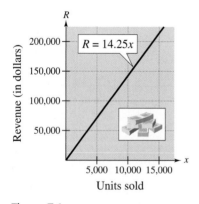

Figure 7.1

Example 2   Direct Variation

Hooke's Law for springs states that the distance a spring is stretched (or compressed) is proportional to the force on the spring. A force of 20 pounds stretches a particular spring 5 inches.

a. Find a mathematical model that relates the distance the spring is stretched to the force applied to the spring.

b. How far will a force of 30 pounds stretch the spring?

Solution

a. For this problem, let d represent the distance (in inches) that the spring is stretched and let F represent the force (in pounds) that is applied to the spring. Because the distance d is proportional to the force F, the model is

$$d = kF.$$

To find the value of the constant k, use the fact that $d = 5$ when $F = 20$. Substituting these values into the given model produces

$$5 = k(20) \qquad \text{Substitute 5 for } d \text{ and 20 for } F.$$

$$\frac{5}{20} = k \qquad \text{Divide both sides by 20.}$$

$$\frac{1}{4} = k. \qquad \text{Simplify.}$$

So, the equation relating distance and force is

$$d = \frac{1}{4}F. \qquad \text{Direct variation model}$$

b. When $F = 30$, the distance is

$$d = \frac{1}{4}(30) = 7.5 \text{ inches.} \qquad \text{See Figure 7.2.}$$

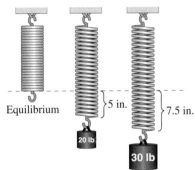

Figure 7.2

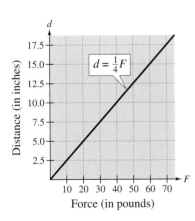

Figure 7.3

In Example 2, you can get a clearer understanding of Hooke's Law by using the model $d = \frac{1}{4}F$ to create a table or a graph (see Figure 7.3). From the table or from the graph, you can see what it means for the distance to be "proportional to the force."

Force, F	10 lb	20 lb	30 lb	40 lb	50 lb	60 lb
Distance, d	2.5 in.	5.0 in.	7.5 in.	10.0 in.	12.5 in.	15.0 in.

In Examples 1 and 2, the direct variations are such that an *increase* in one variable corresponds to an *increase* in the other variable. There are, however, other applications of direct variation in which an increase in one variable corresponds to a *decrease* in the other variable. For instance, in the model $y = -2x$, an increase in x will yield a decrease in y.

A second type of direct variation relates one variable to a *power* of another.

> ▶ Direct Variation as *n*th Power
>
> The following statements are equivalent.
>
> **1.** *y* **varies directly as the *n*th power** of *x*.
>
> **2.** *y* is **directly proportional to the *n*th power** of *x*.
>
> **3.** $y = kx^n$ for some constant *k*.

Example 3 Direction Variation as a Power

The distance a ball rolls down an inclined plane is directly proportional to the square of the time it rolls. Assume that, during the first second, a ball rolls down a particular plane a distance of 6 feet.

a. Find a mathematical model that relates the distance traveled to the time.

b. How far will the ball roll during the first 2 seconds?

Solution

a. Letting *d* be the distance (in feet) that the ball rolls and letting *t* be the time (in seconds), you obtain the model

$$d = kt^2.$$

Because $d = 6$ when $t = 1$, you obtain

$d = kt^2$	Original equation
$6 = k(1)^2$	Substitute 6 for *d* and 1 for *t*.
$6 = k.$	Simplify.

So, the equation relating distance to time is

$$d = 6t^2. \qquad \text{Direct variation as 2nd power model}$$

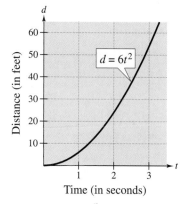

Figure 7.4

The graph of this equation is shown in Figure 7.4.

b. When $t = 2$, the distance traveled is

$$d = 6(2)^2 = 6(4) = 24 \text{ feet.} \qquad \text{See Figure 7.5.}$$

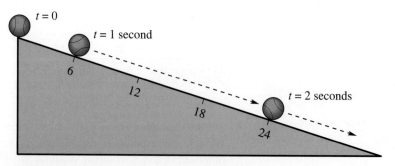

Figure 7.5

2 Solve an application problem involving inverse variation.

Inverse Variation

Another type of variation is called **inverse variation.** For this type of variation, we say that one of the variables is inversely proportional to the other variable.

▶ **Inverse Variation**

    1. The following three statements are equivalent.

        a. y **varies inversely** as x.

        b. y is **inversely proportional** to x.

        c. $y = \dfrac{k}{x}$ for some constant k.

    2. If $y = \dfrac{k}{x^n}$, then y is inversely proportional to the nth power of x.

Example 4   Inverse Variation

The marketing department of a large company has found that the demand for one of its products varies inversely as the price of the product. (When the price is low, more people are willing to buy the product than when the price is high.) When the price of the product is \$7.50, the monthly demand is 50,000 units. Approximate the monthly demand if the price is reduced to \$6.00.

Solution

Let x represent the number of units that are sold each month (the demand), and let p represent the price per unit (in dollars). Because the demand is inversely proportional to the price, the model is

$$x = \frac{k}{p}.$$

By substituting $x = 50{,}000$ when $p = 7.50$, you obtain

$$50{,}000 = \frac{k}{7.50} \qquad \text{Substitute 50,000 for } x \text{ and 7.50 for } p.$$

$$375{,}000 = k. \qquad \text{Multiply both sides by 7.50.}$$

So, the model is

$$x = \frac{375{,}000}{p}. \qquad \text{Inverse variation model}$$

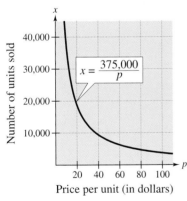

Figure 7.6

The graph of this equation is shown in Figure 7.6. To find the demand that corresponds to a price of \$6.00, substitute $p = 6$ into the equation and obtain

$$x = \frac{375{,}000}{6} = 62{,}500 \text{ units.}$$

So, if the price were lowered from \$7.50 per unit to \$6.00 per unit, the monthly demand could be expected to increase from 50,000 units to 62,500 units.

Some applications of variation involve problems with *both* direct and inverse variation in the same model.

Example 5 Direct and Inverse Variation

A company determines that the demand for one of its products is directly proportional to the amount spent on advertising and inversely proportional to the price of the product. When $40,000 is spent on advertising and the price per unit is $20, the monthly demand is 10,000 units.

a. If the amount of advertising were increased to $50,000, how much could the price be increased to maintain a monthly demand of 10,000 units?

b. If you were in charge of the advertising department, would you recommend this increased expense in advertising?

Solution

a. Let x represent the number of units that are sold each month (the demand), let a represent the amount spent on advertising (in dollars), and let p represent the price per unit (in dollars). Because the demand is directly proportional to the advertising and inversely proportional to the price, the model is

$$x = \frac{ka}{p}.$$

By substituting $x = 10,000$ when $a = 40,000$ and $p = 20$, you obtain

$$10,000 = \frac{k(40,000)}{20} \qquad \text{Substitute 10,000 for } x, 40,000 \text{ for } a, \text{ and 20 for } p.$$

$$200,000 = 40,000k \qquad \text{Multiply both sides by 20.}$$

$$5 = k. \qquad \text{Divide both sides by 40,000.}$$

So, the model is

$$x = \frac{5a}{p}. \qquad \text{Direct and inverse variation model}$$

To find the price that corresponds to a demand of 10,000 and an advertising expense of $50,000, substitute $x = 10,000$ and $a = 50,000$ into the model and solve for p.

$$10,000 = \frac{5(50,000)}{p} \implies p = \frac{5(50,000)}{10,000} = \$25$$

So, the price increase would be $25 - $20 = $5.

b. The total revenue from selling 10,000 units at $20 each is $200,000, and the revenue from selling 10,000 units at $25 each is $250,000. So, increasing the advertising expense from $40,000 to $50,000 would increase the revenue by $50,000. This implies that you should recommend the increased expense in advertising.

Amount of Advertising	Price	Revenue
$40,000	$20.00	10,000 × 20 = $200,000
$50,000	$25.00	10,000 × 25 = $250,000

3 Solve an application problem involving joint variation.

Joint Variation

A third type of variation is called **joint variation.** For this type of variation, we say that one variable varies directly with the product of two variables.

> ▶ Joint Variation
>
> **1.** The following three statements are equivalent.
>
>   **a.** z **varies jointly** as x and y.
>
>   **b.** z is **jointly proportional** to x and y.
>
>   **c.** $z = kxy$ for some constant k.
>
> **2.** If $z = kx^n y^m$, then z is jointly proportional to the nth power of x and the mth power of y.

Example 6   Joint Variation

The *simple interest* for a certain savings account is jointly proportional to the time and the principal. After one quarter (3 months), the interest for a principal of $6000 is $120. How much interest would a principal of $7500 earn in 5 months?

Solution

To begin, let I represent the interest earned (in dollars), let P represent the principal (in dollars), and let t represent the time (in years). Because the interest is jointly proportional to the time and the principal, the model is

$$I = ktP.$$

Because $I = 120$ when $P = 6000$ and $t = \frac{1}{4}$, it follows that

$$k = 120/\left(6000 \cdot \tfrac{1}{4}\right) = 0.08.$$

So, the model that relates interest to time and principal is

$$I = 0.08tP. \qquad \text{\small Joint variation model}$$

To find the interest earned on a principal of $7500 over a 5-month period of time, substitute $P = 7500$ and $t = \frac{5}{12}$ into the model and obtain an interest of

$$I = 0.08\left(\frac{5}{12}\right)(7500) = \$250.$$

Discussing the Concept	Creating Variation Models

For each type of variation, create a problem for which $k = 24$. Sketch a graph of each model and discuss how the graphs are the same and how they differ.

Direct variation: $y = kx$ *Inverse variation:* $y = \dfrac{k}{x}$ *Joint variation:* $z = kxy$

7.1 Exercises

Integrated Review — Concepts, Skills, and Problem Solving

Keep mathematically in shape by doing these exercises *before* the problems of this section.

Properties and Definitions

1. Sketch a curve on the rectangular coordinate system such that *y is not* a function of *x*. Explain.

2. Sketch a curve on the rectangular coordinate system such that *y is* a function of *x*. Explain.

3. Determine the domain of $f(x) = x^2 - 4x + 9$.

4. Determine the domain of $h(x) = \dfrac{x - 1}{x^2(x^2 + 1)}$.

Functions

In Exercises 5–8, consider the function

$$f(x) = 2x^3 - 3x^2 - 18x + 27$$
$$= (2x - 3)(x + 3)(x - 3).$$

5. Use a graphing utility to graph both expressions for the function. Are the graphs the same?

6. Verify the factorization by multiplying the polynomials in the factored form of *f*.

7. Verify the factorization by performing the long division

$$\frac{2x^3 - 3x^2 - 18x + 27}{2x - 3}$$

and then factoring the quotient.

8. Verify the factorization by performing the long division

$$\frac{2x^3 - 3x^2 - 18x + 27}{x^2 - 9}.$$

In Exercises 9 and 10, use the function to find and simplify the expression for

$$\frac{f(2 + h) - f(2)}{h}.$$

9. $f(x) = x^2 - 3$

10. $f(x) = \dfrac{3}{x + 5}$

Modeling

11. The inventor of a new game believes that the variable cost for producing the game is $5.75 per unit and the fixed costs are $12,000. If *x* is the number of games produced, express the total cost *C* as a function of *x*.

12. The length of a rectangle is one and one-half times its width. Express the perimeter *P* of the rectangle as a function of the rectangle's width *w*.

Developing Skills

In Exercises 1–14, write a model for the statement.

1. *I* varies directly as *V*.

2. *C* varies directly as *r*.

3. *V* is directly proportional to *t*.

4. *s* varies directly as the cube of *t*.

5. *u* is directly proportional to the square of *v*.

6. *V* varies directly as the cube root of *x*.

7. *p* varies inversely as *d*.

8. *S* is inversely proportional to the square of *v*.

9. *P* is inversely proportional to the square root of $1 + r$.

10. *A* varies inversely as the fourth power of *t*.

11. *A* varies jointly as *l* and *w*.

12. *V* varies jointly as *h* and the square of *r*.

13. *Boyle's Law* If the temperature of a gas is not allowed to change, its absolute pressure *P* is inversely proportional to its volume *V*.

14. *Newton's Law of Universal Gravitation*   The gravitational attraction F between two particles of masses m_1 and m_2 is directly proportional to the product of the masses and inversely proportional to the square of the distance r between the particles.

In Exercises 15–22, write a verbal sentence using variation terminology to describe the formula.

15. *Area of a Triangle:* $A = \frac{1}{2}bh$

16. *Area of a Circle:* $A = \pi r^2$

17. *Area of a Rectangle:* $A = lw$

18. *Surface Area of a Sphere:* $A = 4\pi r^2$

19. *Volume of a Right Circular Cylinder:* $V = \pi r^2 h$

20. *Volume of a Sphere:* $V = \frac{4}{3}\pi r^3$

21. *Average Speed:* $r = \dfrac{d}{t}$

22. *Height of a Cylinder:* $h = \dfrac{V}{\pi r^2}$

In Exercises 23–36, find the constant of proportionality and write an equation that relates the variables.

23. s varies directly as t, and $s = 20$ when $t = 4$.

24. h is directly proportional to r, and $h = 28$ when $r = 12$.

25. F is directly proportional to the square of x, and $F = 500$ when $x = 40$.

26. v varies directly as the square root of s, and $v = 24$ when $s = 16$.

27. H is directly proportional to u, and $H = 100$ when $u = 40$.

28. M varies directly as the cube of n, and $M = 0.012$ when $n = 0.2$.

29. n varies inversely as m, and $n = 32$ when $m = 1.5$.

30. q is inversely proportional to p, and $q = \frac{3}{2}$ when $p = 50$.

31. g varies inversely as the square root of z, and $g = \frac{4}{5}$ when $z = 25$.

32. u varies inversely as the square of v, and $u = 40$ when $v = \frac{1}{2}$.

33. F varies jointly as x and y, and $F = 500$ when $x = 15$ and $y = 8$.

34. V varies jointly as h and the square of b, and $V = 288$ when $h = 6$ and $b = 12$.

35. d varies directly as the square of x and inversely with r, and $d = 3000$ when $x = 10$ and $r = 4$.

36. z is directly proportional to x and inversely proportional to the square root of y, and $z = 720$ when $x = 48$ and $y = 81$.

Solving Problems

37. *Revenue*   The total revenue R is directly proportional to the number of units sold x. When 500 units are sold, the revenue is $3875.

 (a) Find the revenue when 635 units are sold.

 (b) Interpret the constant of proportionality.

38. *Revenue*   The total revenue R is directly proportional to the number of units sold x. When 25 units are sold, the revenue is $300.

 (a) Find the revenue when 42 units are sold.

 (b) Interpret the constant of proportionality.

39. *Hooke's Law*   A force of 50 pounds stretches a spring 5 inches (see figure).

 (a) How far will a force of 20 pounds stretch the spring?

 (b) What force is required to stretch the spring 1.5 inches?

Equilibrium   5 in.

50 lb

Figure for 39

40. *Hooke's Law*   A force of 50 pounds stretches a spring 3 inches.

 (a) How far will a force of 20 pounds stretch the spring?

 (b) What force is required to stretch the spring 1.5 inches?

41. *Hooke's Law*   A baby weighing $10\frac{1}{2}$ pounds compresses the spring of a baby scale 7 millimeters (see figure). Determine the weight of a baby that will compress the spring 12 millimeters.

42. *Hooke's Law*   A force of 50 pounds stretches the spring of a scale 1.5 inches.

(a) Write the force F as a function of the distance x the spring is stretched.

(b) Graph the function in part (a) where $0 \le x \le 5$. Identify the graph.

43. *Free-Falling Object*   The velocity v of a free-falling object is proportional to the time that the object has fallen. The constant of proportionality is the acceleration due to gravity. Find the acceleration due to gravity if the velocity of a falling object is 96 feet per second after the object has fallen for 3 seconds.

44. *Free-Falling Object*   Neglecting air resistance, the distance d that an object falls varies directly as the square of the time t it has fallen. If an object falls 64 feet in 2 seconds, determine the distance it will fall in 6 seconds.

45. *Stopping Distance*   The stopping distance d of an automobile is directly proportional to the square of its speed s. On a certain road surface, a car requires 75 feet to stop when its speed is 30 miles per hour. Estimate the stopping distance if the brakes are applied when the car is traveling at 50 miles per hour under similar road conditions.

46. *Velocity of a Stream*   The diameter d of a particle that can be moved by a stream is directly proportional to the square of the velocity v of the stream. A stream with a velocity of $\frac{1}{4}$ mile per hour can move coarse sand particles of about 0.02 inch diameter. What must the velocity be to carry particles with a diameter of 0.12 inch?

47. *Frictional Force*   The frictional force F between the tires and the road that is required to keep a car on a curved section of a highway is directly proportional to the square of the speed s of the car. If the speed of the car is doubled, the force will change by what factor?

48. *Power Generation*   The power P generated by a wind turbine varies directly as the cube of the wind speed w. The turbine generates 750 watts of power in a 25-mile-per-hour wind. Find the power it generates in a 40-mile-per-hour wind.

49. *Best Buy*   The prices of 9-inch, 12-inch, and 15-inch diameter pizzas at a certain pizza shop are $6.78, $9.78, and $12.18, respectively. One would expect that the price of a certain size pizza would be directly proportional to its surface area. Is that the case for this pizza shop? If not, which size pizza is the best buy?

50. *Travel Time*   The travel time between two cities is inversely proportional to the average speed. If a train travels between two cities in 3 hours at an average speed of 65 miles per hour, how long would it take at an average speed of 80 miles per hour? What does the constant of proportionality measure in this problem?

51. *Demand Function*   A company has found that the daily demand x for its product is inversely proportional to the price p. When the price is $5, the demand is 800 units. Approximate the demand if the price is increased to $6.

52. *Predator-Prey*   The number N of prey t months after a natural predator is introduced into a test area is inversely proportional to $t + 1$. If $N = 500$ when $t = 0$, find N when $t = 4$.

53. *Weight of an Astronaut*   A person's weight on the moon varies directly with his or her weight on earth. Neil Armstrong, the first man on the moon, weighed 360 pounds on earth, including his heavy equipment. On the moon he weighed only 60 pounds with the equipment. If the first woman in space, Valentina V. Tereshkova, had landed on the moon and weighed 54 pounds with equipment, how much would she have weighed on earth with her equipment?

54. *Weight of an Astronaut*   The gravitational force F with which an object is attracted to the earth is inversely proportional to the square of its distance r from the center of the earth. If an astronaut weighs 190 pounds on the surface of the earth ($r \approx 4000$ miles), what will the astronaut weigh 1000 miles above the earth's surface?

55. *Amount of Illumination*   The illumination *I* from a light source varies inversely as the square of the distance *d* from the light source. If you raise a study lamp from 18 inches to 36 inches above your desk (see figure), the illumination will change by what factor?

56. *Snowshoes*   When a person walks, the pressure *P* on each sole varies inversely with the area *A* of the sole. Denise is trudging through deep snow, wearing boots that have a sole area of 29 square inches each. The sole pressure is 4 pounds per square inch. If Denise were wearing snowshoes, each with an area 11 times that of her boot soles, what would be the pressure on each snowshoe? The constant of variation in this problem is Denise's weight. How much does she weigh?

57. *Oil Spill*   The graph shows the percent *p* of oil that remained in Chedabucto Bay, Nova Scotia, after an oil spill. The cleaning of the spill was left primarily to natural actions such as wave motion, evaporation, photochemical decomposition, and bacterial decomposition. After about a year, the percent that remained varied inversely as time. Find a model that relates *p* and *t*, where *t* is the number of years since the spill. Then use it to find the percent of oil that remained $6\frac{1}{2}$ years after the spill, and compare the result with the graph.

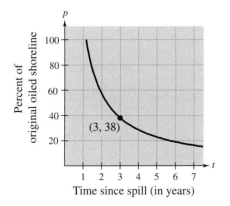

58. *Ocean Temperatures*   The graph shows the temperature of the water in the north central Pacific Ocean. At depths greater than 900 meters, the water temperature varies inversely with the water depth. Find a model that relates the temperature *T* to the depth *d*. Then use it to find the water temperature at a depth of 4385 meters, and compare the result with the graph.

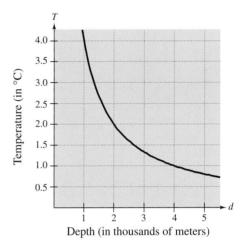

59. *Simple Interest*   Simple interest varies jointly as the product of the interest rate and the time. An investment at 9% for 3 years earns $202.50. How much will the investment earn in 4 years? What does the constant of proportionality measure in this problem?

60. *Engineering*   The load *P* that can be safely supported by a horizontal beam varies jointly as the product of the width *W* of the beam and the square of the depth *D* and inversely as the length *L*.

(a) Write a model for the statement.

(b) How does *P* change when the width and length of the beam are both doubled?

(c) How does *P* change when the width and depth of the beam are doubled?

(d) How does *P* change when all three of the dimensions are doubled?

(e) How does *P* change when the depth of the beam is cut in half?

(f) A beam with width 3 inches, depth 8 inches, and length 10 feet can safely support 2000 pounds. Determine the safe load of a beam made from the same material if its depth is increased to 10 inches.

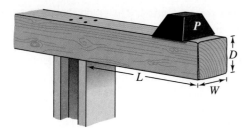

Figure for 60

In Exercises 61–64, complete the table and plot the resulting points.

x	2	4	6	8	10
$y = kx^2$					

61. $k = 1$ **62.** $k = 2$

63. $k = \frac{1}{2}$ **64.** $k = \frac{1}{4}$

In Exercises 65–68, complete the table and plot the resulting points.

x	2	4	6	8	10
$y = \dfrac{k}{x^2}$					

65. $k = 2$ **66.** $k = 5$

67. $k = 10$ **68.** $k = 20$

In Exercises 69–72, determine whether the variation model is of the form $y = kx$ or $y = k/x$, and find k.

69.

x	10	20	30	40	50
y	$\frac{2}{5}$	$\frac{1}{5}$	$\frac{2}{15}$	$\frac{1}{10}$	$\frac{2}{25}$

70.

x	10	20	30	40	50
y	2	4	6	8	10

71.

x	10	20	30	40	50
y	-3	-6	-9	-12	-15

72.

x	10	20	30	40	50
y	60	30	20	15	12

Explaining Concepts

73. Suppose the constant of proportionality is positive and y varies directly as x. If one of the variables increases, how will the other change? Explain.

74. Suppose the constant of proportionality is positive and y varies inversely as x. If one of the variables increases, how will the other change? Explain.

75. If y varies directly as the square of x and x is doubled, how will y change? Use the properties of exponents to explain your answer.

76. If y varies inversely as the square of x and x is doubled, how will y change? Use the properties of exponents to explain your answer.

7.2   Graphs of Linear Inequalities

Objectives

1 Verify a solution to a linear inequality in two variables.

2 Sketch the graph of a linear inequality in two variables.

1 Verify a solution to a linear inequality in two variables.

Linear Inequalities in Two Variables

A **linear inequality** in variables x and y is an inequality that can be written in one of the following forms.

$$ax + by < c, \quad ax + by > c, \quad ax + by \leq c, \quad \text{and} \quad ax + by \geq c$$

Here are some examples.

$$4x - 3y < 7, \quad x - y > -3, \quad x \leq 2, \quad \text{and} \quad y \geq -4$$

An ordered pair (x_1, y_1) is a **solution** of a linear inequality in x and y if the inequality is true when x_1 and y_1 are substituted for x and y, respectively. For instance, the ordered pair $(3, 2)$ is a solution of the inequality $x - y > 0$ because $3 - 2 > 0$ is a true statement.

Example 1   Verifying Solutions of Linear Inequalities

Decide whether each point is a solution of $2x - 3y \geq -2$.

a. $(0, 0)$ **b.** $(2, 2)$ **c.** $(0, 1)$

Solution

a. $2x - 3y \geq -2$ Original inequality

$2(0) - 3(0) \overset{?}{\geq} -2$ Substitute 0 for x and 0 for y.

$0 \geq -2$ Inequality is satisfied. ✓

Because the inequality is satisfied, the point $(0, 0)$ is a solution.

b. $2x - 3y \geq -2$ Original inequality

$2(2) - 3(2) \overset{?}{\geq} -2$ Substitute 2 for x and 2 for y.

$-2 \geq -2$ Inequality is satisfied. ✓

Because the inequality is satisfied, the point $(2, 2)$ is a solution.

c. $2x - 3y \geq -2$ Original inequality

$2(0) - 3(1) \overset{?}{\geq} -2$ Substitute 0 for x and 1 for y.

$-3 \ngeq -2$ Inequality is not satisfied. ✗

Because the inequality is not satisfied, the point $(0, 1)$ is not a solution.

2 Sketch the graph of a linear inequality in two variables.

The Graph of a Linear Inequality

The **graph** of a linear inequality is the collection of all solution points of the inequality. To sketch the graph of a linear inequality such as

$$4x - 3y < 12 \qquad \text{Original inequality}$$

begin by sketching the graph of the *corresponding linear equation*

$$4x - 3y = 12. \qquad \text{Corresponding equation}$$

Use *dashed* lines for the inequalities $<$ and $>$ and *solid* lines for the inequalities $\le$ and $\ge$. The graph of the equation separates the plane into two **half-planes.** In each half-plane, one of the following must be true.

1. All points in the half-plane are solutions of the inequality.

2. No point in the half-plane is a solution of the inequality.

So, you can determine whether the points in an entire half-plane satisfy the inequality by simply testing *one* point in the region.

Study Tip

When the inequality is strictly less than ($<$) or greater than ($>$), the line of the corresponding equation is dashed because the points on the line are *not* solutions of the inequality. When the inequality is less than or equal to ($\le$) or greater than or equal to ($\ge$), the line of the corresponding equation is solid because the points on the line *are* solutions of the inequality. The test point used to determine whether the points in a half-plane satisfy the inequality cannot lie on the line of the corresponding equation.

Example 2   Sketching the Graphs of Linear Inequalities

Sketch the graph of each linear inequality.

a. $x \ge -3$ **b.** $y < 4$ **c.** $x \le 2$

Solution

a. The graph of the corresponding equation $x = -3$ is a vertical line. The points that satisfy the inequality $x \ge -3$ are those lying on or to the right of this line, as shown in Figure 7.7.

b. The graph of the corresponding equation $y = 4$ is a horizontal line. The points that satisfy the inequality $y < 4$ are those lying below this line, as shown in Figure 7.8.

c. The graph of the corresponding equation $x = 2$ is a vertical line. The points that satisfy the inequality $x \le 2$ are those lying on or to the left of this line, as shown in Figure 7.9.

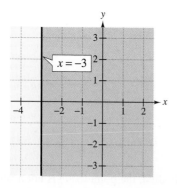

Figure 7.7

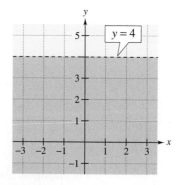

Figure 7.8

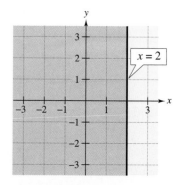

Figure 7.9

We summarize the guidelines for sketching the graph of a linear inequality in two variables as follows.

▶ **Guidelines for Graphing a Linear Inequality**

1. Replace the inequality sign by an equal sign, and sketch the graph of the resulting equation. (Use a dashed line for $<$ or $>$, and a solid line for $\leq$ or $\geq$.)

2. Test one point in one of the half-planes formed by the graph in Step 1.

   a. If the point satisfies the inequality, shade the entire half-plane to denote that every point in the region satisfies the inequality.

   b. If the point does not satisfy the inequality, then shade the other half-plane.

Study Tip

A convenient test point for determining which half-plane contains solutions to the inequality is the origin $(0, 0)$. In Example 3(a), when you substitute 0 for x and 0 for y you can easily see that $(0, 0)$ does not satisfy the inequality.

$$x + y > 3$$

$$0 + 0 > 3$$

Remember that the origin cannot be used as a test point if it lies on the graph of the corresponding equation.

Example 3   Sketching the Graphs of Linear Inequalities

Sketch the graph of each linear inequality.

a. $x + y > 3$ **b.** $2x + y \leq 2$

Solution

a. The graph of the corresponding equation $x + y = 3$ is a line, as shown in Figure 7.10. Because the origin $(0, 0)$ does not satisfy the inequality, the graph consists of the half-plane lying above the line. (Try checking a point above the line. Regardless of which point you choose, you will see that it is a solution.)

b. The graph of the corresponding equation $2x + y = 2$ is a line, as shown in Figure 7.11. Because the origin $(0, 0)$ satisfies the inequality, the graph consists of the half-plane lying on or below the line. (Try checking a point on or below the line. Regardless of which point you choose, you will see that it is a solution.)

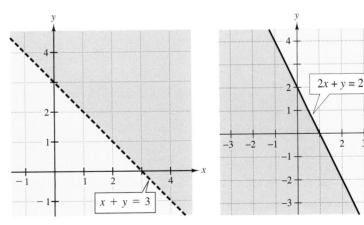

Figure 7.10   Figure 7.11

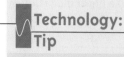

Technology: Tip

Most graphing utilities can graph inequalities in two variables. Consult the user's guide for your graphing utility for keystrokes. The graph of

$$y \leq \frac{1}{2}x - 3$$

is shown below.

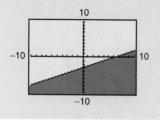

For a linear inequality in two variables, you can sometimes simplify the graphing procedure by writing the inequality in *slope-intercept form*. For instance, by writing $x + y > 1$ in the form

$$y > -x + 1 \qquad \text{Slope-intercept form}$$

you can see that the solution points lie *above* the line $y = -x + 1$, as shown in Figure 7.12. Similarly, by writing the inequality $4x - 3y > 12$ in the form

$$y < \frac{4}{3}x - 4 \qquad \text{Slope-intercept form}$$

you can see that the solutions lie *below* the line $y = \frac{4}{3}x - 4$, as shown in Figure 7.13.

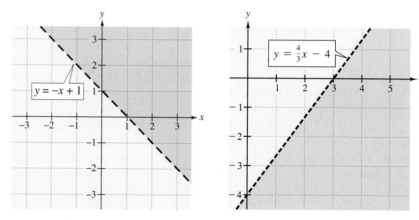

Figure 7.12   Figure 7.13

Example 4   Sketching the Graph of a Linear Inequality

Use the slope-intercept form of a linear equation as an aid in sketching the graph of the inequality $2x - 3y \leq 15$.

Solution

To begin, rewrite the inequality in slope-intercept form.

$$2x - 3y \leq 15 \qquad \text{Original inequality}$$

$$-3y \leq -2x + 15 \qquad \text{Subtract } 2x \text{ from both sides.}$$

$$y \geq \frac{2}{3}x - 5 \qquad \text{Slope-intercept form}$$

From this form, you can conclude that the solution is the half-plane lying *on or above* the line

$$y = \frac{2}{3}x - 5.$$

The graph is shown in Figure 7.14. To verify the solution, test any point in the shaded region.

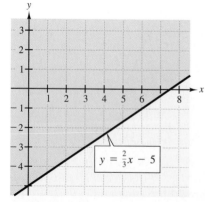

Figure 7.14

Example 5   Working to Meet a Budget

Your budget requires you to earn *at least* $160 per week. You work two part-time jobs. The first job pays $10 per hour and the second job pays $8 per hour. Let x represent the number of hours worked at the first job and let y represent the number of hours worked at the second job.

a. Write an inequality that represents the number of hours worked at each job in order to meet your budget requirements.

b. Graph the inequality and identify at least two ordered pairs (x, y) that identify the number of hours you must work at each job in order to meet your budget requirements.

Solution

a. *Verbal model:*

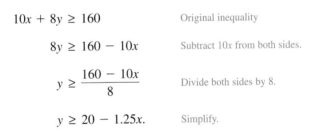

$$10 \cdot \boxed{\text{Number of hours at job 1}} + 8 \cdot \boxed{\text{Number of hours at job 2}} \geq 160$$

Labels: Number of hours at job 1 $= x$ (hours)
                Number of hours at job 2 $= y$ (hours)

Inequality: $10x + 8y \geq 160$

b. Solving the inequality for y, you get

$$10x + 8y \geq 160$$ Original inequality

$$8y \geq 160 - 10x$$ Subtract 10x from both sides.

$$y \geq \frac{160 - 10x}{8}$$ Divide both sides by 8.

$$y \geq 20 - 1.25x.$$ Simplify.

Graph the corresponding equation $y = 20 - 1.25x$ and shade the half-plane lying above the line, as shown in Figure 7.15. From the graph, you can see that two solutions that will yield the desired weekly earnings of at least $160 are $(8, 10)$ and $(12, 5)$. There are many other solutions.

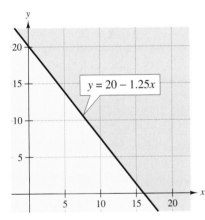

$y = 20 - 1.25x$

Figure 7.15

Discussing the Concept   Using Inequalities

Try the following activity. One person picks a point with whole number coordinates on a grid like the one at the left without revealing the coordinates. A second person writes the equation of a line passing through the grid region. The first person graphs the line on the grid and indicates whether the secret point lies above, below, or on the line. Continue writing and graphing lines until the second person is able to guess the coordinates of the secret point. Switch roles and try again. What is the fewest number of turns your team required to guess the point?

7.2   Exercises

Integrated Review   *Concepts, Skills, and Problem Solving*

Keep mathematically in shape by doing these exercises *before* the problems of this section.

Properties and Definitions

1. Identify the index and the radicand of $\sqrt[4]{6x}$.

2. Write the radical form of $a^{1/n}$.

Solving Inequalities

In Exercises 3–8, solve the inequality.

3. $7 - 3x > 4 - x$ **4.** $2(x + 6) - 20 < 2$

5. $\dfrac{x}{6} + \dfrac{x}{4} < 1$ **6.** $\dfrac{5 - x}{2} \geq 8$

7. $|x - 3| < 2$ **8.** $|x - 5| > 3$

Graphing

In Exercises 9–12, use a graphing utility to graph the function g and identify the transformation of $f(x) = x^5$ represented by g.

9. $g(x) = x^5 - 2$ **10.** $g(x) = (x - 2)^5$

11. $g(x) = -x^5$ **12.** $g(x) = (-x)^5$

Developing Skills

In Exercises 1–8, determine whether the points are solutions of the inequality. See Example 1.

Inequality	*Points*	
1. $x - 2y < 4$	(a) $(0, 0)$	(b) $(2, -1)$
	(c) $(3, 4)$	(d) $(5, 1)$
2. $x + y < 3$	(a) $(0, 6)$	(b) $(4, 0)$
	(c) $(0, -2)$	(d) $(1, 1)$
3. $3x + y \geq 10$	(a) $(1, 3)$	(b) $(-3, 1)$
	(c) $(3, 1)$	(d) $(2, 15)$
4. $-3x + 5y \geq 6$	(a) $(2, 8)$	(b) $(-10, -3)$
	(c) $(0, 0)$	(d) $(3, 3)$
5. $y > 0.2x - 1$	(a) $(0, 2)$	(b) $(6, 0)$
	(c) $(4, -1)$	(d) $(-2, 7)$
6. $y < -3.5x + 7$	(a) $(1, 5)$	(b) $(5, -1)$
	(c) $(-1, 4)$	(d) $\left(0, \tfrac{4}{3}\right)$
7. $y \leq 3 - |x|$	(a) $(-1, 4)$	(b) $(2, -2)$
	(c) $(6, 0)$	(d) $(5, -2)$
8. $y \geq |x - 3|$	(a) $(0, 0)$	(b) $(1, 2)$
	(c) $(4, 10)$	(d) $(5, -1)$

In Exercises 9–14, match the inequality with its graph. [The graphs are labeled (a), (b), (c), (d), (e), and (f).]

(a)
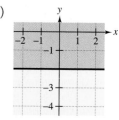

(b)
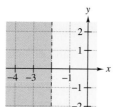

(c)
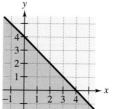

(d)
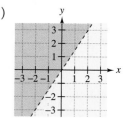

(e)
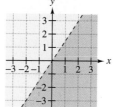

(f)
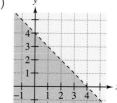

9. $y \geq -2$ **10.** $x < -2$

11. $3x - 2y < 0$ **12.** $3x - 2y > 0$

13. $x + y < 4$ **14.** $x + y \leq 4$

In Exercises 15–38, sketch the graph of the solution of the linear inequality. See Examples 2–4.

15. $x \geq 2$ **16.** $x < -3$

17. $y < 5$ **18.** $y > 2$

19. $y > \frac{1}{2}x$ **20.** $y \leq 2x$

21. $y \geq 5 - x$ **22.** $y > 4 - x$

23. $y \leq x + 2$ **24.** $y \leq x + 1$

25. $x + y \geq 4$ **26.** $x + y \leq 5$

27. $x - 2y \geq 6$ **28.** $3x + y \leq 9$

29. $3x + 2y \geq 2$ **30.** $3x + 5y \leq 15$

31. $3x - 2y \geq 4$ **32.** $4x + 3y \leq 12$

33. $0.2x + 0.3y < 2$ **34.** $0.25x - 0.75y > 6$

35. $y - 1 > -\frac{1}{2}(x - 2)$ **36.** $y - 2 < -\frac{2}{3}(x - 3)$

37. $\dfrac{x}{3} + \dfrac{y}{4} \leq 1$ **38.** $\dfrac{x}{2} + \dfrac{y}{6} \geq 1$

In Exercises 39–46, use a graphing utility to graph (shade) the solution of the inequality.

39. $y \geq \frac{3}{4}x - 1$ **40.** $y \leq 9 - \frac{3}{2}x$

41. $y \leq -\frac{2}{3}x + 6$ **42.** $y \geq \frac{1}{4}x + 3$

43. $x - 2y - 4 \geq 0$ **44.** $2x + 4y - 3 \leq 0$

45. $2x + 3y - 12 \leq 0$ **46.** $x - 3y + 9 \geq 0$

In Exercises 47–52, write an inequality for the shaded region shown in the figure.

47.
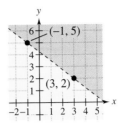

48.
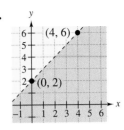

49.
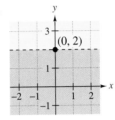

50.
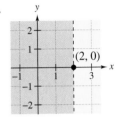

51.
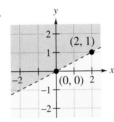

52.
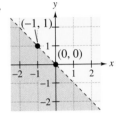

Solving Problems

53. *Geometry*   The perimeter of a rectangle of length x and width y cannot exceed 500 inches. Write a linear inequality for this constraint. Use a graphing utility to graph the solution of the inequality.

54. *Geometry*   The perimeter of a rectangle of length x and width y must be at least 100 centimeters. Write a linear inequality for this constraint. Use a graphing utility to graph the solution of the inequality.

55. *Storage Space*   A warehouse for storing chairs and tables has 1000 square feet of floor space. Each chair requires 10 square feet of floor space and each table requires 15 square feet. Write a linear inequality for this space constraint if x is the number of chairs and y is the number of tables stored. Sketch a graph of the solution of the inequality.

56. *Storage Space*   A warehouse for storing desks and filing cabinets has 2000 square feet of floor space. Each desk requires 15 square feet of floor space and each filing cabinet requires 6 square feet. Write a linear inequality for this space constraint if x is the number of desks and y is the number of filing cabinets stored. Sketch a graph of the solution of the inequality.

57. *Roasting a Turkey*   The time t (in minutes) that it takes to roast a turkey weighing p pounds is given by the following inequalities.

For a turkey up to 6 pounds: $t \geq 20p$

For a turkey over 6 pounds: $t \geq 15p + 30$

Sketch the graphs of these inequalities. What are the coordinates for a 12-pound turkey that has been roasting for 3 hours and 40 minutes? Is this turkey fully cooked?

58. *Pizza and Soda Pop* You and some friends go out for pizza. Together you have $26. You want to order two large pizzas with cheese at $8 each. Each additional topping costs $0.40, and each small soft drink costs $0.80. Write an inequality that represents the number of toppings x and drinks y that your group can afford. Sketch a graph of the solution of the inequality. What are the coordinates for an order of six soft drinks and two large pizzas with cheese, each with three additional toppings? Is this a solution of the inequality? (Assume there is no sales tax.)

59. *Pizza and Soda Pop* You and some friends go out for pizza. Together you have $48. You want to order three large pizzas with cheese at $9 each. Each additional topping costs $1, and each soft drink costs $1.50. Write an inequality that represents the number of toppings x and drinks y that your group can afford. Sketch a graph of the solution of the inequality. What are the coordinates for an order of eight soft drinks and three large pizzas with cheese, each with two additional toppings? Is this a solution of the inequality? (Assume there is no sales tax.)

60. *Diet Supplement* A dietician is asked to design a special diet supplement using two foods. Each ounce of food X contains 30 units of calcium and each ounce of food Y contains 20 units of calcium. The minimum daily requirement in the diet is 300 units of calcium. Write an inequality that represents the different numbers of units of food X and food Y required. Sketch a graph of the solution of the inequality. From the graph, find several ordered pairs with positive integer coordinates that are solutions of the inequality.

61. *Weekly Pay* You have two part-time jobs. One is at a grocery store, which pays $9 per hour, and the other is mowing lawns, which pays $6 per hour. Between the two jobs, you want to earn at least $150 a week. Write an inequality that shows the different numbers of hours you can work at each job, and sketch the graph of the solution of the inequality. From the graph, find several ordered pairs with positive integer coordinates that are solutions of the inequality.

62. *Weekly Pay* You have two part-time jobs. One is at a fast-food restaurant, which pays $6 per hour, and the other is providing childcare, which pays $5 per hour. Between the two jobs, you want to earn at least $120 a week. Write an inequality that shows the different numbers of hours you can work at each job, and sketch the graph of the solution of the inequality. From the graph, find several ordered pairs with positive integer coordinates that are solutions of the inequality.

63. *Getting a Workout* The maximum heart rate r (in beats per minute) of a person in normal health is related to the person's age A (in years). The relationship between r and A is given by $r \leq 220 - A$.

(a) Sketch a graph of the solution of the inequality with A measured along the horizontal axis and r measured along the vertical axis.

(b) Physiologists recommend that during a workout a person strive to increase his or her heart rate to 75% of the maximum rate for the person's age. Sketch the graph of $r = 0.75(220 - A)$ on the same set of coordinate axes used in part (a).

Explaining Concepts

64. List the four forms of a linear inequality in variables x and y.

65. What is meant by saying that (x_1, y_1) is a solution of a linear inequality in x and y?

66. Explain the meaning of the term *half-plane*. Give an example of an inequality whose graph is a half-plane.

67. How does the solution of $x - y > 1$ differ from the solution of $x - y \geq 1$?

68. After graphing the corresponding equation, how do you decide which half-plane is the solution of a linear inequality?

69. Explain the difference between graphing the solution of the inequality $x \leq 3$ on the real number line and graphing it on a rectangular coordinate system.

7.3   Graphs of Quadratic Functions

Objectives

1   Identify the graph of a quadratic function as a parabola and determine its vertex by completing the square.

2   Sketch the graph of a parabola by plotting its vertex, axis, and additional points.

3   Write the equation of a parabola given its vertex and a point on the graph.

4   Use a parabola to solve an application problem.

---

1 Identify the graph of a quadratic function as a parabola and determine its vertex by completing the square.

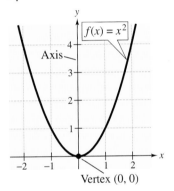

Figure 7.16

Graphs of Quadratic Functions

In this section, you will study graphs of quadratic functions.

$$f(x) = ax^2 + bx + c \qquad \text{Quadratic function}$$

Figure 7.16 shows the graph of a simple quadratic function, $f(x) = x^2$.

> ▶ **Graphs of Quadratic Functions**
>
> The graph of $f(x) = ax^2 + bx + c$, $a \neq 0$, is a **parabola.** The completed-square form
>
> $$f(x) = a(x - h)^2 + k \qquad \text{Standard form}$$
>
> is the **standard form** of the function. The **vertex** of the parabola occurs at the point (h, k), and the vertical line passing through the vertex is the **axis** of the parabola.

Every parabola is *symmetric* about its axis, which means that if it were folded along its axis, the two parts would match.

If a is positive, the graph of $f(x) = ax^2 + bx + c$ opens up, and if a is negative, the graph opens down, as shown in Figure 7.17. Observe in Figure 7.17 that the y-coordinate of the vertex identifies the minimum function value if $a > 0$ and the maximum function value if $a < 0$.

Technology: Discovery

You can use a graphing utility to discover a rule for determining the appearance of a parabola. Graph the equations below.

$$y_1 = x^2 - 3x - 5$$
$$y_2 = 7 - 3x^2$$
$$y_3 = -4 + 6x^2$$
$$y_4 = -x^2 - 6x$$

In your own words, write a rule for determining whether the graph of a parabola opens up or down by just looking at the equation. Does $y = 8 - 2x - 2x^2$ open up or down?

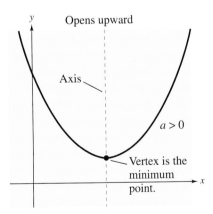

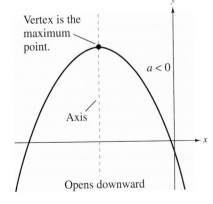

Figure 7.17

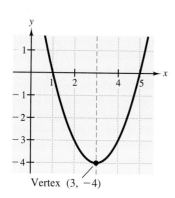

Vertex (3, −4)

Figure 7.18

Example 1   Finding the Vertex by Completing the Square

Find the vertex of the graph of $f(x) = x^2 - 6x + 5$.

Solution

Begin by writing the function in standard form.

$f(x) = x^2 - 6x + 5$	Original function
$f(x) = x^2 - 6x + (-3)^2 - (-3)^2 + 5$	Add and subtract $(-3)^2$.
$f(x) = (x^2 - 6x + 9) - 9 + 5$	Regroup terms.
$f(x) = (x - 3)^2 - 4$	Standard form

From the standard form, you can see that the vertex of the parabola occurs at the point $(3, -4)$, as shown in Figure 7.18. The minimum value of the function is $f(3) = -4$.

Study Tip

When a number is added to a function and then that same number is subtracted from the function, the value of the function remains unchanged. Notice in Example 1 that $(-3)^2$ is added to the function to complete the square and then $(-3)^2$ is subtracted from the function so that the value of the function remains the same.

In Example 1, the vertex of the graph was found by *completing the square*. Another approach to finding the vertex is to complete the square once for a general function and then use the resulting formula for the vertex.

$f(x) = ax^2 + bx + c$	Quadratic function
$= ax^2 + bx + \dfrac{b^2}{4a} - \dfrac{b^2}{4a} + c$	Add and subtract $\dfrac{b^2}{4a}$.
$= a\left[x^2 + \dfrac{b}{a}x + \left(\dfrac{b}{2a}\right)^2\right] + c - \dfrac{b^2}{4a}$	Group terms.
$= a\left(x + \dfrac{b}{2a}\right)^2 + c - \dfrac{b^2}{4a}$	Standard form

From this form you can see that the vertex occurs when $x = -b/2a$.

Example 2   Finding the Vertex with a Formula

Find the vertex of the graph of $f(x) = x^2 + x$.

Solution

From the given function, it follows that $a = 1$ and $b = 1$. So, the x-coordinate of the vertex is

$$x = \frac{-b}{2a} = \frac{-1}{2(1)} = -\frac{1}{2}$$

and the y-coordinate is

$$f\left(-\frac{b}{2a}\right) = f\left(-\frac{1}{2}\right) = \left(-\frac{1}{2}\right)^2 + \left(-\frac{1}{2}\right) = \frac{1}{4} - \frac{1}{2} = -\frac{1}{4}.$$

So, the vertex of the parabola is $\left(-\frac{1}{2}, -\frac{1}{4}\right)$, the minimum value of the function is $f\left(-\frac{1}{2}\right) = -\frac{1}{4}$, and the parabola opens upward, as shown in Figure 7.19.

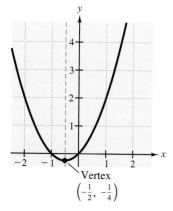

Vertex $\left(-\frac{1}{2}, -\frac{1}{4}\right)$

Figure 7.19

2 Sketch the graph of a parabola by plotting its vertex, axis, and additional points.

Sketching a Parabola by Point-Plotting

To obtain an accurate sketch of a parabola, the following guidelines are useful. Remember that the intercepts are convenient points to plot as well.

> ▶ **Sketching a Parabola**
>
> 1. Determine the vertex and axis of the parabola by completing the square or by formula.
>
> 2. Plot the vertex, axis, and a few additional points on the parabola. (Using the symmetry about the axis can reduce the number of points you need to plot.)
>
> 3. Use the fact that the parabola opens upward if $a > 0$ and opens downward if $a < 0$ to complete the sketch.

Study Tip

The x- and y-intercepts are useful points to plot. Another convenient fact is that the x-coordinate of the vertex lies halfway between the x-intercepts. Keep this in mind as you study the examples and do the exercises in this section.

Example 3   Sketching a Parabola

Sketch the graph of $x^2 - y + 6x + 8 = 0$.

Solution

Begin by writing the equation in standard form.

$x^2 - y + 6x + 8 = 0$	Original equation
$-y = -x^2 - 6x - 8$	Subtract $x^2 + 6x + 8$ from both sides.
$y = x^2 + 6x + 8$	Multiply both sides by -1.
$y = (x^2 + 6x + 3^2 - 3^2) + 8$	Add and subtract 3^2.
$y = (x^2 + 6x + 9) - 9 + 8$	Regroup terms.
$y = (x + 3)^2 - 1$	Standard form

The vertex occurs at the point $(-3, -1)$ and the axis is given by the line $x = -3$. After plotting this information, calculate a few additional points on the parabola, as shown in the table. Note that the y-intercept is $(0, 8)$ and the x-intercepts are solutions to the equation

$$x^2 + 6x + 8 = (x + 4)(x + 2) = 0.$$

The graph of the parabola is shown in Figure 7.20. Note that it opens upward because the leading coefficient (in standard form) is positive. Use your graphing utility to verify the graph shown in Figure 7.20.

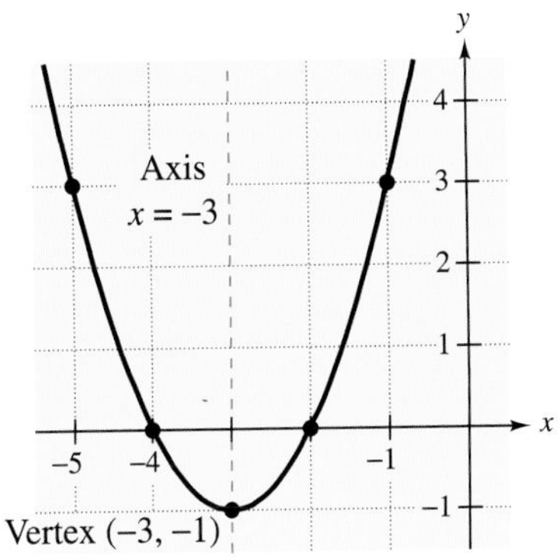

Figure 7.20

x-Value	-5	-4	-3	-2	-1
y-Value	3	0	-1	0	3
Solution point	$(-5, 3)$	$(-4, 0)$	$(-3, -1)$	$(-2, 0)$	$(-1, 3)$

3 Write the equation of a parabola given its vertex and a point on the graph.

Writing the Equation of a Parabola

To write the equation of a parabola with a vertical axis, use the fact that its standard equation has the form $y = a(x - h)^2 + k$, where (h, k) is the vertex.

Example 4   Writing the Equation of a Parabola

Write the equation of the parabola whose vertex is $(-2, 1)$ and whose y-intercept is $(0, -3)$, as shown in Figure 7.21.

Solution

Because the vertex occurs at $(h, k) = (-2, 1)$, you can write the following.

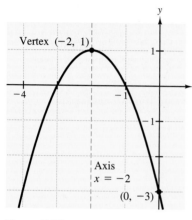

$$y = a(x - h)^2 + k \qquad \text{Standard form}$$

$$y = a[x - (-2)]^2 + 1 \qquad \text{Substitute } -2 \text{ for } h \text{ and } 1 \text{ for } k.$$

$$y = a(x + 2)^2 + 1 \qquad \text{Simplify.}$$

To find the value of a, use the fact that the y-intercept is $(0, -3)$.

$$y = a(x + 2)^2 + 1 \qquad \text{Standard form}$$

$$-3 = a(0 + 2)^2 + 1 \qquad \text{Substitute } 0 \text{ for } x \text{ and } -3 \text{ for } y.$$

$$-1 = a \qquad \text{Simplify.}$$

This implies that the standard form of the equation of the parabola is

$$y = -(x + 2)^2 + 1.$$

Figure 7.21

Example 5   Writing the Equation of a Parabola

Write the equation of the parabola that has a vertex of $(3, -4)$ and contains the point $(5, -2)$, as shown in Figure 7.22.

Solution

Because the vertex occurs at $(h, k) = (3, -4)$, you can write the following.

$$y = a(x - h)^2 + k \qquad \text{Standard form}$$

$$y = a(x - 3)^2 + (-4) \qquad \text{Substitute } 3 \text{ for } h \text{ and } -4 \text{ for } k.$$

$$y = a(x - 3)^2 - 4 \qquad \text{Simplify.}$$

To find the value of a, use the fact that $(5, -2)$ is a point on the parabola.

$$y = a(x - 3)^2 - 4 \qquad \text{Standard form}$$

$$-2 = a(5 - 3)^2 - 4 \qquad \text{Substitute } 5 \text{ for } x \text{ and } -2 \text{ for } y.$$

$$\tfrac{1}{2} = a \qquad \text{Simplify.}$$

This implies that the standard form of the equation is

$$y = \frac{1}{2}(x - 3)^2 - 4.$$

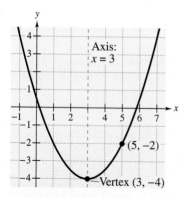

Figure 7.22

4 Use a parabola to solve an application problem.

Application

| Example 6 | An Application Involving a Minimum Point |

A suspension bridge is 100 feet long, as shown in Figure 7.23(a). The bridge is supported by cables attached at the tops of the towers at each end of the bridge. Each cable hangs in the shape of a parabola (see Figure 7.23(b)) given by

$$y = 0.01x^2 - x + 35$$

where x and y are both measured in feet. (a) Find the distance between the lowest point of the cable and the roadbed of the bridge. (b) How tall are the towers?

Solution

a. Because the lowest point occurs at the vertex of the parabola and $a = 0.01$ and $b = -1$, it follows that the vertex of the parabola occurs when

$$x = \frac{-b}{2a} = \frac{1}{0.02} = 50.$$

At this x-value, the value of y is

$$y = 0.01(50)^2 - 50 + 35 = 10.$$

Thus, the minimum distance between the cable and the roadbed is 10 feet.

b. Because the vertex of the parabola occurs at the midpoint of the bridge, the two towers are located at the points where $x = 0$ and $x = 100$. Substituting an x-value of 0, you can find that the corresponding y-value is

$$y = 0.01(0)^2 - 0 + 35 = 35.$$

So, the towers are each 35 feet high. (Try substituting $x = 100$ in the equation to see that you obtain the same y-value.)

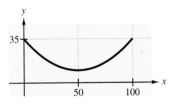

(a)

(b)

Figure 7.23

| Discussing the Concept | Quadratic Modeling |

The data in the table represent the average monthly temperature y in degrees Fahrenheit in Savannah, Georgia for the month x, with $x = 1$ corresponding to November.   (Source: National Climate Data Center) Plot the data. Find a quadratic model for the data and use it to find the average temperatures for December and February. The actual average temperature for both December and February is **52°F**. How well do you think the model fits the data? Use the model to predict the average temperature for June. How useful do you think the model would be for the whole year?

x	1	3	5
y	59	49	59

7.3 Exercises

Integrated Review — Concepts, Skills, and Problem Solving

Keep mathematically in shape by doing these exercises *before* the problems of this section.

Properties and Definitions

1. Fill in the blanks: $(x + b)^2 = x^2 + \quad x +$

_____ .

2. Fill in the blank so that the expression is a perfect square trinomial. Explain how the constant is determined.

$x^2 + 5x +$ _____

Simplifying Expressions

In Exercises 3–10, simplify the expression.

3. $(4x + 3y) - 3(5x + y)$

4. $(-15u + 4v) + 5(3u - 9v)$

5. $2x^2 + (2x - 3)^2 + 12x$

6. $y^2 - (y + 2)^2 + 4y$

7. $\sqrt{24x^2y^3}$ **8.** $\sqrt[3]{9} \cdot \sqrt[3]{15}$

9. $(12a^{-4}b^6)^{1/2}$ **10.** $(16^{1/3})^{3/4}$

Problem Solving

In Exercises 11 and 12, find the time required for an object to reach the ground when it is dropped from a height of s_0 feet. The height h (in feet) is given by $h = -16t^2 + s_0$, where t measures time in seconds from when the object is released.

11. $s_0 = 80$ **12.** $s_0 = 150$

Developing Skills

In Exercises 1–6, match the equation with its graph. [The graphs are labeled (a), (b), (c), (d), (e), and (f).]

1. $y = 4 - 2x$ **2.** $y = \frac{1}{2}x - 4$

3. $y = x^2 - 3$ **4.** $y = -x^2 + 3$

5. $y = (x - 2)^2$ **6.** $y = 2 - (x - 2)^2$

(a)

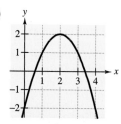

(b)

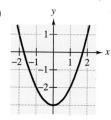

(c)

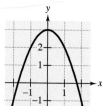

(d)

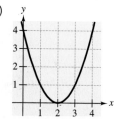

(e)

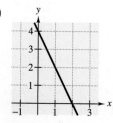

(f)

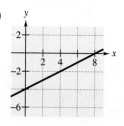

In Exercises 7–18, write the equation in standard form and find the vertex of its graph. See Example 1.

7. $y = x^2 + 2$ **8.** $y = x^2 + 2x$

9. $y = x^2 - 4x + 7$ **10.** $y = x^2 + 6x - 5$

11. $y = x^2 + 6x + 5$ **12.** $y = x^2 - 4x + 5$

13. $y = -x^2 + 6x - 10$ **14.** $y = 4 - 8x - x^2$

15. $y = -x^2 + 2x - 7$ **16.** $y = -x^2 - 10x + 10$

17. $y = 2x^2 + 6x + 2$ **18.** $y = 3x^2 - 3x - 9$

In Exercises 19–24, find the vertex of the graph of the function by formula. See Example 2.

19. $f(x) = x^2 - 8x + 15$ **20.** $f(x) = x^2 + 4x + 1$

21. $g(x) = -x^2 - 2x + 1$

22. $h(x) = -x^2 + 14x - 14$

23. $y = 4x^2 + 4x + 4$ **24.** $y = 9x^2 - 12x$

In Exercises 25–32, state whether the graph opens upward or downward and find the vertex.

25. $y = 2(x - 0)^2 + 2$ **26.** $y = -3(x + 5)^2 - 3$

27. $y = 4 - (x - 10)^2$ **28.** $y = 2(x - 12)^2 + 3$

29. $y = x^2 - 6$ **30.** $y = -(x + 1)^2$

31. $y = -(x - 3)^2$ **32.** $y = x^2 - 6x$

In Exercises 33–40, find the *x*- and *y*-intercepts of the graph.

33. $y = 25 - x^2$ **34.** $y = x^2 - 49$

35. $y = x^2 - 9x$ **36.** $y = x^2 + 4x$

37. $y = 4x^2 - 12x + 9$ **38.** $y = 10 - x - 2x^2$

39. $y = x^2 - 3x + 3$ **40.** $y = x^2 - 3x - 10$

In Exercises 41–64, sketch the graph of the function. Identify the vertex and any *x*-intercepts. Use a graphing utility to verify your results. See Example 3.

41. $g(x) = x^2 - 4$ **42.** $h(x) = x^2 - 9$

43. $f(x) = -x^2 + 4$ **44.** $f(x) = -x^2 + 9$

45. $f(x) = x^2 - 3x$ **46.** $g(x) = x^2 - 4x$

47. $y = -x^2 + 3x$ **48.** $y = -x^2 + 4x$

49. $y = (x - 4)^2$ **50.** $y = -(x + 4)^2$

51. $y = x^2 - 8x + 15$ **52.** $y = x^2 + 4x + 2$

53. $y = -(x^2 + 6x + 5)$ **54.** $y = -x^2 + 2x + 8$

55. $q(x) = -x^2 + 6x - 7$ **56.** $g(x) = x^2 + 4x + 7$

57. $y = 2(x^2 + 6x + 8)$ **58.** $y = 3x^2 - 6x + 4$

59. $y = \frac{1}{2}(x^2 - 2x - 3)$

60. $y = -\frac{1}{2}(x^2 - 6x + 7)$

61. $y = \frac{1}{5}(3x^2 - 24x + 38)$

62. $y = \frac{1}{5}(2x^2 - 4x + 7)$ **63.** $f(x) = 5 - \frac{1}{3}x^2$

64. $f(x) = \frac{1}{3}x^2 - 2$

In Exercises 65–72, identify the transformation of the graph of $f(x) = x^2$ and sketch a graph of h.

65. $h(x) = x^2 + 2$ **66.** $h(x) = x^2 - 4$

67. $h(x) = (x + 2)^2$ **68.** $h(x) = (x - 4)^2$

69. $h(x) = (x - 1)^2 + 3$ **70.** $h(x) = (x + 2)^2 - 1$

71. $h(x) = (x + 3)^2 + 1$ **72.** $h(x) = (x - 3)^2 - 2$

In Exercises 73–76, use a graphing utility to approximate the vertex of the graph. Check the result algebraically.

73. $y = \frac{1}{6}(2x^2 - 8x + 11)$

74. $y = -\frac{1}{4}(4x^2 - 20x + 13)$

75. $y = -0.7x^2 - 2.7x + 2.3$

76. $y = 0.75x^2 - 7.50x + 23.00$

In Exercises 77–82, write an equation of the parabola. See Example 4.

77.

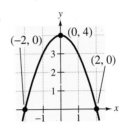

78.

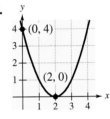

79.

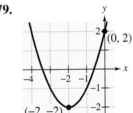

80.

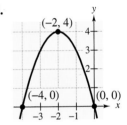

81.

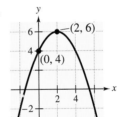

82.

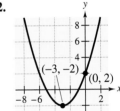

In Exercises 83–90, write an equation of the parabola

$$y = a(x - h)^2 + k$$

that satisfies the conditions. See Example 5.

83. Vertex: $(2, 1)$; $a = 1$

84. Vertex: $(-3, -3)$; $a = 1$

85. Vertex: $(2, -4)$; Point on the graph: $(0, 0)$

86. Vertex: $(-2, -4)$; Point on the graph: $(0, 0)$

87. Vertex: $(3, 2)$; Point on the graph: $(1, 4)$

88. Vertex: $(-1, -1)$; Point on the graph: $(0, 4)$

89. Vertex: $(-1, 5)$; Point on the graph: $(0, 1)$

90. Vertex: $(5, 2)$; Point on the graph: $(10, 3)$

In Exercises 91–94, identify the transformation of $y = x^2$ that will produce the given graph.

91.

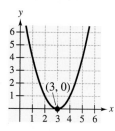

$(3, 0)$

92.

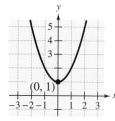

$(0, 1)$

93.

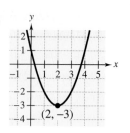

$(2, -3)$

94.

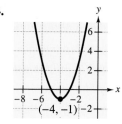

$(-4, -1)$

Solving Problems

95. *Path of a Ball* The height y (in feet) of a ball thrown by a child is given by

$$y = -\frac{1}{12}x^2 + 2x + 4$$

where x is the horizontal distance (in feet) from where the ball was thrown.

(a) How high was the ball when it left the child's hand?

(b) How high was the ball when it reached its maximum height?

(c) How far from the child did the ball strike the ground?

96. *Path of a Ball* Repeat Exercise 95 if the path of the ball is modeled by $y = -\frac{1}{16}x^2 + 2x + 5$.

97. *Maximum Height of a Diver* The path of a diver is given by $y = -\frac{4}{9}x^2 + \frac{24}{9}x + 10$ where y is the height in feet and x is the horizontal distance from the end of the diving board in feet. What is the maximum height of the diver?

98. *Maximum Height of a Diver* Repeat Exercise 97 if the path of the diver is modeled by

$$y = -\frac{4}{3}x^2 + \frac{10}{3}x + 10.$$

99. *Graphical Estimation* The number N (in thousands) of personnel in the Marine Corps reserves in the United States for the years 1990 through 1996 is approximated by the model

$$N = 83.64 + 14.89t - 2.04t^2, \quad 0 \le t \le 6.$$

In this model, t is time in years, with $t = 0$ corresponding to 1990. (Source: U.S. Department of Defense)

(a) Use a graphing utility to graph the function.

(b) Determine the year when the number in the Marine Corps reserves was greatest. Approximate the number that year.

100. *Graphical Estimation* The profit (in thousands of dollars) for a company is given by

$$P = 230 + 20s - \frac{1}{2}s^2$$

where s is the amount (in hundreds of dollars) spent on advertising. Use a graphing utility to graph the profit function and approximate the amount of advertising that yields a maximum profit. Verify the maximum profit algebraically.

101. *Graphical Interpretation* A company manufactures radios that cost the company $60 each. For buyers who purchase 100 or fewer radios, the purchase price is $90 per radio. To encourage large orders, the company will reduce the price *per radio* for orders over 100, as follows. If 101 radios are purchased, the price is $89.85 per unit. If 102 radios are purchased, the price is $89.70 per unit. If $(100 + x)$ radios are purchased, the price per unit is

$$p = 90 - x(0.15)$$

where x is the amount over 100 in the order.

(a) Show algebraically that the profit P for x orders over 100 is

$$P = (100 + x)[90 - x(0.15)] -$$
$$(100 + x)60$$
$$= 3000 + 15x - \frac{3}{20}x^2.$$

(b) Find the vertex of the profit curve and determine the order size for maximum profit.

(c) Would you recommend this pricing scheme? Explain.

102. *Graphical Estimation* The cost of producing x units of a product is given by

$$C = 800 - 10x + \frac{1}{4}x^2, \quad 0 < x < 40.$$

Use a graphing utility to graph this function and use the trace feature to approximate the value of x when C is minimum.

103. *Geometry* The area of a rectangle is given by the function

$$A = \frac{2}{\pi}(100x - x^2), \quad 0 < x < 100$$

where x is the length of the base of the rectangle in feet. Use a graphing utility to graph the function and use the trace feature to approximate the value of x when A is maximum.

104. *Bridge Design* A bridge is to be constructed over a gorge with the main supporting arch being a parabola (see figure). The equation of the parabola is

$$y = 4\left(100 - \frac{x^2}{2500}\right)$$

where x and y are measured in feet.

(a) Find the length of the road across the gorge.

(b) Find the height of the parabolic arch at the center of the span.

(c) Find the lengths of the vertical girders at intervals of 100 feet from the center of the bridge.

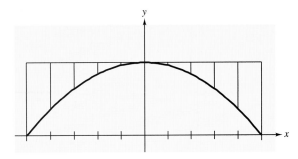

Figure for 104

105. *Highway Design* A highway department engineer must design a parabolic arc to create a turn in a freeway around a city. The vertex of the parabola is placed at the origin, and the parabola must connect with roads represented by the equations

$$y = -0.4x - 100, \quad x < -500$$

and

$$y = 0.4x - 100, \quad x > 500$$

(see figure). Find an equation for the parabolic arc.

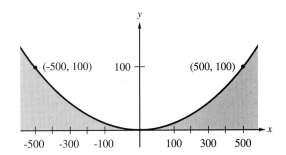

Explaining Concepts

106. Answer parts (a) and (b) of Motivating the Chapter on page 423.

107. In your own words, describe the graph of the quadratic function $f(x) = ax^2 + bx + c$.

108. Explain how to find the vertex of the graph of a quadratic function.

109. Explain how to find any x- or y-intercepts of the graph of a quadratic function.

110. Explain how to determine whether the graph of a quadratic function opens up or down.

111. How is the discriminant related to the graph of a quadratic function?

112. Is it possible for the graph of a quadratic function to have two y-intercepts? Explain.

113. Explain how to determine the maximum (or minimum) value of a quadratic function.

Mid-Chapter Quiz

Take this quiz as you would take a quiz in class. After you are done, check your work against the answers given in the back of the book.

In Exercises 1 and 2, write a model for the statement.

1. *A* varies directly as the square of *r*.

2. *z* varies directly as *x* and inversely as the square of *y*.

In Exercises 3 and 4, write a sentence using variation terminology to describe the formula.

3. Distance: $d = rt$ **4.** Volume of a cube: $V = s^3$

5. Find the constant of proportionality and write an equation that relates the variables. *z* varies directly as the square of *x* and inversely as *y*, and $z = 6$ when $x = 6$ and $y = 4$.

6. Decide whether the points are solutions of the inequality $2x - 3y \leq 4$. Explain your reasoning.

     (a) $(5, 2)$ (b) $(-2, 4)$ (c) $(2, -4)$ (d) $(3, 0)$

In Exercises 7 and 8, write an inequality for the shaded region.

7. See figure at left.   **8.** See figure at left.

In Exercises 9–11, sketch the graph of the solution of the linear inequality.

9. $x > -2$ **10.** $2x + 3y \leq 9$ **11.** $2x - y \leq 4$

In Exercises 12 and 13, write an equation for the indicated parabola.

12. Vertex: $(3, -1)$; passes through the point $(5, 3)$

13. Vertex: $(5, 4)$; passes through the point $(3, 3)$

In Exercises 14 and 15, sketch the graph of the quadratic function. Identify the vertex and the *x*-intercepts.

14. $y = -\frac{1}{4}(x^2 + 6x + 1)$ **15.** $y = 2x^2 - 4x - 7$

16. Methane forms an explosive mixture with air at a concentration of 5% or greater. A steady leak of methane begins in a coal mine so that the concentration of methane gas varies directly as time. Twelve minutes after a leak begins, the concentration of methane in the air is 2%. If the leak continues at the same rate, when could an explosion occur?

17. A store sells two models of computers. The costs to the store of the two models are $900 and $1400, respectively. Management does not want more than $20,000 in computer inventory at any time. Write an inequality that models this constraint.

18. The path of a ball is given by $y = -0.005x^2 + x + 5$. Determine the maximum height of the ball.

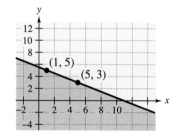

Figure for 7

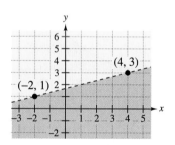

Figure for 8

7.4   Conic Sections

Objectives

1. Find an equation of a circle and sketch its graph.
2. Find an equation of an ellipse and sketch its graph.
3. Find an equation of a hyperbola and sketch its graph.
4. Use the equation of an ellipse to sketch a diagram of an archway.

1. Find an equation of a circle and sketch its graph.

Circles

In Section 7.3, you learned that the graph of a second-degree equation of the form $y = ax^2 + bx + c$ is a parabola. A parabola is one of four types of **conics** or **conic sections.** The other three types are circles, ellipses, and hyperbolas. All four types have equations that are of second degree. As indicated in Figure 7.24, the name "conic" relates to the fact that each of these figures can be obtained by intersecting a plane with a double-napped cone.

   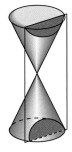

Circle   *Parabola*   *Ellipse*   *Hyperbola*

Figure 7.24

Hypatia

(370–415 A.D.)

One of the first recognized female mathematicians, Hypatia, wrote a textbook entitled *On the Conics of Apollonius*. Her death marked the end of major mathematical discoveries in Europe for several hundred years.

Conic sections occur in many practical applications. Reflective surfaces in satellite dishes, flashlights, and telescopes often are of parabolic shape. The orbits of planets are elliptical, and the orbits of comets are usually elliptical or hyperbolic. Ellipses and parabolas are also used in building archways and bridges.

A **circle** in the rectangular coordinate plane consists of all points (x, y) that are a given positive distance r from a fixed point, called the **center** of the circle. The positive distance r is the **radius** of the circle. If the center of the circle is the origin, as shown in Figure 7.25, the relationship between the coordinates of any point (x, y) on the circle and the radius r is given by

$$\text{Radius} = r = \sqrt{(x - 0)^2 + (y - 0)^2} \qquad \text{Distance Formula, center at } (0, 0)$$
$$= \sqrt{x^2 + y^2}.$$

If the center of the circle is translated to the point (h, k), the relationship between the coordinates of any point (x, y) on the circle and the radius r is given by

$$\text{Radius} = r = \sqrt{(x - h)^2 + (y - k)^2}. \qquad \text{Distance Formula, center at } (h, k)$$

By squaring both sides of this equation, you obtain the **standard form of the equation of a circle.**

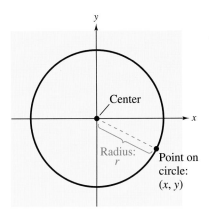

Figure 7.25

▶ Standard Form of the Equation of a Circle

The **standard form of the equation of a circle with center at the origin** is

$$x^2 + y^2 = r^2.$$ Circle with center at (0, 0)

The **standard form of the equation of a circle with center at (h, k)** is

$$(x - h)^2 + (y - k)^2 = r^2.$$ Circle with center at (h, k)

The positive number r is the **radius** of the circle.

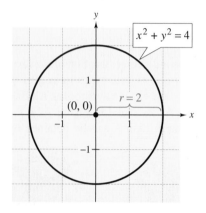

Figure 7.26

Example 1   Finding an Equation of a Circle

Find an equation of the circle whose center is at $(0, 0)$ and whose radius is 2.

Solution

Use the standard form of the equation of a circle with center at the origin.

$$x^2 + y^2 = r^2$$ Standard form with center at (0, 0)

$$x^2 + y^2 = 2^2$$ Substitute 2 for r.

$$x^2 + y^2 = 4$$ Equation of circle

The circle given by this equation is shown in Figure 7.26.

To sketch the circle for a given equation, write the equation in standard form. From the standard form, you can identify the center and radius. For instance, the standard form of the equation $(x - 1)^2 + (y + 2)^2 = 4$ indicates the center to be $(h, k) = (1, -2)$ and the radius to be $r = 2$.

Example 2   Finding the Center and Radius of a Circle

Identify the center and radius of the circle given by the equation, and sketch the circle.

$$x^2 + y^2 + 2x - 6y + 1 = 0$$

Solution

$$x^2 + y^2 + 2x - 6y + 1 = 0$$ Original equation

$$(x^2 + 2x) + (y^2 - 6y) = -1$$ Group terms.

$$(x^2 + 2x + 1) + (y^2 - 6y + 9) = -1 + 1 + 9$$ Complete each square.

$$(x + 1)^2 + (y - 3)^2 = 9$$ Standard form

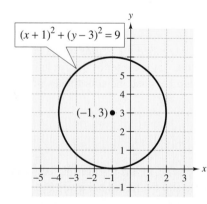

Figure 7.27

From this standard form, you can see that $h = -1$, $k = 3$, and $r = 3$. So, the center of the circle is $(-1, 3)$ and the radius is 3. The graph of the equation of the circle is shown in Figure 7.27.

2 Find an equation of an ellipse and sketch its graph.

Ellipses

An **ellipse** is the set of all points (x, y) such that the sum of the distances between (x, y) and two distinct fixed points is a constant. As shown in Figure 7.28(a), each of the two fixed points is a **focus** of the ellipse. (The plural of focus is *foci*.) In this text, we restrict the study of ellipses to those whose centers are at the origin.

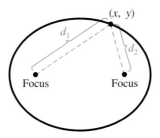

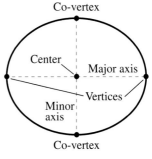

(a) $d_1 + d_2$ is constant.   (b)

Figure 7.28

The line through the foci intersects the ellipse at the **vertices,** as shown in Figure 7.28(b). The line segment joining the vertices is the **major axis,** and its midpoint is the **center** of the ellipse. The line segment perpendicular to the major axis at the center is the **minor axis** of the ellipse, and the points at which the minor axis intersects the ellipse are **co-vertices.**

To trace an ellipse, place two thumbtacks at the foci, as shown in Figure 7.29. If the ends of a fixed length of string are fastened to the thumbtacks and the string is drawn taut with a pencil, the path traced by the pencil will be an ellipse.

Figure 7.29

Study Tip

If the equation of an ellipse is of the form

$$\frac{x^2}{a^2} + \frac{y^2}{b^2} = 1,$$

its major axis is horizontal. Because a is greater than b and its square is the denominator of the x^2 term, you can conclude that the major axis lies along the x-axis. Similarly, if the equation of an ellipse is of the form

$$\frac{x^2}{b^2} + \frac{y^2}{a^2} = 1,$$

its major axis is vertical.

▶ Standard Form of the Equation of an Ellipse

The **standard form of the equation of an ellipse** with center at the origin and major and minor axes of lengths $2a$ and $2b$, respectively, is

$$\frac{x^2}{a^2} + \frac{y^2}{b^2} = 1 \quad \text{or} \quad \frac{x^2}{b^2} + \frac{y^2}{a^2} = 1, \quad 0 < b < a.$$

The vertices lie on the major axis, a units from the center, and the co-vertices lie on the minor axis, b units from the center.

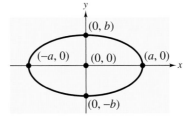

Major axis is horizontal.
Minor axis is vertical.

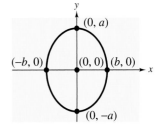

Major axis is vertical.
Minor axis is horizontal.

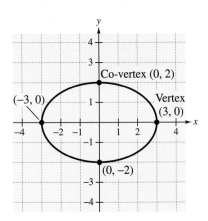

Figure 7.30

Example 3 Finding an Equation of an Ellipse

Find an equation of the ellipse, centered at the origin, whose vertices are $(-3, 0)$ and $(3, 0)$ and whose co-vertices are $(0, -2)$ and $(0, 2)$.

Solution

Begin by plotting the vertices and co-vertices, as shown in Figure 7.30. The center of the ellipse is $(0, 0)$, because it is the point that lies halfway between the vertices (and halfway between the co-vertices). So, the equation of the ellipse has the form

$$\frac{x^2}{a^2} + \frac{y^2}{b^2} = 1.$$

For this ellipse, the major axis is horizontal. So, a is the distance between the center and either vertex, which implies that $a = 3$. Similarly, b is the distance between the center and either co-vertex, which implies that $b = 2$. So, the standard form of the equation of the ellipse is

$$\frac{x^2}{3^2} + \frac{y^2}{2^2} = 1. \qquad \text{Standard form}$$

To sketch an ellipse, it helps first to write its equation in standard form, as shown in the next example.

Example 4 Sketching an Ellipse

Sketch the ellipse given by

$$4x^2 + y^2 = 36$$

and identify the vertices and co-vertices.

Solution

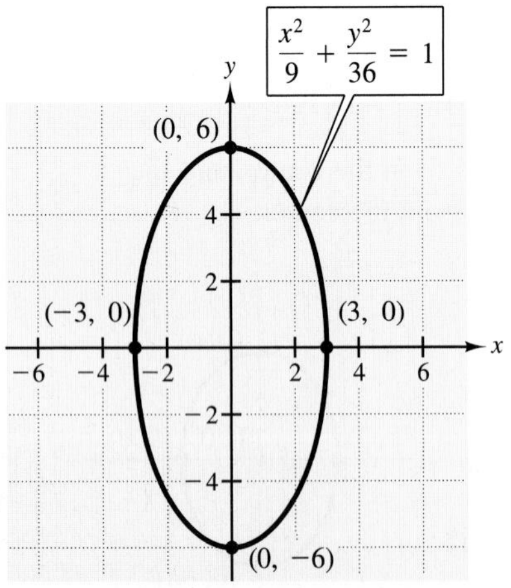

Begin by writing the equation in standard form.

$$4x^2 + y^2 = 36 \qquad \text{Given equation}$$

$$\frac{4x^2}{36} + \frac{y^2}{36} = \frac{36}{36} \qquad \text{Divide both sides by 36.}$$

$$\frac{x^2}{9} + \frac{y^2}{36} = 1 \qquad \text{Simplify.}$$

$$\frac{x^2}{3^2} + \frac{y^2}{6^2} = 1 \qquad \text{Standard form}$$

Because the denominator of the y^2 term is larger than the denominator of the x^2 term, you can conclude that the major axis is vertical. Moreover, because $a = 6$, the vertices are $(0, -6)$ and $(0, 6)$. Finally, because $b = 3$, the co-vertices are $(-3, 0)$ and $(3, 0)$, as shown in Figure 7.31.

Figure 7.31

3 Find an equation of a hyperbola and sketch its graph.

Hyperbolas

A **hyperbola** in the rectangular coordinate system consists of all points (x, y) such that the *difference* of the distances between (x, y) and two fixed points is a constant, as shown in Figure 7.32. The two fixed points are called the **foci** of the hyperbola. We will consider only equations of hyperbolas whose foci lie on the x-axis or on the y-axis. The line on which the foci lie is called the **transverse axis** of the hyperbola.

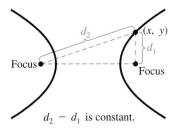

$d_2 - d_1$ is constant.

Figure 7.32

> ▶ Standard Form of the Equation of a Hyperbola
>
> The **standard form of the equation of a hyperbola** whose center is at the origin is given by
>
> $$\frac{x^2}{a^2} - \frac{y^2}{b^2} = 1 \qquad \text{Transverse axis is horizontal.}$$
>
> $$\frac{y^2}{a^2} - \frac{x^2}{b^2} = 1 \qquad \text{Transverse axis is vertical.}$$
>
> where a and b are positive real numbers. The **vertices** of the hyperbola lie on the transverse axis, a units from the center.
>
>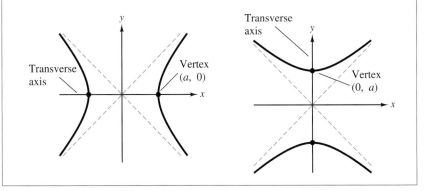

A hyperbola has two disconnected parts, each of which is a **branch** of the hyperbola. The two branches approach a pair of intersecting straight lines called **asymptotes** of the hyperbola. The two asymptotes intersect at the center of the hyperbola.

To sketch a hyperbola, form a **central rectangle** whose center is the origin and whose width and height are $2a$ and $2b$. Note in Figure 7.33 (on page 458) that the asymptotes pass through the corners of the central rectangle and the vertices of the hyperbola lie at the centers of opposite sides of the central rectangle.

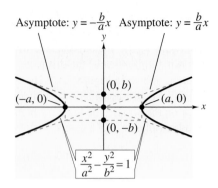

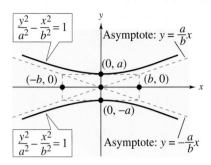

Transverse axis is horizontal.

Figure 7.33

Transverse axis is vertical.

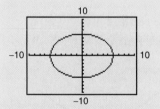
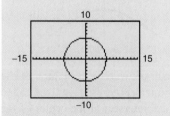
Example 5   Sketching a Hyperbola

Sketch the hyperbola

$$\frac{x^2}{36} - \frac{y^2}{16} = 1.$$

Solution

From the standard form of the equation

$$\frac{x^2}{6^2} - \frac{y^2}{4^2} = 1$$

you can see that the center of the hyperbola is the origin and the transverse axis is horizontal. So, the vertices lie 6 units to the left and right of the center at the points $(-6, 0)$ and $(6, 0)$. Because $a = 6$ and $b = 4$, you can sketch the hyperbola by first drawing a central rectangle whose width is $2a = 12$ and whose height is $2b = 8$, as shown in Figure 7.34(a). Next, draw the asymptotes of the hyperbola through the corners of the central rectangle and plot the vertices. Finally, draw the hyperbola, as shown in Figure 7.34(b).

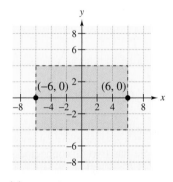

(a)

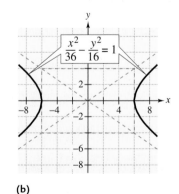

(b)

Figure 7.34

Finding an equation of a hyperbola is a little more difficult than finding equations of the other three types of conics. However, if you know the vertices and the asymptotes, you can find the values of a and b, which enable you to write the equation. Notice in Example 6 that the key to this procedure is knowing that the central rectangle has a width of $2b$ and a height of $2a$.

Example 6   Finding the Equation of a Hyperbola

Find an equation of the hyperbola with a vertical transverse axis whose vertices are $(0, 3)$ and $(0, -3)$ and whose asymptotes are given by $y = \frac{3}{5}x$ and $y = -\frac{3}{5}x$.

Solution

To begin, sketch the lines that represent the asymptotes, as shown in Figure 7.35(a). Note that these two lines intersect at the origin, which implies that the center of the hyperbola is $(0, 0)$. Next, plot the two vertices at the points $(0, 3)$ and $(0, -3)$. Because you know where the vertices are located, you can sketch the central rectangle of the hyperbola, as shown in Figure 7.35(a). Note that the corners of the central rectangle occur at the points

$$(-5, 3), \quad (5, 3), \quad (-5, -3), \quad \text{and} \quad (5, -3).$$

Because the width of the central rectangle is $2b = 10$, it follows that $b = 5$. Similarly, because the height of the central rectangle is $2a = 6$, it follows that $a = 3$. Now that you know the values of a and b, you can use the standard form of the equation of the hyperbola to write the equation.

$$\frac{y^2}{a^2} - \frac{x^2}{b^2} = 1 \qquad \text{Transverse axis is vertical.}$$

$$\frac{y^2}{3^2} - \frac{x^2}{5^2} = 1 \qquad \text{Substitute 3 for } a \text{ and 5 for } b.$$

$$\frac{y^2}{9} - \frac{x^2}{25} = 1 \qquad \text{Equation of the hyperbola}$$

The graph is shown in Figure 7.35(b).

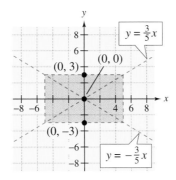

(a)

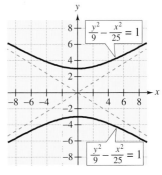

(b)

Figure 7.35

4 Use the equation of an ellipse to sketch a diagram of an archway.

Application

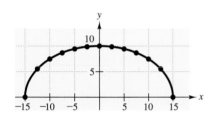

Figure 7.36

Example 7   An Application Involving an Ellipse

You are responsible for designing a semielliptical archway, as shown in Figure 7.36. The height of the archway is 10 feet and its width is 30 feet. Find an equation of the ellipse and use the equation to sketch an accurate diagram of the archway.

Solution

To make the equation simple, place the origin at the center of the ellipse. This means that the standard form of the equation is

$$\frac{x^2}{a^2} + \frac{y^2}{b^2} = 1.$$

Because the major axis is horizontal, it follows that $a = 15$ and $b = 10$, which implies that the equation is

$$\frac{x^2}{15^2} + \frac{y^2}{10^2} = 1. \qquad \text{Standard form}$$

To make an accurate sketch of the ellipse, solve this equation for y as follows.

$$\frac{x^2}{225} + \frac{y^2}{100} = 1 \qquad \text{Simplify denominators.}$$

$$\frac{y^2}{100} = 1 - \frac{x^2}{225} \qquad \text{Subtract } \frac{x^2}{225} \text{ from both sides.}$$

$$y^2 = 100\left(1 - \frac{x^2}{225}\right) \qquad \text{Multiply both sides by 100.}$$

$$y = 10\sqrt{1 - \frac{x^2}{225}} \qquad \text{Take the positive square root of both sides.}$$

Next, calculate several y-values for the archway, as shown in the table. Then use the values in the table to sketch the archway, as shown in Figure 7.37.

Figure 7.37

x-value	± 15	± 12.5	± 10	± 7.5	± 5	± 2.5	0
y-value	0	5.53	7.45	8.66	9.43	9.86	10

Discussing the Concept	Identifying Conic Sections

Cut cone-shaped pieces of styrofoam to demonstrate how to obtain each type of conic section: circle, parabola, ellipse, and hyperbola. Discuss how you could write directions for someone else to form each conic section. Compile a list of real-life situations and/or everyday objects in which conic sections may be seen.

7.4   Exercises

Integrated Review   *Concepts, Skills, and Problem Solving*

Keep mathematically in shape by doing these exercises *before* the problems of this section.

Properties and Definitions

In Exercises 1–4, name the property demonstrated.

1. $(3t + 1) - (3t + 1) = 0$

2. $3x(x - 2) = 3x^2 - 6x$

3. $2(3y) = (2 \cdot 3)y$

4. $-3 + x = x - 3$

Simplifying Expressions

In Exercises 5–10, simplify the expression.

5. $(x^2 \cdot x^3)^4$

6. $4^{-2} \cdot x^2$

7. $\dfrac{15y^{-3}}{10y^2}$

8. $\left(\dfrac{3x^2}{2y}\right)^{-2}$

9. $\dfrac{3x^2 y^3}{18x^{-1} y^2}$

10. $(x^2 + 1)^0$

Problem Solving

11. A service organization paid \$288 for a block of tickets to a ball game. The block contained three more tickets than the organization needed for its members. By inviting three more people to attend (and share in the cost), the organization lowered the price per ticket by \$8. How many people are going to the game?

12. To begin a small business, \$135,000 is needed. The cost will be divided equally among investors. Some have made a commitment to invest. If three more investors could be found, the amount required from each would decrease by \$1500. How many people have made a commitment to invest in the business?

Developing Skills

In Exercises 1–6, match the equation with its graph. [The graphs are labeled (a), (b), (c), (d), (e), and (f).]

(a)
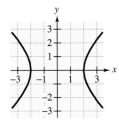

(b)
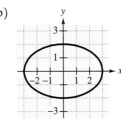

(c)
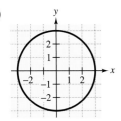

(d)
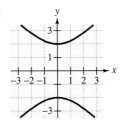

(e)
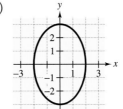

(f)
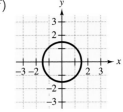

1. $x^2 + y^2 = 9$

2. $4x^2 + 4y^2 = 9$

3. $\dfrac{x^2}{4} + \dfrac{y^2}{9} = 1$

4. $\dfrac{x^2}{9} + \dfrac{y^2}{4} = 1$

5. $x^2 - y^2 = 4$

6. $x^2 - y^2 = -4$

In Exercises 7–14, find an equation of the circle with center at $(0, 0)$ that satisfies the given criterion. See Example 1.

7. Radius: 5

8. Radius: 7

9. Radius: $\frac{2}{3}$

10. Radius: $\frac{5}{2}$

11. Passes through the point $(0, 8)$

12. Passes through the point $(-2, 0)$

13. Passes through the point $(5, 2)$

14. Passes through the point $(-1, -4)$

In Exercises 15–22, find an equation of the circle with center at (h, k) that satisfies the given criteria.

15. Center: $(4, 3)$
   Radius: 10

16. Center: $(-2, 5)$
   Radius: 6

17. Center: $(5, -3)$
   Radius: 9

18. Center: $(-5, -2)$
   Radius: $\frac{5}{2}$

19. Center: $(-2, 1)$
   Passes through the point $(0, 1)$

20. Center: $(8, 2)$
   Passes through the point $(8, 0)$

21. Center: $(3, 2)$
   Passes through the point $(4, 6)$

22. Center: $(-3, -5)$
   Passes through the point $(0, 0)$

In Exercises 23–36, identify the center and radius of the circle and sketch its graph. See Example 2.

23. $x^2 + y^2 = 16$

24. $x^2 + y^2 = 25$

25. $x^2 + y^2 = 36$

26. $x^2 + y^2 = 10$

27. $4x^2 + 4y^2 = 1$

28. $9x^2 + 9y^2 = 64$

29. $(x - 2)^2 + (y - 3)^2 = 4$

30. $(x + 4)^2 + (y - 3)^2 = 25$

31. $\left(x + \frac{5}{2}\right)^2 + (y + 3)^2 = 9$

32. $(x - 5)^2 + \left(y + \frac{3}{4}\right)^2 = 1$

33. $x^2 + y^2 - 4x - 2y + 1 = 0$

34. $x^2 + y^2 + 6x - 4y - 3 = 0$

35. $x^2 + y^2 + 2x + 6y + 6 = 0$

36. $x^2 + y^2 - 2x + 6y - 15 = 0$

In Exercises 37–40, use a graphing utility to graph the circle. (*Note:* Solve for y. Use the square setting so the circles appear correct.)

37. $x^2 + y^2 = 30$

38. $4x^2 + 4y^2 = 45$

39. $(x - 2)^2 + y^2 = 10$

40. $(x + 3)^2 + y^2 = 15$

In Exercises 41–52, find the standard form of the equation of the ellipse, centered at the origin. See Example 3.

Vertices	Co-vertices
41. $(-4, 0)$, $(4, 0)$	$(0, -3)$, $(0, 3)$
42. $(-4, 0)$, $(4, 0)$	$(0, -1)$, $(0, 1)$
43. $(-2, 0)$, $(2, 0)$	$(0, -1)$, $(0, 1)$
44. $(-10, 0)$, $(10, 0)$	$(0, -4)$, $(0, 4)$
45. $(0, -4)$, $(0, 4)$	$(-3, 0)$, $(3, 0)$
46. $(0, -5)$, $(0, 5)$	$(-1, 0)$, $(1, 0)$
47. $(0, -2)$, $(0, 2)$	$(-1, 0)$, $(1, 0)$
48. $(0, -8)$, $(0, 8)$	$(-4, 0)$, $(4, 0)$

49. Major axis (vertical) 10 units, minor axis 6 units

50. Major axis (horizontal) 24 units, minor axis 10 units

51. Major axis (horizontal) 20 units, minor axis 12 units

52. Major axis (horizontal) 50 units, minor axis 30 units

In Exercises 53–62, sketch the ellipse. Identify its vertices and co-vertices. See Example 4.

53. $\dfrac{x^2}{16} + \dfrac{y^2}{4} = 1$

54. $\dfrac{x^2}{25} + \dfrac{y^2}{9} = 1$

55. $\dfrac{x^2}{4} + \dfrac{y^2}{16} = 1$

56. $\dfrac{x^2}{9} + \dfrac{y^2}{25} = 1$

57. $\dfrac{x^2}{25/9} + \dfrac{y^2}{16/9} = 1$

58. $\dfrac{x^2}{1} + \dfrac{y^2}{1/4} = 1$

59. $4x^2 + y^2 - 4 = 0$

60. $4x^2 + 9y^2 - 36 = 0$

61. $10x^2 + 16y^2 - 160 = 0$

62. $16x^2 + 4y^2 - 64 = 0$

In Exercises 63–66, use a graphing utility to graph the ellipse. Identify the vertices. (*Note:* Solve for y.)

63. $x^2 + 2y^2 = 4$

64. $9x^2 + y^2 = 64$

65. $3x^2 + y^2 - 12 = 0$

66. $5x^2 + 2y^2 - 10 = 0$

In Exercises 67–78, sketch the hyperbola. Identify its vertices and asymptotes. See Example 5.

67. $x^2 - y^2 = 9$

68. $y^2 - x^2 = 9$

69. $y^2 - x^2 = 1$

70. $x^2 - y^2 = 1$

71. $\dfrac{x^2}{9} - \dfrac{y^2}{25} = 1$

72. $\dfrac{y^2}{9} - \dfrac{x^2}{25} = 1$

73. $\dfrac{y^2}{4} - \dfrac{x^2}{9} = 1$

74. $\dfrac{x^2}{4} - \dfrac{y^2}{9} = 1$

75. $\dfrac{x^2}{1} - \dfrac{y^2}{9/4} = 1$

76. $\dfrac{y^2}{1/4} - \dfrac{x^2}{25/4} = 1$

77. $4y^2 - x^2 + 16 = 0$

78. $4y^2 - 9x^2 - 36 = 0$

In Exercises 79–86, find an equation of the hyperbola centered at the origin. See Example 6.

Vertices	Asymptotes	
79. $(-4, 0)$, $(4, 0)$	$y = 2x$	$y = -2x$
80. $(-2, 0)$, $(2, 0)$	$y = \frac{1}{3}x$	$y = -\frac{1}{3}x$
81. $(0, -4)$, $(0, 4)$	$y = \frac{1}{2}x$	$y = -\frac{1}{2}x$
82. $(0, -2)$, $(0, 2)$	$y = 3x$	$y = -3x$
83. $(-9, 0)$, $(9, 0)$	$y = \frac{2}{3}x$	$y = -\frac{2}{3}x$
84. $(0, -5)$, $(0, 5)$	$y = x$	$y = -x$
85. $(0, -1)$, $(0, 1)$	$y = 2x$	$y = -2x$
86. $(-1, 0)$, $(1, 0)$	$y = \frac{1}{2}x$	$y = -\frac{1}{2}x$

In Exercises 87–90, use a graphing utility to graph the equation. (*Note:* Solve for y.)

87. $\dfrac{x^2}{16} - \dfrac{y^2}{4} = 1$

88. $\dfrac{y^2}{16} - \dfrac{x^2}{4} = 1$

89. $5x^2 - 2y^2 + 10 = 0$

90. $x^2 - 2y^2 - 4 = 0$

In Exercises 91–100, identify the graph of the equation as a line, circle, parabola, ellipse, or hyperbola.

91. $y = 2x^2 - 8x + 2$

92. $y = 10 - \frac{3}{2}x$

93. $4x^2 + 9y^2 = 36$

94. $4x^2 + 4y^2 = 36$

95. $4x^2 - 9y^2 = 36$

96. $x^2 - 4y + 2x = 0$

97. $x^2 + y^2 - 1 = 0$

98. $2x^2 + 2y^2 = 9$

99. $3x + 2 = 0$

100. $y^2 = x^2 + 2$

Solving Problems

101. *Satellite Orbit*   Find an equation of the circular orbit of a satellite 500 miles above the surface of the earth. Place the origin of the rectangular coordinate system at the center of the earth and assume the radius of the earth is 4000 miles.

102. *Architecture*   The top portion of a stained-glass window is in the form of a pointed Gothic arch (see figure). Each side of the arch is an arc of a circle of radius 12 feet and center at the base of the opposite arch. Find an equation of one of the circles and use it to determine the height of the point of the arch above the horizontal base of the window.

|◄──── 12 ft ────►|

103. *Graphical Estimation*   A rectangle centered at the origin with sides parallel to the coordinate axes is placed in a circle of radius 25 inches centered at the origin (see figure). The length of the rectangle is $2x$ inches.

(a) Show that the height and area are given by $2\sqrt{625 - x^2}$ and $4x\sqrt{625 - x^2}$, respectively.

(b) Use a graphing utility to graph the area function. Approximate the value of x for which the area is maximum.

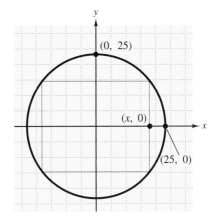

104. *Height of an Arch*   A *semicircular* arch for a tunnel under a river has a diameter of 100 feet (see figure). Find an equation of the circle. Determine the height of the arch 5 feet from the edge of the tunnel.

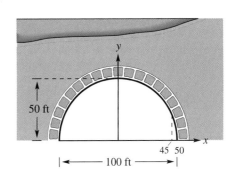

105. *Height of an Arch*   A *semielliptical* arch for a tunnel under a river has a width of 100 feet and a height of 40 feet (see figure). Find an equation of the ellipse. Determine the height of the arch 5 feet from the edge of the tunnel.

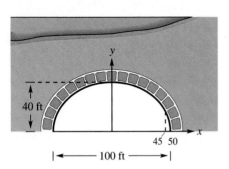

106. *Bicycle Chainwheel*   The pedals of a bicycle drive a chainwheel, which drives a smaller sprocket wheel on the rear axle (see figure). Many chainwheels are circular. Some, however, are slightly elliptical, which tends to make pedaling easier. Find an equation of an elliptical chainwheel that measures 8 inches at its widest point and $7\frac{1}{2}$ inches at its narrowest point.

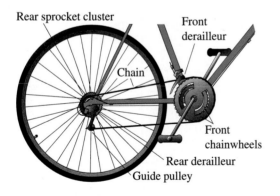

107. *Area*   The area A of the ellipse

$$\frac{x^2}{a^2} + \frac{y^2}{b^2} = 1$$

is given by $A = \pi ab$. Find the equation of an ellipse with area 301.59 square units and $a + b = 20$.

108. (a) Sketch a graph of the ellipse that consists of all points (x, y) such that the sum of the distances between (x, y) and two fixed points is 15 units and for which the foci are located at the centers of the two sets of concentric circles in the figure.

(b) Sketch a graph of the hyperbola that consists of all points (x, y) such that the difference of the distances between (x, y) and two fixed points is 8 units and for which the foci are located at the centers of the two sets of concentric circles in the figure.

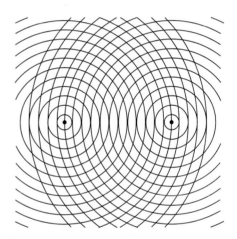

Explaining Concepts

109. Name the four types of conics.

110. Define a circle and give the standard form of the equation of a circle centered at the origin.

111. Define an ellipse and give the standard form of the equation of an ellipse centered at the origin.

112. Define a hyperbola and give the standard form of the equation of a hyperbola centered at the origin.

113. How can you tell if an ellipse is a circle from the equation?

114. From its equation, how can you determine the lengths of the axes of an ellipse?

115. Explain how the central rectangle of a hyperbola can be used to sketch its asymptotes.

Think About It   In Exercises 116 and 117, describe the part of the hyperbola

$$\frac{x^2}{4} - \frac{y^2}{9} = 1$$

given by the equation.

116. $x = -\frac{2}{3}\sqrt{9 + y^2}$ **117.** $y = \frac{3}{2}\sqrt{x^2 - 4}$

7.5   Graphs of Rational Functions

Objectives

1 Use a table of values to sketch the graph of a rational function.

2 Determine horizontal and vertical asymptotes of a rational function.

3 Use asymptotes and intercepts to sketch the graph of a rational function.

4 Use the graph of a rational function to solve an application problem.

1 Use a table of values to sketch the graph of a rational function.

Introduction

Recall that the domain of a rational function consists of all values of x for which the denominator is not zero. For instance, the domain of

$$f(x) = \frac{x + 2}{x - 1}$$

is all real numbers except $x = 1$. When graphing a rational function, pay special attention to the shape of the graph near x-values that are not in the domain.

Example 1   Sketching the Graph of a Rational Function

Sketch the graph of $f(x) = \dfrac{x + 2}{x - 1}$.

Solution

Begin by noticing that the domain is all real numbers except $x = 1$. Next, construct a table of values, including x-values that are close to 1 on the left *and* the right.

x-Values to the Left of 1

x	-3	-2	-1	0	0.5	0.9
$f(x)$	0.25	0	-0.5	-2	-5	-29

x-Values to the Right of 1

x	1.1	1.5	2	3	4	5
$f(x)$	31	7	4	2.5	2	1.75

Plot the points to the left of 1 and connect them with a smooth curve, as shown in Figure 7.38. Do the same for the points to the right of 1. *Do not* connect the two portions of the graph, which are called its **branches.**

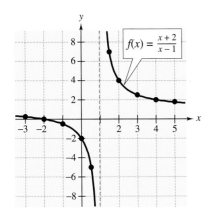

Figure 7.38

In Figure 7.38, as x approaches 1 from the left, the values of $f(x)$ approach negative infinity, and as x approaches 1 from the right, the values of $f(x)$ approach positive infinity.

2 Determine horizontal and vertical asymptotes of a rational function.

Horizontal and Vertical Asymptotes

An **asymptote** of a graph is a line to which the graph becomes arbitrarily close as $|x|$ or $|y|$ increases without bound. In other words, if a graph has an asymptote, it is possible to move far enough out on the graph so that there is almost no difference between the graph and the asymptote.

The graph in Figure 7.39(a) has two asymptotes: the line $x = -2$ is a **vertical asymptote,** and the line $y = 1$ is a **horizontal asymptote.** The graph in Figure 7.39(b) has three asymptotes: the lines $x = -2$ and $x = 2$ are vertical asymptotes, and the line $y = 2$ is a horizontal asymptote.

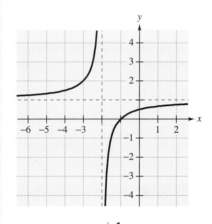

(a) Graph of $y = \dfrac{x + 1}{x + 2}$

Horizontal asymptote: $y = 1$

Vertical asymptote: $x = -2$

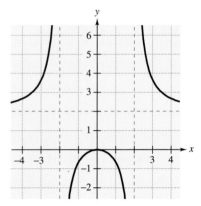

(b) Graph of $y = \dfrac{2x^2}{x^2 - 4}$

Horizontal asymptote: $y = 2$

Vertical asymptotes: $x = \pm 2$

Figure 7.39

The graph of a rational function may have no horizontal or vertical asymptotes, or it may have several.

Technology: Tip

A graphing utility can help you sketch the graph of a rational function. With most graphing utilities, however, there are problems with graphs of rational functions. If you use *connected mode*, the graphing utility will try to connect any branches of the graph. If you use *dot mode*, the graphing utility will draw a dotted (rather than a solid) graph. Both of these options are shown below for the graph of $y = (x - 1)/(x - 3)$.

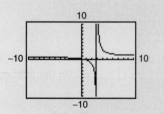

Connected Mode

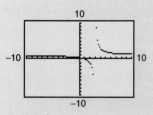

Dot Mode

▶ **Guidelines for Finding Asymptotes**

Let $f(x) = p(x)/q(x)$, where $p(x)$ and $q(x)$ have no common factors.

1. The graph of f has a vertical asymptote at each x-value for which the denominator is zero.

2. The graph of f has at most one horizontal asymptote.
   (a) If the degree of $p(x)$ is less than the degree of $q(x)$, the line $y = 0$ is a horizontal asymptote.
   (b) If the degree of $p(x)$ is equal to the degree of $q(x)$, the line $y = a/b$ is a horizontal asymptote, where a is the leading coefficient of $p(x)$ and b is the leading coefficient of $q(x)$.
   (c) If the degree of $p(x)$ is greater than the degree of $q(x)$, the graph has no horizontal asymptote.

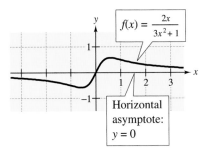

Figure 7.40

Example 2   Finding Horizontal and Vertical Asymptotes

Find all horizontal and vertical asymptotes of the graph of

$$f(x) = \frac{2x}{3x^2 + 1}.$$

Solution

For this rational function, the degree of the numerator is less than the degree of the denominator. This implies that the graph has the line

$$y = 0 \qquad\qquad\qquad \text{Horizontal asymptote}$$

as a horizontal asymptote, as shown in Figure 7.40. To find any vertical asymptotes, set the denominator equal to zero and solve the resulting equation for x.

$$3x^2 + 1 = 0 \qquad\qquad\qquad \text{Set denominator equal to zero.}$$

Because this equation has no real solution, you can conclude that the graph has no vertical asymptote.

Remember that the graph of a rational function can have at most one horizontal asymptote, but it can have several vertical asymptotes. For instance, the graph in Example 3 has two vertical asymptotes.

Example 3   Finding Horizontal and Vertical Asymptotes

Find all horizontal and vertical asymptotes of the graph of

$$f(x) = \frac{2x^2}{x^2 - 1}.$$

Solution

For this rational function, the degree of the numerator is equal to the degree of the denominator. The leading coefficient of the numerator is 2, and the leading coefficient of the denominator is 1. So, the graph has the line

$$y = \frac{2}{1} = 2 \qquad\qquad\qquad \text{Horizontal asymptote}$$

as a horizontal asymptote, as shown in Figure 7.41. To find any vertical asymptotes, set the denominator equal to zero and solve the resulting equation for x.

$$x^2 - 1 = 0 \qquad\qquad\qquad \text{Set denominator equal to zero.}$$
$$(x + 1)(x - 1) = 0 \qquad\qquad\qquad \text{Factor.}$$
$$x + 1 = 0 \quad\Longrightarrow\quad x = -1 \qquad \text{Set 1st factor equal to 0.}$$
$$x - 1 = 0 \quad\Longrightarrow\quad x = 1 \qquad \text{Set 2nd factor equal to 0.}$$

This equation has two real solutions: -1 and 1. So, the graph has two vertical asymptotes: the lines $x = -1$ and $x = 1$.

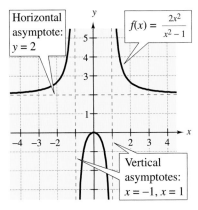

Figure 7.41

3 Use asymptotes and intercepts to sketch the graph of a rational function.

Graphing Rational Functions

To sketch the graph of a rational function, we suggest the following guidelines.

▶ **Guidelines for Graphing Rational Functions**

Let $f(x) = p(x)/q(x)$, where $p(x)$ and $q(x)$ have no common factors.

1. Find and plot the y-intercept (if any) by evaluating $f(0)$.

2. Set the numerator equal to zero and solve the equation for x. The real solutions represent the x-intercepts of the graph. Plot these intercepts.

3. Find and sketch the horizontal and vertical asymptotes of the graph.

4. Plot at least one point both between and beyond each x-intercept and vertical asymptote.

5. Use smooth curves to complete the graph between and beyond the vertical asymptotes.

6. If $p(x)$ and $q(x)$ have a common factor $x - a$, then the graph of $p(x)/q(x)$ has a hole at $x = a$.

Technology: Discovery

The rational function

$$f(x) = \frac{x + 2}{x^2 - 4}$$

has a common factor $x + 2$ in its numerator and denominator. Use the table feature of your graphing utility to examine the values of the function *near* and *at* $x = -2$. What happens at $x = -2$? Is $x = -2$ a vertical asymptote? Explain. Graph this function and use the trace feature to verify your explanation.

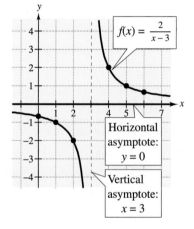

Figure 7.42

> **Example 4** Sketching the Graph of a Rational Function

Sketch the graph of $f(x) = \dfrac{2}{x - 3}$.

Solution

Begin by noting that the numerator and denominator have no common factors. Following the guidelines above produces the following.

- Because $f(0) = \dfrac{2}{0 - 3} = -\dfrac{2}{3}$, the y-intercept is $\left(0, -\dfrac{2}{3}\right)$.

- Because the numerator is never zero, there are no x-intercepts.

- Because the denominator is zero when $x - 3 = 0$ or $x = 3$, the line $x = 3$ is a vertical asymptote.

- Because the degree of the numerator is less than the degree of the denominator, the line $y = 0$ is a horizontal asymptote.

Plot the intercepts, asymptotes, and the additional points from the following table. Then complete the graph by drawing two branches, as shown in Figure 7.42. Note that the two branches are not connected.

x	1	2	4	5	6
$f(x)$	-1	-2	2	1	$\frac{2}{3}$

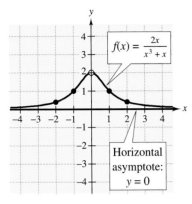

Figure 7.43

Example 5   A Rational Function with a Hole in Its Graph

Sketch the graph of $f(x) = \dfrac{2x}{x^3 + x}$.

Solution

Begin by noting that the numerator and denominator have a common factor x, and so the graph has a hole at $x = 0$.

- The simplified form of the equation is $f(x) = \dfrac{2}{x^2 + 1}$.

- Because the numerator of the simplified form is never zero, there are no x-intercepts.

- Because the denominator is never zero, there are no vertical asymptotes.

- Because the degree of the numerator is less than the degree of the denominator, the line $y = 0$ is a horizontal asymptote.

By identifying the hole, sketching the asymptotes, and plotting the additional points from the following table, you can obtain the graph shown in Figure 7.43.

x	-2	-1	0	1	2
$f(x)$	$\frac{2}{5}$	1	Undefined	1	$\frac{2}{5}$

Example 6   Sketching the Graph of a Rational Function

Sketch the graph of $f(x) = \dfrac{4x^2}{x^2 - 4}$.

Solution

Begin by noting that the numerator and denominator have no common factors.

- Because $f(0) = \dfrac{4(0)^2}{(0)^2 - 4} = 0$, the y-intercept is $(0, 0)$.

- Because the numerator is zero when $4x^2 = 0$ or $x = 0$, the x-intercept is $(0, 0)$.

- Because the denominator is zero when $x^2 - 4 = 0$ or $(x - 2)(x + 2) = 0$, the lines $x = 2$ and $x = -2$ are vertical asymptotes.

- Because the degree of the numerator equals the degree of the denominator, the line $y = a/b = 4/1 = 4$ is a horizontal asymptote.

By plotting the intercept, sketching the asymptotes, and plotting the additional points from the following table, you can obtain the graph shown in Figure 7.44.

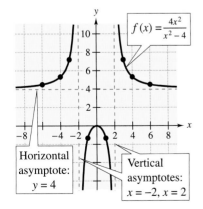

Figure 7.44

x	-6	-4	-3	-1	1	3	4	6
$f(x)$	$\frac{9}{2}$	$\frac{16}{3}$	$\frac{36}{5}$	$-\frac{4}{3}$	$-\frac{4}{3}$	$\frac{36}{5}$	$\frac{16}{3}$	$\frac{9}{2}$

4 Use the graph of a rational function to solve an application problem.

Application

Example 7 Finding the Average Cost

As a fundraising project, a club is publishing a calendar. The cost of photography and typesetting is $850. In addition to these "one-time" charges, the unit cost of printing each calendar is $3.25. Let x represent the number of calendars printed. Write a model that represents the average cost per calendar.

Solution

To begin, you need to find the total cost of printing the calendars. The verbal model for the total cost is

$$\boxed{\text{Total cost}} = \boxed{\text{Unit cost}} \times \boxed{\begin{array}{c}\text{Number of}\\\text{calendars}\end{array}} + \boxed{\begin{array}{c}\text{Cost of photography}\\\text{and typesetting}\end{array}} .$$

The total cost C of printing x calendars is

$$C = 3.25x + 850. \qquad \text{Total cost function}$$

The verbal model for the average cost per calendar is

$$\boxed{\begin{array}{c}\text{Average}\\\text{cost}\end{array}} = \frac{\boxed{\text{Total cost}}}{\boxed{\text{Number of calendars}}} .$$

The average cost per calendar $\overline{C}$ for printing x calendars is

$$\overline{C} = \frac{3.25x + 850}{x}. \qquad \text{Average cost function}$$

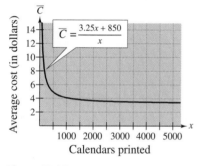

Figure 7.45

From the graph shown in Figure 7.45, notice that the average cost decreases as the number of calendars increases.

Discussing the Concept	More About the Average Cost

In Example 7, what is the horizontal asymptote of the graph of the average cost function? What is the significance of this asymptote in the problem? Is it possible to sell enough calendars to obtain an average cost of $3.00 per calendar? Explain your reasoning.

7.5   Exercises

Integrated Review   *Concepts, Skills, and Problem Solving*

Keep mathematically in shape by doing these exercises *before* the problems of this section.

Properties and Definitions

1. Identify the leading coefficient in $7x^2 + 3x - 4$. Explain.

2. State the degree of the product $(x^4 + 3)(x - 4)$. Explain.

3. Sketch a graph for which y is not a function of x. Explain.

4. Sketch a graph for which y is a function of x. Explain.

Multiplying Expressions

In Exercises 5–10, find the product.

5. $-2x^5(5x^3)$

6. $3x(5 - 2x)$

7. $(2x - 15)^2$

8. $(3x + 2)(7x - 10)$

9. $[(x + 1) - y][(x + 1) + y]$

10. $(x + 3)(x^2 - 3x + 9)$

Problem Solving

11. The height of a triangle is 12 meters less than its base. Find the base and height of the triangle if its area is 80 square meters.

12. An open box with a square base is to be constructed from 825 square inches of material. What should be the dimensions of the base if the height of the box is to be 10 inches?

Developing Skills

Numerical Analysis   In Exercises 1 and 2, (a) complete each table, (b) use the tables to sketch the graph, and (c) determine the domain of the function. See Example 1.

x	0	0.5	0.9	0.99	0.999
y					

x	2	1.5	1.1	1.01	1.001
y					

x	2	5	10	100	1000
y					

1. $f(x) = \dfrac{4}{x - 1}$

2. $f(x) = \dfrac{2x}{x - 1}$

Numerical Analysis   In Exercises 3–6, (a) complete each table, (b) use the tables to sketch the graph, and (c) determine the domain of the function.

x	2	2.5	2.9	2.99	2.999
y					

x	4	3.5	3.1	3.01	3.001
y					

x	4	5	10	100	1000
y					

3. $f(x) = 2 + \dfrac{1}{x - 3}$

4. $f(x) = \dfrac{2}{x - 3}$

5. $f(x) = \dfrac{3x}{x^2 - 9}$

6. $f(x) = \dfrac{5x^2}{x^2 - 9}$

In Exercises 7–24, find the domain of the function and identify any horizontal and vertical asymptotes. See Examples 2 and 3.

7. $f(x) = \dfrac{5}{x^2}$

8. $g(x) = \dfrac{3}{x}$

9. $f(x) = \dfrac{x}{x + 8}$

10. $f(u) = \dfrac{u^2}{u - 10}$

11. $g(t) = \dfrac{2t - 5}{3t - 9}$

12. $h(x) = \dfrac{4x - 3}{2x + 5}$

13. $y = \dfrac{3 - 5x}{1 - 3x}$

14. $y = \dfrac{3x + 2}{2x - 1}$

15. $g(t) = \dfrac{3}{t(t - 1)}$

16. $h(s) = \dfrac{2s + 1}{s(s + 3)}$

17. $y = \dfrac{2x^2}{x^2 + 1}$

18. $g(t) = \dfrac{3t^3}{t^2 + 1}$

19. $y = \dfrac{x^2 - 4}{x^2 - 1}$

20. $y = \dfrac{x^2 - 9}{x^2 - 2x - 8}$

21. $g(z) = 1 - \dfrac{2}{z}$

22. $h(v) = \dfrac{3}{v} - 2$

23. $g(x) = 2x + \dfrac{4}{x}$

24. $f(t) = \dfrac{5}{t} - 4t$

In Exercises 25–28, identify the horizontal and vertical asymptotes of the function. Use the asymptotes to match the rational function with its graph. [The graphs are labeled (a), (b), (c), and (d).]

(a)

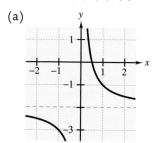

(b)

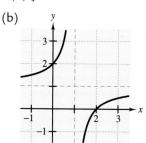

(c)

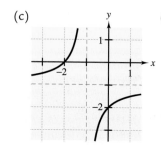

(d)

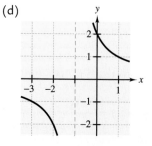

25. $f(x) = \dfrac{2}{x + 1}$

26. $f(x) = \dfrac{1 - 2x}{x}$

27. $f(x) = \dfrac{x - 2}{x - 1}$

28. $f(x) = -\dfrac{x + 2}{x + 1}$

In Exercises 29–32, match the rational function with its graph. [The graphs are labeled (a), (b), (c), and (d).]

(a)

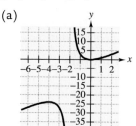

(b)

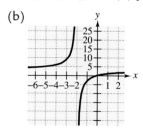

(c)

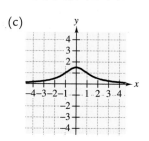

(d)

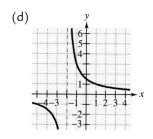

29. $f(x) = \dfrac{3}{x + 2}$

30. $f(x) = \dfrac{3x}{x + 2}$

31. $f(x) = \dfrac{3x^2}{x + 2}$

32. $f(x) = \dfrac{3}{x^2 + 2}$

In Exercises 33–58, sketch the graph of the function. As sketching aids, check for intercepts, vertical asymptotes, and horizontal asymptotes. See Examples 4–6.

33. $g(x) = \dfrac{5}{x}$

34. $f(x) = \dfrac{5}{x^2}$

35. $g(x) = \dfrac{5}{x - 4}$

36. $f(x) = \dfrac{5}{(x - 4)^2}$

37. $f(x) = \dfrac{1}{x - 2}$

38. $f(x) = \dfrac{3}{x + 1}$

39. $g(x) = \dfrac{1}{2 - x}$

40. $g(x) = \dfrac{-3}{x + 1}$

41. $y = \dfrac{3x}{x^2 + 4x}$

42. $y = \dfrac{2x}{x^2 + 4x}$

43. $h(u) = \dfrac{3u^2}{u^2 - 3u}$

44. $g(v) = \dfrac{2v^2}{v^2 + v}$

45. $y = \dfrac{2x + 4}{x}$

46. $y = \dfrac{x - 2}{x}$

47. $y = \dfrac{2x^2}{x^2 + 1}$

48. $y = \dfrac{10}{x^2 + 2}$

49. $y = \dfrac{4}{x^2 + 1}$

50. $y = \dfrac{4x^2}{x^2 + 1}$

51. $g(t) = 3 - \dfrac{2}{t}$

52. $f(x) = \dfrac{4}{x} + 2$

53. $y = -\dfrac{x}{x^2 - 4}$

54. $y = \dfrac{4x + 6}{x^2 - 9}$

55. $f(x) = \dfrac{3x^2}{x^2 - x - 2}$

56. $g(x) = \dfrac{2x^2}{x^2 + 2x - 3}$

57. $f(x) = \dfrac{x^2 - 4}{x^2 - 3x - 10}$

58. $g(t) = \dfrac{t^2 - 9}{t^2 + 6t + 9}$

In Exercises 59–68, use a graphing utility to graph the function. Give the domain of the function and identify any horizontal or vertical asymptotes.

59. $f(x) = \dfrac{3}{x + 2}$

60. $f(x) = \dfrac{3x}{x + 2}$

61. $h(x) = \dfrac{x - 3}{x - 1}$

62. $h(x) = \dfrac{x^2}{x - 2}$

63. $f(t) = \dfrac{6}{t^2 + 1}$

64. $g(t) = 2 + \dfrac{3}{t + 1}$

65. $y = \dfrac{2(x^2 + 1)}{x^2}$

66. $y = \dfrac{2(x^2 - 1)}{x^2}$

67. $y = \dfrac{3}{x} + \dfrac{1}{x - 2}$

68. $y = \dfrac{x}{2} - \dfrac{2}{x}$

Think About It   In Exercises 69 and 70, use a graphing utility to graph the function. Explain why there is no vertical asymptote when a superficial examination of the function seems to indicate that there should be one.

69. $g(x) = \dfrac{4 - 2x}{x - 2}$

70. $h(x) = \dfrac{x^2 - 9}{x + 3}$

Solving Problems

71. *Average Cost*   The cost of producing x units of a product is $C = 2500 + 0.50x$, $x > 0$.

(a) Write the average cost $\overline{C}$ as a function of x.

(b) Find the average cost of producing $x = 1000$ and $x = 10,000$ units.

(c) Use a graphing utility to graph the average cost function. Determine the horizontal asymptote of the graph. Interpret the result.

72. *Average Cost*   The cost of producing x units of a product is $C = 30,000 + 1.25x$, $x > 0$.

(a) Write the average cost $\overline{C}$ as a function of x.

(b) Find the average cost of producing $x = 10,000$ and $x = 100,000$ units.

(c) Use a graphing utility to graph the average cost function. Determine the horizontal asymptote of the graph. Interpret the result.

73. *Medicine*   The concentration of a certain chemical in the bloodstream t hours after injection into the muscle tissue is given by

$$C = \dfrac{2t}{4t^2 + 25}, \quad t \geq 0.$$

(a) Determine the horizontal asymptote of the function and interpret its meaning in the context of the problem.

(b) Use a graphing utility to graph the function. Approximate the time when the concentration is the greatest.

74. *Concentration of a Mixture*   A 25-liter container contains 5 liters of a 25% brine solution. You add x liters of a 75% brine solution to the container. The concentration C of the resulting mixture is

$$C = \dfrac{3x + 5}{4(x + 5)}.$$

(a) Determine the domain of the rational function based on the physical constraints of the problem.

(b) Use a graphing utility to graph the function. As the container is filled, what does the concentration of the brine appear to approach?

75. *Geometry*   A rectangular region of length x and width y has an area of 400 square meters.

(a) Sketch a figure that gives a visual representation of the problem.

(b) Verify that the perimeter P is given by

$$P = 2\left(x + \dfrac{400}{x}\right).$$

(c) Determine the domain of the function based on the physical constraints of the problem.

(d) Use a graphing utility to graph the function. Approximate the dimensions of the rectangle that has a minimum perimeter.

⊞ **76.** *Sales* The cumulative number N (in thousands) of units of a product sold over a period of t years is modeled by

$$N = \frac{150t(1 + 4t)}{1 + 0.15t^2}, \; t \geq 0.$$

(a) Estimate N when $t = 1$, $t = 2$, and $t = 4$.

(b) Use a graphing utility to graph the function. Determine the horizontal asymptote.

(c) Explain the meaning of the horizontal asymptote in the context of the problem.

Think About It In Exercises 77–80, write a rational function satisfying each of the criteria. Use a graphing utility to verify the results.

77. Vertical asymptote: $x = 3$

Horizontal asymptote: $y = 2$

Zero of the function: $x = -1$

78. Vertical asymptote: $x = -2$

Horizontal asymptote: $y = 0$

Zero of the function: $x = 3$

79. Vertical asymptotes: $x = 4$ and $x = -2$

Horizontal asymptote: $y = 0$

Zero of the function: $x = 6$

80. Vertical asymptotes: $x = 1$ and $x = -1$

Horizontal asymptote: $y = 1$

Zero of the function: $x = 0$

Explaining Concepts

81. Answer parts (c)–(e) of Motivating the Chapter on page 423.

82. In your own words, describe how to determine the domain of a rational function. Give an example of a rational function whose domain is all real numbers except 2.

83. In your own words, describe what is meant by an *asymptote* of a graph.

84. *True or False?* If the graph of a rational function f has a vertical asymptote at $x = 3$, it is possible to sketch the graph without lifting your pencil from the paper. Explain.

85. Does every rational function have a vertical asymptote? Explain.

Key Terms

constant of
  proportionality, *p. 424*
varies directly, *p. 424*
varies directly as the *n*th
  power, *p. 426*
varies inversely, *p. 427*
varies jointly, *p. 429*

linear inequality, *p. 435*
graph of a linear inequality,
  p. 436
half-planes, *p. 436*
parabola, *p.443*
vertex of a parabola,
  p. 443

conic sections, *p. 453*
circle, *p. 453*
ellipse, *p. 455*
vertices of an ellipse,
  p. 455
co-vertices of an ellipse,
  p. 455

hyperbola, *p. 457*
vertices of a hyperbola,
  p. 457
asymptotes of a
  hyperbola, *p. 457*
asymptotes of a rational
  function, *p. 466*

Key Concepts

7.1 Direct variation

1. The following three statements are equivalent.

  a. y varies directly as x.

  b. y is directly proportional to x.

  c. $y = kx$ for some constant k.

2. If $y = kx^n$, then y is directly proportional to the nth power of x.

7.1 Inverse variation

1. The following three statements are equivalent.

  a. y varies inversely as x.

  b. y is inversely proportional to x.

  c. $y = k/x$ for some constant k.

2. If $y = k/x^n$, then y is inversely proportional to the nth power of x.

7.1 Joint variation

1. The following three statements are equivalent.

  a. z varies jointly as x and y.

  b. z is jointly proportional to x and y.

  c. $z = kxy$ for some constant k.

2. If $z = kx^n y^m$, then z is jointly proportional to the nth power of x and the mth power of y.

7.2 Guidelines for graphing a linear inequality

1. Replace the inequality sign by an equal sign, and sketch the graph of the resulting equation.

2. Test one point in one of the half-planes formed by the graph in Step 1. If the point satisfies the inequality, shade the entire half-plane to denote that every point in the region satisfies the inequality. If the point does not satisfy the inequality, then shade the other half-plane.

7.3 Sketching a parabola

1. Determine the vertex and axis of the parabola by completing the square or by formula.

2. Plot the vertex, axis, and a few additional points on the parabola.

3. Use the fact that the parabola opens upward if $a > 0$ and opens downward if $a < 0$ to complete the sketch.

7.4 Standard forms of the equations of conics

1. Circle with center at the origin and radius r:
$$x^2 + y^2 = r^2$$

2. Circle with center at (h, k) and radius r:
$$(x - h)^2 + (y - k)^2 = r^2$$

3. Ellipse with center at the origin:
$$\frac{x^2}{a^2} + \frac{y^2}{b^2} = 1 \quad \text{or} \quad \frac{x^2}{b^2} + \frac{y^2}{a^2} = 1, \quad 0 < b < a$$

4. Hyperbola with center at the origin:
$$\frac{x^2}{a^2} - \frac{y^2}{b^2} = 1 \quad \text{or} \quad \frac{y^2}{a^2} - \frac{x^2}{b^2} = 1, \quad a > 0,\ b > 0$$

7.5 Guidelines for graphing rational functions

Let $f(x) = p(x)/q(x)$, where $p(x)$ and $q(x)$ have no common factors.

1. Find and plot the y-intercept and the x-intercept(s).

2. Find and sketch the horizontal and vertical asymptotes of the graph.

3. Plot at least one point both between and beyond each x-intercept and vertical asymptote.

4. Use smooth curves to complete the graph between and beyond the vertical asymptotes.

5. If $p(x)$ and $q(x)$ have a common factor $x - a$, then the graph of $p(x)/q(x)$ has a hole at $x = a$.

REVIEW EXERCISES

Reviewing Skills

7.1 In Exercises 1–4, find a mathematical model for the verbal statement.

1. P varies directly as the cube of t.
2. R varies jointly as u and v.
3. z varies inversely as the square of s.
4. F varies directly as g and inversely as r^2.

In Exercises 5–8, find the constant of proportionality and write an equation that relates the variables.

5. y varies directly as the cube root of x, and $y = 12$ when $x = 8$.
6. r varies inversely as s, and $r = 45$ when $s = \frac{3}{5}$.
7. T varies jointly as r and the square of s, and $T = 5000$ when $r = 0.09$ and $s = 1000$.
8. D is directly proportional to the cube of x and inversely proportional to y, and $D = 810$ when $x = 3$ and $y = 25$.

7.2 In Exercises 9–16, sketch the graph of the solution of the inequality.

9. $y > 4$
10. $x \le 5$
11. $x - 2 \ge 0$
12. $y + 3 < 0$
13. $2x + y < 1$
14. $3x - 4y > 2$
15. $-(x - 1) \le 4y - 2$
16. $(y - 3) \ge 2(x - 5)$

In Exercises 17–20, use a graphing utility to graph (shade) the solution of the inequality.

17. $y \le 12 - \frac{3}{2}x$
18. $y \le \frac{1}{3}x + 1$
19. $x + y \ge 0$
20. $4x - 3y \ge 2$

7.3 In Exercises 21–24, determine the vertex of the graph of the quadratic function.

21. $f(x) = x^2 - 8x + 3$
22. $g(x) = x^2 + 12x - 9$
23. $h(u) = 2u^2 - u + 3$
24. $f(t) = 3t^2 + 2t - 6$

In Exercises 25–28, sketch the graph of the function. Identify the vertex and any x-intercepts. Use a graphing utility to verify your results.

25. $y = x^2 + 8x$
26. $y = -x^2 + 3x$

27. $y = x^2 - 6x + 5$
28. $y = x^2 + 3x - 10$

In Exercises 29–32, identify the transformation of the graph of $f(x) = x^2$ and sketch a graph of h.

29. $h(x) = x^2 + 3$
30. $h(x) = x^2 - 1$
31. $h(x) = (x + 2)^2 - 3$
32. $h(x) = (x - 2)^2 + 4$

In Exercises 33–38, write the standard form of an equation of the parabola that satisfies the conditions.

33. Vertex: $(3, 5)$; $a = -2$
34. Vertex: $(-2, 3)$; $a = 3$
35. Vertex: $(2, -5)$; y-intercept: $(0, 3)$
36. Vertex: $(-4, 0)$; y-intercept: $(0, -6)$
37. Vertex: $(5, 0)$; passes through the point $(1, 1)$
38. Vertex: $(-2, 5)$; passes through the point $(0, 1)$

7.4 In Exercises 39–44, match the equation with the correct graph. [The graphs are labeled (a), (b), (c), (d), (e), and (f).]

(a)
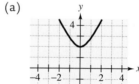

(b)
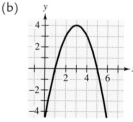

(c)
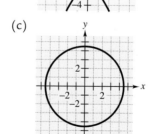

(d)
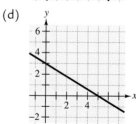

(e)
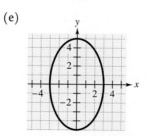

(f)
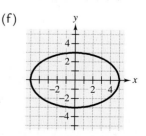

39. $4x^2 + 4y^2 = 81$

40. $3x + 5y = 15$

41. $\dfrac{y^2}{4} - x^2 = 1$

42. $\dfrac{x^2}{25} + \dfrac{y^2}{9} = 1$

43. $y = -x^2 + 6x - 5$

44. $\dfrac{x^2}{9} + \dfrac{y^2}{25} = 1$

In Exercises 45–54, identify and sketch the conic.

45. $x^2 - 2y = 0$

46. $x^2 - y^2 = 64$

47. $x^2 + y^2 = 64$

48. $x^2 + 4y^2 = 64$

49. $y = (x - 6)^2 + 1$

50. $y = 9 - (x - 3)^2$

51. $\dfrac{x^2}{25} + \dfrac{y^2}{4} = 1$

52. $\dfrac{x^2}{25} - \dfrac{y^2}{4} = -1$

53. $4x^2 + 4y^2 - 9 = 0$

54. $x^2 + 9y^2 - 9 = 0$

In Exercises 55–62, find the standard form of the equation of the conic meeting the given criteria.

55. *Parabola:* Vertex: $(5, 0)$

Passes through the point $(1, 1)$

56. *Parabola:* Vertex: $(-2, 5)$

Passes through the point $(0, 1)$

57. *Ellipse:* Vertices: $(0, -5), (0, 5)$

Co-vertices: $(-2, 0), (2, 0)$

58. *Ellipse:* Vertices: $(-10, 0), (10, 0)$

Co-vertices: $(0, -6), (0, 6)$

59. *Circle:* Center: $(0, 0)$

Radius: 20

60. *Circle:* Center: $(3, -2)$

Radius: 4

61. *Hyperbola:* Vertices: $(-3, 0), (3, 0)$

Asymptotes: $y = -\frac{1}{2}x, \ y = \frac{1}{2}x$

62. *Hyperbola:* Vertices: $(0, -4), (0, 4)$

Asymptotes: $y = 2x, \ y = -2x$

 In Exercises 63–66, match the function with its graph. [The graphs are labeled (a), (b), (c), and (d).]

(a)

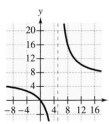

(b)

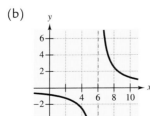

(c)

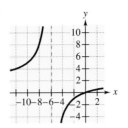

(d)

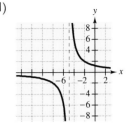

63. $f(x) = \dfrac{5}{x - 6}$

64. $f(x) = \dfrac{6}{x + 5}$

65. $f(x) = \dfrac{6x}{x - 5}$

66. $f(x) = \dfrac{2x}{x + 6}$

In Exercises 67–82, use asymptotes and intercepts to graph the rational function.

67. $f(x) = -\dfrac{5}{x^2}$

68. $f(x) = \dfrac{4}{x}$

69. $P(x) = \dfrac{3x + 6}{x - 2}$

70. $s(x) = \dfrac{2x - 6}{x + 4}$

71. $g(x) = \dfrac{2 + x}{1 - x}$

72. $h(x) = \dfrac{x - 3}{x - 2}$

73. $f(x) = \dfrac{x}{x^2 + 1}$

74. $f(x) = \dfrac{2x}{x^2 + 4}$

75. $h(x) = \dfrac{4}{(x - 1)^2}$

76. $g(x) = \dfrac{-2}{(x + 3)^2}$

77. $y = \dfrac{x}{x^2 - 1}$

78. $y = \dfrac{2x}{x^2 - 4}$

79. $y = \dfrac{2x^2}{x^2 - 4x}$

80. $y = \dfrac{2x}{x^2 + 3x}$

81. $y = \dfrac{x - 4}{x^2 - 3x - 4}$

82. $y = \dfrac{2x + 1}{2x^2 - 5x - 3}$

Think About It In Exercises 83 and 84, write a rational function satisfying each of the criteria. Use a graphing utility to verify the results.

83. Vertical asymptote: $x = 4$

Horizontal asymptote: $y = 3$

Zero of the function: $x = 0$

84. Vertical asymptote: $x = -3$

Horizontal asymptote: $y = 0$

Zero of the function: $x = 2$

Solving Problems

85. *Hooke's Law*   A force of 100 pounds stretches a spring 4 inches. Find the force required to stretch the spring 6 inches.

86. *Stopping Distance*   The stopping distance d of an automobile is directly proportional to the square of its speed s. How will the stopping distance be changed by doubling the speed of the car?

87. *Demand Function*   A company has found that the daily demand x for its product varies inversely as the square root of the price p. When the price is $25, the demand is approximately 1000 units. Approximate the demand if the price is increased to $28.

88. *Weight of an Astronaut*   The gravitational force F with which an object is attracted to the earth is inversely proportional to its distance r from the center of the earth. If an astronaut weighs 200 pounds on the surface of the earth ($r \approx 4000$ miles), what will the astronaut weigh 500 miles above the earth's surface?

89. *Weekly Pay*   You have two part-time jobs. One is at a grocery store, which pays $8 per hour, and the other is mowing lawns, which pays $10 per hour. Between the two jobs, you want to earn at least $200 a week. Write an inequality that shows the different numbers of hours you can work at each job, and sketch the graph of the inequality. From the graph, find several ordered pairs with positive integer coordinates that are solutions of the inequality.

90. *Perimeter of a Rectangle*   The perimeter of a rectangle of length x and width y cannot exceed 800 feet. Write a linear inequality for this constraint and sketch its graph.

91. *Graphical Estimation*   The height y (in feet) of a ball thrown by a child is given by

$$y = -\frac{1}{10}x^2 + 3x + 6$$

where x is the horizontal distance (in feet) from where the ball was thrown.

(a) Use a graphing utility to graph the path of the ball.

(b) How high was the ball when it left the child's hand?

(c) What was the maximum height of the ball?

(d) How far from the child did the ball hit the ground?

92. *Graphical Estimation*   The number N of orders for U.S. civil jet transport aircraft for the years 1990 through 1996 is approximated by the model

$$N = 653.50 - 379.89t + 62.75t^2, \quad 0 \le t \le 6$$

where t represents the calender year, with $t = 0$ corresponding to 1990.   (Source: Aerospace Industries Association of America)

(a) Use a graphing utility to graph the function N.

(b) Use the graph in part (a) to approximate the year when the number of orders was minimum.

(c) Use the model to predict the number of orders in the year 2000.

93. *Satellite Orbit*   Find an equation of the circular orbit of a satellite 1000 miles above the surface of the earth. Place the origin of the rectangular coordinate system at the center of the earth and assume that the radius of the earth is 4000 miles.

94. *Average Cost*   A business produces x units at a cost of $C = 0.5x + 500$. The average cost per unit is

$$\overline{C} = \frac{C}{x} = \frac{0.5x + 500}{x}, \quad x > 0.$$

Find the horizontal asymptote and state what it represents in the model.

95. *Population of Fish*   The Parks and Wildlife Commission introduces 80,000 fish into a lake. The population of the fish (in thousands) is

$$N = \frac{20(4 + 3t)}{1 + 0.05t}, \quad t \ge 0$$

where t is time in years.

(a) Find the population when t is 5, 10, and 25.

(b) Find the horizontal asymptote and state what it represents in the model.

96. *Seizure of Illegal Drugs*   The cost (in millions of dollars) for a government agency to seize $p\%$ of a certain illegal drug as it enters the country is

$$C = \frac{528p}{100 - p}, \quad 0 \le p < 100.$$

(a) Find the cost of seizing 25%.

(b) Find the cost of seizing 75%.

(c) According to this model, would it be possible to seize 100% of the drug? Explain.

Chapter Test

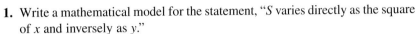

Take this test as you would take a test in class. After you are done, check your work against the answers given in the back of the book.

1. Write a mathematical model for the statement, "S varies directly as the square of x and inversely as y."

2. Find the constant of proportionality if v varies directly as the square root of u, and $v = \frac{3}{2}$ when $u = 36$.

In Exercises 3 and 4, sketch the graph of the solution of the linear inequality.

3. $y < 4$

4. $3x - 2y > 6$

5. Write an inequality for the shaded region shown in the figure.

6. Sketch the graph of $f(x) = -2(x - 2)^2 + 8$. Label its vertex and intercepts.

In Exercises 7–10, identify the conic and sketch its graph.

7. $x^2 + y^2 = 9$

8. $\dfrac{x^2}{9} + \dfrac{y^2}{16} = 1$

9. $\dfrac{x^2}{9} - \dfrac{y^2}{16} = 1$

10. $y = (x - 3)^2$

11. Find an equation of the circle shown in the figure.

12. Find an equation of the parabola shown in the figure.

13. Find the standard form of the equation of the ellipse with vertices $(0, -10)$ and $(0, 10)$ and co-vertices $(-3, 0)$ and $(3, 0)$.

14. Find the standard form of the equation of the hyperbola with vertices $(-3, 0)$ and $(3, 0)$ and asymptotes $y = \frac{1}{2}x$ and $y = -\frac{1}{2}x$.

In Exercises 15 and 16, graph the function and identify any asymptotes.

15. $f(x) = \dfrac{3}{x - 3}$

16. $g(x) = \dfrac{3x}{x^2 - 2x - 15}$

17. If the temperature of a gas is not allowed to change, its absolute pressure P is inversely proportional to its volume V, according to Boyle's Law. A large balloon is filled with 180 cubic meters of helium at atmospheric pressure (1 atm) at sea level. What is the volume of the helium if the balloon rises to an altitude at which the atmospheric pressure is 0.75 atm? (Assume that the temperature does not change.)

18. A warehouse operator has up to 24,000 square feet of floor space in which to store two products. Each unit of product I requires 20 square feet of floor space and each unit of product II requires 30 square feet. Write a linear inequality that models this constraint. Graph the inequality.

19. The revenue R for a chartered bus trip is given by $R = -\frac{1}{20}(n^2 - 240n)$, where n is the number of passengers and $80 \le n \le 160$. How many passengers will produce a maximum revenue? Explain your reasoning.

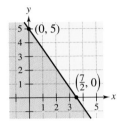

Figure for 5

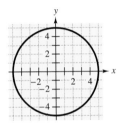

Figure for 11

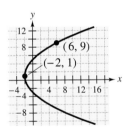

Figure for 12

8 Systems of Equations

Tony Duffy/Allsport

During the 1996–1997 season, there were 691 women's NCAA soccer teams in the United States. (Source: U.S. Bureau of the Census)

Motivating the Chapter

 ## Soccer Club Fundraiser

A collegiate soccer club has a fundraising dinner. Student tickets sell for $8 and nonstudent tickets sell for $15. There are 115 tickets sold and the total revenue is $1445.

See Section 8.2, Exercise 78

a. Set up a system of linear equations that can be used to determine how many tickets of each type were sold.

b. Solve the system in part (a) by the method of substitution.

c. Solve the system in part (a) by the method of elimination.

The soccer club decides to set goals for the next fundraising dinner. To meet these goals, a "major contributor" category is added. A person donating $100 is considered a major contributor to the soccer club and receives a "free" ticket to the dinner. The club's goals are to have 200 people in attendance, with the number of major contributors being one-fourth the number of students, and to raise $4995.

See Section 8.3, Exercise 65

d. Set up a system of linear equations to determine how many of each kind of ticket would need to be sold for the second fundraising dinner.

e. Solve the system in part (d) by Gaussian elimination.

f. Would it be possible for the soccer club to meet its goals if only 18 people donated $100? Explain.

See Section 8.5, Exercise 100

g. Solve the system in part (d) using matrices.

h. Solve the system in part (d) using determinants.

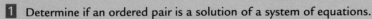

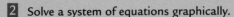

8.1   Systems of Equations

Objectives

1   Determine if an ordered pair is a solution of a system of equations.

2   Solve a system of equations graphically.

3   Solve a system of linear equations algebraically using the method of substitution.

4   Solve a nonlinear system of equations algebraically using the method of substitution.

5   Solve an application problem using a system of equations.

---

1 Determine if an ordered pair is a solution of a system of equations.

Systems of Equations

Many problems in business and science involve **systems of equations.** These systems consist of two or more equations, each containing two or more variables.

$$ax + by = c \qquad\qquad \text{Equation 1}$$

$$dx + ey = f \qquad\qquad \text{Equation 2}$$

A **solution** of such a system is an ordered pair (x, y) of real numbers that satisfies *each* equation in the system. When you find the set of all solutions of the system of equations, you are **solving the system of equations.**

Example 1   Checking Solutions of a System of Equations

Which of the ordered pairs is a solution of the system: (a) $(3, 3)$ or (b) $(4, 2)$?

$$x + \ y = \ \ 6 \qquad\qquad \text{Equation 1}$$

$$2x - 5y = -2 \qquad\qquad \text{Equation 2}$$

Solution

a. To determine whether the ordered pair $(3, 3)$ is a solution of the system of equations, substitute $x = 3$ and $y = 3$ into *each* of the equations. Substituting into Equation 1 produces

$$3 + 3 = 6. \ \checkmark \qquad\qquad \text{Substitute 3 for } x \text{ and 3 for } y.$$

Substituting into Equation 2 produces

$$2(3) - 5(3) \neq -2. \ ✗ \qquad\qquad \text{Substitute 3 for } x \text{ and 3 for } y.$$

Because the ordered pair $(3, 3)$ fails to check in *both* equations, you can conclude that it *is not* a solution of the original system of equations.

b. By substituting $x = 4$ and $y = 2$ into the original equations, you can determine that the ordered pair $(4, 2)$ is a solution of the first equation

$$4 + 2 = 6 \ \checkmark \qquad\qquad \text{Substitute 4 for } x \text{ and 2 for } y.$$

and is also a solution of the second equation

$$2(4) - 5(2) = -2. \ \checkmark \qquad\qquad \text{Substitute 4 for } x \text{ and 2 for } y.$$

So, $(4, 2)$ *is* a solution of the original system of equations.

2 Solve a system of equations graphically.

Solving Systems of Equations by Graphing

A system of two equations in two variables can have exactly one solution, more than one solution, or no solution. In practice, you can gain insight about the location and number of solutions of a system of equations by sketching the graph of each equation in the same coordinate plane. The solutions of the system correspond to the **points of intersection** of the graphs. For instance, Figure 8.1 shows the graphs of three pairs (systems) of equations.

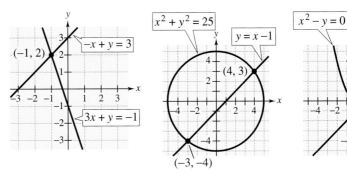

One solution   *Two solutions*   *No solution*

Figure 8.1

For a *general* system of equations, it is possible to have exactly one solution, two or more solutions, or no solution. For a system of *linear* equations, the second of the three possibilities is different.

Specifically, if a system of linear equations has two different solutions, it must have an *infinite* number of solutions. To see why this is true, consider the following graphical interpretations of a system of two linear equations in two variables. (Remember that the graph of a linear equation in two variables is a straight line.)

▶ **Graphical Interpretation of Solutions**

For a system of two linear equations in two variables, the number of solutions is given by one of the following.

Number of Solutions	*Graphical Interpretation*
1. Exactly one solution	The two lines intersect at one point. The system is called **consistent.**
2. Infinitely many solutions	The two lines are identical. The system is called **dependent (consistent).**
3. No solution	The two lines are parallel. The system is called **inconsistent.**

These three possibilities are shown in Figure 8.2 on page 484.

Note that the word *consistent* is used to mean that the system of linear equations has at least one solution, whereas the word *inconsistent* is used to mean that the system of linear equations has no solution.

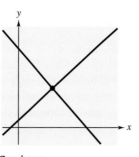

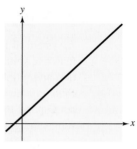

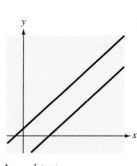

Consistent
Two lines intersect at a single point.
(Slopes are not equal.)

Dependent (consistent)
Two lines coincide and have infinitely many points of intersection.
(Slopes are equal.)

Inconsistent
Two lines are parallel and have no point of intersection.
(Slopes are equal.)

Figure 8.2

Study Tip

Note that for dependent systems, the slopes of the lines and the y-intercepts are equal. For inconsistent systems, the slopes are equal, but the y-intercepts of the two lines are different.

You can see from Figure 8.2 that a comparison of the slopes of two lines gives useful information about the number of solutions of the corresponding system of equations. So, to solve a system of equations graphically, it helps to begin by writing the equations in slope-intercept form.

Example 2 The Graphical Method of Solving a System

Use the graphical method to solve the system of equations.

$$2x + 3y = \ \ 7 \qquad \text{Equation 1}$$
$$2x - 5y = -1 \qquad \text{Equation 2}$$

Solution

Because both equations in the system are linear, you know that they have graphs that are straight lines. To sketch these lines, first write each equation in slope-intercept form, as follows.

$$y = -\frac{2}{3}x + \frac{7}{3} \qquad \text{Equation 1}$$

$$y = \frac{2}{5}x + \frac{1}{5} \qquad \text{Equation 2}$$

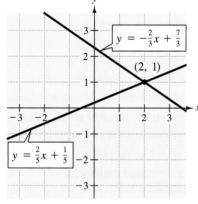

Figure 8.3

Because their slopes are not equal, you can conclude that the graphs will intersect at a single point. The lines corresponding to these two equations are shown in Figure 8.3. From this figure, it appears that the two lines intersect at the point $(2, 1)$. You can check these coordinates as follows.

Substitute into 1st Equation   *Substitute into 2nd Equation*

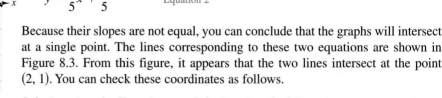

$$2x + 3y = 7 \qquad\qquad 2x - 5y = -1$$
$$2(2) + 3(1) \overset{?}{=} 7 \qquad\qquad 2(2) - 5(1) \overset{?}{=} -1$$
$$7 = 7 \ \checkmark \qquad\qquad\qquad -1 = -1 \ \checkmark$$

Because *both* equations are satisfied, the point $(2, 1)$ *is* a solution of the system.

3 Solve a system of linear equations algebraically using the method of substitution.

Solving Systems of Equations by Substitution

Solving a system of equations graphically is limited by the ability to sketch an accurate graph. It is difficult to obtain an accurate solution if one or both coordinates of a solution point are fractional or irrational. One analytic way to determine an exact solution to a system of two equations in two variables is to convert the system to *one* equation in *one* variable by an appropriate substitution.

> **Example 3**   The Method of Substitution: One-Solution Case

Solve the system of equations.

$$-x + y = 3 \qquad \text{Equation 1}$$
$$3x + y = -1 \qquad \text{Equation 2}$$

Solution

Begin by solving for y in Equation 1.

$$y = x + 3 \qquad \text{Revised Equation 1}$$

Next, substitute this expression for y into Equation 2.

$$3x + y = -1 \qquad \text{Equation 2}$$
$$3x + (x + 3) = -1 \qquad \text{Substitute } x + 3 \text{ for } y.$$
$$4x + 3 = -1 \qquad \text{Combine like terms.}$$
$$4x = -4 \qquad \text{Subtract 3 from both sides.}$$
$$x = -1 \qquad \text{Divide both sides by 4.}$$

At this point, you know that the x-coordinate of the solution is -1. To find the y-coordinate, *back-substitute* the x-value into the revised Equation 1.

$$y = x + 3 \qquad \text{Revised Equation 1}$$
$$y = -1 + 3 \qquad \text{Substitute } -1 \text{ for } x.$$
$$y = 2 \qquad \text{Simplify.}$$

The solution is $(-1, 2)$. Check this in the original system of equations.

Study Tip

The term **back-substitute** implies that you work backwards. After finding a value for one of the variables, substitute that value back into one of the equations in the original (or revised) system to find the value of the other variable.

When you use substitution, it does not matter which variable you solve for first. Whether you solve for y first or x first, you will obtain the same solution. When making your choice, you should choose the variable that is easier to work with. For instance, in the system

$$3x - 2y = 1 \qquad \text{Equation 1}$$
$$x + 4y = 3 \qquad \text{Equation 2}$$

it is easier to begin by solving for x in the second equation. But in the system

$$2x + y = 5 \qquad \text{Equation 1}$$
$$3x - 2y = 11 \qquad \text{Equation 2}$$

it is easier to begin by solving for y in the first equation.

The steps for using substitution are summarized as follows.

> ▶ **The Method of Substitution**
>
> 1. Solve one of the equations for one variable in terms of the other.
> 2. Substitute the expression obtained in Step 1 into the other equation to obtain an equation in one variable.
> 3. Solve the equation obtained in Step 2.
> 4. Back-substitute the solution from Step 3 into the expression obtained in Step 1 to find the value of the other variable.
> 5. Check the solution in the original system.

4 Solve a nonlinear system of equations algebraically using the method of substitution.

Solving Nonlinear Systems

Both of the equations in Example 3 are linear. That is, the variables x and y appear in the first power only. The method of substitution can also be used to solve a system of equations in which one or both of the equations are nonlinear.

Example 4 The Method of Substitution: Nonlinear Case

Solve the system of equations.

$$x^2 + y^2 = 25 \qquad\qquad \text{Equation 1}$$
$$-x + y = -1 \qquad\qquad \text{Equation 2}$$

Solution

In this system it is easier to solve for y in Equation 2.

$$y = x - 1 \qquad\qquad \text{Revised Equation 2}$$

Substitute this expression for y into Equation 1.

$$x^2 + (x - 1)^2 = 25 \qquad\qquad \text{Substitute } x - 1 \text{ for } y \text{ in Equation 1.}$$
$$x^2 + (x^2 - 2x + 1) = 25 \qquad\qquad \text{Use FOIL Method.}$$
$$2x^2 - 2x - 24 = 0 \qquad\qquad \text{Combine like terms.}$$
$$2(x - 4)(x + 3) = 0 \qquad\qquad \text{Factor.}$$
$$(x - 4) = 0 \quad\Longrightarrow\quad x = 4 \quad \text{Set factor equal to 0 and solve for } x.$$
$$(x + 3) = 0 \quad\Longrightarrow\quad x = -3 \quad \text{Set factor equal to 0 and solve for } x.$$

Finally, back-substitute these values of x into the revised second equation to solve for y. For $x = 4$, you have

$$y = 4 - 1 = 3 \qquad\qquad \text{Substitute 4 into revised Equation 2.}$$

which implies that $(4, 3)$ is a solution. For $x = -3$, you have

$$y = -3 - 1 = -4 \qquad\qquad \text{Substitute } -3 \text{ into revised Equation 2.}$$

which implies that $(-3, -4)$ is a solution. Check these in the original system.

5 Solve an application problem using a system of equations.

Applications

To model a real-life situation with a system of equations, you can use the same basic problem-solving strategy that has been used throughout the text.

Write a verbal model. → Assign labels. → Write an algebraic model. → Solve the algebraic model. → Answer the question.

After answering the question, remember to check the answer in the original statement of the problem.

Example 5   An Application

A roofing contractor bought 30 bundles of shingles and four rolls of roofing paper for $528. A second purchase (at the same prices) cost $140 for eight bundles of shingles and one roll of roofing paper. Find the price per bundle of shingles and the price per roll of roofing paper.

Solution

Verbal Model:
$$30\left(\begin{array}{c}\text{Price of}\\\text{a bundle}\end{array}\right) + 4\left(\begin{array}{c}\text{Price of}\\\text{a roll}\end{array}\right) = 528$$

$$8\left(\begin{array}{c}\text{Price of}\\\text{a bundle}\end{array}\right) + 1\left(\begin{array}{c}\text{Price of}\\\text{a roll}\end{array}\right) = 140$$

Labels: Price of bundle of shingles $= x$ (dollars)
Price of roll of roofing paper $= y$ (dollars)

System: $30x + 4y = 528$ Equation 1
$8x + y = 140$ Equation 2

Solving the second equation for y produces $y = 140 - 8x$, and substituting this expression into the first equation produces the following.

$3x + 4(140 - 8x) = 528$ Substitute $140 - 8x$ for y.

$30x + 560 - 32x = 528$ Distributive Property

$-2x = -32$ Simplify.

$x = 16$ Divide both sides by -2.

Back-substituting $x = 16$ into the revised second equation produces

$y = 140 - 8(16)$ Substitute 16 for x.

$= 12.$ Simplify.

So, you can conclude that the price of shingles is $16 per bundle and the price of roofing paper is $12 per roll. Check this in the original statement of the problem.

Check

Equation 1	*Equation 2*	
$30(16) + 4(12) \stackrel{?}{=} 528$	$8(16) + 12 \stackrel{?}{=} 140$	Substitute 16 for x and 12 for y.
$480 + 48 = 528$	$128 + 12 = 140$	Solution checks. ✓

In 1997, there were about 1.1 million newly constructed single-family houses completed. The median floor area was 1975 square feet.

Mark Joseph/Tony Stone Images

The total cost C of producing x units of a product usually has two components—the initial cost and the cost per unit. When enough units have been sold so that the total revenue R equals the total cost, the sales are said to have reached the **break-even point.** You can find this break-even point by setting C equal to R and solving for x. In other words, the break-even point corresponds to the point of intersection of the cost and revenue graphs.

Example 6   An Application: Break-Even Analysis

A small business invests $14,000 in equipment to produce a product. Each unit of the product costs $0.80 to produce and is sold for $1.50. How many items must be sold before the business breaks even?

Solution

Verbal Model:

| Total cost | = | Cost per unit | · | Number of units | + | Initial cost |

| Total revenue | = | Price per unit | · | Number of units |

Labels: Total cost = C (dollars)
            Cost per unit = 0.80   (dollars per unit)
            Number of units = x (units)
            Initial cost = 14,000   (dollars)
            Total revenue = R (dollars)
            Price per unit = 1.50   (dollars per unit)

System: $C = 0.80x + 14,000$ Equation 1

            $R = 1.50x$ Equation 2

Because the break-even point occurs when $R = C$, you have

$$1.5x = 0.8x + 14,000$$

$$0.7x = 14,000$$

$$x = 20,000.$$

So, it follows that the business must sell 20,000 units before it breaks even. Profit P (or loss) for the business can be determined by the equation $P = R - C$. Note in Figure 8.4 that sales less than the break-even point correspond to a loss for the business, whereas sales greater than the break-even point correspond to a profit for the business. The following table helps confirm this conclusion.

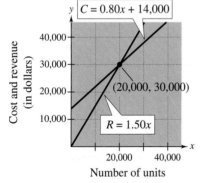

Figure 8.4

Units, x	0	5000	10,000	15,000	20,000	25,000
Revenue, R	$0	$7500	$15,000	$22,500	$30,000	$37,500
Cost, C	$14,000	$18,000	$22,000	$26,000	$30,000	$34,000
Profit, P	−$14,000	−$10,500	−$7000	−$3500	$0	$3500

Example 7   An Interest Rate Problem

A total of \$12,000 was invested in two funds paying 6% and 8% simple interest. If the interest for 1 year is \$880, how much of the \$12,000 was invested in each fund?

Solution

Verbal Model:

$$\boxed{\text{Amount in 6\% fund}} + \boxed{\text{Amount in 8\% fund}} = 12{,}000$$

$$6\% \cdot \boxed{\text{Amount in 6\% fund}} + 8\% \cdot \boxed{\text{Amount in 8\% fund}} = 880$$

Labels: Amount in 6% fund $= x$ (dollars)
         Amount in 8% fund $= y$ (dollars)

System: $x + y = 12{,}000$ Equation 1

          $0.06x + 0.08y = 880$ Equation 2

Begin by solving for x in the first equation.

$x + y = 12{,}000$ Equation 1

$x = 12{,}000 - y$ Revised Equation 1

Substituting this expression for x into the second equation produces the following.

$0.06x + 0.08y = 880$ Equation 2

$0.06(12{,}000 - y) + 0.08y = 880$ Substitute $12{,}000 - y$ for x.

$720 - 0.06y + 0.08y = 880$ Distributive Property

$0.02y = 160$ Simplify.

$y = 8000$ Simplify.

Back-substitute for y in the revised first equation.

$x = 12{,}000 - y$ Revised Equation 1

$x = 12{,}000 - 8000$ Substitute 8000 for y.

$x = 4000$ Simplify.

So, \$4000 was invested in the fund paying 6% and \$8000 was invested in the fund paying 8%. Check this in the original statement of the problem.

Discussing the Concept	Problem Posing

You want to create several systems of equations with relatively simple solutions that students can use for practice. Discuss how to create a system of equations that has a given solution. Illustrate your method by creating a system of linear equations that has one of the following solutions: (1, 4), (−2, 5), (−3, 1), or (4, 2). Trade your system for that of another student, and verify that each other's systems have the desired solutions.

8.1 Exercises

Integrated Review   *Concepts, Skills, and Problem Solving*

Keep mathematically in shape by doing these exercises *before* the problems of this section.

Properties and Definitions

1. Sketch the graph of a line with positive slope.

2. Sketch the graph of a line with undefined slope.

3. The slope of a line is $-\frac{2}{3}$. What is the slope of a line perpendicular to this line?

4. Two lines have slopes $m = -3$ and $m = \frac{3}{2}$. Which line is steeper? Explain.

Solving Equations

In Exercises 5–8, solve the equation and check your answer.

5. $y - 3(4y - 2) = 1$ **6.** $x + 6(3 - 2x) = 4$

7. $\frac{1}{2}x - \frac{1}{5}x = 15$ **8.** $\frac{1}{10}(x - 4) = 6$

In Exercises 9 and 10, solve for y in terms of x.

9. $3x + 4y - 5 = 0$

10. $-2x - 3y + 6 = 0$

Graphing

In Exercises 11–14, sketch the graph of the linear equation.

11. $y = -3x + 2$ **12.** $4x - 2y = -4$

13. $3x + 2y = 8$ **14.** $x + 3 = 0$

Developing Skills

In Exercises 1–8, determine whether each ordered pair is a solution of the system of equations. See Example 1.

1. $x + 2y = 9$
$-2x + 3y = 10$
(a) $(1, 4)$
(b) $(3, -1)$

2. $5x - 4y = 34$
$x - 2y = 8$
(a) $(0, 3)$
(b) $(6, -1)$

3. $-2x + 7y = 46$
$3x + y = 0$
(a) $(-3, 2)$
(b) $(-2, 6)$

4. $-5x - 2y = 23$
$x + 4y = -19$
(a) $(-3, -4)$
(b) $(3, 7)$

5. $4x - 5y = 12$
$3x + 2y = -2.5$
(a) $(8, 4)$
(b) $\left(\frac{1}{2}, -2\right)$

6. $2x - y = 1.5$
$4x - 2y = 3$
(a) $\left(0, -\frac{3}{2}\right)$
(b) $\left(2, \frac{5}{2}\right)$

7. $x^2 + y^2 = 169$
$17x - 7y = 169$
(a) $(5, -12)$
(b) $(-7, 10)$

8. $x^2 + y^2 = 17$
$x + 3y = 7$
(a) $(4, 1)$
(b) $(2, 1)$

In Exercises 9–16, determine whether the system is consistent or inconsistent.

9. $x + 2y = 6$
$x + 2y = 3$

10. $x - 2y = 3$
$2x - 4y = 7$

11. $2x - 3y = -12$
$-8x + 12y = -12$

12. $-5x + 8y = 8$
$7x - 4y = 14$

13. $-x + 4y = 7$
$3x - 12y = -21$

14. $3x + 8y = 28$
$-4x + 9y = 1$

15. $5x - 3y = 1$
$6x - 4y = -3$

16. $9x + 6y = 10$
$-6x - 4y = 3$

In Exercises 17–20, use a graphing utility to graph the equations in the system. Use the graphs to determine whether the system is consistent or inconsistent. If the system is consistent, determine the number of solutions.

17. $\frac{1}{3}x - \frac{1}{2}y = 1$
$-2x + 3y = 6$

18. $x + y = 5$
$x - y = 5$

19. $-2x + 3y = 6$
$x - y = -1$

20. $2x - 4y = 9$
$x - 2y = 4.5$

In Exercises 21–28, use the graphs of the equations to determine whether the system has any solutions. Find any solutions that exist.

21. $x + y = 4$
$\quad\;\; x + y = -1$

22. $-x + \;\; y = 5$
$\qquad\;\; x + 2y = 4$

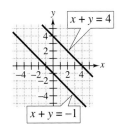

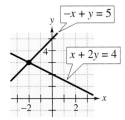

23. $5x - 3y = 4$
$\quad\;\; 2x + 3y = 3$

24. $\quad\;\; 2x - \;\; y = 4$
$\quad -4x + 2y = -12$

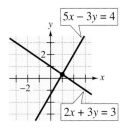

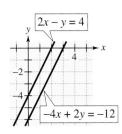

25. $\qquad\;\; x - 2y = -4$
$\quad -0.5x + \;\; y = \;\; 2$

26. $2x - 3y = 6$
$\quad\;\; 4x + 3y = 12$

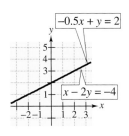

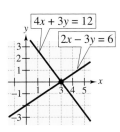

27. $x - 2y = 4$
$\quad\;\; x^2 - \;\; y = 0$

28. $\qquad\;\; y = \sqrt{x - 2}$
$\quad\;\; x - 2y = 1$

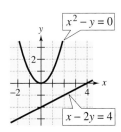

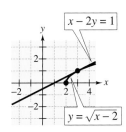

In Exercises 29–42, solve the system of linear equations by graphing. See Example 2.

29. $y = -x + 3$
$\quad\;\; y = \;\; x + 1$

30. $y = \;\; 2x - 4$
$\quad\;\; y = -\frac{1}{2}x + 1$

31. $x - y = 2$
$\quad\;\; x + y = 2$

32. $x - y = 0$
$\quad\;\; x + y = 4$

33. $3x - 4y = 5$
$\qquad\;\; x \qquad = 3$

34. $5x + 2y = 18$
$\qquad\qquad\; y = \;\; 2$

35. $4x + 5y = 20$
$\quad\;\; \frac{4}{5}x + \;\; y = \;\; 4$

36. $-x + 3y = 7$
$\quad\;\; 2x - 6y = 6$

37. $2x - 5y = 20$
$\quad\;\; 4x - 5y = 40$

38. $5x + 3y = 24$
$\quad\;\; x - 2y = 10$

39. $\;\; x + \;\; y = 2$
$\quad\;\; 3x + 3y = 6$

40. $4x - 3y = -3$
$\quad\;\; 8x - 6y = -6$

41. $4x + 5y = 7$
$\quad\;\; 2x - 3y = 9$

42. $7x + 4y = \;\; 6$
$\quad\;\; 5x - 3y = -25$

In Exercises 43–46, use a graphing utility to graph the equations and approximate any solutions of the system of equations.

43. $y = x^2$
$\quad\;\; y = 4x - x^2$

44. $y = 8 - x^2$
$\quad\;\; y = 6 - x$

45. $y = x^3$
$\quad\;\; y = x^3 - 3x^2 + 3x$

46. $y = x^2 - 2x$
$\quad\;\; y = x^3 - 4x$

In Exercises 47–76, solve the given system by substitution. See Examples 3 and 4.

47. $\;\; x - 2y = 0$
$\quad\;\; 3x + 2y = 8$

48. $\;\; x - \;\; y = 0$
$\quad\;\; 5x - 2y = 6$

49. $x \qquad = \;\; 4$
$\quad\;\; x - 2y = -2$

50. $\qquad\;\; y = \;\; 2$
$\quad\;\; x - 6y = -6$

51. $\;\; x + y = 3$
$\quad\;\; 2x - y = 0$

52. $-x + \;\; y = 5$
$\qquad\;\; x - 4y = 0$

53. $x + \;\; y = \;\; 2$
$\quad\;\; x - 4y = 12$

54. $x - 2y = -1$
$\quad\;\; x - 5y = \;\; 2$

55. $x + 6y = \;\; 19$
$\quad\;\; x - 7y = -7$

56. $\;\; x - 5y = -6$
$\quad\;\; 4x - 3y = \;\; 10$

57. $8x + \;\; 5y = 100$
$\quad\;\; 9x - 10y = \;\; 50$

58. $x + 4y = 300$
$\quad\;\; x - 2y = \;\; 0$

59. $-13x + 16y = \;\; 10$
$\qquad\;\; 5x + 16y = -26$

60. $2x + 5y = 29$
$\quad\;\; 5x + 2y = 13$

61. $4x - 14y = -15$
$18x - 12y = 9$

62. $5x - 24y = -12$
$17x - 24y = 36$

63. $\frac{1}{5}x + \frac{1}{2}y = 8$
$\phantom{\frac{1}{5}}x + \phantom{\frac{1}{2}}y = 20$

64. $\frac{1}{2}x + \frac{3}{4}y = 10$
$\phantom{\frac{1}{2}}\frac{3}{2}x - \phantom{\frac{3}{4}}y = 4$

65. $y = 2x^2$
$y = -2x + 12$

66. $y = 5x^2$
$y = -15x - 10$

67. $3x + 2y = 30$
$y = 3x^2$

68. $x + 2y = 16$
$x = 5y^2$

69. $x^2 + y = 9$
$x - y = -3$

70. $x - y^2 = 0$
$x - y = 2$

71. $x^2 + y^2 = 100$
$y + x = 2$

72. $x^2 + y^2 = 169$
$x + y = 7$

73. $x^2 - y = 2$
$3x + y = 2$

74. $x^2 + 2y = 6$
$x - y = -4$

75. $x^2 + y^2 = 25$
$2x - y = -5$

76. $x^2 - y^2 = 16$
$3x - y = 12$

In Exercises 77 and 78, use a graphing utility to graph each equation in the system. The graphs appear parallel. Yet, from the slope-intercept forms of the lines, you can find that the slopes are not equal and so the graphs intersect. Find the point of intersection of the two lines.

77. $x - 100y = -200$
$3x - 275y = 198$

78. $35x - 33y = 0$
$12x - 11y = 92$

Think About It   In Exercises 79–82, write a system of equations having the given solution.

79. $(4, 5)$

80. $(-2, 6)$

81. $(-1, -2)$

82. $\left(\frac{1}{2}, 3\right)$

Solving Problems

83. *Break-Even Analysis*   A small business invests $8000 in equipment to produce a product. Each unit of the product costs $1.20 to produce and is sold for $2.00. How many items must be sold before the business breaks even?

84. *Break-Even Analysis*   A business invests $50,000 in equipment to produce a product. Each unit of the product costs $19.25 to produce and is sold for $35.95. How many items must be sold before the business breaks even?

85. *Break-Even Analysis*   Suppose you are setting up a small business and have invested $10,000 to produce an item that will sell for $3.25. If each unit can be produced for $1.65, how many units must you sell to break even?

86. *Break-Even Analysis*   Suppose you are setting up a small business and have made an initial investment of $30,000. The unit cost of the product is $16.40, and the selling price is $31.40. How many units must you sell to break even?

87. *Simple Interest*   A combined total of $20,000 is invested in two bonds that pay 8% and 9.5% simple interest. The annual interest is $1675. How much is invested in each bond?

88. *Simple Interest*   A combined total of $12,000 is invested in two bonds that pay 8.5% and 10% simple interest. The annual interest is $1140. How much is invested in each bond?

89. *Simple Interest*   A combined total of $25,000 is invested in two funds paying 8% and 8.5% simple interest. The annual interest is $2060. How much is invested in each fund?

Number Problems   In Exercises 90–93, find two positive integers that satisfy the given requirements.

90. The sum of two numbers is 80 and their difference is 18.

91. The sum of the larger number and twice the smaller number is 61 and their difference is 7.

92. The sum of two numbers is 52 and the larger number is 8 less than twice the smaller number.

93. The sum of two numbers is 160 and the larger number is 3 times the smaller number.

Geometry   In Exercises 94–97, find the dimensions of the rectangle meeting the specified conditions.

Perimeter	Condition
94. 50 feet	The length is 5 feet greater than the width.
95. 320 inches	The width is 20 inches less than the length.
96. 68 yards	The width is $\frac{7}{10}$ of the length.
97. 90 meters	The length is $1\frac{1}{2}$ times the width.

98. *Dimensions of a Corral*   You have 250 feet of fencing to enclose two corrals of equal size (see figure). The combined area of the corrals is 2400 square feet. Find the dimensions of each corral.

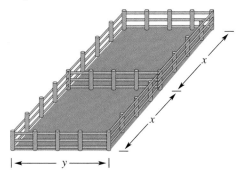

99. *Graphical Analysis*   From 1980 through 1996, the northeastern part of the United States grew at a slower rate than the western part. Models that represent the populations of the two regions are

$$E = 49{,}088.2 + 194.6t - 2.2t^2 \qquad Northeast$$

$$W = 43{,}132.9 + 977.3t \qquad West$$

where E and W are the populations in thousands and t is the calendar year, with $t = 0$ corresponding to 1980. Use a graphing utility to determine when the population of the West overtook the population of the Northeast. (Source: U.S. Bureau of the Census)

Explaining Concepts

100. What is meant by a solution of a system of equations in two variables?

101. List and explain the basic steps in solving a system of equations by substitution.

102. When solving a system of equations by substitution, how do you recognize that the system has no solution?

103. What does it mean to *back-substitute* when solving a system of equations?

104. Give a geometric description of the solution of a system of equations in two variables.

105. Describe any advantages of the method of substitution over the graphical method of solving a system of equations.

106. Is it possible for a consistent system of linear equations to have exactly two solutions? Explain.

8.2   Linear Systems in Two Variables

Objectives

1 Solve a system of linear equations algebraically using the method of elimination.

2 Use a system of equations to solve an application problem.

---

1 Solve a system of linear equations algebraically using the method of elimination.

The Method of Elimination

In Section 8.1, you studied two ways of solving a system of equations—substitution and graphing. In this section, you will study a third way—the **method of elimination.**

The key step in the method of elimination is to obtain, for one of the variables, coefficients that differ only in sign so that when the two equations are *added* this variable is eliminated. Notice how this is accomplished in Example 1—when the two equations are added, the y-terms are eliminated.

Example 1   The Method of Elimination

Solve the system of linear equations.

$$3x + 2y = 4 \qquad \text{Equation 1}$$
$$5x - 2y = 8 \qquad \text{Equation 2}$$

Solution

Begin by noting that the coefficients of y differ only in sign. By adding the two equations, you can eliminate y.

$$3x + 2y = 4 \qquad \text{Equation 1}$$
$$\underline{5x - 2y = 8} \qquad \text{Equation 2}$$
$$8x \phantom{{}- 2y} = 12 \qquad \text{Add equations.}$$

So, $x = \frac{3}{2}$. By back-substituting this value into the first equation, you can solve for y, as follows.

$$3x + 2y = 4 \qquad \text{Equation 1}$$
$$3\left(\frac{3}{2}\right) + 2y = 4 \qquad \text{Substitute } \tfrac{3}{2} \text{ for } x.$$
$$2y = -\frac{1}{2} \qquad \text{Simplify.}$$
$$y = -\frac{1}{4} \qquad \text{Divide both sides by 2.}$$

The solution is $\left(\frac{3}{2}, -\frac{1}{4}\right)$. Check this solution in the original system of linear equations.

The steps for solving a system of linear equations by the method of elimination are summarized as follows.

> ▶ **The Method of Elimination**
>
> 1. Obtain coefficients for x (or y) that differ only in sign by multiplying all terms of one or both equations by suitably chosen constants.
>
> 2. Add the equations to eliminate one variable and solve the resulting equation.
>
> 3. Back-substitute the value obtained in Step 2 into either of the original equations and solve for the other variable.
>
> 4. Check your solution in both of the original equations.

To obtain coefficients (for one of the variables) that differ only in sign, you often need to multiply one or both of the equations by a suitable constant.

Example 2 The Method of Elimination

Solve the system of linear equations.

$$4x - 5y = 13 \qquad \text{Equation 1}$$
$$3x - y = 7 \qquad \text{Equation 2}$$

Solution

To obtain coefficients of y that differ only in sign, multiply Equation 2 by -5.

$4x - 5y = 13$ ⟹	$4x - 5y = 13$	Equation 1
$3x - y = 7$ ⟹	$-15x + 5y = -35$	Multiply Equation 2 by -5.
	$-11x = -22$	Add equations.

So, $x = 2$. Back-substitute this value into Equation 2 and solve for y.

$$3x - y = 7 \qquad \text{Equation 2}$$
$$3(2) - y = 7 \qquad \text{Substitute 2 for } x.$$
$$y = -1 \qquad \text{Solve for } y.$$

The solution is $(2, -1)$. Check this in the original system of equations, as follows.

Substitute into 1st Equation

$$4x - 5y = 13 \qquad \text{Equation 1}$$
$$4(2) - 5(-1) \overset{?}{=} 13 \qquad \text{Substitute 2 for } x \text{ and } -1 \text{ for } y.$$
$$8 + 5 = 13 \qquad \text{Solution checks. } ✓$$

Substitute into 2nd Equation

$$3x - y = 7 \qquad \text{Equation 2}$$
$$3(2) - (-1) \overset{?}{=} 7 \qquad \text{Substitute 2 for } x \text{ and } -1 \text{ for } y.$$
$$6 + 1 = 7 \qquad \text{Solution checks. } ✓$$

Example 3 shows how the method of elimination can be used to determine that a system of linear equations has no solution.

Example 3 The Method of Elimination: No-Solution Case

Solve the system of linear equations.

$$3x + 9y = 8 \qquad \text{Equation 1}$$

$$2x + 6y = 7 \qquad \text{Equation 2}$$

Solution

To obtain coefficients of x that differ only in sign, multiply the first equation by 2 and multiply the second equation by -3.

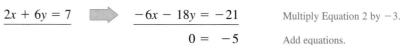

$3x + 9y = 8$	$6x + 18y = 16$ Multiply Equation 1 by 2.
$2x + 6y = 7$	$-6x - 18y = -21$ Multiply Equation 2 by -3.
	$0 = -5$ Add equations.

Because $0 = -5$ is a false statement, you can conclude that the system is inconsistent and has no solution. You can confirm this by graphing the lines corresponding to the two equations given in this system, as shown in Figure 8.5. Because the two lines are parallel, the system is inconsistent.

$2x + 6y = 7$

$3x + 9y = 8$

Figure 8.5

Example 4 shows how the method of elimination works with a system that has infinitely many solutions.

Example 4 The Method of Elimination: Many-Solutions Case

Solve the system of linear equations.

$$-2x + 6y = 3 \qquad \text{Equation 1}$$

$$4x - 12y = -6 \qquad \text{Equation 2}$$

Solution

To obtain coefficients of x that differ only in sign, multiply the first equation by 2.

$-2x + 6y = 3$	$-4x + 12y = 6$ Multiply Equation 1 by 2.
$4x - 12y = -6$	$4x - 12y = -6$ Equation 2
	$0 = 0$ Add equations.

Because $0 = 0$ is a true statement, the system has infinitely many solutions. You can confirm this by graphing the lines corresponding to the two equations, as shown in Figure 8.6. So, the solution set consists of all points (x, y) lying on the line $-2x + 6y = 3$.

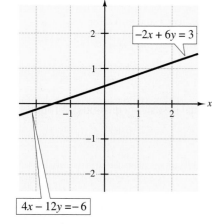

$-2x + 6y = 3$

$4x - 12y = -6$

Figure 8.6

2 Use a system of equations to solve an application problem.

Applications

To determine whether a real-life problem can be solved using a system of linear equations, consider the following. (1) Does the problem involve more than one unknown quantity? (2) Are there two (or more) equations or conditions to be satisfied? If one or both of these conditions occur, the appropriate mathematical model for the problem may be a system of linear equations.

Example 5   A Mixture Problem

A company with two stores buys six large delivery vans and five small delivery vans. The first store receives four of the large vans and two of the small vans for a total cost of $160,000. The second store receives two of the large vans and three of the small vans for a total cost of $128,000. What is the cost of each type of van?

Solution

The two unknowns in this problem are the costs of the two types of vans.

Verbal Model: $4 \left(\dfrac{\text{Cost of}}{\text{large van}} \right) + 2 \left(\dfrac{\text{Cost of}}{\text{small van}} \right) = \$160,000$

$2 \left(\dfrac{\text{Cost of}}{\text{large van}} \right) + 3 \left(\dfrac{\text{Cost of}}{\text{small van}} \right) = \$128,000$

Labels: Cost of large van $= x$ (dollars)
        Cost of small van $= y$ (dollars)

System: $4x + 2y = 160,000$ Equation 1

        $2x + 3y = 128,000$ Equation 2

To solve this system of linear equations, use the method of elimination. To obtain coefficients of x that differ only in sign, multiply the second equation by -2.

$4x + 2y = 160,000$ ⟹ $4x + 2y = 160,000$

$2x + 3y = 128,000$ ⟹ $\underline{-4x - 6y = -256,000}$ Multiply Equation 2 by -2.

$-4y = -96,000$ Add equations.

$y = 24,000$ Divide by -4.

The cost of each small van is $y = \$24,000$. Back-substitute this value into Equation 1 to find the cost of each large van.

$4x + 2y = 160,000$ Equation 1

$4x + 2(24,000) = 160,000$ Substitute 24,000 for y.

$4x = 112,000$ Simplify.

$x = 28,000$ Divide both sides by 4.

The cost of each large van is $x = \$28,000$. Check this solution in the original statement of the problem.

Example 6   An Application Involving Two Speeds

You take a motorboat trip on a river (18 miles upstream and 18 miles downstream). You run the motor at the same speed going up and down the river, but because of the river's current, the trip upstream takes $1\frac{1}{2}$ hours and the trip downstream takes only 1 hour. Determine the speed of the current.

Solution

Verbal Model:

| Boat speed (still water) | − | Speed of current | = | Upstream speed |

| Boat speed (still water) | + | Speed of current | = | Downstream speed |

Labels: Boat speed in still water $= x$ (miles per hour)
Current speed $= y$ (miles per hour)
Upstream speed $= 18/1.5 = 12$ (miles per hour)
Downstream speed $= 18/1 = 18$ (miles per hour)

System: $x - y = 12$ Equation 1
           $x + y = 18$ Equation 2

To solve this system of linear equations, use the method of elimination.

$$\begin{aligned} x - y &= 12 \qquad &\text{Equation 1}\\ \underline{x + y} &= \underline{18} \qquad &\text{Equation 2}\\ 2x \phantom{{}+y} &= 30 \qquad &\text{Add equations.}\end{aligned}$$

So, the speed of the boat in still water is 15 miles per hour. Back-substitute this value into Equation 2 to find the speed of the current.

$$15 + y = 18 \qquad \text{Substitute 15 for } x \text{ in Equation 2.}$$
$$y = 3 \qquad \text{Subtract 15 from both sides.}$$

The speed of the current is 3 miles per hour. Check this solution in the original statement of the problem.

Discussing the Concept   Fitting a Line to Data

The median purchase price of a home for first-time buyers in the United States in 1993 was $121,100, and in 1997 it was $135,400. Suppose you wish to find a linear equation that fits the points representing these data: (3, 121,100) and (7, 135,400). You remember that one form of the equation of a line is $y = mx + b$, but you can't remember how to use the methods for finding the equation of a line that you used in previous chapters. Write a system of equations that could be solved to find the values of m and b. Solve the system and write the linear equation that fits the data. Interpret the slope in the context of the data. By your model, what was the median purchase price of a home for first-time buyers in 1995? The actual median purchase price of a home for first-time buyers in 1995 was $128,300. How does the value obtained from your model compare? (Source: National Association of Realtors)

8.2   Exercises

Integrated Review   *Concepts, Skills, and Problem Solving*

Keep mathematically in shape by doing these exercises *before* the problems of this section.

Properties and Definitions

1. Name the property illustrated by $2(x + y) = 2x + 2y$.

2. What property of equality is demonstrated in solving the following equation?

$$x - 4 = 7$$
$$x - 4 + 4 = 7 + 4$$
$$x = 11$$

Solving Inequalities

In Exercises 3–8, solve the inequality.

3. $1 < 2x + 5 < 9$ **4.** $0 \leq \dfrac{x - 4}{2} < 6$

5. $|6x| > 12$ **6.** $|1 - 2x| < 5$

7. $4x - 12 < 0$ **8.** $4x + 4 \geq 9$

Problem Solving

9. The annual operating cost of a truck is $C = 0.45m + 6200$, where m is the number of miles traveled by the truck in a year. What number of miles will yield an annual operating cost that is less than $15,000?

10. You must select one of two plans of payment when working for a company. One plan pays $2500 per month. The second pays $1500 per month plus a commission of 4% of your gross sales. Write an inequality whose solution is such that the second option gives the greater monthly wage. Solve the inequality.

Developing Skills

In Exercises 1–12, solve the system of linear equations by elimination. Identify and label each line with its equation and label the point of intersection (if any). See Examples 1–4.

1. $2x + y = 4$
$x - y = 2$

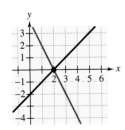

2. $x + 3y = 2$
$-x + 2y = 3$

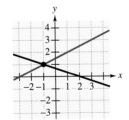

3. $-x + 2y = 1$
$x - y = 2$

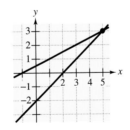

4. $x + y = 0$
$3x - 2y = 10$

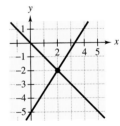

5. $3x + y = 3$
$2x - y = 7$

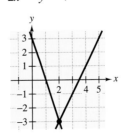

6. $-x + 2y = 2$
$3x + y = 15$

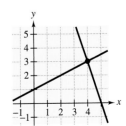

7. $x - y = 1$
$-3x + 3y = 8$

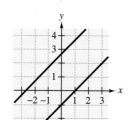

8. $3x + 4y = 2$
$0.6x + 0.8y = 1.6$

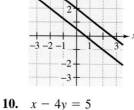

9. $x - 3y = 5$
$-2x + 6y = -10$

10. $x - 4y = 5$
$5x + 4y = 7$

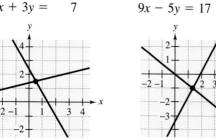

11. $2x - 8y = -11$
$5x + 3y = 7$

12. $3x + 4y = 0$
$9x - 5y = 17$

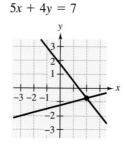

In Exercises 13–40, solve the system of linear equations by the method of elimination. See Examples 1–4.

13. $3x - 2y = 5$
$x + 2y = 7$

14. $-x + 2y = 9$
$x + 3y = 16$

15. $4x + y = -3$
$-4x + 3y = 23$

16. $-3x + 5y = -23$
$2x - 5y = 22$

17. $3x - 5y = 1$
$2x + 5y = 9$

18. $-x + 2y = 12$
$x + 6y = 20$

19. $5x + 2y = 7$
$3x - y = 13$

20. $4x + 3y = 8$
$x - 2y = 13$

21. $x - 3y = 2$
$3x - 7y = 4$

22. $2s - t = 9$
$3s + 4t = -14$

23. $2x + y = 9$
$3x - y = 16$

24. $7r - s = -25$
$2r + 5s = 14$

25. $2u + 3v = 8$
$3u + 4v = 13$

26. $4x - 3y = 25$
$-3x + 8y = 10$

27. $12x - 5y = 2$
$-24x + 10y = 6$

28. $-2x + 3y = 9$
$6x - 9y = -27$

29. $\frac{2}{3}r - s = 0$
$10r + 4s = 19$

30. $x - y = -\frac{1}{2}$
$4x - 48y = -35$

31. $0.05x - 0.03y = 0.21$
$x + y = 9$

32. $0.02x - 0.05y = -0.19$
$0.03x + 0.04y = 0.52$

33. $0.7u - v = -0.4$
$0.3u - 0.8v = 0.2$

34. $0.15x - 0.35y = -0.5$
$-0.12x + 0.25y = 0.1$

35. $5x + 7y = 25$
$x + 1.4y = 5$

36. $12b - 13m = 2$
$-6b + 6.5m = -2$

37. $\frac{3}{2}x - y = 4$
$-x + \frac{2}{3}y = -1$

38. $12x - 3y = 6$
$4x - y = 2$

39. $2x = 25$
$4x - 10y = 0.52$

40. $6x - 6y = 25$
$3y = 11$

In Exercises 41–48, solve the system by any convenient method.

41. $3x + 2y = 5$
$y = 2x + 13$

42. $4x + y = -2$
$-6x + y = 18$

43. $y = 5x - 3$
$y = -2x + 11$

44. $3y = 2x + 21$
$x = 50 - 4y$

45. $2x - y = 20$
$-x + y = -5$

46. $3x - 2y = -20$
$5x + 6y = 32$

47. $\frac{3}{2}x + 2y = 12$
$\frac{1}{4}x + y = 4$

48. $x + 2y = 4$
$\frac{1}{2}x + \frac{1}{3}y = 1$

In Exercises 49–54, decide whether the system is consistent or inconsistent.

49. $4x - 5y = 3$
$-8x + 10y = -6$

50. $4x - 5y = 3$
$-8x + 10y = 14$

51. $-2x + 5y = 3$
$5x + 2y = 8$

52. $x + 10y = 12$
$-2x + 5y = 2$

53. $-10x + 15y = 25$
$2x - 3y = -24$

54. $4x - 5y = 28$
$-2x + 2.5y = -14$

In Exercises 55 and 56, determine the value of k such that the system of linear equations is inconsistent.

55. $5x - 10y = 40$
$-2x + ky = 30$

56. $12x - 18y = 5$
$-18x + ky = 10$

In Exercises 57 and 58, find a system of linear equations that has the given solution. (There are many correct answers.)

57. $\left(3, -\frac{3}{2}\right)$

58. $(-8, 12)$

Solving Problems

59. *Break-Even Analysis*   You are planning to open a small business. You need an initial investment of $85,000. Each week your costs will be about $7400. If your projected weekly revenue is $8100, how many weeks will it take to break even?

60. *Break-Even Analysis*   A small business invests $8000 in equipment to produce a product. Each unit of the product costs $1.20 to produce and is sold for $2.00. How many units must be sold before the business breaks even?

61. *Simple Interest*   A combined total of $20,000 is invested in two bonds that pay 8% and 9.5% simple interest. The annual interest is $1675. How much is invested in each bond?

62. *Simple Interest*   A total of $4500 is invested in two funds paying 4% and 5% simple interest. The annual interest is $210. How much is invested in each fund?

63. *Average Speed*   A van travels for 2 hours at an average speed of 40 miles per hour. How much longer must the van travel at an average speed of 55 miles per hour so that the average speed for the total trip will be 50 miles per hour?

64. *Average Speed*   A truck travels for 4 hours at an average speed of 42 miles per hour. How much longer must the truck travel at an average speed of 55 miles per hour so that the average speed for the total trip will be 50 miles per hour?

65. *Air Speed*   An airplane flying into a headwind travels 1800 miles in 3 hours and 36 minutes. On the return flight, the same distance is traveled in 3 hours. Find the speed of the plane in still air and the speed of the wind, assuming that both remain constant throughout the round trip.

66. *Air Speed*   An airplane flying into a headwind travels the 3000-mile flying distance between two cities in 6 hours and 15 minutes. On the return flight, the distance is traveled in 5 hours. Find the speed of the plane in still air and the speed of the wind, assuming that both remain constant throughout the round trip.

67. *Ticket Sales*   Five hundred tickets were sold for a fundraising dinner. The receipts totaled $3312.50. Adult tickets were $7.50 each and children's tickets were $4.00 each. How many tickets of each type were sold?

68. *Ticket Sales*   A fundraising dinner was held on two consecutive nights. On the first night, 100 adult tickets and 175 children's tickets were sold, for a total of $937.50. On the second night, 200 adult tickets and 316 children's tickets were sold, for a total of $1790.00. Find the price of each type of ticket.

69. *Gasoline Mixture*   Twelve gallons of regular unleaded gasoline plus 8 gallons of premium unleaded gasoline cost $23.08. The price of premium unleaded is 11 cents more per gallon than the price of regular unleaded. Find the price per gallon for each grade of gasoline.

70. *Gasoline Mixture*   The total cost of 8 gallons of regular unleaded gasoline and 12 gallons of premium unleaded gasoline is $27.84. Premium unleaded gasoline costs $0.17 more per gallon than regular unleaded. Find the price per gallon for each grade of gasoline.

71. *Alcohol Mixture*   How many liters of a 40% alcohol solution must be mixed with how many liters of a 65% solution to obtain 20 liters of a 50% solution?

72. *Acid Mixture*   Fifty gallons of 70% acid solution is obtained by mixing an 80% solution with a 50% solution. How many gallons of each solution must be used to obtain the desired mixture?

73. *Nut Mixture* Ten pounds of mixed nuts sell for $6.95 per pound. The mixture is obtained from two kinds of nuts, with one variety priced at $5.65 per pound and the other at $8.95 per pound. How many pounds of each variety of nut were used in the mixture?

74. *Best-Fitting Line* The slope and y-intercept of the line $y = mx + b$ that best fits the three noncollinear points $(0, 0)$, $(1, 1)$, and $(2, 3)$ are given by the solution of the following system of linear equations.

$$5m + 3b = 7$$
$$3m + 3b = 4$$

(a) Solve the system and find the equation of the best-fitting line.

(b) Plot the three points and sketch the graph of the best-fitting line.

75. *Best-Fitting Line* The slope and y-intercept of the line $y = mx + b$ that best fits the three noncollinear points $(0, 4)$, $(1, 2)$, and $(2, 1)$ are given by the solution of the following system of linear equations.

$$3b + 3m = 7$$
$$3b + 5m = 4$$

(a) Solve the system and find the equation of the best-fitting line.

(b) Plot the three points and sketch the graph of the best-fitting line.

76. *U.S. Aircraft Industry* The average hourly wages for those employed in the aircraft industry in the United States from 1994 through 1996 are given in the table. (Source: U.S. Bureau of Labor Statistics)

Year	1994	1995	1996
Wage	$19.50	$19.97	$20.49

(a) Plot the data given in the table, where $x = 0$ corresponds to 1990.

(b) The line $y = mx + b$ that best fits the data is given by the solution of the following system.

$$3b + 15m = 59.96$$
$$15b + 77m = 300.79$$

Solve the system and find the equation of this line. Sketch the graph of the line on the same set of coordinate axes used in part (a).

(c) Explain the meaning of the slope of the line in the context of this problem.

77. *Vietnam Veterans Memorial* "The Wall" in Washington, D.C., designed by Maya Ling Lin when she was a student at Yale University, has two vertical, triangular sections of black granite with a common side (see figure). The top of each section is level with the ground. The bottoms of the two sections can be modeled by the equations $y = \frac{2}{25}x - 10$ and $y = -\frac{5}{61}x - 10$ when the x-axis is superimposed on the top of the wall. Each unit in the coordinate system represents 1 foot. How deep is the memorial at the point where the two sections meet? How long is each section?

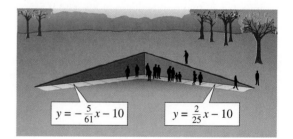

$$y = -\frac{5}{61}x - 10 \qquad y = \frac{2}{25}x - 10$$

Explaining Concepts

78. Answer parts (a)–(c) of Motivating the Chapter on page 481.

79. When solving a system by elimination, how do you recognize that it has infinitely many solutions?

80. Explain what is meant by an *inconsistent* system of linear equations.

81. In your own words, explain how to solve a system of linear equations by elimination.

82. How can you recognize that a system of linear equations has no solution? Give an example.

83. Under what conditions might substitution be better than elimination for solving a system of linear equations?

8.3   Linear Systems in Three Variables

Objectives

1 Solve a system of equations using row-echelon form with back-substitution.

2 Solve a system of linear equations using elimination with back-substitution.

3 Solve an application problem using elimination with back-substitution.

---

1 Solve a system of equations using row-echelon form with back-substitution.

Row-Echelon Form

The method of elimination can be applied to a system of linear equations in more than two variables. In fact, this method easily adapts to computer use for solving systems of linear equations with dozens of variables.

When the method of elimination is used to solve a system of linear equations, the goal is to rewrite the system in a form to which back-substitution can be applied. For instance, consider the following two systems of linear equations.

$$
\begin{aligned}
x - 2y + 2z &= 9 \\
-x + 3y &= -4 \\
2x - 5y + z &= 10
\end{aligned}
\qquad
\begin{aligned}
x - 2y + 2z &= 9 \\
y + 2z &= 5 \\
z &= 3
\end{aligned}
$$

The system on the right is said to be in **row-echelon form,** which means that it has a "stair-step" pattern with leading coefficients of 1. Which of these two systems do you think is easier to solve? After comparing the two systems, it should be clear that it is easier to solve the system on the right because the value of z is already shown and back-substitution will readily yield the values of x and y.

Example 1   Using Back-Substitution

Solve the system of linear equations.

$$
\begin{aligned}
x - 2y + 2z &= 9 \qquad &\text{Equation 1} \\
y + 2z &= 5 \qquad &\text{Equation 2} \\
z &= 3 \qquad &\text{Equation 3}
\end{aligned}
$$

Study Tip

When checking a solution, remember that the solution must satisfy each equation in the original system.

Solution

From Equation 3, you know the value of z. To solve for y, substitute $z = 3$ into Equation 2 to obtain

$$
\begin{aligned}
y + 2(3) &= 5 \qquad &\text{Substitute 3 for } z. \\
y &= -1. \qquad &\text{Solve for } y.
\end{aligned}
$$

Finally, substitute $y = -1$ and $z = 3$ into Equation 1 to obtain

$$
\begin{aligned}
x - 2(-1) + 2(3) &= 9 \qquad &\text{Substitute } -1 \text{ for } y \text{ and 3 for } z. \\
x &= 1. \qquad &\text{Solve for } x.
\end{aligned}
$$

The solution is $x = 1$, $y = -1$, and $z = 3$, which can also be written as the **ordered triple** $(1, -1, 3)$. Check this in the original system of equations.

2 Solve a system of linear equations using elimination with back-substitution.

The Method of Elimination

Two systems of equations are **equivalent** if they have the same solution set. To solve a system that is not in row-echelon form, first convert it to an *equivalent* system that is in row-echelon form. To see how this is done, let's take another look at the method of elimination, as applied to a system of two linear equations.

Example 2 The Method of Elimination

Solve the system of linear equations.

$$3x - 2y = -1 \qquad \text{Equation 1}$$
$$x - y = 0 \qquad \text{Equation 2}$$

Solution

$$x - y = 0 \qquad \text{You can interchange two equations in the system.}$$
$$3x - 2y = -1$$

$$-3x + 3y = 0 \qquad \text{Multiply the first equation by } -3.$$

$$-3x + 3y = 0 \qquad \text{You can add the multiple of the first equation to the}$$
$$\underline{3x - 2y = -1} \qquad \text{second equation to obtain a new second equation.}$$
$$y = -1$$

$$x - y = 0 \qquad \text{New system in row-echelon form}$$
$$y = -1$$

Using back-substitution, you can determine that the solution is $(-1, -1)$. Check the solution in each equation in the original system, as follows.

Equation 1	*Equation 2*
$3x - 2y = -1$	$x - y = 0$
$3(-1) - 2(-1) = -1$	$(-1) - (-1) = 0$

Rewriting a system of linear equations in row-echelon form usually involves a chain of equivalent systems, each of which is obtained by using one of the three basic row operations. This process is called **Gaussian elimination.**

> ▶ **Operations That Produce Equivalent Systems**
>
> Each of the following **row operations** on a system of linear equations produces an *equivalent* system of linear equations.
>
> **1.** Interchange two equations.
>
> **2.** Multiply one of the equations by a nonzero constant.
>
> **3.** Add a multiple of one of the equations to another equation to replace the latter equation.

Example 3   Using Elimination to Solve a System

Solve the system of linear equations.

$$x - 2y + 2z = 9 \qquad \text{Equation 1}$$
$$-x + 3y \qquad\quad = -4 \qquad \text{Equation 2}$$
$$2x - 5y + z = 10 \qquad \text{Equation 3}$$

Solution

There are many ways to begin, but we suggest saving the x in the upper left position, because it has a leading coefficient of 1, and eliminating the other x's from the first column.

$$x - 2y + 2z = 9$$
$$y + 2z = 5$$
$$2x - 5y + z = 10$$

> Adding the first equation to the second equation produces a new second equation.

$$x - 2y + 2z = 9$$
$$y + 2z = 5$$
$$-y - 3z = -8$$

> Adding -2 times the first equation to the third equation produces a new third equation.

Now that all but the first x have been eliminated from the first column, go to work on the second column. (You need to eliminate y from the third equation.)

$$x - 2y + 2z = 9$$
$$y + 2z = 5$$
$$-z = -3$$

> Adding the second equation to the third equation produces a new third equation.

Finally, you need a coefficient of 1 for z in the third equation.

$$x - 2y + 2z = 9$$
$$y + 2z = 5$$
$$z = 3$$

> Multiplying the third equation by -1 produces a new third equation.

This is the same system that was solved in Example 1, and, as in that example, you can conclude by back-substitution that the solution is

$$x = 1, \quad y = -1, \quad \text{and} \quad z = 3.$$

You can check the solution by substituting $x = 1$, $y = -1$, and $z = 3$ into each equation of the original system, as follows.

Equation 1: $\quad x - 2y + 2z = 9$
$$(1) - 2(-1) + 2(3) = 9 \checkmark$$

Equation 2: $\quad -x + 3y = -4$
$$-(1) + 3(-1) = -4 \checkmark$$

Equation 3: $\quad 2x - 5y + z = 10$
$$2(1) - 5(-1) + (3) = 10 \checkmark$$

Example 4 Using Elimination to Solve a System

Solve the following system of linear equations.

$$4x + y - 3z = 11 \qquad \text{Equation 1}$$
$$2x - 3y + 2z = 9 \qquad \text{Equation 2}$$
$$x + y + z = -3 \qquad \text{Equation 3}$$

Solution

$$x + y + z = -3$$
$$2x - 3y + 2z = 9$$
$$4x + y - 3z = 11$$

Interchange the first and third equations.

$$x + y + z = -3$$
$$-5y = 15$$
$$4x + y - 3z = 11$$

Adding -2 times the first equation to the second equation produces a new second equation.

$$x + y + z = -3$$
$$-5y = 15$$
$$-3y - 7z = 23$$

Adding -4 times the first equation to the third equation produces a new third equation.

$$x + y + z = -3$$
$$y = -3$$
$$-3y - 7z = 23$$

Multiplying the second equation by $-\frac{1}{5}$ produces a new second equation.

$$x + y + z = -3$$
$$y = -3$$
$$-7z = 14$$

Adding 3 times the second equation to the third equation produces a new third equation.

$$x + y + z = -3$$
$$y = -3$$
$$z = -2$$

Multiplying the third equation by $-\frac{1}{7}$ produces a new third equation.

Now you can see that $z = -2$ and $y = -3$. Moreover, by back-substituting these values into Equation 1, you can determine that $x = 2$. So, the solution is $x = 2$, $y = -3$, and $z = -2$. You can check this solution as follows.

$$\text{Equation 1:} \quad 4x + y - 3z = 11$$
$$4(2) + (-3) - 3(-2) = 11 \ \checkmark$$

$$\text{Equation 2:} \quad 2x - 3y + 2z = 9$$
$$2(2) - 3(-3) + 2(-2) = 9 \ \checkmark$$

$$\text{Equation 3:} \quad x + y + z = -3$$
$$(2) + (-3) + (-2) = -3 \ \checkmark$$

The next example involves an inconsistent system—one that has no solution. The key to recognizing an inconsistent system is that at some stage in the elimination process, you obtain a false statement such as $0 = -2$. Watch for such statements when you do the exercises for this section.

Example 5   An Inconsistent System

Solve the system of linear equations.

$$
\begin{aligned}
x - 3y + z &= 1 \qquad &\text{Equation 1}\\
2x - y - 2z &= 2 \qquad &\text{Equation 2}\\
x + 2y - 3z &= -1 \qquad &\text{Equation 3}
\end{aligned}
$$

Solution

$$
\begin{aligned}
x - 3y + z &= 1\\
5y - 4z &= 0\\
x + 2y - 3z &= -1
\end{aligned}
$$

Adding -2 times the first equation to the second equation produces a new second equation.

$$
\begin{aligned}
x - 3y + z &= 1\\
5y - 4z &= 0\\
5y - 4z &= -2
\end{aligned}
$$

Adding -1 times the first equation to the third equation produces a new third equation.

$$
\begin{aligned}
x - 3y + z &= 1\\
5y - 4z &= 0\\
0 &= -2
\end{aligned}
$$

Adding -1 times the second equation to the third equation produces a new third equation.

Because the third "equation" is impossible, you can conclude that this system is inconsistent and therefore has no solution. Moreover, because this system is equivalent to the original system, you can conclude that the original system also has no solution.

As with a system of linear equations in two variables, the solution(s) of a system of linear equations in more than two variables must fall into one of three categories.

▶ **The Number of Solutions of a Linear System**

For a system of linear equations, exactly one of the following is true.

1. There is exactly one solution.

2. There are infinitely many solutions.

3. There is no solution.

Example 6 A System with Infinitely Many Solutions

Solve the system of linear equations.

$$
\begin{aligned}
x + y - 3z &= -1 \qquad &\text{Equation 1} \\
y - z &= 0 \qquad &\text{Equation 2} \\
-x + 2y &= 1 \qquad &\text{Equation 3}
\end{aligned}
$$

Solution

Begin by rewriting the system in row-echelon form.

$$
\begin{aligned}
x + y - 3z &= -1 \\
y - z &= 0 \\
3y - 3z &= 0
\end{aligned}
$$

Adding the first equation to the third equation produces a new third equation.

$$
\begin{aligned}
x + y - 3z &= -1 \\
y - z &= 0 \\
0 &= 0
\end{aligned}
$$

Adding -3 times the second equation to the third equation produces a new third equation.

This means that Equation 3 depends on Equations 1 and 2 in the sense that it gives us no additional information about the variables. So, the original system is equivalent to the system

$$
\begin{aligned}
x + y - 3z &= -1 \\
y - z &= 0.
\end{aligned}
$$

In this last equation, solve for y in terms of z to obtain $y = z$. Back-substituting for y in the previous equation produces $x = 2z - 1$. Finally, letting $z = a$, where a is any real number, the solutions to the given system are all of the form

$$x = 2a - 1, \; y = a, \text{ and } z = a.$$

So, every ordered triple of the form

$$(2a - 1, a, a), \qquad a \text{ is a real number}$$

is a solution of the system.

In Example 6, there are other ways to write the same infinite set of solutions. For instance, the solutions could have been written as

$$\left(b, \frac{1}{2}(b + 1), \frac{1}{2}(b + 1)\right), \qquad b \text{ is a real number.}$$

Try convincing yourself of this by substituting $a = 0$, $a = 1$, $a = 2$, and $a = 3$ into the solution listed in Example 6. Then substitute $b = -1$, $b = 1$, $b = 3$, and $b = 5$ into the solution listed above. In both cases, you should obtain the same ordered triples. So, when comparing descriptions of an infinite solution set, keep in mind that there is more than one way to describe the set.

3 Solve an application problem using elimination with back-substitution.

Applications

Example 7   Position Equation

The height at time t of an object that is moving in a (vertical) line with constant acceleration a is given by the **position equation**

$$s = \frac{1}{2}at^2 + v_0 t + s_0.$$

The height s is measured in feet, the acceleration a is measured in feet per second squared, the time t is measured in seconds, v_0 is the initial velocity (at time $t = 0$), and s_0 is the initial height. Find the values of a, v_0, and s_0, if $s = 164$ feet at 1 second, $s = 180$ feet at 2 seconds, and $s = 164$ feet at 3 seconds.

Solution

By substituting the three values of t and s into the position equation, you obtain three linear equations in a, v_0, and s_0.

When $t = 1$, $s = 164$: $\quad \frac{1}{2}a(1)^2 + v_0(1) + s_0 = 164$

When $t = 2$, $s = 180$: $\quad \frac{1}{2}a(2)^2 + v_0(2) + s_0 = 180$

When $t = 3$, $s = 164$: $\quad \frac{1}{2}a(3)^2 + v_0(3) + s_0 = 164$

By multiplying the first and third equations by 2, this system can be rewritten

$$a + 2v_0 + 2s_0 = 328 \qquad \text{Equation 1}$$
$$2a + 2v_0 + s_0 = 180 \qquad \text{Equation 2}$$
$$9a + 6v_0 + 2s_0 = 328 \qquad \text{Equation 3}$$

and you can apply elimination to obtain

$$a + 2v_0 + 2s_0 = 328$$
$$-2v_0 - 3s_0 = -476$$
$$2s_0 = 232.$$

From the third equation, $s_0 = 116$, so that back-substitution into the second equation yields

$$-2v_0 - 3(116) = -476$$
$$-2v_0 = -128$$
$$v_0 = 64.$$

Finally, back-substituting $s_0 = 116$ and $v_0 = 64$ into the first equation yields

$$a + 2(64) + 2(116) = 328$$
$$a = -32.$$

So, the position equation for this object is $s = -16t^2 + 64t + 116$.

Example 8   Data Analysis: Curve-Fitting

Find a quadratic equation

$$y = ax^2 + bx + c$$

whose graph passes through the points $(-1, 3)$, $(1, 1)$, and $(2, 6)$.

Solution

Because the graph of $y = ax^2 + bx + c$ passes through the points $(-1, 3)$, $(1, 1)$, and $(2, 6)$, you can write the following.

When $x = -1$, $y = 3$: $a(-1)^2 + b(-1) + c = 3$

When $x = 1$, $y = 1$: $a(1)^2 + b(1) + c = 1$

When $x = 2$, $y = 6$: $a(2)^2 + b(2) + c = 6$

This produces the following system of linear equations.

$$a - b + c = 3 \qquad \text{Equation 1}$$
$$a + b + c = 1 \qquad \text{Equation 2}$$
$$4a + 2b + c = 6 \qquad \text{Equation 3}$$

The solution of this system is $a = 2$, $b = -1$, and $c = 0$. So, the equation of the parabola is

$$y = 2x^2 - x$$

as shown in Figure 8.7.

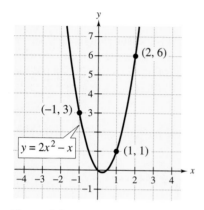

Figure 8.7

Discussing the Concept   Fitting a Quadratic Model

The data in the table represent the United States government's annual net receipts y (in billions of dollars) from individual income taxes for the year x from 1994 through 1996, where $x = 4$ corresponds to 1994. (Source: U.S. Office of Management and Budget)

x	4	5	6
y	543	590	656

Use a system of three linear equations to find a quadratic model that fits the data. According to your model, what were the annual net receipts from individual income taxes in 1997? The actual annual net receipts for 1997 were $737 billion. How does the value obtained from your quadratic model compare? Suppose you had been involved in planning the 1997 federal budget and had used this model to estimate how much federal income could be expected from 1997 individual income taxes. When you review the actual 1997 tax receipts and see that the model wasn't completely accurate, how do you evaluate the model's prediction performance? Are you satisfied with it? Why or why not?

8.3   Exercises

Integrated Review   *Concepts, Skills, and Problem Solving*

Keep mathematically in shape by doing these exercises *before* the problems of this section.

Properties and Definitions

1. A linear equation of the form $2x + 8 = 7$ has how many solutions?

2. What is the usual first step in solving an equation such as

$$\frac{t}{6} + \frac{5}{8} = \frac{7}{4}?$$

Simplifying Expressions

In Exercises 3–6, simplify the expression. (Assume all variables are positive.)

3. $4x^2(x^3)^2$

4. $(2x^2y)^3(xy^3)^4$

5. $\dfrac{8x^{-4}}{2x^7}$

6. $\left(\dfrac{t^4}{3}\right)^{-1}$

Solving Equations

7. $|2x - 4| = 6$

8. $\frac{1}{4}(5 - 2x) = 9x - 7x$

Models and Graphs

9. The speed of a ship is 15 knots. Write the distance d the ship travels as a function of time t. Graph the model.

10. The length of each edge of a cube is s inches. Write the volume V of the cube as a function of s.

11. Express the area A of a circle as a function of its circumference C.

Developing Skills

In Exercises 1 and 2, determine whether each ordered triple is a solution of the system of linear equations.

1.
$$
\begin{aligned}
x + 3y + 2z &= 1 \\
5x - y + 3z &= 16 \\
-3x + 7y + z &= -14
\end{aligned}
$$
   (a) $(0, 3, -2)$ (b) $(12, 5, -13)$
   (c) $(1, -2, 3)$ (d) $(-2, 5, -3)$

2.
$$
\begin{aligned}
3x - y + 4z &= -10 \\
-x + y + 2z &= 6 \\
2x - y + z &= -8
\end{aligned}
$$
   (a) $(-2, 4, 0)$ (b) $(0, -3, 10)$
   (c) $(1, -1, 5)$ (d) $(7, 19, -3)$

In Exercises 3–6, use back-substitution to solve the system of linear equations. See Example 1.

3.
$$
\begin{aligned}
x - 2y + 4z &= 4 \\
3y - z &= 2 \\
z &= -5
\end{aligned}
$$

4.
$$
\begin{aligned}
5x + 4y - z &= 0 \\
10y - 3z &= 11 \\
z &= 3
\end{aligned}
$$

5.
$$
\begin{aligned}
x - 2y + 4z &= 4 \\
y &= 3 \\
y + z &= 2
\end{aligned}
$$

6.
$$
\begin{aligned}
x &= 10 \\
3x + 2y &= 2 \\
x + y + 2z &= 0
\end{aligned}
$$

In Exercises 7 and 8, determine whether the two systems of linear equations are equivalent. Give reasons for your answer.

7.
$$
\begin{aligned}
x + 3y - z &= 6 \\
2x - y + 2z &= 1 \\
3x + 2y - z &= 2
\end{aligned}
\qquad
\begin{aligned}
x + 3y - z &= 6 \\
-7y + 4z &= 1 \\
-7y - 4z &= -16
\end{aligned}
$$

8.
$$
\begin{aligned}
x - 2y + 3z &= 9 \\
-x + 3y &= -4 \\
2x - 5y + 5z &= 17
\end{aligned}
\qquad
\begin{aligned}
x - 2y + 3z &= 9 \\
y + 3z &= 5 \\
-y - z &= -1
\end{aligned}
$$

In Exercises 9–12, perform the row operation and write the equivalent system of linear equations. See Example 2.

9. Add Equation 1 to Equation 2.

$$x - 2y = 8 \qquad \text{Equation 1}$$
$$-x + 3y = 6 \qquad \text{Equation 2}$$

What did this operation accomplish?

10. Add -2 times Equation 1 to Equation 2.

$$2x + 3y = 7 \qquad \text{Equation 1}$$
$$4x - 2y = -2 \qquad \text{Equation 2}$$

What did this operation accomplish?

11. Add Equation 1 to Equation 2.

$$x - 2y + 3z = 5 \qquad \text{Equation 1}$$
$$-x + y + 5z = 4 \qquad \text{Equation 2}$$
$$2x \quad\quad - 3z = 0 \qquad \text{Equation 3}$$

What did this operation accomplish?

12. Add -2 times Equation 1 to Equation 3.

$$x - 2y + 3z = 5 \qquad \text{Equation 1}$$
$$-x + y + 5z = 4 \qquad \text{Equation 2}$$
$$2x \quad\quad - 3z = 0 \qquad \text{Equation 3}$$

What did this operation accomplish?

In Exercises 13–40, solve the system of linear equations. See Examples 3–6.

13.
$$x \quad + z = 4$$
$$y \quad\quad = 2$$
$$4x \quad + z = 7$$

14.
$$x \quad\quad\quad = 3$$
$$-x + 3y \quad\quad = 3$$
$$y + 2z = 4$$

15.
$$x + y + z = 6$$
$$2x - y + z = 3$$
$$3x \quad\quad - z = 0$$

16.
$$x + y + z = 2$$
$$-x + 3y + 2z = 8$$
$$4x + y \quad\quad = 4$$

17.
$$x + y + z = -3$$
$$4x + y - 3z = 11$$
$$2x - 3y + 2z = 9$$

18.
$$x - y + 2z = -4$$
$$3x + y - 4z = -6$$
$$2x + 3y - 4z = 4$$

19.
$$x + 2y + 6z = 5$$
$$-x + y - 2z = 3$$
$$x - 4y - 2z = 1$$

20.
$$x + 6y + 2z = 9$$
$$3x - 2y + 3z = -1$$
$$5x - 5y + 2z = 7$$

21.
$$2x \quad\quad + 2z = 2$$
$$5x + 3y \quad\quad = 4$$
$$3y - 4z = 4$$

22.
$$6y + 4z = -12$$
$$3x + 3y \quad\quad = 9$$
$$2x \quad\quad - 3z = 10$$

23.
$$x + y + 8z = 3$$
$$2x + y + 11z = 4$$
$$x \quad\quad + 3z = 0$$

24.
$$2x - 4y + z = 0$$
$$3x \quad\quad + 2z = -1$$
$$-6x + 3y + 2z = -10$$

25.
$$2x + y + 3z = 1$$
$$2x + 6y + 8z = 3$$
$$6x + 8y + 18z = 5$$

26.
$$3x - y - 2z = 5$$
$$2x + y + 3z = 6$$
$$6x - y - 4z = 9$$

27.
$$y + z = 5$$
$$2x \quad\quad + 4z = 4$$
$$2x - 3y \quad\quad = -14$$

28.
$$5x + 2y \quad\quad = -8$$
$$z = 5$$
$$3x - y + z = 9$$

29.
$$2x + 6y - 4z = 8$$
$$3x + 10y - 7z = 12$$
$$-2x - 6y + 5z = -3$$

30.
$$x + 2y - 2z = 4$$
$$2x + 5y - 7z = 5$$
$$3x + 7y - 9z = 10$$

31.
$$2x \quad\quad + z = 1$$
$$5y - 3z = 2$$
$$6x + 20y - 9z = 11$$

32.
$$2x + y - z = 4$$
$$y + 3z = 2$$
$$3x + 2y \quad\quad = 4$$

33.
$$3x + y + z = 2$$
$$4x \quad\quad + 2z = 1$$
$$5x - y + 3z = 0$$

34.
$$2x \quad\quad + 3z = 4$$
$$5x + y + z = 2$$
$$11x + 3y - 3z = 0$$

35.
$$0.2x + 1.3y + 0.6z = 0.1$$
$$0.1x \quad\quad + 0.3z = 0.7$$
$$2x + 10y + 8z = 8$$

36.
$$0.3x - 0.1y + 0.2z = 0.35$$
$$2x + y - 2z = -1$$
$$2x + 4y + 3z = 10.5$$

37.
$$x + 4y - 2z = 2$$
$$-3x + y + z = -2$$
$$5x + 7y - 5z = 6$$

38.
$$x - 2y - z = 3$$
$$2x + y - 3z = 1$$
$$x + 8y - 3z = -7$$

39.
$$-4x + y + 0.2z = 6$$
$$6x - 3y + 0.5z = -4$$
$$-8x + 2y + 0.6z = 14$$

40.
$$x + 6y + 2z = 9$$
$$3x - 2y + 3z = -1$$
$$5x - 5y + 2z = 7$$

In Exercises 41 and 42, find a system of linear equations in three variables that has the given point as a solution. (*Note:* There are many correct answers.)

41. $(4, -3, 2)$

42. $(5, 7, -10)$

Solving Problems

Vertical Motion   In Exercises 43–46, find the position equation $s = \frac{1}{2}at^2 + v_0t + s_0$ for an object that has the indicated heights at the specified times.

43. $s = 128$ feet at $t = 1$ second

   $s = 80$ feet at $t = 2$ seconds

   $s = 0$ feet at $t = 3$ seconds

44. $s = 48$ feet at $t = 1$ second

   $s = 64$ feet at $t = 2$ seconds

   $s = 48$ feet at $t = 3$ seconds

45. $s = 32$ feet at $t = 1$ second

   $s = 32$ feet at $t = 2$ seconds

   $s = 0$ feet at $t = 3$ seconds

46. $s = 10$ feet at $t = 0$ second

   $s = 54$ feet at $t = 1$ second

   $s = 46$ feet at $t = 3$ seconds

Curve-Fitting   In Exercises 47–52, find a quadratic equation $y = ax^2 + bx + c$ whose graph passes through the points.

47. $(0, -4), (1, 1), (2, 10)$ **48.** $(0, 5), (1, 6), (2, 5)$

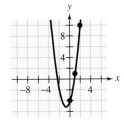
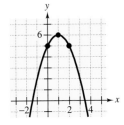

49. $(1, 0), (2, -1), (3, 0)$ **50.** $(1, 2), (2, 1), (3, -4)$

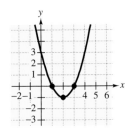
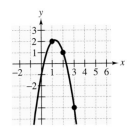

51. $(-1, -3), (1, 1), (2, 0)$

52. $(-1, -1), (1, 1), (2, -4)$

53. *Diagonals of a Polygon*   The total numbers of sides and diagonals of regular polygons with three, four, and five sides are 3, 6, and 10, as shown in the figure. Find a quadratic function $y = ax^2 + bx + c$, where x is the number of sides in the polygon, that fits these data. Does it give the correct answer for a polygon with six sides?

3   6

10   15

54. *Graphical Estimation*   The table gives the numbers y of metric tons of newsprint (in thousands) produced in the years 1993 through 1995 in the United States.   (Source: American Forest and Paper Association)

Year	1993	1994	1995
y	6412	6336	6352

(a) Find a quadratic equation $y = at^2 + bt + c$ whose graph passes through the three points, letting $t = 0$ correspond to 1990.

(b) Use a graphing utility to graph the model found in part (a).

(c) Use the model in part (a) to predict newsprint production in the year 2000 if the trend continues.

Curve-Fitting   In Exercises 55–60, find the equation of the circle $x^2 + y^2 + Dx + Ey + F = 0$ that passes through the points.

55. $(0, 0), (2, -2), (4, 0)$ **56.** $(0, 0), (0, 6), (-3, 3)$

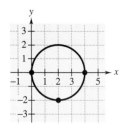   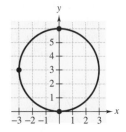

57. $(3, -1), (-2, 4), (6, 8)$ **58.** $(0, 0), (0, 2), (3, 0)$

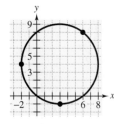   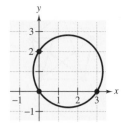

59. $(-3, 5), (4, 6), (5, 5)$

60. $(5, 13), (17, 5), (10, 12)$

61. *Crop Spraying*   A mixture of 12 gallons of chemical A, 16 gallons of chemical B, and 26 gallons of chemical C is required to kill a certain destructive crop insect. Commercial spray X contains 1, 2, and 2 parts, respectively, of these chemicals. Spray Y contains only chemical C. Spray Z contains only chemicals A and B in equal amounts. How much of each type of commercial spray is needed to get the desired mixture?

62. *Chemistry*   A chemist needs 10 liters of a 25% acid solution. It is mixed from three solutions whose concentrations are 10%, 20%, and 50%. How many liters of each solution will satisfy the following?

(a) Use 2 liters of the 50% solution.

(b) Use as little as possible of the 50% solution.

(c) Use as much as possible of the 50% solution.

63. *School Orchestra*   The table shows the percents of each section of the North High School orchestra that were chosen to participate in the city orchestra, the county orchestra, and the state orchestra. Thirty members of the city orchestra, 17 members of the county orchestra, and 10 members of the state orchestra are from North High. How many members are in each section of North High's orchestra?

Orchestra	String	Wind	Percussion
City orchestra	40%	30%	50%
County orchestra	20%	25%	25%
State orchestra	10%	15%	25%

64. *Rewriting a Fraction*   The fraction $1/(x^3 - x)$ can be written as a sum of three fractions as follows.

$$\frac{1}{x^3 - x} = \frac{A}{x} + \frac{B}{x + 1} + \frac{C}{x - 1}.$$

The numbers $A, B,$ and C are the solutions of the system

$$\begin{aligned} A + B + C &= 0 \\ -B + C &= 0 \\ -A \qquad\qquad &= 1. \end{aligned}$$

Solve the system and write the expression as the sum of three fractions.

Explaining Concepts

65. Answer parts (d)–(f) of Motivating the Chapter on page 481.

66. Give an example of a system of linear equations that is in row-echelon form.

67. Show how to use back-substitution to solve the system you found in Exercise 66.

68. Describe the row operations that are performed on a system of linear equations to produce an equivalent system of equations.

69. Write a system of four linear equations in four unknowns, and solve it by elimination.

Mid-Chapter Quiz

Take this quiz as you would take a quiz in class. After you are done, check your work against the answers given in the back of the book.

1. Which is the solution of the system $5x - 12y = 2$ and $2x + 1.5y = 26$: $(1, -2)$ or $(10, 4)$? Explain your reasoning.

In Exercises 2–4, graph the equations in the system. Use the graphs to determine the number of solutions of the system.

2. $-6x + 9y = 9$
$2x - 3y = 6$

3. $x - 2y = -4$
$3x - 2y = 4$

4. $y = x - 1$
$y = 1 + 2x - x^2$

In Exercises 5–8, solve the system of equations graphically.

5. $x \quad = 4$
$2x - y = 6$

6. $y = \frac{1}{3}(1 - 2x)$
$y = \frac{1}{3}(5x - 13)$

7. $2x + 7y = 16$
$3x + 2y = 24$

8. $7x - 17y = -169$
$x^2 + y^2 = 169$

In Exercises 9–12, use substitution to solve the system.

9. $2x - 3y = 4$
$y = 2$

10. $y = 5 - x^2$
$y = 2(x + 1)$

11. $5x - y = 32$
$6x - 9y = 18$

12. $0.2x + 0.7y = 8$
$-x + 2y = 15$

In Exercises 13–16, use Gaussian elimination to solve the linear system.

13. $x + 10y = 18$
$5x + 2y = 42$

14. $3x + 11y = 38$
$7x - 5y = -34$

15. $a + b + c = 1$
$4a + 2b + c = 2$
$9a + 3b + c = 4$

16. $x \quad + 4z = 17$
$-3x + 2y - z = -20$
$x - 5y + 3z = 19$

In Exercises 17 and 18, find a system of linear equations that has the unique solution. (There are many correct answers.)

17. $(10, -12)$

18. $(1, 3, -7)$

19. Twenty gallons of a 30% brine solution is obtained by mixing a 20% solution with a 50% solution. Let x represent the number of gallons of the 20% solution and let y represent the number of gallons of the 50% solution. Write a system of equations that models this problem and solve the system.

20. Find the equation of the parabola $y = ax^2 + bx + c$ that passes through the points $(1, 2)$, $(-1, -4)$, and $(2, 8)$.

8.4 Matrices and Linear Systems

Objectives

1 Form a coefficient and an augmented matrix and form a linear system from the augmented matrix.

2 Perform elementary row operations to solve a system of linear equations.

3 Use matrices and Gaussian elimination with back-substitution to solve a system of linear equations.

1 Form a coefficient and an augmented matrix and form a linear system from the augmented matrix.

Matrices

In this section, you will study a streamlined technique for solving systems of linear equations. This technique involves the use of a rectangular array of real numbers called a **matrix.** (The plural of matrix is *matrices.*) Here is an example of a matrix.

$$
\begin{array}{cccc}
\textit{Column} & \textit{Column} & \textit{Column} & \textit{Column} \\
1 & 2 & 3 & 4
\end{array}
$$

$$
\begin{array}{c}
\textit{Row 1} \\
\textit{Row 2} \\
\textit{Row 3}
\end{array}
\begin{bmatrix}
3 & -2 & 4 & 1 \\
0 & 1 & -1 & 2 \\
2 & 0 & -3 & 0
\end{bmatrix}
$$

This matrix has three rows and four columns, which means that its **order** is 3×4, which is read as "3 by 4." Each number in the matrix is an **entry** of the matrix.

Example 1   Examples of Matrices

The following matrices have the indicated orders.

a. Order: 2×3

$$
\begin{bmatrix}
1 & -2 & 4 \\
0 & 1 & -2
\end{bmatrix}
$$

b. Order: 2×2

$$
\begin{bmatrix}
0 & 0 \\
0 & 0
\end{bmatrix}
$$

c. Order: 3×2

$$
\begin{bmatrix}
1 & -3 \\
-2 & 0 \\
4 & -2
\end{bmatrix}
$$

A matrix with the same number of rows as columns is called a **square matrix.** For instance, the 2×2 matrix in part (b) is square.

Study Tip

Note the use of 0 for the missing y-variable in the third equation, and also note the fourth column of constant terms in the augmented matrix.

A matrix derived from a system of linear equations (each written in standard form) is the **augmented matrix** of the system. Moreover, the matrix derived from the coefficients of the system (but that does not include the constant terms) is the **coefficient matrix** of the system. Here is an example.

$$
\begin{array}{ccc}
\textit{System} & \textit{Coefficient Matrix} & \textit{Augmented Matrix}
\end{array}
$$

$$
\begin{aligned}
x - 4y + 3z &= 5 \\
-x + 3y - z &= -3 \\
2x \quad\;\; - 4z &= 6
\end{aligned}
\qquad
\begin{bmatrix}
1 & -4 & 3 \\
-1 & 3 & -1 \\
2 & 0 & -4
\end{bmatrix}
\qquad
\left[\begin{array}{ccc:c}
1 & -4 & 3 & 5 \\
-1 & 3 & -1 & -3 \\
2 & 0 & -4 & 6
\end{array}\right]
$$

When forming either the coefficient matrix or the augmented matrix of a system, you should begin by vertically aligning the variables in the equations.

Given System	*Align Variables*	*Form Augmented Matrix*

$$\begin{array}{rcr} x + 3y = & 9 \\ -y + 4z = & -2 \\ x - 5z = & 0 \end{array} \qquad \begin{array}{rcr} x + 3y & = & 9 \\ -y + 4z & = & -2 \\ x & -5z & = & 0 \end{array} \qquad \left[\begin{array}{rrr|r} 1 & 3 & 0 & 9 \\ 0 & -1 & 4 & -2 \\ 1 & 0 & -5 & 0 \end{array}\right]$$

Example 2   Forming Coefficient and Augmented Matrices

Form the coefficient matrix and the augmented matrix for each system of linear equations.

a. $-x + 5y = 2$ **b.** $3x + 2y - z = 1$ **c.** $x = 3y - 1$
     $7x - 2y = -6$ $x + 2z = -3$ $2y - 5 = 9x$
                         $-2x - y = 4$

Solution

System	*Coefficient Matrix*	*Augmented Matrix*

a. $\begin{array}{rcr} -x + 5y = & 2 \\ 7x - 2y = & -6 \end{array}$ $\left[\begin{array}{rr} -1 & 5 \\ 7 & -2 \end{array}\right]$ $\left[\begin{array}{rr|r} -1 & 5 & 2 \\ 7 & -2 & -6 \end{array}\right]$

b. $\begin{array}{rcr} 3x + 2y - z = & 1 \\ x + 2z = & -3 \\ -2x - y = & 4 \end{array}$ $\left[\begin{array}{rrr} 3 & 2 & -1 \\ 1 & 0 & 2 \\ -2 & -1 & 0 \end{array}\right]$ $\left[\begin{array}{rrr|r} 3 & 2 & -1 & 1 \\ 1 & 0 & 2 & -3 \\ -2 & -1 & 0 & 4 \end{array}\right]$

c. $\begin{array}{rcr} x - 3y = & -1 \\ -9x + 2y = & 5 \end{array}$ $\left[\begin{array}{rr} 1 & -3 \\ -9 & 2 \end{array}\right]$ $\left[\begin{array}{rr|r} 1 & -3 & -1 \\ -9 & 2 & 5 \end{array}\right]$

Example 3   Forming Linear Systems from Their Matrices

Write systems of linear equations that are represented by the following matrices.

a. $\left[\begin{array}{rr|r} 3 & -5 & 4 \\ -1 & 2 & 0 \end{array}\right]$ **b.** $\left[\begin{array}{rr|r} 1 & 3 & 2 \\ 0 & 1 & -3 \end{array}\right]$ **c.** $\left[\begin{array}{rrr|r} 2 & 0 & -8 & 1 \\ -1 & 1 & 1 & 2 \\ 5 & -1 & 7 & 3 \end{array}\right]$

Solution

a. $3x - 5y = 4$
     $-x + 2y = 0$

b. $x + 3y = 2$
         $y = -3$

c. $2x - 8z = 1$
     $-x + y + z = 2$
     $5x - y + 7z = 3$

2 Perform elementary row operations to solve a system of linear equations.

Elementary Row Operations

In Section 8.3, you studied three operations that can be used on a system of linear equations to produce an equivalent system: (1) interchange two rows, (2) multiply a row by a nonzero constant, and (3) add a multiple of a row to another row. In matrix terminology, these three operations correspond to **elementary row operations.**

Study Tip

Although elementary row operations are simple to perform, they involve a lot of arithmetic. Because it is easy to make a mistake, we suggest that you get in the habit of noting the elementary row operations performed in each step so that you can go back and check your work. People use different schemes to denote which elementary row operations have been performed. The scheme we use is to write an abbreviated version of the row operation to the left of the row that has been changed.

▶ **Elementary Row Operations**

Any of the following **elementary row operations** performed on an augmented matrix will produce a matrix that is row-equivalent to the original matrix. Two matrices are **row-equivalent** if one can be obtained from the other by a sequence of elementary row operations.

1. Interchange two rows.

2. Multiply a row by a nonzero constant.

3. Add a multiple of a row to another row.

> **Example 4**   Elementary Row Operations

a. Interchange the first and second rows.

Original Matrix

$$\begin{bmatrix} 0 & 1 & 3 & 4 \\ -1 & 2 & 0 & 3 \\ 2 & -3 & 4 & 1 \end{bmatrix}$$

New Row-Equivalent Matrix

$$\begin{matrix} R_2 \\ R_1 \end{matrix} \begin{bmatrix} -1 & 2 & 0 & 3 \\ 0 & 1 & 3 & 4 \\ 2 & -3 & 4 & 1 \end{bmatrix}$$

b. Multiply the first row by $\frac{1}{2}$.

Original Matrix

$$\begin{bmatrix} 2 & -4 & 6 & -2 \\ 1 & 3 & -3 & 0 \\ 5 & -2 & 1 & 2 \end{bmatrix}$$

New Row-Equivalent Matrix

$$\frac{1}{2}R_1 \rightarrow \begin{bmatrix} 1 & -2 & 3 & -1 \\ 1 & 3 & -3 & 0 \\ 5 & -2 & 1 & 2 \end{bmatrix}$$

c. Add -2 times the first row to the third row.

Original Matrix

$$\begin{bmatrix} 1 & 2 & -4 & 3 \\ 0 & 3 & -2 & -1 \\ 2 & 1 & 5 & -2 \end{bmatrix}$$

New Row-Equivalent Matrix

$$\begin{bmatrix} 1 & 2 & -4 & 3 \\ 0 & 3 & -2 & -1 \\ -2R_1 + R_3 \rightarrow 0 & -3 & 13 & -8 \end{bmatrix}$$

d. Add 6 times the first row to the second row.

Original Matrix

$$\begin{bmatrix} 1 & 2 & 2 & -4 \\ -6 & -11 & 3 & 18 \\ 0 & 0 & 4 & 7 \end{bmatrix}$$

New Row-Equivalent Matrix

$$6R_1 + R_2 \rightarrow \begin{bmatrix} 1 & 2 & 2 & -4 \\ 0 & 1 & 15 & -6 \\ 0 & 0 & 4 & 7 \end{bmatrix}$$

In Section 8.3, Gaussian elimination was used with back-substitution to solve a system of linear equations. Example 5 demonstrates the matrix version of Gaussian elimination. The two methods are essentially the same. The basic difference is that with matrices you do not need to keep writing the variables.

Example 5   Solving a System of Linear Equations

Linear System

$$x - 2y + 2z = 9$$
$$-x + 3y = -4$$
$$2x - 5y + z = 10$$

Associated Augmented Matrix

$$\begin{bmatrix} 1 & -2 & 2 & \vdots & 9 \\ -1 & 3 & 0 & \vdots & -4 \\ 2 & -5 & 1 & \vdots & 10 \end{bmatrix}$$

Add the first equation to the second equation.

$$x - 2y + 2z = 9$$
$$y + 2z = 5$$
$$2x - 5y + z = 10$$

Add the first row to the second row $(R_1 + R_2)$.

$$R_1 + R_2 \rightarrow \begin{bmatrix} 1 & -2 & 2 & \vdots & 9 \\ 0 & 1 & 2 & \vdots & 5 \\ 2 & -5 & 1 & \vdots & 10 \end{bmatrix}$$

Add -2 times the first equation to the third equation.

$$x - 2y + 2z = 9$$
$$y + 2z = 5$$
$$-y - 3z = -8$$

Add -2 times the first row to the third row $(-2R_1 + R_3)$.

$$-2R_1 + R_3 \rightarrow \begin{bmatrix} 1 & -2 & 2 & \vdots & 9 \\ 0 & 1 & 2 & \vdots & 5 \\ 0 & -1 & -3 & \vdots & -8 \end{bmatrix}$$

Add the second equation to the third equation.

$$x - 2y + 2z = 9$$
$$y + 2z = 5$$
$$-z = -3$$

Add the second row to the third row $(R_2 + R_3)$.

$$R_2 + R_3 \rightarrow \begin{bmatrix} 1 & -2 & 2 & \vdots & 9 \\ 0 & 1 & 2 & \vdots & 5 \\ 0 & 0 & -1 & \vdots & -3 \end{bmatrix}$$

Multiply the third equation by -1.

$$x - 2y + 2z = 9$$
$$y + 2z = 5$$
$$z = 3$$

Multiply the third row by -1.

$$-R_3 \rightarrow \begin{bmatrix} 1 & -2 & 2 & \vdots & 9 \\ 0 & 1 & 2 & \vdots & 5 \\ 0 & 0 & 1 & \vdots & 3 \end{bmatrix}$$

At this point, you can use back-substitution to find that the solution is

$$x = 1, \quad y = -1, \quad \text{and} \quad z = 3.$$

Check this in the original system as follows.

$$(1) - 2(-1) + 2(3) = 9 \qquad \text{Substitute in Equation 1.} \checkmark$$
$$-(1) + 3(-1) = -4 \qquad \text{Substitute in Equation 2.} \checkmark$$
$$2(1) - 5(-1) + (3) = 10 \qquad \text{Substitute in Equation 3.} \checkmark$$

Technology: Tip

Most graphing utilities are capable of performing row operations on matrices. Some graphing utilities have a function that will return the reduced row-echelon form of a matrix. Consult the user's guide of your graphing utility to learn how to perform elementary row operations. Most graphing utilities store the resulting matrix of each step into an answer variable. We suggest that you store the results of each operation into a matrix variable.

Enter the matrix from Example 5 into your graphing utility and perform the indicated row operations.

The last matrix in Example 5 is in **row-echelon form.** The term *echelon* refers to the stair-step pattern formed by the nonzero elements of the matrix.

3 Use matrices and Gaussian elimination with back-substitution to solve a system of linear equations.

Solving a System of Linear Equations

> ▶ **Gaussian Elimination with Back-Substitution**
>
> To use matrices and Gaussian elimination to solve a system of linear equations, use the following steps.
>
> 1. Write the augmented matrix of the system of linear equations.
>
> 2. Use elementary row operations to rewrite the augmented matrix in row-echelon form.
>
> 3. Write the system of linear equations corresponding to the matrix in row-echelon form, and use back-substitution to find the solution.

When you perform Gaussian elimination with back-substitution, we suggest that you operate from *left to right by columns,* using elementary row operations to obtain zeros in all entries directly below the leading 1s.

Example 6 Gaussian Elimination with Back-Substitution

Solve the system of linear equations.

$$2x - 3y = -2$$
$$x + 2y = 13$$

Solution

$$\begin{bmatrix} 2 & -3 & \vdots & -2 \\ 1 & 2 & \vdots & 13 \end{bmatrix}$$ Augmented matrix for system of linear equations

$$\begin{matrix} R_2 \\ R_1 \end{matrix} \begin{bmatrix} 1 & 2 & \vdots & 13 \\ 2 & -3 & \vdots & -2 \end{bmatrix}$$ First column has leading 1 in upper left corner.

$$-2R_1 + R_2 \rightarrow \begin{bmatrix} 1 & 2 & \vdots & 13 \\ 0 & -7 & \vdots & -28 \end{bmatrix}$$ First column has a zero under its leading 1.

$$-\tfrac{1}{7}R_2 \rightarrow \begin{bmatrix} 1 & 2 & \vdots & 13 \\ 0 & 1 & \vdots & 4 \end{bmatrix}$$ Second column has leading 1 in second row.

The system of linear equations that corresponds to the (row-echelon) matrix is

$$x + 2y = 13$$
$$y = 4.$$

Using back-substitution, you can find that the solution of the system is

$$x = 5 \text{ and } y = 4.$$

Check this solution in the original system, as follows.

$$2(5) - 3(4) = -2$$ Substitute in Equation 1. ✓

$$5 - 2(4) = 13$$ Substitute in Equation 2. ✓

Example 7   Gaussian Elimination with Back-Substitution

Solve the system of linear equations.

$$3x + 3y \quad\quad = \quad 9$$
$$2x \quad\quad - 3z = \quad 10$$
$$6y + 4z = -12$$

Solution

$$\begin{bmatrix} 3 & 3 & 0 & \vdots & 9 \\ 2 & 0 & -3 & \vdots & 10 \\ 0 & 6 & 4 & \vdots & -12 \end{bmatrix}$$
Augmented matrix for system of linear equations

$$\tfrac{1}{3}R_1 \rightarrow \begin{bmatrix} 1 & 1 & 0 & \vdots & 3 \\ 2 & 0 & -3 & \vdots & 10 \\ 0 & 6 & 4 & \vdots & -12 \end{bmatrix}$$
First column has leading 1 in upper left corner.

$$-2R_1 + R_2 \rightarrow \begin{bmatrix} 1 & 1 & 0 & \vdots & 3 \\ 0 & -2 & -3 & \vdots & 4 \\ 0 & 6 & 4 & \vdots & -12 \end{bmatrix}$$
First column has zeros under its leading 1.

$$-\tfrac{1}{2}R_2 \rightarrow \begin{bmatrix} 1 & 1 & 0 & \vdots & 3 \\ 0 & 1 & \tfrac{3}{2} & \vdots & -2 \\ 0 & 6 & 4 & \vdots & -12 \end{bmatrix}$$
Second column has leading 1 in second row.

$$-6R_2 + R_3 \rightarrow \begin{bmatrix} 1 & 1 & 0 & \vdots & 3 \\ 0 & 1 & \tfrac{3}{2} & \vdots & -2 \\ 0 & 0 & -5 & \vdots & 0 \end{bmatrix}$$
Second column has zero under its leading 1.

$$-\tfrac{1}{5}R_3 \rightarrow \begin{bmatrix} 1 & 1 & 0 & \vdots & 3 \\ 0 & 1 & \tfrac{3}{2} & \vdots & -2 \\ 0 & 0 & 1 & \vdots & 0 \end{bmatrix}$$
Third column has leading 1 in third row.

The system of linear equations that corresponds to this (row-echelon) matrix is

$$x + y \quad\quad = \quad 3$$
$$y + \frac{3}{2}z = -2$$
$$z = \quad 0.$$

Using back-substitution, you can find that the solution is

$$x = 5, \quad y = -2, \quad \text{and} \quad z = 0.$$

Check this in the original system, as follows.

$$3(5) + 3(-2) \quad\quad = \quad 9$$ Substitute in Equation 1. ✓
$$2(5) \quad\quad - 3(0) = \quad 10$$ Substitute in Equation 2. ✓
$$6(-2) + 4(0) = -12$$ Substitute in Equation 3. ✓

Example 8 A System with No Solution

Solve the system of linear equations.

$$6x - 10y = -4$$
$$9x - 15y = 5$$

Solution

$$\begin{bmatrix} 6 & -10 & \vdots & -4 \\ 9 & -15 & \vdots & 5 \end{bmatrix}$$ Augmented matrix for system of linear equations

$$\tfrac{1}{6}R_1 \rightarrow \begin{bmatrix} 1 & -\tfrac{5}{3} & \vdots & -\tfrac{2}{3} \\ 9 & -15 & \vdots & 5 \end{bmatrix}$$ First column has leading 1 in upper left corner.

$$-9R_1 + R_2 \rightarrow \begin{bmatrix} 1 & -\tfrac{5}{3} & \vdots & -\tfrac{2}{3} \\ 0 & 0 & \vdots & 11 \end{bmatrix}$$ First column has a zero under its leading 1.

The "equation" that corresponds to the second row of this matrix is $0 = 11$. Because this is a false statement, the system of equations has no solution.

Example 9 A System with Infinitely Many Solutions

Solve the system of linear equations.

$$12x - 6y = -3$$
$$-8x + 4y = 2$$

Solution

$$\begin{bmatrix} 12 & -6 & \vdots & -3 \\ -8 & 4 & \vdots & 2 \end{bmatrix}$$ Augmented matrix for system of linear equations

$$\tfrac{1}{12}R_1 \rightarrow \begin{bmatrix} 1 & -\tfrac{1}{2} & \vdots & -\tfrac{1}{4} \\ -8 & 4 & \vdots & 2 \end{bmatrix}$$ First column has leading 1 in upper left corner.

$$8R_1 + R_2 \rightarrow \begin{bmatrix} 1 & -\tfrac{1}{2} & \vdots & -\tfrac{1}{4} \\ 0 & 0 & \vdots & 0 \end{bmatrix}$$ First column has a zero under its leading 1.

Because the second row of the matrix is all zeros, you can conclude that the system of equations has an infinite number of solutions, represented by all points (x, y) on the line

$$x - \frac{1}{2}y = -\frac{1}{4}.$$

Because this line can be written as

$$x = -\frac{1}{4} + \frac{1}{2}y$$

you can write the solution set as

$$\left(-\frac{1}{4} + \frac{1}{2}a, a\right), \quad \text{where } a \text{ is any real number.}$$

Example 10   An Investment Portfolio

You have a portfolio totaling $219,000 and want to invest in municipal bonds, blue-chip stocks, and growth or speculative stocks. The municipal bonds pay 6% annually. Over a 5-year period, you expect blue-chip stocks to return 10% annually and growth stocks to return 15% annually. You want a combined annual return of 8%, and you also want to have only one-fourth of the portfolio invested in stocks. How much should be allocated to each type of investment?

Solution

To solve this problem, let M represent municipal bonds, B represent blue-chip stocks, and G represent growth stocks. These three equations make up the following system.

$$M + B + G = 219,000 \qquad \text{Equation 1: total investment is \$219,000.}$$

$$0.06M + 0.10B + 0.15G = 17,520 \qquad \text{Equation 2: combined annual return is 8\%.}$$

$$B + G = 54,750 \qquad \text{Equation 3: } \tfrac{1}{4} \text{ of investment is allocated to stocks.}$$

The augmented matrix for this system is

$$\begin{bmatrix} 1 & 1 & 1 & \vdots & 219,000 \\ 0.06 & 0.10 & 0.15 & \vdots & 17,520 \\ 0 & 1 & 1 & \vdots & 54,750 \end{bmatrix}.$$

Using elementary row operations, the reduced row-echelon form of this matrix is

$$\begin{bmatrix} 1 & 1 & 1 & \vdots & 219,000 \\ 0 & 1 & 2.25 & \vdots & 109,500 \\ 0 & 0 & 1 & \vdots & 43,800 \end{bmatrix}.$$

From the row-echelon form, you can see that $G = 43,800$. By back-substituting G into the revised second equation, you can determine that $B = 10,950$. By back-substituting B and G into the first equation, you can determine that $M = 164,250$. So, you should invest $164,250 in municipal bonds, $10,950 in blue-chip stocks, and $43,800 in growth or speculative stocks. Check your solution by substituting these values into the original equations of the system.

Discussing the Concept   Analyzing Solutions to Systems of Equations

Use a graphing utility to graph each system of equations given in Example 6, Example 8, and Example 9. Verify the solution given in each example and explain how you can reach the same conclusion by using the graph. Summarize how you can conclude that a system has a unique solution, no solution, or infinitely many solutions when you use Gaussian elimination.

8.4 Exercises

Integrated Review Concepts, Skills, and Problem Solving

Keep mathematically in shape by doing these exercises *before* the problems of this section.

Properties and Definitions

In Exercises 1–4, name the property illustrated.

1. $2ab - 2ab = 0$

2. $8t \cdot 1 = 8t$

3. $b + 3a = 3a + b$

4. $3(2x) = (3 \cdot 2)x$

Algebraic Operations

In Exercises 5–10, plot the points on the rectangular coordinate system. Find the slope of the line passing through the points. If not possible, state why.

5. $(-3, 2), \left(-\frac{3}{2}, -2\right)$

6. $(0, -6), (8, 0)$

7. $\left(\frac{5}{2}, \frac{7}{2}\right), \left(\frac{5}{2}, 4\right)$

8. $\left(-\frac{5}{8}, -\frac{3}{4}\right), \left(1, -\frac{9}{2}\right)$

9. $(3, 1.2), (-3, 2.1)$

10. $(12, 8), (6, 8)$

Problem Solving

11. Through a membership drive, the membership for a public television station was increased by 10%. The current number of members is 8415. How many members did the station have before the membership drive?

12. A sales representative indicates that if a customer waits another month for a new car that currently costs $23,500, the price will increase by 4%. The customer has a certificate of deposit that comes due in 1 month and will pay a penalty for early withdrawal if the money is withdrawn before the due date. Determine the maximum penalty for early withdrawal that would equal the cost increase of waiting to buy the car.

Developing Skills

In Exercises 1–6, determine the order of the matrix. See Example 1.

1. $\begin{bmatrix} 3 & -2 \\ -4 & 0 \\ 2 & -7 \\ -1 & -3 \end{bmatrix}$

2. $\begin{bmatrix} 4 & 0 & -5 \\ -1 & 8 & 9 \\ 0 & -3 & 4 \end{bmatrix}$

3. $\begin{bmatrix} 5 & -8 & 32 \\ 7 & 15 & 28 \end{bmatrix}$

4. $\begin{bmatrix} -2 & 5 \\ 0 & -1 \end{bmatrix}$

5. $\begin{bmatrix} 4 \\ -2 \\ 0 \\ 1 \end{bmatrix}$

6. $\begin{bmatrix} 1 & -1 & 2 & 3 \end{bmatrix}$

In Exercises 7–12, form the augmented matrix for the system of linear equations. See Example 2.

7. $4x - 5y = -2$
$-x + 8y = 10$

8. $8x + 3y = 25$
$3x - 9y = 12$

9. $x + 10y - 3z = 2$
$5x - 3y + 4z = 0$
$2x + 4y = 6$

10. $9x - 3y + z = 13$
$12x - 8z = 5$
$3x + 4y - z = 6$

11. $5x + y - 3z = 7$
$ 2y + 4z = 12$

12. $10x + 6y - 8z = -4$
$-4x - 7y = 9$

In Exercises 13–18, write the system of linear equations represented by the augmented matrix. (Use variables x, y, z, and w.) See Example 3.

13. $\begin{bmatrix} 4 & 3 & \vdots & 8 \\ 1 & -2 & \vdots & 3 \end{bmatrix}$

14. $\begin{bmatrix} 9 & -4 & \vdots & 0 \\ 6 & 1 & \vdots & -4 \end{bmatrix}$

15. $\begin{bmatrix} 1 & 0 & 2 & \vdots & -10 \\ 0 & 3 & -1 & \vdots & 5 \\ 4 & 2 & 0 & \vdots & 3 \end{bmatrix}$

16. $\begin{bmatrix} 4 & -1 & 3 & \vdots & 5 \\ 2 & 0 & -2 & \vdots & -1 \\ -1 & 6 & 0 & \vdots & 3 \end{bmatrix}$

17. $\begin{bmatrix} 5 & 8 & 2 & 0 & \vdots & -1 \\ -2 & 15 & 5 & 1 & \vdots & 9 \\ 1 & 6 & -7 & 0 & \vdots & -3 \end{bmatrix}$

18. $\begin{bmatrix} 7 & 3 & -2 & 4 & \vdots & 2 \\ -1 & 0 & 4 & -1 & \vdots & 6 \\ 8 & 3 & 0 & 0 & \vdots & -4 \\ 0 & 2 & -4 & 3 & \vdots & 12 \end{bmatrix}$

In Exercises 19–24, fill in the blank(s) by using elementary row operations to form a row-equivalent matrix. See Examples 4 and 5.

19. $\begin{bmatrix} 1 & 4 & 3 \\ 2 & 10 & 5 \end{bmatrix}$

$\begin{bmatrix} 1 & 4 & 3 \\ 0 & \blacksquare & -1 \end{bmatrix}$

20. $\begin{bmatrix} 3 & 6 & 8 \\ 4 & -3 & 6 \end{bmatrix}$

$\begin{bmatrix} 3 & 6 & 8 \\ 1 & -9 & \blacksquare \end{bmatrix}$

21. $\begin{bmatrix} 9 & -18 & 6 \\ 2 & 8 & 15 \end{bmatrix}$

$\begin{bmatrix} 1 & \blacksquare & \blacksquare \\ 2 & 8 & 15 \end{bmatrix}$

22. $\begin{bmatrix} 2 & 3 & -5 & 6 \\ 5 & -7 & 12 & 9 \\ -4 & 6 & 9 & 5 \end{bmatrix}$

$\begin{bmatrix} 2 & 3 & -5 & 6 \\ 5 & -7 & 12 & 9 \\ 0 & 12 & \blacksquare & \blacksquare \end{bmatrix}$

23. $\begin{bmatrix} 1 & 1 & 4 & -1 \\ 3 & 8 & 10 & 3 \\ -2 & 1 & 12 & 6 \end{bmatrix}$

$\begin{bmatrix} 1 & 1 & 4 & -1 \\ 0 & 5 & \blacksquare & \blacksquare \\ 0 & 3 & \blacksquare & \blacksquare \end{bmatrix}$

$\begin{bmatrix} 1 & 1 & 4 & -1 \\ 0 & 1 & -\frac{2}{5} & \frac{6}{5} \\ 0 & 3 & \blacksquare & \blacksquare \end{bmatrix}$

24. $\begin{bmatrix} 2 & 4 & 8 & 3 \\ 1 & -1 & -3 & 2 \\ 2 & 6 & 4 & 9 \end{bmatrix}$

$\begin{bmatrix} 1 & \blacksquare & \blacksquare & \blacksquare \\ 1 & -1 & -3 & 2 \\ 2 & 6 & 4 & 9 \end{bmatrix}$

$\begin{bmatrix} 1 & 2 & 4 & \frac{3}{2} \\ 0 & \blacksquare & -7 & \frac{1}{2} \\ 0 & 2 & \blacksquare & \blacksquare \end{bmatrix}$

In Exercises 25–30, convert the matrix to row-echelon form. (*Note:* There is more than one correct answer.)

25. $\begin{bmatrix} 1 & 2 & 3 \\ 2 & -1 & -4 \end{bmatrix}$

26. $\begin{bmatrix} 1 & 3 & 6 \\ -4 & -9 & 3 \end{bmatrix}$

27. $\begin{bmatrix} 4 & 6 & 1 \\ -2 & 2 & 5 \end{bmatrix}$

28. $\begin{bmatrix} 3 & 2 & 6 \\ 2 & 3 & -3 \end{bmatrix}$

29. $\begin{bmatrix} 1 & 1 & 0 & 5 \\ -2 & -1 & 2 & -10 \\ 3 & 6 & 7 & 14 \end{bmatrix}$

30. $\begin{bmatrix} 1 & 2 & -1 & 3 \\ 3 & 7 & -5 & 14 \\ -2 & -1 & -3 & 8 \end{bmatrix}$

In Exercises 31–34, use the matrix capabilities of a graphing utility to write the matrix in row-echelon form. (*Note:* There is more than one correct answer.)

31. $\begin{bmatrix} 1 & -1 & -1 & 1 \\ 4 & -4 & 1 & 8 \\ -6 & 8 & 18 & 0 \end{bmatrix}$

32. $\begin{bmatrix} 1 & -3 & 0 & -7 \\ -3 & 10 & 1 & 23 \\ 4 & -10 & 2 & -24 \end{bmatrix}$

33. $\begin{bmatrix} 1 & 1 & -1 & 3 \\ 2 & 1 & 2 & 5 \\ 3 & 2 & 1 & 8 \end{bmatrix}$

34. $\begin{bmatrix} 1 & -3 & -2 & -8 \\ 1 & 3 & -2 & 17 \\ 1 & 2 & -2 & -5 \end{bmatrix}$

In Exercises 35–40, write the system of linear equations represented by the augmented matrix. Then use back-substitution to find the solution. (Use variables x, y, and z.)

35. $\begin{bmatrix} 1 & -2 & \vdots & 4 \\ 0 & 1 & \vdots & -3 \end{bmatrix}$

36. $\begin{bmatrix} 1 & 5 & \vdots & 0 \\ 0 & 1 & \vdots & -1 \end{bmatrix}$

37. $\begin{bmatrix} 1 & 5 & \vdots & 3 \\ 0 & 1 & \vdots & -2 \end{bmatrix}$

38. $\begin{bmatrix} 1 & 5 & -3 & \vdots & 0 \\ 0 & 1 & 0 & \vdots & 6 \\ 0 & 0 & 1 & \vdots & -5 \end{bmatrix}$

39. $\begin{bmatrix} 1 & -1 & 2 & \vdots & 4 \\ 0 & 1 & -1 & \vdots & 2 \\ 0 & 0 & 1 & \vdots & -2 \end{bmatrix}$

40. $\begin{bmatrix} 1 & 2 & -2 & \vdots & -1 \\ 0 & 1 & 1 & \vdots & 9 \\ 0 & 0 & 1 & \vdots & -3 \end{bmatrix}$

In Exercises 41–62, use matrices to solve the system of linear equations. See Examples 5–9.

41. $x + 2y = 7$
$3x + y = 8$

42. $2x + 6y = 16$
$2x + 3y = 7$

43. $6x - 4y = 2$
$5x + 2y = 7$

44. $x - 3y = 5$
$-2x + 6y = -10$

45. $-x + 2y = 1.5$
$2x - 4y = 3$

46. $2x - y = -0.1$
$3x + 2y = 1.6$

47. $x - 2y - z = 6$
$y + 4z = 5$
$4x + 2y + 3z = 8$

48. $x - 3z = -2$
$3x + y - 2z = 5$
$2x + 2y + z = 4$

49. $x + y - 5z = 3$
$x - 2z = 1$
$2x - y - z = 0$

50. $2y + z = 3$
$-4y - 2z = 0$
$x + y + z = 2$

51. $2x + 4y = 10$
$2x + 2y + 3z = 3$
$-3x + y + 2z = -3$

52. $2x - y + 3z = 24$
$2y - z = 14$
$7x - 5y = 6$

53. $x - 3y + 2z = 8$
$2y - z = -4$
$x + z = 3$

54. $2x + 3z = 3$
$4x - 3y + 7z = 5$
$8x - 9y + 15z = 9$

55. $-2x - 2y - 15z = 0$
$x + 2y + 2z = 18$
$3x + 3y + 22z = 2$

56. $2x + 4y + 5z = 5$
$x + 3y + 3z = 2$
$2x + 4y + 4z = 2$

57. $2x + 4z = 1$
$x + y + 3z = 0$
$x + 3y + 5z = 0$

58. $3x + y - 2z = 2$
$6x + 2y - 4z = 1$
$-3x - y + 2z = 1$

59. $x + 3y = 2$
$2x + 6y = 4$
$2x + 5y + 4z = 3$

60. $2x + 2y + z = 8$
$2x + 3y + z = 7$
$6x + 8y + 3z = 22$

61. $2x + y - 2z = 4$
$3x - 2y + 4z = 6$
$-4x + y + 6z = 12$

62. $3x + 3y + z = 4$
$2x + 6y + z = 5$
$-x - 3y + 2z = -5$

Solving Problems

63. *Simple Interest*   A corporation borrowed $1,500,000 to expand its product line. Some of the money was borrowed at 8%, some at 9%, and the remainder at 12%. The annual interest payment to the lenders was $133,000. If the amount borrowed at 8% was 4 times the amount borrowed at 12%, how much was borrowed at each rate?

64. *Investments*   An inheritance of $16,000 was divided among three investments yielding a total of $990 in simple interest per year. The interest rates for the three investments were 5%, 6%, and 7%. Find the amount placed in each investment if the 5% and 6% investments were $3000 and $2000 less than the 7% investment, respectively.

Investment Portfolio   In Exercises 65 and 66, consider an investor with a portfolio totaling $500,000 that is to be allocated among the following types of investments: certificates of deposit, municipal bonds, blue-chip stocks, and growth or speculative stocks. How much should be allocated to each type of investment?

65. The certificates of deposit pay 10% annually, and the municipal bonds pay 8% annually. Over a 5-year period, the investor expects the blue-chip stocks to return 12% annually and the growth stocks to return 13% annually. The investor wants a combined annual return of 10% and also wants to have only one-fourth of the portfolio invested in stocks.

66. The certificates of deposit pay 9% annually, and the municipal bonds pay 5% annually. Over a 5-year period, the investor expects the blue-chip stocks to return 12% annually and the growth stocks to return 14% annually. The investor wants a combined annual return of 10% and also wants to have only one-fourth of the portfolio invested in stocks.

67. *Nut Mixture*   A grocer wishes to mix three kinds of nuts costing $3.50, $4.50, and $6.00 per pound to obtain 50 pounds of a mixture priced at $4.95 per pound. How many pounds of each variety should the grocer use if half the mixture is composed of the two cheapest varieties?

68. *Nut Mixture*   A grocer wishes to mix three kinds of nuts costing $3.00, $4.00, and $6.00 per pound to obtain 50 pounds of a mixture priced at $4.10 per pound. How many pounds of each variety should the grocer use if three-quarters of the mixture is composed of the two cheapest varieties?

69. *Number Problem*   The sum of three positive numbers is 33. The second number is 3 greater than the first, and the third is 4 times the first. Find the three numbers.

70. *Number Problem*   The sum of three positive numbers is 24. The second number is 4 greater than the first, and the third is 3 times the first. Find the three numbers.

Curve-Fitting   In Exercises 71–74, find a quadratic equation $y = ax^2 + bx + c$ whose graph passes through the given points.

71.
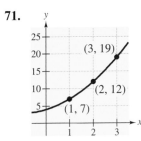

72.
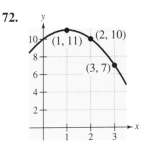

73. $(1, 8), (2, 2), (3, -25)$
74. $(1, 1), (-3, 17), \left(2, -\frac{1}{2}\right)$

Curve-Fitting   In Exercises 75 and 76, find the equation of the circle $x^2 + y^2 + Dx + Ey + F = 0$ that passes through the points.

75. $(1, 1), (3, 3), (4, 2)$
76. $(-1, 2), (2, 3), (3, 2)$

77. *Mathematical Modeling*   A videotape of the path of a ball thrown by a baseball player was analyzed on a television set with a grid covering the screen. The tape was paused three times and the coordinates of the ball were measured each time. The coordinates were approximately $(0, 6)$, $(25, 18.5)$, and $(50, 26)$ (see figure). The x-coordinate was the horizontal distance in feet from the player and the y-coordinate was the height in feet of the ball above the ground.

(a) Find the equation $y = ax^2 + bx + c$ of the graph that passes through the three points.

(b) Use a graphing utility to graph the model in part (a). Use the graph to approximate the maximum height of the ball and the point at which the ball struck the ground.

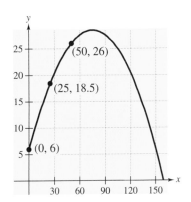

78. *Data Analysis*   The table gives the gross private savings y (in billions of dollars) for the years 1994 through 1996 in the United States. (Source: U.S. Bureau of Economic Analysis)

Year	1994	1995	1996
y	1006.3	1072.3	1161.0

(a) Create a bar graph of the data.

(b) Find a quadratic equation $y = at^2 + bt + c$ whose graph passes through the three points, with $t = 0$ corresponding to 1990.

(c) Use a graphing utility to graph the model in part (b).

(d) Use the model in part (b) to predict gross private savings in the year 2000 if the trend continues.

79. *Rewriting a Fraction*   The fraction

$$\frac{2x^2 - 9x}{(x - 2)^3}$$

can be written as a sum of three fractions, as follows.

$$\frac{2x^2 - 9x}{(x - 2)^3} = \frac{A}{x - 2} + \frac{B}{(x - 2)^2} + \frac{C}{(x - 2)^3}.$$

The numbers $A, B,$ and C are the solutions of the system

$$
\begin{aligned}
4A - 2B + C &= 0 \\
-4A + B &= -9 \\
A &= 2.
\end{aligned}
$$

Write the expression as the sum of three fractions.

80. *Rewriting a Fraction*   The fraction

$$\frac{x + 1}{x(x^2 + 1)}$$

can be written as a sum of two fractions, as follows.

$$\frac{x + 1}{x(x^2 + 1)} = \frac{A}{x} + \frac{Bx + C}{x^2 + 1}$$

The numbers $A, B,$ and C are the solutions of the system

$$
\begin{aligned}
2A + B + C &= 2 \\
2A + B - C &= 0 \\
5A + 4B + 2C &= 3.
\end{aligned}
$$

Solve the system and verify that the sum of the two resulting fractions is the original fraction.

Explaining Concepts

81. Describe the three elementary row operations that can be performed on an augmented matrix.

82. What is the relationship between the three elementary row operations on an augmented matrix and the row operations on a system of linear equations?

83. What is meant by saying that two augmented matrices are *row-equivalent*?

84. Give an example of a matrix in *row-echelon form.*

85. Describe the row-echelon form of an augmented matrix that corresponds to a system of linear equations that is inconsistent.

86. Describe the row-echelon form of an augmented matrix that corresponds to a system of linear equations that has an infinite number of solutions.

8.5   Determinants and Linear Systems

Objectives

1   Find the determinants of a 2×2 matrix and a 3×3 matrix.

2   Use determinants and Cramer's Rule to solve a system of linear equations.

3   Use the determinant to find the area of a triangle, to test for collinear points, and to find the equation of a line.

1 Find the determinants of a 2×2 matrix and a 3×3 matrix.

Arthur Cayley

(1821–1895)

Cayley is credited with creating the theory of matrices. Determinants had been studied as rectangular arrays of numbers since the middle of the 18th century. So, the use and basic properties of matrices were well established when Cayley first published articles introducing matrices as distinct entities.

The Determinant of a Matrix

Associated with each square matrix is a real number called its **determinant.** The use of determinants arose from special number patterns that occur during the solution of systems of linear equations. For instance, the system

$$a_1x + b_1y = c_1$$
$$a_2x + b_2y = c_2$$

has a solution given by

$$x = \frac{c_1b_2 - c_2b_1}{a_1b_2 - a_2b_1} \quad \text{and} \quad y = \frac{a_1c_2 - a_2c_1}{a_1b_2 - a_2b_1}$$

provided that $a_1b_2 - a_2b_1 \neq 0$. Note that the denominator of each fraction is the same. This denominator is called the **determinant** of the coefficient matrix of the system.

Coefficient Matrix   *Determinant*

$$A = \begin{bmatrix} a_1 & b_1 \\ a_2 & b_2 \end{bmatrix} \qquad \det(A) = a_1b_2 - a_2b_1$$

The determinant of the matrix A can also be denoted by vertical bars on both sides of the matrix, as indicated in the following definition.

> ### ▶ Definition of the Determinant of a 2×2 Matrix
>
> $$\det(A) = \begin{vmatrix} a_1 & b_1 \\ a_2 & b_2 \end{vmatrix} = a_1b_2 - a_2b_1$$

A convenient method for remembering the formula for the determinant of a 2×2 matrix is shown in the following diagram.

$$\det(A) = \begin{vmatrix} a_1 & b_1 \\ a_2 & b_2 \end{vmatrix} = a_1b_2 - a_2b_1$$

Note that the determinant is given by the difference of the products of the two diagonals of the matrix.

Example 1 The Determinant of a 2 × 2 Matrix

Find the determinant of each matrix.

a. $A = \begin{bmatrix} 2 & -3 \\ 1 & 4 \end{bmatrix}$ **b.** $B = \begin{bmatrix} -1 & 2 \\ 2 & -4 \end{bmatrix}$ **c.** $C = \begin{bmatrix} 1 & 3 \\ 2 & 5 \end{bmatrix}$

Solution

a. $\det(A) = \begin{vmatrix} 2 & -3 \\ 1 & 4 \end{vmatrix} = 2(4) - 1(-3) = 8 + 3 = 11$

b. $\det(B) = \begin{vmatrix} -1 & 2 \\ 2 & -4 \end{vmatrix} = (-1)(-4) - 2(2) = 4 - 4 = 0$

c. $\det(C) = \begin{vmatrix} 1 & 3 \\ 2 & 5 \end{vmatrix} = 1(5) - 2(3) = 5 - 6 = -1$

Notice in Example 1 that the determinant of a matrix can be positive, zero, or negative.

One way to evaluate the determinant of a 3 × 3 matrix, called **expanding by minors,** allows you to write the determinant of a 3 × 3 matrix in terms of three 2 × 2 determinants. The **minor** of an entry in a 3 × 3 matrix is the determinant of the 2 × 2 matrix that remains after deletion of the row and column in which the entry occurs. Here are two examples.

Given Determinant	*Entry*	*Minor of Entry*	*Value of Minor*
$\begin{vmatrix} 1 & -1 & 3 \\ 0 & 2 & 5 \\ -2 & 4 & -7 \end{vmatrix}$	1	$\begin{vmatrix} 2 & 5 \\ 4 & -7 \end{vmatrix}$	$2(-7) - 4(5) = -34$
$\begin{vmatrix} 1 & -1 & 3 \\ 0 & 2 & 5 \\ -2 & 4 & -7 \end{vmatrix}$	-1	$\begin{vmatrix} 0 & 5 \\ -2 & -7 \end{vmatrix}$	$0(-7) - (-2)(5) = 10$

> ▶ **Expanding by Minors**
>
> $$\det(A) = \begin{vmatrix} a_1 & b_1 & c_1 \\ a_2 & b_2 & c_2 \\ a_3 & b_3 & c_3 \end{vmatrix}$$
>
> $$= a_1(\text{minor of } a_1) - b_1(\text{minor of } b_1) + c_1(\text{minor of } c_1)$$
>
> $$= a_1\begin{vmatrix} b_2 & c_2 \\ b_3 & c_3 \end{vmatrix} - b_1\begin{vmatrix} a_2 & c_2 \\ a_3 & c_3 \end{vmatrix} + c_1\begin{vmatrix} a_2 & b_2 \\ a_3 & b_3 \end{vmatrix}$$
>
> This pattern is called **expanding by minors** along the first row. A similar pattern can be used to expand by minors along any row or column.

The *signs* of the terms used in expanding by minors follow the alternating pattern shown in Figure 8.8. For instance, the signs used to expand by minors along the second row are $-, +, -$, as shown at the top of page 531.

Technology: Tip

A graphing utility with matrix capabilities can be used to evaluate the determinant of a square matrix. Consult the user's guide for your graphing utility to learn how to evaluate a determinant. Use the graphing utility to check the result in Example 1(a). Then try to evaluate the determinant of the 3 × 3 matrix at the right using a graphing utility. Finish the evaluation of the determinant by expanding by minors to check the result.

$$\begin{bmatrix} + & - & + \\ - & + & - \\ + & - & + \end{bmatrix}$$

Figure 8.8 *Sign Pattern for a 3 × 3 Matrix*

$$\det(A) = \begin{vmatrix} a_1 & b_1 & c_1 \\ a_2 & b_2 & c_2 \\ a_3 & b_3 & c_3 \end{vmatrix}$$

$$= -a_2(\text{minor of } a_2) + b_2(\text{minor of } b_2) - c_2(\text{minor of } c_2)$$

Example 2   Finding the Determinant of a 3 × 3 Matrix

Find the determinant of $A = \begin{bmatrix} -1 & 1 & 2 \\ 0 & 2 & 3 \\ 3 & 4 & 2 \end{bmatrix}$.

Solution

By expanding by minors along the *first column*, you obtain the following.

$$\det(A) = \begin{vmatrix} -1 & 1 & 2 \\ 0 & 2 & 3 \\ 3 & 4 & 2 \end{vmatrix}$$

$$= (-1)\begin{vmatrix} 2 & 3 \\ 4 & 2 \end{vmatrix} - (0)\begin{vmatrix} 1 & 2 \\ 4 & 2 \end{vmatrix} + (3)\begin{vmatrix} 1 & 2 \\ 2 & 3 \end{vmatrix}$$

$$= (-1)(4 - 12) - (0)(2 - 8) + (3)(3 - 4)$$

$$= 8 - 0 - 3$$

$$= 5$$

Example 3   Finding the Determinant of a 3 × 3 Matrix

Find the determinant of $A = \begin{bmatrix} 1 & 2 & 1 \\ 3 & 0 & 2 \\ 4 & 0 & -1 \end{bmatrix}$.

Solution

By expanding by minors along the *second column*, you obtain the following.

$$\det(A) = \begin{vmatrix} 1 & 2 & 1 \\ 3 & 0 & 2 \\ 4 & 0 & -1 \end{vmatrix}$$

$$= -(2)\begin{vmatrix} 3 & 2 \\ 4 & -1 \end{vmatrix} + (0)\begin{vmatrix} 1 & 1 \\ 4 & -1 \end{vmatrix} - (0)\begin{vmatrix} 1 & 1 \\ 3 & 2 \end{vmatrix}$$

$$= -(2)(-3 - 8) + 0 - 0$$

$$= 22$$

Note in the expansions in Examples 2 and 3 that a zero entry will always yield a zero term when expanding by minors. Thus, when you are evaluating the determinant of a matrix, you should choose to expand along the row or column that has the most zero entries.

2 Use determinants and Cramer's Rule to solve a system of linear equations.

Cramer's Rule

So far in this chapter, you have studied three methods for solving a system of linear equations: substitution, elimination (with equations), and elimination (with matrices). We now look at one more method, called **Cramer's Rule,** which is named after Gabriel Cramer (1704–1752). This rule uses determinants to write the solution of a system of linear equations.

▶ **Cramer's Rule**

1. For the system of linear equations

$$a_1x + b_1y = c_1$$
$$a_2x + b_2y = c_2$$

the solution is given by

$$x = \frac{D_x}{D} = \frac{\begin{vmatrix} c_1 & b_1 \\ c_2 & b_2 \end{vmatrix}}{\begin{vmatrix} a_1 & b_1 \\ a_2 & b_2 \end{vmatrix}}, \qquad y = \frac{D_y}{D} = \frac{\begin{vmatrix} a_1 & c_1 \\ a_2 & c_2 \end{vmatrix}}{\begin{vmatrix} a_1 & b_1 \\ a_2 & b_2 \end{vmatrix}}$$

provided that $D \neq 0$.

2. For the system of linear equations

$$a_1x + b_1y + c_1z = d_1$$
$$a_2x + b_2y + c_2z = d_2$$
$$a_3x + b_3y + c_3z = d_3$$

the solution is given by

$$x = \frac{D_x}{D} = \frac{\begin{vmatrix} d_1 & b_1 & c_1 \\ d_2 & b_2 & c_2 \\ d_3 & b_3 & c_3 \end{vmatrix}}{\begin{vmatrix} a_1 & b_1 & c_1 \\ a_2 & b_2 & c_2 \\ a_3 & b_3 & c_3 \end{vmatrix}},$$

$$y = \frac{D_y}{D} = \frac{\begin{vmatrix} a_1 & d_1 & c_1 \\ a_2 & d_2 & c_2 \\ a_3 & d_3 & c_3 \end{vmatrix}}{\begin{vmatrix} a_1 & b_1 & c_1 \\ a_2 & b_2 & c_2 \\ a_3 & b_3 & c_3 \end{vmatrix}},$$

$$z = \frac{D_z}{D} = \frac{\begin{vmatrix} a_1 & b_1 & d_1 \\ a_2 & b_2 & d_2 \\ a_3 & b_3 & d_3 \end{vmatrix}}{\begin{vmatrix} a_1 & b_1 & c_1 \\ a_2 & b_2 & c_2 \\ a_3 & b_3 & c_3 \end{vmatrix}}, D \neq 0$$

Example 4   Using Cramer's Rule for a 2 × 2 System

Use Cramer's Rule to solve the system of linear equations.

$$4x - 2y = 10$$

$$3x - 5y = 11$$

Solution

Begin by finding the determinant of the coefficient matrix, $D = -14$.

$$x = \frac{D_x}{D} = \frac{\begin{vmatrix} 10 & -2 \\ 11 & -5 \end{vmatrix}}{-14} = \frac{(-50) - (-22)}{-14} = \frac{-28}{-14} = 2$$

$$y = \frac{D_y}{D} = \frac{\begin{vmatrix} 4 & 10 \\ 3 & 11 \end{vmatrix}}{-14} = \frac{44 - 30}{-14} = \frac{14}{-14} = -1$$

The solution is $(2, -1)$. Check this in the original system of equations.

Example 5   Using Cramer's Rule for a 3 × 3 System

Use Cramer's Rule to solve the system of linear equations.

$$-x + 2y - 3z = 1$$

$$2x + \quad\quad z = 0$$

$$3x - 4y + 4z = 2$$

Solution

The determinant of the coefficient matrix is $D = 10$.

$$x = \frac{D_x}{D} = \frac{\begin{vmatrix} 1 & 2 & -3 \\ 0 & 0 & 1 \\ 2 & -4 & 4 \end{vmatrix}}{10} = \frac{8}{10} = \frac{4}{5}$$

$$y = \frac{D_y}{D} = \frac{\begin{vmatrix} -1 & 1 & -3 \\ 2 & 0 & 1 \\ 3 & 2 & 4 \end{vmatrix}}{10} = \frac{-15}{10} = -\frac{3}{2}$$

$$z = \frac{D_z}{D} = \frac{\begin{vmatrix} -1 & 2 & 1 \\ 2 & 0 & 0 \\ 3 & -4 & 2 \end{vmatrix}}{10} = \frac{-16}{10} = -\frac{8}{5}$$

The solution is $\left(\frac{4}{5}, -\frac{3}{2}, -\frac{8}{5}\right)$. Check this in the original system of equations.

When using Cramer's Rule, remember that the method *does not* apply if the determinant of the coefficient matrix is zero.

3 Use the determinant to find the area of a triangle, to test for collinear points, and to find the equation of a line.

Applications of Determinants

In addition to Cramer's Rule, determinants have many other practical applications. For instance, you can use a determinant to find the area of a triangle whose vertices are given by three points on a rectangular coordinate system.

> ▶ **Area of a Triangle**
>
> The area of a triangle with vertices (x_1, y_1), (x_2, y_2), and (x_3, y_3) is
>
> $$\text{Area} = \pm\frac{1}{2}\begin{vmatrix} x_1 & y_1 & 1 \\ x_2 & y_2 & 1 \\ x_3 & y_3 & 1 \end{vmatrix}$$
>
> where the symbol $(\pm)$ indicates that the appropriate sign should be chosen to yield a positive area.

Example 6 Finding the Area of a Triangle

Find the area of the triangle whose vertices are $(2, 0)$, $(1, 3)$, and $(3, 2)$, as shown in Figure 8.9.

Solution

Choose $(x_1, y_1) = (2, 0)$, $(x_2, y_2) = (1, 3)$, and $(x_3, y_3) = (3, 2)$. To find the area of the triangle, evaluate the determinant

$$\begin{vmatrix} x_1 & y_1 & 1 \\ x_2 & y_2 & 1 \\ x_3 & y_3 & 1 \end{vmatrix} = \begin{vmatrix} 2 & 0 & 1 \\ 1 & 3 & 1 \\ 3 & 2 & 1 \end{vmatrix}$$

$$= 2\begin{vmatrix} 3 & 1 \\ 2 & 1 \end{vmatrix} - 0\begin{vmatrix} 1 & 1 \\ 3 & 1 \end{vmatrix} + 1\begin{vmatrix} 1 & 3 \\ 3 & 2 \end{vmatrix}$$

$$= 2(1) - 0 + 1(-7)$$

$$= -5.$$

Using this value, you can conclude that the area of the triangle is

$$\text{Area} = -\frac{1}{2}\begin{vmatrix} 2 & 0 & 1 \\ 1 & 3 & 1 \\ 3 & 2 & 1 \end{vmatrix}$$

$$= -\frac{1}{2}(-5)$$

$$= \frac{5}{2}.$$

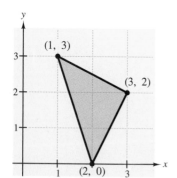

Figure 8.9

To see the benefit of the "determinant formula," try finding the area of the triangle in Example 6 using the standard formula:

$$\text{Area} = \tfrac{1}{2}(\text{base})(\text{height}).$$

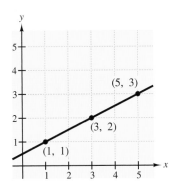

Figure 8.10

Suppose the three points in Example 6 had been on the same line. What would have happened had we applied the area formula to three such points? The answer is that the determinant would have been zero. Consider, for instance, the three collinear points $(1, 1)$, $(3, 2)$, and $(5, 3)$, as shown in Figure 8.10. The area of the "triangle" that has these three points as vertices is

$$\frac{1}{2}\begin{vmatrix} 1 & 1 & 1 \\ 3 & 2 & 1 \\ 5 & 3 & 1 \end{vmatrix} = \frac{1}{2}\left(1\begin{vmatrix} 2 & 1 \\ 3 & 1 \end{vmatrix} - 1\begin{vmatrix} 3 & 1 \\ 5 & 1 \end{vmatrix} + 1\begin{vmatrix} 3 & 2 \\ 5 & 3 \end{vmatrix}\right)$$

$$= \frac{1}{2}[-1 - (-2) + (-1)]$$

$$= 0.$$

This result is generalized as follows.

> **▶ Test for Collinear Points**
>
> Three points (x_1, y_1), (x_2, y_2), and (x_3, y_3) are collinear (lie on the same line) if and only if
>
> $$\begin{vmatrix} x_1 & y_1 & 1 \\ x_2 & y_2 & 1 \\ x_3 & y_3 & 1 \end{vmatrix} = 0.$$

Example 7   Testing for Collinear Points

Determine whether the points $(-2, -2)$, $(1, 1)$, and $(7, 5)$ lie on the same line. (See Figure 8.11.)

Solution

Letting $(x_1, y_1) = (-2, -2)$, $(x_2, y_2) = (1, 1)$, and $(x_3, y_3) = (7, 5)$, you have

$$\begin{vmatrix} x_1 & y_1 & 1 \\ x_2 & y_2 & 1 \\ x_3 & y_3 & 1 \end{vmatrix} = \begin{vmatrix} -2 & -2 & 1 \\ 1 & 1 & 1 \\ 7 & 5 & 1 \end{vmatrix}$$

$$= -2\begin{vmatrix} 1 & 1 \\ 5 & 1 \end{vmatrix} - (-2)\begin{vmatrix} 1 & 1 \\ 7 & 1 \end{vmatrix} + 1\begin{vmatrix} 1 & 1 \\ 7 & 5 \end{vmatrix}$$

$$= -2(-4) - (-2)(-6) + 1(-2)$$

$$= -6.$$

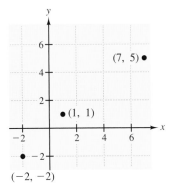

Figure 8.11

Because the value of this determinant *is not* zero, you can conclude that the three points *do not* lie on the same line.

As a good review, look at how the slope can be used to verify the result in Example 7. Label the points $A(-2, -2)$, $B(1, 1)$, and $C(7, 5)$. Because the slopes from A to B and from A to C are different, the points are not collinear.

You can also use determinants to find the equation of a line through two points. In this case the first row consists of the variables x and y and the number 1. By expanding by minors along the first row, the resulting 2×2 determinants are the coefficients of the variables x and y and the constant of the linear equation, as shown in Example 8.

▶ **Two-Point Form of the Equation of a Line**

An equation of the line passing through the distinct points (x_1, y_1) and (x_2, y_2) is given by

$$\begin{vmatrix} x & y & 1 \\ x_1 & y_1 & 1 \\ x_2 & y_2 & 1 \end{vmatrix} = 0.$$

Example 8   Finding an Equation of a Line

Find an equation of the line passing through $(-2, 1)$ and $(3, -2)$.

Solution

$$\begin{vmatrix} x & y & 1 \\ -2 & 1 & 1 \\ 3 & -2 & 1 \end{vmatrix} = 0$$

$$x \begin{vmatrix} 1 & 1 \\ -2 & 1 \end{vmatrix} - y \begin{vmatrix} -2 & 1 \\ 3 & 1 \end{vmatrix} + 1 \begin{vmatrix} -2 & 1 \\ 3 & -2 \end{vmatrix} = 0$$

$$3x + 5y + 1 = 0$$

So, an equation of the line is $3x + 5y + 1 = 0$.

Discussing the Concept   Determinant of a 3×3 Matrix

There is an alternative method for evaluating the determinant of a 3×3 matrix A. (This method works *only* for 3×3 matrices.) To apply this method, copy the first and second columns of A to form fourth and fifth columns. The determinant of A is then obtained by adding the products of three diagonals and subtracting the products of three diagonals.

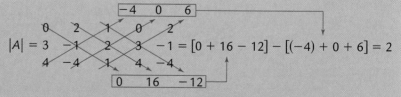

Try using this technique to find the determinants of the matrices in Examples 2 and 3. Do you think this method is easier than expanding by minors?

8.5   Exercises

Integrated Review   *Concepts, Skills, and Problem Solving*

Keep mathematically in shape by doing these exercises *before* the problems of this section.

Properties and Definitions

In Exercises 1–4, use $(px + m)(qx + n) = ax^2 + bx + c$.

1. $a =$

2. $b =$

3. $c =$

4. If $a = 1$, must $p = 1$ and $q = 1$? Explain.

Solving Equations

In Exercises 5–10, solve the equation.

5. $3x^2 + 9x - 12 = 0$ **6.** $x^2 - x - 6 = 0$

7. $4x^2 - 20x + 25 = 0$

8. $x^2 - 16 = 0$

9. $x^3 + 64 = 0$

10. $3x^3 - 6x^2 + 4x - 8 = 0$

Models

In Exercises 11 and 12, translate the phrase into an algebraic expression.

11. The time to travel 320 miles if the average speed is r miles per hour

12. The perimeter of a triangle if the sides are $x + 1, \frac{1}{2}x + 5,$ and $3x + 1$

Developing Skills

In Exercises 1–12, find the determinant of the matrix. See Example 1.

1. $\begin{bmatrix} 2 & 1 \\ 3 & 4 \end{bmatrix}$

2. $\begin{bmatrix} -3 & 1 \\ 5 & 2 \end{bmatrix}$

3. $\begin{bmatrix} 5 & 2 \\ -6 & 3 \end{bmatrix}$

4. $\begin{bmatrix} 2 & -2 \\ 4 & 3 \end{bmatrix}$

5. $\begin{bmatrix} 5 & -4 \\ -10 & 8 \end{bmatrix}$

6. $\begin{bmatrix} 4 & -3 \\ 0 & 0 \end{bmatrix}$

7. $\begin{bmatrix} 2 & 6 \\ 0 & 3 \end{bmatrix}$

8. $\begin{bmatrix} -2 & 3 \\ 6 & -9 \end{bmatrix}$

9. $\begin{bmatrix} -7 & 6 \\ \frac{1}{2} & 3 \end{bmatrix}$

10. $\begin{bmatrix} \frac{2}{3} & \frac{5}{6} \\ 14 & -2 \end{bmatrix}$

11. $\begin{bmatrix} 0.3 & 0.5 \\ 0.5 & 0.3 \end{bmatrix}$

12. $\begin{bmatrix} -1.2 & 4.5 \\ 0.4 & -0.9 \end{bmatrix}$

In Exercises 13–32, evaluate the determinant of the matrix. Expand by minors along the row or column that appears to make the computation easiest. See Examples 2 and 3.

13. $\begin{bmatrix} 2 & 3 & -1 \\ 6 & 0 & 0 \\ 4 & 1 & 1 \end{bmatrix}$

14. $\begin{bmatrix} 10 & 2 & -4 \\ 8 & 0 & -2 \\ 4 & 0 & 2 \end{bmatrix}$

15. $\begin{bmatrix} 1 & 1 & 2 \\ 3 & 1 & 0 \\ -2 & 0 & 3 \end{bmatrix}$

16. $\begin{bmatrix} 2 & 1 & 3 \\ 1 & 4 & 4 \\ 1 & 0 & 2 \end{bmatrix}$

17. $\begin{bmatrix} 2 & 4 & 6 \\ 0 & 3 & 1 \\ 0 & 0 & -5 \end{bmatrix}$

18. $\begin{bmatrix} 2 & 3 & 1 \\ 0 & 5 & -2 \\ 0 & 0 & -2 \end{bmatrix}$

19. $\begin{bmatrix} -2 & 2 & 3 \\ 1 & -1 & 0 \\ 0 & 1 & 4 \end{bmatrix}$

20. $\begin{bmatrix} 3 & 2 & 2 \\ 2 & 2 & 2 \\ -4 & 4 & 3 \end{bmatrix}$

21. $\begin{bmatrix} 1 & 4 & -2 \\ 3 & 6 & -6 \\ -2 & 1 & 4 \end{bmatrix}$

22. $\begin{bmatrix} 2 & -1 & 0 \\ 4 & 2 & 1 \\ 4 & 2 & 1 \end{bmatrix}$

23. $\begin{bmatrix} -3 & 2 & 1 \\ 4 & 5 & 6 \\ 2 & -3 & 1 \end{bmatrix}$

24. $\begin{bmatrix} -3 & 4 & 2 \\ 6 & 3 & 1 \\ 4 & -7 & -8 \end{bmatrix}$

25. $\begin{bmatrix} 1 & 4 & -2 \\ 3 & 2 & 0 \\ -1 & 4 & 3 \end{bmatrix}$

26. $\begin{bmatrix} 6 & 8 & -7 \\ 0 & 0 & 0 \\ 4 & -6 & 22 \end{bmatrix}$

27. $\begin{bmatrix} 2 & -5 & 0 \\ 4 & 7 & 0 \\ -7 & 25 & 3 \end{bmatrix}$

28. $\begin{bmatrix} 8 & 7 & 6 \\ -4 & 0 & 0 \\ 5 & 1 & 4 \end{bmatrix}$

29. $\begin{bmatrix} 0.1 & 0.2 & 0.3 \\ -0.3 & 0.2 & 0.2 \\ 5 & 4 & 4 \end{bmatrix}$ **30.** $\begin{bmatrix} -0.4 & 0.4 & 0.3 \\ 0.2 & 0.2 & 0.2 \\ 0.3 & 0.2 & 0.2 \end{bmatrix}$

31. $\begin{bmatrix} x & y & 1 \\ 3 & 1 & 1 \\ -2 & 0 & 1 \end{bmatrix}$ **32.** $\begin{bmatrix} x & y & 1 \\ -2 & -2 & 1 \\ 1 & 5 & 1 \end{bmatrix}$

In Exercises 33–38, use a graphing utility to evaluate the determinant of the matrix.

33. $\begin{bmatrix} 5 & -3 & 2 \\ 7 & 5 & -7 \\ 0 & 6 & -1 \end{bmatrix}$ **34.** $\begin{bmatrix} -\frac{1}{2} & -1 & 6 \\ 8 & -\frac{1}{4} & -4 \\ 1 & 2 & 1 \end{bmatrix}$

35. $\begin{bmatrix} 3 & -1 & 2 \\ 1 & -1 & 2 \\ -2 & 3 & 10 \end{bmatrix}$ **36.** $\begin{bmatrix} \frac{1}{2} & \frac{3}{2} & \frac{1}{2} \\ 4 & 8 & 10 \\ -2 & -6 & 12 \end{bmatrix}$

37. $\begin{bmatrix} 0.2 & 0.8 & -0.3 \\ 0.1 & 0.8 & 0.6 \\ -10 & -5 & 1 \end{bmatrix}$

38. $\begin{bmatrix} 0.4 & 0.3 & 0.3 \\ -0.2 & 0.6 & 0.6 \\ 3 & 1 & 1 \end{bmatrix}$

In Exercises 39–56, use Cramer's Rule to solve the system of linear equations. (If not possible, state the reason.) See Examples 4 and 5.

39. $\begin{aligned} x + 2y &= 5 \\ -x + y &= 1 \end{aligned}$ **40.** $\begin{aligned} 2x - y &= -10 \\ 3x + 2y &= -1 \end{aligned}$

41. $\begin{aligned} 3x + 4y &= -2 \\ 5x + 3y &= 4 \end{aligned}$ **42.** $\begin{aligned} 18x + 12y &= 13 \\ 30x + 24y &= 23 \end{aligned}$

43. $\begin{aligned} 20x + 8y &= 11 \\ 12x - 24y &= 21 \end{aligned}$ **44.** $\begin{aligned} 13x - 6y &= 17 \\ 26x - 12y &= 8 \end{aligned}$

45. $\begin{aligned} -0.4x + 0.8y &= 1.6 \\ 2x - 4y &= 5 \end{aligned}$

46. $\begin{aligned} -0.4x + 0.8y &= 1.6 \\ 0.2x + 0.3y &= 2.2 \end{aligned}$

47. $\begin{aligned} 3u + 6v &= 5 \\ 6u + 14v &= 11 \end{aligned}$

48. $\begin{aligned} 3x_1 + 2x_2 &= 1 \\ 2x_1 + 10x_2 &= 6 \end{aligned}$

49. $\begin{aligned} 4x - y + z &= -5 \\ 2x + 2y + 3z &= 10 \\ 5x - 2y + 6z &= 1 \end{aligned}$

50. $\begin{aligned} 4x - 2y + 3z &= -2 \\ 2x + 2y + 5z &= 16 \\ 8x - 5y - 2z &= 4 \end{aligned}$

51. $\begin{aligned} 3x + 4y + 4z &= 11 \\ 4x - 4y + 6z &= 11 \\ 6x - 6y &= 3 \end{aligned}$

52. $\begin{aligned} 14x_1 - 21x_2 - 7x_3 &= 10 \\ -4x_1 + 2x_2 - 2x_3 &= 4 \\ 56x_1 - 21x_2 + 7x_3 &= 5 \end{aligned}$

53. $\begin{aligned} 3a + 3b + 4c &= 1 \\ 3a + 5b + 9c &= 2 \\ 5a + 9b + 17c &= 4 \end{aligned}$

54. $\begin{aligned} 2x + 3y + 5z &= 4 \\ 3x + 5y + 9z &= 7 \\ 5x + 9y + 17z &= 13 \end{aligned}$

55. $\begin{aligned} 5x - 3y + 2z &= 2 \\ 2x + 2y - 3z &= 3 \\ x - 7y + 8z &= -4 \end{aligned}$

56. $\begin{aligned} 3x + 2y + 5z &= 4 \\ 4x - 3y - 4z &= 1 \\ -8x + 2y + 3z &= 0 \end{aligned}$

In Exercises 57–60, solve the system of linear equations using a graphing utility and Cramer's Rule. See Examples 4 and 5.

57. $\begin{aligned} -3x + 10y &= 22 \\ 9x - 3y &= 0 \end{aligned}$

58. $\begin{aligned} 3x + 7y &= 3 \\ 7x + 25y &= 11 \end{aligned}$

59. $\begin{aligned} 3x - 2y + 3z &= 8 \\ x + 3y + 6z &= -3 \\ x + 2y + 9z &= -5 \end{aligned}$

60. $\begin{aligned} 6x + 4y - 8z &= -22 \\ -2x + 2y + 3z &= 13 \\ -2x + 2y - z &= 5 \end{aligned}$

In Exercises 61 and 62, solve the equation.

61. $\begin{vmatrix} 5 - x & 4 \\ 1 & 2 - x \end{vmatrix} = 0$

62. $\begin{vmatrix} 4 - x & -2 \\ 1 & 1 - x \end{vmatrix} = 0$

Solving Problems

Area of a Triangle   In Exercises 63–70, use a determinant to find the area of the triangle with the given vertices.

63. $(0, 3), (4, 0), (8, 5)$ **64.** $(2, 0), (0, 5), (6, 3)$

65. $(0, 0), (3, 1), (1, 5)$

66. $(-2, -3), (2, -3), (0, 4)$

67. $(-2, 1), (3, -1), (1, 6)$

68. $(-4, 2), (1, 5), (4, -4)$

69. $\left(0, \frac{1}{2}\right), \left(\frac{5}{2}, 0\right), (4, 3)$ **70.** $\left(\frac{1}{4}, 0\right), \left(0, \frac{3}{4}\right), (8, -2)$

Area of a Region   In Exercises 71–74, find the area of the shaded region of the figure.

71.

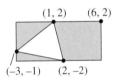

72.

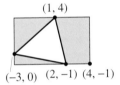

73.

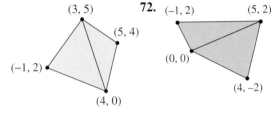

74.

75. *Area of a Region*   A large region of forest has been infested with gypsy moths. The region is roughly triangular, as shown in the figure. Approximate the number of square miles in this region.

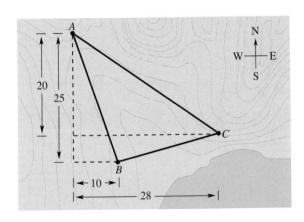

76. *Area of a Region*   You have purchased a triangular tract of land, as shown in the figure. How many square feet are there in the tract of land?

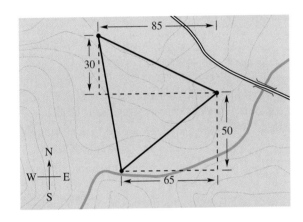

Collinear Points   In Exercises 77–82, determine whether the points are collinear.

77. $(-1, 11), (0, 8), (2, 2)$

78. $(-1, -1), (1, 9), (2, 13)$

79. $(-1, -5), (1, -1), (4, 5)$

80. $(-1, 8), (1, 2), (2, 0)$

81. $\left(-2, \frac{1}{3}\right), (2, 1), \left(3, \frac{1}{5}\right)$ **82.** $\left(0, \frac{1}{2}\right), \left(1, \frac{7}{6}\right), \left(9, \frac{13}{2}\right)$

Equation of a Line   In Exercises 83–90, use a determinant to find the equation of the line through the points.

83. $(0, 0), (5, 3)$ **84.** $(-4, 3), (2, 1)$

85. $(10, 7), (-2, -7)$ **86.** $(-8, 3), (4, 6)$

87. $\left(-2, \frac{3}{2}\right), (3, -3)$ **88.** $\left(-\frac{1}{2}, 3\right), \left(\frac{5}{2}, 1\right)$

89. $(2, 3.6), (8, 10)$ **90.** $(3, 1.6), (5, -2.2)$

⊞ *Curve-Fitting*   In Exercises 91–96, use Cramer's Rule to find a quadratic equation $y = ax^2 + bx + c$ whose graph passes through the points. Use a graphing utility to plot the points and graph the model.

91. $(0, 1), (1, -3), (-2, 21)$

92. $(-1, 0), (1, 4), (4, -5)$

93. $(-2, 6), (2, -2), (4, 0)$

94. $(-2, 6), (1, 9), (3, 1)$

95. $(1, -1), (-1, -5), \left(\frac{1}{2}, \frac{1}{4}\right)$

96. $(2, 3), \left(-1, \frac{9}{2}\right), (-2, 9)$

97. *Mathematical Modeling* The table gives the merchandise exports y_1 and the merchandise imports y_2 (in billions of dollars) for the years 1995 through 1997 in the United States. (Source: U.S. International Trade Administration)

Year	1995	1996	1997
y_1	584.7	624.8	689.2
y_2	743.4	791.4	870.7

(a) Find the quadratic model $y_1 = a_1 t^2 + b_1 t + c_1$ for exports. Let $t = 0$ represent 1990.

(b) Find the quadratic model $y_2 = a_2 t^2 + b_2 t + c_2$ for imports. Let $t = 0$ represent 1990.

(c) Use a graphing utility to graph the models found in parts (a) and (b).

(d) Find a model for the merchandise trade balance $y_1 - y_2$.

(e) Use a graphing utility to graph the model for the merchandise trade balance. What does the graph show concerning this balance?

98. *Mathematical Modeling* The table gives the agricultural products exports y_1 and the agricultural products imports y_2 (in billions of dollars) for the years 1995 through 1997 in the United States. (Source: U.S. International Trade Administration)

Year	1995	1996	1997
y_1	56.0	60.6	57.1
y_2	29.3	32.6	35.2

Table for 98

(a) Find the quadratic model $y_1 = a_1 t^2 + b_1 t + c_1$ for the exports. Let $t = 0$ represent 1990.

(b) Find the quadratic model $y_2 = a_2 t^2 + b_2 t + c_2$ for the imports. Let $t = 0$ represent 1990.

(c) Use a graphing utility to graph the models found in parts (a) and (b).

(d) Find a model for the agricultural products trade balance $y_1 - y_2$.

(e) Use a graphing utility to graph the model for the agricultural products trade balance. What does the graph show concerning this balance?

99. (a) Use Cramer's Rule to solve the following system of linear equations.

$$kx + (1 - k)y = 1$$
$$(1 - k)x + ky = 3$$

(b) For what value(s) of k will the system be inconsistent?

Explaining Concepts

100. Answer parts (g) and (h) of Motivating the Chapter on page 481.

101. Explain the difference between a square matrix and its determinant.

102. Is it possible to find the determinant of a 2×3 matrix? Explain.

103. What is meant by the *minor* of an entry of a square matrix?

104. What conditions must be met in order to use Cramer's Rule to solve a system of linear equations?

Key Terms

system of equations,
  p. 482
solution of a system of
  equations, p. 482
points of intersection,
  p. 483

consistent system, p. 483
dependent system, p. 483
inconsistent system,
  p. 483
row-echelon form, p. 503
equivalent systems, p. 504

Gaussian elimination,
  p. 504
row operations, p. 504
matrix, p. 516
matrix order, p. 516
square matrix, p. 516

augmented matrix, p. 516
coefficient matrix, p. 516
row-equivalent matrices,
  p. 518
minor (of an entry),
  p. 530

Key Concepts

8.1 The method of substitution

1. Solve one of the equations for one variable in terms of the other.

2. Substitute the expression found in Step 1 into the other equation to obtain an equation in one variable.

3. Solve the equation obtained in Step 2.

4. Back-substitute the solution from Step 3 into the expression obtained in Step 1 to find the value of the other variable.

5. Check the solution in the original system.

8.2 The method of elimination

1. Obtain coefficients for x (or y) that differ only in sign by multiplying all of the terms of one or both equations by suitably chosen constants.

2. Add the equations to eliminate one variable and solve the resulting equation.

3. Back-substitute the value obtained in Step 2 into either of the original equations and solve for the other variable.

4. Check your solution in both of the original equations.

8.3 Operations that produce equivalent systems

Each of the following row operations on a system of linear equations produces an equivalent system of linear equations.

1. Interchange two equations.

2. Multiply one of the equations by a nonzero constant.

3. Add a multiple of one of the equations to another equation to replace the latter equation.

8.4 Elementary row operations for matrices

Two matrices are row-equivalent if one can be obtained from the other by a sequence of elementary row operations.

1. Interchange two rows.

2. Multiply a row by a nonzero constant.

3. Add a multiple of a row to another row.

8.4 Gaussian elimination with back-substitution

To use matrices and Gaussian elimination to solve a system of linear equations, use the following steps.

1. Write the augmented matrix of the system of linear equations.

2. Use elementary row operations to rewrite the augmented matrix in row-echelon form.

3. Write the system of linear equations corresponding to the matrix in row-echelon form, and use back-substitution to find the solution.

8.5 Determinant of a matrix

$$\det(A) = \begin{vmatrix} a_1 & b_1 \\ a_2 & b_2 \end{vmatrix} = a_1 b_2 - a_2 b_1$$

8.5 Expanding by minors

The minor of an entry in a 3×3 matrix is the determinant of the 2×2 matrix that remains after the deletion of the row and column in which the entry occurs.

$$\det(A) = \begin{vmatrix} a_1 & b_1 & c_1 \\ a_2 & b_2 & c_2 \\ a_3 & b_3 & c_3 \end{vmatrix}$$

$$= a_1(\text{minor of } a_1) - b_1(\text{minor of } b_1) + c_1(\text{minor of } c_1)$$

$$= a_1 \begin{vmatrix} b_2 & c_2 \\ b_3 & c_3 \end{vmatrix} - b_1 \begin{vmatrix} a_2 & c_2 \\ a_3 & c_3 \end{vmatrix} + c_1 \begin{vmatrix} a_2 & b_2 \\ a_3 & b_3 \end{vmatrix}$$

541

REVIEW EXERCISES

Reviewing Skills

8.1 In Exercises 1–4, determine whether each ordered pair is a solution of the system of equations.

1. $3x + 7y = 2$
$5x + 6y = 9$
(a) $(3, 4)$
(b) $(3, -1)$

2. $-2x + 5y = 21$
$9x - y = 13$
(a) $(2, 5)$
(b) $(-2, 4)$

3. $x^2 + y^2 = 41$
$20x + 10y = 30$
(a) $(4, -5)$
(b) $(7, 12)$

4. $x^2 + 2y = 1$
$2x - 5y = 26$
(a) $(3, -4)$
(b) $(2, 8)$

In Exercises 5–14, solve the system graphically.

5. $x + y = 2$
$x - y = 0$

6. $2x = 3(y - 1)$
$y = x$

7. $x - y = 3$
$-x + y = 1$

8. $x + y = -1$
$3x + 2y = 0$

9. $2x - y = 0$
$-x + y = 4$

10. $x = y + 3$
$x = y + 1$

11. $2x + y = 4$
$-4x - 2y = -8$

12. $3x - 2y = 6$
$-6x + 4y = 12$

13. $3x - 2y = -2$
$-5x + 2y = 2$

14. $2x - y = 4$
$-3x + 4y = -11$

In Exercises 15–18, use a graphing utility to solve the system.

15. $5x - 3y = 3$
$2x + 2y = 14$

16. $8x + 5y = 1$
$3x - 4y = 18$

17. $y = x^2 - 4$
$2x - 3y = 11$

18. $y = 9 - x^2$
$2x + y = 6$

In Exercises 19–28, use substitution to solve the system.

19. $2x + 3y = 1$
$x + 4y = -2$

20. $3x - 7y = 10$
$-2x + y = -14$

21. $-5x + 2y = 4$
$10x - 4y = 7$

22. $5x + 2y = 3$
$2x + 3y = 10$

23. $3x - 7y = 5$
$5x - 9y = -5$

24. $24x - 4y = 20$
$6x - y = 5$

25. $y = 5x^2$
$y = -15x - 10$

26. $y^2 = 16x$
$x - y = -4$

27. $x^2 + y^2 = 1$
$x + y = -1$

28. $x^2 + y^2 = 32$
$x + y = 0$

8.2 In Exercises 29–34, use elimination to solve the system in two variables.

29. $x + y = 0$
$2x + y = 0$

30. $4x + y = 1$
$x - y = 4$

31. $2x - y = 2$
$6x + 8y = 39$

32. $3x + 2y = 11$
$x - 3y = -11$

33. $0.2x + 0.3y = 0.14$
$0.4x + 0.5y = 0.20$

34. $0.1x + 0.5y = -0.17$
$-0.3x - 0.2y = -0.01$

8.3 In Exercises 35–40, use elimination to solve the system in three variables.

35. $-x + y + 2z = 1$
$2x + 3y + z = -2$
$5x + 4y + 2z = 4$

36. $2x + 3y + z = 10$
$2x - 3y - 3z = 22$
$4x - 2y + 3z = -2$

37. $x - y - z = 1$
$-2x + y + 3z = -5$
$3x + 4y - z = 6$

38. $-3x + y + 2z = -13$
$-x - y + z = 0$
$2x + 2y - 3z = -1$

39.
$$x \quad\;\; - 4z = \quad 17$$
$$-2x + 4y + 3z = -14$$
$$5x - \;y + 2z = \;-3$$

40. $2x + 3y - 5z = \;3$
$$-x + 2y \qquad = \;3$$
$$3x + 5y + 2z = 15$$

8.4 In Exercises 41–48, use matrices and elementary row operations to solve the system.

41. $5x + 4y = \quad 2$
$$-x + \;y = -22$$

42. $2x - 5y = 2$
$$3x - 7y = 1$$

43. $0.2x - 0.1y = \quad 0.07$
$$0.4x - 0.5y = -0.01$$

44. $2x + y = \quad 0.3$
$$3x - y = -1.3$$

45. $\quad x + 2y + 6z = \quad 4$
$$-3x + 2y - \;z = -4$$
$$4x + \qquad 2z = 16$$

46. $-x + 3y - \;z = -4$
$$2x \qquad + 6z = \;14$$
$$-3x - \;y + \;z = \;10$$

47. $2x_1 + 3x_2 + \;3x_3 = \;3$
$$6x_1 + 6x_2 + 12x_3 = 13$$
$$12x_1 + 9x_2 - \quad x_3 = \;2$$

48. $-x_1 + 2x_2 + 3x_3 = \quad 4$
$$2x_1 - 4x_2 - \;x_3 = -13$$
$$3x_1 + 2x_2 - 4x_3 = \;-1$$

8.5 In Exercises 49–54, find the determinant of the matrix.

49. $\begin{bmatrix} 7 & 10 \\ 10 & 15 \end{bmatrix}$ **50.** $\begin{bmatrix} -3.4 & 1.2 \\ -5 & 2.5 \end{bmatrix}$

51. $\begin{bmatrix} 8 & 6 & 3 \\ 6 & 3 & 0 \\ 3 & 0 & 2 \end{bmatrix}$ **52.** $\begin{bmatrix} 7 & -1 & 10 \\ -3 & 0 & -2 \\ 12 & 1 & 1 \end{bmatrix}$

53. $\begin{bmatrix} 8 & 3 & 2 \\ 1 & -2 & 4 \\ 6 & 0 & 5 \end{bmatrix}$ **54.** $\begin{bmatrix} 4 & 0 & 10 \\ 0 & 10 & 0 \\ 10 & 0 & 34 \end{bmatrix}$

In Exercises 55–60, solve the system of linear equations by using Cramer's Rule. (If not possible, state the reason.)

55. $7x + 12y = 63$
$$2x + \;3y = 15$$

56. $12x + 42y = -17$
$$30x - 18y = \quad 19$$

57. $3x - 2y = \;16$
$$12x - 8y = -5$$

58. $\quad 4x + 24y = \;20$
$$-3x + 12y = -5$$

59. $-x + \;y + 2z = \quad 1$
$$2x + 3y + \;z = -2$$
$$5x + 4y + 2z = \quad 4$$

60. $2x_1 + \;x_2 + 2x_3 = 4$
$$2x_1 + 2x_2 \qquad = 5$$
$$2x_1 - \;x_2 + 6x_3 = 2$$

In Exercises 61 and 62, create a system of equations having the given solution. (Each problem has many correct answers.)

61. $\left(\frac{2}{3}, -4\right)$ **62.** $(-10, 12)$

Solving Problems

63. *Break-Even Analysis* A small business invests $25,000 in equipment to produce a product. Each unit of the product costs $3.75 to produce and is sold for $5.25. How many units must be sold before the business breaks even?

64. *Break-Even Analysis* A small business invests $33,000 in equipment to produce a product. Each unit of the product costs $1.70 to produce and is sold for $5.00. How many units must be sold before the business breaks even?

65. *Acid Mixture* One hundred gallons of a 60% acid solution is obtained by mixing a 75% solution with a 50% solution. How many gallons of each solution must be used to obtain the desired mixture?

66. *Alcohol Mixture* Fifty gallons of a 90% alcohol solution is obtained by mixing a 100% solution with a 75% solution. How many gallons of each solution must be used to obtain the desired mixture?

67. *Geometry*   The perimeter of a rectangle is 480 meters and its length is 150% of its width. Find the dimensions of the rectangle.

68. *Rope Length*   Suppose that you must cut a rope that is 128 inches long into two pieces such that one piece is three times as long as the other. Find the length of each piece.

69. *Cassette Tape Sales*   You are the manager of a music store and are going over receipts for the previous week's sales. Six hundred and fifty cassette tapes of two different types were sold. One type of cassette sold for $9.95 and the other sold for $14.95. The total cassette receipts were $7717.50. The cash register that was supposed to record the number of each type of cassette sold malfunctioned. Can you recover the information? If so, how many of each type of cassette were sold?

70. *Flying Speeds*   Two planes leave Pittsburgh and Philadelphia at the same time, each going to the other city. Because of the wind, one plane flies 25 miles per hour faster than the other. Find the ground speed of each plane if the cities are 275 miles apart and the planes pass one another (at different altitudes) after 40 minutes of flying time.

71. *Flying Speeds*   One plane flies 450 miles from City A to City B while a second plane, leaving at the same time, flies from City B to City A. Because of the wind, one plane travels 40 miles per hour faster than the other. Find the ground speed of each plane, if the planes pass one another (at different altitudes) after 50 minutes of flying time.

72. *Investments*   An inheritance of $20,000 is divided among three investments yielding $1780 in interest per year. The interest rates for the three investments are 7%, 9%, and 11%. Find the amount placed in each investment if the second and third are $3000 and $1000 less than the first, respectively.

73. *Number Problem*   The sum of three positive numbers is 68. The second number is four greater than the first, and the third is twice the first. Find the three numbers.

Curve-Fitting   In Exercises 74 and 75, find a quadratic equation $y = ax^2 + bx + c$ whose graph passes through the given points.

74. $(0, -6), (1, -3),$
    $(2, 4)$

75. $(-5, 0), (1, -6),$
    $(2, 14)$

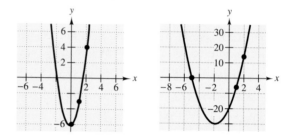

76. *Mathematical Modeling*   A child throws a softball over a garage. The location of the eaves and the peak of the roof are given by $(0, 10), (15, 15),$ and $(30, 10)$.

(a) Find the equation $y = ax^2 + bx + c$ for the path of the ball if the ball follows a path 1 foot over the eaves and the peak of the roof.

(b) Use a graphing utility to graph the path of the ball in part (a).

(c) From the graph, estimate how far from the edge of the garage the child was standing if the ball was at a height of 5 feet when it left his hand.

In Exercises 77–80, use a determinant to find the area of the triangle with the given vertices.

77. $(1, 0), (5, 0), (5, 8)$

78. $(-4, 0), (4, 0), (0, 6)$

79. $(1, 2), (4, -5), (3, 2)$

80. $\left(\frac{3}{2}, 1\right), \left(4, -\frac{1}{2}\right), (4, 2)$

In Exercises 81–84, use a determinant to find the equation of the line through the points.

81. $(-4, 0), (4, 4)$

82. $(2, 5), (6, -1)$

83. $\left(-\frac{5}{2}, 3\right), \left(\frac{7}{2}, 1\right)$

84. $(-0.8, 0.2), (0.7, 3.2)$

Chapter Test

Take this test as you would take a test in class. After you are done, check your work against the answers given in the back of the book.

$2x - 2y = 1$
$-x + 2y = 0$
System for 1

1. Which ordered pair is the solution of the system at the left: $(3, -4)$ or $\left(1, \frac{1}{2}\right)$?

In Exercises 2–13, use the indicated method to solve the system.

2. *Substitution:* $5x - y = 6$
$4x - 3y = -4$

3. *Substitution:* $x + y = 8$
$x^2 + y = 10$

4. *Graphical:* $x - 2y = -1$
$2x + 3y = 12$

5. *Elimination:* $3x - 4y = -14$
$-3x + y = 8$

6. *Elimination:* $8x + 3y = 3$
$4x - 6y = -1$

7. *Elimination:* $x + 2y - 4z = 0$
$3x + y - 2z = 5$
$3x - y + 2z = 7$

8. *Matrices:* $x \qquad - 3z = -10$
$-2y + 2z = 0$
$x - 2y \qquad = -7$

9. *Matrices:* $x - 3y + z = -3$
$3x + 2y - 5z = 18$
$y + z = -1$

10. *Cramer's Rule:* $2x - 7y = 7$
$3x + 7y = 13$

11. *Graphical:* $x - 2y = -3$
$2x + 3y = 22$

12. *Any Method:* $3x - 2y + z = 12$
$x - 3y \qquad = 2$
$-3x \qquad - 9z = -6$

13. *Any Method:* $4x + y + 2z = -4$
$3y + z = 8$
$-3x + y - 3z = 5$

$A = \begin{bmatrix} 3 & -2 & 0 \\ -1 & 5 & 3 \\ 2 & 7 & 1 \end{bmatrix}$

Matrix for 15

$5x - 8y = 3$
$3x + ay = 0$
System for 16

14. Describe the types of possible solutions of a system of linear equations.

15. Evaluate the determinant of A, as shown at the left.

16. Find the value of a such that the system at the left is inconsistent.

17. Find a system of linear equations with integer coefficients that has the solution $(5, -3)$. (The problem has many correct answers.)

18. Two people share the driving on a 200-mile trip. One person drives four times as far as the other. Write a system of linear equations that models the problem. Find the distance each person drives.

19. Find a quadratic equation $y = ax^2 + bx + c$ whose graph passes through the points $(0, 4)$, $(1, 3)$, and $(2, 6)$.

20. An inheritance of $25,000 is divided among three investments yielding $1275 in interest per year. The interest rates for the three investments are 4.5%, 5%, and 8%. Find the amount placed in each investment if the second and third investments are $4000 and $10,000 less than the first, respectively.

21. Find the area of the triangle with vertices $(0, 0)$, $(5, 4)$, and $(6, 0)$.

9 Exponential and Logarithmic Functions

Bruce Ayers/Tony Stone Images

In 1997, the total financial assets held by households in the United States was $27 trillion. Assets included deposits in savings and money market accounts, government securities, corporate equities, pension funds, and mutual fund shares. (Source: Board of Governors of the Federal Reserve System)

Choosing the Best Investment

You receive an inheritance of $5000 and want to invest it.

See Section 9.1, Exercise 104

a. Complete the table by finding the amount A of the $5000 investment after 3 years with an annual interest rate of $r = 6\%$. Which form of compounding gives you the greatest balance?

Compounding	Amount A
Annually	
Quarterly	
Monthly	
Daily	
Hourly	
Continuously	

b. You are considering two different investment options. The first investment option has an interest rate of 7% compounded continuously. The second investment option has an interest rate of 8% compounded quarterly. Which investment would you choose? Explain.

See Section 9.5, Exercise 139

c. What annual percentage rate is needed to obtain a balance of $6200 in 3 years if the interest is compounded monthly?

d. If $r = 6\%$ and the interest is compounded continuously, how long will it take for your inheritance to grow to $7500?

e. What is the *effective yield* on your investment if the interest rate is 8% compounded quarterly?

f. With an interest rate of 6%, compounded continuously, how long will it take your inheritance to double? How long will it take your inheritance to quadruple (reach four times the original amount)?

Objectives

1 Evaluate exponential functions.

2 Use the point-plotting method or a graphing utility to graph an exponential function.

3 Evaluate the natural base e and graph the natural exponential function.

4 Use an exponential function to solve an application problem.

1 Evaluate exponential functions.

Exponential Functions

In this section, you will study a new type of function called an **exponential function.** Whereas polynomial and rational functions have terms with variable bases and constant exponents, exponential functions have terms with *constant bases* and *variable exponents.* Here are some examples.

Polynomial or Rational Function

Constant Exponent

$$x^2, \quad x^{-3}$$

Variable Base

Exponential Function

Variable Exponent

$$2^x, \quad 3^{-x}$$

Constant Base

> ▶ **Definition of Exponential Function**
>
> The **exponential function** f **with base** a is denoted by
>
> $$f(x) = a^x$$
>
> where $a > 0$, $a \neq 1$, and x is any real number.

The base $a = 1$ is excluded from exponential functions because $f(x) = 1^x = 1$ is a constant function, *not* an exponential function.

In Chapter 5, you learned to evaluate a^x for integer and rational values of x. For example, you know that

$$a^3 = a \cdot a \cdot a, \quad a^{-4} = \frac{1}{a^4}, \quad \text{and} \quad a^{5/3} = \left(\sqrt[3]{a}\right)^5.$$

However, to evaluate a^x for any real number x, you need to interpret forms with *irrational* exponents, such as $a^{\sqrt{2}}$ or a^{π}. For the purpose of this text, it is sufficient to think of a number such as

$$a^{\sqrt{2}}$$

where $\sqrt{2} \approx 1.414214$, as the number that has the successively closer approximations

$$a^{1.4}, a^{1.41}, a^{1.414}, a^{1.4142}, a^{1.41421}, a^{1.414214}, \ldots .$$

The rules of exponents that were discussed in Section 5.1 can be extended to cover exponential functions, as described on page 549.

> ▶ **Rules of Exponential Functions**
>
> **1.** $a^x \cdot a^y = a^{x+y}$ Product Rule
>
> **2.** $\dfrac{a^x}{a^y} = a^{x-y}$ Quotient Rule
>
> **3.** $(a^x)^y = a^{xy}$ Power Rule
>
> **4.** $a^{-x} = \dfrac{1}{a^x} = \left(\dfrac{1}{a}\right)^x$ Negative Exponent Rule

To evaluate exponential functions with a calculator, you can use the exponential key $\boxed{y^x}$ (where y is the base and x is the exponent) or $\boxed{\wedge}$. For example, to evaluate $3^{-1.3}$, you can use the following keystrokes.

Keystrokes	*Display*	
3 $\boxed{y^x}$ 1.3 $\boxed{+/-}$ $\boxed{=}$	0.239741	Scientific
3 $\boxed{\wedge}$ $\boxed{(}$ $\boxed{(-)}$ 1.3 $\boxed{)}$ $\boxed{\text{ENTER}}$	0.239741	Graphing

Example 1   Evaluating Exponential Functions

Evaluate each function at the indicated values of x. Use a calculator only if it is necessary or more efficient.

Function	*Values*
a. $f(x) = 2^x$	$x = 3, x = -4, x = \pi$
b. $g(x) = 12^x$	$x = 3, x = -0.1, x = \frac{5}{7}$
c. $j(x) = 200(1.04)^{2x}$	$x = 1, x = -2, x = \sqrt{2}$

Solution

Evaluation	*Comment*
a. $f(3) = 2^3 = 8$	Calculator is not necessary.
$f(-4) = 2^{-4} = \dfrac{1}{2^4} = \dfrac{1}{16}$	Calculator is not necessary.
$f(\pi) = 2^{\pi} \approx 8.825$	Calculator is necessary.
b. $g(3) = 12^3 = 1728$	Calculator is more efficient.
$g(-0.1) = 12^{-0.1} \approx 0.7800$	Calculator is necessary.
$g\left(\dfrac{5}{7}\right) = 12^{5/7} \approx 5.900$	Calculator is necessary.
c. $j(1) = 200(1.04)^{2(1)} = 216.32$	Calculator is more efficient.
$j(-2) = 200(1.04)^{2(-2)} \approx 170.961$	Calculator is more efficient.
$j(\sqrt{2}) = 200(1.04)^{2\sqrt{2}} \approx 223.464$	Calculator is necessary.

2 Use the point-plotting method or a graphing utility to graph an exponential function.

Graphs of Exponential Functions

The basic nature of the graph of an exponential function can be determined by the point-plotting method or by using a graphing utility.

> **Example 2** The Graphs of Exponential Functions

In the same coordinate plane, sketch the graphs of the following functions. Determine the domains and ranges.

a. $f(x) = 2^x$ **b.** $g(x) = 4^x$

Solution

The table lists some values of each function, and Figure 9.1 shows their graphs. From the graphs, you can see that the domain of each function is the set of all real numbers and that the range of each function is the set of all positive real numbers.

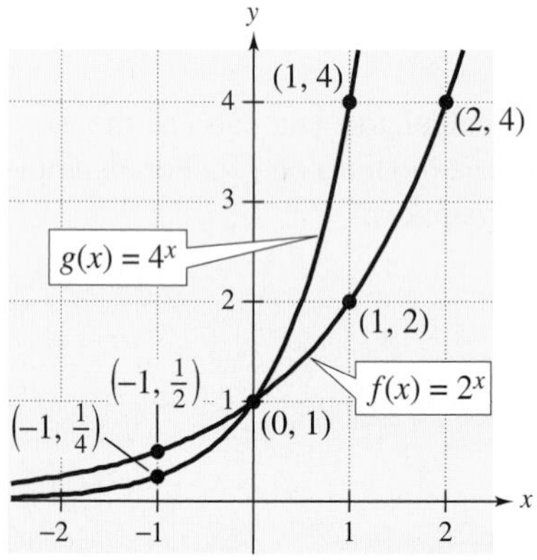

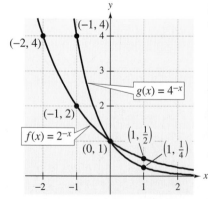

Figure 9.1

x	-2	-1	0	1	2	3
2^x	$\frac{1}{4}$	$\frac{1}{2}$	1	2	4	8
4^x	$\frac{1}{16}$	$\frac{1}{4}$	1	4	16	64

Note in the next example that a graph of the form $f(x) = a^x$ (as shown in Example 2) is a reflection in the y-axis of a graph of the form $g(x) = a^{-x}$.

> **Example 3** The Graphs of Exponential Functions

In the same coordinate plane, sketch the graph of each function.

a. $f(x) = 2^{-x}$ **b.** $g(x) = 4^{-x}$

Solution

The table lists some values of each function, and Figure 9.2 shows their graphs.

x	-3	-2	-1	0	1	2
2^{-x}	8	4	2	1	$\frac{1}{2}$	$\frac{1}{4}$
4^{-x}	64	16	4	1	$\frac{1}{4}$	$\frac{1}{16}$

Figure 9.2

Examples 2 and 3 suggest that for $a > 1$, the graph of $y = a^x$ increases and the graph of $y = a^{-x}$ decreases. Remember that $a^{-x} = (1/a)^x$. So, increasing the power on a fraction $(1/a) < 1$ yields smaller and smaller values. The graphs shown in Figure 9.3 are typical of the graphs of exponential functions. Note that each has a y-intercept at $(0, 1)$ and a horizontal asymptote of $y = 0$.

Graph of $y = a^x$

- Domain: $(-\infty, \infty)$
- Range: $(0, \infty)$
- Intercept: $(0, 1)$
- Increasing

Graph of $y = a^{-x} = \left(\dfrac{1}{a}\right)^x$

- Domain: $(-\infty, \infty)$
- Range: $(0, \infty)$
- Intercept: $(0, 1)$
- Decreasing

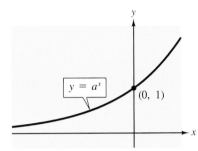

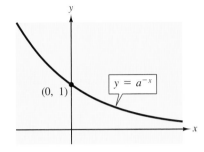

Figure 9.3 *Characteristics of the Exponential Functions a^x and $a^{-x}\, (a > 1)$*

In the next two examples, notice how the graph of $y = a^x$ can be used to sketch the graphs of functions of the form $f(x) = b \pm a^{x+c}$. Also note that the transformation in Example 4(a) keeps the x-axis as a horizontal asymptote, but the transformation in Example 4(b) yields a new horizontal asymptote of $y = -2$. Also, be sure to note how the y-intercept is affected by each transformation.

Example 4   Sketching Graphs of Exponential Functions

Each of the following graphs is a transformation of the graph of $f(x) = 3^x$, as shown in Figure 9.4.

a. Because $g(x) = 3^{x+1} = f(x + 1)$, the graph of g can be obtained by shifting the graph of f 1 unit to the left.

b. Because $h(x) = 3^x - 2 = f(x) - 2$, the graph of h can be obtained by shifting the graph of f down 2 units.

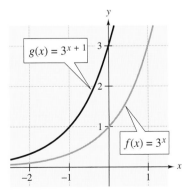

(a)

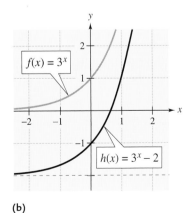

(b)

Figure 9.4

| Example 5 | Reflections of Exponential Functions |

Each of the following graphs is a reflection of the graph of $f(x) = 3^x$, as shown in Figure 9.5.

a. Because $k(x) = -3^x = -f(x)$, the graph of k can be obtained by reflecting the graph of f in the x-axis.

b. Because $j(x) = 3^{-x} = f(-x)$, the graph of j can be obtained by reflecting the graph of f in the y-axis.

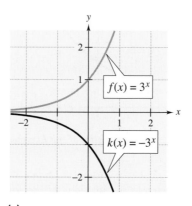

(a)

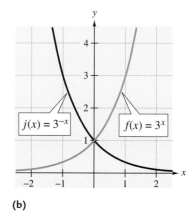

(b)

Figure 9.5

3 Evaluate the natural base e and graph the natural exponential function.

The Natural Exponential Function

So far, we have used integers or rational numbers as bases of exponential functions. In many applications of exponential functions, the convenient choice for a base is the irrational number, denoted by the letter "e."

$$e \approx 2.71828 \ldots \qquad \text{Natural base}$$

This number is called the **natural base.** The function

$$f(x) = e^x \qquad \text{Natural exponential function}$$

is called the **natural exponential function.** Be sure you understand that for this function, e is the constant number $2.71828 \ldots$, and x is a variable. To evaluate the natural exponential function, you need a calculator, preferably one having a natural exponential key $\boxed{e^x}$. Here are some examples of how to use such a calculator to evaluate the natural exponential function.

Value	Keystrokes	Display	
e^2	2 $\boxed{e^x}$	7.3890561	Scientific
e^2	$\boxed{e^x}$ 2 $\boxed{\text{ENTER}}$	7.3890561	Graphing
e^{-3}	3 $\boxed{+/-}$ $\boxed{e^x}$	0.049787	Scientific
e^{-3}	$\boxed{e^x}$ $\boxed{(}$ $\boxed{(-)}$ 3 $\boxed{)}$ $\boxed{\text{ENTER}}$	0.049787	Graphing
$e^{0.32}$	.32 $\boxed{e^x}$	1.3771278	Scientific
$e^{0.32}$	$\boxed{e^x}$ .32 $\boxed{\text{ENTER}}$	1.3771278	Graphing

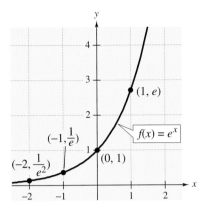

Figure 9.6

Some calculators do not have a key labeled $\boxed{e^x}$. If your calculator does not have this key, but does have a key labeled $\boxed{\ln x}$, you will have to use the two-keystroke sequence $\boxed{\text{INV}}$ $\boxed{\ln x}$ in place of $\boxed{e^x}$.

After evaluating the natural exponential function at several values, as shown in the table, you can sketch its graph, as shown in Figure 9.6.

x	-1.5	-1.0	-0.5	0.0	0.5	1.0	1.5
$f(x) = e^x$	0.223	0.368	0.607	1.000	1.649	2.718	4.482

From the graph, notice the following properties of the natural exponential function.

- The domain is the set of all real numbers.
- The range is the set of positive real numbers.
- The y-intercept is $(0, 1)$.

4 Use an exponential function to solve an application problem.

Applications

A common scientific application of exponential functions is that of **radioactive decay**.

Example 6   Radioactive Decay

Let y represent the mass of a particular radioactive element whose half-life is 25 years. The initial mass is 10 grams. After t years, the mass (in grams) is given by

$$y = 10\left(\frac{1}{2}\right)^{t/25}, \quad t \geq 0.$$

How much of the initial mass remains after 120 years?

Solution

When $t = 120$, the mass is given by

$$y = 10\left(\frac{1}{2}\right)^{120/25} \qquad \text{Substitute 120 for } t.$$

$$= 10\left(\frac{1}{2}\right)^{4.8} \qquad \text{Simplify.}$$

$$\approx 0.359 \text{ gram.} \qquad \text{Use a calculator.}$$

So, after 120 years, the mass has decayed from an initial amount of 10 grams to only 0.359 gram. Note in Figure 9.7 that the graph of the function shows the 25-year half-life. That is, after 25 years the mass is 5 grams (half of the original), after another 25 years the mass is 2.5 grams, and so on.

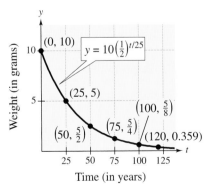

Figure 9.7

One of the most familiar uses of exponential functions involves **compound interest.** A principal P is invested at an annual interest rate r (in decimal form), compounded once a year. If the interest is added to the principal at the end of the year, the balance is

$$A = P + Pr = P(1 + r).$$

This pattern of multiplying the previous principal by $(1 + r)$ is then repeated each successive year, as shown below.

Time in Years	Balance at Given Time
0	$A = P$
1	$A = P(1 + r)$
2	$A = P(1 + r)(1 + r) = P(1 + r)^2$
3	$A = P(1 + r)^2(1 + r) = P(1 + r)^3$
$\vdots$	$\vdots$
t	$A = P(1 + r)^t$

To account for more frequent compounding of interest (such as quarterly or monthly compounding), let n be the number of compoundings per year and let t be the number of years. Then the rate per compounding is r/n and the account balance after t years is

$$A = P\left(1 + \frac{r}{n}\right)^{nt}.$$

Example 7   Finding the Balance for Compound Interest

A sum of $10,000 is invested at an annual interest rate of 7.5%, compounded monthly. Find the balance in the account after 10 years.

Solution

Using the formula for compound interest, with $P = 10{,}000$, $r = 0.075$, $n = 12$ (for monthly compounding), and $t = 10$, you obtain the following balance.

$$A = 10{,}000\left(1 + \frac{0.075}{12}\right)^{12(10)} \approx \$21{,}120.65$$

A second method that banks use to compute interest is called **continuous compounding.** The formula for the balance for this type of compounding is

$$A = Pe^{rt}.$$

The formulas for both types of compounding are summarized on the next page.

▶ **Formulas for Compound Interest**

After t years, the balance A in an account with principal P and annual interest rate r (in decimal form) is given by the following formulas.

1. For n compoundings per year: $A = P\left(1 + \dfrac{r}{n}\right)^{nt}$

2. For continuous compounding: $A = Pe^{rt}$

Example 8   Comparing Three Types of Compounding

A total of \$15,000 is invested at an annual interest rate of 8%. Find the balance after 6 years if the interest is compounded

a. quarterly,   **b.** monthly,   and   **c.** continuously.

Solution

a. Letting $P = 15{,}000$, $r = 0.08$, $n = 4$, and $t = 6$, the balance after 6 years at quarterly compounding is

$$A = 15{,}000\left(1 + \frac{0.08}{4}\right)^{4(6)}$$

$$= \$24{,}126.56.$$

b. Letting $P = 15{,}000$, $r = 0.08$, $n = 12$, and $t = 6$, the balance after 6 years at monthly compounding is

$$A = 15{,}000\left(1 + \frac{0.08}{12}\right)^{12(6)}$$

$$= \$24{,}202.53.$$

c. Letting $P = 15{,}000$, $r = 0.08$, and $t = 6$, the balance after 6 years at continuous compounding is

$$A = 15{,}000e^{0.08(6)}$$

$$= \$24{,}241.12.$$

Note that the balance is greater with continuous compounding than with quarterly and monthly compounding.

Example 8 illustrates the following general rule. For a given principal, interest rate, and time, the more often the interest is compounded per year, the greater the balance will be. Moreover, the balance obtained by continuous compounding is larger than the balance obtained by compounding n times per year.

Technology: Discovery

Use a graphing utility to evaluate

$$A = 15{,}000\left(1 + \frac{0.08}{n}\right)^{n(6)}$$

for $n = 1000$, $10{,}000$, and $100{,}000$. Compare these values with those found in parts (a) and (b) of Example 8.

As n gets larger and larger, do you think that the value of A will ever exceed the value found in Example 8(c)? Explain.

Discussing the Concept   **Finding a Pattern**

Use a graphing utility to investigate the function $f(x) = k^x$ for $0 < k < 1$, $k = 1$, and $k > 1$. Discuss the effect that k has on the shape of the graph.

9.1 Exercises

Keep mathematically in shape by doing these exercises *before* the problems of this section.

Properties and Definitions

1. Explain how to determine the half-plane satisfying $x + y < 5$.

2. Describe the difference between the graphs of $3x - 5y \leq 15$ and $3x - 5y < 15$.

Graphing Inequalities

In Exercises 3–10, graph the inequality.

3. $y > x - 2$

4. $y \leq 5 - \frac{3}{2}x$

5. $y < \frac{2}{3}x - 1$

6. $x > 6 - y$

7. $y \leq -2$

8. $x > 7$

9. $2x + 3y \geq 6$

10. $5x - 2y < 5$

Problem Solving

11. Working together, two people can complete a task in 10 hours. Working alone, one person takes 3 hours longer than the other. How long would it take each to do the task alone?

12. A family is setting up the boundaries for a backyard volleyball court. The court is to be 60 feet long and 30 feet wide. To be assured that the court is rectangular, someone suggests that they measure the diagonals of the court. What should be the length of each diagonal?

Developing Skills

In Exercises 1–8, simplify the expression.

1. $2^x \cdot 2^{x-1}$

2. $10e^{2x} \cdot e^{-x}$

3. $\dfrac{e^{x+2}}{e^x}$

4. $\dfrac{3^{2x+3}}{3^{x+1}}$

5. $(2e^x)^3$

6. $-4e^{-2x}$

7. $\sqrt[3]{-8e^{3x}}$

8. $\sqrt{4e^{6x}}$

In Exercises 9–16, approximate the expression to three decimal places.

9. $4^{\sqrt{3}}$

10. $6^{-\pi}$

11. $e^{1/3}$

12. $e^{-1/3}$

13. $4(3e^4)^{1/2}$

14. $(9e^2)^{3/2}$

15. $\dfrac{4e^3}{12e^2}$

16. $\dfrac{6e^5}{10e^7}$

In Exercises 17–30, evaluate the function as indicated. Use a calculator only if it is more efficient or necessary. (Round to three decimal places.) See Example 1.

17. $f(x) = 3^x$
 (a) $x = -2$
 (b) $x = 0$
 (c) $x = 1$

18. $F(x) = 3^{-x}$
 (a) $x = -2$
 (b) $x = 0$
 (c) $x = 1$

19. $g(x) = 1.07^x$
 (a) $x = -1$
 (b) $x = 3$
 (c) $x = \sqrt{5}$

20. $G(x) = 2.04^{-x}$
 (a) $x = -1$
 (b) $x = 1$
 (c) $x = \sqrt{3}$

21. $f(t) = 500\left(\frac{1}{2}\right)^t$
 (a) $t = 0$
 (b) $t = 1$
 (c) $t = \pi$

22. $g(s) = 1200\left(\frac{2}{3}\right)^s$
 (a) $s = 0$
 (b) $s = 2$
 (c) $s = \sqrt{2}$

23. $f(x) = 1000(1.05)^{2x}$
 (a) $x = 0$
 (b) $x = 5$
 (c) $x = 10$

24. $g(t) = 10,000(1.03)^{4t}$
 (a) $t = 1$
 (b) $t = 3$
 (c) $t = 5.5$

25. $h(x) = \dfrac{5000}{(1.06)^{8x}}$
 (a) $x = 5$
 (b) $x = 10$
 (c) $x = 20$

26. $P(t) = \dfrac{10,000}{(1.01)^{12t}}$
 (a) $t = 2$
 (b) $t = 10$
 (c) $t = 20$

27. $g(x) = 10e^{-0.5x}$
 (a) $x = -4$
 (b) $x = 4$
 (c) $x = 8$

28. $A(t) = 200e^{0.1t}$
 (a) $t = 10$
 (b) $t = 20$
 (c) $t = 40$

29. $g(x) = \dfrac{1000}{2 + e^{-0.12x}}$

(a) $x = 0$

(b) $x = 10$

(c) $x = 50$

30. $f(z) = \dfrac{100}{1 + e^{-0.05z}}$

(a) $z = 0$

(b) $z = 10$

(c) $z = 20$

In Exercises 31–50, graph the function. See Examples 2 and 3.

31. $f(x) = 3^x$

32. $f(x) = 3^{-x} = \left(\frac{1}{3}\right)^x$

33. $h(x) = \frac{1}{2}(3^x)$

34. $h(x) = \frac{1}{2}(3^{-x})$

35. $g(x) = 3^x - 2$

36. $g(x) = 3^x + 1$

37. $f(x) = 4^{x-5}$

38. $f(x) = 4^{x+1}$

39. $g(x) = 4^x - 5$

40. $g(x) = 4^x + 1$

41. $f(t) = 2^{-t^2}$

42. $f(t) = 2^{t^2}$

43. $f(x) = -2^{0.5x}$

44. $h(t) = -2^{-0.5t}$

45. $h(x) = 2^{0.5x}$

46. $g(x) = 2^{-0.5x}$

47. $f(x) = -\left(\frac{1}{3}\right)^x$

48. $f(x) = \left(\frac{3}{4}\right)^x + 1$

49. $g(t) = 200\left(\frac{1}{2}\right)^t$

50. $h(y) = 27\left(\frac{2}{3}\right)^y$

In Exercises 51–58, match the function with its graph. [The graphs are labeled (a), (b), (c), (d), (e), (f), (g), and (h).]

(a)

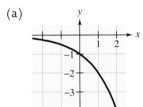

(b)

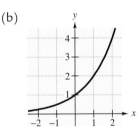

(c)

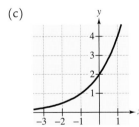

(d)

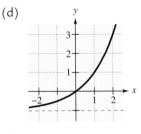

(e)

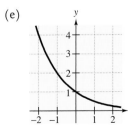

(f)

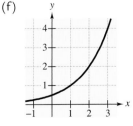

(g)

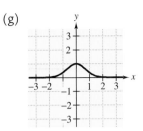

(h)

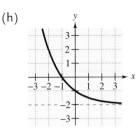

51. $f(x) = 2^x$

52. $f(x) = -2^x$

53. $f(x) = 2^{-x}$

54. $f(x) = 2^x - 1$

55. $f(x) = 2^{x-1}$

56. $f(x) = 2^{x+1}$

57. $f(x) = \left(\frac{1}{2}\right)^x - 2$

58. $f(x) = e^{-x^2}$

In Exercises 59–70, use a graphing utility to graph the function.

59. $y = 5^{x/3}$

60. $y = 5^{-x/3}$

61. $y = 5^{(x-2)/3}$

62. $y = 5^{-x/3} + 2$

63. $y = 500(1.06)^t$

64. $y = 100(1.06)^{-t}$

65. $y = 3e^{0.2x}$

66. $y = 50e^{-0.05x}$

67. $P(t) = 100e^{-0.1t}$

68. $A(t) = 1000e^{0.08t}$

69. $y = 6e^{-x^2/3}$

70. $g(x) = 7e^{(x+1)/2}$

In Exercises 71–76, identify the transformation of the graph of $f(x) = 4^x$ and sketch a graph of h. See Examples 4 and 5.

71. $h(x) = 4^x - 1$

72. $h(x) = 4^x + 2$

73. $h(x) = 4^{x+2}$

74. $h(x) = 4^{x-4}$

75. $h(x) = -4^x$

76. $h(x) = -4^x + 2$

77. *Think About It*   What type of function does each equation represent?

(a) $f(x) = 2x$

(b) $f(x) = \sqrt{2}x$

(c) $f(x) = 2^x$

(d) $f(x) = 2x^2$

78. *Identifying Graphs*   Identify the graphs of

$$y_1 = e^{0.2x}, \quad y_2 = e^{0.5x}, \quad \text{and} \quad y_3 = e^x$$

in the figure. Describe the effect on the graph of $y = e^{kx}$ when $k > 0$ is changed.

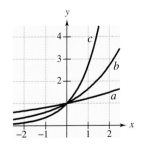

Solving Problems

79. *Radioactive Decay* After t years, the initial mass of 16 grams of a radioactive element whose half-life is 30 years is given by

$$y = 16\left(\frac{1}{2}\right)^{t/30}, \quad t \geq 0.$$

How much of the initial mass remains after 80 years?

80. *Radioactive Decay* After t years, the initial mass of 23 grams of a radioactive element whose half-life is 45 years is given by

$$y = 23\left(\frac{1}{2}\right)^{t/45}, \quad t \geq 0.$$

How much of the initial mass remains after 150 years?

In Exercises 81–86, complete the table to determine the balance A for P dollars invested at rate r for t years, compounded n times per year.

n	1	4	12	365	Continuous compounding
A					

	Principal	Rate	Time
81.	$P = \$100$	$r = 8\%$	$t = 20$ years
82.	$P = \$400$	$r = 8\%$	$t = 50$ years
83.	$P = \$2000$	$r = 9\%$	$t = 10$ years
84.	$P = \$1500$	$r = 7\%$	$t = 2$ years
85.	$P = \$5000$	$r = 10\%$	$t = 40$ years
86.	$P = \$10,000$	$r = 9.5\%$	$t = 30$ years

In Exercises 87–90, complete the table to determine the principal P that will yield a balance of A dollars when invested at rate r for t years, compounded n times per year.

n	1	4	12	365	Continuous compounding
P					

	Balance	Rate	Time
87.	$A = \$5000$	$r = 7\%$	$t = 10$ years
88.	$A = \$100,000$	$r = 9\%$	$t = 20$ years
89.	$A = \$1,000,000$	$r = 10.5\%$	$t = 40$ years
90.	$A = \$2500$	$r = 7.5\%$	$t = 2$ years

91. *Price and Demand* The daily demand x and the price p for a certain product are related by

$$p = 25 - 0.4e^{0.02x}.$$

Find the prices for demands of (a) $x = 100$ units and (b) $x = 125$ units.

92. *Population Growth* The population of the United States (in recent years) can be approximated by the exponential function

$$P(t) = 205.7(1.0098)^t$$

where P is the population (in millions) and t is the time in years, with $t = 0$ corresponding to 1970. Use the model to estimate the population in the years (a) 2000 and (b) 2010.

93. *Property Value* Suppose that the value of a piece of property doubles every 15 years. If you buy the property for $64,000, its value t years after the date of purchase should be

$$V(t) = 64,000(2)^{t/15}.$$

Use the model to approximate the value of the property (a) 5 years and (b) 20 years after it is purchased.

94. *Inflation Rate* Suppose that the annual rate of inflation averages 5% over the next 10 years. With this rate of inflation, the approximate cost C of goods or services during any year in that decade will be given by

$$C(t) = P(1.05)^t, \quad 0 \leq t \leq 10$$

where t is time in years and P is the present cost. If the price of an oil change for your car is presently $24.95, estimate the price 10 years from now.

95. *Depreciation* After t years, the value of a car that originally cost $16,000 depreciates so that each year it is worth $\frac{3}{4}$ of its value for the previous year. Find a model for $V(t)$, the value of the car after t years. Sketch a graph of the model and determine the value of the car 2 years after it was purchased.

96. *Depreciation* Suppose straight-line depreciation is used to determine the value of the car in Exercise 95. Assume that the car depreciates $3000 per year.

(a) Write a linear equation for $V(t)$, the value of the car for year t.

(b) Sketch the graph of the model in part (a) on the same coordinate axes used for the graph in Exercise 95.

(c) If you were selling the car after owning it for 2 years, which depreciation model would you prefer?

(d) If you were selling the car after 4 years, which model would be to your advantage?

97. *Graphical Interpretation* An investment of $500 in two different accounts with respective interest rates of 6% and 8% is compounded continuously.

(a) Write an exponential function to represent the balance after t years for each account.

(b) Use a graphing utility to graph each of the models in the same viewing rectangle.

(c) Use a graphing utility to graph the function $A_2 - A_1$ on the same screen used for the graphs in part (b).

(d) Use the graphs to discuss the rates of increase of the balances in the two accounts.

98. *Savings Plan* Suppose you decide to start saving pennies according to the following pattern. You save 1 penny the first day, 2 pennies the second day, 4 the third day, 8 the fourth day, and so on. Each day you save twice the number of pennies as the previous day. Write an exponential function that models this problem. How many pennies do you save on the thirtieth day? (In the next chapter you will learn how to find the total number saved.)

99. *Parachute Drop* A parachutist jumps from a plane and opens the parachute at a height of 2000 feet (see figure). The distance between the parachutist and the ground is

$h = 1950 + 50e^{-1.6t} - 20t$

where h is the distance in feet and t is the time in seconds. (The time $t = 0$ corresponds to the time when the parachute is opened.)

(a) Use a graphing utility to graph the function.

(b) Find the distance between the parachutist and the ground when $t = 0, 25, 50,$ and 75.

(c) Approximate the time when the parachutist reaches the ground.

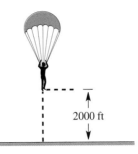

2000 ft

Figure for 99

100. *Parachute Drop* A parachutist jumps from a plane and opens the parachute at a height of 3000 feet. The distance between the parachutist and the ground is

$h = 2940 + 60e^{-1.7t} - 22t$

where h is the distance in feet and t is the time in seconds. (The time $t = 0$ corresponds to the time when the parachute is opened.)

(a) Use a graphing utility to graph the function.

(b) Find the distance between the parachutist and the ground when $t = 0, 50,$ and 100.

(c) Approximate the time when the parachutist reaches the ground.

101. *Data Analysis* A meteorologist measures the atmospheric pressure P (in kilograms per square meter) at altitude h (in kilometers). The data are shown in the table.

h	0	5	10	15	20
P	10,332	5583	2376	1240	517

(a) Use a graphing utility to plot the data points.

(b) A model for the data is given by

$P = 10{,}958e^{-0.15h}$.

Use a graphing utility to graph the model in the same viewing rectangle as in part (a). How well does the model fit the data?

(c) Use a graphing utility to create a table comparing the model with the data points.

(d) Estimate the atmospheric pressure at an altitude of 8 kilometers.

(e) Use the graph to estimate the altitude at which the atmospheric pressure is 2000 kilograms per square meter.

102. *Data Analysis* For the years 1991 through 1996, the median price of a one-family home sold in the United States is given in the table. (Source: National Association of Realtors)

Year	1991	1992	1993
Price	$100,300	$103,700	$106,800

Year	1994	1995	1996
Price	$109,900	$113,100	$118,200

A model for these data is given by

$$y = 97,107e^{0.0317t}$$

where t is time in years, with $t = 0$ representing 1990.

(a) Use the model to complete the table and compare the results with the actual data.

Year	1991	1992	1993	1994	1995	1996
Price						

(b) Use a graphing utility to graph the model.

(c) If the model were used to predict home prices in the years ahead, would the predictions be increasing at a faster rate or a slower rate with increasing t?

(d) Compare the model with the continuous compound interest model. What does the coefficient of t represent?

103. *Calculator Experiment*

(a) Use a calculator to complete the table.

x	1	10	100	1000	10,000
$\left(1 + \dfrac{1}{x}\right)^x$					

(b) Use the table to sketch the graph of the function

$$f(x) = \left(1 + \frac{1}{x}\right)^x.$$

Does this graph appear to be approaching a horizontal asymptote?

(c) From parts (a) and (b), what conclusions can you make about the value of

$$\left(1 + \frac{1}{x}\right)^x$$

as x gets larger and larger?

Explaining Concepts

104. Answer parts (a) and (b) of Motivating the Chapter on page 547.

105. Describe the differences between exponential functions and polynomial or rational functions.

106. Explain why 1^x is not an exponential function.

107. Compare the graphs of $f(x) = 3^x$ and $g(x) = \left(\frac{1}{3}\right)^x$.

108. Describe some applications of the exponential functions $f(x) = a^x$ and $g(x) = a^{-x}$.

109. *True or False?* $e = \dfrac{271,801}{99,990}$. Explain.

110. Without using a calculator, how do you know that $2^{\sqrt{2}}$ is greater than 2, but less than 4?

9.2   Inverse Functions

Objectives

1 Form the composition of two functions and find the domain of the composite function.

2 Use the Horizontal Line Test to determine if a function has an inverse.

3 Find the inverse of a function algebraically.

4 Graphically show that two functions are inverses of each other.

1 Form the composition of two functions and find the domain of the composite function.

Composition of Functions

Two functions can be combined to form another function called the **composition** of the two functions. For instance, if $f(x) = 2x^2$ and $g(x) = x - 1$, the composition of f with g is denoted by $f \circ g$ and is evaluated as

$$f(g(x)) = f(x - 1) = 2(x - 1)^2.$$

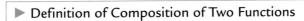 $f \circ g$

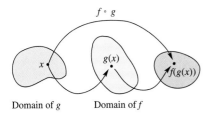

$g(x)$ $\quad$ $f(g(x))$

x

Domain of g $\qquad$ Domain of f

Figure 9.8

> ▶ **Definition of Composition of Two Functions**
>
> The **composition** of the functions f and g is given by
>
> $$(f \circ g)(x) = f(g(x)).$$
>
> The domain of the **composite function** $(f \circ g)$ is the set of all x in the domain of g such that $g(x)$ is in the domain of f. (See Figure 9.8.)

Study Tip

A composite function can be viewed as a *function within a function,* where the composition

$$(f \circ g)(x) = f(g(x))$$

has f as the "outer" function and g as the "inner" function. This is reversed in the composition

$$(g \circ f)(x) = g(f(x)).$$

Example 1   Forming the Composition of Two Functions

Find the composition of f with g. Evaluate the composite function when $x = 1$ and when $x = -3$.

$$f(x) = 2x + 4 \quad \text{and} \quad g(x) = 3x - 1$$

Solution

The composition of f with g is given by

$$\begin{aligned}
(f \circ g)(x) &= f(g(x)) && \text{Composite form} \\
&= f(3x - 1) && g(x) = 3x - 1 \text{ is the inner function.} \\
&= 2(3x - 1) + 4 && \text{Input } 3x - 1 \text{ into the outer function } f. \\
&= 6x - 2 + 4 && \text{Simplify.} \\
&= 6x + 2. && \text{Simplify.}
\end{aligned}$$

When $x = 1$, the value of this function is

$$(f \circ g)(1) = 6(1) + 2 = 8.$$

When $x = -3$, the value of this function is

$$(f \circ g)(-3) = 6(-3) + 2 = -16.$$

The composition of f with g is generally *not* the same as the composition of g with f. This is illustrated in Example 2.

Example 2   Comparing the Compositions of Functions

Given $f(x) = 2x - 3$ and $g(x) = x^2 + 1$, find each of the following.

a. $(f \circ g)(x)$ **b.** $(g \circ f)(x)$

Solution

a. The composition of f with g is as follows.

$$(f \circ g)(x) = f(g(x)) \qquad \text{Composite form}$$

$$= f(x^2 + 1) \qquad g(x) = x^2 + 1 \text{ is the inner function.}$$

$$= 2(x^2 + 1) - 3 \qquad \text{Input } x^2 + 1 \text{ into the outer function } f.$$

$$= 2x^2 + 2 - 3 \qquad \text{Distributive Property}$$

$$= 2x^2 - 1 \qquad \text{Simplify.}$$

b. The composition of g with f is as follows.

$$(g \circ f)(x) = g(f(x)) \qquad \text{Composite form}$$

$$= g(2x - 3) \qquad f(x) = 2x - 3 \text{ is the inner function.}$$

$$= (2x - 3)^2 + 1 \qquad \text{Input } 2x - 3 \text{ into the outer function } g.$$

$$= 4x^2 - 12x + 9 + 1 \qquad \text{Expand } (2x - 3)^2.$$

$$= 4x^2 - 12x + 10 \qquad \text{Simplify.}$$

Note that $(f \circ g)(x) \neq (g \circ f)(x)$.

Study Tip

To determine the domain of a composite function, first write the composite function in simplest form. Then use the fact that its domain *either is equal to or is a restriction of the domain of the "inner" function.* Check to see if the domain of $g \circ f$ in Example 3 is the same as or is a restriction of the domain of f.

Example 3   Finding the Domain of a Composite Function

Find the domain of the composition of $(f \circ g)(x)$ when $f(x) = x^2$ and $g(x) = \sqrt{x}$.

Solution

The composition of f with g is given by

$$(f \circ g)(x) = f(g(x)) \qquad \text{Composite form}$$

$$= f(\sqrt{x}) \qquad g(x) = \sqrt{x} \text{ is the inner function.}$$

$$= (\sqrt{x})^2 \qquad \text{Input } \sqrt{x} \text{ into the outer function } f.$$

$$= x, \quad x \geq 0. \qquad \text{Domain of } f \circ g \text{ is all } x \geq 0.$$

The domain of the inner function $g(x) = \sqrt{x}$ is the interval $[0, \infty)$. The simplified form of $f \circ g$ has no restriction on this set of numbers. So, the restriction $x \geq 0$ must be added to the composition of this function. The domain of $f \circ g$ is $[0, \infty)$.

2 Use the Horizontal Line Test to determine if a function has an inverse.

Inverse and One-to-One Functions

In Section 2.5, you learned that a function can be represented by a set of ordered pairs. For instance, the function $f(x) = x + 2$ from the set $A = \{1, 2, 3, 4\}$ to the set $B = \{3, 4, 5, 6\}$ can be written as follows.

$$f(x) = x + 2: \quad \{(1, 3), (2, 4), (3, 5), (4, 6)\}$$

By interchanging the first and second coordinates of each of these ordered pairs, you can form another function that is called the **inverse function** of f. The inverse function is denoted by f^{-1}. It is a function from the set B to the set A, and can be written as follows.

$$f^{-1}(x) = x - 2: \quad \{(3, 1), (4, 2), (5, 3), (6, 4)\}$$

A: Domain of f B: Range of f

1
2
3
4
f
3
4
5
6
f^{-1}

Range of f^{-1} Domain of f^{-1}

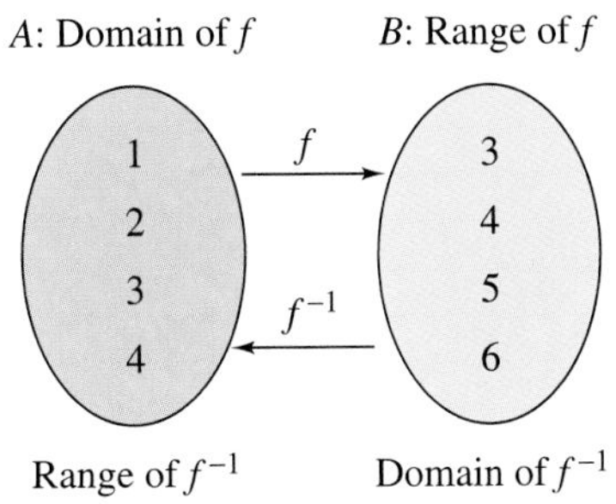

Figure 9.9 f *is one-to-one and has inverse* f^{-1}.

Interchanging the ordered pairs for a function f will only produce another function when f is one-to-one. A function f is **one-to-one** if each value of the dependent variable corresponds to exactly one value of the independent variable. Figure 9.9 shows that the domain of f is the range of f^{-1} and the range of f is the domain of f^{-1}.

> ▶ **Horizontal Line Test for Inverse Functions**
>
> A function f has an inverse function if and only if f is one-to-one. Graphically, a function f has an inverse function if and only if no *horizontal* line intersects the graph of f at more than one point.

Example 4   Applying the Horizontal Line Test

a. The function $f(x) = x^3 - 1$ *has* an inverse because no horizontal line intersects its graph at more than one point, as shown in Figure 9.10.

b. The function $f(x) = x^2 - 1$ *does not have* an inverse because it is possible to find a horizontal line that intersects the graph of f at more than one point, as shown in Figure 9.11.

c. The function f: $\{(1, 3), (2, 0), (3, -1), (4, 3)\}$ *does not have* an inverse function because the horizontal line $y = 3$ intersects two points of the graph of f, as shown in Figure 9.12.

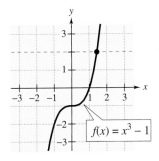

Figure 9.10

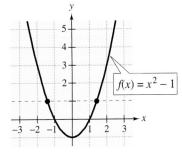

Figure 9.11

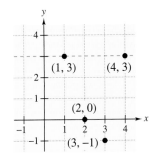

Figure 9.12

Don't be confused by the use of -1 to denote the inverse function f^{-1}. Whenever we write f^{-1}, we will *always* be referring to the inverse of the function f and *not* the reciprocal of $f(x)$.

The formal definition of the inverse of a function is given as follows.

▶ **Definition of the Inverse of a Function**

Let f and g be two functions such that

$$f(g(x)) = x \quad \text{for every } x \text{ in the domain of } g$$

and

$$g(f(x)) = x \quad \text{for every } x \text{ in the domain of } f.$$

The function g is the **inverse** of the function f, and is denoted by f^{-1} (read "f-inverse"). So, $f(f^{-1}(x)) = x$ and $f^{-1}(f(x)) = x$. The domain of f is equal to the range of f^{-1}, and vice versa.

If the function g is the inverse of the function f, it must also be true that the function f is the inverse of the function g. For this reason, you can refer to the functions f and g as being *inverses of each other*.

Example 5   Verifying Inverse Functions

Show that each function is an inverse of the other.

$$f(x) = x^3 + 1 \quad \text{and} \quad g(x) = \sqrt[3]{x - 1}$$

Solution

Begin by noting that the domain and range of both functions are the entire set of real numbers. To show that f and g are inverses of each other, you need to show that $f(g(x)) = x$ and $g(f(x)) = x$, as follows.

$$f(g(x)) = f\left(\sqrt[3]{x - 1}\right) \qquad \text{$g(x) = \sqrt[3]{x - 1}$ is the inner function.}$$
$$= \left(\sqrt[3]{x - 1}\right)^3 + 1 \qquad \text{Input $\sqrt[3]{x - 1}$ into the outer function f.}$$
$$= (x - 1) + 1 \qquad \text{Simplify.}$$
$$= x \qquad \text{Simplify.}$$

$$g(f(x)) = g(x^3 + 1) \qquad \text{$f(x) = x^3 + 1$ is the inner function.}$$
$$= \sqrt[3]{(x^3 + 1) - 1} \qquad \text{Input $x^3 + 1$ into the outer function g.}$$
$$= \sqrt[3]{x^3} \qquad \text{Simplify.}$$
$$= x \qquad \text{Simplify.}$$

Note that the two functions f and g "undo" each other in the following verbal sense. The function f first cubes the input x and then adds 1, whereas the function g first subtracts 1, and then takes the cube root of the result.

3 Find the inverse of a function algebraically.

Finding Inverse Functions

You can find the inverse of a simple function by inspection. For instance, the inverse of $f(x) = 10x$ is $f^{-1}(x) = x/10$. For more complicated functions, however, it is best to use the following steps for finding the inverse of a function. The key step in these guidelines is switching the roles of x and y. This step corresponds to the fact that inverse functions have ordered pairs with the coordinates reversed.

Study Tip

You can graph a function and use the Horizontal Line Test to see if the function is one-to-one before trying to find its inverse.

▶ **Finding the Inverse of a Function**

1. In the equation for $f(x)$, replace $f(x)$ by y.

2. Interchange the roles of x and y, and solve for y.

3. If the new equation does not represent y as a function of x, the function f does not have an inverse function.

4. If the new equation represents y as a function of x, replace y by $f^{-1}(x)$.

Example 6   Finding the Inverse of a Function

Find the inverse, if it exists.

a. $f(x) = 2x + 3$ **b.** $f(x) = x^2$ **c.** $f(x) = x^3 + 3$

Technology: Discovery

Use a graphing utility to graph $f(x) = x^3 + 1$, $f^{-1}(x) = \sqrt[3]{x - 1}$, and $y = x$ in the same viewing window.

a. Relative to the line $y = x$, how do the graphs of f and f^{-1} compare?

b. For the graph of f, complete the table.

x	-1	0	1
f			

For the graph of f^{-1}, complete the table.

x	0	1	2
f^{-1}			

What can you conclude about the coordinates of the points on the graph of f compared with those on the graph of f^{-1}?

Solution

a. $f(x) = 2x + 3$ Original function

$y = 2x + 3$ Replace $f(x)$ by y.

$x = 2y + 3$ Interchange x and y.

$y = \dfrac{x - 3}{2}$ Solve for y.

$f^{-1}(x) = \dfrac{x - 3}{2}$ Replace y by $f^{-1}(x)$.

b. $f(x) = x^2$ Original function

$y = x^2$ Replace $f(x)$ by y.

$x = y^2$ Interchange x and y.

$\pm\sqrt{x} = y$ Solve for y.

Because y is not a function of x, the original function f has no inverse.

c. $f(x) = x^3 + 3$ Original function

$y = x^3 + 3$ Replace $f(x)$ by y.

$x = y^3 + 3$ Interchange x and y.

$\sqrt[3]{x - 3} = y$ Solve for y.

$f^{-1}(x) = \sqrt[3]{x - 3}$ Replace y by $f^{-1}(x)$.

Graphically show that two
functions are inverses of each other.

Graphs of Inverse Functions

The graphs of a function f and its inverse f^{-1} are related to each other in the following way. If the point (a, b) lies on the graph of f, the point (b, a) must lie on the graph of f^{-1}, and vice versa. This means that the graph of f^{-1} is a reflection of the graph of f in the line $y = x$, as shown in Figure 9.13.

**Technology:
Tip**

A graphing utility program that graphs the function f and its reflection in the line $y = x$ can be found at our website *www.hmco.com*. Programs are available for several current models of graphing utilities.

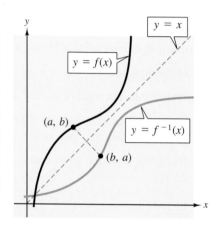

Figure 9.13

Example 7   The Graphs of f and f^{-1}

Sketch the graphs of the inverse functions $f(x) = 2x - 3$ and $f^{-1}(x) = \frac{1}{2}(x + 3)$ on the same rectangular coordinate system and show that the graphs are reflections of each other in the line $y = x$.

Solution

The graphs of f and f^{-1} are shown in Figure 9.14. Visually, it appears that the graphs are reflections of each other. You can further verify this by testing a few points on each graph. Note in the following list that if the point (a, b) is on the graph of f, the point (b, a) is on the graph of f^{-1}.

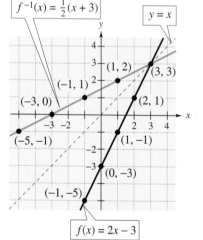

Figure 9.14

$f(x) = 2x - 3$	$f^{-1}(x) = \frac{1}{2}(x + 3)$
$(-1, -5)$	$(-5, -1)$
$(0, -3)$	$(-3, 0)$
$(1, -1)$	$(-1, 1)$
$(2, 1)$	$(1, 2)$
$(3, 3)$	$(3, 3)$

You can sketch the graph of the inverse of a function without knowing the equation of the inverse function. Simply find the coordinates of points that lie on the original function. By interchanging the x- and y-coordinates you have points that lie on the graph of the inverse function. Plot these points and sketch the graph of the inverse.

In Example 6(b) we said that the function $f(x) = x^2$ has no inverse. What we mean is that *assuming the domain of f is the entire real line*, the function $f(x) = x^2$ has no inverse. If, however, we restrict the domain of f to the nonnegative real numbers, f does have an inverse.

Example 8   The Graphs of f and f^{-1}

Graphically show that each function is an inverse of the other.

$$f(x) = x^2, \quad x \geq 0, \quad \text{and} \quad f^{-1}(x) = \sqrt{x}$$

Solution

The graphs of f and f^{-1} are shown in Figure 9.15. Visually, it appears that the graphs are reflections of each other in the line $y = x$. You can further verify this by testing a few points on each graph. Note in the following list that if the point (a, b) is on the graph of f, the point (b, a) is on the graph of f^{-1}.

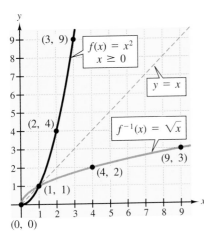

Figure 9.15

$f(x) = x^2, \quad x \geq 0$	$f^{-1}(x) = \sqrt{x}$
$(0, 0)$	$(0, 0)$
$(1, 1)$	$(1, 1)$
$(2, 4)$	$(4, 2)$
$(3, 9)$	$(9, 3)$

Discussing the Concept   **Error Analysis**

Suppose you are an algebra instructor and one of your students hands in the following solutions. Find and correct the errors and discuss how you can help your student avoid such errors in the future.

a. If $f(x) = 2x - 1$ and $g(x) = x^3 + 1$, find $(f \circ g)(2)$.

$$(f \circ g)(2) = (2 \cdot 2 - 1)(2^3 + 1)$$
$$= (4 - 1)(8 + 1)$$
$$= (3)(9)$$
$$= 27$$

b. If $f(x) = 3x^2 + x$ and $g(x) = x - 2$, find $(f \circ g)(1)$.

$$(f \circ g)(1) = f(1) - 2$$
$$= [3(1)^2 + 1] - 2$$
$$= (3 + 1) - 2$$
$$= 2$$

9.2 Exercises

Integrated Review   Concepts, Skills, and Problem Solving

Keep mathematically in shape by doing these exercises *before* the problems of this section.

Properties and Definitions

1. Decide whether $x - y^2 = 0$ represents y as a function of x. Explain.

2. Decide whether $|x| - 2y = 4$ represents y as a function of x. Explain.

3. Explain why the domains of f and g are not the same.

$$f(x) = \sqrt{4 - x^2} \qquad g(x) = \frac{6}{\sqrt{4 - x^2}}$$

4. Determine the range of $h(x) = 8 - \sqrt{x}$ over the domain $\{0, 4, 9, 16\}$.

Simplifying Expressions

In Exercises 5–10, perform the operations and simplify.

5. $-(5x^2 - 1) + (3x^2 - 5)$

6. $(-2x)(-5x)(3x + 4)$

7. $(u - 4v)(u + 4v)$

8. $(3a - 2b)^2$

9. $(t - 2)^3$

10. $\dfrac{6x^3 - 3x^2}{12x}$

Problem Solving

11. The velocity of a free-falling body is given by $v = \sqrt{2gh}$, where v is the velocity measured in feet per second, $g = 32$ ft/sec^2, and h is the distance (in feet) the object has fallen. Find the distance an object has fallen if its velocity is 80 feet per second.

12. The cost for a long-distance telephone call is \$0.95 for the first minute and \$0.35 for each additional minute. The total cost of a call is \$5.15. Find the length of the call.

Developing Skills

In Exercises 1–10, find the compositions. See Examples 1 and 2.

1. $f(x) = x - 3$, $g(x) = 2x - 4$
   (a) $(f \circ g)(x)$
   (b) $(g \circ f)(x)$
   (c) $(f \circ g)(4)$
   (d) $(g \circ f)(7)$

2. $f(x) = x + 1$, $g(x) = 5 - 3x$
   (a) $(f \circ g)(x)$
   (b) $(g \circ f)(x)$
   (c) $(f \circ g)(3)$
   (d) $(g \circ f)(3)$

3. $f(x) = x + 5$, $g(x) = 2x^2 - 6$
   (a) $(f \circ g)(x)$
   (b) $(g \circ f)(x)$
   (c) $(f \circ g)(2)$
   (d) $(g \circ f)(-3)$

4. $f(x) = x^2 - 3x$, $g(x) = 3 - 2x$
   (a) $(f \circ g)(x)$
   (b) $(g \circ f)(x)$
   (c) $(f \circ g)(-1)$
   (d) $(g \circ f)(3)$

5. $f(x) = |x - 3|$, $g(x) = 3x$
   (a) $(f \circ g)(x)$
   (b) $(g \circ f)(x)$
   (c) $(f \circ g)(1)$
   (d) $(g \circ f)(2)$

6. $f(x) = |x|$, $g(x) = 2x + 5$
   (a) $(f \circ g)(x)$
   (b) $(g \circ f)(x)$
   (c) $(f \circ g)(-2)$
   (d) $(g \circ f)(-4)$

7. $f(x) = \sqrt{x - 4}$, $g(x) = x + 5$
   (a) $(f \circ g)(x)$
   (b) $(g \circ f)(x)$
   (c) $(f \circ g)(3)$
   (d) $(g \circ f)(8)$

8. $f(x) = \sqrt{x + 6}$, $g(x) = 2x - 3$
   (a) $(f \circ g)(x)$
   (b) $(g \circ f)(x)$
   (c) $(f \circ g)(3)$
   (d) $(g \circ f)(-2)$

9. $f(x) = \dfrac{1}{x-3}$, $g(x) = \dfrac{2}{x^2}$

  (a) $(f \circ g)(x)$ (b) $(g \circ f)(x)$

  (c) $(f \circ g)(-1)$ (d) $(g \circ f)(2)$

10. $f(x) = \dfrac{4}{x^2-4}$, $g(x) = \dfrac{1}{x}$

  (a) $(f \circ g)(x)$ (b) $(g \circ f)(x)$

  (c) $(f \circ g)(-2)$ (d) $(g \circ f)(1)$

In Exercises 11–14, use the functions f and g to find the indicated values.

$f = \{(-2, 3), (-1, 1), (0, 0), (1, -1), (2, -3)\}$,

$g = \{(-3, 1), (-1, -2), (0, 2), (2, 2), (3, 1)\}$

11. (a) $f(1)$ **12.** (a) $g(0)$

    (b) $g(-1)$ (b) $f(2)$

    (c) $(g \circ f)(1)$ (c) $(f \circ g)(0)$

13. (a) $(f \circ g)(-3)$ **14.** (a) $(f \circ g)(2)$

    (b) $(g \circ f)(-2)$ (b) $(g \circ f)(2)$

In Exercises 15–18, use the functions f and g to find the indicated values.

$f = \{(0, 1), (1, 2), (2, 5), (3, 10), (4, 17)\}$,

$g = \{(5, 4), (10, 1), (2, 3), (17, 0), (1, 2)\}$

15. (a) $f(3)$ **16.** (a) $g(2)$

    (b) $g(10)$ (b) $f(0)$

    (c) $(g \circ f)(3)$ (c) $(f \circ g)(10)$

17. (a) $(g \circ f)(4)$ **18.** (a) $(f \circ g)(1)$

    (b) $(f \circ g)(2)$ (b) $(g \circ f)(0)$

In Exercises 19–26, find the domain of the composition (a) $f \circ g$ and (b) $g \circ f$. See Example 3.

19. $f(x) = x + 1$ **20.** $f(x) = 2 - 3x$

    $g(x) = 2x - 5$ $g(x) = 5x + 3$

21. $f(x) = \sqrt{x}$ **22.** $f(x) = \sqrt{x - 5}$

    $g(x) = x - 2$ $g(x) = x + 3$

23. $f(x) = x^2 - 1$ **24.** $f(x) = \sqrt{2x - 1}$

    $g(x) = \sqrt{x + 3}$ $g(x) = x^2 + 1$

25. $f(x) = \dfrac{x}{x + 5}$ **26.** $f(x) = \dfrac{x}{x - 4}$

    $g(x) = \sqrt{x - 1}$ $g(x) = \sqrt{x}$

In Exercises 27–32, use the Horizontal Line Test to determine whether the function has an inverse. See Example 4.

27. $f(x) = x^2 - 2$

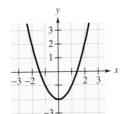

28. $f(x) = \frac{1}{5}x$

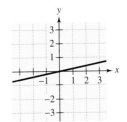

29. $f(x) = x^2$, $x \geq 0$

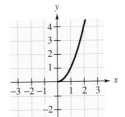

30. $f(x) = \sqrt{-x}$

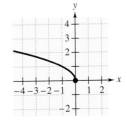

31. $g(x) = \sqrt{25 - x^2}$

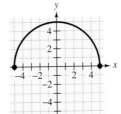

32. $g(x) = |x - 4|$

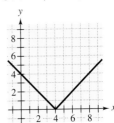

In Exercises 33–42, use a graphing utility to graph the function and determine whether the function is one-to-one.

33. $f(x) = x^3 - 1$ **34.** $f(x) = (2 - x)^3$

35. $f(t) = \sqrt[3]{5 - t}$ **36.** $h(t) = 4 - \sqrt[3]{t}$

37. $g(x) = x^4 - 6$ **38.** $f(x) = (x + 2)^5$

39. $h(t) = \dfrac{5}{t}$ **40.** $g(t) = \dfrac{5}{t^2}$

41. $f(s) = \dfrac{4}{s^2 + 1}$ **42.** $f(x) = \dfrac{1}{x^2 - 2}$

In Exercises 43–54, verify algebraically that the functions f and g are inverses of each other. See Example 5.

43. $f(x) = 10x$ **44.** $f(x) = \frac{2}{3}x$

    $g(x) = \frac{1}{10}x$ $g(x) = \frac{3}{2}x$

45. $f(x) = x + 15$ **46.** $f(x) = 3 - x$

    $g(x) = x - 15$ $g(x) = 3 - x$

47. $f(x) = 1 - 2x$
$g(x) = \frac{1}{2}(1 - x)$

48. $f(x) = 2x - 1$
$g(x) = \frac{1}{2}(x + 1)$

49. $f(x) = 2 - 3x$
$g(x) = \frac{1}{3}(2 - x)$

50. $f(x) = -\frac{1}{4}x + 3$
$g(x) = -4(x - 3)$

51. $f(x) = \sqrt[3]{x + 1}$
$g(x) = x^3 - 1$

52. $f(x) = x^7$
$g(x) = \sqrt[7]{x}$

53. $f(x) = \frac{1}{x}$

$g(x) = \frac{1}{x}$

54. $f(x) = \frac{1}{x - 3}$

$g(x) = 3 + \frac{1}{x}$

In Exercises 55–66, find the inverse of the function f. Verify that $f(f^{-1}(x))$ and $f^{-1}(f(x))$ are equal to the identity function. See Example 6.

55. $f(x) = 5x$

56. $f(x) = 2x$

57. $f(x) = \frac{1}{2}x$

58. $f(x) = \frac{1}{3}x$

59. $f(x) = x + 10$

60. $f(x) = x - 5$

61. $f(x) = 3 - x$

62. $f(x) = 8 - x$

63. $f(x) = x^7$

64. $f(x) = x^5$

65. $f(x) = \sqrt[3]{x}$

66. $f(x) = x^{1/5}$

In Exercises 67–82, find the inverse of the function. See Example 6.

67. $f(x) = 8x$

68. $f(x) = \frac{1}{10}x$

69. $g(x) = x + 25$

70. $f(x) = 7 - x$

71. $g(x) = 3 - 4x$

72. $g(t) = 6t + 1$

73. $g(t) = \frac{1}{4}t + 2$

74. $h(s) = 5 - \frac{3}{2}s$

75. $h(x) = \sqrt{x}$

76. $h(x) = \sqrt{x + 5}$

77. $f(t) = t^3 - 1$

78. $h(t) = t^5 + 8$

79. $g(s) = \dfrac{5}{s + 4}$

80. $f(s) = \dfrac{2}{3 - s}$

81. $f(x) = \sqrt{x + 3}, \quad x \geq -3$

82. $f(x) = \sqrt{x^2 - 4}, \quad x \geq 2$

In Exercises 83–88, sketch the graphs of f and f^{-1} on the same rectangular coordinate system. Show that the graphs are reflections of each other in the line $y = x$. See Example 7.

83. $f(x) = x + 4, \ f^{-1}(x) = x - 4$

84. $f(x) = x - 7, \ f^{-1}(x) = x + 7$

85. $f(x) = 3x - 1, \ f^{-1}(x) = \frac{1}{3}(x + 1)$

86. $f(x) = 5 - 4x, \ f^{-1}(x) = -\frac{1}{4}(x - 5)$

87. $f(x) = x^2 - 1, \ x \geq 0,$
$f^{-1}(x) = \sqrt{x + 1}$

88. $f(x) = (x + 2)^2, \ x \geq -2,$
$f^{-1}(x) = \sqrt{x} - 2$

In Exercises 89–92, match the graph with the graph of its inverse. [The graphs of the inverse functions are labeled (a), (b), (c), and (d).]

(a)
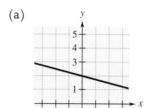

(b)
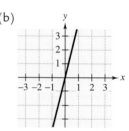

(c)
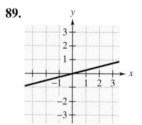

(d)
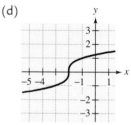

89.
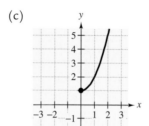

90.
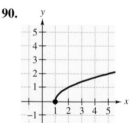

91.
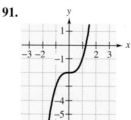

92.
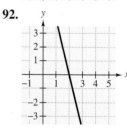

In Exercises 93–100, use a graphing utility to verify that the functions are inverses of each other.

93. $f(x) = \frac{1}{3}x$
$g(x) = 3x$

94. $f(x) = \frac{1}{5}x - 1$
$g(x) = 5x + 5$

95. $f(x) = \sqrt{x + 1}$
$g(x) = x^2 - 1, \ x \geq 0$

96. $f(x) = \sqrt{4 - x}$
$g(x) = 4 - x^2, \ x \geq 0$

97. $f(x) = \frac{1}{8}x^3$
$g(x) = 2\sqrt[3]{x}$

98. $f(x) = \sqrt[3]{x + 2}$
$g(x) = x^3 - 2$

99. $f(x) = 3x + 4$
$g(x) = \frac{1}{3}(x - 4)$

100. $f(x) = |x - 2|, \ x \geq 2$
$g(x) = x + 2, \ x \geq 0$

In Exercises 101–104, delete part of the graph of the function so that the remaining part is one-to-one. Find the inverse of the remaining part and give the domain of the inverse. (*Note:* There is more than one correct answer.) See Example 8.

101. $f(x) = (x - 2)^2$ **102.** $f(x) = 9 - x^2$

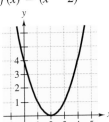   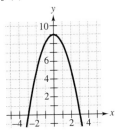

103. $f(x) = |x| + 1$ **104.** $f(x) = |x - 2|$

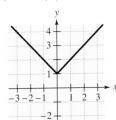   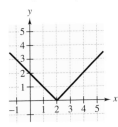

In Exercises 105–108, use the graph of f to sketch the graph of f^{-1}.

105.   **106.**

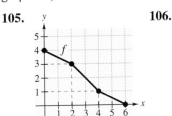

107.   **108.**

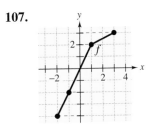

109. Consider the function $f(x) = 3 - 2x$.
(a) Find $f^{-1}(x)$.   (b) Find $(f^{-1})^{-1}(x)$.

Solving Problems

110. *Hourly Wage*   Your wage is $9.00 per hour plus $0.65 for each unit produced per hour. Thus, your hourly wage y in terms of the number of units produced is $y = 9 + 0.65x$.
(a) Determine the inverse of the function.
(b) What does each variable represent in the inverse function?
(c) Determine the number of units produced when your hourly wage averages $14.20.

111. *Cost*   You need 100 pounds of two commodities that cost $0.50 and $0.75 per pound.
(a) Verify that your total cost is $y = 0.50x + 0.75(100 - x)$, where x is the number of pounds of the less expensive commodity.
(b) Find the inverse of the function. What does each variable represent in the inverse function?
(c) Use the context of the problem to determine the domain of the inverse function.
(d) Determine the number of pounds of the less expensive commodity purchased if the total cost is $60.

112. *Geometry*   You are standing on a bridge over a calm pond and drop a pebble, causing ripples of concentric circles in the water (see figure). The radius (in feet) of the outer ripple is given by $r(t) = 0.6t$, where t is time in seconds after the pebble hits the water. The area of the circle is given by the function $A(r) = \pi r^2$. Find an equation for the composition $A(r(t))$. What are the input and output of this composite function?

113. *Sales Bonus*   You are a sales representative for a clothing manufacturer. You are paid an annual salary plus a bonus of 2% of your sales over $200,000. Consider the two functions $f(x) = x - 200,000$ and $g(x) = 0.02x$. If x is greater than $200,000$, which of the following represents your bonus? Explain.
(a) $f(g(x))$ (b) $g(f(x))$

114. *Daily Production Cost* The daily cost of producing x units in a manufacturing process is $C(x) = 8.5x + 300$. The number of units produced in t hours during a day is given by $x(t) = 12t$, $0 \le t \le 8$. Find, simplify, and interpret $(C \circ x)(t)$.

115. *Rebate and Discount* The suggested retail price of a new car is p dollars. The dealership advertised a factory rebate of $2000 and a 5% discount.

(a) Write a function R in terms of p, giving the cost of the car after receiving the factory rebate.

(b) Write a function S in terms of p, giving the cost of the car after receiving the dealership discount.

(c) Form the composite functions $(R \circ S)(p)$ and $(S \circ R)(p)$ and interpret each.

(d) Find $(R \circ S)(26{,}000)$ and $(S \circ R)(26{,}000)$. Which yields the smaller cost for the car? Explain.

116. *Exploration and Conjecture* Consider the functions $f(x) = 4x$ and $g(x) = x + 6$.

(a) Find $(f \circ g)(x)$.

(b) Find $(f \circ g)^{-1}(x)$.

(c) Find $f^{-1}(x)$ and $g^{-1}(x)$.

(d) Find $(g^{-1} \circ f^{-1})(x)$ and compare the result with that of part (b).

(e) Repeat parts (a) through (d) for $f(x) = x^3 + 1$ and $g(x) = 2x$.

(f) Write two one-to-one functions f and g, and repeat parts (a) through (d) for these functions.

(g) Make a conjecture about $(f \circ g)^{-1}(x)$ and $(g^{-1} \circ f^{-1})(x)$.

Explaining Concepts

True or False? In Exercises 117–120, decide whether the statement is true or false. If true, explain your reasoning. If false, give an example.

117. If the inverse of f exists, the y-intercept of f is an x-intercept of f^{-1}. Explain.

118. There exists no function f such that $f = f^{-1}$.

119. If the inverse of f exists, the domains of f and f^{-1} are the same.

120. If the inverse of f exists and its graph passes through the point $(2, 2)$, the graph of f^{-1} also passes through the point $(2, 2)$.

121. Give an example showing that the composite functions $(f \circ g)(x)$ and $(g \circ f)(x)$ are not necessarily the same.

122. Describe how to find the inverse of a function given by a set of ordered pairs. Give an example.

123. Describe how to find the inverse of a function given by an equation in x and y. Give an example.

124. Give an example of a function that does not have an inverse.

125. Explain the Horizontal Line Test. What is the relationship between this test and a function being one-to-one?

126. Describe the relationship between the graph of a function and its inverse.

9.3   Logarithmic Functions

Objectives

1 Evaluate a logarithmic function.

2 Sketch the graph of a logarithmic function using its inverse exponential function.

3 Use the natural base e to define the natural logarithmic function.

4 Use the change-of-base formula to evaluate logarithms.

---

1 Evaluate a logarithmic function.

Logarithmic Functions

In Section 9.2, you were introduced to the concept of the inverse of a function. Moreover, you saw that if a function has the property that no horizontal line intersects the graph of the function more than once, the function must have an inverse. By looking back at the graphs of the exponential functions introduced in Section 9.1, you will see that every function of the form

$$f(x) = a^x \qquad \text{Exponential functions have inverses.}$$

passes the Horizontal Line Test, and so must have an inverse. To describe the inverse of $f(x) = a^x$, we follow the steps used in Section 9.2.

$$y = a^x \qquad \text{Replace } f(x) \text{ by } y.$$

$$x = a^y \qquad \text{Interchange } x \text{ and } y.$$

At this point we have no way to solve for y. A verbal description of y in the equation $x = a^y$ is "y equals the exponent needed on base a to get x." This inverse of $f(x) = a^x$ is denoted by

$$f^{-1}(x) = \log_a x.$$

> ▶ **Definition of Logarithmic Function**
>
> Let a and x be positive real numbers such that $a \neq 1$. The **logarithm of x with base a** is denoted by $\log_a x$ and is defined as follows.
>
> $$y = \log_a x \quad \text{if and only if} \quad a^y = x$$
>
> The function $f(x) = \log_a x$ is the **logarithmic function with base a.**

From the definition it is clear that

Logarithmic Equation   *Exponential Equation*

$$y = \log_a x \qquad \text{is equivalent to} \qquad a^y = x.$$

So, to find the value of $\log_a x$, *think*

$$\text{"}\log_a x = \text{the exponent needed on base } a \text{ to get } x.\text{"}$$

For instance,

$$y = \log_2 8 \qquad \text{Think: "The exponent needed on 2 to get 8."}$$

$$y = 3.$$

That is,

$$3 = \log_2 8. \qquad \text{This is equivalent to } 2^3 = 8.$$

By now it should be clear that *a logarithm is an exponent.*

The Granger Collection

John Napier

(1550–1617)

John Napier, a Scottish mathematician, developed logarithms as a way to simplify some of the tedious calculations of his day. Beginning in 1594, Napier worked about 20 years on logarithms, but he was only partially successful in his quest. Nonetheless, the development of logarithms was a step forward and received immediate recognition.

Study Tip

Study the results in Example 2 carefully. Each of the logarithms illustrates an important special property of logarithms that you should know.

Example 1 Evaluating Logarithms

Evaluate each logarithm.

a. $\log_2 16$ **b.** $\log_3 9$ **c.** $\log_4 2$

Solution

In each case you should answer the question, "To what power must the base be raised to obtain the given number?"

a. The power to which 2 must be raised to obtain 16 is 4. That is,

$$2^4 = 16 \quad \Longrightarrow \quad \log_2 16 = 4.$$

b. The power to which 3 must be raised to obtain 9 is 2. That is,

$$3^2 = 9 \quad \Longrightarrow \quad \log_3 9 = 2.$$

c. The power to which 4 must be raised to obtain 2 is $\frac{1}{2}$. That is,

$$4^{1/2} = 2 \quad \Longrightarrow \quad \log_4 2 = \frac{1}{2}.$$

Example 2 Evaluating Logarithms

Evaluate each logarithm.

a. $\log_5 1$ **b.** $\log_{10} \dfrac{1}{10}$ **c.** $\log_3(-1)$ **d.** $\log_4 0$

Solution

a. The power to which 5 must be raised to obtain 1 is 0. That is,

$$5^0 = 1 \quad \Longrightarrow \quad \log_5 1 = 0.$$

b. The power to which 10 must be raised to obtain $\frac{1}{10}$ is -1. That is,

$$10^{-1} = \frac{1}{10} \quad \Longrightarrow \quad \log_{10} \frac{1}{10} = -1.$$

c. There is no power to which 3 can be raised to obtain -1. The reason for this is that for any value of x, 3^x is a positive number. So, $\log_3(-1)$ is undefined.

d. There is no power to which 4 can be raised to obtain 0. So, $\log_4 0$ is undefined.

The following properties of logarithms follow directly from the definition of the logarithmic function with base a.

▶ **Properties of Logarithms**

Let a and x be positive real numbers such that $a \neq 1$. Then the following properties are true.

1. $\log_a 1 = 0$ because $a^0 = 1$.

2. $\log_a a = 1$ because $a^1 = a$.

3. $\log_a a^x = x$ because $a^x = a^x$.

The logarithmic function with base 10 is called the **common logarithmic function.** On most calculators, this function can be evaluated with the common logarithmic key $\boxed{\text{LOG}}$, as illustrated in the next example.

Example 3   Evaluating Common Logarithms

Evaluate each logarithm. Use a calculator only if necessary.

a. $\log_{10} 100$

b. $\log_{10} 0.01$

c. $\log_{10} 5$

Solution

a. Because $10^2 = 100$, it follows that

$$\log_{10} 100 = 2.$$

b. Because $10^{-2} = \frac{1}{100} = 0.01$, it follows that

$$\log_{10} 0.01 = -2.$$

c. There is no simple power to which 10 can be raised to obtain 5, so you should use a calculator to evaluate $\log_{10} 5$.

Keystrokes	Display	
5 $\boxed{\text{LOG}}$	0.69897	Scientific
$\boxed{\text{LOG}}$ 5 $\boxed{\text{ENTER}}$	0.69897	Graphing

So, rounded to three decimal places, $\log_{10} 5 \approx 0.699$.

Be sure you see that the value of a logarithm can be zero or negative, as in Example 3(b), *but* you cannot take the logarithm of zero or a negative number. This means that the logarithms $\log_{10}(-10)$ and $\log_5 0$ are not valid.

2 Sketch the graph of a logarithmic function using its inverse exponential function.

Graphs of Logarithmic Functions

To sketch the graph of $y = \log_a x$, we can use the fact that the graphs of inverse functions are reflections of each other in the line $y = x$.

Example 4   Graphs of Exponential and Logarithmic Functions

On the same rectangular coordinate system, sketch the graphs of the following.

a. $f(x) = 2^x$ **b.** $g(x) = \log_2 x$

Solution

a. Begin by making a table of values for $f(x) = 2^x$.

x	-2	-1	0	1	2	3
$f(x) = 2^x$	$\frac{1}{4}$	$\frac{1}{2}$	1	2	4	8

By plotting these points and connecting them with a smooth curve, you obtain the graph shown in Figure 9.16.

b. Because $g(x) = \log_2 x$ is the inverse function of $f(x) = 2^x$, the graph of g is obtained by reflecting the graph of f in the line $y = x$, as shown in Figure 9.16.

Figure 9.16   *Inverse Functions*

Study Tip

In Example 4, the inverse nature of logarithmic functions is used to sketch the graph of $g(x) = \log_2 x$. You could also use a standard point-plotting approach or a graphing utility.

Notice from the graph of $g(x) = \log_2 x$, shown in Figure 9.16, that the domain of the function is the set of positive numbers and the range is the set of all real numbers. The basic characteristics of the graph of a logarithmic function are summarized in Figure 9.17. In this figure, note that the graph has one x-intercept at $(1, 0)$. Also note that the y-axis is a vertical asymptote of the graph.

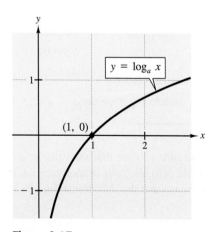

Graph of $y = \log_a x$, $a > 1$

- Domain: $(0, \infty)$
- Range: $(-\infty, \infty)$
- Intercept: $(1, 0)$
- Vertical asymptote: $x = 0$
- Increasing

Figure 9.17

In the following example, the graph of $\log_a x$ is used to sketch the graphs of functions of the form $y = b \pm \log_a(x + c)$. Notice how each transformation affects the vertical asymptote.

Example 5   Sketching the Graphs of Logarithmic Functions

The graph of each of the following functions is similar to the graph of $f(x) = \log_{10} x$, as shown in Figure 9.18. From the graph you can determine the domain of the function.

a. Because $g(x) = \log_{10}(x - 1) = f(x - 1)$, the graph of g can be obtained by shifting the graph of f 1 unit to the right. The vertical asymptote of the graph of g is $x = 1$. The domain of g is $x > 1$.

b. Because $h(x) = 2 + \log_{10} x = 2 + f(x)$, the graph of h can be obtained by shifting the graph of f 2 units up. The vertical asymptote of the graph of h is $x = 0$. The domain of h is $x > 0$.

c. Because $k(x) = -\log_{10} x = -f(x)$, the graph of k can be obtained by reflecting the graph of f in the x-axis. The vertical asymptote of the graph of k is $x = 0$. The domain of k is $x > 0$.

d. Because $j(x) = \log_{10}(-x) = f(-x)$, the graph of j can be obtained by reflecting the graph of f in the y-axis. The vertical asymptote of the graph of j is $x = 0$. The domain of j is $x < 0$.

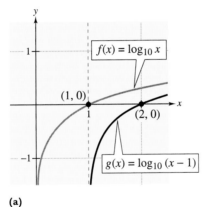

(a)

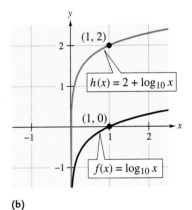

(b)

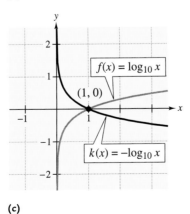

(c)

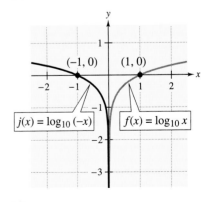

(d)

Figure 9.18

③ Use the natural base e to define the natural logarithmic function.

The Natural Logarithmic Function

As with exponential functions, the most widely used base for logarithmic functions is the number e. The logarithmic function with base e is the **natural logarithmic function** and is denoted by the special symbol $\ln x$, which is read as "el en of x."

▶ **The Natural Logarithmic Function**

The function defined by

$$f(x) = \log_e x = \ln x$$

where $x > 0$, is called the **natural logarithmic function.**

The three properties of logarithms listed earlier in this section are also valid for natural logarithms.

▶ **Properties of Natural Logarithms**

Let x be a positive real number. Then the following properties are true.

1. $\ln 1 = 0$ because $e^0 = 1$.

2. $\ln e = 1$ because $e^1 = e$.

3. $\ln e^x = x$ because $e^x = e^x$.

Example 6   Evaluating the Natural Logarithmic Function

Evaluate each of the following. Then incorporate the results into a graph of the natural logarithmic function.

a. $\ln e^2$ **b.** $\ln \dfrac{1}{e}$

Solution

Using the property that $\ln e^x = x$, you obtain the following.

a. $\ln e^2 = 2$

b. $\ln \dfrac{1}{e} = \ln e^{-1} = -1$

Using the points $(1/e, -1)$, $(1, 0)$, $(e, 1)$, and $(e^2, 2)$, you can sketch the graph of the natural logarithmic function, as shown in Figure 9.19.

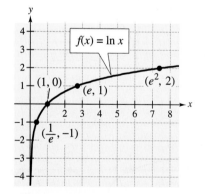

Figure 9.19

On most calculators, the natural logarithm key is denoted by ⎡LN⎤. For instance, on a scientific calculator, you can evaluated $\ln 2$ as 2 ⎡LN⎤ and on a graphing calculator, you can evaluate it as ⎡LN⎤ 2 ⎡ENTER⎤. In either case, you should obtain a display of 0.6931472.

4 Use the change-of-base formula to evaluate logarithms.

Change of Base

Although 10 and e are the most frequently used bases, you occasionally need to evaluate logarithms with other bases. In such cases the following **change-of-base formula** is useful.

> ▶ Change-of-Base Formula
>
> Let a, b, and x be positive real numbers such that $a \neq 1$ and $b \neq 1$. Then $\log_a x$ is given as follows.
>
> $$\log_a x = \frac{\log_b x}{\log_b a} \qquad \text{or} \qquad \log_a x = \frac{\ln x}{\ln a}$$

Technology: Tip

You can use a graphing utility to graph logarithmic functions that do not have a base of 10 by using the change-of-base formula. Use the change-of-base formula to rewrite $g(x) = \log_2 x$ in Example 4 on page 576. Use the trace feature to estimate $g(x) = \log_2 x$ when $x = 3$. Verify your estimate arithmetically using a calculator.

The usefulness of this change-of-base formula is that you can use a calculator that has only the common logarithm key [LOG] and the natural logarithm key [LN] to evaluate logarithms to any base.

Example 7   Changing the Base to Evaluate Logarithms

a. Use *common* logarithms to evaluate $\log_3 5$.

b. Use *natural* logarithms to evaluate $\log_6 2$.

Solution

Using the change-of-base formula, you can convert to common and natural logarithms by writing

$$\log_3 5 = \frac{\log_{10} 5}{\log_{10} 3} \qquad \text{and} \qquad \log_6 2 = \frac{\ln 2}{\ln 6}.$$

Now, use the following keystrokes.

a.

Keystrokes	Display	
5 [LOG] [÷] 3 [LOG] [=]	1.4649735	Scientific
[LOG] 5 [÷] [LOG] 3 [ENTER]	1.4649735	Graphing

So, $\log_3 5 \approx 1.465$.

b.

Keystrokes	Display	
2 [LN] [÷] 6 [LN] [=]	0.3868528	Scientific
[LN] 2 [÷] [LN] 6 [ENTER]	0.3868528	Graphing

So, $\log_6 2 \approx 0.387$.

In Example 7(a), $\log_3 5$ could have been evaluated using natural logarithms in the change-of-base formula.

$$\log_3 5 = \frac{\ln 5}{\ln 3} \approx 1.465$$

Notice that you get the same answer whether you use natural logarithms or common logarithms in the change-of-base formula.

At this point, you have been introduced to all the basic types of functions that are covered in this course: polynomial functions, radical functions, rational functions, exponential functions, and logarithmic functions. The only other common types of functions are *trigonometric functions*, which you will study if you go on to take a course in trigonometry or precalculus.

Discussing the Concept   Comparing Models

Suppose you work for a research and development firm that deals with a wide variety of disciplines. Your supervisor has asked you to give a presentation to your department on four basic kinds of mathematical models. Identify each of the models shown below. Develop a presentation describing the types of data sets that each model would best represent. Include distinctions in domain, range, intercepts, and a discussion of the types of applications to which each model is suited.

a.

b.

c.

d.

9.3   Exercises

Integrated Review   *Concepts, Skills, and Problem Solving*

Keep mathematically in shape by doing these exercises *before* the problems of this section.

Properties and Definitions

In Exercises 1–4, identify the transformation of $f(x) = x^2$ needed to sketch the graph of the function.

1. $g(x) = (x - 4)^2$ **2.** $h(x) = -x^2$

3. $j(x) = x^2 + 1$ **4.** $k(x) = (x + 3)^2 - 5$

Factoring

In Exercises 5–8, factor the expression completely.

5. $2x^3 - 6x$ **6.** $16 - (y + 2)^2$

7. $t^2 + 10t + 25$ **8.** $5 - u + 5u^2 - u^3$

Graphing

In Exercises 9–12, graph the equation.

9. $y = 3 - \frac{1}{2}x$ **10.** $3x - 4y = 6$

11. $y = x^2 - 6x + 5$ **12.** $y = -(x - 2)^2 + 1$

Developing Skills

In Exercises 1–12, write the logarithmic equation in exponential form.

1. $\log_5 25 = 2$ **2.** $\log_6 36 = 2$

3. $\log_4 \frac{1}{16} = -2$ **4.** $\log_8 \frac{1}{8} = -1$

5. $\log_3 \frac{1}{243} = -5$ **6.** $\log_{10} 10{,}000 = 4$

7. $\log_{36} 6 = \frac{1}{2}$ **8.** $\log_{32} 4 = \frac{2}{5}$

9. $\log_8 4 = \frac{2}{3}$ **10.** $\log_{16} 8 = \frac{3}{4}$

11. $\log_2 2.462 \approx 1.3$ **12.** $\log_3 1.179 \approx 0.15$

In Exercises 13–24, write the exponential equation in logarithmic form.

13. $7^2 = 49$ **14.** $6^4 = 1296$

15. $3^{-2} = \frac{1}{9}$ **16.** $5^{-4} = \frac{1}{625}$

17. $8^{2/3} = 4$ **18.** $81^{3/4} = 27$

19. $25^{-1/2} = \frac{1}{5}$ **20.** $6^{-3} = \frac{1}{216}$

21. $4^0 = 1$ **22.** $6^1 = 6$

23. $5^{1.4} \approx 9.518$ **24.** $10^{0.12} \approx 1.318$

In Exercises 25–48, evaluate the logarithm without using a calculator. (If not possible, state the reason.) See Examples 1 and 2.

25. $\log_2 8$ **26.** $\log_3 27$

27. $\log_{10} 10$ **28.** $\log_8 8$

29. $\log_{10} 1000$ **30.** $\log_{10} 0.00001$

31. $\log_2 \frac{1}{4}$ **32.** $\log_3 \frac{1}{9}$

33. $\log_4 \frac{1}{64}$ **34.** $\log_5 \frac{1}{125}$

35. $\log_{10} \frac{1}{10{,}000}$ **36.** $\log_{10} \frac{1}{100}$

37. $\log_2(-3)$ **38.** $\log_4(-4)$

39. $\log_4 1$ **40.** $\log_3 1$

41. $\log_5(-6)$ **42.** $\log_2 0$

43. $\log_9 3$ **44.** $\log_{25} 125$

45. $\log_{16} 8$ **46.** $\log_{144} 12$

47. $\log_7 7^4$ **48.** $\log_5 5^3$

In Exercises 49–54, use a calculator to evaluate the common logarithm. Round to four decimal places. See Example 3.

49. $\log_{10} 31$ **50.** $\log_{10} 5310$

51. $\log_{10} 0.85$ **52.** $\log_{10} 0.345$

53. $\log_{10}(\sqrt{2} + 4)$ **54.** $\log_{10} \frac{\sqrt{3}}{2}$

In Exercises 55–58, sketch the graphs of f and g on the same set of coordinate axes. What can you conclude about the relationship between f and g? See Example 4.

55. $f(x) = \log_3 x$
$\quad g(x) = 3^x$

56. $f(x) = \log_4 x$
$\quad g(x) = 4^x$

57. $f(x) = \log_6 x$
$\quad g(x) = 6^x$

58. $f(x) = \log_{1/2} x$
$\quad g(x) = \left(\frac{1}{2}\right)^x$

In Exercises 59–62, state the relationship between the functions f and g.

59. $f(x) = 7^x$
$\quad g(x) = \log_7 x$

60. $f(x) = 5^x$
$\quad g(x) = \log_5 x$

61. $f(x) = e^x$
$\quad g(x) = \ln x$

62. $f(x) = 10^x$
$\quad g(x) = \log_{10} x$

In Exercises 63–68, identify the transformation of the graph of $f(x) = \log_2 x$ and sketch the graph of h. See Example 5.

63. $h(x) = 3 + \log_2 x$

64. $h(x) = -4 + \log_2 x$

65. $h(x) = \log_2(x - 2)$

66. $h(x) = \log_2(x + 4)$

67. $h(x) = \log_2(-x)$

68. $h(x) = -\log_2(x)$

In Exercises 69–74, match the function with its graph. [The graphs are labeled (a), (b), (c), (d), (e), and (f).]

(a)
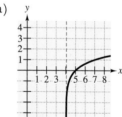

(b)
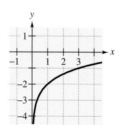

(c)
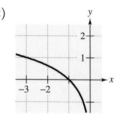

(d)
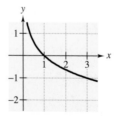

(e)
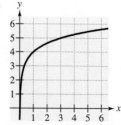

(f)
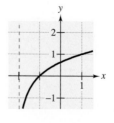

69. $f(x) = 4 + \log_3 x$

70. $f(x) = -2 + \log_3 x$

71. $f(x) = -\log_3 x$

72. $f(x) = \log_3(-x)$

73. $f(x) = \log_3(x - 4)$

74. $f(x) = \log_3(x + 2)$

In Exercises 75–84, sketch the graph of the function.

75. $f(x) = \log_5 x$

76. $g(x) = \log_8 x$

77. $g(t) = -\log_2 t$

78. $h(s) = -2 \log_3 s$

79. $f(x) = 3 + \log_2 x$

80. $f(x) = -2 + \log_3 x$

81. $g(x) = \log_2(x - 3)$

82. $h(x) = \log_3(x + 1)$

83. $f(x) = \log_{10}(10x)$

84. $g(x) = \log_4(4x)$

In Exercises 85–90, find the domain and vertical asymptote of the logarithmic function. Sketch its graph.

85. $f(x) = \log_4 x$

86. $g(x) = \log_6 x$

87. $h(x) = \log_4(x - 3)$

88. $f(x) = -\log_6(x + 2)$

89. $y = -\log_3 x + 2$

90. $y = \log_5(x - 1) + 4$

 In Exercises 91–96, use a graphing utility to graph the function. Determine the domain and identify any vertical asymptotes.

91. $y = 5 \log_{10} x$

92. $y = 5 \log_{10}(x - 3)$

93. $y = -3 + 5 \log_{10} x$

94. $y = 5 \log_{10}(3x)$

95. $y = \log_{10}\left(\frac{x}{5}\right)$

96. $y = \log_{10}(-x)$

In Exercises 97–102, use a calculator to evaluate the natural logarithm. Round to four decimal places. See Example 6.

97. $\ln 25$

98. $\ln 6.57$

99. $\ln 0.75$

100. $\ln(\sqrt{3} + 1)$

101. $\ln\left(\dfrac{1 + \sqrt{5}}{3}\right)$

102. $\ln\left(1 + \dfrac{0.10}{12}\right)$

In Exercises 103–108, match the function with its graph. [The graphs are labeled (a), (b), (c), (d), (e), and (f).]

(a)
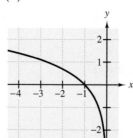

(b)
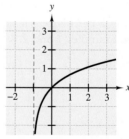

(c)

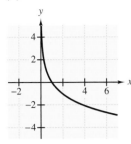

(d)

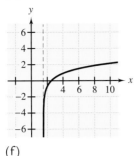

(e)

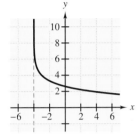

(f)

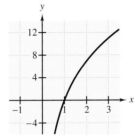

103. $f(x) = \ln(x + 1)$ **104.** $f(x) = 4 - \ln(x + 4)$

105. $f(x) = \ln\left(x - \frac{3}{2}\right)$ **106.** $f(x) = -\frac{3}{2} \ln x$

107. $f(x) = 10 \ln x$ **108.** $f(x) = \ln(-x)$

In Exercises 109–116, sketch the graph of the function.

109. $f(x) = -\ln x$ **110.** $f(x) = -2 \ln x$

111. $f(x) = 3 \ln x$ **112.** $h(t) = 4 \ln t$

113. $f(x) = 1 + \ln x$ **114.** $h(x) = 2 + \ln x$

115. $g(t) = 2 \ln(t - 4)$ **116.** $g(x) = -3 \ln(x + 3)$

In Exercises 117–120, use a graphing utility to graph the function. Determine the domain and identify any vertical asymptotes.

117. $g(x) = \ln(x + 6)$ **118.** $h(x) = -\ln(x - 2)$

119. $f(t) = 3 + 2 \ln t$ **120.** $g(t) = \ln(3 - t)$

In Exercises 121–134, use a calculator to evaluate the logarithm by means of the change-of-base formula. Use (a) the common logarithm key and (b) the natural logarithm key. See Example 7.

121. $\log_8 132$ **122.** $\log_5 510$

123. $\log_3 7$ **124.** $\log_7 4$

125. $\log_2 0.72$ **126.** $\log_{12} 0.6$

127. $\log_{15} 1250$ **128.** $\log_{20} 125$

129. $\log_{1/2} 4$ **130.** $\log_{1/3} 18$

131. $\log_4 \sqrt{42}$ **132.** $\log_3 \sqrt{26}$

133. $\log_2(1 + e)$ **134.** $\log_4(2 + e^3)$

Solving Problems

135. *American Elk*   The antler spread a (in inches) and shoulder height h (in inches) of an adult male American elk are related by the model

$$h = 116 \log_{10}(a + 40) - 176.$$

Approximate the shoulder height of a male American elk with an antler spread of 55 inches.

136. *Intensity of Sound*   The relationship between the number of decibels B and the intensity of a sound I in watts per centimeter squared is given by

$$B = 10 \log_{10}\left(\frac{I}{10^{-16}}\right).$$

Determine the number of decibels of a sound with an intensity of 10^{-4} watts per centimeter squared.

137. *Creating a Table*   The time t in years for an investment to double in value when compounded continuously at the rate r is given by

$$t = \frac{\ln 2}{r}.$$

Complete the following table, which shows the "doubling times" for several annual percentage rates.

r	0.07	0.08	0.09	0.10	0.11	0.12
t						

138. *Tornadoes*   Most tornadoes last less than 1 hour and travel less than 20 miles. The speed of the wind S (in miles per hour) near the center of the tornado is related to the distance the tornado travels d (in miles) by the model

$$S = 93 \log_{10} d + 65.$$

On March 18, 1925, a large tornado struck portions of Missouri, Illinois, and Indiana, covering a distance of 220 miles. Approximate the speed of the wind near the center of this tornado.

139. *Tractrix* A person walking along a dock (the y-axis) drags a boat by a 10-foot rope (see figure). The boat travels along a path known as a *tractrix*. The equation of the path is

$$y = 10 \ln\left(\frac{10 + \sqrt{100 - x^2}}{x}\right) - \sqrt{100 - x^2}.$$

(a) Use a graphing utility to graph the function. What is the domain of the function?

(b) Identify any asymptotes of the graph.

(c) Determine the position of the person when the x-coordinate of the position of the boat is $x = 2$.

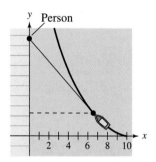

140. *Home Mortgage* The model

$$t = 10.042 \ln\left(\frac{x}{x - 1250}\right), \quad 1250 < x$$

approximates the length t (in years) of a home mortgage of $150,000 at 10% in terms of the monthly payment x.

(a) Use a graphing utility to graph the model. Describe the change in the length of the mortgage as the monthly payment increases.

(b) Use the graph in part (a) to approximate the length of the mortgage if the monthly payment is $1316.35.

(c) Use the result of part (b) to find the total amount paid over the term of the mortgage. What amount of the total is interest costs?

141. *Data Analysis* The table gives the prices x of a half-gallon of milk and the prices y of a half-gallon of ice cream in the United States for the years 1990 through 1996. The data were collected in December of each year. (Source: U.S. Bureau of Labor Statistics)

Year	1990	1991	1992	1993	1994	1995	1996
x	$1.39	$1.40	$1.39	$1.43	$1.44	$1.48	$1.65
y	$2.54	$2.63	$2.49	$2.59	$2.62	$2.68	$2.94

A model for the data is

$$y = 435.33 - 527.72x + 396.68 \ln x + 88.05x^2.$$

(a) Use a graphing utility to plot the data and graph the model on the same screen.

(b) Use the model to estimate the price of a half-gallon of ice cream if the price of a half-gallon of milk was $1.59.

Explaining Concepts

142. Write "logarithm of x with base 5" symbolically.

143. Explain the relationship between the functions $f(x) = 2^x$ and $g(x) = \log_2 x$.

144. Explain why $\log_a a = 1$.

145. Explain why $\log_a a^x = x$.

146. What are common logarithms and natural logarithms?

147. Describe how to use a calculator to find the logarithm of a number if the base is not 10 or e.

Think About It In Exercises 148–153, answer the question for the function $f(x) = \log_{10} x$. (Do not use a calculator.)

148. What is the domain of f?

149. Find the inverse function of f.

150. Describe the values of $f(x)$ for $1000 \le x \le 10,000$.

151. Describe the values of x, given that $f(x)$ is negative.

152. By what amount will x increase, given that $f(x)$ is increased by 1 unit?

153. Find the ratio of a to b, given that $f(a) = 3 + f(b)$.

Mid-Chapter Quiz

Take this quiz as you would take a quiz in class. After you are done, check your work against the answers given in the back of the book.

1. Given $f(x) = \left(\frac{4}{3}\right)^x$, find (a) $f(2)$, (b) $f(0)$, (c) $f(-1)$, and (d) $f(1.5)$.
2. Find the domain and range of $g(x) = 2^{-0.5x}$.

In Exercises 3–6, sketch the graph of the function.

3. $y = \frac{1}{2}(4^x)$
4. $y = 5(2^{-x})$
5. $f(t) = 12e^{-0.4t}$
6. $g(x) = 100(1.08)^x$

7. You deposit \$750 at $7\frac{1}{2}\%$ interest, compounded n times per year or continuously. Complete the table, which shows the balance A after 20 years for several types of compounding.

n	1	4	12	365	Continuous compounding
A					

8. A gallon of milk costs \$2.23 now. If the price increases by 4% each year, what will the price be after 5 years?
9. Given $f(x) = 2x - 3$ and $g(x) = x^3$, find the indicated composition.
   (a) $(f \circ g)(x)$ (b) $(g \circ f)(x)$ (c) $(f \circ g)(-2)$ (d) $(g \circ f)(4)$
10. Verify algebraically and graphically that $f(x) = 3 - 5x$ and $g(x) = \frac{1}{5}(3 - x)$ are inverses of each other.

In Exercises 11 and 12, find the inverse of the function.

11. $h(x) = 10x + 3$
12. $g(t) = \frac{1}{2}t^3 + 2$

13. Write the logarithmic equation $\log_4\left(\frac{1}{16}\right) = -2$ in exponential form.
14. Write the exponential equation $3^4 = 81$ in logarithmic form.
15. Evaluate $\log_5 125$ without the aid of a calculator.
16. Write a paragraph comparing the graphs of $f(x) = \log_5 x$ and $g(x) = 5^x$.

In Exercises 17 and 18, use a graphing utility to sketch the graph of the function.

17. $f(t) = \frac{1}{2} \ln t$
18. $h(x) = 3 - \ln x$

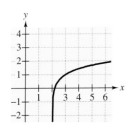

Figure for 19

19. Use the graph of f at the left to determine h and k if $f(x) = \log_5(x - h) + k$.
20. Use a calculator and the change-of-base formula to evaluate $\log_6 450$.

9.4 Properties of Logarithms

Objectives

1 Use the properties of logarithms to evaluate a logarithm.

2 Use the properties of logarithms to rewrite a logarithmic expression.

3 Use the properties of logarithms to solve an application problem.

1 Use the properties of logarithms to evaluate a logarithm.

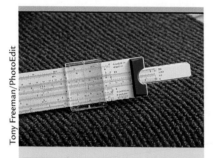

Slide Rule

Before electronic hand-held calculators became available in the 1970's, mathematicians, engineers, and scientists relied on a tool called the slide rule. Created by Edmund Gunter, (1581–1626) the slide rule uses logarithms to quickly multiply and divide numbers.

Tony Freeman/PhotoEdit

Properties of Logarithms

You know from the preceding section that the logarithmic function with base a is the *inverse* of the exponential function with base a. So, it makes sense that each property of exponents should have a corresponding property of logarithms. For instance, the exponential property

$$a^0 = 1 \qquad \text{Exponential property}$$

has the corresponding logarithmic property

$$\log_a 1 = 0. \qquad \text{Corresponding logarithmic property}$$

In this section you will study the logarithmic properties that correspond to the following three exponential properties:

Base a	*Natural Base*	
1. $a^n a^m = a^{n+m}$	$e^n e^m = e^{n+m}$	Product Rule
2. $\dfrac{a^n}{a^m} = a^{n-m}$	$\dfrac{e^n}{e^m} = e^{n-m}$	Quotient Rule
3. $(a^n)^m = a^{nm}$	$(e^n)^m = e^{nm}$	Power Rule

▶ **Properties of Logarithms**

Let a be a positive real number such that $a \neq 1$, and let n be a real number. If u and v are real numbers, variables, or algebraic expressions such that $u > 0$ and $v > 0$, the following properties are true.

Logarithm with Base a	*Natural Logarithm*	
1. $\log_a(uv) = \log_a u + \log_a v$	$\ln(uv) = \ln u + \ln v$	Product Rule
2. $\log_a \dfrac{u}{v} = \log_a u - \log_a v$	$\ln \dfrac{u}{v} = \ln u - \ln v$	Quotient Rule
3. $\log_a u^n = n \log_a u$	$\ln u^n = n \ln u$	Power Rule

There is no general property of logarithms that can be used to simplify $\log_a(u + v)$. Specifically,

$$\log_a(u + v) \text{ } does \text{ } not \text{ } equal \text{ } \log_a u + \log_a v.$$

| Example 1 | Using Properties of Logarithms |

Use the fact that $\ln 2 \approx 0.693$, $\ln 3 \approx 1.099$, and $\ln 5 \approx 1.609$ to approximate each of the following.

a. $\ln \dfrac{2}{3}$ **b.** $\ln 10$ **c.** $\ln 30$

Solution

a. $\ln \dfrac{2}{3} = \ln 2 - \ln 3$ 　　　　　　　Quotient Rule

$\approx 0.693 - 1.099$ 　　　　　　Substitute for $\ln 2$ and $\ln 3$.

$= -0.406$ 　　　　　　　　Simplify.

b. $\ln 10 = \ln(2 \cdot 5)$ 　　　　　　　Factor.

$= \ln 2 + \ln 5$ 　　　　　　　Product Rule

$\approx 0.693 + 1.609$ 　　　　　　Substitute for $\ln 2$ and $\ln 5$.

$= 2.302$ 　　　　　　　　Simplify.

c. $\ln 30 = \ln(2 \cdot 3 \cdot 5)$ 　　　　　　Factor.

$= \ln 2 + \ln 3 + \ln 5$ 　　　　Product Rule

$\approx 0.693 + 1.099 + 1.609$ 　Substitute for $\ln 2$, $\ln 3$, and $\ln 5$.

$= 3.401$ 　　　　　　　　Simplify.

When using the properties of logarithms, it helps to state the properties *verbally.* For instance, the verbal form of the Product Property $\ln(uv) = \ln u + \ln v$ is: *The log of a product is the sum of the logs of the factors.* Similarly, the verbal form of the Quotient Property $\ln\left(\dfrac{u}{v}\right) = \ln u - \ln v$ is: *The log of a quotient is the difference of the logs of the numerator and denominator.*

Study Tip

Remember that you can verify results such as those given in Example 2 with a calculator.

| Example 2 | Using Properties of Logarithms |

Use the properties of logarithms to verify that $-\ln 2 = \ln \frac{1}{2}$.

Solution

Using the Power Rule, you can write the following.

$-\ln 2 = (-1) \ln 2$ 　　　　　　　Rewrite coefficient as -1.

$= \ln 2^{-1}$ 　　　　　　　　Power Rule

$= \ln \dfrac{1}{2}$ 　　　　　　　　Rewrite 2^{-1} as $\frac{1}{2}$.

Rewriting Logarithmic Expressions

In Examples 1 and 2, the properties of logarithms were used to rewrite logarithmic expressions involving the log of a *constant*. A more common use of the properties is to rewrite the log of a *variable expression*.

Example 3 Rewriting the Logarithm of a Product

Use the properties of logarithms to rewrite $\log_{10} 7x^3$.

Solution

$$\log_{10} 7x^3 = \log_{10} 7 + \log_{10} x^3 \qquad \text{Product Rule}$$

$$= \log_{10} 7 + 3 \log_{10} x \qquad \text{Power Rule}$$

When you rewrite a logarithmic expression as in Example 3, you are **expanding** the expression. The reverse procedure is demonstrated in Example 4, and is called **condensing** a logarithmic expression.

Example 4 Condensing a Logarithmic Expression

Use the properties of logarithms to condense each expression.

a. $\ln x - \ln 3$

b. $2 \log_3 x + \log_3 5$

Solution

a. Using the Quotient Rule, you can write

$$\ln x - \ln 3 = \ln \frac{x}{3}. \qquad \text{Quotient Rule}$$

b. $2 \log_3 x + \log_3 5 = \log_3 x^2 + \log_3 5 \qquad \text{Power Rule}$

$$= \log_3 5x^2 \qquad \text{Product Rule}$$

Confirm these results by graphing both sides of the equation on a graphing utility.

Example 5 Expanding a Logarithmic Expression

Expand the logarithmic expression.

$$\log_2 3xy^2, \quad x > 0, \ y > 0$$

Solution

$$\log_2 3xy^2 = \log_2 3 + \log_2 x + \log_2 y^2 \qquad \text{Product Rule}$$

$$= \log_2 3 + \log_2 x + 2 \log_2 y \qquad \text{Power Rule}$$

Sometimes expanding or condensing logarithmic expressions involves several steps. In the next example, be sure that you can justify each step in the solution. Also, notice how different the expanded expression is from the original.

Example 6   Expanding a Logarithmic Expression

Expand the logarithmic expression.

$$\ln \sqrt{x^2 - 1}, \quad x > 1$$

Solution

$$
\begin{aligned}
\ln \sqrt{x^2 - 1} &= \ln(x^2 - 1)^{1/2} && \text{Rewrite using fractional exponent.} \\
&= \tfrac{1}{2}\ln(x^2 - 1) && \text{Power Rule} \\
&= \tfrac{1}{2}\ln[(x - 1)(x + 1)] && \text{Factor.} \\
&= \tfrac{1}{2}[\ln(x - 1) + \ln(x + 1)] && \text{Product Rule} \\
&= \tfrac{1}{2}\ln(x - 1) + \tfrac{1}{2}\ln(x + 1) && \text{Distributive Property}
\end{aligned}
$$

Example 7   Condensing a Logarithmic Expression

Use the properties of logarithms to condense the expression.

a. $\ln 2 - 2 \ln x$ **b.** $3(\ln 4 + \ln x)$

Solution

$$
\begin{aligned}
\textbf{a.} \ \ln 2 - 2 \ln x &= \ln 2 - \ln x^2, \quad x > 0 && \text{Power Rule} \\
&= \ln \frac{2}{x^2}, \quad x > 0 && \text{Quotient Rule}
\end{aligned}
$$

$$
\begin{aligned}
\textbf{b.} \ 3(\ln 4 + \ln x) &= 3(\ln 4x) && \text{Product Rule} \\
&= \ln(4x)^3 && \text{Power Rule} \\
&= \ln 64x^3, \quad x \geq 0 && \text{Simplify.}
\end{aligned}
$$

When you expand or condense a logarithmic expression, it is possible to change the domain of the expression. For instance, the domain of the function

$$f(x) = 2 \ln x \qquad \text{Domain is the set of positive real numbers.}$$

is the set of positive real numbers, whereas the domain of

$$g(x) = \ln x^2 \qquad \text{Domain is the set of nonzero real numbers.}$$

is the set of nonzero real numbers. So, when you expand or condense a logarithmic expression, you should check to see whether the rewriting has changed the domain of the expression. In such cases, you should restrict the domain appropriately. For instance, you can write

$$f(x) = 2 \ln x = \ln x^2, \quad x > 0.$$

3 Use the properties of logarithms to solve an application problem.

Application

Example 8   Human Memory Model

Students participating in a psychological experiment attended several lectures on a subject. Every month for a year after that, the students were tested to see how much of the material they remembered. The average scores for the group are given by the **human memory model**

$$f(t) = 80 - \ln(t + 1)^9, \quad 0 \le t \le 12$$

where t is the time in months. Find the average scores for the group after 2 months and 8 months.

Solution

To make the calculations easier, rewrite the model as

$$f(t) = 80 - 9\ln(t + 1), \quad 0 \le t \le 12.$$

After 2 months, the average score will be

$$f(2) = 80 - 9\ln 3 \approx 70.1 \qquad \text{Average score after 2 months}$$

and after 8 months, the average score will be

$$f(8) = 80 - 9\ln 9 \approx 60.2. \qquad \text{Average score after 8 months}$$

The graph of the function is shown in Figure 9.20.

Figure 9.20   *Human Memory Model*

(graph showing points (0, 80), (2, 70.1), (8, 60.2); y-axis labeled "Average score" from 20 to 80; x-axis labeled "Time (in months)" from 1 to 12; curve labeled $f(t) = 80 - 9\ln(t + 1)$)

Discussing the Concept	Mathematical Modeling

The data in the table represent the annual number y of new AIDS cases in males reported in the United States for the year x from 1993 through 1997, with $x = 3$ corresponding to 1993.   (Source: U.S. Centers for Disease Control and Prevention)

x	3	4	5	6	7
y	85,781	63,357	57,512	53,009	45,738

a. Plot the data. Would a linear model fit the points well?

b. Add a third row to the table giving the values of $\ln y$.

c. Plot the coordinate pairs $(x, \ln y)$. Would a linear model fit these points well? If so, draw the best-fitting line and find its equation.

d. Describe the shapes of the two scatter plots. Using your knowledge of logarithms, explain why the second scatter plot is so different from the first.

9.4   Exercises

Integrated Review   *Concepts, Skills, and Problem Solving*

Keep mathematically in shape by doing these exercises *before* the problems of this section.

Properties and Definitions

In Exercises 1 and 2, use the property of radicals to fill in the blank.

1. Multiplication Property: $\sqrt[n]{u}\ \sqrt[n]{v} =$

2. Division Property: $\dfrac{\sqrt[n]{u}}{\sqrt[n]{v}} =$

3. Explain why the radicals $\sqrt{2x}$ and $\sqrt[3]{2x}$ cannot be added.

4. Is $1/\sqrt{2x}$ in simplest form? Explain.

Simplifying Expressions

In Exercises 5-10, perform the operations and simplify. (Assume all variables are positive.)

5. $25\sqrt{3x} - 3\sqrt{12x}$ 6. $\left(\sqrt{x} + 3\right)\left(\sqrt{x} - 3\right)$

7. $\sqrt{u}\left(\sqrt{20} - \sqrt{5}\right)$

8. $\left(2\sqrt{t} + 3\right)^2$

9. $\dfrac{50x}{\sqrt{2}}$

10. $\dfrac{12}{\sqrt{t+2} + \sqrt{t}}$

Problem Solving

11. The demand equation for a product is given by $p = 30 - \sqrt{0.5(x - 1)}$, where x is the number of units demanded per day and p is the price per unit. Find the demand if the price is $26.76.

12. The sale price of a computer is $1955. The discount is 15% of the list price. Find the list price.

Developing Skills

In Exercises 1–24, use properties of logarithms to evaluate the expression without using a calculator. (If not possible, state the reason.)

1. $\log_5 5^2$

2. $\log_3 9$

3. $\log_2\left(\tfrac{1}{8}\right)^3$

4. $\log_8\left(\tfrac{1}{64}\right)^5$

5. $\log_6 \sqrt{6}$

6. $\ln \sqrt[3]{e}$

7. $\ln 8^0$

8. $\log_4 4^2$

9. $\ln e^4$

10. $\ln e^{-4}$

11. $\log_4 8 + \log_4 2$

12. $\log_6 2 + \log_6 3$

13. $\log_8 4 + \log_8 16$

14. $\log_{10} 5 + \log_{10} 20$

15. $\log_4 8 - \log_4 2$

16. $\log_5 50 - \log_5 2$

17. $\log_6 72 - \log_6 2$

18. $\log_3 324 - \log_3 4$

19. $\log_2 5 - \log_2 40$

20. $\log_3\left(\tfrac{2}{3}\right) + \log_3\left(\tfrac{1}{2}\right)$

21. $\ln e^8 + \ln e^4$

22. $\ln e^5 - \ln e^2$

23. $\ln \dfrac{e^3}{e^2}$

24. $\ln(e^2 \cdot e^4)$

In Exercises 25–36, use $\log_4 2 = 0.5000$, $\log_4 3 \approx 0.7925$, and the properties of logarithms to approximate the value of the logarithm. Do not use a calculator. See Example 1.

25. $\log_4 4$

26. $\log_4 8$

27. $\log_4 6$

28. $\log_4 24$

29. $\log_4 \tfrac{3}{2}$

30. $\log_4 \tfrac{9}{2}$

31. $\log_4 \sqrt{2}$

32. $\log_4 \sqrt[3]{9}$

33. $\log_4(3 \cdot 2^4)$

34. $\log_4 \sqrt{3 \cdot 2^5}$

35. $\log_4 3^0$

36. $\log_4 4^3$

In Exercises 37–42, use $\log_{10} 3 \approx 0.477$, $\log_{10} 12 \approx 1.079$, and the properties of logarithms to approximate the value of the logarithm. Use a calculator to verify your result.

37. $\log_{10} 9$

38. $\log_{10} \tfrac{1}{4}$

39. $\log_{10} 36$

40. $\log_{10} 144$

41. $\log_{10} \sqrt{36}$

42. $\log_{10} 5^0$

In Exercises 43–76, use the properties of logarithms to expand the given expression. See Examples 3, 5, and 6.

43. $\log_3 11x$

44. $\log_2 3x$

45. $\log_7 x^2$

46. $\log_3 x^3$

47. $\log_5 x^{-2}$

48. $\log_2 s^{-4}$

49. $\log_4 \sqrt{3x}$

50. $\log_3 \sqrt[3]{5y}$

51. $\ln 3y$

52. $\ln 5x$

53. $\log_2 \dfrac{z}{17}$

54. $\log_{10} \dfrac{7}{y}$

55. $\ln \dfrac{5}{x-2}$

56. $\log_4 \dfrac{1}{\sqrt{t}}$

57. $\ln x^2(y-2)$

58. $\ln y(y-1)^2$

59. $\log_4[x^6(x-7)^2]$

60. $\log_8[(x-y)^4 z^6]$

61. $\log_3 \sqrt[3]{x+1}$

62. $\log_5 \sqrt{xy}$

63. $\ln \sqrt{x(x+2)}$

64. $\ln \sqrt[3]{x(x+5)}$

65. $\ln\left(\dfrac{x+1}{x-1}\right)^2$

66. $\log_2\left(\dfrac{x^2}{x-3}\right)^3$

67. $\ln \sqrt[3]{\dfrac{x^2}{x+1}}$

68. $\ln \sqrt{\dfrac{3x}{x-5}}$

69. $\ln \dfrac{a^3(b-4)}{c^2}$

70. $\log_3 \dfrac{x^2 y}{z^7}$

71. $\ln \dfrac{x\sqrt[3]{y}}{(wz)^4}$

72. $\log_4 \dfrac{\sqrt[3]{a+1}}{(ab)^4}$

73. $\log_6[a\sqrt{b}(c-d)^3]$

74. $\ln[(xy)^2(x+3)^4]$

75. $\ln\left[(x+y)\dfrac{\sqrt[5]{w+2}}{3t}\right]$

76. $\ln\left[(u-v)\dfrac{\sqrt[3]{u-4}}{3v}\right]$

In Exercises 77–108, use the properties of logarithms to condense the expression. See Examples 4 and 7.

77. $\log_{12} x - \log_{12} 3$

78. $\log_6 12 - \log_6 y$

79. $\log_2 3 + \log_2 x$

80. $\log_5 2x + \log_5 3y$

81. $\log_{10} 4 - \log_{10} x$

82. $\ln 10x - \ln z$

83. $4 \ln b$

84. $10 \log_4 z$

85. $-2 \log_5 2x$

86. $-5 \ln(x+3)$

87. $\frac{1}{3} \ln(2x+1)$

88. $-\frac{1}{2} \log_3 5y$

89. $\log_3 2 + \frac{1}{2} \log_3 y$

90. $\ln 6 - 3 \ln z$

91. $2 \ln x + 3 \ln y - \ln z$

92. $4 \ln 3 - 2 \ln x - \ln y$

93. $5 \ln 2 - \ln x + 3 \ln y$

94. $4 \ln 2 + 2 \ln x - \frac{1}{2} \ln y$

95. $4(\ln x + \ln y)$

96. $\frac{1}{2}(\ln 8 + \ln 2x)$

97. $2[\ln x - \ln(x+1)]$

98. $5\left[\ln x - \frac{1}{2} \ln(x+4)\right]$

99. $\log_4(x+8) - 3 \log_4 x$

100. $5 \log_3 x + \log_3(x-6)$

101. $\frac{1}{2} \log_5(x+2) - \log_5(x-3)$

102. $\frac{1}{4} \log_6(x+1) - 5 \log_6(x-4)$

103. $5 \log_6(c+d) - \frac{1}{2} \log_6(m-n)$

104. $2 \log_5(x+y) + 3 \log_5 w$

105. $\frac{1}{5}(3 \log_2 x - 4 \log_2 y)$

106. $\frac{1}{3}[\ln(x-6) - 4 \ln y - 2 \ln z]$

107. $\frac{1}{5} \log_6(x-3) - 2 \log_6 x - 3 \log_6(x+1)$

108. $3\left[\frac{1}{2} \log_9(a+6) - 2 \log_9(a-1)\right]$

In Exercises 109–114, simplify the expression.

109. $\ln 3e^2$

110. $\log_3(3^2 \cdot 4)$

111. $\log_5 \sqrt{50}$

112. $\log_2 \sqrt{22}$

113. $\log_4 \dfrac{4}{x^2}$

114. $\ln \dfrac{6}{e^5}$

In Exercises 115–118, use a graphing utility to graph the two equations on the same screen. Use the graphs to verify that the expressions are equivalent. Assume $x > 0$.

115. $y_1 = \ln\left(\dfrac{10}{x^2+1}\right)^2$

$y_2 = 2[\ln 10 - \ln(x^2+1)]$

116. $y_1 = \ln \sqrt{x(x+1)}$

$y_2 = \frac{1}{2}[\ln x + \ln(x+1)]$

117. $y_1 = \ln[x^2(x+2)]$

$y_2 = 2 \ln x + \ln(x+2)$

118. $y_1 = \ln\left(\dfrac{\sqrt{x}}{x-3}\right)$

$y_2 = \frac{1}{2} \ln x - \ln(x-3)$

119. *Think About It* Explain how you can show that

$$\dfrac{\ln x}{\ln y} \neq \ln \dfrac{x}{y}.$$

120. *Think About It* Without a calculator, approximate the natural logarithms of as many integers as possible between 1 and 20 using $\ln 2 \approx 0.6931$, $\ln 3 \approx 1.0986$, $\ln 5 \approx 1.6094$, and $\ln 7 \approx 1.9459$. Explain the method you used. Then verify your results with a calculator and explain any differences in the results.

Solving Problems

121. *Intensity of Sound*   The relationship between the number of decibels B and the intensity of a sound I in watts per centimeter squared is given by

$$B = 10 \log_{10}\left(\frac{I}{10^{-16}}\right).$$

Use properties of logarithms to write the formula in simpler form, and determine the number of decibels of a sound with an intensity of 10^{-10} watts per centimeter squared.

122. *Human Memory Model*   Students participating in a psychological experiment attended several lectures on a subject. Every month for a year after that, the students were tested to see how much of the material they remembered. The average scores for the group are given by the human memory model

$$f(t) = 80 - \log_{10}(t + 1)^{12}, \quad 0 \leq t \leq 12$$

where t is the time in months.

(a) Find the average scores for the group after 2 months and 8 months.

(b) Use a graphing utility to graph the function.

Biology   In Exercises 123 and 124, use the following information. The energy E (in kilocalories per gram molecule) required to transport a substance from the outside to the inside of a living cell is given by

$$E = 1.4(\log_{10} C_2 - \log_{10} C_1)$$

where C_1 and C_2 are the concentrations of the substance outside and inside the cell, respectively.

123. Condense the equation.

124. The concentration of a particular substance inside a cell is twice the concentration outside the cell. How much energy is required to transport the substance from outside to inside the cell?

Explaining Concepts

True or False?   In Exercises 125–132, use properties of logarithms to determine whether the equation is true or false. If false, state why or give an example to show that it is false.

125. $\ln e^{2-x} = 2 - x$

126. $\log_2 8x = 3 + \log_2 x$

127. $\log_8 4 + \log_8 16 = 2$

128. $\log_3(u + v) = \log_3 u + \log_3 v$

129. $\log_3(u + v) = \log_3 u \cdot \log_3 v$

130. $\dfrac{\log_6 10}{\log_6 3} = \log_6 10 - \log_6 3$

131. If $f(x) = \log_a x$, then $f(ax) = 1 + f(x)$.

132. If $f(x) = \log_a x$, then $f(a^n) = n$.

True or False?   In Exercises 133–138, determine whether the statement is true or false given that $f(x) = \ln x$. If false, state why or give an example to show that the statement is false.

133. $f(0) = 0$

134. $f(2x) = \ln 2 + \ln x$

135. $f(x - 3) = \ln x - \ln 3, \quad x > 3$

136. $\sqrt{f(x)} = \frac{1}{2} \ln x$

137. If $f(u) = 2f(v)$, then $v = u^2$.

138. If $f(x) > 0$, then $x > 1$.

9.5 Solving Exponential and Logarithmic Equations

Objectives

1 Solve an exponential or a logarithmic equation in which both sides have the same base.

2 Use inverse properties to solve an exponential equation.

3 Use inverse properties to solve a logarithmic equation.

4 Use an exponential or a logarithmic equation to solve an application problem.

1 Solve an exponential or a logarithmic equation in which both sides have the same base.

Exponential and Logarithmic Equations

In this section, you will study procedures for *solving equations* that involve exponential or logarithmic expressions. As a simple example, consider the exponential equation $2^x = 16$. By rewriting this equation in the form $2^x = 2^4$, you can see that the solution is $x = 4$. To solve this equation, you can use one of the following properties, which result from the fact that exponential and logarithmic functions are one-to-one.

> ▶ **Properties of Exponential and Logarithmic Equations**
>
> Let a be a positive real number such that $a \neq 1$, and let x and y be real numbers. Then the following properties are true.
>
> **1.** $a^x = a^y$ $\qquad$ if and only if $x = y$.
>
> **2.** $\log_a x = \log_a y$ $\quad$ if and only if $x = y$ $(x > 0, y > 0)$.

Example 1 Solving Exponential and Logarithmic Equations

Solve each equation.

a. $4^{x+2} = 64$ $\qquad$ **b.** $\ln(2x - 3) = \ln 11$

Solution

a. $\quad 4^{x+2} = 64$ $\qquad$ Original equation

$\qquad 4^{x+2} = 4^3$ $\qquad$ Rewrite with like bases.

$\qquad x + 2 = 3$ $\qquad$ Property of exponential equations

$\qquad\qquad x = 1$ $\qquad$ Subtract 2 from both sides.

The solution is 1. Check this in the original equation.

b. $\ln(2x - 3) = \ln 11$ $\qquad$ Original equation

$\qquad\qquad 2x - 3 = 11$ $\qquad$ Property of logarithmic equations

$\qquad\qquad\qquad 2x = 14$ $\qquad$ Add 3 to both sides.

$\qquad\qquad\qquad\quad x = 7$ $\qquad$ Divide both sides by 2.

The solution is 7. Check this in the original equation.

2 Use inverse properties to solve an exponential equation.

Solving Exponential Equations

In Example 1(a), you were able to solve the given equation because both sides of the equation could be written in exponential form (with the same base). However, if only one side of the equation can be written in exponential form or if both sides cannot be written with the same base, it is more difficult to solve the equation. For example, how would you solve the following equation?

$$2^x = 7$$

To solve this equation, you must find the power to which 2 can be raised to obtain 7. To do this, rewrite the exponential form into logarithmic form by taking the logarithm of both sides and use one of the following inverse properties of exponents and logarithms.

▶ **Inverse Properties of Exponents and Logarithms**

Base a	*Natural Base e*
1. $\log_a(a^x) = x$	$\ln(e^x) = x$
2. $a^{(\log_a x)} = x$	$e^{(\ln x)} = x$

Technology: Discovery

Use a graphing utility to graph both sides of each equation. What does this tell you about the inverse properties of exponents and logarithms?

1. (a) $\log_{10}(10^x) = x$

(b) $10^{(\log_{10} x)} = x$

2. (a) $\ln(e^x) = x$

(b) $e^{(\ln x)} = x$

Example 2 Solving an Exponential Equation

Solve the exponential equation.

a. $2^x = 7$ **b.** $4^{x-3} = 9$ **c.** $2e^x = 10$

Solution

a. To isolate the x, take the $\log_2$ of both sides of the equation, as follows.

$2^x = 7$	Original equation
$\log_2 2^x = \log_2 7$	Take the logarithm of both sides.
$x = \log_2 7$	Inverse property

The solution is $x = \log_2 7 \approx 2.807$. Check this in the original equation.

b. To isolate the x, take the $\log_4$ of both sides of the equation, as follows.

$4^{x-3} = 9$	Original equation
$\log_4 4^{x-3} = \log_4 9$	Take the logarithm of both sides.
$x - 3 = \log_4 9$	Inverse property
$x = \log_4 9 + 3$	Add 3 to both sides.

The solution is $x = \log_4 9 + 3 \approx 4.585$. Check this in the original equation.

c.

$2e^x = 10$	Original equation
$e^x = 5$	Divide both sides by 2.
$\ln e^x = \ln 5$	Take the logarithm of both sides.
$x = \ln 5$	Inverse property

The solution is $x = \ln 5 \approx 1.609$. Check this in the original equation.

Study Tip

Remember that to evaluate a logarithm like $\log_2 7$ you need to use the change of base formula.

$$\log_2 7 = \frac{\ln 7}{\ln 2} \approx 2.807$$

Similarly,

$$\log_4 9 + 3 = \frac{\ln 9}{\ln 4} + 3 \approx$$

$$1.585 + 3 \approx 4.585$$

Technology: Tip

Graphical Check of Solutions
Remember that you can use a graphing utility to solve equations graphically or check solutions that are obtained algebraically. For instance, to check the solutions in Examples 2(a) and 2(c), graph both sides of the equations, as shown below.

Graph $y = 2^x$ and $y = 7$. Then approximate the intersection of the two graphs to be $x \approx 2.807$.

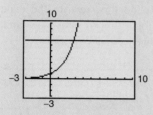

Graph $y = 2e^x$ and $y = 10$. Then approximate the intersection of the two graphs to be $x \approx 1.609$.

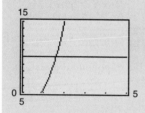

Example 3 Solving an Exponential Equation

Solve $5 + e^{x+1} = 20$.

Solution

$5 + e^{x+1} = 20$	Original equation
$e^{x+1} = 15$	Subtract 5 from both sides.
$\ln e^{x+1} = \ln 15$	Take the logarithm of both sides.
$x + 1 = \ln 15$	Inverse property
$x = -1 + \ln 15$	Subtract 1 from both sides.

The solution is $x = -1 + \ln 15 \approx 1.708$.

Check

$5 + e^{x+1} = 20$	Original equation
$5 + e^{1.708+1} \overset{?}{=} 20$	Substitute 1.708 for x.
$5 + e^{2.708} \overset{?}{=} 20$	Simplify.
$5 + 14.999 \approx 20$	Solution checks. ✓

3 Use inverse properties to solve a logarithmic equation.

Solving Logarithmic Equations

You know how to solve an exponential equation by *taking the logarithms of both sides*. To solve a logarithmic equation, you need to **exponentiate** both sides. For instance, to solve a logarithmic equation such as

$$\ln x = 2$$

you can exponentiate both sides of the equation as follows.

$\ln x = 2$	Original equation
$e^{\ln x} = e^2$	Exponentiate both sides.
$x = e^2$	Inverse property

This procedure is demonstrated in the next three examples. We suggest the following guidelines for solving exponential and logarithmic equations.

▶ **Solving Exponential and Logarithmic Equations**

1. To solve an exponential equation, first isolate the exponential expression, then **take the logarithm of both sides of the equation** and solve for the variable.

2. To solve a logarithmic equation, first isolate the logarithmic expression, then **exponentiate both sides of the equation** and solve for the variable.

Example 4   Solving a Logarithmic Equation

Solve $2 \log_4 x = 5$.

Solution

$2 \log_4 x = 5$	Original equation
$\log_4 x = \dfrac{5}{2}$	Divide both sides by 2.
$4^{\log_4 x} = 4^{5/2}$	Exponentiate both sides.
$x = 4^{5/2}$	Inverse property
$x = 32$	Simplify.

The solution is $x = 32$. Check this in the original equation, as follows.

Check

$2 \log_4 x = 5$	Original equation
$2 \log_4(32) \overset{?}{=} 5$	Substitute 32 for x.
$2(2.5) \overset{?}{=} 5$	Use a calculator.
$5 = 5$	Solution checks. ✓

Example 5   Solving a Logarithmic Equation

Solve $3 \log_{10} x = 6$.

Solution

$$3 \log_{10} x = 6 \qquad \text{Original equation}$$
$$\log_{10} x = 2 \qquad \text{Divide both sides by 3.}$$
$$10^{\log_{10} x} = 10^2 \qquad \text{Exponentiate both sides.}$$
$$x = 100 \qquad \text{Inverse property}$$

The solution is $x = 100$.   Check this in the original equation.

Example 6   Solving a Logarithmic Equation

Solve $20 \ln 0.2x = 30$.

Solution

$$20 \ln 0.2x = 30 \qquad \text{Original equation}$$
$$\ln 0.2x = 1.5 \qquad \text{Divide both sides by 20.}$$
$$e^{\ln 0.2x} = e^{1.5} \qquad \text{Exponentiate both sides.}$$
$$0.2x = e^{1.5} \qquad \text{Inverse property}$$
$$x = 5e^{1.5} \qquad \text{Divide both sides by 0.2.}$$

The solution is $x = 5e^{1.5} \approx 22.408$. Check this in the original equation.

The next example uses logarithmic properties as part of the solution.

Example 7   Solving a Logarithmic Equation

Solve $\log_3 2x - \log_3(x - 3) = 1$.

Solution

$$\log_3 2x - \log_3(x - 3) = 1 \qquad \text{Original equation}$$
$$\log_3 \frac{2x}{x - 3} = 1 \qquad \text{Condense the left side.}$$
$$3^{\log[2x/(x-3)]} = 3^1 \qquad \text{Exponentiate both sides.}$$
$$\frac{2x}{x - 3} = 3 \qquad \text{Inverse property}$$
$$2x = 3x - 9 \qquad \text{Multiply both sides by } x - 3.$$
$$-x = -9 \qquad \text{Subtract } 3x \text{ from both sides.}$$
$$x = 9 \qquad \text{Divide both sides by } -1.$$

The solution is $x = 9$. Check this in the original equation.

4 Use an exponential or a logarithmic equation to solve an application problem.

Application

Example 8   Compound Interest

A deposit of $5000 is placed in a savings account for 2 years. The interest on the account is compounded continuously. At the end of 2 years, the balance in the account is $5867.55. What is the annual interest rate for this account?

Solution

Using the formula for continuously compounded interest, $A = Pe^{rt}$, you have the following solution.

Formula: $A = Pe^{rt}$

Labels: Principal $= P = 5000$ (dollars)
             Amount $= A = 5867.55$ (dollars)
             Time $= t = 2$ (years)
             Annual interest rate $= r$ (percent in decimal form)

Equation: $5867.55 = 5000e^{2r}$

$$\frac{5867.55}{5000} = e^{2r}$$ Divide both sides by 5000.

$$1.1735 \approx e^{2r}$$ Simplify.

$$\ln 1.1735 \approx \ln(e^{2r})$$ Logarithmic form

$$0.16 \approx 2r$$ Inverse property

$$0.08 \approx r$$ Divide both sides by 2.

The rate is 8%. Check this in the original statement of the problem, as follows.

Check

$A = Pe^{rt}$ Original equation

$A = 5000e^{(0.08)(2)}$ Substitute 5000 for P, 0.08 for r, and 2 for t.

$A = \$5867.55$ Solution checks. ✓

Discussing the Concept   Solving Equations

Solve each equation.

a. $x^2 - 5x - 14 = 0$
b. $e^{2x} - 5e^x - 14 = 0$
c. $(\ln x)^2 - 5\ln x - 14 = 0$

Explain your strategy. What are the similarities among the three equations? One of the equations has only one solution. Explain why.

9.5 Exercises

Integrated Review — Concepts, Skills, and Problem Solving

Keep mathematically in shape by doing these exercises *before* the problems of this section.

Properties and Definitions

1. Is it possible for the system

$$7x - 2y = 8$$
$$x + y = 4$$

to have exactly two solutions? Explain.

2. Explain why the following system has no solution.

$$8x - 4y = 5$$
$$-2x + y = 1$$

Solving Equations

In Exercises 3–8, solve the equation.

3. $\frac{2}{3}x + \frac{2}{3} = 4x - 6$

4. $x^2 - 10x + 17 = 0$

5. $\frac{5}{2x} - \frac{4}{x} = 3$

6. $\frac{1}{x} + \frac{2}{x - 5} = 0$

7. $|x - 4| = 3$

8. $\sqrt{x + 2} = 7$

Models and Graphing

9. A train is traveling at 73 miles per hour. Write the distance d the train travels as a function of the time t. Graph the function.

10. The diameter of a right circular cylinder is 10 centimeters. Write the volume V of the cylinder as a function of its height h if the formula for its volume is $V = \pi r^2 h$. Graph the model.

11. The height of a right circular cylinder is 10 centimeters. Write the volume V of the cylinder as a function of its radius r if the formula for its volume is $V = \pi r^2 h$. Graph the model.

12. A force of 100 pounds stretches a spring 4 inches. Write the force F as a function of the distance x that the spring is stretched. Graph the model.

Developing Skills

In Exercises 1–6, determine whether the x-values are solutions of the equation.

1. $3^{2x-5} = 27$
  (a) $x = 1$
  (b) $x = 4$

2. $4^{x+3} = 16$
  (a) $x = -1$
  (b) $x = 0$

3. $e^{x+5} = 45$
  (a) $x = -5 + \ln 45$
  (b) $x = -5 + e^{45}$

4. $2^{3x-1} = 324$
  (a) $x \approx 3.1133$
  (b) $x \approx 2.4327$

5. $\log_9(6x) = \frac{3}{2}$
  (a) $x = 27$
  (b) $x = \frac{9}{2}$

6. $\ln(x + 3) = 2.5$
  (a) $x = -3 + e^{2.5}$
  (b) $x \approx 9.1825$

In Exercises 7–34, solve the equation. (Do not use a calculator.) See Example 1.

7. $2^x = 2^5$

8. $5^x = 5^3$

9. $3^{x+4} = 3^{12}$

10. $10^{1-x} = 10^4$

11. $3^{x-1} = 3^7$

12. $4^{x+4} = 4^3$

13. $4^{3x} = 16$

14. $3^{2x} = 81$

15. $6^{2x-1} = 216$

16. $5^{3-2x} = 625$

17. $5^x = \frac{1}{125}$

18. $3^x = \frac{1}{243}$

19. $2^{x+2} = \frac{1}{16}$

20. $3^{2-x} = 9$

21. $4^{x+3} = 32^x$

22. $9^{x-2} = 243^{x+1}$

23. $\ln 5x = \ln 22$

24. $\ln 3x = \ln 24$

25. $\log_6 3x = \log_6 18$

26. $\log_5 2x = \log_5 36$

27. $\ln(2x - 3) = \ln 15$

28. $\ln(2x - 3) = \ln 17$

29. $\log_2(x + 3) = \log_2 7$

30. $\log_4(x - 4) = \log_4 12$

31. $\log_5(2x - 3) = \log_5(4x - 5)$

32. $\log_3(4 - 3x) = \log_3(2x + 9)$

33. $\log_3(2 - x) = 2$

34. $\log_2(3x - 1) = 5$

In Exercises 35–38, simplify the expression.

35. $\ln e^{2x-1}$

36. $\log_3 3^{x^2}$

37. $10^{\log_{10} 2x}$

38. $e^{\ln(x+1)}$

In Exercises 39–82, solve the exponential equation. Round the result to two decimal places. See Examples 2 and 3.

39. $2^x = 45$

40. $5^x = 21$

41. $3^x = 3.6$

42. $2^x = 1.5$

43. $10^{2y} = 52$

44. $8^{4x} = 20$

45. $7^{3y} = 126$

46. $5^{5y} = 305$

47. $3^{x+4} = 6$

48. $5^{3-x} = 15$

49. $10^{x+6} = 250$

50. $12^{x-1} = 324$

51. $3e^x = 42$

52. $6e^{-x} = 3$

53. $\frac{1}{4}e^x = 5$

54. $\frac{2}{3}e^x = 1$

55. $\frac{1}{2}e^{3x} = 20$

56. $4e^{-3x} = 6$

57. $250(1.04)^x = 1000$

58. $32(1.5)^x = 640$

59. $300e^{x/2} = 9000$

60. $6000e^{-2t} = 1200$

61. $1000^{0.12x} = 25{,}000$

62. $10{,}000e^{-0.1t} = 4000$

63. $\frac{1}{5}(4^{x+2}) = 300$

64. $3(2^{t+4}) = 350$

65. $6 + 2^{x-1} = 1$

66. $5^{x+6} - 4 = 12$

67. $7 + e^{2-x} = 28$

68. $9 + e^{5-x} = 32$

69. $8 - 12e^{-x} = 7$

70. $4 - 2e^x = -23$

71. $4 + e^{2x} = 10$

72. $10 + e^{4x} = 18$

73. $32 + e^{7x} = 46$

74. $50 - e^{x/2} = 35$

75. $23 - 5e^{x+1} = 3$

76. $2e^x + 5 = 115$

77. $4(1 + e^{x/3}) = 84$

78. $50(3 - e^{2x}) = 125$

79. $\dfrac{8000}{(1.03)^t} = 6000$

80. $\dfrac{5000}{(1.05)^x} = 250$

81. $\dfrac{300}{2 - e^{-0.15t}} = 200$

82. $\dfrac{500}{1 + e^{-0.1t}} = 400$

In Exercises 83–118, solve the logarithmic equation. Round the result to two decimal places. See Examples 4–7.

83. $\log_{10} x = 3$

84. $\log_{10} x = -2$

85. $\log_2 x = 4.5$

86. $\log_4 x = 2.1$

87. $4 \log_3 x = 28$

88. $6 \log_2 x = 18$

89. $16 \ln x = 30$

90. $12 \ln x = 20$

91. $\log_{10} 4x = 2$

92. $\log_3 6x = 4$

93. $\ln 2x = 3$

94. $\ln(0.5t) = \frac{1}{4}$

95. $\ln x^2 = 6$

96. $\ln \sqrt{x} = 6.5$

97. $2 \log_4(x + 5) = 3$

98. $5 \log_{10}(x + 2) = 15$

99. $2 \log_8(x + 3) = 3$

100. $\frac{2}{3} \ln(x + 1) = -1$

101. $1 - 2 \ln x = -4$

102. $5 - 4 \log_2 x = 2$

103. $-1 + 3 \log_{10} \dfrac{x}{2} = 8$

104. $-5 + 2 \ln 3x = 5$

105. $\log_4 x + \log_4 5 = 2$

106. $\log_5 x - \log_5 4 = 2$

107. $\log_6(x + 8) + \log_6 3 = 2$

108. $\log_7(x - 1) - \log_7 4 = 1$

109. $\log_5(x + 3) - \log_5 x = 1$

110. $\log_3(x - 2) + \log_3 5 = 3$

111. $\log_{10} x + \log_{10}(x - 3) = 1$

112. $\log_{10} x + \log_{10}(x + 1) = 0$

113. $\log_2(x - 1) + \log_2(x + 3) = 3$

114. $\log_6(x - 5) + \log_6 x = 2$

115. $\log_4 3x + \log_4(x - 2) = \frac{1}{2}$

116. $\log_{10}(25x) - \log_{10}(x - 1) = 2$

117. $\log_2 x + \log_2(x + 2) - \log_2 3 = 4$

118. $\log_3 2x + \log_3(x - 1) - \log_3 4 = 1$

In Exercises 119–122, use a graphing utility to approximate the x-intercept of the graph.

119. $y = 10^{x/2} - 5$

120. $y = 2e^x - 21$

121. $y = 6 \ln(0.4x) - 13$

122. $y = 5 \log_{10}(x + 1) - 3$

In Exercises 123–126, use a graphing utility to approximate the point of intersection of the graphs.

123. $y_1 = 2$
$y_2 = e^x$

124. $y_1 = 2$
$y_2 = \ln x$

125. $y_1 = 3$
$y_2 = 2 \ln(x + 3)$

126. $y_1 = 200$
$y_2 = 1000e^{-x/2}$

Solving Problems

127. *Compound Interest*   A deposit of $10,000 is placed in a savings account for 2 years. The interest for the account is compounded continuously. At the end of 2 years, the balance in the account is $11,972.17. What is the annual interest rate for this account?

128. *Compound Interest*   A deposit of $2500 is placed in a savings account for 2 years. The interest for the account is compounded continuously. At the end of 2 years, the balance in the account is $2847.07. What is the annual interest rate for this account?

129. *Doubling Time*   Solve the exponential equation

$$5000 = 2500e^{0.09t}$$

for t to determine the number of years for an investment of $2500 to double in value when compounded continuously at the rate of 9%.

130. *Doubling Rate*   Solve the exponential equation

$$10,000 = 5000e^{10r}$$

for r to determine the interest rate required for an investment of $5000 to double in value when compounded continuously for 10 years.

131. *Intensity of Sound*   The relationship between the number of decibels B and the intensity of a sound I in watts per centimeter squared is given by

$$B = 10 \log_{10}\left(\frac{I}{10^{-16}}\right).$$

Determine the intensity of a sound I if it registers 75 decibels on a decibel meter.

132. *Intensity of Sound*   The relationship between the number of decibels B and the intensity of a sound I in watts per centimeter squared is given by

$$B = 10 \log_{10}\left(\frac{I}{10^{-16}}\right).$$

Determine the intensity of a sound I if it registers 90 decibels on a decibel meter.

133. *Muon Decay*   A muon is an elementary particle that is similar to an electron, but much heavier. Muons are unstable—they quickly decay to form electrons and other particles. In an experiment conducted in 1943, the number of muon decays m (of an original 5000 muons) was related to the time T (in microseconds) by the model

$$T = 15.7 - 2.48 \ln m.$$

How many decays were recorded when $T = 2.5$?

134. *Friction*   In order to restrain an untrained horse, a person partially wraps the rope around a cylindrical post in a corral (see figure). If the horse is pulling on the rope with a force of 200 pounds, the force F in pounds required by the person is

$$F = 200e^{-0.5\pi\theta/180}$$

where *theta* is the angle of wrap in degrees. Find the smallest value of *theta* if F cannot exceed 80 pounds.

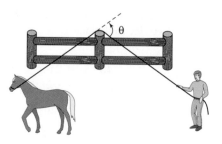

Figure for 134

135. *Human Memory Model*   The average score A for a group of students who took a test t months after the completion of a course is given by the memory model

$$A = 80 - \log_{10}(t + 1)^{12}.$$

How long after completing the course will the average score fall to $A = 72$?

(a) Answer the question algebraically by letting $A = 72$ and solving the resulting equation.

(b) Answer the question graphically by using a graphing utility to graph the equations $y_1 = 80 - \log_{10}(t + 1)^{12}$ and $y_2 = 72$, and finding their point(s) of intersection.

(c) Which strategy works better for this problem? Explain.

136. *Military Personnel*   The number N (in thousands) of United States military personnel on active duty in foreign countries for the years 1990 through 1996 is modeled by the equation

$$N = 273.1 + 355.8e^{-t}, \quad 0 \le t \le 6$$

where t is time in years, with $t = 0$ corresponding to 1990.   (Source: U.S. Department of Defense)

(a) Use a graphing utility to graph the equation over the specified domain.

(b) Use the graph in part (a) to estimate the value of t when $N = 500$.

137. *Making Ice Cubes*   You place a tray of 60°F water into a freezer that is set at 0°F. The water cools according to Newton's Law of Cooling

$$kt = \ln\frac{T - S}{T_0 - S}$$

where T is the temperature of the water (in °F), t is the number of hours the tray is in the freezer, S is the temperature of the surrounding air, and T_0 is the original temperature of the water.

(a) If the water freezes in 4 hours, what is the constant k? (*Hint:* Water freezes at 32°F.)

(b) Suppose you lower the temperature in the freezer to $-10°F$. At this temperature, how long will it take for the ice cubes to form?

(c) Suppose the initial temperature of the water is 50°F. If the freezer temperature is 0°F, how long will it take for the ice cubes to form?

138. *Oceanography*   Oceanographers use the density d (in grams per cubic centimeter) of seawater to obtain information about the circulation of water masses and the rates at which waters of different densities mix. For water with a salinity of 30%, the water temperature T (in °C) is related to the density by

$T = 7.9 \ln(1.0245 - d) + 61.84.$

Find the densities of the subantarctic water and the antarctic bottom water shown in the figure.

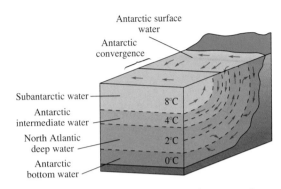

Figure for 138   *This cross section shows complex currents at various depths in the South Atlantic Ocean off Antarctica.*

Explaining Concepts

139. Answer parts (c)–(f) of Motivating the Chapter on page 547.

140. State the three basic properties of logarithms.

141. Which equation requires logarithms for its solution?

$2^{x-1} = 32$ or $2^{x-1} = 30$?

142. Explain how to solve $10^{2x-1} = 5316$.

143. In your own words, state the guidelines for solving exponential and logarithmic equations.

9.6 Applications

Objectives

1 Use an exponential equation to solve a compound interest problem.

2 Use an exponential equation to solve a growth or decay problem.

3 Use a logarithmic equation to solve an intensity problem.

---

1 Use an exponential equation to solve a compound interest problem.

Compound Interest

In Section 9.1, you were introduced to the following two exponential formulas for compound interest. In these formulas, A is the balance, P is the principal, r is the annual interest rate (in decimal form), and t is the time in years.

$$n \text{ Compoundings per Year} \qquad \qquad Continuous\ Compounding$$

$$A = P\left(1 + \frac{r}{n}\right)^{nt} \qquad\qquad\qquad A = Pe^{rt}$$

Example 1 Finding the Annual Interest Rate

An investment of $50,000 is made in an account that compounds interest quarterly. After 4 years, the balance in the account is $71,381.07. What is the annual interest rate for this account?

Solution

Formula: $\quad A = P\left(1 + \dfrac{r}{n}\right)^{nt}$

Labels: Principal $= P = 50{,}000$ (dollars)
              Amount $= A = 71{,}381.07$ (dollars)
              Time $= t = 4$ (years)
              Number of compoundings per year $= n = 4$
              Annual interest rate $= r$ (percent in decimal form)

Equation: $\quad 71{,}381.07 = 50{,}000\left(1 + \dfrac{r}{4}\right)^{(4)(4)}$

$\dfrac{71{,}381.07}{50{,}000} = \left(1 + \dfrac{r}{4}\right)^{16}$ Divide both sides by 50,000.

$1.42762 \approx \left(1 + \dfrac{r}{4}\right)^{16}$ Simplify.

$(1.42762)^{1/16} \approx 1 + \dfrac{r}{4}$ Raise both sides to $\frac{1}{16}$ power.

$1.0225 \approx 1 + \dfrac{r}{4}$ Simplify.

$0.09 \approx r$ Subtract 1 and then multiply both sides by 4.

The annual interest rate is approximately 9%. Check this in the original problem.

| Example 2 | Doubling Time for Continuous Compounding |

An investment is made in a trust fund at an annual interest rate of 8.75%, compounded continuously. How long will it take for the investment to double?

Solution

$$A = Pe^{rt}$$ Formula for continuous compounding

$$2P = Pe^{0.0875t}$$ Substitute known values.

$$2 = e^{0.0875t}$$ Divide both sides by P.

$$\ln 2 = 0.0875t$$ Inverse property

$$\frac{\ln 2}{0.0875} = t$$ Divide both sides by 0.0875.

$$7.92 \approx t$$

It will take approximately 7.92 years for the investment to double.

Check

$$A = Pe^{rt}$$ Formula for continuous compounding

$$2P \overset{?}{=} Pe^{0.0875(7.92)}$$ Substitute $2P$ for A, 0.0875 for r, and 7.92 for t.

$$2P \overset{?}{=} Pe^{0.693}$$ Simplify.

$$2P \approx 1.9997P$$ Solution checks. ✓

| Example 3 | Finding the Type of Compounding |

You deposit $1000 in an account. At the end of 1 year your balance is $1077.63. If the bank tells you that the annual interest rate for the account is 7.5%, how was the interest compounded?

Solution

If the interest had been compounded continuously at 7.5%, the balance would have been

$$A = 1000e^{(0.075)(1)} = \$1077.88.$$

Because the actual balance is slightly less than this, you should use the formula for interest that is compounded n times per year.

$$A = 1000\left(1 + \frac{0.075}{n}\right)^n = 1077.63$$

At this point, it is not clear what you should do to solve the equation for n. However, by completing a table, you can see that $n = 12$. So, the interest was compounded monthly.

n	1	4	12	365
$\left(1 + \dfrac{0.075}{n}\right)^n$	1.075	1.07714	1.07763	1.07788

In Example 3, notice that an investment of $1000 compounded monthly produced a balance of $1077.63 at the end of 1 year. Because $77.63 of this amount is interest, the **effective yield** for the investment is

$$\text{Effective yield} = \frac{\text{year's interest}}{\text{amount invested}} = \frac{77.63}{1000} = 0.07763 = 7.763\%.$$

In other words, the effective yield for an investment collecting compound interest is the *simple interest rate* that would yield the same balance at the end of 1 year.

> **Example 4**   Finding the Effective Yield

An investment is made in an account that pays 6.75% interest, compounded continuously. What is the effective yield for this investment?

Solution

Notice that you do not have to know the principal or the time that the money will be left in the account. Instead, you can choose an arbitrary principal, such as $1000. Then, because effective yield is based on the balance at the end of 1 year, you can use the following formula.

$$A = Pe^{rt}$$

$$= 1000e^{0.0675(1)}$$

$$= 1069.83$$

Now, because the account would earn $69.83 in interest after 1 year for a principal of $1000, you can conclude that the effective yield is

$$\text{Effective yield} = \frac{69.83}{1000}$$

$$= 0.06983$$

$$= 6.983\%.$$

Growth and Decay

2 Use an exponential equation to solve a growth or decay problem.

The balance in an account earning *continuously* compounded interest is one example of a quantity that increases over time according to the **exponential growth model** $y = Ce^{kt}$.

> ▶ **Exponential Growth and Decay**
>
> The mathematical model for exponential growth or decay is given by
>
> $$y = Ce^{kt}.$$
>
> For this model, t is the time, C is the original amount of the quantity, and y is the amount after time t. The number k is a constant that is determined by the rate of growth. If $k > 0$, the model represents **exponential growth,** and if $k < 0$, it represents **exponential decay.**

One common application of exponential growth is in modeling the growth of a population. Example 5 illustrates the use of the growth model

$$y = Ce^{kt}, \quad k > 0.$$

Example 5   Population Growth

A country's population was 2 million in 1990 and 3 million in 2000. What would you predict the population of the country to be in 2010?

Solution

If you assumed a *linear growth model*, you would simply predict the population in the year 2010 to be 4 million. If, however, you assumed an *exponential growth model*, the model would have the form

$$y = Ce^{kt}.$$

In this model, let $t = 0$ represent the year 1990. The given information about the population can be described by the following table.

t (years)	0	10	20
Ce^{kt} (million)	$Ce^{k(0)} = 2$	$Ce^{k(10)} = 3$	$Ce^{k(20)} = ?$

To find the population when $t = 20$, you must first find the values of C and k. From the table, you can use the fact that $Ce^{k(0)} = Ce^{0} = 2$ to conclude that $C = 2$. Then, using this value of C, you can solve for k as follows.

$Ce^{k(10)} = 3$	From table
$2e^{10k} = 3$	Substitute value of C.
$e^{10k} = \dfrac{3}{2}$	Divide both sides by 2.
$10k = \ln \dfrac{3}{2}$	Inverse property
$k = \dfrac{1}{10} \ln \dfrac{3}{2}$	Divide both sides by 10.
$k \approx 0.0405$	Simplify.

Finally, you can use this value of k to conclude that the population in the year 2010 is given by

$$2e^{0.0405(20)} \approx 2(2.25) = 4.5 \text{ million.}$$

Figure 9.21 graphically compares the exponential growth model with a linear growth model.

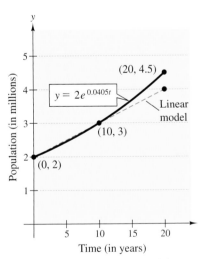

Figure 9.21   *Population Models*

Example 6 Radioactive Decay

Radioactive iodine is a by-product of some types of nuclear reactors. Its **half-life** is 60 days. That is, after 60 days, a given amount of radioactive iodine will have decayed to half the original amount. Suppose a nuclear accident occurs and releases 20 grams of radioactive iodine. How long will it take for the radioactive iodine to decay to a level of 1 gram?

Solution

To solve this problem, use the model for exponential decay.

$$y = Ce^{kt}$$

Next, use the information given in the problem to set up the following table.

t (days)	0	60	?
Ce^{kt} (grams)	$Ce^{k(0)} = 20$	$Ce^{k(60)} = 10$	$Ce^{k(t)} = 1$

Because $Ce^{k(0)} = Ce^0 = 20$, you can conclude that $C = 20$. Then, using this value of C, you can solve for k, as follows.

$$Ce^{k(60)} = 10 \qquad \text{From table}$$

$$20e^{60k} = 10 \qquad \text{Substitute value of } C.$$

$$e^{60k} = \frac{1}{2} \qquad \text{Divide both sides by 20.}$$

$$60k = \ln \frac{1}{2} \qquad \text{Inverse property}$$

$$k = \frac{1}{60} \ln \frac{1}{2} \qquad \text{Divide both sides by 60.}$$

$$\approx -0.01155 \qquad \text{Simplify.}$$

Finally, you can use this value of k to find the time when the amount is 1 gram, as follows.

$$Ce^{kt} = 1 \qquad \text{From table}$$

$$20e^{-0.01155t} = 1 \qquad \text{Substitute values of } C \text{ and } k.$$

$$e^{-0.01155t} = \frac{1}{20} \qquad \text{Divide both sides by 20.}$$

$$-0.01155t = \ln \frac{1}{20} \qquad \text{Inverse property}$$

$$t = \frac{1}{-0.01155} \ln \frac{1}{20} \qquad \text{Divide both sides by } -0.01155.$$

$$\approx 259.4 \text{ days} \qquad \text{Simplify.}$$

So, 20 grams of radioactive iodine will have decayed to 1 gram after about 259.4 days. This solution is shown graphically in Figure 9.22.

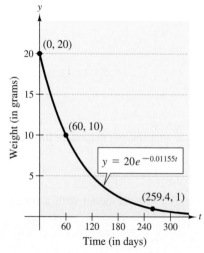

Figure 9.22 *Radioactive Decay*

3 Use a logarithmic equation to solve an intensity problem.

Intensity Models

On the **Richter scale,** the magnitude R of an earthquake can be measured by the **intensity model**

$$R = \log_{10} I$$

where I is the intensity of the shock wave.

Example 7 Earthquake Intensity

In 1906, San Francisco experienced an earthquake that measured 8.6 on the Richter scale. In 1989, another earthquake, which measured 7.7 on the Richter scale, struck the same area. Compare the intensities of these two earthquakes.

Earthquakes take place along faults in the earth's crust. The 1989 earthquake in California took place along the San Andreas Fault.

Solution

The intensity of the 1906 earthquake is given as follows.

$$8.6 = \log_{10} I \qquad \text{Given}$$
$$10^{8.6} = I \qquad \text{Inverse property}$$

The intensity of the 1989 earthquake can be found in a similar way.

$$7.7 = \log_{10} I \qquad \text{Given}$$
$$10^{7.7} = I \qquad \text{Inverse property}$$

The ratio of these two intensities is

$$\frac{I \text{ for } 1906}{I \text{ for } 1989} = \frac{10^{8.6}}{10^{7.7}}$$
$$= 10^{8.6 - 7.7}$$
$$= 10^{0.9}$$
$$\approx 7.94.$$

Thus, the 1906 earthquake had an intensity that was about eight times greater than the 1989 earthquake.

Discussing the Concept   Problem Posing

Write a problem that could be answered by investigating the exponential growth model

$$y = 10e^{0.08t}$$

or the exponential decay model

$$y = 5e^{-0.25t}.$$

Exchange your problem for that of another class member, and solve one another's problems.

9.6 Exercises

Integrated Review   *Concepts, Skills, and Problem Solving*

Keep mathematically in shape by doing these exercises *before* the problems of this section.

Properties and Definitions

In Exercises 1–4, identify the type of variation given in the model.

1. $y = kx^2$

2. $y = \dfrac{k}{x}$

3. $z = kxy$

4. $z = \dfrac{kx}{y}$

Simplifying Expressions

In Exercises 5–10, solve the system of equations.

5. $x - y = 0$
$x + 2y = 9$

6. $2x + 5y = 15$
$3x + 6y = 20$

7. $y = x^2$
$-3x + 2y = 2$

8. $x - y^3 = 0$
$x - 2y^2 = 0$

9. $x - y \quad = -1$
$x + 2y - 2z = 3$
$3x - y + 2z = 3$

10. $2x + y - 2z = 1$
$x \quad - z = 1$
$3x + 3y + z = 12$

Graphs

In Exercises 11 and 12, use the function $y = -x^2 + 4x$.

11. (a) Does the graph open up or down? Explain.

(b) Find the x-intercepts algebraically.

(c) Find the coordinates of the vertex of the parabola.

12. Use a graphing utility to graph the function and geometrically verify the results of Exercise 11.

Solving Problems

Annual Interest Rate   In Exercises 1–8, find the annual interest rate. See Example 1.

	Principal	Balance	Time	Compounding
1.	$500	$1004.83	10 years	Monthly
2.	$3000	$21,628.70	20 years	Quarterly
3.	$1000	$36,581.00	40 years	Daily
4.	$200	$314.85	5 years	Yearly
5.	$750	$8267.38	30 years	Continuously
6.	$2000	$4234.00	10 years	Continuously
7.	$5000	$22,405.68	25 years	Daily
8.	$10,000	$110,202.78	30 years	Daily

Doubling Time   In Exercises 9–16, find the time for an investment to double. Use a graphing utility to check the result graphically. See Example 2.

	Principal	Rate	Compounding
9.	$6000	8%	Quarterly
10.	$500	$5\frac{1}{4}\%$	Monthly
11.	$2000	10.5%	Daily
12.	$10,000	9.5%	Yearly
13.	$1500	7.5%	Continuously
14.	$100	6%	Continuously
15.	$300	5%	Yearly
16.	$12,000	4%	Continuously

Type of Compounding   In Exercises 17–20, determine the type of compounding. Solve the problem by trying the common types of compounding. See Example 3.

	Principal	Balance	Time	Rate
17.	$750	$1587.75	10 years	7.5%
18.	$10,000	$73,890.56	20 years	10%
19.	$100	$141.48	5 years	7%
20.	$4000	$4788.76	2 years	9%

Effective Yield   In Exercises 21–28, find the effective yield. See Example 4.

	Rate	Compounding
21.	8%	Continuously
22.	9.5%	Daily
23.	7%	Monthly
24.	8%	Yearly
25.	6%	Quarterly
26.	9%	Quarterly
27.	8%	Monthly
28.	$5\frac{1}{4}$%	Daily

29. *Think About It*   Is it necessary to know the principal P to find the doubling time in Exercises 9–16? Explain.

30. *Effective Yield*

   (a) Is it necessary to know the principal P to find the effective yield in Exercises 21–28? Explain.

   (b) When the interest is compounded more frequently, what inference can you make about the difference between the effective yield and the stated annual percentage rate?

Principal   In Exercises 31–38, find the principal that must be deposited in an account to obtain the given balance.

	Balance	Rate	Time	Compounding
31.	$10,000	9%	20 years	Continuously
32.	$5000	8%	5 years	Continuously
33.	$750	6%	3 years	Daily
34.	$3000	7%	10 years	Monthly
35.	$25,000	7%	30 years	Monthly
36.	$8000	6%	2 years	Monthly
37.	$1000	5%	1 year	Daily
38.	$100,000	9%	40 years	Daily

Balance After Monthly Deposits   In Exercises 39–42, you make monthly deposits of P dollars into a savings account at an annual interest rate r, compounded continuously. Find the balance A after t years given that

$$A = \frac{P(e^{rt} - 1)}{e^{r/12} - 1}.$$

	Principal	Rate	Time
39.	$P = 30$	$r = 8\%$	$t = 10$ years

	Principal	Rate	Time
40.	$P = 100$	$r = 9\%$	$t = 30$ years
41.	$P = 50$	$r = 10\%$	$t = 40$ years
42.	$P = 20$	$r = 7\%$	$t = 20$ years

Balance After Monthly Deposits   In Exercises 43 and 44, you make monthly deposits of $30 into a savings account at an annual interest rate of 8%, compounded continuously (see figure).

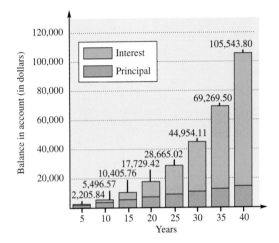

43. Find the total amount that has been deposited into the account in 20 years and the total interest earned.

44. Find the total amount that has been deposited into the account in 40 years and the total interest earned.

Exponential Growth and Decay   In Exercises 45–48, find the constant k such that the graph of $y = Ce^{kt}$ passes through the given points.

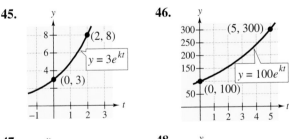

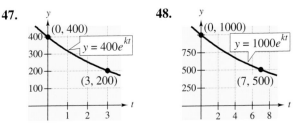

Population of a Region   In Exercises 49–56, the population (in millions) of an urban region for 1994 and the predicted population (in millions) for the year 2015 are given. Find the constants C and k to obtain the exponential growth model $y = Ce^{kt}$ for the population. (Let $t = 0$ correspond to the year 1994.) Use the model to predict the population of the region in the year 2020. See Example 5.   (Source: United Nations)

Region	1994	2015
49. Los Angeles	12.2	14.3
50. New York	16.3	17.6
51. Shanghai, China	14.7	23.4
52. Jakarta, Indonesia	11.0	21.2
53. Osaka, Japan	10.5	10.6
54. Seoul, Korea	11.5	13.1
55. Mexico City, Mexico	15.5	18.8
56. Sao Paulo, Brazil	16.1	20.8

57. *Rate of Growth*

(a) Compare the values of k in Exercises 51 and 53. Which is larger? Explain.

(b) What variable in the continuous compound interest formula is equivalent to k in the model for population growth? Use your answer to give an interpretation of k.

58. *World Population*   The figure in the next column shows the population P (in billions) of the world as projected by the Population Reference Bureau. The bureau's projection can be modeled by

$$P = \frac{11.14}{1 + 1.101e^{-0.051t}}$$

where $t = 0$ represents 1990. Use the model to estimate the population in 2020.

Radioactive Decay   In Exercises 59–64, complete the table for the radioactive isotopes. See Example 6.

Isotope	Half-Life (Years)	Initial Quantity	Amount After 1000 Years
59. Ra^{226}	1620	6 g	
60. Ra^{226}	1620		0.25 g
61. C^{14}	5730		4.0 g
62. C^{14}	5730	10 g	
63. Pu^{230}	24,360	4.2 g	
64. Pu^{230}	24,360		1.5 g

65. *Radioactive Decay*   Radioactive radium (Ra^{226}) has a half-life of 1620 years. If you start with 5 grams of the isotope, how much remains after 1000 years?

66. *Radioactive Decay*   The isotope Pu^{230} has a half-life of 24,360 years. If you start with 10 grams of the isotope, how much remains after 10,000 years?

67. *Radioactive Decay*   Carbon 14 (C^{14}) has a half-life of 5730 years. If you start with 5 grams of this isotope, how much remains after 1000 years?

68. *Carbon 14 Dating* C^{14} dating assumes that the carbon dioxide on earth today has the same radioactive content as it did centuries ago. If this is true, the amount of C^{14} absorbed by a tree that grew several centuries ago should be the same as the amount of C^{14} absorbed by a tree growing today. A piece of ancient charcoal contains only 15% as much of the radioactive carbon as a piece of modern charcoal. How long ago did the tree burn to make the ancient charcoal if the half-life of C^{14} is 5730 years? (Round your answer to the nearest 100 years.)

69. *Depreciation*   A car that cost $22,000 new has a depreciated value of $16,500 after 1 year. Find the value of the car when it is 3 years old by using the exponential model $y = Ce^{kt}$.

70. *Depreciation*   After x years, the value y of a truck that cost $32,000 new is given by

$$y = 32,000(0.8)^x.$$

(a) Use a graphing utility to graph the model.

(b) Graphically approximate the value of the truck after 1 year.

(c) Graphically approximate the time when the truck's value will be $16,000.

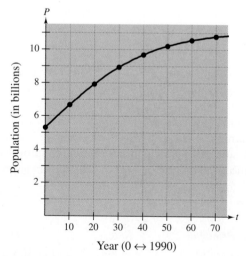

Figure for 58

Earthquake Intensity   In Exercises 71–74, compare the intensities of the two earthquakes. See Example 7.

	Location	Date	Magnitude
71.	Alaska	3/27/64	8.4
	San Fernando Valley	2/9/71	6.6
72.	Long Beach, California	3/10/33	6.2
	Morocco	2/29/60	5.8
73.	Mexico City, Mexico	9/19/85	8.1
	Nepal	8/20/88	6.5
74.	Chile	8/16/06	8.6
	Armenia, USSR	12/7/88	6.8

Acidity Model   In Exercises 75–78, use the acidity model

$$pH = -\log_{10}[H^+]$$

where acidity (pH) is a measure of the hydrogen ion concentration $[H^+]$ (measured in moles of hydrogen per liter) of a solution.

75. Find the pH of a solution that has a hydrogen ion concentration of 9.2×10^{-8}.

76. Compute the hydrogen ion concentration if the pH of a solution is 4.7.

77. A certain fruit has a pH of 2.5 and an antacid tablet has a pH of 9.5. The hydrogen ion concentration of the fruit is how many times the concentration of the tablet?

78. If the pH of a solution is decreased by 1 unit, the hydrogen ion concentration is increased by what factor?

79. *Population Growth*   The population p of a certain species t years after it is introduced into a new habitat is given by

$$p(t) = \frac{5000}{1 + 4e^{-t/6}}.$$

(a) Use a graphing utility to graph the population function.

(b) Determine the population size that was introduced into the habitat.

(c) Determine the population size after 9 years.

(d) After how many years will the population be 2000?

80. *Sales Growth*   Annual sales y of a product x years after it is introduced are approximated by

$$y = \frac{2000}{1 + 4e^{-x/2}}.$$

(a) Use a graphing utility to graph the model.

(b) Use the graph in part (a) to approximate annual sales when $x = 4$.

(c) Use the graph in part (a) to approximate the time when annual sales are $y = 1100$ units.

(d) Use the graph in part (a) to estimate the maximum level that annual sales will approach.

81. *Advertising Effect*   The sales S (in thousands of units) of a product after spending x hundred dollars in advertising are given by

$$S = 10(1 - e^{kx}).$$

(a) Find S as a function of x if 2500 units are sold when $500 is spent on advertising.

(b) How many units will be sold if advertising expenditures are raised to $700?

Explaining Concepts

82. If the equation $y = Ce^{kt}$ models exponential growth, what must be true about k?

83. If the equation $y = Ce^{kt}$ models exponential decay, what must be true about k?

84. The formulas for periodic and continuous compounding have the four variables $A, P, r,$ and t in common. Explain what each variable measures.

85. What is meant by the effective yield of an investment? Explain how it is computed.

86. In your own words, explain what is meant by the half-life of a radioactive isotope.

87. If the reading on the Richter scale is increased by 1, the intensity of the earthquake is increased by what factor?

Key Terms

exponential function, p. 548
natural base e, p. 552
natural exponential function, p. 552

composition, p. 561
inverse function, p. 563
one-to-one, p. 563
logarithmic function, p. 573

common logarithmic function, p. 575
natural logarithmic function, p. 578

exponentiate, p. 597
exponential growth, p. 606
exponential decay, p. 606

Key Concepts

9.1 Rules of exponential functions

1. $a^x \cdot a^y = a^{x+y}$
2. $\dfrac{a^x}{a^y} = a^{x-y}$
3. $(a^x)^y = a^{xy}$
4. $a^{-x} = \dfrac{1}{a^x} = \left(\dfrac{1}{a}\right)^x$

9.2 Composition of two functions

The composition of two functions f and g is given by $(f \circ g)(x) = f(g(x))$. The domain of the composite function $(f \circ g)$ is the set of all x in the domain of g such that $g(x)$ is in the domain of f.

9.2 Horizontal Line Test for inverse functions

A function f has an inverse function if and only if f is one-to-one. Graphically, a function f has an inverse if and only if no horizontal line intersects the graph of f at more than one point.

9.2 Finding the inverse of a function

1. In the equation for $f(x)$, replace $f(x)$ by y.
2. Interchange the roles of x and y, and solve for y.
3. If the new equation does not represent y as a function of x, the function f does not have an inverse function.
4. If the new equation represents y as a function of x, replace y by $f^{-1}(x)$.

9.3 Properties of logarithms and natural logarithms

Let a and x be positive real numbers such that $a \neq 1$. Then the following properties are true.

1. $\log_a 1 = 0$ because $a^0 = 1$.
   $\ln 1 = 0$ because $e^0 = 1$.
2. $\log_a a = 1$ because $a^1 = a$.
   $\ln e = 1$ because $e^1 = e$.
3. $\log_a a^x = x$ because $a^x = a^x$.
   $\ln e^x = x$ because $e^x = e^x$.

9.3 Change-of-base formula

Let a, b, and x be positive real numbers such that $a \neq 1$ and $b \neq 1$. Then $\log_a x = \dfrac{\log_b x}{\log_b a}$ or $\log_a x = \dfrac{\ln x}{\ln a}$.

9.4 Properties of logarithms

Let a be a positive real number such that $a \neq 1$, and let n be a real number. If u and v are real numbers, variables, or algebraic expressions such that $u > 0$ and $v > 0$, the following properties are true.

Logarithm with base a	Natural logarithm
1. $\log_a(uv) = \log_a u + \log_a v$	$\ln(uv) = \ln u + \ln v$
2. $\log_a \dfrac{u}{v} = \log_a u - \log_a v$	$\ln \dfrac{u}{v} = \ln u - \ln v$
3. $\log_a u^n = n \log_a u$	$\ln u^n = n \ln u$

9.5 Properties of exponential and logarithmic equations

Let a be a positive real number such that $a \neq 1$, and let x and y be real numbers. Then the following properties are true.

1. $a^x = a^y$ if and only if $x = y$.
2. $\log_a x = \log_a y$ if and only if $x = y$ $(x > 0, y > 0)$.

9.5 Inverse properties of exponents and logarithms

Base a	Natural base e
1. $\log_a(a^x) = x$	$\ln(e^x) = x$
2. $a^{(\log_a x)} = x$	$e^{(\ln x)} = x$

9.5 Solving exponential and logarithmic equations

1. To solve an exponential equation, first isolate the exponential expression, then take the logarithm of both sides of the equation and solve for the variable.
2. To solve a logarithmic equation, first isolate the logarithmic expression, then exponentiate both sides of the equation and solve for the variable.

REVIEW EXERCISES

Reviewing Skills

9.1 In Exercises 1–4, evaluate the exponential function as indicated.

1. $f(x) = 2^x$
   (a) $x = -3$ (b) $x = 1$ (c) $x = 2$

2. $g(x) = 2^{-x}$
   (a) $x = -2$ (b) $x = 0$ (c) $x = 2$

3. $g(t) = e^{-t/3}$
   (a) $t = -3$ (b) $t = \pi$ (c) $t = 6$

4. $h(s) = 1 - e^{0.2s}$
   (a) $s = 0$ (b) $s = 2$ (c) $s = \sqrt{10}$

In Exercises 5–8, match the function with the sketch of its graph. [The graphs are labeled (a), (b), (c), and (d).]

(a)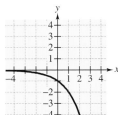

(b)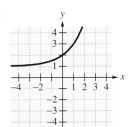

(c)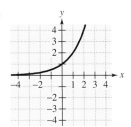

(d)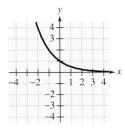

5. $f(x) = 2^x$ **6.** $f(x) = 2^{-x}$
7. $f(x) = -2^x$ **8.** $f(x) = 2^x + 1$

In Exercises 9–18, use the point-plotting method to sketch the graph of the exponential function.

9. $f(x) = 3^x$ **10.** $f(x) = 3^{-x}$
11. $f(x) = 3^x - 1$ **12.** $f(x) = 3^x + 2$
13. $f(x) = 3^{(x+1)}$ **14.** $f(x) = 3^{(x-1)}$
15. $y = 3^{x/2}$ **16.** $f(x) = 3^{-x/2}$
17. $y = 3^{x/2} - 2$ **18.** $f(x) = 3^{x/2} + 3$

In Exercises 19–22, use a graphing utility to graph the exponential function.

19. $y = 5e^{-x/4}$
20. $y = 6 - e^{x/2}$
21. $f(x) = e^{x+2}$
22. $h(t) = \dfrac{8}{1 + e^{-t/5}}$

9.2 In Exercises 23–26, form $f \circ g$ and $g \circ f$ and evaluate the composite functions.

23. $f(x) = x + 2$, $g(x) = x^2$
   (a) $(f \circ g)(2)$ (b) $(g \circ f)(-1)$

24. $f(x) = \sqrt[3]{x}$, $g(x) = x + 2$
   (a) $(f \circ g)(6)$ (b) $(g \circ f)(64)$

25. $f(x) = \sqrt{x + 1}$, $g(x) = x^2 - 1$
   (a) $(f \circ g)(5)$ (b) $(g \circ f)(-1)$

26. $f(x) = \dfrac{1}{x - 5}$, $g(x) = \dfrac{5x + 1}{x}$
   (a) $(f \circ g)(1)$ (b) $(g \circ f)\left(\dfrac{1}{5}\right)$

In Exercises 27 and 28, form the compositions (a) $f \circ g$ and (b) $g \circ f$, and find the domains of the composites.

27. $f(x) = \sqrt{x - 4}$, $g(x) = 2x$

28. $f(x) = \dfrac{2}{x - 4}$, $g(x) = x^2$

In Exercises 29–32, use the Horizontal Line Test to determine whether the function has an inverse.

29. $f(x) = x^2 - 25$ **30.** $f(x) = \frac{1}{4}x^3$

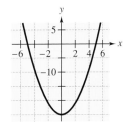

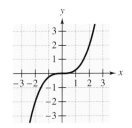

31. $h(x) = 4\sqrt[3]{x}$

32. $g(x) = \sqrt{9 - x^2}$

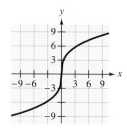

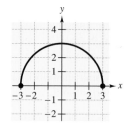

In Exercises 33–38, find the inverse of the function algebraically. (If not possible, state the reason.)

33. $f(x) = 3x + 4$

34. $f(x) = 2x - 3$

35. $h(x) = \sqrt{x}$

36. $g(x) = x^2 + 2,\ x \geq 0$

37. $f(t) = t^3 + 4$

38. $h(t) = \sqrt[3]{t - 1}$

9.3 In Exercises 39 and 40, write the exponential equation in logarithmic form.

39. $4^3 = 64$

40. $25^{3/2} = 125$

In Exercises 41 and 42, write the logarithmic equation in exponential form.

41. $\ln e = 1$

42. $\log_3 \frac{1}{9} = -2$

In Exercises 43–50, evaluate the logarithm.

43. $\log_{10} 1000$

44. $\log_9 3$

45. $\log_3 \frac{1}{9}$

46. $\log_4 \frac{1}{16}$

47. $\ln e^7$

48. $\log_a \frac{1}{a}$

49. $\ln 1$

50. $\ln e^{-3}$

In Exercises 51–56, evaluate the logarithmic function as indicated.

51. $f(x) = \log_3 x$
  (a) $x = 1$ (b) $x = 27$ (c) $x = 0.5$

52. $g(x) = \log_{10} x$
  (a) $x = 0.01$ (b) $x = 0.1$ (c) $x = 30$

53. $f(x) = \ln x$
  (a) $x = e$ (b) $x = \frac{1}{3}$ (c) $x = 10$

54. $h(x) = \ln x$
  (a) $x = e^2$ (b) $x = \frac{5}{4}$ (c) $x = 1200$

55. $g(x) = \ln e^{3x}$
  (a) $x = -2$ (b) $x = 0$ (c) $x = 7.5$

56. $f(x) = \log_2 \sqrt{x}$
  (a) $x = 4$ (b) $x = 64$ (c) $x = 5.2$

In Exercises 57–66, use the point-plotting method to graph the logarithmic function.

57. $f(x) = \log_3 x$

58. $f(x) = -\log_3 x$

59. $f(x) = -2 + \log_3 x$

60. $f(x) = 2 + \log_3 x$

61. $y = \log_2(x - 4)$

62. $y = \log_4(x + 1)$

63. $y = \ln(x - 3)$

64. $y = -\ln(x + 2)$

65. $y = 5 - \ln x$

66. $y = 3 + \ln x$

In Exercises 67–70, use the change-of-base formula to evaluate the logarithm. Round the result to three decimal places.

67. $\log_4 9$

68. $\log_{1/2} 5$

69. $\log_{12} 200$

70. $\log_3 0.28$

9.4 In Exercises 71–76, approximate the logarithm given that $\log_5 2 \approx 0.43068$ and $\log_5 3 \approx 0.68261$.

71. $\log_5 18$

72. $\log_5 \sqrt{6}$

73. $\log_5 \frac{1}{2}$

74. $\log_5 \frac{2}{3}$

75. $\log_5 (12)^{2/3}$

76. $\log_5 (5^2 \cdot 6)$

In Exercises 77–84, use the properties of logarithms to expand the expression.

77. $\log_4 6x^4$

78. $\log_{10} 2x^{-3}$

79. $\log_5 \sqrt{x + 2}$

80. $\ln \sqrt[3]{\dfrac{x}{5}}$

81. $\ln \dfrac{x + 2}{x - 2}$

82. $\ln x(x - 3)^2$

83. $\ln\left[\sqrt{2x}(x + 3)^5\right]$

84. $\log_3 \dfrac{a^2\sqrt{b}}{cd^5}$

In Exercises 85–94, use the properties of logarithms to condense the expression.

85. $-\frac{2}{3} \ln 3y$

86. $5 \log_2 y$

87. $\log_8 16x + \log_8 2x^2$

88. $\log_4 6x - \log_4 10$

89. $-2(\ln 2x - \ln 3)$

90. $4(1 + \ln x + \ln x)$

91. $4[\log_2 k - \log_2(k - t)]$

92. $\frac{1}{3}(\log_8 a + 2 \log_8 b)$

93. $3 \ln x + 4 \ln y + \ln z$

94. $\ln(x + 4) - 3 \ln x - \ln y$

True or False? In Exercises 95–100, use the properties of logarithms to determine whether the equation is true or false. If false, state why or give an example to show that it is false.

95. $\log_2 4x = 2 \log_2 x$

96. $\dfrac{\ln 5x}{\ln 10x} = \ln \dfrac{1}{2}$

97. $\log_{10} 10^{2x} = 2x$ **98.** $e^{\ln t} = t$

99. $\log_4 \dfrac{16}{x} = 2 - \log_4 x$

100. $6 \ln x + 6 \ln y = \ln(xy)^6$

9.5 In Exercises 101–110, solve the equation.

101. $2^x = 64$ **102.** $5^x = 25$

103. $4^{x-3} = \frac{1}{16}$ **104.** $3^{x-2} = 81$

105. $\log_3 x = 5$ **106.** $\log_4 x = 3$

107. $\log_2 2x = \log_2 100$ **108.** $\ln(x + 4) = \ln 7$

109. $\log_3(2x + 1) = 2$ **110.** $\log_5(x - 10) = 2$

In Exercises 111–124, solve the equation.

111. $3^x = 500$ **112.** $8^x = 1000$

113. $\ln x = 7.25$ **114.** $\ln x = -0.5$

115. $2e^{0.5x} = 45$ **116.** $100e^{-0.6x} = 20$

117. $12(1 - 4^x) = 18$ **118.** $25(1 - e^t) = 12$

119. $\log_{10} 2x = 1.5$ **120.** $\log_2 2x = -0.65$

121. $\frac{1}{3} \log_2 x + 5 = 7$ **122.** $4 \log_5(x + 1) = 4.8$

123. $\log_2 x + \log_2 3 = 3$

124. $2 \log_4 x - \log_4(x - 1) = 1$

9.6 *Annual Interest Rate*   In Exercises 125–130, find the annual interest rate.

	Principal	Balance	Time	Compounding
125.	$250	$410.90	10 years	Quarterly
126.	$1000	$1348.85	5 years	Monthly
127.	$5000	$15,399.30	15 years	Daily

	Principal	Balance	Time	Compounding
128.	$10,000	$35,236.45	20 years	Yearly
129.	$1500	$24,666.97	40 years	Continuously
130.	$7500	$15,877.50	15 years	Continuously

Effective Yield   In Exercises 131–136, find the effective yield.

	Rate	Compounding
131.	5.5%	Daily
132.	6%	Monthly
133.	7.5%	Quarterly
134.	8%	Yearly
135.	7.5%	Continuously
136.	4%	Continuously

Radioactive Decay   In Exercises 137–142, complete the table for the radioactive isotopes.

	Isotope	Half-Life (Years)	Initial Quantity	Amount After 1000 Years
137.	Ra^{226}	1620	3.5 g	
138.	Ra^{226}	1620		0.5 g
139.	C^{14}	5730		2.6 g
140.	C^{14}	5730	10 g	
141.	Pu^{230}	24,360	5 g	
142.	Pu^{230}	24,360		2.5 g

Solving Problems

143. *Inflation Rate*   If the annual rate of inflation averages 5% over the next 10 years, the approximate cost C of goods or services during any year in that decade will be given by

$$C(t) = P(1.05)^t, \quad 0 \le t \le 10$$

where t is time in years and P is the present cost. If the price of an oil change for your car is presently $24.95, when will it cost $30.00?

144. *Doubling Time*   Find the time for an investment of $1000 to double in value when invested at 8% compounded monthly.

145. *Doubling Time*   Find the time for an investment of $750 to double in value when invested at 5.5% compounded continuously.

146. *Product Demand*   The daily demand x and price p for a product are related by

$$p = 25 - 0.4e^{0.02x}.$$

Approximate the demand if the price is $16.97.

147. *Sound Intensity*   The relationship between the number of decibels B and the intensity of a sound I in watts per centimeter squared is given by

$$B = 10 \log_{10}\left(\frac{I}{10^{-16}}\right).$$

Determine the intensity of a sound in watts per centimeter squared if the decibel level is 125.

148. *Sound Intensity*   The relationship between the number of decibels B and the intensity of a sound I in watts per centimeter squared is given by

$$B = 10 \log_{10}\left(\frac{I}{10^{-16}}\right).$$

Determine the intensity of a sound in watts per centimeter squared if the decibel level is 150.

149. *Population Limit*   The population p of a certain species t years after it is introduced into a new habitat is given by

$$p(t) = \frac{600}{1 + 2e^{-0.2t}}.$$

Use a graphing utility to graph the function. Use the graph to determine the limiting size of the population in this habitat.

150. *Deer Herd*   The state Parks and Wildlife Department releases 100 deer into a wilderness area. The population P of the herd can be modeled by

$$P = \frac{500}{1 + 4e^{-0.36t}}$$

where t is measured in years.

(a) Find the population after 5 years.

(b) After how many years will the population be 250?

151. *Ventilation*   The table gives the required ventilation rate V (in cubic feet per minute per person) for a room in a public building with an air space of x cubic feet per person.

x	100	200	300	400
V	25	17	12	9

A model for the data is $V = 78.56 - 11.6314 \ln x$.

(a) Use a graphing utility to plot the data and graph the model.

(b) Use the model to determine V if $x = 250$.

152. *Comparing Models*   The figure gives the assets (in billions of dollars) in client accounts for the years 1993 through 1997 for Merrill Lynch. A list of models ($t = 3$ represents 1993) for the data is also given. For each of the models, (a) use a graphing utility to obtain its graph, and (b) use the graphs of part (a) to determine which model "best fits" the data.   (Source: Merrill Lynch 1997 Annual Report)

Linear: $A = 156.5t - 8.3$

Quadratic: $A = 50.6t^2 - 349.9t + 1156.5$

Exponential: $A = 282.4e^{0.193t}$

Logarithmic: $A = 1133.3 + 620.8t - 2210.9 \ln t$

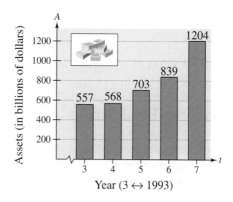

Year (3 ↔ 1993)

Chapter Test

Take this test as you would take a test in class. After you are done, check your work against the answers given in the back of the book.

1. Evaluate $f(t) = 54\left(\frac{2}{3}\right)^t$ when $t = -1, 0, \frac{1}{2}$, and 2.

2. Sketch a graph of the function $f(x) = 2^{x/3}$.

3. Form the composition of (a) $f \circ g$ and (b) $g \circ f$, and find the domains of the composites.

$$f(x) = 3x - 4, \qquad g(x) = x^2 + 1$$

4. Find the inverse of the function $f(x) = 5x + 6$.

5. Verify algebraically that the two functions are inverses of each other.

$$f(x) = -\frac{1}{2}x + 3, \qquad g(x) = -2x + 6$$

6. Describe the relationship between the graphs of $f(x) = \log_5 x$ and $g(x) = 5^x$.

7. Use the properties of logarithms to expand $\log_4\left(5x^2/\sqrt{y}\right)$.

8. Use the properties of logarithms to condense $\ln x - 4 \ln y$.

9. Simplify $\log_5 5^3 \cdot 6$.

In Exercises 10–17, solve the equation.

10. $\log_4 x = 3$

11. $10^{3y} = 832$

12. $400e^{0.08t} = 1200$

13. $3 \ln(2x - 3) = 10$

14. $8(2 - 3^x) = -56$

15. $\log_2 x + \log_2 4 = 5$

16. $\ln x - \ln 2 = 4$

17. $30(e^x + 9) = 300$

18. Determine the balance after 20 years if $2000 is invested at 7% compounded (a) quarterly and (b) continuously.

19. Determine the principal that will yield $100,000 when invested at 9% compounded quarterly for 25 years.

20. A principal of $500 yields a balance of $1006.88 in 10 years when the interest is compounded continuously. What is the annual interest rate?

21. A car that cost $18,000 new has a depreciated value of $14,000 after 1 year. Find the value of the car when it is 3 years old by using the exponential model $y = Ce^{kt}$.

In Exercises 22–24, the population of a certain species t years after it is introduced into a new habitat is given by

$$p(t) = \frac{2400}{1 + 3e^{-t/4}}.$$

22. Determine the population size that was introduced into the habitat.

23. Determine the population after 4 years.

24. After how many years will the population be 1200?

Cumulative Test: Chapters 7–9

Take this test as you would take a test in class. After you are done, check your work against the answers given in the back of the book.

1. Write a mathematical model for the statement, "V varies directly as the square root of x and inversely as y."

2. Find the constant of proportionality if v varies directly as the square of t and $v = -64$ when $t = 2$.

3. The stopping distance d of a car is directly proportional to the square of its speed s. On a certain type of pavement, a car requires 50 feet to stop when its speed is 25 miles per hour. Estimate the stopping distance when the speed of the car is 40 miles per hour. Explain your reasoning.

4. The number N of prey t months after a predator is introduced into an area is inversely proportional to $t + 1$. If $N = 300$ when $t = 0$, find N when $t = 5$.

5. Sketch a graph of the solution of the linear inequality $5x + 2y > 10$.

6. Find an inequality for the graph shown in the figure.

7. Find an equation of the parabola shown in the figure.

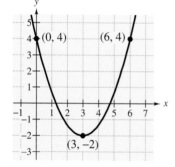

Figure for 6

In Exercises 8–11, graph the equation.

8. $x^2 + y^2 = 8$

9. $x^2 + 2y = 0$

10. $\dfrac{x^2}{1} + \dfrac{y^2}{4} = 1$

11. $\dfrac{x^2}{1} - \dfrac{y^2}{4} = 1$

Figure for 7

12. A semicircular arch is positioned over the roadway onto the grounds of an estate. The roadway is 10 feet wide and the arch is sitting on pillars that are 8 feet tall. Find the maximum height of a truck that can be driven onto the estate if the truck's width is 8 feet.

In Exercises 13 and 14, graph the rational function.

13. $y = \dfrac{4}{x - 2}$

14. $y = \dfrac{4x^2}{x^2 + 1}$

15. Find a rational function with a vertical asymptote at $x = 3$ and a horizontal asymptote at $y = 2$.

16. The cost of producing x units of a product is $C = 10x + 13$, and therefore the average cost per unit is

$$\overline{C} = \frac{C}{x} = \frac{10x + 13}{x}, \quad x > 0.$$

Identify the horizontal asymptote of $\overline{C}$ and explain its meaning in the context of the problem.

In Exercises 17–20, solve the system of equations by the specified method.

17. *Graphical:* $x - y = 1$
$\qquad\qquad\quad 2x + y = 5$

18. *Substitution:* $4x + 2y = 8$
$\qquad\qquad\qquad\quad x - 5y = 13$

19. *Elimination:*

$$\begin{aligned} 4x - 3y + 2z &= -2 \\ -2x + y + z &= 1 \\ x - 2y - 6z &= -12 \end{aligned}$$

20. *Cramer's Rule:*

$$\begin{aligned} 2x - y &= 4 \\ 3x + y &= -5 \end{aligned}$$

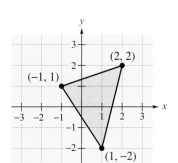

Figure for 22

21. Solve the system of linear equations using matrices.

$$\begin{aligned} x + 5y \quad\ &= 29 \\ 2x \quad - 3z &= 4 \\ -4y + z &= -26 \end{aligned}$$

22. Find the area of a triangle whose vertices are $(-1, 1)$, $(2, 2)$, and $(1, -2)$, as shown in the figure.

23. Find the value of k such that the system is inconsistent.

$$\begin{aligned} 4x - 8y &= -3 \\ 2x + ky &= 16 \end{aligned}$$

24. Graph $g(x) = \log_3(x - 1)$.

25. Evaluate $\log_4 \frac{1}{16}$ without using a calculator.

26. Describe the relationship between the graphs of $f(x) = e^x$ and $g(x) = \ln x$.

27. Use the properties of logarithms to condense $3(\log_2 x + \log_2 y) - \log_2 z$.

28. Use the properties of logarithms to expand $\ln \dfrac{5x}{(x + 1)^2}$.

29. Solve each equation.

(a) $\log_x\left(\frac{1}{9}\right) = -2$ (b) $4 \ln x = 10$

(c) $500(1.08)^t = 2000$ (d) $3(1 + e^{2x}) = 20$

30. If the inflation rate averages 3.5% over the next 5 years, the approximate cost C of goods and services t years from now is given by

$$C(t) = P(1.035)^t, \quad 0 \le t \le 5$$

where P is the present cost. If the price of an oil change is presently $24.95, estimate the price 5 years from now.

31. Determine the effective yield of an 8% interest rate compounded continuously.

32. Determine the length of time for an investment of $1000 to quadruple in value if the investment earns 9% compounded continuously.

10 Sequences, Series, and Probability

Lawrence Migdale/Stock Boston

Interest in genealogy, which is the study of family history and ancestry, is growing. Today there are many resources and organizations available to assist both the amateur and professional genealogist.

622

Motivating the Chapter

 Ancestors and Descendants

See Section 10.3, Exercise 123

a. Your ancestors consist of your two parents (first generation), your four grandparents (second generation), your eight great-grandparents (third generation), and so on. Write a geometric sequence that will describe the number of ancestors for each generation.

b. If your ancestry could be traced back 66 generations (approximately 2000 years), how many different ancestors would you have?

c. A common ancestor is one to whom you are related in more than one way. (See figure.) From the results of part (b), do you think that you have had no common ancestors in the past 2000 years?

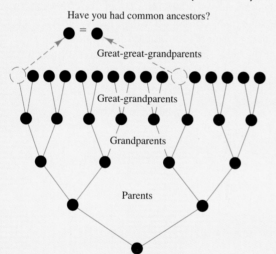

Have you had common ancestors?

Great-great-grandparents

Great-grandparents

Grandparents

Parents

See Section 10.6, Exercise 55

d. One set of your great-grandparents had eight children (of which one is your grandparent). Each of those children had five children (of which one is your parent). And each of those children had three children. How many direct descendants do your great-grandparents have?

e. One hundred fifty people were able to attend your family reunion, to which everyone was asked to bring an exchange gift. All the names were put into a bowl to be drawn randomly to determine the order for receiving a gift. Will the drawing be done *with* or *without* replacement? Explain.

f. What is the probability that your name will be chosen first to receive a gift? After 45 names are drawn, your name is still in the bowl. What is the probability that your name will be chosen next?

10.1   Sequences and Series

Objectives

1 Write the terms of a sequence given its nth term.

2 Write the terms of a sequence involving factorials.

3 Find the apparent nth term of a sequence.

4 Sum the terms of a sequence to obtain a series.

---

1 Write the terms of a sequence given its nth term.

Sequences

Suppose you were given the following choice of contract offers for the next 5 years of employment.

Contract A $20,000 the first year and a $2200 raise each year

Contract B $20,000 the first year and a 10% raise each year

Which contract offers the greater salary over the 5-year period? The salaries for each contract are shown at the left. The salaries for contract A represent the first five terms of an **arithmetic sequence,** and the salaries for contract B represent the first five terms of a **geometric sequence.** Notice that after 5 years the geometric sequence represents a better contract offer than the arithmetic sequence.

A mathematical **sequence** is simply an ordered list of numbers. Each number in the list is a **term** of the sequence. A sequence can have a finite number of terms or an infinite number of terms. For instance, the sequence of positive odd integers that are less than 15 is a *finite* sequence

Year	*Contract A*	*Contract B*
1	$20,000	$20,000
2	$22,200	$22,000
3	$24,400	$24,200
4	$26,600	$26,620
5	$28,800	$29,282
Total	$122,000	$122,102

$$1, 3, 5, 7, 9, 11, 13 \qquad \text{Finite sequence}$$

whereas the sequence of positive odd integers is an *infinite* sequence.

$$1, 3, 5, 7, 9, 11, 13, . . . \qquad \text{Infinite sequence}$$

Note that the three dots indicate that the sequence continues and has an infinite number of terms.

Because each term of a sequence is matched with its location, a sequence can be defined as a **function** whose domain is a subset of positive integers.

> ▶ **Sequences**
>
> An **infinite sequence** $a_1, a_2, a_3, . . . , a_n, . . .$ is a function whose domain is the set of positive integers.
>
> A **finite sequence** $a_1, a_2, a_3, . . . , a_n$ is a function whose domain is the finite set $\{1, 2, 3, . . ., n\}$.

In some cases it is convenient to begin subscripting a sequence with 0 instead of 1. Then the domain of the infinite sequence is the set of nonnegative integers and the domain of the finite sequence is the set $\{0, 1, 2, . . . , n\}$.

$$a_{(\ )} = 2(\ ) + 1$$
$$a_{(1)} = 2(1) + 1 = 3$$
$$a_{(2)} = 2(2) + 1 = 5$$
$$\vdots$$
$$a_{(51)} = 2(51) + 1 = 103$$

The subscripts of a sequence are used in place of function notation. For instance, if parentheses replaced the n in $a_n = 2n + 1$, the notation would be similar to function notation, as shown at the left.

Example 1   Finding the Terms of a Sequence

Write the first six terms of the sequence whose nth term is

$$a_n = n^2 - 1.$$ Begin sequence with $n = 1$.

Solution

$$a_1 = (1)^2 - 1 = 0 \qquad a_2 = (2)^2 - 1 = 3 \qquad a_3 = (3)^2 - 1 = 8$$
$$a_4 = (4)^2 - 1 = 15 \qquad a_5 = (5)^2 - 1 = 24 \qquad a_6 = (6)^2 - 1 = 35$$

To represent the entire sequence, you can write the following.

$$0, 3, 8, 15, 24, 35, \ldots, n^2 - 1, \ldots$$

Example 2   Finding the Terms of a Sequence

Write the first six terms of the sequence whose nth term is

$$a_n = 3(2^n).$$ Begin sequence with $n = 0$.

Solution

$$a_0 = 3(2^0) = 3 \cdot 1 = 3 \qquad a_1 = 3(2^1) = 3 \cdot 2 = 6$$
$$a_2 = 3(2^2) = 3 \cdot 4 = 12 \qquad a_3 = 3(2^3) = 3 \cdot 8 = 24$$
$$a_4 = 3(2^4) = 3 \cdot 16 = 48 \qquad a_5 = 3(2^5) = 3 \cdot 32 = 96$$

The entire sequence can be written as follows.

$$3, 6, 12, 24, 48, 96, \ldots, 3(2^n), \ldots$$

Example 3   A Sequence Whose Terms Alternate in Sign

Write the first six terms of the sequence whose nth term is

$$a_n = \frac{(-1)^n}{2n - 1}.$$ Begin sequence with $n = 1$.

Solution

$$a_1 = \frac{(-1)^1}{2(1) - 1} = -\frac{1}{1} \qquad a_2 = \frac{(-1)^2}{2(2) - 1} = \frac{1}{3} \qquad a_3 = \frac{(-1)^3}{2(3) - 1} = -\frac{1}{5}$$
$$a_4 = \frac{(-1)^4}{2(4) - 1} = \frac{1}{7} \qquad a_5 = \frac{(-1)^5}{2(5) - 1} = -\frac{1}{9} \qquad a_6 = \frac{(-1)^6}{2(6) - 1} = \frac{1}{11}$$

The entire sequence can be written as follows.

$$-1, \frac{1}{3}, -\frac{1}{5}, \frac{1}{7}, -\frac{1}{9}, \frac{1}{11}, \ldots, \frac{(-1)^n}{2n - 1}, \ldots$$

Technology: Tip

Most graphing utilities have a "sequence graphing mode" that allows you to plot the terms of a sequence as points on a rectangular coordinate system. For instance, the graph of the first six terms of the sequence given by

$$a_n = n^2 - 1$$

is shown below.

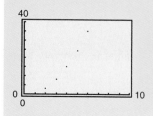

2 Write the terms of a sequence involving factorials.

Factorial Notation

Some very important sequences in mathematics involve terms that are defined with special types of products called **factorials.**

> ▶ **Definition of Factorial**
>
> If n is a positive integer, n **factorial** is defined as
>
> $$n! = 1 \cdot 2 \cdot 3 \cdot 4 \cdot \cdot \cdot \cdot \cdot (n-1) \cdot n.$$
>
> As a special case, zero factorial is defined as $0! = 1$.

The first several factorial values are as follows.

$$0! = 1 \qquad\qquad 1! = 1$$
$$2! = 1 \cdot 2 = 2 \qquad\qquad 3! = 1 \cdot 2 \cdot 3 = 6$$
$$4! = 1 \cdot 2 \cdot 3 \cdot 4 = 24 \qquad 5! = 1 \cdot 2 \cdot 3 \cdot 4 \cdot 5 = 120$$

Many calculators have a factorial key, denoted by $\boxed{n!}$. If your calculator has such a key, try using it to evaluate $n!$ for several values of n. You will see that the value of n does not have to be very large before the value of $n!$ is huge. For instance,

$$10! = 3,628,800.$$

Example 4   A Sequence Involving Factorials

Write the first six terms of the sequence with the given nth term.

a. $a_n = \dfrac{1}{n!}$ **b.** $a_n = \dfrac{2^n}{n!}$ Begin both sequences with $n = 0$.

Solution

a. $a_0 = \dfrac{1}{0!} = \dfrac{1}{1} = 1$ $\qquad\qquad a_1 = \dfrac{1}{1!} = \dfrac{1}{1} = 1$

$a_2 = \dfrac{1}{2!} = \dfrac{1}{2}$ $\qquad\qquad a_3 = \dfrac{1}{3!} = \dfrac{1}{1 \cdot 2 \cdot 3} = \dfrac{1}{6}$

$a_4 = \dfrac{1}{4!} = \dfrac{1}{1 \cdot 2 \cdot 3 \cdot 4} = \dfrac{1}{24}$ $\qquad a_5 = \dfrac{1}{5!} = \dfrac{1}{1 \cdot 2 \cdot 3 \cdot 4 \cdot 5} = \dfrac{1}{120}$

b. $a_0 = \dfrac{2^0}{0!} = \dfrac{1}{1} = 1$ $\qquad\qquad a_1 = \dfrac{2^1}{1!} = \dfrac{2}{1} = 2$

$a_2 = \dfrac{2^2}{2!} = \dfrac{4}{2} = 2$ $\qquad\qquad a_3 = \dfrac{2^3}{3!} = \dfrac{8}{1 \cdot 2 \cdot 3} = \dfrac{8}{6} = \dfrac{4}{3}$

$a_4 = \dfrac{2^4}{4!} = \dfrac{2 \cdot 2 \cdot 2 \cdot 2}{1 \cdot 2 \cdot 3 \cdot 4} = \dfrac{2}{3}$ $\qquad a_5 = \dfrac{2^5}{5!} = \dfrac{2 \cdot 2 \cdot 2 \cdot 2 \cdot 2}{1 \cdot 2 \cdot 3 \cdot 4 \cdot 5} = \dfrac{4}{15}$

3 Find the apparent *n*th term of a sequence.

Finding the *n*th Term of a Sequence

Sometimes you will have the first several terms of a sequence and need to find a formula (the *n*th term) that will generate those terms. *Pattern recognition* is crucial in finding a form for the *n*th term.

Study Tip

Simply listing the first few terms is not sufficient to define a unique sequence—the *n*th term *must be given*. Consider the sequence

$$\frac{1}{2}, \frac{1}{4}, \frac{1}{8}, \frac{1}{15}, \cdots$$

The first three terms are identical to the first three terms of the sequence in Example 5(a). However, the *n*th term of this sequence is defined as

$$a_n = \frac{6}{(n+1)(n^2 - n + 6)}.$$

Example 5   Finding the *n*th Term of a Sequence

Write an expression for the *n*th term of each sequence.

a. $\dfrac{1}{2}, \dfrac{1}{4}, \dfrac{1}{8}, \dfrac{1}{16}, \dfrac{1}{32}, \cdots$ **b.** $1, -4, 9, -16, 25, \ldots$

Solution

a.

n:	1	2	3	4	5	$\cdots$	*n*
Terms:	$\dfrac{1}{2}$	$\dfrac{1}{4}$	$\dfrac{1}{8}$	$\dfrac{1}{16}$	$\dfrac{1}{32}$	$\cdots$	a_n

Pattern: The numerator is 1 and each denominator is an increasing power of 2.

$$a_n = \frac{1}{2^n}$$

b.

n:	1	2	3	4	5	$\cdots$	*n*
Terms:	1	-4	9	-16	25	$\cdots$	a_n

Pattern: The terms have alternating signs with those in the even positions being negative. Each term is the square of *n*.

$$a_n = (-1)^{n+1} n^2$$

4 Sum the terms of a sequence to obtain a series.

Series

In the opening illustration of this section, the terms of the finite sequence were *added*. If you add all the terms of an infinite sequence, you obtain a **series.**

> ▶ Definition of a Series
>
> For an infinite sequence
>
> $$a_1, a_2, a_3, \ldots, a_n, \ldots$$
>
> **1.** the sum of all the terms
>
> $$a_1 + a_2 + a_3 + \cdots + a_n + \cdots$$
>
> is called an **infinite series,** or simply a **series,** and
>
> **2.** the sum of the first *n* terms
>
> $$S_n = a_1 + a_2 + a_3 + \cdots + a_n$$
>
> is called a **partial sum.**

Example 6 Finding Partial Sums

Find the indicated partial sums for each sequence.

a. Find S_1, S_2, and S_5 for $a_n = 3n - 1$.

b. Find S_2, S_3, and S_4 for $a_n = \dfrac{(-1)^n}{n+1}$.

Solution

a. The first five terms of the sequence $a_n = 3n - 1$ are

$$a_1 = 2, a_2 = 5, a_3 = 8, a_4 = 11, \text{ and } a_5 = 14.$$

So, the partial sums are

$$S_1 = 2, S_2 = 2 + 5 = 7, \text{ and } S_5 = 2 + 5 + 8 + 11 + 14 = 40.$$

b. The first four terms of the sequence $a_n = \dfrac{(-1)^n}{n+1}$ are

$$a_1 = -\frac{1}{2}, a_2 = \frac{1}{3}, a_3 = -\frac{1}{4}, \text{ and } a_4 = \frac{1}{5}.$$

So, the partial sums are

$$S_2 = -\frac{1}{2} + \frac{1}{3} = -\frac{1}{6}, S_3 = -\frac{1}{2} + \frac{1}{3} - \frac{1}{4} = -\frac{5}{12}, \text{ and}$$

$$S_4 = -\frac{1}{2} + \frac{1}{3} - \frac{1}{4} + \frac{1}{5} = -\frac{13}{60}.$$

Technology: Tip

Most graphing utilities have a built-in program that will calculate the partial sum of a sequence. Consult the user's guide for your graphing utility.

A convenient shorthand notation for denoting a partial sum is called **sigma notation.** This name comes from the use of the uppercase Greek letter sigma, written as Σ.

Study Tip

If a_i is a constant in the partial sum

$$\sum_{i=m}^{n} a_i$$

then the partial sum is the product $(n - m + 1)(a_i)$.

For instance,

$$\sum_{i=1}^{4} 5 = 5 + 5 + 5 + 5$$

$$= 4(5)$$

$$= 20.$$

▶ Definition of Sigma Notation

The sum of the first n terms of the sequence whose nth term is a_n is

$$\sum_{i=1}^{n} a_i = a_1 + a_2 + a_3 + a_4 + \cdots + a_n$$

where i is the **index of summation,** n is the **upper limit of summation,** and 1 is the **lower limit of summation.**

Example 7 Sigma Notation for Sums

$$\sum_{i=1}^{6} 2i = 2(1) + 2(2) + 2(3) + 2(4) + 2(5) + 2(6)$$

$$= 2 + 4 + 6 + 8 + 10 + 12$$

$$= 42$$

Example 8   Sigma Notation for Sums

Study Tip

In Example 7, the index of summation is i and the summation begins with $i = 1$. Any letter can be used as the index of summation, and the summation can begin with any integer. For instance, in Example 8, the index of summation is k and the summation begins with $k = 0$.

Find the sum $\sum\limits_{k=0}^{8} \dfrac{1}{k!}$.

Solution

$$\sum_{k=0}^{8} \frac{1}{k!} = \frac{1}{0!} + \frac{1}{1!} + \frac{1}{2!} + \frac{1}{3!} + \frac{1}{4!} + \frac{1}{5!} + \frac{1}{6!} + \frac{1}{7!} + \frac{1}{8!}$$

$$= 1 + 1 + \frac{1}{2} + \frac{1}{6} + \frac{1}{24} + \frac{1}{120} + \frac{1}{720} + \frac{1}{5040} + \frac{1}{40,320}$$

$$\approx 2.71828$$

Note that this sum is approximately $e = 2.71828. \ldots$

Example 9   Writing a Sum in Sigma Notation

Write the following sums in sigma notation.

a. $\dfrac{2}{2} + \dfrac{2}{3} + \dfrac{2}{4} + \dfrac{2}{5} + \dfrac{2}{6}$ **b.** $1 - \dfrac{1}{3} + \dfrac{1}{9} - \dfrac{1}{27} + \dfrac{1}{81}$

Solution

a. The pattern has numerators of 2 and denominators that range over the integers from 2 to 6. So, one possible sigma notation is

$$\sum_{i=1}^{5} \frac{2}{i + 1} = \frac{2}{2} + \frac{2}{3} + \frac{2}{4} + \frac{2}{5} + \frac{2}{6}.$$

b. The numerators alternate in sign and the denominators are integer powers of 3, starting with 3^0 and ending with 3^4. So, one possible sigma notation is

$$\sum_{i=0}^{4} \frac{(-1)^i}{3^i} = \frac{1}{3^0} + \frac{-1}{3^1} + \frac{1}{3^2} + \frac{-1}{3^3} + \frac{1}{3^4}.$$

Discussing the Concept   Finding a Pattern

You learned in this section that a sequence is an ordered list of numbers. Study the following sequence and see if you can guess what its next term should be.

Z, O, T, T, F, F, S, S, E, N, T, E, T, . . .

(*Hint:* You might try to figure out what numbers the letters represent.) Construct another sequence with letters. Can the other members of your class guess the next term?

10.1   Exercises

Integrated Review   *Concepts, Skills, and Problem Solving*

Keep mathematically in shape by doing these exercises *before* the problems of this section.

Properties and Definitions

1. Demonstrate the Multiplicative Property of Equality for the equation $-7x = 35$.

2. Demonstrate the Additive Property of Equality for the equation $7x + 63 = 35$.

3. How do you determine whether $t = -3$ is a solution of the equation $t^2 + 4t + 3 = 0$?

4. What is the usual first step in solving an equation such as

$$\frac{3}{x} - \frac{1}{x+1} = 10?$$

Simplifying Expressions

In Exercises 5–10, simplify the expression.

5. $(x + 10)^{-2}$

6. $\dfrac{18(x-3)^5}{(x-3)^2}$

7. $(a^2)^{-4}$

8. $(8x^3)^{1/3}$

9. $\sqrt{128x^3}$

10. $\dfrac{5}{\sqrt{x}-2}$

Graphs and Models

In Exercises 11 and 12, (a) write a function for the area of the region, (b) use a graphing utility to graph the function, and (c) approximate the value of x if the area of the region is 200 square units.

11.

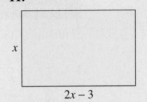

x

$2x - 3$

12.

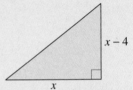

$x - 4$

x

Developing Skills

In Exercises 1–22, write the first five terms of the sequence. (Begin with $n = 1$.) See Examples 1–4.

1. $a_n = 2n$

2. $a_n = 3n$

3. $a_n = (-1)^n 2n$

4. $a_n = (-1)^{n+1} 3n$

5. $a_n = \left(\frac{1}{2}\right)^n$

6. $a_n = \left(\frac{1}{3}\right)^n$

7. $a_n = \left(-\frac{1}{2}\right)^{n+1}$

8. $a_n = \left(\frac{2}{3}\right)^{n-1}$

9. $a_n = (-0.2)^{n-1}$

10. $a_n = \left(-\frac{2}{3}\right)^{n-1}$

11. $a_n = \dfrac{1}{n+1}$

12. $a_n = \dfrac{3}{2n+1}$

13. $a_n = \dfrac{2n}{3n+2}$

14. $a_n = \dfrac{5n}{4n+3}$

15. $a_n = \dfrac{(-1)^n}{n^2}$

16. $a_n = \dfrac{1}{\sqrt{n}}$

17. $a_n = 5 - \dfrac{1}{2^n}$

18. $a_n = 7 + \dfrac{1}{3^n}$

19. $a_n = \dfrac{(n+1)!}{n!}$

20. $a_n = \dfrac{n!}{(n-1)!}$

21. $a_n = \dfrac{2 + (-2)^n}{n!}$

22. $a_n = \dfrac{1 + (-1)^n}{n^2}$

In Exercises 23–26, find the indicated term of the sequence.

23. $a_n = (-1)^n (5n - 3)$

$a_{15} = $

24. $a_n = (-1)^{n-1}(2n + 4)$

$a_{14} = $

25. $a_n = \dfrac{n^2 - 2}{(n-1)!}$

$a_8 = $

26. $a_n = \dfrac{n^2}{n!}$

$a_{12} = $

In Exercises 27–38, simplify the expression.

27. $\dfrac{5!}{4!}$

28. $\dfrac{18!}{17!}$

29. $\dfrac{10!}{12!}$

30. $\dfrac{5!}{8!}$

31. $\dfrac{25!}{20!5!}$

32. $\dfrac{20!}{15! \cdot 5!}$

33. $\dfrac{n!}{(n+1)!}$

34. $\dfrac{(n+2)!}{n!}$

35. $\dfrac{(2n+1)!}{(n-1)!}$

36. $\dfrac{(3n)!}{(3n+2)!}$

37. $\dfrac{(2n)!}{(2n-1)!}$

38. $\dfrac{(2n+2)!}{(2n)!}$

In Exercises 39–42, match the sequence with the graph of its first 10 terms. [The graphs are labeled (a), (b), (c), and (d).]

(a)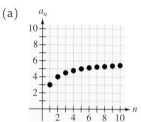

(b)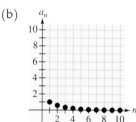

(c)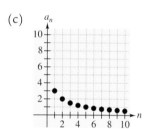

(d)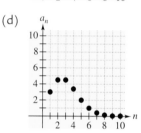

39. $a_n = \dfrac{6}{n+1}$

40. $a_n = \dfrac{6n}{n+1}$

41. $a_n = (0.6)^{n-1}$

42. $a_n = \dfrac{3^n}{n!}$

In Exercises 43–48, use a graphing utility to graph the first 10 terms of the sequence.

43. $a_n = (-0.8)^{n-1}$

44. $a_n = \dfrac{2n^2}{n^2+1}$

45. $a_n = \dfrac{1}{2}n$

46. $a_n = \dfrac{n+2}{n}$

47. $a_n = 3 - \dfrac{4}{n}$

48. $a_n = 10\left(\dfrac{3}{4}\right)^{n-1}$

In Exercises 49–66, write an expression for the nth term of the sequence. (Assume that n begins with 1.) See Example 5.

49. 3, 6, 9, 12, 15, . . .

50. 5, 10, 15, 20, 25, . . .

51. 1, 4, 7, 10, 13, . . .

52. 3, 7, 11, 15, 19, . . .

53. 0, 3, 8, 15, 24, . . .

54. 1, 8, 27, 64, 125, . . .

55. 2, −4, 6, −8, 10, . . .

56. 1, −1, 1, −1, 1, . . .

57. $\dfrac{2}{3}, \dfrac{3}{4}, \dfrac{4}{5}, \dfrac{5}{6}, \dfrac{6}{7}, \ldots$

58. $\dfrac{2}{1}, \dfrac{3}{3}, \dfrac{4}{5}, \dfrac{5}{7}, \dfrac{6}{9}, \ldots$

59. $\dfrac{1}{2}, \dfrac{-1}{4}, \dfrac{1}{8}, \dfrac{-1}{16}, \ldots$

60. $1, \dfrac{1}{4}, \dfrac{1}{9}, \dfrac{1}{16}, \dfrac{1}{25}, \ldots$

61. $1, \dfrac{1}{2}, \dfrac{1}{4}, \dfrac{1}{8}, \ldots$

62. $\dfrac{1}{3}, \dfrac{2}{9}, \dfrac{4}{27}, \dfrac{8}{81}, \ldots$

63. $1 + \dfrac{1}{1}, 1 + \dfrac{1}{2}, 1 + \dfrac{1}{3}, 1 + \dfrac{1}{4}, 1 + \dfrac{1}{5}, \ldots$

64. $1 + \dfrac{1}{2}, 1 + \dfrac{3}{4}, 1 + \dfrac{7}{8}, 1 + \dfrac{15}{16}, 1 + \dfrac{31}{32}, \ldots$

65. $1, \dfrac{1}{2}, \dfrac{1}{6}, \dfrac{1}{24}, \dfrac{1}{120}, \ldots$

66. $1, 2, \dfrac{2^2}{2}, \dfrac{2^3}{6}, \dfrac{2^4}{24}, \dfrac{2^5}{120}, \ldots$

In Exercises 67–82, find the partial sum. See Examples 6–8.

67. $\displaystyle\sum_{k=1}^{6} 3k$

68. $\displaystyle\sum_{k=1}^{4} 5k$

69. $\displaystyle\sum_{i=0}^{6} (2i+5)$

70. $\displaystyle\sum_{i=0}^{4} (2i+3)$

71. $\displaystyle\sum_{j=3}^{7} (6j-10)$

72. $\displaystyle\sum_{i=2}^{7} (4i-1)$

73. $\displaystyle\sum_{j=1}^{5} \dfrac{(-1)^{j+1}}{j^2}$

74. $\displaystyle\sum_{j=0}^{3} \dfrac{1}{j^2+1}$

75. $\displaystyle\sum_{m=2}^{6} \dfrac{2m}{2(m-1)}$

76. $\displaystyle\sum_{k=1}^{5} \dfrac{10k}{k+2}$

77. $\displaystyle\sum_{k=1}^{6} (-8)$

78. $\displaystyle\sum_{n=3}^{12} 10$

79. $\displaystyle\sum_{i=1}^{8} \left(\dfrac{1}{i} - \dfrac{1}{i+1}\right)$

80. $\displaystyle\sum_{k=1}^{5} \left(\dfrac{2}{k} - \dfrac{2}{k+2}\right)$

81. $\displaystyle\sum_{n=0}^{5} \left(-\dfrac{1}{3}\right)^n$

82. $\displaystyle\sum_{n=0}^{6} \left(\dfrac{3}{2}\right)^n$

In Exercises 83–90, use a graphing utility to find the sum.

83. $\displaystyle\sum_{n=1}^{6} 3n^2$

84. $\displaystyle\sum_{n=0}^{5} 2n^2$

85. $\displaystyle\sum_{j=2}^{6} (j!-j)$

86. $\displaystyle\sum_{i=0}^{4} (i!+4)$

87. $\displaystyle\sum_{j=0}^{4} \dfrac{6}{j!}$

88. $\displaystyle\sum_{k=1}^{6} \left(\dfrac{1}{2k} - \dfrac{1}{2k-1}\right)$

89. $\displaystyle\sum_{k=1}^{6} \ln k$

90. $\displaystyle\sum_{k=2}^{4} \dfrac{k}{\ln k}$

In Exercises 91-108, write the sum using sigma notation. (Begin with $k = 0$ or $k = 1$.) See Example 9.

91. $1 + 2 + 3 + 4 + 5$

92. $8 + 9 + 10 + 12 + 13 + 14$

93. $2 + 4 + 6 + 8 + 10$

94. $24 + 30 + 36 + 42$

95. $\dfrac{1}{2(1)} + \dfrac{1}{2(2)} + \dfrac{1}{2(3)} + \dfrac{1}{2(4)} + \cdots + \dfrac{1}{2(10)}$

96. $\dfrac{3}{1 + 1} + \dfrac{3}{1 + 2} + \dfrac{3}{1 + 3} + \cdots + \dfrac{3}{1 + 50}$

97. $\dfrac{1}{1^2} + \dfrac{1}{2^2} + \dfrac{1}{3^2} + \dfrac{1}{4^2} + \cdots + \dfrac{1}{20^2}$

98. $\dfrac{1}{2^0} + \dfrac{1}{2^1} + \dfrac{1}{2^2} + \dfrac{1}{2^3} + \cdots + \dfrac{1}{2^{12}}$

99. $\dfrac{1}{3^0} - \dfrac{1}{3^1} + \dfrac{1}{3^2} - \dfrac{1}{3^3} + \cdots - \dfrac{1}{3^9}$

100. $\left(-\dfrac{2}{3}\right)^0 + \left(-\dfrac{2}{3}\right)^1 + \left(-\dfrac{2}{3}\right)^2 + \cdots + \left(-\dfrac{2}{3}\right)^{20}$

101. $\dfrac{4}{1 + 3} + \dfrac{4}{2 + 3} + \dfrac{4}{3 + 3} + \cdots + \dfrac{4}{20 + 3}$

102. $\dfrac{1}{2^3} - \dfrac{1}{4^3} + \dfrac{1}{6^3} - \dfrac{1}{8^3} + \cdots + \dfrac{1}{14^3}$

103. $\dfrac{1}{2} + \dfrac{2}{3} + \dfrac{3}{4} + \dfrac{4}{5} + \dfrac{5}{6} + \cdots + \dfrac{11}{12}$

104. $\dfrac{2}{4} + \dfrac{4}{7} + \dfrac{6}{10} + \dfrac{8}{13} + \dfrac{10}{16} + \cdots + \dfrac{20}{31}$

105. $\dfrac{2}{4} + \dfrac{4}{5} + \dfrac{6}{6} + \dfrac{8}{7} + \cdots + \dfrac{40}{23}$

106. $\left(2 + \dfrac{1}{1}\right) + \left(2 + \dfrac{1}{2}\right) + \left(2 + \dfrac{1}{3}\right) + \cdots + \left(2 + \dfrac{1}{25}\right)$

107. $1 + 1 + 2 + 6 + 24 + 120 + 720$

108. $1 + 1 + \dfrac{1}{2} + \dfrac{1}{6} + \dfrac{1}{24} + \dfrac{1}{120} + \dfrac{1}{720}$

Arithmetic Mean In Exercises 109–112, find the arithmetic mean of the set. The *arithmetic mean* of a set of n measurements $x_1, x_2, x_3, \ldots, x_n$ is

$$\bar{x} = \dfrac{1}{n}\sum_{i=1}^{n} x_i$$

109. 3, 7, 2, 1, 5

110. 84, 69, 66, 96

111. 0.5, 0.8, 1.1, 0.8, 0.7, 0.7, 1.0

112. -1.0, 4.2, 5.4, -3.2, 3.6

Solving Problems

113. *Compound Interest* A deposit of $500 is made in an account that earns 7% interest compounded yearly. The balance in the account after N years is given by

$$A_N = 500(1 + 0.07)^N, \quad N = 1, 2, 3, \ldots$$

(a) Compute the first eight terms of the sequence.

(b) Find the balance in this account after 40 years by computing A_{40}.

(c) Use a graphing utility to graph the first 40 terms of the sequence.

(d) The terms are increasing. Is the rate of growth of the terms increasing? Explain.

114. *Depreciation* At the end of each year, the value of a car with an initial cost of $26,000 is three-fourths what it was at the beginning of the year. Thus, after n years, its value is given by

$$a_n = 26,000\left(\dfrac{3}{4}\right)^n, \quad n = 1, 2, 3, \ldots$$

(a) Find the value of the car 3 years after it was purchased by computing a_3.

(b) Find the value of the car 6 years after it was purchased by computing a_6. Is this value half of what it was after 3 years? Explain.

115. *Soccer Ball* The number of degrees a_n in each angle of a regular n-sided polygon is

$$a_n = \dfrac{180(n - 2)}{n}, \quad n \geq 3.$$

The surface of a soccer ball is made of regular hexagons and pentagons. If a soccer ball is taken apart and flattened, as shown in the figure, the sides don't meet each other. Use the terms a_5 and a_6 to explain why there are gaps between adjacent hexagons.

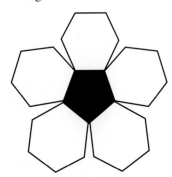

116. *Stars*   The number of degrees d_n in each tip of each n-pointed star in the figure is given by

$$d_n = \frac{180(n-4)}{n}, \quad n \geq 5.$$

Write the first six terms of this sequence.

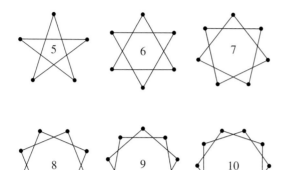

117. *Stars*   The stars in Exercise 116 were formed by placing n equally spaced points on a circle and connecting each point with the second point from it on the circle. The stars in the figure for this exercise were formed in a similar way except that each point was connected with the third point from it. For these stars, the number of degrees in a tip is given by

$$d_n = \frac{180(n-6)}{n}, \quad n \geq 7.$$

Write the first five terms of this sequence.

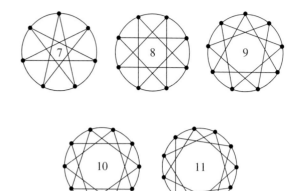

Figure for 117

118. *Outpatient Surgery*   The number a_n (in thousands) of outpatient surgeries performed in hospitals in the United States for the years 1990 through 1995 is modeled by

$$a_n = 11{,}791 + 436n, \quad n = 1, 2, \ldots, 5$$

where n is the year, with $n = 0$ corresponding to 1990. Find the terms of this finite sequence and use a graphing utility to construct a bar graph that represents the sequence. (Source: American Hospital Association)

Explaining Concepts

119. Give an example of an infinite sequence.

120. State the definition of n factorial.

121. The nth term of a sequence is $a_n = (-1)^n n$. Which terms of the sequence are negative?

In Exercises 122–124, decide whether the statement is true. Explain your reasoning.

122. $\displaystyle\sum_{i=1}^{4}(i^2 + 2i) = \sum_{i=1}^{4}i^2 + \sum_{i=1}^{4}2i$

123. $\displaystyle\sum_{k=1}^{4}3k = 3\sum_{k=1}^{4}k$

124. $\displaystyle\sum_{j=1}^{4}2^j = \sum_{j=3}^{6}2^{j-2}$

10.2   Arithmetic Sequences

Objectives

1 Find the common difference and the nth term of an arithmetic sequence.

2 Find the nth partial sum of an arithmetic sequence.

3 Use an arithmetic sequence to solve an application problem.

1 Find the common difference and the nth term of an arithmetic sequence.

Arithmetic Sequences

A sequence whose consecutive terms have a common difference is called an **arithmetic sequence.**

▶ **Definition of an Arithmetic Sequence**

A sequence is called **arithmetic** if the differences between consecutive terms are the same. So, the sequence

$$a_1, a_2, a_3, a_4, \ldots, a_n, \ldots$$

is arithmetic if there is a number d such that

$$a_2 - a_1 = d, \quad a_3 - a_2 = d, \quad a_4 - a_3 = d$$

and so on. The number d is the **common difference** of the sequence.

Example 1   Examples of Arithmetic Sequences

a. The sequence whose nth term is $3n + 2$ is arithmetic. For this sequence, the common difference between consecutive terms is 3.

$$5, 8, 11, 14, \ldots, 3n + 2, \ldots$$

$8 - 5 = 3$

b. The sequence whose nth term is $7 - 5n$ is arithmetic. For this sequence, the common difference between consecutive terms is -5.

$$2, -3, -8, -13, \ldots, 7 - 5n, \ldots$$

$-3 - 2 = -5$

c. The sequence whose nth term is $\frac{1}{4}(n + 3)$ is arithmetic. For this sequence, the common difference between consecutive terms is $\frac{1}{4}$.

$$1, \frac{5}{4}, \frac{3}{2}, \frac{7}{4}, \ldots, \frac{n + 3}{4}, \ldots$$

$\frac{5}{4} - 1 = \frac{1}{4}$

Study Tip

The nth term of an arithmetic sequence can be derived from the following pattern.

$a_1 = a_1$ 1st term

$a_2 = a_1 + d$ 2nd term

$a_3 = a_1 + 2d$ 3rd term

$a_4 = a_1 + 3d$ 4th term

$a_5 = a_1 + 4d$ 5th term

      1 less

$\vdots$ $\vdots$

$a_n = a_1 + (n-1)d$ nth term

      1 less

▶ The nth Term of an Arithmetic Sequence

The nth term of an arithmetic sequence has the form

$$a_n = a_1 + (n-1)d$$

where d is the common difference between the terms of the sequence, and a_1 is the first term.

Example 2   Finding the nth Term of an Arithmetic Sequence

Find a formula for the nth term of the arithmetic sequence whose common difference is 2 and whose first term is 5.

Solution

You know that the formula for the nth term is of the form $a_n = a_1 + (n-1)d$. Moreover, because the common difference is $d = 2$, and the first term is $a_1 = 5$, the formula must have the form

$$a_n = 5 + 2(n-1).$$

So, the formula for the nth term is

$$a_n = 2n + 3.$$

The sequence therefore has the following form.

$$5, 7, 9, 11, 13, \ldots, 2n + 3, \ldots$$

If you know the nth term and the common difference of an arithmetic sequence, you can find the $(n+1)$th term by using the following **recursion formula.**

$$a_{n+1} = a_n + d$$

Example 3   Using a Recursion Formula

The 12th term of an arithmetic sequence is 52 and the common difference is 3.

a. What is the 13th term of the sequence?   **b.** What is the first term?

Solution

a. Using the recursion formula $a_{13} = a_{12} + d$, you know that $a_{12} = 52$ and $d = 3$. So, the 13th term of the sequence is

$$a_{13} = 52 + 3 = 55.$$

b. Using $n = 12$, $d = 3$, and $a_{12} = 52$ in the formula $a_n = a_1 + (n-1)d$ yields

$$52 = a_1 + (12 - 1)(3)$$

$$19 = a_1.$$

2 Find the *n*th partial sum of an arithmetic sequence.

The Partial Sum of an Arithmetic Sequence

The sum of the first n terms of an arithmetic sequence is called the **nth partial sum** of the sequence. For instance, the 5th partial sum of the arithmetic sequence whose nth term is $3n + 4$ is

$$\sum_{i=1}^{5} (3i + 4) = 7 + 10 + 13 + 16 + 19 = 65.$$

To find a formula for the nth partial sum S_n of an arithmetic sequence, write out S_n forwards and backwards and then add the two forms, as follows.

$$S_n = a_1 + (a_1 + d) + (a_1 + 2d) + \cdots + [a_1 + (n - 1)d] \qquad \text{Forwards}$$

$$S_n = a_n + (a_n - d) + (a_n - 2d) + \cdots + [a_n - (n - 1)d] \qquad \text{Backwards}$$

$$2S_n = (a_1 + a_n) + (a_1 + a_n) + (a_1 + a_n) + \cdots + [a_1 + a_n] \qquad \begin{array}{l}\text{Sum of two} \\ \text{equations}\end{array}$$

$$= n(a_1 + a_n) \qquad \begin{array}{l} n \text{ groups of} \\ (a_1 + a_n)\end{array}$$

Dividing both sides by 2 yields the following formula.

Study Tip

You can use the formula for the nth partial sum of an arithmetic sequence to find the sum of consecutive numbers. For instance, the sum of the integers from 1 to 100 is

$$\sum_{i=1}^{100} i = 100\left(\frac{1 + 100}{2}\right)$$

$$= 50(101)$$

$$= 5050.$$

▶ **The *n*th Partial Sum of an Arithmetic Sequence**

The nth partial sum of the arithmetic sequence whose nth term is a_n is

$$\sum_{i=1}^{n} a_i = a_1 + a_2 + a_3 + a_4 + \cdots + a_n$$

$$= n\left(\frac{a_1 + a_n}{2}\right).$$

In other words, to find the sum of the first n terms of an arithmetic sequence, find the average of the first and nth terms, and multiply by n.

Example 4   Finding the *n*th Partial Sum

Find the sum of the first 20 terms of the arithmetic sequence whose nth term is $4n + 1$.

Solution

The first term of this sequence is $a_1 = 4(1) + 1 = 5$ and the 20th term is $a_{20} = 4(20) + 1 = 81$. So, the sum of the first 20 terms is given by

$$\sum_{i=1}^{n} a_i = n\left(\frac{a_1 + a_n}{2}\right) \qquad \text{nth partial sum formula}$$

$$\sum_{i=1}^{20} (4i + 1) = 20\left(\frac{a_1 + a_{20}}{2}\right) \qquad \text{Substitute 20 for } n.$$

$$= 10(5 + 81) \qquad \text{Substitute 5 for } a_1 \text{ and 81 for } a_{20}.$$

$$= 10(86) \qquad \text{Simplify.}$$

$$= 860. \qquad \text{nth partial sum}$$

3 Use an arithmetic sequence to solve an application problem.

Application

Example 5   Total Sales

Your business sells $100,000 worth of products during its first year. You have a goal of increasing annual sales by $25,000 each year for 9 years. If you meet this goal, how much will you sell during your first 10 years of business?

Solution

The annual sales during the first 10 years form the following arithmetic sequence.

$100,000, $125,000, $150,000, $175,000, $200,000,
$225,000, $250,000, $275,000, $300,000, $325,000

Using the formula for the nth partial sum of an arithmetic sequence, you find the total sales during the first 10 years as follows.

$$\text{Total sales} = n\left(\frac{a_1 + a_n}{2}\right)$$ nth partial sum formula

$$= 10\left(\frac{100,000 + 325,000}{2}\right)$$ Substitute for n, a_1, and a_n.

$$= 10(212,500)$$ Simplify.

$$= \$2,125,000$$ Simplify.

From the bar graph shown in Figure 10.1, notice that the annual sales for this company follows a *linear growth* pattern. In other words, saying that a quantity increases arithmetically is the same as saying that it increases linearly.

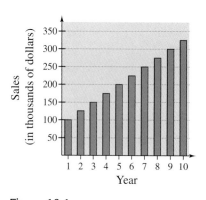

Figure 10.1

Discussing the Concept   Using Arithmetic Sequences

A magic square is a square table of positive integers in which each row, column, and diagonal adds up to the same number. One example is shown below. In addition, the values in the middle row, in the middle column, and along both diagonals form arithmetic sequences. See if you can complete the following magic squares.

6	1	8
7	5	3
2	9	4

a.

	11	14
	10	
		15

b.

8		
	9	
	13	

c.

		20
	13	
6		

10.2 Exercises

Integrated Review — Concepts, Skills, and Problem Solving

Keep mathematically in shape by doing these exercises *before* the problems of this section.

Properties and Definitions

1. In your own words, state the definition of an algebraic expression.

2. State the definition of a term of an algebraic expression.

3. Write a trinomial of degree 3.

4. Write a monomial of degree 4.

Domain

In Exercises 5–10, find the domain of the function.

5. $f(x) = x^3 - 2x$

6. $g(x) = \sqrt[3]{x}$

7. $h(x) = \sqrt{16 - x^2}$

8. $A(x) = \dfrac{3}{36 - x^2}$

9. $g(t) = \ln(t - 2)$

10. $f(s) = 630e^{-0.2s}$

Problem Solving

11. Determine the balance when $10,000 is invested at $7\frac{1}{2}\%$ compounded daily for 15 years.

12. Determine the amount after 5 years if $4000 is invested in an account earning 6% compounded monthly.

Developing Skills

In Exercises 1–10, find the common difference of the arithmetic sequence. See Example 1.

1. $2, 5, 8, 11, \ldots$

2. $-8, 0, 8, 16, \ldots$

3. $100, 94, 88, 82, \ldots$

4. $3200, 2800, 2400, 2000, \ldots$

5. $10, -2, -14, -26, -38, \ldots$

6. $4, \frac{9}{2}, 5, \frac{11}{2}, 6, \ldots$

7. $1, \frac{5}{3}, \frac{7}{3}, 3, \ldots$

8. $\frac{1}{2}, \frac{5}{4}, 2, \frac{11}{4}, \ldots$

9. $\frac{7}{2}, \frac{9}{4}, 1, -\frac{1}{4}, -\frac{3}{2}, \ldots$

10. $\frac{5}{2}, \frac{11}{6}, \frac{7}{6}, \frac{1}{2}, -\frac{1}{6}, \ldots$

In Exercises 11–26, determine whether the sequence is arithmetic. If so, find the common difference.

11. $2, 4, 6, 8, \ldots$

12. $1, 2, 4, 8, 16, \ldots$

13. $10, 8, 6, 4, 2, \ldots$

14. $2, 6, 10, 14, \ldots$

15. $32, 16, 0, -16, \ldots$

16. $32, 16, 8, 4, \ldots$

17. $3.2, 4, 4.8, 5.6, \ldots$

18. $8, 4, 2, 1, 0.5, 0.25, \ldots$

19. $2, \frac{7}{2}, 5, \frac{13}{2}, \ldots$

20. $3, \frac{5}{2}, 2, \frac{3}{2}, 1, \ldots$

21. $\frac{1}{3}, \frac{2}{3}, \frac{4}{3}, \frac{8}{3}, \frac{16}{3}, \ldots$

22. $\frac{9}{4}, 2, \frac{7}{4}, \frac{3}{2}, \frac{5}{4}, \ldots$

23. $1, \sqrt{2}, \sqrt{3}, 2, \sqrt{5}, \ldots$

24. $1, 4, 9, 16, 25, \ldots$

25. $\ln 4, \ln 8, \ln 12, \ln 16, \ldots$

26. $e, e^2, e^3, e^4, \ldots$

In Exercises 27–36, write the first five terms of the arithmetic sequence. (Begin with $n = 1$.)

27. $a_n = 3n + 4$

28. $a_n = 5n - 4$

29. $a_n = -2n + 8$

30. $a_n = -10n + 100$

31. $a_n = \frac{5}{2}n - 1$

32. $a_n = \frac{2}{3}n + 2$

33. $a_n = \frac{3}{5}n + 1$

34. $a_n = \frac{3}{4}n - 2$

35. $a_n = -\frac{1}{4}(n - 1) + 4$

36. $a_n = 4(n + 2) + 24$

In Exercises 37–54, find a formula for the nth term of the arithmetic sequence. See Example 2.

37. $a_1 = 3, \quad d = \frac{1}{2}$ **38.** $a_1 = -1, \quad d = 1.2$

39. $a_1 = 1000, \quad d = -25$

40. $a_1 = 64, \quad d = -8$

41. $a_3 = 20, \quad d = -4$ **42.** $a_1 = 12, \quad d = -3$

43. $a_1 = 3, \quad d = \frac{3}{2}$ **44.** $a_6 = 5, \quad d = \frac{3}{2}$

45. $a_1 = 5, \quad a_5 = 15$ **46.** $a_2 = 93, \quad a_6 = 65$

47. $a_3 = 16, \quad a_4 = 20$ **48.** $a_5 = 30, \quad a_4 = 25$

49. $a_1 = 50, \quad a_3 = 30$ **50.** $a_{10} = 32, \quad a_{12} = 48$

51. $a_2 = 10, \quad a_6 = 8$ **52.** $a_7 = 8, \quad a_{13} = 6$

53. $a_1 = 0.35, \quad a_2 = 0.30$

54. $a_1 = 0.08, \quad a_2 = 0.082$

In Exercises 55–62, write the first five terms of the arithmetic sequence defined recursively. See Example 3.

55. $a_1 = 25$
    $a_{k+1} = a_k + 3$

56. $a_1 = 12$
    $a_{k+1} = a_k - 6$

57. $a_1 = 9$
    $a_{k+1} = a_k - 3$

58. $a_1 = 8$
    $a_{k+1} = a_k + 7$

59. $a_1 = -10$
    $a_{k+1} = a_k + 6$

60. $a_1 = -20$
    $a_{k+1} = a_k - 4$

61. $a_1 = 100$
    $a_{k+1} = a_k - 20$

62. $a_1 = 4.2$
    $a_{k+1} = a_k + 0.4$

In Exercises 63–72, find the nth partial sum. See Example 4.

63. $\sum_{k=1}^{20} k$ **64.** $\sum_{k=1}^{30} 4k$

65. $\sum_{k=1}^{50} (k + 3)$ **66.** $\sum_{n=1}^{30} (n + 2)$

67. $\sum_{k=1}^{10} (5k - 2)$ **68.** $\sum_{k=1}^{100} (4k - 1)$

69. $\sum_{n=1}^{500} \frac{n}{2}$ **70.** $\sum_{n=1}^{600} \frac{2n}{3}$

71. $\sum_{n=1}^{30} \left(\frac{1}{3}n - 4\right)$ **72.** $\sum_{n=1}^{75} (0.3n + 5)$

In Exercises 73–84, find the nth partial sum of the arithmetic sequence.

73. $5, 12, 19, 26, 33, \ldots, \quad n = 12$

74. $2, 12, 22, 32, 42, \ldots, \quad n = 20$

75. $2, 8, 14, 20, \ldots, \quad n = 25$

76. $500, 480, 460, 440, \ldots, \quad n = 20$

77. $200, 175, 150, 125, 100, \ldots, \quad n = 8$

78. $800, 785, 770, 755, 740, \ldots, \quad n = 25$

79. $-50, -38, -26, -14, -2, \ldots, \quad n = 50$

80. $-16, -8, 0, 8, 16, \ldots, \quad n = 30$

81. $1, 4.5, 8, 11.5, 15, \ldots, \quad n = 12$

82. $2.2, 2.8, 3.4, 4.0, 4.6, \ldots, \quad n = 12$

83. $a_1 = 0.5, \quad a_4 = 1.7, \ldots, \quad n = 10$

84. $a_1 = 15, \quad a_{100} = 307, \ldots, \quad n = 100$

In Exercises 85–90, match the arithmetic sequence with its graph. [The graphs are labeled (a), (b), (c), (d), (e), and (f).]

(a) (b)

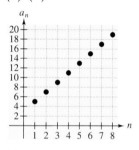

(c) (d)

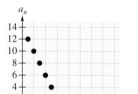 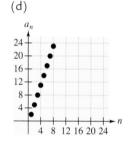

(e) (f)

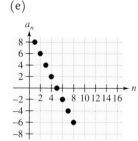 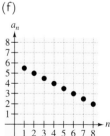

85. $a_n = \frac{1}{2}n + 1$ **86.** $a_n = -\frac{1}{2}n + 6$

87. $a_n = -2n + 10$ **88.** $a_n = 2n + 3$

89. $a_1 = 12$ **90.** $a_1 = 2$
    $a_{n+1} = a_n - 2$ $a_{n+1} = a_n + 3$

In Exercises 91–96, use a graphing utility to graph the first 10 terms of the sequence.

91. $a_n = -2n + 21$

92. $a_n = -25n + 500$

93. $a_n = \frac{3}{5}n + \frac{3}{2}$

94. $a_n = \frac{3}{2}n + 1$

95. $a_n = 2.5n - 8$

96. $a_n = 6.2n + 3$

In Exercises 97–102, use a graphing utility to find the sum.

97. $\displaystyle\sum_{j=1}^{25} (750 - 30j)$

98. $\displaystyle\sum_{n=1}^{40} (1000 - 25n)$

99. $\displaystyle\sum_{i=1}^{60} \left(300 - \frac{8}{3}i\right)$

100. $\displaystyle\sum_{n=1}^{20} \left(500 - \frac{1}{10}n\right)$

101. $\displaystyle\sum_{n=1}^{50} (2.15n + 5.4)$

102. $\displaystyle\sum_{n=1}^{60} (200 - 3.4n)$

Solving Problems

103. Find the sum of the first 75 positive integers.

104. Find the sum of the integers from 35 to 100.

105. Find the sum of the first 50 positive even integers.

106. Find the sum of the first 100 positive odd integers.

107. *Salary Increases* In your new job you are told that your starting salary will be $36,000 with an increase of $2000 at the end of each of the first 5 years. How much will you be paid through the end of your first six years of employment with the company?

108. *Would You Accept This Job?* Suppose that you receive 25 cents on the first day of the month, 50 cents on the second day, 75 cents on the third day, and so on. Determine the total amount that you will receive during a 30-day month.

109. *Ticket Prices* There are 20 rows of seats on the main floor of a concert hall: 20 seats in the first row, 21 seats in the second row, 22 seats in the third row, and so on (see figure). How much should you charge per ticket in order to obtain $15,000 for the sale of all the seats on the main floor?

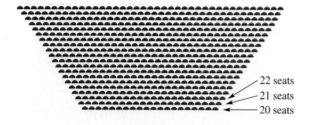

— 22 seats
— 21 seats
— 20 seats

110. *Pile of Logs* Logs are stacked in a pile as shown in the figure. The top row has 15 logs and the bottom row has 21 logs. How many logs are in the pile?

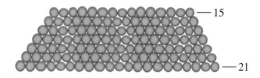

— 15

— 21

111. *Baling Hay* In the first two trips baling hay around a large field (see figure), a farmer obtains 93 bales and 89 bales, respectively. The farmer estimates that the same pattern will continue. Estimate the total number of bales made if there are another six trips around the field.

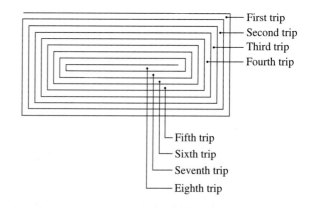

First trip
Second trip
Third trip
Fourth trip

Fifth trip
Sixth trip
Seventh trip
Eighth trip

112. *Baling Hay*   In the first two trips baling hay around a field (see figure), a farmer obtains 64 bales and 60 bales, respectively. The farmer estimates that the same pattern will continue. Estimate the total number of bales made if there are another four trips around the field.

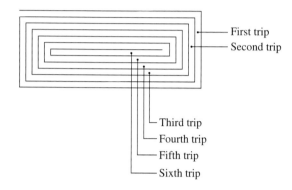

113. *Clock Chimes*   A clock chimes once at 1:00, twice at 2:00, three times at 3:00, and so on. The clock also chimes once at 15-minute intervals that are not on the hour. How many times does the clock chime in a 12-hour period?

114. *Clock Chimes*   A clock chimes once at 1:00, twice at 2:00, three times at 3:00, and so on. The clock also chimes once on the half-hour. How many times does the clock chime in a 12-hour period?

115. *Free-Falling Object*   A free-falling object will fall 16 feet during the first second, 48 more feet during the second, 80 more feet during the third, and so on. What is the total distance the object will fall in 8 seconds if this pattern continues?

116. *Free-Falling Object*   A free-falling object will fall 4.9 meters during the first second, 14.7 more meters during the second, 24.5 more meters during the third, and so on. What is the total distance the object will fall in 5 seconds if this pattern continues?

117. *Pattern Recognition*

(a) Compute the sums of positive odd integers.

$$1 + 3 = $$

$$1 + 3 + 5 = $$

$$1 + 3 + 5 + 7 = $$

$$1 + 3 + 5 + 7 + 9 = $$

$$1 + 3 + 5 + 7 + 9 + 11 = $$

(b) Use the sums in part (a) to make a conjecture about the sums of positive odd integers. Check your conjecture for the sum

$$1 + 3 + 5 + 7 + 9 + 11 + 13 = \underline{\qquad} .$$

(c) Verify your conjecture in part (b) analytically.

Explaining Concepts

118. In your own words, explain what makes a sequence arithmetic.

119. The second and third terms of an arithmetic sequence are 12 and 15, respectively. What is the first term?

120. Explain how the first two terms of an arithmetic sequence can be used to find the nth term.

121. Explain what is meant by a recursion formula.

122. Explain what is meant by the nth partial sum of a sequence.

123. Explain how to find the sum of the integers from 100 to 200.

124. Each term of an arithmetic sequence is multiplied by a constant C. Is the resulting sequence arithmetic? If so, how does the common difference compare with the common difference of the original sequence?

10.3 Geometric Sequences and Series

Objectives

1 Find the common ratio and the nth term of a geometric sequence.

2 Find the nth partial sum of a geometric sequence.

3 Find the sum of an infinite geometric series.

4 Use a geometric sequence to solve an application problem.

1 Find the common ratio and the nth term of a geometric sequence.

Geometric Sequences

In Section 10.2, you studied sequences whose consecutive terms have a common *difference*. In this section, you will study sequences whose consecutive terms have a common *ratio*.

> ▶ **Definition of a Geometric Sequence**
>
> A sequence is called **geometric** if the ratios of consecutive terms are the same. So, the sequence $a_1, a_2, a_3, a_4, \ldots, a_n, \ldots$ is geometric if there is a number r, $r \neq 0$, such that
>
> $$\frac{a_2}{a_1} = r, \quad \frac{a_3}{a_2} = r, \quad \frac{a_4}{a_3} = r$$
>
> and so on. The number r is the **common ratio** of the sequence.

Example 1 Examples of Geometric Sequences

a. The sequence whose nth term is 2^n is geometric. For this sequence, the common ratio between consecutive terms is 2.

$$2, 4, 8, 16, \ldots, 2^n, \ldots$$
$$\underbrace{}$$
$$\tfrac{4}{2} = 2$$

b. The sequence whose nth term is $4(3^n)$ is geometric. For this sequence, the common ratio between consecutive terms is 3.

$$12, 36, 108, 324, \ldots, 4(3^n), \ldots$$
$$\underbrace{}$$
$$\tfrac{36}{12} = 3$$

c. The sequence whose nth term is $\left(-\tfrac{1}{3}\right)^n$ is geometric. For this sequence, the common ratio between consecutive terms is $-\tfrac{1}{3}$.

$$-\frac{1}{3}, \frac{1}{9}, -\frac{1}{27}, \frac{1}{81}, \ldots, \left(-\frac{1}{3}\right)^n, \ldots$$
$$\underbrace{\phantom{-\frac{1}{3}, \frac{1}{9}}}$$
$$\frac{1/9}{-1/3} = -\frac{1}{3}$$

Study Tip

If you know the nth term of a geometric sequence, the $(n + 1)$th term can be found by multiplying by r. That is,

$$a_{n+1} = ra_n.$$

> ▶ **The nth Term of a Geometric Sequence**
>
> The nth term of a geometric sequence has the form
>
> $$a_n = a_1 r^{n-1}$$
>
> where r is the common ratio of consecutive terms of the sequence. So, every geometric sequence can be written in the following form.
>
> $$a_1, a_1 r, a_1 r^2, a_1 r^3, a_1 r^4, \ldots, a_1 r^{n-1}, \ldots$$

Example 2   Finding the nth Term of a Geometric Sequence

Find a formula for the nth term of the geometric sequence whose common ratio is 3 and whose first term is 1. What is the eighth term of this sequence?

Solution

The formula for the nth term is of the form $a_1 r^{n-1}$. Moreover, because the common ratio is $r = 3$, and the first term is $a_1 = 1$, the formula must have the form

$$a_n = a_1 r^{n-1} \qquad \text{Formula for geometric sequence}$$

$$= (1)(3)^{n-1} \qquad \text{Substitute 1 for } a_1 \text{ and 3 for } r.$$

$$= 3^{n-1}. \qquad \text{Simplify.}$$

The sequence therefore has the following form.

$$1, 3, 9, 27, 81, \ldots, 3^{n-1}, \ldots$$

The eighth term of the sequence is $a_8 = 3^{8-1} = 3^7 = 2187$.

Example 3   Finding the nth Term of a Geometric Sequence

Find a formula for the nth term of the geometric sequence whose first two terms are 4 and 2.

Solution

Because $a_1 = 4$ and $a_2 = 2$, the common ratio is

$$\frac{a_2}{a_1} = \frac{2}{4} = \frac{1}{2} \qquad \text{So, } r = \tfrac{1}{2}.$$

the formula for the nth term must be

$$a_n = a_1 r^{n-1} \qquad \text{Formula for geometric sequence}$$

$$= 4\left(\frac{1}{2}\right)^{n-1}. \qquad \text{Substitute 4 for } a_1 \text{ and } \tfrac{1}{2} \text{ for } r.$$

The sequence therefore has the following form.

$$4, 2, 1, \frac{1}{2}, \frac{1}{4}, \ldots, 4\left(\frac{1}{2}\right)^{n-1}, \ldots$$

2 Find the *n*th partial sum of a geometric sequence.

The Partial Sum of a Geometric Sequence

▶ **The *n*th Partial Sum of a Geometric Sequence**

The *n*th partial sum of the geometric sequence whose *n*th term is $a_n = a_1 r^{n-1}$ is given by

$$\sum_{i=1}^{n} a_1 r^{i-1} = a_1 + a_1 r + a_1 r^2 + a_1 r^3 + \cdots + a_1 r^{n-1} = a_1\left(\frac{r^n - 1}{r - 1}\right).$$

Joseph Fourier

(1768–1830)

Some of the early work in representing functions by series was done by the French mathematician Joseph Fourier. Fourier's work is important in the history of calculus, partly because it forced 18th-century mathematicians to question the then-prevailing narrow concept of a function. Both Cauchy and Dirichlet were motivated by Fourier's work in series, and in 1837 Dirichlet published the general defini-tion of a function that is used today.

> **Example 4**   Finding the *n*th Partial Sum

Find the sum $1 + 2 + 4 + 8 + 16 + 32 + 64 + 128$.

Solution

This is a geometric sequence whose common ratio is $r = 2$. Because the first term of the sequence is $a_1 = 1$, it follows that the sum is

$$\sum_{i=1}^{8} 2^{i-1} = (1)\left(\frac{2^8 - 1}{2 - 1}\right) = \frac{256 - 1}{2 - 1} = 255. \qquad \text{Substitute 1 for } a_1 \text{ and 2 for } r.$$

> **Example 5**   Finding the *n*th Partial Sum

Find the sum of the first five terms of the geometric sequence whose *n*th term is $a_n = \left(\frac{2}{3}\right)^n$.

Solution

$$\sum_{i=1}^{5} \left(\frac{2}{3}\right)^i = \frac{2}{3}\left[\frac{(2/3)^5 - 1}{(2/3) - 1}\right] \qquad \text{Substitute } \tfrac{2}{3} \text{ for } a_1 \text{ and } \tfrac{2}{3} \text{ for } r.$$

$$= \frac{2}{3}\left[\frac{(32/243) - 1}{-1/3}\right] \qquad \text{Simplify.}$$

$$= \frac{2}{3}\left(-\frac{211}{243}\right)(-3) \qquad \text{Simplify.}$$

$$= \frac{422}{243} \qquad \text{Simplify.}$$

$$\approx 1.737$$

3 Find the sum of an infinite geometric series.

Geometric Series

Suppose that in Example 5, you were to find the sum of all the terms of the *infinite* geometric sequence

$$\frac{2}{3}, \frac{4}{9}, \frac{8}{27}, \frac{16}{81}, \cdots, \left(\frac{2}{3}\right)^n, \cdots.$$

A summation of all the terms of an infinite geometric sequence is called an **infinite geometric series,** or simply a **geometric series.**

Technology: Discovery

Evaluate $\left(\frac{1}{2}\right)^n$ for $n = 1, 10, 100,$ and 1000. What happens to the value of $\left(\frac{1}{2}\right)^n$ as n increases? Make a conjecture about the value of $\left(\frac{1}{2}\right)^n$ as n approaches infinity.

In your mind, would this sum be infinitely large or would it be a finite number? Consider the formula for the nth partial sum of a geometric sequence.

$$S_n = a_1\left(\frac{r^n - 1}{r - 1}\right) = a_1\left(\frac{1 - r^n}{1 - r}\right)$$

Suppose that $|r| < 1$ and you let n become larger and larger. It follows that r^n gets closer and closer to 0, so that the term r^n drops out of the formula above. You then get the sum

$$S = a_1\left(\frac{1}{1 - r}\right) = \frac{a_1}{1 - r}.$$

Notice that this sum is not dependent on the nth term of the sequence. In the case of Example 5, $r = \left(\frac{2}{3}\right) < 1$, and so the sum of the infinite geometric sequence is

$$S = \sum_{i=1}^{\infty} \left(\frac{2}{3}\right)^i = \frac{a_1}{1 - r} = \frac{2/3}{1 - (2/3)} = \frac{2/3}{1/3} = 2.$$

▶ **Sum of an Infinite Geometric Series**

If $a_1, a_1r, a_1r^2, \ldots, a_1r^n, \ldots$ is an infinite geometric sequence, then for $|r| < 1$, the sum of the terms is

$$S = \sum_{i=0}^{\infty} a_1 r^i = \frac{a_1}{1 - r}.$$

Example 6   Evaluating a Geometric Series

Find the value of each sum.

a. $\displaystyle\sum_{i=1}^{\infty} 5\left(\frac{3}{4}\right)^{i-1}$ **b.** $\displaystyle\sum_{n=0}^{\infty} 4\left(\frac{3}{10}\right)^n$ **c.** $\displaystyle\sum_{i=0}^{\infty} \left(-\frac{3}{5}\right)^i$

Solution

a. The series is geometric with $a_1 = 5\left(\frac{3}{4}\right)^{1-1} = 5$ and $r = \frac{3}{4}$. So,

$$\sum_{i=1}^{\infty} 5\left(\frac{3}{4}\right)^{i-1} = \frac{5}{1 - (3/4)}$$

$$= \frac{5}{1/4} = 20.$$

b. The series is geometric with $a_1 = 4\left(\frac{3}{10}\right)^0 = 4$ and $r = \frac{3}{10}$. So,

$$\sum_{n=0}^{\infty} 4\left(\frac{3}{10}\right)^n = \frac{4}{1 - (3/10)}$$

$$= \frac{4}{7/10} = \frac{40}{7}.$$

c. The series is geometric with $a_1 = \left(-\frac{3}{5}\right)^0 = 1$ and $r = \frac{3}{5}$. So,

$$\sum_{i=0}^{\infty} \left(-\frac{3}{5}\right)^i = \frac{1}{1 - (-3/5)} = \frac{1}{1 + (3/5)} = \frac{5}{8}.$$

4 Use a geometric sequence to solve an application problem.

Applications

| Example 7 | A Lifetime Salary

You have accepted a job that pays a salary of $28,000 the first year. During the next 39 years, suppose you receive a 6% raise each year. What will your total salary be over the 40-year period?

Solution

Using a geometric sequence, your salary during the first year will be

$$a_1 = 28,000.$$

Then, with a 6% raise, your salary during the next 2 years will be as follows.

$$a_2 = 28,000 + 28,000(0.06)$$

$$= 28,000(1.06)^1$$

$$a_3 = 28,000(1.06) + 28,000(1.06)(0.06)$$

$$= 28,000(1.06)^2$$

From this pattern, you can see that the common ratio of the geometric sequence is $r = 1.06$. Using the formula for the nth partial sum of a geometric sequence, you will find that the total salary over the 40-year period is given by

$$\text{Total salary} = \sum_{i=1}^{n} a_1 r^{i-1}$$

$$= a_1 \left(\frac{r_n - 1}{r - 1} \right)$$

$$= 28,000 \left[\frac{(1.06)^{40} - 1}{1.06 - 1} \right]$$

$$= 28,000 \left[\frac{(1.06)^{40} - 1}{0.06} \right]$$

$$\approx \$4,333,335.$$

The bar graph in Figure 10.2 illustrates your salary during the 40-year period.

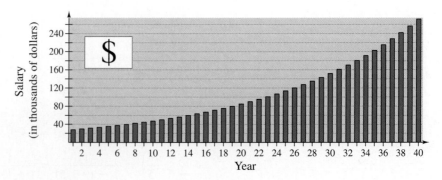

Figure 10.2

Example 8   Increasing Annuity

You deposit $100 in an account each month for 2 years. The account pays an annual interest rate of 9%, compounded monthly. What is your balance at the end of 2 years? (This type of savings plan is called an **increasing annuity.**)

Solution

The first deposit would earn interest for the full 24 months, the second deposit would earn interest for 23 months, the third deposit would earn interest for 22 months, and so on. Using the formula for compound interest,

$$a_n = P\left(1 + \frac{r}{12}\right)^n \qquad \text{\small } n \text{ \small is months}$$

$$= 100\left(1 + \frac{0.09}{12}\right)^n$$

$$= 100(1 + 0.0075)^n$$

you can see that the total of the 24 deposits would be

$$\text{Total} = a_1 + a_2 + \cdots + a_{24}$$

$$= 100(1.0075)^1 + 100(1.0075)^2 + \cdots + 100(1.0075)^{24}$$

$$= 100(1.0075)\left(\frac{1.0075^{24} - 1}{1.0075 - 1}\right) \qquad a_1\left(\frac{r^n - 1}{r - 1}\right)$$

$$= \$2638.49.$$

Discussing the Concept   Annual Revenue

The two bar graphs below show the annual revenues for two companies. One company's revenue grew at an arithmetic rate, whereas the other grew at a geometric rate. Which company had the greatest revenue during the 10-year period? Which company would you rather own? Explain.

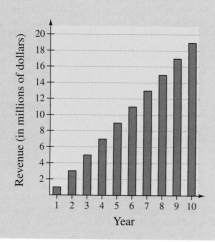

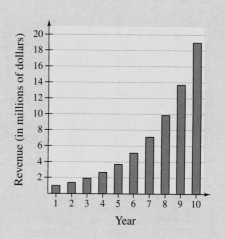

10.3 Exercises

Integrated Review   Concepts, Skills, and Problem Solving

Keep mathematically in shape by doing these exercises *before* the problems of this section.

Properties and Definitions

1. Relative to the x- and y-axes, explain the meaning of each coordinate of the point $(-6, 4)$.

2. A point lies 5 units from the x-axis and 10 units from the y-axis. Give the ordered pair for such a point in each quadrant.

3. In your own words, define the graph of the function $y = f(x)$.

4. Describe the procedure for finding the x- and y-intercepts of the graph of $f(x) = 2\sqrt{x} + 4$.

Solving Inequalities

In Exercises 5–10, solve the inequality.

5. $3x - 5 > 0$

6. $\frac{3}{2}y + 11 < 20$

7. $100 < 2x + 30 < 150$

8. $-5 < -\dfrac{x}{6} < 2$

9. $2x^2 - 7x + 5 > 0$

10. $2x - \dfrac{5}{x} > 3$

Problem Solving

11. A television set is advertised as having a 19-inch screen. Determine the dimensions of the square screen if its diagonal is 19 inches.

12. A construction worker is building the forms for the rectangular foundation of a home that is 25 feet wide and 40 feet long. To make sure that the corners are square the worker measures the diagonal of the foundation. What should that measurement be?

Developing Skills

In Exercises 1–12, find the common ratio of the geometric sequence. See Example 1.

1. $2, 6, 18, 54, \ldots$

2. $5, -10, 20, -40, \ldots$

3. $1, -3, 9, -27, \ldots$

4. $54, 18, 6, 2, \ldots$

5. $12, -6, 3, -\frac{3}{2}, \ldots$

6. $9, 6, 4, \frac{8}{3}, \ldots$

7. $1, -\frac{3}{2}, \frac{9}{4}, -\frac{27}{8}, \ldots$

8. $5, -\frac{5}{2}, \frac{5}{4}, -\frac{5}{8}, \ldots$

9. $1, \pi, \pi^2, \pi^3, \ldots$

10. $e, e^2, e^3, e^4, \ldots$

11. $500(1.06), 500(1.06)^2, 500(1.06)^3, 500(1.06)^4, \ldots$

12. $1.1, (1.1)^2, (1.1)^3, (1.1)^4, \ldots$

In Exercises 13–24, determine whether the sequence is geometric. If so, find the common ratio.

13. $64, 32, 16, 8, \ldots$

14. $64, 32, 0, -32, \ldots$

15. $10, 15, 20, 25, \ldots$

16. $10, 20, 40, 80, \ldots$

17. $5, 10, 20, 40, \ldots$

18. $54, -18, 6, -2, \ldots$

19. $1, 8, 27, 64, 125, \ldots$

20. $12, 7, 2, -3, -8, \ldots$

21. $1, -\frac{2}{3}, \frac{4}{9}, -\frac{8}{27}, \ldots$

22. $\frac{1}{3}, -\frac{2}{3}, \frac{4}{3}, -\frac{8}{3}, \ldots$

23. $10(1 + 0.02), 10(1 + 0.02)^2, 10(1 + 0.02)^3, \ldots$

24. $1, 0.2, 0.04, 0.008, \ldots$

In Exercises 25–38, write the first five terms of the geometric sequence.

25. $a_1 = 4, \quad r = 2$

26. $a_1 = 3, \quad r = 4$

27. $a_1 = 6, \quad r = \frac{1}{3}$

28. $a_1 = 4, \quad r = \frac{1}{2}$

29. $a_1 = 1, \quad r = -\frac{1}{2}$

30. $a_1 = 32, \quad r = -\frac{3}{4}$

31. $a_1 = 4, \quad r = -\frac{1}{2}$

32. $a_1 = 4, \quad r = \frac{3}{2}$

33. $a_1 = 1000, \quad r = 1.01$

34. $a_1 = 200, \quad r = 1.07$

35. $a_1 = 4000, \quad r = 1/1.01$

36. $a_1 = 1000, \quad r = 1/1.05$

37. $a_1 = 10, \quad r = \frac{3}{5}$

38. $a_1 = 36, \quad r = \frac{2}{3}$

In Exercises 39–52, find the specified nth term of the geometric sequence.

39. $a_1 = 6, \quad r = \frac{1}{2}, \quad a_{10} = $

40. $a_1 = 8, \quad r = \frac{3}{4}, \quad a_8 = $

41. $a_1 = 3, \quad r = \sqrt{2}, \quad a_{10} = $

42. $a_1 = 5, \quad r = \sqrt{3}, \quad a_9 = $

43. $a_1 = 200, \quad r = 1.2, \quad a_{12} = $

44. $a_1 = 500, \quad r = 1.06, \quad a_{40} = $

45. $a_1 = 120$, $r = -\frac{1}{3}$, $a_{10} =$

46. $a_1 = 240$, $r = -\frac{1}{4}$, $a_{13} =$

47. $a_1 = 4$, $a_2 = 3$, $a_5 =$

48. $a_1 = 1$, $a_2 = 9$, $a_7 =$

49. $a_1 = 1$, $a_3 = \frac{9}{4}$, $a_6 =$

50. $a_3 = 6$, $a_5 = \frac{8}{3}$, $a_6 =$

51. $a_2 = 12$, $a_3 = 16$, $a_4 =$

52. $a_4 = 100$, $a_5 = -25$, $a_7 =$

In Exercises 53–66, find the formula for the nth term of the geometric sequence. (Begin with $n = 1$.) See Examples 2 and 3.

53. $a_1 = 2$, $r = 3$

54. $a_1 = 5$, $r = 4$

55. $a_1 = 1$, $r = 2$

56. $a_1 = 25$, $r = 4$

57. $a_1 = 1$, $r = -\frac{1}{5}$

58. $a_1 = 12$, $r = -\frac{4}{3}$

59. $a_1 = 4$, $r = -\frac{1}{2}$

60. $a_1 = 9$, $r = \frac{2}{3}$

61. $a_1 = 8$, $a_2 = 2$

62. $a_1 = 18$, $a_2 = 8$

63. $a_1 = 14$, $a_2 = \frac{21}{2}$

64. $a_1 = 36$, $a_2 = \frac{27}{2}$

65. $4, -6, 9, -\frac{27}{2}, \ldots$

66. $1, \frac{3}{2}, \frac{9}{4}, \frac{27}{8}, \ldots$

In Exercises 67–70, match the geometric sequence with its graph. [The graphs are labeled (a), (b), (c), and (d).]

(a)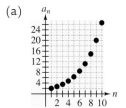

(b)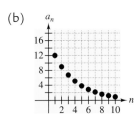

(c)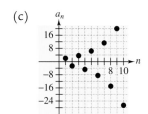

(d)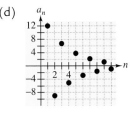

67. $a_n = 12\left(\frac{3}{4}\right)^{n-1}$

68. $a_n = 12\left(-\frac{3}{4}\right)^{n-1}$

69. $a_n = 2\left(\frac{4}{3}\right)^{n-1}$

70. $a_n = 2\left(-\frac{4}{3}\right)^{n-1}$

In Exercises 71–80, find the nth partial sum. See Examples 4 and 5.

71. $\displaystyle\sum_{i=1}^{10} 2^{i-1}$

72. $\displaystyle\sum_{i=1}^{6} 3^{i-1}$

73. $\displaystyle\sum_{i=1}^{12} 3\left(\frac{3}{2}\right)^{i-1}$

74. $\displaystyle\sum_{i=1}^{20} 12\left(\frac{2}{3}\right)^{i-1}$

75. $\displaystyle\sum_{i=1}^{15} 3\left(-\frac{1}{3}\right)^{i-1}$

76. $\displaystyle\sum_{i=1}^{8} 8\left(-\frac{1}{4}\right)^{i-1}$

77. $\displaystyle\sum_{i=1}^{12} 4(-2)^{i-1}$

78. $\displaystyle\sum_{i=1}^{20} 16\left(\frac{1}{2}\right)^{i-1}$

79. $\displaystyle\sum_{i=1}^{8} 6(0.1)^{i-1}$

80. $\displaystyle\sum_{i=1}^{24} 1000(1.06)^{i-1}$

In Exercises 81–92, find the nth partial sum of the geometric sequence.

81. $1, -3, 9, -27, 81, \ldots, n = 10$

82. $3, -6, 12, -24, 48, \ldots, n = 12$

83. $8, 4, 2, 1, \frac{1}{2}, \ldots, n = 15$

84. $9, 6, 4, \frac{8}{3}, \frac{16}{9}, \ldots, n = 10$

85. $4, 12, 36, 108, \ldots, n = 8$

86. $\frac{1}{36}, -\frac{1}{12}, \frac{1}{4}, -\frac{3}{4}, \ldots, n = 20$

87. $60, -15, \frac{15}{4}, -\frac{15}{16}, \ldots, n = 12$

88. $40, -10, \frac{5}{2}, -\frac{5}{8}, \frac{5}{32}, \ldots, n = 10$

89. $30, 30(1.06), 30(1.06)^2, 30(1.06)^3, \ldots, n = 20$

90. $100, 100(1.08), 100(1.08)^2, 100(1.08)^3, \ldots, n = 40$

91. $500, 500(1.04), 500(1.04)^2, 500(1.04)^3, \ldots, n = 18$

92. $1, \sqrt{2}, 2, 2\sqrt{2}, 4, \ldots, n = 12$

In Exercises 93–100, find the sum. See Example 6.

93. $\displaystyle\sum_{n=0}^{\infty} \left(\frac{1}{2}\right)^n$

94. $\displaystyle\sum_{n=0}^{\infty} 2\left(\frac{2}{3}\right)^n$

95. $\displaystyle\sum_{n=0}^{\infty} \left(-\frac{1}{2}\right)^n$

96. $\displaystyle\sum_{n=0}^{\infty} \left(\frac{1}{10}\right)^n$

97. $\displaystyle\sum_{n=0}^{\infty} 2\left(-\frac{2}{3}\right)^n$

98. $\displaystyle\sum_{n=0}^{\infty} 4\left(\frac{1}{4}\right)^n$

99. $8 + 6 + \frac{9}{2} + \frac{27}{8} + \cdots$

100. $3 - 1 + \frac{1}{3} - \frac{1}{9} + \cdots$

In Exercises 101–104, use a graphing utility to graph the first 10 terms of the sequence.

101. $a_n = 20(-0.6)^{n-1}$

102. $a_n = 4(1.4)^{n-1}$

103. $a_n = 15(0.6)^{n-1}$

104. $a_n = 8(-0.6)^{n-1}$

Solving Problems

105. *Depreciation* A company buys a machine for $250,000. During the next 5 years, the machine depreciates at the rate of 25% per year. (That is, at the end of each year, the depreciated value is 75% of what it was at the beginning of the year.)

(a) Find a formula for the nth term of the geometric sequence that gives the value of the machine n full years after it was purchased.

(b) Find the depreciated value of the machine at the end of 5 full years.

(c) During which year did the machine depreciate the most?

106. *Population Increase* A city of 500,000 people is growing at the rate of 1% per year. (That is, at the end of each year, the population is 1.01 times the population at the beginning of the year.)

(a) Find a formula for the nth term of the geometric sequence that gives the population n years from now.

(b) Estimate the population 20 years from now.

107. *Salary Increases* You accept a job that pays a salary of $30,000 the first year. During the next 39 years, you receive a 5% raise each year. What would your total salary be over the 40-year period?

108. *Salary Increases* You accept a job that pays a salary of $30,000 the first year. During the next 39 years, you receive a 5.5% raise each year.

(a) What would your total salary be over the 40-year period?

(b) How much more income did the extra 0.5% provide than the result in Exercise 107?

Increasing Annuity In Exercises 109–114, find the balance A in an increasing annuity in which a principal of P dollars is invested each month for t years, compounded monthly at rate r.

109. $P = \$100$ $t = 10$ years $r = 9\%$

110. $P = \$50$ $t = 5$ years $r = 7\%$

111. $P = \$30$ $t = 40$ years $r = 8\%$

112. $P = \$200$ $t = 30$ years $r = 10\%$

113. $P = \$75$ $t = 30$ years $r = 6\%$

114. $P = \$100$ $t = 25$ years $r = 8\%$

115. *Would You Accept This Job?* You start work at a company that pays $0.01 for the first day, $0.02 for the second day, $0.04 for the third day, and so on. If the daily wage keeps doubling, what would your total income be for working (a) 29 days and (b) 30 days?

116. *Would You Accept This Job?* You start work at a company that pays $0.01 for the first day, $0.03 for the second day, $0.09 for the third day, and so on. If the daily wage keeps tripling, what would your total income be for working (a) 25 days and (b) 26 days?

117. *Power Supply* The electrical power for an implanted medical device decreases by 0.1% each day.

(a) Find a formula for the nth term of the geometric sequence that gives the percent of the initial power n days after the device is implanted.

(b) What percent of the initial power is still available 1 year after the device is implanted?

(c) The power supply needs to be changed when half the power is depleted. Use a graphing utility to graph the first 750 terms of the sequence and estimate when the power source should be changed.

118. *Cooling* The temperature of water in an ice cube tray is 70°F when it is placed in a freezer. Its temperature n hours after being placed in the freezer is 20% less than 1 hour earlier.

(a) Find a formula for the nth term of the geometric sequence that gives the temperature of the water n hours after being placed in the freezer.

(b) Find the temperature of the water 6 hours after it is placed in the freezer.

(c) Use a graphing utility to estimate the time when the water freezes. Explain your reasoning.

119. *Area* A square has 12-inch sides. A new square is formed by connecting the midpoints of the sides of the square. Then two of the triangles are shaded (see figure). This process is repeated five more times. What is the total area of the shaded region?

120. *Area* A square has 12-inch sides. The square is divided into nine smaller squares and the center square is shaded (see figure). Each of the eight unshaded squares is then divided into nine smaller squares and each center square is shaded. This process is repeated four more times. What is the total area of the shaded region?

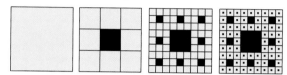

121. *Bungee Jumping* A bungee jumper jumps from a bridge and stretches a cord 100 feet. Successive bounces stretch the cord 75% of each previous length (see figure). Find the total distance traveled by the bungee jumper during 10 bounces.

$$100 + 2(100)(0.75) + \cdots + 2(100)(0.75)^{10}$$

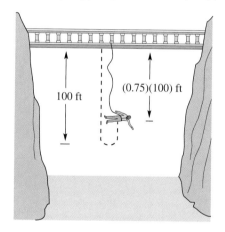

100 ft

(0.75)(100) ft

122. *Distance* A ball is dropped from a height of 16 feet. Each time it drops h feet, it rebounds 0.81*h* feet.

(a) Find the total distance traveled by the ball.

(b) The ball takes the following time for each fall.

$$s_1 = -16t^2 + 16, \qquad s_1 = 0 \text{ if } t = 1$$
$$s_2 = -16t^2 + 16(0.81), \qquad s_2 = 0 \text{ if } t = 0.9$$
$$s_3 = -16t^2 + 16(0.81)^2, \quad s_3 = 0 \text{ if } t = (0.9)^2$$
$$s_4 = -16t^2 + 16(0.81)^3, \quad s_4 = 0 \text{ if } t = (0.9)^3$$
$$\vdots \qquad\qquad\qquad \vdots$$
$$s_n = -16t^2 + 16(0.81)^{n-1}, s_n = 0 \text{ if } t = (0.9)^{n-1}$$

Beginning with s_2, the ball takes the same amount of time to bounce up as it does to fall, and thus the total time elapsed before it comes to rest is

$$t = 1 + 2\sum_{n=1}^{\infty} (0.9)^n.$$

Find this total.

Explaining Concepts

123. Answer parts (a)–(c) of Motivating the Chapter on page 623.

124. In your own words, explain what makes a sequence geometric.

125. What is the general formula for the *n*th term of a geometric sequence?

126. The second and third terms of a geometric sequence are 6 and 3, respectively. What is the first term?

127. Give an example of a geometric sequence whose terms alternate in sign.

128. Explain why the terms of a geometric sequence decrease when $a_1 > 0$ and $0 < r < 1$.

129. In your own words, describe an increasing annuity.

130. Explain what is meant by the nth partial sum of a sequence.

Mid-Chapter Quiz

Take this quiz as you would take a quiz in class. After you are done, check your work against the answers given in the back of the book.

In Exercises 1 and 2, write the first five terms of the sequence.

1. $a_n = 32\left(\dfrac{1}{4}\right)^{n-1}$ (Begin with $n = 1$.)   **2.** $a_n = \dfrac{(-3)^n n}{n + 4}$ (Begin with $n = 1$.)

In Exercises 3–6, find the sum.

3. $\displaystyle\sum_{k=1}^{4} 10k$ **4.** $\displaystyle\sum_{i=1}^{10} 4$ **5.** $\displaystyle\sum_{j=1}^{5} \dfrac{60}{j + 1}$ **6.** $\displaystyle\sum_{n=1}^{8} 8\left(-\dfrac{1}{2}\right)$

In Exercises 7 and 8, write the sum using sigma notation.

7. $\dfrac{2}{3(1)} + \dfrac{2}{3(2)} + \dfrac{2}{3(3)} + \cdots + \dfrac{2}{3(20)}$ **8.** $\dfrac{1}{1^3} - \dfrac{1}{2^3} + \dfrac{1}{3^3} - \cdots + \dfrac{1}{25^3}$

In Exercises 9 and 10, find the common difference of the arithmetic sequence.

9. $1, \frac{3}{2}, 2, \frac{5}{2}, 3, \ldots$ **10.** $100, 94, 88, 82, 76, \ldots$

In Exercises 11 and 12, find the common ratio of the geometric sequence.

11. $2, 6, 18, 54, 162, \ldots$ **12.** $2, 1, \frac{1}{2}, \frac{1}{4}, \frac{1}{8}, \ldots$

In Exercises 13 and 14, find a formula for a_n.

13. Arithmetic, $a_1 = 20$, $a_4 = 11$ **14.** Geometric, $a_1 = 32$, $r = -\dfrac{1}{4}$

In Exercises 15–20, find the sum.

15. $\displaystyle\sum_{n=1}^{50} (3n + 5)$ **16.** $\displaystyle\sum_{n=1}^{300} \dfrac{n}{5}$ **17.** $\displaystyle\sum_{i=1}^{8} 9\left(\dfrac{2}{3}\right)^{i-1}$

18. $\displaystyle\sum_{j=1}^{20} 500(1.06)^{j-1}$ **19.** $\displaystyle\sum_{i=0}^{\infty} 3\left(\dfrac{2}{3}\right)^i$ **20.** $\displaystyle\sum_{i=0}^{\infty} \dfrac{4}{5}\left(\dfrac{1}{4}\right)^i$

21. Find the 12th term of $625, -250, 100, -40, \ldots$.

22. Match $a_n = 10\left(\frac{1}{2}\right)^{n-1}$ and $b_n = 10\left(-\frac{1}{2}\right)^{n-1}$ with the graphs at the left.

23. The temperature of a coolant decreases by 25.75°F the first hour. For each subsequent hour, the temperature decreases by 2.25°F less than it decreased the previous hour. How much does the temperature decrease during the 10th hour?

24. The sequence given by $a_n = 2^{n-1}$ is geometric. Describe the sequence given by $b_n = \ln a_n$.

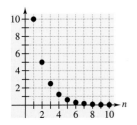

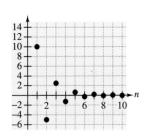

Figure for 22

10.4   The Binomial Theorem

Objectives

1 Determine the coefficients of a binomial raised to a power.

2 Arrange the binomial coefficients in a triangular pattern known as Pascal's Triangle.

3 Expand a binomial raised to a power using Pascal's Triangle and the Binomial Theorem.

1 Determine the coefficients of a binomial raised to a power.

Binomial Coefficients

Recall that a **binomial** is a polynomial that has two terms. In this section, you will study a formula that provides a quick method of raising a binomial to a power. To begin, let's look at the expansion of $(x + y)^n$ for several values of n.

$$(x + y)^0 = 1$$

$$(x + y)^1 = x + y$$

$$(x + y)^2 = x^2 + 2xy + y^2$$

$$(x + y)^3 = x^3 + 3x^2y + 3xy^2 + y^3$$

$$(x + y)^4 = x^4 + 4x^3y + 6x^2y^2 + 4xy^3 + y^4$$

$$(x + y)^5 = x^5 + 5x^4y + 10x^3y^2 + 10x^2y^3 + 5xy^4 + y^5$$

There are several observations you can make about these expansions.

1. In each expansion, there are $n + 1$ terms.

2. In each expansion, x and y have symmetrical roles. The powers of x decrease by 1 in successive terms, whereas the powers of y increase by 1.

3. The sum of the powers of each term is n. For instance, in the expansion of $(x + y)^5$, the sum of the powers of each term is 5.

$$4 + 1 = 5 \quad 3 + 2 = 5$$

$$(x + y)^5 = x^5 + 5x^4y^1 + 10x^3y^2 + 10x^2y^3 + 5xy^4 + y^5$$

4. The coefficients increase and then decrease in a symmetric pattern.

The coefficients of a binomial expansion are called **binomial coefficients.** To find them, you can use the following theorem.

▶ **The Binomial Theorem**

In the expansion of $(x + y)^n$

$$(x + y)^n = x^n + nx^{n-1}y + \cdots + {}_nC_r x^{n-r}y^r + \cdots + nxy^{n-1} + y^n$$

the coefficient of $x^{n-r}y^r$ is given by

$${}_nC_r = \frac{n!}{(n - r)!r!}.$$

| Example 1 | Finding Binomial Coefficients |

Find each binomial coefficient.

a. $_8C_2$ **b.** $_{10}C_3$ **c.** $_7C_0$ **d.** $_8C_8$ **e.** $_9C_6$

Solution

a. $_8C_2 = \dfrac{8!}{6! \cdot 2!} = \dfrac{(8 \cdot 7) \cdot 6!}{6! \cdot 2!} = \dfrac{8 \cdot 7}{2 \cdot 1} = 28$

b. $_{10}C_3 = \dfrac{10!}{7! \cdot 3!} = \dfrac{(10 \cdot 9 \cdot 8) \cdot 7!}{7! \cdot 3!} = \dfrac{10 \cdot 9 \cdot 8}{3 \cdot 2 \cdot 1} = 120$

c. $_7C_0 = \dfrac{7!}{7! \cdot 0!} = 1$

d. $_8C_8 = \dfrac{8!}{0! \cdot 8!} = 1$

e. $_9C_6 = \dfrac{9!}{3! \cdot 6!} = \dfrac{(9 \cdot 8 \cdot 7) \cdot 6!}{3! \cdot 6!} = \dfrac{9 \cdot 8 \cdot 7}{3 \cdot 2 \cdot 1} = 84$

Technology: Tip

As you will learn later in this text, the formula for the binomial coefficient is the same as the formula for combinations in the study of probability. Most graphing utilities have a function called *combination* that can be used to evaluate a binomial coefficient. Consult the user's guide for your graphing utility.

When $r \neq 0$ and $r \neq n$, as in parts (a) and (b) above, there is a simple pattern for evaluating binomial coefficients.

$$_8C_2 = \overset{\overset{\text{2 factors}}{\frown}}{\underset{\underset{\text{2 factorial}}{\smile}}{\dfrac{8 \cdot 7}{2 \cdot 1}}} \qquad \text{and} \qquad _{10}C_3 = \overset{\overset{\text{3 factors}}{\frown}}{\underset{\underset{\text{3 factorial}}{\smile}}{\dfrac{10 \cdot 9 \cdot 8}{3 \cdot 2 \cdot 1}}}$$

| Example 2 | Finding Binomial Coefficients |

Find each binomial coefficient.

a. $_7C_3$ **b.** $_7C_4$ **c.** $_{12}C_1$ **d.** $_{12}C_{11}$

Solution

a. $_7C_3 = \dfrac{7 \cdot 6 \cdot 5}{3 \cdot 2 \cdot 1} = 35$

b. $_7C_4 = \dfrac{7 \cdot 6 \cdot 5 \cdot 4}{4 \cdot 3 \cdot 2 \cdot 1} = 35$

c. $_{12}C_1 = \dfrac{12!}{11!1!} = \dfrac{(12) \cdot 11!}{11! \cdot 1!} = \dfrac{12}{1} = 12$

d. $_{12}C_{11} = \dfrac{12!}{1! \cdot 11!} = \dfrac{(12) \cdot 11!}{1! \cdot 11!} = \dfrac{12}{1} = 12$

In Example 2, it is not a coincidence that the answers to parts (a) and (b) are the same, and that those in parts (c) and (d) are the same. In general, it is true that

$$_nC_r = {_nC_{n-r}}.$$

This shows the symmetric property of binomial coefficients.

2 Arrange the binomial coefficients in a triangular pattern known as Pascal's Triangle.

Pascal's Triangle

There is a convenient way to remember a pattern for binomial coefficients. By arranging the coefficients in a triangular pattern, you obtain the following array, which is called **Pascal's Triangle.** This triangle is named after the famous French mathematician Blaise Pascal (1623–1662).

$$
\begin{array}{ccccccccccccccc}
&&&&&&& 1 \\
&&&&&& 1 && 1 \\
&&&&& 1 && 2 && 1 &&&&& 1+2=3 \\
&&&& 1 && 3 && 3 && 1 \\
&&& 1 && 4 && 6 && 4 && 1 \\
&& 1 && 5 && 10 && 10 && 5 && 1 &&& 10+5=15 \\
& 1 && 6 && 15 && 20 && 15 && 6 && 1 \\
1 && 7 && 21 && 35 && 35 && 21 && 7 && 1
\end{array}
$$

The first and last numbers in each row of Pascal's Triangle are 1. As shown above, every other number in each row is formed by adding the two numbers immediately above the number. Pascal noticed that numbers in this triangle are precisely the same numbers that are the coefficients of binomial expansions.

Study Tip

The top row in Pascal's Triangle is called the *zero row* because it corresponds to the binomial expansion

$$(x + y)^0 = 1.$$

Similarly, the next row is called the *first row* because it corresponds to the binomial expansion

$$(x + y)^1 = 1(x) + 1(y).$$

In general, the *nth row* in Pascal's Triangle gives the coefficients of $(x + y)^n$.

$$
\begin{aligned}
(x + y)^0 &= 1 & \text{0th row} \\
(x + y)^1 &= 1x + 1y & \text{1st row} \\
(x + y)^2 &= 1x^2 + 2xy + 1y^2 & \text{2nd row} \\
(x + y)^3 &= 1x^3 + 3x^2y + 3xy^2 + 1y^3 & \text{3rd row} \\
(x + y)^4 &= 1x^4 + 4x^3y + 6x^2y^2 + 4xy^3 + 1y^4 \\
(x + y)^5 &= 1x^5 + 5x^4y + 10x^3y^2 + 10x^2y^3 + 5xy^4 + 1y^5 \\
(x + y)^6 &= 1x^6 + 6x^5y + 15x^4y^2 + 20x^3y^3 + 15x^2y^4 + 6xy^5 + 1y^6 \\
(x + y)^7 &= 1x^7 + 7x^6y + 21x^5y^2 + 35x^4y^3 + 35x^3y^4 + 21x^2y^5 + 7xy^6 + 1y^7
\end{aligned}
$$

You can use the seventh row of Pascal's Triangle to find the eighth row.

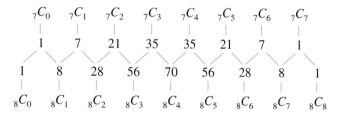

$$
\begin{array}{ccccccccc}
_7C_0 & _7C_1 & _7C_2 & _7C_3 & _7C_4 & _7C_5 & _7C_6 & _7C_7 \\
1 & 7 & 21 & 35 & 35 & 21 & 7 & 1 \\
1 & 8 & 28 & 56 & 70 & 56 & 28 & 8 & 1 \\
_8C_0 & _8C_1 & _8C_2 & _8C_3 & _8C_4 & _8C_5 & _8C_6 & _8C_7 & _8C_8
\end{array}
$$

Example 3   Using Pascal's Triangle

To evaluate $_5C_2$, use the fifth row of Pascal's Triangle

$$
\begin{array}{cccccc}
1 & 5 & 10 & 10 & 5 & 1 \\
_5C_0 & _5C_1 & _5C_2 & _5C_3 & _5C_4 & _5C_5
\end{array}
$$

to obtain $_5C_2 = 10$.

3 Expand a binomial raised to a power using Pascal's Triangle and the Binomial Theorem.

Binomial Expansions

As mentioned at the beginning of this section, when you write out the coefficients for a binomial that is raised to a power, you are **expanding a binomial.** The formulas for binomial coefficients give you an easy way to expand binomials, as demonstrated in the next three examples.

Example 4   Expanding a Binomial

Write the expansion for the expression.

$$(x + 1)^5$$

Solution

The binomial coefficients from the fifth row of Pascal's Triangle are

$$1, 5, 10, 10, 5, 1.$$

So, the expansion is as follows.

$$(x + 1)^5 = (1)x^5 + (5)x^4(1) + (10)x^3(1^2) + (10)x^2(1^3) + (5)x(1^4) + (1)(1^5)$$

$$= x^5 + 5x^4 + 10x^3 + 10x^2 + 5x + 1$$

To expand binomials representing *differences*, rather than sums, you alternate signs. Here are two examples.

$$(x - 1)^3 = x^3 - 3x^2 + 3x - 1$$

$$(x - 1)^4 = x^4 - 4x^3 + 6x^2 - 4x + 1$$

Example 5   Expanding a Binomial

Write the expansion for the expression.

a. $(x - 3)^4$ **b.** $(2x - 1)^3$

Solution

a. The binomial coefficients from the fourth row of Pascal's Triangle are

$$1, 4, 6, 4, 1.$$

So, the expansion is as follows.

$$(x - 3)^4 = (1)x^4 - (4)x^3(3) + (6)x^2(3^2) - (4)x(3^3) + (1)(3^4)$$

$$= x^4 - 12x^3 + 54x^2 - 108x + 81$$

b. The binomial coefficients from the third row of Pascal's Triangle are

$$1, 3, 3, 1.$$

So, the expansion is as follows.

$$(2x - 1)^3 = (1)(2x)^3 - (3)(2x)^2(1) + (3)(2x)(1^2) - (1)(1^3)$$

$$= 8x^3 - 12x^2 + 6x - 1$$

Example 6   Expanding a Binomial

Write the expansion for $(x - 2y)^4$.

Solution

Use the fourth row of Pascal's Triangle, as follows.

$$(x - 2y)^4 = (1)x^4 - (4)x^3(2y) + (6)x^2(2y)^2 - (4)x(2y)^3 + (1)(2y)^4$$

$$= x^4 - 8x^3y + 24x^2y^2 - 32xy^3 + 16y^4$$

Example 7   Finding a Term in the Binomial Expansion

a. Find the sixth term of $(a + 2b)^8$.   **b.** Find the 12th term of $(2a - b)^{15}$.

Solution

a. From the Binomial Theorem, you can see that the $(r + 1)$th term is $_nC_r x^{n-r}y^r$. So in this case, $6 = r + 1$ means that $r = 5$. Because $n = 8$, $x = a$, and $y = 2b$, the sixth term in the binomial expansion is

$$_8C_5 a^{8-5}(2b)^5 = 56 \cdot a^3 \cdot (2b)^5$$

$$= 56(2^5)a^3b^5$$

$$= 1792\, a^3b^5.$$

b. From the Binomial Theorem, you can see that the $(r + 1)$th term is $_nC_r x^{n-r}y^r$. So in this case, $12 = r + 1$ means that $r = 11$. Because $n = 15$, $x = 2a$, and $y = -b$, the 12th term in the binomial expansion is

$$_{15}C_{11}(2a)^{15-11}(-b)^{11} = 1365 \cdot (2a)^4 \cdot (-b)^{11}$$

$$= 1365(2^4)(-1)^{11}\, a^4b^{11}$$

$$= -21{,}840a^4b^{11}.$$

Discussing the Concept   **Finding a Pattern**

By adding the terms in each of the rows of Pascal's Triangle, you obtain the following.

Row 0: $1 = 1$

Row 1: $1 + 1 = 2$

Row 2: $1 + 2 + 1 = 4$

Row 3: $1 + 3 + 3 + 1 = 8$

Row 4: $1 + 4 + 6 + 4 + 1 = 16$

Find a pattern for this sequence. Then use the pattern to find the sum of the terms in the 10th row of Pascal's Triangle. Finally, check your answer by actually adding the terms of the 10th row.

10.4 Exercises

Integrated Review — Concepts, Skills, and Problem Solving

Keep mathematically in shape by doing these exercises *before* the problems of this section.

Properties and Definitions

1. Is it possible to find the determinant of the following matrix? Explain.

$$\begin{bmatrix} 3 & 2 & 6 \\ 1 & -4 & 7 \end{bmatrix}$$

2. State the three elementary row operations that can be used to transform a matrix into a second row-equivalent matrix.

3. Is the following matrix in row-echelon form? Explain.

$$\begin{bmatrix} 1 & 2 & 6 \\ 0 & 1 & 7 \end{bmatrix}$$

Determinants

In Exercises 4–7, evaluate the determinant of the matrix.

4. $\begin{bmatrix} 10 & 25 \\ 6 & -5 \end{bmatrix}$

5. $\begin{bmatrix} 3 & 7 \\ -2 & 6 \end{bmatrix}$

6. $\begin{bmatrix} 3 & -2 & 1 \\ 0 & 5 & 3 \\ 6 & 1 & 1 \end{bmatrix}$

7. $\begin{bmatrix} 4 & 3 & 5 \\ 3 & 2 & -2 \\ 5 & -2 & 0 \end{bmatrix}$

Problem Solving

8. Use a determinant to find the area of the triangle with vertices $(-5, 8)$, $(10, 0)$, and $(3, -4)$.

9. The path of a ball passes through the points $(0, 2)$, $(10, 8)$, and $(20, 0)$. Use Cramer's Rule to find the equation of the parabola that models the path of the ball.

10. Use determinants to find the equation of the line through $(2, -1)$ and $(4, 7)$.

Developing Skills

In Exercises 1–12, evaluate the binomial coefficient $_nC_r$. See Examples 1 and 2.

1. $_6C_4$
2. $_7C_3$
3. $_{10}C_5$
4. $_{12}C_9$
5. $_{20}C_{20}$
6. $_{15}C_0$
7. $_{18}C_{18}$
8. $_{200}C_1$
9. $_{50}C_{48}$
10. $_{75}C_1$
11. $_{25}C_4$
12. $_{18}C_5$

In Exercises 13–22, use a graphing utility to evaluate $_nC_r$.

13. $_{30}C_6$
14. $_{25}C_{10}$
15. $_{12}C_7$
16. $_{40}C_5$
17. $_{52}C_5$
18. $_{100}C_6$
19. $_{200}C_{195}$
20. $_{500}C_4$
21. $_{25}C_{12}$
22. $_{1000}C_2$

In Exercises 23–28, use Pascal's Triangle to evaluate $_nC_r$. See Example 3.

23. $_6C_2$
24. $_9C_3$
25. $_7C_3$
26. $_9C_5$
27. $_8C_4$
28. $_{10}C_6$

In Exercises 29–38, use Pascal's Triangle to expand the expression. See Examples 4–6.

29. $(a + 2)^3$
30. $(x + 3)^5$
31. $(x + y)^8$
32. $(r - s)^7$
33. $(2x - 1)^5$
34. $(4 - 3y)^3$
35. $(2y + z)^6$
36. $(2t - s)^5$
37. $(x^2 + 2)^4$
38. $(3 - y^4)^5$

In Exercises 39–50, use the Binomial Theorem to expand the expression.

39. $(x + 3)^6$

40. $(x - 5)^4$

41. $(x - 4)^6$

42. $(x - 8)^4$

43. $(x + y)^4$

44. $(u + v)^6$

45. $(u - 2v)^3$

46. $(2x + y)^5$

47. $(3a + 2b)^4$

48. $(4u - 3v)^3$

49. $(2x^2 - y)^5$

50. $(x - 4y^3)^4$

In Exercises 51–60, find the coefficient of the given term of the expression.

Expression	Term
51. $(x + 1)^{10}$	x^7
52. $(x + 3)^{12}$	x^9
53. $(x - y)^{15}$	$x^4 y^{11}$
54. $(x - 3y)^{14}$	$x^3 y^{11}$

Expression	Term
55. $(2x + y)^{12}$	$x^3 y^9$
56. $(x + y)^{10}$	$x^7 y^3$
57. $(x^2 - 3)^4$	x^4
58. $(3 - y^3)^5$	y^9
59. $(\sqrt{x} + 1)^8$	x^2
60. $\left(\dfrac{1}{u} + 2\right)^6$	$\dfrac{1}{u^4}$

In Exercises 61–64, use the Binomial Theorem to approximate the quantity accurate to three decimal places. For example,

$$(1.02)^{10} \approx 1 + 10(0.02) + 45(0.02)^2.$$

61. $(1.02)^8$

62. $(2.005)^{10}$

63. $(2.99)^{12}$

64. $(1.98)^9$

Solving Problems

Probability   In Exercises 65–68, use the Binomial Theorem to expand the expression. In the study of probability, it is sometimes necessary to use the expansion $(p + q)^n$, where $p + q = 1$.

65. $\left(\frac{1}{2} + \frac{1}{2}\right)^5$

66. $\left(\frac{2}{3} + \frac{1}{3}\right)^4$

67. $\left(\frac{1}{4} + \frac{3}{4}\right)^4$

68. $\left(\frac{2}{5} + \frac{3}{5}\right)^3$

69. *Patterns in Pascal's Triangle*   Describe the pattern.

```
              1
           1     1
        1     2     1
     1     3     3     1
  1     4     6     4     1
1     5    10    10     5     1
1  6   15   20   15    6     1
```

70. *Patterns in Pascal's Triangle*   Use each encircled group of numbers to form a 2×2 matrix. Find the determinant of each matrix. Describe the pattern.

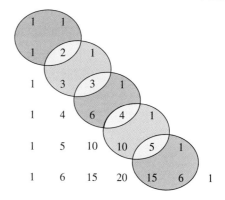

Explaining Concepts

71. How many terms are in the expansion of $(x + y)^n$?

72. Describe the pattern for the exponents with base x in the expansion of $(x + y)^n$.

73. How do the expansions of $(x + y)^n$ and $(x - y)^n$ differ?

74. Which of the following is equal to $_{11}C_5$? Explain.

(a) $\dfrac{11 \cdot 10 \cdot 9 \cdot 8 \cdot 7}{5 \cdot 4 \cdot 3 \cdot 2 \cdot 1}$

(b) $\dfrac{11 \cdot 10 \cdot 9 \cdot 8 \cdot 7}{6 \cdot 5 \cdot 4 \cdot 3 \cdot 2 \cdot 1}$

75. What is the relationship between $_nC_r$ and $_nC_{n-r}$? Explain.

76. In your own words, explain how to form the rows in Pascal's Triangle.

10.5 Counting Principles

Objectives

1 Count the number of ways an event can occur.

2 Determine the number of ways two or three events can occur using the Fundamental Counting Principle.

3 Determine the number of ways n elements can be arranged.

4 Determine the number of ways n elements can be taken r at a time.

---

1 Count the number of ways an event can occur.

Simple Counting Problems

The last two sections of this chapter contain a brief introduction to some of the basic counting principles and their application to probability. In the next section, you will see that much of probability has to do with counting the number of ways an event can occur. Examples 1, 2, and 3 describe some simple cases.

Example 1 A Random Number Generator

A random number generator (on a computer) selects an integer from 1 to 30. Find the number of ways each event can occur.

a. An even integer is selected.

b. A number that is less than 12 is selected.

c. A prime number is selected.

d. A perfect square is selected.

Solution

a. Because half of the numbers from 1 to 30 are even, this event can occur in 15 different ways.

b. The positive integers that are less than 12 are as follows.

$$\{1, 2, 3, 4, 5, 6, 7, 8, 9, 10, 11\}$$

Because this set has 11 members, you can conclude that there are 11 different ways this event can happen.

c. The prime numbers between 1 and 30 are as follows.

$$\{2, 3, 5, 7, 11, 13, 17, 19, 23, 29\}$$

Because this set has 10 members, you can conclude that there are 10 different ways this event can happen.

d. The perfect square numbers between 1 and 30 are as follows.

$$\{1, 4, 9, 16, 25\}$$

Because this set has five members, you can conclude that there are five different ways this event can happen.

Example 2   Selecting Pairs of Numbers at Random

Eight pieces of paper are numbered from 1 to 8 and placed in a box. One piece of paper is drawn from the box, its number is written down, and the piece of paper is replaced in the box. Then, a second piece of paper is drawn from the box, and its number is written down. Finally, the two numbers are added together. How many different ways can a total of 12 be obtained?

Solution

To solve this problem, count the different ways that a total of 12 can be obtained using two numbers between 1 and 8.

First number + Second number = 12

After considering the various possibilities, you can see that this equation can be satisfied in the following five ways.

First Number: 4 5 6 7 8
Second Number: 8 7 6 5 4

So, a total of 12 can be obtained in *five* different ways.

Solving counting problems can be tricky. Often, seemingly minor changes in the statement of a problem can affect the answer. For instance, compare the counting problem in the next example with that given in Example 2.

Example 3   Selecting Pairs of Numbers at Random

Eight pieces of paper are numbered from 1 to 8 and placed in a box. Two pieces of paper are drawn from the box, and the numbers on them are written down and totaled. How many different ways can a total of 12 be obtained?

Solution

To solve this problem, count the different ways that a total of 12 can be obtained *using two different numbers* between 1 and 8.

First number + Second number = 12

After considering the various possibilities, you can see that this equation can be satisfied in the following four ways.

First Number: 4 5 7 8
Second Number: 8 7 5 4

So, a total of 12 can be obtained in *four* different ways.

Examples 2 and 3 differ in that the random selection in Example 2 occurs *with replacement*, whereas the random selection in Example 3 occurs *without replacement*, which eliminates the possibility of choosing two 6s. When doing such exercises, be sure to note if selection is *with* or *without* replacement.

2 Determine the number of ways two or three events can occur using the Fundamental Counting Principle.

Counting Principles

The first three examples in this section are considered simple counting problems in which you can *list* each possible way that an event can occur. When it is possible, this is always the best way to solve a counting problem. However, some events can occur in so many different ways that it is not feasible to write out the entire list. In such cases, you must rely on formulas and counting principles. The most important of these is called the **Fundamental Counting Principle.**

> ▶ **Fundamental Counting Principle**
>
> Let E_1 and E_2 be two events. The first event E_1 can occur in m_1 different ways. After E_1 has occurred, E_2 can occur in m_2 different ways. The number of ways that the two events can occur is
>
> $$m_1 \cdot m_2.$$

The Fundamental Counting Principle can be extended to three or more events. For instance, the number of ways that three events E_1, E_2, and E_3 can occur is

$$m_1 \cdot m_2 \cdot m_3.$$

Example 4 Applying the Fundamental Counting Principle

How many "two-letter words" can be made from the English alphabet?

Solution

The English alphabet contains 26 letters. So, the number of possible "two-letter words" is $26 \cdot 26 = 676$.

Example 5 Applying the Fundamental Counting Principle

Telephone numbers in the United States have 10 digits. The first three are the *area code* and the next seven are the *local telephone number*. How many different telephone numbers are possible within each area code? (A telephone number cannot have 0 or 1 as its first or second digit.)

Solution

There are only eight choices for the first and second digits because neither can be 0 or 1. For each of the other digits, there are 10 choices.

In 1996, there were 178 million active telephone numbers in use in the United States.

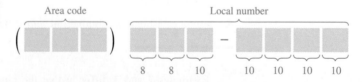

So, by the Fundamental Counting Principle, the number of local telephone numbers that are possible within each area code is

$$8 \cdot 8 \cdot 10 \cdot 10 \cdot 10 \cdot 10 \cdot 10 = 6{,}400{,}000.$$

3 Determine the number of ways *n* elements can be arranged.

Permutations

One important application of the Fundamental Counting Principle is in determining the number of ways that *n* elements can be arranged (in order). An ordering of *n* elements is called a **permutation** of the elements.

> ▶ Definition of Permutation
>
> A **permutation** of *n* different elements is an ordering of the elements such that one element is first, one is second, one is third, and so on.

Example 6   Listing Permutations

How many permutations are possible for the letters A, B, and C?

Solution

The possible permutations of the letters A, B, and C are as follows.

> A, B, C   B, A, C   C, A, B
>
> A, C, B   B, C, A   C, B, A

So, six permutations are possible.

Example 7   Finding the Number of Permutations of *n* Elements

How many permutations are possible for the letters A, B, C, D, E, and F?

Solution

1st position:	Any of the *six* letters.
2nd position:	Any of the remaining *five* letters.
3rd position:	Any of the remaining *four* letters.
4th position:	Any of the remaining *three* letters.
5th position:	Any of the remaining *two* letters.
6th position:	The *one* remaining letter.

So, the number of choices for the six positions are as follows.

Permutations of six letters

By the Fundamental Counting Principle, the total number of permutations of the six letters is

$$6 \cdot 5 \cdot 4 \cdot 3 \cdot 2 \cdot 1 = 6!$$
$$= 720.$$

The result obtained in Example 7 is generalized below.

▶ **Number of Permutations of *n* Elements**

The number of permutations of *n* elements is given by

$$n \cdot (n - 1) \cdots \cdots 4 \cdot 3 \cdot 2 \cdot 1 = n!.$$

So, there are *n*! different ways that *n* elements can be ordered.

Example 8   Finding the Number of Permutations

How many ways can you form a four-digit number using each of the digits 1, 3, 5, and 7 exactly once?

Solution

One way to solve this problem is simply to list the number of ways.

1357, 1375, 1537, 1573, 1735, 1753
3157, 3175, 3517, 3571, 3715, 3751
5137, 5173, 5317, 5371, 5713, 5731
7135, 7153, 7315, 7351, 7513, 7531

Another way to solve the problem is to use the formula for the number of permutations of four elements. By that formula, there are 4! = 24 permutations.

Example 9   Finding the Number of Permutations

How many ways can you form a six-digit number using each of the digits 1, 2, 3, 4, 5, and 6 exactly once?

Solution

By the formula for the number of permutations of six elements, there are 6! = 720 permutations.

Example 10   Finding the Number of Permutations

You are a supervisor for 11 different employees. One of your responsibilities is to perform an annual evaluation for each employee, and then rank the 11 different performances. How many different rankings are possible?

Solution

Because there are 11 different employees, you have 11 choices for first ranking. After choosing the first ranking, you can choose any of the remaining 10 for second ranking, and so on.

Rankings of 11 Employees

So, the number of different rankings is 11! = 39,916,800.

---

4 Determine the number of ways
n elements can be taken *r* at a time.

Combinations

When counting the number of possible permutations of a set of elements, order is important. The final topic in this section is a method of selecting subsets of a larger set in which order is *not important*. Such subsets are called **combinations of *n* elements taken *r* at a time.** For instance, the combinations

{A, B, C}   and   {B, A, C}

are equivalent because both sets contain the same three elements, and the order in which the elements are listed is *not important*. So, you would count only one of the two sets. A common example of how a combination occurs is a card game in which the player is free to reorder the cards after they have been dealt.

Do you remember this riddle? How many different ways can two of the three (wolf, goat, and cabbage) be left on the shore?

A farmer wants to take a wolf, a goat, and cabbage across a river. Unattended, the wolf will eat the goat, and the goat will eat the cabbage. The farmer can only take one of the three on each trip across the river. How can the farmer take all three safely across the river?

There are three possible combinations that can be left on the shore.

{wolf, goat}, {wolf, cabbage}, {goat, cabbage}

Of the three combinations, the farmer can only leave the wolf alone with the cabbage when he crosses the river.

Example 11   Combination of *n* Elements Taken *r* at a Time

In how many different ways can three letters be chosen from the letters A, B, C, D, and E? (The order of the three letters is not important.)

Solution

The following subsets represent the different combinations of three letters that can be chosen from five letters.

{A, B, C}   {A, B, D}   {A, B, E}   {A, C, D}   {A, C, E}
{A, D, E}   {B, C, D}   {B, C, E}   {B, D, E}   {C, D, E}

From this list, you can conclude that there are 10 different ways that three letters can be chosen from five letters. Because order is not important, the set {B, C, A} is not chosen. It is represented by the set {A, B, C}.

The formula for the number of combinations of *n* elements taken *r* at a time is as follows.

> ▶ **Number of Combinations of *n* Elements Taken *r* at a Time**
>
> The number of combinations of *n* elements taken *r* at a time is
>
> $$_nC_r = \frac{n!}{(n-r)!r!}.$$

Study Tip

When solving problems involving counting principles, you need to be able to distinguish among the various counting principles in order to determine which is necessary to solve the problem correctly. To do this, ask yourself the following questions.

1. Is the order of the elements important? *Permutation*

2. Are the chosen elements a subset of a larger set in which order is not important? *Combination*

3. Does the problem involve two or more separate events? *Fundamental Counting Principle*

Note that the formula for $_nC_r$ is the same one given for binomial coefficients. To see how this formula is used, consider the counting problem given in Example 11. In that problem, you need to find the number of combinations of five elements taken three at a time. Thus, $n = 5$, $r = 3$, and the number of combinations is

$$_5C_3 = \frac{5!}{2!3!}$$

$$= \frac{5 \cdot 4 \cdot 3}{3 \cdot 2 \cdot 1}$$

$$= 10$$

which is the same as the answer obtained in Example 11.

Example 12   Combinations of n Elements Taken r at a Time

A standard poker hand consists of five cards dealt from a deck of 52. How many different poker hands are possible? (After the cards are dealt, the player may reorder them, and therefore order is not important.)

Solution

Use the formula for the number of combinations of 52 elements taken five at a time, as follows.

$$_{52}C_5 = \frac{52!}{47!5!}$$

$$= \frac{52 \cdot 51 \cdot 50 \cdot 49 \cdot 48}{5 \cdot 4 \cdot 3 \cdot 2 \cdot 1}$$

$$= 2,598,960$$

So, there are almost 2.6 million different hands.

Discussing the Concept   **Applying Counting Methods**

The Boston Market restaurant chain offers individual rotisserie chicken meals with two side items. Customers can choose either dark or white meat and can select side items from a list of 15. How many different meals are available if two different side items are to be ordered? Of the 15 side items, nine are hot and six are cold. How many different meals are available if a customer wishes to order one hot and one cold side item? (Source: Boston Market, Inc.)

10.5   Exercises

Integrated Review   *Concepts, Skills, and Problem Solving*

Keep mathematically in shape by doing these exercises *before* the problems of this section.

Properties and Definitions

1. Which of the following functions are exponential? Explain.

$$f(x) = 5x^2 \quad g(x) = 2(5^x)$$

2. Explain why $e^{2-x^2} = e^2 \cdot e^{-x^2}$.

Logarithms and Exponents

In Exercises 3–6, rewrite the equation in exponential form.

3. $\log_4 64 = 3$ **4.** $\log_3 \frac{1}{81} = -4$

5. $\ln 1 = 0$ **6.** $\ln 5 \approx 1.6094 \ldots$

In Exercises 7–10, solve the equation. (Round the result to two decimal places.)

7. $3^x = 50$ **8.** $e^{x/2} = 8$

9. $\log_2(x - 5) = 6$ **10.** $\ln(x + 3) = 10$

Problem Solving

11. After t years, the value of a car that cost $22,000 is given by

$$V(t) = 22,000(0.8)^t.$$

Sketch a graph of the function and determine when the value of the car is $15,000.

12. Carbon 14 has a half-life of 5730 years. If you start with 10 grams of this isotope, how much remains after 3000 years?

Solving Problems

Random Selection   In Exercises 1–6, find the number of ways the specified event can occur when one or more marbles are selected from a bowl containing 10 marbles numbered 0 through 9. See Examples 1–3.

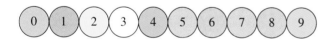

1. One marble is drawn and its number is even.

2. One marble is drawn and its number is prime.

3. Two marbles are drawn one after the other. The first is replaced before the second is drawn. The sum of the numbers is 10.

4. Two marbles are drawn one after the other. The first is replaced before the second is drawn. The sum of the numbers is 7.

5. Two marbles are drawn without replacement. The sum of the numbers is 10.

6. Two marbles are drawn without replacement. The sum of the numbers is 7.

Random Selection   In Exercises 7–16, find the number of ways the specified event can occur when one or more marbles are selected from a bowl containing 20 marbles numbered 1 through 20.

7. One marble is drawn and its number is odd.

8. One marble is drawn and its number is even.

9. One marble is drawn and its number is prime.

10. One marble is drawn and its number is greater than 12.

11. One marble is drawn and its number is divisible by 3.

12. One marble is drawn and its number is divisible by 6.

13. Two marbles are drawn one after the other. The first is replaced before the second is drawn. The sum of the numbers is 8.

14. Two marbles are drawn one after the other. The first is replaced before the second is drawn. The sum of the numbers is 15.

15. Two marbles are drawn without replacement. The sum of the numbers is 8.

16. Three marbles are drawn one after the other. Each marble is replaced before the next is drawn. The sum of the numbers is 15.

17. *Staffing Choices*   A small grocery store needs to open another checkout line. Three people who can run the cash register are available and two people are available to bag groceries. In how many different ways can the additional checkout line be staffed?

18. *Computer System*   You are in the process of purchasing a new computer system. You must choose one of three monitors, one of two computers, and one of two keyboards. How many different configurations of the system are available to you?

19. *Identification Numbers*   In a statistical study, each participant was given an identification label consisting of a letter of the alphabet followed by a single digit (0 is a digit). How many distinct identification labels can be made in this way?

20. *Identification Numbers*   How many identification labels (see Exercise 19) can be made by one letter of the alphabet followed by a two-digit number?

21. *License Plates*   How many distinct automobile license plates can be formed by using a four-digit number followed by two letters?

22. *License Plates*   How many distinct license plates can be formed by using three letters followed by a three-digit number?

23. *Three-Digit Numbers*   How many three-digit numbers can be formed in each of the following situations?

(a) The hundreds digit cannot be 0.

(b) No repetition of digits is allowed.

(c) The number cannot be greater than 400.

24. *Toboggan Ride*   Five people line up on a toboggan at the top of a hill. In how many ways can they be seated if only two of the five are willing to sit in the front seat?

25. *Task Assignment*   Four people are assigned to four different tasks. In how many ways can the assignments be made if one of the four is not qualified for the first task?

26. *Taking a Trip*   Five people are taking a long trip in a car. Two sit in the front seat and three in the back seat. Three of the people agree to share the driving. In how many different arrangements can the five people sit?

27. *Aircraft Boarding*   Eight people are boarding an aircraft. Three have tickets for first class and board before those in economy class. In how many different ways can the eight people board the aircraft?

28. *Permutations*   List all the permutations of the letters X, Y, and Z.

29. *Permutations*   List all the permutations of the letters A, B, C, and D.

30. *Permutations*   List all the permutations of two letters selected from the letters A, B, and C.

31. *Permutations*   List all the permutations of two letters selected from the letters A, B, C, and D.

32. *Seating Arrangement*   In how many ways can five children be seated in a single row of five chairs?

33. *Seating Arrangement*   In how many ways can six people be seated in a six-passenger car?

34. *Posing for a Photograph*   In how many ways can four children line up in one row to have their picture taken?

35. *Combination Lock*   A combination lock will open when the right choice of three numbers (from 1 to 40, inclusive) is selected. How many different lock combinations are possible?

36. *Access Code*   An access code will unlock a door when the right choice of five numbers (from 0 to 9, inclusive) is selected. How many different access codes are possible?

37. *Work Assignments*   Eight workers are assigned to eight different tasks. In how many ways can this be done assuming there are no restrictions in making the assignments?

38. *Work Assignments*   Out of eight workers, five are selected and assigned to different tasks. In how many ways can this be done if there are no restrictions in making the assignments?

39. *Choosing Offers*   From a pool of 10 candidates, the offices of president, vice-president, secretary, and treasurer will be filled. In how many ways can the offices be filled, if each of the 10 candidates can hold any office?

40. *Time Management Study*   There are eight steps in accomplishing a certain task and these steps can be performed in any order. Management wants to test each possible order to determine which is the least time-consuming.

(a) How many different orders will have to be tested?

(b) How many different orders will have to be tested if one step in accomplishing the task must be done first? (The other seven steps can be performed in any order.)

41. *Number of Subsets*   List all the subsets with two elements that can be formed from the set of letters

{A, B, C, D, E, F}.

42. *Number of Subsets*   List all the subsets with three elements that can be formed from the set of letters

{A, B, C, D, E, F}.

43. *Committee Selection*   Three students are selected from a class of twenty to form a fundraising committee. In how many ways can the committee be formed?

44. *Committee Selection*   In how many ways can a committee of five be formed from a group of 30 people?

45. *Menu Selection*   A group of four people go out to dinner at a restaurant. There are nine entrees on the menu and the four people decide that no two will order the same thing. In how many ways can the four people order from the nine entrees?

46. *Menu Selection*   A group of six people go out to dinner at a restaurant. There are 12 entrees on the menu and the six people decide that no two will order the same thing. In how many ways can the six people order from the 12 entrees?

47. *Test Questions*   A student may answer any nine questions from the 12 questions on an exam. Determine the number of ways the student can select the nine questions.

48. *Test Questions*   A student may answer any three questions from the 10 questions on an exam. Determine the number of ways the student can select the three questions.

49. *Basketball Lineup*   A high school basketball team has 15 players. Use a graphing utility to determine the number of ways the coach can choose 5 players in the starting lineup. (Assume each player can play each position.)

50. *Softball League*   Six churches form a softball league. If each team must play every other team twice during the season, what is the total number of league games played?

51. *Job Applicants*   An employer interviews six people for four openings in the company. Four of the six people are women. If all six are qualified, in how many ways could the employer fill the four positions if

(a) the selection is random?

(b) exactly two women are selected?

52. *Defective Units*   A shipment of 10 microwave ovens contains two defective units. In how many ways can a vending company purchase three of these units and receive

(a) all good units?   (b) two good units?

(c) one good unit?

53. *Group Selection*   Four people are to be selected from four couples. In how many ways can this be done if

(a) there are no restrictions?

(b) one person from each couple must be selected?

54. *Geometry*   Eight points are located in the coordinate plane such that no three lie on the same line. How many different triangles can be formed having their vertices as three of the eight points?

55. *Geometry*   Three points that are not on the same line determine three lines. How many lines are determined by seven points, no three of which are on a line?

Geometry   In Exercises 56–59, find the number of diagonals of the polygon. (A line segment connecting any two nonadjacent vertices of a polygon is called a *diagonal* of the polygon.)

56. Pentagon

57. Hexagon

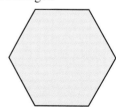

58. Octagon

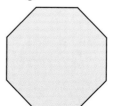

59. Decagon

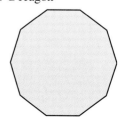

60. *Relationships*   As the size of a group increases, the number of relationships increases dramatically (see figure). Determine the number of two-person relationships in a group that has the following numbers.

(a) 3   (b) 4   (c) 6   (d) 8   (e) 10   (f) 12

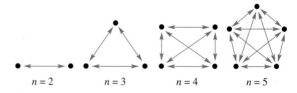

$n = 2$ $n = 3$ $n = 4$ $n = 5$

Figure for 60

Explaining Concepts

61. State the Fundamental Counting Principle.

62. When you use the Fundamental Counting Principle, what are you counting?

63. Give examples of a permutation and a combination.

64. Without calculating the numbers, determine which is greater: the combination of 10 elements taken six at a time or the permutation of 10 elements taken six at a time. Explain.

10.6   Probability

Objectives

1 Determine the probability that an event will occur.

2 Use counting principles to determine the probability that an event will occur.

1 Determine the probability that an event will occur.

The Probability of an Event

The **probability of an event** is a number from 0 to 1 that indicates the likelihood that the event will occur. An event that is certain to occur has a probability of 1. An event that cannot occur has a probability of 0. An event that is equally likely to occur or not occur has a probability of $\frac{1}{2}$ or 0.5.

Probability of 0: Event cannot occur.	Probability of 0.5: Event is equally likely to occur or not occur.	Probability of 1: Event must occur.

$$
\begin{array}{ccc}
\rule{0pt}{0pt} & \rule{0pt}{0pt} & \rule{0pt}{0pt} \\
0 & 0.5 & 1
\end{array}
$$

> ▶ **The Probability of an Event**
>
> Consider a **sample space** S that is composed of a finite number of outcomes, each of which is equally likely to occur. A subset E of the sample space is an **event.** The probability that an outcome E will occur is
>
> $$P = \frac{\text{number of outcomes in event}}{\text{number of outcomes in sample space}}.$$

Example 1   Finding the Probability of an Event

a. You select a card from a standard deck of 52 cards. What is the probability that the card is an ace?

$$P = \frac{\text{number of aces in the deck}}{\text{number of cards in the deck}} = \frac{4}{52} = \frac{1}{13}$$

b. You are dialing a friend's phone number but cannot remember the last digit. If you choose a digit at random, the probability that it is correct is

$$P = \frac{\text{number of correct digits}}{\text{number of possible digits}} = \frac{1}{10}.$$

c. On a multiple-choice test, you know that the answer to a question is not (a) or (d), but you are not sure about (b), (c), and (e). If you guess, the probability that you are wrong is

$$P = \frac{\text{number of wrong answers}}{\text{number of possible answers}} = \frac{2}{3}.$$

Example 2   Conducting a Poll

The Centers for Disease Control took a survey of 11,631 high school students. The students were asked whether they considered themselves to be a good weight, underweight, or overweight. The results are shown in Figure 10.3.

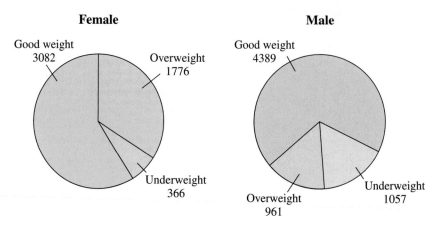

Figure 10.3

a. If you choose a female at random from those surveyed, the probability that she said she was underweight is

$$P = \frac{\text{number of females who answered "underweight"}}{\text{number of females in survey}}$$

$$= \frac{366}{3082 + 366 + 1776}$$

$$= \frac{366}{5224}$$

$$\approx 0.07.$$

b. If you choose a person who answered "underweight" from those surveyed, the probability that the person is female is

$$P = \frac{\text{number of females who answered "underweight"}}{\text{number in survey who answered "underweight"}}$$

$$= \frac{366}{366 + 1057}$$

$$= \frac{366}{1423}$$

$$\approx 0.26.$$

Polls such as the one described in Example 2 are often used to make inferences about a population that is larger than the sample. For instance, from Example 2, you might infer that 7% of *all* high school girls consider themselves to be underweight. When you make such an inference, it is important that those surveyed are representative of the entire population.

Example 3   Using Area to Find Probability

You have just stepped into the tub to take a shower when one of your contact lenses falls out. (You have not yet turned on the water.) Assuming the lens is equally likely to land anywhere on the bottom of the tub, what is the probability that it lands in the drain? Use the dimensions in Figure 10.4 to answer the question.

Solution

Because the area of the tub bottom is $(26)(50) = 1300$ square inches and the area of the drain is

$$\pi(1^2) = \pi \qquad \text{Area of drain}$$

square inches, the probability that the lens lands in the drain is about

$$P = \frac{\pi}{1300} \approx 0.0024.$$

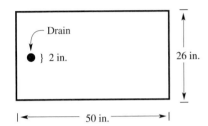

Figure 10.4

Example 4   The Probability of Inheriting Certain Genes

Common parakeets have genes that can produce any one of four feather colors:

Green:	BBCC, BBCc, BbCC, BbCc
Blue:	BBcc, Bbcc
Yellow:	bbCC, bbCc
White:	bbcc

Use the *Punnett square* in Figure 10.5 to find the probability that an offspring to two green parents (both with BbCc feather genes) will be yellow. Note that each parent passes along a B or b gene and a C or c gene.

Solution

The probability that an offspring will be yellow is

$$P = \frac{\text{number of yellow possibilities}}{\text{number of possibilities}}$$

$$= \frac{3}{16}.$$

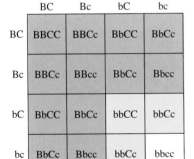

Figure 10.5

2 Use counting principles to determine the probability that an event will occur.

Standard 52-Card Deck

A♠	A♥	A♦	A♣
K♠	K♥	K♦	K♣
Q♠	Q♥	Q♦	Q♣
J♠	J♥	J♦	J♣
10♠	10♥	10♦	10♣
9♠	9♥	9♦	9♣
8♠	8♥	8♦	8♣
7♠	7♥	7♦	7♣
6♠	6♥	6♦	6♣
5♠	5♥	5♦	5♣
4♠	4♥	4♦	4♣
3♠	3♥	3♦	3♣
2♠	2♥	2♦	2♣

Figure 10.6

Using Counting Methods to Find Probabilities

In the following examples, you will see how the basic counting principles you learned in Section 10.5 are used to determine the probability that an event will occur.

Example 5 The Probability of a Royal Flush

Five cards are dealt at random from a standard deck of 52 playing cards (see Figure 10.6). What is the probability that the cards are 10-J-Q-K-A of the same suit?

Solution

On page 666, you saw that the number of possible five-card hands from a deck of 52 cards is $_{52}C_5 = 2,598,960$. Because only four of these five-card hands are 10-J-Q-K-A of the same suit, the probability that the hand contains these cards is

$$P = \frac{4}{2,598,960} = \frac{1}{649,740}.$$

Example 6 Conducting a Survey

A survey was conducted of 500 adults who had worn Halloween costumes. Each person was asked how he or she acquired a Halloween costume: created it, rented it, bought it, or borrowed it. The results are shown in Figure 10.7. What is the probability that the first four people who were polled all created their costumes?

Solution

To answer this question, you need to use the formula for the number of combinations *twice*. First, find the number of ways to choose four people from 360 who created their own costumes.

$$_{360}C_4 = \frac{360 \cdot 359 \cdot 358 \cdot 357}{4 \cdot 3 \cdot 2 \cdot 1} = 688,235,310$$

Next, find the number of ways to choose four people from the 500 who were surveyed.

$$_{500}C_4 = \frac{500 \cdot 499 \cdot 498 \cdot 497}{4 \cdot 3 \cdot 2 \cdot 1} = 2,573,031,125$$

The probability that all of the first four people surveyed created their own costumes is the ratio of these two numbers.

$$P = \frac{\text{number of ways to choose 4 from 360}}{\text{number of ways to choose 4 from 500}}$$

$$= \frac{688,235,310}{2,573,031,125}$$

$$\approx 0.267$$

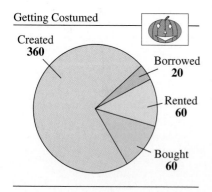

Figure 10.7

Example 7   Forming a Committee

To obtain input from 200 company employees, the management of a company selected a committee of five. Of the 200 employees, 56 were from minority groups. None of the 56, however, was selected to be on the committee. Does this indicate that the management's selection was biased?

Solution

Part of the solution is similar to that of Example 6. If the five committee members were selected at random, the probability that all five would be nonminority is

$$P = \frac{\text{number of ways to choose 5 from 144 nonminority employees}}{\text{number of ways to choose 5 from 200 employees}}$$

$$= \frac{{}_{144}C_5}{{}_{200}C_5}$$

$$= \frac{481{,}008{,}528}{2{,}535{,}650{,}040}$$

$$\approx 0.19.$$

So, if the committee were chosen at random (that is, without bias), the likelihood that it would have no minority members is about 0.19. Although this does not *prove* that there was bias, it does suggest it.

Discussing the Concept   **Probability of Guessing Correctly**

You are taking a chemistry test and are asked to arrange the first 10 elements in the order in which they appear on the periodic table of elements. Suppose that you have no idea of the correct order and simply guess. Does the following computation represent the probability that you guess correctly? Explain.

Solution

You have 10 choices for the first element, nine choices for the second, eight choices for the third, and so on. The number of different orders is $10! = 3{,}628{,}800$, which means that your probability of guessing correctly is

$$P = \frac{1}{3{,}628{,}800}.$$

10.6 Exercises

Integrated Review — Concepts, Skills, and Problem Solving

Keep mathematically in shape by doing these exercises *before* the problems of this section.

Properties and Definitions

In Exercises 1–6, complete the property of logarithms.

1. $\log_a 1 = $ ▢

2. $\log_a a = $ ▢

3. $\log_a a^x = $ ▢

4. $\log_a (uv) = $ ▢

5. $\log_a \dfrac{u}{v} = $ ▢

6. $\log_a u^n = $ ▢

Rewriting Expressions

In Exercises 7–10, use the properties of logarithms to expand the expression as a sum, difference, or multiple of logarithms.

7. $\log_2(x^2 y)$

8. $\log_2 \sqrt{x^2 + 1}$

9. $\ln \dfrac{7}{x - 3}$

10. $\ln\left(\dfrac{u + 2}{u - 2}\right)^2$

Graphs and Models

11. Annual sales y of a product x years after it is introduced are approximated by

$$y = \frac{10,000}{1 + 4e^{-x/3}}.$$

(a) Use a graphing utility to graph the equation.

(b) Use the graph in part (a) to approximate the time when annual sales are $y = 5000$ units.

(c) Use the graph in part (a) to estimate the maximum level of annual sales.

12. An investment is made in an account that pays 5.5% interest, compounded continuously. What is the effective yield for this investment?

Solving Problems

Sample Space In Exercises 1–4, determine the number of outcomes in the sample space for the experiment.

1. One letter from the alphabet is chosen.

2. A six-sided die is tossed twice and the sum is recorded.

3. Two county supervisors are selected from five supervisors, A, B, C, D, and E, to study a recycling plan.

4. A salesperson makes a presentation about a product in three homes per day. In each home there may be a sale (denote by Y) or there may be no sale (denote by N).

Sample Space In Exercises 5–8, list the outcomes in the sample space for the experiment.

5. A taste tester must taste and rank three brands of yogurt, A, B, and C, according to preference.

6. A coin and a die are tossed.

7. A basketball tournament between two teams consists of three games. For each game, your team may win (denote by W) or lose (denote by L).

8. Two students are randomly selected from four students, A, B, C, and D.

In Exercises 9 and 10, you are given the probability that an event *will* occur. Find the probability that the event will *not* occur.

9. $p = 0.35$

10. $p = 0.8$

In Exercises 11 and 12, you are given the probability that an event *will not* occur. Find the probability that the event *will* occur.

11. $p = 0.82$

12. $p = 0.13$

Coin Tossing In Exercises 13–16, a coin is tossed three times. Find the probability of the specified event. Use the sample space

$S = \{HHH, HHT, HTH, HTT, THH, THT, TTH, TTT\}$.

13. The event of getting exactly two heads

14. The event of getting a tail on the second toss

15. The event of getting at least one head

16. The event of getting no more than two heads

Playing Cards   In Exercises 17–20, a card is drawn from a standard deck of 52 playing cards. Find the probability of drawing the indicated event.

17. A red card   **18.** A queen

19. A face card   **20.** A black face card

Tossing a Die   In Exercises 21–24, a six-sided die is tossed. Find the probability of the specified event.

21. The number is a 5.

22. The number is a 7.

23. The number is no more than 5.

24. The number is at least 1.

United States Blood Types   In Exercise 25 and 26, use the circle graph, which shows the percent of people in the United States in 1996 with each blood type. See Example 2.   (Source: America's Blood Centers)

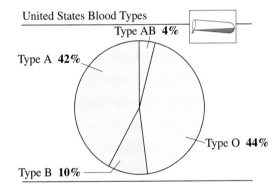

25. A person is selected at random from the United States population. What is the probability that the person *does not* have blood type B?

26. What is the probability that a person selected at random from the United States population *does* have blood type B? How is this probability related to the probability found in Exercise 25?

Charitable Giving   In Exercises 27–30, use the circle graph, which shows the percent of households in the United States by dollar amount given for the year 1995. Answer the question for a household selected at random from the population.   (Source: U.S. Bureau of the Census)

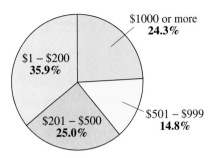

Figure for 27–30

27. What is the probability that the household gave at least $1000?

28. What is the probability that the household gave no more than $200?

29. What is the probability that the household gave no more than $500?

30. What is the probability that the household gave more than $500?

31. *Multiple-Choice Test*   A student takes a multiple-choice test in which there are five choices for each question. Find the probability that the first question is answered correctly given the following conditions.

   (a) The student has no idea of the answer and guesses at random.

   (b) The student can eliminate two of the choices and guesses from the remaining choices.

   (c) The student knows the answer.

32. *Multiple-Choice Test*   A student takes a multiple-choice test in which there are four choices for each question. Find the probability that the first question is answered correctly given the following conditions.

   (a) The student has no idea of the answer and guesses at random.

   (b) The student can eliminate two of the choices and guesses from the remaining choices.

   (c) The student knows the answer.

33. *Class Election*   Three people are running for class president. It is estimated that the probability that Candidate A will win is 0.5 and the probability that Candidate B will win is 0.3.

   (a) What is the probability that *either* Candidate A *or* Candidate B will win?

   (b) What is the probability that the third candidate will win?

34. *Winning an Election*   Jones, Smith, and Thomas are candidates for public office. It is estimated that Jones and Smith have about the same probability of winning, and Thomas is believed to be twice as likely to win as either of the others. Find the probability of each candidate winning the election.

35. *Continuing Education*   In a high school graduating class of 325 students, 255 are going to continue their education. What is the probability that a student selected at random from the class will not be furthering his or her education?

36. *Study Questions*   An instructor gives the class a list of four study questions for the next exam. Two of the four study questions will be on the exam. Find the probability that a student who only knows the material relating to three of the four questions will be able to correctly answer both questions selected for the exam.

Geometry   In Exercises 37–40, the specified probability is the ratio of two areas. See Example 3.

37. *Meteorites*   The largest meteorite in the United States was found in the Willamette Valley of Oregon in 1902. Earth contains 57,510,000 square miles of land and 139,440,000 square miles of water. What is the probability that a meteorite that hits the earth will fall onto land? What is the probability that a meteorite that hits the earth will fall into water?

38. *Game*   A child uses a spring-loaded device to shoot a marble into the square box shown in the figure. The base of the square is horizontal and the marble has an equal likelihood of coming to rest at any point on the base. In parts (a)–(d), find the probability that the marble comes to rest in the specified region.

(a) The red center   (b) The blue ring

(c) The purple border   (d) Not in the yellow ring

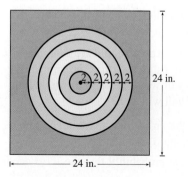

39. *Meeting Time*   You and a friend agree to meet at a favorite restaurant between 5:00 and 6:00 P.M. The one who arrives first will wait 15 minutes for the other, after which the first person will leave (see figure). What is the probability that the two of you actually meet, assuming that your arrival times are random within the hour?

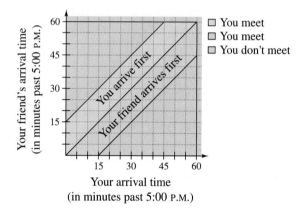

40. *Estimating* π A coin of diameter d is dropped onto a paper that contains a grid of squares d units on a side (see figure).

(a) Find the probability that the coin covers a vertex of one of the squares in the grid.

(b) Repeat the experiment 100 times and use the results to approximate π.

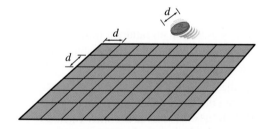

In Exercises 41 and 42, complete the Punnett square and answer the questions. See Example 4.

41. *Girl or Boy?*   The genes that determine the sex of humans are denoted by XX (female) and XY (male). (See figure.)

(a) What is the probability that a newborn will be a girl? a boy?

(b) Explain why it is equally likely that a newborn baby will be a boy or a girl.

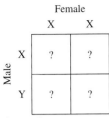

Female

	X	X
X	?	?
Y	?	?

Male

Figure for 41

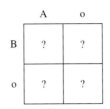

	A	o
B	?	?
o	?	?

Figure for 42

42. *Blood Types* There are four basic human blood types: A (AA or Ao), B (BB or Bo), AB (AB), and O (oo) (see figure).

(a) What is the blood type of each parent?

(b) What is the probability that their offspring will have blood type A? B? AB? O?

In Exercises 43–54, some of the sample spaces are large. Therefore, you should use the counting principles discussed in Section 10.5. See Examples 5 and 6.

43. *Game Show* On a game show, you are given five digits to arrange in the proper order for the price of a car. If you arrange them correctly, you win the car. Find the probability of winning if you know the correct position of only one digit and must guess the positions of the other digits.

44. *Game Show* On a game show you are given four digits to arrange in the proper order for the price of a grandfather clock. What is the probability of winning given the following conditions?

(a) You guess the position of each digit.

(b) You know the first digit, but must guess the remaining three.

45. *Lottery* You buy a lottery ticket inscribed with a five-digit number. On the designated day, five digits (from 0 to 9 inclusive) are randomly selected. What is the probability that you have a winning ticket?

46. *Shelving Books* A parent instructs a young child to place a five-volume set of books on a bookshelf.

Find the probability that the books are in the correct order if the child places them at random.

47. *Preparing for a Test* An instructor gives her class a list of 10 study problems, from which she will select eight to be answered on an exam. If you know how to solve eight of the problems, what is the probability that you will be able to correctly answer all eight questions on the exam?

48. *Committee Selection* A committee of three students is to be selected from a group of three girls and five boys. Find the probability that the committee is composed entirely of girls.

49. *Defective Units* A shipment of 10 food processors to a certain store contains two defective units. If you purchase two of these food processors as birthday gifts for friends, determine the probability that you get both defective units.

50. *Defective Units* A shipment of 12 compact disc players contains two defective units. A husband and wife buy three of these compact disc players to give to their children as Christmas gifts.

(a) What is the probability that none of the three units is defective?

(b) What is the probability that at least one of the units is defective?

51. *Book Selection* Four books are selected at random from a shelf containing six novels and four autobiographies. Find the probability that the four autobiographies are selected.

52. *Card Selection* Five cards are selected from a standard deck of 52 cards. Find the probability that four aces are selected.

53. *Card Selection* Five cards are selected from a standard deck of 52 cards. Find the probability that all hearts are selected.

54. *Card Selection* Five cards are selected from a standard deck of 52 cards. Find the probability that two aces and three queens are selected.

Explaining Concepts

55. Answer parts (d)–(f) of Motivating the Chapter on page 623.

56. The probability of an event must be a real number in what interval? Is the interval open or closed?

57. The probability of an event is $\frac{3}{4}$. What is the probability that the event *does not* occur? Explain.

58. What is the sum of the probabilities of all the occurrences of outcomes in a sample space? Explain.

59. The weather forecast indicates that the probability of rain is 40%. Explain what this means.

Key Terms

arithmetic sequence, *pp. 624, 634*

geometric sequence, *pp. 624, 642*

sequence, *p. 624*

term of a sequence, *p. 624*

infinite sequence, *p. 624*

finite sequence, *p. 624*

factorials, *p. 626*

series, *p. 627*

infinite series, *p. 627*

partial sum, *p. 627*

sigma notation, *p. 628*

index of summation, *p. 628*

upper limit of summation, *p. 628*

lower limit of summation, *p. 628*

common difference, *p. 634*

recursion formula, *p. 635*

*n*th partial sum, *p. 636*

common ratio, *p. 642*

infinite geometric series, *p. 644*

binomial coefficients, *p. 653*

Pascal's Triangle, *p. 655*

expanding a binomial, *p. 656*

permutation, *p. 663*

combination, *p. 665*

probability of an event, *p. 671*

sample space, *p. 671*

event, *p. 671*

Key Concepts

10.2 The *n*th term of an arithmetic sequence

The *n*th term of an arithmetic sequence has the form $a_n = a_1 + (n - 1)d$, where d is the common difference of the sequence, and a_1 is the first term.

10.2 The *n*th partial sum of an arithmetic sequence

The *n*th partial sum of the arithmetic sequence whose *n*th term is a_n is

$$\sum_{i=1}^{n} a_i = a_1 + a_2 + a_3 + a_4 + \cdots + a_n = n\left(\frac{a_1 + a_n}{2}\right).$$

10.3 The *n*th term of a geometric sequence

The *n*th term of a geometric sequence has the form $a_n = a_1 r^{n-1}$, where r is the common ratio of consecutive terms of the sequence. So, every geometric sequence can be written in the following form.

$$a_1, a_1 r, a_1 r^2, a_1 r^3, a_1 r^4, \ldots, a_1 r^{n-1}, \ldots$$

10.3 The *n*th partial sum of a geometric sequence

The *n*th partial sum of the geometric sequence whose *n*th term is $a_n = a_1 r^{n-1}$ is given by

$$\sum_{i=1}^{n} a_1 r^{i-1} = a_1 + a_1 r + a_1 r^2 + a_1 r^3 + \cdots + a_1 r^{n-1}$$

$$= a_1\left(\frac{r^n - 1}{r - 1}\right).$$

10.3 Sum of an infinite geometric series

If $a_1, a_1 r, a_1 r^2, \ldots, a_1 r^n, \ldots$ is an infinite geometric sequence, then for $|r| < 1$, the sum of the terms is

$$\sum_{i=1}^{\infty} a_1 r^i = \frac{a_1}{1 - r}.$$

10.4 The Binomial Theorem

In the expansion of $(x + y)^n$

$$(x + y)^n = x^n + nx^{n-1}y + \cdots + {}_nC_r x^{n-r}y^r + \cdots + nxy^{n-1} + y^n$$

the coefficient of $x^{n-r}y^r$ is given by

$${}_nC_r = \frac{n!}{(n - r)!r!}.$$

10.5 Fundamental Counting Principle

Let E_1 and E_2 be two events. The first event E_1 can occur in m_1 different ways. After E_1 has occurred, E_2 can occur in m_2 different ways. The number of ways that the two events can occur is $m_1 \cdot m_2$.

10.5 Number of permutations of *n* elements

The number of permutations of *n* elements is given by $n \cdot (n - 1) \cdots \cdots 4 \cdot 3 \cdot 2 \cdot 1 = n!$.

10.5 Number of combinations of *n* elements taken *r* at a time

The number of combinations of *n* elements taken *r* at a time is ${}_nC_r = \dfrac{n!}{(n - r)!r!}$.

10.6 The probability of an event

The probability that an outcome E will occur is

$$P = \frac{\text{number of outcomes in event}}{\text{number of outcomes in sample space}}.$$

REVIEW EXERCISES

Reviewing Skills

10.1 In Exercises 1–4, write the first five terms of the sequence. (Begin with $n = 1$.)

1. $a_n = 3n + 5$

2. $a_n = \frac{1}{2}n - 4$

3. $a_n = \frac{1}{2^n} + \frac{1}{2}$

4. $a_n = (n + 1)!$

In Exercises 5–8, find the nth term of the sequence.

5. $1, 3, 5, 7, 9, \ldots$

6. $3, -6, 9, -12, 15, \ldots$

7. $\frac{1}{4}, \frac{2}{9}, \frac{3}{16}, \frac{4}{25}, \frac{5}{36}, \ldots$

8. $\frac{0}{2}, \frac{1}{3}, \frac{2}{4}, \frac{3}{5}, \frac{4}{6}, \ldots$

In Exercises 9–14, match the sequence with its graph. [The graphs are labeled (a), (b), (c), (d), (e), and (f).]

(a)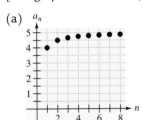

(b)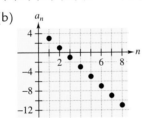

(c)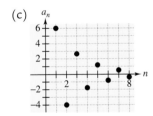

(d)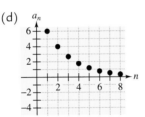

(e)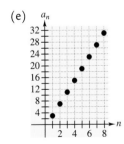

(f)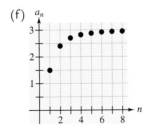

9. $a_n = 5 - \dfrac{1}{n}$

10. $a_n = \dfrac{3n^2}{n^2 + 1}$

11. $a_n = 5 - 2n$

12. $a_n = 4n - 1$

13. $a_n = 6\left(\frac{2}{3}\right)^{n-1}$

14. $a_n = 6\left(-\frac{2}{3}\right)^{n-1}$

In Exercises 15–18, evaluate the sum.

15. $\displaystyle\sum_{k=1}^{4} 7$

16. $\displaystyle\sum_{k=1}^{4} \frac{(-1)^k}{k}$

17. $\displaystyle\sum_{n=1}^{4} \left(\frac{1}{n} - \frac{1}{n+1}\right)$

18. $\displaystyle\sum_{n=1}^{4} \left(\frac{1}{n} - \frac{1}{n+2}\right)$

In Exercises 19–22, use sigma notation to write the sum.

19. $[5(1) - 3] + [5(2) - 3] + [5(3) - 3] + [5(4) - 3]$

20. $[9 - 10(1)] + [9 - 10(2)] + [9 - 10(3)] + [9 - 10(4)]$

21. $\dfrac{1}{3(1)} + \dfrac{1}{3(2)} + \dfrac{1}{3(3)} + \dfrac{1}{3(4)} + \dfrac{1}{3(5)} + \dfrac{1}{3(6)}$

22. $\left(-\frac{1}{3}\right)^0 + \left(-\frac{1}{3}\right)^1 + \left(-\frac{1}{3}\right)^2 + \left(-\frac{1}{3}\right)^3 + \left(-\frac{1}{3}\right)^4$

10.2 In Exercises 23 and 24, find the common difference of the arithmetic sequence.

23. $30, 27.5, 25, 22.5, 20, \ldots$

24. $9, 12, 15, 18, 21, \ldots$

In Exercises 25–32, write the first five terms of the arithmetic sequence. (Begin with $n = 1$.)

25. $a_n = 132 - 5n$

26. $a_n = 2n + 3$

27. $a_n = \frac{3}{4}n + \frac{1}{2}$

28. $a_n = -\frac{3}{5}n + 1$

29. $a_1 = 5$

$a_{k+1} = a_k + 3$

30. $a_1 = 12$

$a_{k+1} = a_k + 1.5$

31. $a_1 = 80$

$a_{k+1} = a_k - \frac{5}{2}$

32. $a_1 = 25$

$a_{k+1} = a_k - 6$

In Exercises 33–36, find a formula for the nth term of the arithmetic sequence.

33. $a_1 = 10, \quad d = 4$

34. $a_1 = 32, \quad d = -2$

35. $a_1 = 1000, \quad a_2 = 950$

36. $a_1 = 12, \quad a_2 = 20$

In Exercises 37–40, find the nth partial sum of the arithmetic sequence.

37. $\displaystyle\sum_{k=1}^{12} (7k - 5)$

38. $\displaystyle\sum_{k=1}^{10} (100 - 10k)$

39. $\displaystyle\sum_{j=1}^{100} \frac{j}{4}$

40. $\displaystyle\sum_{j=1}^{50} \frac{3j}{2}$

In Exercises 41 and 42, use a graphing utility to evaluate the sum.

41. $\displaystyle\sum_{i=1}^{60} (1.25i + 4)$

42. $\displaystyle\sum_{i=1}^{100} (5000 - 3.5i)$

10.3 In Exercises 43 and 44, find the common ratio of the geometric sequence.

43. $8, 12, 18, 27, \frac{81}{2}, \ldots$

44. $27, -18, 12, -8, \frac{16}{3}, \ldots$

In Exercises 45–50, write the first five terms of the geometric sequence.

45. $a_1 = 10, \quad r = 3$

46. $a_1 = 2, \quad r = -5$

47. $a_1 = 100, \quad r = -\frac{1}{2}$

48. $a_1 = 12, \quad r = \frac{1}{6}$

49. $a_1 = 3$
$\quad a_{k+1} = 2a_k$

50. $a_1 = 36$
$\quad a_{k+1} = \frac{1}{2}a_k$

In Exercises 51–56, find a formula for the nth term of the geometric sequence.

51. $a_1 = 1, \quad r = -\frac{2}{3}$

52. $a_1 = 100, \quad r = 1.07$

53. $a_1 = 24, \quad a_2 = 48$

54. $a_1 = 16, \quad a_2 = -4$

55. $a_1 = 12, \quad a_4 = -\frac{3}{2}$

56. $a_2 = 1, \quad a_3 = \frac{1}{3}$

In Exercises 57–64, find the nth partial sum of the geometric sequence.

57. $\displaystyle\sum_{n=1}^{12} 2^n$

58. $\displaystyle\sum_{n=1}^{12} (-2)^n$

59. $\displaystyle\sum_{k=1}^{8} 5\left(-\frac{3}{4}\right)^k$

60. $\displaystyle\sum_{k=1}^{10} 4\left(\frac{3}{2}\right)^k$

61. $\displaystyle\sum_{i=1}^{8} (1.25)^{i-1}$

62. $\displaystyle\sum_{i=1}^{8} (-1.25)^{i-1}$

63. $\displaystyle\sum_{n=1}^{120} 500(1.01)^n$

64. $\displaystyle\sum_{n=1}^{40} 1000(1.1)^n$

In Exercises 65–68, find the sum of the infinite geometric series.

65. $\displaystyle\sum_{i=1}^{\infty} \left(\frac{7}{8}\right)^{i-1}$

66. $\displaystyle\sum_{i=1}^{\infty} \left(\frac{1}{3}\right)^{i-1}$

67. $\displaystyle\sum_{k=1}^{\infty} 4\left(\frac{2}{3}\right)^{k-1}$

68. $\displaystyle\sum_{k=1}^{\infty} 1.3\left(\frac{1}{10}\right)^{k-1}$

In Exercises 69 and 70, use a graphing utility to evaluate the sum.

69. $\displaystyle\sum_{k=1}^{50} 50(1.2)^{k-1}$

70. $\displaystyle\sum_{j=1}^{60} 25(0.9)^{j-1}$

10.4 In Exercises 71–74, find the binomial coefficient.

71. $_8C_3$

72. $_{12}C_2$

73. $_{12}C_0$

74. $_{100}C_1$

In Exercises 75–78, use a graphing utility to evaluate $_nC_r$.

75. $_{40}C_4$

76. $_{15}C_9$

77. $_{25}C_6$

78. $_{32}C_2$

In Exercises 79–84, use the Binomial Theorem to expand the binomial expression. Simplify the result.

79. $(x + 1)^{10}$

80. $(y - 2)^6$

81. $(3x - 2y)^4$

82. $(2u + 5v)^4$

83. $(u^2 + v^3)^9$

84. $(x^4 - y^5)^8$

In Exercises 85–88, find the coefficient of the given term of the expression.

Expression	Term
85. $(x - 3)^{10}$	x^5
86. $(x + 4)^9$	x^6
87. $(x + 2y)^7$	x^4y^3
88. $(2x - 3y)^5$	x^2y^3

Solving Problems

89. Find the sum of the first 50 positive integers that are multiples of 4.

90. Find the sum of the integers from 225 to 300.

91. *Auditorium Seating*   Each row in a small auditorium has three more seats than the preceding row. Find the seating capacity of the auditorium if the front row seats 22 people and there are 12 rows of seats.

92. *Depreciation*   A company pays $120,000 for a machine. During the next 5 years, the machine depreciates at the rate of 30% per year. (That is, at the end of each year, the depreciated value is 70% of what it was at the beginning of the year.)

(a) Find a formula for the nth term of the geometric sequence that gives the value of the machine n full years after it was purchased.

(b) Find the depreciated value of the machine at the end of 5 full years.

93. *Population Increase*   A city of 85,000 people is growing at the rate of 1.2% per year. (That is, at the end of each year, the population is 1.012 times what it was at the beginning of the year.)

(a) Find a formula for the nth term of the geometric sequence that gives the population n years from now.

(b) Estimate the population 50 years from now.

94. *Salary Increase*   You accept a job that pays a salary of $32,000 the first year. During the next 39 years, you receive a 5.5% raise each year. What would your total salary be over the 40-year period?

95. *Morse Code*   In Morse code, all characters are transmitted using a sequence of *dots* and *dashes*. How many different characters can be formed by using a sequence of three dots and dashes? (These can be repeated. For example, dash-dot-dot represents the letter *d*.)

96. *Random Selection*   Ten marbles numbered 0 through 9 are placed in a bag. List the ways in which two marbles having a sum of 8 can be drawn without replacement.

97. *Committee Selection*   Determine the number of ways a committee of five people can be formed from a group of 15 people.

98. *Program Listing*   There are seven participants in a piano recital. In how many orders can their names be listed in the program?

99. *Rolling a Die*   Find the probability of obtaining a number greater than 4 when a six-sided die is rolled.

100. *Coin Tossing*   Find the probability of obtaining at least one head when a coin is tossed four times.

101. *Book Selection*   A child who does not know how to read carries a four-volume set of books to a bookshelf. Find the probability that the child will put the books on the shelf in the correct order.

102. *Rolling a Die*   Are the chances of rolling a 3 with one six-sided die the same as the chances of rolling a total of 6 with two six-sided dice? If not, which has the greater probability of occurring?

103. *Hospital Inspection*   As part of a monthly inspection at a hospital, the inspection team randomly selects reports from eight of the 84 nurses who are on duty. What is the probability that none of the reports selected will be from the ten most experienced nurses?

104. *Target Shooting*   An archer shoots an arrow at the target shown in the figure. Suppose that the arrow is equally likely to hit any point on the target. What is the probability that the arrow hits the bull's-eye? What is the probability that the arrow hits the blue ring?

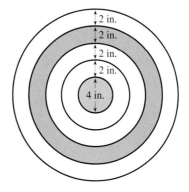

Chapter Test

Take this test as you would take a test in class. After you are done, check your work against the answers given in the back of the book.

1. Write the first five terms of the sequence $a_n = \left(-\frac{2}{3}\right)^{n-1}$. (Begin with $n = 1$.)

2. Evaluate: $\displaystyle\sum_{j=0}^{4} (3j + 1)$

3. Evaluate: $\displaystyle\sum_{n=1}^{5} (3 - 4n)$

4. Use sigma notation to write $\dfrac{2}{3(1) + 1} + \dfrac{2}{3(2) + 1} + \cdots + \dfrac{2}{3(12) + 1}$.

5. Write the first five terms of the arithmetic sequence whose first term is $a_1 = 12$ and whose common difference is $d = 4$.

6. Find a formula for the nth term of the arithmetic sequence whose first term is $a_1 = 5000$ and whose common difference is $d = -100$.

7. Find the sum of the first 50 positive integers that are multiples of 3.

8. Find the common ratio of the geometric sequence: $2, -3, \frac{9}{2}, -\frac{27}{4}, \ldots$.

9. Find a formula for the nth term of the geometric sequence whose first term is $a_1 = 4$ and whose common ratio is $r = \frac{1}{2}$.

10. Evaluate: $\displaystyle\sum_{n=1}^{8} 2(2^n)$

11. Evaluate: $\displaystyle\sum_{n=1}^{10} 3\left(\frac{1}{2}\right)^n$

12. Evaluate: $\displaystyle\sum_{i=1}^{\infty} \left(\frac{1}{2}\right)^i$

13. Evaluate: $\displaystyle\sum_{i=1}^{\infty} 4\left(\frac{2}{3}\right)^{i-1}$

14. Fifty dollars is deposited each month in an increasing annuity that pays 8%, compounded monthly. What is the balance after 25 years?

15. Evaluate: $_{20}C_3$

16. Explain how to use Pascal's Triangle to expand $(x - 2)^5$.

17. Find the coefficient of the term x^3y^5 in the expansion of $(x + y)^8$.

18. How many license plates can consist of one letter followed by three digits?

19. Four students are randomly selected from a class of 25 to answer questions from a reading assignment. In how many ways can the four be selected?

20. The weather report indicates that the probability of snow tomorrow is 0.75. What is the probability that it will not snow?

21. A card is drawn from a standard deck of playing cards. Find the probability that it is a red face card.

22. Suppose two spark plugs require replacement in a four-cylinder engine. If the mechanic randomly removes two plugs, find the probability that they are the two defective plugs.

Appendix A

Introduction to Graphing Utilities

Introduction ▪ Using a Graphing Utility ▪ Using Special Features of a Graphing Utility

Introduction

In Section 2.2 you studied the point-plotting method for sketching the graph of an equation. One of the disadvantages of the point-plotting method is that to get a good idea about the shape of a graph you need to plot *many* points. By plotting only a few points, you can badly misrepresent the graph.

For instance, consider the equation $y = x^3$. To graph this equation, suppose you calculated only the following three points.

x	-1	0	1
$y = x^3$	-1	0	1

By plotting these three points, as shown in Figure A.1(a), you might assume that the graph of the equation is a straight line. This, however, is not correct. By plotting several more points, as shown in Figure A.1(b), you can see that the actual graph is not straight at all.

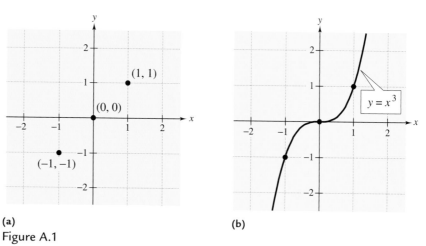

(a)   (b)

Figure A.1

So, the point-plotting method leaves you with a dilemma. On the one hand, the method can be very inaccurate if only a few points are plotted. But, on the other hand, it is very time-consuming to plot a dozen (or more) points. Technology can help you solve this dilemma. Plotting several points (or even hundreds of points) on a rectangular coordinate system is something that a computer or graphing calculator can do easily.

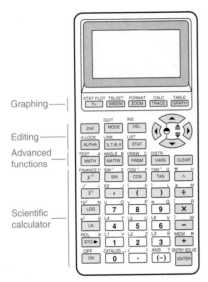

Graphing

Editing

Advanced functions

Scientific calculator

Figure A.2 *Keypad of a TI-83 Graphics Calculator*

Using a Graphing Utility

There are many different graphing utilities: some are graphing packages for computers and some are hand-held graphing calculators. In this section we describe the steps used to graph an equation with a *TI-83* graphing utility. (See Figure A.2.) We will often give keystroke sequences for illustration; however, these may not agree precisely with the steps required by *your* calculator.[*]

▶ Graphing an Equation with a *TI-83* Graphing Calculator

Before performing the following steps, set your calculator so that all of the standard defaults are active. For instance, all of the options at the left of the [MODE] screen should be highlighted.

1. Set the viewing window for the graph. (See Example 3.) To set the standard viewing window, press [ZOOM] 6.

2. Rewrite the equation so that *y* is isolated on the left side of the equation.

3. Press the [Y=] key. Then enter the right side of the equation on the first line of the display. (The first line is labeled $Y_1 = .$)

4. Press the [GRAPH] key.

Example 1 Graphing a Linear Equation

Sketch the graph of $2y + x = 4$.

Solution

To begin, solve the given equation for *y* in terms of *x*.

$$2y + x = 4 \qquad \text{Original equation}$$

$$2y = -x + 4 \qquad \text{Subtract } x \text{ from both sides.}$$

$$y = -\frac{1}{2}x + 2 \qquad \text{Divide both sides by 2.}$$

Press the [Y=] key, and enter the following keystrokes.

[(−)] [X,T,θ,n] [÷] 2 [+] 2

The top row of the display should now be as follows.

$$Y_1 = \text{-X/2} + 2$$

Press the [GRAPH] key, and the screen should look like that shown in Figure A.3.

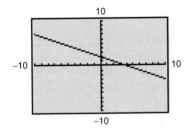

Figure A.3

---

[*]The graphing calculator keystrokes given in this section correspond to the *TI-83* graphing utility by Texas Instruments. For other graphing utilities, the keystrokes may differ. Consult your user's guide.

In Figure A.3, notice that the calculator screen does not label the tick marks on the *x*-axis or the *y*-axis. To see what the tick marks represent, you can press [WINDOW]. If you set your calculator to the standard graphing defaults before working Example 1, the screen should show the following values.

Xmin = -10   The minimum *x*-value is −10.
Xmax = 10   The maximum *x*-value is 10.
Xscl = 1   The *x*-scale is 1 unit per tick mark.
Ymin = -10   The minimum *y*-value is −10.
Ymax = 10   The maximum *y*-value is 10.
Yscl = 1   The *y*-scale is 1 unit per tick mark.
Xres = 1   Sets the pixel resolution.

These settings are summarized visually in Figure A.4.

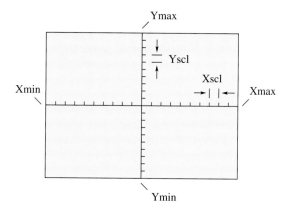

Figure A.4

Example 2   Graphing an Equation Involving Absolute Value

Sketch the graph of $y = |x - 3|$.

Solution

This equation is already written so that *y* is isolated on the left side of the equation. Press the [Y=] key, and enter the following keystrokes.

[ABS] [(] [X,T,θ,n] [−] 3 [)]

The top row of the display should now be as follows.

$$Y_1 = \text{abs}(X - 3)$$

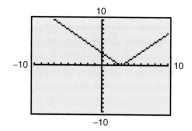

Figure A.5

Press the [GRAPH] key, and the screen should look like that shown in Figure A.5.

Using the Special Features of a Graphing Utility

To use your graphing utility to its best advantage, you must learn to set the viewing window, as illustrated in the next example.

Example 3 Setting the Viewing Window

Sketch the graph of $y = x^2 + 12$.

Solution

Press Y= and enter $x^2 + 12$ on the first line.

X,T,θ,n x^2 + 12

Press the GRAPH key. If your calculator is set to the standard viewing window, nothing will appear on the screen. The reason for this is that the lowest point on the graph of $y = x^2 + 12$ occurs at the point $(0, 12)$. Using the standard viewing window, you obtain a screen whose largest y-value is 10. In other words, none of the graph is visible on a screen whose y-values vary between -10 and 10, as shown in Figure A.6(a). To change these settings, press WINDOW and enter the following values.

Xmin = -10	The minimum x-value is -10.
Xmax = 10	The maximum x-value is 10.
Xscl = 1	The x-scale is 1 unit per tick mark.
Ymin = -10	The minimum y-value is -10.
Ymax = 30	The maximum y-value is 30.
Yscl = 5	The y-scale is 5 units per tick mark.
Xres = 1	Sets the pixel resolution.

Press GRAPH and you will obtain the graph shown in Figure A.6(b). On this graph, note that each tick mark on the y-axis represents 5 units because you changed the y-scale to 5. Also note that the highest point on the y-axis is now 30 because you changed the maximum value of y to 30.

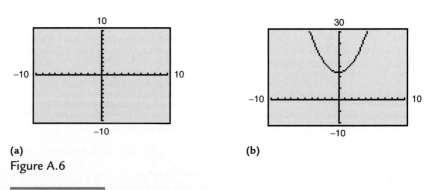

(a)   (b)

Figure A.6

If you changed the y-maximum and y-scale on your utility as indicated in Example 3, you should return to the standard settings before working Example 4. To do this, press ZOOM 6.

Example 4   Using a Square Setting

Sketch the graph of $y = x$. The graph of this equation is a straight line that makes a 45° angle with the x-axis and with the y-axis. From the graph on your utility, does the angle appear to be 45°?

Solution

Press $\boxed{Y=}$ and enter x on the first line.

$$Y_1 = X$$

Press the $\boxed{\text{GRAPH}}$ key and you will obtain the graph shown in Figure A.7(a). Notice that the angle the line makes with the x-axis doesn't appear to be 45°. The reason for this is that the screen is wider than it is tall. This makes the tick marks on the x-axis farther apart than the tick marks on the y-axis. To obtain the same distance between tick marks on both axes, you can change the graphing settings from "standard" to "square." To do this, press the following keys.

$\boxed{\text{ZOOM}}$ 5   Square setting

The screen should look like that shown in Figure A.7(b). Note in this figure that the square setting has changed the viewing window so that the x-values vary between -15 and 15.

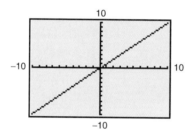

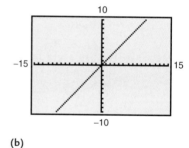

(a)   (b)

Figure A.7

There are many possible square settings on a graphing utility. To create a square setting, you need the following ratio to be $\frac{2}{3}$.

$$\frac{\text{Ymax} - \text{Ymin}}{\text{Xmax} - \text{Xmin}}$$

For instance, the setting in Example 4 is square because $(\text{Ymax} - \text{Ymin}) = 20$ and $(\text{Xmax} - \text{Xmin}) = 30$.

Example 5   Sketching More than One Graph on the Same Screen

Sketch the graphs of the following equations on the same screen.

$$y = -x + 4, \quad y = -x, \quad \text{and} \quad y = -x - 4$$

Solution

To begin, press $\boxed{Y=}$ and enter all three equations on the first three lines. The display should now be as follows.

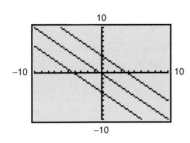

Figure A.8

$$Y_1 = \text{-}X + 4 \qquad \boxed{(\text{-})} \; \boxed{X,T,\theta,n} \; \boxed{+} \; 4$$

$$Y_2 = \text{-}X \qquad\qquad \boxed{(\text{-})} \; \boxed{X,T,\theta,n}$$

$$Y_3 = \text{-}X - 4 \qquad \boxed{(\text{-})} \; \boxed{X,T,\theta,n} \; \boxed{-} \; 4$$

Press the $\boxed{\text{GRAPH}}$ key and you will obtain the graph shown in Figure A.8. Note that the graph of each equation is a straight line, and that the lines are parallel to each other.

_____

Another special feature of a graphing utility is the trace feature. This feature is used to find solution points of an equation. For example, you can approximate the x- and y-intercepts of $y = 3x + 6$ by first graphing the equation, then pressing the $\boxed{\text{TRACE}}$ key, and finally pressing the $\boxed{\blacktriangleleft}\boxed{\blacktriangleright}$ keys. To get a better approximation of a solution point, you can use the following keystrokes repeatedly.

$$\boxed{\text{ZOOM}} \; 2 \; \boxed{\text{ENTER}}$$

Check to see that you get an x-intercept of $(-2, 0)$ and a y-intercept of $(0, 6)$. Use the trace feature to find the x- and y-intercepts of $y = \frac{1}{2}x - 4$.

Appendix A   Exercises

In Exercises 1–12, use a graphing utility to graph the equation. (Use the standard setting.)

1. $y = -3x$
2. $y = x - 4$
3. $y = \frac{3}{4}x - 6$
4. $y = -3x + 2$
5. $y = \frac{1}{2}x^2$
6. $y = -\frac{2}{3}x^2$
7. $y = x^2 - 4x + 2$
8. $y = -0.5x^2 - 2x + 2$
9. $y = |x - 3|$
10. $y = |x + 4|$
11. $y = |x^2 - 4|$
12. $y = |x - 2| - 5$

In Exercises 13–16, use a graphing utility to graph the equation using the given window settings.

13. $y = 27x + 100$
14. $y = 50,000 - 6000x$

Xmin = 0
Xmax = 5
Xscl = .5
Ymin = 75
Ymax = 250
Yscl = 25
Xres = 1

Xmin = 0
Xmax = 7
Xscl = .5
Ymin = 0
Ymax = 50000
Yscl = 5000
Xres = 1

15. $y = 0.001x^2 + 0.5x$
16. $y = 100 - 0.5|x|$

Xmin = -500
Xmax = 200
Xscl = 50
Ymin = -100
Ymax = 100
Yscl = 20
Xres = 1

Xmin = -300
Xmax = 300
Xscl = 60
Ymin = -100
Ymax = 100
Yscl = 20
Xres = 1

In Exercises 17–20, find a viewing window that shows the important characteristics of the graph.

17. $y = 15 + |x - 12|$
18. $y = 15 + (x - 12)^2$
19. $y = -15 + |x + 12|$
20. $y = -15 + (x + 12)^2$

In Exercises 21–24, graph both equations on the same screen. Are the graphs identical? If so, what rule of algebra is being illustrated?

21. $y_1 = 2x + (x + 1)$
   $y_2 = (2x + x) + 1$
22. $y_1 = \frac{1}{2}(3 - 2x)$
   $y_2 = \frac{3}{2} - x$
23. $y_1 = 2\left(\frac{1}{2}\right)$
   $y_2 = 1$
24. $y_1 = x(0.5x)$
   $y_2 = (0.5x)x$

In Exercises 25–32, use the trace feature of a graphing utility to approximate the *x*- and *y*-intercepts of the graph.

25. $y = 9 - x^2$

26. $y = 3x^2 - 2x - 5$

27. $y = 6 - |x + 2|$

28. $y = |x - 2|^2 - 3$

29. $y = 2x - 5$

30. $y = 4 - |x|$

31. $y = x^2 + 1.5x - 1$

32. $y = x^3 - 4x$

Geometry   In Exercises 33–36, graph the equations on the same display. Using a "square setting," determine the geometrical shape bounded by the graphs.

33. $y = -4, \quad y = -|x|$

34. $y = |x|, \quad y = 5$

35. $y = |x| - 8, \quad y = -|x| + 8$

36. $y = -\frac{1}{2}x + 7, \quad y = \frac{8}{3}(x + 5), \quad y = \frac{2}{7}(3x - 4)$

Modeling Data   In Exercises 37 and 38, use the following models, which give the number of pieces of first-class mail and the number of periodicals handled by the U.S. Postal Service.

First Class

$y = 0.07x^2 + 1.06x + 88.97, \quad 0 \le x \le 7$

Periodicals

$y = 0.02x^2 - 0.23x + 10.70, \quad 0 \le x \le 7$

In these models, *y* is the number of pieces handled (in billions) and *x* is the year, with $x = 0$ corresponding to 1990.   (Source: U.S. Postal Service)

37. Use the following setting to graph both models on the same display of a graphing utility.

Xmin = 0
Xmax = 7
Xscl = 1
Ymin = -5
Ymax = 115
Yscl = 10
Xres = 1

38. (a) Were the numbers of pieces of first-class mail and periodicals increasing or decreasing over time?

(b) Is the distance between the graphs increasing or decreasing over time? What does this mean to the U.S. Postal Service?

Appendix B
Further Concepts in Geometry

B.1 Exploring Congruence and Similarity

Identifying Congruent Figures ■ Identifying Similar Figures ■
Reading and Using Definitions ■ Congruent Triangles ■ Classifying Triangles

Identifying Congruent Figures

Two figures are *congruent* if they have the same shape and the same size. Each of the triangles in Figure B.1 is congruent to each of the other triangles. The triangles in Figure B.2 are not congruent to each other.

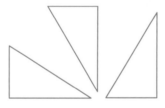

Congruent
Figure B.1

Not Congruent
Figure B.2

Notice that two figures can be congruent without having the same orientation. If two figures are congruent, then either one can be moved (and turned or flipped if necessary) so that it coincides with the other figure.

Example 1 Dividing Regions into Congruent Parts

Divide the region into two congruent parts.

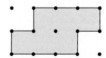

Solution
There are many solutions to this problem. Some of the solutions are shown in Figure B.3. Can you think of others?

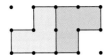  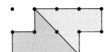 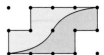

Figure B.3

Identifying Similar Figures

Two figures are *similar* if they have the same shape. (They may or may not have the same size.) Each of the quadrilaterals in Figure B.4 is similar to the others. The quadrilaterals in Figure B.5 are not similar to each other.

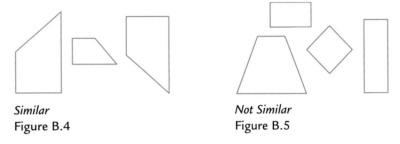

Similar
Figure B.4

Not Similar
Figure B.5

Example 2   Determining Similarity

Two of the figures are similar. Which two are they?

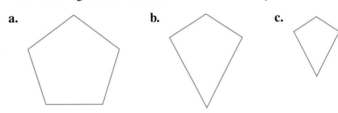

a.   **b.**   **c.**

Solution

The first figure has five sides and the other two figures have four sides. Because similar figures must have the same shape, the first figure is not similar to either of the others. Because you are told that two figures are similar, it follows that the second and third figures are similar.

Example 3   Determining Similarity

You wrote an essay on Euclid, the Greek mathematician who is famous for writing a geometry book titled *Elements of Geometry*. You are making a copy of the essay using a photocopier that is set at 75% reduction. Is each image on the copied pages similar to its original?

Solution

Every image on a copied page *is* similar to its original. The copied pages are smaller, but that doesn't matter because similar figures do not have to be the same size.

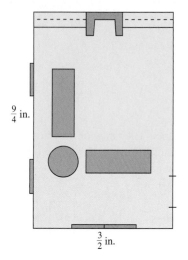

$\frac{9}{4}$ in.

$\frac{3}{2}$ in.

Figure B.6

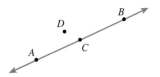

C is between A and B. D is not between A and B.

Figure B.7

Example 4   Drawing an Object to Scale

You are drawing a floor plan of a building. You choose a scale of $\frac{1}{8}$ inch to 1 foot. That is, $\frac{1}{8}$ inch of the floor plan represents 1 foot of the actual building. What dimensions should you draw for a room that is 12 feet wide and 18 feet long?

Solution

Because each foot is represented as $\frac{1}{8}$ inch, the width of the room should be

$$12\left(\tfrac{1}{8}\right) = \tfrac{12}{8} = \tfrac{3}{2} = 1\tfrac{1}{2}$$

and the length of the room should be

$$18\left(\tfrac{1}{8}\right) = \tfrac{18}{8} = \tfrac{9}{4} = 2\tfrac{1}{4}.$$

The scale dimensions of the room are $1\frac{1}{2}$ inches by $2\frac{1}{4}$ inches. See Figure B.6.

Reading and Using Definitions

A definition uses *known* words to describe a *new* word. If no words were known, then no new words could be defined. Hence, some words such as **point, line,** and **plane** must be commonly understood without being defined. Some statements such as "a point lies on a line" and "point *C* lies between points *A* and *B*" are also not defined. See Figure B.7.

▶ **Segments and Rays**

Consider the line $\overleftrightarrow{AB}$ that contains the points *A* and *B*. (In geometry, the word *line* means a *straight line*.)

The **line segment** (or simply **segment**) $\overline{AB}$ consists of the *endpoints A* and *B* and all points on the line $\overleftrightarrow{AB}$ that lie between *A* and *B*.

The **ray** $\overrightarrow{AB}$ consists of the *initial point A* and all points on the line $\overleftrightarrow{AB}$ that lie on the same side of *A* as *B* lies. If *C* is between *A* and *B*, then $\overrightarrow{CA}$ and $\overrightarrow{CB}$ are **opposite** rays.

Points, segments, or rays that lie on the same line are **collinear.**

Lines are drawn with two arrowheads, line segments are drawn with no arrowhead, and rays are drawn with a single arrowhead. See Figure B.8.

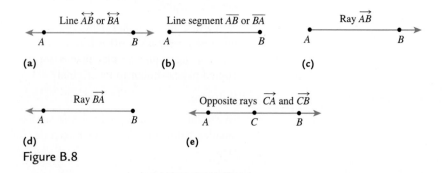

Figure B.8

It follows that $\overline{AB}$ and $\overline{BA}$ denote the same segment, but $\overrightarrow{AB}$ and $\overrightarrow{BA}$ do not denote the same ray. No length is given to lines or rays because each is infinitely long. The *length* of the line segment $\overline{AB}$ is denoted by AB.

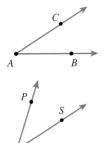

▶ **Angles**

An **angle** consists of two different rays that have the same initial point. The rays are the *sides* of the angle. The angle that consists of the rays $\overrightarrow{AB}$ and $\overrightarrow{AC}$ is denoted by $\angle BAC$, $\angle CAB$, or $\angle A$. The point A is the **vertex** of the angle. See Figure B.9.

The measure of $\angle A$ is denoted by $m\angle A$. Angles are classified as **acute, right, obtuse,** and **straight.**

Acute	$0° < m\angle A < 90°$
Right	$m\angle A = 90°$
Obtuse	$90° < m\angle A < 180°$
Straight	$m\angle A = 180°$

The top angle can be denoted by $\angle A$ or by $\angle BAC$. In the lower figure, the angle $\angle ROS$ should not be denoted by $\angle O$ because the figure contains three angles whose vertex is O.
Figure B.9

In geometry, *unless specifically stated otherwise*, angles are assumed to have a measure that is greater than 0° and less than or equal to 180°.

Every nonstraight angle has an **interior** and an **exterior.** A point D is in the interior of $\angle A$ if it is between points that lie on each side of the angle. Two angles (such as $\angle ROS$ and $\angle SOP$ shown in Figure B.9) are **adjacent** if they share a common vertex and side, but have no common interior points. In Figure B.10, $\angle 1$ and $\angle 3$ share a vertex, but not a common side, so $\angle 1$ and $\angle 3$ are not adjacent.

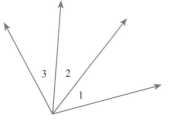

$\angle 1$ and $\angle 2$ are adjacent.
$\angle 1$ and $\angle 3$ are not adjacent.
Figure B.10

▶ **Segment and Angle Congruence**

Two segments are **congruent,** $\overline{AB} \cong \overline{CD}$, if they have the same length. Two angles are **congruent,** $\angle P \cong \angle Q$, if they have the same measure.

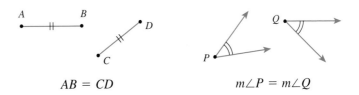

$$AB = CD \qquad\qquad m\angle P = m\angle Q$$

Definitions can always be interpreted "forward" and "backward." For instance, the definition of congruent segments means (1) if two segments have the same measure, then they are congruent, and (2) if two segments are congruent, then they have the same measure. You learned that two figures are congruent if they have the same shape and size.

Congruent Triangles

If $\triangle ABC$ is **congruent** to $\triangle PQR$, then there is a correspondence between their angles and sides such that corresponding angles are congruent and corresponding sides are congruent. The notation $\triangle ABC \cong \triangle PQR$ indicates the congruence *and* the correspondence, as shown in Figure B.11. If two triangles are congruent, then you know that they share many properties.

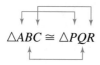

$$\triangle ABC \cong \triangle PQR$$

Corresponding angles are	Corresponding sides are
$\angle A \cong \angle P$	$\overline{AB} \cong \overline{PQ}$
$\angle B \cong \angle Q$	$\overline{BC} \cong \overline{QR}$
$\angle C \cong \angle R$	$\overline{CA} \cong \overline{RP}$

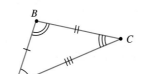

Figure B.11

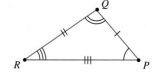

Example 5   Naming Congruent Parts

You and a friend have identical drafting triangles, as shown in Figure B.12. Name all congruent parts.

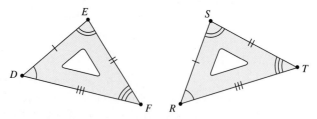

Figure B.12

Solution

Given that $\triangle DEF \cong \triangle RST$, the congruent angles and sides are as follows.

Angles: $\angle D \cong \angle R$, $\qquad \angle E \cong \angle S$, $\qquad \angle F \cong \angle T$

Sides: $\overline{DE} \cong \overline{RS}$, $\qquad \overline{EF} \cong \overline{ST}$, $\qquad \overline{FD} \cong \overline{TR}$

Classifying Triangles

A triangle can be classified by relationships among its sides or among its angles, as shown in the following definitions.

▶ **Classification by Sides**

1. An **equilateral triangle** has three congruent sides.

2. An **isosceles triangle** has at least two congruent sides.

3. A **scalene triangle** has no sides congruent.

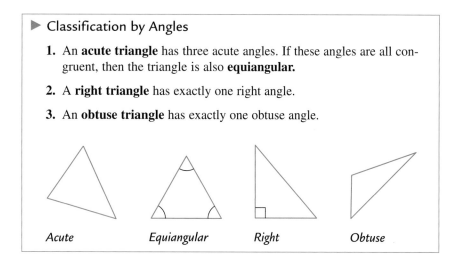

Equilateral   *Isosceles*   *Scalene*

▶ **Classification by Angles**

1. An **acute triangle** has three acute angles. If these angles are all congruent, then the triangle is also **equiangular.**

2. A **right triangle** has exactly one right angle.

3. An **obtuse triangle** has exactly one obtuse angle.

Acute   *Equiangular*   *Right*   *Obtuse*

In △*ABC*, each of the points *A*, *B*, and *C* is a **vertex** of the triangle. (The plural of vertex is *vertices*.) The side $\overline{BC}$ is the side *opposite* ∠*A*. Two sides that share a common vertex are *adjacent sides* (see Figure B.13).

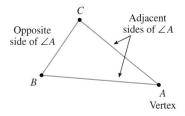

Figure B.13

The sides of right triangles and isosceles triangles are given special names. In a right triangle, the sides adjacent to the right angle are the **legs** of the triangle. The side opposite the right angle is the **hypotenuse** of the triangle (see Figure B.14).

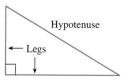

Figure B.14

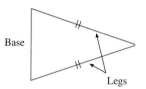

Figure B.15

An isosceles triangle can have three congruent sides. If it has only two, then the two congruent sides are the **legs** of the triangle. The third side is the **base** of the triangle (see Figure B.15).

B.1   Exercises

In Exercises 1–3, copy the region on a piece of dot paper. Then divide the region into two congruent parts. How many different ways can you do this?

1.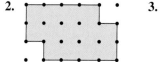

2. 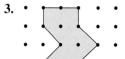   3.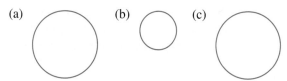

4. Two of the figures are congruent. Which are they?

(a)   (b)   (c)

5. Two of the figures are similar. Which are they?

(a)   (b)   (c)

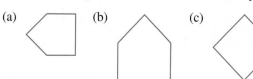

In Exercises 6 and 7, copy the region on a piece of paper. Then divide the region into four congruent parts.

6.    7.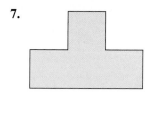

In Exercises 8–11, use the triangular grid below. In the grid, each small triangle has sides of 1 unit.

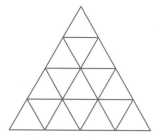

8. How many congruent triangles with 1-unit sides are in the grid?

9. How many congruent triangles with 2-unit sides are in the grid?

10. How many congruent triangles with 3-unit sides are in the grid?

11. Does the grid contain triangles that are not similar to each other?

12. *True or False?*   If two figures are congruent, then they are similar.

13. *True or False?*   If two figures are similar, then they are congruent.

14. *True or False?*   A triangle can be similar to a square.

15. *True or False?*   Any two squares are similar.

In Exercises 16–19, match the description with its correct notation.

(a) $\overline{PQ}$ (b) PQ (c) $\overleftrightarrow{PQ}$ (d) $\overrightarrow{PQ}$

16. The line through P and Q

17. The ray from P through Q

18. The segment between P and Q

19. The length of the segment between P and Q

20. The point R is between points S and T. Which of the following are true?

(a) R, S, and T are collinear.

(b) $\overrightarrow{SR}$ is the same as $\overrightarrow{ST}$.

(c) $\overline{ST}$ is the same as $\overline{TS}$.

(d) $\overrightarrow{ST}$ is the same as $\overrightarrow{TS}$.

In Exercises 21–23, use the figure below.

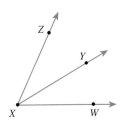

21. The figure shows three angles whose vertex is X. Write two names for each angle. Which two angles are adjacent?

22. Is Y in the interior or exterior of $\angle WXZ$?

23. Which is the best estimate for $m\angle WXY$?

(a) $15°$ (b) $30°$ (c) $45°$

In Exercises 24–29, match the triangle with its name.

(a) Equilateral   (b) Scalene   (c) Obtuse

(d) Equiangular   (e) Isosceles   (f) Right

24. Side lengths: 2 cm, 3 cm, 4 cm

25. Angle measures: $60°$, $60°$, $60°$

26. Side lengths: 3 cm, 2 cm, 3 cm

27. Angle measures: $30°$, $60°$, $90°$

28. Side lengths: 4 cm, 4 cm, 4 cm

29. Angle measures: $20°$, $145°$, $15°$

In Exercises 30–32, use the figure, in which $\triangle LMP \cong \triangle ONQ$.

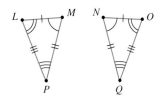

30. Name three pairs of congruent angles.

31. Name three pairs of congruent sides.

32. If $\triangle LMP$ is isosceles, explain why $\triangle ONQ$ must be isosceles.

In Exercises 33–36, use the definition of congruence to complete the statement.

33. If $\triangle ABC \cong \triangle TUV$, then $m\angle C = $ _____ .

34. If $\triangle PQR \cong \triangle XYZ$, then $\angle P \cong$ _____ .

35. If $\triangle LMN \cong \triangle TUV$, then $\overline{LN} \cong$ _____ .

36. If $\triangle DEF \cong \triangle NOP$, then $DE = $ _____ .

37. Copy and complete the table. Write *Yes* if it is possible to sketch a triangle with both characteristics. Write *No* if it is not possible. Illustrate your results with sketches. (The first is done for you.)

	Scalene	Isosceles	Equilateral
Acute	Yes	?	?
Obtuse	?	?	?
Right	?	?	?

Acute and Scalene

In Exercises 38–41, △ABC is isosceles with $\overline{AC} \cong \overline{BC}$. Solve for x. Then decide whether the triangle is equilateral. (The figures are not necessarily drawn to scale.)

38.

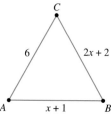

39.

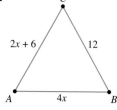

40.

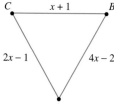

41.

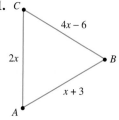

Coordinate Geometry   In Exercises 44 and 45, use the following figure.

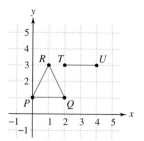

44. Find a location of S such that △PQR ≅ △PQS.

45. Find two locations of V such that △PQR ≅ △TUV.

46. *Logical Reasoning*   Arrange 16 toothpicks as shown below. What is the least number of toothpicks you must remove to create four congruent triangles? (Each toothpick must be the side of at least one triangle.) Sketch your result.

47. *Logical Reasoning*   Show how you could arrange six toothpicks to form four congruent triangles. Each triangle has one toothpick for each side, and you cannot bend, break, or overlap the toothpicks. (*Hint:* The figure can be three-dimensional.)

42. *Landscape Design*   You are designing a patio. Your plans use a scale of $\frac{1}{8}$ inch to 1 foot. The patio is 24 feet by 36 feet. What are its dimensions on the plans?

43. *Architecture*   The Pentagon, near Washington D.C., covers a region that is about 1200 feet by 1200 feet. About how large would a $\frac{1}{8}$-inch to 1-foot scale drawing of the Pentagon be? Would such a scale be reasonable?

B.2   Angles

Identifying Special Pairs of Angles   ■   Angles Formed by a Transversal   ■   Angles of a Triangle

Identifying Special Pairs of Angles

You have been introduced to several definitions concerning angles. For instance, you know that two angles are *adjacent* if they share a common vertex and side but have no common interior points. Here are some other definitions for pairs of angles. See Figure B.16.

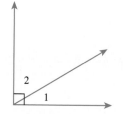

∠1 *and* ∠2 *are complementary angles.*
Figure B.16

> ### Definitions for Pairs of Angles
>
> Two angles are **vertical angles** if their sides form two pairs of opposite rays.
>
> Two adjacent angles are a **linear pair** if their noncommon sides are opposite rays.
>
> Two angles are **complementary** if the sum of their measures is 90°. Each angle is the *complement* of the other.
>
> Two angles are **supplementary** if the sum of their measures is 180°. Each angle is the *supplement* of the other.

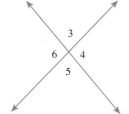

Figure B.17

| Example 1 | Identifying Special Pairs of Angles |

Use the terms defined above to describe relationships between the labeled angles in Figure B.17.

Solution

a. $\angle 3$ and $\angle 5$ are vertical angles. So are $\angle 4$ and $\angle 6$.

b. There are four sets of linear pairs:

$\angle 3$ and $\angle 4$, $\angle 4$ and $\angle 5$, $\angle 5$ and $\angle 6$, and $\angle 3$ and $\angle 6$.

The angles in each of these pairs are also supplementary angles.

In Example 1(b), note that the linear pairs are also supplementary. This result is stated in the following postulate.

> ### Linear Pair Postulate
>
> If two angles form a linear pair, then they are supplementary—i.e., the sum of their measures is 180°.

The relationship between vertical angles is stated in the following theorem.

> ### Vertical Angles Theorem
>
> If two angles are vertical angles, then they are congruent.

Angles Formed by a Transversal

A **transversal** is a line that intersects two or more coplanar lines at different points. The angles that are formed when the transversal intersects the lines have the following names.

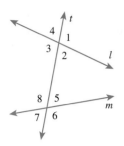

Figure B.18

> ▶ **Angles Formed by a Transversal**
>
> In Figure B.18, the transversal t intersects the lines l and m.
>
> Two angles are **corresponding angles** if they occupy corresponding positions, such as $\angle 1$ and $\angle 5$.
>
> Two angles are **alternate interior angles** if they lie between l and m on opposite sides of t, such as $\angle 2$ and $\angle 8$.
>
> Two angles are **alternate exterior angles** if they lie outside l and m on opposite sides of t, such as $\angle 1$ and $\angle 7$.
>
> Two angles are **consecutive interior angles** if they lie between l and m on the same side of t, such as $\angle 2$ and $\angle 5$.

Example 2   Naming Pairs of Angles

In Figure B.19, how is $\angle 9$ related to the other angles?

Solution

You can consider that $\angle 9$ is formed by the transversal l as it intersects m and n, or you can consider $\angle 9$ to be formed by the transversal m as it intersects l and n. Considering one or the other of these, you have the following.

a. $\angle 9$ and $\angle 10$ are a linear pair. So are $\angle 9$ and $\angle 12$.

b. $\angle 9$ and $\angle 11$ are vertical angles.

c. $\angle 9$ and $\angle 7$ are alternate exterior angles. So are $\angle 9$ and $\angle 3$.

d. $\angle 9$ and $\angle 5$ are corresponding angles. So are $\angle 9$ and $\angle 1$.

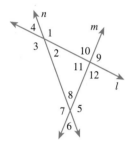

Figure B.19

To help build understanding regarding angles formed by a transversal, consider relationships between two lines. **Parallel lines** are coplanar lines that do not intersect. (Recall from Section 2.3 that two nonvertical lines are parallel if and only if they have the same slope.) **Intersecting lines** are coplanar and have exactly one point in common. If intersecting lines meet at right angles, they are perpendicular; otherwise, they are **oblique.** See Figure B.20.

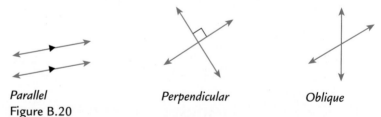

Parallel   *Perpendicular*   *Oblique*

Figure B.20

Many of the angles formed by a transversal that intersects *parallel* lines are congruent. The following postulate and theorems list useful results.

> ▶ **Corresponding Angles Postulate**
>
> If two parallel lines are cut by a transversal, then the pairs of corresponding angles are congruent.

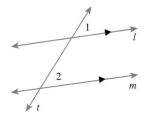

$l \parallel m$, $\angle 1 \cong \angle 2$
Figure B.21

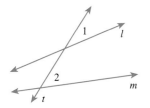

$l \nparallel m$, $\angle 1 \ncong \angle 2$
Figure B.22

Note that the hypothesis of this postulate states that the lines must be parallel, as shown in Figure B.21. If the lines are not parallel, then the corresponding angles are not congruent, as shown in Figure B.22.

▶ **Angle Theorems**

Alternate Interior Angles Theorem If two parallel lines are cut by a transversal, then the pairs of alternate interior angles are congruent.

Consecutive Interior Angles Theorem If two parallel lines are cut by a transversal, then the pairs of consecutive interior angles are supplementary.

Alternate Exterior Angles Theorem If two parallel lines are cut by a transversal, then the pairs of alternate exterior angles are congruent.

Perpendicular Transversal Theorem If a transversal is perpendicular to one of two parallel lines, then it is perpendicular to the second.

Example 3   Using Properties of Parallel Lines

In Figure B.23, lines r and s are parallel lines cut by a transversal, l. Find the measure of each labeled angle.

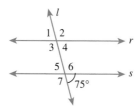

Figure B.23

Solution

$\angle 1$ and the given angle are alternate exterior angles and are congruent. So $m\angle 1 = 75°$. $\angle 5$ and the given angle are vertical angles. Because vertical angles are congruent, they have the same measure. So, $m\angle 5 = 75°$. Similarly, $\angle 1 \cong \angle 4$ and $m\angle 1 = m\angle 4 = 75°$. There are several sets of linear pairs, including

$\angle 1$ and $\angle 2$; $\angle 3$ and $\angle 4$; $\angle 5$ and $\angle 6$; $\angle 5$ and $\angle 7$.

The angles in each of these pairs are also supplementary angles; the sum of the measures of each pair of angles is 180°. Because one angle of each pair measures 75°, the supplements each measure 105°. So $\angle 2$, $\angle 3$, $\angle 6$, and $\angle 7$ each measure 105°.

Angles of a Triangle

The word "triangle" means "three angles." When the sides of a triangle are extended, however, other angles are formed. The original three angles of the triangle are the **interior angles.** The angles that are adjacent to the interior angles are the **exterior angles** of the triangle. Each vertex has a pair of exterior angles, as shown in Figure B.24 on the following page.

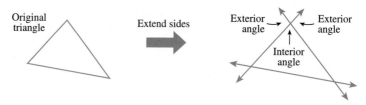

Figure B.24

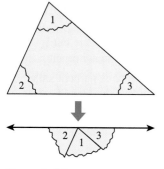

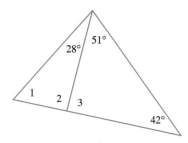

Figure B.25

You could cut a triangle out of a piece of paper. Tear off the three angles and place them adjacent to each other, as shown in Figure B.25. What do you observe? (You could perform this investigation by measuring with a protractor or using a computer drawing program.) You should arrive at the conclusion given in the following theorem.

▶ **Triangle Sum Theorem**

The sum of the measures of the interior angles of a triangle is 180°.

Example 4 Using the Triangle Sum Theorem

In the triangle in Figure B.26, find $m\angle 1$, $m\angle 2$, and $m\angle 3$.

Solution

To find the measure of $\angle 3$, use the Triangle Sum Theorem, as follows.

$$m\angle 3 = 180° - (51° + 42°) = 87°$$

Knowing the measure of $\angle 3$, you can use the Linear Pair Postulate to write $m\angle 2 = 180° - 87° = 93°$. Using the Triangle Sum Theorem, you have

$$m\angle 1 = 180° - (28° + 93°) = 59°.$$

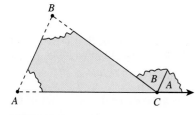

Figure B.26

The next theorem is one that you might have anticipated from the investigation earlier. As shown in Figure B.27, if you had torn only two of the angles from the paper triangle, you could put them together to cover exactly one of the exterior angles.

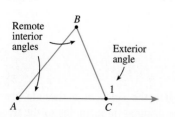

Figure B.27

▶ **Exterior Angle Theorem**

The measure of an exterior angle of a triangle is equal to the sum of the measures of the two remote (nonadjacent) interior angles. (See Figure B.28.)

Figure B.28

B.2   Exercises

In Exercises 1–6, sketch a pair of angles that fits the description. Label the angles as $\angle 1$ and $\angle 2$.

1. A linear pair of angles

2. Supplementary angles for which $\angle 1$ is acute

3. Acute vertical angles

4. Adjacent congruent complementary angles

5. Obtuse vertical angles

6. Adjacent congruent supplementary angles

In Exercises 7–12, use the figure to determine relationships between the given angles.

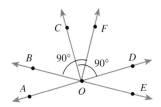

7. $\angle AOC$ and $\angle COD$

8. $\angle AOB$ and $\angle BOC$

9. $\angle BOC$ and $\angle COE$

10. $\angle AOB$ and $\angle EOD$

11. $\angle BOC$ and $\angle COF$

12. $\angle AOB$ and $\angle AOE$

In Exercises 13–18, use the following information to decide whether the statement is true or false. (*Hint:* Make a sketch.)

Vertical angles: $\angle 1$ and $\angle 2$;
Linear pairs: $\angle 1$ and $\angle 3$, $\angle 1$ and $\angle 4$.

13. If $m\angle 3 = 30°$, then $m\angle 4 = 150°$.

14. If $m\angle 1 = 150°$, then $m\angle 4 = 30°$.

15. $\angle 2$ and $\angle 3$ are congruent.

16. $m\angle 3 + m\angle 1 = m\angle 4 + m\angle 2$

17. $\angle 3 \cong \angle 4$

18. $m\angle 3 = 180° - m\angle 2$

In Exercises 19–24, find the value of x.

19.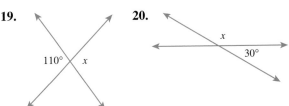

20.

21.

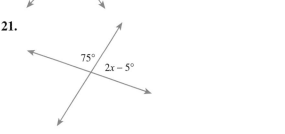

22. 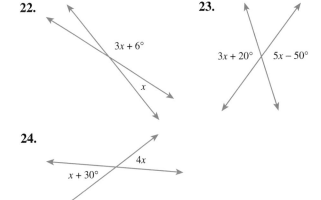   **23.**

24.

25. In the figure, P, S, and T are collinear. If $m\angle P = 40°$ and $m\angle QST = 110°$, what is $m\angle Q$? (*Hint:* $m\angle P + m\angle Q + m\angle PSQ = 180°$)

(a) $40°$ (b) $55°$ (c) $70°$ (d) $110°$ (e) $140°$

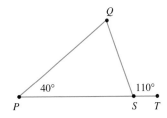

In Exercises 26–29, use the figure below.

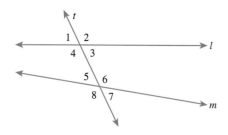

26. Name two corresponding angles.

27. Name two alternate interior angles.

28. Name two alternate exterior angles.

29. Name two consecutive interior angles.

In Exercises 30–33, $l_1 \parallel l_2$. Find the measures of $\angle 1$ and $\angle 2$. Explain your reasoning.

30.

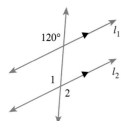

31.

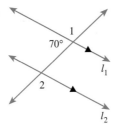

32.

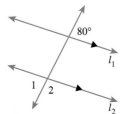

33.

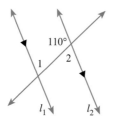

In Exercises 34–36, $m \parallel n$ and $k \parallel l$. Determine the values of a and b.

34.

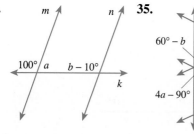

35.

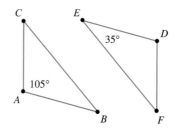

36.

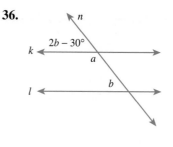

In Exercises 37–39, use the following figure.

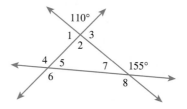

37. Name the interior angles of the triangle.

38. Name the exterior angles of the triangle.

39. Two angle measures are given in the figure. Find the measure of the eight labeled angles.

In Exercises 40–43, use the following figure, in which $\triangle ABC \cong \triangle DEF$.

40. What is the measure of $\angle D$?

41. What is the measure of $\angle B$?

42. What is the measure of $\angle C$?

43. What is the measure of $\angle F$?

44. *True or False?*   A right triangle can have an obtuse angle.

45. *True or False?*   A triangle that has two 60° angles must be equiangular.

46. *True or False?*   If a right triangle has two congruent angles, then it must have two 45° angles.

In Exercises 47 and 48, find the measure of each labeled angle.

47.

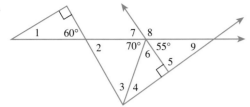

48.

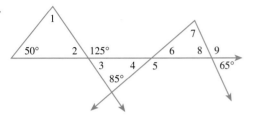

In Exercises 49 and 50, draw and label a right triangle, $\triangle ABC$, for which the right angle is $\angle C$. What is $m\angle B$?

49. $m\angle A = 13°$ **50.** $m\angle A = 47°$

In Exercises 51 and 52, draw two noncongruent, isosceles triangles that have an exterior angle with the given measure.

51. 130°   **52.** 145°

In Exercises 53–56, find the measures of the interior angles.

53.

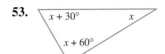

54.

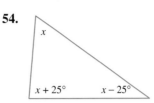

55.

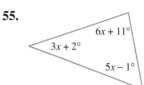

56.

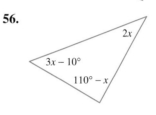

Appendix C

Further Concepts in Statistics

Stem-and-Leaf Plots   ■   Histograms and Frequency Distributions   ■
Line Graphs   ■   Choosing an Appropriate Graph   ■   Scatter Plots   ■
Fitting a Line to Data   ■   Measures of Central Tendency

Stem-and-Leaf Plots

Statistics is the branch of mathematics that studies techniques for collecting, organizing, and interpreting data. In this section, you will study several ways to organize and interpret data.

One type of plot that can be used to organize sets of numbers by hand is a **stem-and-leaf plot.** A set of test scores and the corresponding stem-and-leaf plot are shown below.

Test Scores	Stems	Leaves
93, 70, 76, 58, 86, 93, 82, 78, 83, 86,	5	8
64, 78, 76, 66, 83, 83, 96, 74, 69, 76,	6	4 4 6 9
64, 74, 79, 76, 88, 76, 81, 82, 74, 70	7	0 0 4 4 4 6 6 6 6 8 8 9
	8	1 2 2 3 3 3 6 6 8
	9	3 3 6

Note that the *leaves* represent the units digits of the numbers and the *stems* represent the tens digits. Stem-and-leaf plots can also be used to compare two sets of data, as shown in the following example.

Example 1   Comparing Two Sets of Data

Use a stem-and-leaf plot to compare the test scores given above with the following test scores. Which set of test scores is better?

90, 81, 70, 62, 64, 73, 81, 92, 73, 81, 92, 93, 83, 75, 76,
83, 94, 96, 86, 77, 77, 86, 96, 86, 77, 86, 87, 87, 79, 88

Solution

Begin by ordering the second set of scores.

62, 64, 70, 73, 73, 75, 76, 77, 77, 77, 79, 81, 81, 81, 83,
83, 86, 86, 86, 87, 87, 88, 90, 92, 92, 93, 94, 96, 96

Now that the data have been ordered, you can construct a *double* stem-and-leaf plot by letting the leaves to the right of the stems represent the units digits for the first group of test scores and letting the leaves to the left of the stems represent the units digits for the second group of test scores.

Leaves (2nd Group)	Stems	Leaves (1st Group)
	5	8
4 2	6	4 4 6 9
9 7 7 7 6 5 3 3 0	7	0 0 4 4 4 6 6 6 6 6 8 8 9
8 7 7 6 6 6 6 3 3 1 1 1	8	1 2 2 3 3 3 6 6 8
6 6 4 3 2 2 0	9	3 3 6

By comparing the two sets of leaves, you can see that the second group of test scores is better than the first group.

Example 2   Using a Stem-and-Leaf Plot

The table below shows the percent of the population of each state and the District of Columbia that was at least 65 years old in 1997. Use a stem-and-leaf plot to organize the data.   (Source: U.S. Bureau of the Census)

AK	5.3	AL	13.0	AR	14.3	AZ	13.2	CA	11.1
CO	10.1	CT	14.4	DC	13.9	DE	12.9	FL	18.5
GA	9.9	HI	13.2	IA	15.0	ID	11.3	IL	12.5
IN	12.5	KS	13.5	KY	12.5	LA	11.4	MA	14.1
MD	11.5	ME	13.9	MI	12.4	MN	12.3	MO	13.7
MS	12.2	MT	13.2	NC	12.5	ND	14.4	NE	13.7
NH	12.1	NJ	13.7	NM	11.2	NV	11.5	NY	13.4
OH	13.4	OK	13.4	OR	13.3	PA	15.8	RI	15.8
SC	12.1	SD	14.3	TN	12.5	TX	10.1	UT	8.7
VA	11.2	VT	12.3	WA	11.5	WI	13.2	WV	15.1
WY	11.3								

Solution

Begin by ordering the numbers, as shown below.

5.3, 8.7, 9.9, 10.1, 10.1, 11.1, 11.2, 11.2, 11.3, 11.3, 11.4,
11.5, 11.5, 11.5, 12.1, 12.1, 12.2, 12.3, 12.3, 12.4, 12.5,
12.5, 12.5, 12.5, 12.5, 12.9, 13.0, 13.2, 13.2, 13.2, 13.2,
13.3, 13.4, 13.4, 13.4, 13.5, 13.7, 13.7, 13.7, 13.9, 13.9,
14.1, 14.3, 14.3, 14.4, 14.4, 15.0, 15.1, 15.8, 15.8, 18.5

Next construct the stem-and-leaf plot using the leaves to represent the digits to the right of the decimal points.

Stems	Leaves	
5.	3	Alaska has the lowest percent.
6.		
7.		
8.	7	
9.	9	
10.	1 1	
11.	1 2 2 3 3 4 5 5 5	
12.	1 1 2 3 3 4 5 5 5 5 5 9	
13.	0 2 2 2 2 3 4 4 4 5 7 7 7 9 9	
14.	1 3 3 4 4	
15.	0 1 8 8	
16.		
17.		
18.	5	Florida has the highest percent.

Histograms and Frequency Distributions

With data such as those given in Example 2, it is useful to group the numbers into intervals and plot the frequency of the data in each interval. For instance, the **frequency distribution** and **histogram** shown in Figure C.1 represent the data given in Example 2.

Frequency Distribution

Interval	Tally
$[5, 7)$	I
$[7, 9)$	I
$[9, 11)$	III
$[11, 13)$	LHT LHT LHT LHT I
$[13, 15)$	LHT LHT LHT LHT
$[15, 17)$	IIII
$[17, 19)$	I

Histogram

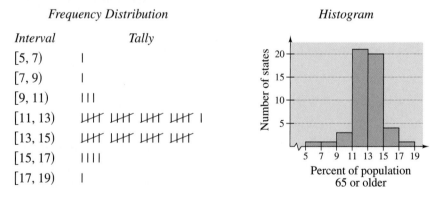

Figure C.1

A histogram has a portion of a real number line as its horizontal axis. A **bar graph** is similar to a histogram, except that the rectangles (bars) can be either horizontal or vertical and the labels of the bars are not necessarily numbers.

Another difference between a bar graph and a histogram is that the bars in a bar graph are usually separated by spaces, whereas the bars in a histogram are not separated by spaces.

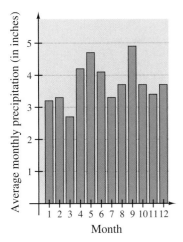

Figure C.2

The y-axis is labeled "Average monthly precipitation (in inches)" with values 1, 2, 3, 4, 5. The x-axis is labeled "Month" with values 1 2 3 4 5 6 7 8 9 10 11 12.

Example 3   Constructing a Bar Graph

The data below show the average monthly precipitation (in inches) in Houston, Texas. Construct a bar graph for these data. What can you conclude?   (Source: PC USA)

January	3.2	February	3.3	March	2.7
April	4.2	May	4.7	June	4.1
July	3.3	August	3.7	September	4.9
October	3.7	November	3.4	December	3.7

Solution

To create a bar graph, begin by drawing a vertical axis to represent the precipitation and a horizontal axis to represent the months. The bar graph is shown in Figure C.2. From the graph, you can see that Houston receives a fairly consistent amount of rain throughout the year—the driest month tends to be March and the wettest month tends to be September.

Line Graphs

A **line graph** is similar to a standard coordinate graph. Line graphs are usually used to show trends over periods of time.

Example 4   Constructing a Line Graph

The following data show the number of immigrants (in thousands) to the United States for the years 1970 through 1996. Construct a line graph of the data. What can you conclude?   (Source: U.S. Immigration and Naturalization Service)

Year	Number	Year	Number	Year	Number
1970	373	1971	370	1972	385
1973	400	1974	395	1975	386
1976	399	1977	462	1978	601
1979	460	1980	531	1981	597
1982	594	1983	560	1984	544
1985	570	1986	602	1987	602
1988	643	1989	1091	1990	1536
1991	1827	1992	974	1993	904
1994	804	1995	720	1996	916

Solution

Begin by drawing a vertical axis to represent the number of immigrants in thousands. Then label the horizontal axis with years and plot the points shown in the table. Finally, connect the points with line segments, as shown on the next page in Figure C.3. From the line graph, you can see that the number of immigrants steadily increased until 1989, when there was a sharp increase followed by a sudden decrease in 1992.

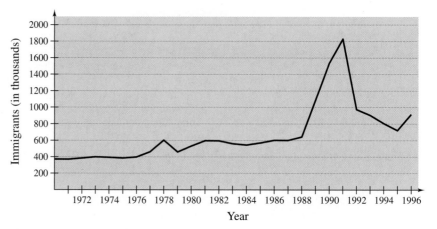

Figure C.3

Choosing an Appropriate Graph

Line graphs and bar graph are commonly used for displaying data. When you are using a graph to organize and present data, you must first decide which type of graph to use.

Example 5 Organizing Data with a Graph

Listed below are the daily average numbers of miles walked by people while working at their jobs. Organize the data graphically. (Source: American Podiatry Association)

Occupation	Miles Walked per Day
Mail Carrier	4.4
Medical Doctor	3.5
Nurse	3.9
Police Officer	6.8
Television Reporter	4.2

Solution

You can use a bar graph because the data fall into distinct categories, and it would be useful to compare totals. The bar graph shown in Figure C.4 is horizontal. This makes it easier to label each bar. Also notice that the occupations are listed in order of the number of miles walked.

Study Tip

Here are some guidelines to use when you must decide which type of graph to use.

1. Use a bar graph when the data fall into distinct categories and you want to compare totals.

2. Use a line graph when you want to show the relationship between consecutive amounts or data over time.

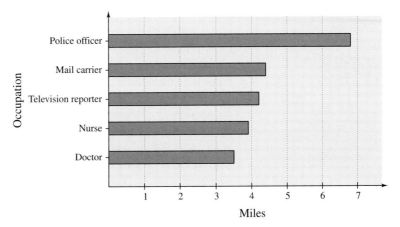

Figure C.4

Scatter Plots

Many real-life situations involve finding relationships between two variables, such as the year and the number of people in the labor force. In a typical situation, data are collected and written as a set of ordered pairs. The graph of such a set is called a **scatter plot.**

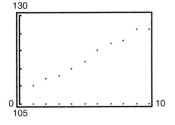

Figure C.5

From the scatter plot in Figure C.5 that relates the year t with the number of people in the labor force P, it appears that the points describe a relationship that is nearly linear. (The relationship is not *exactly* linear because the labor force did not increase by precisely the same amount each year.) A mathematical equation that approximates the relationship between t and P is called a *mathematical model.* When developing a mathematical model, you strive for two (often conflicting) goals—accuracy and simplicity.

Consider a collection of ordered pairs of the form (x, y). If y tends to increase as x increases, the collection is said to have a **positive correlation.** If y tends to decrease as x increases, the collection is said to have a **negative correlation.** Figure C.6 shows three examples: one with a positive correlation, one with a negative correlation, and one with no (discernible) correlation.

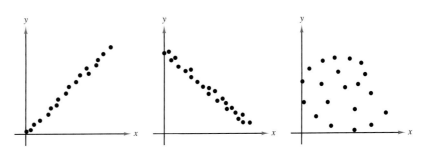

Figure C.6

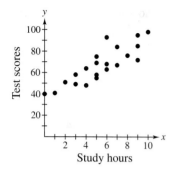

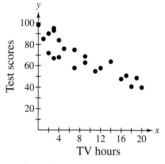

Figure C.7

Example 6   Interpreting Scatter Plots

On a Friday, 22 students in a class were asked to keep track of the numbers of hours they spent studying for a test on Monday and the numbers of hours they spent watching television. The numbers are shown below. Construct a scatter plot for each set of data. Then determine whether the points are positively correlated, are negatively correlated, or have no discernible correlation. What can you conclude? (The first coordinate is the number of hours and the second coordinate is the score obtained on Monday's test.)

Study Hours: (0, 40), (1, 41), (2, 51), (3, 58), (3, 49), (4, 48), (4, 64), (5, 55), (5, 69), (5, 58), (5, 75), (6, 68), (6, 63), (6, 93), (7, 84), (7, 67), (8, 90), (8, 76), (9, 95), (9, 72), (9, 85), (10, 98)

TV Hours: (0, 98), (1, 85), (2, 72), (2, 90), (3, 67), (3, 93), (3, 95), (4, 68), (4, 84), (5, 76), (7, 75), (7, 58), (9, 63), (9, 69), (11, 55), (12, 58), (14, 64), (16, 48), (17, 51), (18, 41), (19, 49), (20, 40)

Solution

Scatter plots for the two sets of data are shown in Figure C.7. The scatter plot relating study hours and test scores has a positive correlation. This means that the more a student studied, the higher his or her score tended to be. The scatter plot relating television hours and test scores has a negative correlation. This means that the more time a student spent watching television, the lower his or her score tended to be.

Fitting a Line to Data

Finding a linear model that represents the relationship described by a scatter plot is called **fitting a line to data.** You can do this graphically by simply sketching the line that appears to fit the points, finding two points on the line, and then finding the equation of the line that passes through the two points.

Example 7   Fitting a Line to Data

Find a linear model that relates the year and the number of people P (in millions) who were part of the United States labor force from 1987 through 1997. In the table below, t represents the year, with $t = 0$ corresponding to 1987.   (Source: U.S. Bureau of Labor Statistics)

t	0	1	2	3	4	5	6	7	8	9	10
P	120	122	124	126	126	128	129	131	132	134	136

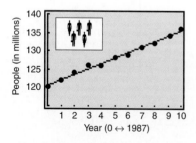

Figure C.8

Solution

After plotting the data from the table, draw the line that you think best represents the data, as shown in Figure C.8. Two points that lie on this line are (0, 120) and (9, 134). Using the point-slope form, you can find the equation of the line to be $P = 14t/9 + 120$.

Once you have found a model, you can measure how well the model fits the data by comparing the actual values with the values given by the model, as shown in the following table for the data and model in Example 7.

t	0	1	2	3	4	5	6	7	8	9	10
Actual P	120	122	124	126	126	128	129	131	132	134	136
Model P	120	121.6	123.1	124.7	126.2	127.8	129.3	130.9	132.4	134	135.6

The sum of the squares of the differences between the actual values and the model's values is the **sum of the squared differences.** The model that has the least sum is called the **least squares regression line** for the data. For the model in Example 7, the sum of the squared differences is 3.16. The least squares regression line for the data is

$$P = 1.5t + 120.5. \qquad \text{\small Best-fitting linear model}$$

The sum of the squared differences is 2.5.

Many graphing utilities have "built-in" least squares regression programs. If your graphing utility has such a program, enter the data in the table and use it to find the least squares regression line.

Measures of Central Tendency

In many real-life situations, it is helpful to describe data by a single number that is most representative of the entire collection of numbers. Such a number is called a **measure of central tendency.** The most commonly used measures are as follows.

1. The **mean,** or **average,** of n numbers is the sum of the numbers divided by n.
2. The **median** of n numbers is the middle number when the numbers are written in order. If n is even, the median is the average of the two middle numbers.
3. The **mode** of n numbers is the number that occurs most frequently. If two numbers tie for most frequent occurrence, the collection has two modes and is called **bimodal.**

Example 8   Comparing Measures of Central Tendency

You are interviewing for a job. The interviewer tells you that the average income of the company's 25 employees is $60,849. The actual annual incomes of the 25 employees are shown below. What are the mean, median, and mode of the incomes? Was the person telling you the truth?

$17,305, $478,320, $45,678, $18,980, $17,408,
$25,676, $28,906, $12,500, $24,540, $33,450,
$12,500, $33,855, $37,450, $20,432, $28,956,
$34,983, $36,540, $250,921, $36,853, $16,430,
$32,654, $98,213, $48,980, $94,024, $35,671

Solution

The mean of the incomes is

$$\text{Mean} = \frac{17{,}305 + 478{,}320 + 45{,}678 + 18{,}980 + \cdots + 35{,}671}{25}$$

$$= \frac{1{,}521{,}225}{25} = \$60{,}849.$$

To find the median, order the incomes as follows.

$12,500,	$12,500,	$16,430,	$17,305,	$17,408,
$18,980,	$20,432,	$24,540,	$25,676,	$28,906,
$28,956,	$32,654,	$33,450,	$33,855,	$34,983,
$35,671,	$36,540,	$36,853,	$37,450,	$45,678,
$48,980,	$94,024,	$98,213,	$250,921,	$478,320

From this list, you can see that the median (the middle number) is $33,450. From the same list, you can see that $12,500 is the only income that occurs more than once. Thus, the mode is $12,500. Technically, the person was telling the truth because the average is (generally) defined to be the mean. However, of the three measures of central tendency

Mean: $60,849   *Median:* $33,450   *Mode:* $12,500

it seems clear that the median is most representative. The mean is inflated by the two highest salaries.

Which of the three measures of central tendency is the most representative? The answer is that it depends on the distribution of the data *and* the way in which you plan to use the data.

For instance, in Example 8, the mean salary of $60,849 does not seem very representative to a potential employee. To a city income tax collector who wants to estimate 1% of the total income of the 25 employees, however, the mean is precisely the right measure.

Example 9   Choosing a Measure of Central Tendency

Which measure of central tendency is the most representative of the data given in each of the following frequency distributions?

a. *Number*	*Tally*	b. *Number*	*Tally*	c. *Number*	*Tally*
1	7	1	9	1	6
2	20	2	8	2	1
3	15	3	7	3	2
4	11	4	6	4	3
5	8	5	5	5	5
6	3	6	6	6	5
7	2	7	7	7	4
8	0	8	8	8	3
9	15	9	9	9	0

Solution

a. For these data, the mean is 4.23, the median is 3, and the mode is 2. Of these, the mode is probably the most representative.

b. For these data, the mean and median are each 5 and the modes are 1 and 9 (the distribution is bimodal). Of these, the mean or median is the most representative.

c. For these data, the mean is 4.59, the median is 5, and the mode is 1. Of these, the mean or median is the most representative.

Appendix C   Exercises

1. Construct a stem-and-leaf plot for the following exam scores for a class of 30 students. The scores are for a 100-point exam.

   77, 100, 77, 70, 83, 89, 87, 85, 81, 84, 81, 78, 89, 78, 88, 85, 90, 92, 75, 81, 85, 100, 98, 81, 78, 75, 85, 89, 82, 75

2. *Insurance Coverage*   The following table shows the total number of persons (in thousands) without health insurance coverage in the 50 states and the District of Columbia in 1996. Use a stem-and-leaf plot to organize the data.   (Source: U.S. Bureau of the Census)

AK	89	AL	550	AR	566	AZ	1159	CA	6514
CO	644	CT	368	DC	80	DE	98	FL	2722
GA	1319	HI	101	IA	335	ID	196	IL	1337
IN	600	KS	292	KY	601	LA	890	MA	766
MD	581	ME	146	MI	857	MN	480	MO	700
MS	518	MT	124	NC	1160	ND	62	NE	190
NH	109	NJ	1317	NM	412	NV	255	NY	3132
OH	1292	OK	570	OR	496	PA	1133	RI	93
SC	634	SD	67	TN	841	TX	4680	UT	240
VA	811	VT	65	WA	761	WI	438	WV	261
WY	66								

In Exercises 3 and 4, use the following set of data, which lists students' scores on a 100-point exam.

93, 84, 100, 92, 66, 89, 78, 52, 71, 85, 83, 95, 98, 99, 93, 81, 80, 79, 67, 59, 90, 55, 77, 62, 90, 78, 66, 63, 93, 87, 74, 96, 72, 100, 70, 73

3. Use a stem-and-leaf plot to organize the data.

4. Draw a histogram to represent the data.

5. Complete the following frequency distribution table and draw a histogram to represent the data.

   44, 33, 17, 23, 16, 18, 44, 47, 18, 20, 25, 27, 18, 29, 29, 28, 27, 18, 36, 22, 32, 38, 33, 41, 49, 48, 45, 38, 49, 15

Interval	Tally
[15, 22)	
[22, 29)	
[29, 36)	
[36, 43)	
[43, 50)	

6. *Snowfall*   The data below show the seasonal snowfall (in inches) in Lincoln, Nebraska for the years 1968 through 1997 (the amounts are listed in order by year). How would you organize these data? Explain your reasoning.   (Source: University of Nebraska-Lincoln)

   39.8, 26.2, 49.0, 21.6, 29.2, 33.6, 42.1, 21.1, 21.8, 31.0, 34.4, 23.3, 13.0, 32.3, 38.0, 47.5, 21.5, 18.9, 15.7, 13.0, 19.1, 18.7, 25.8, 23.8, 32.1, 21.3, 21.8, 30.7, 29.0, 44.6

7. *Travel to the United States*   The data below give the places of origin and the numbers of travelers (in millions) to the United States in 1995. Construct a bar graph for these data.   (Source: U.S. Department of Commerce)

Canada	14.7	Mexico	8.0
Europe	8.8	Far East	6.6
Other	5.2		

8. *Fruit Crops* The data below show the cash receipts (in millions of dollars) from fruit crops of farmers in 1996. Construct a bar graph for these data. (Source: U.S. Department of Agriculture)

Apples	1846	Peaches	380
Cherries	264	Pears	292
Grapes	2334	Plums and Prunes	295
Lemons	228	Strawberries	770
Oranges	1798		

Handling Garbage In Exercises 9–14, use the line graph given below. (Source: Franklin Associates)

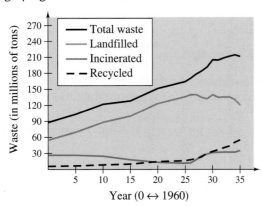

9. Estimate the total waste in 1985 and 1995.

10. Estimate the amount of incinerated garbage in 1990.

11. Which quantities increased every year?

12. During which time period did the amount of incinerated garbage decrease?

13. What is the relationship among the four quantities in the line graph?

14. Why do you think landfill garbage is decreasing?

15. *College Attendance* The following table shows the enrollment in a liberal arts college. Construct a line graph for the data.

Year	1993	1994	1995	1996
Enrollment	1675	1704	1710	1768

Year	1997	1998	1999	2000
Enrollment	1833	1918	1967	1972

16. *Oil Imports* The table shows the crude oil imports into the United States in millions of barrels for the years 1988 through 1997. Construct a line graph for the data and state what information it reveals. (Source: U.S. Energy Information Administration)

Year	1988	1989	1990	1991	1992
Oil imports	1864	2133	2151	2110	2220

Year	1993	1994	1995	1996	1997
Oil imports	2477	2578	2643	2748	2918

Table for 16

17. *Stock Market* The list below shows stock prices for selected companies in April of 1999. Draw a graph that best represents the data. Explain why you chose that type of graph. (Source: Value Line)

Company	Stock Price
Sears, Roebuck	$46
Wal-Mart Stores	$98
JC Penney	$40
K Mart Corp.	$16
The Gap, Inc.	$68

18. *Net Profit* The table shows the net profits (in millions of dollars) of Callaway Golf Co. for the years 1992 through 1997. Draw a graph that best represents the net profit and explain why you chose that type of graph. (Source: Value Line)

Year	1992	1993	1994	1995	1996	1997
Net profit	19.3	41.2	78.0	97.7	122.3	139.9

19. *Camcorders* The factory sales (in millions of dollars) of camcorders for the years 1990 through 1996 are given in the table. Organize the data graphically. Explain your reasoning. (Source: Electronic Industries Association)

Year	1990	1991	1992	1993	1994	1995	1996
Number	2260	2013	1841	1958	1985	2135	2084

20. *Owning Cats* The average numbers (out of 100) of cat owners who state various reasons for owning a cat are listed below. Organize the data graphically. Explain your reasoning. (Source: Gallup Poll)

Reason for Owning a Cat	Number
Have a pet to play with	93
Companionship	84
Help children learn responsibility	78
Have a pet to communicate with	62
Security	51

Interpreting a Scatter Plot   In Exercises 21–24, use the scatter plot shown. The scatter plot compares the number of hits x made by 30 softball players during the first half of the season with the number of runs batted in y.

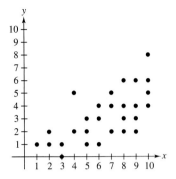

21. Do x and y have a positive correlation, a negative correlation, or no correlation?

22. Why does the scatter plot show only 28 points?

23. From the scatter plot, does it appear that players with more hits tend to have more runs batted in?

24. Can a player have more runs batted in than hits? Explain.

In Exercises 25–28, decide whether a scatter plot relating the two quantities would tend to have a positive, a negative, or no correlation. Explain.

25. The age and value of a car

26. A student's study time and test scores

27. The height and age of a pine tree

28. A student's height and test scores

In Exercises 29–32, use the data in the table, which shows the relationship between the altitude A (in thousands of feet) and the air pressure P (in pounds per square foot).

A	0	5	10	15	20	25
P	14.7	12.3	10.2	8.4	6.8	5.4

A	30	35	40	45	50
P	4.5	3.5	2.8	2.1	1.8

29. Sketch a scatter plot of the data.

30. How are A and P related?

31. Estimate the air pressure at 42,500 feet.

32. Estimate the altitude at which the air pressure is 5.0 pounds per square foot.

Crop Yield   In Exercises 33–36, use the data in the table, where x is the number of units of fertilizer applied to sample plots and y is the yield (in bushels) of a crop.

x	0	1	2	3	4	5	6	7	8
y	58	60	59	61	63	66	65	67	70

33. Sketch a scatter plot of the data.

34. Determine whether the points are positively correlated, are negatively correlated, or have no discernible correlation.

35. Sketch a linear model that you think best represents the data. Find an equation of the line you sketched. Use the line to predict the yield if 10 units of fertilizer are used.

36. Can the model found in Exercise 35 be used to predict yields for arbitrarily large values of x? Explain.

Speed of Sound   In Exercises 37–40, use the data in the table, where h is altitude in thousands of feet and v is the speed of sound in feet per second.

h	0	5	10	15	20	25	30	35
v	1116	1097	1077	1057	1036	1015	995	973

37. Sketch a scatter plot of the data.

38. Determine whether the points are positively correlated, are negatively correlated, or have no discernible correlation.

39. Sketch a linear model that you think best represents the data. Find an equation of the line you sketched. Use the line to predict the speed of sound at an altitude of 27,000 feet.

40. The speed of sound at an altitude of 70,000 feet is approximately 971 feet per second. What does this suggest about the validity of using the model in Exercise 39 to extrapolate beyond the data given in the table?

In Exercises 41–44, use a graphing utility to find the least squares regression line for the data. Sketch a scatter plot and the regression line.

41. (0, 23), (1, 20), (2, 19), (3, 17), (4, 15), (5, 11), (6, 10)

42. (4, 52.8), (5, 54.7), (6, 55.7), (7, 57.8), (8, 60.2), (9, 63.1), (10, 66.5)

43. (−10, 5.1), (−5, 9.8), (0, 17.5), (2, 25.4), (4, 32.8), (6, 38.7), (8, 44.2), (10, 50.5)

44. $(-10, 213.5)$, $(-5, 174.9)$, $(0, 141.7)$, $(5, 119.7)$, $(8, 102.4)$, $(10, 87.6)$

45. *School Enrollment* The table gives the preprimary school enrollments y (in millions) for the years 1990 through 1995, where $t = 0$ corresponds to 1990. (Source: U.S. Bureau of the Census)

t	0	1	2	3	4	5
y	11.21	11.37	11.54	11.95	12.33	12.52

(a) Use a graphing utility to find the least squares regression line. Use the equation to estimate enrollment in 1996.

(b) Make a scatter plot of the data and sketch the graph of the regression line.

(c) Use a graphing utility to determine the correlation coefficient.

46. *Advertising* The management of a department store ran an experiment to determine if a relationship existed between sales S (in thousands of dollars) and the amount spent on advertising x (in thousands of dollars). The following data were collected.

x	1	2	3	4	5	6	7	8
S	405	423	455	466	492	510	525	559

(a) Use a graphing utility to find the least squares regression line. Use the equation to estimate sales if $4500 is spent on advertising.

(b) Make a scatter plot of the data and sketch the graph of the regression line.

(c) Use a graphing utility to determine the correlation coefficient.

In Exercises 47–50, find the mean, median, and mode of the set of measurements.

47. 5, 12, 7, 14, 8, 9, 7   **48.** 30, 37, 32, 39, 33, 34, 32

49. 5, 12, 7, 24, 8, 9, 7   **50.** 20, 37, 32, 39, 33, 34, 32

51. *Electric Bills* A person had the following monthly bills for electricity. What are the mean and median of the collection of bills?

Jan.	$67.92	Feb.	$59.84	Mar.	$52.00
Apr.	$52.50	May	$57.99	June	$65.35
July	$81.76	Aug.	$74.98	Sept.	$87.82
Oct.	$83.18	Nov.	$65.35	Dec.	$57.00

52. *Car Rental* A car rental company kept the following record of the numbers of miles driven by a car that was rented. What are the mean, median, and mode of these data?

Monday	410	Tuesday	260
Wednesday	320	Thursday	320
Friday	460	Saturday	150

53. *Six-Child Families* A study was done on families having six children. The table gives the number of families in the study with the indicated number of girls. Determine the mean, median, and mode of this set of data.

Number of girls	0	1	2	3	4	5	6
Frequency	1	24	45	54	50	19	7

54. *Baseball* A baseball fan examined the records of a favorite baseball player's performance during his last 50 games. The number of games in which the player had 0, 1, 2, 3, and 4 hits are recorded in the table.

Number of hits	0	1	2	3	4
Frequency	14	26	7	2	1

(a) Determine the average number of hits per game.

(b) Determine the player's batting average if he had 200 at bats during the 50-game series.

55. *Think About It* Construct a collection of numbers that has the following properties. If this is not possible, explain why it is not.

Mean $= 6$, median $= 4$, mode $= 4$

56. *Think About It* Construct a collection of numbers that has the following properties. If this is not possible, explain why it is not.

Mean $= 6$, median $= 6$, mode $= 4$

57. *Test Scores* A professor records the following scores for a 100-point exam.

99, 64, 80, 77, 59, 72, 87, 79, 92, 88,
90, 42, 20, 89, 42, 100, 98, 84, 78, 91

Which measure of central tendency best describes these test scores?

58. *Shoe Sales* A salesperson sold eight pairs of a certain style of men's shoes. The sizes of the eight pairs were as follows: $10\frac{1}{2}$, 8, 12, $10\frac{1}{2}$, 10, $9\frac{1}{2}$, 11, and $10\frac{1}{2}$. Which measure (or measures) of central tendency best describes the typical shoe size for these data?

Appendix D

Introduction to Logic

D.1   Statements and Truth Tables

Statements   ■   Truth Tables

Statements

In everyday speech and in mathematics we make inferences that adhere to common **laws of logic.** These laws (or methods of reasoning) allow us to build an algebra of statements by using logical operations to form compound statements from simpler ones. One of the primary goals of logic is to determine the truth value (true or false) of a compound statement knowing the truth values of its simpler component statements. For instance, the compound statement "The temperature is below freezing and it is snowing" is true only if both component statements are true.

▶ **Definition of a Statement**

1. A **statement** (or proposition) is a sentence to which only one truth value (either true or false) can be meaningfully assigned.

2. An **open statement** is a sentence that contains one or more variables and becomes a statement when each variable is replaced by a specific item from a designated set.

Example 1   Statements, Nonstatements, and Open Statements

Statement	*Truth Value*
A square is a rectangle.	T
-3 is less than -5.	F

Nonstatement	*Truth Value*
Do your homework.	No truth value can be meaningfully assigned.
Did you call the police?	No truth value can be meaningfully assigned.

Open Statement	*Truth Value*
x is an irrational number.	We need a value of x.
She is a computer science major.	We need a specific person.

Symbolically, statements are represented by lowercase letters p, q, r, and so on. Statements can be changed or combined to form **compound statements** by means of the three logical operations **and, or,** and **not,** which are represented by $\wedge$ (and), $\vee$ (or), and $\sim$ (not). In logic, the word *or* is used in the *inclusive* sense (meaning "and/or" in everyday language). That is, the statement "p or q" is true if p is true, q is true, or both p and q are true. The following list summarizes the terms and symbols used with these three operations of logic.

▶ Operations of Logic

Operation	Verbal Statement	Symbolic Form	Name of Operation
$\sim$	not p	$\sim p$	**Negation**
$\wedge$	p and q	$p \wedge q$	**Conjunction**
$\vee$	p or q	$p \vee q$	**Disjunction**

Compound statements can be formed using more than one logical operation, as demonstrated in Example 2.

Example 2   Forming Negations and Compound Statements

The statements p and q are as follows.

    p: The temperature is below freezing.
    q: It is snowing.

Write the verbal form for each of the following.

a. $p \wedge q$ **b.** $\sim p$ **c.** $\sim (p \vee q)$ **d.** $\sim p \wedge \sim q$

Solution

a. The temperature is below freezing and it is snowing.

b. The temperature is not below freezing.

c. It is not true that the temperature is below freezing or it is snowing.

d. The temperature is not below freezing and it is not snowing.

Example 3   Forming Compound Statements

The statements p and q are as follows.

    p: The temperature is below freezing.
    q: It is snowing.

a. Write the symbolic form for: *The temperature is not below freezing or it is not snowing.*

b. Write the symbolic form for: *It is not true that the temperature is below freezing and it is snowing.*

Solution

a. The symbolic form is: $\sim p \vee \sim q$ **b.** The symbolic form is: $\sim (p \wedge q)$

Truth Tables

To determine the truth value of a compound statement, it is helpful to construct charts called **truth tables.** The following tables represent the three basic logical operations.

Negation					Conjunction				Disjunction		
p	q	$\sim p$	$\sim q$		p	q	$p \wedge q$		p	q	$p \vee q$
T	T	F	F		T	T	T		T	T	T
T	F	F	T		T	F	F		T	F	T
F	T	T	F		F	T	F		F	T	T
F	F	T	T		F	F	F		F	F	F

For the sake of uniformity, all truth tables with two component statements will have **T** and **F** values for p and q assigned in the order shown in the first columns of these three tables. Truth tables for several operations can be combined into one chart by using the same two first columns. For each operation, a new column is added. Such an arrangement is especially useful with compound statements that involve more than one logical operation and for showing that two statements are logically equivalent.

> ▶ Logical Equivalence
>
> Two compound statements are **logically equivalent** if they have identical truth tables. Symbolically, the equivalence of the statements p and q is denoted by writing $p \equiv q$.

Example 4   Logical Equivalence

Use a truth table to show the logical equivalence of the statements $\sim p \wedge \sim q$ and $\sim (p \vee q)$.

Solution

p	q	$\sim p$	$\sim q$	$\sim p \wedge \sim q$	$p \vee q$	$\sim (p \vee q)$
T	T	F	F	F	T	F
T	F	F	T	F	T	F
F	T	T	F	F	T	F
F	F	T	T	T	F	T

↑——— Identical ———↑

Because the fifth and seventh columns in the table are identical, the two given statements are logically equivalent.

The equivalence established in Example 4 is one of two well-known rules in logic called **DeMorgan's Laws.** Verification of the second of DeMorgan's Laws is left as an exercise.

> ▶ **DeMorgan's Laws**
>
> **1.** $\sim(p \vee q) \equiv \sim p \wedge \sim q$
>
> **2.** $\sim(p \wedge q) \equiv \sim p \vee \sim q$

Compound statements that are true, no matter what the truth values of the component statements, are called **tautologies.** One simple example is the statement "p or not p," as shown in the table.

$p \vee \sim p$ is a tautology

p	$\sim p$	$p \vee \sim p$
T	F	T
F	T	T

D.1   Exercises

In Exercises 1–12, classify the sentence as a statement, a nonstatement, or an open statement.

1. All dogs are brown.

2. Can I help you?

3. That figure is a circle.

4. Substitute 4 for x.

5. x is larger than 4.

6. 8 is larger than 4.

7. $x + y = 10$

8. $12 + 3 = 14$

9. Hockey is fun to watch.

10. One mile is greater than 1 kilometer.

11. It is more than 1 mile to the school.

12. Come to the party.

In Exercises 13–20, determine whether the open statement is true for the given values of x.

13. $x^2 - 5x + 6 = 0$
   (a) $x = 2$
   (b) $x = -2$

14. $x^2 - x - 6 = 0$
   (a) $x = 2$
   (b) $x = -2$

15. $x^2 \le 4$
   (a) $x = -2$
   (b) $x = 0$

16. $|x - 3| = 4$
   (a) $x = -1$
   (b) $x = 7$

17. $4 - |x| = 2$
   (a) $x = 0$
   (b) $x = 1$

18. $\sqrt{x^2} = x$
   (a) $x = 3$
   (b) $x = -3$

19. $\dfrac{x}{x} = 1$
   (a) $x = -4$
   (b) $x = 0$

20. $\sqrt[3]{x} = -2$
   (a) $x = 8$
   (b) $x = -8$

In Exercises 21–24, write the verbal form for each of the following.

(a) $\sim p$ (b) $\sim q$ (c) $p \wedge q$ (d) $p \vee q$

21. p: The sun is shining.
   q: It is hot.

22. p: The car has a radio.
   q: The car is red.

23. p: Lions are mammals.
   q: Lions are carnivorous.

24. p: Twelve is less than 15.
   q: Seven is a prime number.

In Exercises 25–28, write the verbal form for each of the following.

(a) $\sim p \wedge q$ (b) $\sim p \vee q$

(c) $p \wedge \sim q$ (d) $p \vee \sim q$

25. p: The sun is shining.
   q: It is hot.

26. p: The car has a radio.
   q: The car is red.

27. p: Lions are mammals.
   q: Lions are carnivorous.

28. p: Twelve is less than 15.
   q: Seven is a prime number.

In Exercises 29–32, write the symbolic form of the given compound statement. In each case let p represent the statement "It is four o'clock," and let q represent the statement "It is time to go home."

29. It is four o'clock and it is not time to go home.

30. It is not four o'clock or it is not time to go home.

31. It is not four o'clock or it is time to go home.

32. It is four o'clock and it is time to go home.

In Exercises 33–36, write the symbolic form of the given compound statement. In each case let p represent the statement "The dog has fleas," and let q represent the statement "The dog is scratching."

33. The dog does not have fleas or the dog is not scratching.

34. The dog has fleas and the dog is scratching.

35. The dog does not have fleas and the dog is scratching.

36. The dog has fleas or the dog is not scratching.

In Exercises 37–42, write the negation of the given statement.

37. The bus is not blue.

38. Frank is not 6 feet tall.

39. x is equal to 4.

40. x is not equal to 4.

41. The earth is not flat.

42. The earth is flat.

In Exercises 43–48, construct a truth table for the given compound statement.

43. $\sim p \wedge q$ **44.** $\sim p \vee q$

45. $\sim p \vee \sim q$ **46.** $\sim p \wedge \sim q$

47. $p \vee \sim q$ **48.** $p \wedge \sim q$

In Exercises 49–54, use a truth table to determine whether the given statements are logically equivalent.

49. $\sim p \wedge q, \quad p \vee \sim q$

50. $\sim(p \wedge \sim q), \quad \sim p \vee q$

51. $\sim(p \vee \sim q), \quad \sim p \wedge q$

52. $\sim(p \vee q), \quad \sim p \vee \sim q$

53. $p \wedge \sim q, \quad \sim(\sim p \vee q)$

54. $p \wedge \sim q, \quad \sim(\sim p \wedge q)$

In Exercises 55–58, determine whether the statements are logically equivalent.

55. (a) The house is red and it is not made of wood.

   (b) The house is red or it is not made of wood.

56. (a) It is not true that the tree is not green.

   (b) The tree is green.

57. (a) The statement that the house is white or blue is not true.

   (b) The house is not white and it is not blue.

58. (a) I am not 25 years old and I am not applying for this job.

   (b) The statement that I am 25 years old and I am applying for this job is not true.

In Exercises 59–62, use a truth table to determine whether the given statement is a tautology.

59. $\sim p \wedge p$ **60.** $\sim p \vee p$

61. $\sim(\sim p) \vee \sim p$ **62.** $\sim(\sim p) \wedge \sim p$

63. Use a truth table to verify the second of DeMorgan's Laws:

$$\sim(p \wedge q) \equiv \sim p \vee \sim q$$

D.2 Implications, Quantifiers, and Venn Diagrams

Implications ■ Logical Quantifiers ■ Venn Diagrams

Implications

A statement of the form "If p, then q," is called an **implication** (or a conditional statement) and is denoted by

$$p \rightarrow q.$$

The statement p is called the **hypothesis** and the statement q is called the **conclusion.** There are many different ways to express the implication $p \rightarrow q$, as shown in the following list.

▶ **Different Ways of Stating Implications**

The implication $p \rightarrow q$ has the following equivalent verbal forms.

1. If p, then q.　　　　　**2.** p implies q.

3. p, only if q.　　　　　**4.** q follows from p.

5. q is necessary for p.　　**6.** p is sufficient for q.

Normally, we think of the implication $p \rightarrow q$ as having a cause-and-effect relationship between the hypothesis p and the conclusion q. However, you should be careful not to confuse the truth value of the component statements with the truth value of the implication. The following truth table should help you keep this distinction in mind.

Implication

p	q	$p \rightarrow q$
T	T	T
T	F	F
F	T	T
F	F	T

Note in the table that the implication $p \rightarrow q$ is false only when p is true and q is false. This is like a promise. Suppose you promise a friend that "If the sun shines, I will take you fishing." The only way you can break your promise is if the sun shines (p is true) and you do not take your friend fishing (q is false). If the sun doesn't shine (p is false), you have no obligation to go fishing, and so the promise cannot be broken.

Example 1   Finding Truth Values of Implications

Give the truth value of each implication.

a. If 3 is odd, then 9 is odd.

b. If 3 is odd, then 9 is even.

c. If 3 is even, then 9 is odd.

d. If 3 is even, then 9 is even.

Solution

	Hypothesis	*Conclusion*	*Implication*
a.	T	T	T
b.	T	F	F
c.	F	T	T
d.	F	F	T

The next example shows how to write an implication as a disjunction.

Example 2   Identifying Equivalent Statements

Use a truth table to show the logical equivalence of the following statements.

a. If I get a raise, I will take my family on a vacation.

b. I will not get a raise *or* I will take my family on a vacation.

Solution

Let p represent the statement "I will get a raise," and let q represent the statement "I will take my family on a vacation." Then, you can represent the statement in part (a) as $p \rightarrow q$ and the statement in part (b) as $\sim p \vee q$. The logical equivalence of these two statements is shown in the following truth table.

$$p \rightarrow q \equiv \sim p \vee q$$

p	q	$\sim p$	$\sim p \vee q$	$p \rightarrow q$
T	T	F	T	T
T	F	F	F	F
F	T	T	T	T
F	F	T	T	T

↑ Identical ↑

Because the fourth and fifth columns of the truth table are identical, you can conclude that the two statements $p \rightarrow q$ and $\sim p \vee q$ are equivalent.

From the table in Example 2 and the fact that $\sim(\sim p) \equiv p$, we can write the **negation of an implication.** That is, because $p \rightarrow q$ is equivalent to $\sim p \vee q$, it follows that the negation of $p \rightarrow q$ must be $\sim(\sim p \vee q)$, which by DeMorgan's Laws can be written as follows.

$$\sim(p \rightarrow q) \equiv p \wedge \sim q$$

For the implication $p \rightarrow q$, there are three important associated implications.

1. The **converse** of $p \rightarrow q$: $q \rightarrow p$
2. The **inverse** of $p \rightarrow q$: $\sim p \rightarrow \sim q$
3. The **contrapositive** of $p \rightarrow q$: $\sim q \rightarrow \sim p$

From the table below, you can see that these four statements yield two pairs of logically equivalent implications. The connective "$\rightarrow$" is used to determine the truth values in the last three columns of the table.

p	q	$\sim p$	$\sim q$	$p \rightarrow q$	$\sim q \rightarrow \sim p$	$q \rightarrow p$	$\sim p \rightarrow \sim q$
T	T	F	F	T	T	T	T
T	F	F	T	F	F	T	T
F	T	T	F	T	T	F	F
F	F	T	T	T	T	T	T

⎿ Identical ⏋   ⎿ Identical ⏋

Example 3   Writing the Converse, Inverse, and Contrapositive

Write the converse, inverse, and contrapositive for the implication "If I get a B on my test, then I will pass the course."

Solution

a. *Converse:* If I pass the course, then I got a B on my test.

b. *Inverse:* If I do not get a B on my test, then I will not pass the course.

c. *Contrapositive:* If I do not pass the course, then I did not get a B on my test.

In Example 3, be sure you see that neither the converse nor the inverse is logically equivalent to the original implication. To see this, consider that the original implication simply states that if you get a B on your test, then you will pass the course. The converse is not true because knowing that you passed the course does not imply that you got a B on the test. (After all, you might have gotten an A on the test.)

A **biconditional statement,** denoted by $p \leftrightarrow q$, is the conjunction of the implications $p \rightarrow q$ and $q \rightarrow p$. Often a biconditional statement is written as "p if and only if q," or in shorter form as "p iff q." A biconditional statement is true when both components are true and when both components are false, as shown in the following truth table.

Biconditional Statement: *p if and only if q*

p	q	$p \to q$	$q \to p$	$p \leftrightarrow q$	$(p \to q) \wedge (q \to p)$
T	T	T	T	T	T
T	F	F	T	F	F
F	T	T	F	F	F
F	F	T	T	T	T

The following list summarizes some of the laws of logic that we have discussed up to this point.

▶ **Laws of Logic**

1. For every statement p, either p is true or p is false. Law of Excluded Middle

2. $\sim(\sim p) \equiv p$ Law of Double Negation

3. $\sim(p \vee q) \equiv \sim p \wedge \sim q$ DeMorgan's Law

4. $\sim(p \wedge q) \equiv \sim p \vee \sim q$ DeMorgan's Law

5. $p \to q \equiv \sim p \vee q$ Law of Implication

6. $p \to q \equiv \sim q \to \sim p$ Law of Contraposition

Logical Quantifiers

Logical quantifiers are words such as *some*, *all*, *every*, *each*, *one*, and *none*. Here are some examples of statements with quantifiers.

Some isosceles triangles are right triangles.

Every painting on display is for sale.

Not all corporations have male chief executive officers.

All squares are parallelograms.

Being able to recognize the negation of a statement involving a quantifier is one of the most important skills in logic. For instance, consider the statement "All dogs are brown." In order for this statement to be false, you do not have to show that *all* dogs are not brown, you must simply find at least one dog that is not brown. So, the negation of the statement is "Some dogs are brown."

Next we list some of the more common negations involving quantifiers.

▶ **Negating Statements with Quantifiers**

Statement	*Negation*
1. All p are q.	Some p are not q.
2. Some p are q.	No p is q.
3. Some p are not q.	All p are q.
4. No p is q.	Some p are q.

When using logical quantifiers, the word *all* can be replaced by the words *each* or *every*. For instance, the following are equivalent.

All *p* are *q*.   Each *p* is *q*.   Every *p* is *q*.

Similarly, the word *some* can be replaced by the words *at least one*. For instance, the following are equivalent.

Some *p* are *q*.   At least one *p* is *q*.

> **Example 4** Negating Quantifying Statements

Write the negation of each statement.

a. All students study.

b. Not all prime numbers are odd.

c. At least one mammal can fly.

d. Some bananas are not yellow.

Solution

a. Some students do not study.

b. All prime numbers are odd.

c. No mammals can fly.

d. All bananas are yellow.

Venn Diagrams

Venn diagrams are figures that are used to show relationships between two or more sets of objects. They can help us to interpret quantifying statements. Study the following Venn diagrams in which the circle marked *A* represents people more than 6 feet tall and the circle marked *B* represents the basketball players.

1. All basketball players are more than 6 feet tall.

2. Some basketball players are more than 6 feet tall.

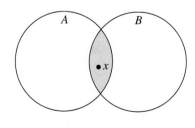

3. Some basketball players are not more than 6 feet tall.

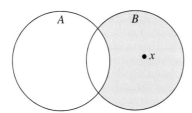

4. No basketball player is more than 6 feet tall.

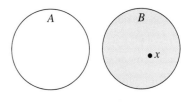

D.2   Exercises

In Exercises 1–4, write the verbal form for each of the following.

(a) $p \rightarrow q$ (b) $q \rightarrow p$ (c) $\sim q \rightarrow \sim p$ (d) $p \rightarrow \sim q$

1. p: The engine is running.
   q: The engine is wasting gasoline.

2. p: The student is at school.
   q: It is nine o'clock.

3. p: The integer is even.
   q: It is divisible by 2.

4. p: The person is generous.
   q: The person is rich.

In Exercises 5–10, write the symbolic form of the compound statement. Let p represent the statement "The economy is expanding," and let q represent the statement "Interest rates are low."

5. If interest rates are low, then the economy is expanding.

6. If interest rates are not low, then the economy is not expanding.

7. An expanding economy implies low interest rates.

8. Low interest rates are sufficient for an expanding economy.

9. Low interest rates are necessary for an expanding economy.

10. The economy will expand only if interest rates are low.

In Exercises 11–20, give the truth value of the implication.

11. If 4 is even, then 12 is even.

12. If 4 is even, then 2 is odd.

13. If 4 is odd, then 3 is odd.

14. If 4 is odd, then 2 is odd.

15. If $2n$ is even, then $2n + 2$ is odd.

16. If $2n + 1$ is even, then $2n + 2$ is odd.

17. $3 + 11 > 16$ only if $2 + 3 = 5$.

18. $\frac{1}{6} < \frac{2}{3}$ is necessary for $\frac{1}{2} > 0$.

19. $x = -2$ follows from $2x + 3 = x + 1$.

20. If $2x = 224$, then $x = 10$.

In Exercises 21–26, write the converse, inverse, and contrapositive of the statement.

21. If the sky is clear, then you can see the eclipse.

22. If the person is nearsighted, then he is ineligible for the job.

23. If taxes are raised, then the deficit will increase.

24. If wages are raised, then the company's profits will decrease.

25. It is necessary to have a birth certificate to apply for the visa.

26. The number is divisible by 3 only if the sum of its digits is divisible by 3.

In Exercises 27–40, write the negation of the statement.

27. Paul is a junior or a senior.

28. Jack is a senior and he plays varsity basketball.

29. If the temperature increases, then the metal rod will expand.

30. If the test fails, then the project will be halted.

31. We will go to the ocean only if the weather forecast is good.

32. Completing the pass on this play is necessary if we are going to win the game.

33. Some students are in extracurricular activities.

34. Some odd integers are not prime numbers.

35. All contact sports are dangerous.

36. All members must pay their dues prior to June 1.

37. No child is allowed at the concert.

38. No contestant is over the age of 12.

39. At least one of the $20 bills is counterfeit.

40. At least one unit is defective.

In Exercises 41–48, construct a truth table for the compound statement.

41. $\sim(p \to \sim q)$

42. $\sim q \to (p \to q)$

43. $\sim(q \to p) \wedge q$

44. $p \to (\sim p \vee q)$

45. $[(p \vee q) \wedge (\sim p)] \to q$

46. $[(p \to q) \wedge (\sim q)] \to p$

47. $(p \leftrightarrow \sim q) \to \sim p$

48. $(p \vee \sim q) \leftrightarrow (q \to \sim p)$

In Exercises 49–56, use a truth table to show the logical equivalence of the two statements.

49. $q \to p$ $\qquad$ $\sim p \to \sim q$

50. $\sim p \to q$ $\qquad$ $p \vee q$

51. $\sim(p \to q)$ $\qquad$ $p \wedge \sim q$

52. $(p \vee q) \to q$ $\qquad$ $p \to q$

53. $(p \to q) \vee \sim q$ $\qquad$ $p \vee \sim p$

54. $q \to (\sim p \vee q)$ $\qquad$ $q \vee \sim q$

55. $p \to (\sim p \wedge q)$ $\qquad$ $\sim p$

56. $\sim(p \wedge q) \to \sim q$ $\qquad$ $p \vee \sim q$

57. Select the statement that is logically equivalent to the statement "If a number is divisible by 6, then it is divisible by 2."

   (a) If a number is divisible by 2, then it is divisible by 6.

   (b) If a number is not divisible by 6, then it is not divisible by 2.

   (c) If a number is not divisible by 2, then it is not divisible by 6.

   (d) Some numbers are divisible by 6 and not divisible by 2.

58. Select the statement that is logically equivalent to the statement "It is not true that Pam is a conservative and a Democrat."

   (a) Pam is a conservative and a Democrat.

   (b) Pam is not a conservative and not a Democrat.

   (c) Pam is not a conservative or she is not a Democrat.

   (d) If Pam is not a conservative, then she is a Democrat.

59. Select the statement that is *not* logically equivalent to the statement "Every citizen over the age of 18 has the right to vote."

   (a) Some citizens over the age of 18 have the right to vote.

   (b) Each citizen over the age of 18 has the right to vote.

   (c) All citizens over the age of 18 have the right to vote.

   (d) No citizen over the age of 18 can be restricted from voting.

60. Select the statement that is *not* logically equivalent to the statement "It is necessary to pay the registration fee to take the course."

   (a) If you take the course, then you must pay the registration fee.

   (b) If you do not pay the registration fee, then you cannot take the course.

   (c) If you pay the registration fee, then you may take the course.

   (d) You may take the course only if you pay the registration fee.

In Exercises 61–70, sketch a Venn diagram and shade the region that illustrates the given statement. Let A be a circle that represents people who are happy, and let B be a circle that represents college students.

61. All college students are happy.

62. All happy people are college students.

63. No college students are happy.

64. No happy people are college students.

65. Some college students are not happy.

66. Some happy people are not college students.

67. At least one college student is happy.

68. At least one happy person is not a college student.

69. Each college student is sad.

70. Each sad person is not a college student.

In Exercises 71–74, state whether the statement follows from the given Venn diagram. Assume that each area shown in the Venn diagram is non-empty. (*Note:* Use only the information given in the diagram. Do not be concerned with whether the statement is actually true or false.)

71. (a) All toads are green.

(b) Some toads are green.

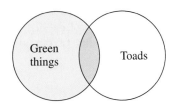

72. (a) All men are company presidents.

(b) Some company presidents are women.

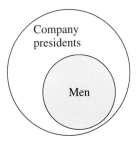

73. (a) All blue cars are old.

(b) Some blue cars are not old.

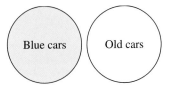

74. (a) No football players are more than 6 feet tall.

(b) Every football player is more than 6 feet tall.

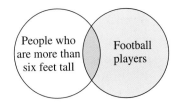

D.3   Logical Arguments

Arguments   ■   Venn Diagrams and Arguments   ■   Proofs

Arguments

An **argument** is a collection of statements, listed in order. The last statement is called the **conclusion** and the other statements are called the **premises.** An argument is **valid** if the conjunction of all the premises implies the conclusion. The most common type of argument takes the following form.

Premise #1: $p \to q$

Premise #2: p

Conclusion: q

This form of argument is called the **Law of Detachment** or *Modus Ponens.* It is illustrated in the following example.

Example 1	A Valid Argument

Show that the following argument is valid.

Premise #1:   If Sean is a freshman, then he is taking algebra.

Premise #2:   Sean is a freshman.

Conclusion: So, Sean is taking algebra.

Solution

Let p represent the statement "Sean is a freshman," and let q represent the statement "Sean is taking algebra." Then the argument fits the Law of Detachment, which can be written as $[(p \rightarrow q) \wedge p] \rightarrow q$. The validity of this argument is shown in the following truth table.

Law of Detachment

p	q	$p \rightarrow q$	$(p \rightarrow q) \wedge p$	$[(p \rightarrow q) \wedge p] \rightarrow q$
T	T	T	T	T
T	F	F	F	T
F	T	T	F	T
F	F	T	F	T

Keep in mind that the validity of an argument has nothing to do with the truthfulness of the premises or conclusion. For instance, the following argument is valid—the fact that it is fanciful does not alter its validity.

Premise #1:   If I snap my fingers, elephants will stay out of my house.

Premise #2:   I am snapping my fingers.

Conclusion: So, elephants will stay out of my house.

▶ **Four Types of Valid Arguments**

Name		*Pattern*	
1. Law of Detachment or *Modus Ponens*	Premise #1: Premise #2: Conclusion:	$p \rightarrow q$ p q	
2. Law of Contraposition or *Modus Tollens*	Premise #1: Premise #2: Conclusion:	$p \rightarrow q$ $\sim q$ $\sim p$	
3. Law of Transitivity or *Syllogism*	Premise #1: Premise #2: Conclusion:	$p \rightarrow q$ $q \rightarrow r$ $p \rightarrow r$	
4. Law of Disjunctive Syllogism	Premise #1: Premise #2: Conclusion:	$p \vee q$ $\sim p$ q	

Example 2   An Invalid Argument

Determine whether the following argument is valid.

Premise #1:   If John is elected, the income tax will be increased.

Premise #2:   The income tax was increased.

Conclusion: So, John was elected.

Solution

This argument has the following form.

	Pattern	*Implication*
Premise #1:	$p \rightarrow q$	$[(p \rightarrow q) \wedge q] \rightarrow p$
Premise #2:	q	
Conclusion:	p	

This is not one of the four valid forms of arguments that were listed. You can construct a truth table to verify that the argument is invalid, as follows.

An Invalid Argument

p	q	$p \rightarrow q$	$(p \rightarrow q) \wedge q$	$[(p \rightarrow q) \wedge q] \rightarrow p$
T	T	T	T	T
T	F	F	F	T
F	T	T	T	F
F	F	T	F	T

An invalid argument, such as the one in Example 2, is called a **fallacy.** Other common fallacies are given in the following example.

Example 3   Common Fallacies

Each of the following arguments is invalid.

a. *Arguing from the Converse:* If the football team wins the championship, then students will skip classes. The students skipped classes. So, the football team won the championship.

b. *Arguing from the Inverse:* If the football team wins the championship, then students will skip classes. The football team did not win the championship. So, the students did not skip classes.

c. *Arguing from False Authority:* Wheaties are best for you because Joe Montana eats them.

d. *Arguing from an Example:* Beta brand products are not reliable because my Beta brand snowblower does not start in cold weather.

e. *Arguing from Ambiguity:* If an automobile carburetor is modified, the automobile will pollute. Brand X automobiles have modified carburetors. So, Brand X automobiles pollute.

f. *Arguing by False Association:* Joe was running through the alley when the fire alarm went off. So, Joe started the fire.

Example 4 A Valid Argument

Determine whether the following argument is valid.

Premise #1:   You like strawberry pie or you like chocolate pie.

Premise #2:   You do not like strawberry pie.

Conclusion: So, you like chocolate pie.

Solution

This argument has the following form.

Premise #1: $p \lor q$

Premise #2: $\sim p$

Conclusion: q

This argument is a disjunctive syllogism, which is one of the four common types of valid arguments.

In a valid argument, the conclusion drawn from the premise is called a **valid conclusion.**

Example 5 Making Valid Conclusions

Given the following two premises, which of the conclusions are valid?

Premise #1:   If you like boating, then you like swimming.

Premise #2:   If you like swimming, then you are a scholar.

a. Conclusion: If you like boating, then you are a scholar.

b. Conclusion: If you do not like boating, then you are not a scholar.

c. Conclusion: If you are not a scholar, then you do not like boating.

Solution

a. This conclusion is valid. It follows from the Law of Transitivity or Syllogism.

b. This conclusion is invalid. The fallacy stems from arguing from the inverse.

c. This conclusion is valid. It follows from the Law of Contraposition.

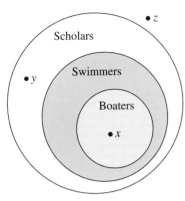

Figure D.1

Venn Diagrams and Arguments

Venn diagrams can be used to test informally the validity of an argument. For instance, a Venn diagram for the premises in Example 5 is shown in Figure D.1. In this figure, the validity of Conclusion (a) is seen by choosing a boater x in all three sets. Conclusion (b) is seen to be invalid by choosing a person y who is a scholar but does not like boating. Finally, person z indicates the validity of Conclusion (c).

Venn diagrams work well for testing arguments that involve quantifiers, as shown in the next two examples.

Example 6   Using a Venn Diagram to Show That an Argument Is Not Valid

Use a Venn diagram to test the validity of the following argument.

Premise #1:	Some plants are green.
Premise #2:	All lettuce is green.
Conclusion:	So, lettuce is a plant.

Solution

From the Venn diagram shown in Figure D.2, you can see that this is not a valid argument. Remember that even though the conclusion is true (lettuce is a plant), this does not imply that the argument is true.

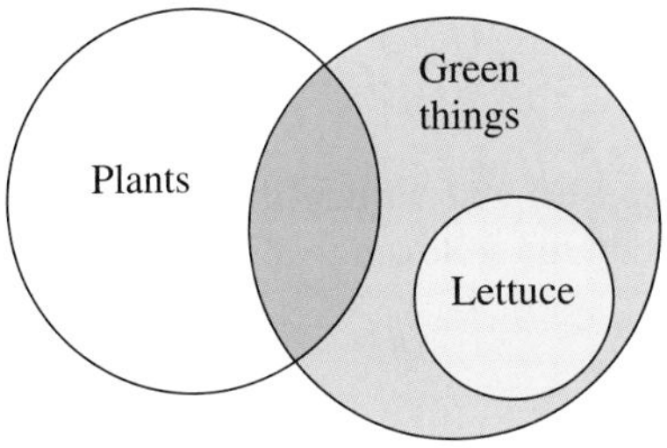

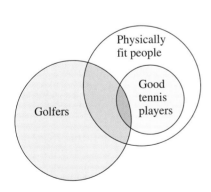

Figure D.2

When you are using Venn diagrams, you must remember to draw the most general case. For example, in Figure D.2 the circle representing plants is not drawn entirely within the circle representing green things because you are told that only *some* plants are green.

Example 7   Using a Venn Diagram to Show That an Argument Is Valid

Use a Venn diagram to test the validity of the following argument.

Premise #1:	All good tennis players are physically fit.
Premise #2:	Some golfers are good tennis players.
Conclusion:	So, some golfers are physically fit.

Solution

Because the set of golfers intersects the set of good tennis players, as shown in Figure D.3, the set of golfers must also intersect the set of physically fit people. So, the argument is valid.

Figure D.3

Proofs

What does the word *proof* mean to you? In mathematics, the word *proof* is used to mean simply a valid argument. Many proofs involve more than two premises and a conclusion. For instance, the proof in Example 8 involves three premises and a conclusion.

> **Example 8** A Proof by Contraposition

Use the following three premises to prove that "It is not snowing today."

Premise #1: If it is snowing today, Greg will go skiing.

Premise #2: If Greg is skiing today, then he is not studying.

Premise #3: Greg is studying today.

Solution

Let p represent the statement "It is snowing today," let q represent "Greg is skiing," and let r represent "Greg is studying today." So, the given premises have the following form.

Premise #1: $p \rightarrow q$

Premise #2: $q \rightarrow \sim r$

Premise #3: r

By noting that $r \equiv \sim(\sim r)$, reordering the premises, and writing the contrapositives of the first and second premises, you obtain the following valid argument.

Premise #3: r

Contrapositive of Premise #2: $r \rightarrow \sim q$

Contrapositive of Premise #1: $\sim q \rightarrow \sim p$

Conclusion: $\sim p$

So, you can conclude $\sim p$. That is, "It is not snowing today."

D.3 Exercises

In Exercises 1–4, use a truth table to show that the given argument is valid.

1. Premise #1: $p \rightarrow \sim q$
   Premise #2: q
   Conclusion: $\sim p$

2. Premise #1: $p \leftrightarrow q$
   Premise #2: p
   Conclusion: q

3. Premise #1: $p \vee q$
   Premise #2: $\sim p$
   Conclusion: q

4. Premise #1: $p \wedge q$
   Premise #2: $\sim p$
   Conclusion: q

In Exercises 5–8, use a truth table to show that the given argument is invalid.

5. Premise #1: $\sim p \rightarrow q$
   Premise #2: p
   Conclusion: $\sim q$

6. Premise #1: $p \rightarrow q$
   Premise #2: $\sim p$
   Conclusion: $\sim q$

7. Premise #1: $p \vee q$
   Premise #2: q
   Conclusion: p

8. Premise #1: $\sim(p \wedge q)$
   Premise #2: q
   Conclusion: p

In Exercises 9–22, determine whether the argument is valid or invalid.

9. Premise #1: If taxes are increased, then businesses will leave the state.
Premise #2: Taxes are increased.
Conclusion: So, businesses will leave the state.

10. Premise #1: If a student does the homework, then a good grade is certain.
Premise #2: Liza does the homework.
Conclusion: So, Liza will receive a good grade for the course.

11. Premise #1: If taxes are increased, then businesses will leave the state.
Premise #2: Businesses are leaving the state.
Conclusion: So, taxes were increased.

12. Premise #1: If a student does the homework, then a good grade is certain.
Premise #2: Liza received a good grade for the course.
Conclusion: So, Liza did her homework.

13. Premise #1: If the doors are kept locked, then the car will not be stolen.
Premise #2: The car was stolen.
Conclusion: So, the car doors were unlocked.

14. Premise #1: If Jan passes the exam, she is eligible for the position.
Premise #2: Jan is not eligible for the position.
Conclusion: So, Jan did not pass the exam.

15. Premise #1: All cars manufactured by the Ford Motor Company are reliable.
Premise #2: Lincolns are manufactured by Ford.
Conclusion: So, Lincolns are reliable cars.

16. Premise #1: Some cars manufactured by the Ford Motor Company are reliable.
Premise #2: Lincolns are manufactured by Ford.
Conclusion: So, Lincolns are reliable .

17. Premise #1: All federal income tax forms are subject to the Paperwork Reduction Act of 1980.
Premise #2: The 1040 Schedule A form is subject to the Paperwork Reduction Act of 1980.
Conclusion: So, the 1040 Schedule A form is a federal income tax form.

18. Premise #1: All integers divisible by 6 are divisible by 3.
Premise #2: Eighteen is divisible by 6.
Conclusion: So, 18 is divisible by 3.

19. Premise #1: Eric is at the store or the handball court.
Premise #2: He is not at the store.
Conclusion: So, he must be at the handball court.

20. Premise #1: The book must be returned within 2 weeks or you pay a fine.
Premise #2: The book was not returned within 2 weeks.
Conclusion: So, you must pay a fine.

21. Premise #1: It is not true that it is a diamond and it sparkles in the sunlight.
Premise #2: It does sparkle in the sunlight.
Conclusion: So, it is a diamond.

22. Premise #1: Either I work tonight or I pass the mathematics test.
Premise #2: I'm going to work tonight.
Conclusion: So, I will fail the mathematics test.

In Exercises 23–30, determine which conclusion is valid from the given premises.

23. Premise #1: If 7 is a prime number, then 7 does not divide evenly into 21.
Premise #2: Seven divides evenly into 21.
(a) Conclusion: So, 7 is a prime number.
(b) Conclusion: So, 7 is not a prime number.
(c) Conclusion: So, 21 divided by 7 is 3.

24. Premise #1: If the fuel is shut off, then the fire will be extinguished.
Premise #2: The fire continues to burn.
(a) Conclusion: So, the fuel was not shut off.
(b) Conclusion: So, the fuel was shut off.
(c) Conclusion: So, the fire becomes hotter.

25. Premise #1: It is necessary that interest rates be lowered for the economy to improve.
Premise #2: Interest rates were not lowered.
(a) Conclusion: So, the economy will improve.
(b) Conclusion: So, interest rates are irrelevant to the performance of the economy.
(c) Conclusion: So, the economy will not improve.

26. Premise #1: It will snow only if the temperature is below 32° at some level of the atmosphere.
Premise #2: It is snowing.
(a) Conclusion: So, the temperature is below 32° at ground level.
(b) Conclusion: So, the temperature is above 32° at some level of the atmosphere.
(c) Conclusion: So, the temperature is below 32° at some level of the atmosphere.

27. Premise #1:   Smokestack emissions must be reduced or acid rain will continue as an environmental problem.

Premise #2:   Smokestack emissions have not decreased.

(a) Conclusion: So, the ozone layer will continue to be depleted.

(b) Conclusion: So, acid rain will continue as an environmental problem.

(c) Conclusion: So, stricter automobile emission standards must be enacted.

28. Premise #1:   The library must upgrade its computer system or service will not improve.

Premise #2:   Service at the library has improved.

(a) Conclusion: So, the computer system was upgraded.

(b) Conclusion: So, more personnel were hired for the library.

(c) Conclusion: So, the computer system was not upgraded.

29. Premise #1:   If Rodney studies, then he will make good grades.

Premise #2:   If he makes good grades, then he will get a good job.

(a) Conclusion: So, Rodney will get a good job.

(b) Conclusion: So, if Rodney doesn't study, then he won't get a good job.

(c) Conclusion: So, if Rodney doesn't get a good job, then he didn't study.

30. Premise #1:   It is necessary to have a ticket and an ID card to get into the arena.

Premise #2:   Janice entered the arena.

(a) Conclusion: So, Janice does not have a ticket.

(b) Conclusion: So, Janice has a ticket and an ID card.

(c) Conclusion: So, Janice has an ID card.

In Exercises 31–34, use a Venn diagram to test the validity of the argument.

31. Premise #1: All numbers divisible by 10 are divisible by 5.

Premise #2: Fifty is divisible by 10.

Conclusion: So, 50 is divisible by 5.

32. Premise #1: All human beings require adequate rest.

Premise #2: All infants are human beings.

Conclusion: So, all infants require adequate rest.

33. Premise #1: No person under the age of 18 is eligible to vote.

Premise #2: Some college students are eligible to vote.

Conclusion: So, some college students are under the age of 18.

34. Premise #1: Every amateur radio operator has a radio license.

Premise #2: Jackie has a radio license.

Conclusion: So, Jackie is an amateur radio operator.

In Exercises 35–38, use the premises to prove the given conclusion.

35. Premise #1: If Sue drives to work, then she will stop at the grocery store.

Premise #2: If she stops at the grocery store, then she will buy milk.

Premise #3: Sue drove to work today.

Conclusion: So, Sue will buy milk.

36. Premise #1: If Bill is patient, then he will succeed.

Premise #2: Bill will get bonus pay if he succeeds.

Premise #3: Bill did not get bonus pay.

Conclusion: So, Bill is not patient.

37. Premise #1: If this is a good product, then we should buy it.

Premise #2: Either it was made by XYZ Corporation, or we will not buy it.

Premise #3: It is not made by XYZ Corporation.

Conclusion: So, it is not a good product.

38. Premise #1: If it is raining today, Pam will clean her apartment.

Premise #2: If Pam is cleaning her apartment today, then she is not riding her bike.

Premise #3: Pam is riding her bike today.

Conclusion: It is not raining today.

Answers to Integrated Reviews, Odd-Numbered Exercises, Quizzes, and Tests

Chapter P
Section P.1   *(page 9)*

1. (a) $1, 4, 6$

(b) $-10, 0, 1, 4, 6$

(c) $-10, -\frac{2}{3}, -\frac{1}{4}, 0, \frac{5}{8}, 1, 4, 6$

(d) $-\sqrt{5}, \sqrt{3}, 2\pi$

3. (a) 3

(b) $-\sqrt{4}, 0, 3$

(c) $-3.5, -\sqrt{4}, -\frac{1}{2}, -0.\overline{3}, 0, 3, 25.2$

(d) $\sqrt{5}, 3\pi$

5. $-5, -4, -3, -2, -1, 0, 1, 2, 3$

7. $1, 3, 5, 7, 9$

9. (a)   (b)

(c)   (d)

11. $-1 < 3$ **13.** $-\frac{9}{2} < -2$ **15.** $2 < 5$

17. $10 > 4$ **19.** $-7 < -2$ **21.** $-5 < -2$

23. $\frac{1}{3} > \frac{1}{4}$ **25.** $-\frac{5}{8} < \frac{1}{2}$ **27.** $-\frac{2}{3} > -\frac{10}{3}$

29. $2.75 < \pi$ **31.** 6 **33.** 19 **35.** 50 **37.** 8

39. 35 **41.** 3 **43.** 10 **45.** 225 **47.** -85

49. -16 **51.** $-\frac{3}{4}$ **53.** -3.5 **55.** π

57. $|-6| > |2|$ **59.** $|47| > |-27|$

61. $-|-16.8| = -|16.8|$ **63.** $\left|-\frac{3}{4}\right| > -\left|\frac{4}{5}\right|$

65. $-34, 34$ **67.** $160, 160$ **69.** $\frac{3}{11}, \frac{3}{11}$ **71.** $-\frac{5}{4}, \frac{5}{4}$

73. $-4.7, 4.7$ **75.** 7

77. 5 **79.** $\frac{3}{5}$

81. $\frac{5}{3}$ **83.** 4.25

85. 0.7 **87.** $x < 0$

89. $x \geq 0$ **91.** $2 < z \leq 10$ **93.** $p < 225$

95. True   **97.** False. $\frac{2}{3}$ is not an integer.

99. False. $|3 + (-2)| = 1 \neq 5 = |3| + |-2|$

101. The set of integers includes the natural numbers, zero, and the negative integers.

103. Yes. The nonnegative real numbers include 0.

105. Place them on the real number line. The number on the right is greater.

Section P.2   *(page 18)*

1. 45 **3.** 19 **5.** -19 **7.** -22 **9.** -21

11. -20 **13.** 0.7 **15.** $\frac{5}{4}$ **17.** $\frac{1}{2}$ **19.** $\frac{1}{10}$

21. $\frac{1}{24}$ **23.** $\frac{105}{8}$ **25.** 60 **27.** 45.95 **29.** -28

31. $5 + 5 + 5 + 5$ **33.** $(-4) + (-4) + (-4)$

35. $6\left(\frac{1}{4}\right)$ **37.** $4(-15)$ **39.** -30 **41.** 48

43. -72 **45.** $\frac{1}{2}$ **47.** $-\frac{12}{5}$ **49.** $\frac{1}{12}$ **51.** $-\frac{1}{3}$

53. 6 **55.** -3 **57.** -9 **59.** $-\frac{5}{2}$ **61.** $\frac{2}{5}$

63. $\frac{46}{17}$ **65.** $\frac{11}{12}$ **67.** $(4)(4)(4)$

69. $\left(-\frac{3}{4}\right)\left(-\frac{3}{4}\right)\left(-\frac{3}{4}\right)\left(-\frac{3}{4}\right)$

71. $(-0.8)(-0.8)(-0.8)(-0.8)(-0.8)(-0.8)$ **73.** $(-7)^3$

75. $(-5)^4$ **77.** -7^3 **79.** 16 **81.** -8

83. -64 **85.** $\frac{64}{125}$ **87.** $\frac{1}{32}$ **89.** 0.027

91. -0.32 **93.** 0 **95.** 4 **97.** 22 **99.** 6

101. 57 **103.** 3 **105.** 27 **107.** 135 **109.** -5

111. -6 **113.** 1 **115.** 161 **117.** $14,425$

119. 171.36 **121.** $\frac{17}{180}$ **123.** $\$2533.56$

125. (a) $\$5, \$8, -\$5, \$16$

(b) The stock gained $24 in value during the week. Find the difference between the first bar (Monday) and the last bar (Friday).

127. (a) $\$10,800$

(b) $\$27,018.72$

(c) $\$16,218.72$

129. 15 square meters   **131.** 20 square inches

133. 6.125 cubic feet

135. (a) $20 \times 3 + 18 = 60 + 18 = 78$

    (b) Yes

    (c) No. The check digit should be 4.

137. True

139. False. A negative number raised to an odd power is negative.

141. Yes. $-3 + (-4) = -7$

143. If the numbers have like signs, the product or quotient is positive. If the numbers have unlike signs, the product or quotient is negative.

145. Evaluate additions and subtractions from left to right. For example, $6 - 5 - 2 = (6 - 5) - 2 = 1 - 2 = -1$. If this order were not understood, one might *incorrectly* write $6 - (5 - 2) = 6 - 3 = 3$.

147. Only common factors (not terms) of the numerator and denominator can be canceled.

149. $3 \cdot 4^2 = 3 \cdot 16 = 48$; Exponents must be evaluated before multiplication is done.

Section P.3   *(page 27)*

1. Commutative Property of Addition

3. Additive Inverse Property

5. Commutative Property of Multiplication

7. Multiplicative Identity Property

9. Commutative Property of Addition

11. Associative Property of Addition

13. Distributive Property

15. Associative Property of Addition

17. Associative Property of Multiplication

19. Multiplicative Identity Property

21. Additive Identity Property

23. Associative Property of Addition

25. Distributive Property   **27.** Distributive Property

29. $(3 \cdot 6)y$ **31.** $-3(15)$ **33.** $5 \cdot 6 + 5 \cdot z$

35. $-x + 25$ **37.** $x + 8$

39. (a) -10 (b) $\frac{1}{10}$ **41.** (a) 16   (b) $-\frac{1}{16}$

43. (a) $-6z$ (b) $\dfrac{1}{6z}$

45. (a) $-x - 1$ or $-(x + 1)$ (b) $\dfrac{1}{x + 1}$

47. $x + (5 - 3)$ **49.** $(32 + 4) + y$ **51.** $(3 \cdot 4)5$

53. $(6 \cdot 2)y$ **55.** $20 \cdot 2 + 20 \cdot 5$

57. $5(3x) + 5(-4)$ or $15x - 20$

59. $x(-2) + 6(-2)$ or $-2x - 12$

61. $-6(2y) + (-6)(-5)$ or $-12y + 30$

63. $3x + 15$ **65.** $-2x - 16$ **67.** Answers will vary.

69. Original equation
Addition Property of Equality
Associative Property of Addition
Additive Inverse Property
Additive Identity Property

71. Original equation
Addition Property of Equality
Associative Property of Addition
Additive Inverse Property
Additive Identity Property
Multiplication Property of Equality
Associative Property of Multiplication
Multiplicative Inverse Property
Multiplicative Identity Property

73. 28   **75.** 434   **77.** 62.82

79. $a(b + c) = ab + ac$ **81.** \$0.60   **83.** \$1.01

85. Given two real numbers a and b, the sum a plus b is the same as the sum b plus a.

87. The multiplicative inverse of a real number a $(a \neq 0)$ is the number $1/a$. The product of a number and its multiplicative inverse is the multiplicative identity 1. For example, $8 \cdot \frac{1}{8} = 1$.

89. $0 \cdot a = 0$

91. $4 \odot 7 = 15 \neq 18 = 7 \odot 4$

    $3 \odot (4 \odot 7) = 21 \neq 27 = (3 \odot 4) \odot 7$

Mid-Chapter Quiz   *(page 30)*

1.   **2.**

    $-4.5 > -6$ $\frac{3}{4} < \frac{3}{2}$

3. 3.2   **4.** -5.75 **5.** 22   **6.** 3.75   **7.** 14

8. -22 **9.** $\frac{5}{2}$ **10.** $\frac{1}{2}$ **11.** 48   **12.** $-\frac{3}{8}$

13. $\frac{7}{10}$ **14.** $-\frac{27}{8}$ **15.** 4   **16.** 2

17. (a) Distributive Property

    (b) Additive Inverse Property

18. (a) Associative Property of Addition

    (b) Multiplicative Identity Property

19. \$1492.28   **20.** \$3600   **21.** $\frac{7}{24}$

Section P.4   *(page 37)*

1. $10x$, 5   **3.** $-3y^2$, $2y$, -8 **5.** $4x^2$, $-3y^2$, $-5x$, $2y$

7. x^2, $-2.5x$, $-\dfrac{1}{x}$ **9.** 5   **11.** $-\dfrac{3}{4}$

13. Commutative Property of Addition

15. Associative Property of Multiplication

17. $5x + 5 \cdot 6$ or $5x + 30$ **19.** $6(x + 1)$

21. $x \cdot x \cdot x \cdot x \cdot x \cdot x \cdot x$ **23.** $z \cdot z \cdot z \cdot z \cdot z \cdot z \cdot z$

25. $(-5x)^4$ **27.** $x^3 y^3$ **29.** -2^7 **31.** x^{13}

33. $27y^6$ **35.** $16x^2$ **37.** $-16x^2$ **39.** $-125z^6$

41. $6x^3 y^4$ **43.** $-10y^9$ **45.** $-3125z^8$ **47.** $64a^7$

49. $-54u^5 v^3$ **51.** x^{4n} **53.** x^{n+4} **55.** $7x$

57. $8y$ **59.** $8x + 18y$ **61.** $6x^2 - 2x$

63. $-2z^4 + 5z + 8$ **65.** $4u^2 v^2 + uv$

67. $8x^2 + 4x - 12$ **69.** $-18y^2 + 3y + 6$

71. $12x - 35$ **73.** $-7y - 7$ **75.** $y^3 + 6$

77. $2y^3 + y^2 + y$ **79.** $-6x + 96$ **81.** $12x^2 + 2x$

83. $-2b^2 + 4b - 36$ **85.** $-4x^4 - 50x^3$

87. (a) 3   (b) -10 **89.** (a) 6   (b) 9

91. (a) 0   (b) $\frac{3}{10}$ **93.** (a) 13   (b) -36

95. (a) 7   (b) 7   **97.** (a) 3   (b) 0

99. (a) 210   (b) 140

101. $\frac{1}{2}b(b - 3)$; 90   **103.** $x(2x + 3) = 2x^2 + 3x$

105. $2800 million, $2800.34 million

107. $134 thousand   **109.** 1440 square feet

111. (e) $(0 + 3 + 1 + 2 + 6 + 7) \times 3 +$
$(4 + 8 + 9 + 3 + a)$

(f) 5, no

113. Add (or subtract) their respective coefficients and attach the common variable factor.
$5x^4 - 3x^4 = (5 - 3)x^4 = 2x^4$

115. $5x + 3x = (5 + 3)x = 8x$

117. No. When $y = 3$, the expression is undefined.

Section P.5   *(page 46)*

1. $8 + n$ **3.** $12 + 2n$ **5.** $n - 6$ **7.** $4n - 3$

9. $\frac{1}{3}n$ **11.** $\frac{x}{6}$ **13.** $8 \cdot \frac{N}{5}$ **15.** $4c + 10$

17. $0.30L$ **19.** $\frac{n + 5}{10}$ **21.** $|n - 5|$ **23.** $3x^2 - 4$

25. A number decreased by 2

27. A number increased by 50

29. The sum of three times a number and 2

31. The ratio of a number to 2

33. Four-fifths of a number

35. Eight times the difference of a number and 5

37. The sum of a number and 10 divided by 3

39. The product of a number and the sum of the number and 7

41. $0.25n$ **43.** $\frac{m}{10}$ **45.** $5m + 10n$ **47.** $55t$

49. $\frac{100}{r}$ **51.** $0.45y$ **53.** $0.0125I$

55. $L - 0.20L = 0.80L$ **57.** $8.25 + 0.60q$

59. $n + 3n = 4n$ **61.** $(2n + 1) + (2n + 3) = 4n + 4$

63. $\frac{2n(2n + 2)}{4}$ **65.** s^2 **67.** $0.375b^2$

69. Perimeter: $2(2w) + 2w = 6w$
Area: $2w(w) = 2w^2$

71. Perimeter: $6 + 2x + 3 + x + 3 + x = 4x + 12$
Area: $3x + 6x = 9x$

73. $l(l - 6) = l^2 - 6l$ square feet

75. $-3, 2, 7, 12, 17, 22$ **77.** a **79.** Subtraction
5, 5, 5, 5, 5

81. a and c

83. Using a specific case may make it easier to see the form of the expression for the general case.

Review Exercises   *(page 51)*

1.
$-5 < 3$

3.
$-\frac{8}{5} < -\frac{2}{5}$

5. 11   **7.** 7.3   **9.** 5   **11.** -7.2 **13.** 11

15. 230   **17.** -41.8 **19.** $\frac{11}{21}$ **21.** $\frac{1}{6}$ **23.** $\frac{17}{8}$

25. -28 **27.** -4200 **29.** $-\frac{1}{20}$ **31.** 14

33. 2   **35.** -216 **37.** -16 **39.** $\frac{1}{8}$ **41.** 20

43. 98   **45.** Additive Inverse Property

47. Distributive Property

49. Associative Property of Addition

51. Commutative Property of Multiplication

53. Distributive Property   **55.** $u - 3v$

57. $-3y^2 + 10y$ **59.** x^6 **61.** $-3x^3 y^4$ **63.** $125a^4 b$

$5v$ **69.** $5x - 10$ **71.** $5x - y$

75. (a) 0   (b) -3 **77.** $200 - 3n$

81. The sum of two times a number and 7

83. The difference of a number and 5 divided by 4

85. $0.18I$ **87.** $I(I - 5)$ **89.** $111.0 billion

91. 15.6 million

93.

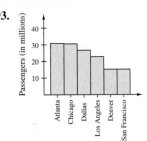

95. $644

Chapter Test   *(page 53)*

1. (a) $-\frac{5}{2} > -|-3|$ (b) $-\frac{2}{3} > -\frac{3}{2}$ **2.** 11.9

3. -20 **4.** $-\frac{1}{2}$ **5.** -150 **6.** 60   **7.** $\frac{1}{6}$

8. $\frac{4}{27}$ **9.** $-\frac{27}{125}$ **10.** 15

11. (a) Associative Property of Multiplication

(b) Multiplicative Inverse Property

12. $5(2x) - 5 \cdot 3$ **13.** $3x^4y^3$ **14.** $-2x^2 + 5x - 1$

15. a^2 **16.** $11t + 7$ **17.** (a) 4   (b) -12

18. $5n - 8$ **19.** Perimeter: $3.2l$; Area: $0.6l^2$

20. $4n + 2$ **21.** 16 feet   **22.** 640 cubic feet

Chapter 1

Section 1.1   *(page 63)*

Integrated Review   *(page 63)*

1. Commutative Property of Addition

2. Inverse Property of Multiplication

3. Distributive Property

4. Associative Property of Addition

5. 1   **6.** 4   **7.** $\frac{9}{2}$ **8.** 5   **9.** $\frac{4}{5}$ **10.** $\frac{28}{15}$

11. $18,000   **12.** 9 feet

1. (a) Not a solution   (b) Solution

3. (a) Solution   (b) Not a solution

5. (a) Solution   (b) Not a solution

7. (a) Not a solution   (b) Solution

9. No solution   **11.** Identity   **13.** Linear

15. Not linear, because the variable is in the denominator

17. Original equation
Subtract 15 from both sides.
Combine like terms.
Divide both sides by 3.
Simplify.

19. Original equation
Subtract 5 from both sides.
Combine like terms.
Divide both sides by -2.
Simplify.

21. 3   **23.** 4   **25.** -0.7 **27.** $-\frac{2}{3}$ **29.** -1

31. -1 **33.** 2   **35.** $-\frac{10}{3}$ **37.** -2 **39.** 2

41. $\frac{1}{3}$ **43.** No solution, because $-3 \neq 0$.   **45.** 0

47. No solution, because $-4 \neq 0$.   **49.** 11

51. -2 **53.** $-\frac{9}{2}$ **55.** $\frac{6}{5}$ **57.** -3 **59.** -3

61. $\frac{25}{3}$ **63.** 50   **65.** $\frac{19}{10}$ **67.** $-\frac{10}{3}$ **69.** $-\frac{20}{9}$

71. -20 **73.** $-\frac{8}{31}$ **75.** 23   **77.** 12   **79.** $\frac{1}{5}$

81. 125, 126   **83.** 82, 84   **85.** 1.5 hours

87. 1.5 seconds   **89.** 6 hours

91. (a)

t	1	1.5	2	3	4	5
Width	300	240	200	150	120	100
Length	300	360	400	450	480	500
Area	90,000	86,400	80,000	67,500	57,600	50,000

(b) Because the length is t times the width and the perimeter is fixed, as t gets larger the length gets larger and the area gets smaller. The maximum area occurs when the length and width are equal.

93. 1996

95. An equation whose solution set is not the entire set of real numbers is called a conditional equation. The solution set of an identity is all real numbers.

97. Evaluating an expression means finding its value when its variables are replaced by real numbers. Solving an equation means finding all values of the variable for which the equation is true.

99. Equivalent equations have the same solution set. For example, $3x + 4 = 10$ and $3x - 6 = 0$ are equivalent.

101. False

Section 1.2 *(page 73)*

Integrated Review *(page 73)*

1. A collection of letters (called variables) and real numbers (called constants) combined using the operations of addition, subtraction, multiplication, and division is called an algebraic expression.

2. The terms of an algebraic expression are those parts separated by addition or subtraction.

3. $a^m \cdot a^n = a^{m+n}$ 4. $(ab)^m = a^m \cdot b^m$

5. -240 6. 120 7. $-\frac{1}{4}$ 8. $\frac{6}{5}$ 9. -27

10. $\frac{25}{64}$ 11. $6x + 1$ 12. $14x + 2$

---

1. (A number) $+ 30 = 82$

   Equation: $x + 30 = 82$

   Solution: 52

3. 26(biweekly pay) $+$ (bonus) $= 30,500$

   Equation: $26x + 2300 = 30,500$

   Solution: $1084.62

	Percent	Parts out of 100	Decimal	Fraction
5.	30%	30	0.30	$\frac{3}{10}$
7.	7.5%	7.5	0.075	$\frac{3}{40}$
9.	12.5%	12.5	0.125	$\frac{1}{8}$

11. 87.5   13. 69.36   15. 600   17. 350

19. 400   21. 12,000   23. 33%   25. 175%

27. $\frac{2}{3}$ 29. $\frac{3}{4}$ 31. $\frac{1}{25}$ 33. $\frac{10}{3}$ 35. 4   37. 6

39. $\frac{15}{2}$ 41. 6   43. 4   45. 1140   47. 2

49. 15.625%   51. 18%   53. 7%   55. 200

57. 177.77%, 56.25%

59. Monroe: 30.84%   61. 8 pounds, 11.1%
   Washington: 6.01%
   Howard: 11.30%
   Spring: 17.44%
   West: 12.77%
   Clark: 21.66%

63. 11,750 million pounds   65. $\frac{1}{50}$ 67. $\frac{85}{4}$ 69. $\frac{4}{9}$ or $\frac{9}{4}$

71. $0.0475   73. $0.0845   75. $14\frac{1}{2}$-ounce bag

77. 6-ounce tube   79. $5\frac{1}{11}$ 81. 3

83. $\frac{516}{11} \approx 46.9$ feet   85. 17.1 gallons   87. $2400

89. 2667   91. Percent means parts out of 100.

93. No. $\frac{1}{2}\% = 0.5\% = 0.005$

95. No. It is necessary to know one of the following: the total number of students in the class, the number of boys in the class, or the number of girls in the class.

97. Mathematical modeling is using mathematics to solve problems that occur in real-life situations. For examples, review the real-life problems in the exercise set.

Section 1.3 *(page 86)*

Integrated Review *(page 86)*

1. Negative. To add two real numbers with like signs, add their absolute values and attach the common sign to the result.

2. Negative. To add two real numbers with unlike signs, subtract the smaller absolute value from the greater absolute value and attach the sign of the number with the greater absolute value.

3. Positive. To multiply two real numbers with like signs, find the product of their absolute values. The product is positive.

4. Negative. To multiply two real numbers with unlike signs, find the product of their absolute values. The product is negative.

5. 14   6. 0   7. 0   8. $-10,000$

9. -200 10. 40   11. 0.7 mile

12. $78\frac{11}{30}$ tons

---

1. $18.36, 40%   3. $152.00, 65%

5. $22,250.00, 21%   7. $416.70, $191.70

9. $24.21, 48.5%   11. $111.00, 63%

13. $33.25, $61.75   15. $1145.00, 22%

17. $22.05   19. $66\frac{2}{3}\%$ 21. $25   23. 20%

25. 9 minutes, $2.06   27. $54.15

29. Tax: $267   31. 2.5 hours
   Total bill: $4717
   Amount financed: $3717

33. 3 hours   35. 50 gallons at 20%; 50 gallons at 60%

37. 8 quarts at 15%; 16 quarts at 60%

39. 75 pounds at $12 per pound; 25 pounds at $20 per pound

41. 100   43. $\frac{5}{6}$ gallon   45. 2275 miles   47. $\frac{100}{11}$ hour

49. $\frac{2000}{3}$ feet per second   51. $2\frac{1}{2}$ hours   53. 1440 miles

55. 17.65 minutes   57. 3 hours at 58 mph

                    $2\frac{3}{4}$ hours at 52 mph

59. (a) 8 pages per minute   (b) $\frac{15}{4}$ units per hour

61. (a) $\frac{1}{3}, \frac{1}{4}$ (b) $\frac{12}{7} = 1\frac{5}{7}$ hours

63. $R = \dfrac{E}{I}$　**65.** $L = \dfrac{S}{1 - r}$　**67.** $a = \dfrac{2h - 96t}{t^2}$

69. 24 cubic units　**71.** $147\pi \approx 461.8$ cubic centimeters

73. 0.926 foot

75. 43 centimeters

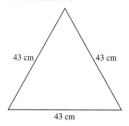

77. \$2850　**79.** \$3571.43　**81.** \$15,000

83. (a) 1993, yes　(b) \$0.307　**85.** Bus drivers

87. Markup is the difference between the cost a retailer pays for a product and the price at which the retailer sells the product. Markup rate is the percent increase of the markup.

89. $\dfrac{1}{t}$

91. No, it quadruples. The area of a square of side s is s^2. If the length of the sides is $2s$, the area is $(2s)^2 = 4s^2$.

Mid-Chapter Quiz　*(page 91)*

1. 2　**2.** 2　**3.** 2　**4.** Identity　**5.** $\frac{28}{5}$　**6.** $\frac{12}{7}$

7. $\frac{33}{2}$　**8.** 6　**9.** -0.20　**10.** 1.41　**11.** $\frac{9}{20}$, 45%

12. 200　**13.** \$0.1958 per ounce　**14.** 2000

15. Catalog　**16.** 7 hours

17. 25% solution: 40 gallons

50% solution: 10 gallons

18. 1.5 hours, 4.5 hours

19. 3.43 hours　**20.** 169 square inches

Section 1.4　*(page 101)*

Integrated Review　*(page 101)*

1. Commutative Property of Multiplication

2. Additive Inverse Property

3. Distributive Property

4. Additive Identity Property　**5.** 7　**6.** 0

7. 0　**8.** 1　**9.** 4　**10.** 9

11. 19.8 square meters　**12.** 104 square feet

1. (a) Yes　(b) No　(c) Yes　(d) No

3. (a) No　(b) Yes　(c) Yes　(d) No

5. d　**7.** a　**9.** f　**11.** a　**13.** d

15.

17.

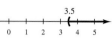

19.

21.

23.

25.

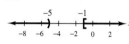

27.

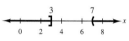

29. $-15 + x < -24$

31. $x \geq 4$

33. $x \leq 2$

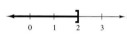

35. $x < 4$

37. $x \leq -4$

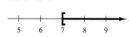

39. $x > 8$

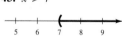

41. $x \geq 7$

43. $x > 7$

45. $x > -\frac{2}{3}$

47. $x > \frac{9}{2}$

49. $x > \frac{20}{11}$

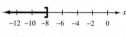

51. $x > \frac{8}{3}$

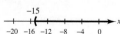

53. $x \leq -8$

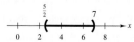

55. $x > -15$

57. $\frac{5}{2} < x < 7$

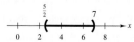

59. $-3 \le x < -1$

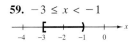

61. $-6 < x < 6$

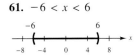

63. $-\frac{3}{2} < x < \frac{9}{2}$

65. $1 < x < 10$

67. $-1 < x \le 4$

69. $x \le -6$

71. $-\infty < x < \infty$

73. $x < -\frac{8}{3}$ or $x \ge \frac{5}{2}$

75. $y \le -10$

77. $-5 < x \le 0$

79. $x < -3$ or $x \ge 2$, $\{x | x < -3\} \cup \{x | x \ge 2\}$

81. $-5 \le x < 4$, $\{x | x \ge -5\} \cap \{x | x < 4\}$

83. $x \le -2.5$ or $x \ge -0.5$, $\{x | x \le -2.5\} \cup \{x | x \ge -0.5\}$

85. $\{x | x \ge -7\} \cap \{x | x < 0\}$

87. $\{x | x < -5\} \cup \{x | x > 3\}$

89. $\{x | x > -\frac{9}{2}\} \cap \{x | x \le -\frac{3}{2}\}$ **91.** $x \ge 0$ **93.** $z \ge 2$

95. $10 \le n \le 16$ **97.** x is at least $\frac{5}{2}$.

99. y is at least 3 and less than 5.

101. z is more than 0 and no more than π.   **103.** \$2600

105. The average temperature in Miami is greater than the average temperature in New York.

107. 26,000 miles   **109.** $x \ge 31$

111. The call must be less than 6.38 minutes. If a portion of a minute is billed as a full minute, the call must be less than or equal to 6 minutes.

113. $2 \le x \le 16$ **115.** $3 \le n \le \frac{15}{2}$

117. $12.50 < 8 + 0.75n$; $n > 6$ **119.** 1987, 1988, 1989

121. (f) 1   (g) 4   **123.** Yes.

125. The multiplication and division properties differ. The inequality symbol is reversed if both sides of the inequality are multiplied or divided by a negative real number.

127. $-8 < -t \le 5$

Section 1.5   *(page 111)*

Integrated Review   *(page 111)*

1. $2n$ is an even integer and $2n + 1$ is an odd integer.

2. No. $-2x^4 \ne 16x^4 = (-2x)^4$

3. $\frac{35}{14} = \frac{7 \cdot 5}{7 \cdot 2} = \frac{5}{2}$ **4.** $\frac{4}{5} \div \frac{z}{3} = \frac{4}{5} \cdot \frac{3}{z} = \frac{12}{5z}$

5. $-3.2 < 2$ **6.** $-3.2 > -4.1$ **7.** $-\frac{3}{4} > -5$

8. $-\frac{1}{5} > -\frac{1}{3}$ **9.** $\pi > -3$ **10.** $6 < \frac{13}{2}$

11. More than \$500   **12.** Less than \$500

1. Not a solution   **3.** Solution

5. $x - 10 = 17$; $x - 10 = -17$

7. $4x + 1 = \frac{1}{2}$; $4x + 1 = -\frac{1}{2}$ **9.** $4, -4$

11. No solution   **13.** 0   **15.** $3, -3$ **17.** 21, 11

19. $11, -14$ **21.** $\frac{16}{3}, 16$ **23.** No solution   **25.** $\frac{4}{3}$

27. $\frac{15}{2}, -\frac{39}{2}$ **29.** $18.75, -6.25$ **31.** $\frac{17}{5}, -\frac{11}{5}$

33. $-\frac{5}{3}, -\frac{13}{3}$ **35.** $\frac{15}{4}, -\frac{1}{4}$ **37.** 2, 3   **39.** $7, -3$

41. $\frac{3}{2}, -\frac{1}{4}$ **43.** 11, 13   **45.** $\frac{1}{2}$ **47.** $|x - 5| = 3$

49. (a) Solution   **51.** (a) Not a solution

　(b) Not a solution   　(b) Solution

　(c) Not a solution   　(c) Solution

　(d) Solution   　(d) Not a solution

53. $-3 < y + 5 < 3$ **55.** $7 - 2h \ge 9$ or $7 - 2h \le -9$

57.

59.

61. $-4 < y < 4$ **63.** $x \le -6$ or $x \ge 6$

65. $-7 < x < 7$ **67.** $-9 \le y \le 9$ **69.** $-2 \le y \le 6$

71. $x < -16$ or $x > 4$ **73.** $-3 \le x \le 4$

75. $t \le -\frac{15}{2}$ or $t \ge \frac{5}{2}$ **77.** $-\infty < x < \infty$

79. No solution   **81.** $-82 \le x \le 78$

83. $-104 < y < 136$ **85.** $z < -50$ or $z > 110$

87. $-5 < x < 35$ **89.** $\frac{28}{3} \le x \le \frac{32}{3}$

91. $-\infty < x < \infty$ **93.** $-4 \le x \le 40$

95. $-2 < x < \frac{2}{3}$ **97.** $x < -6$ or $x > 3$

99. $3 \le x \le 7$ **101.** d   **103.** b   **105.** $|x| \le 2$

107. $|x - 19| < 3$ **109.** $|x| < 3$ **111.** $|x - 5| > 6$

113.

115. (a) $|s - x| \leq \frac{3}{16}$ (b) $\frac{79}{16} \leq x \leq \frac{85}{16}$

117. The absolute value of a real number measures the distance of the real number from zero.

119. The solutions of $|x| = a$ are $x = a$ and $x = -a$. $|x - 3| = 5$ means $x - 3 = 5$ or $x - 3 = -5$. Thus, $x = 8$ or $x = -2$.

121. All real numbers less than 1 unit from 4 **123.** 3

Review Exercises *(page 115)*

1. (a) Not a solution (b) Solution

3. (a) Solution (b) Not a solution **5.** 3 **7.** -12

9. -24 **11.** 3 **13.** 2 **15.** 4 **17.** 4

19. 14 **21.** No solution **23.** 2 **25.** -4.2

27. $87, 0.87, \frac{87}{100}$ **29.** 65 **31.** 3000 **33.** 125%

35. $\frac{4}{3}$ **37.** $\frac{3}{20}$ **39.** $\frac{7}{2}$ **41.** $\frac{10}{3}$ **43.** $\$49.98, 50\%$

45. $\$81.72, 54\%$ **47.** $\$17.99, 25\%$

49. $\$1396.85, 30\%$ **51.** $x = \frac{1}{2}(7y - 4)$ **53.** $h = \dfrac{V}{\pi r^2}$

55. $x \leq 4$

57. $x > -6$

59. $x > 3$

61. $x \leq 6$

63. $y > -\frac{70}{3}$

65. $-7 \leq x < -2$

67. $-16 < x < -1$

69. $-3 < x < 2$

71. $x \leq -3$

73. $z \leq 10$

75. $7 \leq y < 14$ **77.** ± 6 **79.** $4, -\frac{4}{3}$ **81.** $0, -\frac{8}{5}$

83. $\frac{1}{2}, 3$ **85.** $x < 1$ or $x > 7$ **87.** $x < -3$ or $x > 3$

89. $-4 < x < 11$ **91.** $b < -9$ or $b > 5$

93. $(-\infty, 2], [3, \infty)$ **95.** $|x - 3| < 2$ **97.** 73, 74

99. 6% **101.** Department store **103.** $\$2.47$

105. $3\frac{3}{4}$ cups **107.** 25 pints or 3.125 gallons

109. $\frac{14}{3}$ **111.** 80 feet **113.** $\$27,166.25$

115. 84.21% **117.** Department store

119. 30% solution: $3\frac{1}{3}$ liters **121.** 2800 miles
60% solution: $6\frac{2}{3}$ liters

123. 43.6 miles per hour **125.** $\frac{18}{7} \approx 2.57$ hours

127. $\$340$ **129.** $\$210,526.32$ **131.** $\$30,000$

133. 8 inches **135.** $2 \leq x \leq 27$

137.

Chapter Test *(page 119)*

1. 4 **2.** 3 **3.** 4 **4.** 24 **5.** 864 **6.** 150%

7. $\$8000$ **8.** 15-ounce package **9.** $\$1466.67$

10. $2\frac{1}{2}$ hours **11.** $33\frac{1}{3}$ liters of 10% solution
$66\frac{2}{3}$ liters of 40% solution

12. 40 minutes **13.** $\$2000$

14. (a) $5, -11$ (b) $\frac{2}{3}, -\frac{4}{3}$ (c) $5, -\frac{1}{2}$

15. (a) $x \geq -6$ (b) $x > 2$

(c) $-7 < x \leq 1$ (d) $-1 \leq x < \frac{5}{4}$

16. (a) $1 \leq x \leq 5$ **17.** $t \geq 8$ **18.** 25,000
(b) $x < -\frac{9}{5}$ or $x > 3$
(c) $-\frac{44}{5} < x < -\frac{36}{5}$

Chapter 2

Section 2.1 *(page 130)*

Integrated Review *(page 130)*

1. $3x = 7$ is a linear equation because it can be written in the form $ax + b = 0$. Since $x^2 + 3x = 2$ cannot be written in the form $ax + b = 0$, it is not a linear equation.

2. Substitute 3 for x in the equation. If the result is true, $x = 3$ is a solution.

3. $12x^3$ **4.** $3t^6$ **5.** $54x^{10}$ **6.** $-8x^4y^5$

7. $1 - 6x$ **8.** $-3x + 22$ **9.** $12y$

10. $0.02x + 100$ **11.** $\frac{1}{4}, \frac{1}{5}; \frac{20}{9} \approx 2.2$ hours

12. 45.65 miles per hour

1.
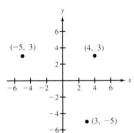

3.
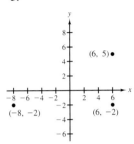

43.
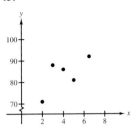

45.
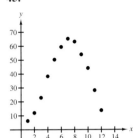

47. $(-3, -4) \Rightarrow (-1, 1)$
$(1, -3) \Rightarrow (3, 2)$
$(-2, -1) \Rightarrow (0, 4)$

5.
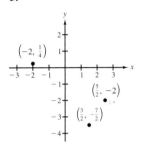

7.
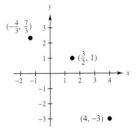

49.

x	-2	0	2	4	6
$y = 5x - 1$	-11	-1	9	19	29

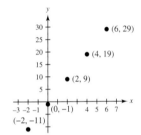

9. *A*: $(-2, 4)$
  B: $(0, -2)$
  C: $(4, -2)$

11. *A*: $(4, -2)$
  B: $\left(-3, -2\frac{1}{2}\right)$
  C: $\left(3, \frac{1}{2}\right)$

13.
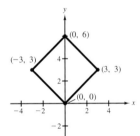

15.
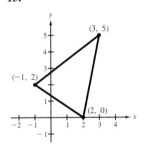

51.

x	-4	$\frac{2}{5}$	4	8	12
$y = -\frac{5}{2}x + 4$	14	3	-6	-16	-26

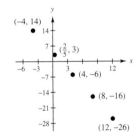

17.
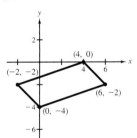

19.
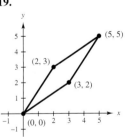

53.

x	-2	0	2	4	6
$y = 4x^2 + x - 2$	12	-2	16	66	148

55. (a) Not a solution
   (b) Solution
   (c) Not a solution
   (d) Solution

57. (a) Not a solution
   (b) Solution
   (c) Solution
   (d) Not a solution

21. $(-5, 2)$ **23.** $(3, -2)$ **25.** $(-10, -10)$
27. $(10, 0)$ **29.** Quadrant III   **31.** Quadrant IV
33. Quadrant I   **35.** Quadrant IV   **37.** Quadrants I or II
39. Quadrants II or III   **41.** Quadrants I or III

59. (a) Solution

(b) Solution

(c) Not a solution

(d) Not a solution

61.

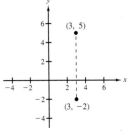

Distance: 7
Vertical line

63.

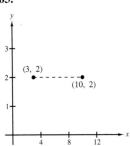

Distance: 7
Horizontal line

65.

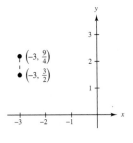

Distance: $\frac{3}{4}$
Vertical line

67.

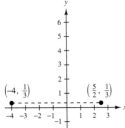

Distance: $\frac{13}{2}$
Horizontal line

69. $\sqrt{5}$　　**71.** 5

73. $\sqrt{58}$　　**75.** $\sqrt{61}$　　**77.** $\sqrt{29}$

79. $\left(\sqrt{13}\right)^2 + \left(\sqrt{13}\right)^2 = \left(\sqrt{26}\right)^2$

81. $\left(2\sqrt{5}\right)^2 + \left(2\sqrt{5}\right)^2 = \left(2\sqrt{10}\right)^2$　　**83.** Not collinear

85. Collinear

87.

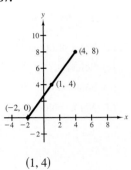

$(1, 4)$

89.

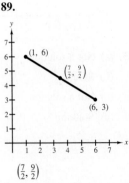

$\left(\frac{7}{2}, \frac{9}{2}\right)$

91.

x	100	150	200	250	300
$C = 28x + 3000$	5800	7200	8600	10,000	11,400

93. 18.55 feet　　**95.** $3 + \sqrt{26} + \sqrt{29} \approx 13.48$

97. The word *ordered* is significant because each number in the pair has a particular interpretation. The first measures horizontal distance and the second measures vertical distance.

99. The x-coordinate of any point on the y-axis is 0. The y-coordinate of any point on the x-axis is 0.

101. No. The scales on the x- and y-axes are determined by the magnitudes of the quantities being measured by x and y.

103.

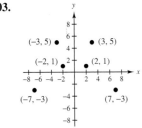

Reflection in the y-axis

Section 2.2　*(page 139)*

Integrated Review　*(page 139)*

1. $t - 3 + c > 7 + c$　　**2.** $(t - 3)c > 7c$

3. $y\left(\dfrac{1}{y}\right) = 1$

4. Commutative Property of Addition　　**5.** $x \geq 1$

6. $x < -3$　　**7.** $-\frac{1}{2} < x < \frac{1}{2}$

8. $-\frac{1}{2} \leq x \leq \frac{3}{2}$　　**9.** $-6 \leq x \leq 6$

10. $20 < x < 30$　　**11.** \$29,018　　**12.** \$108.50

1. e　　**3.** f　　**5.** d

7.

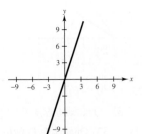

9.

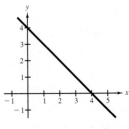

11.

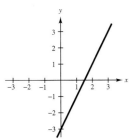

13.

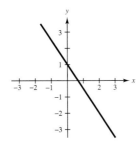

31. $\left(\frac{1}{2}, 0\right), (0, -3)$ **33.** $(-20, 0), (0, 15)$

35. $(10, 0), (0, 5)$ **37.** $\left(-\frac{3}{4}, 0\right), (0, 3)$

39. $(\pm 1, 0), (0, -1)$ **41.** $(-2, 0), (0, 2)$

43. $(-2, 0), (4, 0), (0, -2)$ **45.** $(3, 0), (0, 2)$

47. $(0, 3)$

49.

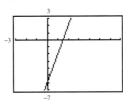

$(1.5, 0), (0, -6)$

51.

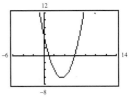

$(1, 0), (6, 0), (0, 6)$

15.

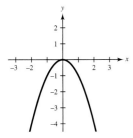

17.

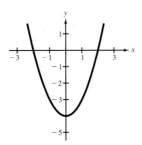

53.

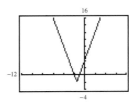

$(-2, 0), (-1, 0), (0, 4)$

55.

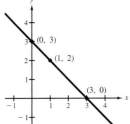

19.

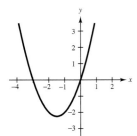

21.

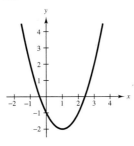

57.

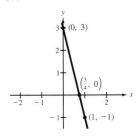

59.

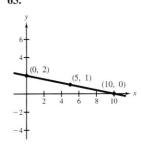

23.

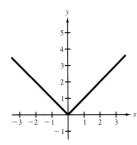

25.

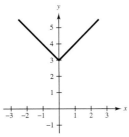

61.

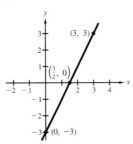

63.

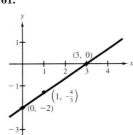

27.

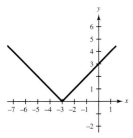

29.

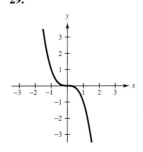

65.

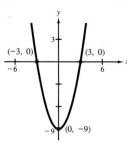

67.

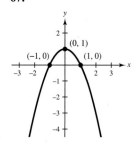

81. (a)

x	0	3	6	9	12
F	0	4	8	12	16

(b)

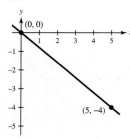

(c) *F* doubles.

69.

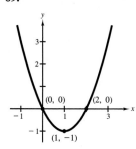

71.

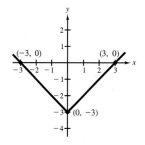

83. The scales on the *y*-axes are different. From graph (a) it appears that sales have not increased. From graph (b) it appears that sales have increased dramatically.

85. The graph of an equation is the set of all solutions of the equation plotted on a rectangular coordinate system.

73.

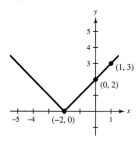

75.

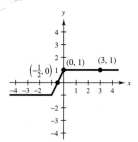

87. To find the *x*-intercepts, let $y = 0$ and solve the equation for *x*. To find the *y*-intercepts, let $x = 0$ and solve the equation for *y*. To find any *x*-intercepts of the graph of $2x - y = 4$, solve $2x - 0 = 4$ to obtain the intercept $(2, 0)$. To find any *y*-intercepts of the graph of $2x - y = 4$, solve $2(0) - y = 4$ to obtain the intercept $(0, -4)$.

89. (a) 6 miles

    (b) Stopped

    (c) $6 \leq t \leq 10$. The graph is steepest.

77.

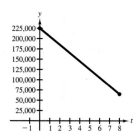

Section 2.3   (page 150)

Integrated Review   *(page 150)*

1. Equivalent   **2.** 5   **3.** $\frac{8}{3}$ **4.** 27   **5.** 5

6. $\frac{10}{3}$ **7.** 18   **8.** -19 **9.** No solution

10. 0.1   **11.** $0 < t \leq 23$ **12.** $m < 23{,}846$

79. $y = 40{,}000 - 5000t, \ 0 \leq t \leq 7$

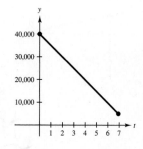

1. $\frac{2}{3}$ **3.** -2 **5.** Undefined

7. (a) L_3 (b) L_2 (c) L_1

9.

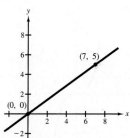

$m = \frac{5}{7}$; rises

11.

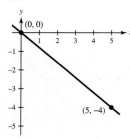

$m = -\frac{4}{5}$; falls

13.

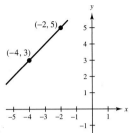

$m = 1$; rises
m is undefined; vertical

15.

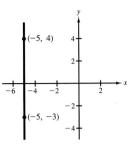

17.

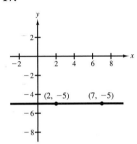

$m = 0$; horizontal

19.

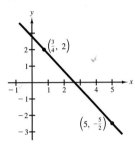

$m = -\frac{18}{17}$; falls

21.

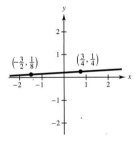

$m = \frac{1}{18}$; rises

23.

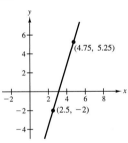

$m = \frac{29}{9}$; rises

25.

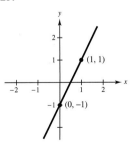

$m = 2$

27.

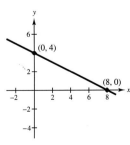

$m = -\frac{1}{2}$

29.

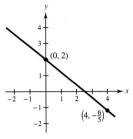

$m = -\frac{4}{5}$

31. $x = 1$ **33.** $y = -15$

35. $(6, 2), (10, 2)$ **37.** $(4, -1), (5, 2)$ **39.** $(1, 2), (2, 1)$

41. $(-2, 4), (1, 8)$ **43.** $y = 2x - 3$

45. $y = \frac{1}{4}x - 1$ **47.** $y = -\frac{2}{5}x + \frac{3}{5}$

49. $y = \frac{1}{2}x + 2$ **51.** $m = 3$; $(0, -2)$

53. $m = \frac{2}{3}$; $(0, 1)$ **55.** $m = -\frac{5}{3}$; $\left(0, \frac{2}{3}\right)$

57. $y = 3x - 2$ **59.** $y = -x$

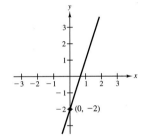   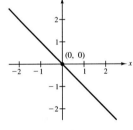

61. $y = -\frac{3}{2}x + 1$ **63.** $y = \frac{1}{4}x + \frac{1}{2}$

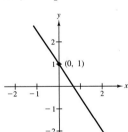   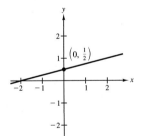

65.

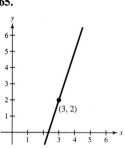

67.

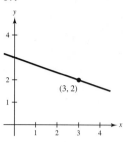

69.

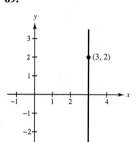

71.

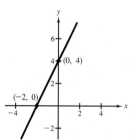

73.

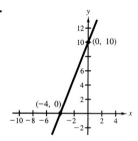

75. Parallel

77. Perpendicular   **79.** 16,667 feet   **81.** $\frac{45}{4}$ feet

83. (a)

t	0	1	2	3
y	\$2015.79	\$2208.43	\$2401.07	\$2593.71

t	4	5	6
y	\$2786.35	\$2978.99	\$3171.63

(b)

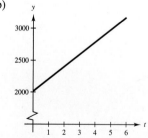

(c) \$192.64; the increase each year is the slope of the graph.

(d) \$4905.39

85. (a)

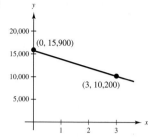

(b) -1900

(c) Annual depreciation

87. Negative slope: line falls to the right.

Zero slope: line is horizontal.

Positive slope: line rises to the right.

89. m is the slope; b is the y-intercept.

91. No. Their slopes must be negative reciprocals of each other.

Mid-Chapter Quiz   *(page 153)*

1. Quadrants I or II   **2.** $(10, -3)$

3. (a) Not a solution   (b) Solution

(c) Solution   (d) Solution

4.

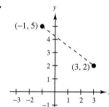

Distance: 5

5.

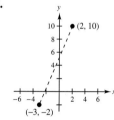

Distance: 13

6. $(-8, 0), (0, 6)$

7.

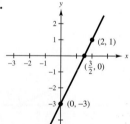

8.

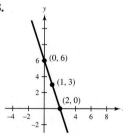

9.

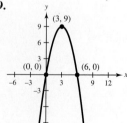

10.

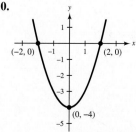

11.

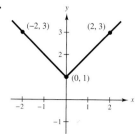

12.

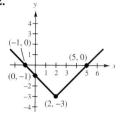

13. m is undefined; Vertical   **14.** $m = 0$; Horizontal

15. $m = \frac{5}{3}$; Rises   **16.** $m = -\frac{5}{3}$; Falls

17. $y = -\frac{1}{2}x + 1$ **18.** $y = 2x + 8$

$\quad m = -\frac{1}{2}$ $\quad m = 2$

$\quad (0, 1)$ $\quad (0, 8)$

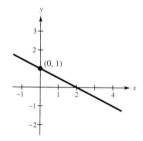

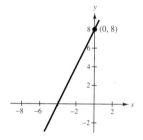

19. $y = \frac{1}{2}x - 2$ **20.** Perpendicular

$\quad m = \frac{1}{2}$

$\quad (0, -2)$

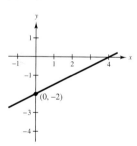

21. Neither   **22.** Parallel

23. $V = 85,000 - 8100t, \quad 0 \le t \le 10$

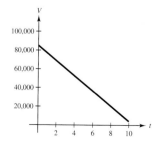

Section 2.4 *(page 160)*

Integrated Review *(page 160)*

1. $\frac{a}{b}$ **2.** A proportion   **3.** 1.875   **4.** 9000

5. 150%   **6.** 15.5%   **7.** $66\frac{2}{3}\%$ **8.** 350

9. 80,000   **10.** 275   **11.** 72 pounds

12. 3 seconds

1. b   **3.** a   **5.** $3x - y = 1$ **7.** $x + 2y = -1$

9. $4x - 5y = 8$ **11.** $y = -\frac{1}{2}x$ **13.** $y + 4 = 3x$

15. $y - 6 = -\frac{3}{4}x$ **17.** $y - 8 = -2(x + 2)$

19. $y + 7 = \frac{5}{4}(x + 4)$ **21.** $y - \frac{7}{2} = -4(x + 2)$

23. $y - \frac{5}{2} = \frac{4}{3}\left(x - \frac{3}{4}\right)$ **25.** $y + 1 = 0$

27. $3x - 2y = 0$ **29.** $x + y - 4 = 0$

31. $x + 2y - 4 = 0$ **33.** $2x + 5y = 0$

35. $2x - 6y + 15 = 0$ **37.** $5x + 34y - 67 = 0$

39. $52x + 15y - 395 = 0$ **41.** $4x + 5y - 11 = 0$

43. $y = \frac{1}{2}x + 3$ **45.** $y = 3$ **47.** $x = -1$

49. $y = 6$ **51.** $x = -7$

53. (a) $y = 3x - 5$ (b) $y = -\frac{1}{3}x + \frac{5}{3}$

55. (a) $y = -\frac{5}{4}x - \frac{9}{4}$ (b) $y = \frac{4}{5}x + 8$

57. (a) $y = 4x - 5$ (b) $y = -\frac{1}{4}x + \frac{31}{4}$

59. (a) $x = \frac{2}{3}$ (b) $y = \frac{4}{3}$ **61.** (a) $y = 2$ (b) $x = -1$

63. $\frac{x}{3} + \frac{y}{2} = 1$ **65.** $-\frac{6x}{5} - \frac{3y}{7} = 1$

67. $C = 20x + 5000$; $13,000

69. $S = 100,000t$; $600,000   **71.** $S = 0.03M + 1500$; 3%

73. (a) $S = 0.70L$ (b) $94.50

75. (a) $V = 7400 - 1475t$ (b) $4450

77. (a) $N = 60t + 1500$ (b) 2400   (c) 1800

79. (a) and (b)

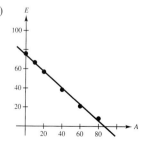

(c) $E = 74.56 - 0.86A$ (d) 48.8 years

81. $x - 8y = 0$

Distance from the deep end	0	8	16	24	32	40
Depth of water	9	8	7	6	5	4

83. Yes. When different pairs of points are selected, the change in y and the change in x are the lengths of the sides of similar triangles. Corresponding sides of similar triangles are proportional.

85. 3 is the slope; 5 is the y-intercept.

Section 2.5 *(page 172)*

Integrated Review *(page 172)*

1. $a < c$; Transitive Property

2. $9x = 36$
$\frac{1}{9}(9x) = \frac{1}{9}(36)$
$x = 4$

3. $y \le 45$ **4.** $x \ge 15$

5. $-4y$ **6.** $5(x - 2)$ **7.** $\frac{3}{2}t - \frac{5}{8}$

8. $\frac{7}{24}x + 8$ **9.** $-30x^2 + 23x + 3$

10. $4x^3 + 12x^2y + 4xy^2 + y^3$ **11.** $8\frac{3}{4}$

12. 16 pints or 2 gallons

---

1. Domain: $\{-2, 0, 1\}$
Range: $\{-1, 0, 1, 4\}$

3. Domain: $\{0, 2, 4, 5, 6\}$
Range: $\{-3, 0, 5, 8\}$

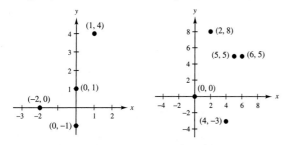

5. $(3, 150), (2, 100), (8, 400), (6, 300), \left(\frac{1}{2}, 25\right)$

7. $(1, 1), (2, 8), (3, 27), (4, 64), (5, 125), (6, 216), (7, 343)$

9. (1995, Atlanta Braves), (1996, New York Yankees), (1997, Florida Marlins), (1998, New York Yankees)

11. Not a function   **13.** Function   **15.** Not a function

17. Not a function   **19.** Function   **21.** Not a function

23. (a) Function from A to B
(b) Not a function from A to B
(c) Function from A to B
(d) Not a function from A to B

25. There are two values of y associated with one value of x.

27. There are two values of y associated with one value of x.

29. There is one value of y associated with one value of x.

31. There is one value of y associated with one value of x.

33. There is one value of y associated with one value of x.

35. (a) $3(2) + 5 = 11$
(b) $3(-2) + 5 = -1$
(c) $3(k) + 5$
(d) $3(k + 1) + 5 = 3k + 8$

37. (a) $3 - 0^2 = 3$
(b) $3 - (-3)^2 = -6$
(c) $3 - m^2$
(d) $3 - (2t)^2 = 3 - 4t^2$

39. (a) $\frac{3}{3 + 2} = \frac{3}{5}$ (b) $\frac{-4}{-4 + 2} = 2$
(c) $\frac{s}{s + 2}$ (d) $\frac{s - 2}{s - 2 + 2} = \frac{s - 2}{s}$

41. (a) 29 (b) 11 (c) $12a - 2$ (d) $12a + 5$

43. (a) 2 (b) 2 (c) $4y^2 - 8y + 2$ (d) 16

45. (a) 2 (b) 3 (c) $\sqrt{z}$ (d) $\sqrt{5z + 5}$

47. (a) 4 (b) 4 (c) -7 (d) $8 - |x - 6|$

49. (a) 0 (b) $-\frac{3}{2}$ (c) $-\frac{5}{2}$ (d) $\frac{3x + 12}{x - 1}$

51. (a) 2 (b) -2 (c) 10 (d) -8

53. (a) 0 (b) $\frac{7}{4}$ (c) 3 (d) 0

55. (a) 2 (b) $\frac{2x - 12}{x}$ **57.** All real numbers x

59. All real numbers x such that $x \ne 3$

61. All real numbers t such that $t \ne 0, -2$

63. All real numbers x such that $x \ge -4$

65. All real numbers x such that $x \ge \frac{1}{2}$

67. All real numbers t **69.** Domain: $\{0, 2, 4, 6\}$
Range: $\{0, 1, 8, 27\}$

71. Domain: $\{-3, -1, 4, 10\}$
Range: $\left\{-\frac{17}{2}, -\frac{5}{2}, 2, 11\right\}$

73. Domain: All real numbers r such that $r > 0$
Range: All real numbers C such that $C > 0$

75. Domain: All real numbers r such that $r > 0$
Range: All real numbers A such that $A > 0$

77. $P = 4x$ **79.** $V = x^3$ **81.** $d = 230t$

83. $V = x(24 - 2x)^2$ **85.** $A = (32 - x)^2$
$\quad\;\; = 4x(12 - x)^2$

87. (a) 10,680 pounds (b) 8010 pounds

89. Yes. Yes. For each year there is associated one public school enrollment and one private school enrollment.

91. (g) $y(x) = 15,900 - 1900x$; $2600

(h) Depreciates more slowly as the car ages

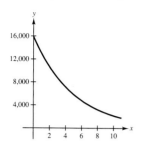

(i) Domain: All real numbers x such that $0 < x \leq 8.37$

Range: All real numbers y such that $0 < y \leq 15,900$

93. (a) Not correct (b) Correct

95. No. $\{(4, 3), (4, -2)\}$ is a relation, but not a function.

97. You can name the function $(f, g, \text{etc.})$ that is convenient when there is more than one function used in solving a problem. The values of the independent and dependent variables are easily seen in function notation.

Section 2.6 *(page 183)*

Integrated Review *(page 183)*

1. Multiplicative Inverse Property

2. Additive Identity Property

3. Distributive Property

4. Associative Property of Addition **5.** $5x^6$

6. $3(x + 1)^5$ **7.** $-64t^3$ **8.** $-16x^4$ **9.** u^8v^4

10. $18a^4b^5$ **11.** Department store **12.** $960.70

1.

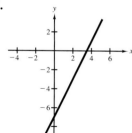

Domain: $-\infty < x < \infty$
Range: $-\infty < y < \infty$

3.

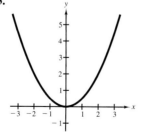

Domain: $-\infty < x < \infty$
Range: $0 \leq y < \infty$

5.

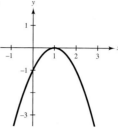

Domain: $-\infty < x < \infty$
Range: $-\infty < y \leq 0$

7.

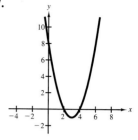

Domain: $-\infty < x < \infty$
Range: $-1 \leq y < \infty$

9.

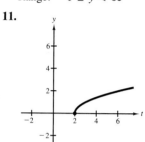

Domain: $0 \leq x < \infty$
Range: $-1 \leq y < \infty$

11.

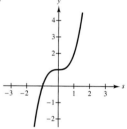

Domain: $2 \leq t < \infty$
Range: $0 \leq y < \infty$

13.

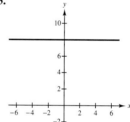

Domain: $-\infty < x < \infty$
Range: $y = 8$

15.

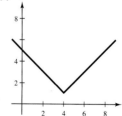

Domain: $-\infty < s < \infty$
Range: $-\infty < y < \infty$

17.

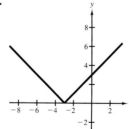

Domain: $-\infty < x < \infty$
Range: $0 \leq y < \infty$

19.

Domain: $-\infty < s < \infty$
Range: $1 \leq y < \infty$

21.

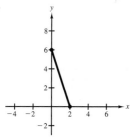

Domain: $0 \le x \le 2$
Range: $0 \le y \le 6$

23.

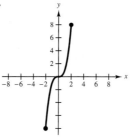

Domain: $-2 \le x \le 2$
Range: $-8 \le y \le 8$

41.

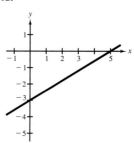

y is a function of x.

43.

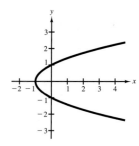

y is not a function of x.

25.

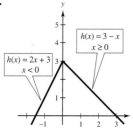

Domain: $-\infty < x < \infty$
Range: $-\infty < y \le 3$

27.

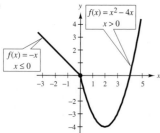

Domain: $-\infty < x < \infty$
Range: $-4 \le y < \infty$

29.

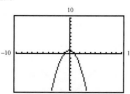

Domain: $-\infty < x < \infty$
Range: $-\infty < y \le 1$

31.

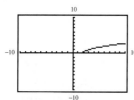

Domain: $2 \le x < \infty$
Range: $0 \le y < \infty$

33. Function   **35.** Function   **37.** Not a function

39. Not a function

45. b   **47.** a   **49.**

Xmin = 0
Xmax = 20
Xscl = 2
Ymin = -10
Ymax = 60
Yscl = 6

51. (a) Vertical shift
        2 units upward

(b) Vertical shift
     4 units downward

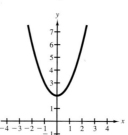

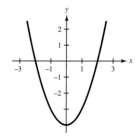

(c) Horizontal shift
     2 units to the left

(d) Horizontal shift
     4 units to the right

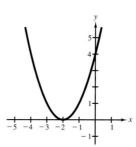

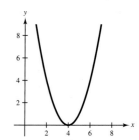

(e) Reflection in the
x-axis

(f) Reflection in the x-axis
and a vertical shift
4 units upward

73. (a)

(b)

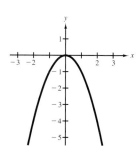

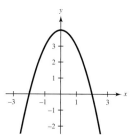

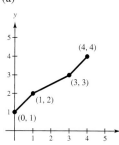

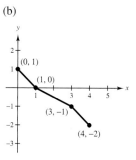

(g) Horizontal shift 3 units
to the right and a verti-
cal shift 1 unit upward

(h) Reflection in the x-axis,
a horizontal shift 2 units
to the left, and a vertical
shift 3 units downward

(c)

(d)

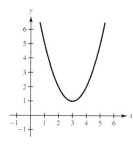

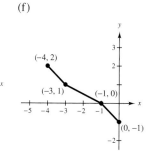

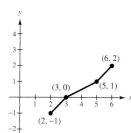

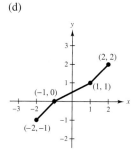

53. Horizontal shift 5 units
to the right

55. Vertical shift 5 units
downward

(e)

(f)

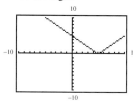

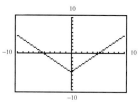

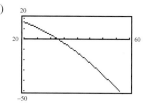

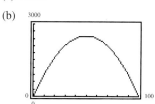

75. (a)

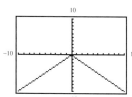

(b) 46%

57. Reflection in the
x-axis

59. $h(x) = (x + 3)^2$

77. (a) Proof

(b)

61. $h(x) = -x^2$ **63.** $h(x) = -(x + 3)^2$
65. $h(x) = -x^2 + 2$ **67.** $h(x) = -\sqrt{x}$
69. $h(x) = \sqrt{x + 2}$ **71.** $h(x) = \sqrt{-x}$

(c) $x = 50$. The figure is a square.

79. (a) 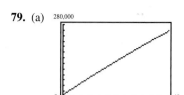   (b) 1970

(c)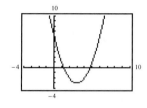

81. The range changes from $[0, 4]$ to $[0, 8]$.

83. Vertical shift upward, vertical shift downward, horizontal shift to the left, horizontal shift to the right

85. Reflection in the y-axis

Review Exercises   *(page 189)*

1. 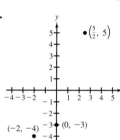   **3.**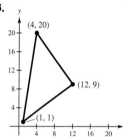

5. Quadrant IV   **7.** Quadrants I or IV

9. (a) Yes   (b) No   (c) No   (d) Yes

11. 5   **13.** $3\sqrt{5}$ **15.** c   **17.** a

19. 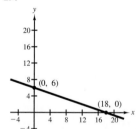   **21.**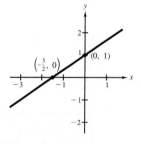

23. 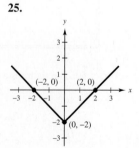   **25.**

27. $\left(\frac{3}{2}, 0\right)$, $(0, -6)$ **29.** $(-2, 0)$, $(0, 7)$

31. $(5, 0)$, $(0, 5)$ **33.** $(-3, 0)$, $(2, 0)$, $(0, -4)$

35. 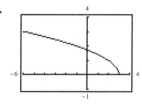   **37.**

$(1.27, 0)$, $(4.73, 0)$, $(0, 6)$ $(0, -11)$

39.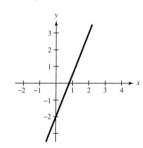

$(3, 0)$, $(0, 1.73)$

41. $\frac{2}{7}$ **43.** 0   **45.** $-\frac{3}{4}$ **47.** $\frac{3}{2}$

49. $(1, -1)$, $(0, 2)$ **51.** $(7, 6)$, $(11, 11)$

53. $(3, 0)$, $(3, 5)$

55. $y = \frac{5}{2}x - 2$ **57.** $y = -\frac{1}{2}x + 1$

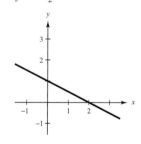

59. Neither   **61.** Perpendicular   **63.** Neither

65. $2x - y - 6 = 0$ **67.** $4x + y = 0$

69. $2x + 3y - 17 = 0$ **71.** $y - 5 = 0$

73. $y = -\frac{1}{2}(x + 6)$ **75.** $y - 6 = \frac{3}{2}(x - 4)$

77. $y - \frac{7}{6} = \frac{3}{8}(x - 4)$

79. (a) $3x + y - 1 = 0$ (b) $x - 3y - 3 = 0$

81. (a) $x - 12 = 0$ (b) $y - 1 = 0$

83. Not a function   **85.** Function

87. (a) 29   (b) 3   (c) $\dfrac{36 - 5t}{2}$ (d) $4 - \dfrac{5}{2}(x + h)$

89. (a) 3   (b) 0   (c) $\sqrt{2}$ (d) $\sqrt{5 - 5z}$

91. (a) -3 (b) 2   (c) 0   (d) -7

93. (a) -2 (b) $\dfrac{12 - 2x}{x}$ **95.** $-\infty < x < \infty$

97. $-\infty < x \le \frac{5}{2}$

99.

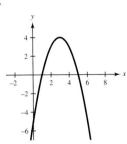

101.

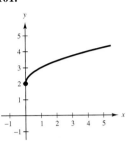

103.

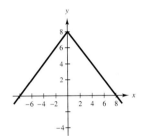

105.

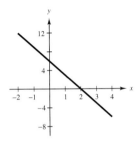

107.

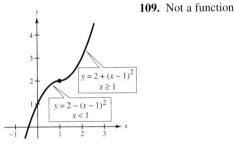

109. Not a function

111. Function

113. Reflection in the x-axis

115. Horizontal shift
1 unit to the right

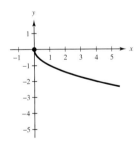

117. $y = x^2 - 2$ **119.** $y = -(x + 3)^2$

121. $3\sqrt{145} \approx 36.12$ feet

123. $V = 20{,}000 - 2000t$ **125.** $y = 3.87 - x$
$0 \le t \le 7$

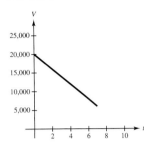

127. $A = x(75 - x)$
$0 < x < \frac{75}{2}$

129. (a) 16 feet per second

(b) 2.5 seconds

(c) -16 feet per second (The ball is on the way down.)

Chapter Test   *(page 193)*

1. Quadrant IV   **2.**

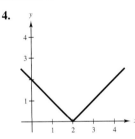

Distance: 5

3. $(-1, 0),\ (0, -3)$ **4.**

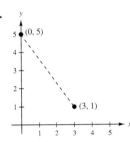

5. (a) $-\frac{2}{3}$ (b) Undefined

6.

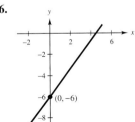

7.

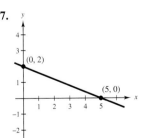

8. $y = -\frac{5}{3}x + 3$ **9.** $x - 2y - 55 = 0$
$\frac{3}{5}$

10. $2x + y = 0$ **11.** $x + 2 = 0$ **12.** No

13. (a) Function (b) Not a function

14. (a) -2 (b) 7 (c) $\dfrac{x+2}{x-1}$

15. (a) All real values of t such that $t \le 9$

(b) All real values of x such that $x \ne 4$

16.

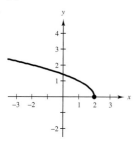

17. Reflection in the x-axis, horizontal shift 2 units to the right, vertical shift 1 unit upward

18. $V = 26{,}000 - 4000t$; $t = 2.5$

19. (a) $y = |x - 2|$ (b) $y = |x| - 2$ (c) $y = 2 - |x|$

Chapter 3

Section 3.1 *(page 201)*

Integrated Review *(page 201)*

1. Negative   **2.** 3 and 9   **3.** False. -36

4. True. 36   **5.** $x \ge 6$ **6.** $x > \frac{3}{2}$

7. $-3 < x < 3$ **8.** $-1 \le x < 3$

9. $1 < x < 5$ **10.** $x < 2$ or $x > 8$

11. $1489.66   **12.** $11\frac{3}{8}$ gallons

---

1. $10x - 4$; 1; 10   **3.** $3x^2 - x + 2$; 2; 3

5. $y^5 - 3y^4 - 2y^3 + 5$; 5; 1   **7.** $-3x^3 - 2x^2 - 3$; 3; -3

9. -4; 0; -4 **11.** $-16t^2 + v_0t$; 2; -16

13. Binomial   **15.** Trinomial   **17.** Monomial

19. $3x^3$ **21.** $8x^2 + 5$

23. The first term is not of the form ax^k (k must be nonnegative).

25. The term is not of the form ax^k (k must be nonnegative).

27. $7 + 3x$ **29.** $7x^2 + 3$ **31.** $4y^2 - y + 3$

33. $-2y^4 - 5y + 4$ **35.** 13   **37.** $6x^2 - 7x + 8$

39. $\frac{1}{6}x^3 + 3x + \frac{2}{5}$ **41.** $2.69t^2 + 7.35t - 4.2$

43. $2x^2 - 3x$ **45.** $4x^3 + 2x^2 + 9x - 6$

47. $2p^2 - 2p - 5$ **49.** $0.6b^2 - 0.6b + 7.1$ **51.** $-2y^3$

53. $x^2 - 3x + 2$ **55.** $7t^3 - t - 10$ **57.** $\frac{7}{4}y^2 - 9y - 12$

59. $9.37t^5 + 10.4t^4 - 5.4t^2 + 7.35t - 2.6$

61. $-2x^3 + x^2 + 2x$ **63.** $x^2 - 2x + 5$

65. $-4x^3 - 2x + 13$ **67.** $-11x^7 - 10x^5 + 8x^4 + 16$

69. $2x^3 - 2x + 3$ **71.** $-2x^3 - x^2 + 6x - 11$

73. $7y^2 - 9y + 2$ **75.** $7x^3 + 2x$ **77.** $3x^3 + 5x^2 + 2$

79. $3t^2 + 29$ **81.** $3v^2 + 78v + 27$ **83.** $29s + 8$

85.

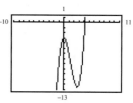

$y_1 = y_2$

87. $-x^3 - 4x^2 - x + 16$

89. (a) 64 feet (b) 60 feet (c) 48 feet (d) 0 feet

91. (a) 50 feet (b) 146 feet (c) 114 feet (d) 50 feet

93. Dropped; 100 feet   **95.** Thrown downward; 50 feet

97. 224 feet; 216 feet; 176 feet   **99.** $15,000

101. $14x + 8$ **103.** $36x$ **105.** $5x + 72$

107. (a) $y = -0.42t^2 + 11.424t + 59.89$

(b)

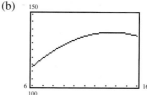

No, increasing over the interval $6 \le t \le 13.6$

109. The degree of the term ax^k is k. The term of highest degree in a polynomial has the same degree as the polynomial.

111. $8x^2 - 3x^2 = (8 - 3)x^2 = 5x^2$ **113.** No. $x^3 + 2x + 3$

Section 3.2 *(page 211)*

Integrated Review *(page 211)*

1. The point is 2 units to the left of the y-axis and 3 units above the x-axis.

2. $(4, 3), (-4, 3), (-4, -3), (4, -3)$ **3.** 13

4. $-\frac{11}{3}$ **5.** 6.84   **6.** 7   **7.** (a) 12 (b) $\frac{3}{16}$

8. (a) -7 (b) $-2x$ **9.** (a) $\dfrac{1}{3}$ (b) $\dfrac{c-6}{c+4}$

10. (a) $2\sqrt{3}$ (b) $\sqrt{t-1}$

11.

12.

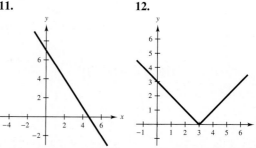

---

1. $t^{3+4} = t^7$ **3.** $(-5)^5 x^5 = -3125x^5$

5. $u^{4 \cdot 2} = u^8$ **7.** $x^{6-4} = x^2$ **9.** $\dfrac{y^4}{5^4} = \dfrac{y^4}{625}$

11. (a) $-3x^8$ (b) $9x^7$ **13.** (a) $-125z^6$ (b) $25z^8$

15. (a) $2u^3v^3$ (b) $-4u^9v$ **17.** (a) $-15u^8$ (b) $64u^5$

19. (a) $-m^{19}n^7$ (b) $-m^7n^3$ **21.** (a) $3m^4n^3$ (b) $3m^2n^3$

23. (a) $\dfrac{9x^2}{16y^2}$ (b) $\dfrac{125u^3}{27v^3}$ **25.** (a) $3x^4y$ (b) $-\dfrac{2}{3}x^2y^4$

27. (a) $\frac{25}{4}u^8v^2$ (b) $\frac{1}{4}u^8v^2$

29. (a) $x^{2n-1}y^{2n-1}$ (b) $x^{2n-2}y^{n-12}$ **31.** $16a^3$

33. $10y - 2y^2$ **35.** $8x^3 - 12x^2 + 20x$

37. $-10x^2 - 6x^4 + 14x^5$ **39.** $-x^7 + 2x^6 - 5x^4 + 6x^3$

41. $75x^3 + 30x^2$ **43.** $3u^6v - 5u^4v^2 + 6u^3v^4$

45. $x^2 + 6x + 8$ **47.** $x^2 - x - 30$ **49.** $x^2 - 8x + 16$

51. $2x^2 + 7x - 15$ **53.** $10x^2 - 34x + 12$

55. $-12x^3 - 3x^2 + 32x + 8$ **57.** $48y^2 + 32y - 3$

59. $6x^2 + 7xy + 2y^2$ **61.** $4t^2 - 6t + 4$

63. $x^3 - 5x^2 + 10x - 6$ **65.** $3a^3 + 11a^2 + 9a + 2$

67. $8u^3 + 22u^2 - u - 20$ **69.** $x^4 - 2x^3 - 3x^2 + 8x - 4$

71. $5x^4 + 20x^3 - 3x^2 + 8x - 2$ **73.** $t^4 - t^2 + 4t - 4$

75. $28x^5 - 56x^4 + 36x^3 + 21x^2 - 42x + 27$

77. $2u^3 + u^2 - 7u - 6$ **79.** $-2x^3 + 3x^2 - 1$

81. $t^4 - t^2 + 4t - 4$ **83.** $x^2 - 4$ **85.** $x^2 - 49$

87. $4 - 49y^2$ **89.** $36 - 16x^2$ **91.** $4a^2 - 25b^2$

93. $36x^2 - 81y^2$ **95.** $4x^2 - \frac{1}{16}$ **97.** $0.04t^2 - 0.25$

99. $x^2 + 10x + 25$ **101.** $x^2 - 20x + 100$

103. $4x^2 + 20x + 25$ **105.** $36x^2 - 12x + 1$

107. $4x^2 - 28xy + 49y^2$

109. $x^2 + 2xy + y^2 + 4x + 4y + 4$

111. $u^2 - v^2 + 6v - 9$ **113.** $x^3 + 9x^2 + 27x + 27$

115. $u^3 + 3u^2v + 3uv^2 + v^3$

117.
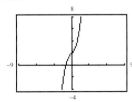
$y_1 = y_2$

119.
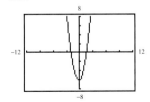
$y_1 = y_2$

121. (a) $t^2 - 8t + 15$ (b) $h^2 + 2h$

123. (a) $V(n) = n^3 + 6n^2 + 8n$ (b) 48 cubic inches

  (c) $A(n) = n^2 + 2n$ (d) $A(n + 4) = n^2 + 10n + 24$

125. $8x^2 + 26x$ **127.** $1.2x^2$ **129.** (a) $5w$ (b) $\frac{3}{2}w^2$

131. $1000 + 2000r + 1000r^2$

133. $(x + a)(x + b) = x^2 + ax + bx + ab$; FOIL method

135. (a) $x^2 - 1$ (b) $x^3 - 1$ (c) $x^4 - 1$, $x^5 - 1$

137. (a) $(x) \times (x + 5) \times (3x - 2)$

  $V_B(x) = 3x^3 + 13x^2 - 10x$

  (b) $(x - 3) \times (x - 1) \times 2(x - 3)$

  $V_P(x) = \frac{2}{3}x^3 - \frac{14}{3}x^2 + 10x - 6$

  (c) $V_S = \frac{7}{3}x^3 + \frac{53}{3}x^2 - 20x + 6$

139. $(2x)^3 = 2^3 \cdot x^3 = 8x^3 \neq 2x^3$

141. First, Outer, Inner, Last

143. (a) True (b) False. $(x + 2)(x - 3) = x^2 - x - 6$

Section 3.3   *(page 221)*

Integrated Review   *(page 221)*

1. A function f from a set A to a set B is a rule of correspondence that assigns to each element x in the set A exactly one element y in the set B.

2. The set A (see Exercise 1) is called the domain (or set of inputs) of the function f, and the set B (see Exercise 1) contains the range (or set of outputs) of the function.

3.
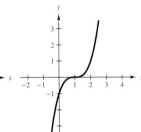

4.

5.
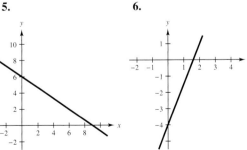

6.

Function   Function

7.

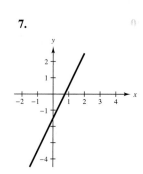

Function

8.

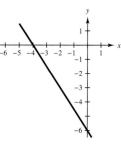

Function

9.

Not a function

10.

Not a function

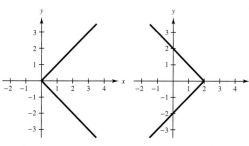

11.

12. (a) $A = x(250 - x)$

(b)

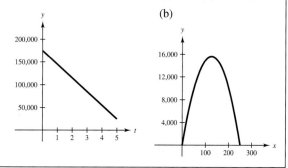

1. 6 **3.** $3x$ **5.** $6z^2$ **7.** $14b^2$ **9.** $21(x + 8)^2$

11. $8(z - 1)$ **13.** $2(2u + 5)$ **15.** $6(4x^2 - 3)$

17. $x(2x + 1)$ **19.** $7u(3u - 2)$

21. No common factor other than 1 **23.** $4(7x^2 + 4x - 2)$

25. $3y(x^2y - 5)$ **27.** $3xy(5y - x + 3)$

29. $x^2(14x^2y^3 + 21xy^2 + 9)$ **31.** $-(x - 10)$

33. $-7(2x - 1)$ **35.** $-2(3x^2 - 2x - 8)$

37. $-y(3y^2 + 2y - 1)$ **39.** $10y - 3$ **41.** $6x + 5$

43. $(y - 3)(2y + 5)$ **45.** $(3x + 2)(5x - 3)$

47. $(7a + 6)(2 - 3a^2)$ **49.** $(4t - 1)^2(8t^3 + 3)$

51. $(4x + 9)(-2x - 9)$ **53.** $(x + 25)(x + 1)$

55. $(y - 6)(y + 2)$ **57.** $(x + 2)(x^2 + 1)$

59. $(a - 4)(3a^2 - 2)$ **61.** $(z + 3)(z^3 - 2)$

63. $(x - 2y)(5x^2 + 7y^2)$ **65.** $(x + 8)(x - 8)$

67. $(1 + a)(1 - a)$ **69.** $(4y + 3)(4y - 3)$

71. $(9 + 2x)(9 - 2x)$ **73.** $(2z + y)(2z - y)$

75. $(6x + 5y)(6x - 5y)$ **77.** $\left(u + \frac{1}{4}\right)\left(u - \frac{1}{4}\right)$

79. $\left(\frac{2}{3}x + \frac{4}{5}y\right)\left(\frac{2}{3}x - \frac{4}{5}y\right)$ **81.** $(x + 3)(x - 5)$

83. $(14 + z)(4 - z)$ **85.** $(x + 9)(3x + 1)$

87. $(x - 2)(x^2 + 2x + 4)$ **89.** $(y + 4)(y^2 - 4y + 16)$

91. $(2t - 3)(4t^2 + 6t + 9)$ **93.** $(3u + 1)(9u^2 - 3u + 1)$

95. $(4a + b)(16a^2 - 4ab + b^2)$

97. $(x + 3y)(x^2 - 3xy + 9y^2)$ **99.** $2(2 - 5x)(2 + 5x)$

101. $8(x + 2)(x^2 - 2x + 4)$

103. $(y - 3)(y + 3)(y^2 + 9)$ **105.** $3x^2(x + 10)(x - 10)$

107. $6(x^2 - 2y^2)(x^4 + 2x^2y^2 + 4y^4)$

109. $(2x^n + 5)(2x^n - 5)$

111.

$y_1 = y_2$

113.

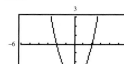

$y_1 = y_2$

115. $x^2(3x + 4) - (3x + 4) = (3x + 4)(x + 1)(x - 1)$

$3x(x^2 - 1) + 4(x^2 - 1) = (x + 1)(x - 1)(3x + 4)$

117. $p = 800 - 0.25x$ **119.** $P(1 + rt)$

121. Width $= 45 - l$ **123.** $S = 2x(x + 2h)$

125. $\pi(R - r)(R + r)$

127. The polynomial is written as a product of polynomials.

129. Determine the prime factorization of each integer. The greatest common factor is the product of each common prime factor raised to its lowest power in either one of the integers.

131. $x^2 + 2x = x(x + 2)$

Mid-Chapter Quiz   (page 224)

1. $4; -2$

2. The exponent in the term $-3x^{1/2}$ is not an integer.

3. $3t^3 + 3t^2 + 7$ **4.** $7y^2 - 5y$ **5.** $9x^3 - 4x^2 + 1$

6. $2u^2 - u + 1$ **7.** $10n^5$ **8.** $-8x^{10}$ **9.** $-\frac{3}{4}x$

10. $\dfrac{16y^4}{25x^2}$ **11.** $28y - 21y^2$ **12.** $x^2 - 4x - 21$

13. $24x^2 - 26xy + 5y^2$ **14.** $2z^2 + 3z - 35$

15. $36r^2 - 25$ **16.** $4x^2 - 12x + 9$ **17.** $x^3 + 1$

18. $x^4 + 2x^3 - 23x^2 + 40x - 20$ **19.** $7a(4a - 3)$

20. $(5 + 2x)(5 - 2x)$ **21.** $(z + 3)^2(z - 3)$

22. $4(y - 2x)(y^2 + 2xy + 4x^2)$

23. $(5x + 10)(2x + 1)$ $(5x - 10)(2x - 1)$

   $(5x + 1)(2x + 10)$ $(5x - 1)(2x - 10)$

   $(5x + 2)(2x + 5)$ $(5x - 2)(2x - 5)$

   $(5x + 5)(2x + 2)$ $(5x - 5)(2x - 2)$

24. $\frac{1}{2}(x + 2)^2 - \frac{1}{2}x^2 = 2(x + 1)$

25. 79 feet; 26 feet   **26.** \$12,000

Section 3.4   *(page 233)*

Integrated Review   *(page 233)*

1. A function can have only one value of y corresponding to $x = 0$.

2. 6   **3.** $|x| < 5$ **4.** $|x - 6| > 3$

5.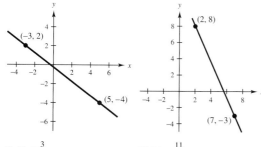

6.

$m = -\frac{3}{4}$ $m = -\frac{11}{5}$

7.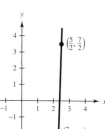

8.

$m = 33$ $m = -\frac{19}{3}$

9.   **10.**

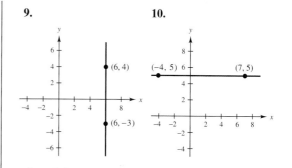

m is undefined. $m = 0$

11. \$12,720   **12.** 49.1 miles per hour

1. $(x + 2)^2$ **3.** $(a - 6)^2$ **5.** $(5y - 1)^2$

7. $(3b + 2)^2$ **9.** $(u + 4v)^2$ **11.** $(6x - 5y)^2$

13. $5(x + 3)^2$ **15.** $2x(x + 6)^2$ **17.** $5v^2(2v - 3)^2$

19. $\frac{1}{36}(3x - 4)^2$ **21.** ± 18 **23.** ± 12 **25.** 16

27. 9   **29.** $x + 1$ **31.** $y - 5$ **33.** $x - 6$

35. $z - 2$ **37.** $(x + 3)(x + 1)$ **39.** $(x - 3)(x - 2)$

41. $(y + 10)(y - 3)$ **43.** $(t - 7)(t + 3)$

45. $(x - 8)(x - 12)$ **47.** $(x - 7y)(x + 5y)$

49. $(x + 12y)(x + 18y)$ **51.** $\pm 9, \pm 11, \pm 19$

53. $\pm 4, \pm 20$ **55.** $\pm 12, \pm 36$ **57.** $-16, 8$

59. $-18, 2$ **61.** $5x + 3$ **63.** $5a - 3$ **65.** $2y - 9$

67. $(3x + 1)(x + 1)$ **69.** $(7x + 1)(x + 2)$

71. $(2x - 3)(x - 3)$ **73.** $(2x - 3)(3x - 1)$

75. Prime   **77.** $(3b - 1)(2b + 7)$

79. $(2y + 3)(9y + 4)$ **81.** $(2 + x)(3 - 2x)$

83. $(1 + 4x)(1 - 15x)$ **85.** $3(x - 4)(2x + 7)$

87. $5y(3y - 2)(4y + 5)$ **89.** $(a + 2b)(10a + 3b)$

91. $(4x - 3y)(6x + y)$ **93.** $(3x + 4)(x + 2)$

95. $(2x - 1)(3x + 2)$ **97.** $(3x - 1)(5x - 2)$

99. $3x^3(x - 4)$ **101.** $2t(5t - 9)(t + 2)$

103. $2(3x - 1)(9x^2 + 3x + 1)$

105. $9ab^2(3ab + 2)(ab - 1)$

107. $(x + 2)(x + 4)(x - 4)$ **109.** $(3 - z)(9 + z)$

111. $(x - 5 + y)(x - 5 - y)$

113. $(x^4 + 1)(x^2 + 1)(x + 1)(x - 1)$

115.

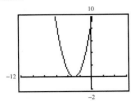

117.

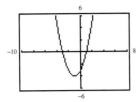

$y_1 = y_2$

$y_1 = y_2$

119. c   **121.** b   **123.** $4(6 + x)(6 - x)$

125. (a) $(2n - 2)(2n)(2n + 2)$ (b) 18, 20, 22

127. Begin by finding the factors of 6 whose sum is -5. They are -2 and -3. The factorization is $(x - 2)(x - 3)$.

129. Multiply the factors. The factors of $x^2 - 5x + 6$ are $x - 2$ and $x - 3$ because $(x - 2)(x - 3) = x^2 - 5x + 6$.

131. No. $x(x + 2) - 2(x + 2) = (x + 2)(x - 2)$

Section 3.5 *(page 243)*

Integrated Review *(page 243)*

1. Additive Inverse Property

2. Multiplicative Identity Property

3. Distributive Property

4. Associative Property of Multiplication   **5.** -4

6. 353.33   **7.** No solution   **8.** -19 **9.** 40

10. 24

11. (a) $P = -\frac{1}{4}x^2 + 8x - 12$ **12.** 6 seconds

(b)

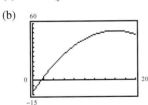

(c) 52

1. 0, 8   **3.** $-10, 3$ **5.** $-4, 2$ **7.** $-\frac{5}{2}, -\frac{1}{3}$

9. $-\frac{25}{2}, 0, \frac{3}{2}$ **11.** $-4, -\frac{1}{2}, 3$ **13.** 0, 5   **15.** $-\frac{5}{3}, 0$

17. $-2, 10$ **19.** ± 3 **21.** ± 5 **23.** ± 4

25. $-2, 5$ **27.** 4, 6   **29.** $-5, \frac{5}{4}$ **31.** $-\frac{1}{2}, 7$

33. 4   **35.** -8 **37.** $\frac{3}{2}$ **39.** $-4, 9$ **41.** $-12, 6$

43. $-\frac{7}{2}, 5$ **45.** $-7, 0$ **47.** $-6, 5$ **49.** $-2, 6$

51. $-5, 1$ **53.** 0, 7, 12   **55.** $-\frac{1}{3}, 0, \frac{1}{2}$ **57.** ± 2

59. $\pm 3, -2$ **61.** ± 3 **63.** $\pm 1, 0, 3$ **65.** $\pm 2, -\frac{3}{2}, 0$

67. $(-3, 0), (3, 0)$; the x-intercepts are solutions of the polynomial equation.

69. $(-1, 0), (3, 0)$; the x-intercepts are solutions of the polynomial equation.

71.

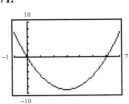

$(0, 0), (6, 0)$

73.

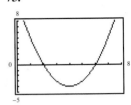

$(2, 0), (6, 0)$

75.

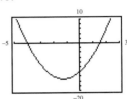

$(-4, 0), \left(\frac{3}{2}, 0\right)$

77.

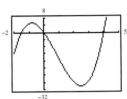

$\left(-\frac{3}{2}, 0\right), (0, 0), (4, 0)$

79. $-\frac{b}{a}, 0$ **81.** $x^2 - 2x - 15 = 0$ **83.** 15

85. 11, 12   **87.** 15 feet $\times$ 22 feet

89. Base: 8 inches; Height: 12 inches

91. (a) Length $= 5 - 2x$

Width $= 4 - 2x$

Height $= x$

Volume $=$ (length)(width)(height)

$V = (5 - 2x)(4 - 2x)(x)$

(b) $0, 2, \frac{5}{2}$

$0 < x < 2$

(c)

x	0.25	0.50	0.75	1.00	1.25	1.50	1.75
V	3.94	6	6.56	6	4.69	3	1.31

(d) $\frac{3}{2}$

(e) 0.74

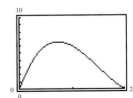

93. 20 seconds   **95.** 10 units, 20 units

97. (a) and (b): $-6, -\frac{1}{2}$ (c) Answers will vary.

99. (d) 3 feet $\times$ 5 feet $\times$ 6 feet   (e) 1026 cubic feet

(f) $x = 6$

101. False. This is not an application of the Zero-Factor Property, because there are an unlimited number of factors whose product is 1.

103. Maximum number: n. The third-degree equation $(x + 1)^3 = 0$ has only one solution: $x = -1$.

Review Exercises *(page 247)*

1. The third term is not of the form ax^k (k must be a nonnegative integer).

3. $-x^4 + 6x^3 + 5x^2 - 4x$

Leading coefficient: -1

Degree: 4

5. $-7x^3 + 3x^2 - 6x + 14$

Leading coefficient: -7

Degree: 3

7. $3x^4 - 2$ **9.** $5x^3$ **11.** $6 + 4x - x^2$

13. $-3x^3 - x^2 + 16$ **15.** $-t^2 + 4t$

17. $x^5 - 4x^3 + 7x^2 - 9x + 3$ **19.** $-9x^3 + 9x - 4$

21. $-2y - 15$ **23.** x^5 **25.** u^6 **27.** $-8z^3$

29. $4u^7v^3$ **31.** $2z^3$ **33.** $8u^2v^2$ **35.** $144x^4$

37. $-8x^4 - 32x^3$ **39.** $6x^3 - 15x^2 + 9x$

41. $x^2 + 5x - 14$ **43.** $15x^2 - 11x - 12$

45. $24x^4 + 22x^2 + 3$ **47.** $4x^3 - 5x + 6$

49. $u^2 - 8u + 7$ **51.** $16x^2 - 56x + 49$

53. $4x^2 + 12xy + 9y^2$ **55.** $25u^2 - 64$ **57.** $4u^2 - v^2$

59. $u^2 - v^2 - 6u + 9$ **61.** $3x^2(2 + 5x)$

63. $-14(x + 5)(5x + 23)$ **65.** $(v + 1)(v - 1)(v - 2)$

67. $(t + 3)(t^2 + 3)$ **69.** $(x + 6)(x - 6)$

71. $(3a + 10)(3a - 10)$ **73.** $(u - 3)(u + 15)$

75. $(u - 1)(u^2 + u + 1)$ **77.** $(2x + 3)(4x^2 - 6x + 9)$

79. $(x - 9)^2$ **81.** $(2s + 10t)^2$ **83.** $(x + 7)(x - 5)$

85. $(2x - 3)(x - 2)$ **87.** $(3x + 2)(6x + 5)$

89. $4a(1 + 4a)(1 - 4a)$ **91.** $4(2x - 3)(2x - 1)$

93. $\left(\frac{1}{2}x + y\right)^2$ **95.** $(2u - 7)^2$

97. $(x + y - 5)(x - y - 5)$ **99.** $-\frac{4}{3}, 2$ **101.** ± 10

103. $0, 3$ **105.** $-4, 9$ **107.** $-4, 0, 3$

109.

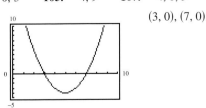

$(3, 0), (7, 0)$

111. $x^2 + 4x - 45 = 0$ **113.** $2x^3 - x^2 - 6x = 0$

115. (a)

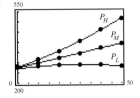

(b) $0.003t^2 + 2.365t + 274.445$

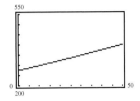

P_M; yes; average of the high and low projections

(c) $0.05t^2 + 2.07t + 7.47$

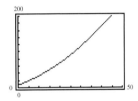

The difference between the high and low projections is increasing.

117. \$2000   **119.** $P = 8x + 8$

$A = 15x$

121. $14x + 3$ **123.** (a) $P = 4l - 10$ (b) $A = l^2 - 5l$

125. 100 feet $\times$ 300 feet   **127.** 2 seconds   **129.** 14, 16

Chapter Test *(page 250)*

1. Degree: 3; leading coefficient: -5.2

2. The variable appears in the denominator.

3. (a) $6a^2 - 3a$ (b) $-2y^2 - 2y$

4. (a) $8x^2 - 4x + 10$ (b) $11t + 7$

5. (a) $-24u^6v^5$ (b) $60x^3y^2$ **6.** (a) $\frac{1}{8}y^3$ (b) $\frac{27}{2}x^6y^4$

7. (a) $-3x^2 + 12x$ (b) $2x^2 + 7xy - 15y^2$

8. (a) $3x^2 - 6x + 3$ (b) $6s^3 - 17s^2 + 26s - 21$

9. (a) $16x^2 - 24x + 9$ (b) $16 - a^2 - 2ab - b^2$

10. $6y(3y - 2)$ **11.** $\left(v - \frac{4}{3}\right)\left(v + \frac{4}{3}\right)$

12. $(x + 2)(x - 2)(x - 3)$ **13.** $(3u - 1)^2$

14. $2(x - 5)(3x + 2)$ **15.** $(x + 3)(x^2 - 3x + 9)$

16. $1, -5$ **17.** $3, -\frac{4}{3}$ **18.** $x^2 + 26x$

19. 6 centimeters $\times$ 9 centimeters   **20.** 2 seconds

21. Base: 5 feet

Height: 14 feet

Cumulative Test: Chapters P–3 *(page 251)*

1. (a) $-2 < 5$ (b) $\frac{1}{3} < \frac{1}{2}$
(c) $|-5| > -5$ (d) $|2.3| > -|-4.5|$

2. $3n - 8$ **3.** (a) $t^2 - 9t$ (b) $2x^3 - 11x$

4. (a) $8a^8b^7$ (b) $\frac{1}{4}x^2y^2$

5. (a) $2x^2 - 9x - 5$ (b) $x^2 - 2xy + y^2 + 4x - 4y + 4$

6. (a) $\frac{3}{2}$ (b) $-\frac{3}{2}$ **7.** (a) $-\frac{2}{3}, 4$ (b) $-\frac{1}{2}, 3$

8. (a) $x < -1$ (b) $-\frac{3}{2} \le x < 4$ **9.** $1408.75

10. $\frac{13}{2}$ **11.** $x \le -1$ or $x \ge 5$ **12.** $x \ge 103$

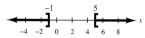

13. Function **14.** $2 \le x < \infty$

15. (a) 4 (b) $c^2 + 3c$ **16.** (a) $m = \frac{3}{4}$ (b) Distance: 10

17. (a) $2x - y + 5 = 0$ (b) $2x - 3y + 7 = 0$

18. (a) $(x - 5)(3x + 7)$ (b) $9(x + 4)(x - 4)$

19. (a) $(y + 3)(y - 3)^2$ (b) $2t(2t - 5)^2$

20. **21.**

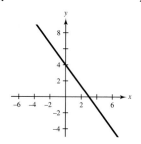 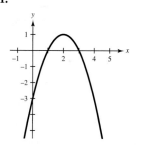

Chapter 4

Section 4.1 *(page 259)*

Integrated Review *(page 259)*

1. The graph of an equation is the set of solution points of the equation on a rectangular coordinate system.

2. Create a table of solution points of the equation, plot those points on a rectangular coordinate system, and connect the points with a smooth curve or line.

3. $(2, 0), (6, 2)$

4. To find the x-intercept, let $y = 0$ and solve the equation for x. To find the y-intercept, let $x = 0$ and solve the equation for y.

5. $14x^5$ **6.** y^2z^{11} **7.** a^3 **8.** $x + 2$

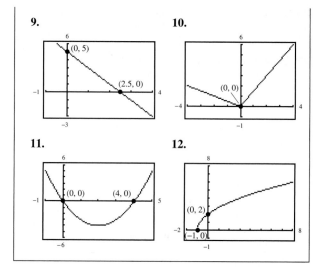

1. $\frac{1}{25}$ **3.** $-\frac{1}{1000}$ **5.** 1 **7.** 64 **9.** -32

11. $\frac{3}{2}$ **13.** 1 **15.** 1 **17.** 729 **19.** $100,000$

21. $\frac{1}{16}$ **23.** $\frac{1}{64}$ **25.** $\frac{3}{16}$ **27.** $\frac{64}{121}$ **29.** $\frac{16}{15}$

31. y^2 **33.** z^2 **35.** $\frac{7}{x^4}$ **37.** $\frac{1}{64x^3}$ **39.** x^6

41. $\frac{4}{3}a$ **43.** t^2 **45.** $\frac{1}{4x^4}$ **47.** $-\frac{12}{xy^3}$ **49.** $\frac{y^4}{9x^4}$

51. $\frac{10}{x}$ **53.** $\frac{x^5}{2y^4}$ **55.** $\frac{81v^8}{u^6}$ **57.** $\frac{b^5}{a^5}$ **59.** $\frac{1}{2x^8y^3}$

61. $6u$ **63.** x^8y^{12} **65.** $\frac{2b^{11}}{25a^{12}}$ **67.** $\frac{v^2}{uv^2 + 1}$

69. $\frac{ab}{b - a}$ **71.** 3.6×10^6 **73.** 4.762×10^7

75. 3.1×10^{-4} **77.** 3.81×10^{-8} **79.** 5.75×10^7

81. 9.461×10^{15} **83.** 8.99×10^{-5} **85.** $60,000,000$

87. 0.0000001359 **89.** $31,700,000,000$

91. $13,000,000$ **93.** 0.00000000048 **95.** 6.8×10^5

97. 2.5×10^9 **99.** 6×10^6 **101.** 9×10^{15}

103. 1.6×10^{12} **105.** 3.46×10^{10} **107.** 4.70×10^{11}

109. 1.67×10^{14} **111.** 2.74×10^{20} **113.** 9.3×10^7

115. $1.58 \times 10^{-5} \approx 8.3$ minutes **117.** 3.33×10^5

119. $20,393 **121.** $3x$ is the base and 4 is the exponent.

123. Change the sign of the exponent of the factor.

125. When the numbers are very large or very small

Section 4.2   *(page 268)*

Integrated Review   *(page 268)*

1. $m = \dfrac{y_2 - y_1}{x_2 - x_1}$ **2.** (a) $m > 0$ **3.** 10

(b) $m < 0$

(c) $m = 0$

(d) m is undefined.

4. 12   **5.** $-8x - 10$ **6.** $-2x^2 + 14x$ **7.** $\dfrac{25}{x^4}$

8. $\dfrac{4u^3}{3}$ **9.** 30% solution: $13\frac{1}{3}$ gallons   **10.** $500

60% solution: $6\frac{2}{3}$ gallons

---

1. $(-\infty, 8) \cup (8, \infty)$ **3.** $(-\infty, -4) \cup (-4, \infty)$

5. $(-\infty, \infty)$ **7.** $(-\infty, \infty)$ **9.** $(-\infty, \infty)$

11. $(-\infty, -3) \cup (-3, 0) \cup (0, \infty)$

13. $(-\infty, -4) \cup (-4, 4) \cup (4, \infty)$

15. $(-\infty, 0) \cup (0, 3) \cup (3, \infty)$

17. $(-\infty, 2) \cup (2, 3) \cup (3, \infty)$

19. $\left(-\infty, -1\right) \cup \left(-1, \frac{5}{3}\right) \cup \left(\frac{5}{3}, \infty\right)$

21. (a) 1

(b) -8

(c) Undefined (division by 0)

(d) 0

23. (a) 0

(b) 0

(c) Undefined (division by 0)

(d) Undefined (division by 0)

25. (a) $\frac{25}{22}$ **27.** $(0, \infty)$

(b) 0

(c) Undefined (division by 0)

(d) Undefined (division by 0)

29. $\{1, 2, 3, 4, \ldots\}$ **31.** $[0, 100]$ **33.** $x + 3$

35. $(3)(x + 16)^2$ **37.** $(x)(x - 2)$ **39.** $x + 2$

41. $\dfrac{x}{5}$ **43.** $6y, \ y \neq 0$ **45.** $\dfrac{6x}{5y^3}, \ x \neq 0$ **47.** $\dfrac{x - 3}{4x}$

49. $x, \ x \neq 8, x \neq 0$ **51.** $\dfrac{1}{2}, \ x \neq \dfrac{3}{2}$ **53.** $-\dfrac{1}{3}, \ x \neq 5$

55. $\dfrac{1}{a + 3}$ **57.** $\dfrac{x}{x - 7}$ **59.** $\dfrac{y(y + 2)}{y + 6}, \ y \neq 2$

61. $\dfrac{x(x + 2)}{x - 3}, \ x \neq 2$ **63.** $-\dfrac{3x + 5}{x + 3}, \ x \neq 4$

65. $\dfrac{x + 8}{x - 3}, \ x \neq -\dfrac{3}{2}$ **67.** $\dfrac{3x - 1}{5x - 4}, \ x \neq -\dfrac{4}{5}$

69. $\dfrac{3y^2}{y^2 + 1}, \ x \neq 0$ **71.** $\dfrac{y - 8x}{15}, \ y \neq -8x$

73. $\dfrac{5 + 3xy}{y^2}, \ x \neq 0$ **75.** $\dfrac{u - 2v}{u - v}, \ u \neq -2v$

77. $\dfrac{3(m - 2n)}{m + 2n}$

79. Evaluating both sides when $x = 10$ yields $\frac{3}{2} \neq 9$.

81. Evaluating both sides when $x = 0$ yields $1 \neq \frac{3}{4}$.

83.

x	-2	-1	0	1	2	3	4
$\dfrac{x^2 - x - 2}{x - 2}$	-1	0	1	2	Undefined	4	5
$x + 1$	-1	0	1	2	3	4	5

$$\dfrac{x^2 - x - 2}{x - 2} = \dfrac{(x - 2)(x + 1)}{x - 2} = x + 1, \ x \neq 2$$

85. $\dfrac{x}{x + 3}, \ x > 0$

87. (a) $C = 2500 + 9.25x$ **89.** (a) Van: $45(t + 3)$;
Car: $60t$

(b) $\overline{C} = \dfrac{2500 + 9.25x}{x}$ (b) $d = |15(9 - t)|$

(c) $\{1, 2, 3, 4, \ldots\}$ (c) $\dfrac{4t}{3(t + 3)}$

(d) $34.25

91. π **93.** $\dfrac{1000(10{,}730 + 1509t)}{3426 + 65t}$

95. Let u and v be polynomials. The algebraic expression u/v is a rational expression.

97. The rational expression is in simplified form if the numerator and denominator have no factors in common (other than ± 1).

99. You can cancel only common factors.

Section 4.3   *(page 277)*

Integrated Review   *(page 277)*

1. $u^2 - v^2 = (u + v)(u - v)$

$9t^2 - 4 = (3t + 2)(3t - 2)$

2. $u^2 - 2uv + v^2 = (u - v)^2$

$4x^2 - 12x + 9 = (2x - 3)^2$

3. $u^3 + v^3 = (u + v)(u^2 - uv + v^2)$

$8x^3 + 64 = (2x + 4)(4x^2 - 8x + 16)$

4. $(3x - 2)(x + 5)$. Multiply   **5.** $5x(1 - 4x)$

6. $(2 + x)(14 - x)$ **7.** $(3x - 5)(5x + 3)$

8. $(4t + 1)^2$ **9.** $(y - 4)(y^2 + 4y + 16)$

10. $(2x + 1)(4x^2 - 2x + 1)$

11.

12.

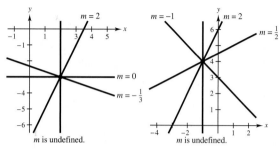

m is undefined. m is undefined.

1. (a) 0 (b) Undefined   (c) $\frac{3}{2}$ (d) $\frac{1}{24}$

3. x^2 **5.** $(x + 2)^2$ **7.** $u + 1$ **9.** $(-1)(2 + x)$

11. $\dfrac{99}{40}$ **13.** $\dfrac{9}{2}$ **15.** $\dfrac{s^3}{6}$, $s \neq 0$ **17.** $24u^2$, $u \neq 0$

19. 24, $x \neq -\dfrac{3}{4}$ **21.** $\dfrac{2uv(u + v)}{3(3u + v)}$, $u \neq 0$

23. -1, $r \neq 12$ **25.** $-\dfrac{x + 8}{x^2}$, $x \neq \dfrac{3}{2}$

27. $4(r + 2)$, $r \neq 3$, $r \neq 2$ **29.** $2t + 5$, $t \neq 3$, $t \neq -2$

31. $\dfrac{xy(x + 2y)}{x - 2y}$ **33.** $\dfrac{(x - y)^2}{x + y}$, $x \neq -3y$

35. $\dfrac{(x - 1)(2x + 1)}{(3x - 2)(x + 2)}$, $x \neq \pm 5$, $x \neq -1$

37. $\dfrac{x^2(x^2 - 9)(2x + 5)(3x - 1)}{2(2x + 1)(2x + 3)(3 - 2x)}$, $x \neq 0$, $x \neq \dfrac{1}{2}$

39. $\dfrac{(x + 3)^2}{x}$, $x \neq 3$, $x \neq 4$ **41.** $-\dfrac{8}{27}$

43. $\dfrac{4x}{3}$, $x \neq 0$ **45.** $\dfrac{3y^2}{2ux^2}$, $v \neq 0$ **47.** $\dfrac{3}{2(a + b)}$

49. $x^4 y(x + 2y)$, $x \neq 0$, $y \neq 0$, $x \neq -2y$

51. $\dfrac{3x}{10}$, $x \neq 0$ **53.** $-\dfrac{5x(x + 1)}{2}$, $x \neq 0$, $x \neq 5$, $x \neq -1$

55. $\dfrac{(x + 3)(4x + 1)}{(3x - 1)(x - 1)}$, $x \neq -3$, $x \neq -\dfrac{1}{4}$

57. $x + 2$, $x \neq \pm 2$, $x \neq -3$

59. $-\dfrac{(x + 2)(x^2 - 3x - 10)}{(x + 3)(x^2 - 4x + 4)}$, $x \neq \pm 2$, $x \neq 7$

61. $\dfrac{x + 4}{3}$, $x \neq -2$, $x \neq 0$ **63.** $\dfrac{1}{4}$, $x \neq -1$, $x \neq 0$, $y \neq 0$

65. $\dfrac{(x + 1)(2x - 5)}{x}$, $x \neq -1$, $x \neq -5$, $x \neq -\dfrac{2}{3}$

67. $\dfrac{x^4}{(x^n + 1)^2}$, $x^n \neq -3$, $x^n \neq 3$, $x \neq 0$

69.

71.

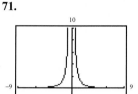

73. $\dfrac{2w^2 + 3w}{6}$ **75.** $\dfrac{x}{4(2x + 1)}$ **77.** $\dfrac{x}{2(2x + 1)}$

79. (a) $\dfrac{1}{20}$ minute   (b) $\dfrac{x}{20}$ minutes   (c) $\dfrac{7}{4}$ minutes

81. (a)

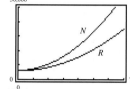

(b) $\dfrac{6{,}115{,}200 + 590{,}700t^2}{12(6357 + 1070t^2)}$

(c)

Year, t	0	2	4	6
Monthly bill	\$80.16	\$66.42	\$55.25	\$50.84

(d) The number of subscribers was increasing at a faster rate than the revenue.

83. Multiply the rational expression by the reciprocal of the polynomial.

85. Invert the divisor, not the dividend.

Mid-Chapter Quiz   *(page 281)*

1. $\dfrac{3}{t^{-9}}$ **2.** $\dfrac{3x^6}{16y^3}$ **3.** $\dfrac{2}{3u^3}$ **4.** 1

5. (a) 1.34×10^7 **6.** (a) 8.1×10^{13}

    (b) 7.5×10^{-4} (b) 2×10^{-4}

7. $(-\infty, 0) \cup (0, 4) \cup (4, \infty)$

8. (a) 0 (b) $\dfrac{9}{2}$ (c) Undefined   (d) $\dfrac{8}{9}$

9. $\dfrac{3}{2}y$ **10.** $\dfrac{2u^2}{9v}$ **11.** $-\dfrac{2x + 1}{x}$ **12.** $\dfrac{z + 3}{2z - 1}$

13. $\dfrac{7 + 3ab}{a}$ **14.** $\dfrac{n^2}{m + n}$ **15.** $\dfrac{t}{2}$ **16.** $\dfrac{5x}{x - 2}$

17. $\dfrac{8x}{3(x - 1)(x^2 + 2x - 3)}$ **18.** $\dfrac{4(u - v)^2}{5uv}$

19. $-\dfrac{3t}{2}$ **20.** $\dfrac{2(x + 1)}{3x}$

21. (a) $\dfrac{6000 + 10.50x}{x}$ (b) $22.50

Section 4.4   *(page 288)*

> ### Integrated Review   *(page 288)*
>
> **1.** (a) $y = \frac{3}{5}x + \frac{4}{5}$
>
> (b) $y - 2 = \frac{3}{5}(x - 2)$
>
> **2.** If $m > 0$, the line rises from left to right.
> If $m < 0$, the line falls from left to right.
>
> **3.** $42x^2 - 60x$ **4.** $6 + y - 2y^2$ **5.** $121 - x^2$
>
> **6.** $16 - 25z^2$ **7.** $x^2 + 2x + 1$ **8.** $2t$
>
> **9.** $x^3 - 8$ **10.** $2t^3 - 5t^2 - 12t$
>
> **11.** $P = 12x + 6$ **12.** $P = 12x$
>
> $A = 5x^2 + 9x$ $A = 6x^2$

1. $\dfrac{3}{2}$ **3.** $-\dfrac{x}{4}$ **5.** $-\dfrac{3}{a}$ **7.** $-\dfrac{2}{9}$ **9.** $\dfrac{2z^2 - 2}{3}$

11. $\dfrac{x + 6}{3}$ **13.** $-\dfrac{4}{3}$ **15.** $1,\ y \neq 6$

17. $\dfrac{1}{x - 3},\ x \neq 0$ **19.** $20x^3$ **21.** $36y^3$

23. $15x^2(x + 5)$ **25.** $126z^2(z + 1)^4$

27. $56t(t + 2)(t - 2)$ **29.** $6x(x + 2)(x - 2)$

31. x^2 **33.** $(u + 1)$ **35.** $-(x + 2)$

37. $\dfrac{2n^2(n + 8)}{6n^2(n - 4)},\ \dfrac{10(n - 4)}{6n^2(n - 4)}$

39. $\dfrac{2(x + 3)}{x^2(x + 3)(x - 3)},\ \dfrac{5x(x - 3)}{x^2(x + 3)(x - 3)}$

41. $\dfrac{3v^2}{6v^2(v + 1)},\ \dfrac{8(v + 1)}{6v^2(v + 1)}$

43. $\dfrac{(x - 8)(x - 5)}{(x + 5)(x - 5)^2},\ \dfrac{9x(x + 5)}{(x + 5)(x - 5)^2}$ **45.** $\dfrac{25 - 12x}{20x}$

47. $\dfrac{7(a + 2)}{a^2}$ **49.** $0,\ x \neq 4$ **51.** $\dfrac{3(x + 2)}{x - 8}$

53. $\dfrac{5(5x + 22)}{x + 4}$ **55.** $1,\ x \neq \frac{2}{3}$ **57.** $\dfrac{1}{2x(x - 3)}$

59. $\dfrac{x^2 - 7x - 15}{(x + 3)(x - 2)}$ **61.** $\dfrac{x - 2}{x(x + 1)}$

63. $\dfrac{5(x + 1)}{(x + 5)(x - 5)}$ **65.** $\dfrac{4}{x^2(x^2 + 1)}$

67. $\dfrac{x^2 + x + 9}{(x - 2)(x - 3)(x + 3)}$ **69.** $\dfrac{4x}{(x - 4)^2}$

71. $\dfrac{y - x}{xy},\ x \neq -y$ **73.** $\dfrac{2(4x^2 + 5x - 3)}{x^2(x + 3)}$

75. $-\dfrac{u^2 - uv - 5u + 2v}{(u - v)^2}$ **77.** $\dfrac{x}{x - 1},\ x \neq -6$

79.

81. $\dfrac{x}{2(3x + 1)},\ x \neq 0$

83. $\dfrac{4 + 3x}{4 - 3x},\ x \neq 0$ **85.** $-4x - 1,\ x \neq 0,\ x \neq \frac{1}{4}$

87. $\dfrac{3}{4},\ x \neq 0,\ x \neq 3$ **89.** $\dfrac{5(x + 3)}{2x(5x - 2)}$

91. $y - x,\ x \neq 0,\ y \neq 0,\ x \neq -y$ **93.** $\dfrac{-(y - 1)(y - 3)}{y(4y - 1)}$

95. $\dfrac{x(x + 6)}{3x^3 + 10x - 30},\ x \neq 0,\ x \neq 3$ **97.** $-\dfrac{1}{2(h + 2)}$

99.

x	-3	-2	-1	0	1	2	3
$\dfrac{\left(1 - \frac{1}{x}\right)}{\left(1 - \frac{1}{x^2}\right)}$	$\frac{3}{2}$	2	Undef.	Undef.	Undef.	$\frac{2}{3}$	$\frac{3}{4}$
$\dfrac{x}{x + 1}$	$\frac{3}{2}$	2	Undef.	0	$\frac{1}{2}$	$\frac{2}{3}$	$\frac{3}{4}$

Domain of the complex fraction: $(-\infty, -1) \cup (-1, 0) \cup (0, 1) \cup (1, \infty)$. Domain of the simplified fraction: $(-\infty, -1) \cup (-1, \infty)$. The two expressions are equivalent except at $x = 0$ and $x = 1$.

101. $\dfrac{5t}{12}$ **103.** $\dfrac{5x}{24}$ **105.** $\dfrac{11x}{45},\ \dfrac{13x}{45}$ **107.** $\dfrac{R_1 R_2}{R_1 + R_2}$

109. (a) Upstream: $\dfrac{10}{5 - x}$; Downstream: $\dfrac{10}{5 + x}$

(b) $f(x) = \dfrac{10}{5 - x} + \dfrac{10}{5 + x}$ (c) $f(x) = \dfrac{100}{(5 + x)(5 - x)}$

111. Rewrite each fraction in terms of the lowest common denominator, combine the numerators, and place the result over the lowest common denominator.

113. When the numerators are subtracted, the result should be $(x - 1) - (4x - 11) = x - 1 - 4x + 11$.

Section 4.5 (page 298)

Integrated Review (page 298)

1. $\dfrac{120y}{90} = \dfrac{30 \cdot 4y}{30 \cdot 3} = \dfrac{4y}{3}$

2. $(2n + 1)(2n + 3) = 4n^2 + 8n + 3$

3. $(2n + 1) + (2n + 3) = 4n + 4$

4. $2n(2n + 2) = 4n^2 + 4n$ **5.** $\frac{3}{4}$ **6.** $\frac{5}{2}$

7. $\pm\frac{5}{2}$ **8.** $0, 8$ **9.** $-7, 6$ **10.** 5

11. $y = 1500 + 0.12x$

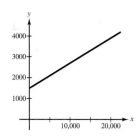

12. $N = 3500 + 60t$

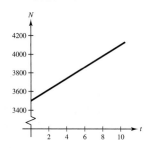

1. $3z + 5$ **3.** $\frac{5}{2}z^2 + z - 3$ **5.** $7x^2 - 2x,\ x \neq 0$

7. $m^3 + 2m - \dfrac{7}{m}$ **9.** $-10z^2 - 6,\ z \neq 0$

11. $4z^2 + \frac{3}{2}z - 1,\ z \neq 0$

13. $\frac{5}{2}x - 4 + \frac{7}{2}y,\ x \neq 0,\ y \neq 0$

15. $x - 5,\ x \neq 3$ **17.** $x + 10,\ x \neq -5$

19. $x - 3 + \dfrac{2}{x - 2}$ **21.** $x + 7,\ x \neq 3$

23. $5x - 8 + \dfrac{19}{x + 2}$ **25.** $4x + 3 - \dfrac{11}{3x + 2}$

27. $6t - 5,\ t \neq \frac{5}{2}$ **29.** $y + 3,\ y \neq -\frac{1}{2}$

31. $x^2 + 4,\ x \neq 2$ **33.** $2x^2 + x + 4 + \dfrac{6}{x - 3}$

35. $2 + \dfrac{5}{x + 2}$ **37.** $x - 4 + \dfrac{32}{x + 4}$

39. $\dfrac{6}{5}z + \dfrac{41}{25} + \dfrac{41}{25(5z - 1)}$ **41.** $4x - 1,\ x \neq -\dfrac{1}{4}$

43. $x^2 - 5x + 25,\ x \neq -5$ **45.** $x + 2$

47. $4x^2 + 12x + 25 + \dfrac{52x - 55}{x^2 - 3x + 2}$

49. $x^5 + x^4 + x^3 + x^2 + x + 1,\ x \neq 1$

51. $x^3 - x + \dfrac{x}{x^2 + 1}$ **53.** $2x,\ x \neq 0$

55. $7uv,\ u \neq 0,\ v \neq 0$ **57.** $x^2 - 3x - 3 - \dfrac{10}{x - 2}$

59. $x^2 - x + 4 - \dfrac{17}{x + 4}$

61. $x^3 - 2x^2 - 4x - 7 - \dfrac{4}{x - 2}$

63. $5x^2 + 14x + 56 + \dfrac{232}{x - 4}$

65. $10x^3 + 10x^2 + 60x + 360 + \dfrac{1360}{x - 6}$

67. $0.1x + 0.82 + \dfrac{1.164}{x - 0.2}$ **69.** $(x - 3)(x - 1)(x + 4)$

71. $(x - 1)(2x - 1)(3x - 2)$

73. $(x + 5)(3x + 2)(3x - 2)$ **75.** $(x + 3)^2(x - 3)(x + 4)$

77. $5\left(x - \frac{4}{5}\right)(3x + 2)$ **79.** -8

81.

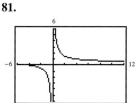

83.

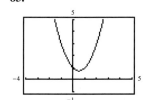

85. $x^{2n} + x^n + 4,\ x^n \neq -2$ **87.** $x^3 - 5x^2 - 5x - 10$

89.

x-Values	Polynomial values	Divisors	Remainders
-2	-8	$x + 2$	-8
-1	0	$x + 1$	0
0	0	x	0
$\frac{1}{2}$	$-\frac{9}{8}$	$x - \frac{1}{2}$	$-\frac{9}{8}$
1	-2	$x - 1$	-2
2	0	$x - 2$	0

$f(k)$ equals the remainder when dividing by $(x - k)$.

91. $x^2 - 3$ **93.** $2x + 8$

95. x is not a factor of the numerator.

97. The remainder is 0 and the divisor is a factor of the dividend.

99. True. If $\dfrac{n(x)}{d(x)} = q(x)$, then $n(x) = d(x) \cdot q(x)$.

Section 4.6 *(page 308)*

> ### Integrated Review *(page 308)*
>
> **1.** Quadrants II or III   **2.** Quadrants I or II
> **3.** x-axis   **4.** $(9, -6)$ **5.** $x < \frac{3}{2}$ **6.** $x < 5$
> **7.** $1 < x < 5$ **8.** $x < 2$ or $x > 8$
> **9.** $x \le -8$ or $x \ge 16$ **10.** $-24 \le x \le 36$
> **11.** 15 minutes, 2 miles   **12.** 7.5%: $15,000
> 9%: $9000

1. (a) Not a solution   **3.** (a) Not a solution
  (b) Not a solution   (b) Solution
  (c) Not a solution   (c) Solution
  (d) Solution   (d) Not a solution

5. 10 **7.** 8 **9.** $-\frac{9}{32}$ **11.** 10 **13.** $-\frac{2}{9}$
15. $\frac{7}{4}$ **17.** $\frac{43}{8}$ **19.** 61 **21.** $\frac{18}{5}$ **23.** $-\frac{26}{5}$
25. 3 **27.** 3 **29.** $-\frac{11}{5}$ **31.** $\frac{4}{3}$ **33.** ± 6
35. ± 4 **37.** $-9, 8$ **39.** 3, 13 **41.** No solution
43. -5 **45.** 5 **47.** $-\frac{11}{10}, 2$ **49.** 20
51. $\frac{3}{2}$ **53.** 3, -1 **55.** No solution **57.** 2, 3
59. (a) and (b) $(-2, 0)$ **61.** (a) and (b) $(-1, 0), (1, 0)$
63. (a)   **65.** (a)

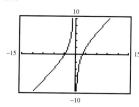

  (b) $(4, 0)$ (b) $(1, 0)$
67.   **69.**

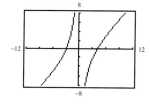

  $(-3, 0), (2, 0)$ $(-3, 0), (4, 0)$
71. 8, $\frac{1}{8}$ **73.** 40 miles per hour

75. 8 miles per hour   **77.** 4 miles per hour   **79.** 10
  10 miles per hour
81. 12   **83.** (a)

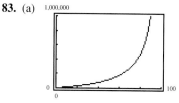

  (b) 85%

85. 3 hours; $\frac{15}{8}$ minutes; $\frac{5}{3}$ hours   **87.** 15 hours; $22\frac{1}{2}$ hours
89. $11\frac{1}{4}$ hours

91.

Year	1990	1991	1992	1993	1994	1995
Revenue	87.7	95.3	104.7	116.5	131.8	152.4

93. (a) $\{4, 6, 8, 10, \ldots\}$
  (b)

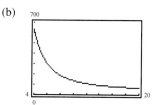

  (c) 10d
95. (d) 3 miles per hour, obtained by solving $\dfrac{10}{5-x} + \dfrac{10}{5+x}$
   $= 6.25$
  (e) Yes. 11.1 hours

97. Multiply both sides of the equation by the lowest common denominator, solve the resulting equation, and check the result. It is important to check the result for any errors or extraneous solutions.

99. (a) Simplify each side by removing symbols of grouping, combining like terms, and reducing fractions on one or both sides.
  (b) Add (or subtract) the same quantity to (from) both sides of the equation.
  (c) Multiply (or divide) both sides of the equation by the same nonzero real number.
  (d) Interchange the two sides of the equation.

101. When the equation involves only two fractions, one on each side of the equation, the equation can be solved by cross-multiplication.

Review Exercises *(page 313)*

1. $\frac{1}{72}$ **3.** $\frac{125}{8}$ **5.** 3.6×10^7 **7.** 500
9. 5.38×10^{-5} **11.** 483,300,000 **13.** $12y$
15. $\frac{2}{x^3}$ **17.** $\frac{x^6}{y^8}$ **19.** $\frac{1}{t^3}$ **21.** $\frac{27}{y^3}$

23. $(-\infty, 8) \cup (8, \infty)$ **25.** $(-\infty, 1) \cup (1, 6) \cup (6, \infty)$

27. $\dfrac{2x^3}{5}$, $x \neq 0$, $y \neq 0$ **29.** $\dfrac{b-3}{6(b-4)}$

31. -9, $x \neq y$ **33.** $\dfrac{x}{2(x+5)}$, $x \neq 5$ **35.** $3x^5y^2$

37. $\dfrac{8}{5}x^3$ **39.** $\dfrac{y}{8x}$, $y \neq 0$ **41.** $12z(z-6)$, $z \neq -6$

43. $-\dfrac{1}{4}$, $u \neq 0$, $u \neq 3$ **45.** $3x^2$, $x \neq 0$

47. $\dfrac{125y}{x}$, $y \neq 0$ **49.** $\dfrac{x(x-1)}{x-7}$, $x \neq -1$, $x \neq 1$

51. $\dfrac{6(x+5)}{x(x+7)}$, $x \neq \pm 5$ **53.** $-\dfrac{7}{9}$ **55.** $-\dfrac{13}{48}$

57. $\dfrac{4x+3}{(x+5)(x-12)}$ **59.** $\dfrac{5x^3 - 5x^2 - 31x + 13}{(x+2)(x-3)}$

61. $\dfrac{x+24}{x(x^2+4)}$ **63.** $\dfrac{6(x-9)}{(x+3)^2(x-3)}$

65. $\dfrac{3t^2}{5t-2}$, $t \neq 0$

67. $\dfrac{-a^2 + a + 16}{(4a^2 + 16a + 1)(a-4)}$, $a \neq 0$, $a \neq -4$

69.   **71.**

73. $2x^2 - \dfrac{1}{2}$, $x \neq 0$ **75.** $2x^2 + \dfrac{4}{3}x - \dfrac{8}{9} + \dfrac{10}{9(3x-1)}$

77. $x^2 - 2$, $x \neq \pm 1$

79. $x^2 - x - 3 - \dfrac{3x^2 - 2x - 3}{x^3 - 2x^2 + x - 1}$

81. $x^2 + 5x - 7$, $x \neq -2$

83. $x^3 + 3x^2 + 6x + 18 + \dfrac{29}{x-3}$

85. -120 **87.** $\dfrac{36}{23}$ **89.** 5 **91.** $-4, 6$

93. $-\dfrac{16}{3}, 3$ **95.** $-\dfrac{5}{2}, 1$ **97.** $-2, 2$ **99.** $-\dfrac{9}{5}, 3$

101. (a)   (b) -3

$(-3, 0)$

103. $(0, 6]$ **105.** 56 miles per hour   **107.** 25

109. 4   **111.** $6\dfrac{2}{3}$ minutes

113. (a) 304,000, 453,333, 702,222   (b) 29.8 years

Chapter Test   *(page 317)*

1. $\dfrac{3}{8}$ **2.** 3.0×10^{-5} **3.** $\dfrac{10}{a}$ **4.** $\dfrac{1}{r^3 s^5}$

5. $\dfrac{x^8}{y^{12}}$ **6.** $108x^2y^8$ **7.** 3.2×10^{-5} **8.** 30,400,000

9. $(-\infty, -5) \cup (-5, 5) \cup (5, \infty)$

10. $x^3(x+3)(x-3)$

11. (a) $-\dfrac{1}{3}$, $x \neq 2$ (b) $\dfrac{2a+3}{5}$, $a \neq 4$

12. $\dfrac{5z}{3}$, $z \neq 0$ **13.** $\dfrac{4}{y+4}$, $y \neq 2$

14. $\dfrac{(2x+3)^2}{x+1}$, $x \neq \dfrac{3}{2}$ **15.** $\dfrac{14y^6}{15}$, $x \neq 0$

16. $\dfrac{x^3}{4}$, $x \neq 0$, $x \neq -2$

17. $-(3x+1)$, $x \neq 0$, $x \neq \dfrac{1}{3}$ **18.** $\dfrac{-2x^2 + 2x + 1}{x+1}$

19. $\dfrac{5x^2 - 15x - 2}{(x-3)(x+2)}$ **20.** $\dfrac{5x^3 + x^2 - 7x - 5}{x^2(x+1)^2}$

21. 4, $x \neq -1$ **22.** $t^2 + 3 - \dfrac{6t-6}{t^2-2}$

23. $2x^3 + 6x^2 + 3x + 9 + \dfrac{20}{x-3}$ **24.** 22

25. $-1, -\dfrac{15}{2}$ **26.** No solution

27. $6\dfrac{2}{3}$ hours, 10 hours

Chapter 5

Section 5.1   *(page 327)*

Integrated Review   *(page 327)*

1. a^{m+n} **2.** $a^m b^m$ **3.** a^{mn} **4.** a^{m-n}

5. $y = 4 - 3x$ **6.** $y = \dfrac{2}{3}(1-x)$

7. $y = \dfrac{1}{3}(4 - x^2)$ **8.** $y = 4 - x^2$

9. $y = \dfrac{1}{3}\left(2\sqrt{x} - 15\right)$ **10.** $y = \dfrac{1}{5}(6|x| + 10)$

11. $\dfrac{12}{5}$ hours   **12.** $\dfrac{189}{4} = 47.25$ miles per hour

---

1. 8   **3.** -7 **5.** -2 **7.** Not a real number

9. 7   **11.** 4.2   **13.** Square root   **15.** 8   **17.** 10

19. Not a real number   **21.** $-\dfrac{2}{3}$ **23.** Not a real number

25. 5   **27.** -23 **29.** 5   **31.** 10   **33.** 6

35. $-\dfrac{1}{4}$ **37.** 11   **39.** -24 **41.** 3

43. Not a real number   **45.** Irrational   **47.** Rational

49. $16^{1/2} = 4$ **51.** $27^{2/3} = 9$ **53.** $\sqrt[4]{256^3} = 64$

55. 5   **57.** -6 **59.** -8 **61.** $\dfrac{1}{4}$ **63.** $\dfrac{1}{9}$

65. $\frac{4}{9}$ **67.** $\frac{3}{11}$ **69.** 9   **71.** -64 **73.** 25

75. $t^{1/2}$ **77.** $x^{7/4}$ **79.** $u^{7/3}$ **81.** $s^{13/2}$

83. $x^{-1} = \dfrac{1}{x}$ **85.** $t^{-9/4} = \dfrac{1}{t^{9/4}}$ **87.** x^3 **89.** $y^{13/12}$

91. $x^{3/4}y^{1/4}$ **93.** $y^{5/2}z^4$ **95.** 3   **97.** $\sqrt[3]{2}$ **99.** $\frac{1}{2}$

101. $\sqrt{c}$ **103.** $\dfrac{3y^2}{4z^{4/3}}$ **105.** $\dfrac{9y^{3/2}}{x^{2/3}}$ **107.** $x^{1/4}$

109. $\sqrt[8]{y}$ **111.** $x^{3/8}$ **113.** $\sqrt{x+y}$

115. $\dfrac{1}{(3u-2v)^{5/6}}$ **117.** 8.5440   **119.** 9.9845

121. 0.0038   **123.** 4.3004   **125.** 66.7213

127. 1.0420   **129.** 0.7915   **131.** $[0, \infty)$

133. $(0, \infty)$ **135.** $(-\infty, 0]$

137.

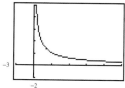

Domain: $(0, \infty)$

139.

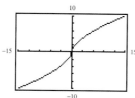

Domain: $(-\infty, \infty)$

141. $2x^{3/2} - 3x^{1/2}$ **143.** $1 + 5y$ **145.** 0.128

147. 23 feet $\times$ 23 feet   **149.** 10.49 centimeters

151. (a) 15.0 feet

(b) $h = \sqrt{15.0^2 - \left(\dfrac{a}{2}\right)^2}$

(c) 8.29 feet

(d) 25.38 feet

153. Given $\sqrt[n]{a}$, a is the radicand and n is the index.

155. No. $\sqrt{2}$ is an irrational number. Its decimal representation is a nonterminating, nonrepeating decimal.

157. 1, 4, 5, 6, 9; Yes

Section 5.2   *(page 335)*

Integrated Review   *(page 335)*

1. Replace the inequality sign with an equal sign and sketch the graph of the resulting equation. (Use a dashed line for $<$ or $>$, and a solid line for $\leq$ and $\geq$.) Test one point in each of the half-planes formed by the graph. If the point satisfies the inequality, shade the entire half-plane to denote that every point in the region satisfies the inequality.

2. The first includes the points on the line $3x + 4y = 4$ and the second does not.

3. $-(x-3)(x^2+1)$ **4.** $(2t+13)(2t-13)$

5. $(x-1)(x-2)$ **6.** $(x-1)(2x+7)$

7. $(x+1)(11x-5)$ **8.** $(2x-7)^2$

9. 816 adults   **10.** 267 units
384 students

1. $2\sqrt{5}$ **3.** $5\sqrt{2}$ **5.** $4\sqrt{6}$ **7.** $6\sqrt{6}$

9. $13\sqrt{7}$ **11.** 0.2   **13.** $0.06\sqrt{2}$ **15.** $1.1\sqrt{2}$

17. $\dfrac{\sqrt{15}}{2}$ **19.** $\dfrac{\sqrt{13}}{5}$ **21.** $3x^2\sqrt{x}$ **23.** $4y^2\sqrt{3}$

25. $3\sqrt{13}|y^3|$ **27.** $2|x|y\sqrt{30y}$ **29.** $8a^2b^3\sqrt{3ab}$

31. $2\sqrt[3]{6}$ **33.** $2\sqrt[3]{14}$ **35.** $2x\sqrt[3]{5x^2}$ **37.** $3|y|\sqrt{2y}$

39. $xy\sqrt[3]{x}$ **41.** $|x|\sqrt[4]{3y^2}$ **43.** $2xy\sqrt[5]{y}$ **45.** $\dfrac{\sqrt[3]{35}}{4}$

47. $\dfrac{\sqrt[5]{15}}{3}$ **49.** $\dfrac{2\sqrt[5]{x^2}}{y}$ **51.** $\dfrac{3a\sqrt[3]{2a}}{b^3}$ **53.** $\dfrac{4a^2\sqrt{2}}{|b|}$

55. $3x^2$ **57.** $\dfrac{\sqrt{3}}{3}$ **59.** $\dfrac{\sqrt{7}}{7}$ **61.** $4\sqrt{3}$

63. $\dfrac{\sqrt[4]{20}}{2}$ **65.** $\dfrac{3\sqrt[3]{2}}{2}$ **67.** $\dfrac{\sqrt{y}}{y}$ **69.** $\dfrac{2\sqrt{x}}{x}$

71. $\dfrac{\sqrt{2x}}{2x}$ **73.** $\dfrac{2\sqrt{3b}}{b^2}$ **75.** $\dfrac{\sqrt[3]{18xy^2}}{3y}$

77. $\dfrac{a^2\sqrt[3]{a^2b}}{b}$ **79.** $2\sqrt{2}$ **81.** $24\sqrt{2} - 6$

83. $-11\sqrt[4]{3} - 5\sqrt[4]{7}$ **85.** $30\sqrt[3]{2}$ **87.** $12\sqrt{x}$

89. $13\sqrt{y}$ **91.** $(10-z)\sqrt[3]{z}$ **93.** $\dfrac{2\sqrt{5}}{5}$ **95.** $\dfrac{9\sqrt{5}}{5}$

97. $\dfrac{\sqrt{2x}(2x-3)}{2x}$ **99.** $\sqrt{7} + \sqrt{18} > \sqrt{7+18}$

101. $5 > \sqrt{3^2 + 2^2}$ **103.** $3\sqrt{5}$ **105.** $3\sqrt{13}$

107. (a) $5\sqrt{10}$ (b) $400\sqrt{10} \approx 1264.9$ square feet

109. 89.44 cycles per second

111. (a)

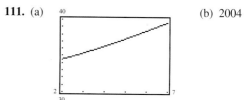

(b) 2004

113. No. $\dfrac{5\sqrt{3}}{3} \neq \dfrac{25}{3}$

115. $\sqrt{6} \cdot \sqrt{15} = \sqrt{9 \cdot 10} = 3\sqrt{10}$

117. No. $\sqrt{2} + \sqrt{18} = \sqrt{2} + 3\sqrt{2} = 4\sqrt{2}$

119. $x < 0$. $\sqrt{(-8)^2} = \sqrt{64} = 8$

Section 5.3 *(page 342)*

> ### Integrated Review *(page 342)*
>
> **1.** $mn = c$ **2.** The signs are the same.
>
> **3.** The signs are different.   **4.** $m + n = b$
>
> **5.** $2x - y = 0$ **6.** $x + y - 6 = 0$
>
> **7.** $y - 3 = 0$ **8.** $x - 4 = 0$
>
> **9.** $6x + 11y - 96 = 0$ **10.** $x + y - 11 = 0$
>
> **11.** $\dfrac{360}{r}$ **12.** $2L + 2\left(\dfrac{L}{3}\right)$

1. 4   **3.** $3\sqrt{2}$ **5.** $2\sqrt[3]{9}$ **7.** $2\sqrt[4]{3}$

9. $2\sqrt{5} - \sqrt{15}$ **11.** $2\sqrt{10} + 8\sqrt{2}$ **13.** $3\sqrt{2}$

15. $6 - 2\sqrt{5}$ **17.** $y + 4\sqrt{y}$ **19.** $4\sqrt{a} - a$

21. $2 - 7\sqrt[3]{4}$ **23.** -1

25. $\sqrt{15} + 3\sqrt{3} - 5\sqrt{5} - 15$ **27.** $8\sqrt{5} + 24$

29. $2\sqrt[3]{3} - 3\sqrt[3]{4} + 3\sqrt[3]{6} - 9$ **31.** $2x + 20\sqrt{2x} + 100$

33. $45x - 17\sqrt{x} - 6$ **35.** $9x - 25$

37. $\sqrt[3]{4x^2} + 10\sqrt[3]{2x} + 25$ **39.** $y - 5\sqrt[3]{y} + 2\sqrt[3]{y^2} - 10$

41. $t + 5\sqrt[3]{t^2} + \sqrt[3]{t} - 3$ **43.** $(x + 3)$ **45.** $(4 - 3x)$

47. $\left(2u + \sqrt{2u}\right)$ **49.** $2 - \sqrt{5}, -1$

51. $\sqrt{11} + \sqrt{3}, 8$ **53.** $\sqrt{15} - 3, 6$

55. $\sqrt{x} + 3, x - 9$ **57.** $\sqrt{2u} + \sqrt{3}, 2u - 3$

59. $2\sqrt{2} - 2, 4$ **61.** $\sqrt{x} - \sqrt{y}, x - y$

63. $\dfrac{1 - 2\sqrt{x}}{3}$ **65.** $\dfrac{-1 + \sqrt{3y}}{4}$ **67.** (a) $2\sqrt{3} - 4$

      (b) 0

69. (a) 0   **71.** $-3\left(\sqrt{2} + 2\right)$ **73.** $\dfrac{7\left(5 - \sqrt{3}\right)}{22}$

   (b) -1

75. $\dfrac{5 + 2\sqrt{10}}{5}$ **77.** $\dfrac{\sqrt{6} - \sqrt{2}}{2}$ **79.** $\dfrac{-9\left(\sqrt{3} + \sqrt{7}\right)}{4}$

81. $\dfrac{4\sqrt{7} + 11}{3}$ **83.** $\dfrac{2x - 9\sqrt{x} - 5}{4x - 1}$

85. $\dfrac{\left(\sqrt{15} + \sqrt{3}\right)x}{4}$ **87.** $\dfrac{2t^2\left(\sqrt{5} + \sqrt{t}\right)}{5 - t}$

89. $4\left(\sqrt{3a} - \sqrt{a}\right)$ **91.** $\dfrac{3(x - 4)\left(x^2 + \sqrt{x}\right)}{x(x - 1)(x^2 + x + 1)}$

93. $-\dfrac{\sqrt{u + v}\left(\sqrt{u - v} + \sqrt{u}\right)}{v}$

95.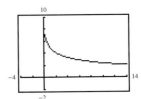

97.

99. $\dfrac{2}{7\sqrt{2}}$ **101.** $\dfrac{4}{5\left(\sqrt{7} - \sqrt{3}\right)}$

103. $192\sqrt{2}$ square inches   **105.** $\dfrac{500k\sqrt{k^2 + 1}}{k^2 + 1}$

107. $\sqrt{3}\left(1 - \sqrt{6}\right)$

$\qquad = \sqrt{3} - \sqrt{3} \cdot \sqrt{6}$ Distributive Property

$\qquad = \sqrt{3} - \sqrt{9 \cdot 2}$ Multiplication Property
of Radicals

$\qquad = \sqrt{3} - 3\sqrt{2}$ Simplify radicals.

109. $\left(3 - \sqrt{2}\right)\left(3 + \sqrt{2}\right) = 9 - 2 = 7$

Multiplying the number by its conjugate yields the difference of two squares. Squaring a square root eliminates the radical.

Mid-Chapter Quiz *(page 345)*

1. 15   **2.** $\frac{3}{2}$ **3.** 8   **4.** 9   **5.** $3|x|\sqrt{3}$

6. $3|x|\sqrt{x}$ **7.** $\dfrac{2|u|\sqrt{u}}{3}$ **8.** $\dfrac{2\sqrt[3]{2}}{u^2}$ **9.** $4\sqrt{2y}$

10. $6x\sqrt[3]{5x^2} + 4x\sqrt[3]{5x}$ **11.** $16 + 6\sqrt{2}$ **12.** $10 - 4\sqrt{2}$

13. $3 + 5\sqrt{6}$ **14.** $60 + 67\sqrt{3}$ **15.** $\dfrac{\sqrt{21} - \sqrt{7}}{2}$

16. $-3 - 3\sqrt{2}$ **17.** $\dfrac{4}{3}\left(3 - \sqrt{6}\right)$

18. $\frac{1}{2}\left(4\sqrt{3} - 3\sqrt{2} + \sqrt{6} - 4\right)$ **19.** $1 - \sqrt{4}, -3$

20. $\sqrt{10} + 5, -15$ **21.** $\sqrt{5^2 + 12^2} = \sqrt{169} = 13$

22. $23 + 8\sqrt{2}$ inches

Section 5.4 *(page 352)*

> ### Integrated Review *(page 352)*
>
> **1.** The function is undefined when the denominator is zero. The domain is all real numbers x such that $x \neq -2$ and $x \neq 3$.
>
> **2.** $\dfrac{2x^2 + 5x - 3}{x^2 - 9}$ is undefined if $x = -3$.
>
> **3.** $36x^5y^8$ **4.** 1   **5.** $4rs^2$ **6.** $\dfrac{9x^2}{16y^6}$

7. $-\dfrac{x + 13}{5x^2}$ **8.** $\dfrac{x^2 - 4}{25(x^2 - 9)}$ **9.** $\dfrac{2x + 5}{x - 5}$

10. $-\dfrac{5x - 8}{x - 1}$

11. 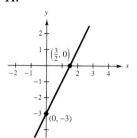 **12.**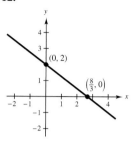

1. (a) Not a solution **3.** (a) Not a solution

(b) Not a solution (b) Solution

(c) Not a solution (c) Not a solution

(d) Solution (d) Not a solution

5. 400 **7.** 9 **9.** 27 **11.** 49 **13.** No solution

15. 64 **17.** 90 **19.** -27 **21.** $\frac{4}{5}$ **23.** 5

25. No solution **27.** $\frac{44}{3}$ **29.** $\frac{14}{25}$ **31.** No solution

33. 4 **35.** No solution **37.** 7 **39.** -15

41. $-\frac{2}{3}$ **43.** 8 **45.** 1, 3 **47.** $\frac{1}{4}$ **49.** $\frac{1}{2}$ **51.** 4

53. 4 **55.** 216 **57.** 4, -12 **59.** -16

61. 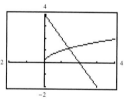 **63.**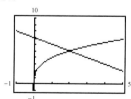

1.407 1.569

65. 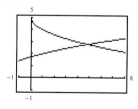 **67.**

4.840 1.978

69. 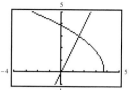 **71.** c **73.** d **75.** f

1.500

77. 9.00 **79.** 12.00

81.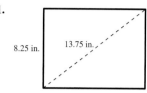

8.25 in. 13.75 in.

11 inches

83. $10\sqrt{17} \approx 41.23$ feet **85.** 15 feet

87. 30 inches $\times$ 16 inches **89.** $h = \dfrac{\sqrt{s^2 - \pi^2 r^4}}{\pi r}$

91. 64 feet **93.** 56.57 feet per second **95.** 56.25 feet

97. 1.82 feet **99.** 500 units

101. (a) 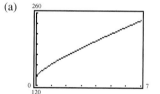 (b) 1995

103. \$12,708.73

105. No. It is not an operation that necessarily yields an equivalent equation. There may be extraneous solutions.

107. $\left(\sqrt{x} + \sqrt{6}\right)^2 \neq \left(\sqrt{x}\right)^2 + \left(\sqrt{6}\right)^2$

Section 5.5 *(page 362)*

Integrated Review *(page 362)*

1. $\dfrac{u}{v} \cdot \dfrac{w}{z} = \dfrac{uw}{vz}$

$\dfrac{3t}{5} \cdot \dfrac{8t^2}{15} = \dfrac{3t(8t^2)}{5(15)} = \dfrac{8t^3}{25}$

2. $\dfrac{u}{v} \div \dfrac{w}{z} = \dfrac{u}{v} \cdot \dfrac{z}{w}$

$\dfrac{3t}{5} \div \dfrac{8t^2}{15} = \dfrac{3t}{5} \cdot \dfrac{15}{8t^2} = \dfrac{9}{8t}$

3. Rewrite the fractions so they have like denominators and then use the rule

$$\frac{u}{w} + \frac{v}{w} = \frac{u+v}{w}.$$

$$\frac{3t}{5} + \frac{8t^2}{15} = \frac{9t}{15} + \frac{8t^2}{15} = \frac{9t + 8t^2}{15}$$

4. $\dfrac{t-5}{5-t} = \dfrac{-1(5-t)}{5-t} = -1$ **5.** $\dfrac{x}{5}$ **6.** $\dfrac{x}{5(x+y)}$

7. $\dfrac{9}{2(x+3)}$ **8.** $\dfrac{1}{x-2}$ **9.** $\dfrac{x^2 + 2x - 13}{x(x-2)}$

10. $\dfrac{(x+1)(x+3)}{3}$ **11.** $\dfrac{7x}{9}, \dfrac{19x}{18}$ **12.** $\dfrac{C_1 C_2}{C_1 + C_2}$

---

1. $2i$ **3.** $-12i$ **5.** $\frac{2}{5}i$ **7.** $0.3i$ **9.** $2\sqrt{2}i$

11. $3\sqrt{3}i$ **13.** $\sqrt{7}i$ **15.** 2 **17.** $\sqrt{5}i$

19. $\dfrac{3\sqrt{2}}{8}i$ **21.** $10i$ **23.** $3\sqrt{2}i$ **25.** $3\sqrt{3}i$

27. -4 **29.** $-3\sqrt{6}$ **31.** -0.44 **33.** $-2\sqrt{3} - 3$

35. $5\sqrt{2} - 4\sqrt{5}$ **37.** $4 + 3\sqrt{2}i$ **39.** -16

41. $-8i$ **43.** $a = 3, b = -4$ **45.** $a = 2, b = -3$

47. $a = -4, b = -2\sqrt{2}$ **49.** $a = 2, b = -2$

51. $10 + 4i$ **53.** $-14 - 40i$ **55.** $-14 + 20i$

57. $9 - 7i$ **59.** $3 + 6i$ **61.** $\frac{13}{6} + \frac{3}{2}i$

63. $-3 + 49i$ **65.** $7 + (3\sqrt{7} - 5)i$ **67.** -36

69. 24 **71.** $-36i$ **73.** $27i$ **75.** -9

77. $-65 - 10i$ **79.** $20 - 12i$ **81.** $4 + 18i$

83. $-20 + 12\sqrt{5}i$ **85.** $-40 - 5i$ **87.** $-14 + 42i$

89. 9 **91.** $-7 - 24i$ **93.** $-21 + 20i$

95. $2 + 11i$ **97.** 5 **99.** 68 **101.** 31

103. 100 **105.** 4 **107.** 2.5 **109.** $-10i$

111. $2 + 2i$ **113.** $-\frac{24}{53} + \frac{84}{53}i$ **115.** $-\frac{6}{5} + \frac{2}{5}i$

117. $\frac{8}{5} - \frac{1}{5}i$ **119.** $1 - \frac{6}{5}i$ **121.** $-\frac{53}{25} + \frac{29}{25}i$

123. (a) Solution   **125.** (a) Solution
   (b) Solution   (b) Solution

127. (a) $\left(\dfrac{-5 + 5\sqrt{3}i}{2}\right)^3 = 125$

   (b) $\left(\dfrac{-5 - 5\sqrt{3}i}{2}\right)^3 = 125$

129. (a) $1, \dfrac{-1 + \sqrt{3}i}{2}, \dfrac{-1 - \sqrt{3}i}{2}$

   (b) $2, \dfrac{-2 + 2\sqrt{3}i}{2} = -1 + \sqrt{3}i, \dfrac{-2 - 2\sqrt{3}i}{2}$
   $= -1 - \sqrt{3}i$

(c) $4, \dfrac{-4 + 4\sqrt{3}i}{2} = -2 + 2\sqrt{3}i, \dfrac{-4 - 4\sqrt{3}i}{2}$
   $= -2 - 2\sqrt{3}i$

131. $2a$ **133.** $2bi$ **135.** $i = \sqrt{-1}$

137. $\sqrt{-3}\sqrt{-3} = (\sqrt{3}i)(\sqrt{3}i)$ **139.** 13
   $= 3i^2 = -3$

Review Exercises   *(page 366)*

1. 7 **3.** -9 **5.** -2 **7.** -4 **9.** 1.2 **11.** $\frac{5}{6}$

13. $-\frac{1}{5}$ **15.** $2i$ **17.** $49^{1/2} = 7$ **19.** $\sqrt[3]{216} = 6$

21. 81 **23.** -125 **25.** $\frac{1}{16}$ **27.** $-\frac{9}{16}$ **29.** $x^{7/12}$

31. $z^{5/3}$ **33.** $\dfrac{1}{x^{5/4}}$ **35.** $a\sqrt[3]{b^2}$ **37.** $\sqrt[8]{x}$

39. $\sqrt[3]{3x + 2}$ **41.** 0.04 **43.** 10.63

45. 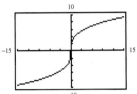   **47.**

$(-\infty, \infty)$ $[0, \infty)$

49. $6\sqrt{10}$ **51.** $5u^2v^2\sqrt{3u}$ **53.** $0.5x^2\sqrt{y}$

55. $2b\sqrt[4]{4a^2b}$ **57.** $2ab\sqrt[3]{6b}$ **59.** $\dfrac{\sqrt{30}}{6}$ **61.** $\dfrac{\sqrt{3x}}{2x}$

63. $\dfrac{\sqrt[3]{4x^2}}{x}$ **65.** $\sqrt{7}$ **67.** $-24\sqrt{10}$

69. $14\sqrt{x} - 9\sqrt[3]{x}$ **71.** $7\sqrt[4]{y} + 3$ **73.** $12\sqrt{x} - 2\sqrt[3]{x}$

75. $\dfrac{2\sqrt{5}}{5}$ **77.** $10\sqrt{3}$ **79.** $5\sqrt{2} + 3\sqrt{5}$

81. $5\sqrt{2} + 2\sqrt{5}$ **83.** $7\sqrt{6} + 6\sqrt{2} - 4\sqrt{3} - 14$

85. $12\sqrt{5} + 41$ **87.** $3 - x$ **89.** $-3(1 + \sqrt{2})$

91. $\dfrac{6(4 - \sqrt{6})}{5}$ **93.** $-\dfrac{(\sqrt{2} - 1)(\sqrt{3} + 4)}{13}$

95. $\dfrac{(\sqrt{x} + 10)^2}{x - 100}$

97. 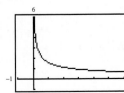   **99.**

101. 225 **103.** No real solution   **105.** 105 **107.** 3

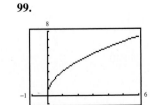

109. 5   **111.** $-5, -3$ **113.** 6   **115.** $\frac{3}{32}$

117. $4\sqrt{3}\,i$ **119.** $10 - 9\sqrt{3}\,i$ **121.** $\frac{3}{4} - \sqrt{3}\,i$

123. $15i$ **125.** $\left(11 - 2\sqrt{21}\right)i$ **127.** -5

129. $\sqrt{70} - 2\sqrt{10}$ **131.** $x = 2, y = 3$

133. $x = 4, y = 125$ **135.** $8 - 3i$ **137.** $8 + 4i$

139. 25   **141.** $11 - 60i$ **143.** $-\frac{7}{3}i$

145. $-\frac{8}{17} + \frac{2}{17}i$ **147.** $\frac{13}{37} - \frac{33}{37}i$

149. $21 + 12\sqrt{2}$ inches   **151.** 1.37 feet

153. 500 watts   **155.** 9000 watts   **157.** 9.77 feet

Chapter Test *(page 369)*

1. (a) 64   **2.** (a) $\frac{1}{9}$ **3.** (a) $x^{1/3}$

    (b) 10   (b) 6   (b) 25

4. (a) $\frac{4}{3}\sqrt{2}$ **5.** (a) $2x\sqrt{6x}$

    (b) $2\sqrt[3]{3}$ (b) $2xy^2\sqrt[4]{x}$

6. Multiply the numerator and denominator of a fraction by a factor such that no radical contains a fraction and no denominator of a fraction contains a radical.

$$\frac{\sqrt{6}}{2}$$

7. $-10\sqrt{3x}$ **8.** $5\sqrt{3x} + 3\sqrt{5}$ **9.** $16 - 8\sqrt{2x} + 2x$

10. $7\sqrt{3}(3 + 4y)$ **11.** 27   **12.** No solution   **13.** 9

14. $x = 4, y = 400$ **15.** $x = 3, y = 1$ **16.** $2 - 2i$

17. $-5 - 12i$ **18.** $-8 + 4i$ **19.** $13 + 13i$

20. $-2 - 5i$ **21.** 100 feet

Chapter 6

Section 6.1 *(page 377)*

Integrated Review *(page 377)*

1. -3. Coefficient of the term of highest degree

2. 5. $(y^2 - 2)(y^3 + 7) = y^5 - 2y^3 + 7y^2 - 14$

3.   **4.**

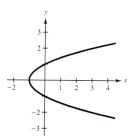   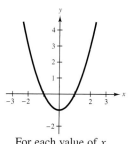

For some values of x there correspond two values of y.

For each value of x there corresponds exactly one value of y.

---

   5. $\dfrac{1}{x^3}$ **6.** $-\dfrac{15y^4}{x^2}$ **7.** $\dfrac{9y^2}{4x^2}$ **8.** $\dfrac{v^4}{u^5}$

   9. $\dfrac{2u}{9v^6}$ **10.** $\dfrac{r^3}{7}$ **11.** 100   **12.** $\dfrac{29}{18} \approx 1.6$ hours

                                           Distance

1. 5, 7   **3.** $-9, 8$ **5.** $-9, 5$ **7.** 6   **9.** $-\frac{4}{3}$

11. 0, 3   **13.** 9, 12   **15.** $\frac{5}{3}, 6$ **17.** 1, 6   **19.** $-\frac{5}{6}, \frac{1}{2}$

21. ± 8 **23.** ± 3 **25.** $\pm\frac{4}{5}$ **27.** ± 8 **29.** $\pm\frac{5}{2}$

31. $\pm\frac{15}{2}$ **33.** $-17, 9$ **35.** 2.5, 3.5   **37.** $2 \pm \sqrt{7}$

39. $\dfrac{-1 \pm 5\sqrt{2}}{2}$ **41.** $\dfrac{3 \pm 7\sqrt{2}}{4}$ **43.** $\pm 6i$

45. $\pm 2i$ **47.** $\pm\dfrac{\sqrt{17}}{3}i$ **49.** $3 \pm 5i$ **51.** $-\dfrac{4}{3} \pm 4i$

53. $-\dfrac{3}{2} \pm \dfrac{3\sqrt{6}}{2}i$ **55.** $-6 \pm \dfrac{11}{3}i$ **57.** $1 \pm 3\sqrt{3}i$

59. $-1 \pm 0.2i$ **61.** $\dfrac{2}{3} \pm \dfrac{1}{3}i$ **63.** $-\dfrac{7}{3} \pm \dfrac{\sqrt{38}}{3}i$

65. $0, \frac{5}{2}$ **67.** $-4, \frac{3}{2}$ **69.** ± 30 **71.** $\pm 30i$ **73.** ± 3

75. $-5, 15$ **77.** $5 \pm 10i$ **79.** $-2 \pm 3\sqrt{2}i$

81.   **83.**

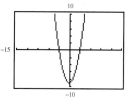   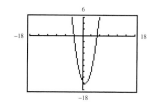

$(-3, 0), (3, 0)$ $(-3, 0), (5, 0)$

85.   **87.**

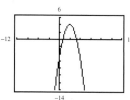   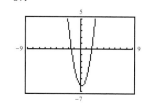

$(1, 0), (5, 0)$ $(2, 0), \left(-\frac{3}{2}, 0\right)$

89.   **91.**

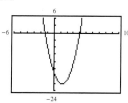   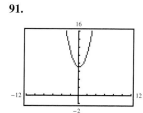

$\left(-\frac{4}{3}, 0\right), (4, 0)$ $\pm\sqrt{7}i$

93.

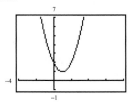

95.

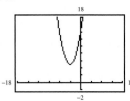

$$1 \pm i$$

$$-3 \pm \sqrt{5}i$$

97. $f(x) = \sqrt{4 - x^2}$
$\quad g(x) = -\sqrt{4 - x^2}$

99. $f(x) = \frac{1}{2}\sqrt{4 - x^2}$
$\quad g(x) = -\frac{1}{2}\sqrt{4 - x^2}$

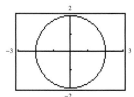

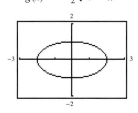

101. $\pm 1, \pm 2$ **103.** $\pm\sqrt{2}, \pm\sqrt{3}$ **105.** $\pm 2, \pm i$

107. $\pm 1, \pm\sqrt{5}$ **109.** $4, 25$ **111.** $-8, 27$

113. $1, \frac{125}{8}$ **115.** $1, 32$ **117.** $\frac{1}{32}, 243$ **119.** $\frac{1}{2}, 1$

121. $(x - 5)(x + 2) = x^2 - 3x - 10 = 0$

123. $\left[x - \left(1 + \sqrt{2}\right)\right]\left[x - \left(1 - \sqrt{2}\right)\right] = x^2 - 2x - 1 = 0$

125. $(x - 5i)(x + 5i) = x^2 + 25 = 0$

127. 4 seconds   **129.** $2\sqrt{2} \approx 2.83$ seconds

131. 9 seconds   **133.** 6%   **135.** 1993

137. (a) 2.5 seconds. Extracting square roots

(b) 0 seconds, 2 seconds. Factoring

139. Factoring and the Zero-Factor Property allow you to solve a quadratic equation by converting it into two linear equations that you already know how to solve.

141. False. The solutions are $x = 5$ and $x = -5$.

143. To solve an equation of quadratic form, determine an algebraic expression u such that substitution yields the quadratic equation $au^2 + bu + c = 0$. Solve this quadratic equation for u and then, through back-substitution, find the solution of the original equation.

Section 6.2   (page 384)

Integrated Review   (page 384)

1. a^4b^4 **2.** a^{rs} **3.** $\dfrac{b^r}{a^r}$ **4.** $\dfrac{1}{a^r}$ **5.** 6

6. 3   **7.** $-\frac{2}{3}, 5$ **8.** $-5, 8$

9.

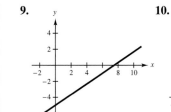

10.

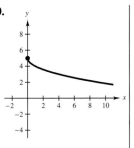

11.

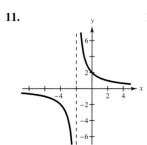

12.

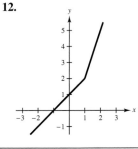

1. 16   **3.** 100   **5.** 64   **7.** $\frac{25}{4}$ **9.** $\frac{81}{4}$ **11.** $\frac{1}{36}$

13. $\frac{9}{100}$ **15.** 0.04   **17.** 0, 20   **19.** $-6, 0$

21. 0, 5   **23.** 1, 7   **25.** $-6, 4$ **27.** $-4, -3$

29. $-3, 6$ **31.** 1, 6   **33.** $-\frac{5}{2}, \frac{3}{2}$

35. $2 + \sqrt{7} \approx 4.65$
$\quad\ 2 - \sqrt{7} \approx -0.65$

37. $-2 + \sqrt{7} \approx 0.65$
$\quad\ -2 - \sqrt{7} \approx -4.65$

39. $2 + \sqrt{3} \approx 3.73$
$\quad\ 2 - \sqrt{3} \approx 0.27$

41. $-1 + \sqrt{2}i \approx -1 + 1.41i$
$\quad\ -1 - \sqrt{2}i \approx -1 - 1.41i$

43. $5 + 3\sqrt{3} \approx 10.20$
$\quad\ 5 - 3\sqrt{3} \approx -0.20$

45. $-10 + 3\sqrt{10} \approx -0.51$
$\quad\ -10 - 3\sqrt{10} \approx -19.49$

47. $\dfrac{-5 + \sqrt{13}}{2} \approx -0.70$
$\quad\ \dfrac{-5 - \sqrt{13}}{2} \approx -4.30$

49. $\dfrac{-3 + \sqrt{17}}{2} \approx 0.56$
$\quad\ \dfrac{-3 - \sqrt{17}}{2} \approx -3.56$

51. $\dfrac{1}{2} + \dfrac{\sqrt{3}}{2}i \approx 0.5 + 0.87i$
$\quad\ \dfrac{1}{2} - \dfrac{\sqrt{3}}{2}i \approx 0.5 - 0.87i$

53. 3, 4

55. $\dfrac{1 + 2\sqrt{7}}{3} \approx 2.10$
$\quad\ \dfrac{1 - 2\sqrt{7}}{3} \approx -1.43$

57. $\dfrac{-3 + \sqrt{137}}{8} \approx 1.09$
$\quad\ \dfrac{-3 - \sqrt{137}}{8} \approx -1.84$

59. $\dfrac{-4 + \sqrt{10}}{2} \approx -0.42$
$\quad\ \dfrac{-4 - \sqrt{10}}{2} \approx -3.58$

61. $\dfrac{-9 + \sqrt{21}}{6} \approx -0.74$
$\quad\ \dfrac{-9 - \sqrt{21}}{6} \approx -2.26$

63. $\dfrac{-1 + \sqrt{10}}{2} \approx 1.08$

$\dfrac{-1 - \sqrt{10}}{2} \approx -2.08$

65. $\dfrac{3}{10} + \dfrac{\sqrt{191}}{10}i \approx 0.30 + 1.38i$

$\dfrac{3}{10} - \dfrac{\sqrt{191}}{10}i \approx 0.30 - 1.38i$

67. $\dfrac{7 + \sqrt{57}}{2} \approx 7.27$

$\dfrac{7 - \sqrt{57}}{2} \approx -0.27$

69. $-1 + \sqrt{3}i \approx -1 + 1.73i$ **71.** $-1 \pm 2i$

$-1 - \sqrt{3}i \approx -1 - 1.73i$

73. $1 \pm \sqrt{3}$ **75.** $1 \pm \sqrt{3}$ **77.** $4 \pm 2\sqrt{2}$

79.

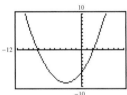

$\left(-2 \pm \sqrt{5}, 0\right)$

81.

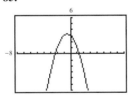

$\left(1 \pm \sqrt{6}, 0\right)$

83.

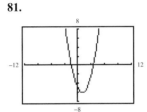

$\left(-3 \pm 3\sqrt{3}, 0\right)$

85.

$\left(\dfrac{-1 \pm \sqrt{13}}{2}, 0\right)$

87. (a) $x^2 + 8x$ (b) $x^2 + 8x + 16$ (c) $(x + 4)^2$

89. 4 centimeters, 6 centimeters

91. 15 feet $\times$ $46\frac{2}{3}$ feet or 20 feet $\times$ 35 feet

93. 271 meters, 129 meter

95. 139 units, 861 units

97. $\frac{25}{4}$. Divide the coefficient of the first-degree term by 2, and square the result to obtain $\left(\frac{5}{2}\right)^2 = \frac{25}{4}$.

99. Yes. $x^2 + 1 = 0$

101. True. Given the solutions $x = r_1$ and $x = r_2$, the quadratic equation can be written as $(x - r_1)(x - r_2) = 0$.

Section 6.3   (*page 392*)

Integrated Review   (*page 392*)

1. $\sqrt{a}\sqrt{b}$ **2.** $\dfrac{\sqrt{a}}{\sqrt{b}}$

3. No. $\sqrt{72} = \sqrt{36 \cdot 2} = 6\sqrt{2}$

4. No. $\dfrac{10}{\sqrt{5}} = \dfrac{10\sqrt{5}}{(\sqrt{5})^2} = 2\sqrt{5}$ **5.** $23\sqrt{2}$

6. 150   **7.** 7   **8.** $11 + 6\sqrt{2}$ **9.** $\dfrac{4\sqrt{10}}{5}$

10. $\dfrac{5(1 + \sqrt{3})}{4}$ **11.** 10 inches $\times$ 15 inches

12. 200 units

1. $2x^2 + 2x - 7 = 0$ **3.** $-x^2 + 10x - 5 = 0$

5. $4, 7$ **7.** $-2, -4$ **9.** $-\frac{1}{2}$ **11.** $-\frac{3}{2}$ **13.** $-\frac{1}{2}, \frac{2}{3}$

15. $-15, 20$ **17.** $1 \pm \sqrt{5}$ **19.** $-2 \pm \sqrt{3}$

21. $-3 \pm 2\sqrt{3}$ **23.** $5 \pm \sqrt{2}$ **25.** $-\dfrac{3}{4} \pm \dfrac{\sqrt{15}}{4}i$

27. $-\dfrac{1}{3}, 1$ **29.** $\dfrac{-2 \pm \sqrt{10}}{2}$ **31.** $\dfrac{-1 \pm \sqrt{5}}{3}$

33. $\dfrac{-3 \pm \sqrt{21}}{4}$ **35.** $\dfrac{3}{8} \pm \dfrac{\sqrt{7}}{8}i$ **37.** $\dfrac{5 \pm \sqrt{73}}{4}$

39. $\dfrac{3 \pm \sqrt{13}}{6}$ **41.** $\dfrac{-3 \pm \sqrt{57}}{6}$ **43.** $\dfrac{1 \pm \sqrt{5}}{5}$

45. $\dfrac{-1 \pm \sqrt{10}}{5}$

47. Two distinct imaginary solutions

49. Two distinct irrational solutions

51. Two distinct imaginary solutions

53. One (repeated) rational solution

55. Two distinct imaginary solutions   **57.** ± 13

59. $-3, 0$ **61.** $\frac{9}{5}, \frac{21}{5}$ **63.** $-\frac{3}{2}, 18$ **65.** $-4 \pm 3i$

67. $8, 16$ **69.** $\dfrac{13}{6} \pm \dfrac{13\sqrt{11}}{6}i$ **71.** $\dfrac{-5 \pm 5\sqrt{17}}{12}$

73. $-\frac{11}{6}, \frac{5}{2}$

75.

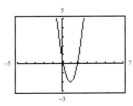

$(0.18, 0), (1.82, 0)$

77.

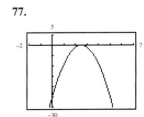

$(2.50, 0)$

79.

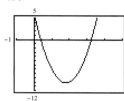

81.

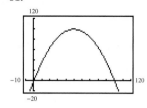

$(3.23, 0), (0.37, 0)$ $(99.80, 0), (0.20, 0)$

83. No real solutions   **85.** Two real solutions

87. $\dfrac{5 \pm \sqrt{185}}{8}$ **89.** $\dfrac{3 + \sqrt{17}}{2}$

91. (a) $c < 9$ (b) $c = 9$ (c) $c > 9$

93. (a) $c < 16$ (b) $c = 16$ (c) $c > 16$

95. 5.1 inches $\times$ 11.4 inches

97. (a) 2.5 seconds (b) $\dfrac{5 + 5\sqrt{3}}{4} \approx 3.4$ seconds

99. (a)

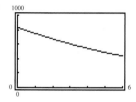

(b) 1991

(c) 400,500

101.

	x_1	x_2	$x_1 + x_2$	$x_1 x_2$
(a)	-2	3	1	-6
(b)	-3	$\frac{1}{2}$	$-\frac{5}{2}$	$-\frac{3}{2}$
(c)	$-\frac{3}{2}$	$\frac{3}{2}$	0	$-\frac{9}{4}$
(d)	$5 + 3i$	$5 - 3i$	10	34

103. (c) $\dfrac{4 + \sqrt{66}}{4} \approx 3.0$ seconds; Quadratic Formula

(d) 84-foot level: 3.5 seconds
100-foot level: 3.7 seconds

105. $b^2 - 4ac$. If the discriminant is positive, the quadratic equation has two real solutions; if it is zero, the equation has one (repeated) real solution; and if it is negative, the equation has no real solutions.

107. The four methods are factoring, extracting square roots, completing the square, and the Quadratic Formula.

Mid-Chapter Quiz   *(page 395)*

1. ± 6 **2.** $-4, \frac{5}{2}$ **3.** $\pm 2\sqrt{3}$ **4.** $-1, 7$

5. $-5 \pm 2\sqrt{6}$ **6.** $\dfrac{-3 \pm \sqrt{19}}{2}$ **7.** $-2 \pm \sqrt{10}$

8. $\dfrac{3 \pm \sqrt{105}}{12}$ **9.** $-\dfrac{5}{2} \pm \dfrac{\sqrt{3}}{2}i$ **10.** $-2, 10$

11. $-3, 10$ **12.** $-2, 5$ **13.** $\dfrac{3}{2}$ **14.** $\dfrac{-5 \pm \sqrt{10}}{3}$

15. 36   **16.** $\pm 2i, \pm \sqrt{3}\,i$

17.

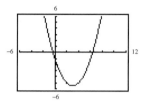

18.

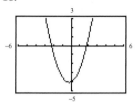

$(-0.32, 0), (6.32, 0)$ $(-2.24, 0), (1.79, 0)$

19. 50 units   **20.** 35 meters $\times$ 65 meters

Section 6.4   *(page 402)*

Integrated Review   *(page 402)*

1. $m = \dfrac{y_2 - y_1}{x_2 - x_1}$ **2.** (a) $y = mx + b$

(b) $y - y_1 = m(x - x_1)$

(c) $Ax + By + C = 0$

(d) $y - b = 0$

3. $2x + 4y = 0$ **4.** $3x - 4y = 0$

5. $2x - y = 0$ **6.** $x + y - 6 = 0$

7. $22x + 16y - 161 = 0$

8. $134x - 73y + 146 = 0$ **9.** $y - 8 = 0$

10. $x + 3 = 0$ **11.** 8 people

12. 3 miles per hour

---

1. 18 dozen, \$1.20 per dozen   **3.** 16, \$30

5. 108 square inches   **7.** 70 feet   **9.** 64 inches

11. 180 square kilometers   **13.** 440 meters

15. 12 inches $\times$ 16 inches   **17.** Base: 24 inches
Height: 16 inches

19. 50 feet $\times$ 250 feet or 100 feet $\times$ 125 feet

21. No.

Area $= \frac{1}{2}(b_1 + b_2)h = \frac{1}{2}x[x + (550 - 2x)] = 43{,}560$

This equation has no real solution.

23. Height: 12 inches   **25.** 8%   **27.** 6%
Width: 24 inches

29. 2.59%   **31.** 48   **33.** 5

35. 15.86 miles or 2.14 miles

37. (a) $d = \sqrt{(3 + x)^2 + (4 + x)^2}$

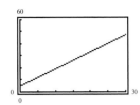

$x \approx 3.5$ when $d = 10$

(b) $\dfrac{-7 + \sqrt{199}}{2} \approx 3.55$ meters

39. 9.1 hours, 11.1 hours   **41.** 6.8 days, 9.8 days

43. 3 seconds   **45.** 9.5 seconds   **47.** 4.7 seconds

49. (a) 3 seconds, 7 seconds   **51.** 15, 16

(b) 10 seconds

53. 14, 16   **55.** 21, 23   **57.** 400 miles per hour

59. 46 miles per hour or 65 miles per hour

61. (a) $b = 20 - a$

$A = \pi ab$

$A = \pi a(20 - a)$

(b)

a	4	7	10	13	16
A	201.1	285.9	314.2	285.9	201.1

(c) 7.9, 12.1   (d)

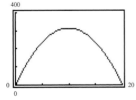

63. (a) Write a verbal model that will describe what you need to know.

(b) Assign labels to each part of the verbal model—numbers to the known quantities and letters to the variable quantities.

(c) Use the labels to write an algebraic model based on the verbal model.

(d) Solve the resulting algebraic equation and check your solution.

65. Dollars

67. $(x + 4)^2 = 0$

Section 6.5   *(page 413)*

Integrated Review   *(page 413)*

1. No. 3.682×10^9 **2.** $[10^6, 10^8)$

3. $6v(u^2 - 32v)$ **4.** $5x^{1/3}(x^{1/3} - 2)$

5. $(x - 10)(x - 4)$ **6.** $(x + 2)(x - 2)(x + 3)$

7. $(4x + 11)(4x - 11)$ **8.** $4x(x^2 - 3x + 4)$

9. $\frac{3}{2}h^2$ **10.** $\frac{1}{3}b^2$ **11.** $5x^2$ **12.** $x^2 + 8x$

1. $0, \frac{5}{2}$ **3.** $\pm\frac{9}{2}$ **5.** $-3, 5$ **7.** $1, 3$ **9.** $\frac{5}{2}$

11. Negative: $(-\infty, 4)$
Positive: $(4, \infty)$

13. Negative: $(6, \infty)$
Positive: $(-\infty, 6)$

15. Negative: $(0, 4)$
Positive: $(-\infty, 0) \cup (4, \infty)$

17. Negative: $(-\infty, -2) \cup (2, \infty)$
Positive: $(-2, 2)$

19. Negative: $(-1, 5)$
Positive: $(-\infty, -1) \cup (5, \infty)$

21. $[-3, \infty)$ **23.** $(8, \infty)$

   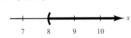

25. $(0, 2)$ **27.** $[0, 2]$

29. $(-\infty, -2) \cup (2, \infty)$ **31.** $[-5, 2]$

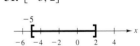

33. $(-\infty, -3) \cup (1, \infty)$ **35.** No solution

37. $(-\infty, \infty)$

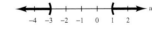

39. $\left(-\infty, 2 - \sqrt{2}\right) \cup \left(2 + \sqrt{2}, \infty\right)$

41. $(-\infty, \infty)$ **43.** No solution

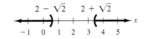

45. $\left[-2, \frac{4}{3}\right]$

47. $\left(\frac{2}{3}, \frac{5}{2}\right)$

81. $(-6, 4)$

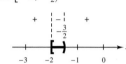

83. $(-\infty, 3] \cup \left(\frac{11}{2}, \infty\right)$

49. $\left(-\infty, -\frac{1}{2}\right) \cup (4, \infty)$

51. $-\frac{7}{2}$

85. $\left[-2, -\frac{3}{2}\right)$

87. $(-1, 3)$

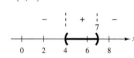

53. No solution   **55.** $\left(-\infty, 5 - \sqrt{6}\right) \cup \left(5 + \sqrt{6}, \infty\right)$

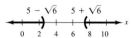

89. $(4, 7)$

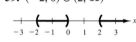

91. $\left(-2, -\frac{2}{5}\right)$

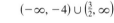

57. $(-\infty, -9] \cup [-1, \infty)$

59. $(-2, 0) \cup (2, \infty)$

93. $(-\infty, 3) \cup [5, \infty)$ **95.** $(-\infty, -1) \cup (0, 1)$

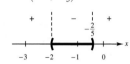

61.

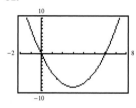

$(0, 6)$

63.

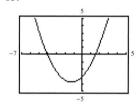

$(-\infty, -4) \cup \left(\frac{3}{2}, \infty\right)$

97. $(-\infty, -1) \cup (4, \infty)$ **99.** $\left(-5, \frac{13}{4}\right)$

101. $(0, 0.382) \cup (2.618, \infty)$

103.

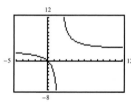

(a) $[0, 2)$

(b) $(2, 4]$

105.

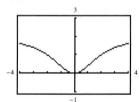

(a) $(-\infty, -2] \cup [2, \infty)$

(b) $(-\infty, \infty)$

65.

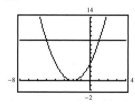

$(-\infty, -5] \cup [1, \infty)$

67.

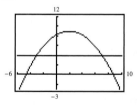

$(-\infty, -3) \cup (7, \infty)$

107. $(3, 5)$ **109.** $r > 7.24\%$

111. $90,000 \le x \le 100,000$ **113.** $(12, 20)$

115. (a)

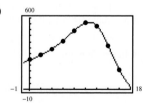

69. 3

71. $0, -5$

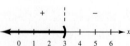

(b) $5.7 \le t \le 13.7$

117. The direction of the inequality is reversed.

73. $(3, \infty)$

75. $(-\infty, 3)$

119. A polynomial can change signs only at the x-values that make the polynomial zero. The zeros of the polynomial are called the critical numbers, and they are used to determine the test intervals in solving polynomial inequalities.

77. $(0, 3)$

79. $[-3, 4)$

121. $x^2 + 1 < 0$

Review Exercises *(page 417)*

1. $-12, 0$ **3.** $\pm\frac{1}{2}$ **5.** $-\frac{5}{2}$ **7.** $-9, 10$

9. $-\frac{3}{2}, 6$ **11.** ± 50 **13.** $\pm 2\sqrt{3}$ **15.** $-4, 36$

17. $\pm 11i$ **19.** $\pm 5\sqrt{2}\,i$ **21.** $-4 \pm 3\sqrt{2}\,i$

23. $\pm\sqrt{5}, \pm i$ **25.** $1, 9$ **27.** $1, 1 \pm \sqrt{6}$

29. 64 **31.** $3 \pm 2\sqrt{3}$ **33.** $\frac{3}{2} \pm \frac{\sqrt{3}}{2}i$

35. $\frac{1}{3} \pm \frac{\sqrt{17}}{3}i$ **37.** $\frac{-5 \pm \sqrt{19}}{2}$ **39.** $-6, 5$

41. $-\frac{7}{2}, 3$ **43.** $\frac{8 \pm 3\sqrt{6}}{5}$ **45.** $\frac{10}{3} \pm \frac{5\sqrt{2}}{3}i$

47. One repeated rational solution

49. Two distinct rational solutions

51. Two distinct rational solutions

53. Two distinct imaginary solutions

55. $(0, 7)$ **57.** $(-\infty, -2] \cup [6, \infty)$

   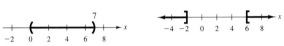

59. $\left(-4, \frac{5}{2}\right)$ **61.** $\left(-\infty, -3\right] \cup \left(\frac{7}{2}, \infty\right)$

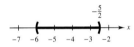   

63. $\left(-6, -\frac{5}{2}\right)$ **65.** $16;\ \$5000$

67. 6 inches $\times$ 18 inches   **69.** 3.5%   **71.** 48

73. 15   **75.** 60 feet, 80 feet   **77.** 19 hours, 21 hours

79. (a) 2 seconds   (b) 6 seconds   **81.** $x > 13{,}158$

83. $(5.3, 14.2)$

Chapter Test *(page 420)*

1. $-5, 10$ **2.** $-\frac{3}{8}, 3$ **3.** $1.7, 2.3$ **4.** $-3 \pm 9i$

5. $\frac{3 \pm \sqrt{3}}{2}$ **6.** $\frac{2 \pm 3\sqrt{2}}{2}$ **7.** $1, 16$

8. $\pm\sqrt{2}, \pm 2\sqrt{2}\,i$ **9.** -56; two imaginary solutions

10. $(x + 4)(x - 5) = x^2 - x - 20 = 0$

11. $(-\infty, -2] \cup [6, \infty)$ **12.** $(0, 3)$

13. $[-8, 3)$ **14.** $\left(2, \frac{11}{4}\right)$

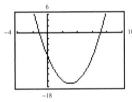

15. 12 feet $\times$ 20 feet   **16.** 40

17. $\frac{\sqrt{10}}{2} \approx 1.58$ seconds   **18.** 120   **19.** $(5, 13)$

Cumulative Test: Chapters 4–6 *(page 421)*

1. $\frac{9x^{18}}{4y^{12}}$ **2.** 1.6×10^7 **3.** $2x^3 - 2x^2 - x - \frac{4}{2x - 1}$

4. $\frac{x(x + 2)(x + 4)}{9(x - 4)}$ **5.** $\frac{3x + 5}{x(x + 3)}$ **6.** $x + y$

7. $-4 + 3\sqrt{2}\,i$ **8.** $-7 - 24i$ **9.** $t^{1/2}$

10. $35\sqrt{5x}$ **11.** $2x - 6\sqrt{2x} + 9$ **12.** $\sqrt{10} + 2$

13. $\frac{2}{17} - \frac{9}{17}i$ **14.** $2, 5$ **15.** $2, 9$ **16.** 16

17. 4   **18.** $5 \pm 5\sqrt{2}\,i$ **19.** $\frac{-3 \pm \sqrt{3}}{3}$

20. $r_2 = \frac{\sqrt{15}\,r_1}{5}$ **21.** $16(1 + \sqrt{2}) \approx 38.6$ inches

22.   **23.** $x^2 - 4x - 12 = 0$

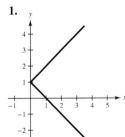

$(-1.12, 0), (7.12, 0)$

Chapter 7

Section 7.1 *(page 430)*

Integrated Review *(page 430)*

1.   **2.**

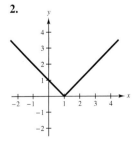

For some x there corre-   For each x there corresponds more than one   sponds exactly one
value of y.   value of y.

3. $(-\infty, \infty)$ **4.** $(-\infty, 0) \cup (0, \infty)$

5.

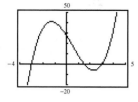

Yes, the graphs are the same.

6. Proof　　**7.** Proof　　**8.** Proof　　**9.** $h + 4$

10. $-\dfrac{3}{7(h + 7)}$　　**11.** $C = 12{,}000 + 5.75x$

12. $P = 5w$

---

1. $I = kV$　　**3.** $V = kt$　　**5.** $u = kv^2$　　**7.** $p = \dfrac{k}{d}$

9. $P = \dfrac{k}{\sqrt{1 + r}}$　　**11.** $A = klw$　　**13.** $P = \dfrac{k}{V}$

15. A varies jointly as the base and height.

17. A varies jointly as the length and the width.

19. V varies jointly as the square of the radius and the height.

21. r varies directly as the distance and inversely as the time.

23. $s = 5t$　　**25.** $F = \frac{5}{16}x^2$　　**27.** $H = \frac{5}{2}u$

29. $n = \dfrac{48}{m}$　　**31.** $g = \dfrac{4}{\sqrt{z}}$　　**33.** $F = \dfrac{25}{6}xy$

35. $d = \dfrac{120x^2}{r}$　　**37.** (a) \$4921.25　(b) Price per unit

39. (a) 2 inches　(b) 15 pounds　　**41.** 18 pounds

43. 32 feet per second per second　　**45.** $208\frac{1}{3}$ feet　　**47.** 4

49. No, k is different for each pizza. The 15-inch pizza is the best buy.

51. 667 units　　**53.** 324 pounds　　**55.** $\frac{1}{4}$

57. $p = \dfrac{114}{t}$, 17.5%　　**59.** \$270; the amount invested

61.

x	2	4	6	8	10
$y = kx^2$	4	16	36	64	100

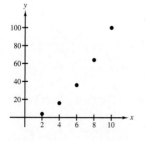

63.

x	2	4	6	8	10
$y = kx^2$	2	8	18	32	50

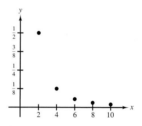

65.

x	2	4	6	8	10
$y = \dfrac{k}{x^2}$	$\frac{1}{2}$	$\frac{1}{8}$	$\frac{1}{18}$	$\frac{1}{32}$	$\frac{1}{50}$

67.

x	2	4	6	8	10
$y = \dfrac{k}{x^2}$	$\frac{5}{2}$	$\frac{5}{8}$	$\frac{5}{18}$	$\frac{5}{32}$	$\frac{1}{10}$

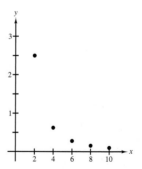

69. $y = \dfrac{k}{x}$ with $k = 4$　　**71.** $y = kx$ with $k = -\dfrac{3}{10}$

73. Increase. Because $y = kx$ and $k > 0$, the variables increase or decrease together.

75. y will quadruple. If $y = kx^2$ and x is replaced with $2x$, you have $y = k(2x)^2 = 4kx^2$.

Section 7.2 *(page 440)*

Integrated Review *(page 440)*

1. Index: 4; radicand: $6x$ **2.** $\sqrt[n]{a}$ **3.** $x < \frac{3}{2}$

4. $x < 5$ **5.** $x < \frac{12}{5}$ **6.** $x \le -11$

7. $1 < x < 5$ **8.** $x < 2$ or $x > 8$

9. Vertical shift 2 units downward **10.** Horizontal shift 2 units to the right

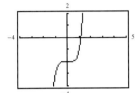
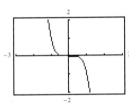

11. Reflection in the x-axis **12.** Reflection in the y-axis

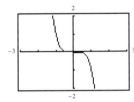

1. (a) Solution
(b) Not a solution
(c) Solution
(d) Solution

3. (a) Not a solution
(b) Not a solution
(c) Solution
(d) Solution

5. (a) Solution
(b) Not a solution
(c) Not a solution
(d) Solution

7. (a) Not a solution
(b) Solution
(c) Not a solution
(d) Solution

9. b **11.** d **13.** f

15.

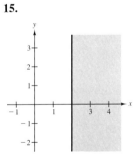

17.

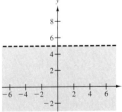

19.

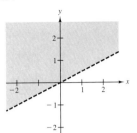

21.

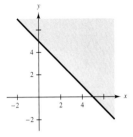

23.

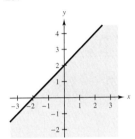

25.

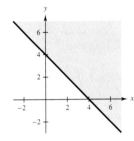

27.

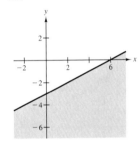

29.

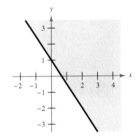

31.

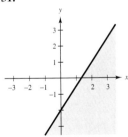

33.

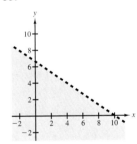

35.

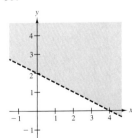

37.

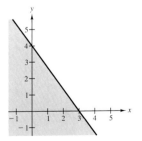

39.

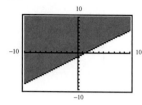

41.

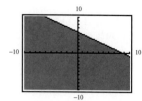

43.

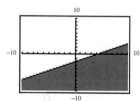

45.

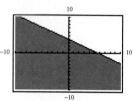

47. $3x + 4y > 17$ **49.** $y < 2$ **51.** $x - 2y < 0$

53. $2x + 2y \le 500$ or

$y \le -x + 250$

(*Note:* x and y cannot be negative.)

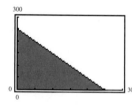

55. $10x + 15y \le 1000$ or

$y \le -\frac{2}{3}x + \frac{200}{3}$

(*Note:* x and y cannot be negative.)

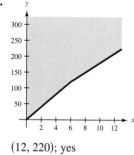

57.

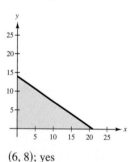

$(12, 220)$; yes

59. $x + 1.5y \le 21$

(*Note:* x and y cannot be negative.)

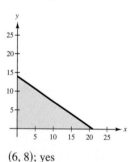

$(6, 8)$; yes

61. $9x + 6y \ge 150$

(*Note:* x and y cannot be negative.)

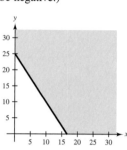

$(10, 10), (8, 14), (20, 0)$

63. (a) and (b)

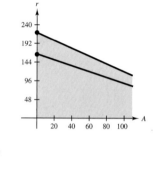

65. The inequality is true when x_1 and y_1 are substituted for x and y, respectively.

67. The solution of $x - y > 1$ does not include the points on the line $x - y = 1$. The solution of $x - y \ge 1$ does include the points on the line $x - y = 1$.

69. On the real number line, the solution of $x \le 3$ is an unbounded interval.

On a rectangular coordinate system, the solution is a half-plane.

Section 7.3 *(page 448)*

Integrated Review *(page 448)*

1. $(x + b)^2 = x^2 + 2bx + b^2$ **2.** $x^2 + 5x + \frac{25}{4}$

3. $-11x$ **4.** $-41v$ **5.** $6x^2 + 9$ **6.** -4

7. $2|x|y\sqrt{6y}$ **8.** $3\sqrt[3]{5}$ **9.** $\dfrac{2\sqrt{3}b^3}{a^2}$ **10.** 2

11. $\sqrt{5} \approx 2.24$ seconds **12.** $\frac{5}{4}\sqrt{6} \approx 3.06$ seconds

1. e **3.** b **5.** d

7. $y = (x - 0)^2 + 2, \ (0, 2)$ **9.** $y = (x - 2)^2 + 3, \ (2, 3)$

11. $y = (x + 3)^2 - 4, \ (-3, -4)$

13. $y = -(x - 3)^2 - 1, \ (3, -1)$

15. $y = -(x - 1)^2 - 6, \ (1, -6)$

17. $y = 2\left(x + \frac{3}{2}\right)^2 - \frac{5}{2}, \left(-\frac{3}{2}, -\frac{5}{2}\right)$ **19.** $(4, -1)$

21. $(-1, 2)$ **23.** $\left(-\frac{1}{2}, 3\right)$ **25.** Upward, $(0, 2)$

27. Downward, $(10, 4)$ **29.** Upward, $(0, -6)$

31. Downward, $(3, 0)$ **33.** $(\pm 5, 0), (0, 25)$

35. $(0, 0), (9, 0)$ **37.** $\left(\frac{3}{2}, 0\right), (0, 9)$ **39.** $(0, 3)$

41.

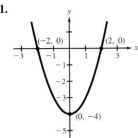

43.

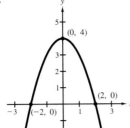

61.

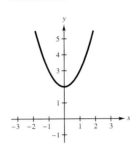

63.

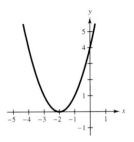

65. Vertical shift

67. Horizontal shift

45.

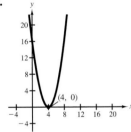

47.

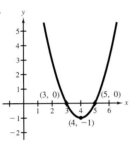

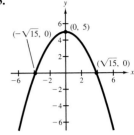

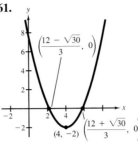

69. Horizontal and vertical shifts

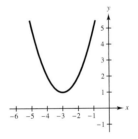

49.

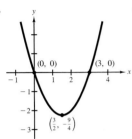

51.

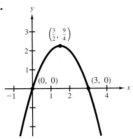

53.

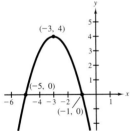

55.

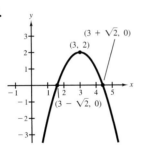

71. Horizontal and vertical shifts

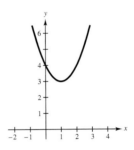

57.

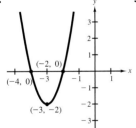

59.

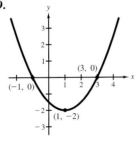

73.

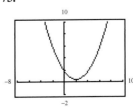

Vertex: $(2, 0.5)$

75.

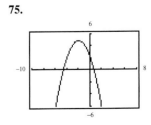

Vertex: $(-1.9, 4.9)$

77. $y = -x^2 + 4$ **79.** $y = x^2 + 4x + 2$

81. $y = -\frac{1}{2}x^2 + 2x + 4$ **83.** $y = x^2 - 4x + 5$

85. $y = x^2 - 4x$ **87.** $y = \frac{1}{2}x^2 - 3x + \frac{13}{2}$

89. $y = -4x^2 - 8x + 1$

91. Horizontal shift 3 units to the right

93. Horizontal shift 2 units to the right and vertical shift 3 units downward

95. (a) 4 feet

(b) 16 feet

(c) $12 + 8\sqrt{3} \approx 25.9$ feet

97. 14 feet

99. (a)

(b) 1993, 110,800

101. (a) Answers will vary.

(b) $(50, 3375)$, 150 radios

(c) Recommend for orders of 150 radios or less

103.

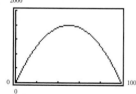

105. $y = \frac{1}{2500}x^2$ **107.** Parabola

109. To find any x-intercepts, set $y = 0$ and solve the resulting equation for x. To find any y-intercepts, set $x = 0$ and solve the resulting equation for y.

111. If the discriminant is positive, the parabola has two x-intercepts; if it is zero, the parabola has one x-intercept; and if it is negative, the parabola has no x-intercepts.

113. Find the y-coordinate of the vertex of the graph of the function.

Mid-Chapter Quiz   *(page 452)*

1. $A = kr^2$ **2.** $z = \dfrac{kx}{y^2}$

3. Distance varies jointly proportional to rate and time.

4. The volume of a cube varies directly as the cube of the length of the sides.

5. $k = \dfrac{2}{3}$, $z = \dfrac{2x^2}{3y}$

6. (a) Solution

(b) Solution

(c) Not a solution

(d) Not a solution

7. $x + 2y \le 11$ **8.** $x - 3y > -5$

9. 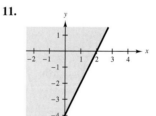   **10.**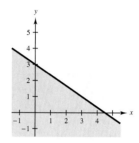

11.

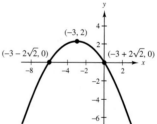

12. $y = (x - 3)^2 - 1$ **13.** $y = -\frac{1}{4}(x - 5)^2 + 4$

14.

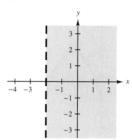

15.

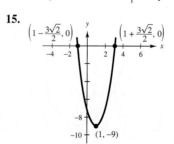

16. 30 minutes after the leak began   **17.** $9x + 14y \le 200$

18. 55 feet

Section 7.4 *(page 461)*

Integrated Review *(page 461)*

1. Additive Inverse Property
2. Distributive Property
3. Associative Property of Multiplication
4. Commutative Property of Addition
5. x^{20} 6. $\dfrac{x^2}{16}$ 7. $\dfrac{3}{2y^5}$ 8. $\dfrac{4y^2}{9x^4}$
9. $\dfrac{x^3y}{6}$ 10. 1 11. 12 12. 15

1. c 3. e 5. a 7. $x^2 + y^2 = 25$
9. $x^2 + y^2 = \dfrac{4}{9}$ 11. $x^2 + y^2 = 64$
13. $x^2 + y^2 = 29$ 15. $(x - 4)^2 + (y - 3)^2 = 100$
17. $(x - 5)^2 + (y + 3)^2 = 81$
19. $(x + 2)^2 + (y - 1)^2 = 4$
21. $(x - 3)^2 + (y - 2)^2 = 17$
23. Center: $(0, 0)$
$r = 4$

25. Center: $(0, 0)$
$r = 6$

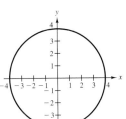

27. Center: $(0, 0)$
$r = \dfrac{1}{2}$

29. Center: $(2, 3)$
$r = 2$

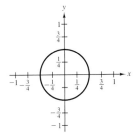
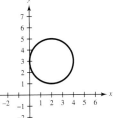

31. Center: $\left(-\dfrac{5}{2}, -3\right)$
$r = 3$

33. Center: $(2, 1)$
$r = 2$

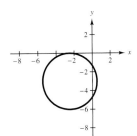
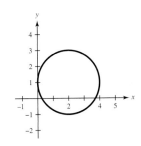

35. Center: $(-1, -3)$
$r = 2$

37.

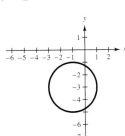
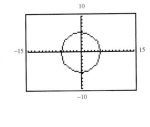

39.

41. $\dfrac{x^2}{16} + \dfrac{y^2}{9} = 1$

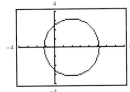

43. $\dfrac{x^2}{4} + \dfrac{y^2}{1} = 1$ 45. $\dfrac{x^2}{9} + \dfrac{y^2}{16} = 1$ 47. $\dfrac{x^2}{1} + \dfrac{y^2}{4} = 1$
49. $\dfrac{x^2}{9} + \dfrac{y^2}{25} = 1$ 51. $\dfrac{x^2}{100} + \dfrac{y^2}{36} = 1$
53. Vertices: $(\pm 4, 0)$
Co-vertices: $(0, \pm 2)$

55. Vertices: $(0, \pm 4)$
Co-vertices: $(\pm 2, 0)$

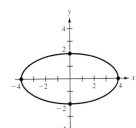
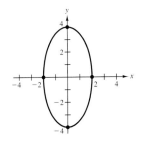

57. Vertices: $\left(\pm\frac{5}{3}, 0\right)$
Co-vertices: $\left(0, \pm\frac{4}{3}\right)$

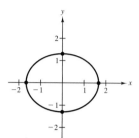

59. Vertices: $(0, \pm 2)$
Co-vertices: $(\pm 1, 0)$

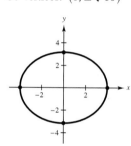

73. Vertices: $(0, \pm 2)$
Asymptotes: $y = \pm\frac{2}{3}x$

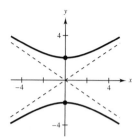

75. Vertices: $(\pm 1, 0)$
Asymptotes: $y = \pm\frac{3}{2}x$

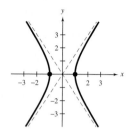

61. Vertices: $(\pm 4, 0)$
Co-vertices: $\left(0, \pm\sqrt{10}\right)$

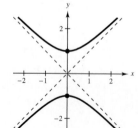

63. Vertices: $(\pm 2, 0)$

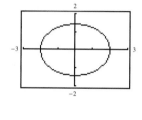

77. Vertices: $(\pm 4, 0)$
Asymptotes: $y = \pm\frac{1}{2}x$

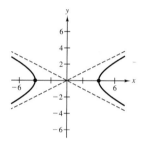

65. Vertices: $\left(0, \pm 2\sqrt{3}\right)$

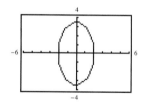

67. Vertices: $(\pm 3, 0)$
Asymptotes: $y = \pm x$

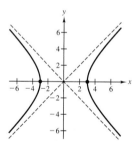

79. $\dfrac{x^2}{16} - \dfrac{y^2}{64} = 1$ **81.** $\dfrac{y^2}{16} - \dfrac{x^2}{64} = 1$

83. $\dfrac{x^2}{81} - \dfrac{y^2}{36} = 1$ **85.** $\dfrac{y^2}{1} - \dfrac{x^2}{1/4} = 1$

87.

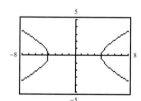

89.

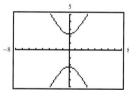

91. Parabola **93.** Ellipse **95.** Hyperbola

97. Circle **99.** Line **101.** $x^2 + y^2 = 4500^2$

103. (a) Proof

(b)

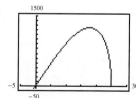

Maximum when $x \approx 17.68$

69. Vertices: $(0, \pm 1)$
Asymptotes: $y = \pm x$

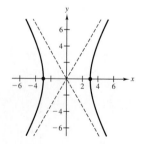

71. Vertices: $(\pm 3, 0)$
Asymptotes: $y = \pm\frac{5}{3}x$

105. $\dfrac{x^2}{2500} + \dfrac{y^2}{1600} = 1$; $\sqrt{304} \approx 17.4$ feet

107. $\dfrac{x^2}{144} + \dfrac{y^2}{64} = 1$

109. Circles, parabolas, ellipses, and hyperbolas

111. An ellipse is the set of all points (x, y) such that the sum of the distances between (x, y) and two distinct fixed points is a constant.

$$\dfrac{x^2}{a^2} + \dfrac{y^2}{b^2} = 1 \quad \text{or} \quad \dfrac{x^2}{b^2} + \dfrac{y^2}{a^2} = 1$$

113. It is a circle if the coefficients of the second-degree terms are equal.

115. The asymptotes are the extended diagonals of the central rectangle.

117. Top half

Section 7.5 *(page 471)*

Integrated Review *(page 471)*

1. 7. Coefficient of the term of highest degree

2. 5. $(x^4 + 3)(x - 4) = x^5 - 4x^4 + 3x - 12$

3.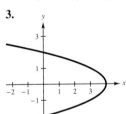

4.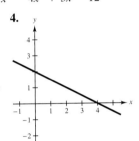

5. $-10x^8$ **6.** $15x - 6x^2$ **7.** $4x^2 - 60x + 225$

8. $21x^2 - 16x - 20$ **9.** $x^2 - y^2 + 2x + 1$

10. $x^3 + 27$ **11.** Base: 20 meters
Height: 8 meters

12. 15 inches × 15 inches

1. (a)

x	0	0.5	0.9	0.99	0.999
y	-4	-8	-40	-400	-4000

x	2	1.5	1.1	1.01	1.001
y	4	8	40	400	4000

x	2	5	10	100	1000
y	4	1	0.4444	0.0404	0.0040

(b)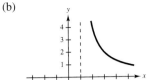

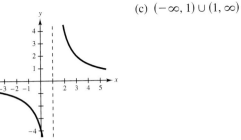

(c) $(-\infty, 1) \cup (1, \infty)$

3. (a)

x	2	2.5	2.9	2.99	2.999
y	1	0	-8	-98	-998

x	4	3.5	3.1	3.01	3.001
y	3	4	12	102	1002

x	4	5	10	100	1000
y	3	2.5	2.143	2.010	2.001

(b)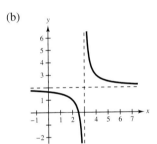

(c) $(-\infty, 3) \cup (3, \infty)$

5. (a)

x	2	2.5	2.9	2.99	2.999
y	-1.2	-2.727	-14.75	-149.7	-1500

x	4	3.5	3.1	3.01	3.001
y	1.714	3.231	15.246	150.25	1500.2

x	4	5	10	100	1000
y	1.714	0.938	0.330	0.030	0.003

(b)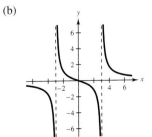

(c) $(-\infty, -3) \cup (-3, 3) \cup (3, \infty)$

7. Domain: $(-\infty, 0) \cup (0, \infty)$
Horizontal asymptote: $y = 0$
Vertical asymptote: $x = 0$

9. Domain: $(-\infty, -8) \cup (-8, \infty)$
Horizontal asymptote: $y = 1$
Vertical asymptote: $x = -8$

11. Domain: $(-\infty, 3) \cup (3, \infty)$
Horizontal asymptote: $y = \frac{2}{3}$
Vertical asymptote: $t = 3$

13. Domain: $\left(-\infty, \frac{1}{3}\right) \cup \left(\frac{1}{3}, \infty\right)$
Horizontal asymptote: $y = \frac{5}{3}$
Vertical asymptote: $x = \frac{1}{3}$

15. Domain: $(-\infty, 0) \cup (0, 1) \cup (1, \infty)$
Horizontal asymptote: $y = 0$
Vertical asymptotes: $t = 0, t = 1$

17. Domain: $(-\infty, \infty)$
Horizontal asymptote: $y = 2$
Vertical asymptote: None

19. Domain: $(-\infty, -1) \cup (-1, 1) \cup (1, \infty)$
Horizontal asymptote: $y = 1$
Vertical asymptotes: $x = -1, x = 1$

21. Domain: $(-\infty, 0) \cup (0, \infty)$
Horizontal asymptote: $y = 1$
Vertical asymptote: $z = 0$

23. Domain: $(-\infty, 0) \cup (0, \infty)$
Horizontal asymptote: None
Vertical asymptote: $x = 0$

25. d   **27.** b   **29.** d   **31.** a

33.   **35.**

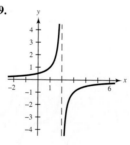

37.   **39.**

41.

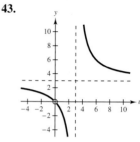

43.

45.

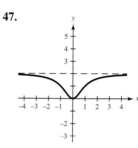

47.

49.

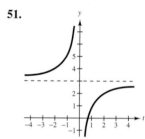

51.

53.

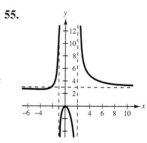

55.

57.

59.

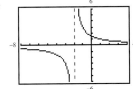

Domain: $(-\infty, -2) \cup (-2, \infty)$
Horizontal asymptote: $y = 0$
Vertical asymptote: $x = -2$

61.

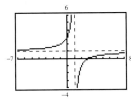

Domain: $(-\infty, 1) \cup (1, \infty)$
Horizontal asymptote: $y = 1$
Vertical asymptote: $x = 1$

63.

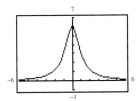

Domain: $(-\infty, \infty)$
Horizontal asymptote: $y = 0$
Vertical asymptote: None

65.

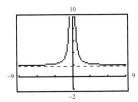

Domain: $(-\infty, 0) \cup (0, \infty)$
Horizontal asymptote: $y = 2$
Vertical asymptote: $x = 0$

67.

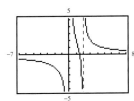

Domain: $(-\infty, 0) \cup (0, 2) \cup (2, \infty)$
Horizontal asymptote: $y = 0$
Vertical asymptotes: $x = 0, \quad x = 2$

69.

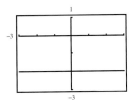

The fraction is not reduced to lowest terms.

71. (a) $\overline{C} = \dfrac{2500 + 0.50x}{x}$ (b) \$3, \$0.75

(c)

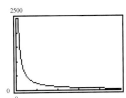

$\overline{C} = \$0.50$. The average cost for many units will be \$0.50.

73. (a) $C = 0$. The chemical is eliminated from the body.

(b)

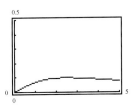

$t \approx 2.5$ hours

75. (a) Answers will vary.   (b) Proof   (c) $(0, \infty)$

(d)

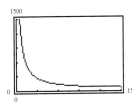

20 units $\times$ 20 units

77. $y = \dfrac{2(x + 1)}{x - 3}$ **79.** $y = \dfrac{x - 6}{(x - 4)(x + 2)}$

81. (c) Domain: All real numbers except $x \approx 7.69$
x-intercept: $x \approx 10.10$
Horizontal asymptote: $y \approx 36.85$
Vertical asymptote: $x \approx 7.69$

(d)

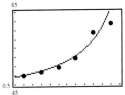

The model appears to be accurate for the restricted domain.

(e) The models are not accurate for the years before 1991 and after 1996. Use the quadratic model to estimate the value of the shipment in 1998, because the rational model has an asymptote at $x \approx 7.69$. Answers will vary.

83. An asymptote of a graph is a line to which the graph becomes arbitrarily close as $|x|$ or $|y|$ increases without bound.

85. No. $f(x) = \dfrac{1}{x^2 + 1}$ has no vertical asymptote.
Vertical asymptotes: $x = 0, x = 2$

Review Exercises *(page 476)*

1. $P = kt^3$ **3.** $z = \dfrac{k}{s^2}$ **5.** $k = 6, \; y = 6\sqrt[3]{x}$

7. $k = \frac{1}{18}, \; T = \frac{1}{18}rs^2$ **9.**

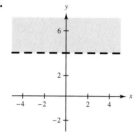

11.

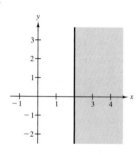

13.

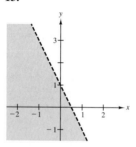

15.

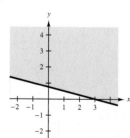

17.

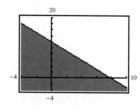

19.

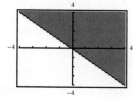

21. $(4, -13)$ **23.** $\left(\frac{1}{4}, \frac{23}{8}\right)$

25.

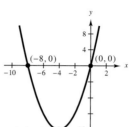

27.

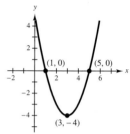

29. Vertical shift

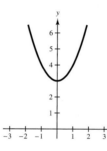

31. Horizontal shift and vertical shift

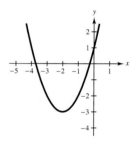

33. $y = -2(x - 3)^2 + 5$ **35.** $y = 2(x - 2)^2 - 5$

37. $y = \frac{1}{16}(x - 5)^2$ **39.** c   **41.** a   **43.** b

45. Parabola

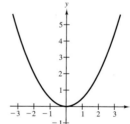

47. Circle

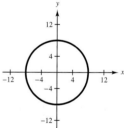

49. Parabola

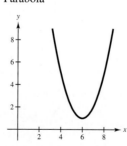

51. Ellipse

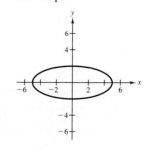

53. Circle

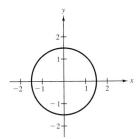

55. $y = \frac{1}{16}x^2 - \frac{5}{8}x + \frac{25}{16}$ **57.** $\frac{x^2}{4} + \frac{y^2}{25} = 1$

59. $x^2 + y^2 = 400$ **61.** $\frac{x^2}{9} - \frac{y^2}{9/4} = 1$ **63.** b   **65.** a

67.

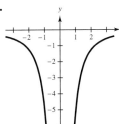

69.

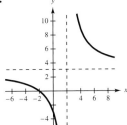

71.

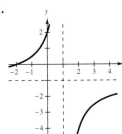

73.

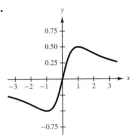

75.

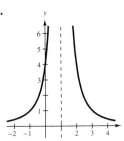

77.

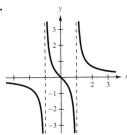

79.

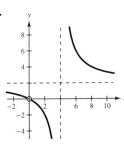

81.

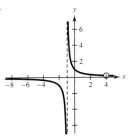

83. $f(x) = \frac{3x}{x - 4}$ **85.** 150 pounds   **87.** 945 units

89. $4x + 5y \geq 100$

(*Note:* x and y cannot be negative.)

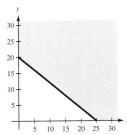

$(10, 12), (12, 11), (8, 15)$

91. (a)

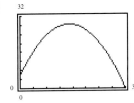

(b) 6 feet
(c) 28.5 feet
(d) 31.9 feet

93. $x^2 + y^2 = 5000^2$

95. (a) $N(5) = 304$ thousand fish; $N(10) \approx 4533$ thousand fish; $N(25) \approx 702.2$ thousand fish

(b) The population is limited by the horizontal asymptote $N = 1200$ thousand fish.

Chapter Test   (*page 479*)

1. $S = \frac{kx^2}{y}$ **2.** $v = \frac{1}{4}\sqrt{u}$

3.

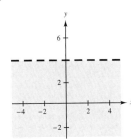

4.

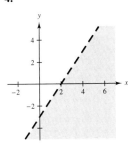

5. $10x + 7y \leq 35$ **6.**

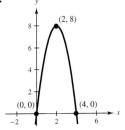

7. Circle

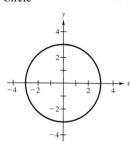

8. Ellipse

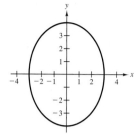

Chapter 8

Section 8.1 *(page 490)*

9. Hyperbola

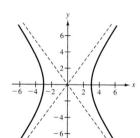

10. Parabola

Integrated Review *(page 490)*

1.

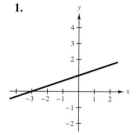

2.

11. $x^2 + y^2 = 25$ **12.** $(y - 1)^2 = 8(x + 2)$

13. $\dfrac{x^2}{9} + \dfrac{y^2}{100} = 1$ **14.** $\dfrac{x^2}{9} - \dfrac{y^2}{9/4} = 1$

3. $\frac{3}{2}$ **4.** $m = -3$ has the greater absolute value.

5. $\frac{5}{11}$ **6.** $\frac{14}{11}$ **7.** 50   **8.** 64

9. $y = \frac{1}{4}(5 - 3x)$ **10.** $y = \frac{2}{3}(3 - x)$

15.

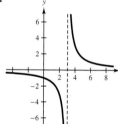

16.

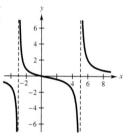

11.

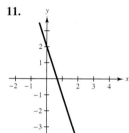

12.

Asymptotes: $x = 3, y = 0$ Asymptotes: $x = -3, x = 5,$
$y = 0$

17. 240 cubic meters

18. $2x + 3y \le 2400$

(*Note:* x and y cannot be negative.)

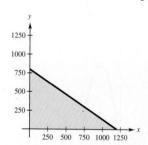

19. 120

13.

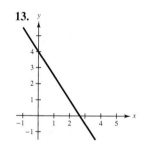

14.

1. (a) Solution   **3.** (a) Not a solution
    (b) Not a solution   (b) Solution

5. (a) Not a solution   **7.** (a) Solution
    (b) Solution   (b) Not a solution

9. Inconsistent   **11.** Inconsistent

13. Consistent   **15.** Consistent

17.

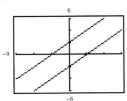

19.

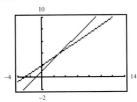

Inconsistent   One solution

21. No solution   **23.** $\left(1, \frac{1}{3}\right)$

25. Infinite number of solutions   **27.** No solution

29. $(1, 2)$ **31.** $(2, 0)$ **33.** $(3, 1)$

35. Infinitely many solutions   **37.** $(10, 0)$

39. Infinitely many solutions   **41.** $(3, -1)$

43. $(0, 0), (2, 4)$ **45.** $(0, 0), (1, 1)$

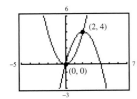
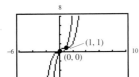

47. $(2, 1)$ **49.** $(4, 3)$ **51.** $(1, 2)$ **53.** $(4, -2)$

55. $(7, 2)$ **57.** $(10, 4)$ **59.** $(-2, -1)$

61. $\left(\frac{3}{2}, \frac{3}{2}\right)$ **63.** $\left(\frac{20}{3}, \frac{40}{3}\right)$ **65.** $(-3, 18), (2, 8)$

67. $(2, 12), \left(-\frac{5}{2}, \frac{75}{4}\right)$ **69.** $(2, 5), (-3, 0)$

71. $(8, -6), (-6, 8)$ **73.** $(1, -1), (-4, 14)$

75. $(0, 5), (-4, -3)$ **77.**

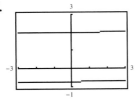

$\left(2992, \frac{798}{25}\right)$

79. $2x - 3y = -7$ **81.** $7x + y = -9$
     $x + y = 9$ $-x + 3y = -5$

83. 10,000 units   **85.** 6250 units   **87.** $15,000 at 8%
                                                      $5000 at 9.5%

89. $13,000 at 8%   **91.** 18, 25   **93.** 40, 120
     $12,000 at 8.5%

95. 70 inches $\times$ 90 inches   **97.** 27 meters $\times$ 18 meters

99. 1987

101. (a) Solve one of the equations for one variable in terms of the other.

(b) Substitute the expression found in Step (a) into the other equation to obtain an equation in one variable.

(c) Solve the equation obtained in Step (b).

(d) Back-substitute the solution from Step (c) into the expression obtained in Step (a) to find the value of the other variable.

(e) Check the solution in each of the original equations of the system.

103. After finding a value for one of the variables, substitute this value back into one of the original equations. This is called back-substitution.

105. The graphical method usually yields approximate solutions.

Section 8.2   *(page 499)*

Integrated Review   *(page 499)*

1. Distributive Property

2. Addition Property of Equality   **3.** $-2 < x < 2$

4. $4 \le x < 16$ **5.** $x < -2$ or $x > 2$

6. $-2 < x < 3$ **7.** $x < 3$ **8.** $x \ge \frac{5}{4}$

9. $m < 19,555.56$ **10.** $1500 + 0.04x > 2500$

$x > \$25,000$

---

1. $(2, 0)$ **3.** $(5, 3)$

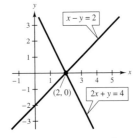
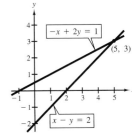

5. $(2, -3)$ **7.** Inconsistent

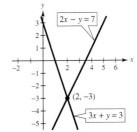
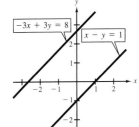

9. Infinitely many solutions **11.** $\left(\frac{1}{2}, \frac{3}{2}\right)$

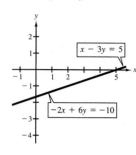

13. $(3, 2)$ **15.** $(-2, 5)$ **17.** $(2, 1)$ **19.** $(3, -4)$

21. $(-1, -1)$ **23.** $(5, -1)$ **25.** $(7, -2)$

27. Inconsistent **29.** $\left(\frac{3}{2}, 1\right)$ **31.** $(6, 3)$

33. $(-2, -1)$ **35.** Infinitely many solutions

37. Inconsistent **39.** $(12.5, 4.948)$ **41.** $(-3, 7)$

43. $(2, 7)$ **45.** $(15, 10)$ **47.** $(4, 3)$ **49.** Consistent

51. Consistent **53.** Inconsistent **55.** $k = 4$

57. $x + 2y = 0$
$4x + 2y = 9$

59. 122 weeks

61. 8% bond: $15,000 **63.** 4 hours
9.5% bond: $5000

65. Speed of plane in still air: 550 miles per hour
Wind speed: 50 miles per hour

67. Adult: 375 **69.** Regular unleaded: $1.11
Children: 125 Premium unleaded: $1.22

71. 40% solution: 12 liters **73.** $5.65 variety: 6.1 pounds
65% solution: 8 liters $8.95 variety: 3.9 pounds

75. (a) $y = -1.5x + 3\frac{5}{6}$

(b)

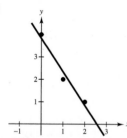

77. Depth: 10 feet; Lengths of sections: 122 feet, 125 feet

79. When adding a nonzero multiple of one equation to another equation to eliminate a variable, you get $0 = 0$ for the second equation.

81. (a) Obtain coefficients for x or y that differ only in sign by multiplying all terms of one or both equations by suitably chosen constants.

(b) Add the equations to eliminate one variable, and solve the resulting equation.

(c) Back-substitute the value obtained in Step (b) into either of the original equations and solve for the other variable.

(d) Check your solution in both of the original equations.

83. When it is easy to solve for one of the variables in one of the equations of the system, it might be better to use substitution.

Section 8.3 (page 511)

Integrated Review (page 511)

1. One solution

2. Multiply both sides of the equation by the lowest common denominator, 24.

3. $4x^8$ **4.** $8x^{10}y^{15}$ **5.** $\dfrac{4}{x^{11}}$ **6.** $\dfrac{3}{t^4}$

7. $-1, 5$ **8.** $\frac{1}{2}$

9. $d = 15t$ **10.** $V = s^3$ **11.** $A = \dfrac{C^2}{4\pi}$

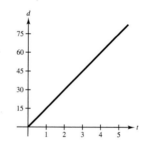

1. (a) Not a solution **3.** $(22, -1, -5)$

(b) Solution

(c) Solution

(d) Not a solution

5. $(14, 3, -1)$

7. No. When the first equation was multiplied by -2 and added to the second equation, the constant term should have been -11.

9. $x - 2y = 8$ **11.** $x - 2y + 3z = 5$
$y = 14$ $-y + 8z = 9$
Eliminated the x-term $2x - 3z = 0$
from the second equation Eliminated the x-term in
Equation 2

13. $(1, 2, 3)$ **15.** $(1, 2, 3)$ **17.** $(2, -3, -2)$

19. No solution **21.** $(-4, 8, 5)$ **23.** No solution

25. $\left(\frac{3}{10}, \frac{2}{5}, 0\right)$ **27.** $(-4, 2, 3)$ **29.** $(-1, 5, 5)$

31. $\left(-\frac{1}{2}a + \frac{1}{2}, \frac{3}{5}a + \frac{2}{5}, a\right)$ **33.** $\left(-\frac{1}{2}a + \frac{1}{4}, \frac{1}{2}a + \frac{5}{4}, a\right)$

35. $(1, -1, 2)$ **37.** $\left(\frac{6}{13}a + \frac{10}{13}, \frac{5}{13}a + \frac{4}{13}, a\right)$

39. $\left(-\frac{1}{2}, 2, 10\right)$ **41.** $\begin{aligned} x + 2y - z &= -4 \\ y + 2z &= 1 \\ 3x + y + 3z &= 15 \end{aligned}$

43. $s = -16t^2 + 144$ **45.** $s = -16t^2 + 48t$

47. $y = 2x^2 + 3x - 4$ **49.** $y = x^2 - 4x + 3$

51. $y = -x^2 + 2x$ **53.** $y = \frac{1}{2}x^2 - \frac{1}{2}x$; Yes

55. $x^2 + y^2 - 4x = 0$ **57.** $x^2 + y^2 - 6x - 8y = 0$

59. $x^2 + y^2 - 2x - 4y - 20 = 0$

61. 20 gallons of spray X   **63.** Strings: 50
18 gallons of spray Y   Winds: 20
16 gallons of spray Z   Percussion: 8

65. (d) $\begin{aligned} x + y + z &= 200 \\ 8x + 15y + 100z &= 4995 \\ x - 4z &= 0 \end{aligned}$

(e) Students: 140
Nonstudents: 25
Major contributors: 35

(f) No

67. Substitute $y = 3$ into the first equation to obtain $x + 2(3) = 2$ or $x = 2 - 6 = -4$.

69. Answers will vary.

Mid-Chapter Quiz   *(page 515)*

1. $(10, 4)$

2.

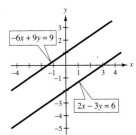

No solution

3.

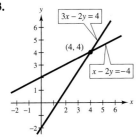

One solution

4.

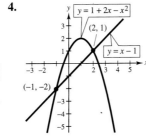

Two solutions

5.

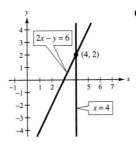

$(4, 2)$

6.

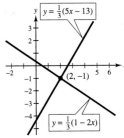

$(2, -1)$

7.

$(8, 0)$

8.

$(5, 12), (-12, 5)$

9. $(5, 2)$ **10.** $(1, 4), (-3, -4)$ **11.** $\left(\frac{90}{13}, \frac{34}{13}\right)$

12. $(5, 10)$ **13.** $(8, 1)$ **14.** $(-2, 4)$ **15.** $\left(\frac{1}{2}, -\frac{1}{2}, 1\right)$

16. $(5, -1, 3)$ **17.** $\begin{aligned} x + y &= -2 \\ 2x - y &= 32 \end{aligned}$

18. $\begin{aligned} x + y - z &= 11 \\ x + 2y - z &= 14 \\ -2x + y + z &= -6 \end{aligned}$ **19.** $\begin{aligned} x + y &= 20 \\ 0.2x + 0.5y &= 6 \end{aligned}$

20% solution: $13\frac{1}{3}$ gallons

50% solution: $6\frac{2}{3}$ gallons

20. $y = x^2 + 3x - 2$

Section 8.4   *(page 524)*

Integrated Review   *(page 524)*

1. Additive Inverse Property

2. Multiplicative Identity Property

3. Commutative Property of Addition

4. Associative Property of Multiplication

5.

6.

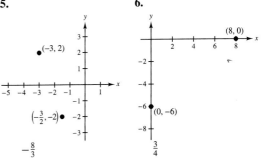

7.

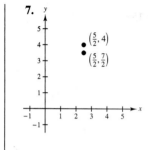

Undefined

8.

$-\dfrac{30}{13}$

9.

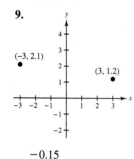

-0.15

10.

0

11. 7650   **12.** $940

---

1. 4×2 **3.** 2×3 **5.** 4×1

7. $\begin{bmatrix} 4 & -5 & \vdots & -2 \\ -1 & 8 & \vdots & 10 \end{bmatrix}$ **9.** $\begin{bmatrix} 1 & 10 & -3 & \vdots & 2 \\ 5 & -3 & 4 & \vdots & 0 \\ 2 & 4 & 0 & \vdots & 6 \end{bmatrix}$

11. $\begin{bmatrix} 5 & 1 & -3 & \vdots & 7 \\ 0 & 2 & 4 & \vdots & 12 \end{bmatrix}$ **13.** $\begin{cases} 4x + 3y = 8 \\ x - 2y = 3 \end{cases}$

15. $\begin{cases} x \quad\;\; + 2z = -10 \\ \quad 3y - z = \quad 5 \\ 4x + 2y \quad\;\; = \quad 3 \end{cases}$

17. $\begin{cases} 5x + 8y + 2z \quad\quad = -1 \\ -2x + 15y + 5z + w = \quad 9 \\ x + 6y - 7z \quad\quad = -3 \end{cases}$

19. $\begin{bmatrix} 1 & 4 & 3 \\ 0 & 2 & -1 \end{bmatrix}$ **21.** $\begin{bmatrix} 1 & -2 & \frac{2}{3} \\ 2 & 8 & 15 \end{bmatrix}$

23. $\begin{bmatrix} 1 & 1 & 4 & -1 \\ 0 & 5 & -2 & 6 \\ 0 & 3 & 20 & 4 \end{bmatrix}$ **25.** $\begin{bmatrix} 1 & 2 & 3 \\ 0 & 1 & 2 \end{bmatrix}$

$\begin{bmatrix} 1 & 1 & 4 & -1 \\ 0 & 1 & -\frac{2}{5} & \frac{6}{5} \\ 0 & 3 & 20 & 4 \end{bmatrix}$

27. $\begin{bmatrix} 1 & 0 & -\frac{7}{5} \\ 0 & 1 & \frac{11}{10} \end{bmatrix}$ **29.** $\begin{bmatrix} 1 & 1 & 0 & 5 \\ 0 & 1 & 2 & 0 \\ 0 & 0 & 1 & -1 \end{bmatrix}$

31. $\begin{bmatrix} 1 & -1 & -1 & 1 \\ 0 & 1 & 6 & 3 \\ 0 & 0 & 1 & \frac{4}{5} \end{bmatrix}$ **33.** $\begin{bmatrix} 1 & 1 & -1 & 3 \\ 0 & 1 & -4 & 1 \\ 0 & 0 & 0 & 0 \end{bmatrix}$

35. $\begin{cases} x - 2y = \quad 4 \\ \quad\quad y = -3 \end{cases}$ **37.** $\begin{cases} x + 5y = \quad 3 \\ \quad\quad y = -2 \end{cases}$

$(-2, -3)$ $(13, -2)$

39. $\begin{cases} x - y + 2z = \quad 4 \\ \quad y - z = \quad 2 \\ \quad\quad z = -2 \end{cases}$ **41.** $\left(\frac{9}{5}, \frac{13}{5}\right)$ **43.** $(1, 1)$

$(8, 0, -2)$

45. No solution   **47.** $(2, -3, 2)$

49. $(2a + 1, 3a + 2, a)$ **51.** $(1, 2, -1)$

53. $(1, -1, 2)$ **55.** $(34, -4, -4)$ **57.** No solution

59. $(-12a - 1, 4a + 1, a)$ **61.** $\left(2, 5, \frac{5}{2}\right)$

63. 8%: $800,000
9%: $500,000
12%: $200,000

65. Certificates of deposit: $250{,}000 - \frac{1}{2}s$
Municipal bonds: $125{,}000 + \frac{1}{2}s$
Blue-chip stocks: $125{,}000 - s$
Growth stocks: s
If $s = \$100{,}000$, then
　Certificates of deposit: $200,000
　Municipal bonds: $175,000
　Blue-chip stocks: $25,000
　Growth stocks: $100,000

67. $3.50: 15 pounds   **69.** 5, 8, 20
$4.50: 10 pounds
$6.00: 25 pounds

71. $y = x^2 + 2x + 4$ **73.** $y = -10.5x^2 + 25.5x - 7$

75. $x^2 + y^2 - 5x - 3y + 6 = 0$

77. (a) $y = -\frac{1}{250}x^2 + \frac{3}{5}x + 6$

(b)

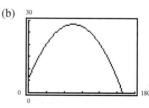

Maximum height: 28.5 ft
The ball struck the ground at approximately $(159.4, 0)$.

79. $\dfrac{2x^2 - 9x}{(x - 2)^3} = \dfrac{2}{x - 2} - \dfrac{1}{(x - 2)^2} - \dfrac{10}{(x - 2)^3}$

81. (a) Interchange two rows.

(b) Multiply a row by a nonzero constant.

(c) Add a multiple of a row to another row.

83. The one matrix can be obtained from the other by using the elementary row operations.

85. There will be a row in the matrix with all zero entries except in the last column.

Section 8.5   *(page 537)*

Integrated Review   *(page 537)*

1. pq **2.** $mq + np$ **3.** mn

4. No. $p = -1$, $q = -1$ is another solution.

5. $-4, 1$ **6.** $-2, 3$ **7.** $\dfrac{5}{2}$ **8.** ± 4

9. $-4, 2 \pm 2\sqrt{3}i$ **10.** $2, \pm \dfrac{2\sqrt{3}}{3}i$

11. $\dfrac{320}{r}$ **12.** $\dfrac{9}{2}x + 7$

---

1. 5   **3.** 27   **5.** 0   **7.** 6   **9.** -24

11. -0.16 **13.** -24 **15.** -2 **17.** -30 **19.** 3

21. 0   **23.** -75 **25.** -58 **27.** 102   **29.** -0.22

31. $x - 5y + 2$ **33.** 248   **35.** -32 **37.** -6.37

39. $(1, 2)$ **41.** $(2, -2)$ **43.** $\left(\dfrac{3}{4}, -\dfrac{1}{2}\right)$

45. Not possible, $D = 0$ **47.** $\left(\dfrac{2}{3}, \dfrac{1}{2}\right)$ **49.** $(-1, 3, 2)$

51. $\left(1, \dfrac{1}{2}, \dfrac{3}{2}\right)$ **53.** $(1, -2, 1)$ **55.** Not possible, $D = 0$

57. $\left(\dfrac{22}{27}, \dfrac{22}{9}\right)$ **59.** $\left(\dfrac{51}{16}, -\dfrac{7}{16}, -\dfrac{13}{16}\right)$ **61.** 1, 6   **63.** 16

65. 7   **67.** $\dfrac{31}{2}$ **69.** $\dfrac{33}{8}$ **71.** 16   **73.** $\dfrac{53}{2}$

75. 250 square miles   **77.** Collinear   **79.** Collinear

81. Not collinear   **83.** $3x - 5y = 0$

85. $7x - 6y - 28 = 0$ **87.** $9x + 10y + 3 = 0$

89. $32x - 30y + 44 = 0$

91. $y = 2x^2 - 6x + 1$ **93.** $y = \dfrac{1}{2}x^2 - 2x$

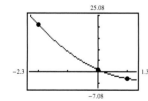   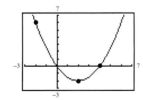

95. $y = -3x^2 + 2x$

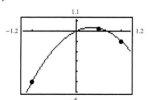

97. (a) $y_1 = 574.95x^2 - 6847.15x + 20{,}446.7$

(b) $y_2 = 728.05x^2 - 8672.95x + 25{,}906.9$

(c)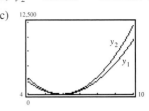

(d) $y_1 - y_2 = -153.1x^2 + 1825.8x - 5460.2$

(e)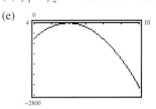

The trade deficit is increasing.

99. (a) $\left(\dfrac{4k - 3}{2k - 1}, \dfrac{4k - 1}{2k - 1}\right)$ (b) $\dfrac{1}{2}$

101. A determinant is a real number associated with a square matrix.

103. The minor of an entry of a square matrix is the determinant of the matrix that remains after deleting the row and column in which the entry occurs.

Review Exercises   *(page 542)*

1. (a) Not a solution   **3.** (a) Solution

(b) Solution   (b) Not a solution

5. $(1, 1)$ **7.** No solution

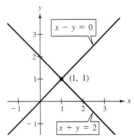   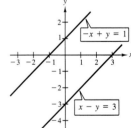

9. $(4, 8)$ **11.** Infinite number of solutions

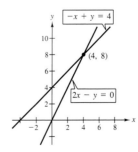   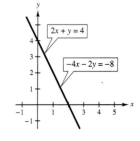

13. $(0, 1)$ **15.** $(3, 4)$

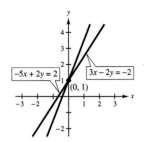

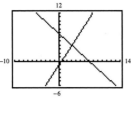

17. $\left(1, -3\right), \left(-\frac{1}{3}, -\frac{35}{9}\right)$ **19.** $(2, -1)$

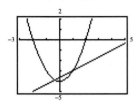

21. No solution   **23.** $(-10, -5)$ **25.** $(-2, 20), (-1, 5)$

27. $(-1, 0), (0, -1)$ **29.** $(0, 0)$ **31.** $\left(\frac{5}{2}, 3\right)$

33. $(-0.5, 0.8)$ **35.** $(2, -3, 3)$ **37.** $(0, 1, -2)$

39. $(1, 0, -4)$ **41.** $(10, -12)$ **43.** $(0.6, 0.5)$

45. $\left(\frac{24}{5}, \frac{22}{5}, -\frac{8}{5}\right)$ **47.** $\left(\frac{1}{2}, -\frac{1}{3}, 1\right)$ **49.** 5   **51.** -51

53. 1   **55.** $(-3, 7)$ **57.** Not possible, $D = 0$

59. $(2, -3, 3)$ **61.** $3x + y = -2$ **63.** 16,667 units
                         $6x + y = 0$

65. 75% solution: 40 gallons   **67.** 96 meters $\times$ 144 meters
     50% solution: 60 gallons

69. \$9.95 tapes: 400   **71.** 250 miles per hour
     \$14.95 tapes: 250   290 miles per hour

73. 16, 20, 32   **75.** $y = 3x^2 + 11x - 20$ **77.** 16

79. 7   **81.** $x - 2y + 4 = 0$ **83.** $2x + 6y - 13 = 0$

Chapter Test   *(page 545)*

1. $\left(1, \frac{1}{2}\right)$ **2.** $(2, 4)$ **3.** $(2, 6), (-1, 9)$ **4.** $(3, 2)$

5. $(-2, 2)$ **6.** $\left(\frac{1}{4}, \frac{1}{3}\right)$ **7.** $(2, 2z - 1, z)$ **8.** $(-1, 3, 3)$

9. $(2, 1, -2)$ **10.** $\left(4, \frac{1}{7}\right)$ **11.** $(5, 4)$ **12.** $(5, 1, -1)$

13. $\left(-\frac{11}{5}, \frac{56}{25}, \frac{32}{25}\right)$ **14.** No solution
                            One solution
                            Infinitely many solutions

15. -62 **16.** $-\frac{24}{5}$

17. $x + 2y = -1$ **18.** $x + y = 200$
     $x + y = 2$ $4x - y = 0$
                       40 miles, 160 miles

19. $y = 2x^2 - 3x + 4$ **20.** \$13,000 at 4.5%   **21.** 12
                             \$9000 at 5%
                             \$3000 at 8%

Chapter 9

Section 9.1   *(page 556)*

Integrated Review   *(page 556)*

1. Test one point in each of the half-planes formed by the graph of $x + y = 5$. If the point satisfies the inequality, shade the entire half-plane to denote that every point in the region satisfies the inequality.

2. The first contains the boundary and the second does not.

3. 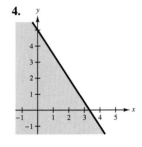   **4.**

5. 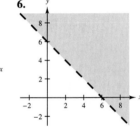   **6.**

7. 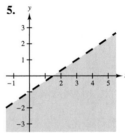   **8.**

9. 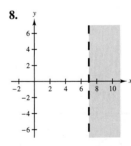   **10.**

11. 18.6 hours, 21.6 hours   **12.** $30\sqrt{5} \approx 67.1$ feet

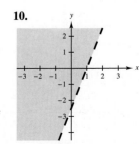

1. 2^{2x-1} **3.** e^2 **5.** $8e^{3x}$ **7.** $-2e^x$ **9.** 11.036

11. 1.396   **13.** 51.193   **15.** 0.906

17. (a) $\frac{1}{9}$ **19.** (a) 0.935   **21.** (a) 500

   (b) 1   (b) 1.225   (b) 250

   (c) 3   (c) 1.163   (c) 56.657

23. (a) 1000   **25.** (a) 486.111   **27.** (a) 73.891

   (b) 1628.895   (b) 47.261   (b) 1.353

   (c) 2653.298   (c) 0.447   (c) 0.183

29. (a) 333.333

   (b) 434.557

   (c) 499.381

31.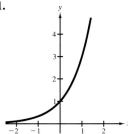

33.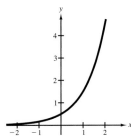

35.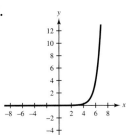

37.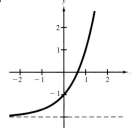

39.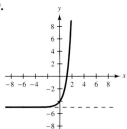

41.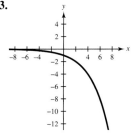

43.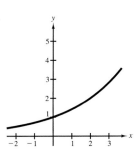

45.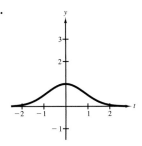

47.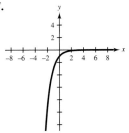

49.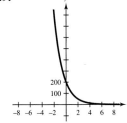

51. b   **53.** e   **55.** f   **57.** h

59.

61.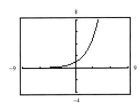

63.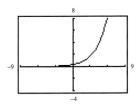

65.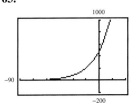

67.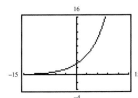

69.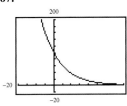

71. Vertical shift

73. Horizontal shift

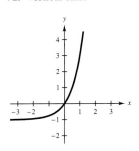

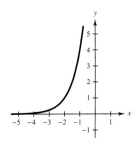

75. Reflection in the *x*-axis

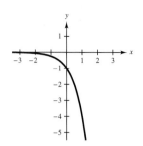

77. (a) Algebraic
   (b) Algebraic
   (c) Exponential
   (d) Algebraic

95. $V(t) = 16,000\left(\frac{3}{4}\right)^t$

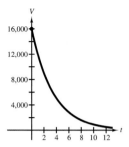

$9000

79. 2.520 grams

81.

n	1	4	12	365	Continuous
A	$466.10	$487.54	$492.68	$495.22	$495.30

83.

n	1	4	12
A	$4734.73	$4870.38	$4902.71

n	365	Continuous
A	$4918.66	$4919.21

85.

n	1	4	12
A	$226,296.28	$259,889.34	$268,503.32

n	365	Continuous
A	$272,841.23	$272,990.75

87.

n	1	4	12
P	$2541.75	$2498.00	$2487.98

n	365	Continuous
P	$2483.09	$2482.93

89.

n	1	4	12
P	$18,429.30	$15,830.43	$15,272.04

n	365	Continuous
P	$15,004.64	$14,995.58

91. (a) $22.04
   (b) $20.13

93. (a) $80,634.95
   (b) $161,269.89

97. (a) $A_1 = 500e^{0.06t}, A_2 = 500e^{0.08t}$
   (b)   (c)

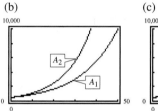

   (d) The difference between the functions increases at an increasing rate.

99. (a)

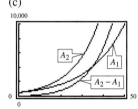

   (b)

t	0	25	50	75
h	2000 ft	1450 ft	950 ft	450 ft

   (c) Ground level: 97.5 seconds

101. (a) and (b)

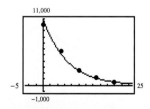

   (c)

h	0	5	10	15	20
P		10,332	5583	2376	1240	517
Approx.		10,958	5176	2445	1155	546

   (d) 3300 kilograms per square meter
   (e) 11.3 kilometers

103. (a)

x	1	10	100	1000	10,000
$\left(1+\dfrac{1}{x}\right)^x$	2	2.5937	2.7048	2.7169	2.7181

(b)

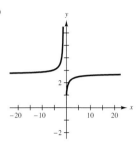

The graph appears to be approaching a horizontal asymptote.

(c) The value approaches e.

105. Polynomials have terms with variable bases and constant exponents. Exponential functions have terms with constant bases and variable exponents.

107. f is an increasing function and g is a decreasing function.

109. False. e is an irrational number.

Section 9.2 *(page 568)*

Integrated Review *(page 568)*

1. y is not a function of x because for some values of x there correspond two values of y. For example, $(4, 2)$ and $(4, -2)$ are solution points.

2. y is a function of x because for each value of x there corresponds exactly one value of y.

3. The domain of f is $-2 \le x \le 2$ and the domain of g is $-2 < x < 2$. g is undefined at $x = \pm 2$.

4. $\{4, 5, 6, 8\}$ **5.** $-2x^2 - 4$ **6.** $30x^3 + 40x^2$

7. $u^2 - 16v^2$ **8.** $9a^2 - 12ab + 4b^2$

9. $t^3 - 6t^2 + 12t - 8$ **10.** $\frac{1}{2}x^2 - \frac{1}{4}x$

11. 100 feet   **12.** 13 minutes

---

1. (a) $2x - 7$ **3.** (a) $2x^2 - 1$

(b) $2x - 10$ (b) $2x^2 + 20x + 44$

(c) 1   (c) 7

(d) 4   (d) 2

5. (a) $|3x - 3|$ **7.** (a) $\sqrt{x + 1}$

(b) $3|x - 3|$ (b) $\sqrt{x - 4} + 5$

(c) 0   (c) 2

(d) 3   (d) 7

9. (a) $\dfrac{x^2}{2 - 3x^2}$ **11.** (a) -1 **13.** (a) -1

(b) $2(x - 3)^2$ (b) -2 (b) 1

(c) -1 (c) -2

(d) 2

15. (a) 10   **17.** (a) 0   **19.** (a) $(-\infty, \infty)$

(b) 1   (b) 10   (b) $(-\infty, \infty)$

(c) 1

21. (a) $[2, \infty)$ **23.** (a) $[-3, \infty)$ **25.** (a) $[1, \infty)$

(b) $[0, \infty)$ (b) $(-\infty, \infty)$ (b) $(-\infty, -5)$

27. No   **29.** Yes   **31.** No

33.   **35.**

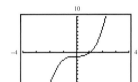   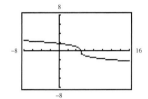

Yes   Yes

37.   **39.**

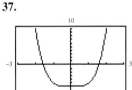   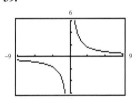

No   Yes

41.

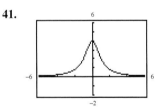

No

43. $f(g(x)) = f\left(\frac{1}{10}x\right) = 10\left(\frac{1}{10}x\right) = x$

$g(f(x)) = g(10x) = \frac{1}{10}(10x) = x$

45. $f(g(x)) = f(x - 15) = (x - 15) + 15 = x$

$g(f(x)) = g(x + 15) = (x + 15) - 15 = x$

47. $f(g(x)) = f\left[\frac{1}{2}(1 - x)\right] = 1 - 2\left[\frac{1}{2}(1 - x)\right]$

$= 1 - (1 - x) = x$

$g(f(x)) = g(1 - 2x) = \frac{1}{2}[1 - (1 - 2x)] = \frac{1}{2}(2x) = x$

49. $f(g(x)) = f\left[\frac{1}{3}(2 - x)\right] = 2 - 3\left[\frac{1}{3}(2 - x)\right]$
$= 2 - (2 - x) = x$
$g(f(x)) = g(2 - 3x) = \frac{1}{3}[2 - (2 - 3x)] = \frac{1}{3}(3x) = x$

51. $f(g(x)) = f(x^3 - 1) = \sqrt[3]{(x^3 - 1) + 1} = \sqrt[3]{x^3} = x$
$g(f(x)) = g(\sqrt[3]{x + 1}) = (\sqrt[3]{x + 1})^3 - 1$
$= x + 1 - 1 = x$

53. $f(g(x)) = f\left(\frac{1}{x}\right) = \frac{1}{(1/x)} = x$

$g(f(x)) = g\left(\frac{1}{x}\right) = \frac{1}{(1/x)} = x$

55. $f^{-1}(x) = \frac{1}{5}x$ **57.** $f^{-1}(x) = 2x$

59. $f^{-1}(x) = x - 10$ **61.** $f^{-1}(x) = 3 - x$

63. $f^{-1}(x) = \sqrt[7]{x}$ **65.** $f^{-1}(x) = x^3$ **67.** $f^{-1}(x) = \frac{x}{8}$

69. $g^{-1}(x) = x - 25$ **71.** $g^{-1}(x) = \dfrac{3 - x}{4}$

73. $g^{-1}(t) = 4t - 8$ **75.** $h^{-1}(x) = x^2, \ x \geq 0$

77. $f^{-1}(t) = \sqrt[3]{t + 1}$ **79.** $g^{-1}(s) = \dfrac{5}{s} - 4$

81. $f^{-1}(x) = x^2 - 3, \ x \geq 0$

83.
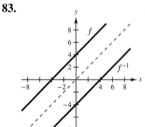

85.
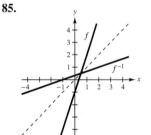

87.
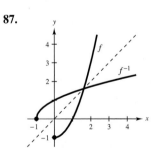

89. b **91.** d

93.
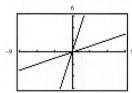

95.
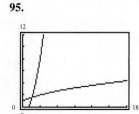

97.
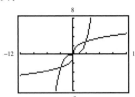

99.
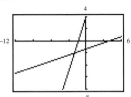

101. $x \geq 2$; $f^{-1}(x) = \sqrt{x} + 2$, domain of f^{-1}: $x \geq 0$

103. $x \geq 0$; $f^{-1}(x) = x - 1$, domain of f^{-1}: $x \geq 1$

105.

x	0	1	3	4
f^{-1}	6	4	2	0

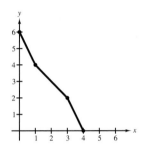

107.

x	-4	-2	2	3
f^{-1}	-2	-1	1	3

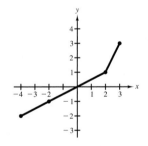

109. (a) $f^{-1}(x) = \frac{1}{2}(3 - x)$ (b) $(f^{-1})^{-1}(x) = 3 - 2x$

111. (a) Total cost = cost of $0.50 compound
    + cost of $0.75 compound
$y = 0.50x + 0.75(100 - x)$

(b) $y = 4(75 - x)$
  x: total cost
  y: number of pounds at $0.50 per pound

(c) $50 \leq x \leq 75$ (d) 60 pounds

113. (a) $f(g(x)) = 0.02x - 200,000$

(b) $g(f(x)) = 0.02(x - 200,000)$

$g(f(x))$ represents the bonus, because it gives 2% of sales over $200,000.

115. (a) $R = p - 2000$

(b) $S = 0.95p$

(c) $(R \circ S)(p) = 0.95p - 2000$;
5% discount followed by the $2000 rebate

$(S \circ R)(p) = 0.95(p - 2000)$;
5% discount after the price is reduced by the rebate

(d) $(R \circ S)(26,000) = 22,700$

$(S \circ R)(26,000) = 22,800$
$R \circ S$ yields the smaller cost because the dealer discount is calculated on a larger base.

117. True

119. False

$f(x) = \sqrt{x - 1}$; Domain: $[1, \infty)$

$f^{-1}(x) = x^2 + 1$, $x \geq 0$; Domain: $[0, \infty)$

121. If $f(x) = 2x$ and $g(x) = x^2$, then $(f \circ g)(x) = 2x^2$ and $(g \circ f)(x) = 4x^2$.

123. • Interchange the roles of x and y.

• If the new equation represents y as a function of x, solve the new equation for y.

• Replace y by $f^{-1}(x)$.

125. Graphically, a function f has an inverse function if and only if no horizontal line intersects the graph of f at more than one point. This is equivalent to saying that the function f is one-to-one.

Section 9.3 *(page 581)*

Integrated Review *(page 581)*

1. Horizontal shift 4 units to the right

2. Reflection in the x-axis

3. Vertical shift 1 unit upward

4. Horizontal shift 3 units to the left and a vertical shift 5 units downward

5. $2x(x^2 - 3)$ **6.** $(2 - y)(6 + y)$

7. $(t + 5)^2$ **8.** $(5 - u)(1 + u^2)$

9. 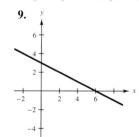   **10.**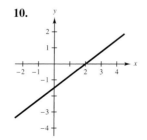

11. 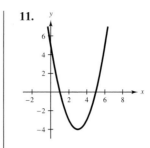   **12.**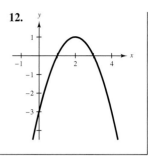

1. $5^2 = 25$ **3.** $4^{-2} = \frac{1}{16}$ **5.** $3^{-5} = \frac{1}{243}$

7. $36^{1/2} = 6$ **9.** $8^{2/3} = 4$ **11.** $2^{1.3} \approx 2.462$

13. $\log_7 49 = 2$ **15.** $\log_3 \frac{1}{9} = -2$ **17.** $\log_8 4 = \frac{2}{3}$

19. $\log_{25} \frac{1}{5} = -\frac{1}{2}$ **21.** $\log_4 1 = 0$

23. $\log_5 9.518 \approx 1.4$ **25.** 3   **27.** 1   **29.** 3

31. -2 **33.** -3 **35.** -4

37. There is no power to which 2 can be raised to obtain -3.

39. 0

41. There is no power to which 5 can be raised to obtain -6.

43. $\frac{1}{2}$ **45.** $\frac{3}{4}$ **47.** 4   **49.** 1.4914   **51.** -0.0706

53. 0.7335

55. f and g are inverse functions.   **57.** f and g are inverse functions.

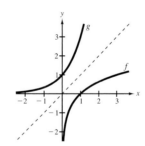   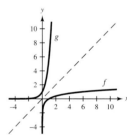

59. f and g are inverse functions.

61. f and g are inverse functions.

63. The graph is shifted 3 units upward.   **65.** The graph is shifted 2 units to the right.

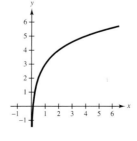   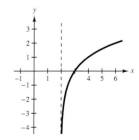

67. The graph is reflected in the *y*-axis.

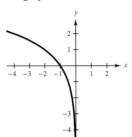

69. e   **71.** d   **73.** a

75.

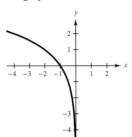

77.

79.

81.

83.

85. Domain: $(0, \infty)$
Vertical asymptote: $x = 0$

87. Domain: $(3, \infty)$
Vertical asymptote:
$x = 3$

89. Domain: $(0, \infty)$
Vertical asymptote:
$x = 0$

91. Domain: $(0, \infty)$
Vertical asymptote:
$x = 0$

93. Domain: $(0, \infty)$
Vertical asymptote:
$x = 0$

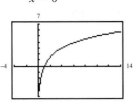

95. Domain: $(0, \infty)$

Vertical asymptote: $x = 0$

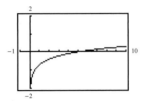

97. 3.2189   **99.** -0.2877 **101.** 0.0757   **103.** b

105. d   **107.** f

109.

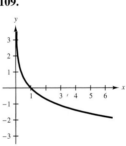

111.

113.

115.

117. Domain: $(-6, \infty)$
Vertical asymptote:
$x = -6$

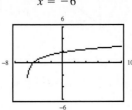

119. Domain: $(0, \infty)$
Vertical asymptote:
$t = 0$

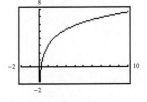

121. 2.3481   **123.** 1.7712   **125.** -0.4739

127. 2.6332   **129.** -2 **131.** 1.3481   **133.** 1.8946

135. 53.4 inches

137.

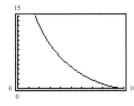

r	0.07	0.08	0.09	0.10	0.11	0.12
t	9.9	8.7	7.7	6.9	6.3	5.8

139. (a)

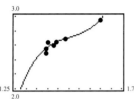

(b) $x = 0$

(c) $(2, y) \approx (2, 13.1)$

Domain: $(0, 10]$

141. (a)

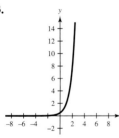

(b) $2.81

143. $g = f^{-1}$ **145.** $a^x = a^x$

147. $\log_b x = \dfrac{\log_{10} x}{\log_{10} b} = \dfrac{\ln x}{\ln b}$ **149.** $f^{-1}(x) = 10^x$

151. $0 < x < 1$ **153.** b^2

Mid-Chapter Quiz   *(page 585)*

1. (a) $\dfrac{16}{9}$ (b) 1 (c) $\dfrac{3}{4}$ (d) $\dfrac{8\sqrt{3}}{9}$

2. Domain: $(-\infty, \infty)$
Range: $(0, \infty)$

3.

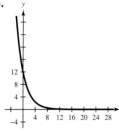

4.

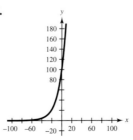

5.

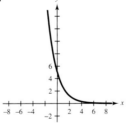

6.

7.

n	1	4	12	365	Continuous
A	\$3185.89	\$3314.90	\$3345.61	\$3360.75	\$3361.27

8. \$2.71

9. (a) $2x^3 - 3$ (b) $(2x - 3)^3$ (c) -19 (d) 125

10. $f(g(x)) = 3 - 5\left[\frac{1}{5}(3 - x)\right] = 3 - 3 + x = x$

$g(f(x)) = \frac{1}{5}[3 - (3 - 5x)] = \frac{1}{5}(5x) = x$

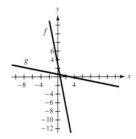

11. $h^{-1}(x) = \frac{1}{10}(x - 3)$ **12.** $g^{-1}(t) = \sqrt[3]{2(t - 2)}$

13. $4^{-2} = \frac{1}{16}$ **14.** $\log_3 81 = 4$ **15.** 3

16. $f^{-1}(x) = g(x)$

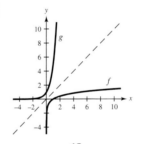

17.

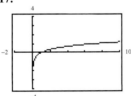

18.

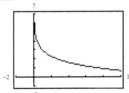

19. $h = 2, k = 1$ **20.** 3.4096

Section 9.4   *(page 591)*

Integrated Review   *(page 591)*

1. $\sqrt[n]{uv}$ **2.** $\sqrt[n]{\dfrac{u}{v}}$ **3.** Different indices

4. No. $\dfrac{1}{\sqrt{2x}} = \dfrac{\sqrt{2x}}{2x}$ **5.** $19\sqrt{3x}$ **6.** $x - 9$

7. $\sqrt{5u}$ **8.** $4t + 12\sqrt{t} + 9$ **9.** $25\sqrt{2}x$

10. $6\left(\sqrt{t + 2} - \sqrt{t}\right)$ **11.** 22 units   **12.** \$2300

1. 2   **3.** -9 **5.** $\frac{1}{2}$ **7.** 0   **9.** 4   **11.** 2

13. 2   **15.** 1   **17.** 2   **19.** -3 **21.** 12   **23.** 1

25. 1   **27.** 1.2925   **29.** 0.2925   **31.** 0.2500

33. 2.7925   **35.** 0   **37.** 0.954   **39.** 1.556

41. 0.778   **43.** $\log_3 11 + \log_3 x$ **45.** $2 \log_7 x$

47. $-2 \log_5 x$ **49.** $\frac{1}{2}(\log_4 3 + \log_4 x)$ **51.** $\ln 3 + \ln y$

53. $\log_2 z - \log_2 17$ **55.** $\ln 5 - \ln(x - 2)$

57. $2 \ln x + \ln(y - 2)$ **59.** $6 \log_4 x + 2 \log_4(x - 7)$

61. $\frac{1}{3} \log_3(x + 1)$ **63.** $\frac{1}{2}[\ln x + \ln(x + 2)]$

65. $2[\ln(x + 1) - \ln(x - 1)]$ **67.** $\frac{1}{3}[2 \ln x - \ln(x + 1)]$

69. $3 \ln a + \ln(b - 4) - 2 \ln c$

71. $\ln x + \frac{1}{3} \ln y - 4(\ln w + \ln z)$

73. $\log_6 a + \frac{1}{2} \log_6 b + 3 \log_6(c - d)$

75. $\ln(x + y) + \frac{1}{5} \ln(w + 2) - (\ln 3 + \ln t)$ **77.** $\log_{12} \dfrac{x}{3}$

79. $\log_2 3x$ **81.** $\log_{10} \dfrac{4}{x}$ **83.** $\ln b^4, \; b > 0$

85. $\log_5(2x)^{-2}, \; x > 0$ **87.** $\ln \sqrt[3]{2x + 1}$ **89.** $\log_3 2\sqrt{y}$

91. $\ln \dfrac{x^2 y^3}{z}, \; x > 0, y > 0, z > 0$

93. $\ln \dfrac{2^5 y^3}{x}, \; x > 0, y > 0$ **95.** $\ln(xy)^4, \; x > 0, y > 0$

97. $\ln\left(\dfrac{x}{x + 1}\right)^2, \; x > 0$ **99.** $\log_4 \dfrac{x + 8}{x^3}, \; x > 0$

101. $\log_5 \dfrac{\sqrt{x + 2}}{x - 3}$ **103.** $\log_6 \dfrac{(c + d)^5}{\sqrt{m - n}}$

105. $\log_2 \sqrt[5]{\dfrac{x^3}{y^4}}, \; y > 0$ **107.** $\log_6 \dfrac{\sqrt[5]{x - 3}}{x^2(x + 1)^3}, \; x > 3$

109. $2 + \ln 3$ **111.** $1 + \frac{1}{2} \log_5 2$ **113.** $1 - 2 \log_4 x$

115.

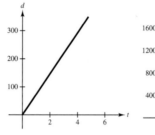

117.

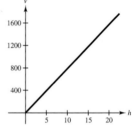

119. Evaluate when $x = e$ and $y = e$

121. $B = 10(\log_{10} I + 16)$; 60 decibels

123. $E = 1.4 \log_{10} \dfrac{C_2}{C_1}$ **125.** True   **127.** True

129. False. $\log_3(u + v)$ does not simplify.   **131.** True

133. False. 0 is not in the domain of f.

135. False. $f(x - 3) = \ln(x - 3)$

137. False. If $v = u^2$, then $f(v) = \ln u^2 = 2 \ln u = 2f(u)$.

Section 9.5   (page 600)

Integrated Review   (page 600)

1. No. A system of linear equations has no solutions, one solution, or an infinite number of solutions.

2. The equations represent parallel lines and therefore have no point of intersection.

3. 2   **4.** $5 \pm 2\sqrt{2}$ **5.** $-\frac{1}{2}$ **6.** $\frac{5}{3}$

7. 1, 7   **8.** 47

9. $d = 73t$ **10.** $V = 25\pi h$

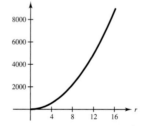

11. $V = 10\pi r^2$ **12.** $F = 25x$

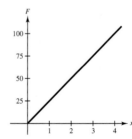

1. (a) Not a solution   **3.** (a) Solution

     (b) Solution   (b) Not a solution

5. (a) Not a solution   **7.** 5   **9.** 8   **11.** 8

     (b) Solution

13. $\frac{2}{3}$ **15.** 2   **17.** -3 **19.** -6 **21.** 2

23. $\frac{22}{5}$ **25.** 6   **27.** 9   **29.** 4   **31.** No solution

33. -7 **35.** $2x - 1$ **37.** $2x, \; x > 0$

39. 5.49   **41.** 1.17   **43.** 0.86   **45.** 0.83

47. -2.37 **49.** -3.60 **51.** 2.64

53. 3.00   **55.** 1.23   **57.** 35.35   **59.** 6.80

61. 12.22   **63.** 3.28   **65.** No solution

67. -1.04 **69.** 2.48   **71.** 0.90   **73.** 0.38

75. 0.39   **77.** 8.99   **79.** 9.73   **81.** 4.62

83. 1000.00   **85.** 22.63   **87.** 2187.00   **89.** 6.52

91. 25.00   **93.** 10.04   **95.** ± 20.09 **97.** 3.00

99. 19.63   **101.** 12.18   **103.** 2000.00   **105.** 3.20

107. 4.00   **109.** 0.75   **111.** 5.00   **113.** 2.46

115. 2.29   **117.** 6.00

119.

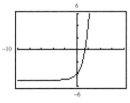

$(1.40, 0)$

121.

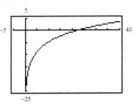

$(21.82, 0)$

123.

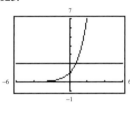

$(0.69, 2)$

125.

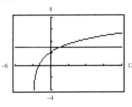

$(1.48, 3)$

127. 9%   **129.** 7.70 years

131. $10^{-8.5}$ watts per square centimeter   **133.** 205

135. (a) 3.64 months

(b)

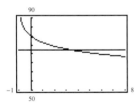

(c) Answers will vary.

137. (a) $k = \frac{1}{4} \ln \frac{8}{15} \approx -0.1572$

(b) ≈ 3.25 hours

(c) ≈ 2.84 hours

139. (c) 7.2%

(d) $6\frac{3}{4}$ years

(e) 8.24%

(f) Double: 11.6 years; Quadruple: 23.1 years

141. $2^{x-1} = 30$

143. To solve an exponential equation, first isolate the exponential expression, then take the logarithms of both sides of the equation, and solve for the variable.

To solve a logarithmic equation, first isolate the logarithmic expression, then exponentiate both sides of the equation, and solve for the variable.

Section 9.6   *(page 610)*

Integrated Review   *(page 610)*

1. Direct variation as nth power

2. Inverse variation   **3.** Joint variation

4. Joint variation   **5.** $(3, 3)$ **6.** $\left(\frac{10}{3}, \frac{5}{3}\right)$

7. $(2, 4), \left(-\frac{1}{2}, \frac{1}{4}\right)$ **8.** $(0, 0), (8, 2)$ **9.** $(1, 2, 1)$

10. $(4, -1, 3)$ **11.** (a) Down

(b) $(0, 0), (4, 0)$

(c) $(2, 4)$

12.

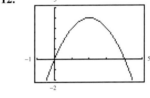

1. 7%   **3.** 9%   **5.** 8%   **7.** 6%   **9.** 8.75 years

11. 6.60 years   **13.** 9.24 years   **15.** 14.21 years

17. Continuous   **19.** Quarterly   **21.** 8.33%

23. 7.23%   **25.** 6.136%   **27.** 8.30%   **29.** No

31. $1652.99   **33.** $626.46   **35.** $3080.15

37. $951.23   **39.** $5496.57   **41.** $320,250.81

43. Total deposits: $7200.00; Total interest: $10,529.42

45. $k = \frac{1}{2} \ln \frac{8}{3} \approx 0.4904$ **47.** $k = \frac{1}{3} \ln \frac{1}{2} \approx -0.2310$

49. $y = 12.2e^{0.0076t}$; 14.9   **51.** $y = 14.7e^{0.0221t}$; 26.1

53. $y = 10.5e^{0.0005t}$; 10.6   **55.** $y = 15.5e^{0.0092t}$; 19.7

57. (a) k is larger in Exercise 51, because the population of Shanghai is increasing faster than the population of Osaka.

(b) k corresponds to r; k gives the annual percentage rate of growth.

59. 3.91 grams   **61.** 4.51 grams   **63.** 4.08 grams

65. 3.3 grams   **67.** 4.43 grams   **69.** $9281

71. The one in Alaska is 63 times as great.

73. The one in Mexico is 40 times as great.

75. 7.04 **77.** 10^7 times

79. (a)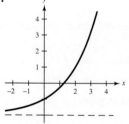

(b) 1000

(c) 2642

(d) 5.88 years

81. (a) $S = 10(1 - e^{-0.0575x})$ **83.** $k < 0$

(b) 3300 units

85. The effective yield of an investment collecting compound interest is the simple interest rate that would yield the same balance at the end of 1 year. To compute the effective yield, divide the interest earned in 1 year by the amount invested.

87. 10

Review Exercises *(page 615)*

1. (a) $\frac{1}{8}$ **3.** (a) 2.718 **5.** c **7.** a

(b) 2 (b) 0.351

(c) 4 (c) 0.135

9. 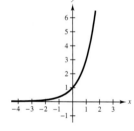 **11.**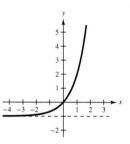

13. 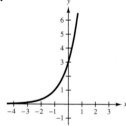 **15.**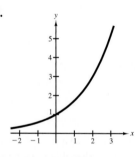

17. **19.**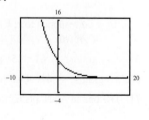

21.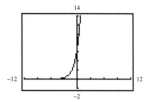

23. (a) 6 (b) 1 **25.** (a) 5 (b) -1

27. (a) $(f \circ g)(x) = \sqrt{2x - 4}$ (b) $(g \circ f)(x) = 2\sqrt{x - 4}$
Domain: $[2, \infty)$ Domain: $[4, \infty)$

29. No **31.** Yes **33.** $f^{-1}(x) = \frac{1}{3}(x - 4)$

35. $h^{-1}(x) = x^2, \ x \geq 0$ **37.** $f^{-1}(t) = \sqrt[3]{t - 4}$

39. $\log_4 64 = 3$ **41.** $e^1 = e$ **43.** 3 **45.** -2

47. 7 **49.** 0 **51.** (a) 0 **53.** (a) 1

(b) 3 (b) -1.099

(c) -0.631 (c) 2.303

55. (a) -6 **57.**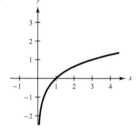

(b) 0

(c) 22.5

59. 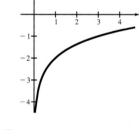 **61.**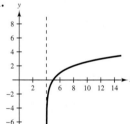

63. 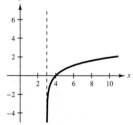 **65.**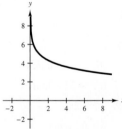

67. 1.585 **69.** 2.132 **71.** 1.7959 **73.** -0.43068

75. 1.02931 **77.** $\log_4 6 + 4 \log_4 x$ **79.** $\frac{1}{2} \log_5(x + 2)$

81. $\ln(x + 2) - \ln(x - 2)$

83. $\frac{1}{2}(\ln 2 + \ln x) + 5\ln(x + 3)$ **85.** $\ln\left(\frac{1}{3y}\right)^{2/3}$

87. $\log_8 32x^3$ **89.** $\ln\frac{9}{4x^2}$, $x > 0$

91. $\log_2\left(\frac{k}{k - t}\right)^4$, $t > k$

93. $\ln(x^3 y^4 z)$, $x > 0$, $y > 0$, $z > 0$

95. False. $\log_2 4x = 2 + \log_2 x$ **97.** True   **99.** True

101. 6   **103.** 1   **105.** 243   **107.** 50   **109.** 4

111. 5.66   **113.** 1408.10   **115.** 6.23

117. No solution   **119.** 15.81   **121.** 64   **123.** 2.67

125. 5%   **127.** 7.5%   **129.** 7%   **131.** 5.65%

133. 7.71%   **135.** 7.79%   **137.** 2.282 grams

139. 2.934 grams   **141.** 4.860 grams

143. 3.8 years   **145.** 12.6 years

147. 3.16×10^{-4} watts per square centimeter

149.

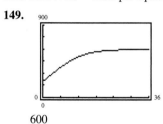

151. (a)

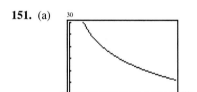

(b) 14.3 cubic feet per minute per person

Chapter Test   *(page 619)*

1. $f(-1) = 81$ **2.**

$f(0) = 54$

$f\left(\frac{1}{2}\right) = 18\sqrt{6} \approx 44.09$

$f(2) = 24$

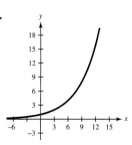

3. (a) $(f \circ g)(x) = 3x^2 - 1$
    Domain: $(-\infty, \infty)$

   (b) $(g \circ f)(x) = 9x^2 - 24x + 17$
     Domain: $(-\infty, \infty)$

4. $f^{-1}(x) = \frac{1}{5}(x - 6)$

5. $(f \circ g)(x) = -\frac{1}{2}(-2x + 6) + 3 = (x - 3) + 3 = x$

    $(g \circ f)(x) = -2\left(-\frac{1}{2}x + 3\right) + 6 = (x - 6) + 6 = x$

6. $g = f^{-1}$

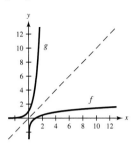

7. $\log_4 5 + 2\log_4 x - \frac{1}{2}\log_4 y$ **8.** $\ln\frac{x}{y^4}$, $y > 0$

9. $3 + \log_5 6$ **10.** 64   **11.** 0.973   **12.** 13.733

13. 15.516   **14.** 2   **15.** 8   **16.** 109.196   **17.** 0

18. (a) $8012.78   **19.** $10,806.08   **20.** 7%

   (b) $8110.40

21. $8469.14   **22.** 600   **23.** 1141   **24.** 4.4 years

Cumulative Test: Chapters 7–9   *(page 620)*

1. $V = \frac{k\sqrt{x}}{y}$ **2.** $k = -16$ **3.** 128 feet   **4.** 50

5.   **6.** $y \geq 2x + 2$

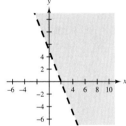

7. $y = \frac{2}{3}(x - 3)^2 - 2$

8.

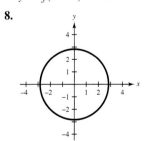

9.

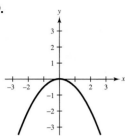

10.

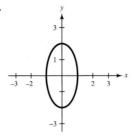

11.

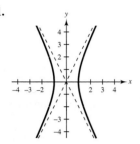

12. 11 feet

13.

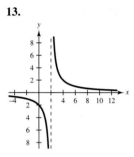

14.

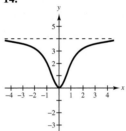

15. $f(x) = \dfrac{2x}{x-3}$

16. Horizontal asymptote: $\overline{C} = 10$; As x increases, the average cost approaches \$10.

17. $(2, 1)$ **18.** $(3, -2)$ **19.** $(2, 4, 1)$

20. $\left(-\frac{1}{5}, -\frac{22}{5}\right)$ **21.** $(-1, 6, -2)$ **22.** $\frac{11}{2}$ **23.** -4

24.

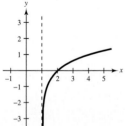

25. -2

26. Reflections in the line $y = x$ because $f^{-1}(x) = g(x)$.

27. $\log_2\left(\dfrac{x^3 y^3}{z}\right)$ **28.** $\ln 5 + \ln x - 2\ln(x+1)$

29. (a) 3

(b) 12.18

(c) 18.01

(d) 0.87

30. \$29.63   **31.** 8.329%   **32.** 15.40 years

Chapter 10

Section 10.1   *(page 630)*

Integrated Review   *(page 630)*

1. $-7x = 35$ **2.** $7x + 63 = 35$

$\dfrac{-7x}{-7} = \dfrac{35}{-7}$ $7x + 63 - 63 = 35 - 63$

$7x = -28$

$x = -5$ $x = -4$

3. It is a solution if the equation is true when -3 is substituted for t.

4. Multiply both sides of the equation by the lowest common denominator $x(x+1)$.

5. $\dfrac{1}{(x+10)^2}$ **6.** $18(x-3)^3$ **7.** $\dfrac{1}{a^8}$ **8.** $2x$

9. $8x\sqrt{2x}$ **10.** $\dfrac{5(\sqrt{x}+2)}{x-4}$

11. (a) $A = x(2x-3)$

(b)

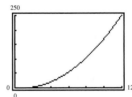

(c) $\dfrac{3 + \sqrt{1609}}{4} \approx 10.8$

12. (a) $A = \frac{1}{2}x(x-4)$

(b)

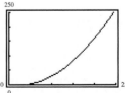

(c) $2\left(1 + \sqrt{101}\right) \approx 22.1$

---

1. $2, 4, 6, 8, 10$ **3.** $-2, 4, -6, 8, -10$

5. $\frac{1}{2}, \frac{1}{4}, \frac{1}{8}, \frac{1}{16}, \frac{1}{32}$ **7.** $\frac{1}{4}, -\frac{1}{8}, \frac{1}{16}, -\frac{1}{32}, \frac{1}{64}$

9. $1, -0.2, 0.04, -0.008, 0.0016$ **11.** $\frac{1}{2}, \frac{1}{3}, \frac{1}{4}, \frac{1}{5}, \frac{1}{6}$

13. $\frac{2}{5}, \frac{1}{2}, \frac{6}{11}, \frac{4}{7}, \frac{10}{17}$ **15.** $-1, \frac{1}{4}, -\frac{1}{9}, \frac{1}{16}, -\frac{1}{25}$

17. $\frac{9}{2}, \frac{19}{4}, \frac{39}{8}, \frac{79}{16}, \frac{159}{32}$ **19.** $2, 3, 4, 5, 6$

21. $0, 3, -1, \frac{3}{4}, -\frac{1}{4}$ **23.** -72 **25.** $\frac{31}{2520}$ **27.** 5

29. $\frac{1}{132}$ **31.** 53,130 **33.** $\frac{1}{n+1}$ **35.** $n(n+1)$

37. $2n$ **39.** c **41.** b

43. 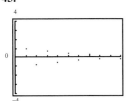 **45.**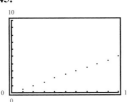

47. 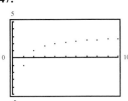 **49.** $a_n = 3n$

51. $a_n = 3n - 2$ **53.** $a_n = n^2 - 1$

55. $a_n = (-1)^{n+1}2n$ **57.** $a_n = \frac{n+1}{n+2}$

59. $a_n = \frac{(-1)^{n+1}}{2^n}$ **61.** $a_n = \frac{1}{2^{n-1}}$ **63.** $a_n = 1 + \frac{1}{n}$

65. $a_n = \frac{1}{n!}$ **67.** 63 **69.** 77 **71.** 100

73. $\frac{3019}{3600}$ **75.** $\frac{437}{60}$ **77.** -48 **79.** $\frac{8}{9}$ **81.** $\frac{182}{243}$

83. 273 **85.** 852 **87.** $\frac{65}{4}$ **89.** 6.5793 **91.** $\sum_{k=1}^{5} k$

93. $\sum_{k=1}^{5} 2k$ **95.** $\sum_{k=1}^{10} \frac{1}{2k}$ **97.** $\sum_{k=1}^{20} \frac{1}{k^2}$ **99.** $\sum_{k=0}^{9} \frac{1}{(-3)^k}$

101. $\sum_{k=1}^{20} \frac{4}{k+3}$ **103.** $\sum_{k=1}^{11} \frac{k}{k+1}$ **105.** $\sum_{k=1}^{20} \frac{2k}{k+3}$

107. $\sum_{k=0}^{6} k!$ **109.** 3.6 **111.** 0.8

113. (a) $535, $572.45, $612.52, $655.40, $701.28, $750.37, $802.89, $859.09

(b) $7487.23 (c)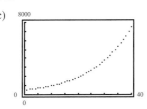

(d) Yes. Investment earning compound interest increases at an increasing rate.

115. $a_5 = 108°, a_6 = 120°$

At the point where any two hexagons and a pentagon meet, the sum of the three angles is $a_5 + 2a_6 = 348° < 360°$. Therefore, there is a gap of 12°.

117. 25.7°, 45°, 60°, 72°, 81.8°

119. $a_n = 3n$: 3, 6, 9, 12, . . .

121. Terms in which n is odd **123.** True

Section 10.2 *(page 638)*

Integrated Review *(page 638)*

1. A collection of letters (called variables) and real numbers (called constants) combined with the operations of addition, subtraction, multiplication, and division is called an algebraic expression.

2. The terms of an algebraic expression are those parts separated by addition or subtraction.

3. $2x^3 - 3x^2 + 2$ **4.** $7x^4$ **5.** $(-\infty, \infty)$

6. $(-\infty, \infty)$ **7.** $[-4, 4)$

8. $(-\infty, -6) \cup (-6, 6) \cup (6, \infty)$ **9.** $(2, \infty)$

10. $(-\infty, \infty)$ **11.** $30,798.61 **12.** $5395.40

1. 3 **3.** -6 **5.** -12 **7.** $\frac{2}{3}$ **9.** $-\frac{5}{4}$

11. Arithmetic, 2 **13.** Arithmetic, -2

15. Arithmetic, -16 **17.** Arithmetic, 0.8

19. Arithmetic, $\frac{3}{2}$ **21.** Not arithmetic

23. Not arithmetic **25.** Not arithmetic

27. 7, 10, 13, 16, 19 **29.** 6, 4, 2, 0, -2

31. $\frac{3}{2}, 4, \frac{13}{2}, 9, \frac{23}{2}$ **33.** $\frac{8}{5}, \frac{11}{5}, \frac{14}{5}, \frac{17}{5}, 4$ **35.** $4, \frac{15}{4}, \frac{7}{2}, \frac{13}{4}, 3$

37. $a_n = \frac{1}{2}n + \frac{5}{2}$ **39.** $a_n = -25n + 1025$

41. $a_n = -4n + 32$ **43.** $a_n = \frac{3}{2}n + \frac{3}{2}$

45. $a_n = \frac{5}{2}n + \frac{5}{2}$ **47.** $a_n = 4n + 4$

49. $a_n = -10n + 60$ **51.** $a_n = -\frac{1}{2}n + 11$

53. $a_n = -0.05n + 0.40$ **55.** 25, 28, 31, 34, 37

57. 9, 6, 3, 0, -3 **59.** $-10, -4, 2, 8, 14$

61. 100, 80, 60, 40, 20 **63.** 210 **65.** 1425

67. 255 **69.** 62,625 **71.** 35 **73.** 522

75. 1850 **77.** 900 **79.** 12,200 **81.** 243

83. 23 **85.** b **87.** e **89.** c

91. 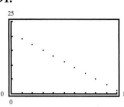   **93.**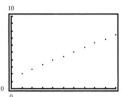

95.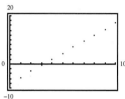

97. 9000   **99.** 13,120   **101.** 3011.25   **103.** 2850

105. 2550   **107.** $246,000   **109.** $25.43

111. 632 bales   **113.** 114   **115.** 1024 feet

117. (a) 4, 9, 16, 25, 36 (b) 49 (c) $\sum_{k=1}^{n} (2k - 1) = n^2$

119. 9

121. A recursion formula gives the relationship between the terms a_{n+1} and a_n.

123. $\frac{101}{2}(100 + 200)$

Section 10.3   *(page 648)*

Integrated Review   *(page 648)*

1. The point is 6 units to the left of the y-axis and 4 units above the x-axis.

2. $(10, 5), (-10, 5), (-10, -5), (10, -5)$

3. The graph of f is the set of ordered pairs $(x, f(x))$, where x is in the domain of f.

4. To find the x-intercept, set $y = 0$ and solve the equation for x. To find the y-intercept, set $x = 0$ and solve the equation for y.

5. $x > \frac{5}{3}$ **6.** $y < 6$ **7.** $35 < x < 60$

8. $-12 < x < 30$ **9.** $x < 1$ or $x > \frac{5}{2}$

10. $-1 < x < 0$ or $x > \frac{5}{2}$

11. $\frac{19\sqrt{2}}{2} \approx 13.4$ inches   **12.** $5\sqrt{89} \approx 47.2$ feet

1. 3   **3.** -3 **5.** $-\frac{1}{2}$ **7.** $-\frac{3}{2}$ **9.** π

11. 1.06   **13.** Geometric, $\frac{1}{2}$ **15.** Not geometric

17. Geometric, 2   **19.** Not geometric

21. Geometric, $-\frac{2}{3}$ **23.** Geometric, 1.02

25. 4, 8, 16, 32, 64   **27.** 6, 2, $\frac{2}{3}, \frac{2}{9}, \frac{2}{27}$

29. 1, $-\frac{1}{2}, \frac{1}{4}, -\frac{1}{8}, \frac{1}{16}$ **31.** 4, $-2, 1, -\frac{1}{2}, \frac{1}{4}$

33. 1000, 1010, 1020.1, 1030.30, 1040.60

35. 4000, 3960.40, 3921.18, 3882.36, 3843.92

37. 10, 6, $\frac{18}{5}, \frac{54}{25}, \frac{162}{125}$ **39.** $\frac{3}{256}$ **41.** $48\sqrt{2}$

43. 1486.02   **45.** -0.00610 **47.** $\frac{81}{64}$ **49.** $\pm\frac{243}{32}$

51. $\frac{64}{3}$ **53.** $a_n = 2(3)^{n-1}$ **55.** $a_n = 2^{n-1}$

57. $a_n = \left(-\frac{1}{5}\right)^{n-1}$ **59.** $a_n = 4\left(-\frac{1}{2}\right)^{n-1}$

61. $a_n = 8\left(\frac{1}{4}\right)^{n-1}$ **63.** $a_n = 14\left(\frac{3}{4}\right)^{n-1}$

65. $a_n = 4\left(-\frac{3}{2}\right)^{n-1}$ **67.** b   **69.** a

71. 1023   **73.** 772.48   **75.** 2.25   **77.** -5460

79. 6.06   **81.** $-14,762$ **83.** 16   **85.** 13,120

87. 48   **89.** 1103.57   **91.** 12,822.71   **93.** 2

95. $\frac{2}{3}$ **97.** $\frac{6}{5}$ **99.** 32

101. **103.**

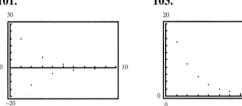

105. (a) $250,000(0.75)^n$ (b) $59,326.17 (c) The first year

107. $3,623,993   **109.** $19,496.56   **111.** $105,428.44

113. $75,715.32

115. (a) $5,368,709.11 (b) $10,737,418.23

117. (a) $P = (0.999)^n$ (b) 69.4%

(c) 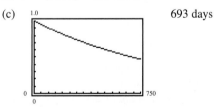   693 days

119. 70.875 square inches   **121.** 666.21 feet

123. (a) $a_n = 2^n$ (b) $2 + 2^2 + 2^3 + 2^4 + \cdots + 2^{66}$

(c) It is likely that you have had common ancestors in the last 2000 years.

125. $a_n = a_1 r^{n-1}$ **127.** $a_n = \left(-\frac{2}{3}\right)^{n-1}$

129. An increasing annuity is an investment plan where equal deposits are made in an account at equal time intervals.

Mid-Chapter Quiz   *(page 652)*

1. $32, 8, 2, \frac{1}{2}, \frac{1}{8}$ **2.** $-\frac{3}{5}, 3, -\frac{81}{7}, \frac{81}{2}, -135$ **3.** 100

4. 40 **5.** 87 **6.** -32 **7.** $\displaystyle\sum_{k=1}^{20} \frac{2}{3k}$

8. $\displaystyle\sum_{k=1}^{25} \frac{(-1)^{k-1}}{k^3}$ **9.** $\frac{1}{2}$ **10.** -6 **11.** 3 **12.** $\frac{1}{2}$

13. $20 - 3(n - 1)$ **14.** $32\left(-\frac{1}{4}\right)^{n-1}$ **15.** 4075

16. 9030 **17.** 25.947 **18.** $18{,}392.796$ **19.** 9

20. $\frac{16}{15}$ **21.** -0.026 **22.** a_n: upper graph
b_n: lower graph

23. $5.5°$ **24.** Arithmetic

Section 10.4   *(page 658)*

Integrated Review   *(page 658)*

1. No. The matrix must be square.

2. Interchange two rows.
Multiply a row by a nonzero constant.
Add a multiple of one row to another row.

3. Yes   **4.** -200 **5.** 32 **6.** -60

7. -126 **8.** 58 **9.** $y = -0.07x^2 + 1.3x + 2$

10. $y = 4x - 9$

1. 15 **3.** 252 **5.** 1 **7.** 1 **9.** 1225

11. $12{,}650$ **13.** $593{,}775$ **15.** 792 **17.** $2{,}598{,}960$

19. $2{,}535{,}650{,}040$ **21.** $5{,}200{,}300$ **23.** 15 **25.** 35

27. 70 **29.** $a^3 + 6a^2 + 12a + 8$

31. $x^8 + 8x^7y + 28x^6y^2 + 56x^5y^3 + 70x^4y^4 + 56x^3y^5$
    $+ 28x^2y^6 + 8xy^7 + y^8$

33. $32x^5 - 80x^4 + 80x^3 - 40x^2 + 10x - 1$

35. $64y^6 + 192y^5z + 240y^4z^2 + 160y^3z^3 + 60y^2z^4$
    $+ 12yz^5 + z^6$

37. $x^8 + 8x^6 + 24x^4 + 32x^2 + 16$

39. $x^6 + 18x^5 + 135x^4 + 540x^3 + 1215x^2 + 1458x + 729$

41. $x^6 - 24x^5 + 240x^4 - 1280x^3 + 3840x^2 - 6144x + 4096$

43. $x^4 + 4x^3y + 6x^2y^2 + 4xy^3 + y^4$

45. $u^3 - 6u^2v + 12uv^2 - 8v^3$

47. $81a^4 + 216a^3b + 216a^2b^2 + 96ab^3 + 16b^4$

49. $32x^{10} - 80x^8y + 80x^6y^2 - 40x^4y^3 + 10x^2y^4 - y^5$

51. 120 **53.** -1365 **55.** 1760 **57.** 54 **59.** 70

61. 1.172 **63.** $510{,}568.785$

65. $\frac{1}{32} + \frac{5}{32} + \frac{10}{32} + \frac{10}{32} + \frac{5}{32} + \frac{1}{32}$

67. $\frac{1}{256} + \frac{12}{256} + \frac{54}{256} + \frac{108}{256} + \frac{81}{256}$

69. The difference between consecutive entries increases by 1.

2, 3, 4, 5

71. $n + 1$

73. The signs of the terms alternate in the expansion of $(x - y)^n$.

75. They are the same.

Section 10.5   *(page 667)*

Integrated Review   *(page 667)*

1. $g(x) = 2(5^x)$ is exponential since it has a constant base and variable exponent.

2. Using the law of exponents $a^m \cdot a^n = a^{m+n}$, you have $e^2 \cdot e^{-x^2} = e^{2+(-x^2)} = e^{2-x^2}$.

3. $4^3 = 64$ **4.** $3^{-4} = \frac{1}{81}$ **5.** $e^0 = 1$

6. $e^{1.6094\cdots} \approx 5$ **7.** $\dfrac{\ln 50}{\ln 3} \approx 3.56$

8. $6\ln 2 \approx 4.16$ **9.** 69 **10.** $e^{10} - 3 \approx 22{,}023.47$

11.

$t = 1.7$

12. 6.96 grams

1. 5 **3.** 9 **5.** 8 **7.** 10 **9.** 8 **11.** 6

13. 7 **15.** 6 **17.** 6 **19.** 260 **21.** $6{,}760{,}000$

23. (a) 900 (b) 720 (c) 400 **25.** 18 **27.** 720

29. ABCD, ABDC, ACBD, ACDB, ADBC, ADCB, BACD, BADC, BCAD, BCDA, BDAC, BDCA, CABD, CADB, CBAD, CBDA, CDAB, CDBA, DABC, DACB, DBAC, DBCA, DCAB, DCBA

31. AB, BA, AC, CA, AD, DA, BC, CB, BD, DB, CD, DC

33. 720 **35.** $64{,}000$ **37.** $40{,}320$ **39.** 5040

41. $\{A, B\}, \{A, C\}, \{A, D\}, \{A, E\}, \{A, F\}, \{B, C\},$
$\{B, D\}, \{B, E\}, \{B, F\}, \{C, D\}, \{C, E\}, \{C, F\},$
$\{D, E\}, \{D, F\}, \{E, F\}$

43. 1140 **45.** 126 **47.** 220 **49.** 3003

51. (a) 15 (b) 6 **53.** (a) 70 (b) 16

55. 21 **57.** 9 **59.** 35

61. Let E_1 and E_2 be two events that can occur in m_1 ways and m_2 ways, respectively. The number of ways the two events can occur is $m_1 \cdot m_2$.

63. Permutation: The ordering of five students for a picture

Combination: The selection of three students from a group of five students for a class project

Section 10.6 *(page 676)*

Integrated Review *(page 676)*

1. 0   **2.** 1   **3.** x **4.** $\log_a u + \log_a v$

5. $\log_a u - \log_a v$ **6.** $n \log_a u$

7. $2 \log_2 x + \log_2 y$ **8.** $\frac{1}{2} \log_2(x^2 + 1)$

9. $\ln 7 - \ln(x - 3)$ **10.** $2[\ln(u + 2) - \ln(u - 2)]$

11. (a)   **12.** 5.65%

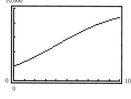

(b) $x \approx 4$ (c) 10,000

---

1. 26   **3.** 10

5. {ABC, ACB, BAC, BCA, CAB, CBA}

7. {WWW, WWL, WLW, WLL, LWW, LWL, LLW, LLL}

9. 0.65   **11.** 0.18   **13.** $\frac{3}{8}$ **15.** $\frac{7}{8}$ **17.** $\frac{1}{2}$

19. $\frac{3}{13}$ **21.** $\frac{1}{6}$ **23.** $\frac{5}{6}$ **25.** 0.9   **27.** 0.243

29. 0.609   **31.** (a) $\frac{1}{5}$ (b) $\frac{1}{3}$ (c) 1

33. (a) 0.8 (b) 0.2   **35.** $\frac{14}{65}$ **37.** $\frac{1917}{6565}, \frac{4648}{6565}$

39. 0.4375

41.

	Female	
	X	X
Male X	XX	XX
Y	XY	XY

(a) Probability of a girl $= \frac{2}{4} = \frac{1}{2}$

Probability of a boy $= \frac{2}{4} = \frac{1}{2}$

(b) Because the probabilities are the same, it is equally likely that a newborn will be a boy or a girl.

43. $\frac{1}{24}$ **45.** $\frac{1}{100,000}$ **47.** $\frac{1}{45}$

49. $\frac{1}{45}$ **51.** $\frac{1}{210}$ **53.** $\frac{33}{66,640}$

55. (d) 120

(e) Without replacement, since each person receives 1 gift

(f) $\frac{1}{150}, \frac{1}{105}$

57. The probability that the event does not occur is $1 - \frac{3}{4} = \frac{1}{4}$.

59. Over an extended period, it will rain 40% of the time under the given weather conditions.

Review Exercises *(page 681)*

1. 8, 11, 14, 17, 20   **3.** $1, \frac{3}{4}, \frac{5}{8}, \frac{9}{16}, \frac{17}{32}$ **5.** $a_n = 2n - 1$

7. $a_n = \dfrac{n}{(n+1)^2}$ **9.** a   **11.** b   **13.** d   **15.** 28

17. $\frac{4}{5}$ **19.** $\displaystyle\sum_{k=1}^{4}(5k - 3)$ **21.** $\displaystyle\sum_{k=1}^{6}\frac{1}{3k}$ **23.** -2.5

25. 127, 122, 117, 112, 107   **27.** $\frac{5}{4}, 2, \frac{11}{4}, \frac{7}{2}, \frac{17}{4}$

29. 5, 8, 11, 14, 17   **31.** $80, \frac{155}{2}, 75, \frac{145}{2}, 70$ **33.** $4n + 6$

35. $-50n + 1050$ **37.** 486   **39.** $\frac{2525}{2}$ **41.** 2527.5

43. $\frac{3}{2}$ **45.** 10, 30, 90, 270, 810

47. $100, -50, 25, -12.5, 6.25$ **49.** 3, 6, 12, 24, 48

51. $a_n = \left(-\frac{2}{3}\right)^{n-1}$ **53.** $a_n = 24(2)^{n-1}$

55. $a_n = 12\left(-\frac{1}{2}\right)^{n-1}$ **57.** 8190   **59.** -1.928

61. 19.842   **63.** 116,169.54   **65.** 8   **67.** 12

69. 2.275×10^6 **71.** 56   **73.** 1

75. 91,390   **77.** 177,100

79. $x^{10} + 10x^9 + 45x^8 + 120x^7 + 210x^6 + 252x^5$
$+ 210x^4 + 120x^3 + 45x^2 + 10x + 1$

81. $81x^4 - 216x^3y + 216x^2y^2 - 96xy^3 + 16y^4$

83. $u^{18} + 9u^{16}v^3 + 36u^{14}v^6 + 84u^{12}v^9 + 126u^{10}v^{12}$
$+ 126u^8v^{15} + 84u^6v^{18} + 36u^4v^{21} + 9u^2v^{24} + v^{27}$

85. $-61,236$ **87.** 280   **89.** 5100   **91.** 462

93. (a) $a_n = 85,000(1.012)^n$ (b) 154,328

95. 8   **97.** 3003   **99.** $\frac{1}{3}$ **101.** $\frac{1}{24}$ **103.** 0.346

Chapter Test *(page 684)*

1. $1, -\frac{2}{3}, \frac{4}{9}, -\frac{8}{27}, \frac{16}{81}$ **2.** 35   **3.** -45

4. $\displaystyle\sum_{k=1}^{12}\frac{2}{3k + 1}$ **5.** 12, 16, 20, 24, 28

6. $a_n = -100n + 5100$ **7.** 3825   **8.** $-\frac{3}{2}$

9. $a_n = 4\left(\frac{1}{2}\right)^{n-1}$ **10.** 1020   **11.** $\frac{3069}{1024}$ **12.** 1

13. 12   **14.** $47,868.33   **15.** 1140

16. $x^5 - 10x^4 + 40x^3 - 80x^2 + 80x - 32$ **17.** 56

18. 26,000   **19.** 12,650   **20.** 0.25   **21.** $\frac{3}{26}$ **22.** $\frac{1}{6}$

Appendix A *(page A6)*

1.
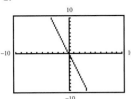

3.
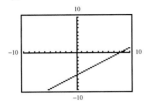

5.
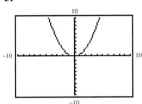

7.
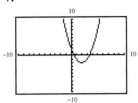

9.
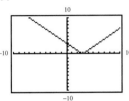

11.
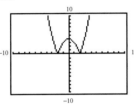

13.
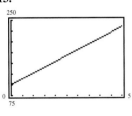

15.
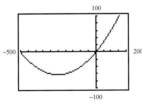

17.
```
Xmin = 4
Xmax = 20
Xscl = 1
Ymin = 14
Ymax = 22
Yscl = 1
```

19.
```
Xmin = -20
Xmax = -4
Xscl = 1
Ymin = -16
Ymax = -8
Yscl = 1
```

21.
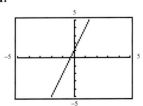

Associative Property of Addition

23.
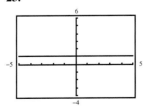

Multiplicative Inverse Property

25. $(-3, 0), (3, 0), (0, 9)$ **27.** $(-8, 0), (4, 0), (0, 4)$

29. $\left(\frac{5}{2}, 0\right), (0, -5)$ **31.** $(-2, 0), \left(\frac{1}{2}, 0\right), (0, -1)$

33. Triangle **35.** Square

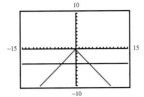 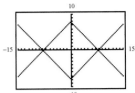

37.
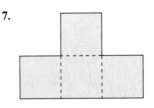

Appendix B

Section B.1 *(page A14)*

1. Answers will vary.

3.
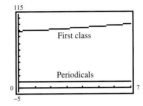

5. a and b **7.**
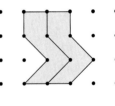
  9. 7

11. No **13.** False **15.** True **17.** d **19.** b

21. $\angle ZXW$ or $\angle WXZ$, $\angle ZXY$ or $\angle YXZ$, **23.** b **25.** d
$\angle YXW$ or $\angle WXY$; $\angle ZXY$ and $\angle YXW$

27. f **29.** c **31.** $\overline{LM} \cong \overline{NO}$, $\overline{MP} \cong \overline{NQ}$, $\overline{LP} \cong \overline{OQ}$

33. $m\angle V$ **35.** $\overline{TV}$

37.

	Scalene	Isosceles	Equilateral
Acute	Yes	Yes	Yes
Obtuse	Yes	Yes	No
Right	Yes	Yes	No

39. 3; Yes **41.** 3; Yes **43.** 12.5 feet by 12.5 feet; No

45. $(3, 5), (3, 1)$ **47.** Form a tetrahedron.

Section B.2 *(page A21)*

1.

3.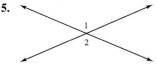

5.

7. Adjacent ≅ suppl. ∠   **9.** Adjacent suppl. ∠

11. Adjacent compl. ∠   **13.** False

15. False   **17.** True   **19.** 110°

21. 55°   **23.** 35°   **25.** c

27. ∠3 and ∠5 *or* ∠4 and ∠6

29. ∠4 and ∠5 *or* ∠3 and ∠6

31. $m\angle 1 = 110°$ because it forms a linear pair with the given angle; $m\angle 2 = 110°$ by the Alternate Exterior Angles Theorem

33. $m\angle 1 = 70°$ by the Consecutive Interior Angles Theorem; $m\angle 2 = 70°$ because it forms a linear pair with the given angle, or by the Alternate Interior Angles Theorem

35. $a = 30°, b = 20°$ **37.** ∠2, ∠5, ∠7

39. $m\angle 1 = m\angle 3 = 70°, m\angle 4 = m\angle 6 = 135°,$
$m\angle 2 = 110°, m\angle 5 = 45°, m\angle 7 = 25°, m\angle 8 = 155°$

41. 35°   **43.** 40°

45. True. The third angle must be $180° - 2(60°) = 60°.$

47. $m\angle 1 = 30°, m\angle 2 = 60°, m\angle 3 = 50°, m\angle 4 = 35°,$
$m\angle 5 = 90°, m\angle 6 = 55°, m\angle 7 = 55°, m\angle 8 = 125°,$
$m\angle 9 = 35°$

49.

$m\angle B = 77°$

51.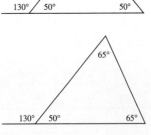

53. 30°, 60°, 90°   **55.** 38°, 59°, 83°

Appendix C   *(page A33)*

1.

Stems	Leaves
7	0 5 5 5 7 7 8 8 8
8	1 1 1 1 2 3 4 5 5 5 5 7 8 9 9 9
9	0 2 8
10	0 0

3.

Stems	Leaves
5	2 5 9
6	2 3 6 6 7
7	0 1 2 3 4 7 8 8 9
8	0 1 3 4 5 7 9
9	0 0 2 3 3 3 5 6 8 9
10	0 0

5. Frequency Distribution

Interval	Tally
[15, 22)	ⅢⅢ |||
[22, 29)	ⅢⅢ	
[29, 36)	ⅢⅢ
[36, 43)	||||
[43, 50)	ⅢⅢ ||

Histogram

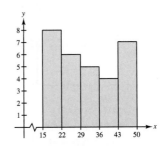

7.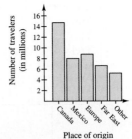

9. 1985: 165 million tons
1995: 210 million tons

11. Recycled waste

13. Total waste equals the sum of the other three quantities.

15.

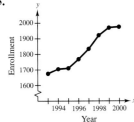

17.

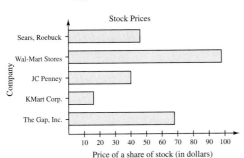

19.

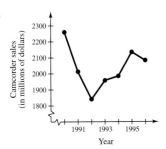

21. Positive correlation   **23.** Yes

25. Negative correlation   **27.** Positive correlation

29.

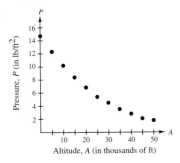

31. 2.45 pounds per square inch

33.

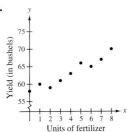

35. $y = 57.49 + 1.43x$; 71.8

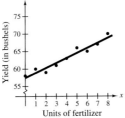

37.

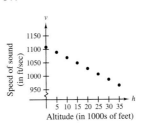

39. $v = 1117.3 - 4.1h$; 1006.6

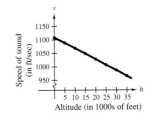

41. $y = -2.179x + 22.964$ **43.** $y = 2.378x + 23.546$

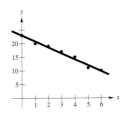

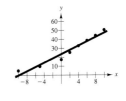

45. (a) $y = 11.1 + 0.28t$; 12.78

(b)

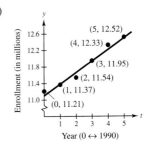

(c) $r \approx 0.987$

47. Mean: 8.86; median: 8; mode: 7

49. Mean: 10.29; median: 8; mode: 7

51. Mean: \$67.14; median: \$65.35

53. Mean: 3.065; median: 3; mode: 3

55. One possibility: {4, 4, 10}

57. The median gives the most representative description.

Appendix D

Section D.1   *(page A40)*

1. Statement   **3.** Open statement   **5.** Open statement

7. Open statement   **9.** Nonstatement

11. Open statement   **13.** (a) True   **15.** (a) True

(b) False   (b) True

17. (a) False   (b) False   **19.** (a) True   (b) False

21. (a) The sun is not shining.

(b) It is not hot.

(c) The sun is shining and it is hot.

(d) The sun is shining or it is hot.

23. (a) Lions are not mammals.

(b) Lions are not carnivorous.

(c) Lions are mammals and lions are carnivorous.

(d) Lions are mammals or lions are carnivorous.

25. (a) The sun is not shining and it is hot.

(b) The sun is not shining or it is hot.

(c) The sun is shining and it is not hot.

(d) The sun is shining or it is not hot.

27. (a) Lions are not mammals and lions are carnivorous.

(b) Lions are not mammals or lions are carnivorous.

(c) Lions are mammals and lions are not carnivorous.

(d) Lions are mammals or lions are not carnivorous.

29. $p \wedge \sim q$ **31.** $\sim p \vee q$ **33.** $\sim p \vee \sim q$

35. $\sim p \wedge q$ **37.** The bus is blue.

39. x is not equal to 4.   **41.** The earth is flat.

43.

p	q	$\sim p$	$\sim p \wedge q$
T	T	F	F
T	F	F	F
F	T	T	T
F	F	T	F

45.

p	q	$\sim p$	$\sim q$	$\sim p \vee \sim q$
T	T	F	F	F
T	F	F	T	T
F	T	T	F	T
F	F	T	T	T

47.

p	q	$\sim q$	$p \vee \sim q$
T	T	F	T
T	F	T	T
F	T	F	F
F	F	T	T

49. Not logically equivalent   **51.** Logically equivalent

53. Logically equivalent   **55.** Not logically equivalent

57. Logically equivalent   **59.** Not a tautology

61. A tautology

63.

p	q	$\sim p$	$\sim q$	$p \wedge q$	$\sim(p \wedge q)$	$\sim p \vee \sim q$
T	T	F	F	T	F	F
T	F	F	T	F	T	T
F	T	T	F	F	T	T
F	F	T	T	F	T	T

Identical: $\sim(p \wedge q)$ and $\sim p \vee \sim q$

Section D.2   *(page A47)*

1. (a) If the engine is running, then the engine is wasting gasoline.

(b) If the engine is wasting gasoline, then the engine is running.

(c) If the engine is not wasting gasoline, then the engine is not running.

(d) If the engine is running, then the engine is not wasting gasoline.

3. (a) If the integer is even, then it is divisible by 2.

(b) If it is divisible by 2, then the integer is even.

(c) If it is not divisible by 2, then the integer is not even.

(d) If the integer is even, then it is not divisible by 2.

5. $q \rightarrow p$ **7.** $p \rightarrow q$ **9.** $p \rightarrow q$ **11.** True

13. True   **15.** False   **17.** True   **19.** True

21. Converse:
If you can see the eclipse, then the sky is clear.

Inverse:
If the sky is not clear, then you cannot see the eclipse.

Contrapositive:
If you cannot see the eclipse, then the sky is not clear.

23. Converse:
If the deficit increases, then taxes were raised.

Inverse:
If taxes are not raised, then the deficit will not increase.

Contrapositive:
If the deficit does not increase, then taxes were not raised.

25. Converse:

It is necessary to apply for the visa to have a birth certificate.

Inverse:
It is not necessary to have a birth certificate to not apply for the visa.

Contrapositive:
It is not necessary to apply for the visa to not have a birth certificate.

27. Paul is not a junior and not a senior.

29. The temperature will increase and the metal rod will not expand.

31. We will go to the ocean and the weather forecast is not good.

33. No students are in extracurricular activities.

35. Some contact sports are not dangerous.

37. Some children are allowed at the concert.

39. None of the $20 bills are counterfeit.

41.

p	q	$\sim q$	$p \rightarrow \sim q$	$\sim(p \rightarrow \sim q)$
T	T	F	F	T
T	F	T	T	F
F	T	F	T	F
F	F	T	T	F

43.

p	q	$q \rightarrow p$	$\sim(q \rightarrow p)$	$\sim(q \rightarrow p) \wedge q$
T	T	T	F	F
T	F	T	F	F
F	T	F	T	T
F	F	T	F	F

45.

p	q	$\sim p$	$p \vee q$	$(p \vee q) \wedge (\sim p)$
T	T	F	T	F
T	F	F	T	F
F	T	T	T	T
F	F	T	F	F

$[(p \vee q) \wedge (\sim p)] \rightarrow q$
T
T
T
T

47.

p	q	$\sim p$	$\sim q$	$p \leftrightarrow (\sim q)$	$(p \leftrightarrow \sim q) \rightarrow \sim p$
T	T	F	F	F	T
T	F	F	T	T	F
F	T	T	F	T	T
F	F	T	T	F	T

49.

p	q	$\sim p$	$\sim q$	$q \rightarrow p$	$\sim p \rightarrow \sim q$
T	T	F	F	T	T
T	F	F	T	T	T
F	T	T	F	F	F
F	F	T	T	T	T

Identical

51.

p	q	$\sim q$	$p \rightarrow q$	$\sim(p \rightarrow q)$	$p \wedge \sim q$
T	T	F	T	F	F
T	F	T	F	T	T
F	T	F	T	F	F
F	F	T	T	F	F

Identical

53.

p	q	$\sim p$	$\sim q$	$p \rightarrow q$	$(p \rightarrow q) \vee \sim q$
T	T	F	F	T	T
T	F	F	T	F	T
F	T	T	F	T	T
F	F	T	T	T	T

$p \vee \sim p$
T
T
T
T

Identical

55.

p	q	$\sim p$	$\sim p \wedge q$	$p \rightarrow (\sim p \wedge q)$
T	T	F	F	F
T	F	F	F	F
F	T	T	T	T
F	F	T	F	T

└──── Identical ────┘

57. c   **59.** a   **61.**

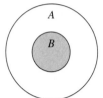

63.   **65.**

67.   **69.**

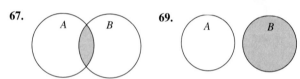

71. (a) Statement does not follow.
   (b) Statement follows.
73. (a) Statement does not follow.
   (b) Statement does not follow.

Section D.3   *(page A54)*

1.

p	q	$\sim p$	$\sim q$	$p \rightarrow \sim q$	$(p \rightarrow \sim q) \wedge q$
T	T	F	F	F	F
T	F	F	T	T	F
F	T	T	F	T	T
F	F	T	T	T	F

$[(p \rightarrow \sim q) \wedge q] \rightarrow \sim p$
T
T
T
T

3.

p	q	$\sim p$	$p \vee q$	$(p \vee q) \wedge \sim p$
T	T	F	T	F
T	F	F	T	F
F	T	T	T	T
F	F	T	F	F

$[(p \vee q) \wedge \sim p] \rightarrow q$
T
T
T
T

5.

p	q	$\sim p$	$\sim q$	$\sim p \rightarrow q$	$(\sim p \rightarrow q) \wedge p$
T	T	F	F	T	T
T	F	F	T	T	T
F	T	T	F	T	F
F	F	T	T	F	F

$[(\sim p \rightarrow q) \wedge p] \rightarrow \sim q$
F
T
T
T

7.

p	q	$p \vee q$	$(p \vee q) \wedge q$	$[(p \vee q) \wedge q] \rightarrow p$
T	T	T	T	T
T	F	T	F	T
F	T	T	T	F
F	F	F	F	T

9. Valid   **11.** Invalid   **13.** Valid   **15.** Valid
17. Invalid   **19.** Valid   **21.** Invalid   **23.** b
25. c   **27.** b   **29.** c

31. Valid

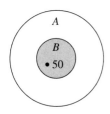

A: All numbers divisible by 5
B: All numbers divisible by 10

33. Invalid

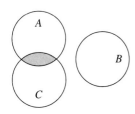

A: People eligible to vote
B: People under the age of 18
C: College students

35. Let p represent the statement "Sue drives to work," let q represent "Sue will stop at the grocery store," and let r represent "Sue will buy milk."

First write:

  Premise #1: $p \rightarrow q$
  Premise #2: $q \rightarrow r$
  Premise #3: p

Reorder the premises:

  Premise #3: p
  Premise #1: $p \rightarrow q$
  Premise #2: $q \rightarrow r$
  Conclusion: r

Then we can conclude r. That is, "Sue will buy milk."

37. Let p represent "This is a good product," let q represent "We will buy it," and let r represent "The product was made by XYZ Corporation."

First write:

  Premise #1: $p \rightarrow q$
  Premise #2: $r \vee \sim q$
  Premise #3: $\sim r$

Note that $p \rightarrow q \equiv \sim q \rightarrow \sim p$, and reorder the premises:

  Premise #2: $r \vee \sim q$
  Premise #3: $\sim r$
  (Conclusion from Premise #2, Premise #3: $\sim q$)
  Premise #1: $\sim q \rightarrow \sim p$
  Conclusion: $\sim p$

Then we can conclude $\sim p$. That is, "It is not a good product."

Index of Applications

TIME AND DISTANCE APPLICATIONS

Index